KU-216-495

GEOLOGY

Thomas Moran's *The Chasm of the Colorado,* 1873–74, oil on canvas, 84 3/8″ × 144 3/4″ (214.3 × 267.6 cm). Lent by the U.S. Department of the Interior, Office of the Secretary. National Museum of American Art, Washington, D.C./Art Resource, N.Y.

GEOLOGY

An Introduction to Physical Geology

Stanley Chernicoff
University of Washington, Seattle

Ramesh Venkatakrishnan
Consulting Geologist/Illustrator

WORTH PUBLISHERS

GEOLOGY

Manufactured in the United States of America

Library of Congress Catalog Card Number: 94-060651

ISBN: 0-87901-451-2

Printing: 1 2 3 4 5—99 98 97 96 95

Development editor: Marjorie P.K. Weiser

Design: Malcolm Grear Designers

Art director: George Touloumes

Project editor: Elizabeth Geller

Production supervisor: Stacey B. Alexander

Layout: Matthew Dvorozniak

Picture researcher: Inge King

Picture research assistants: Siva Bonatti, Silvia Dinale

Composition and separations: York Graphic Services, Inc.

Printing and binding: Von Hoffmann Press, Inc.

Also Available for Students

Carl Shellenberger: *Study Guide to Accompany Geology*
ISBN: 0-87901-452-0

Cover: Thomas Moran, *The Chasm of the Colorado* (detail), 1873–74.
Lent by the U.S. Department of the Interior, Office of the Secretary.
National Museum of American Art, Washington, D.C./Art
Resource, N.Y.

Illustration credits begin on page IC-1 and constitute an extension
of the copyright page.

Worth Publishers
33 Irving Place
New York, New York 10003

Moran (center), with journalist J.E. Colburn and a Kaibab Paiute boy, photographed by Jack Hillers, 1873. National Anthropological Archives, Smithsonian Institution (photo #1592-B).

Cover Artist

> *You cannot see the Grand Canyon in one view, as if it were a changeless spectacle from which a curtain might be lifted, but to see it you have to toil from month to month through its labyrinths.*
>
> John Wesley Powell

Thomas Moran (1837-1926), who was already well known for an impressive painting of the cliffs along the Yellowstone River, was a guest artist accompanying John Wesley Powell's 1873 expedition to the Grand Canyon. Powell, who in 1869 had led the first expedition ever to explore the entire Grand Canyon, returned there from 1870 to 1874 with larger and better prepared explorers, the Congressionally chartered Geographical and Geological Survey of the Rocky Mountains (also known as the Powell Survey and the Colorado Exploration Expedition). Powell would be a founder of the United States Geological Survey in 1879, and its second director. He would also be a founder and director of the Smithsonian Institution's Bureau of American Ethnology.

After viewing the Canyon from a mountain overlook, Moran wrote:

> The whole gorge for miles lay beneath us and it was by far the most awfully grand and impressive scene that I have ever yet seen. . . . Above and around us rose a wall of 2000 feet and below us a vast chasm 2500 feet in perpendicular depth and 1/2 mile wide. . . . The color of the Grand Canyon itself is red, a light Indian Red, and the material sandstone and red marble and is in terraces all the way down. All above the canyon is variously colored sandstone mainly a light flesh or cream color and worn into very fine forms. . . .

Powell described Moran's painting:

> Mr. Moran has represented depths and magnitudes and distances and forms and color and clouds with the greatest fidelity. But his picture not only tells the truth, it displays the beauty of the truth. The painting is called "The Chasm of the Colorado," and rightly. The Grand Canyon of the Colorado is yet to be painted. The view selected is from a point many miles away from the canyon itself and at a place where the side of the plateau had been deeply eroded by a cataract stream. . . .

Moran returned to and painted views of the Grand Canyon numerous times. His paintings were widely exhibited in cities back east. He also made prints of Grand Canyon scenes, which were published in various magazines and journals, thus making the grandeur of the continent's western scenery accessible everywhere.

About the Author

Born in Brooklyn, New York, **Stan Chernicoff** began his academic career as a political science major at Brooklyn College of the City University of New York. Upon graduation, he intended to enter law school and pursue a career in constitutional law. He had, however, the good fortune to take geology as his last requirement for graduation in the spring of his senior year, and he was so thoroughly captivated by it that his plans were immediately changed.

After an intensive post-baccalaureate program of physics, calculus, chemistry, and geology, Stan entered the University of Minnesota-Twin Cities, where he received his doctorate in Glacial and Quaternary Geology under the guidance of one of North America's preeminent glacial geologists, Dr. H. E. Wright. Stan launched his career as a purveyor of geological knowledge as a senior graduate student teaching physical geology to hundreds of bright Minnesotans.

Stan has been a member of the faculty of the Department of Geological Sciences at the University of Washington in Seattle since 1981, where he has won several teaching awards. At Washington, he has taught Physical Geology, the Great Ice Ages, and the Geology of the Pacific Northwest to more than 20,000 students, and he has trained hundreds of graduate teaching assistants in the art of bringing geology alive for non-science majors. Stan studies the glacial history of the Puget Sound region and pursues his true passion, coaching Little League baseball. He lives in Seattle with his wife, Dr. Julie Stein, a professor of archaeology, and their two sons, Matthew (the shortstop) and David (the left fielder).

About the Illustrator

Born in New Delhi, India, **Ramesh Venkatakrishnan** came to the United States after receiving his M.S. in Applied Geology from the University of Delhi in 1975. Ramesh first met Stan Chernicoff at the University of Minnesota-Twin Cities, where he became a laboratory teaching assistant for Physical Geology and in collaboration with Stan developed a series of charts and drawings for the laboratories. In 1979, Ramesh moved to the University of Idaho in Moscow, where he received his doctorate in Geology in 1982.

Ramesh began teaching at Old Dominion University in Norfolk, Virginia, in 1981, teaching physical geology, structural geology, tectonics, remote sensing, and field geology. During the summers he taught the Geology Field Camp for Virginia Tech. In 1986 he and Stan began developing the basic ideas for the illustration program for this textbook. Ramesh left academia in 1990 to pursue a career in consulting with Golder Associates, Inc., where he was a Senior Geologist. Ramesh became an independent consulting geologist in 1993. He has authored or coauthored more than 30 papers on a variety of topics, ranging from mineral exploration to tectonics, and remote sensing to site assessments. The illustrations in GEOLOGY are a natural outgrowth of Ramesh's passion for geologic field investigations and for representing geology in all its color, texture, and three-dimensionality. In addition to geology, his interests include watercolor painting, bicycling, and growing indoor plants. Ramesh works from his home in Mt. Laurel, New Jersey, where he lives with his wife, Kalpana, and daughter, Priya.

Contents in Brief

Contents

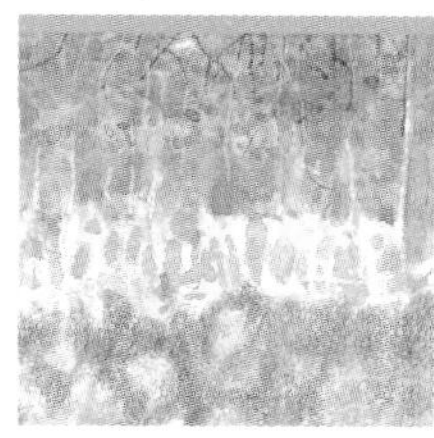

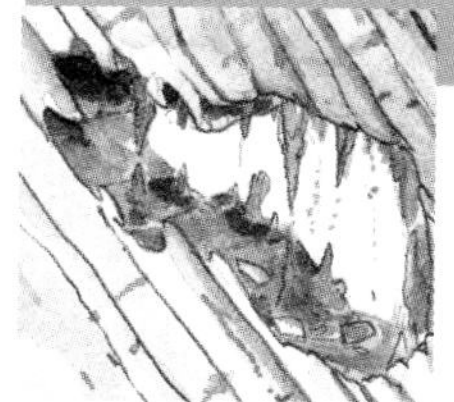

Chapter 16

Caves and Karst 445

Chapter 17

Glaciers and Ice Ages 467

Chapter 18

Deserts and Wind Action 503

Chapter 19

Shores and Coastal Processes 533

Chapter 20

Human Use of the Earth's Resources 563

Preface

The introductory course in physical geology, taken predominantly by non-science majors, may be the only science course some students will take in their college years. What a wonderful opportunity this provides us to introduce students to the field we love and show them how fascinating and useful it is. Indeed, much of what students will learn in the course will be recalled throughout their lives, as they travel across this and other continents, dig in backyards, walk along a beach, or sit by a mountain stream. For this reason, Ramesh Venkatakrishnan, the book's illustrator, and I have expended the best of our abilities to craft an exciting, stimulating, and enduring introduction to the field.

The Book's Goal

The goal is basic—to teach "what everyone should know about geology" in a way that will engage and stimulate. The book embodies the view that this is the most useful college-level science class a non-science student can take, and one that all students should take. Physical geology can show students the essence of how science and scientists work at the same time as it nurtures their interest in understanding, appreciating, and protecting their surroundings. In this course they can learn to prepare for any number of geologic and environmental threats, and see how our Earth can continue providing all of our needs for food, shelter, and material well-being as long as we don't squander these resources.

Content and Organization

The unifying themes of plate tectonics, environmental geology and natural resources, and planetary geology are introduced in Chapter 1 and discussed in their proper context within nearly every chapter. Chapter 1 also presents the important groups of rocks, the rock cycle, and geological time—building a foundation for the succeeding chapters. Chapters 1 through 8 introduce the "basics"—minerals, rocks, and time—to prepare the reader for the in-depth discussions of structural geology, geophysics, and tectonics that follow in Chapters 9 through 12. After the Earth's first-order features—ocean basins, continents, mountain systems—have been discussed, the processes that sculpt these large-scale features are addressed. Chapters 13 through 19 present the principal geomorphic processes of mass movement, streams, groundwater, glacial flow, arid region and eolian processes, and coastal evolution. Chapter 16, Caves and Karst, is a brief separate chapter covering material that is usually embedded in or appended to groundwater chapters. Because caves are among the natural settings that many students visit, and because karst environments are particularly sensitive to environmental damage, we have expanded the discussion as a separate chapter. The final chapter, Human Use of the Earth's Resources, ties together earlier discussions from throughout the book. It reinforces principles that relate to the origins of resources, especially energy-producing ones, and stresses our responsibility to manage them wisely.

Pedagogy

To help readers learn and retain the important principles, key terms are in boldface type and listed at the chapter's end. They also appear in boldface in the end-of-chapter Summary, a narrative discussion that recalls all the key chapter concepts. Also at the end of every chapter are two question sets: Questions for Review to help students retain the facts presented and those For Further Thought to challenge readers to think more deeply about the implications of the material studied.

The author and illustrator have tried to introduce readers to world geology. This book emphasizes, however, the geology of North America (including the "offshore" state, Hawaii), while acknowledging that geological processes do not stop at national boundaries or at the continent's coasts. Wherever data are available—from the distribution of coal to the survey of seismic hazards—we have tried to show our readers as much of this continent, and beyond, as feasible. Photos and examples have been selected from throughout the United States and Canada and from many other regions of the world.

The metric system is used for all numerical units, with their English equivalents in parentheses, so that U.S. students can become more familiar with the units of measurement used by every other country in the world.

The Artwork

The drawings in this book are unique. Ramesh Venkatakrishnan is an experienced and respected geology professor and consultant. He is also a gifted artist. His drawings evolved along with the earliest drafts of the manuscript, sometimes leading the way for the text discussions. We have worked together since we were graduate teaching assistants at the University of Minnesota-Twin Cities. The compulsion for "illuminating" what we want introductory students to know is shared by both of us.

As you will see when you leaf through this book, the art explains, describes, stimulates, and teaches. It is not schematic; it shows how the Earth and its geological features actually look. It is also not static; it shows geological processes in action, allowing students to see how geological features evolve through time. Every effort has been made to illustrate accurately a wide range of geologic and geomorphic settings, including vegetation and wildlife, weathering patterns, even the shadows cast by the Sun at various latitudes. The artistic style is consistent throughout, so that students may become familiar with the appearance of some features even before reading about them in subsequent chapters. For example, the stream drainage patterns appearing on volcanoes in Chapter 4, Volcanoes and Volcanism, set the stage for the discussion of drainage patterns in Chapter 14, Streams and Floods. The colors used and the map symbols keyed to various rock types follow international conventions and are consistent throughout.

The Supplements Package

GEOLOGY is accompanied by an array of materials to enhance teaching and learning.

For students who wish additional help mastering the text, there is the Study Guide by W. Carl Shellenberger (Montana State University—Northern). For each chapter, the Guided Study section helps students to focus on and review in writing the key ideas of each section of the chapter as they read. The Chapter Review, arranged by section and composed of fill-in statements, enables them to see if they have retained the ideas and terminology introduced in the chapter. The Practice Tests and the Challenge Test, which consist of multiple-choice, true/false, and brief essay questions test their mastery of the material. All answers are accompanied by page references for easy review.

The Instructor's Resources by Chip Fox (Clemson University) features chapter objectives, an outline lecture guide with teaching suggestions embedded in it, student activities and classroom demonstrations, and suggestions for further reading and for films, software, slide sets, and model kits. Answers to the end-of-chapter questions in the textbook are also provided.

The comprehensive Test Bank, also by Chip Fox, contains over a thousand questions. There are at least 40 multiple-choice questions per chapter, classified as either factual or conceptual/analytical. There are also 10 short essay questions, complete with answers, for each chapter. A

Computerized Version of the Test Bank is available in both IBM and Macintosh formats.

More than 130 of the text's diagrams and photographs are available for classroom use as full-color slides or transparencies.

Acknowledgments

Some remarkably talented, dedicated people have helped us to accomplish far more than we could have done alone. Worth Publishers assembled a "committee" of top-flight geologists who clarified definitions and explanations, eliminated ambiguities, corrected factual errors, and, in general, helped the author hone the manuscript in countless ways and helped the illustrator to select what to show and how best to do it. Special thanks must go to Jim McClurg at University of Wyoming, Howard Mooers at University of Minnesota-Duluth, and Carl Shellenberger at Montana State University—Northern for their thoughtful advice throughout the development of this book. In addition, for their constructive criticism at various stages along the way, we wish to thank these excellent reviewers:

Gail M. Ashley, *Rutgers University, Piscataway*
David M. Best, *Northern Arizona University*
David P. Bucke, Jr., *University of Vermont*
Michael E. Campana, *University of New Mexico*
Joseph V. Chernosky, Jr., *University of Maine, Orono*
G. Michael Clark, *University of Tennessee, Knoxville*
W. R. Danner, *University of British Columbia*
Paul Frederick Edinger, *Coker College (SC)*
Robert L. Eves, *Southern Utah University*
Stanley C. Finney, *California State University, Long Beach*
Roberto Garza, *San Antonio College*
Charles W. Hickcox, *Emory University*
Kenneth M. Hinkel, *University of Cincinnati*
Darrel Hoff, *Luther College (IA)*
David T. King, Jr., *Auburn University*
Peter T. Kolesar, *Utah State University*
Albert M. Kudo, *University of New Mexico*
Martin B. Lagoe, *University of Texas, Austin*
Lauretta A. Miller, *Fairleigh Dickinson University*
Robert E. Nelson, *Colby College (ME)*
David M. Patrick, *University of Southern Mississippi*
Terry L. Pavlis, *University of New Orleans*
John J. Renton, *West Virginia University*
Vernon P. Scott, *Oklahoma State University*
Dorothy Stout, *Cypress College (CA)*
Daniel A. Sundeen, *University of Southern Mississippi*
Allan M. Thompson, *University of Delaware*
Charles P. Thornton, *Pennsylvania State University*

At Worth Publishers, developmental editor Marjorie Weiser performed the herculean task of reining in the author's long-windedness with extraordinary grace and intelligence and brought organization wherever she found disorder. Project editor Elizabeth Geller polished each chapter of prose and every rough sketch, working with all the elements of the book until they formed a coherent whole. Photo research was handled masterfully by picture editor Inge King and photo research assistants Siva Bonatti and Silvia Dinale. The book's pleasing appearance was created under the supervision of George Touloumes and his gifted staff, including Demetrios Zangos and Matthew Dvorozniak. Production supervisor Stacey Alexander oversaw all scheduling and quality control throughout the production process.

I very much appreciate Managing Editor Anne Vinnicombe's support and behind-the-scenes hard work. Thanks are due also to Carol Bullock, who coordinated and edited the supplements; editorial assistants Paul Hamilton, Jeannie Jhun, and Marian Turk who coordinated communications and numerous other tasks efficiently; and Maja Lorkovic, who provided invaluable editing and production assistance at all stages of the project.

Finally, I also wish to acknowledge with deep appreciation the role of Ron Pullins (formerly of Little, Brown and now of West Educational Publishing) and Worth Publishers' Kerry Baruth who championed the cause of GEOLOGY with their respective companies.

After they leave our classrooms, students may well forget specific facts and terminology of geology, but they will still retain the general impressions and attitudes they formed during the course. We hope that our words and illustrations will help to advance the goals of those teaching this course and contribute to their classes. We have used our teaching experiences to craft a textbook that we think our own students will learn from and enjoy. We hope your students will too. We invite your comments; please send them to the author, whose e-mail address is: sechern@u.washington.edu.

Stan Chernicoff

To The Student

Geology is the scientific study of the structure and origin of the Earth, and the processes that have formed it over time. This book was created to bring you some of the excitement of that study through words and illustrations. We, the author and illustrator, are geologists who love the discipline and have derived much pleasure from introducing it to thousands of our own students. We have also been students and understand that some topics will be more interesting to you than others. We have worked to make every aspect of the book as fascinating and useful to you as possible.

Unlike some subjects you might study, geology is not over when your course is completed. Wherever you live or may travel, geology is far more than "the scenery"—although your appreciation for the landscape will be much enhanced by a basic knowledge of geology. When you drink from a kitchen tap, dig in garden soil, see a forest—or see it being cut down for some construction project—geology has a role. If you gaze at a waterfall, swim in coastal currents, endure an earthquake—or read about those who did—you will understand more about the experience after taking this course. As you will learn, geology is everywhere—in the products you buy, the food you eat, the quality of your environment. The materials and processes encountered in the study of geology supply all of our needs for shelter, food, and warmth. This course will help you to understand why earthquakes occur in some places and not in others, where we can live to avoid floods and landslides, where we can find safe drinking water, and more. Knowledge of geology can also help to make us better citizens as we learn how to prevent further damage to our environment.

You will also learn about more distant matters: the age of the Earth, how the planet has changed over its long lifetime, and how some of its creatures have changed along with it. You will explore some geological ideas about our moon and the planets with which we share our solar system. Highlighted discussions provide additional information about some particularly interesting topics.

The language geologists use helps us to describe natural phenomena with precision and accuracy. We have minimized the new terminology you will need to learn, but to do well in the course you will need to master some technical terms. The key terms are in bold type when introduced. They are also listed at the end of each chapter and defined in the glossary at the end of the book.

The drawings in this book are unique, showing you how the Earth really looks both on and below the surface. As you read and examine each illustration, you will find that the words, the photographs, and the drawings are all important in giving you a full picture of geology. The text and illustrations, the key terms, and even the chapter summaries work together to help you learn the concepts and terms. When you read the summary, ask yourself if you know these key points; can you cite examples beyond those given in the brief summary?

Each chapter also includes two sets of questions, one testing your retention of the facts and one challenging you to think more deeply about some of their implications. Test yourself, and then go back and reread any material you are not sure you understand.

As the author of this book, I hope you will enjoy it and gain an appreciation for geology that will enrich your life. If you have any comments, complaints, or compliments, please send them to me or to Worth Publishers. My e-mail address is: sechern@u.washington.edu.

GEOLOGY

Part 1

Forming the Earth

Geology and the Methods of Science

Highlight 1-1 **What Caused the Extinction of the Dinosaurs?**

The Earth in Space

Plate Tectonics

Proving Plate Tectonics: A Theory Develops

A Preview of Things to Come

Figure 1-1 The San Andreas fault, as seen from the air over Carrizo Plain in California.

Chapter 1

A First Look at Planet Earth

At 5:03 P.M. Pacific daylight time on October 17, 1989, baseball fans across North America were settling down in front of their television sets to watch Game Three of the World Series from San Francisco. Minutes later, violent movement along a small segment of California's San Andreas fault (Fig. 1-1) had caused widespread destruction, taking the lives of scores of Bay Area residents and injuring hundreds more. Instead of baseball, millions viewed grim scenes of collapsed buildings and freeways (Fig. 1-2), broadcast live. People in other earthquake-prone regions wondered whether their own towns and cities might be the next to suffer a life-threatening quake. Four years later, at 4:31 A.M. PST on January 17, 1994, millions of Southern Californians were jolted awake by a powerful earthquake that took 57 lives, buckled numerous freeways, and, with estimated cleanup and repair costs of more than 15 billion dollars, proved to be one of the United States' most expensive natural disasters ever. Yet, the 1989 Loma Prieta earthquake and the 1994 Northridge earthquake may only be preludes to the long-awaited, much feared "Big One" that looms in the minds of 30 million Californians.

Figure 1-2 The collapse of the Nimitz Freeway in Oakland, California, during a major earthquake along the San Andreas fault on October 17, 1989. The San Andreas fault, a fracture in the Earth's crust that cuts north–south across much of California, is responsible for some of North America's most powerful earthquakes.

Do the citizens of Boston, Massachusetts, Memphis, Tennessee, and Charleston, South Carolina, need to worry quite as much about earthquakes as do the citizens of San Francisco and Los Angeles? Considering only recent events, perhaps not. But destructive tremors struck the Boston area in 1755, Memphis in 1811, and Charleston in 1886, and could well do so again. Most North Americans would be wise to understand what geological forces cause phenomena such as earthquakes so that they can prepare for them.

(a)

(b)

(c)

Figure 1-3 Geologists at work. (**a**) Sampling 2000°C lava from Hawaiian volcanoes; (**b**) unearthing dinosaur remains in Montana; (**c**) collecting rocks from the lunar surface.

Geology is the scientific study of the Earth's processes and materials. Geologists examine the origin and evolution of the Earth, the wide variety of materials it contains, and the processes, such as earthquakes, that act at or near its surface. The subjects of their investigations range from the smallest atoms to entire continents and ocean basins. Geologists study erupting volcanoes and windswept sand dunes, raging floods and creeping glaciers, titanic meteorite impacts and the slow, silent enlargement of underground caverns. Some collect and interpret evidence of past life on Earth—from the vestiges of the earliest known single-celled algae found in the 3-billion-year-old rocks of the Australian outback to the bones of our earliest human ancestors excavated from 3-million-year-old volcanic ash layers in East African valleys. Geologists today are not even Earth-bound: They study lunar rocks collected by astronauts and analyze data, gathered by space probes, from the farthest planets in our solar system (Fig. 1-3).

Everything we use comes from the Earth, and so geology has enormous practical effects on our daily lives. The natural resources that shape modern life have come to us, in part, through geological knowledge. Geologists help locate the ingredients for cement, concrete, and asphalt, with which we build our cities and highways, as well as the oil, gas, and coal that fuel our cars and light and heat our homes. In recent years, many geologists have been engaged in finding a rapidly dwindling resource—fresh, uncontaminated underground water for industrial, agricultural, and domestic use. Geological study also helps us to predict and avoid some of nature's life-threatening hazards. Geologists identify slopes that may produce landslides and sites that are too unstable to safely contain dams or nuclear power plants. They detect potential earthquake locations and recommend ways to avoid damage from flooding rivers. They warn us away from eroding shorelines and devise ways to clean up the mess we've made through decades of environmental ignorance or negligence.

Other geologists probe the most fundamental mysteries of our planet: How old is the Earth? How did it form? When did life first appear? Why do some areas suffer from devastating earthquakes, while others are spared? Why are some regions endowed with breathtaking mountains and others with fertile plains? The questions that attract geologists are some of the same questions that have engaged human curiosity for millennia. (You may have pondered some of them as you walked into the lecture hall for your first geology class.) We'll begin to answer these questions by describing how the science of geology operates, introducing some of the basic concepts and standards that underpin this discipline. We will discuss how our planet may have formed and speculate about some changes it has gone through since. Finally, we will examine some of the Earth's large-scale geological processes, and determine how the nature of these processes was deduced.

Geology and the Methods of Science

The principal objective of science is to discover the fundamental patterns of the natural world. Scientists make one basic assumption: The world works in an orderly fashion, in which natural phenomena will recur given the same set of conditions. As scientists understand it, every effect has a cause. Geologists, like other scientists, begin their investigations with a question or set of questions about how some part of the natural world functions.

Scientists have a distinctive strategy for investigating unexplained phenomena. Classical **scientific methods** require investigators to be receptive to whatever they discover about the question being studied. They must gather all available data bearing on their subject, such as measurements, descriptions of naturally occurring phenomena, and the results of laboratory experiments, and derive a logical explanation to support them all. A tentative explanation that fits all the data collected, and is expected to account for future observations as well, is called an **hypothesis**. Often, a number of hypotheses are proposed to explain the same set of data. For example, within the past 20 years, more than fifty hypotheses have been proposed to explain why the Earth's climate periodically plunges into ice ages.

An hypothesis is tested as scientists conduct further experiments or make further observations. Hypotheses that do not explain such subsequent findings may need to be modified or abandoned. Some hypotheses undergo numerous modifications and evolve over many years until they fully account for all observations and/or experiments; others simply don't survive the testing process and are cast aside. New information and more up-to-date technologies capable of more accurate observations often reveal flaws in previously viable hypotheses. The history of science is littered with disproven hypotheses that were quite popular at one time. (The suggestion that the Earth is the center of our solar system is one such failed hypothesis; it could not withstand close scientific scrutiny once astronomic telescopes were invented.) A flawed hypothesis is usually replaced by one that encompasses more of the available data.

An hypothesis that is repeatedly confirmed by observation and experimentation gains credibility and wider acceptance within the scientific community. Ultimately, a comprehensive hypothesis that consistently explains accumulating data may become a **theory**, a generally accepted explanation for a given set of data or observations. An example of how multiple hypotheses are developed and tested, possibly to become theories, is given in Highlight 1-1 (p. 6).

Even when an hypothesis survives testing and becomes a theory, new technology may produce new data that necessitate updated hypotheses, modifying or completely replacing the established theory. (For example, we may someday be able to drill more deeply into the Earth than we can now and discover evidence that will cause geologists to revise current hypotheses regarding deep-Earth processes.) A theory that continues to meet rigorous testing over a long period of time may be declared a **scientific law**. For example, an unfettered object always falls toward the Earth's surface, proving the immutability of the law of gravitational attraction.

Geologists apply scientific methods in ways that are unique to their science. Unlike chemistry or physics, which are experimental sciences, geology is largely an historical science, confronting problems of vast time and scale. Some chemistry or physics experiments involve reactions a millionth of a second in duration and occurring on an atomic scale. Many geological processes, however, take millions of years and occur on a continent-sized scale. Because most geological processes are imperceptibly slow and their scale unimaginably large from a human perspective, geologists can't always test hypotheses through direct observation or experimentation. To supplement their field and laboratory work, they sometimes use scaled-down models to study large-scale geological phenomena, or rely on the power of computers to create mathematical models.

Highlight 1-1 *What Caused the Extinction of the Dinosaurs?*

For years, paleontologists (geologists who study ancient life forms) have wondered what might have caused more than 75% of all the forms of life then on Earth to vanish about 65 million years ago. The most dramatic loss was the extinction of the dinosaurs, a group of animals that had roamed the planet for 150 million years, but numerous other life forms vanished as well—large and small, water- and land-dwelling, plant and animal. Many species, whether living in freshwater lakes or rivers, in salt-water oceans, or on land, all became extinct at about that same time.

Some early hypotheses focused on only one kind of organism to explain these extinctions. Some proposed that epidemic diseases eliminated dinosaur populations or that the rise of egg-stealing mammals ravaged dinosaur nests. But neither of these accounted for the loss of two-thirds of all marine animal species, which led some scientists to propose that the oceans became lethally salty (though this did not explain why some marine creatures survived). To explain the extinction of gigantic terrestrial reptiles, tiny marine organisms, and many life forms in between, a number of hypotheses invoked global environmental change. Did the Earth suffer from a period of drastic cooling 65 million years ago? Did a shift in the planet's protective magnetic field allow harmful solar radiation to reach land and sea, eliminating a wide variety of life forms? Did a nearby star explode, bathing the Earth in cosmic radiation? Surely, each of these events would have affected all life on Earth simultaneously. Why, then, were 25% of the planet's species unaffected?

Several hypotheses agree that wholesale extinction followed some catastrophic disruption of the global food chain, though they disagree on what that disruption was. (The food chain refers to nature's succession of predator–prey relationships, specifying "who eats whom." In this chain, a meat-eating dinosaur such as *Tyrannosaurus rex* would perish if its prey, typically a plant-eating dinosaur, became scarce, which would occur as a result of the prey's own food supply having diminished.) Some scientists looking for indications of a global event that could have caused such a disruption in the food chain noted evidence of widespread fires followed by very rapid cooling at this period, but disagreed on what might have caused this. One group has cited evidence that massive volcanic eruptions of India's Deccan plateau sent a cloud of volcanic ash and gas around the Earth, blocking out sunlight, cooling the planet, and leading to a worldwide decline in vegetation, including microscopic marine plants. These scientists reason that without the plants their diets were based on, many plant-eating animals would have died out, and their extinction would in turn have wiped out the meat-eaters who were their predators.

(a)

Another group of scientists, led by geologist Walter Alvarez and his father, Nobel prize-winning physicist Luis Alvarez, have proposed this scenario: A meteorite at least 10 kilometers (6 miles) in diameter plowed into the Earth, releasing a shower of pulverized rock into the atmosphere. The resulting dust veil would have blocked out sunlight (in much the same way as volcanic ash would have), cooled the planet, and led to an "impact winter" that may have lasted for decades—long enough to devastate the global food chain. The strongest evidence to support this impact hypothesis is a 2.5-centimeter (1 inch)-thick layer of clay found around the world in rocks that date from about 65 million years ago (Fig. 1-4a). The clay contains iridium, an element that is extremely rare in rocks of terrestrial origin, but is quite common in meteorites. The Alvarezes and their associates contend that the iridium-rich layer resulted from the global fallout of pulverized meteorite dust. Fossils of numerous species, including many now extinct, have been found in the rocks that formed just before the iridium-rich layer was deposited, whereas only about a fourth as many species are represented in the rocks that formed just after this layer was deposited, suggesting that many extinctions occurred during the time of deposition. Further evidence of a meteorite impact includes the presence of microscopic glassy spheres called *microtektites* in

(b)

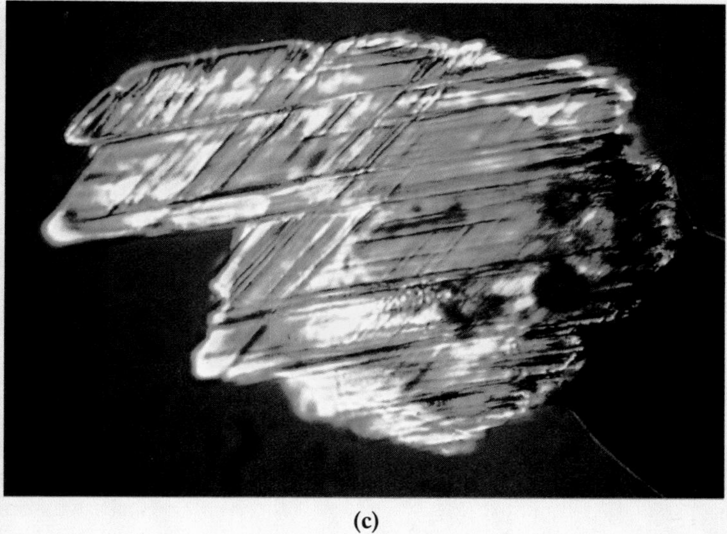

(c)

Figure 1-4 (**a**) An iridium-containing layer of clay (marked by coin) found by Walter Alvarez in Gubbio, Italy. The Alvarezes believe that this clay, which is found around the world in rock of this age, may have been deposited after a meteor impact about 65 million years ago. (**b**) Microtektites from Thailand. Such aerodynamic spheres, found around the world in sediments of about 65 million years of age, are believed to have been formed when rock melted by meteor impact was hurled into the air by the impact and cooled and solidified before falling back to Earth. (**c**) A magnified photograph of a mineral grain that has been shattered by the shock of a meteorite impact.

sediment layers around the world dating from this period (Fig. 1-4b). These may have been formed when superheated rocks at an impact site were hurled into the air in a molten state, dispersed in the atmosphere, and cooled rapidly as they fell back to Earth. Mineral grains shattered by very high pressures—as would occur if they had been struck by a meteorite—have also been found at proposed impact sites (Fig. 1-4c). Finally, the high concentration of carbon soot (a product of burned vegetation) found within the iridium layer could be evidence of global wildfires, which may have been touched off as countless glowing bits of falling debris ignited the Earth's vegetation.

As yet, no extinction hypothesis has achieved theory status. Analysis of the Earth's 65-million-year-old deposits continues today, to document further the proportions of organisms that became extinct at that time, and to search for additional evidence of a meteorite strike or a catastrophic volcanic eruption. The impact hypothesis awaits confirmation by a key piece of evidence—definite location of an impact site. A meteorite capable of such vast environmental disruption would have left behind a crater at least 160 kilometers (100 miles) across. No possible site has been identified on land, although a large impact structure from a meteorite that may be about 65 million years old is being studied near Manson, Iowa. Since most of the Earth's surface is ocean-covered, however, it is possible that the crater may be lying undetected in the sea floor. Two huge submarine craters, one off the coast of Mexico's Yucatán peninsula and the other off the coast of Colombia, are also being investigated. Because of the age of its rocks and the presence of microtektites in its vicinity, the Yucatán crater (called Chicxulub), 180 kilometers (112 miles) in diameter, is considered particularly promising. Smaller impact craters of about the target age have also been widely reported, but the cumulative effect of many small and scattered impacts could not have produced the same effect as a single large impact. (Skeptics point out that there have been numerous proven meteor impacts that did *not* cause radical changes to plant or animal life.) The debate over the cause of this mass extinction is expected to continue for years to come.

The Development of Geological Concepts

Almost two centuries of observation and hypothesis formation and testing have contributed to our current understanding of how our planet developed. People long believed that the Earth's geological evolution was marked by episodes of dramatic and rapid change interspersed with long periods of relative stability and little change. Since natural scientists understood best those processes that they could directly observe, such as volcanic eruptions, monumental earthquakes, and raging floods, they believed that these violent events alone explained the origin of such apparently inactive Earth features as mountains, valleys, and fossils. **Catastrophism,** the hypothesis that the Earth evolved through a series of immense worldwide upheavals, was the prevalent view until the mid-eighteenth century.

During the latter part of the eighteenth century, the Scottish naturalist James Hutton (1726–1797) recognized that slow processes, such as rivers cutting through valley floors and loose soil creeping down gentle slopes, acting over a vast amount of time, may have had a greater effect on the Earth than did occasional catastrophic events. (This idea met with great resistance, because it implied that the Earth was a lot older than was generally believed at the time. Most Christians—encompassing most Europeans and North Americans, including scientists—accepted more or less literally the Biblical account of creation, which was generally interpreted as meaning that the Earth was only a few thousand years old.) Hutton proposed that the physical, chemical, and biological processes that anyone could see changing the Earth in small ways during his or her lifetime must have worked in a similar manner throughout Earth's long history. His hypothesis, called **uniformitarianism,** was that current geological processes could be used to explain long-past geological events. By the 1830s, after much debate, uniformitarianism prevailed over catastrophism. Its acceptance has been hailed as the birth of modern geology.

James Hutton said, "the present is the key to the past," and geologists today recognize that the Earth's present appearance is the result of millions of years of the same physical processes, although probably at varying rates. They also know, however, that some geological events are indeed catastrophic, and that much geological change does occur during brief spectacular events. A great earthquake may shift a land area more than 6 meters (20 feet) in a single moment. The 1989 storm known as Hurricane Hugo eroded more of the Carolina coast in one day than had the preceding century of slow, steady wave action. Thus, slow but consistent processes as well as catastrophic events are continuously reshaping our planet.

The Earth in Space

The Earth is a slightly flattened sphere with an average radius of 6371 kilometers (3957 miles), orbiting approximately 150 million kilometers (93 million miles) from the medium-sized star we call the Sun. Our Sun is only one of about 100 billion stars in the Milky Way galaxy, a pancake-shaped cluster of stars which itself is only one of about 100 billion such galaxies in the observable universe. Despite Earth's relative smallness in the universe, it is perfectly positioned to receive just the right amount of the Sun's radiant energy to support life. Because of Earth's composition and geologic past, it has manufactured a watery envelope and protective atmosphere on which countless living species have relied for billions of years. But how did the Earth come to be the only known life-sustaining planet in the universe?

The Probable Origin of the Sun and Its Planets

Cosmologists (scientists who study the origin of the universe) have proposed that the universe began as a very small, very hot volume of space containing an enormous amount of energy. The birth of all the matter in the universe is believed by many scientists to have occurred when this space expanded rapidly with a "Big Bang" 15 to 20 billion years ago. (The timing of the hypothetical Big Bang has been estimated by observing the current positions and speeds of the visible galaxies, which are moving away from each other, and tracing their paths backward.)

Immediately after the Big Bang, the universe began to expand and cool, which it continues to do today. A few minutes after the explosion, it had cooled to a temperature of about 1 billion degrees Celsius. At this time, the universe consisted only of three kinds of subatomic particles (discussed further in Chapter 2)—remnants of its original composition. These particles would eventually combine to form atoms, the building blocks of all the matter in the universe today. At this early stage, however, although the particles hurled about and collided, any atoms they formed immediately broke apart in violent collisions with other particles.

Only after about a million years had passed, when the universe had cooled to approximately 3000° Celsius (6000° Fahrenheit), could atoms of hydrogen and helium—the simplest, lightest elements—begin to exist without being torn apart. At that time the universe consisted of about 75% hydrogen gas and 25% helium gas by weight, a composition that has not changed much to this day. As the universe continued to expand, it became less uniform, with pockets of relatively high gas concentrations that began to attract more particles of gas by the force of gravity. Where enough gas gathered, the resulting gas clouds collapsed inward by gravity. These accumulations became galaxies (the large disk-shaped structures that are spread fairly uniformly throughout the universe) and clusters of galaxies. Although cosmologists have amassed a great deal of evidence to support these details of the Big Bang hypothesis, galaxy formation is still not well understood.

Within each galaxy, such as our own Milky Way, some gas clouds collapsed further to form stars. Stars continue to be born in this way in all galaxies, including our own. (Through a telescope, one can see the "star nursery" that makes up the belt of the constellation Orion.) The heat released by the collision of infalling gas particles causes the particles in the core of each star to move faster and faster. Eventually they move fast enough to collide and engage in nuclear reactions, which release the energy that keeps the star hot and glowing brightly in the sky. In these nuclear reactions, the particles fuse together, forming larger particles that will become the nuclei of helium and other, heavier elements.

Stars are not only born but they die as well, some slowly and some rapidly. A star that is dying very rapidly is called a *nova* (from the Latin for "new") because it appears as a very bright new star in the heavens. Dying stars are important because they heat up so much that new nuclear reactions occur, producing the nuclei of ever more complex, heavier elements. In this way dying stars act as manufacturing plants. They are also distribution centers, dispersing these nuclei over the space surrounding each dying star. It is believed that all the matter in and on the Earth—including that in our bodies—comes from such dying stars.

Our own Sun is a star, believed to have formed in the same way as any other star, that came into being well after our galaxy was created. When our Sun was born, it was late enough that earlier stars had already died, and their remnants contributed to the gas cloud, or *nebula,* that eventually developed

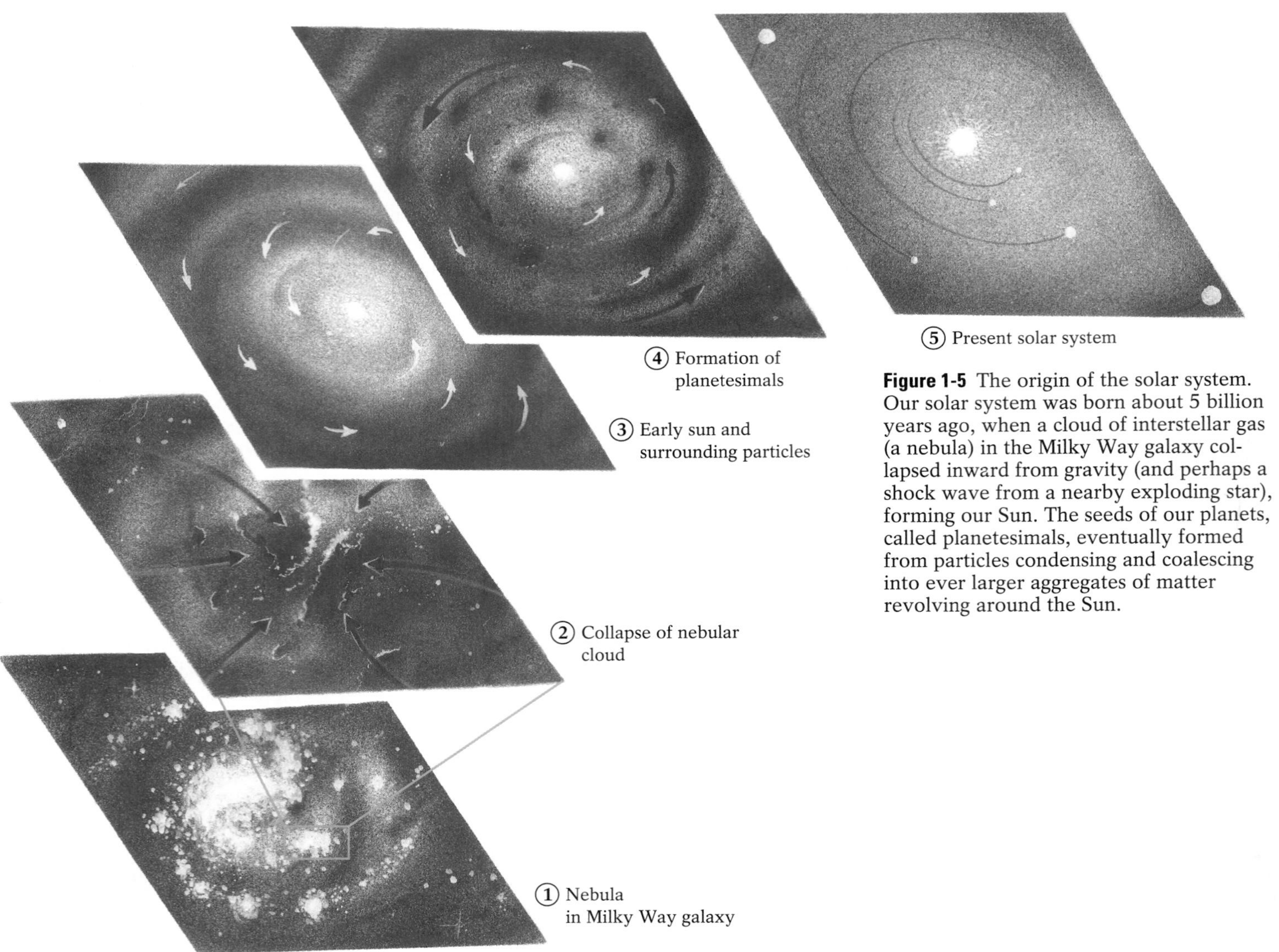

Figure 1-5 The origin of the solar system. Our solar system was born about 5 billion years ago, when a cloud of interstellar gas (a nebula) in the Milky Way galaxy collapsed inward from gravity (and perhaps a shock wave from a nearby exploding star), forming our Sun. The seeds of our planets, called planetesimals, eventually formed from particles condensing and coalescing into ever larger aggregates of matter revolving around the Sun.

into our entire solar system (Fig. 1-5). (A nebula is actually any "fuzzy-looking" area in the sky, such as a distant galaxy or the gas surrounding stars.) The interstellar ("between stars") nebula that would become our Sun and planets was probably originally dispersed across a vast area of space, extending beyond what would become the outermost planet. About 5 billion years ago, however, this nebula began to collapse inward, perhaps due to a shock wave from a nearby exploding star. As its component materials were drawn by gravity toward its center, they collided in nuclear reactions and generated heat in the same way as other stars, forming the infant Sun.

As heat became concentrated in the center of this new star, some material in the outer nebula surrounding it began to cool and condense into infinitesimally small grains of matter. Uncondensed substances nearby were swept outward by strong solar winds, streams of matter and energy that flowed from the infant Sun. In this way, the first solid materials to form in our solar system became separated into a hot inner zone of heavier substances, such as iron and nickel, and a cold outer zone of lighter gases such as hydrogen and helium.

As the first bits of matter condensed, they continued to collide and coalesce, forming aggregates that grew to a few kilometers or larger in diameter. These planetary seeds, or *planetesimals,* formed the cores of the developing planets. They attracted additional material and continued to grow by gradual accumulation, or *accretion,* of particles, sweeping clean the space around them.

Intense solar radiation warmed the four planets closest to the Sun, causing their surface temperatures to rise. Nearly all of their light elements—hydrogen, helium, ammonia, and methane—vaporized and were carried away by solar winds. The matter remaining in these four small, dense, inner planets consisted primarily of iron, nickel, and silicates (a mixture of silicon and oxygen), the heavier substances that remained in the inner zone of the solar nebula. These planets are Mercury, Venus, Earth, and Mars.

The outer planets developed so far from the Sun, and are consequently so frigid, that none of their matter vaporized. These planets—Jupiter, Saturn, Uranus, Neptune, and Pluto—probably consist predominantly of a low-density mass of icy condensed gases (hydrogen, helium, methane, and ammonia).

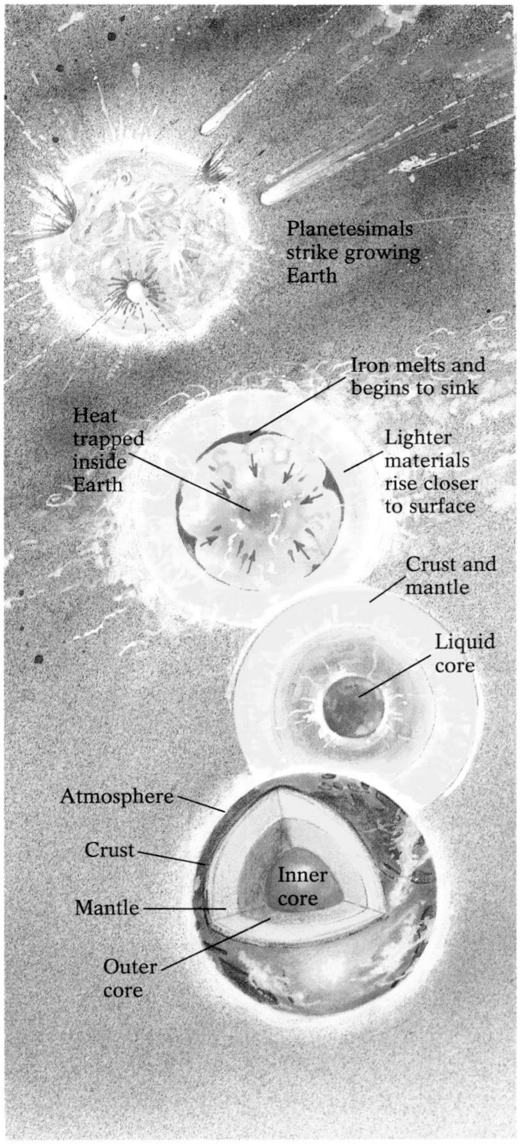

Figure 1-6 Heating and differentiation of the early Earth. Increased temperatures in the planet's interior, caused in part by planetesimal impacts, initiated the so-called iron catastrophe —the event believed responsible for the layering of the Earth's interior.

The Earth's Earliest History

The largest of the four inner planets is our Earth. When it finally accreted to its present size about 4.6 billion years ago, it was a lifeless ball of dust with no surface water and no atmosphere, bearing no resemblance to its appearance today. This ball was probably an homogeneous mixture of iron, magnesium, silicon, and oxygen, with small amounts of gases trapped in its interior. The Earth was probably relatively cool, at least for a while, but it didn't stay that way for long.

During its period of accretion, the proto-Earth had attracted additional planetesimals whose enormous energy of motion was converted to heat upon impact. Some of this heat radiated back into space, but much was retained by the new planet as succeeding planetesimals struck its surface and buried the heat from earlier strikes. In addition, atoms of radioactive substances such as uranium, whose nuclei disintegrate and release heat, also contributed to the planet's internal heat reservoir. (Radioactive elements are discussed in Chapter 2.) As more heat accumulated and became trapped within the Earth's interior, the internal temperature of the proto-Earth began to rise.

The Iron Catastrophe The combined heat from infalling accreted masses and from uranium-type heating eventually set in motion the next significant event in the Earth's ancient past—its evolution from homogeneous planet to one whose constituent substances are sorted into distinct internal layers (Fig. 1-6). Sometime between a few hundred million and one billion years after the Earth formed, the temperature at depths of 400 to 800 kilometers (250–500 miles) below its surface increased to the melting point of iron. As a result, much of the iron there liquefied and began to sink to even greater depths, drawn by gravity toward the planet's center. This movement of molten iron produced additional heat, raising the Earth's temperature to an estimated 2000°C (3600°F), hot enough to melt the other substances in the interior. This critical moment in Earth's history is called the **iron catastrophe.** In the molten state, denser materials began to sink toward the center, while lighter materials began to rise closer to the surface. Matter that had originally made up an homogeneous Earth now became separated, or *differentiated,* into concentric zones of differing densities. The densest materials, probably iron and nickel, formed what may have been a totally liquid core at the planet's center. Lighter materials, composed largely of silicon and oxygen, floated upward, ultimately cooling and solidifying to form the Earth's outer layers. Even lighter materials—gases that had been created or trapped in the interior—escaped, with some hydrogen and oxygen combining to form the first oceans. The lightest gases of all rose to form the Earth's early atmosphere.

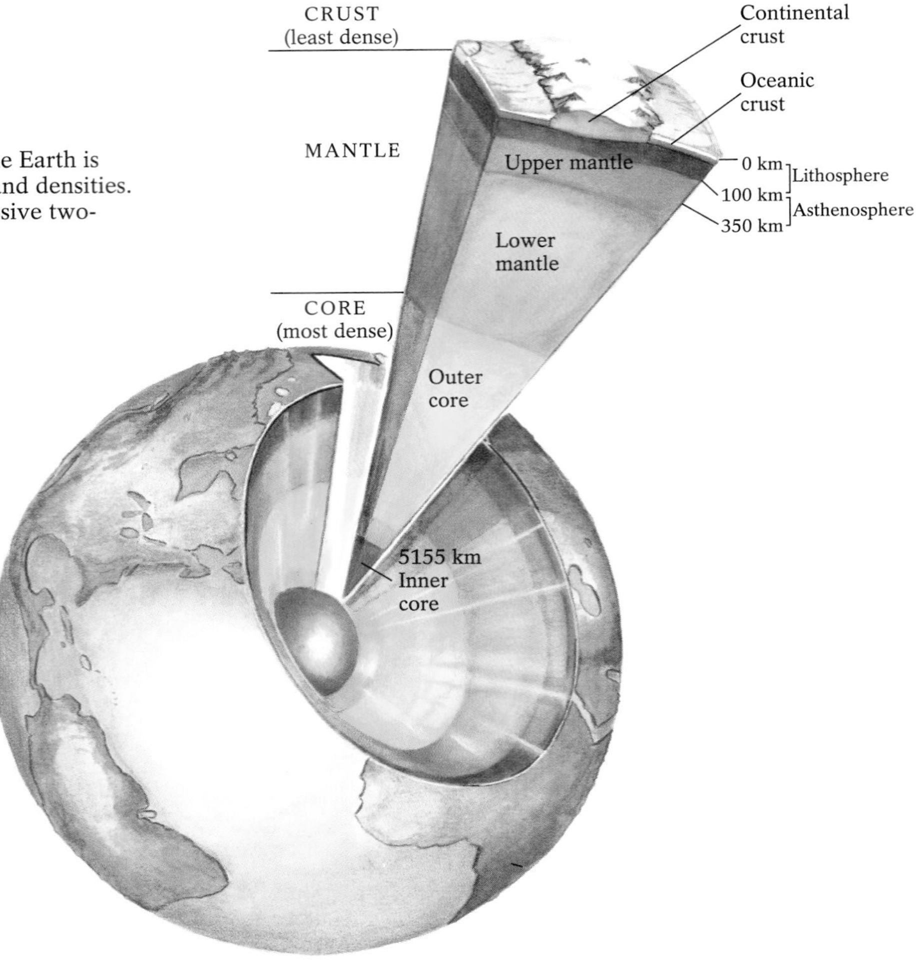

Figure 1-7 A simplified model of the Earth's interior. The Earth is composed of concentric layers of differing thicknesses and densities. A slice of the Earth's interior reveals a thin crust, a massive two-part mantle, and a two-part core.

Gases continued to escape over time, primarily during volcanic eruptions; this ongoing degassing of the Earth's interior led to the creation of more and more surface water, and to progressive changes in the atmosphere.

The Earth's interior probably did not become completely molten during the iron catastrophe. If it had, the planet might have lost all its gases, and its life-supporting atmosphere could never have developed. It did, however, lose its earliest atmosphere, which is believed to have been composed mainly of hydrogen, ammonia, and methane. Because of that loss—and fortunately for us—today's atmosphere is largely free of those gases.

A Glimpse of the Earth's Interior Contrary to those memorable but misleading scenes in old Hollywood movies, the Earth is not a hollow ball filled with jungles and dinosaurs. Scientists have determined that the Earth's interior consists of three principal concentric layers, each with a different density (Fig. 1-7). (*Density* expresses the quantity of matter in a given volume of a substance. Because its chemical structure is more compact, one cubic centimeter of iron is far denser than one cubic centimeter of glass.) The outermost layer of the Earth is a thin **crust** of relatively low-density rocks. Underlying it is the **mantle**, a thick layer of denser rocks. At the Earth's center is its **core,** the densest layer of all, consisting primarily of pure metals such as iron and nickel. The arrangement of these three layers is somewhat like that of a hard-boiled egg with its thin shell, extensive white, and small central yolk, but the egg model does not show a number of important sublayers that are fundamental to our understanding of our dynamic Earth.

The outer 100 kilometers (60 miles) of the Earth, encompassing both the Earth's crust and the uppermost portion of the mantle, is a solid, brittle layer known as the **lithosphere** ("rock layer," from the Greek *lithos*, rock).

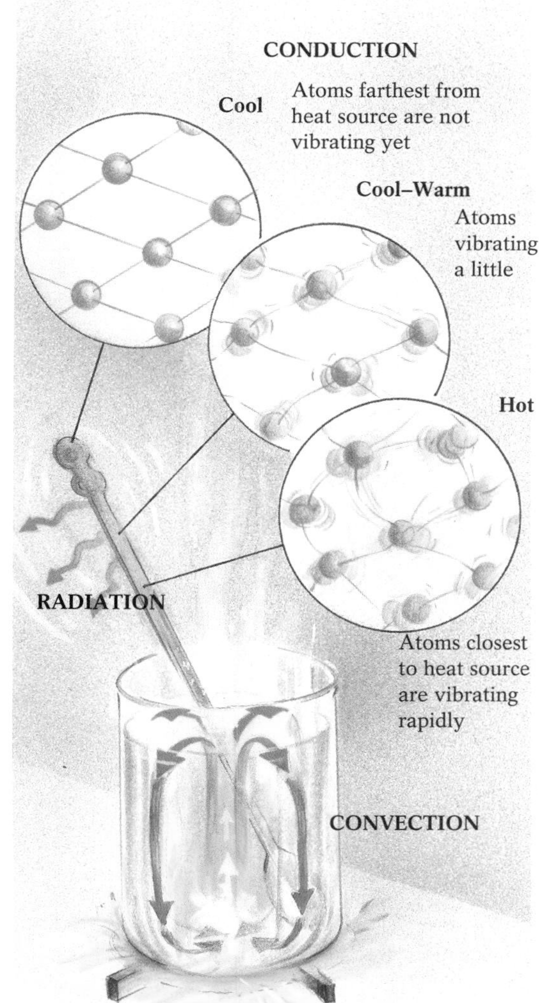

Figure 1-8 Three ways in which heat is transmitted; these phenomena, occurring on a much larger scale within the Earth, are responsible for many geological processes. *Conduction* of heat involves the passage of thermal energy from atom to neighboring atom. In the diagram, the atoms in the part of the spoon that is immersed in the hot soup have begun to vibrate, passing thermal energy to adjoining atoms. The end of the spoon farthest from the heat source is still cool, but eventually the heat will be conducted throughout the length of the spoon. *Convection* involves the movement of heat from place to place by a flowing medium. Because the soup at the bottom of the pot is closer to the flame, it is the first part of the soup to become hot. As it heats up, it expands (becoming less dense), and the cooler (more dense) soup above it sinks to the bottom, displacing it and forcing it upward. When the warm soup arrives at the top, it encounters the relative coolness of the air and contracts in volume (becomes more dense) as it begins to cool. The cooled soup then sinks back toward the bottom, to be reheated and then to rise again. In the diagram heat is also being *radiated* from the warm spoon to the cooler air surrounding it.

Underlying the lithosphere is the **asthenosphere** ("weak layer," from the Greek *aesthenos,* weak), a zone of heat-softened, slow-flowing yet still-solid rock located in the upper mantle from about 100 to 350 kilometers (60–215 miles) beneath the Earth's surface. The lithosphere and asthenosphere are where such large-scale geological processes as mountain building, volcanism, earthquake activity, and the creation of ocean basins originate. Below the mantle, the core is divided into a liquid outer core and a solid inner core. We will explore these layers in detail in Chapter 11.

Heat in the Earth We have seen how the buildup of heat in the Earth's interior caused some of its component substances to melt and sink or rise, depending on their density. The phenomenon of heat, which is a form of energy, is closely tied to numerous processes that continue to shape our planet. Heat energy is transferred from place to place within the Earth, always moving from warmer to cooler areas. There are three primary methods by which heat energy is transmitted through the Earth: conduction, convection, and radiation (Fig. 1-8).

In *conduction,* minute particles such as atoms (discussed in Chapter 2) become energized by heat energy from an outside source until they vibrate rapidly, colliding with neighboring particles and setting them in motion, generating a chain reaction of vibration. However, considerable heat energy must be applied before the atoms in a rock will begin to vibrate enough to nudge their neighboring atoms. Thus, rocks are generally poor conductors of heat. In fact, if the internal heat produced during the Earth's creation had flowed by conduction alone, it would not yet have come to the surface.

When heat is transferred by *convection,* vibrating particles actually move from one place to another, carrying heat with them instead of vibrating in place and passing their heat only to neighboring particles. When the temperature in the Earth's interior became high enough to melt some of its components, heat began to be transported by moving particles of the fluid. Eventually this heat melted surrounding substances as well, and particles of hot low-density materials rose toward the surface, carrying heat with them. Convection is a much faster, more efficient way of transmitting heat than is conduction. The heat and molten material carried by this process probably caused the planet's first volcanic eruptions.

All heated objects radiate energy in one form or another. Energy transmitted by *radiation* moves in the form of one or more different types of electromagnetic waves, such as radio waves, microwaves, infrared waves, visible light waves, ultraviolet light waves, and X-rays. These forms of energy are then converted into heat energy when they strike and are absorbed by an object. In this way, a microwave oven's radiant energy is converted to heat energy to make a frozen pizza palatable, and light radiated by the sun is transformed into heat energy when it is absorbed by the Earth's atmosphere. Similarly, radioactive substances within the Earth emit different forms of electromagnetic radiation that are converted to heat energy when they warm surrounding rocks.

The Origin of the Earth's Crust As the Earth differentiated into its major concentric layers some upwelling material reached the surface, where it cooled and solidified, forming the Earth's earliest crust. Among these low-density substances were oxygen and silicon, which combined to form the silicate minerals that abound in the Earth's crust and upper mantle. Some heat-producing radioactive substances such as uranium and thorium also moved toward the surface; because of the heat radiating from these elements, crustal rocks are repeatedly remelted and reformed into the wide variety of rock types.

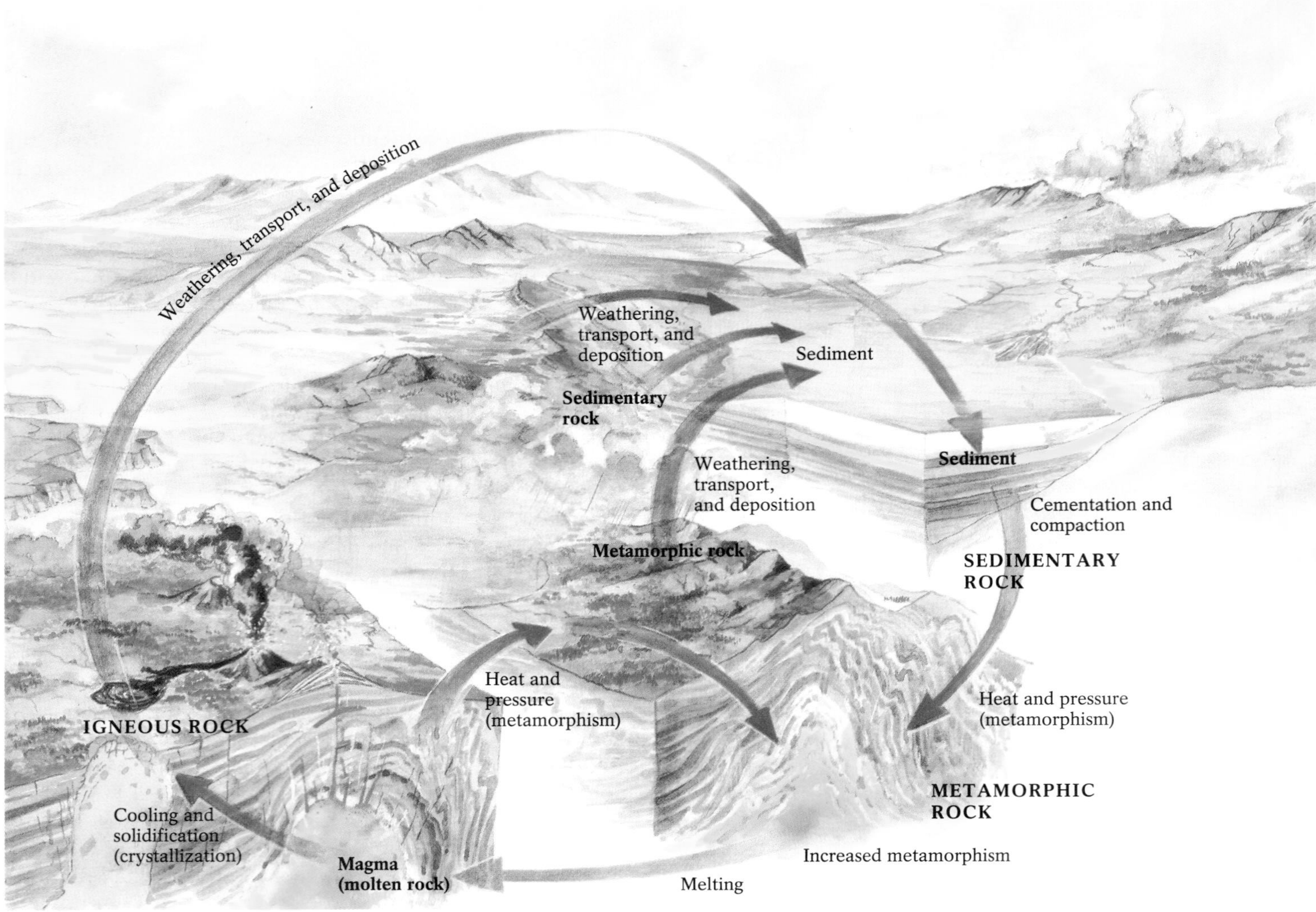

Figure 1-9 The rock cycle—an idealized scheme illustrating the variety of ways that the Earth's rocks may evolve into other types of rocks. For example, an igneous rock may weather away and its particles eventually become a sedimentary rock. The same igneous rock may remain buried deep beneath the Earth's surface, where heat and pressure might convert it into a metamorphic rock. The same igneous rock, if it is buried even deeper, may actually melt and form a new igneous rock once it has recooled and solidified. There is no prescribed sequence to the rock cycle. A given rock's evolution may be altered at any time by a change in the geological conditions around it.

Rock Types and the Rock Cycle

A **rock** is a naturally formed aggregate of inorganic materials that originated within the Earth. Three types of rocks exist in the Earth's crust and at its surface, each type reflecting a different process of origin. **Igneous rocks** are made of molten material from the Earth's interior that has cooled and solidified, either at or beneath the Earth's surface. **Sedimentary rocks** form when preexisting rocks are broken down into fragments that accumulate and become compacted or cemented together, or when materials dissolved from preexisting rocks are left behind as solid rock after water evaporates. **Metamorphic rocks** form when the chemical composition and structure of any type of rock are changed in the Earth's interior by heat, pressure, or chemical reactions with circulating fluids.

Over the great extent of geologic time and through the dynamism of Earth's processes, rocks of any one of these basic types may eventually evolve into either of the other types, or into a different form of the same type. Rocks of any type exposed at the Earth's surface can be worn away (or *weathered*) by rain, wind, crashing waves, flowing glaciers, or other means, and the resulting fragments carried (or *transported*) elsewhere to be deposited as new *sediment*; this sediment might eventually become new sedimantary rock. Sedimentary rocks may become buried so deeply in the Earth's hot interior that they may be changed into metamorphic rocks, or they may melt and eventually form igneous rocks. Under heat and pressure, igneous rocks can also become metamorphic rocks. The processes by which the various rock types can evolve and change over time are illustrated in the **rock cycle** (Fig. 1-9).

Time and Geology

The Earth is believed to be about 4.6 billion years old. Over so vast a period of time, even processes that operate at imperceptibly slow rates have changed the Earth's appearance dramatically (Fig. 1-10). Fossil evidence of ancient marine creatures in some rocks of the Grand Canyon, for example, show that these rocks, today near the top of a hot, dry plateau 2300 meters (7500 feet) above sea level, once lay at the bottom of an ocean; over millions of years, the mud at the bottom of this ocean—still containing the remains of sea creatures that died there—gradually solidified into rock, was uplifted to its present elevation, and was cut through and exposed by the Colorado River. As James Hutton knew, although the short-term effects of such processes go largely unrecognized by us, rocks bear witness to their long-term effects on the Earth's features.

Geologists think of time in both relative and absolute terms. *Relative dating* of rocks determines which rocks are older or younger than others by referring to spatial relationships between the rocks. For example, the *principle of superposition* states that where layers of sedimentary rocks have not been disturbed since deposition, younger rocks overlie older rocks. Rock-dating techniques developed during the twentieth century, based on the constant decay of radioactive elements, now enable geologists to specify the *absolute* ages of rocks in years. The oldest rocks yet found on Earth, near Yellowknife Lake in Canada's Northwest Territories, have been dated by the known decay rate of uranium at 3.96 billion years. (We will discuss the various dating methods in detail in Chapter 8.)

(a)

(b)

Figure 1-10 Two photos of the Grand Canyon, taken from the same perspective (looking down the Colorado River, one-half kilometer below Lees Ferry) about 100 years apart. Other than signs of encroaching vegetation, there is no perceptible geological difference between the earlier scene (**a**), photographed in 1873, and the later scene (**b**), photographed in 1972. Although geological processes in the arid Southwest are very slow on a human time scale, the Grand Canyon was formed by these very processes over the course of millions of years. Before gradually solidifying, becoming uplifted, and being incised by the Colorado River, the rocks of the canyon originated as sediment on the floor of an ancient inland sea.

If Earth's history were viewed as a great textbook, its rocks would be the pages on which most of the events are written. Almost everything we know about our past we know because it was preserved, in some way, in rock. Geologists initially divided Earth's history into four major *eras* of varying length and a dozen smaller *periods,* based largely on fossil evidence showing the existence of various key organisms in sedimentary rocks and on spatial relationships such as superposition. With the advent of modern rock-dating technology, however, the geologic time scale now includes absolute ages for its eras and periods (Fig. 1-11).

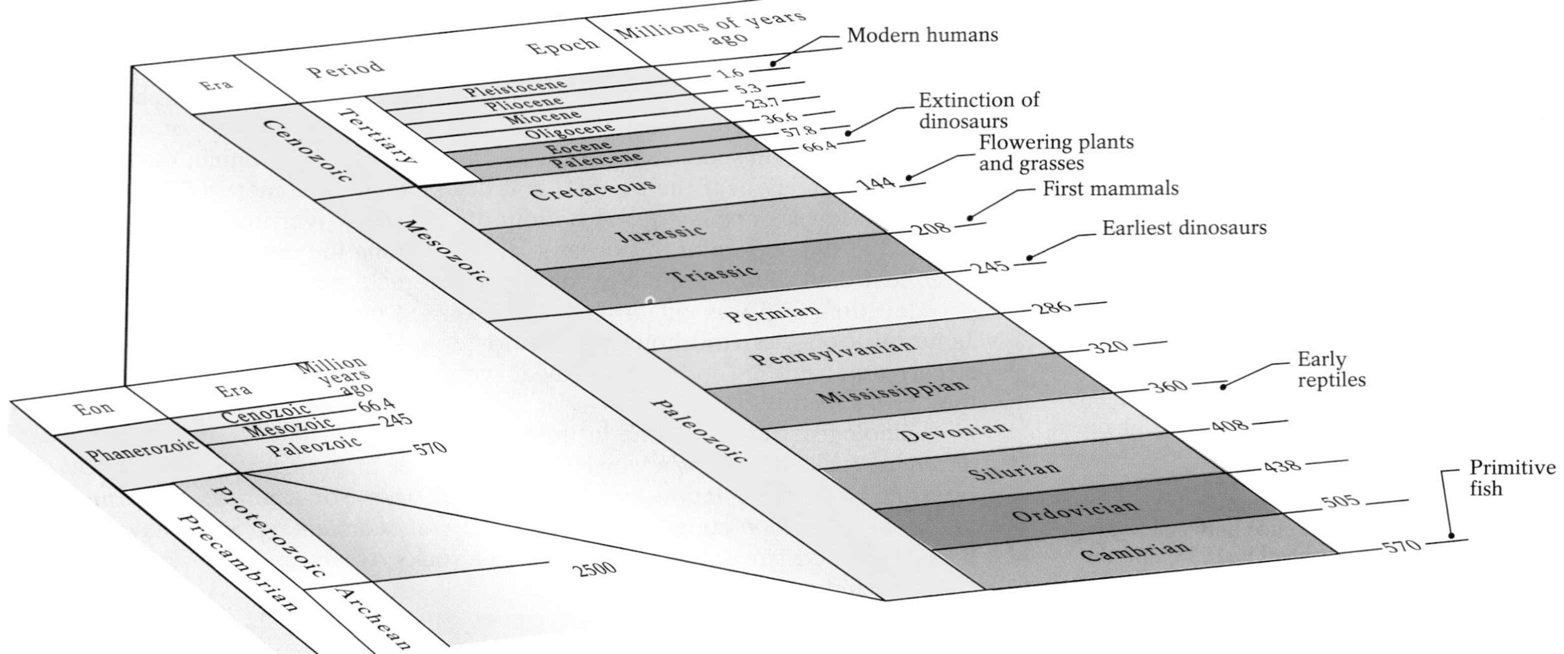

Figure 1-11 A simplified version of the geologic time scale. (See also Figure 8-27.) The different eras of the Phanerozoic Eon derive their names from the nature of the life forms associated with those eras: Paleozoic, "ancient life"; Mesozoic, "middle life"; Cenozoic, "recent life." The names of the various periods of the Paleozoic and Mesozoic Eras are taken primarily from places at which their rocks were originally identified. For example, the Devonian Period is named for the rocks at Devon, England. Note that the Precambrian Eon comprises about 87% of all of the Earth's history. The most recent era, the Cenozoic, is more finely divided because more rocks of this age are exposed and geologists have more precise ways to date them.

Plate Tectonics

Certain regions of the world are periodically devastated by earthquakes, while others are unscathed . . . a chain of volcanoes stretches along the west coast of North America from Alaska to Northern California, but no volcanic peaks can be found in such mid-continent locations as Minnesota, Iowa, or Missouri . . . the slopes of Ohio and Indiana are consistently shunned by skiers, who flock instead to Sun Valley, Idaho . . . Each of these examples highlights a definite pattern in the Earth's functioning—*it is not random.*

For centuries, geologists sought to explain such large-scale patterns with a multitude of independent hypotheses, each tailored to a specific location. For instance, geologists in the Swiss Alps proposed that mountain building occurs when rocks are pushed together, whereas those studying the Grand Tetons of Wyoming proposed that mountains form when rocks are pulled apart. Such hypotheses might have given reasonable pictures of small-scale or regional processes, but failed to explain the common underlying cause of *all* mountain ranges.

In the 1960s, an exciting new hypothesis called **plate tectonics** (from the Greek adjective *tektonikos*, or "built," as in archi*tect*ure) revolutionized our understanding of how the Earth functions. This hypothesis changed the way geologists view the world as dramatically as, a century earlier, the theory of evolution changed how biologists thought about living things. After only a few decades of observation and testing, the hypothesis of plate tectonics has become a widely accepted theory because it provides answers to questions that earlier hypotheses could not resolve. It has given us a way to understand processes such as mountain building, predict such potential catastrophes as earthquakes and volcanic eruptions, and find underground reservoirs of oil, natural gas, and precious metals. Finally, the plate tectonic theory enables us to fit our observations about the ancient past into the same conceptual framework as our understanding of the geological phenomena occurring today.

According to the theory of plate tectonics, the Earth's lithosphere consists of large rigid plates that move in response to the flow of the heat-softened asthenosphere beneath them. Over millions of years, plate movements have created the Earth's ocean basins, continents, and mountains, all of which essentially ride as passengers on the shifting plates.

Basic Plate Tectonic Concepts

The wide-ranging theory of plate tectonics comprises only four basic concepts.

1. The outer portion of the Earth—its crust and uppermost segment of mantle (i.e., its lithosphere)—is composed of rigid units called plates.
2. The plates move.
3. Most of the world's large-scale geological activity, such as earthquakes and volcanic eruptions, occurs at or near plate boundaries.
4. The interiors of plates are relatively quiet geologically, with far fewer and milder earthquakes than occur at plate boundaries and little volcanic activity.

Figure 1-12 shows the Earth's seven large, or major, plates and a number of its smaller ones. Note that the continents generally are not independent plates, but instead are usually parts of composite plates that contain both continental and oceanic segments. For example, the North American plate includes the North American continent and the adjacent western half of the Atlantic Ocean. Continental portions of plates are composed of thicker, lower-density lithosphere, whereas oceanic portions of plates are thinner, and of higher density. Plates are more than 120 kilometers (75 miles) thick in continental regions, but only about 80 kilometers (50 miles) thick at ocean basins.

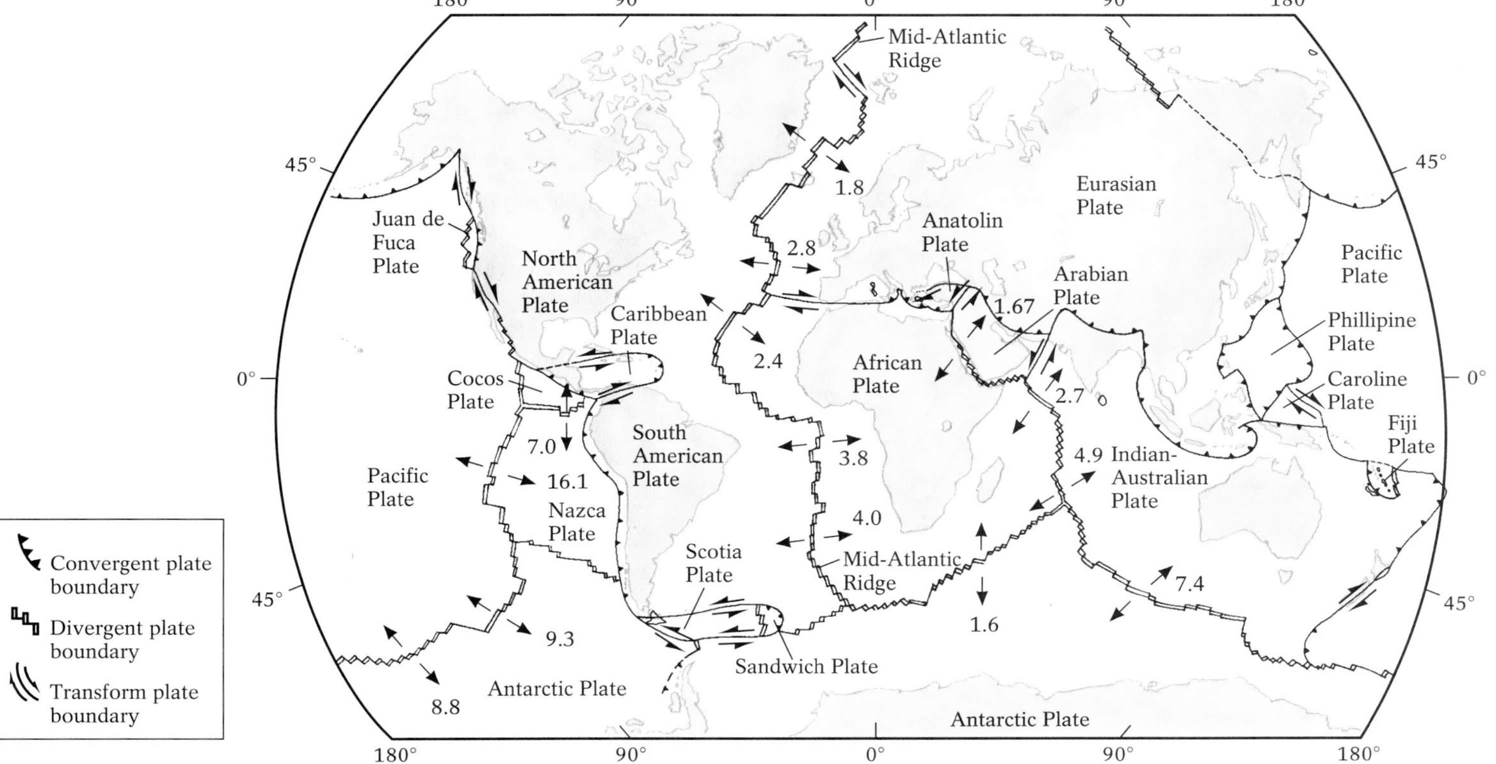

Figure 1-12 A world map showing the Earth's tectonic plates. The arrows show the current direction of plate motion. Note that some plates, such as the North American plate, are composed of both continental and oceanic lithosphere. The Pacific plate is made up almost exclusively of oceanic lithosphere.

As the Earth's plates move, everything on them, even features as large as continents and oceans, moves with them. As you read these words, most of you in North America are moving westward at about 4 centimeters (1½ inches) per year. (Those in Los Angeles and the surrounding communities of westernmost California live on a different plate, the Pacific plate, which is headed northward.) The North American plate is moving westward away from the Eurasian plate, its neighbor to the east, while the Eurasian plate is moving eastward. The increasing distance between cities on these two plates, however, will not cause airfares between Baltimore and London to increase—the length of a trans-Atlantic flight is extended by only about 5 to 10 centimeters (2–4 inches) per year as a result of plate movements. (This is about two times the rate at which your toenails grow.) If Columbus were crossing the Atlantic today, 500 years after his famous voyage, he would have to sail only an extra 30 to 50 meters (100–160 feet) to reach shore.

Although this rate of plate motion may seem insignificant, over the vast course of geologic time it can have large consequences. In just a few tens of millions of years, a thousand extra kilometers of Atlantic Ocean have been created. This does not mean that the Earth's overall size is increasing—our planet's size, it is generally believed, has remained essentially constant since its early formative years. Thus, if plates move apart at certain boundaries, they must converge at others.

When moving plates come in contact with one another, heat and pressure build at their boundaries. Depending on the direction of their movement, plate edges may break and generate earthquakes, buckle to form mountains, melt and erupt volcanically, or do all three. In fact, the locations of earthquakes and volcanoes are used to map the outlines of plates in a kind of geological connect-the-dots game (Fig. 1-13).

Plate interiors are generally placid geologically, typically unaffected by distant plate-edge activity. San Francisco has frequent earthquakes because it is located at a moving plate boundary, whereas Chicago has very few because it is located within a relatively inactive plate interior.

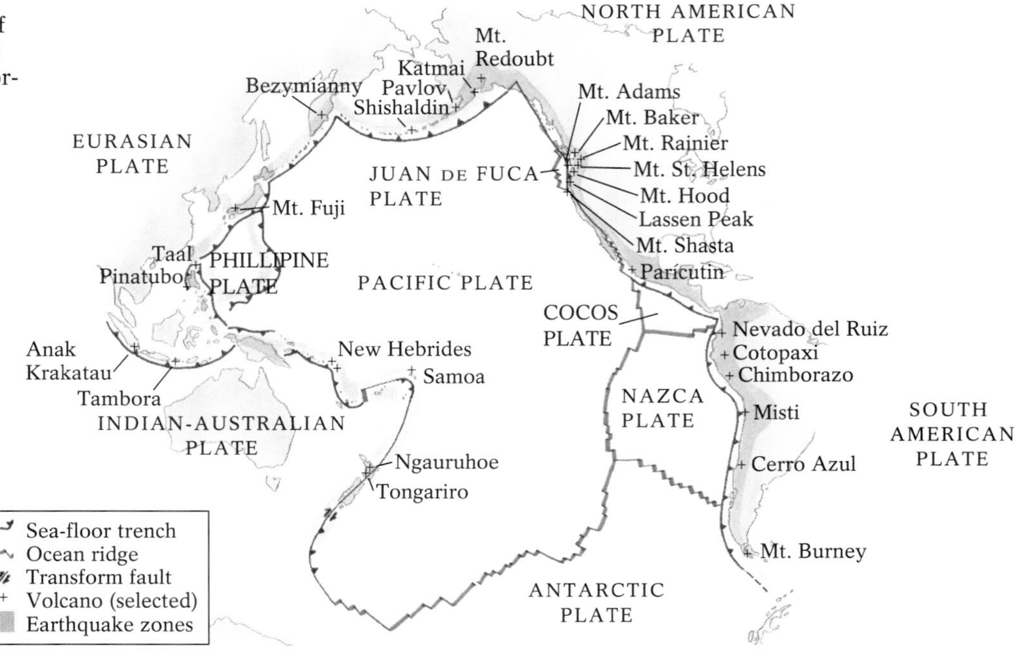

Figure 1-13 A map of the Pacific Ocean and surrounding continents, indicating the sites of earthquakes and volcanic eruptions recorded over the last 100 years or so. Note the proportion of volcanoes and earthquakes that have occurred at plate boundaries.

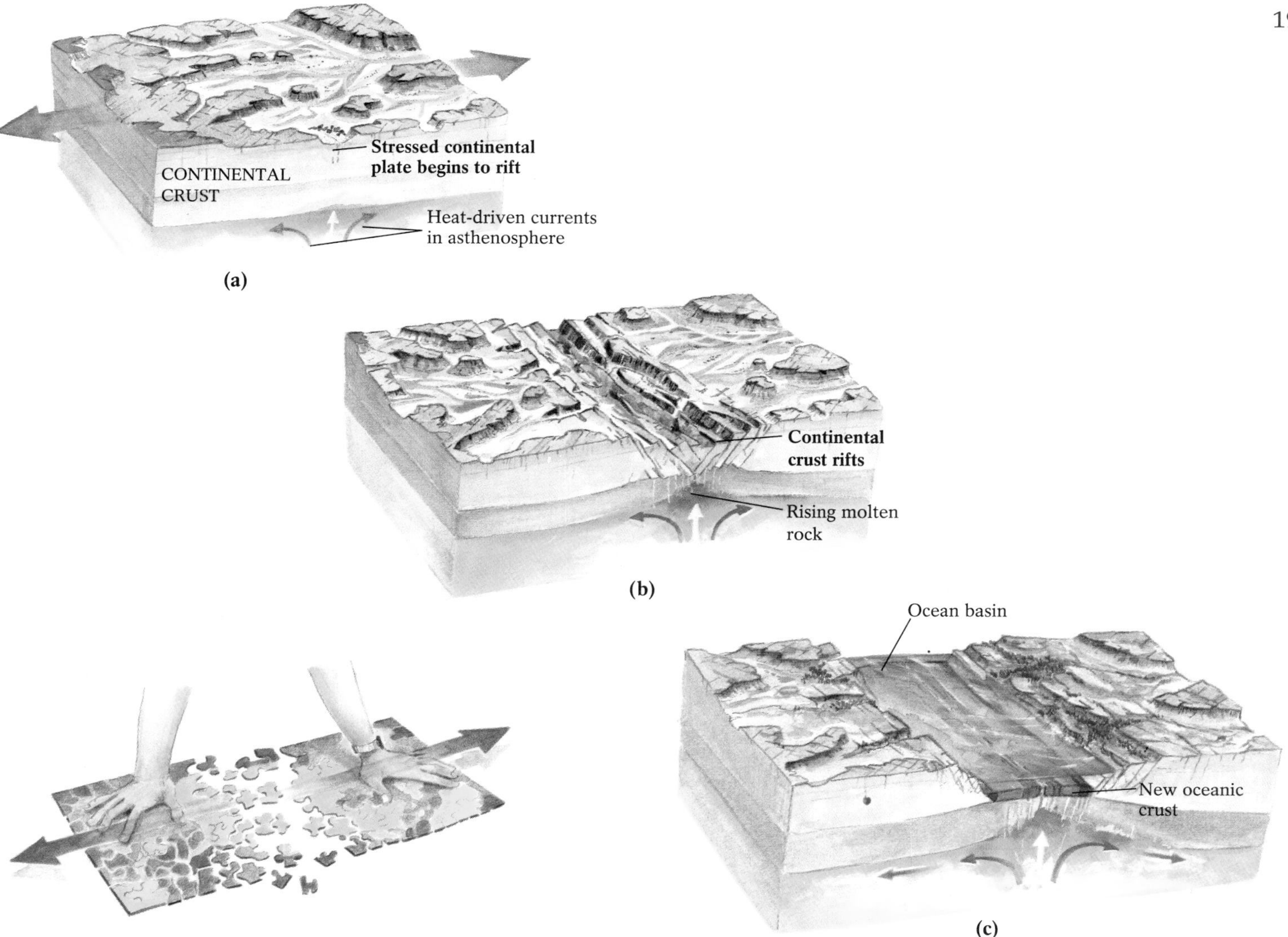

Figure 1-14 Plate rifting and divergence. When currents in the underlying asthenosphere pull one of the Earth's plates in opposite directions (**a**), the plate is stressed and eventually rifts (**b**), much as a jigsaw puzzle being pulled in opposite directions will break apart. As the plate fragments continue to move (diverge) farther from one another, molten rock from the mantle rises into the gap and solidifies along the edges of the plates (**c**), forming new oceanic crust which is eventually covered with water to form a new ocean basin.

Plate Movements and Boundaries

The Earth's plates move in several ways, and plate boundaries are categorized according to which type of movement they demonstrate. There are three major types of plate movements and corresponding boundaries: divergent plate boundaries, where plates move apart; convergent plate boundaries, where plates move together; and transform plate boundaries, where plates move past one another in opposite directions.

Rifting and Divergent Plate Boundaries Plate interiors may become geologically active if slow-flowing currents in the Earth's asthenosphere generate a pulling-apart motion that tears a preexisting plate into two or more smaller plates. This process, discussed in detail in Chapter 12, is known as **rifting.** The Great Rift Valley of East Africa, where the African plate has been coming apart, is a prime example of early rifting. If this rifting continues, there may be two Africas on the world map in the very distant future. Farther north, such rifting completely separated the Arabian plate from the African plate over the last 20 million years, forming the crustal depression occupied by the Red Sea and the Gulf of Aden; as their complementary coastlines illustrate (see Fig. 1-12), these two plates were once attached.

Once a plate has been rifted, the resulting smaller plates may continue separating from one another by a type of plate motion described as **divergence** (Fig. 1-14). Divergence proceeds, typically at a rate of about 1 to 9 centimeters (0.5–4 inches) per year, as molten rock rises into the thousands of fractures between the rifted plates. The molten rock cools and solidifies, becoming attached to the edges of the rifted plates. Meanwhile divergence

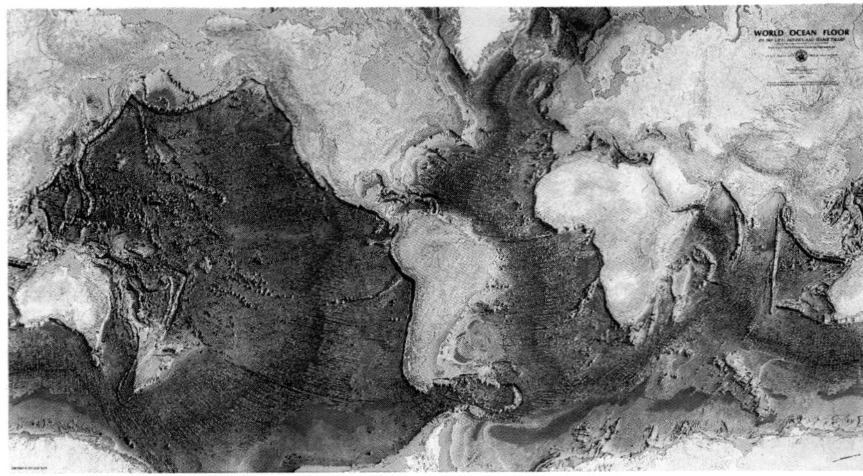

Figure 1-15 A map of the Earth's ocean basins (based on the studies of sea-floor mapping pioneers Marie Tharp and Bruce Heezen), showing sea-floor spreading at mid-ocean ridges. These mark the positions of the world's divergent plate boundaries, particularly those in the Atlantic and eastern Pacific Oceans.

continues, further separating the older rifted segments and eventually forming an ocean basin. The new ocean basin fills with seawater as further rifting opens new connections to other oceans, as well as by the addition of water runoff from land-based streams.

Throughout the period of divergence, ocean basins continue to expand as molten rock erupts, creating new oceanic lithosphere that builds up to form the *mid-ocean ridge*. The mid-ocean ridge is a continuous chain of submarine mountains that bisects the world's oceans, meandering around the globe like the stitches on a baseball. The process of plate growth at mid-ocean ridges is known as **sea-floor spreading** (Fig. 1-15). If the young Red Sea between the African and Arabian plates continues to grow at its present rate, it may someday become a full-blown ocean like the Atlantic or Pacific.

Plate Convergence and Subducting Boundaries Plates move toward each other in a phenomenon known as **convergence**. Plate convergence may involve two continental plates, two oceanic plates, or one of each. Most often, when one oceanic plate and one continental plate or two oceanic plates converge, the denser of the two plates sinks beneath the other and is reabsorbed into the Earth's interior, a process known as **subduction** (Fig. 1-16). Because plates of

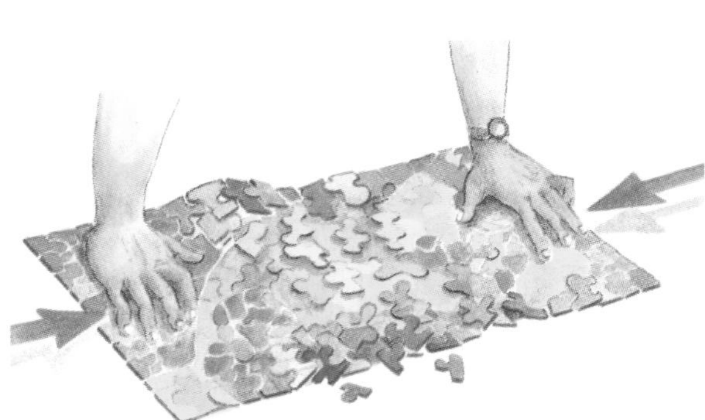

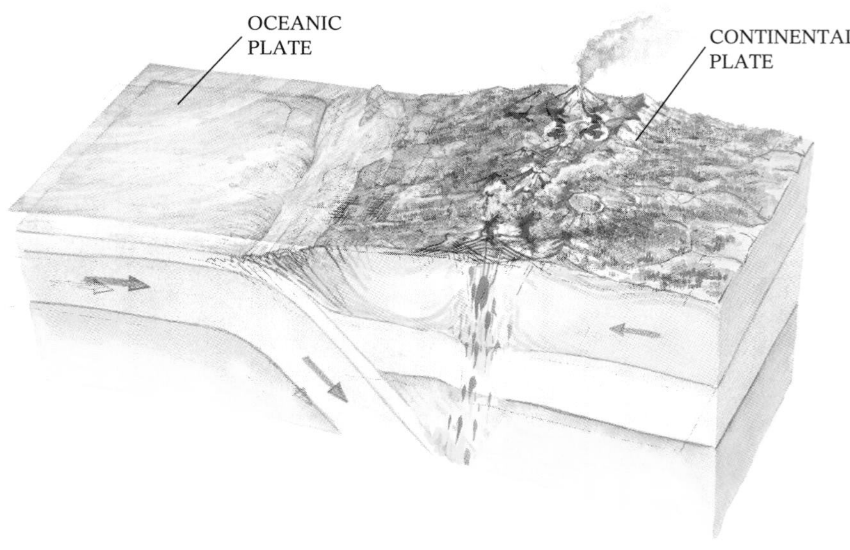

Figure 1-16 Oceanic plate subduction. Like a jigsaw puzzle that buckles from having its edges forced together, converging plate boundaries push against one another and crumble, with one plate often sinking, or subducting, below the other.

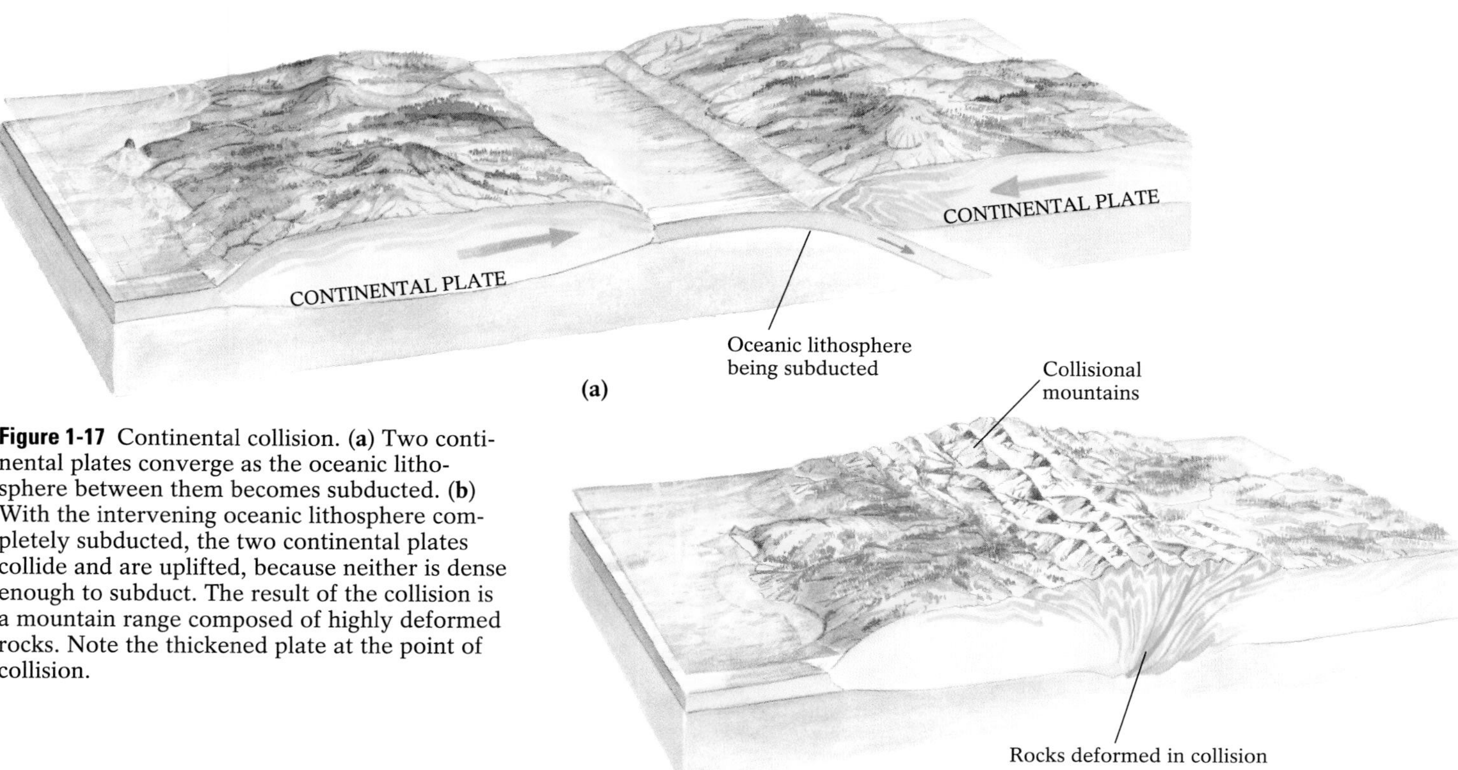

Figure 1-17 Continental collision. (**a**) Two continental plates converge as the oceanic lithosphere between them becomes subducted. (**b**) With the intervening oceanic lithosphere completely subducted, the two continental plates collide and are uplifted, because neither is dense enough to subduct. The result of the collision is a mountain range composed of highly deformed rocks. Note the thickened plate at the point of collision.

oceanic lithosphere are always denser than those of continental lithosphere, when these two types of plates converge, the oceanic plate always subducts. Some oceanic plates, however, are more dense than others; thus, when two oceanic plates converge, the denser one subducts beneath the less-dense one.

Unlike oceanic plates, all continental plates are too buoyant to subduct into the denser underlying mantle—when two continental plates converge, the only place either plate edge can go is up. Such convergence of continental plates, called **continental collision** (Fig. 1-17), has created many of the Earth's largest mountain ranges. Northern Africa's Atlas Mountains and southern Europe's Alps were formed by past collisions of the African and Eurasian plates, the Himalayas from the ongoing collision of the Indian and Eurasian plates, and North America's Appalachians from a three-way collision of the African, Eurasian, and North American plates that took place between about 400 and 250 million years ago.

The convergence of two oceanic plates is often accompanied by earthquakes, caused by the friction of the plates colliding, and the growth of volcano chains. The volcano chains, which are fed largely by molten material produced as the subducting plate sinks deep into the Earth's mantle and is melted, initially form under water. Eventually, however, after millions of years, the volcanoes grow high enough above the sea floor to emerge as a chain of volcanic islands. This has happened in the northern Pacific, where the Pacific plate descends beneath the oceanic edges of the North American plate to form the earthquake-wracked volcanic Aleutian Islands of Alaska. The 1990 eruption of Mount Redoubt in Alaska is evidence of ongoing subduction in this region (Fig. 1-18a).

Mount Redoubt, Alaska

Mount St. Helens, Washington

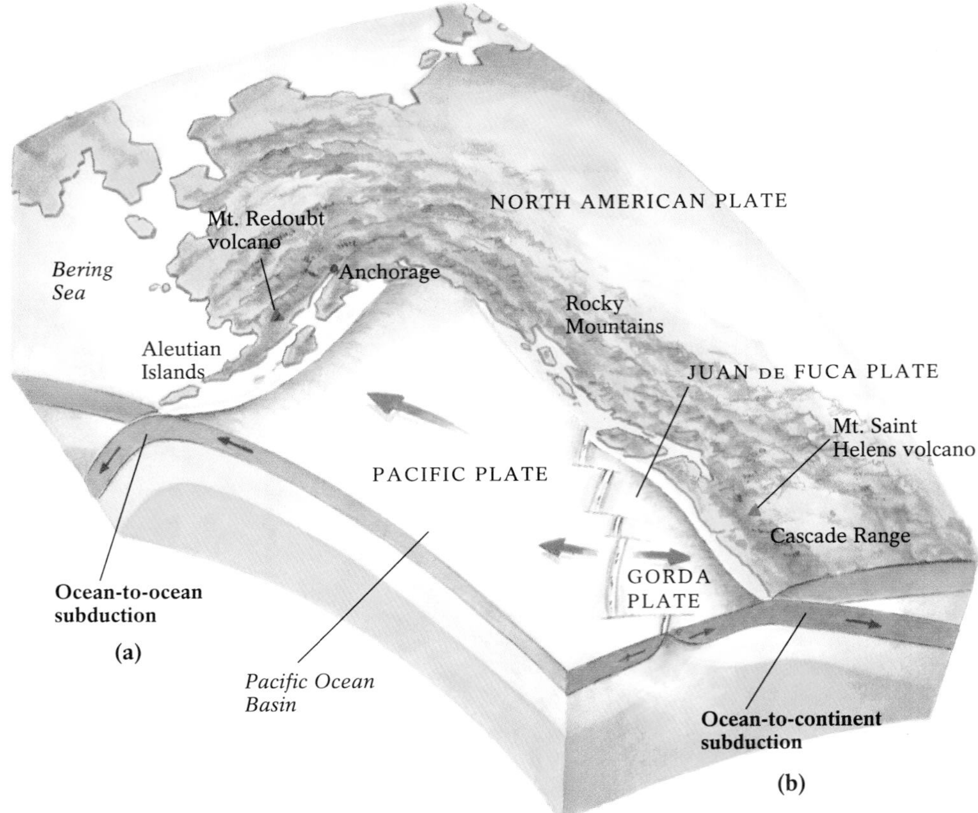

Figure 1-18 (**a**) Subduction between two oceanic plates. The northern portion of the oceanic Pacific plate is currently subducting beneath the oceanic portion of the North American plate that underlies the Bering Sea. Alaska's Mount Redoubt, a product of this subduction, has been disrupting the lives of Anchorage residents intermittently since 1990. (**b**) Subduction between an oceanic plate and a continental plate. Along the coasts of southern British Columbia, Washington, Oregon, and northern California, several small plates of the Pacific Ocean basin are subducting beneath the continental edge of the North American plate. Mount St. Helens, in the subduction-produced Cascade Range mountains of Washington, has erupted periodically since March 1980.

When an oceanic plate subducts beneath a continental plate, earthquakes may also occur and a chain of volcanoes may also grow—but on land. For example, at the western shores of North America, the edges of the oceanic Juan de Fuca and Gorda plates are subducting beneath the advancing continental edge of the lighter North American plate (Fig. 1-18b). As the subducting plates descend to warmer depths, they melt, forming the molten material that fuels the chain of volcanoes of the Cascade Range, which stretches from southern British Columbia to northern California.

Subduction rates can be as high as 15 to 20 centimeters per year (about 6–10 inches/year). This is considerably faster than the average rate of divergence, which might lead us to expect that, over time, oceanic plates would eventually disappear and the Earth itself would shrink. This does not happen, however, because less of the Earth's surface is devoted to subduction than to divergence. As maps of the sea floor show, only about 40,000 kilometers (25,000 miles) are subducting while about 65,000 kilometers (40,000 miles) of the world's plate boundaries are diverging. It is this balance between the creation and the destruction of plates that maintains the Earth's size.

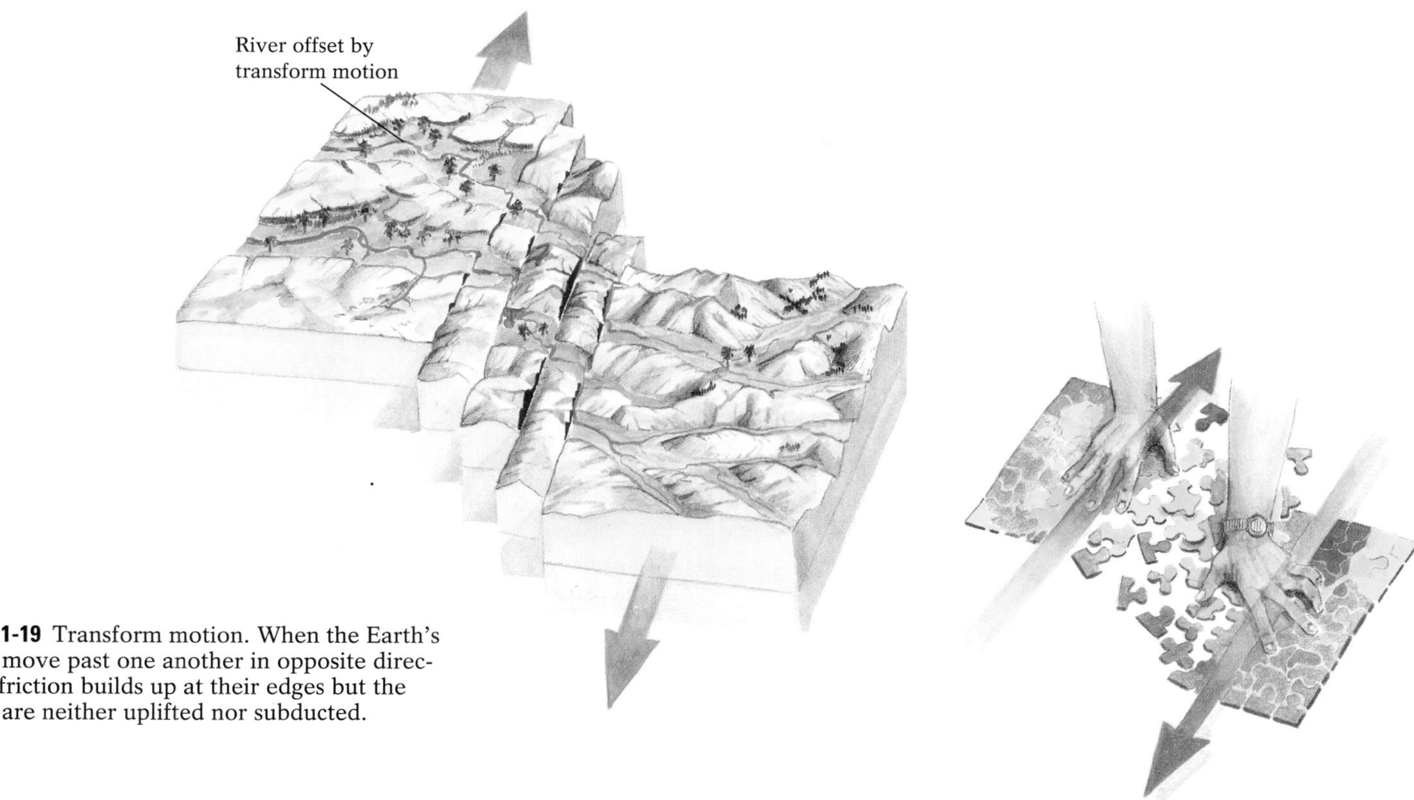

Figure 1-19 Transform motion. When the Earth's plates move past one another in opposite directions, friction builds up at their edges but the plates are neither uplifted nor subducted.

Transform Motion and Transform Plate Boundaries The third major type of plate boundary occurs where two plates, either oceanic or continental, move past one another in opposite directions, a process known as **transform motion** (Fig. 1-19). At transform boundaries, plates neither grow (as at divergent boundaries) nor shrink (as at convergent boundaries). Because no mantle material wells upward, and no plates are subducting to melt in the Earth's interior, transform boundaries do not produce volcanoes. Because great friction results as the moving plates grind past each other, however, these boundaries do produce earthquakes, and any community caught between two plates at a transform boundary may be periodically devastated. Such is the case with San Francisco and Los Angeles, which are located within the San Andreas transform zone between the North American and Pacific plates. Any feature, natural or human-made, that traverses a transform boundary will be broken and displaced by this type of plate motion (Fig. 1-20).

Figure 1-20 An orange grove along the San Andreas transform plate boundary, with its tree rows offset by an earlier earthquake there. Since any feature that cuts across an active transform plate boundary will be broken and displaced by plate movement (as are the rivers in Figure 1-19), precise planting in orchards is wasted effort in the neighborhood of a transform boundary.

Proving Plate Tectonics: A Theory Develops

In 1782, the American scientist–philosopher–statesman Benjamin Franklin hypothesized: "The crust of the Earth must be a shell floating on a fluid interior. Thus the surface of the globe would be capable of being broken and disordered by the violent movements of the fluids on which it rested." Almost 200 years later, geologists would accept that remarkable insight by one of history's finest scientific thinkers as the key concept of the theory of plate tectonics. But how was the comprehensive plate tectonic theory developed? What evidence supports a theory that evokes images of opening and closing oceans and movable continents? And what causes all this to happen?

Alfred Wegener and Continental Drift

The first conception of the revolutionary theory of plate tectonics accompanied by supporting evidence was proposed by the German geophysicist–meteorologist Alfred Wegener (1880–1930). Despite blistering ridicule from the leading geologists of his day, Wegener proposed that the continents float on the denser underlying interior of the Earth, and periodically break up and drift apart. He asserted that all of the Earth's continents had been joined together about 200 million years ago as a supercontinent he called *Pangaea* ("all lands") (Fig. 1-21). Wegener hypothesized that Pangaea had covered about 40% of the Earth's surface, most of it in the southern hemisphere. While Pangaea existed, what is now the area of New York City sweltered in the lush tropics near the equator, and most of Eastern Africa and India shivered under a dome of glacial ice near the South Pole. Pangaea was surrounded by a single ocean, which Wegener called *Panthalassa* (named for the Greek goddess of the sea). Over a vast period of time, Pangaea broke up, forming a number of continents that migrated to all regions of the globe. Wegener called this dispersal **continental drift**.

What evidence led Wegener's controversial hypothesis of continental drift to evolve into the widely accepted theory of plate tectonics? Our discussion begins with Wegener's observations on the shapes of continental margins, patterns of present-day animal life, similarities among far-distant fossils and rocks, and evidence of past climates that are inconsistent with their modern locations. It concludes with more recent, technologically enhanced studies establishing the ages of ocean-floor rocks, which provide continuing support for the plate tectonic theory.

Figure 1-21 A reconstruction of the proposed supercontinent Pangaea, with present-day coastlines and continent names shown for reference. This configuration of plates dates from about 225 million years ago.

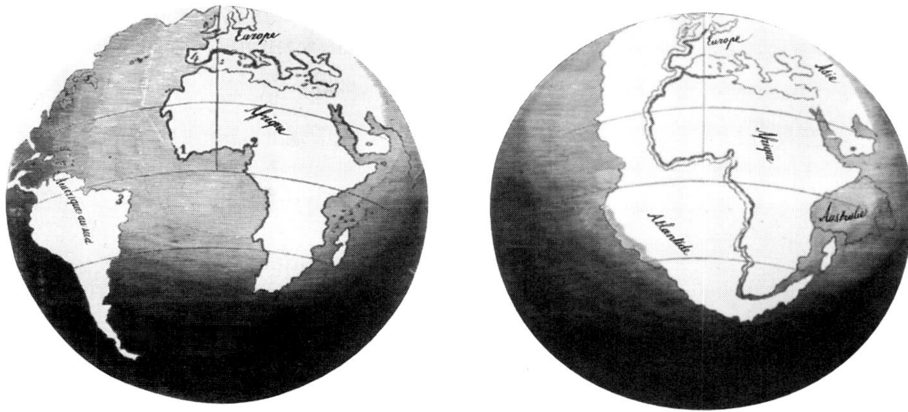

Figure 1-22 An early observation of the continental fit of South America and Africa, which was first proposed by Sir Francis Bacon in 1620. French naturalist Antonio Snider-Pelligrini sketched this diagram in 1858 for his work, *La Création et ses Mystères Dévoiles*, which suggested that Noah's Deluge was responsible for the movement of the continents.

Continental Fit The English philosopher Sir Francis Bacon was the first to note that the outlines of the continents of the world could be pieced together jigsaw-puzzle style. Bacon wrote of this in 1620, soon after seeing the new maps drafted following the global explorations of the sixteenth century, and the concept reappeared periodically for centuries following (Fig. 1-22). Almost 300 years later, Wegener tinkered with the puzzle of continental shapes until he fit them together, forming a model of the landmass he called Pangaea. Today, precise computer fitting has confirmed Bacon's and Wegener's hypotheses, and we can see that the reunited continents would indeed fit together remarkably well. For example, when the borders of South America and Africa are juxtaposed, South America tucks perfectly into the niche in the western coast of Africa (Fig. 1-23).

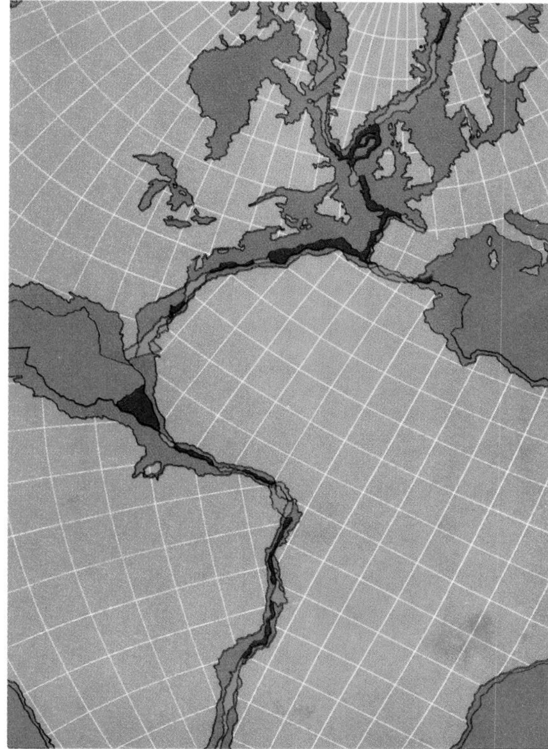

Figure 1-23 Precise matching of the continental shelves of circum-Atlantic continents by computer analysis. (From "The Origin of the Oceans," by Sir Edward Bullard, published 1969. Copyright © 1969 by *Scientific American, Inc.* All rights reserved.) Because the seaward (below sea level) edges of the continental shelves, shaded dark brown, represent the true edges of the continents, their fitting is more precise than simple shoreline fitting. The only places at which the plates appear to overlap, shaded red, are marked by geologic materials that have been deposited *since* the breakup of Pangaea. The precision of this fitting is one of the compelling lines of evidence that led to the wide acceptance of the theory of plate tectonics.

Habitats of Living Animals Studies of the distribution patterns of certain modern animals helped to convince Wegener that now-separated landmasses were once united into a supercontinent. He noted, for example, that the hippopotamus is today found only in Africa and 500 kilometers (300 miles) to the east on the island of Madagascar. How did a creature that is clearly not built for long-distance, open-ocean swimming get from the mainland to the island? Could a landbridge have once existed between Africa and Madagascar in the Mozambique channel between these two landmasses? Oceanographic surveys conducted during the 1940s found no evidence of such a feature. Wegener considered it likely that there was once a single landmass that divided, leaving ancestral hippos on both parts.

Wegener proposed a similar explanation for the unusual wildlife native to Australia, the only continent with (for example) kangaroos, wallabies, and koalas. This continent was once part of a large southern-hemisphere landmass that rifted about 40 million years ago. Drifting alone on their island continent, isolated from the life forms on the larger landmass from which they had separated, Australia's fauna began to evolve along their own distinctive path, gradually becoming the unique animals they are today.

Habitats of Ancient Animals Wegener also examined the fossil record for rare occurrences of past life forms (Fig. 1-24). Fossils of *Mesosaurus*, a small reptile that lived 240 million years ago, have been found only in Brazil and South Africa, which are separated today by 5000 kilometers (3000 miles) of

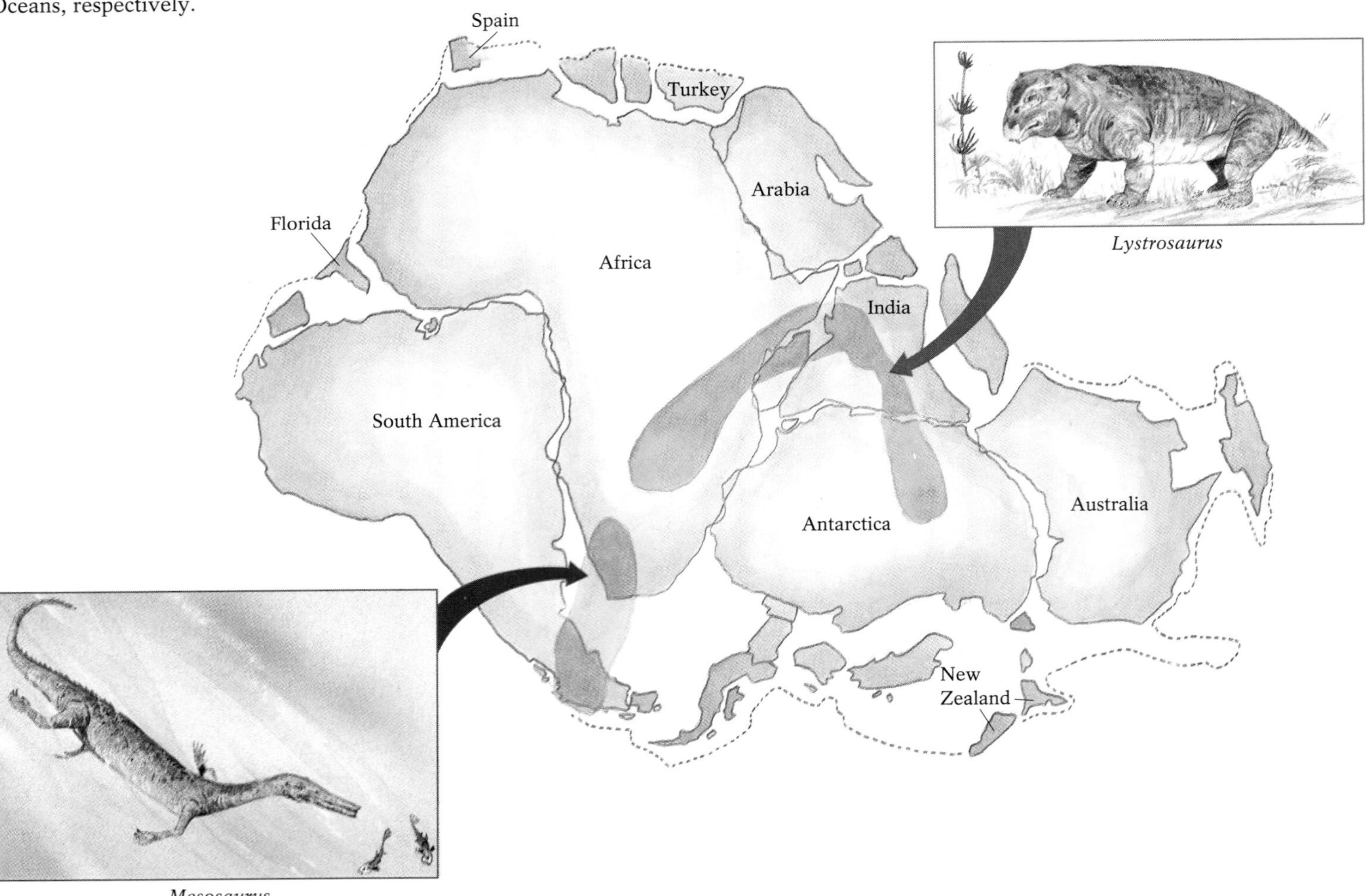

Figure 1-24 The distribution of *Mesosaurus* and *Lystrosaurus* on southern-hemisphere continents. Fossils of these reptiles date from the Triassic Period (208–245 million years ago), a time when the southern-hemisphere continents were joined together as part of Pangaea. The less-than-sleek *Lystrosaurus*, a sheep-sized reptile that lived about 225 million years ago in what is now Antarctica, Africa, Madagascar, and India, and *Mesosaurus* (discussed in the text), neither of which were long-distance swimmers, simply strolled to the spots where we find their remains today. That was, of course, long before these once-contiguous lands rifted and diverged, leaving spreading sea floors that were to become the Indian and the southern Atlantic Oceans, respectively.

the southern Atlantic Ocean. The skeletal structure of *Mesosaurus* and the composition of the deposits in which its fossil remains have been found show that it paddled around in shallow lakes and estuaries (the part of a river that empties into an ocean) and did not swim in open oceans. Paleontologists conclude from this that *Mesosaurus* lived on what was then a single landmass, and simply traveled overland between the sites of what are now Africa and South America before the landmass rifted. A similar explanation accounts for the occurrence in Antarctica, Africa, Madagascar, and India of the fossil remains of another non-ocean-swimming prehistoric creature, *Lystrosaurus*.

Related Rocks In the northern hemisphere, the 390-million-year-old rocks of the mountains of eastern North America are remarkably similar in mineral composition, structure, and fossil content to the same-aged rocks of eastern Greenland, western Europe and western Africa (Fig. 1-25a). Wegener recognized that if North America, Africa, and Europe had been joined in the past, there would have been a continuous chain of mountains from Alabama all the way to Scandinavia (Fig. 1-25b). It might seem that the island of Iceland would have gotten in the way; we now know, however, that Iceland, located atop the diverging mid-Atlantic ridge and only about 20 million years old, originated long after Pangaea broke up to form the separate continents of Europe and North America.

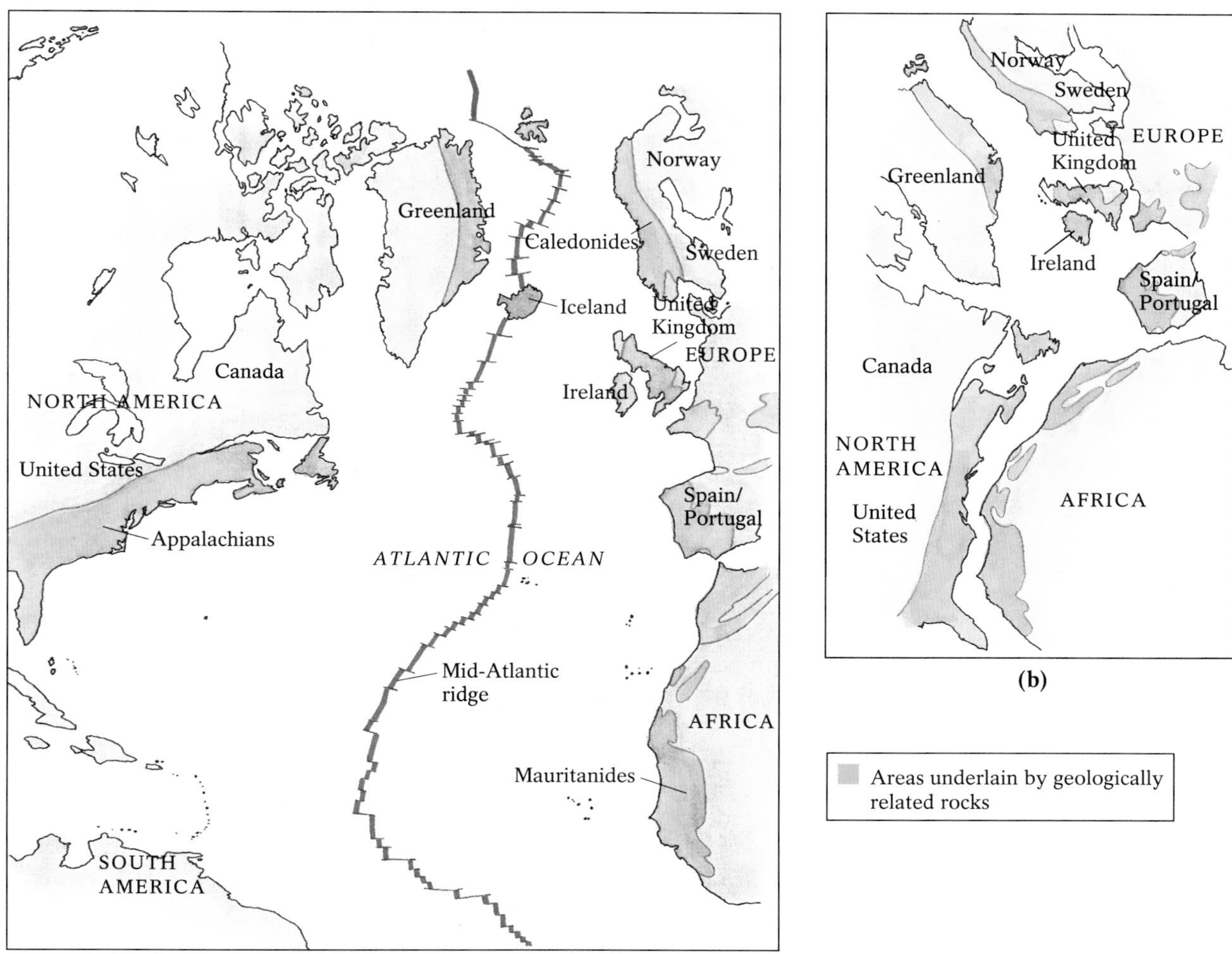

Figure 1-25 The current positions of the northern-hemisphere continents surrounding the Atlantic Ocean (**a**), and their pre-rifting positions as part of Pangaea (**b**). Note the absence of Iceland in the Pangaea reconstruction—the origin of Iceland from volcanic eruptions at the mid-Atlantic ridge post-dates the existence of Pangaea.

Ancient Climates In his attempt to prove that the continents drifted, Wegener also pointed out that geological evidence often suggests a past climate very different from the current climate of a given location. As glaciers slowly flow across a landscape, they pick up rocks, boulders, and sand grains and, as they melt, deposit a jumble of debris of all sizes (Fig. 1-26). The bouldery material shown in Figure 1-26b is glacial debris found on the west coast of South Africa, the presence of which tells us that South Africa was glaciated about 225 million years ago (the estimated age of that deposit). An icy past for this region would be plausible only if South Africa once occupied the space close to the South Pole now claimed by the continent of Antarctica.

(a)

(b)

Figure 1-26 (**a**) Glacial debris at the margin of the Columbia Icefield in Jasper National Park, Alberta, Canada. The debris is in the foreground and on the slope left of the ice flow. (**b**) Glacial debris from the Dwyka area, Cape Province, South Africa.

As further evidence of continental drift, a distinctive pattern of scratches that is found today in the ancient bedrock of such warm places as India, Australia, and South America, typical of the striations produced by glaciers as they drag debris along, indicates that these lands, too, were once located in colder climes (Fig. 1-27).

Similar reasoning explains the discovery of coal in presently cold climates. Coal forms after a great accumulation of vegetation has been buried, compressed, and heated in a warm, moist environment. So, upon learning that a cliff in arctic Spitsbergen, Norway, contains a seam of coal surrounded by kilometers of ice, we can infer that the coal must have formed in a very different climate some 300 million years ago and then journeyed northward a great distance.

The Age of Ocean Rocks Over the past several decades, geologists and oceanographers have used late-twentieth-century technology to add to Wegener's body of evidence in support of continental drift. They have crisscrossed the world's oceans with drilling rigs, taking samples of material from numerous locations on the sea floor. The igneous rocks they have collected from directly atop the mid-ocean ridges are always the youngest in their samples, whereas rocks collected near continents are always the oldest. The sediments covering the older ocean rocks are also older and usually form a thicker layer than those over the younger rocks, indicating that they have accumulated over a longer period of time. These findings are exactly what we would expect if oceanic plates grow at divergent plate boundaries.

(a)

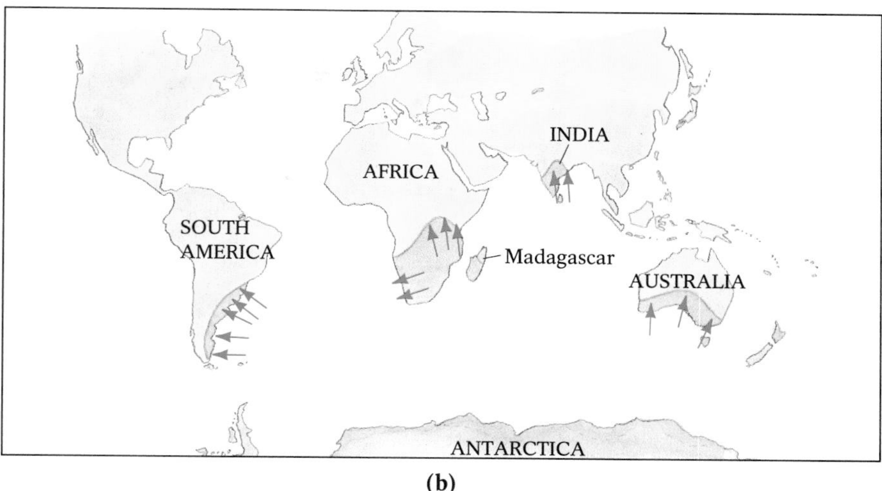

(b)

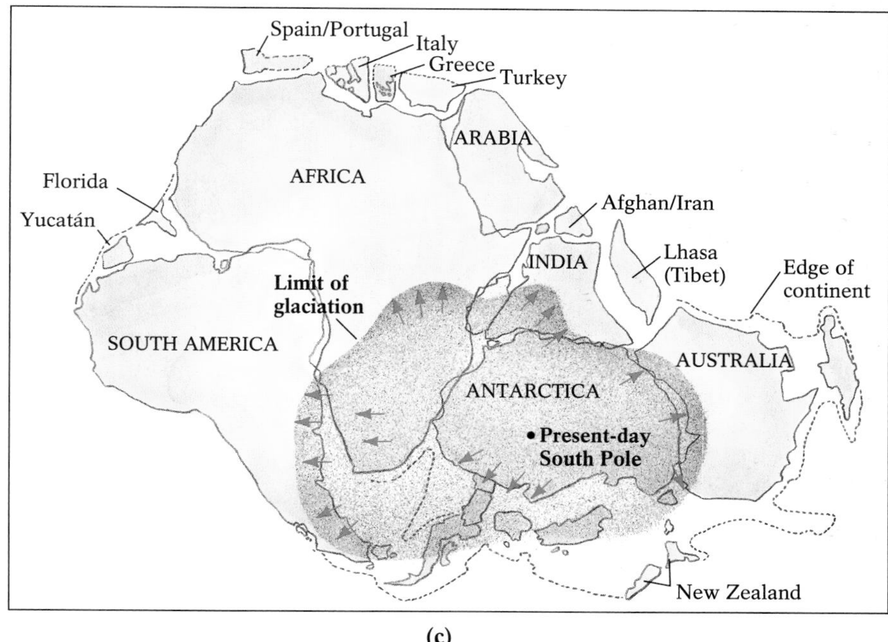

(c)

Figure 1-27 (**a**) Glacial striations in bedrock, in Victoria, British Columbia. (**b**) Bedrock striation patterns as they appear today on separated southern-hemisphere continents: The striations are oriented in what appears to be a random pattern. (**c**) Striation patterns as they would appear on Pangaea, with "reunited" southern-hemisphere continents occupying the South Pole: The striations form a systematic pattern resembling the spokes of a bicycle wheel, a pattern that typifies the outward flow of a large glacier.

The Driving Force Behind Plate Motion

With such strong evidence to support it, why was Wegener's hypothesis of continental drift initially rejected? The best answer is that Wegener's proposal that the continents, driven by the Earth's rotation, plowed through denser oceanic rocks did not seem physically plausible. At the time, nobody knew of the existence of lithospheric plates and the heat-softened underlying asthenosphere, and consequently Wegener was unable to propose a scientifically acceptable mechanism for continental drift. That mechanism would become clear only later, and the scientific community of Wegener's day could not accept so radical an idea as contesting the long-held notion of immovable continents and ancient, featureless sea floors. In November of 1928, only two years before he died, Wegener endured his final rejection when a meeting of esteemed American geologists concluded, "if we are to believe Wegener's hypothesis, we must forget everything which has been learned in the last 70 years and start all over again."

That's where Wegener's hypothesis of continental drift stood for the next three decades. With the advent of deep-sea drilling and the development of other ways to study the Earth's interior (discussed in Chapter 11), scientists discovered the existence of the lithospheric plates and identified several ways in which they might move. During the 1960s and 1970s, a great deal of evidence appeared in support of Wegener's lines of reasoning, enhancing the acceptability of the Pangaea hypothesis and replacing the notion of continental drift with the theory of plate tectonics. Finally, in the 1980s and since, new research technologies have been increasing our knowledge of the physics of the Earth's interior, enabling geologists to propose convincing hypotheses to explain what drives plate tectonics.

Geologists now believe that heat-driven currents in the Earth's mantle are responsible for plate movements. These currents, known as **convection cells**, develop when portions of a heated substance become less dense and rise toward the surface, displaced by cooler (more dense) portions, which are pulled down by gravity (Fig. 1-28). As convection cells move heated mantle rocks toward the surface in this way, the convecting mantle material encounters the solid lithosphere. Unable to reach the surface, the rising material moves laterally beneath the lithosphere, dragging it along as it flows. Sometimes the drag produced by neighboring convection cells pulling the lithosphere in opposite directions is great enough to cause the lithosphere to rift, and the rifted plates are then carried along in opposite directions by the slowly convecting currents. A relatively small amount of mantle-derived material does escape to the surface at such divergent plate boundaries, gradually cooling to form the new outer boundaries of the plates. The plates continue to diverge from the site of the rising warm mantle over millions of years, becoming cooler and more dense. Eventually, their non-diverging boundaries, now far removed from the spreading mid-ocean ridge, become dense enough to sink back into the Earth's interior at a subduction zone, where they will be reheated, eventually to reenter the cycle.

Some geologists believe that convection alone drives plate tectonic processes. Others believe that gravity assists convection, literally dragging the plates, or *slabs*, down into the interior at subduction zones. This process is known as *slab pull.* Heat-driven divergence and gravity-driven subduction, then, may work together to cycle the Earth's plates.

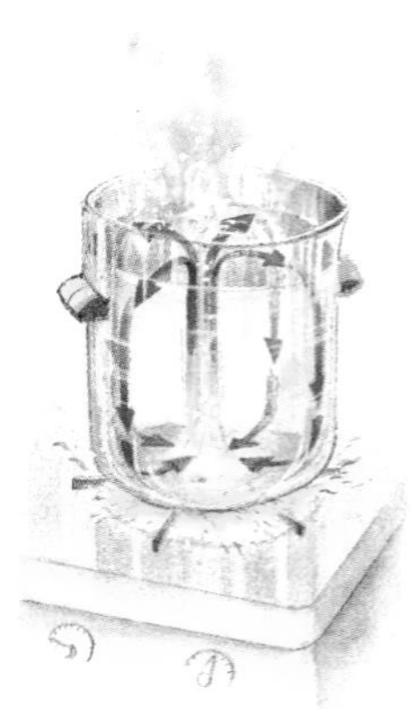

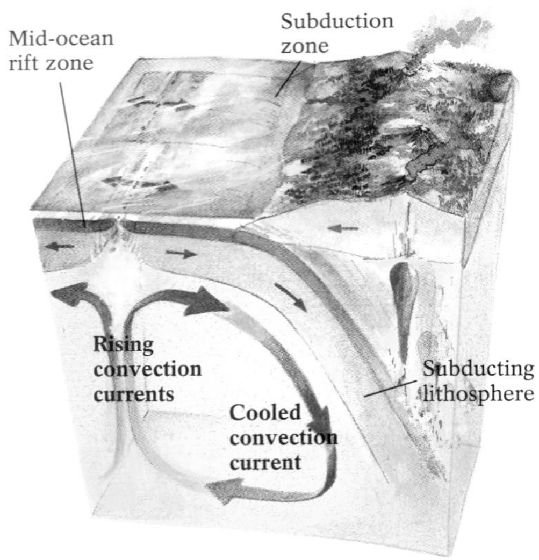

Figure 1-28 Convection cells and plate motion. Heat within the Earth's mantle results in rising ("convecting") currents of warm mantle material, which drag the lighter lithospheric plates along with them as they flow beneath the Earth's surface. As rising mantle material spreads beneath the lithospheric plates, it cools, becomes more dense, and begins to sink back to the deeper interior, where it is reheated to rise again. (A similar dynamic is what causes a pot of soup to boil.) Such a cycle, known as a convection cell, is believed by many scientists to be the principal driving mechanism of plate tectonics.

The Earth's Plate Tectonic Future

Our knowledge of current rates and directions of plate motion enables us to speculate about the Earth's continental configuration millions of years into the future, assuming that those rates and directions remain relatively constant. In Figure 1-29, we see the Earth as it may look 100 million years from now. This scenario, of course, assumes that the rates and directions of all plate movement will remain constant. Plate motions have, however, changed in the past, and may well do so again. Nevertheless, the reunion of the continents to form another supercontinent at some time in the very distant future is a strong possibility.

A Preview of Things to Come

This brief introduction to the principles of plate tectonics should prepare you for the next eleven chapters, which constitute the first two major parts of this text. As Part 1 continues, Chapters 2 through 7 focus on the origins of various Earth materials—from minerals to igneous, sedimentary, and metamor-

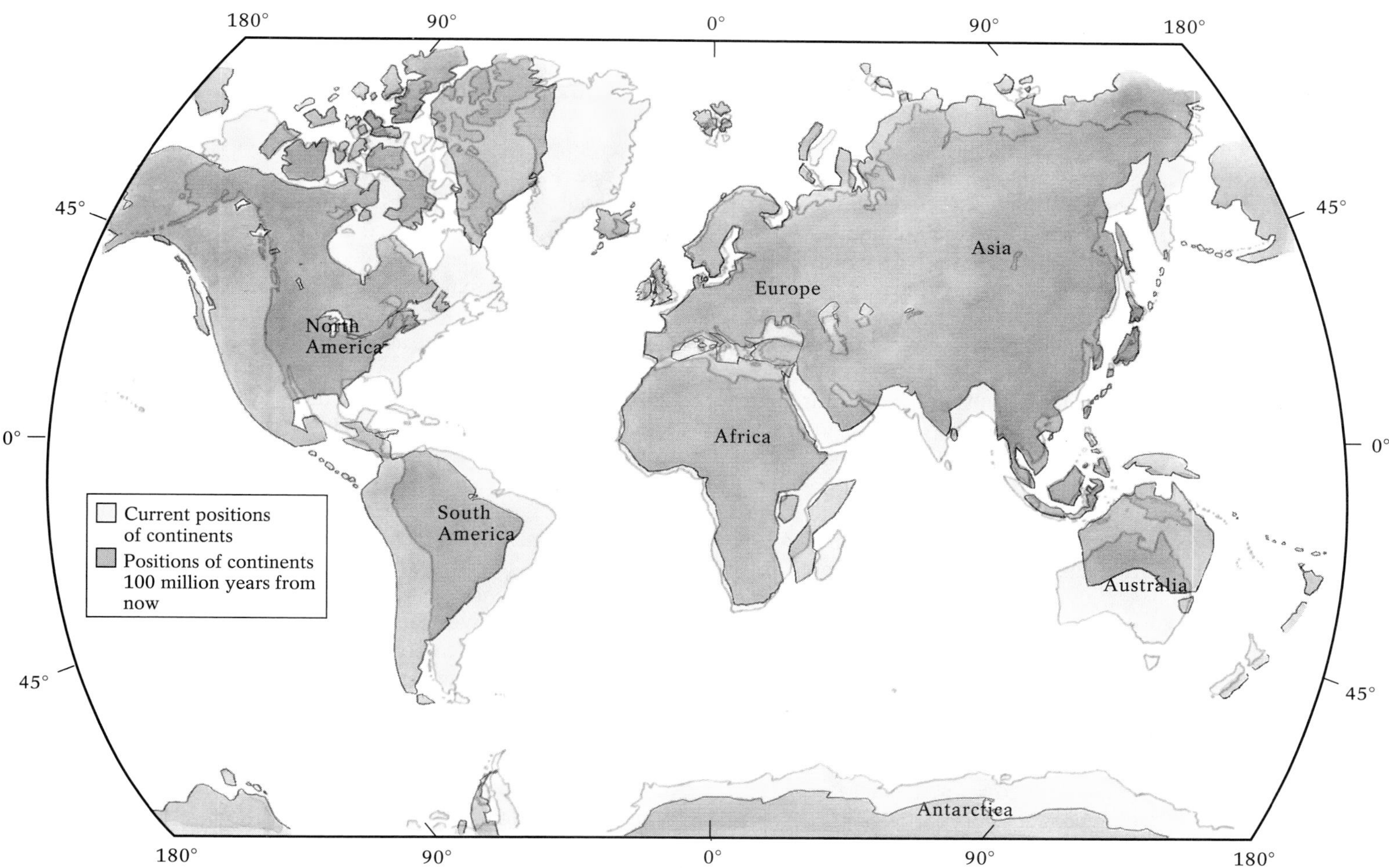

Figure 1-29 The hypothetical position of the Earth's continents 100 million years from now. These positions are based on the assumption that plate velocities and directions will remain as they are today. Note the possible collisions of Africa and Europe (with the loss of the Mediterranean Sea) and Australia and southeast Asia, and the movement of western California toward Alaska along a continent-long transform plate boundary.

phic rocks—and explain some important processes that produce those rocks, such as volcanism and weathering. Chapter 8 explains how geologists interpret the rock record to unravel the Earth's long history. In Part 2, Chapters 9 through 11 are devoted to the large- and small-scale geologic structures at the Earth's surface and the dynamics of the Earth's interior. Chapter 12 provides a more detailed look at the global effects of plate tectonics.

Finally, Part 3 describes how Earth's surface processes sculpt its large-scale features created by tectonic forces. In Chapters 13 through 19 we will explore how solid rock is turned to dust by the Earth's climate; why some slopes are more stable than others (and precautions to take when building a house in a landslide-prone region); how streams modify the landscape (and why they occasionally flood our homes); where to find clean, drinkable, underground water; how caves form; why glaciers once covered much of North America and when they will be back; where sand dunes form and why the world's deserts grow larger every day; and why crashing waves are eroding some of our coasts at an alarming rate. The concluding chapter focuses on the ways humans use and manage the Earth's natural resources—from the oil that heats our homes and fuels our cars to the fertilizers that enrich our crops and guarantee our next meal.

The relevance of geology will be apparent in every chapter of this text. Pay close attention—what you learn will affect you in one way or another every day of the rest of your life.

Chapter Summary

Geology is the scientific study of the Earth. Geologists, like other scientists, systematically collect data derived from experiments and observations. They analyze and interpret their findings and develop **hypotheses** to explain how the forces of nature work. The hypotheses that are consistently supported by further study and investigation may be elevated to the status of a widely accepted explanation, or **theory.** A theory that withstands rigorous testing over a long period of time may be declared a **scientific law**. In order to be accepted, all **scientific methods** must conform to this research strategy. The hypotheses that have undergone such scrutiny include two which attempted to explain the evolution of the Earth's geologic features. **Catastrophism,** which was popular until the mid-eighteenth century, held that the Earth had evolved through a series of immense worldwide upheavals; **uniformitarianism,** which after much debate replaced catastrophism as the hypothesis of choice by the 1830s (and remains so today, though somewhat modified), proposed that the Earth has evolved slowly and gradually by small-scale processes that can still be seen operating today.

The Earth is one of nine planets in our solar system, which formed some 10 to 15 million years after the present universe began with the Big Bang. The Sun, which is a star, formed about 5 billion years ago from the collapse of a gas cloud, the center of which heated up as particles drawn inward by gravity collided and produced nuclear reactions. As the outer region of the gas cloud cooled, the Earth and other planets developed (the Earth about 4.6 billion years ago) by accretion of colliding bits of matter. During the Earth's first few hundred million years of existence, the impact of these accreted masses warmed its interior until the accumulated heat was sufficient to melt most of the planet's constituents. At this time, an event known as the **iron catastrophe** took place, during which the Earth's densest elements, primarily iron, sank toward its interior while its lightest elements moved upward to its surface. Today, the Earth has three concentric layers of different densities: the thin, least dense outer layer, called the **crust;** a thick, more dense underlying layer, called the **mantle;** and a much smaller **core,** which is the most dense of Earth's layers. Over the last 4 billion years, the Earth has cooled slowly and its layers have largely solidified. However, enough heat remains in its interior to generate rising currents of warm mantle material that have kept the outer portion of the Earth mobile. The Earth's **lithosphere,** a composite layer made up of the crust and the outermost segment of the mantle, is solid and brittle and forms large rocky plates; these plates are pulled along at the Earth's surface by the warm, flowing **asthenosphere** beneath them.

Time plays an important role in the evolution of the Earth's geologic features and materials. The Earth is believed to be about 4.6 billion years old; over such a vast amount of time, many gradual geological changes can take place which occur too slowly to be perceived on a human time scale. The Earth's three principal types of rocks undergo such changes, actually turning from one type into another, depending on the environmental forces acting on them. **Rocks,** which are defined as naturally occurring aggregates of inorganic materials, are categorized according to the way in which they form. The three basic rock groups are **igneous rocks**, which solidify from molten material; **sedimentary rocks**, which are compacted and cemented aggregates of fragments of preexisting rocks of any type; and **metamorphic rocks**, which form from any type of rock when its chemical composition is altered by heat, pressure, or chemical reactions in the Earth's interior. The continual transformation of the Earth's rocks from one type into another over time is called the **rock cycle**.

For centuries, scientists have tried to decipher the origin of the Earth's largest geologic features, such as oceans and mountain ranges, and to learn the reasons for geologic catastrophes such as volcanoes and earthquakes. In the late twentieth century, the theory of **plate tectonics** has provided an explanation for all of these. According to this theory, the outermost portion of the Earth (the lithosphere) is composed of seven major and a dozen or more minor plates. The plates consist of relatively dense oceanic lithosphere, relatively light continental lithosphere, or a combination of both types. The Earth's continents and oceans drift from place to place atop the moving plates. The movement of the continents, called **continental drift**, was first recognized by the German geophysicist–meteorologist Alfred Wegener (1880–1930) in what is considered the first step toward the development of the comprehensive plate tectonic theory. The plates move in three ways: away from one another, by **divergence**; toward one another, by **convergence**; or past one another in opposite directions, by **transform motion**.

Divergence is preceded by **rifting,** the process in which a large preexisting plate is torn into a number of smaller plates. New oceanic lithosphere forms between rifted plates as molten rock from the Earth's mantle wells upward, cools, and solidifies. In this manner, new oceanic rock is continuously added to the rifted edges of the diverging plates, a process known as **sea-floor spreading**. Convergence involving either two oceanic plates or one oceanic plate and one continental plate results in **subduction,** in which the denser of the two plates sinks below the other and is reabsorbed into the Earth's mantle; a **continental collision** occurs when both converging plates are continental, in which case neither are dense enough to subduct and their edges are instead uplifted by the pressure of the collision. Because of the hot mantle material that rises into divergent plate boundaries, the pressure that develops at the edges of colliding plates, and the friction that occurs between plates moving past one another at transform boundaries, rocks at the boundaries of such plates may melt (fueling volcanic eruptions), buckle (to form mountains), or break (generating earthquakes). By comparison, plate interiors are relatively quiet geologically.

The force that drives the plates appears to be **convection cells** that circulate the Earth's internal heat, continuously heating the deeper rocks so that they become lighter and rise. Near the surface, rising currents of flowing material spread laterally beneath the Earth's lithosphere, pulling the plates along atop the flowing asthenosphere. The effect of gravity on dense oceanic plates may also contribute to plate motion, by pulling the plates down into the Earth's interior at subduction zones.

Key Terms

geology (p. 4)
scientific methods (p. 5)
hypothesis (p. 5)
theory (p. 5)
scientific law (p. 5)
catastrophism (p. 8)
uniformitarianism (p. 8)
iron catastrophe (p. 11)
crust (p. 12)
mantle (p. 12)
core (p. 12)
lithosphere (p. 12)
asthenosphere (p. 13)
rock (p. 14)
igneous rocks (p. 14)
sedimentary rocks (p. 14)
metamorphic rocks (p. 14)
rock cycle (p. 14)
plate tectonics (p. 16)
rifting (p. 19)
divergence (p. 19)
sea-floor spreading (p. 20)
convergence (p. 20)
subduction (p. 20)
continental collision (p. 21)
transform motion (p. 23)
continental drift (p. 24)
convection cells (p. 30)

Questions for Review

1. Briefly explain the differences between a scientific hypothesis, a scientific theory, and scientific law.
2. Contrast the principles of catastrophism and uniformitarianism.
3. Draw a simple sketch of the major layers that make up the Earth's interior. What parts of the Earth's internal structure form the Earth's plates?
4. Describe the three major types of rocks in the Earth's rock cycle.
5. Draw simple sketches of divergent plate boundaries, two kinds of convergent plate boundaries, and transform plate boundaries.
6. At which type of plate boundary do oceanic plates grow? At which type of plate boundary are oceanic plates consumed?
7. Why are there so few earthquakes in Minneapolis and Indianapolis?
8. How could one use the distribution of modern and ancient animals to prove that continents drift?
9. Briefly describe how oceanic sediments vary in age and thickness with reference to a diverging mid-ocean ridge.
10. Draw a simple sketch of a convection cell and explain how it works. Explain the difference between convection and conduction of heat.

For Further Thought

1. When geologists find ancient glacial deposits in equatorial Africa, they usually interpret them as polar deposits that have drifted from a cold place to a warm place. Formulate another hypothesis to explain this phenomenon.
2. Why is there no current volcanic activity along the east coast of North America?
3. Describe how plate tectonic activity might affect the rock cycle.
4. Find the Ural Mountains on a map of eastern Europe. Briefly explain how they might have formed.
5. As recently as 5 million years ago, South America and North America were completely separated, unattached by an intervening Central America. Using Figure 1-12, speculate about how the Central American connection that binds the western hemisphere might have formed.

Figure 2-1 Crystals of quartz, showing the beauty and symmetry for which many minerals are valued.

Chapter 2

Minerals

Minerals have long been valued in many cultures for their sheer visual appeal—for their stunning colors or luster, or their often-perfect symmetry (Fig. 2-1). But the importance of minerals is hardly just aesthetic, and is not restricted to the "pretty" specimens found in museum or jewelers' showcases. From the simple flint hand scrapers made by our ancestors hundreds of thousands of years ago, to the silicon microchip with which this text was written, minerals have always helped us improve our lot in life (Fig. 2-2). Every day a vast array of minerals, all derived from the Earth, play a role in our existence.

Certain types of minerals are fundamental to our physical well-being. For instance, many of the nutrients our bodies need to survive and grow come from minerals in the Earth's soil; when the minerals are broken down and absorbed by plants, these nutrients are made available to us in the fruits and vegetables that we eat (and in the animals that have eaten the plants).

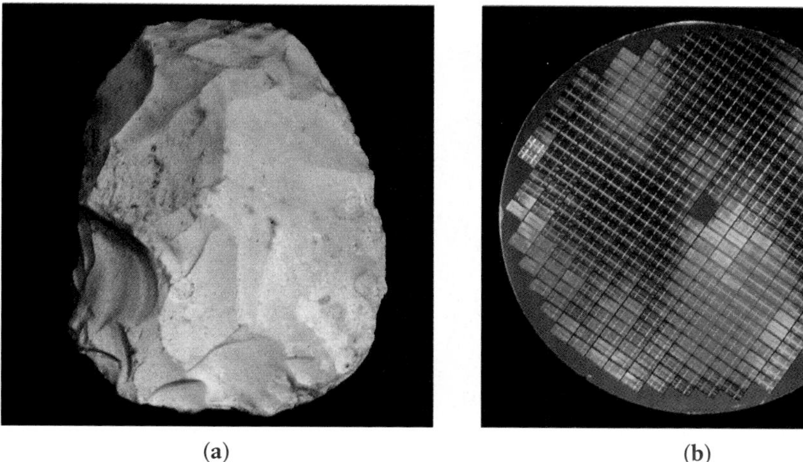

(a) (b)

Figure 2-2 Minerals have been providing us with essential tools for hundreds of thousands of years. (**a**) This flint hand scraper was used by Native Americans more than a thousand years ago, probably to scrape flesh and hair from animal skins. (**b**) This computer "wafer," containing 1-million-bit silicon memory chips, is used today to store large amounts of information.

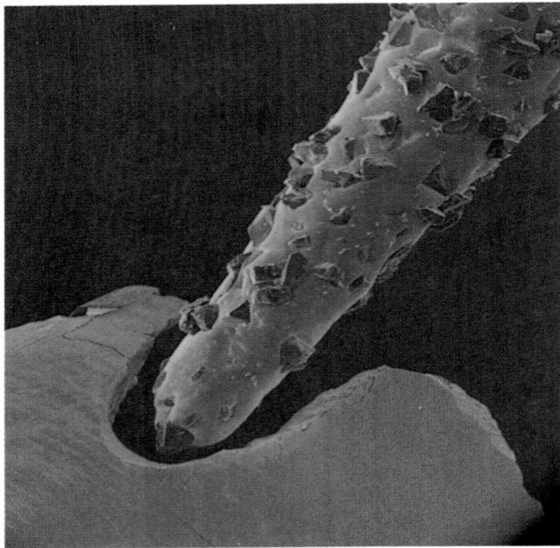

Figure 2-3 Today, through technological advances and the existence of diamonds, a visit to the dentist is not nearly as long or as painful as it used to be. Here we see a drill bit studded with diamonds (the hardest known substance) cutting swiftly through the relatively soft enamel of a human molar composed largely of the calcium-phosphorus mineral, apatite. (Magnified 25×)

Some nutrients we get from this process include calcium, phosphorus, and fluorine, which give our bones and teeth their hardness; sodium and potassium, which help regulate blood pressure; magnesium, which, along with calcium, is required for muscle contraction and normal nerve-impulse transmission; and iron, a major component of the hemoglobin in blood, which carries life-sustaining oxygen to our cells.

We use minerals in a remarkable number of ways. A common absorbent mineral called talc (in talcum powder) is used to dry and soothe our skin. Minerals provide the sulfur used to manufacture fertilizers, paints, dyes, detergents, explosives, synthetic fibers, or a simple book of matches, and the fluorine that helps to refrigerate food and cool homes and offices. Aluminum, commonly derived from the mineral bauxite, is an ideal component for space shuttles, garden furniture, window frames, and beer and soft-drink cans because of its lightness, strength, and resistance to corrosion. When you're laid low by a common intestinal problem, you may run for a spoonful of Kaopectate®, a remedy whose active ingredient is the mineral kaolinite. (And see, for another example, Figure 2-3.)

In this chapter, we will examine what minerals are and how they are formed, and discuss methods of identifying the different kinds of minerals. We will also look at the distinctive structures of some important minerals and see how their characteristics determine our uses for them.

What Is a Mineral?

Minerals are naturally occurring inorganic solids consisting of chemical elements in specific proportions, whose atoms are arranged in a systematic internal pattern. For example, diamonds, emeralds, and quartz are minerals. Because minerals are *naturally* occurring solids, the thousands of synthetic compounds produced in laboratories do not qualify as minerals. Because minerals are *inorganic*, substances such as coal, which is composed of heated and compressed remnants of plants, are not considered to be minerals. Because minerals have a *systematic internal organization*, substances such as the gemstone opal, which lacks systematic internal organization of its atoms, are not considered minerals. **Rocks** are naturally occurring aggregates, or combinations, of one or more minerals, with each mineral retaining its own discrete characteristics. For example, the rock granite contains minerals such as quartz and plagioclase (Fig. 2-4).

Figure 2-4 One can clearly see the individual mineral components in this piece of granite, an igneous rock.

Minerals are composed of one or more **elements** in specific proportions. An element is a form of matter that cannot be broken down into a simpler form by heating, cooling, or reacting with other chemical elements. Aluminum and oxygen are two common elements. **Atoms** (named from the Greek *atomos,* or "indivisible") are the smallest particles of an element that retain all its chemical **properties**. A property is a characteristic of a substance that enables us to distinguish it from other substances. All atoms of a given element are essentially identical, and the atoms of one element differ in fundamental ways from the atoms of every other element. There are 106 known elements, of which 92 form naturally and 14 have been created in laboratories. Every element can be represented by its chemical symbol, a one- or two-letter abbreviation. These symbols are usually the first letter or letters of the English or Latin name of the element, such as O for oxygen, Al for aluminum, and Na (from the Latin *natrium*) for sodium. The 106 elements have been arranged by chemists into the Periodic Table of Elements (Fig. 2-5).

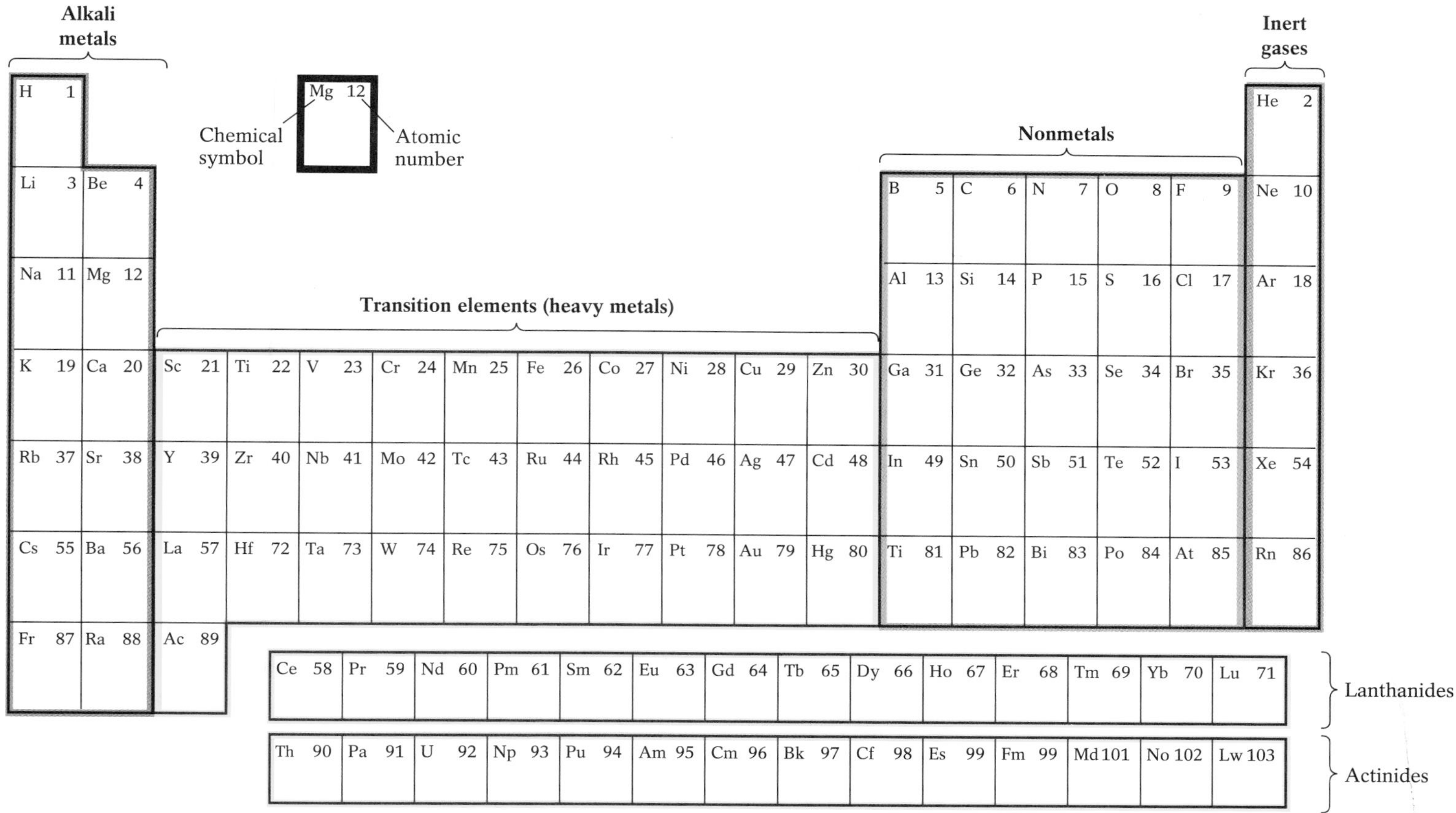

Figure 2-5 The Periodic Table of Elements. The periodic table groups all the elements by similarities in their atomic structures, which result, in turn, in similarities in their chemical properties.

Atoms of one or more elements may combine in specific proportions to form chemical **compounds**. For example, the mineral quartz is a chemical compound that always contains one silicon atom for every two oxygen atoms. The fixed proportions of atoms that make up a compound are expressed by combinations of symbols in a *chemical formula;* for example, the silicon and oxygen combination that makes up quartz has the chemical formula SiO_2. (Subscript numerals denote the ratio of atoms of each element in the chemical compound; thus, one can tell from its formula that quartz contains two atoms of oxygen for every atom of silicon.)

Geology students sometimes wonder why so much discussion about minerals concerns chemistry. The principal reason is that all minerals are chemical compounds, and it is their chemical structures that determine their distinctive characteristics. These characteristics in turn determine their uses and value to society. To know why diamonds are hard, why gold can be pounded into wafer-thin leaves, and why we build skyscrapers with skeletons of titanium steel, we must understand the chemical makeup of minerals.

The Structure of Atoms

An atom is incredibly small, approximately 0.00000001 centimeter (one hundred-millionth of a centimeter) in diameter. This line of type would contain about 800,000,000 atoms laid side by side. As small as an atom is, it con-

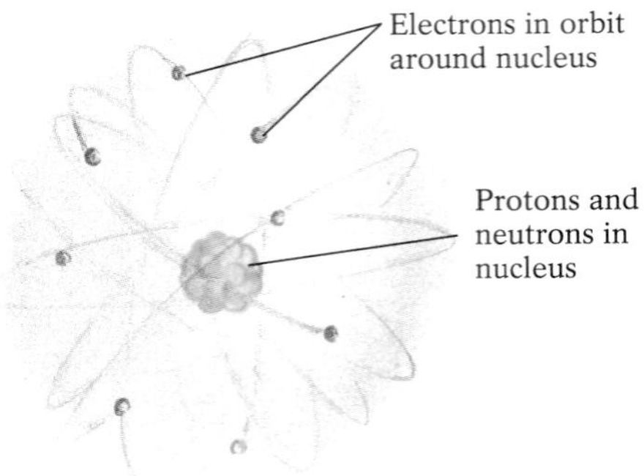

Figure 2-6 A simplified model of an atom. Protons and neutrons compose the nucleus; electrons orbit the nucleus at high speed.

sists of even smaller particles: **protons** and **neutrons**, located in the central region, or **nucleus**, of an atom; and **electrons**, moving about outside the nucleus. A simplified model of an atom is shown in Figure 2-6.

Each proton carries a single positive charge, expressed as +1, and has a mass of 1.67×10^{-24} gram, which for convenience is referred to as an *atomic mass unit* (AMU) of 1. Neutrons are nearly identical to protons in size and mass, but, as their name suggests, they have no charge—they are neutral. They do, however, contribute to the **atomic mass** of the atom, which is the sum of protons and neutrons within an atom's nucleus. Thus, an atom containing one proton and one neutron has an atomic mass of 2 AMU. The third type of atomic particle, the electron, orbits the nucleus at speeds so great that if one were orbiting the Earth, it would do so in less than a single second. Each electron carries a single negative charge, expressed as −1, and has a mass about 1/1836 that of a proton or neutron; thus, an electron's contribution to an atom's mass is negligible.

The number of protons in an atom's nucleus is its **atomic number**. Every atom of an element has the same number of protons in its nucleus; this number differs from the number of protons in the nuclei of all other elements. For example, every magnesium atom has 12 protons in its nucleus, or an atomic number of 12. An atom with 13 protons in its nucleus is an entirely different element, aluminum. Thus, the number of protons determines an atom's identity.

The number of neutrons in an atom's nucleus can vary without causing the atom's identity to change. **Isotopes** are atoms of the same element that have the same number of protons but different numbers of neutrons in their nucleus, and which consequently differ in atomic mass. For example, the element oxygen (atomic number = 8) always contains eight protons in its nucleus, but has three isotopes: $^{16}_{8}O$, with eight neutrons in its nucleus; $^{17}_{8}O$, with nine neutrons; and $^{18}_{8}O$, with ten neutrons. (The subscript numeral in these notations is the atomic number and the superscript numeral is the atomic mass.) Some isotopes of certain elements contain nuclei that break down spontaneously and emit some of their particles. Such isotopes are described as *radioactive*. Two common radioactive isotopes are $^{235}_{92}U$ (uranium-235) and $^{14}_{6}C$ (carbon-14).

The number of electrons in an atom is always the same as the number of protons. For example, hydrogen (atomic number = 1) has one proton and one electron, whereas iron (atomic number = 26) has 26 protons and 26 electrons. Because the number of an atom's protons equals the number of its electrons, its positive charge exactly balances its negative charge and it has no net charge. However, all of an atom's positive charge is concentrated in the protons in its nucleus, whereas its negative charge is distributed among the electrons in its periphery.

An atom's negatively charged electrons are attracted by the positively charged protons in its nucleus, and tend to congregate about the nucleus in ever-changing orbits, forming an *electron cloud*. They do not crash into the nucleus because the momentum associated with their high speeds keeps them in orbit. At the same time, however, the electrons in the cloud repel one another. Most of the time, each electron moves within a specific region of space around the nucleus, called an **energy level**, the position of which maximizes the force of attraction between the electron and the nucleus while minimizing the force of repulsion between the electron and all the other electrons.

Electrons fill an atom's lowest, or first, energy level before any enter the higher energy levels more distant from the nucleus. The lowest energy level in any atom always has a maximum capacity of two electrons. The second

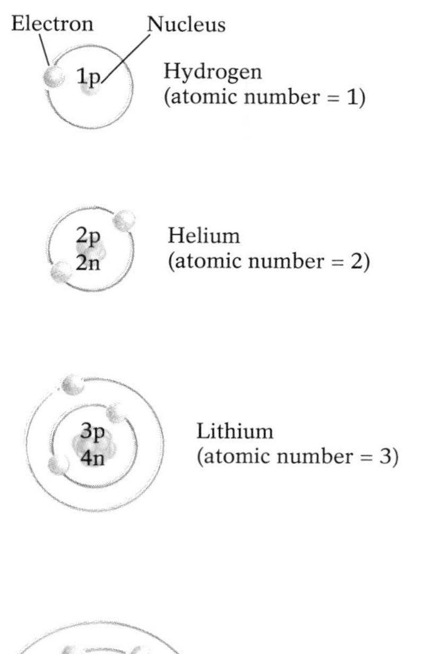

Figure 2-7 Energy-level diagrams of various elements. The nucleus contains the protons (p) and neutrons (n); electrons are shown as balls orbiting the nucleus along concentric circular tracks representing energy levels. (Note that electron size is exaggerated for clarity; electrons are actually much smaller than protons and neutrons.) Hydrogen and helium, because they have two or fewer electrons, have only one energy level; sodium, because it has 11 electrons, has three energy levels (2 electrons in the first level, 8 electrons in the second level, and 1 electron in the third level). Note that the number of electrons in an atom always equals the number of protons.

energy level can hold a maximum of eight electrons, and succeeding energy levels can each hold eight or more electrons. The number and energy-level positions of some atoms' electrons are shown schematically in Figure 2-7.

How Atoms Bond

To form chemical compounds, atoms combine, or **bond**, by losing, gaining, or sharing electrons during a chemical reaction. The transfer or sharing of electrons between bonded atoms changes the electron configuration of each. Two key factors determine which atoms will unite with which others to form compounds: Each atom should achieve chemical stability, and the resulting compound should be electrically neutral.

An atom achieves chemical stability when its outermost energy level is filled with electrons. Thus, atoms bond by transferring or sharing electrons to have full outermost energy levels. (In the case of hydrogen and helium atoms, which have only the lowest energy level, this requires two electrons in the outermost energy level; for all other atoms, it requires eight.) Because chemical stability almost always requires eight electrons in the outermost energy level, this is known as the **octet rule**. Atoms of some elements tend to lose their outer electrons, whereas others tend to gain electrons, depending largely on the number of electrons in their outermost energy levels. Elements with one or two electrons in their outermost energy levels have a strong tendency to give up those electrons. Elements with six or seven electrons in their outermost energy levels tend to acquire electrons. Elements with three, four, or five electrons in their outermost energy levels tend to share electrons with other atoms instead of transferring or receiving them. Elements whose outer energy levels are filled with eight electrons (or, in the case of helium, which has only the first energy level, with two electrons) are very chemically stable, or *inert*; they do not lose, gain, or share electrons.

Due to the diversity of electron configurations among elements, various types of bonding are possible. The atoms that make up the vast majority of the Earth's minerals are most often linked by ionic bonding, covalent bonding, or metallic bonding.

Ionic Bonding An atom does not change its identity when it loses or gains an electron, but it does lose its electrical neutrality and become positively or negatively charged. For example, when a potassium atom loses one electron, it becomes a positively charged particle because it has one more proton than electrons. When a chlorine atom gains a single electron, it becomes a negatively charged particle because it has one less proton than electrons. An atom that has lost or gained one or more electrons to become a charged particle is called an **ion.** A positively charged ion is indicated by the symbol for the element followed by a superscript "+" (for example, K^+). A negatively charged ion is indicated by the symbol for the element followed by a superscript "–" (for example, Cl^-). Symbols for atoms that have gained or lost more than one electron show the number lost or gained as a superscript before the charge (for example, Ca^{2+}).

When an atom with a strong tendency to lose electrons comes in contact with an atom with a strong tendency to gain electrons, electrons are generally transferred so that each atom achieves the chemical stability of a full outer energy level. The donor atom loses one or more electrons and becomes a positively charged ion, and the receiving atom gains one or more electrons and becomes a negatively charged ion. These oppositely charged ions then attract each other to form an **ionic bond**. The result is an electrically neutral compound.

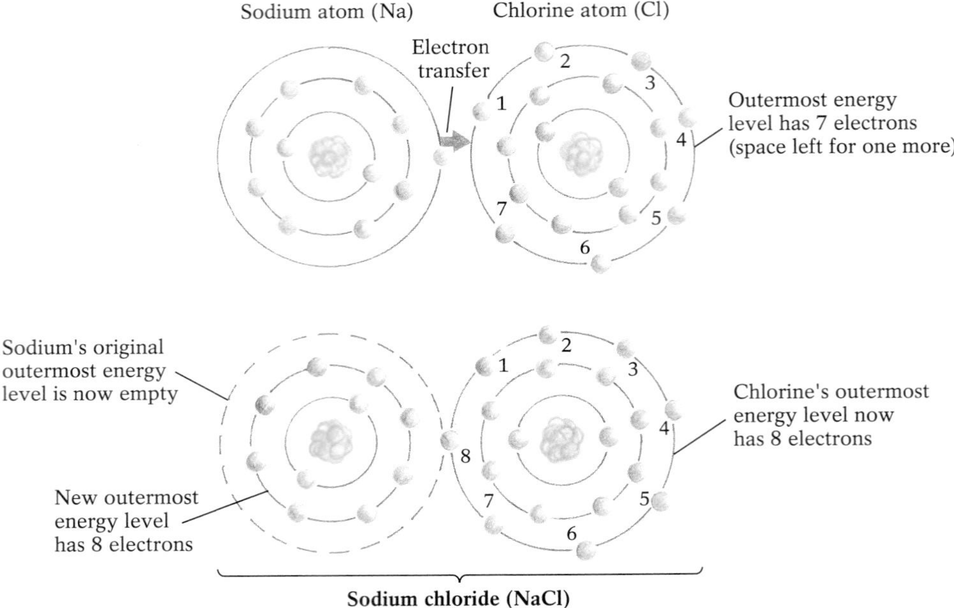

Figure 2-8 Ionic bonding of sodium (Na) and chlorine (Cl). When a sodium atom (with one electron in its outer energy level) donates its outermost electron to a chlorine atom (with seven electrons in its outer energy level), the outer energy level in the new configuration of each atom has eight electrons. The two resulting ions (Na^+ and Cl^-) unite to form sodium chloride (NaCl), a neutral ionic compound.

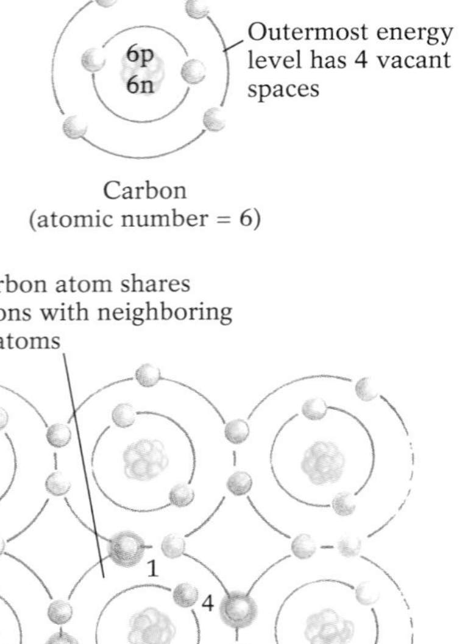

Figure 2-9 Covalent bonding in diamond, a mineral which consists entirely of the element carbon. Each carbon atom in diamond is bonded covalently to four neighboring carbon atoms; the great strength of these bonds accounts for the fact that diamond is the hardest known substance on Earth.

Figure 2-8 shows the ionic bonding of sodium and chlorine atoms to form a neutral compound. Sodium has a strong tendency to lose the lone electron in its third, or outer, energy level, thereby completely eliminating that energy level. Bonding turns the sodium atom into a positively charged ion with only two energy levels; the second energy level, now the outer one, has eight electrons. Chlorine has a strong tendency to gain a single electron to add to the seven in its outer energy level, which then becomes filled with eight electrons. Bonding turns the chlorine atom into a negatively charged ion. The compound that results from ionic bonding of sodium (Na) and chlorine (Cl) is sodium chloride (NaCl).

The physical and chemical properties of ionic compounds differ from those of their component elements. Pure sodium, for instance, is a soft silvery metal that reacts vigorously when mixed with water and may even burst into flame; chlorine usually occurs as Cl_2, a green poisonous gas that was used as a weapon during World War I. Sodium chloride neither ignites nor poisons—it is the white crystalline mineral halite (table salt), a substance that regulates some of the biochemical processes essential to all life.

Covalent Bonding Atoms whose outer energy levels are about half full, containing three, four, or five electrons, tend to achieve chemical stability by sharing electrons with other similarly equipped atoms. In such a case both atoms fill their outer energy levels with the shared electrons rather than transferring electrons from one to the other. Sharing electrons produces a **covalent bond**, in which the outer energy levels of the atoms overlap. Covalent bonds are generally stronger than any other type of bond. In some cases, two or more atoms of a single element may bond covalently with each other. For example, overlapping carbon atoms bond covalently in the all-carbon mineral, diamond (Fig. 2-9).

Metallic Bonding The atoms of some electron-donating elements tend to pack closely together, with each typically surrounded by either eight or twelve others. This produces a cloud of electrons that roam independently among the positively charged nuclei, unattached to any specific nucleus. This phenomenon of a negatively charged electron cloud attracted to a cluster of positively charged nuclei is called **metallic bonding**, and is responsible for the properties that define metals. For example, metals are efficient conductors of electricity, which requires freely moving negatively charged particles. The mobile electrons of metallically bonded substances explain (among other phenomena) why copper wiring can be used to transmit an electrical current to appliances.

Intermolecular Bonding Compounds often exist as *molecules*, stable groups of bonded atoms having the proportions of the compound, which are the smallest particles to exhibit the chemical and physical properties of the compound. (For example, one molecule of water, H_2O, consisting of exactly two atoms of hydrogen stably bonded to one atom of oxygen, has all the properties of water, whereas its component elements alone do not.) Molecules such as those of water, which are held together internally by strong covalent or ionic bonds, are often weakly attached to other molecules by **intermolecular bonds**. Intermolecular bonding results from the relatively weak positive or negative charges that develop at different locations within a molecule or group of atoms, due to the uneven distribution of their moving electrons.

Because of the circumstances under which they are produced (discussed in Chapter 3), minerals do not exist as molecules; however, because the groups of atoms that do compose minerals have many of the same qualities as molecules (specifically in their uneven distribution of charges), they are also subject to intermolecular bonding.

For geologists, the most important example of intermolecular bonding involves water. Each water molecule consists of two hydrogen atoms covalently bonded to one oxygen atom. Because the positive charge of the oxygen atom's nucleus is greater than that of the hydrogen atoms', the shared electrons are more attracted to, and tend to spend more time near, the oxygen nucleus; the oxygen atom subsequently develops a weak negative charge from the presence of the negatively charged electrons. Because the electrons spend less time near the hydrogen nuclei, a weak positive charge develops on the hydrogen side of the molecule (from the relative absence of electrons). These charged regions attract oppositely charged regions of nearby molecules, forming weak **hydrogen bonds** with these molecules.

The formation of hydrogen bonds explains why so many substances, even those that are strongly bonded, dissolve in water. For example, when halite (NaCl) and water are combined, the positively charged sides of the water molecules attract the chlorine ions and the negatively charged sides attract the sodium ions. Although hydrogen bonds are weak compared to the ionic bonds that hold the Na and Cl together in halite, when NaCl is surrounded by many water molecules their combined effect is strong enough to overcome the ionic bonds and split halite into its component Na^+ and Cl^- ions (Fig. 2-10).

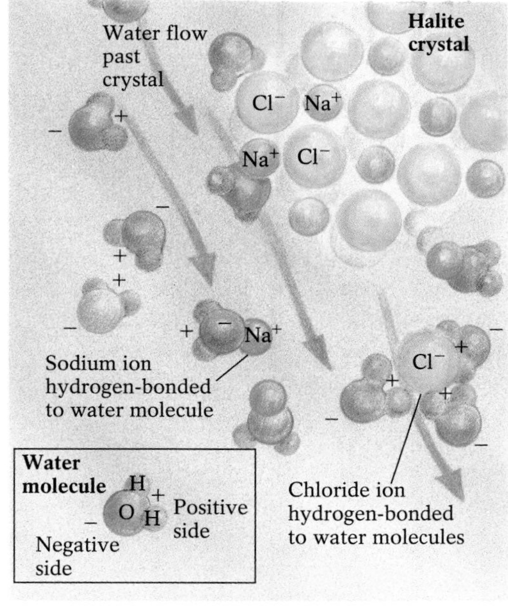

Figure 2-10 Water molecules each have a positively charged side (the hydrogen side) and a negatively charged side (the oxygen side). These regions attract oppositely charged ions from other compounds, forming hydrogen bonds with them. The attraction of sodium ions to the negative side of water molecules and the attraction of chlorine ions to the positive side is what causes halite (table salt) to dissolve in water.

In a different type of intermolecular bond, a number of electrons are momentarily grouped on the same side of an atom's nucleus, giving that side of the atom a slight negative charge and the electron-poor side a slight positive charge. The positive side may briefly attract electrons of neighboring atoms, and the negatively charged side may briefly attract the nuclei of neighboring atoms. This type of weak intermolecular attraction is called a **van der Waals bond**, after Johannes van der Waals, the Dutch physicist who discovered it in the late nineteenth century. Although weak, van der Waals bonds can be sufficient to bond atoms or layers of atoms together in certain minerals, such as graphite.

Hydrogen and van der Waals bonds are much weaker than ionic, covalent, and metallic bonds, and can be broken by the addition of small amounts of heat. For example, the intermolecular bonds between water molecules in ice break when the temperature is raised to 0°C (32°F)—the melting point of ice—changing the water from a solid to a liquid. The strong covalent bonds that form the individual water molecules, however, are unaffected.

Mineral Structure

When a mineral grows unrestricted in an open space, it develops into a regular geometric shape known as a **crystal** (from the Greek *kryos*, or "ice"). This shape is the external expression of the mineral's microscopic internal **crystal structure**, the orderly arrangement of its ions or atoms into a latticework of repeated three-dimensional units. Every crystal of a given mineral has the same crystal structure, as can be seen by the consistent orientation of the planar surfaces, or *faces*, among all well-formed crystals of a given mineral. A mineral's crystalline form is generally a combination of simple geometric shapes such as cubes, pyramids, and prisms.

More often than not in nature, minerals grow in restricted regions, often competing with other growing minerals for the available space. In this circumstance, they form interlocking masses of small irregular grains that bear little resemblance to well-formed mineral crystals. Although these grains do not develop the precise geometric shape of a crystal, their microscopic crystal structure remains characteristically regular.

Molten rock may cool too rapidly for its atoms to form even a semblance of an orderly arrangement. The resulting solids, called **mineraloids**, lack a specific crystal structure and thus do not qualify as true minerals. Natural glass (such as obsidian) is one type of mineraloid. Like manufactured glass, it is made by rapid cooling of molten silica (SiO_2). Glass is essentially frozen liquid, a fact that can be seen in the window panes of very old dwellings; over time the window glass gradually flows until the lower portion of each pane is noticeably thicker than the top.

Table 2-1 **The Most Abundant Elements in the Earth's Continental Crust**

Element	Proportion of Earth's Weight (%)
Oxygen (O)	45.20
Silicon (Si)	27.20
Aluminum (Al)	8.00
Iron (Fe)	5.80
Calcium (Ca)	5.06
Magnesium (Mg)	2.77
Sodium (Na)	2.32
Potassium (K)	1.68
	98.03
Other elements	1.97
Total	100.00

Table 2-2 **The Most Abundant Elements in the Whole Earth**

Element	Proportion of Earth's Weight (%)
Iron (Fe)	34.8
Oxygen (O)	29.3
Silicon (Si)	14.7
Magnesium (Mg)	11.3
Sulfur (S)	3.3
Nickel (Ni)	2.4
Calcium (Ca)	1.4
Aluminum (Al)	1.2
	98.4
Other elements	1.6
Total	100.00

Determinants of Mineral Formation

The kinds of minerals that form in a particular time and place depend on the relative abundances of the available elements, the relative sizes and other characteristics of those elements' atoms and ions, and the temperature and pressure at the time of formation.

Only eight of the 92 naturally occurring elements in the Earth's continental crust are relatively abundant, with oxygen and silicon dominating (Table 2-1). In the entire Earth, iron and oxygen dominate, silicon is fairly abundant, and magnesium is more significant than it is in the crust (Table 2-2). Most minerals in the crust are oxygen-and-silicon–based compounds, and most in the upper mantle are iron–oxygen–silicon–magnesium–based compounds.

In addition to the types and relative abundances of available elements, mineral formation depends on how readily these elements interact with one another to form a crystalline structure: Given two elements of equal abundance, the element that will contribute more readily to mineral formation is that which "fits" better with the other elements present. Atoms and ions in minerals tend to become packed together as closely as their sizes permit. In an ionically bonded mineral, each positive ion attracts as many negative ions as can fit around it, and each negative ion does the same with positive ions. Thus, their relative sizes determine how many negative ions will surround a positive ion and vice versa. In the mineral halite (NaCl), for example, because of the relative sizes of the positive sodium ions (Na^+) and the negative chlorine ions (Cl^-), one sodium ion is always surrounded by six chlorine ions (Fig. 2-11a). Positive ions are generally smaller than negative ions (Fig. 2-11b) because they lose electrons from their outermost energy level, and so lose that energy level (see Fig. 2-8). In a crystal structure, therefore, smaller positive ions usually occupy the spaces between larger negative ions.

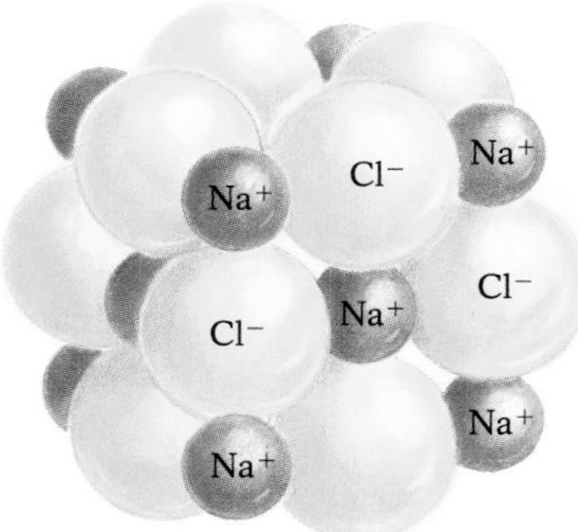

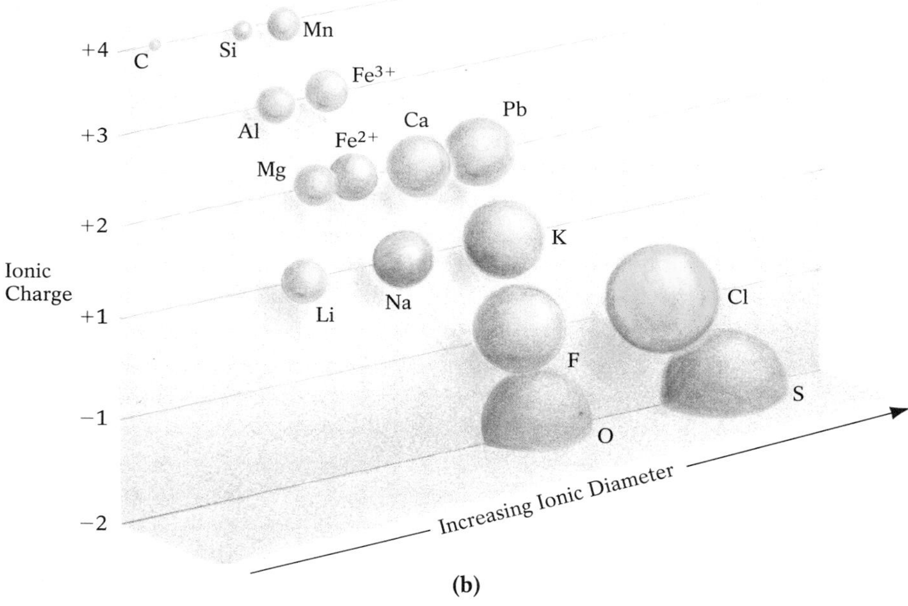

Figure 2-11 (**a**) The crystalline structure of halite (NaCl), in which small sodium ions (Na^+) are tucked in between larger chlorine ions (Cl^-). (**b**) The relative ionic diameters of the Earth's most common elements. Note that negatively charged ions are generally larger than positively charged ions.

Ionic Substitution Certain ions of similar size and charge can replace one another within a crystal structure, depending on which is most available during the mineral's formation. As a result of such **ionic substitution**, some minerals which have the same internal arrangement of ions may have minor variations in composition.

Imagine building a wall of red bricks but running short of them partway through. If you substitute yellow bricks of the same size, the appearance of the wall will be altered, but its stability will not be affected. The same is true of ionic substitution. Iron (Fe^{2+}), and magnesium (Mg^{2+}), which are nearly identical in size and charge, substitute freely for one another in the mineral olivine ($(Fe,Mg)_2SiO_4$). (Note: In a mineral's chemical formula, the elements that can substitute for one another in a mineral's crystal structure appear in parentheses and are separated by commas.) The color, melting point, and other physical characteristics of olivine differ depending on whether Fe^{2+} or Mg^{2+} is predominant, but its chemical stability and crystal structure are unaffected.

Polymorphism Two minerals may have the same chemical composition but different crystal structures because they formed under different temperature and pressure conditions. Such minerals are known as **polymorphs** ("many forms"). Graphite (the "lead" in your pencil) and diamond, for example, are polymorphs that consist entirely of carbon. Graphite's structure forms under the low pressure prevalent at shallow depths (only a few kilometers below the Earth's surface), whereas diamond's structure results from intense pressure at depths greater than 150 kilometers (about 90 miles). From the moment it reaches the surface, however, diamond begins to change into graphite, the carbon polymorph which is more stable in the Earth's low temperature/low pressure surface environment. (Fortunately for owners of diamond jewelry, a great number of strong carbon–carbon bonds must be broken before a diamond is completely converted to graphite. At room temperature, this conversion is too slow to be measured. In terms of human longevity, then, diamonds *are* "forever.")

(a)

(b)

Figure 2-12 Variations in the color of minerals. (**a**) These two mineral samples have one thing in common, although you'd never guess it from their colors—they are both quartz, composed of silicon dioxide (SiO_2). Their colors differ because they each contain minute traces of different impurities. (**b**) Variations in color within a single elbaite crystal. Crystals of this mineral always have one pink end and one green end, due to impurities incorporated into their structures during formation.

How Minerals Are Identified

A mineral's chemical composition and crystal structure give it a unique combination of chemical and physical properties by which geologists can distinguish it from all other minerals. Many of these properties can be observed readily in the field, whereas others require some bulky or sophisticated equipment and can be studied only in laboratories. A mineral can seldom be identified accurately on the basis of only one of its properties; usually several must be established before an identification is considered conclusive.

In the Field

Many of a mineral's properties are instantly apparent or can be ascertained with a minimum of effort or technology. Most geologists and dedicated rockhounds can identify a great many minerals in the field by examining them with the naked eye and performing some very simple tests.

Color Color may be the first thing you notice about a mineral, but it is perhaps the least reliable identifying characteristic. Different-colored minerals may contain the same elements (as is the case with the polymorphs diamond and graphite), whereas minerals that are similar in color may be completely different in composition. As a result, the identity of an unknown mineral—though it may be narrowed down to three or four possibilities—cannot be determined on the basis of its color alone.

A mineral's color depends on how much light its chemical makeup and its crystal structure cause it to absorb. When white light, which contains all the colors of the spectrum, strikes a mineral, part of the spectrum is absorbed and the remainder is reflected back from the mineral and perceived by the eye. For example, when we see an emerald, only the green part of the spectrum is reflected; the rest of the colors of the spectrum are absorbed. Certain elements, such as iron, manganese, chromium, and nickel, absorb a great deal of light and reflect little, and minerals containing these elements are commonly dark in color or black. Minerals with loosely bonded ions and atoms also tend to absorb most incoming light, and thus appear dark or black. (It is the weakly bound layers of carbon atoms in graphite that account for its black color.) Elements such as sodium, potassium, calcium, and silicon absorb little light, and minerals containing them are characteristically light colored.

A few alien atoms in a mineral's structure may completely change its color. For example, pure corundum (Al_2O_3) is generally white or light gray, but even a trace of the element chromium produces a brilliant red variety of corundum called *ruby*. The addition of a little titanium produces a deep blue variety of corundum called *sapphire*. Depending on the types of impurities they contain, different samples of a given mineral may exhibit any number of different colors (Fig. 2-12).

Heating or exposure to radiation displaces some atoms or ions from their designated sites in a crystal structure and can cause significant color variations. Unprincipled jewel merchants have been known to "improve" the colors of poor-quality gems with heat or X-rays. Those colors may fade, however, as the displaced ions and atoms gradually return to their original sites within the crystal structure.

Because a mineral's color is rarely unique to that mineral, and because a mineral's true color can be "disguised" in so many ways, color is not by itself a reliable diagnostic tool for mineral identification.

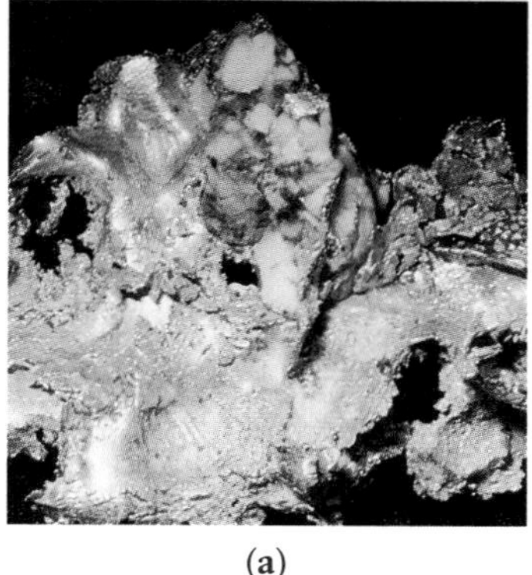

(a)

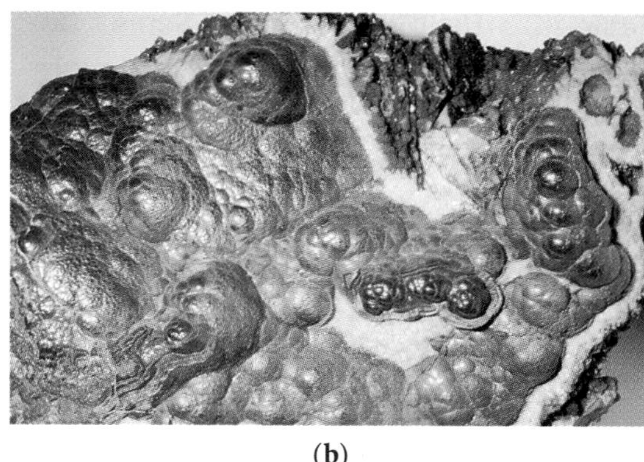

(b)

Figure 2-13 Gold (**a**) and hematite (**b**) each exhibit the shiny luster characteristic of metals.

Luster **Luster** describes how a mineral's surface reflects light. Minerals can exhibit metallic or nonmetallic luster. Car bumpers glisten with a metallic luster in part because of the metallic bonding of the metal chromium. When any light shines on a metal, its energy stimulates the metal's loosely held electrons to vibrate. The vibrating electrons emit a diffuse light, giving metallic surfaces their characteristic shiny luster (Fig. 2-13). Nonmetallic lusters are more varied; they can be vitreous (glassy), pearly, silky, adamantine ("like a diamond"), dull, or earthy (Fig. 2-14).

(a)

(b)

(c)

(d)

(e)

Figure 2-14 Nonmetallic lusters can be: (**a**) vitreous (glassy), as in rose quartz; (**b**) adamantine, as in anglesite; (**c**) silky, as in asbestos; (**d**) pearly, as in gypsum; or (**e**) earthy, as in realgar.

Figure 2-15 Though samples of hematite (Fe_2O_3) are usually steel-gray, hematite's streak is always reddish brown.

Streak **Streak** is the color of a mineral in its powdered form, obtained when the mineral's surface is pulverized by rubbing it across an unglazed porcelain slab known as a streak plate. The color of a mineral's streak often differs from that of the intact mineral sample, and is often a more accurate indicator of identity, because streak color is not affected by trace impurities in the sample. The steel-gray mineral hematite (Fe_2O_3), for instance, has a distinctive reddish brown streak (Fig. 2-15).

Streak can distinguish between similar-appearing minerals that may have considerably different economic values. Pyrite (FeS_2) is often called "fool's gold" because of its brassy yellow color; its streak, however, is black, in contrast with the golden yellow streak of true gold. Several similar-appearing minerals have similar streaks as well, however, and additional tests are necessary to positively identify these.

Table 2-3 **The Mohs Hardness Scale**

Mineral	Hardness	Hardness of Some Common Objects
Talc	1	
Gypsum	2	
		Human fingernail (2.5)
Calcite	3	
		Copper penny (3.5)
Fluorite	4	
Apatite	5	
		Glass (5–6), Pocketknife blade (5–6)
Orthoclase (potassium feldspar)	6	
		Steel file (6.5)
Quartz	7	
Topaz	8	
Corundum	9	
Diamond	10	

Hardness Geologists define **hardness** as a mineral's resistance to scratching or abrasion, which is tested by scratching an unknown mineral with a series of minerals or other substances the hardnesses of which are known. (Geological hardness is *not* a function of how easily a mineral breaks—a solid rap with a hammer will easily shatter a diamond, the world's hardest natural substance.) Because every scratch mark represents the removal of atoms from the surface of the mineral, and thus the breakage of the bonds holding these atoms, a mineral's hardness indicates the relative strength of its bonds.

The Mohs Hardness Scale, named for its developer, German mineralogist Friedrich Mohs (1773–1839), assigns relative hardnesses to several common and a few rare and precious minerals. An unknown mineral that can be scratched by topaz but not quartz has a hardness between 7 and 8 on the Mohs scale. Table 2-3, which lists the minerals and common testing standards used in the Mohs scale, explains why geologists are often found with a few copper pennies, a pocketknife, and well-worn fingernails.

The hardness of a mineral depends on the strength of its weakest bonds. Thus, minerals with many covalent bonds are generally harder than those with ionic or other bonds. Because every atom of carbon in a diamond forms a strong covalent bond with four neighboring carbons (see Fig. 2-9), diamond is exceptionally hard and durable. Graphite, diamond's polymorph, is one of the softest minerals. Its sturdy layers of covalently bonded carbon atoms are only weakly bonded to one another by van der Waals forces (Fig. 2-16). When you write with a pencil, pressure on the point breaks these weak bonds, leaving a trail of carbon layers on the paper.

Cleavage **Cleavage** is the tendency of some minerals, when hammered or struck, to break consistently along distinct planes in their crystal structures where the bonds are weakest. Such breaks form smooth flat surfaces, called *cleavage planes*, on the mineral. When a mineral that tends to cleave is struck along a plane of cleavage, every fragment that breaks off will have the same general shape. For example, if a cube of halite is shattered, numerous smaller cubes of halite will be produced.

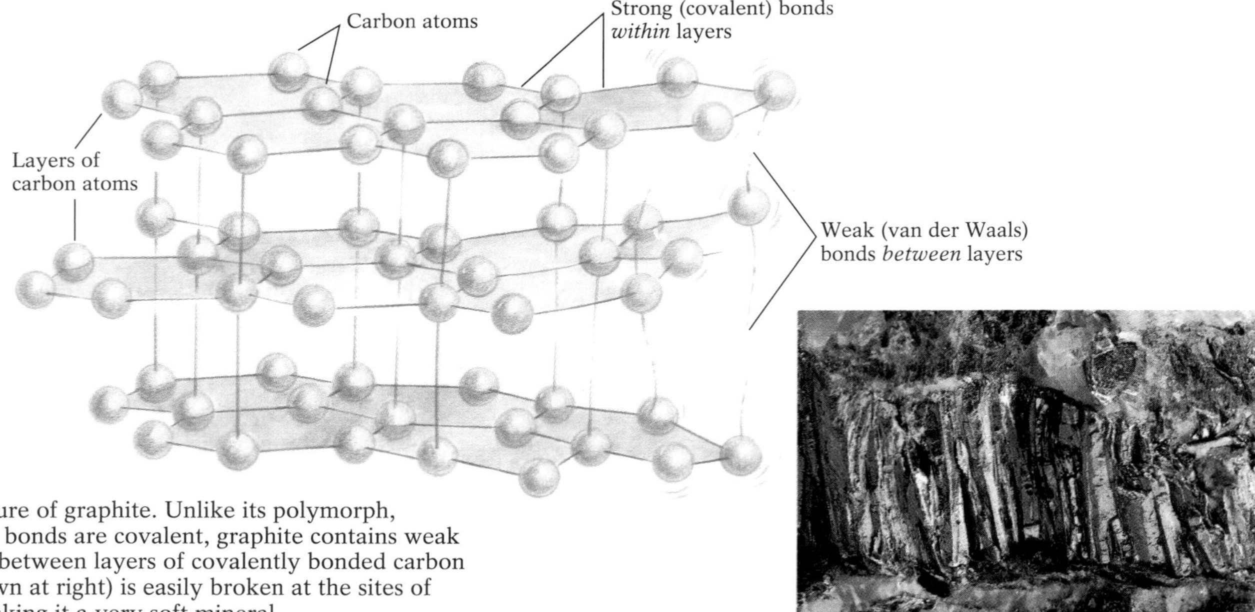

Figure 2-16 The structure of graphite. Unlike its polymorph, diamond, in which all bonds are covalent, graphite contains weak van der Waals bonds between layers of covalently bonded carbon atoms. Graphite (shown at right) is easily broken at the sites of these weak bonds, making it a very soft mineral.

(a)

(b) (c) (d) (e)

Figure 2-17 Distinguishing minerals by their cleavage. (**a**) Halite has three mutually perpendicular cleavage planes, forming cubes. (**b**) Mica has one perfect cleavage plane, forming sheets. (**c**) Calcite's three cleavage planes are not mutually perpendicular; calcite cleavage produces a geometrical shape called a rhomb. (**d**) A photomicrograph of cleavage in augite shows its planes intersecting at 90° angles. (**e**) A photomicrograph of cleavage in hornblende shows its planes intersecting at angles of about 60° and 120°.

In characterizing cleavage, geologists consider the number of cleavage surfaces produced and the angles between adjacent surfaces. Halite, for example, cleaves in three mutually perpendicular directions, forming cubes (Fig. 2-17a), whereas mica cleaves in only one direction, forming sheets (Fig. 2-17b). Calcite's ($CaCO_3$) three cleavage planes are not perpendicular to each other (Fig. 2-17c). Minerals that appear similar by other diagnostic criteria can have different cleavage-plane angles. For example, augite and hornblende are two common black minerals that are similar in external form, hardness, and other characteristics. When a crystal of each is broken, however, the two prominent cleavage planes of augite intersect at an angle of about 90°, whereas the two cleavage planes of hornblende intersect at about 120° and 60° angles. The differing cleavage angles reflect the different arrangement of atoms and strength of bonds within the crystal structures of these minerals (Figs. 2-17d and 2-17e).

The cleavage planes in diamonds and certain other minerals enable gem cutters to fashion beautiful jewels from raw uncut stones. Highlight 2-1 explains how valuable specimens of the Earth's hardest substance can be cut without destroying them.

Highlight 2-1 *Cutting Diamonds*

(a)

(b)

Figure 2-18 Diamond before and after cutting. **(a)** A raw uncut diamond. **(b)** A cut and faceted diamond.

Not long ago, a television commercial for a luxury automobile featured a diamond cutter seated in the rear of a large sedan, anxiously preparing to cut a huge diamond. One false move, or a jarring bump from inferior shock absorbers, and a precious stone would be shattered. The diamond cutter's anxiety would certainly be warranted in such a situation—diamond cutting is a difficult and exacting skill, requiring steady hands and vision and finely honed tools, as well as a precise knowledge of diamonds' crystalline structure.

Actually, diamonds have traditionally been cleaved (not cut) along the planes of their weakest bonds. Knowing the sites of these weaker bonds, an experienced diamond cutter can cleave a raw, irregularly shaped diamond into a perfectly symmetrical jewel. A master cutter may analyze a large stone for days or even weeks to determine which cleavage planes would maintain the largest gem while minimizing any visible imperfections.

Diamond can be cleaved in four directions. Thus, a diamond cutter can choose among many complex combinations of cleavage faces, or *facets* (Fig. 2-18). The cutter scratches the locations of the crucial planes on the raw stone with a diamond-tipped stylus (what else could scratch a diamond?), and then a sharp blade is positioned at each scratch and struck forcefully with a mallet. The stone cleaves smoothly, exposing the chosen planes. If a plane has been marked inaccurately, or if the blade is off the mark, the stone will shatter instead of cleaving. (Because of the great value of diamonds and the unpleasant financial consequences of a cleavage error, diamonds today are also sawed, a slower but safer way to cut a stone; by using ultrafine saw blades with cutting edges impregnated with diamond dust and rotating at high speeds, a cutter can gradually saw a diamond's surface into the desired shape.)

X-rays can locate cleavage planes with precision, and they are sometimes used, but the trained eye of an experienced diamond cutter is usually sufficient. After a diamond has been cut, its flat cleavage surfaces are polished with the only substance hard enough to abrade a diamond, a paste of pulverized non–gem-quality diamonds mixed with olive oil.

During cutting, a diamond may lose as much as half its size and weight. It will, however, gain in value: A well-cut diamond will have scores of polished surfaces that admit light, disperse it into the various colors of the spectrum, and then redirect the light to the viewer's eye as brilliant flashes of color.

Fracture Minerals that do not cleave—because all of their bonds are equally strong—will break at random, or **fracture**. Unlike a straight, smooth-faced cleavage plane, a fracture appears as a jagged irregular surface or as a curved, shell-shaped (*conchoidal*) surface. In the mineral quartz, composed exclusively of silicon and oxygen, all of the atoms are bonded covalently in a three-dimensional framework with equal bond strengths in all directions. Thus, when a crystal of quartz is struck, it fractures (Fig. 2-19). Knowing that quartz fractures instead of cleaving enables us to distinguish it from similar-looking minerals that cleave.

Figure 2-19 A conchoidal fracture surface on a quartz crystal. Quartz, with equally strong covalent bonds in all directions, has no planes of weakness. It therefore fractures irregularly instead of cleaving.

Smell and Taste Experienced geologists occasionally sniff and lick rocks to help identify minerals that have a distinctive smell or taste. Some sulfur-containing minerals emit the familiar rotten-egg stench associated with hydrogen sulfide gas (H_2S). Halite's salty taste distinguishes it from similar-looking minerals such as quartz and calcite; sylvite (KCl) is distinctively bitter. Kaolinite absorbs liquid rapidly—when licked, it absorbs saliva, causing it to stick to the tongue. Novices should be wary of tasting unknown minerals: Realgar and orpiment, which smell like garlic, especially when heated, have as their major element the poisonous metal arsenic.

Effervescence Certain minerals, particularly those that contain carbonate ions (CO_3^{2-}), effervesce, or fizz, when mixed with an acid. A few drops of dilute hydrochloric acid (HCl) on calcite ($CaCO_3$) produce a rapid chemical reaction that releases carbon dioxide gas (in the form of bubbles) and water; this helps to distinguish calcite from similar-looking minerals, such as quartz and halite, which do not effervesce.

Crystal Form Because the three-dimensional geometric form of a crystal is an external expression of a mineral's internal structure, the shape of a well-formed crystal is itself usually distinctive enough to be used to identify the mineral. The angles between adjacent faces of a given mineral crystal are the same in every well-formed, unbroken sample of that mineral. Adjacent faces on a perfect quartz crystal from Herkimer, New York (a good place to find them), will always be at an angle of 120°, the same as on perfect quartz crystals from Hot Springs, Arkansas (another good source).

The crystals of some minerals are not geometrically shaped. Among these are some whose crystals grow together into forms resembling distinctive nonmineral objects, making them easy to identify. A rosette shape characterizes barite ($BaSO_4$); botryoidal malachite ($Cu_2(OH)_2CO_3$) looks like "a bunch of grapes," from which it gets its name (derived from the Greek *botruoeides*); and stellate pyrite (FeS_2) resembles a collection of stars (Fig. 2-20).

Figure 2-20 Unusual crystal aggregates. (**a**) Barite rosettes; (**b**) botryoidal malachite (green); (**c**) stellate ("star-like") pyrite; (**d**) stibnite needles.

(**a**)

(**b**)

(**c**)

(**d**)

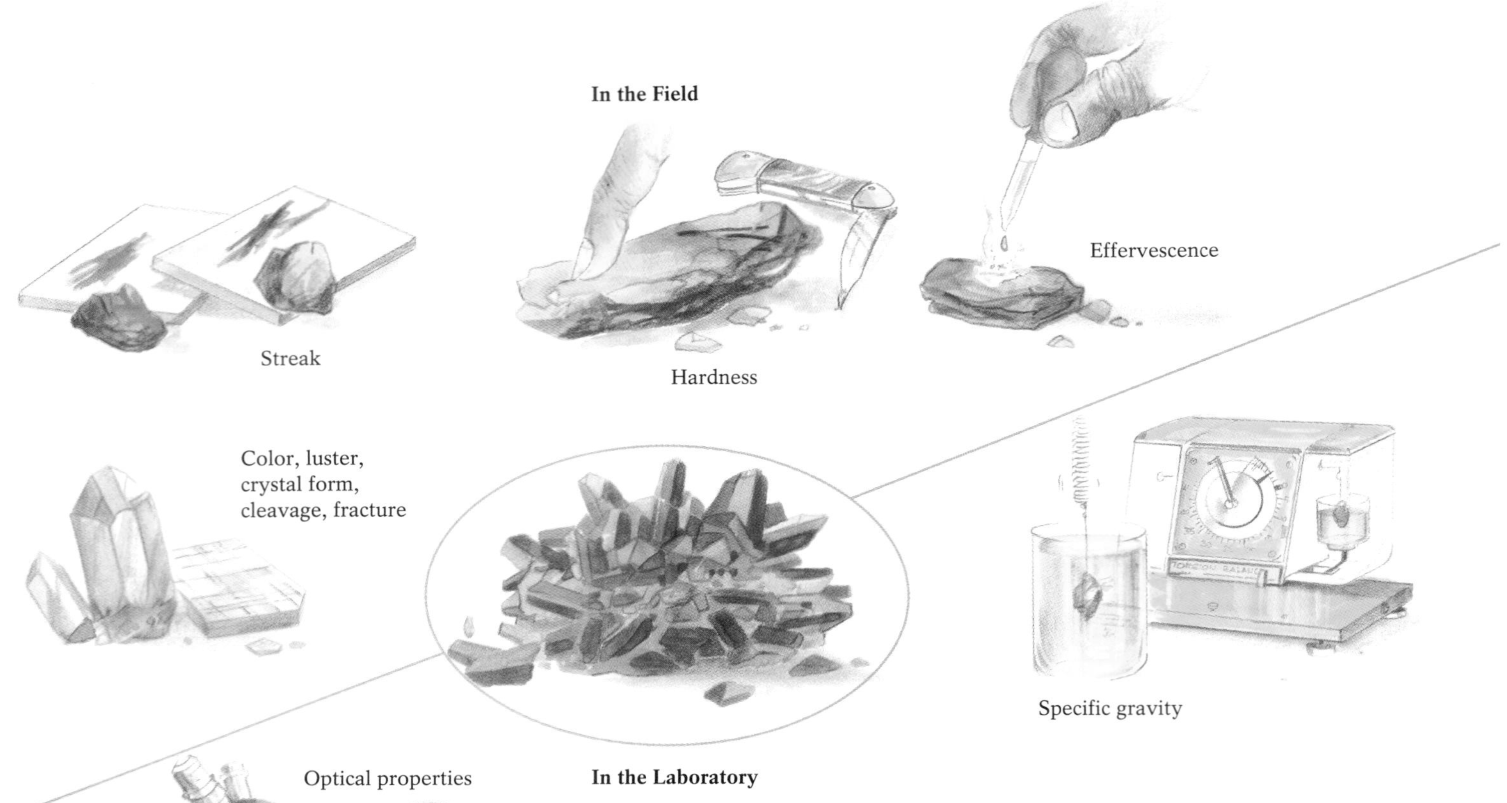

In the Field

Effervescence

Streak

Hardness

Color, luster, crystal form, cleavage, fracture

Specific gravity

Optical properties

In the Laboratory

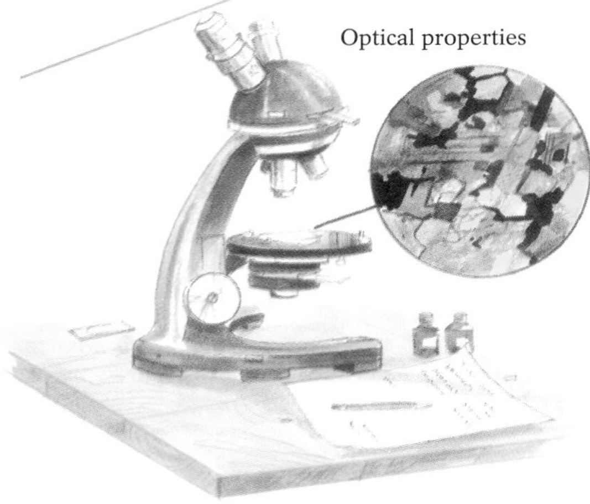

Fluorescence and phosphorescence

Figure 2-21 Any number of analyses—some simple and some complicated—may help determine a mineral's identity.

In the Laboratory

So many minerals have similar physical characteristics that even experienced geologists will make an educated guess as to their identity and then bring samples back from the field to test them in the laboratory. There, special equipment can analyze a variety of physical and chemical properties with greater precision than is ever possible in the field. Figure 2-21 shows a sampling of the methods geologists use to identify minerals in the field and in the laboratory.

Specific Gravity **Specific gravity** is the ratio of a substance's weight to the weight of an equal volume of pure water. For example, a mineral that weighs four times as much as an equal volume of water has a specific gravity of 4. A cubic centimeter of quartz (SiO_2), with a specific gravity of 2.65, is markedly lighter in weight than an equal volume of galena (PbS), with a specific gravity of 7.5.

Specific gravity can help distinguish between two apparently similar minerals. The specific gravity of 24-karat (pure) gold is 19.3, whereas the specific gravity of pyrite ("fool's gold") is 5. Gold's high specific gravity is the reason prospectors can "pan" for it. When they slosh river sand about in a pan, the heavier gold particles sink to the bottom, while lighter particles, usually of such common minerals as quartz and feldspar, readily float away.

Mineral polymorphs, with their different atomic arrangements, usually have different specific gravities. Even though both consist exclusively of carbon atoms, for instance, graphite has a specific gravity of 2.3, whereas the specific gravity of diamond is 3.5, due to its highly compressed crystal structure.

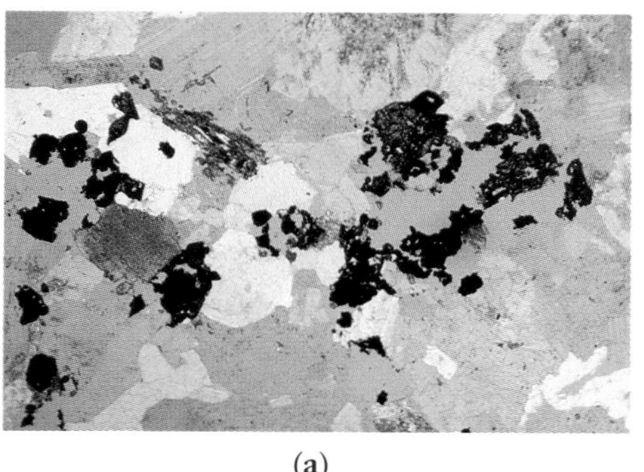

(a)

(b)

Figure 2-22 The colors that appear in a mineral sample while it is magnified under polarized light can be used to identify the mineral. (**a**) A thin section of granite as seen in regular light (magnified 70×). (**b**) The same section as seen under polarized light.

Other Laboratory Tests Different kinds of light can be transmitted through an ultra-thin (0.03-millimeter) slice of a mineral sample while the sample is simultaneously magnified using a *petrographic microscope*. Most minerals can be identified by their colors under these specific light conditions (such as under polarized light, as in Figure 2-22).

When exposed to ultraviolet light, certain minerals glow in distinctive colors. This property, called **fluorescence**, characterizes fluorite, calcite, scheelite ($CaWO_4$), and willemite (Zn_2SiO_4), among other minerals (Fig. 2-23). A mineral that continues to glow after the ultraviolet light has been removed is said to exhibit **phosphorescence**.

In **X-ray diffraction**, X-rays passed through a mineral sample become scattered, or *diffracted*, producing distinctive patterns. These patterns are determined by the arrangement of the atoms and ions in a mineral's crystal structure, and so are unique to each mineral (Fig. 2-24).

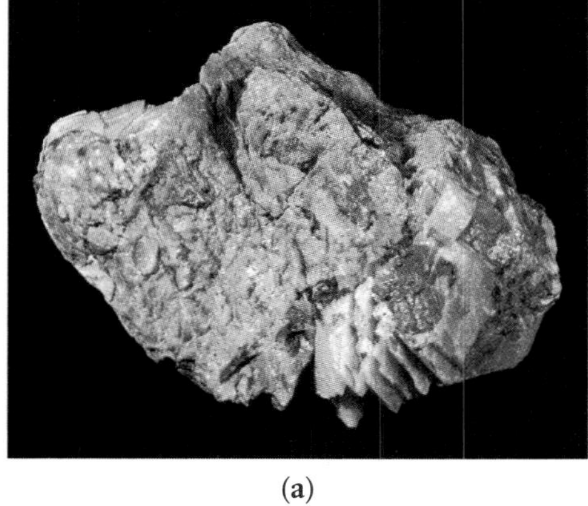

(a)

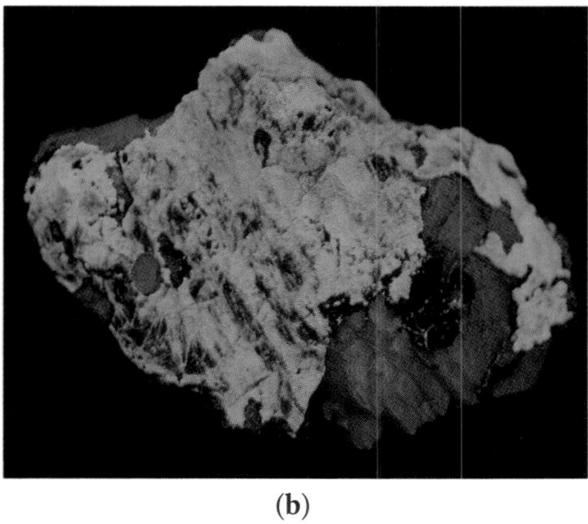

(b)

Figure 2-23 Minerals that glow in distinctive colors when exposed to ultraviolet light are said to be *fluorescent*. The minerals willemite and calcite in this rock specimen (**a**) glow bright green and red, respectively, while exposed to UV light (**b**).

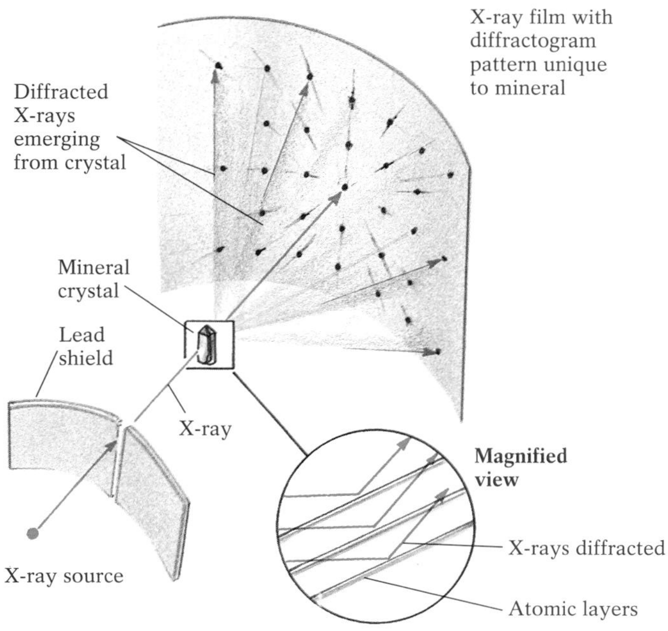

Figure 2-24 Mineral identification by X-ray diffraction. X-rays passed through a mineral scatter in a manner that is unique to each mineral. A *diffractogram* is the photographic record of the pattern of X-rays produced by a mineral.

Some Common Rock-Forming Minerals

Rock-forming minerals are those that compose the common rocks of the Earth's crust and mantle. Five groups of minerals predominate. Most are silicates, which contain silicon, oxygen, and usually one or a few other common elements. Important nonsilicates include carbonates, containing carbon, oxygen, calcium, and perhaps other atoms; oxides, containing oxygen and other atoms; sulfates, containing sulfur, oxygen, and various other atoms; and sulfides, containing sulfur and other atoms but no oxygen.

The Silicates and Their Structures

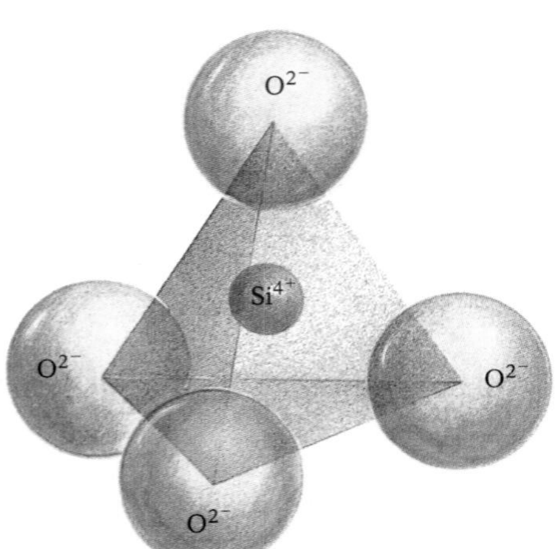

Figure 2-25 The silicon-oxygen tetrahedron. Four oxygen ions occupy the corners of this structure, with a lone silicon ion embedded in the open space at the center. (If you have access to four basketballs and a racquetball, you can easily re-create the SiO_4 tetrahedron.) The silicon-oxygen tetrahedron has an overall charge of −4.

Because silicon and oxygen are so abundant and unite so readily, **silicates** make up more than 90% of the weight of the Earth's crust and almost 100% of the weight of the Earth's mantle (extending from the crust to a depth of about 2900 kilometers). Silicates are the dominant component of most rocks, whether igneous, sedimentary, or metamorphic.

Why oxygen and silicon? Oxygen, the primary element in the Earth's crust, is readily available, and the only common crustal element whose atoms readily accept electrons to form negative ions. Silicon, the second most abundant element in the Earth's crust, is very compatible with oxygen: Its small, positively charged ion fits snugly in the niches among large, closely packed oxygen ions. The crystal structure of all silicates contains repeating groupings of four negative oxygen ions congregated around a single positive silicon ion to form a four-faced structure called a *tetrahedron* (Fig. 2-25). The four oxygen ions, each with a –2 charge, have a combined negative charge of –8; the one silicon ion has a +4 charge. The result is a **silicon-oxygen tetrahedron** (SiO_4^{4-}) with a –4 charge. To form an electrically neutral compound, a silicon-oxygen tetrahedron must either acquire four positive charges or disperse its negative charge. Other, positive ions may bond with the tetrahedron to balance its charge, or adjacent tetrahedra may share its oxygen ions, claiming half of their negative charge.

The number of oxygen ions shared by its tetrahedra, expressed as a silicon-to-oxygen ratio, determines the type of silicate and its crystal structure. A tetrahedron that does not share its oxygen ions has a silicon-to-oxygen ratio of 1:4, whereas a tetrahedron that shares all four of its oxygen ions has a silicon-to-oxygen ratio of 1:2 (see Fig. 2-26).

Because the SiO_4 tetrahedron is so abundant, many geologists believe that it was the first compound to form as the molten Earth began to cool and solidify. Since then, more than a thousand different silicate minerals have crystallized, testimony to the ease with which SiO_4 tetrahedra combine with various positive ions, and with each other, in nature. There are five principal silicate crystal structures: independent tetrahedra, single chains, double chains, sheets, and three-dimensional frameworks (Fig. 2-26a–e). Each structure represents a different means of sharing oxygen atoms, and each has its own silicon-to-oxygen ratio. Consequently, each displays distinctive physical characteristics and properties (Table 2-4).

Independent Tetrahedra Independent tetrahedra, bonded with positive ions of other elements and sharing no oxygen ions (Fig. 2-26a), form several prominent silicates. Olivine, for example, contains positive iron and/or magnesium ions whose +2 charges balance the −4 charge of the tetrahedron. Silicates with independent tetrahedra are noted for their hardness (6.5–8 on the Mohs scale), a consequence of the strong ionic bonds between the tetrahedra and the interspersed positive ions. Independent tetrahedra always maintain a silicon-to-oxygen ratio of 1:4.

Table 2-4 **Common Rock-Forming Silicates**

Silicate	Formula	Silicon: Oxygen Ratio* (silicate structure)	Properties
Quartz	SiO_2	1:2 (framework)	Hardness of 7; breaks by fracture; six-sided prismatic crystals; specific gravity 2.65
Alkali feldspars	$KAlSi_3O_8$	1:2 (framework)	Hardness of 6.0–6.5; strong cleavage in two directions at right angles; pink or white in color; specific gravity 2.5–2.6
Plagioclase feldspars	$(Ca,Na)AlSi_3O_8$	1:2 (framework)	Hardness of 6.0–6.5; strong cleavage in two directions at right angles; white to bluish-gray in color; specific gravity 2.6–2.7
Muscovite mica	$K_2Al_4(Si_6Al_2O_{20})(OH,F)_2$	1:2.5 (sheet)	Hardness of 2–3; perfect cleavage in one direction; colorless and transparent to light green-gray; specific gravity 2.8–3.0
Biotite mica	$K_2(Mg,Fe)_6Si_3O_{10}(OH)_2$	1:2.5 (sheet)	Hardness of 2.5–3.0; perfect cleavage in one direction; black to dark brown in color; specific gravity 2.7–3.2
Amphiboles	$(Na,Ca)_2(Mg,Al,Fe)_5(Si,Al)_8O_{22}(OH)_2$	1:2.75 (double chain)	Hardness of 5–6; cleaves in two directions at 56° and 124°; black to dark green in color; specific gravity 3.0–3.3
Pyroxenes	$(Mg,Fe,Ca,Na)(Mg,Fe,Al)Si_2O_6$	1:3 (single chain)	Hardness of 5–6; cleaves in two directions at about 90°; black to dark green in color; specific gravity 3.1–3.5
Olivine	$(Mg,Fe)_2SiO_4$	1:4 (independent tetrahedra)	Hardness of 6.5–7.0; green in color; breaks by fracture; specific gravity 3.2–3.6

*Aluminum ions may substitute for some silicon ions in certain minerals. These count as silicon ions for purposes of the silicon:oxygen ratio.

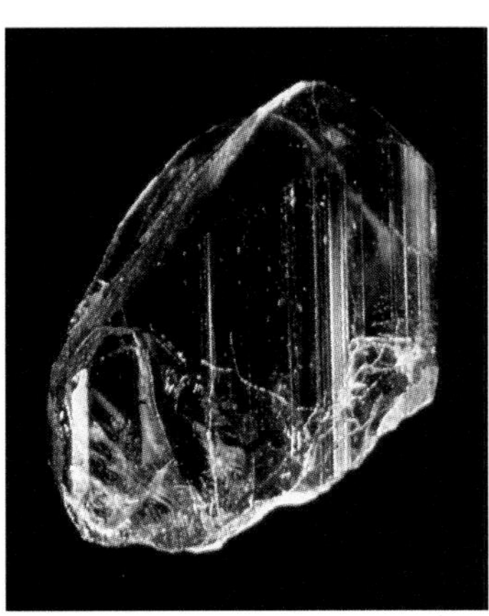

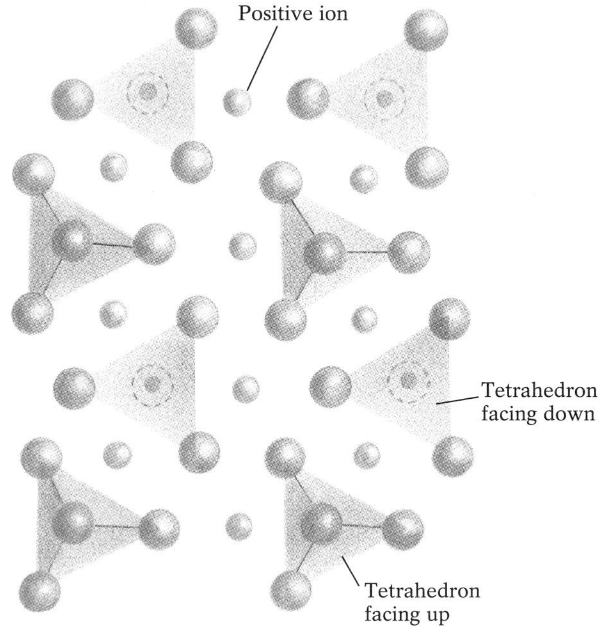

Figure 2-26 Types of silicate structures. **(a)** Independent tetrahedra. Positive ions are positioned between tetrahedra such that each tetrahedron, with a –4 charge, bonds to two positive ions, each with a +2 charge, thereby neutralizing their combined charge. No oxygen ions are shared between the tetrahedra; therefore the silicon-to-oxygen ratio is 1:4. Photo: Olivine. As is typical of silicates with independent tetrahedral structures, this mineral fractures rather than cleaving.

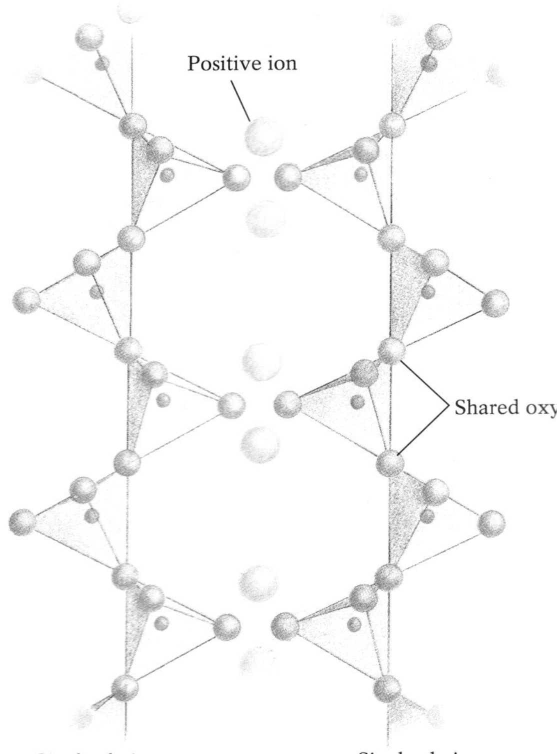

Single Chains
(b)

Figure 2-26 **(b)** Single chains. Each tetrahedron shares two of its corner oxygen ions with adjacent tetrahedra, forming a linear chain of tetrahedra with a silicon-to-oxygen ratio of 1:3. Because each tetrahedron still has a –2 charge after sharing two of its oxygen ions, the cumulative negative charge on such chains attracts a variety of positive ions which bind between them, neutralizing the negative charge of the chains and also joining them loosely together. Photo: Pyroxene, showing the 90° cleavage angles characteristic of single-chain silicates.

Single Chains When each tetrahedron shares two corner oxygens and still has an excess –2 charge, a linear chain of tetrahedra is formed (Fig. 2-26b). The accumulated negative charges cause the chain itself to act as a negative ion complex, attracting positive ions that neutralize its negative charge and bind adjacent chains. Because the bonds *within* chains are strong and those *between* chains are relatively weak, single-chain silicates tend to cleave parallel to the chains. A prominent group of single-chain silicates are the *pyroxenes*, which typically contain iron and/or magnesium ions that bind the negatively charged chains together. This pattern of oxygen-sharing produces a silicon-to-oxygen ratio of 1:3.

Double Chains A double chain forms when adjacent tetrahedra share two corner oxygens, and in addition some share a third oxygen with a tetrahedron in a neighboring chain (Fig. 2-26c). This shared oxygen binds the two chains together. With more oxygen sharing than in single chains, double chains have a silicon-to-oxygen ratio of 1:2.75. Any of several positive ions may link double chains to adjacent double chains, producing the complex silicates known as the *amphiboles*. The most common amphibole is hornblende, in which aluminum substitutes for some of the silicon within the tetrahedron (because their ions are similar in size—see Fig. 2-11b—and charge).

Hornblende's relatively high iron and/or magnesium content gives it a dark green-to-black color, which is similar to that of the pyroxenes. These two similar-looking types of silicates have different crystal structures with distinctive internal planes of weakness, however, and can be readily distinguished by their cleavage angles.

Sheet Silicates When all three oxygens at the base of a tetrahedron are shared with other tetrahedra, a sheet is formed that can extend indefinitely in the two dimensions of a plane. The fourth oxygen, at the peak of each tetrahedron, projects outward from the sheet, free to bond with available positive ions and thus bind two adjacent sheets together. The two-sheet layers are in turn bound to adjacent two-sheet layers by other positive ions (Fig. 2-26d). The bonds between the two-sheet pairs are far weaker than the bonds within each sheet or those that hold each two-sheet pair together, and are easily broken, producing the sheet silicates' characteristic cleavage. The *micas* are sheet silicates with a silicon (and sometimes substituted aluminum)-to-oxygen ratio of 1:2.5.

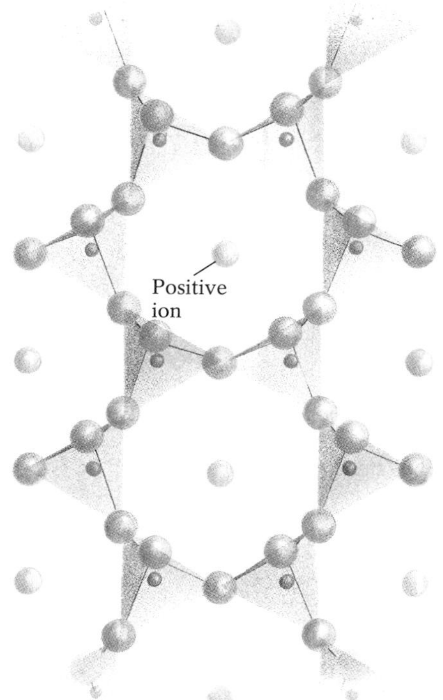

Double Chains
(c)

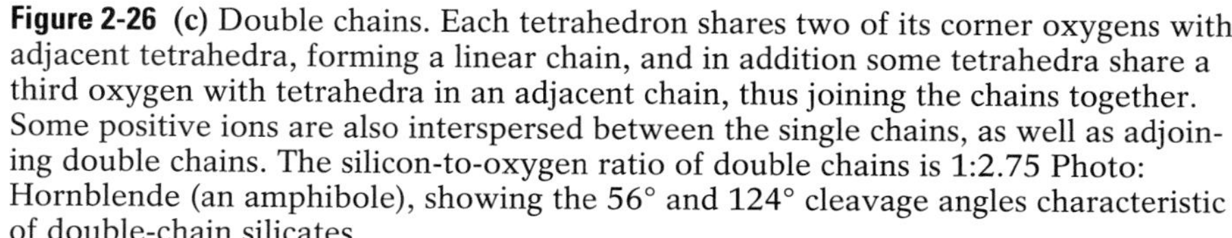

Figure 2-26 (c) Double chains. Each tetrahedron shares two of its corner oxygens with adjacent tetrahedra, forming a linear chain, and in addition some tetrahedra share a third oxygen with tetrahedra in an adjacent chain, thus joining the chains together. Some positive ions are also interspersed between the single chains, as well as adjoining double chains. The silicon-to-oxygen ratio of double chains is 1:2.75 Photo: Hornblende (an amphibole), showing the 56° and 124° cleavage angles characteristic of double-chain silicates.

In the common mica muscovite, adjacent sheets of tetrahedra are securely bonded by aluminum ions to form a two-sheet pair, but the two-sheet pairs are weakly bonded by intervening potassium ions. These weak bonds readily break when we attempt to peel the layers of muscovite, causing sheets of this mica to separate easily. To visualize the structure of a sheet silicate, consider a stack of peanut-butter sandwiches separated by pieces of wax paper. Each sandwich is analogous to a two-sheet layer of mica; the peanut butter represents the strong aluminum bonds between the sheets and the wax paper represents the weak potassium bonds between the two-sheet layers. If you pick up the top sandwich by grasping its upper slice of bread, the whole sandwich, bound strongly by the peanut butter, remains intact. However, each sandwich separates readily from the one below, as does each two-sheet pair in a sheet-silicate mineral such as muscovite mica.

Framework Silicates When all four tetrahedral oxygen ions are shared with adjacent tetrahedra, a three-dimensional framework structure results. This highest degree of oxygen sharing results in the lowest silicon-to-oxygen ratio, 1:2. This is the structure of the two most abundant minerals of the Earth's crust, quartz and feldspar.

Figure 2-26 (d) Sheet silicates. Each tetrahedron shares all three of its corner oxygens, forming a sheet of adjoined tetrahedra. The fourth oxygen in each tetrahedron extends upward to bond with positive ions and, subsequently, another sheet. (Other positive ions bind two-sheet pairs to adjacent two-sheet pairs.) The silicon-to-oxygen ratio of sheet silicates is 1:2.5. Photo: Muscovite mica, showing the planar cleavage of the sheet silicates.

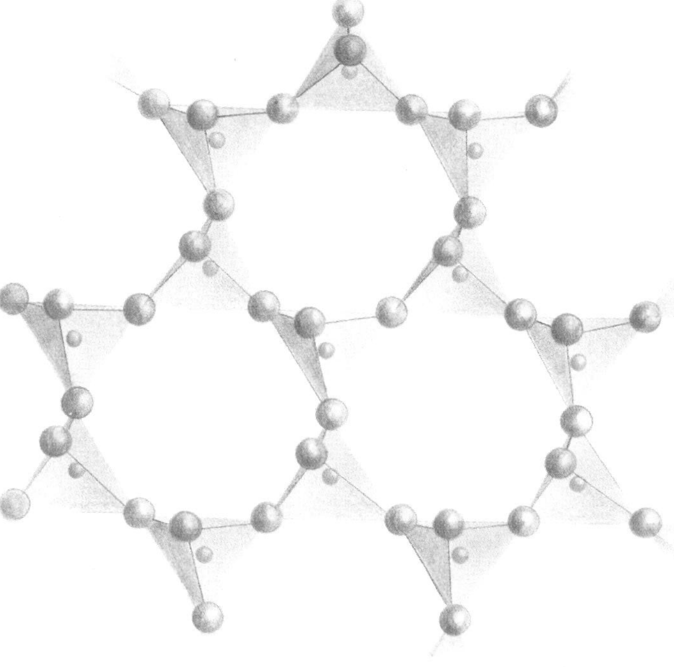

Sheet Silicates
(d)

Figure 2-26 (e) Framework silicates. Each tetrahedron shares all four of its oxygens, forming a three-dimensional framework structure. Because the charge on each tetrahedron is neutralized by the sharing of all its negative charges, no positive ions bond with the structure. The silicon-to-oxygen ratio of framework silicates is 1:2. Photo: Quartz crystals. Due to its sturdy framework structure, quartz does not cleave.

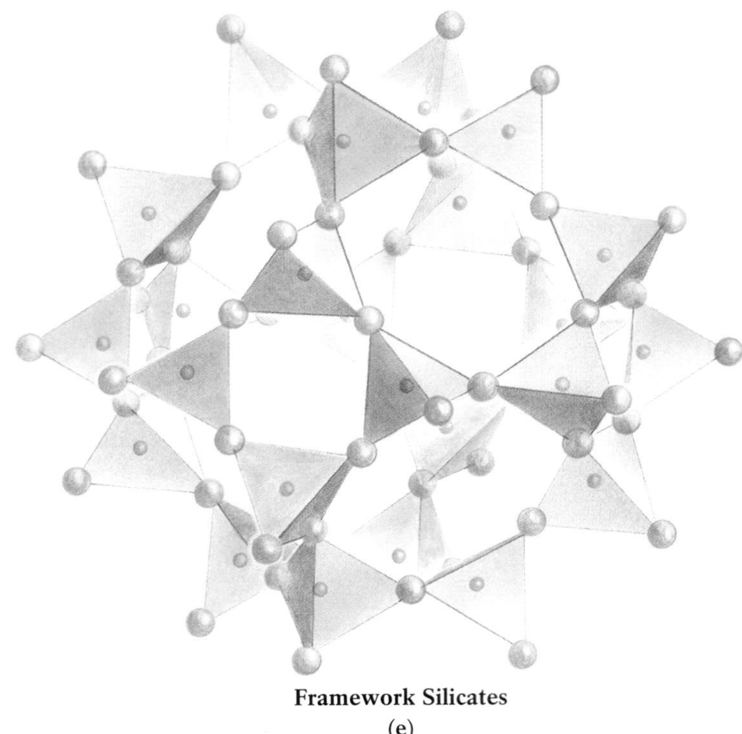

Framework Silicates
(e)

Quartz, the second most abundant mineral in the continental crust, is the only silicate composed entirely of silicon and oxygen. Every oxygen ion in a tetrahedron is shared with an adjacent tetrahedron; quartz achieves electrical neutrality without other ions (Fig. 2-26e). Its strong covalent bonds make quartz the hardest of the common rock-forming minerals (7 on the Mohs scale). Because all of its bonds are equally strong, quartz has no weak planes and therefore does not cleave, but breaks by fracturing. Pure quartz is transparent and colorless, due to the absence of other ions that would absorb light. Because of its open structure, however, quartz often traps a few stray ions when it crystallizes, and these impurities give quartz a wonderful range of colors. When quartz has room to grow undisturbed, it assumes its characteristic crystal shape—perfect six-sided prisms with pyramids on top and bottom.

The *feldspars*, which account for about 60% by volume of continental crust, crystallize from molten rock across a wide range of temperature and pressure. In feldspars, as in quartz, every oxygen ion in a tetrahedron is shared with an adjacent tetrahedron. Unlike quartz, however, which consists exclusively of silicon and oxygen, in the feldspars silicon atoms are often replaced with atoms of aluminum, and potassium, sodium, and calcium ions occupy open spaces between the tetrahedra. There are two types of feldspars, based on their chemical composition. The plagioclase feldspars contain positive sodium and/or calcium ions, which substitute for one another, in varying proportions; the all-sodium plagioclase feldspar albite is at one end of the range, and the all-calcium plagioclase feldspar anorthite is at the other. The alkali feldspars are potassium-rich silicates whose slightly different internal structure is required by the large size of the potassium ion. All feldspars have two prominent cleavage planes that intersect at a 90° angle. They are relatively hard, about 6 on the Mohs scale.

Nonsilicates

Because the common rock-forming silicates are so abundant, no one gets very excited about the discovery of a fresh supply of plagioclase feldspar—we've got plenty already. But nonsilicate minerals constitute only about 5% of the Earth's continental crust. They include native elements—those that are not combined in nature with other elements—such as the precious metals gold, silver, and platinum, and such useful metals as iron, aluminum, copper, nickel, zinc, lead, and tin. Nonsilicates also include a few noteworthy gems, such as diamonds, rubies, and sapphires. Nonsilicates are constructed from a variety of negative-ion groups, most containing oxygen. The most common nonsilicates are the carbonates, oxides, and sulfates, as well as a few groups without oxygen, such as the sulfides (Table 2-5).

Carbonates The **carbonate** ion complex (CO_3^{2-}) contains one central carbon atom with strong covalent bonds to three neighboring oxygen atoms. It typically combines with one or more positive ions, such as calcium to form calcite ($CaCO_3$) or calcium and magnesium to form dolomite ($CaMg(CO_3)_2$). Because the ionic bonds between the negative carbonate ion group and the positive ions are relatively weak, calcite and dolomite are relatively soft, registering 3–4 on the Mohs scale. They also dissolve readily in acidic water, with important geological consequences that we will examine in Chapters 5, 6, and 16.

Oxides Mineral **oxides** are produced when negative oxygen ions combine with one or more positive metallic ions. The resulting minerals include some of our major sources of iron—hematite (Fe_2O_3) and magnetite (Fe_3O_4)—as well as tin, titanium, and uranium.

Table 2-5 **Common Nonsilicate Minerals**

Mineral Type	Composition	Examples	Uses
Carbonates	Metallic ion(s) plus carbonate ion complex (CO_3^{2-})	Calcite ($CaCO_3$)	Cement
		Dolomite ($CaMg(CO_3)_2$)	Cement
Oxides	Metallic ion(s) plus oxygen ion (O^{2-})	Hematite (Fe_2O_3)	Iron ore
		Magnetite (Fe_3O_4)	Iron ore
		Corundum (Al_2O_3)	Gems, abrasives
		Cassiterite (SnO_2)	Tin ore
		Rutile (TiO_2)	Titanium ore
		Ilmenite ($FeTiO_3$)	Titanium ore
		Uraninite (UO_2)	Uranium ore
Sulfides	Metallic ion(s) plus sulfur (S^{2-})	Galena (PbS)	Lead ore
		Pyrite (FeS_2)	Sulfur ore
		Cinnabar (HgS)	Mercury ore
		Sphalerite (ZnS)	Zinc ore
		Molybdenite (MoS_2)	Molybdenum ore
		Chalcopyrite ($CuFeS_2$)	Copper ore
Sulfates	Metallic ion(s) plus sulfate ion (SO_4^{2-})	Gypsum ($CaSO_4{\cdot}2H_2O$)	Plaster
		Anhydrite ($CaSO_4$)	Plaster
		Barite ($BaSO_4$)	Drilling mud
Native elements	Minerals consisting of a single element	Gold (Au)	Jewelry, coins, electronics
		Silver (Ag)	Jewelry, coins, photography
		Platinum (Pt)	Jewelry, catalyst for gasoline production
		Diamond (C)	Jewelry, drill bits, cutting tools

Sulfides and Sulfates With six electrons in its outer energy level, sulfur can either accept or donate electrons. When sulfur accepts electrons, it becomes a negative ion that bonds with various positive ions to form the **sulfides**. This valuable group of minerals, including copper sulfides such as chalcocite (Cu_2S) and lead sulfides such as galena (PbS), is the source of a number of important metals.

When sulfur donates electrons, it becomes a positive ion that bonds with oxygen to form the negative **sulfate** ion complex (SO_4^{2-}). Gypsum ($CaSO_4 \cdot 2H_2O$) is a common and useful sulfate mineral used to manufacture sheetrock and plaster of Paris.

Naming Minerals

There are more than 3000 known minerals, and new ones are identified and named each year. The names of some challenge even the most eloquent among us—try pronouncing yugawaralite, ammoniojarosite, or anabohitsite. Why do some minerals get such tongue-twisting names? Why not simply refer to them in a systematic way based on the elements that form them?

One reason is that atoms of the same chemical compound can be arranged to form more than one mineral. For example, the atoms in calcium carbonate ($CaCO_3$) form several different crystal structures, including those of calcite and its polymorph aragonite. Another reason is that some minerals' chemical compositions are so complex that a name based on their formulas would not be possible. Consider pumpellyite, a mineral found in rocks that have been slightly heated and compressed during certain types of mountain-building events—its chemical formula is $Ca_4(Mg,Fe^{2+})(Al,Fe^{3+})_5O(OH)_3(Si_2O_7)_2(SiO_4)_2 \cdot 2H_2O$. Finally, many minerals were known by common names for centuries before their chemical compositions were identified, and few of us would choose to ask for "sodium chloride" at the dinner table, or wish to receive a two-carat "high-pressure carbon" engagement ring.

An international commission approves the names proposed for new minerals as they are identified. Some names acknowledge the geographical location where a mineral was discovered, such as labradorite; some refer to a distinctive physical characteristic, such as citrine (yellow quartz), named for its color's resemblance to that of a citron, a lemon-like fruit; other mineral names honor heroes, such as the Moon mineral armalcolite, named for the first lunar explorers Neil *Arm*strong, Edwin *Al*drin, and Michael *Col*lins.

Gemstones

Several minerals lead a glamorous life as gemstones—precious or semi-precious minerals that display particularly appealing color, luster, or crystal form and can be cut or polished for ornamental purposes. Some gems, such as diamonds and emeralds, are quite rare. Others are unusually perfect crystals of common minerals such as the rock-forming silicates: Amethyst is a common gemstone the appealing purple tones of which are produced by small numbers of iron atoms scattered throughout the crystal structure of some quartz; the valued gemstone amazonite is an abundant alkali feldspar that owes its vivid green color to a few chromium atoms in its crystal structure. Common nonsilicates also can be gems. The aluminum oxide corundum, with its hardness of 9, is a popular abrasive used in emery cloth and sandpaper—but when its crystals are perfectly formed, and have a few other ions trapped in their structures for color, ordinary corundum becomes not-so-ordinary sapphires and rubies (Fig. 2-27).

(a)

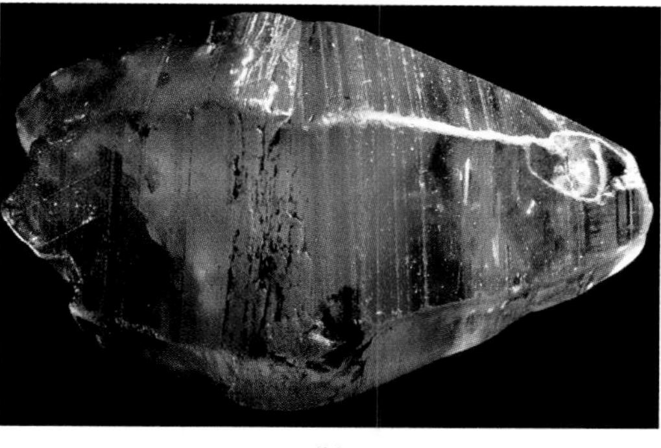

(b)

(c)

Figure 2-27 The minerals corundum, sapphire, and ruby each have the chemical formula Al_2O_3. Their different colors are produced by trace impurities in their crystal structure. (**a**) Ruby (red) embedded in corundum (colorless). (**b**) Sapphire. (**c**) Ruby.

How Gemstones Form

Minerals of gemstone quality form under conditions that promote the development of perfect, large crystals. This happens most often in two ways: when molten rock cools and crystallizes deep underground, or when preexisting rock is subjected to extraordinary pressure and heat. Molten rock often migrates into fractures in surrounding cooler rocks, where its ions and atoms crystallize in reasonably large spaces, producing perfect crystals and, if the space is big enough, enormous crystals. A single pyroxene crystal excavated in South Dakota was more than 12 meters (40 feet) long, 2 meters (6.5 feet) wide, and weighed more than 8000 kilograms (8 tons). This process also produces such complex silicate gemstones as topaz, tourmaline, and beryl.

Gemstones may also form when the heat and pressure applied along the edges of colliding plates cause the ions and atoms in their rocks to migrate and recombine, creating new minerals that are more stable under the new conditions. For example, heating and compression of carbonate rocks that contain aluminum ions can cause the aluminum to combine with oxygen liberated from calcite, forming the aluminum oxides we know as rubies and sapphires.

And then there is that rare stone, the diamond, which is transformed from unspectacular carbon into a brilliant crystal by the unique geological conditions of extremely high pressures at great depths (greater than 150 kilometers, or 90 miles). Diamonds most often are found where superheated gas has propelled molten rock rapidly from great depth to the surface, carrying the diamonds along with it. The resulting diamond-rich structures, known as kimberlite pipes (named for Kimberley, South Africa), are typically a few hundred meters to a kilometer across and many kilometers deep. Kimberlite pipes are found in Siberia, India, Australia, Brazil, and in southern and central Africa, and scattered throughout the Rockies of Colorado and Wyoming. Most apparently formed between 70 and 150 million years ago, when the supercontinent of Pangaea was rifting into the continents that we know today. (See Chapter 1 for a discussion of Pangaea.) As the rifts opened, they apparently provided pathways through which diamond-bearing molten rock could rise from great depths to the surface. Thus, the value of diamonds is in part related to the unique historical circumstances under which they were propelled to the surface. Moreover, only a few kimberlite bodies produce gem-quality diamonds. (In southern Africa, where most of the world's diamonds are mined, only 1 in 200 kimberlite pipes yield enough diamonds to make it worth the high cost of mining them.)

Synthetic Gems: Can We Imitate Nature?

Because the most valuable gemstones are rare, people have tried for centuries to duplicate nature's feat and produce synthetic gems. In the twentieth century they have had some success, and have even managed to surpass nature in some cases. Artificial emerald crystals, first created in the 1930s, are more transparent, richer in color, and more perfect in shape than natural ones, which are often marred by gas bubbles and other impurities. The high quality of synthetic emeralds contributes to their great market value, which, at several hundred dollars per carat, is still far less than the price of the extremely rare natural ones.

Even diamonds can now be made in the laboratory, by subjecting carbon to extreme heat and pressure. (Almost any carbon-rich substance will do as a starting point, even sugar or peanuts.) On December 12, 1954, scientists in the General Electric research lab in Schenectady, New York, created the first tiny synthetic diamonds by subjecting carbon to great pressure and temperatures exceeding 3000°C (5400°F). Today, more than 20 tons of industrial-grade synthetic diamonds are produced each year, destined for such practical uses as to be drill bits in oil-well drilling and modern dentistry.

In 1970, the first gem-quality diamonds were synthesized. Now, some synthetics even outshine the original. Strontium titanate, a synthetic mineral sold under the trade names of Fabulite and Wellington Diamond, glitters four times more vividly than a real diamond. Being only moderately hard, between 5 and 6 on the Mohs scale, it is not very durable. And because its creators can manufacture tons of it, this synthetic gem is only as rare as they choose. Another synthetic, cubic zirconia, can be manufactured in batches of 50 kilograms (110 pounds) and sold wholesale for a few cents per carat. Its optical qualities are virtually indistinguishable from those of nature's diamonds, and it is quite durable. Only the fact that its specific gravity is higher than that of natural diamond reveals its identity.

Minerals as Clues to the Past

After mineral samples are identified, geologists use their knowledge of how different minerals form to determine the geological events and environmental conditions that may have produced them. For example, large deposits of halite, such as those in Michigan, Kansas, and Louisiana, suggest evaporation of an ancient saltwater sea. Glaucophane, a blue variety of amphibole, is known to form only in high-pressure, low-temperature conditions found together only in deep-sea trenches where oceanic plates are subducting at convergent plate boundaries. (See the discussion of subduction in Chapter 1.) Thus, a rock containing glaucophane must have formed in an ancient subduction zone.

Polymorphs, with identical chemical compositions but different crystal structures, form under different heat and pressure conditions. Ordinary quartz, with a specific gravity of 2.65, crystallizes at relatively low temperature and pressure. One of its polymorphs, stishovite, with a specific gravity of 4.28, crystallizes at temperatures higher than 1200°C (2200°F) and pressures in excess of 130,000 times that at sea level. Geologists believe that the dense compressed structure of stishovite probably results from extreme conditions produced by geological events of enormous magnitude, such as meteorite impacts. This hypothesis is supported by the presence of stishovite in the fractured rocks of Meteor Crater near Winslow, Arizona, and at Manicouagan

Figure 2-28 (**a**) Meteor Crater, near Winslow, Arizona, is believed to have been produced by a relatively small meteorite impact about 25,000 years ago. (**b**) The circular basin that forms Manicouagan Lake (75 kilometers, or 40 miles, in diameter), in Quebec, Canada, is also thought to be the product of an ancient meteorite strike, dating from about 210 million years ago. Studies of rocks near Meteor Crater and Manicouagan Lake have revealed the presence of the extremely high-pressure variety of SiO_2, stishovite; given our knowledge of the conditions that form stishovite, hypotheses that propose impact origins for Meteor Crater and Manicouagan Lake are thus reinforced.

(a)

(b)

Lake, Quebec, which have been proposed as ancient meteorite-impact sites (Fig. 2-28).

Now that we have introduced the basic structure of minerals, and how they form and are identified, we can began to examine more closely the common types of rocks that make up the Earth's crust. In the next five chapters, we will discuss how igneous, sedimentary, and metamorphic rocks form (introduced in Chapter 1 with the rock cycle), and how we often use the minerals in these rocks to interpret past geological events.

As a geology student, you may be tempted to browse from time to time through the bins of mineral specimens at rock shops and museums. A few years ago a man in Tucson, Arizona, was doing just that. From a barrel of unspectacular stones, he pulled a potato-sized lavender specimen, paid the proprietor $10, and went home with what may have been the largest sapphire (1905 carats) ever found. Although the value and quality of the stone have been hotly debated, a quote from the keen-eyed rockhound still rings true: "The only reason more gems aren't found in this country is that no one is looking for them." Now that you know something about minerals, you too can start looking.

Chapter Summary

Minerals are naturally occurring inorganic solids with specific chemical compositions and definite internal structures. **Rocks** are naturally occurring aggregates of minerals. Minerals are composed of one or more chemical **elements**, the form of matter that cannot be broken down to a simpler form by heating, cooling, or reacting with other elements. Each element, in turn, is made up of **atoms**, infinitesimally small particles that retain all of an element's chemical **properties**, the characteristics of a substance that enable us to distinguish it from other substances. When atoms of two or more elements combine in specific proportions, they form chemical **compounds**, which have different properties than any of their constituent elements individually. All minerals are chemical compounds.

The center of an atom is occupied by its **nucleus**, which contains both positively charged particles called **protons** and uncharged particles of equal mass called **neutrons**. An element's **atomic mass** is the sum of its protons and neutrons. An element's **atomic number** is determined by the number of protons in its nucleus; every atom of a given element always has the same number of protons, and thus the same atomic number. However, atoms of a given element may differ in atomic mass because the number of neutrons in their nuclei may vary. Atoms of a given element that contain different numbers of neutrons are **isotopes** of that element.

An atom's nucleus is surrounded by a cloud of negatively charged particles called **electrons** that move about the nucleus at high speed. Each electron occupies a specific region of space called an **energy level**; an atom may have one or more energy levels. The electrons occupying the outermost energy level can usually be donated, acquired, or shared with other atoms to achieve the most chemically stable electron configuration, one in which eight electrons fill an atom's outer energy level. The tendency of most atoms to fill their outer energy level in this manner is known as the **octet rule**.

Atoms combine to form chemical compounds in a variety of ways known as **bonding**. When an atom donates or acquires electrons, it becomes an electrically charged particle called an **ion**. An atom that donates electrons becomes positively charged; an atom that acquires electrons becomes negatively charged. Positive and negative ions are attracted to one another by one of several types of chemical bonds to form electrically neutral chemical compounds.

When electrons are acquired or donated between atoms, an **ionic bond** is formed. When the electrons in the outer energy levels of atoms are shared between atoms, a **covalent bond** is formed. In **metallic bonding**, electrons move continually among numerous closely packed nuclei. **Intermolecular bonds**, including **hydrogen bonds** and **van der Waals bonds**, form from weak attractions between groups of atoms, resulting from temporary charges caused by the movement of orbiting electrons.

As a mineral forms through chemical bonding, all of its ions or atoms occupy specific positions to form its **crystal structure**, a three-dimensional pattern that repeats throughout the mineral. When mineral growth is not limited by space, a **crystal** may form with a regular geometric shape that reflects the mineral's internal crystal structure. Naturally occurring inorganic solids that lack systematic crystal structures are **mineraloids**. Volcanic glass, which cools too rapidly from molten rock to develop a crystal structure, is an example of a mineraloid.

The types of minerals that will form at a given time and place are determined by which elements are available to bond, the charges and sizes of their ions, and the temperature and pressure under which the minerals form. Ions and atoms of similar size and charge are able to replace one another within a crystal structure; this **ionic substitution** produces minerals whose crystal structures are the same but which have a few minor variations in composition. Because they form under different environmental conditions, two minerals of the same chemical composition, such as the carbon-based minerals diamond and graphite, may have different crystal structures; such minerals are called **polymorphs**.

Geologists identify minerals out in the field by noting their external characteristics and measuring certain physical properties, including color, **luster**, **streak**, **hardness**, **cleavage**, **fracture**, and crystal form. Methods used to identify minerals in the laboratory include determining their **specific gravity**, examining them under a petrographic microscope, assessing whether they **fluoresce** or **phosphoresce** under ultraviolet light, and analyzing them by **X-ray diffraction**.

Although there are scores of substances that may have orderly internal patterns, the number of naturally occurring minerals is limited by the availability of the Earth's elements, their electrical charges, and the relative sizes of their ions. The Earth's crust is composed primarily of only eight common elements, which combine to produce the **rock-forming minerals**. The two most prominent elements, oxygen and silicon, readily combine to form the **silicon-oxygen tetrahedron**. This is the basic building block of the Earth's most abundant group of minerals, the **silicates**. The silicates, which make up more than 90% of the Earth's crust and nearly 100% of the Earth's mantle, include more than a thousand different minerals. Silicon-oxygen tetrahedra may be linked in a variety of crystal structures: independent tetrahedra (olivine is an example); single chains of tetrahedra (typified by the pyroxenes); double chains (the amphiboles); sheet structures (the micas); and framework structures (feldspars and quartz).

A number of nonsilicates are also common rock-forming minerals. Among them are the **carbonates** (such as calcite and dolomite); the **oxides** (such as iron-rich magnetite and hematite); and the **sulfides** (such as lead-based galena and iron-based pyrite) and **sulfates** (such as calcium-based gypsum).

Gemstones are minerals that are valued for their particularly appealing color, luster, or crystal form. Some gemstones, such as diamonds and emeralds, are quite rare; others are unusually well-formed specimens of relatively commonplace minerals, usually containing trace impurities in their structures which impart distinctive colors to the crystals. Although naturally formed gemstones are considered the most valuable, gemstones can be synthesized in the laboratory, as well.

Minerals that form under unique geological circumstances, when identified in ancient rocks, can provide clues to the past. The presence of the quartz polymorph stishovite, for instance, because it is known to form under high temperature/high pressure conditions, is considered to be evidence of meteoric impact.

Key Terms

minerals (p. 36)
rocks (p. 36)
element (p. 36)
atom (p. 36)
property (p. 36)
compound (p. 37)
proton (p. 38)
neutron (p. 38)
nucleus (p. 38)
electron (p. 38)
atomic mass (p. 38)
atomic number (p. 38)
isotope (p. 38)
energy level (p. 38)
bond (p. 39)
octet rule (p. 39)
ion (p. 39)
ionic bonding (p. 39)
covalent bonding (p. 40)
metallic bonding (p. 40)
intermolecular bonding (p. 41)
hydrogen bonding (p. 41)
van der Waals bonding (p. 41)
crystal (p. 42)
crystal structure (p. 42)
mineraloid (p. 42)
ionic substitution (p. 43)
polymorph (p. 43)
luster (p. 45)
streak (p. 45)
hardness (p. 46)
cleavage (p. 46)
fracture (p. 48)
specific gravity (p. 50)
fluorescence (p. 51)
phosphorescence (p. 51)
X-ray diffraction (p. 51)
rock-forming minerals (p. 52)
silicates (p. 52)
silicon-oxygen tetrahedron (p. 52)
carbonates (p. 57)
oxides (p. 57)
sulfides (p. 58)
sulfates (p. 58)

Questions for Review

1. What is a mineral? How does a mineral differ from a rock?
2. Briefly describe the structure of an atom. What is an isotope? What is an ion?
3. How does an atom achieve chemical stability? How does a chemical compound achieve electrical neutrality?
4. Describe three types of chemical bonding.
5. What is a mineral crystal? Describe two circumstances under which a mineral probably will *not* form into a crystal.
6. Define and give an example of a mineral polymorph.
7. Describe four properties of minerals that you could use to identify an unknown mineral.
8. Briefly discuss why silicon and oxygen are so compatible in nature.
9. List four different silicate structures and give a specific mineral example of each.
10. List two types of common nonsilicates and give a specific mineral example of each.
11. Describe the connections between at least two different geological environments and the formation of gemstones.

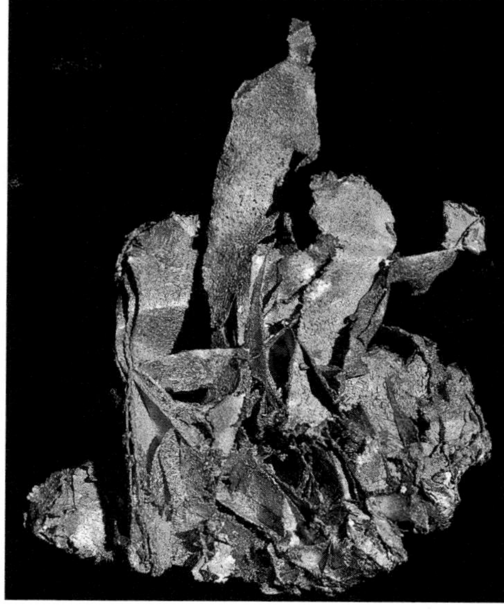

For Further Thought

1. If some sodium (Na^+) substitutes for calcium (Ca^{2+}) in the plagioclase feldspars, why must some aluminum (Al^{3+}) replace some silicon (Si^{4+}) in the mineral's structure?
2. Sulfur forms a small ion with a high positive charge. Why then doesn't sulfur unite universally with oxygen to form the basic building blocks of most crustal minerals?
3. If you were trying to prove that a meteorite struck a particular place on the Earth at some time in the past, what mineralogical evidence would you look for at the proposed impact site?
4. Why doesn't the mineral quartz exhibit the diagnostic property of cleavage? Considering its physical beauty, why isn't a quartz crystal a more valuable gemstone?
5. In the photos at right you can see crystals of real gold and "fool's gold" (pyrite). How would you distinguish these two similar-looking minerals?

Melting Rocks and Crystallizing Magma

Intrusive Rock Formations

Highlight 3-1 **Tabular Plutons Save the Union**

Plate Tectonics and Igneous Rock

Igneous Rocks of the Mcon

The Economic Value of Igneous Rocks

Figure 3-1 Lava fountains from Hawaii's Halemaumau Crater. This red-hot, liquid rock will cool and harden into solid rock within hours.

Chapter 3

Igneous Processes and Igneous Rocks

On the island of Hawaii, you can watch a volcano erupt and later touch the warm rock that passed through the volcano in a molten state just hours or days before (Fig. 3-1). In the Sierra Nevada mountains of California, you can walk on rocks that cooled 80 million years ago from molten rock 20 kilometers below the Earth's surface. The remains of ancient volcanic eruptions and vast uplifted regions of formerly subsurface rocks can be found in almost every state and province of North America. As we saw in Chapter 1, rocks such as these, which cooled and crystallized directly from molten rock, either at the surface or deep underground, are called **igneous rocks** (from the Latin *ignis,* or "fire"). More than 95% of the Earth's outer 50 kilometers (30 miles) consists of igneous rocks.

We know relatively little of the processes that produce igneous rocks, as they are generally shielded from our view by the loose sediment or sedimentary rocks that cover most of the Earth's surface. Moreover, almost all igneous rock solidifies well beneath the surface—although geologists can observe molten material spewing from a volcano, they cannot see it moving underground. (Consider how little you know about the maze of cables and pipelines concealed under the soil and concrete of your community.)

To investigate igneous processes, geologists look for a region where surface rock layers have been removed by erosion (that is, physically worn away by such agents as rivers, glaciers, winds, ocean waves, and gravity), opening up windows through which the subsurface can be seen. Igneous rocks that solidified underground can be seen in some of North America's most scenic places, from Mount Katahdin in northern Maine to the Yosemite Valley of eastern California. Some of our continent's most ancient rocks are the igneous roots of one-time mountains that have long since been eroded by downcutting rivers and rasping glaciers to expose the underlying rocks. Such rocks can be seen in northern Minnesota, Ontario, and Quebec. Some geologists also investigate igneous processes by simulating them in laboratories, observing the effects of pressure, temperature, composition, and other factors on the melting and crystallization points of sample rocks and minerals.

Our discussion of the Earth's igneous processes and rocks describes first some characteristics of molten rock, and then the types of rocks that form when molten material cools and solidifies. We will examine how and why rocks melt and crystallize, paying special attention to how plate tectonics affects the origin and distribution of igneous rocks on Earth (and how Moon rocks differ in this respect). Finally, we will see what economically valuable materials originate from igneous processes.

Melting Rocks and Crystallizing Magma

Geologists distinguish two forms of molten rock. **Magma** is molten rock that flows within the Earth. It may be completely liquid or, as is more common, it may be a fluid mixture of liquid, solid crystals, and dissolved gases. The moment magma reaches the Earth's surface, it becomes **lava**, molten rock that flows above ground.

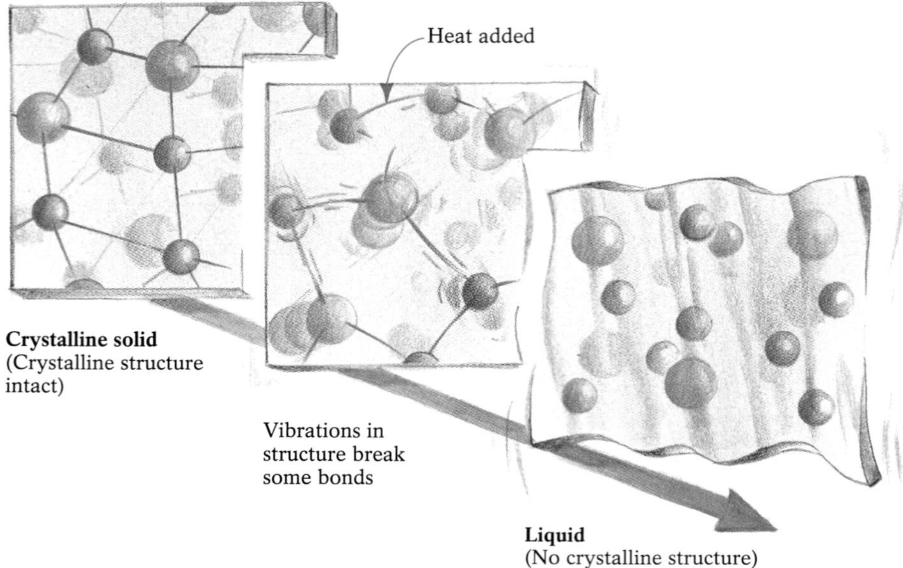

Figure 3-2 Heat causes a solid's atoms to vibrate until some of its chemical bonds weaken and break, causing melting.

A crystalline solid melts when some of the bonds between its ions break, allowing the charged particles to move freely (Fig. 3-2). When underground temperatures become high enough, bonds in minerals are broken, so that eventually heated rock is no longer a crystalline solid but rather a liquid containing some still-solid fragments—a magma. Since the various types of minerals melt at different temperatures, different minerals melt out of the rock as the heat gradually increases; as each newly molten mineral enters and enriches it, the composition of the magma changes. Meanwhile, the mobile ions in the magma continue to move, forming temporary bonds that periodically break and reform.

As the heat dissipates, particles in the molten mass begin to slow down; their bonds cease to break and more bonds start to form, and tiny crystals begin to appear. Additional ions and atoms bond at prescribed sites on the crystal structure, as described in Chapter 2. In this way, crystals grow until they touch the edges of adjacent crystals. As cooling progresses, different minerals crystallize from the magma as its temperature changes, with the magma's composition changing as each crystallized mineral is removed from the fluid mix. Ultimately, if cooling continues, the entire body of magma becomes solidified.

Igneous rocks are classified by their two most obvious properties: their texture, which is determined by the size and shape of their mineral crystals and the manner in which these grew together during cooling; and their composition, which is determined by their mineral content.

Igneous Textures

A rock's *texture* refers to the appearance of its surface, specifically the size, shape, and arrangement of the mineral components of the rock (Fig. 3-3a–e). The most important factor controlling these features in igneous rocks is the rate at which magma or lava cools. When a magma's minerals crystallize slowly underground over thousands of years, there is ample time for crystals to grow large enough to be seen by the unaided eye, producing rock textures in which the crystals can be seen clearly. These are called *phaneritic* (from the Greek *phaneros,* or "visible") textures (Fig. 3-3a). Slow cooling occurs when magmas enter, or *intrude,* preexisting solid rocks; thus, rocks with phaneritic textures are called **intrusive rocks**. They are also known as **plutonic rocks** (for Pluto, the Greek god of the underworld).

Figure 3-3 Igneous textures. (**a**) Rocks that solidify slowly underground, as did this granite in Yosemite National Park, California, have *phaneritic* (coarse-grained) textures.

(**a**)

(**b**)

Figure 3-3 (**b**) Extremely coarse-grained *pegmatites,* such as the one shown here, form from ion-rich magmas having a high water content.

Some igneous rocks develop from ion-rich magmas with a high proportion of water. Under these conditions ions move quite readily to bond to growing crystals, enabling them to become unusually large (sometimes several meters long). Rocks with such exceptionally large crystals are called **pegmatites** (from the Greek *pēgma*, or "fastened together") (Fig. 3-3b). In western Maine, near the towns of Bethel and Rumford, rocks with pegmatitic textures contain 5-meter (17-foot)-long crystals of the mineral beryl. Most pegmatites consist primarily of such common minerals as quartz, feldspar, and mica. Some rare elements, such as beryllium, also occur in pegmatites; pegmatite outcrops that contain beryllium in the form of the gemstones emerald and aquamarine are popular destinations for amateur and professional mineral hunters.

Figure 3-3 (c) Volcanic rocks such as this basalt, because they solidify rapidly above ground, have typically *aphanitic* (very small-grained) textures.

(c)

Some igneous rocks solidify from lava so quickly that their crystals have little time to grow. These *aphanitic* (from the Greek *a phaneros*, or "not visible") rocks have crystals so small they can barely be seen by the naked eye (Fig. 3-3c). Rocks with aphanitic textures are called **extrusive rocks** because they form from lava that has flowed out, or been *extruded*, onto the Earth's surface. They are also known as **volcanic rocks,** because lava is a product of volcanoes (named for Vulcan, the Roman god of fire).

In some igneous rocks, large, often perfect, crystals are surrounded by regions with much smaller or even invisible grains. These *porphyritic* textures are believed to form as a result of slow cooling followed abruptly by rapid cooling. First, gradual underground cooling produces large crystals (Fig. 3-3d) that grow slowly within a magma. Then the mixture of remaining liquid magma and the early-formed crystals it contains erupts to the surface, where the liquid cools rapidly to produce the enveloping body of smaller grains.

A given magma can produce igneous rocks having any of the full range of igneous textures—the amount of time available for cooling and crystal growth determines whether the rock textures will be aphanitic, phaneritic, or porphyritic. Aphanitic igneous rocks that cooled rapidly at the Earth's surface, for example, are commonly underlain by phaneritic rock that derived from the same magma but cooled slowly underground.

(d)

Figure 3-3 (d) Some rocks have a *porphyritic* texture, marked by large crystals surrounded by an aphanitic matrix.

Volcanic Glass When lava from a volcano erupts into the air or flows into a body of water, much of it cools so quickly that its ions don't have enough time to form any crystals at all. The ions are essentially frozen in place randomly, bonded to any available ions nearby. The texture of the resulting rock is described as *glassy* (Fig. 3-3e). There are two common types of volcanic glass. *Pumice* (from the Latin *spuma*, or "foam") forms when bubbling, highly gaseous, silica-rich magma cools instantaneously. Some pumice has so many tiny cavities that it can float. Large rafts of pumice blown from coastal and island volcanoes have been known to float out to sea for hundreds of kilometers before they finally became waterlogged and sank.

(e)

Figure 3-3 (e) The volcanic solid obsidian has a *glassy* texture (containing no crystals) because it solidifies instantaneously.

The second type of volcanic glass is *obsidian*. Obsidian forms when very silica-rich magmas, containing less gas than those that produce pumice, cool instantaneously. Because of the disordered arrangement of its ions, obsidian lacks an organized crystal structure and therefore does not exhibit the systematic internal planes of weakness that characterize most crystals. For this reason, it breaks by fracture rather than by cleavage. Early humans worked obsidian to fashion projectile points and sharp-edged cutting tools. (To demonstrate the quality of such ancient tools, in 1986, David Pokotylo, the curator of archaeology at the University of British Columbia's Museum of Anthropology, underwent hand surgery with an obsidian microblade scalpel prepared in the ancient style of obsidian-tool making. To the surprise of the surgical team, Dr. Pokotylo's incisions healed more rapidly and more cleanly than those made with conventional steel blades.)

Igneous Compositions

The Earth's magmas consist largely of the most common elements: oxygen, silicon, aluminum, iron, calcium, magnesium, sodium, potassium, and sulfur. The relative proportions of these components at any given time within a body of magma give the magma its distinctive characteristics, and ultimately determine the mineral content of the rocks it will form. Water vapor (H_2O), carbon dioxide (CO_2), and sulfur dioxide (SO_2) are the major dissolved gases in molten rock, accounting for a small percentage of a magma's total volume.

We saw in Chapter 2 that the Earth consists primarily of silicon-and-oxygen–based minerals. These silicates are the major constituents of igneous rocks, which are divided into four main compositional groups based on the proportion of their *silica* content (the amount of silicon and oxygen) to the other elements and ions bonded to the silicon-oxygen tetrahedra: Igneous rocks and magmas are classified as ultramafic, mafic, intermediate, or felsic (Table 3-1), with ultramafic material having the smallest proportion of silica and felsic material having the largest. Figure 3-4 illustrates how the mineralogical composition of the common igneous rocks varies in each of these categories. (See also Table 2-5 for chemical compositions of minerals.)

Table 3-1 **Common Igneous Compositions**

Composition Type	Percentage of Silica	Other Major Elements	Relative Viscosity of Magma	Temperature at Which First Crystals Solidify	Igneous Rocks Produced
Felsic	>70%	Al, K, Na	High	~600–800°C (1100–1475°F)	Granite (plutonic) Rhyolite (volcanic)
Intermediate	60%	Al, Ca, Na, Fe, Mg	Medium	~800–1000°C (1475–1830°F)	Diorite (plutonic) Andesite (volcanic)
Mafic	40–50%	Al, Ca, Fe, Mg	Low	~1000–1200°C (1830–2200°F)	Gabbro (plutonic) Basalt (volcanic)
Ultramafic	<40%	Mg, Fe, Al, Ca	Very low	>1200°C (2200°F)	Peridotite (plutonic) Komatiite (volcanic)

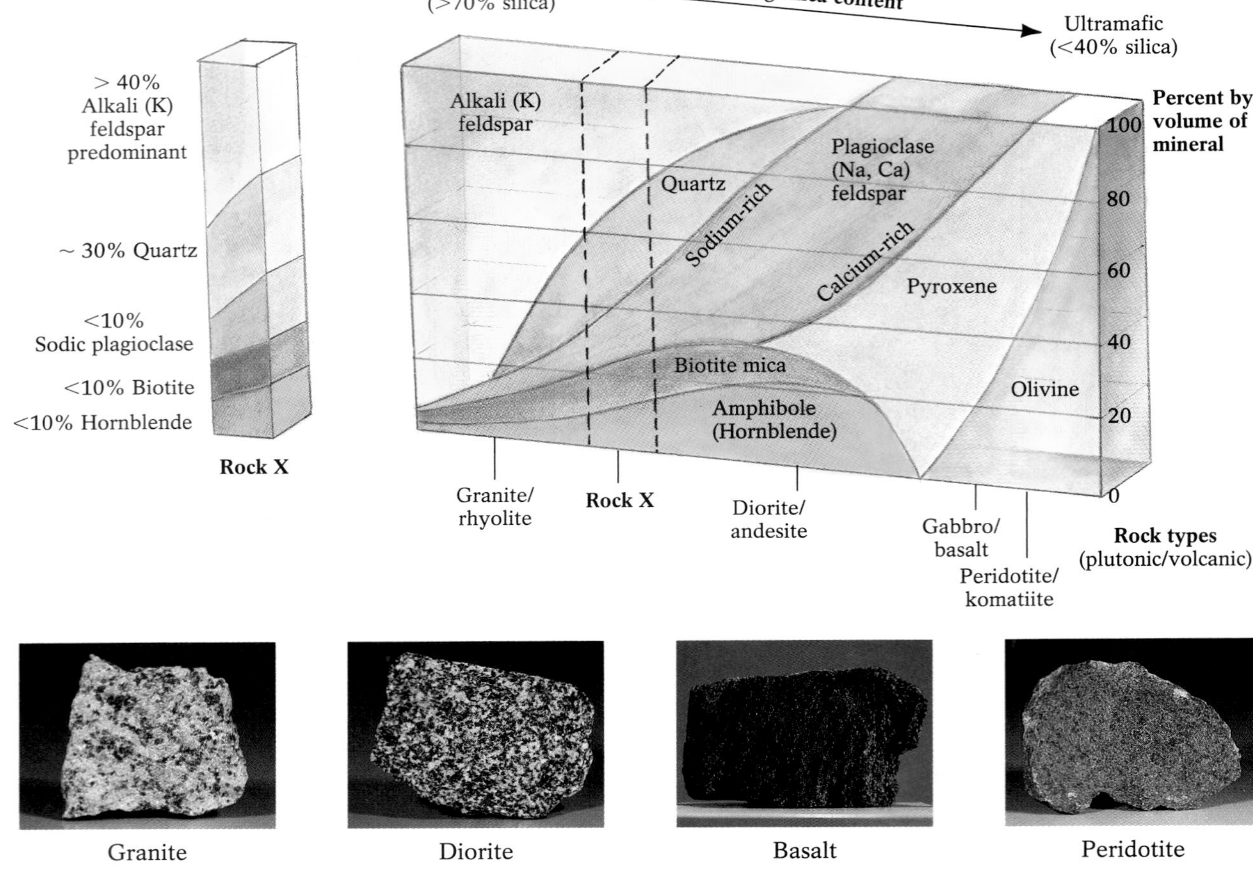

Figure 3-4 An igneous rock classification chart, showing the range of compositional types among the igneous rocks, from felsic to ultramafic. The mineral components of the rocks are indicated by colored areas in the body of the chart. (The sample segment shows how to interpret the chart, using as an example a rock falling between granite and diorite in composition.)

Ultramafic Igneous Rocks The term "mafic" is derived from *ma*gnesium and *f*errum (Latin for "iron"). Ultramafic igneous rocks are dominated by the iron-magnesium silicate minerals olivine and pyroxene and contain relatively little silica (less than 40%) and virtually no feldspars or free quartz. The most common ultramafic rock, **peridotite,** contains 70% to 90% olivine. Ultramafic rocks generally crystallize slowly deep in the Earth's interior, developing their typically coarse-grained phaneritic texture. (The relatively rare extrusive equivalent of peridotite is called komatiite.) These deep-forming rocks, dark in color and very dense, appear at the Earth's surface only where extensive erosion has removed overlying crustal rocks. They are most likely to be found where converging continental tectonic plates have collided and been uplifted, bringing deep rocks closer to the surface.

Mafic Igneous Rocks Mafic igneous rocks have a silica content between 40% and 50%. These are the most abundant rocks of the Earth's crust, and the mafic rock **basalt** is the single most abundant of them. Basalt, whose principal minerals include pyroxene, calcium feldspar, and a minor amount of olivine, is dark in color and relatively dense; because it forms from molten rock which has cooled fairly rapidly at or near the surface, its texture is aphanitic. It is the dominant rock of the world's oceanic plates, making up most of the ocean floor and many islands, including the entire Hawaiian chain, Samoa and Tahiti in the Pacific, and Iceland in the Atlantic. Basalt also constitutes vast areas of our continents, being found in Brazil, India, South Africa, Siberia, and the Pacific Northwest of North America (Fig. 3-5). When a magma containing the same mix of minerals cools more slowly underground, basalt's plutonic equivalent, the coarse-grained phaneritic **gabbro,** is produced. Since it is a deep-forming rock, outcrops of gabbro are generally seen only where extensive erosion has removed surface rocks. Geologists believe that gabbro lies beneath the basalts of the ocean floor.

Intermediate Igneous Rocks Intermediate igneous rocks contain more silica than mafic rocks—about 60%. They typically consist of iron and magnesium silicates, such as pyroxene and amphibole, along with sodium- and aluminum-rich minerals, such as sodium plagioclase and mica, and a small amount of quartz. They are generally lighter in color than mafic rocks. The aphanitic intermediate igneous rock **andesite,** named for the Andes mountains of South America, where it often dominates the local geology, is the world's second most abundant volcanic rock. Andesites contain pyroxene, the amphibole hornblende, quartz, and an abundance of andesine, a plagioclase mineral with about 60% sodium and 40% calcium ions. Andesite's plutonic equivalent is **diorite,** which can be recognized by its coarse-grained, salt-and-pepper appearance.

Felsic Igneous Rocks The term "felsic" is derived from *fel*dspar and *si*lica. Felsic igneous rocks contain more silica—70% or higher—than mafic or intermediate igneous rocks. They are generally poor in iron, magnesium, and calcium silicates, and rich in potassium feldspar, aluminum-rich micas, and quartz. The most common felsic igneous rock is the plutonic rock **granite.** Potassium feldspars impart granite's pinkish cast. Sodium feldspars are often present, contributing a porcelain-like look to these rocks. Small quartz grains and flakes of biotite and muscovite mica are also scattered throughout granite. The magma from which granite crystallizes flows and crystallizes slowly

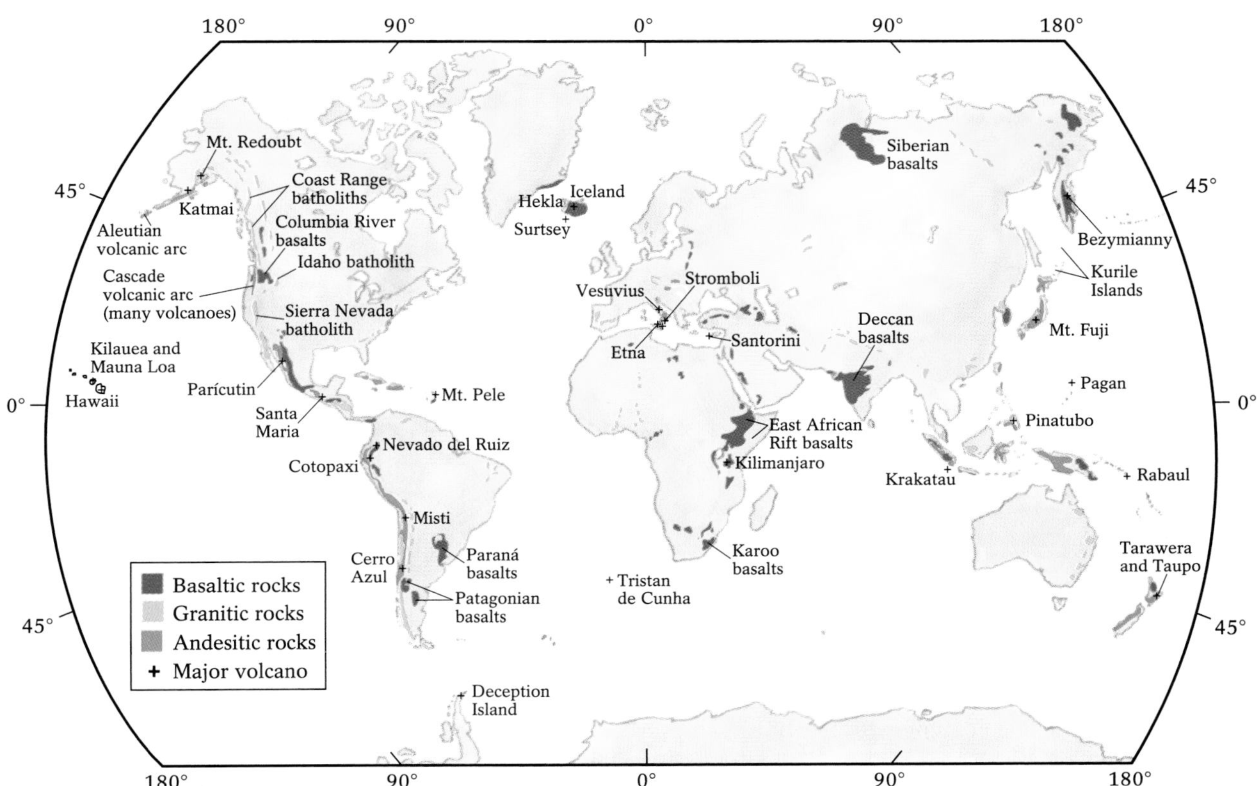

Figure 3-5 Distribution of the Earth's major igneous rock provinces. As well as being the most abundant igneous rock of the ocean floors, the mafic igneous rock basalt also composes several large areas of the continents.

underground, seldom reaching the surface. (**Rhyolite,** granite's volcanic equivalent, is relatively uncommon.) Granitic rocks can be seen at the surface only after erosion removes overlying rocks. As you can see in Figure 3-5, these rocks are quite common on the Earth's continents, but virtually absent from the ocean basins.

Rocks of felsic composition have a greater variety of textures than any other igneous rock. They range from aphanitic rhyolites and several glassy rocks to ultra-coarse pegmatitic rocks.

The Creation of Magma

Numerous factors control the melting of rocks and the creation of magmas. Heat, pressure, the amount of water in the rocks, and the process of partial melting combine to determine the point at which particular rocks melt and enter the molten mass.

Heat As we saw in Chapter 1, the heat in the Earth's interior comes from three primary sources: the heat produced during the formation of the planet, still rising from the Earth's core; the heat liberated continuously by decay of radioactive isotopes; and the frictional heat produced as the Earth's plates move against one another and over the underlying asthenosphere. Temperatures in the Earth increase with depth, at a rate referred to as the *geothermal gradient.* As you can see in Figure 3-6, the geothermal gradient is steepest from about 50 to 250 kilometers (30–150 miles) of depth. This great heat, thought to be due to a high concentration of radioactive isotopes in the rocks of the lower crust and upper mantle, melts a portion of the rocks at these depths.

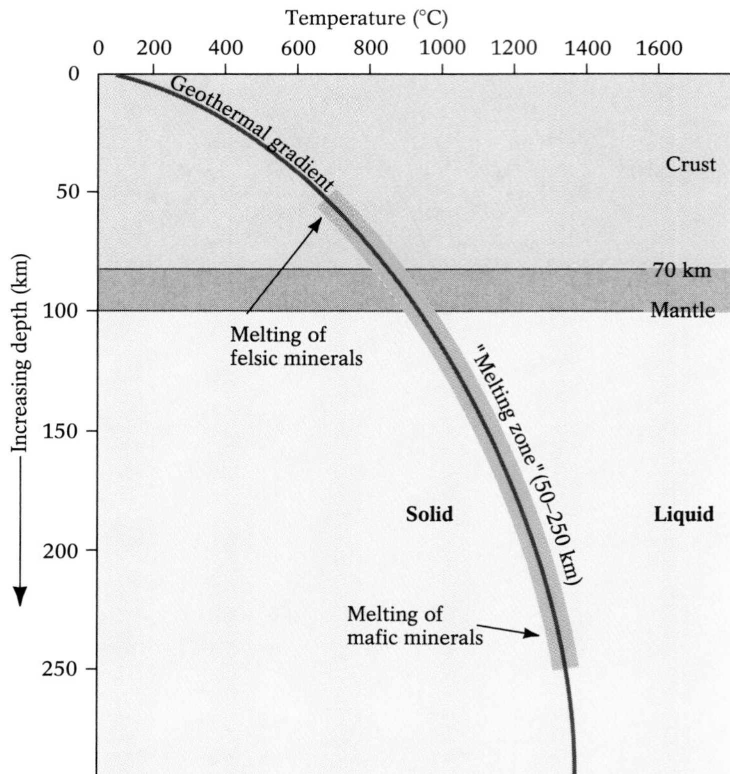

Figure 3-6 A graph showing the geothermal gradient—the rate at which the Earth's internal temperature increases with depth. The temperatures between 50 to 250 kilometers (30–150 miles) in depth exceed the 700°C (1300°F) melting point of felsic minerals and the 1300°C (2400°F) melting point of mafic minerals. Thus, rocks tend to melt at the temperatures found in the Earth's lower crust and upper mantle.

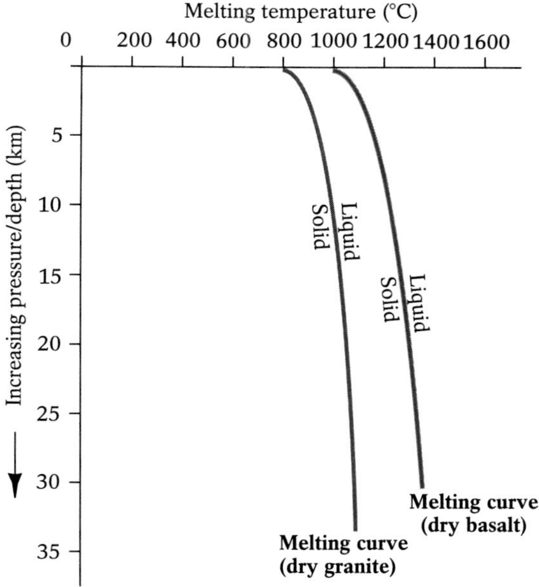

Figure 3-7 Melting-temperature curves for dry basalt and dry granite. (As we'll see in Figure 3-9, adding water to rock changes its melting curve.) For both, melting temperatures increase with increasing depth, because the pressure at greater depths stabilizes rock's crystal structure, raising its melting point.

Pressure When high pressure holds the ions and atoms in a crystalline solid firmly in place, more heat energy is required to vibrate, weaken, and break the bonds of its ions and atoms. Because rocks located far beneath the Earth's surface experience great pressure from the weight of overlying rocks, higher temperatures are needed to melt them. In general, as pressure increases, the temperature at which a mineral melts also increases (Fig. 3-7). At the Earth's surface, for example, a crystal of the sodium feldspar albite melts at 1118°C (2050°F). At a depth of 100 kilometers (60 miles), however, where the pressure is 35,000 times that on the surface, a temperature of 1440°C (2650°F) is required to melt albite.

If the pressure on rock is somehow reduced or removed—as happens when tectonic plates rift and diverge (Fig. 3-8)—its melting point drops below its current temperature and it begins to melt.

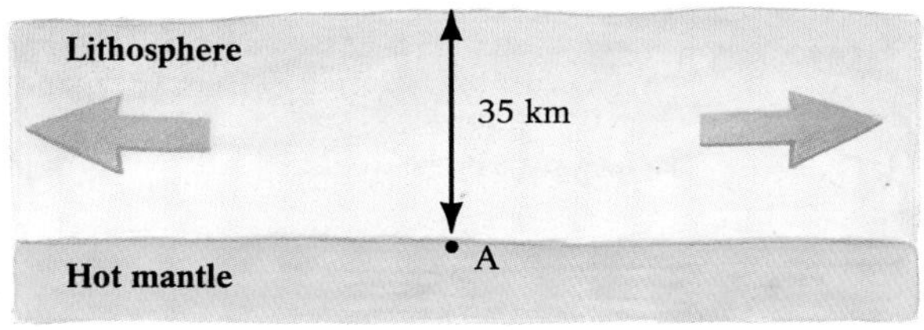

① Divergence begins.

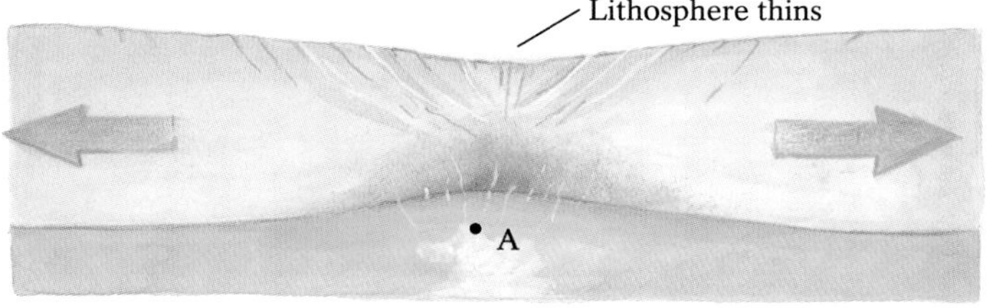

② Pressure at point A is relaxed as crust thins. Hot mantle rock at point A begins to melt.

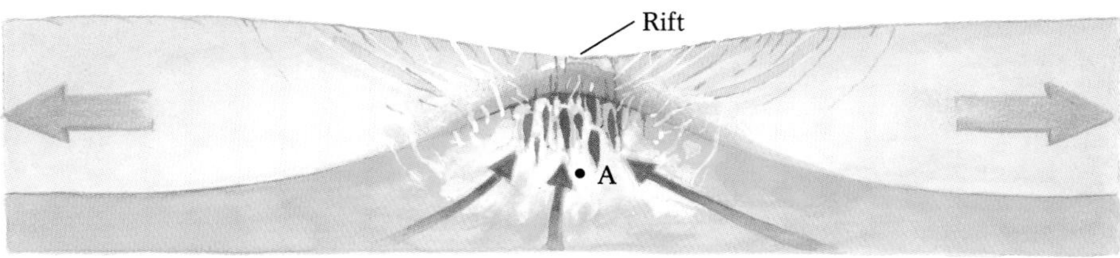

③ As divergence continues, rock at point A has melted and begins to rise into rift.

Figure 3-8 The effect of lithospheric thinning on rock in the Earth's upper mantle. Hot mantle rocks remain solid primarily because of the pressure applied by the weight of overlying rock. This pressure may be removed suddenly through the plate tectonic process of rifting. Directly below the rift zone, where pressure from overlying rocks is reduced, the hot mantle rocks begin to melt and produce new mantle-derived magmas.

Water Water, even a small amount, lowers the melting point of rocks. Recall from Chapter 2 that every water molecule has a positive and a negative side, which attract oppositely charged regions or ions in other compounds. As water molecules tug on charged ions at the surface of a mineral's crystal structure, the mineral's bonds weaken so that less heat is required to vibrate and then completely break them. Under high pressure, water has an even greater effect on the melting point of a mineral: Whereas dry rocks become

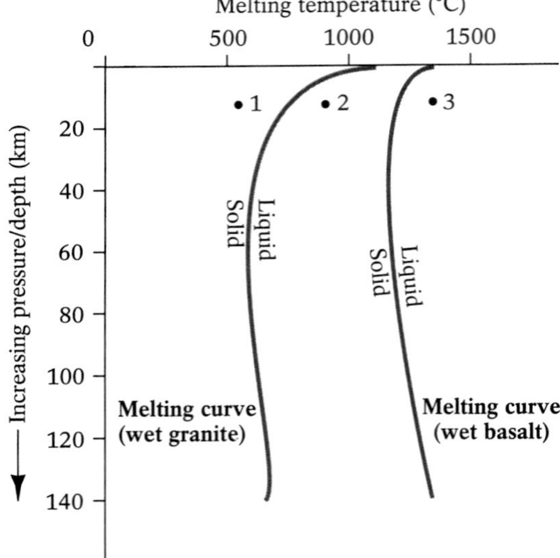

Figure 3-9 Melting-temperature curves for wet basalt and wet granite. (At point 1, both granite and basalt would be solid; at point 3, both would be liquid; at point 2, granite would be liquid but basalt would still be solid.) For both, melting temperatures decrease with increasing depth, because highly pressured water destabilizes rock's crystal structure, lowering its melting point.

more resistant to melting with greater depth, wet rocks become less resistant (Fig. 3-9). This is because high pressure drives more water into the rocks. As we will see, this combination of high pressure and high water content is an important factor in the production of magmas at subducting plate boundaries (Fig. 3-10).

Partial Melting **Partial melting**—the incomplete melting of rock—occurs because different minerals melt at different temperatures. When rocks melt to produce new magma, the process is not quite as simple as the warm midday sun melting last night's snowfall. Snow consists of a single solid mineral, ice (H_2O), and therefore it all melts at one temperature, 0°C (32°F). Most rocks are composed of several minerals, each having a melting point reflecting the strength of its own chemical bonds. For example, the bonds between sodium and oxygen in the plagioclase albite are weaker than those between calcium and oxygen in the plagioclase anorthite. Consequently, albite's bonds break (and albite melts) at 1118°C (2050°F), whereas anorthite's bonds can withstand a temperature of 1553°C (2800°F) before they break (and anorthite melts). A multimineral rock melts completely only when the temperature is higher than the melting points of all of its component minerals.

A rock that is heated to the melting points of some but not all of its component minerals yields a magma of molten minerals that contains still-solid chunks of minerals with higher melting points. Thus, partial melting of mafic rocks may produce intermediate magmas, and partial melting of intermediate rocks may produce felsic magmas. If temperatures are subsequently raised, the minerals with higher melting points will progressively melt and be added to the magma, changing its composition.

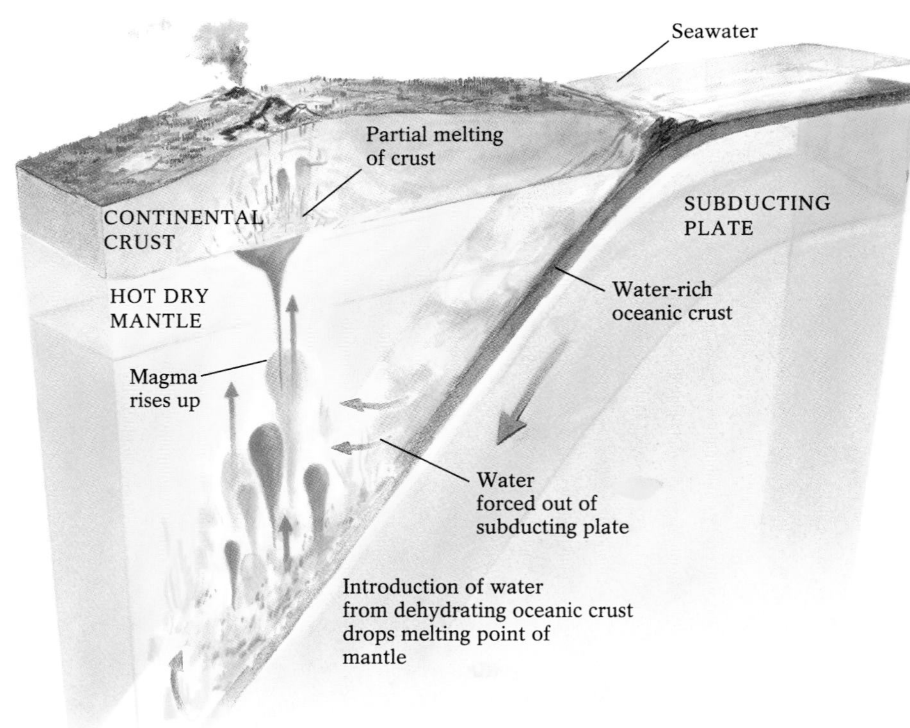

Figure 3-10 As a water-rich oceanic plate subducts, it descends into the Earth's warmer, higher-pressure interior. Increasing pressure during subduction drives water from the plate's sediments and basalts into the dry hot mantle rocks above it. Before this water enters the mantle rocks, high pressure prevents them from melting; when water is driven into these rocks, their melting point drops and they begin to melt, producing the new magmas that fuel subduction zone volcanism.

The Crystallization of Magma

Eventually, every magma cools and solidifies. As it does so, it generally undergoes a number of significant changes in composition.

The temperature at which a mineral melts is the same as the temperature at which it crystallizes. Minerals that melt first (at the lowest temperatures) during heating are thus the last to crystallize during cooling; minerals that melt last (at the highest temperatures) during heating are the first to crystallize during cooling. A partially cooled body of magma contains solid crystals of minerals that crystallize at higher temperatures, along with a liquid containing the atoms and ions of minerals that will not crystallize until the temperature is further lowered. As the magma continues to cool, additional ions and atoms crystallize out of the melt, leaving progressively less liquid. At each stage of cooling, the proportion of crystal to liquid changes, as does the chemical interaction between them.

Bowen's Reaction Series In 1922, Canadian geochemist Norman Levi Bowen and his colleagues at the Geophysical Laboratory of the Carnegie Institution in Washington, D.C., determined the sequence in which silicate minerals crystallize as magma cools. Their work made it possible to summarize a complex set of geochemical relationships, **Bowen's reaction series,** in a single diagram (Fig. 3-11), and demonstrated that a full range of igneous rocks, from mafic to felsic, could be produced from the same, originally mafic, magma. Based on the fact that early-forming crystals remaining in contact with the still-liquid parent magma continue to react with it and so evolve into new minerals, Bowen's reaction series shows that the silicate minerals can crystallize from mafic magmas in two ways—in a discontinuous series or in a continuous series.

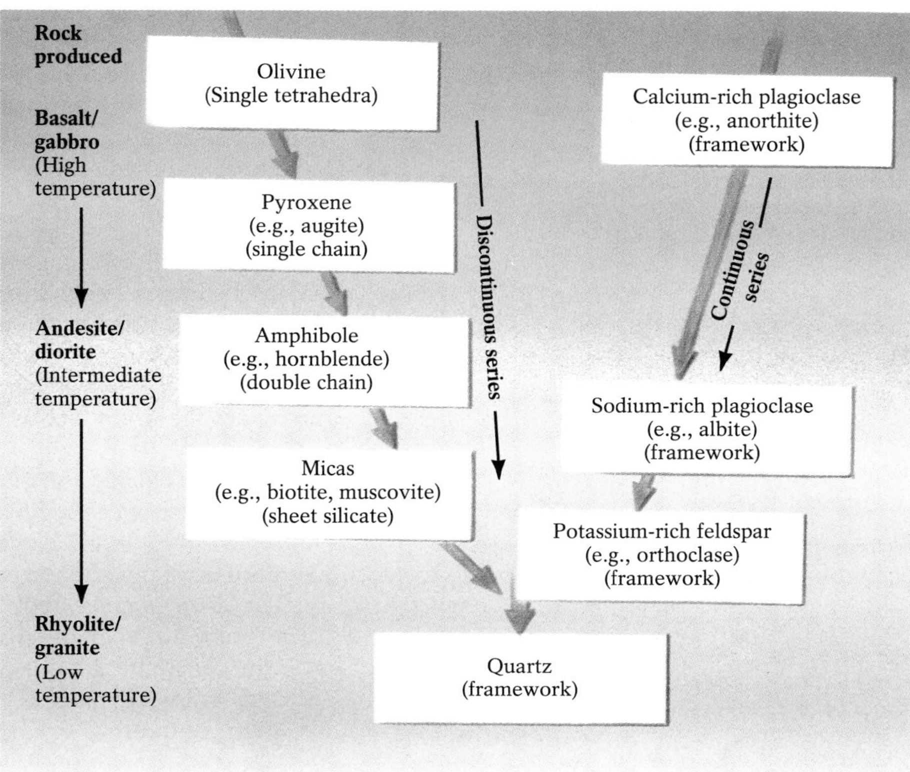

Figure 3-11 Bowen's reaction series, showing the sequence of minerals that crystallize as an initially mafic magma cools under ideal conditions (such that the early-forming crystals remain in contact with the still-liquid magma, enabling them to exchange ions with it).

Bowen showed that ferromagnesian minerals (the iron- and magnesium-rich silicates) crystallize one after another in a specific sequence. Because each successive type of ferromagnesian mineral crystallized differs in both composition and internal structure from the one before, Bowen called this the *discontinuous series*. As mafic magma cools, the first ferromagnesian mineral to crystallize is olivine, which has a low silica content and a relatively simple structure of independent tetrahedra. As olivine crystallizes, some iron and magnesium are removed from the parent magma, increasing the proportion of the other major ions in the magma. Meanwhile, the scattered olivine crystals continue to incorporate silica from the remaining magma, and their tetrahedra begin to become linked in the single-chain structure of the pyroxenes. The discontinuous evolution of the ferromagnesian minerals continues as pyroxene crystals acquire more silica and are transformed into the double-chained amphiboles. Eventually, the series culminates in the formation of the complex sheet silicate biotite mica, the last ferromagnesian mineral to form. By then, all of the magma's original iron and magnesium ions and atoms have been crystallized. Any minerals that crystallize after biotite will contain no iron or magnesium.

Figure 3-12 A zoned plagioclase crystal. The crystal's outer layer is dominated by sodium ions, which replaced earlier-bonding calcium ions as the crystal continued to react with ions in the surrounding magma as it cooled.

Meanwhile, at the same high temperatures at which olivine and the pyroxenes are crystallizing, calcium plagioclase is also crystallizing. As in the case of those ferromagnesian minerals, the early-forming calcium plagioclase crystals continue to interact with the remaining liquid. Gradually the calcium ions are replaced by sodium ions from the liquid magma and the growing crystals are completely converted to sodium plagioclase. Because one type of ion is being replaced by a very similar ion (recall discussion of ionic substitution in Chapter 2), there is no change in the internal structure of the plagioclases; therefore Bowen called this the *continuous series*. The resulting sequence of plagioclase feldspars ranges from anorthite, in which calcium accounts for 90% to 100% of the positive ions, through a variety of intermediate calcium–sodium mixtures, to albite, in which sodium accounts for 90% to 100% of the positive ions.

When magma cools slowly, sodium ions invade anorthite crystals gradually, starting at the surface and then dispersing through the entire crystal. When magma cools rapidly, sodium ions penetrate the surface but do not have time to invade the crystal interior, which retains its calcium ions. In this way, rapid cooling results in the formation of *zoned* plagioclase crystals, which have sodium-rich rinds and calcium-rich cores (Fig. 3-12).

After the ferromagnesian minerals and the plagioclase feldspars have crystallized completely from an initially mafic magma, less than 10% of the original liquid remains. Depending on its initial composition, this liquid may now contain high concentrations of silica, potassium, and aluminum. In such a case potassium feldspar, potassium-aluminum mica (typically muscovite), and quartz are the last minerals to crystallize.

How Magma Changes as It Cools Bowen's reaction series, developed in the laboratory, assumes the ideal condition of early-forming crystals remaining in contact with the liquid magma, enabling them to interact and evolve continually until crystallization is complete. In nature, however, this condition rarely applies. As a magma cools, several things may happen to the crystals that form. Some do remain suspended, continuing to exchange ions and atoms with the remaining liquid. But early-forming crystals might also be physically removed from the magma and have no further chemical interaction with the remaining liquid (Fig. 3-13). Crystals that are denser than the surrounding liquid may sink to the bottom of the magma chamber and become buried by later-settling crystals. Crystals may be plastered against the walls or ceiling of

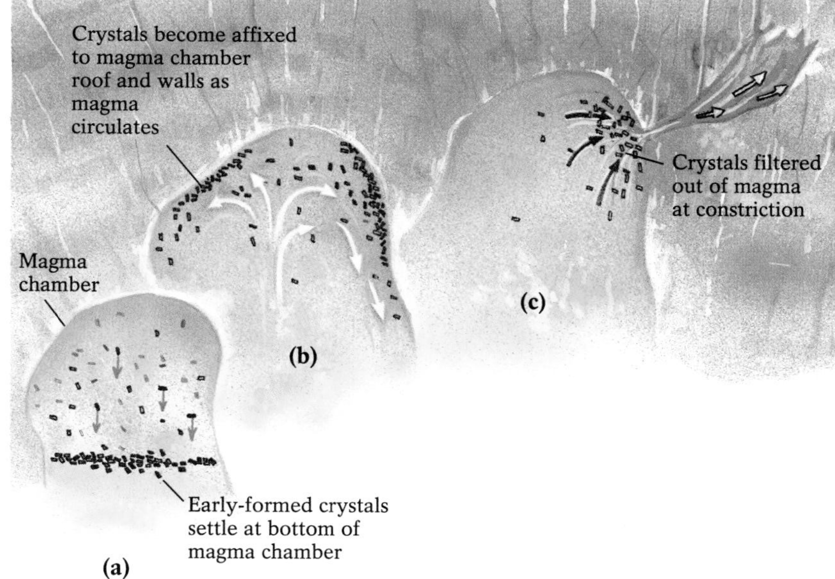

Figure 3-13 Early-forming crystals do not always remain in contact with the liquid magma, as Bowen's reaction series assumes. Instead, the crystals may: (**a**) settle to the bottom of the magma chamber; (**b**) become affixed to the walls and roof of the magma chamber as magma circulates within the chamber; or (**c**) be filtered out of the magma as it is pressed through small fractures in the surrounding rock.

the magma chamber by the hot rising liquid. The largest crystals may even be filtered out of the melt entirely as the remaining liquid component of the magma flows into fractures too narrow for them to pass. Because their ions are no longer available to interact with the remaining magma, removal of crystals in any of these ways substantially affects the composition of the remaining magma, and thereby the composition of any rocks which may later form from it (Fig. 3-14).

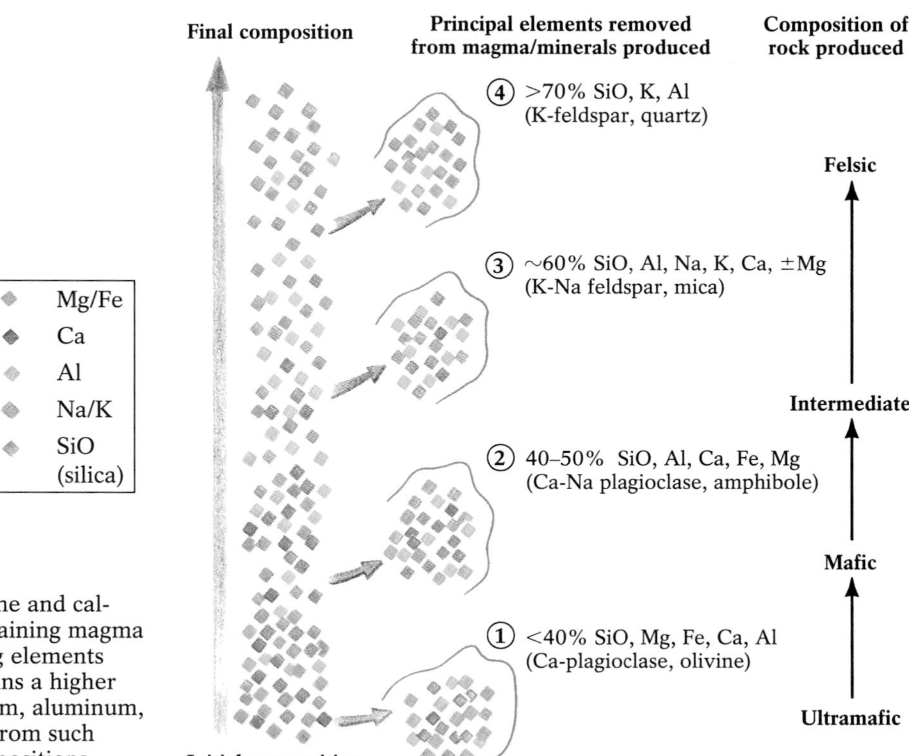

Figure 3-14 When early-forming minerals (such as olivine and calcium plagioclase) are removed from a magma, the remaining magma is depleted of a significant portion of early-crystallizing elements such as magnesium, iron, and calcium; in turn, it contains a higher proportion of later-crystallizing elements such as sodium, aluminum, and potassium. The igneous rocks that crystallize late from such evolving magma therefore have entirely different compositions (progressively more felsic) than rocks that crystallize earlier.

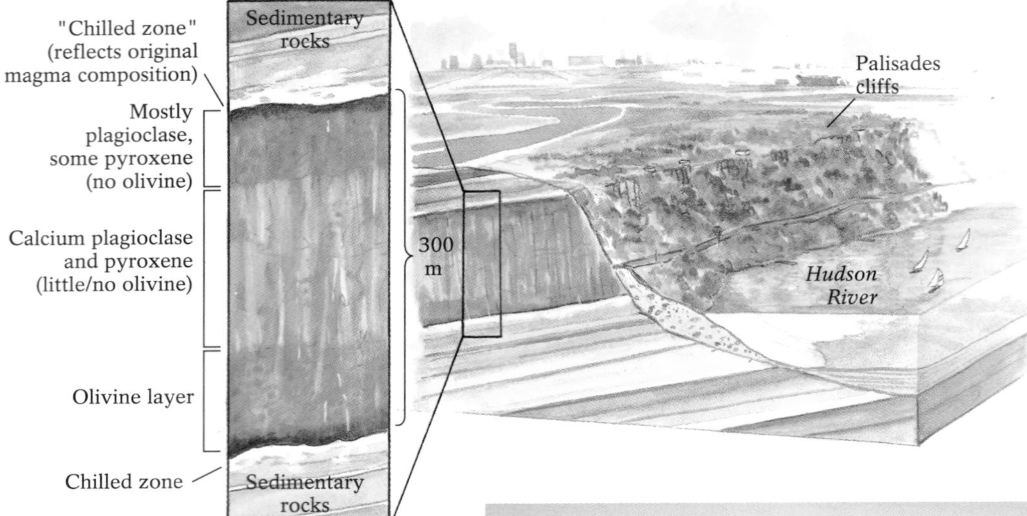

Figure 3-15 The Palisades cliffs, in northeastern New Jersey, demonstrate the result of fractional crystallization. The rocks of the Palisades crystallized from a 300-meter (1000-foot)-thick body of magma that intruded preexisting rocks at temperatures of at least 1200°C (2200°F). The top and bottom of the Palisades solidified very rapidly without undergoing fractional crystallization, probably because the magma came into contact with cold surrounding rocks, and thus provide us with a glimpse of the magma's original composition. The bottom third of the Palisades has a high concentration of olivine crystals, the central third is a mixture of calcium plagioclase and pyroxene with no appreciable olivine, and the upper third consists largely of plagioclase with no olivine and little pyroxene. It appears that early-forming olivine crystallized and then settled to the bottom of the magma body; pyroxene and plagioclase crystallized next, with the denser pyroxenes settling and concentrating in the center, and the lighter plagioclases occupying the uppermost section. Since the entire Palisades magma cooled and solidified fairly quickly, it left no residual magma from which later-forming minerals could crystallize.

A magma from which crystals have been removed at various stages of its cooling has in effect become separated into a number of independently crystallizing bodies; the rocks that form from such a magma are different in composition both from each other and from the original magma, with each successive body crystallized being more silica-rich than the last (see Fig. 3-14). By this process, called **fractional crystallization,** a single parent magma can produce a variety of igneous rocks of different compositions. The Palisades cliffs of northern New Jersey, on the west bank of the Hudson River, are a classic example of this phenomenon (Fig. 3-15).

Other Magma Crystallization Processes Bowen believed that all igneous rocks, even felsic ones, form by fractional crystallization of mafic magmas evolving according to his reaction series. Geologists building on Bowen's work, however, realized that these processes alone could not account for all igneous rocks. For example, although felsic igneous rocks can crystallize from mafic magmas, too little magma remains after the mafic minerals crystallize to produce large bodies of felsic and intermediate rocks. Thus, the 1000-kilometer (600-mile) stretches of felsic igneous rock on our continents must have been produced by processes other than fractional crystallization of mafic magma.

We have seen that partial melting of preexisting rocks can produce intermediate or felsic magmas. As magma moves, blocks of rock from the walls of the magma chamber may break free and be wholly or partially melted by the surrounding hot magma. Assimilation of such rock bodies can significantly alter a magma's composition. In addition, two or more different bodies of magma may flow together and form a magma of hybrid composition. For example, the 1912 volcanic eruption in Alaska's Aleutian Islands produced rocks containing both felsic and mafic minerals, suggesting that two distinct bodies of magma had combined to fuel the eruption.

Intrusive Rock Formations

Magmas tend, for several reasons, to rise. When two materials of different densities occupy a space together, the denser is pulled down by gravity and the lighter is forced to rise. All fluid magma rises because it is less dense than the solid rock that surrounds it. In addition, the gases in magma expand outward as it rises, helping to drive the magma upward. Finally, magma rises when surrounding rocks press on it and squeeze it upward (see Chapter 7), much as toothpaste oozes out when the tube is squeezed.

Because no one has ever actually seen magma move underground, we can only speculate about how it does from the igneous formations that we can see. For example, in some areas erosion has exposed thousands of square kilometers of solidified intrusive magma. Such vast regions could not have resulted from magma flowing into preexisting subterranean cavities because the weight of overlying rocks at depths below about 10 kilometers (6 miles) would have collapsed and destroyed such large underground spaces—the magma must have moved forcefully into cracks in preexisting rock, actually pushing the rock aside to create its own space. In a similar way, rising magma may force overlying rocks to bulge upward. The resulting igneous rock appears as a domed intrusion within other rocks; this distinctive igneous structure is known as a *diapir.*

Figure 3-16 A dioritic xenolith within granite. The granitic magma encompassed the preexisting dioritic rock (seen here as a dark gray mass within the lighter gray granite) but was not hot enough to melt it; as a result, the diorite was preserved as a discrete mass when the magma eventually solidified.

When moving magma incorporates some preexisting rock as it rises, some of the incorporated rocks melt and become assimilated into the magma; unmelted rocks are carried within the magma. When this magma eventually solidifies, such "foreign" rocks can be seen as distinctly different rock masses called **xenoliths** (from the Greek *xenos,* or "stranger," and *lithos,* or "stone") (Fig. 3-16).

Magma moves underground in ways that produce distinctive subsurface igneous forms, which may later be exposed by subsequent erosion of overlying surface rocks. The term **pluton** refers to all such intrusive igneous bodies. Plutons may be classified by their position relative to the preexisting rock, called *country rock,* surrounding it: *Concordant* plutons lie parallel to layers of country rock; *discordant* plutons cut across layers of country rock. Plutons of both varieties come in a range of shapes and sizes, as shown in Figure 3-17.

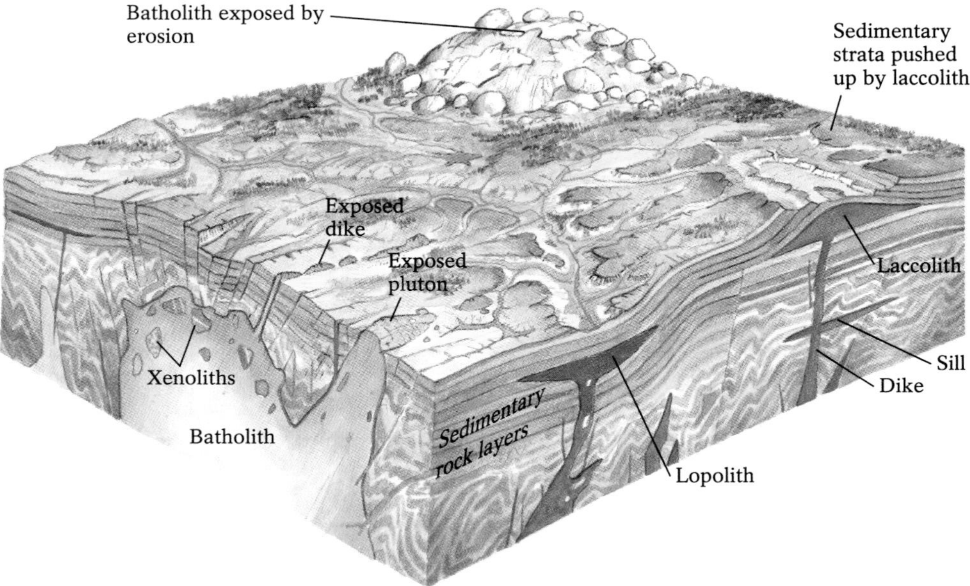

Figure 3-17 Plutonic igneous features. Sills are concordant tabular plutons; dikes are discordant tabular plutons. Laccoliths and lopoliths are larger concordant plutons, and batholiths are even larger discordant plutons.

Tabular Plutons

Tabular plutons are slablike intrusions of igneous rock that are broader than they are thick, like a table top. If magma flows into a relatively thin fracture in country rock, or pushes between sedimentary rock layers, a tabular pluton will result when the magma cools. Tabular plutons may be as small as a few centimeters thick to several hundred meters thick.

A **dike** is a discordant tabular pluton, cutting across preexisting rocks. Dikes are generally steeply inclined or nearly vertical, suggesting that they formed from rising magma, which tends to follow the most direct route upward. Dikes often occur in clusters where magma apparently infiltrated and solidified in a network of fractures.

Many dikes result from magma that rises into volcanoes and then solidifies; we can see these erosion-resistant rocks when the less resistant volcanic material at the surface has been worn away. Some dikes diverge, like the spokes of a bicycle wheel, from a *volcanic neck,* a vertical pluton remaining in what was once a volcano's central magma pathway. Along Route 64, running through the Navajo and Hopi lands of the Four Corners (the intersection of Colorado, New Mexico, Arizona, and Utah), more than a hundred such volcanic necks can be seen, remnants of ancient volcanic plumbing (Fig. 3-18).

A **sill** is a concordant tabular pluton, lying parallel to layers of preexisting rocks. Sills are produced when intruding magma enters a space between layers of rock, melting and incorporating adjacent sedimentary material. Sills can form only within a few kilometers of the Earth's surface, because at greater depths overlying rocks would compress and close off any spaces into which magma might flow.

A sill and a dike in the south-central Pennsylvania town of Gettysburg provided the setting for an event that affected the course of American history. This event is recounted in Highlight 3-1.

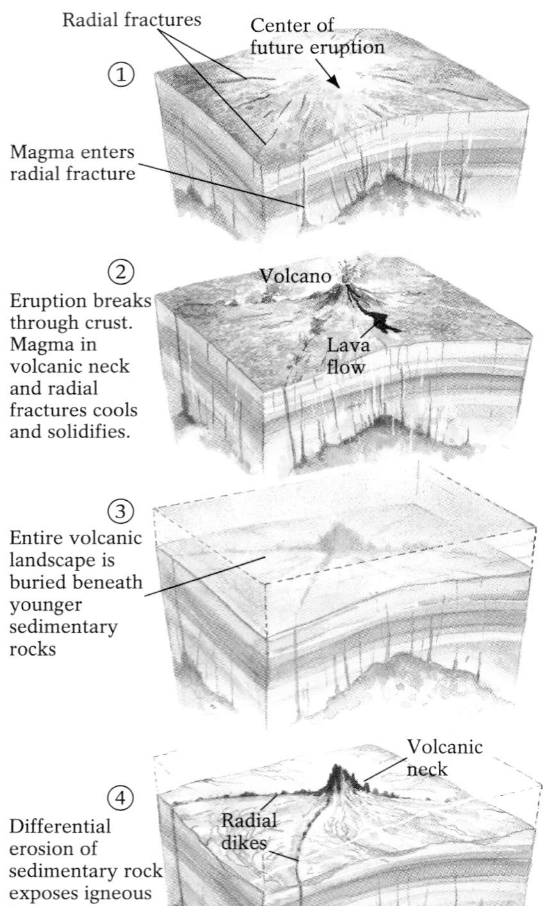

Figure 3-18 Shiprock Peak, in New Mexico, is believed to be a volcanic neck, the congealed lava from the interior of a former volcanic cone. Erosion of the surrounding sedimentary rock and the cone itself has exposed this volcanic neck and the radial dikes that once fed magma to the volcano.

Highlight 3-1 *Tabular Plutons Save the Union*

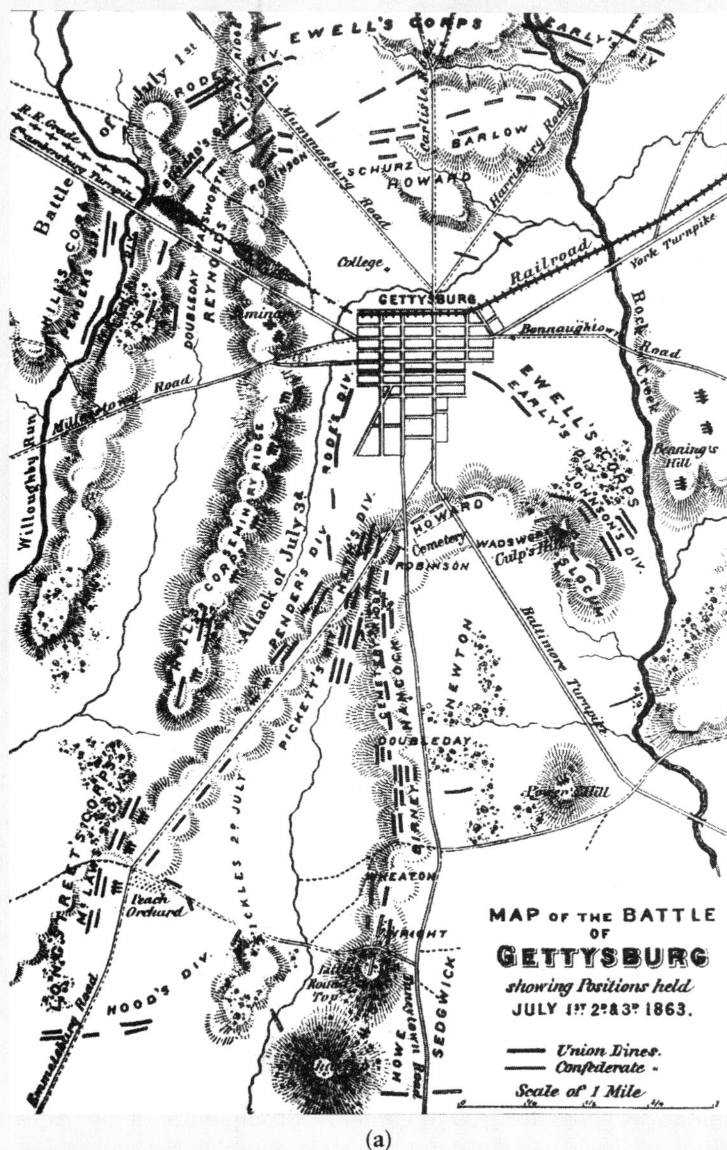

(a)

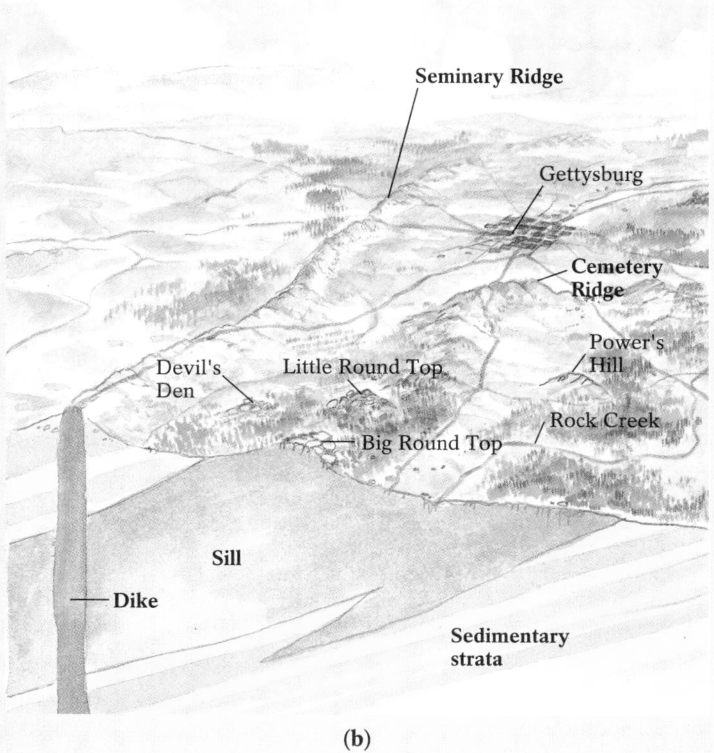

(b)

Figure 3-19 (**a**) A Civil War-era map of the site of the battle of Gettysburg. Seminary Ridge appears in the upper left; Cemetery Ridge is to its lower right. (**b**) A contemporary artist's rendering of the relevant topographic features.

The battle of Gettysburg, which lasted three days and took the lives of tens of thousands of Civil War soldiers, was effectively won by the Union on a hot July 3 in 1863. On this day, Confederate troops ventured forth from their outpost on a narrow dike of resistant basalt called Seminary Ridge to charge against the Union stronghold on the equally resistant but thicker basaltic sill called Cemetery Ridge (Fig. 3-19). (This offensive would become known as "Pickett's charge.") The steep forward slope of the Cemetery Ridge sill impeded the Confederate charge, and a protective wall constructed from basaltic boulders by Union troops concealed them and repelled Confederate shots. Thus, with an assist from a well-placed basaltic sill, Union forces defeated the Confederate offensive at Gettysburg, a turning point in the American Civil War.

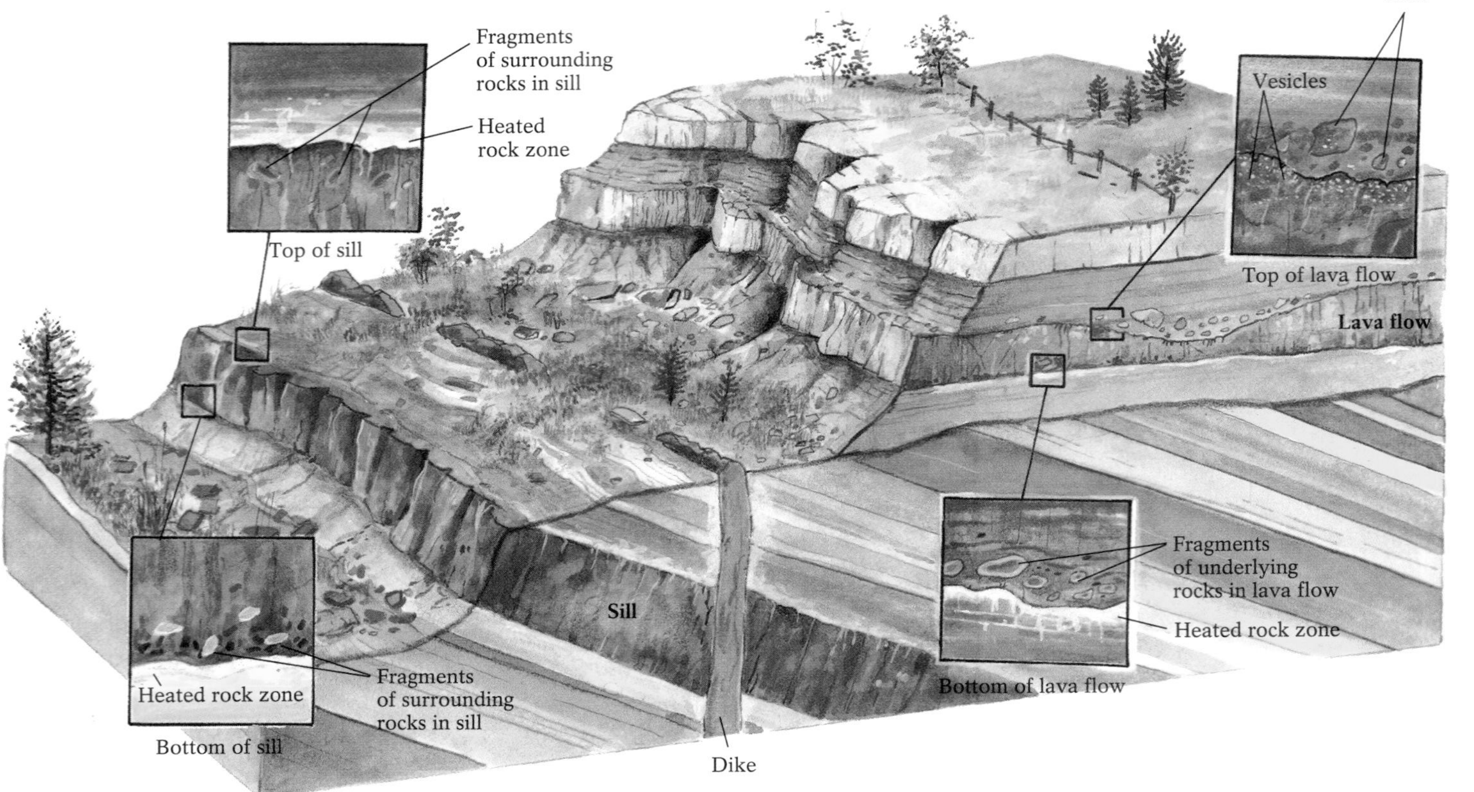

Figure 3-20 Sills and lava flows can be distinguished by determining their relationship to the rocks surrounding them. Because sills result from magma intruding preexisting rock layers, rocks *both above and below* a sill show evidence of having been heated by the magma; only the rock *below* a lava flow is affected, as overlying layers are not deposited until after the flow. In addition, because they occur at the surface, lava flows show evidence of having been exposed to air, whereas sills do not. (See text.)

Sills pose an interesting challenge to geological detectives. How can a sill between layers of preexisting rock be distinguished from a lava flow buried under subsequent flows or sedimentary rocks? There are several critical clues (Fig. 3-20):

1. Look at the adjacent surfaces of the surrounding rocks. When lava is extruded, there are no overlying rocks and so only the top surface of rocks beneath it will show evidence of heating (described in Chapter 7). When hot magma intrudes between two layers of rock to form a sill, it heats the adjacent surfaces of both layers before cooling.
2. Look at the top and bottom surface of the igneous layer. Both surfaces of a sill will contain fragments of the surrounding rock that were pried loose as magma intruded, whereas only the bottom of a lava flow incorporates preexisting rock.
3. Look at the top surface of the igneous layer. Because the top of a lava flow is exposed to the air for some time before being overlain by other lava or sediment, its gases are free to escape. Consequently, a lava flow surface is often pockmarked by cavities called *vesicles* that were formerly occupied by escaping gas bubbles. The tops of most sills, which were never exposed to the air, display few if any vesicles.
4. Look for signs of weathering (discussed in Chapter 5). The upper surface of a lava flow would appear somewhat weathered from its exposure to the atmosphere before being overlain, whereas a sill would not show signs of weathering because it was never exposed.

Batholiths and Other Large Plutons

Large concordant plutons are commonly several kilometers thick and tens or even hundreds of kilometers across. They may be mushroom-shaped or saucer-shaped, close to the surface or deep beneath it. When thick, viscous

Figure 3-21 The best laccolith-viewing in North America is along Route 95, through the Henry Mountains of southeastern Utah, from which this photo was taken. The laccolith is the massive gray structure in the background.

magma intrudes between two parallel layers of rock and lifts the overlying one, it cools to form a mushroom-shaped or domed concordant pluton, or **laccolith** (from the Greek *lakkos,* or "reservoir") (see Fig. 3-17). Laccoliths tend to form at relatively shallow depths, where there is little pressure to keep the overlying rock in place. They are typically granitic, formed from felsic magma that flows so slowly that it tends to bulge upward instead of spreading outward, raising the overlying rock to form a dome; this overlying dome is often eroded away, exposing the igneous rock below (Fig. 3-21). Sills form in a similar way, but they are usually basaltic and relatively flat because they form from faster-flowing mafic magmas that can enter small spaces readily.

Unlike upward-bulging laccoliths, saucer-shaped concordant plutons called **lopoliths** (from the Greek *lopas,* or "saucer") sag downward (see Fig. 3-17). These are probably produced when mafic magma is so dense that it sinks as it intrudes, depressing the country rocks below to create a magma-filled basin. One such structure is evident along the western shore of Lake Superior, where the surrounding country rock has been eroded away (Fig. 3-22). Lopoliths often contain mineralogically distinct layers, with the densest early-forming crystals that settled to the floor of the magma chamber on the bottom. Several layers near the base of the Bushveld lopolith of South Africa contain the Earth's richest concentration of the dense metal platinum.

Figure 3-22 The gabbroic Duluth lopolith, a classic North American lopolith on the shores of the western end of Lake Superior in Minnesota and Wisconsin, is more than 250 kilometers (150 miles) in diameter and about 15 kilometers (10 miles) thick. This photo shows only a portion of the exposed rock.

Figure 3-23 Eighty million years of uplift and erosion have exposed plutons such as this one, El Capitan, in Yosemite Valley, California. This valley, in Yosemite National Park, is part of the Sierra Nevada batholith, a huge plutonic mass believed to have been formed by at least five separate igneous episodes spanning 130 million years.

Some igneous intrusions can be even vaster than these large structures. **Batholiths** (from the Greek *bathos,* or "deep") are massive discordant plutons with surface areas (when exposed) of 100 square kilometers (40 square miles) or more (see Fig. 3-17). Most batholiths appear to be about 30 kilometers (20 miles) deep. They are generally found in elongated mountain ranges where overlying rocks have eroded to expose deep cores of plutonic rocks, as in the Yosemite Valley of California's Sierra Nevada and the White Mountains of New Hampshire. The Coast Range batholith of western British Columbia stretches more than 2000 kilometers (1200 miles) and is as much as about 290 kilometers (180 miles) wide. Batholiths are also found in the ancient centers of continents, the deep, erosion-resistant rocks of long-gone mountains; 2 to 3 billion years ago, such mountains may have towered over mid-continent landscapes in what are now Minnesota, Wisconsin, Michigan, and Ontario.

The texture of batholith rock is relatively aphanitic at the exterior and coarsens gradually to a phaneritic interior. The fine texture develops at the margins of the magma, where it comes in contact with the cooler country rock into which it intrudes, crystallizing relatively quickly. The coarse-grained interior texture develops where the magma cooled more slowly. Recent studies using earthquake waves passing through batholiths underground suggest that their shape may resemble that of a human tooth, the incisor—increasing with depth to its widest point and then tapering.

Some batholiths are a complex of many large plutons that have intruded one another in stages, often over tens of millions of years. The Sierra Nevada batholith in east-central California, which consists of numerous plutons packed so closely as to appear to be a continuous mass, probably formed over a span of 130 million years (Fig. 3-23).

Plate Tectonics and Igneous Rock

The worldwide distribution of igneous structures and rocks is not random. Certain structures and rocks are found consistently in some geological settings but not in others. Plutonic igneous structures, for example, tend to form at or near the boundaries of diverging or converging plates. These plate movements provide openings and opportunities for magma to intrude older rocks. Smaller plutonic features, such as dikes and sills, are generally found in divergent or rifting zones, where mafic magmas move as the Earth's brittle outer layers are stretched and pulled apart. Where oceanic plates have subducted, intermediate and granitic batholiths are found, marking many modern and ancient plate boundaries. Oceanic rocks carried by subduction down into the asthenosphere are partially melted, generating the vast quantities of magma that form coastal batholiths. The chain of western batholiths in North America, which stretches from British Columbia through the California Sierras to Baja California, developed through more than 200 million years of oceanic-plate subduction.

The Origin of Basalts and Gabbros

Basalt and gabbro, the intrusive equivalent of basalt, are the only igneous rocks in oceanic crust. Because the thin oceanic crust lies directly over the mantle, the source for basaltic/gabbroic magma must be the ultramafic mantle. But mantle material is not homogeneous in composition, and therefore neither are the world's basalts and gabbros. The varied compositions of these

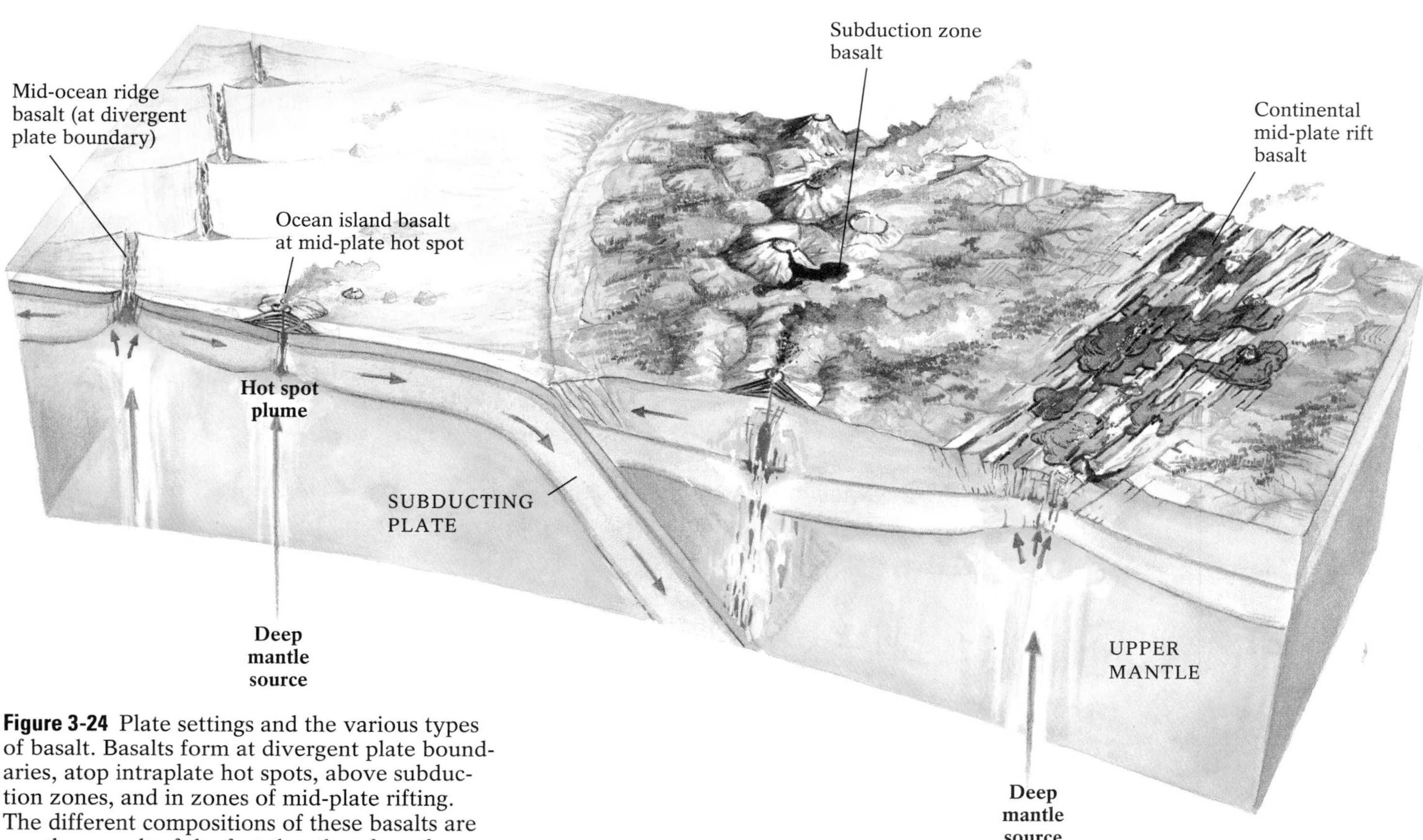

Figure 3-24 Plate settings and the various types of basalt. Basalts form at divergent plate boundaries, atop intraplate hot spots, above subduction zones, and in zones of mid-plate rifting. The different compositions of these basalts are mostly a result of the fact that they form from magmas originating at varying depths in the Earth's mantle.

rocks seem to depend on whether they derived from deep- or shallow-mantle magma sources (Fig. 3-24).

During the Earth's earliest millenia, the uppermost segment of the mantle apparently melted partially and its lighter components rose to become the Earth's earliest crust. The remaining upper mantle thus lacks such light elements as sodium, potassium, and aluminum. Small amounts of these elements may still exist below the depleted upper mantle, however. Thus, a basalt or gabbro that contains these elements probably derived from the mantle's deeper, undepleted zone, whereas one lacking these elements probably derived from the depleted uppermost mantle.

Because gabbros are rarely seen at the Earth's surface, most of what we know about the origin of mafic rock is from the Earth's numerous basalts. These are grouped into two main categories—oceanic and continental—based on the general environmental setting in which they formed. Within each of the categories, basalts can be further distinguished by the related factors of composition, magma source, and plate tectonic setting.

Oceanic Basalts *M*id-ocean *r*idge *b*asalts, or MORBs, the most abundant volcanic rock, account for about 65% of the Earth's surface area. Eruptions at oceanic divergent boundaries produce these rocks. Because they have low concentrations of sodium, potassium, and aluminum, MORBs probably formed from partial melting of the upper mantle, which is depleted of these elements.

*O*cean *i*sland *b*asalts, or OIBs, are found not at divergent plate boundaries, but atop *hot spots,* volcanic zones (generally intraplate) that lie over deep-mantle heat sources. Because they contain small but significant amounts of sodium, potassium, and aluminum, OIBs probably originated from a deep part of the mantle that was not depleted of those elements. Unlike MORBs, which generally erupt unwitnessed beneath thousands of meters of seawater, eruptions of OIBs can be readily seen in such places as the Hawaiian Islands.

Continental Basalts Basalts in continental settings vary more than their oceanic counterparts. They form both where new rifts tear at old continental plates and where oceanic plates subduct. The compositions of basalts associated with continental rifting show that they most likely derived from deep-mantle sources, whereas those at subduction zones tap shallower sources. Both form, however, as hot basaltic magma rises through tens of kilometers of continental crust, incorporating many of the materials in its path. Thus, the varied composition of continental basalts may result from melting and assimilation of continental rocks as well as from partial melting of deep and shallow mantle rocks.

The Origin of Andesites and Diorites

Whereas basalts and gabbros are the most common igneous rocks of the ocean basins, the less mafic andesites and diorites are commonly found along the geologically active subductive margins of continents and on oceanic islands that rose from the sea floor when oceanic plates subducted. Nearly continuous regions of andesitic rock are found on virtually all the lands that border the Pacific Ocean. This **andesite line** follows the nearly continuous pattern of subduction zones surrounding the Pacific Ocean basin (Fig. 3-25).

Figure 3-25 The andesite line. Subduction-produced andesitic and dioritic rocks make up most of the surface geology surrounding the Pacific Ocean basin.

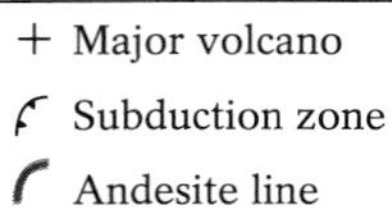

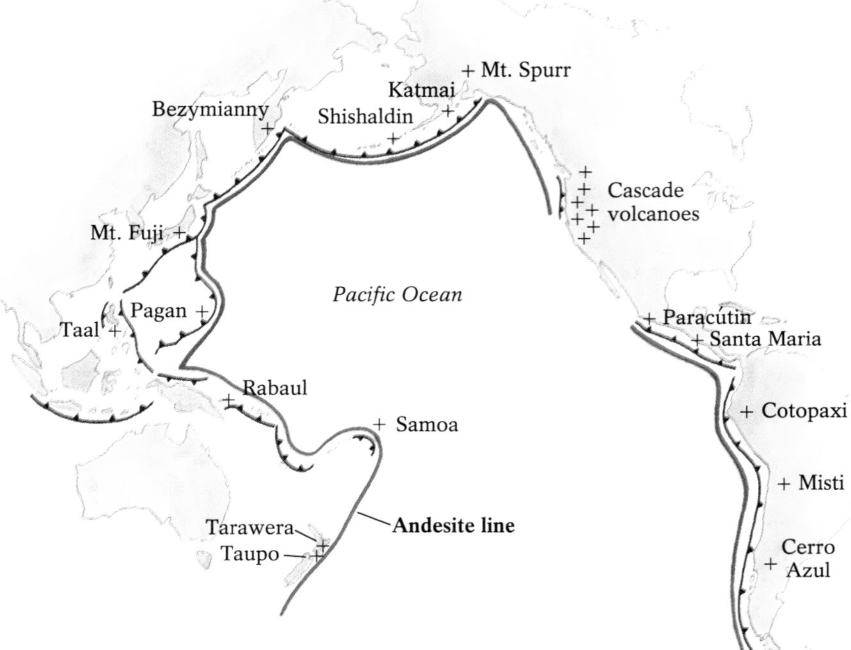

Geologists believe that a number of processes combine to produce rocks of intermediate composition from a subducting oceanic plate. One likely factor is water, which may be trapped in spaces within the sediments blanketing the descending plate, within the crystal structures of minerals in the sedimentary muds, and within fractures in the oceanic basalt. As an oceanic plate subducts, its water is pressed out into the overlying mantle, which, although it is quite warm, has been kept solid by the prevailing pressure. As we saw earlier in this chapter, water lowers the melting point of rocks under pressure, promoting their melting. Thus, water driven from the subducting plate enters the warm mantle and lowers its melting point so that partial melting occurs. Because partial melting of the ultramafic rocks in the upper mantle generally produces mafic magma, a new batch of basaltic magma may form above a subducting oceanic slab.

Because more andesite than basalt erupts from subduction zone volcanoes, however, some process other than water-induced partial melting of the mantle must add felsic material to these basaltic magmas to produce the intermediate composition of andesite. A likely source of this felsic component is the 200 meters (about 650 feet) of sediment that, on average, covers oceanic plates. Oceanic sediment comes principally from the airborne debris of continental volcanism (usually intermediate to felsic in composition), the felsic minerals transported to the oceans by continental rivers, and the silica-based shells and skeletons of microscopic marine organisms. Carried into the mantle on subducting plates, some of these felsic materials, because of their relatively low melting points, melt and then mix with mantle-derived basalt to produce an intermediate magma that cools to form andesite or diorite. Partial melting of the subducting plate may also contribute an intermediate component to subduction zone magmas, or subduction zone magmas may assimilate felsic materials as they rise through overlying rocks. The various factors that combine to produce andesite and diorite from the subduction of wet oceanic lithosphere are summarized in Figure 3-26.

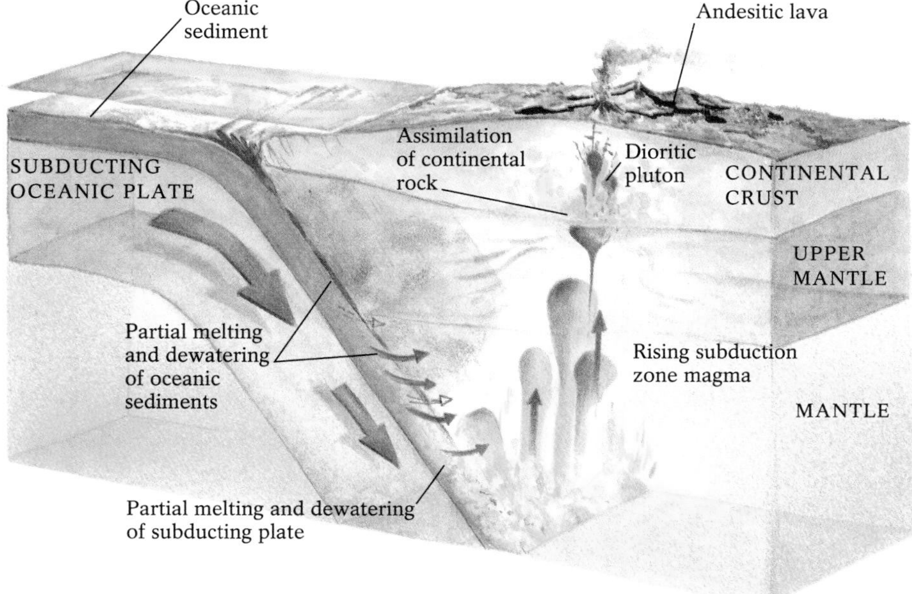

Figure 3-26 The factors involved in the origin of andesite and diorite. Water pressed out of the subducting plate and its associated oceanic sediment enters the mantle rock and lowers its melting point, causing it to melt and rise as basaltic magma. The composition of this initially mafic magma is made more intermediate by its mixing with partially melted felsic oceanic sediment and oceanic crust from the subducting plate, as well as with felsic country rock assimilated by the magma as it rises up through the continental crust of the non-subducting plate.

The Origin of Rhyolites and Granites

Nearly all rhyolitic and granitic rocks are found on continents. Many geologists believe that these rocks originate principally from partial melting of lower continental crust, and there is convincing laboratory evidence for this. When pressures and temperatures comparable to those at the lower crustal depth of 35 to 40 kilometers (21–24 miles) are applied to wet rocks of typical continental compositions (similar in composition to andesite/diorite), the rocks melt partially to yield the felsic rhyolites and granites.

Most granitic intrusions appear at or near modern or ancient subduction plate margins. Apparently, rising hot mafic and intermediate magmas and the frictional heat that accompanies subduction cause partial melting of dioritic rocks at the base of plate-edge mountain belts (Fig. 3-27). Partial melting of intermediate rocks produces a new magma that is predominantly felsic; thus, the resulting rocks (which are almost always plutonic) are typically granitic.

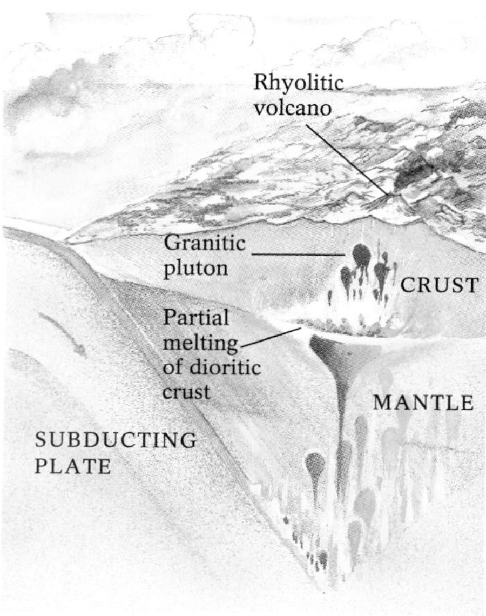

Figure 3-27 The origin of felsic rocks at convergent plate boundaries. Hot rising mafic and intermediate magmas partially melt dioritic rocks in the lower continental crust, producing granitic plutons (and, occasionally, granite's rare volcanic equivalent, rhyolite).

Igneous Rocks of the Moon

Since 1969, geologists who study igneous rocks have extended the reach of their rock hammers some 400,000 kilometers (240,000 miles) to the Moon. Moon rocks collected by Apollo astronauts in the 1970s indicate that the Moon's surface contains at least two distinct types of geological/geographical provinces—the highlands and the *maria* (plural of Latin *mare,* or "sea") (Fig. 3-28). The rocks of the lunar highlands date from about 4.0 to 4.5 billion years ago, when the early Moon's interior was apparently hot enough to develop a multilayered internal structure. (As we saw when we described the early development of the Earth in Chapter 1, a planetary body that is entirely or partially molten separates into distinct layers as its lighter materials rise and denser materials sink toward the interior.) The rocks of the lunar highlands, which probably crystallized as the Moon's earliest crust, consist principally of anorthosite, a type of coarse-grained plutonic igneous rock composed almost exclusively of the calcium plagioclase anorthite.

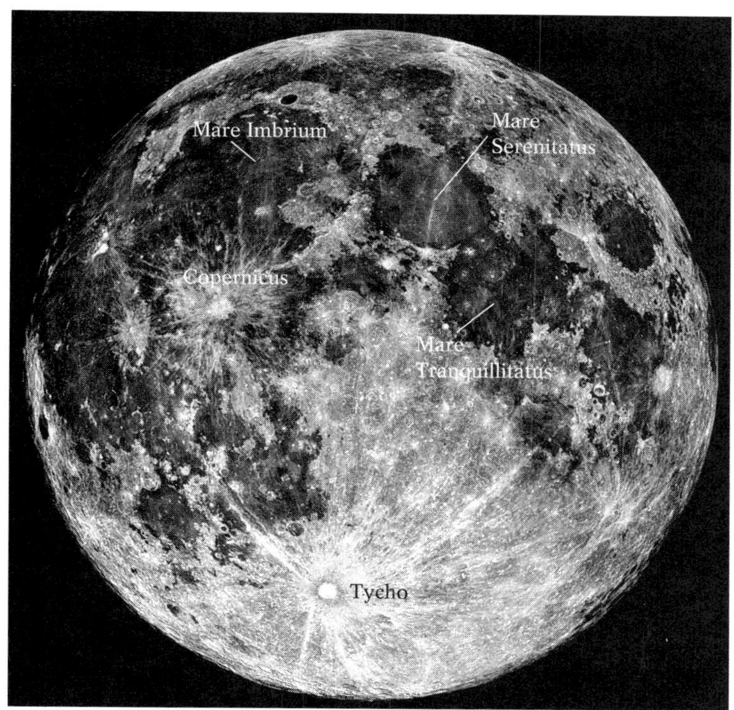

Figure 3-28 A telescopic view of the near side of the Moon, showing its highlands and maria.

The Moon's anorthosites, and its other rocks as well, differ subtly from the Earth's rocks. These differences demonstrate some fundamental geological distinctions between the Earth and the Moon. First, Moon rocks are totally waterless. If there ever was any water on the Moon, it became heated, vaporized, and escaped the Moon's weak gravitational field very early in the Moon's evolution. The lack of water on the Moon explains the absence of such prominent water-containing Earth minerals as the amphiboles and micas.

Another difference between the Moon and the Earth is that the origin of the Moon's igneous rocks seems unrelated to plate tectonics. Most lunar geologists believe that the Moon has never had moving plates. (Unlike most of the Earth's mountains, lunar mountain ranges apparently did not form from plate convergence.) The Moon's igneous activity also seems unrelated to the Moon's internal heat, much of which was probably lost long ago because the Moon's small size enables its heat to be readily conducted from its interior to its surface and then into space.

Some lunar igneous rocks appear to be produced by a process that is rare on Earth—the generation of new magmas by meteorite impact. (Most of the Earth's incoming meteorites are incinerated as they pass through the atmosphere.) The crushing force of a large impact on the Moon first produces an enormous quantity of shattered rock, which collects as angular fragments interspersed with bits of glass fused by the heat of the impact. The heat may raise the temperature of the stricken rocks to their melting points, thus generating new magma. The impact also fractures the lunar crust, providing subsurface magmas with an easy path to the surface.

Unlike the Earth's watery oceans, the lunar maria, or "seas," are actually vast solidified basalt flows. They were named when Galileo and other early astronomers, using the crude telescopes of the time, believed that they were true seas. From about 4 to 3.85 billion years ago, intense meteorite activity gouged numerous craters in the Moon's predominantly anorthositic

Figure 3-29 The formation of lunar maria—the vast solidified basalt flows that are prominent features of our Moon—from meteor impacts. Above: A photo of lunar craters, mountains, and maria taken during the Apollo 10 mission of May 1969.

Incoming meteor

Lunar surface

Meteor impact

Lunar crust broken and melted by impact. Upper mantle melts and wells up into fractured crust.

Lunar maria

Basaltic lava fills and overflows impact craters.

surface. Partial melting in the Moon's ultramafic upper mantle, accelerated by instantaneous reductions in pressure as impacts removed great volumes of overlying rock, produced basaltic magma that rose to the surface through impact-induced fractures in the crust. Basaltic lava flowed into and filled the craters, forming the lunar maria (Fig. 3-29). Although minor cratering continues today, the eruptions of the mare basalts apparently marked the final episodes of major igneous activity on the Moon. By the close of this period, meteorite concentrations in our solar system had been reduced, because much of the debris remaining after the origin of the solar system had already been swept up during the system's first billion years. The lack of younger igneous rocks on the Moon's surface today suggests that its interior is no longer hot enough to produce new magmas.

The Economic Value of Igneous Rocks

The practical uses of igneous materials range from the glittering (gemstones and precious metals) to the utilitarian (crushed basalt for road construction). Any urban center displays one of the principal uses we have found for plutonic igneous rocks—the decorative building stone that adorns the exteriors and lobbies of many banks and office buildings. The same appealing polished granites and diorites can be found in cemeteries, where they serve as durable tombstones. On a smaller scale, glassy pumice is the abrasive in grease-removing cleansers, and is also used to remove callouses from hands and feet. Until recently, pumice was an ingredient in toothpaste because of its ability to remove dental stains and plaque; since it also claimed its share of tooth enamel, however, milder abrasives have now taken its place. A few other familiar and useful minerals, such as the diamonds found in ultramafic rocks and the emeralds and topazes in felsic pegmatitic rocks, are also of igneous origin.

Gold and silver are often found in or around granitic rocks, as are less shiny but equally valuable ores of copper, lead, and zinc. These late-crystallizing metallic ions often become concentrated in the hot magmatic fluids that remain after most other minerals have already crystallized. These fluids may enter fractures in adjoining rocks, where these elements then crystallize as mineral-rich veins.

After being deposited by magmatic fluids, valuable minerals may be redeposited by groundwater that percolates down through an igneous region. When such water comes in contact with a magma chamber or a body of still-warm plutonic rock, it is heated and some or all of it may be converted to steam, which then invades surrounding rocks and dissolves any soluble minerals in them. When the steam eventually cools and condenses, its dissolved load recrystallizes to form mineral-rich deposits. The gold of the Homestake Mine in the Black Hills of South Dakota, the lead, silver, and zinc of northern Idaho, the copper of northern Michigan and Bingham Canyon in northern Utah (Fig. 3-30), and the silver of the Comstock Lode of Nevada all accumulated from the action of heat-driven subterranean fluids.

We can now build on your general knowledge of the Earth's igneous processes and rocks and expand our discussion of igneous activity to that which occurs above ground. The following chapter focuses on igneous phenomena we can observe—the volcanic eruptions and rocks produced when magma reaches the Earth's surface and escapes.

Figure 3-30 The Bingham Canyon copper mine near Salt Lake City, Utah, where plutonic igneous rocks are exposed over a 3 by 5 mile area. The abundant copper found throughout these rocks was deposited by subterranean magmas and groundwater.

Chapter Summary

Igneous rocks are the most abundant type of rock in the Earth's crust and mantle. They form when molten rock cools and crystallizes. Molten rock contained beneath the Earth's surface is called **magma;** when it erupts onto the surface, it is called **lava.** The texture of an igneous rock reflects the rate at which its parent magma or lava cooled, and its mineral content reflects the composition and evolution of the molten rock from which it formed.

Intrusive, or **plutonic,** igneous rocks form from magma that cools slowly underground. These rocks are generally coarse-grained, or phaneritic, because ample cooling time allows crystals to grow to visible sizes. Igneous rocks with exceptionally large crystals are called **pegmatites. Extrusive,** or **volcanic,** igneous rocks form when lava cools quickly at the Earth's surface. These rocks are generally fine-grained, or aphanitic, because rapid cooling limits crystal growth.

The most common igneous rocks include ultramafic **peridotite,** an iron-and-magnesium–rich plutonic rock containing less than 40% silica; mafic **basalt,** an iron-magnesium-and-calcium–rich volcanic rock (45–50% silica), and its intrusive equivalent, **gabbro;** intermediate **andesite,** an iron-aluminum-and-sodium–rich volcanic rock (about 60% silica), and its intrusive equivalent, **diorite;** and felsic **rhyolite,** a potassium-and-aluminum–rich volcanic rock (70% or more silica), and its intrusive equivalent, **granite.**

Magmas are produced when rocks in the Earth's interior melt. The factors that affect a rock's melting point include local heat and pressure conditions and the water content and composition of the rock. Most magmas are created when preexisting rocks **partially melt**—that is, the minerals with lower melting points liquefy first and start to flow as a molten mass, which carries within it still-solid crystals of minerals that melt at higher temperatures.

As magma cools, different minerals crystallize from it at different temperatures. The silicate minerals as a group crystallize in two specific sequences, known together as **Bowen's reaction series:** In the discontinuous series, mafic silicate minerals evolve in distinct steps, with both their compositions and their internal crystal structures changing at each step; in the continuous series, the plagioclase feldspars evolve as sodium ions gradually replace calcium ions in the developing crystals, without any accompanying change in the minerals' internal crystal structures.

As early-forming minerals crystallize, they may remain suspended in the magma, continue to exchange ions with it, and ultimately evolve into later-forming minerals as predicted by Bowen's reaction series. However, some early-forming crystals may separate from the liquid magma by settling to the bottom of the magma chamber (if the crystals are heavier than the magma), becoming plastered to the walls of the magma chamber, or being filtered out as the still-liquid magma moves into fractures too narrow for the crystals to pass. When early-forming crystals are removed from a magma, its composition is changed by the loss of the ions in those crystals. As a result of this process, called **fractional crystallization,** the rocks ultimately produced by the magma will be different in composition from that which would have been produced by the original, unseparated magma. Recent studies of igneous rocks suggest that in addition to fractional crystallization, the composition of a magma may also be modified by assimilation of preexisting rocks or by mixing with another body of magma having a different composition. When masses of preexisting rock are only partly assimilated in a magma, they appear in the solidified rock as distinct bodies known as **xenoliths.**

Bodies of magma that cool underground form **plutons,** igneous features that are distinct from surrounding rocks. Plutons are classified by their shapes, sizes, and orientation relative to the rocks they intrude. Concordant plutons are parallel to the preexisting rock layers; discordant plutons cut across the preexisting rock layers. Tabular plutons are igneous formations that are relatively thin, like a table top. Discordant tabular plutons are called **dikes;** concordant tabular plutons are called **sills.** Large concordant plutons include mushroom-shaped **laccoliths** and saucer-shaped **lopoliths;** large discordant plutons are called **batholiths.**

The principal igneous rock types are typically associated with specific plate tectonic settings. Basalts and gabbros are most often found at divergent plate boundaries (the mid-ocean ridge basalts, or MORBs), atop intraplate hot spots (the ocean island basalts, or OIBs), and where a continental plate is rifting. Andesites and diorites are found where oceanic plates have subducted to form volcanic mountains. The nearly continuous ring of subduction-produced andesites that surrounds the Pacific Ocean is called the **andesite line.** Rhyolites and granites are formed within continents by partial melting of the lower portions of the continental crust. They are often associated with subduction-produced mountains, and also occur where continental rifting has taken place.

Igneous rocks also exist on the Moon. Lunar rocks collected by the Apollo astronauts in the 1970s indicate that the Moon's surface rocks consist largely of anorthosite, a coarse-grained plutonic igneous rock composed almost exclusively of the calcium plagioclase anorthite, and vast areas of basalt. Igneous rocks on the Moon differ fundamentally from those on Earth in that they contain no water, and their formation involves neither subsurface heat nor plate tectonics.

Igneous rocks are valued for the gemstones and precious metals they contain as well as for the variety of practical purposes for which they are used (including the use of crushed basalt in road construction).

Key Terms

igneous rocks (p. 65)
magma (p. 66)
lava (p. 66)
intrusive rocks (p. 67)
plutonic rocks (p. 67)
pegmatites (p. 67)
extrusive rocks (p. 68)
volcanic rocks (p. 68)
peridotite (p. 70)
basalt (p. 70)
gabbro (p. 70)
andesite (p. 71)
diorite (p. 71)
granite (p. 71)
rhyolite (p. 72)
partial melting (p. 74)
Bowen's reaction series (p. 75)
fractional crystallization (p. 78)
xenolith (p. 79)
pluton (p. 79)
dike (p. 80)
sill (p. 80)
laccolith (p. 83)
lopolith (p. 83)
batholith (p. 84)
andesite line (p. 86)

Questions for Review

1. Briefly describe the textural difference between phaneritic and aphanitic rocks. Why do these rocks have different textures?

2. Some igneous rocks contain large visible crystals surrounded by microscopically small crystals. What is the term for these rocks? How does such a texture form?

3. What elements would you expect to predominate in a mafic igneous rock? In a felsic igneous rock?

4. Name the common *extrusive* igneous rocks in which you would expect to find each of the following mineral types: calcium feldspar; potassium feldspar; muscovite mica; olivine; amphiboles; sodium feldspars. Which *plutonic* igneous rock contains abundant quartz and muscovite mica, but virtually no olivine or pyroxene?

5. What factors, in addition to heat, control the melting of rocks to generate magma?

6. What is the basic difference between the continuous and discontinuous series of Bowen's reaction series?

7. Briefly describe three things that might happen to an early-crystallized mineral surrounded by liquid magma.

8. What is the difference between a sill and a dike? Between a batholith and a lopolith?

9. Briefly discuss two specific types of plate tectonic boundaries and the igneous rocks that are associated with them.

10. What is the basic difference between a MORB and an OIB?

For Further Thought

1. What type of igneous feature is shown in the photo below?

2. Felsic rocks such as rhyolite often occur together with basaltic rocks at locations where continents are undergoing rifting. Give one possible explanation for this.

3. Why do we rarely find batholiths made of gabbro?

4. Speculate about how the distribution of the Earth's igneous rocks will change when the Earth's internal heat is exhausted and plate tectonic movement stops.

5. Why are there virtually no granites or diorites on the Moon? How might small volumes of such felsic rock form under the geological conditions believed to be responsible for the Moon's igneous rocks?

The Nature and Origin of Volcanoes

The Products of Volcanism

Eruptive Styles and Associated Landforms

Highlight 4-1 In the Shadow of Mount Pelée

Coping with Volcanic Hazards

Highlight 4-2 Mount St. Helens

Extraterrestrial Volcanism

Figure 4-1 A satellite image of Japan's Mount Fuji (Fujiyama).

Chapter 4

Volcanoes and Volcanism

Volcanism is the set of geological processes resulting in the expulsion of molten rock, as lava, at the Earth's surface. The most notable products of volcanism are **volcanoes**, the landforms created when lava and hot particles escape from the Earth's interior through openings, or **vents**, in the Earth's surface and then cool and solidify around the vents. Volcanoes provide some of the world's most breathtaking scenery. Each year, millions are drawn to the slopes of Mount Rainier in Washington state, Mount Fuji in Japan (shown in Figure 4-1), and Mount Vesuvius in Italy by the mystery and beauty of these volcanoes. Yet the ever-present danger associated with volcanoes worries nearby communities, and for good reason. A recent survey by the Smithsonian Institution found that about 600 volcanoes have erupted in the past 2000 years, most of them more than once. In a single year, there are approximately 50 eruptions around the world.

Powerful volcanic eruptions and their aftereffects can be among the Earth's most destructive natural events. Consider what happened to the uninhabited Indonesian island of Krakatoa one summer day in 1883. Krakatoa, a volcanic island in the Sunda Straits of the southwest Pacific Ocean, had for many years been a landmark for clipper ships carrying tea from China to England. The volcano, which had been inactive for more than two hundred years, stood 792.5 meters (2601 feet) high. On the morning of August 27, the entire island all but vanished in one of history's most explosive eruptions, leaving a hole 304.8 meters (1006.5 feet) below sea level. As far as we know, no one perished directly from the destruction of Krakatoa; however, between 36,000 and 100,000 lives were lost as the resulting waves, up to 37 meters (121 feet) high, pounded coastal villages on the nearby islands of Java and Sumatra.

The eruption of Krakatoa, the force of which was equivalent to the explosion of 100 millions tons of TNT, jostled the atmosphere around the globe. Sharp barometric changes were recorded as far away as San Francisco and London. The sound of the explosion was heard 4802 kilometers (2983 miles) away at Alice Spring in central Australia (which is akin to the residents of San Diego hearing an explosion in Boston). The eruption produced a black cloud of volcanic debris that rose to an altitude of 80 kilometers (about 50 miles), blocked out all sunlight, and plunged the region into darkness for three days. The cloud's finest particles, swept aloft by wind currents, reduced incoming solar radiation by as much as 10% worldwide, causing a drop of

Figure 4-2 Anak Krakatau ("Child of Krakatoa"), the small island that emerged from the remains of the volcanic island Krakatoa during eruptions in the 1920s. The original Krakatoa volcano was demolished in a monumental eruption in 1883.

more than 1°C (1.8°F) in global temperatures. The suspended particles also caused years of spectacular crimson sunsets. A few months after the eruption, on October 30, 1883, residents of Poughkeepsie, New York, and New Haven, Connecticut, summoned fire brigades to douse blazes that were, in fact, only the fiery glow of the brilliant evening sky. Today, a new young volcano is growing where Krakatoa used to be (Fig. 4-2).

Volcanism does not mean unmitigated disaster; it results in some notable benefits as well as destruction. For instance, we owe the air we breathe and even some of the water we drink to volcanic eruptions, which throughout Earth's existence have released useful gases from the planet's interior. Some of the hydrogen and oxygen liberated by volcanoes combine to form the waters of the Earth's *hydrosphere*—its oceans, lakes, rivers, underground waters, glaciers, and clouds. Nitrogen and oxygen combine with other components to produce the Earth's gaseous *atmosphere*.

Volcanoes are like windows into the Earth, providing us with information that would otherwise be inaccessible. Ascending magma carries subterranean rock fragments to the surface, giving us a glimpse of actual rocks from the Earth's interior. In addition, volcanoes can show us the past: Volcanic deposits have provided much of our knowledge of extinct life forms, including our own evolutionary ancestors. Footprints of the earliest upright-walking hominids were preserved in fresh volcanic ash in East Africa 3.6 million years ago (Fig. 4-3).

Volcanic activity also adds to the Earth's inventory of habitable real estate. Iceland, Japan, Hawaii, Tahiti, many islands of the Pacific and Caribbean, and nearly all of Central America are products of volcanism. Volcanic terrains often become prime agricultural lands. The rich coffees grown in South and Central America sprout from fertile volcanic soils. The value of some types of volcanic soils can be seen on the Indonesian island of Java, where fine volcanic ash retains water and abundant nutrients (such as potassium, calcium, and sodium) that nourish plants. Java's population density is about 200 times that of neighboring Borneo, the soils of which are derived from solid extrusive rock and so are markedly less fertile.

Volcanic landscapes often contain a source of inexpensive, clean energy. During the dark polar winter in Reykjavic, Iceland, the world's northernmost capital, people stroll in their shirt-sleeves through warm shopping malls. Iceland has no oil, no coal, no natural gas, and very few trees for fuel, but it does have an abundant underground hot water supply, thanks to the molten rock that fuels Iceland's constant volcanic activity. By tapping the scalding water just meters beneath their feet, Icelander's can heat more than 80% of their homes and businesses. Similar *geothermal* ("Earth heat") resources are being tapped in other current and recently active volcanic settings, among them the Geysers area, 150 kilometers (100 miles) north of San Francisco, California.

Thus, volcanoes and volcanism are at once both a great hazard and a great boon to humankind. In this chapter, we will explain the causes and characteristics of the different types of volcanoes. We will describe the threats they pose to us, and how we have learned to cope with them. Finally, we will examine volcanism on some of our neighboring planets in the solar system.

Figure 4-3 A footprint in volcanic ash from East Africa. Anthropologists have learned a great deal about human evolution from imprints such as this, recording the passage of some of our early ancestors more than 3.5 million years ago.

The Nature and Origin of Volcanoes

Volcanic eruptions range from the quiet oozing of molten basalt from Kilauea volcano in Hawaii to the cataclysmic explosion of Krakatoa in Indonesia. The current status of a volcano is the key to its threat to human life and property. Is it active, dormant, or extinct? An *active* volcano is one that is currently erupting or has erupted recently (in geological terms). Certain active volcanoes, such as Kilauea or Stromboli in the eastern Mediterranean, erupt almost continuously. Others erupt periodically, such as Lassen Peak in northern California, which erupted last in 1917. Active volcanoes can be found on all the continents except Australia, and on the floors of all the major ocean basins. Indonesia, with 76 active volcanoes, Japan, with 60, and the United States, with 53, are the world's most volcanically active nations.

A *dormant* volcano is one that has not erupted recently but is considered likely to do so in the future. A number of signs may suggest that a dormant volcano is stirring to wakefulness. A shallow heat source, such as rising magma or heated rocks, might convert surface water that seeps into the ground into hot water springs and steam. The discovery in 1975 of new hot springs on the slopes of Mount Baker near the Washington–British Columbia border, for example, has raised anxiety in nearby communities about the prospect of a future eruption there. If magma pushes aside fractured rocks as it rises toward the surface, numerous small earthquakes may be set off near a volcano. The presence of relatively fresh (within 1000 years old) volcanic rocks in a volcano's vicinity also suggests that it is capable of a repeat performance (Fig. 4-4).

Figure 4-4 Which of these two volcanoes would you think is more likely to erupt? (**a**) The slopes of Washington state's Mount Rainier remain deeply scored by repeated episodes of glacial erosion that occurred over hundreds of thousands of years and appear not to have received a fresh covering of lava in thousands of years. (**b**) The slopes of Mount St. Helens (shown here before its 1980 eruption), believed to be the youngest of Washington's Cascade Range volcanoes, are relatively uneroded. Although intermittently dormant for hundreds or thousands of years, St. Helens' volcanic cone has continued to grow.

(a)

(b)

A volcano is classified as *extinct* if it has not erupted for a very long time and is considered unlikely to do so in the future; one indication that a volcano is probably extinct is that extensive erosion has taken place since its last eruption. A truly extinct volcano is no longer fueled by a magma source. It is generally wise, however, to view all volcanoes with caution. Residents of the Icelandic island of Heimaey believed their Mount Helgafjell to be extinct until it came to life in a spectacular eruption in 1973, its first in 5000 years.

The Causes of Volcanism

Volcanism begins with the creation of magma by the melting of preexisting rock (discussed in Chapter 3) and culminates with the ascent of this magma to the Earth's surface through fractures, faults, and other cracks in the lithosphere. (The distribution of the Earth's lithospheric cracks, usually associated with tectonic plate boundaries and with intraplate hot spots, determines where most volcanoes will form.) A magma will erupt if it flows upward rapidly enough to reach the surface before it can cool and solidify. Two characteristics of a magma determine its potential to do this: its gas content and its viscosity.

Gas in Volcanic Magma Magmatic gases make up from 1% to 9% of most magmas. The principal gases are water vapor and carbon dioxide, with smaller quantities of nitrogen, sulfur dioxide, chlorine, and a few others. Tens of kilometers underground, gases remain dissolved in magma, held there by the pressure of the surrounding rocks. This pressure decreases as magma rises toward the surface, and the gases, less dense than the surrounding magma, migrate upward, pushing any overlying magma before them.

Gases become concentrated near the top of a rising magma body and press against the overlying rock. When these volcanic gases are completely prevented from escaping, perhaps by a plug of congealed lava blocking the passage to the surface, they accumulate and exert even greater pressure against the overlying rock, until ultimately it shatters. As soon as the overlying rock is removed, the pent-up gases instantaneously expand, much as the reduction in pressure allows the gases in an agitated bottle of a soft drink to fizz and bubble out when the bottle is opened. The initial blast removes the overlying pressure, hurling masses of older rock skyward, and sprays shreds of lava into the air as the gas-charged magma is expelled. The eruption may then settle down to a relatively placid outpouring of degassed magma, or it may cease altogether.

Magma Viscosity A magma's *fluidity*—its ability to flow—is governed by its temperature and composition. Increased heat invariably increases the fluidity of any substance, because when the temperature of a substance increases, its ions and atoms move about more rapidly, thus breaking the temporary bonds that inhibit flow. **Viscosity** is a fluid's *resistance* to flow (the opposite of fluidity): Viscous fluids move quite sluggishly; less viscous fluids move more easily. Viscosity increases with *decreasing* temperature.

The viscosity of magma generally increases with silica content (see Table 3-1), because the oxygen ions at the corners of the unbonded silicon-oxygen tetrahedra (discussed in Chapter 2) form temporary bonds with other ions in the magma. Because felsic magma tends to be relatively cool (because it crystallizes at low temperatures) and has a high silica content, it is very viscous. Conversely, because mafic magma is hot and has a low silica content, it is much less viscous and flows easily. Therefore, mafic magmas are more likely to rise to the surface and erupt than are felsic magmas, which tend to cool underground into plutonic rocks.

The viscosity of magma also has a direct effect on the explosiveness of a volcanic eruption. Gases do not escape from all magmas with equal ease. In more fluid magmas, migrating gases meet with little resistance and therefore escape readily when the magma reaches the surface; these gases do not accumulate to build up the high pressure that causes explosive eruptions. Thus, low-viscosity mafic magmas tend to erupt quietly, with a relatively gentle outpouring of degassed lava. In highly viscous felsic magma, on the other hand, the movement of gases is impeded and gas pressures build within the molten material. Thus, felsic magmas tend to erupt explosively.

The Products of Volcanism

A volcanic eruption can produce a flowing stream of red-hot lava, a shower of ash particles as fine as talcum powder, a hail of volcanic blocks the size of automobiles, or any number of intermediate-sized products. The quantity of lava produced by volcanoes ranges from small spurts to vast floods. (An immense submarine lava flow, believed to have been extruded within the last 25 years, was recently discovered in the vicinity of the East Pacific rise, a divergent plate boundary off the coast of South America. The flow contains approximately 15 cubic kilometers of basalt, enough to pave over the entire U.S. interstate-highway system to a depth of 10 meters, or 35 feet.) Both the type and the amount of material produced by a volcano depend largely on the composition of its lava. In considering the products of volcanism, we will first examine the different types of lava and their properties and then describe the various forms in which volcanic materials are deposited on the surface.

Types of Lava

The composition of a lava is similar to that of its parent magma. (The lava, however, contains less dissolved gas, as most of it escapes into the atmosphere during an eruption.) The most common type of lava is basalt, lava having a mafic composition. As we've learned, mafic magmas are the most likely to become volcanic because they tend to be hot and highly fluid, moving readily to the surface before solidifying. This explains why we find much more basalt than gabbro in the Earth's crust. Magmas of felsic composition tend to be cooler and much more viscous, only rarely reaching the surface—as rhyolite lava—before solidifying. For this reason, there is much more granite than rhyolite in crustal rocks. Andesitic lavas are intermediate between basaltic and rhyolitic lavas in both composition and fluidity; they erupt much more frequently than rhyolitic lavas but are less common than basalt.

Because of their distinct natures, lavas of different compositions act differently upon being extruded; as a result, each is associated with certain characteristic volcanic styles and products.

Basaltic Lava For nearly a century, the Hawaiian Volcano Observatory on the big island of Hawaii has been observing eruptions, producing most of what we know about subaerial ("under air," as opposed to under water) basaltic lava flows and the resulting rocks. The observatory has found the temperature of Hawaiian flows to be as high as 1175°C (2150°F). Such hot, low-viscosity lava cools to produce two principal types of basalt, *pahoehoe* (pronounced "pa-hoy-hoy") and *aa* (pronounced "ah-ah"). Pahoehoe, which means "ropy" in a Polynesian dialect, is aptly named. Highly fluid basaltic lava moves swiftly down a steep slope at speeds that may exceed 30 kilometers per hour (20 mph), spreading out rapidly into sheets about 1 meter thick (Fig. 4-5). The surface of such a flow cools to form an elastic skin that is then dragged into ropelike folds by the continuing movement of the still-fluid lava beneath it. The ropy surface of pahoehoe basalt is generally quite smooth. Native islanders refer to it as "ground you can walk on barefoot," and most old Hawaiian foot trails follow ancient pahoehoe flows.

Figure 4-5 This fast-flowing basaltic lava stream from Hawaii's Kilauea volcano cools to form a ropy, pahoehoe-type surface texture.

As basaltic lava flows farther from the vent, it cools and becomes more viscous. A thick brittle crust develops at its surface and slowly continues to move forward, carried along by the warmer, more fluid lava below it. The molten interior of the flow advances more rapidly than the cooler outer region, breaking it up to produce a rough surface having numerous jagged projections sharp enough to cut animals' hoofs. Flows having these features are called aa flows (Fig. 4-6). (*Aa* is a local term of unknown origin that may recall the cries of a barefoot islander who strayed onto its surface. Ancient foot trails meticulously avoid aa fields.) Aa is often found downstream from pahoehoe, the product of the same flow.

Figure 4-6 This relatively slow-moving basaltic lava cools to form a blocky, jagged, aa-type surface texture.

Figure 4-7 Vesicular basalt, or scoria. These pores remain in the cooled rock after gases burst from bubbles on the surface of highly fluid lava.

Subaerial basalt flows may produce several other distinctive features as they cool. The gas still contained in the lava often migrates to its surface and escapes, leaving small pea-sized vesicles (discussed in Chapter 3) which are preserved at the top of the basalt when it cools. Vesicle-rich basalt is known as *scoria* (Fig. 4-7). As basaltic lava cools, it often shrinks in volume, producing a pattern of cracks known as *columnar jointing*. The cracks extend inward from the top and bottom surfaces of the flow into its interior as cooling proceeds, creating five- and six-sided polygonal columns of basaltic rock. A side view of these columns suggests a large bunch of pencils; an aerial view suggests oversized ceramic bathroom floor tiles (Fig. 4-8). In North America, the Devil's Postpile in California's Sierra Nevada and Devil's Tower in northeastern Wyoming (site of the climax of the film *Close Encounters of the Third Kind*) are spectacular examples of basaltic columns.

Basalt flows may also contain *lava tubes*, cavities that form when lava solidifies into a crust at its surface, but continues to flow underneath. Eventually, as the eruption wanes, the still-molten lava drains from the cooled tubes, leaving them hollow. Lava Beds National Monument in northeastern California is rife with 300 or more tubes in the area's pahoehoe flows. This natural labyrinth sheltered the Modoc Indians in 1872 as they battled several hundred troops from the United States Army in an attempt to reclaim their native lands. (The fewer than 60 Modoc warriors, who knew every crevice in the lava field, inflicted serious harm on their pursuers from their sanctuary within the lava tubes—but eventually were overcome and forced to leave their territory.)

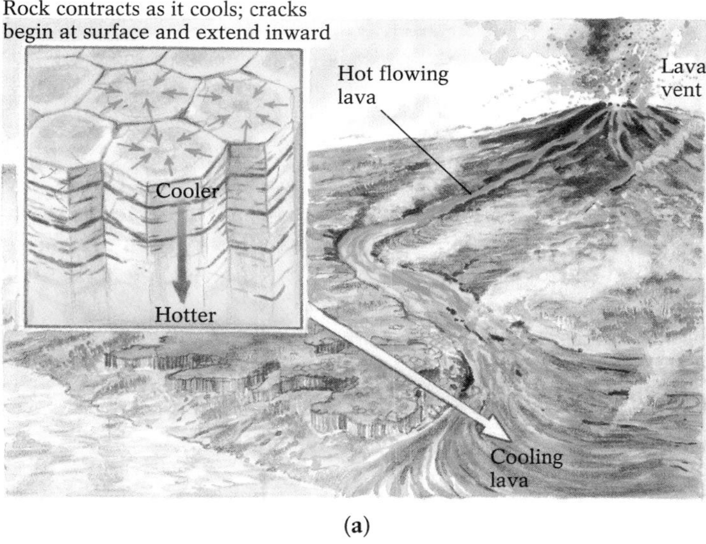

Figure 4-8 Contraction of basaltic lava flows as they cool (**a**) sometimes produces geometrically patterned joints such as the columnar structures of the Giant's Causeway in Northern Ireland (**b**). Similar structures can be found in North America in eastern Washington, eastern Oregon, southern Idaho, eastern California, Yellowstone National Park, and eastern Wyoming.

Subaqueous ("under water") basaltic eruptions, though much more common than subaerial eruptions, are less well understood because they usually take place at inaccessible depth, often beneath thousands of meters of seawater. We do know that when basaltic lava erupts beneath the sea, it develops a distinctive *pillow structure*. Upon contact with cold water, the extruded lava is instantly chilled, immediately forming a thin deformable skin that stretches to resemble an elongated pillow as additional hot lava enters under it. As the pillow shape expands, its surface cracks, allowing some lava to flow from it and form another pillow, from which yet another pillow might grow, and so on (Fig. 4-9).

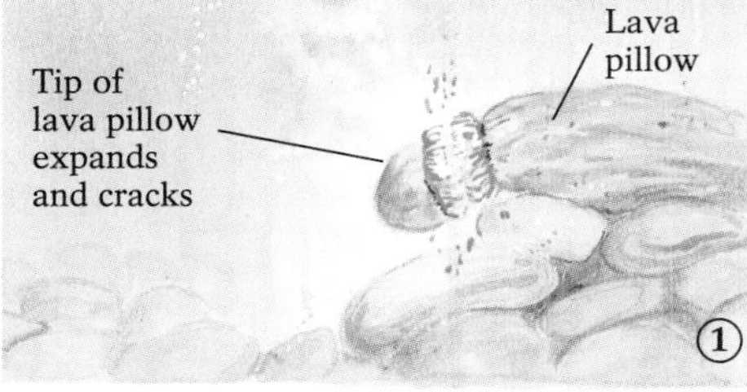

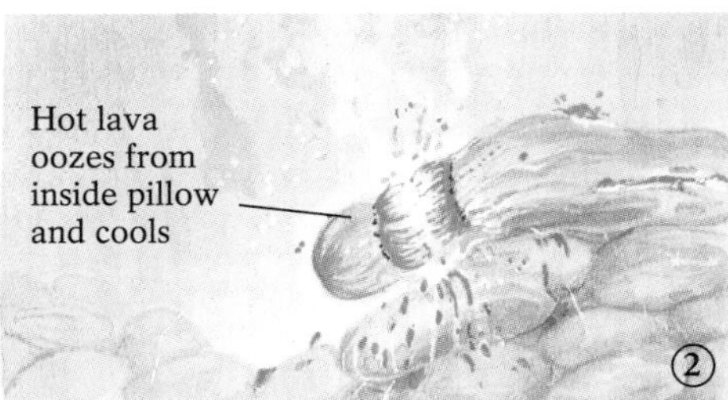

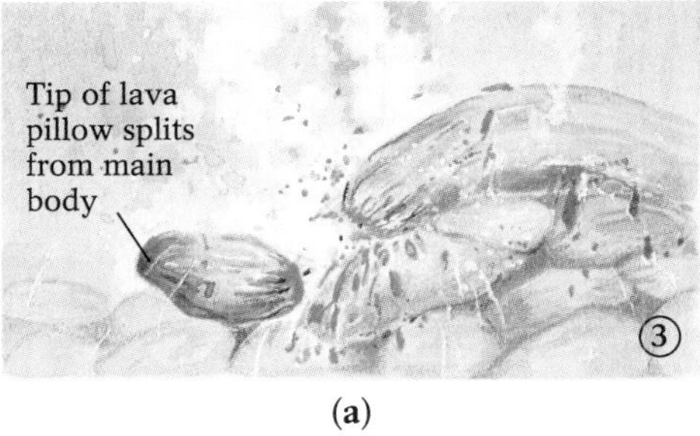

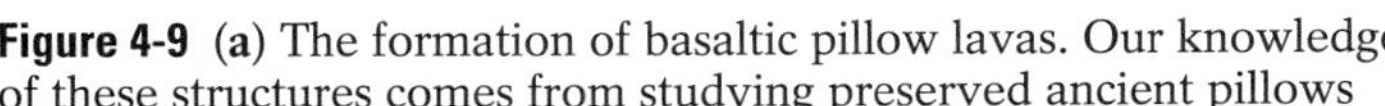

Figure 4-9 (**a**) The formation of basaltic pillow lavas. Our knowledge of these structures comes from studying preserved ancient pillows (**b**), and from observing modern pillows, such as these in the Galapagos (**c**), from deep-sea submersibles.

Andesitic Lava Andesitic lava, which is intermediate in composition between mafic basaltic lava and felsic rhyolitic lava, is also intermediate in viscosity. Consequently, it flows more slowly than basaltic lava, and solidifies before traveling as far from its vent. Like basaltic lavas, andesitic lavas may develop vesicles because their fairly low viscosity allows some trapped gases to bubble out; they may also develop aa-type surface textures. We rarely see pahoehoe-type andesitic flows, however, because these lavas are too viscous to stretch into a ropy structure. Andesitic lavas, particularly the more mafic ones, can also produce columnar jointing and pillow structures. The more felsic andesitic lavas, however, can be viscous enough to impede the passage of rising gases and erupt in major volcanic explosions.

Rhyolitic Lava Rhyolitic magma, being the most felsic and highly viscous, moves so slowly that it tends to cool and solidify underground as plutonic granite, rarely erupting as lava at the Earth's surface. When rhyolite does erupt, it usually explodes violently, producing an enormous volume of solid airborne fragments instead of a lava flow. Because it is so viscous, rhyolitic lava never flows far from the vent, and does not produce the structures that typify less viscous lavas.

Felsic magmas with high water and gas content may bubble out of a vent as a froth of lava that quickly solidifies into an extremely porous volcanic rock, *pumice*. Pumice may contain so many vesicles from which gas escaped that it can be lighter than water. When Krakatoa erupted in 1883, sailing ships were trapped for three days in the waters of the Sunda Straits by huge rafts of floating pumice. The pumice finally became waterlogged and sank, allowing the ships to escape the continuing shower of hot ash.

(a)

(b)

Figure 4-10 Tephra particles range in size from fine dust to large boulders. (**a**) This range can sometimes be found within a single deposit, as here at Mono Craters, California. (**b**) A volcanic bomb, the largest type of tephra.

Pyroclastics

An explosive eruption expels lava forcefully into the atmosphere, where it cools rapidly, solidifying into countless fragments of various sizes and shapes. Such an eruption might also shatter some of the preexisting rock composing the volcano. *Volcanic blocks*, for instance, are chunks of igneous rock ripped from the throat of a volcano during an eruption; they tend to be angular, and range from the size of a baseball to that of a house. Blocks weighing as much as 100 tons have been found as far as 10 kilometers (6 miles) from the volcano that spewed them out. All such fragmental volcanic products are known as **pyroclastics** (from the Greek *pyro*, meaning "fire," and *klastos*, meaning "fragments"). Pyroclastic materials may travel through the air as dispersed particles or they may hug the ground as dense flows.

Tephra Those pyroclastic particles that cool and solidify from lava as it is propelled through the air are called **tephra**. Tephra particles are classified by size, which ranges from a fine dust to massive chunks (Fig. 4-10). *Volcanic dust* particles are only about one thousandth of a millimeter in diameter and have the consistency of cake flour. Because it is so fine, volcanic dust can travel great distances downwind from an erupting volcano and remain in the upper atmosphere for as long as a year. Somewhat grittier is volcanic *ash*, particles of which are less than 2 millimeters in diameter, ranging from the size of a grain of fine sand to that of rice. Ash generally stays in the air for only a few hours or days. During the 1980 eruption of Mount St. Helens in Washington state (see Highlight 4-2, pp. 118–119), ash clogged automobile carburetors, fouled the bearings of farm machinery, and forced commercial air flights to be diverted to avoid abrasion damage to jet engines (as well as reduced visibility). People downwind of the eruption were advised to wear surgical masks outdoors because of the health hazard of inhaling ash.

Coarser tephra, pulled to Earth by gravity, fall sooner and closer to the volcanic vent. *Cinders*, or *lapilli* (Italian for "little stones") range from about the size of peas to that of walnuts (2 to 64 millimeters in diameter). *Volcanic bombs* are large (64 millimeters or more), streamlined chunks of rock formed when sizable blobs of lava solidify in mid-air while being propelled by the force of an eruption.

Tephra of all sizes accumulate where they are deposited, falling into layers that record the frequency and intensity of past volcanic activity. The first step in "reading" a tephra layer is to associate it with a specific volcano, usually by its unique chemical composition; together the layers constitute the volcano's record of past eruptive activity. In general, the coarser-grained and thicker a tephra layer is, the closer it is to its source (Fig. 4-11). Eventually tephra become compacted or cemented into a solid rock known as **volcanic tuff.**

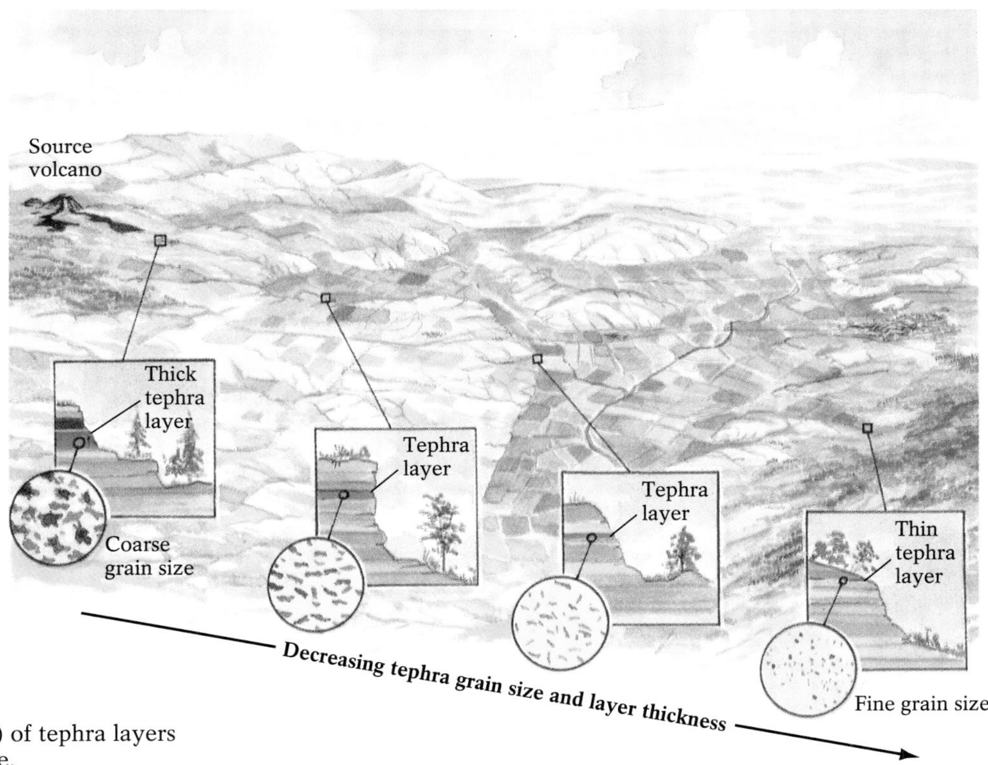

Figure 4-11 The thickness and texture (grain size) of tephra layers decrease with distance from their volcanic source.

Pyroclastic Flows When the amount of pyroclastic material expelled by a volcano is so great that gravity almost immediately pulls it down onto the volcano slope, this material rushes downslope as a **pyroclastic flow**, or **nuée ardente** (French for "glowing cloud"). Because the flow also contains trapped air and magmatic gases, it is buoyant: The gases create a frictionless barrier between the flow and the ground. With little resistance between the flow and

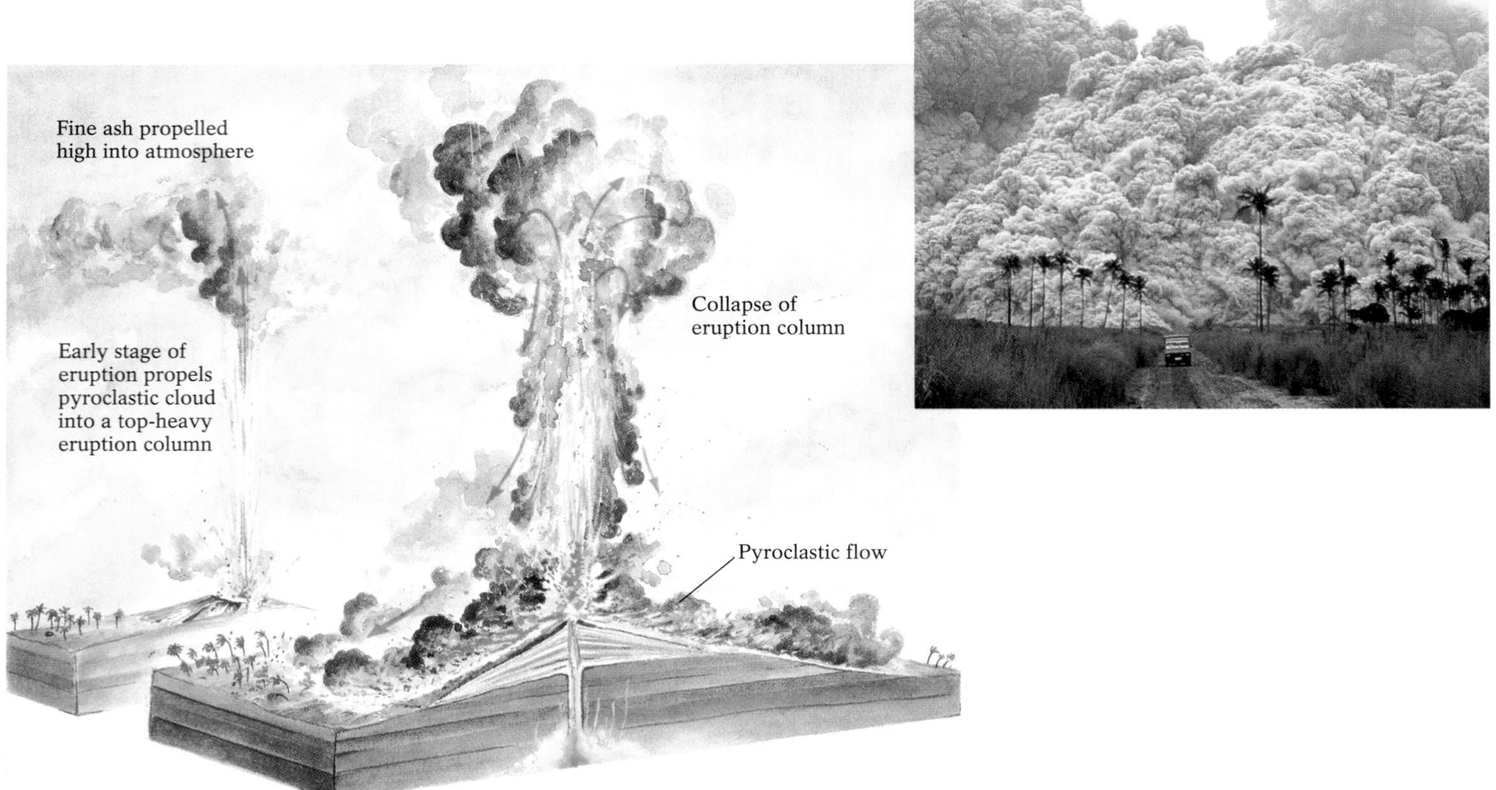

Figure 4-12 Pyroclastic flows, or nuée ardentes, are produced when a massive amount of airborne pyroclastic material is pulled to Earth by gravity and rushes downslope. Inset: A pyroclastic flow from the May 1991 eruption of Mount Pinatubo, in the Philippines.

the slope, pyroclastic flows may reach speeds in excess of 150 kilometers per hour (100 mph), even on gentle slopes (Fig. 4-12).

Particles in a pyroclastic flow travel through the air so briefly that they do not cool significantly, and may still be red-hot when they settle to the ground. With temperatures of 800°C (1475°F) or more, they are capable of melting glass or burning an apple into a smear of carbon. As the flow travels downslope, its gases escape and the warm particles finally come to rest, but they may still be soft enough to fuse with one another, forming a welded *tuff*.

Volcanic Mudflows Pyroclastic material that accumulates on the slope of a volcano may become mixed with water to form a volcanic mudflow, or **lahar** (a term coined on the Indonesian island of Java, where explosive eruptions and abundant loose, moist soil must have frequently caused disastrous lahars). A lahar may contain a range of particle sizes from the finest ash to enormous 100-ton boulders. A lahar is often produced when an explosive eruption occurs on a snow-capped volcano, and hot pyroclastic material melts a large volume of snow or glacial ice. This was the cause of the devastating lahar that buried the highland town of Armero, on the slopes of Colombia's Andes mountains, when that country's Nevado del Ruiz erupted on November 13, 1985 (Fig. 4-13).

The fine dust and ash sprayed from explosive eruptions may actually produce rain clouds, as atmospheric moisture, augmented by water vapor from the eruption, condenses around the cooling tephra particles. The clouds soon result in torrential rains, causing lahars to carry loose ash and soil down a volcano's slope. A mudflow whose water content is particularly high may reach speeds as high as tens of kilometers per hour.

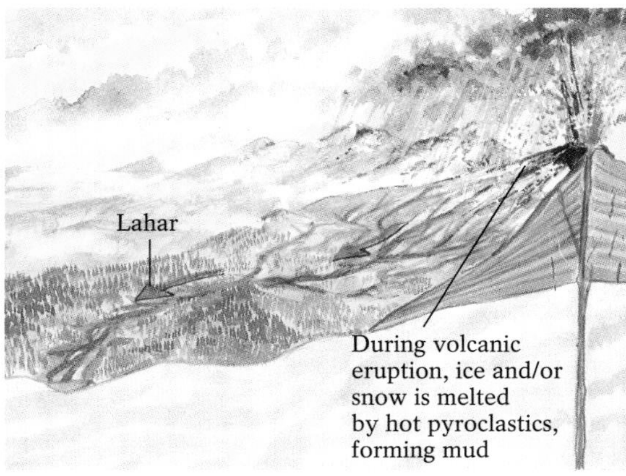

Figure 4-13 Some lahars occur when pyroclastic eruptions melt snow and ice on volcanic slopes, producing torrents of mud. Inset: This lahar resulted when Colombia's Nevado del Ruiz volcano erupted in 1985, melting about 10% of the snow on the slopes of the volcano and producing a 40-meter (137-foot)-high wall of mud that buried the town of Armero and approximately 23,000 of its residents.

Secondary Volcanic Effects

In addition to such primary effects as suffocating ash falls, searing nuée ardentes, and enveloping hot lahars, volcanic eruptions have secondary effects that can alter the environment and affect human, animal, and plant life, changing the composition of the atmosphere, and even, in some cases, the global climate. For example, the magmatic gases that escape during and after an eruption may include sulfur dioxide (SO_2), which combines rapidly with atmospheric water vapor and oxygen to form sulfuric acid (H_2SO_4). Sulfuric acid droplets can remain in the atmosphere for years, producing acidic precipitation and increasing the acidity of local, regional, and global waters. Volcanologists who have analyzed samples of the 2-kilometer (1.2-mile)-thick glacial ice on Greenland have identified areas showing evidence of past increased acidity; presumably, this ice formed when the Earth's atmosphere (and hence its snowflakes) contained an unusually high amount of SO_2, possibly due to volcanic activity.

Volcanic gas and ash emissions also affect worldwide climate. The dust and ash from a large tephra column can rise into the stratosphere and remain suspended there for more than four years. The particles reflect incoming sunlight back into space, lowering the amount of radiation that can reach and warm the Earth's atmosphere. Droplets of emitted SO_2 in the atmosphere also reflect and absorb radiation. The combination of dust and gas from a single eruption can lower the Earth's temperature by as much as 2° to 3° Celsius (4–6°F), an effect that may last for more than a decade. The first scientific demonstration of such an effect took place in May of 1784, when a blue haze and dry fog hovered over Europe and the weather was unusually cool. Scientist–philosopher–statesman Benjamin Franklin hypothesized that the unusual climate was caused by the 1783 flood-basalt eruption in Iceland. He tested his hypothesis by using a magnifying glass to focus the sun's rays on a sheet of paper. Normally, this would burn a hole in the paper, but in the spring of 1784, it did not. Franklin concluded correctly that less sunlight was reaching Europe as a result of the Icelandic eruption.

The greater the eruption, the longer its effects will linger. The spectacular eruption of Indonesia's Mount Tambora in 1815, which took an estimated 50,000 lives and decapitated the peak, was followed by what became known as the "Year without a Summer." In all, 150 to 200 cubic kilometers (60–80 cubic miles) of pyroclastics were ejected into the atmosphere (about 200 times that expelled from Mount St. Helens in 1980). Snow fell in upstate New York the following June; in Connecticut July 4th was celebrated in overcoats; and August frosts destroyed crops from the Midwest to Maine. Tambora's dust and gas reduced sunlight by more than 10%, and produced a long succession of dreary days. The depressing weather that summer at Lake Geneva in Switzerland has been credited with creating the dark mood that prompted author Mary Shelley to write her morbid classic *Frankenstein*.

Eruptive Styles and Associated Landforms

(a)

(b)

Figure 4-14 Volcanoes come in various shapes and sizes. (**a**) Lofty, symmetrical Mount Augustine, in Alaska's Cook Inlet; (**b**) small, stubby volcanoes from the Flagstaff area of northern Arizona.

Every volcano has its own eruptive style, depending on the composition, viscosity, and gas content of its magma. These factors also determine its shape. When we think of volcanoes, highly photogenic, snow-capped peaks such as the one in Figure 4-14a may come to mind. We do not usually picture scruffy little hills such as the ones in northern Arizona (Fig. 4-14b), which are also a common type of volcano. Despite their different appearances, nearly all volcanoes have the same two major components: a mountain, or **volcanic cone**, constructed by the products of numerous eruptions over time; and a steep-walled, bowl-shaped depression, or **volcanic crater**, surrounding the vent from which those volcanic products emanate. A volcano's crater forms following an eruption, when lava and pyroclastics that have accumulated in the area around the vent are left somewhat unsupported and subside to form a depression. Volcanic cones and craters have a variety of shapes and dimensions, depending on the type of eruption and the composition of the volcanic products.

Effusive Eruptions

Effusive eruptions are relatively quiet, nonexplosive events that generally involve basaltic lava. They are seldom accompanied by powerful volcanic blasts and rarely produce large volumes of pyroclastics. Basaltic lava, which is highly fluid, flows freely from central volcanic vents, as well as from elongated cracks on land and from fractures at submarine plate boundaries.

Central-Vent Eruptions In central-vent eruptions, basaltic lava flows out in all directions from one main vent, solidifying in more or less the same volume all around. Over time, succeeding flows accumulate to form a discernible low, broad, cone-shaped structure, known as a **shield volcano** because it resembles a warrior's shield lying on the ground (Fig. 4-15). The gentle slopes and broad summits of shield volcanoes are due to the fluidity of basaltic lava, which flows swiftly and too far from the vent to develop into steeper mountains. A vertical slice through a shield volcano would reveal hundreds of layers of basalt recording eruptions throughout the volcano's life.

The classic location for the study of central-vent eruptions and shield volcanoes is on the big island of Hawaii, whose shield volcanoes are the largest single objects on Earth. The island is located near one end of the 2590-kilometer (1600-mile)-long chain of islands and undersea mountains in the north-central Pacific Ocean that developed as the Pacific plate moved over a stationary hot spot in the Earth's mantle. (We will look more closely at the formation of volcanic island chains in Chapter 12.) The island of Hawaii itself consists of five major shields, the two largest being Mauna Kea and Mauna Loa, the summits of which are 4205 meters (13,792 feet) and 4170 meters (13,678 feet) above sea level, respectively. The bases of the volcanoes are on the floor of the Pacific Ocean, 5000 meters (17,000 feet) below sea level.

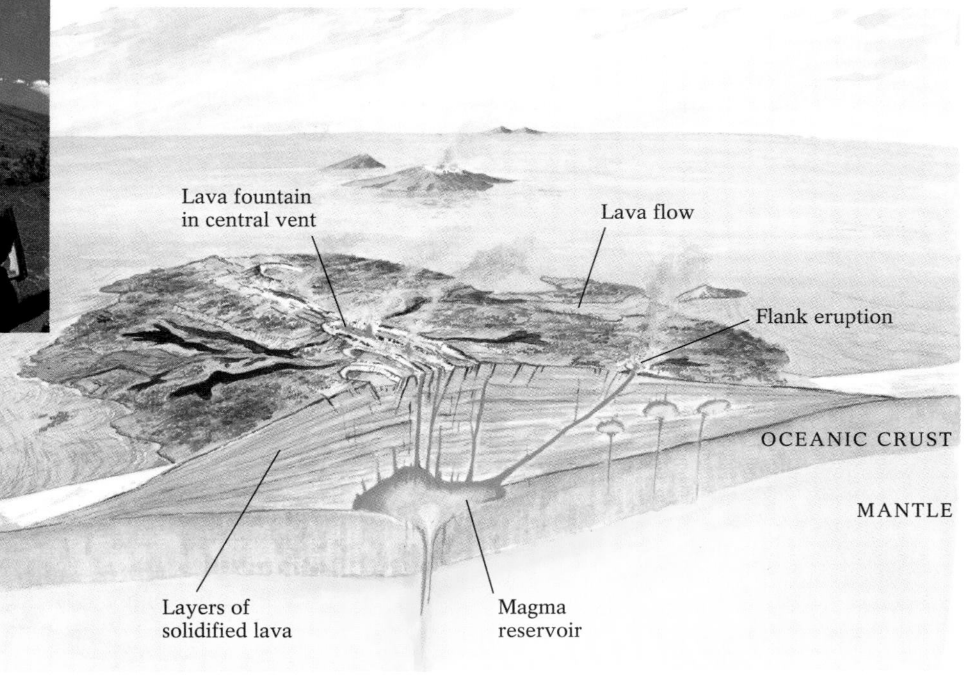

Figure 4-15 Low, broad shield volcanoes form by the gradual accumulation of gently sloping basalt lava flows. Inset: The summit of Hawaii's Kilauea volcano, a shield volcano that actually developed on the flank of an even larger shield volcano, Mauna Loa.

Thus, the total height—more than 9000 meters, or about 30,000 feet—of each of these volcanoes is greater than that of the Earth's highest continental peak, Mount Everest. Mauna Loa, with a circumference of 600 kilometers (400 miles), is composed of thousands of layers of lava flows that have erupted over the past 750,000 years.

At the start of an effusive eruption, lava begins to accumulate in the volcanic crater, forming a lava lake which may eventually overflow the rim of the crater. If enough lava has erupted to empty or partially empty the volcano's subterranean reservoir of magma, the unsupported summit of the volcanic cone may collapse inward, forming a much larger summit depression known as a **caldera**. Whereas an initial summit crater may be 300 meters (1000 feet) in diameter and several hundred meters deep, a caldera may be a few kilometers to more than 15 kilometers (10 miles) in diameter and more than 1000 meters deep.

The weight of such a collapsed summit may close off the central vent and divert any remaining underlying magma laterally, producing a *flank eruption* from the side of the volcano. Flank eruptions also occur when the central vent becomes plugged by congealed lava, or when the volcanic cone has grown so high that rising magma seeks a lower, more direct route to the surface. Pressurized lava spewing from a secondary vent on a volcano's flank often forms spectacular lava fountains 100 to 200 meters (330–660 feet) high. Hawaii's Kilauea is a volcanic cone built up by flank eruptions on Mauna Loa's southeastern slope. One of the world's most active volcanoes, Kilauea's supply of fresh basaltic lava is replenished continuously because it is located directly above the Hawaiian hot spot.

Fissure Eruptions Rising basaltic magma will exploit any available route to the surface. That route may be through a series of linear fractures, or *fissures*, in the Earth's crust, which develop most often where plates are diverging.

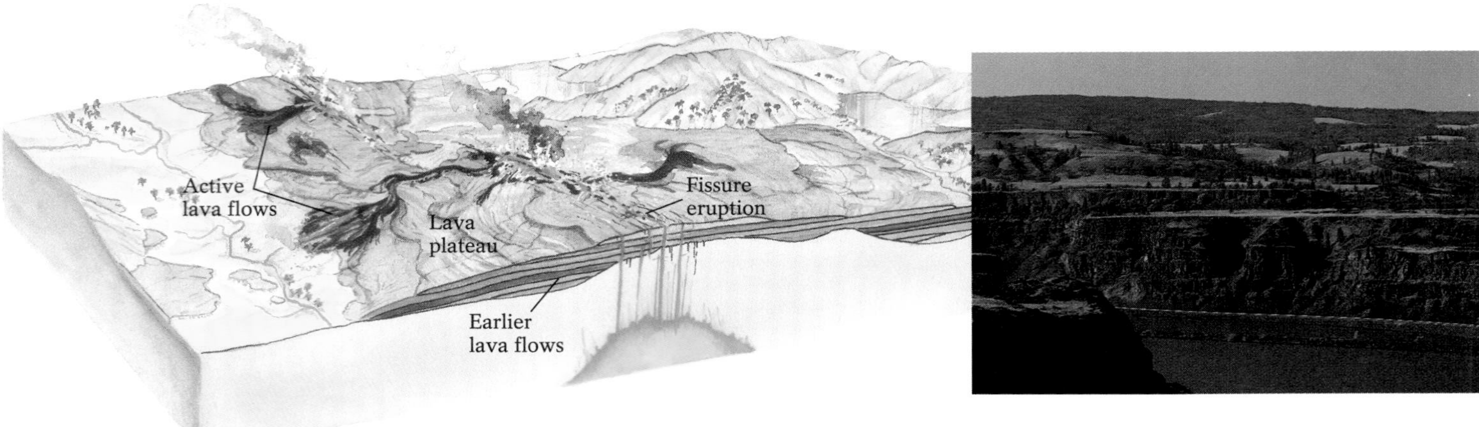

Figure 4-16 Lava-plateau formation. Inset: The 15-million-year-old basaltic Columbia River Plateau of eastern Washington state. Geologists have identified more than 60 individual lava flows here, in some places totaling more than 2 kilometers, or 1 mile, in thickness. The lava from just one of these flows could pave Interstate 90 from Boston to Seattle to a depth of 175 meters (575 feet), about the height of the Washington Monument in Washington, D.C.

(Fissure eruptions are common in Iceland, which is located directly above the mid-Atlantic ridge, the divergent plate boundary that separates North America from Europe.) Highly fluid basaltic lava may gush at speeds of 20 kilometers per hour (12 mph) or more from kilometer-long cracks in the crust. As lava flows away from a fissure it may spread out over thousands of square kilometers; successive flows build up immensely thick *lava plateaus*, or *flood basalts* (Fig. 4-16). A disastrous basaltic flood, the only one in recorded history, began in Iceland on June 8, 1783, as a series of lava flows spread out from one 32-kilometer (20-mile)-long fissure. The lava blocked and diverted rivers, melted snow and ice into raging torrents, and destroyed much of Iceland's scarce agricultural lands. Livestock were poisoned after grazing on grass contaminated by fluorine gas emitted during the eruption. As a result, 10,000 people—20% of Iceland's population—perished from starvation.

Subaqueous Eruptions Most subaqueous eruptions—especially oceanic, or *submarine*, eruptions—are quiet and effusive because below about 300 meters (1000 feet) of water, water pressure prevents the gases and water vapor in lava from expanding or escaping. Effusive submarine eruptions of basalt typically produce pillow structures such as those shown in Figure 4-9. However, there are occasional shallower submarine eruptions during which jets of steam propel lava above the ocean's surface, where it shatters and cools rapidly to form a substantial tephra cloud. Another type of submarine eruption may occur when seawater enters a magma chamber through the ruptured walls of an island volcano. As the cold water comes in contact with the red-hot magma, a cloud of superheated steam is produced that may expand violently, shatter the volcanic cone, and propel magma and cone fragments skyward (Fig. 4-17).

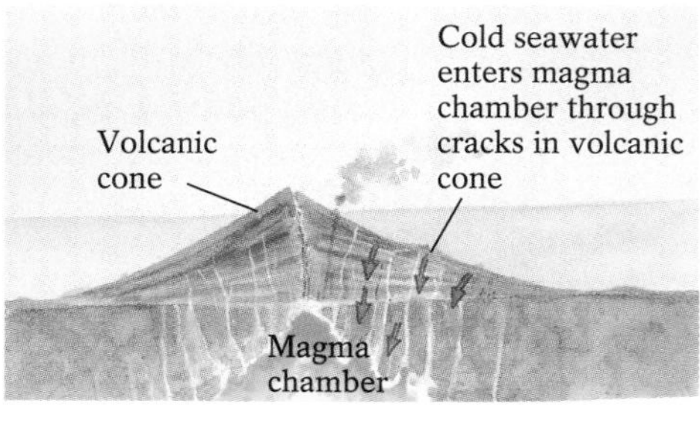

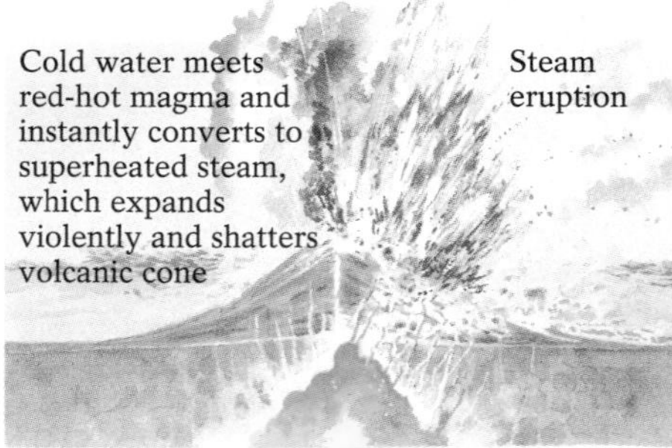

Figure 4-17 When seawater enters fractures in a submarine volcano and makes contact with the underlying magma chamber, it is converted to steam. Pressure from the pent-up steam may build until there is an explosive eruption, such as this one of a submarine volcano near Japan in 1986.

Figure 4-18 When Mount Vesuvius erupted in A.D. 79, the people in nearby Pompeii were trapped and suffocated beneath a layer of volcanic ash up to 8 meters (26 feet) thick. When Pompeii was excavated, archaeologists found cavities lined with an exact imprint of the decomposed bodies; by pouring plaster into the cavities, they were able to make casts that displayed the victims' musculature, agonized facial expressions, and in some cases even the folds in their clothing.

Pyroclastic Eruptions

Pyroclastic eruptions, which involve viscous, gas-rich magmas, vary from moderately to spectacularly explosive and tend to produce a great deal of solid volcanic fragments. Moderately explosive eruptions on the Italian island of Stromboli in the eastern Mediterranean occur almost continuously, producing a constant muffled thumping that is among the daily sounds of nature in the region. The blasts cover the region with a cloud of ash and hurl sizzling bombs, some as heavy as 2000 kilograms (4400 pounds), as far as 3 kilometers (2 miles) from their crater.

Pyroclastic eruptions are common in the eastern Mediterranean. Fed by large quantities of extremely gas-rich magma, they periodically produce spectacular vertical columns of tephra rising tens of kilometers into the atmosphere. The surrounding countryside is simultaneously smothered by suffocating showers of hot ash and covered with nuée ardentes and hot swirling lahars. Residents of the prosperous Roman resort town of Pompeii, which lay below the craggy peak of Mount Vesuvius looming 1220 meters (4002 feet) above the Bay of Naples, did not realize that this vine-clad mountain was a volcano until Vesuvius exploded on August 24, A.D. 79, sending a tephra column into the atmosphere, obscuring the midday sun, and shrouding the region in darkness. Most Pompeiians, inhaling the still-warm particles, perished by asphyxiation; others were crushed by falling statues. Those who tried to escape by sea were trapped by towering waves. The entire city was entombed beneath a layer of more than 6 meters (greater than 20 feet) of volcanic ash that preserved the most minute details of Roman life, not to be seen again until uncovered by archaeologists in the mid-eighteenth century (Fig. 4-18).

If it is fed by a magma containing less gas, a pyroclastic eruption may produce dense clouds of superheated ash and pumice instead of towering tephra columns. These pyroclastics might fall back to the surface almost immediately and race downslope as a nuée ardente.

A felsic magma, if it contains fairly little gas, may become a lava so viscous that it does not even flow out of the volcano's crater. Instead, it cools and hardens within to form a **volcanic dome** which caps the vent, trapping the volcano's gases and building pressure toward another eruption; this pressure might be relieved by the escape of gases laterally through the flank of the volcano, or it might build to a point that it finally shatters the volcanic dome in a particularly explosive eruption (Fig. 4-19). Highlight 4-1 recounts one such violent pyroclastic eruption.

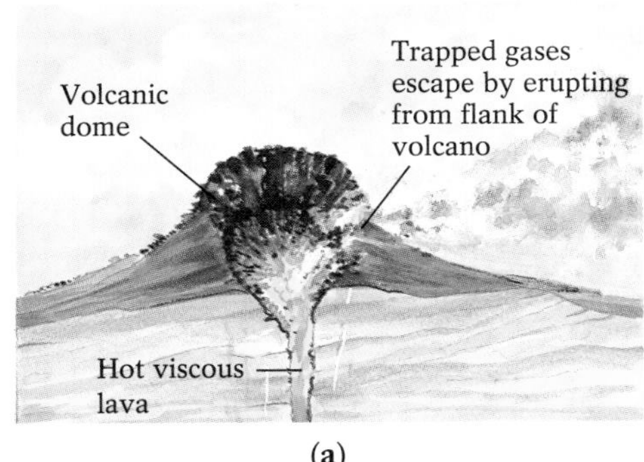

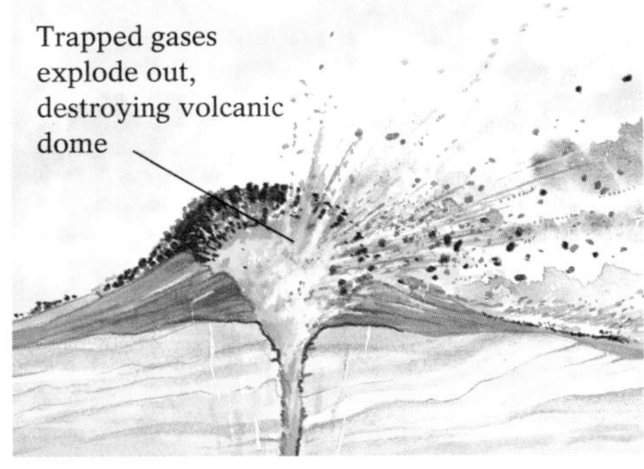

Figure 4-19 Volcanic domes are caused by extremely viscous lavas solidifying before they leave their crater, plugging the volcanic vent. Buildup of pressure beneath a dome may result in gases escaping from the flank of the volcano (**a**), or eventual explosive destruction of the dome (**b**).

Highlight 4-1 *In the Shadow of Mount Pelée*

(a)

Figure 4-20 The city of St. Pierre, on the Caribbean island of Martinique, before (**a**) and after (**b**) its destruction from the 1902 eruption of Mount Pelée.

(b)

On the morning of May 8, 1902, 30,000 residents of the beautiful French colonial city of St. Pierre on the Caribbean island of Martinique perished horribly in recent history's most devastating pyroclastic flow. In a flash, the summit of Mount Pelée vanished and a nuée ardente with a temperature of 700°C (1250°F) raced at 160 kilometers per hour (100 mph) down the mountainside, incinerating St. Pierre and capsizing and burning most of the ships anchored in its harbor (Fig. 4-20).

Only two of those in the city survived—shoemaker Léon Compère-Léandre, whose home was at the very edge of the pyroclastic flow, and condemned murderer Auguste Cyparis, confined to a dungeon whose door and window were blocked by an accumulated of tephra that had fallen in the preceding weeks. He was discovered by rescue teams four days after the blast. Later pardoned, Cyparis emigrated to the United States, where he joined Barnum and Bailey's circus and enjoyed considerable fame and fortune as the "Prisoner of St. Pierre.")

Pelée had given sufficient warning to save the city's inhabitants. Days earlier, a ship's captain from Naples became alarmed. "I know nothing about Mount Pelée, but if Vesuvius were looking the way your volcano looks this morning, I'd get out of Naples," he told everyone, before fleeing with his crew to safety. But the mayor of St. Pierre, engaged in a tight race for reelection, refused to evacuate his city-dwelling supporters. When Mt. Pelée began to spew hot ash across the countryside on April 23, the local newspaper, a loyal instrument of the administration, neglected to report the growing danger. On May 4, a lahar surged down Pelée's slope and engulfed a sugar refinery, killing 24 people. On May 7, the governor met with the mayor and they decided to order the evacuation immediately *after* the election on May 10. Unfortunately, Pelée erupted two days earlier, and the order never came.

The eruption continued until a stiff plug of hardened lava, an extreme example of a volcanic dome, became wedged like a cork in Pelée's vent. Pushed up by the eruption's residual gases, the "Tower of Pelée" grew to a height of 340 meters (1100 feet) before it crumbled. The government of Martinique rebuilt the city at the same ill-fated location, causing the townspeople great concern when Pelée became active again in 1929 and erupted, less explosively, in 1934. Volcanic pressures that have been building ever since may someday result in a replay of the 1902 eruption.

Culminating Eruptions The life of a volcano may consist of repeated eruptions marked by towering tephra columns and fiery pyroclastic flows and culminate in an extremely energetic eruption including a full range of explosive styles. The final explosion often begins with a titanic blast, followed by formation of a massive vertical tephra column. As the escaping gases lose force, the tephra column collapses and pyroclastic flows may follow. Finally, the summit of the cone may subside into the emptied magma chamber, forming a large caldera. A caldera in an oceanic volcano may immediately fill with seawater; a caldera in a land-based volcano might eventually fill with fresh water to form a sizeable lake.

Such a culminating eruption took place in the southern Oregon Cascades about 6700 years ago, when Mount Mazama buried thousands of square kilometers with tephra and pumice before its summit collapsed to form the caldera we know as Crater Lake. Before this eruption, the mountain is believed to have been 3700 meters (12,136 feet) high and glacier-covered, comparable in majesty to Mount Rainier or Mount Shasta. Today, it is a sawed-off goblet-shaped mountain only 1836 meters (6058 feet) above sea level, containing North America's deepest, bluest lake (Fig. 4-21).

This event may not have ended Mount Mazama's excitement, however. Eruptions over the past 1000 years have produced three smaller volcanic cones within the lake, two of which are still below water level. The third, Wizard Island, rises above the surface at the western shore of the lake. Recent dives to the lake bottom reveal that hot water and steam are being vented continuously, suggesting that Mount Mazama may be entering a new eruptive sequence.

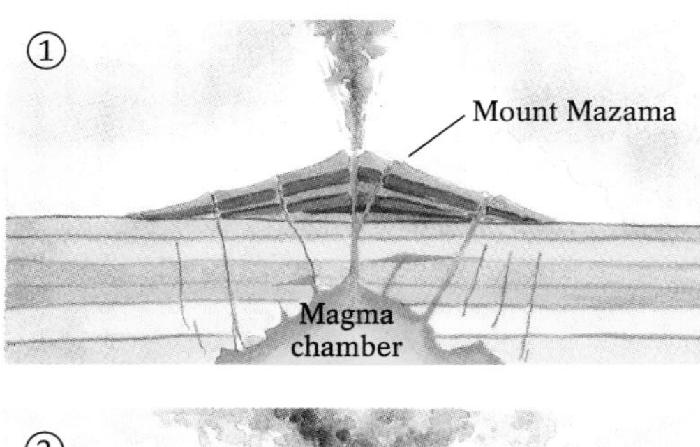

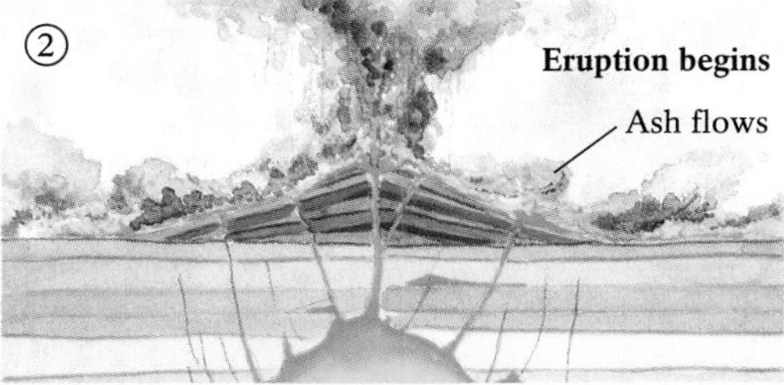

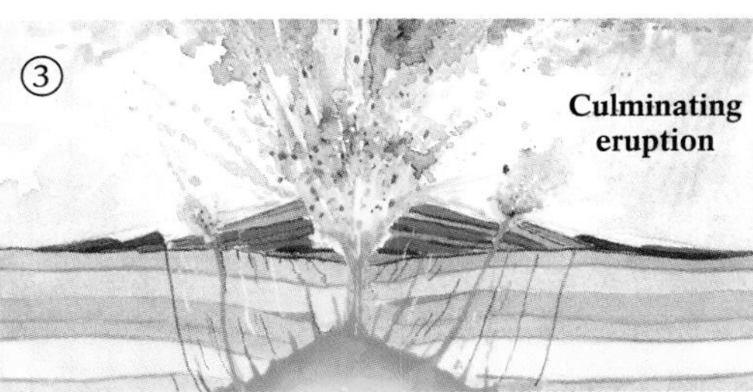

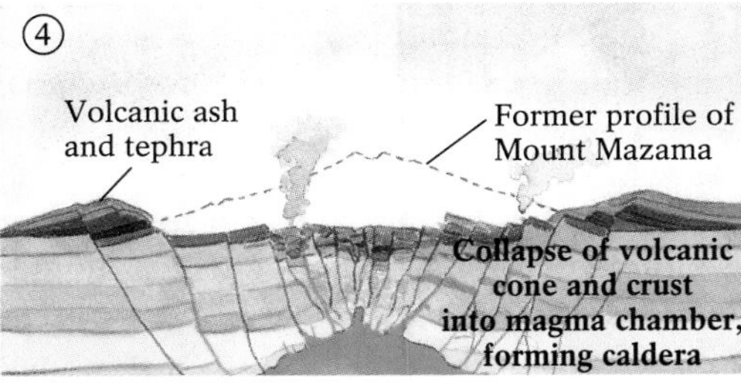

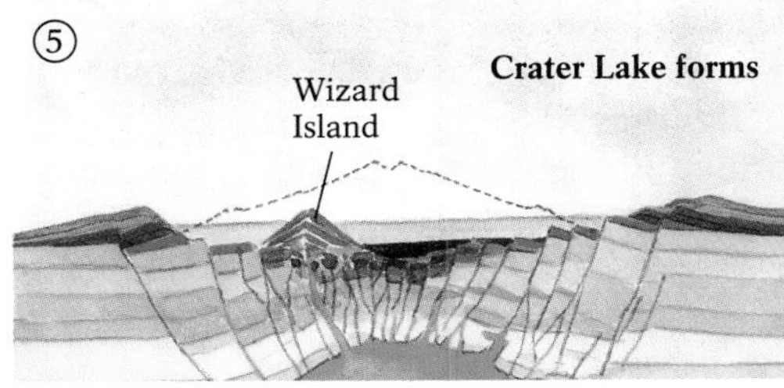

Figure 4-21 The last eruption of Mount Mazama, a volcano in the southern Oregon Cascades, occurred about 6700 years ago and lowered its height by more than a mile. Because very little rock from the shattered cone of the volcano has been found in the areas surrounding Mount Mazama—including Oregon, Washington, and British Columbia—geologists have concluded that the summit of the volcano must have collapsed inward during its last eruption, rather than exploding outward, and remains to this day buried beneath the caldera produced by its collapse. This caldera (above) now contains Crater Lake, North America's deepest lake (more than 600 meters, or 1900 feet, deep).

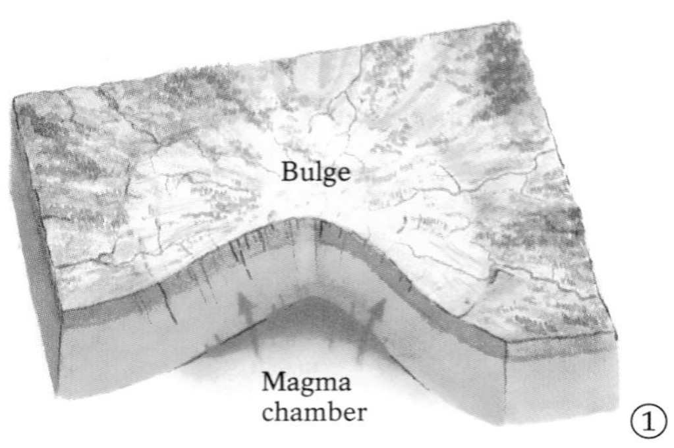

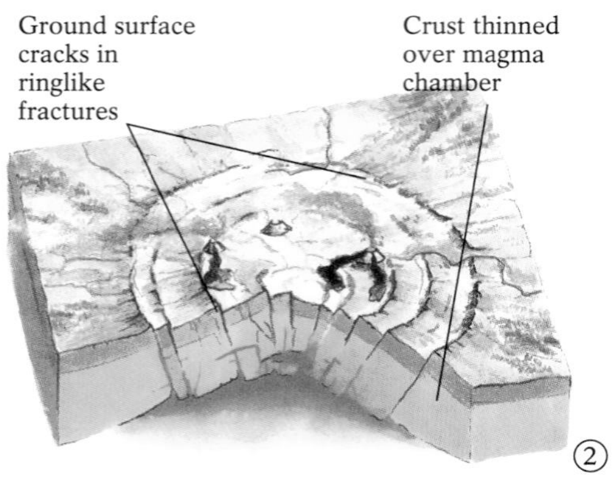

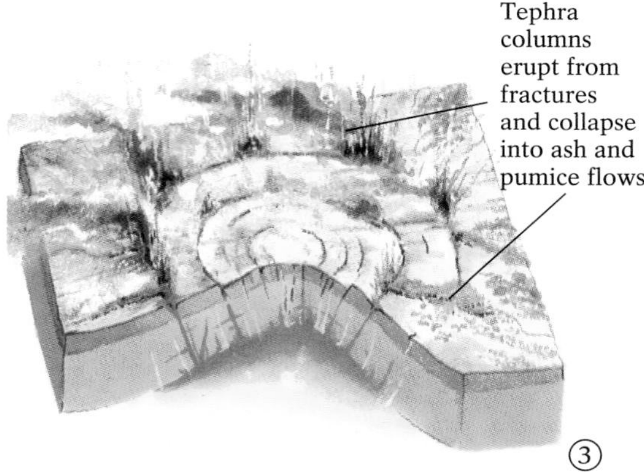

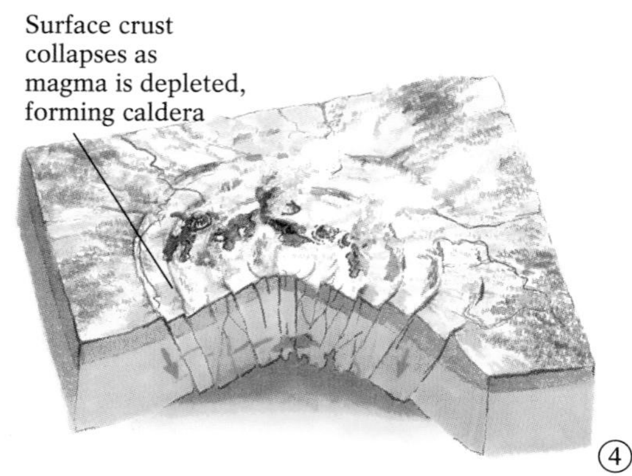

Ash-Flow Eruptions Some spectacular and extremely dangerous pyroclastic eruptions occur when there is no recognizable volcanic cone. An *ash-flow eruption* results when a large reservoir of extremely viscous, gas-rich magma rises to just below the Earth's surface, stretching the crust and forcing it up into a bulge marked by a series of ringlike fractures (Fig. 4-22). The thinned bedrock over such a magma reservoir may collapse, forcing magma into the fractures to produce a circular pattern of tremendous tephra columns. These soon collapse to form numerous flows of hot swirling ash and pumice. As the magma reservoir continues to empty, the surface crust is further undermined until it collapses to form a caldera of monumental proportions.

The Toba ash-flow eruption, which took place approximately 75,000 years ago in the central highlands of the Indonesian island of Sumatra, may have been the Earth's greatest single volcanic event. It produced a caldera 50 kilometers by 100 kilometers (30 miles by 60 miles) and buried an area of 25,000 square kilometers (about 10,000 square miles) beneath a billion-ton blanket of pyroclastics averaging 300 meters (1000 feet) in thickness.

Ash-flow eruptions may recur when a collapsed magma chamber becomes filled with a new magma, the volume and pressure of which force the caldera floor to arch upward again. At least three times in the last 2 million years, the Yellowstone plateau in Wyoming has erupted in devastating ash flows after being pushed up like a blister by an enormous mass of felsic magma. One of these events showered thousands of square kilometers with hot ash and pumice and created the caldera, 80 kilometers by 50 kilometers (50 miles by 30 miles), that now contains beautiful Yellowstone Lake. Today, only a few kilometers below Yellowstone's caldera, there is a huge mass of granitic magma, accumulated since Yellowstone's last great ash-flow eruption 600,000 years ago. Yellowstone National Park's thermal features—its geysers (periodic steam blasts), hot springs (quieter flows of hot water), and gurgling mudpots (pools of boiling mud)—are all heated by this shallow subterranean magma reservoir. Another ash-flow eruption may not be imminent here, but geologists monitor the area continually for signs of increasing volcanic activity.

The most immediate threat of an ash-flow eruption in North America is at the 30-kilometer (19-mile)-long Long Valley caldera near Mammoth Lakes, the popular ski resort in eastern California. Over the past 20 years, the United States Geological Survey has watched the floor of the caldera rise more than 25 centimeters (9 inches), and taken measurements suggesting that magma rose from a depth of about 8 kilometers (5 miles) to about 3.2 kilometers (2 miles) beneath the surface. In 1982, the area was designated a potential volcanic hazard, requiring heightened scientific vigilance and a regional plan for coping with an eruption.

Types of Pyroclastic Volcanic Cones Because pyroclastic lavas are usually intermediate to felsic in composition and thus fairly viscous, they solidify relatively close to the vent. The composition of such a magma, and subsequently

Figure 4-22 Ash-flow eruptions begin when rising viscous magma causes the surface crust to bulge. Cracks develop in the crust, allowing gas-rich tephra columns to erupt and then collapse into hot swirling ash and pumice flows. The partial emptying of the magma chamber weakens its roof and causes the surface to collapse, expelling more pumice in spectacular ash flows and forming a caldera. Later, new magma may refill the chamber and force the caldera upward again until another eruption takes place.

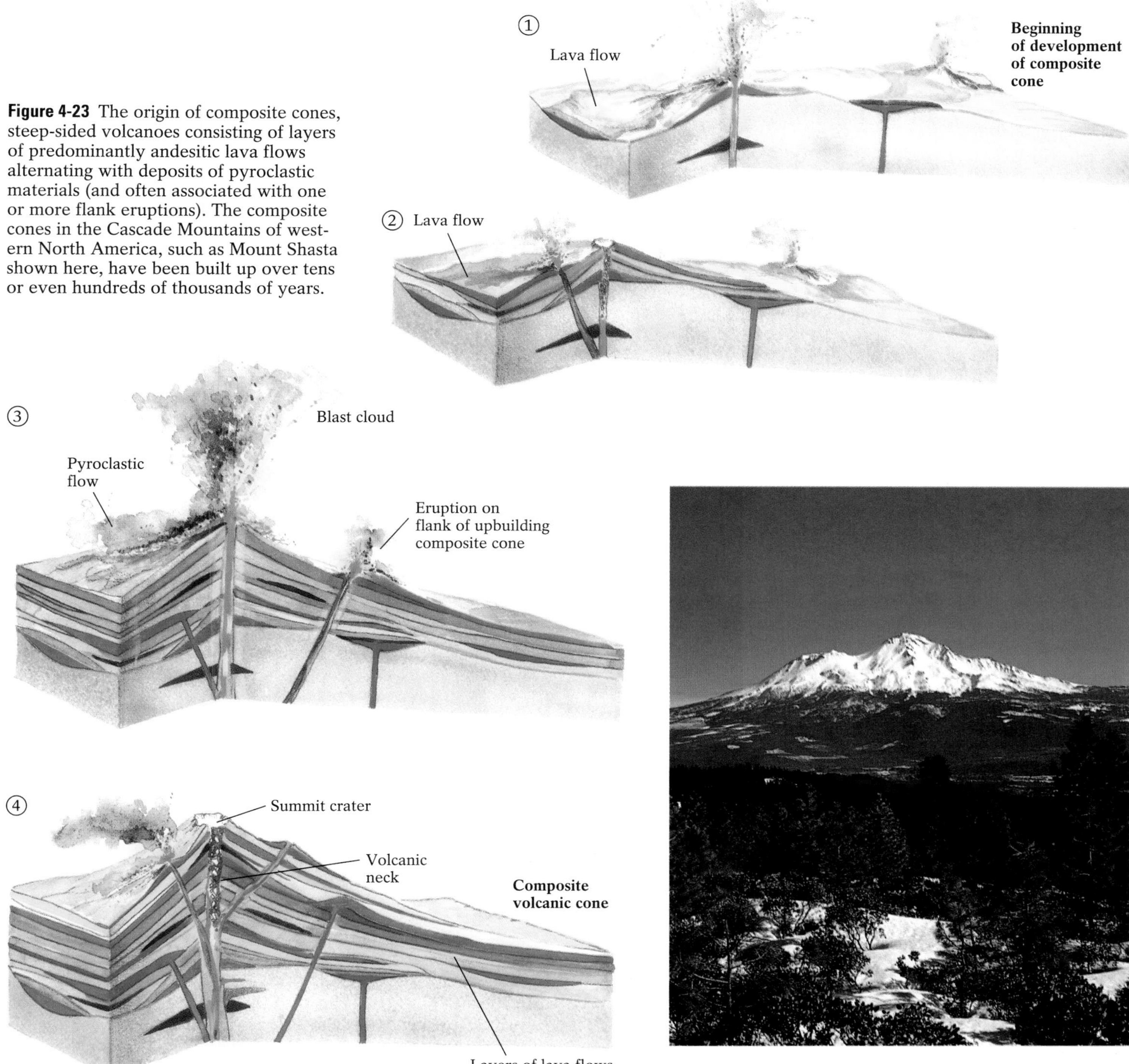

Figure 4-23 The origin of composite cones, steep-sided volcanoes consisting of layers of predominantly andesitic lava flows alternating with deposits of pyroclastic materials (and often associated with one or more flank eruptions). The composite cones in the Cascade Mountains of western North America, such as Mount Shasta shown here, have been built up over tens or even hundreds of thousands of years.

the volcano's eruptive style, may change periodically over time so that the cone intermittently ejects a large quantity of tephra instead of lava. The larger tephra particles from these eruptions fall near the summit to form steep cinder piles, which are then covered by the next lava flow. This produces the characteristic landform of pyroclastic eruptions—the **composite cone**, or **stratovolcano** ("strato" means layered), built up from alternating layers of lava and pyroclastics (Fig. 4-23). Because each pyroclastic deposit produces a steep slope that is then protected from erosion by a successive layer of lava, composite cones grow to be very large and have 10° to 25° slopes. These are the Earth's most picturesque volcanoes, among them Mount Fuji in Japan and Mounts Rainier and Shasta in the western United States.

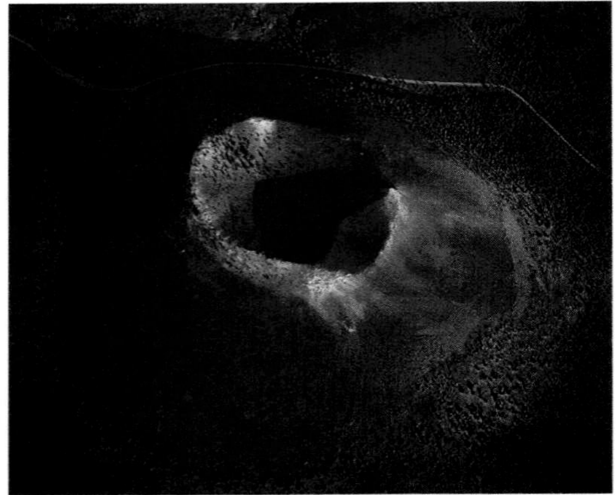

Figure 4-24 Sunset Crater, north of Flagstaff, Arizona. This cinder cone grew to a height of about 300 meters (1000 feet) sometime around A.D. 1065, toward the end of the region's volcanic activity.

Unlike composite cones, **pyroclastic cones** are built up almost entirely from accumulation of loose pyroclastic material. When the dominant pyroclastics are cinders, the structures formed are **cinder cones** (Fig. 4-24). Pyroclastic cones are typically the smallest volcanoes (less than 450 meters, or 1500 feet, high), and generally steep-sided, because pyroclastics can be piled up stably to form slopes between 30° and 40°. Because no intervening lava flows bind their loose pyroclastics, however, pyroclastic cones are readily eroded.

Pyroclastic cones may form from a magma of any composition that contains enough gas to shower lava into the air and create a cone of pyroclastic material around a volcanic vent. Because of the greater gas content of felsic magmas, there are many felsic cinder cones, but mafic cinder cones, which develop where basaltic lava meets subterranean water to form steam-driven lava fountains, are also common.

Eruptive Style, Volcanic Landforms, and Plate Tectonics

About 80% of the Earth's volcanoes surround the Pacific Ocean, where several of the Pacific basin's oceanic plates are subducting beneath adjacent landmasses. Another 15% are similarly situated above subduction zones in the Mediterranean and Caribbean seas. The remainder are scattered along the ridges of divergent plate boundaries (as in the case of Iceland's 22 active volcanoes) and atop continental and oceanic intraplate hot spots (as in the case of Yellowstone National Park and the Hawaiian volcanoes). Each type of plate tectonic setting determines the eruptive style and physical appearance of its volcanoes (Fig. 4-25).

Subduction zones foster explosive pyroclastic volcanism because it is there that intermediate and felsic magmas are produced by partial melting of both the subducting plates and their overlying silica-rich sediments. Subduction zones tend to produce steep-sided composite cones composed primarily of andesite.

Figure 4-25 The worldwide distribution of volcanoes, illustrating the relationship between plate tectonics and volcanism. Note the coincidence of volcanic activity with plate boundaries—especially those surrounding the Pacific Ocean basin, which are often called "the ring of fire."

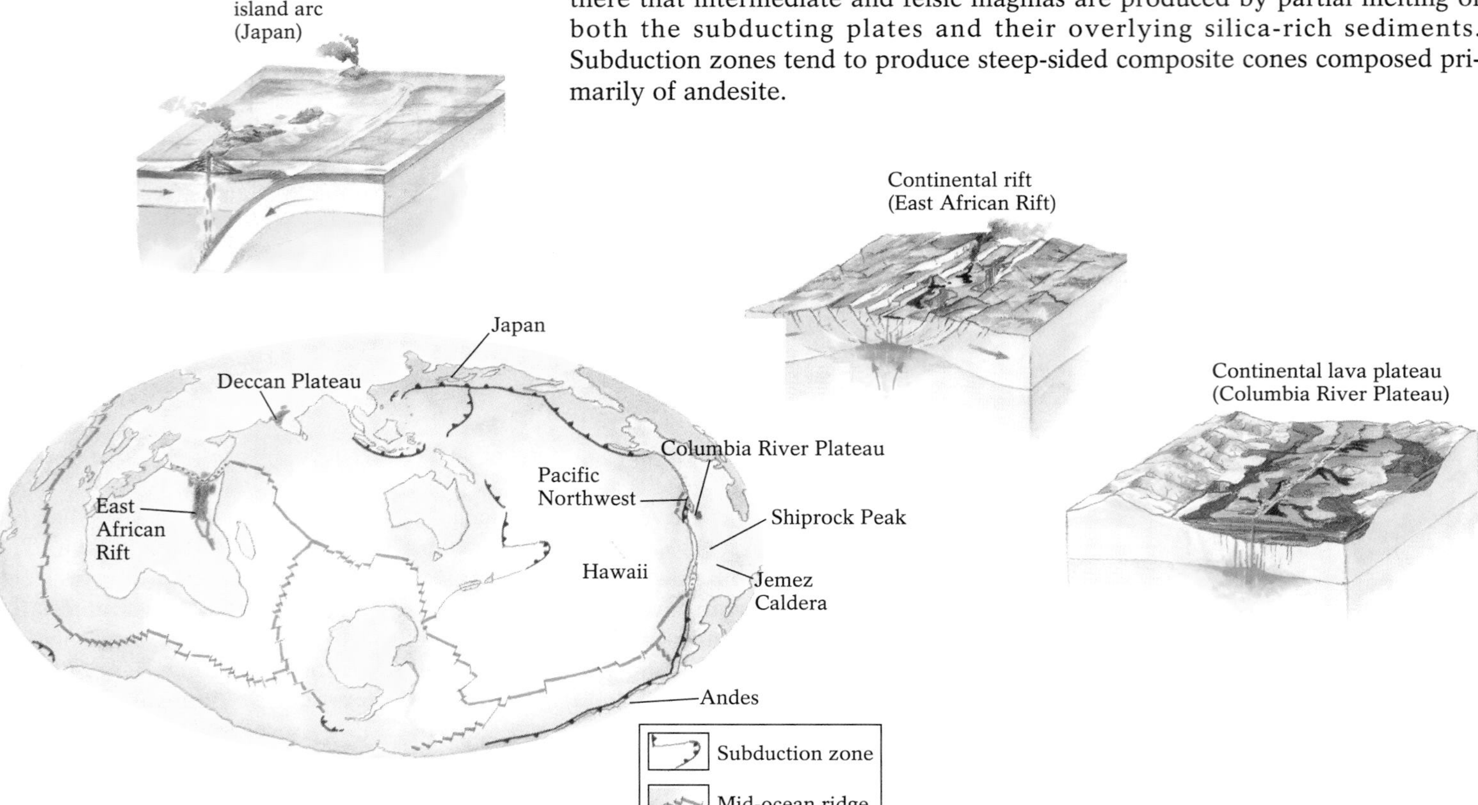

Table 4-1 **Lava Types, Associated Volcanic Features, and Plate Tectonic Settings**

Lava Type	Eruptive Style	Typical Volcanic Landforms	Common Volcanic Products and Effects	Common Plate Tectonic Setting	North American Example(s)
Basaltic (mafic composition)	Quiet, effusive	Lava plateaus, shield volcanoes, occasional cinder cones	Aa lava, pahoehoe lava, vesicular basalts, pillow lavas, columnar basalts	Divergent plate boundaries (such as the mid-Atlantic ridge), oceanic intraplate hot spots (such as underlies Hawaii), intraplate rifts (such as the East African rift)	Columbia River lava plateau (Washington and Oregon), Belknap Crater (eastern Oregon), Craters of the Moon (Idaho)
Andesitic (intermediate composition)	Fairly explosive, pyroclastic	Composite cones, cinder cones	Relatively viscous lava, lahars, tuffs (from airborne ash), welded tuffs (from pyroclastic flows)	Subduction zones	Cascades (British Columbia, Washington, Oregon, northern California), Aleutians (Alaska)
Rhyolitic (felsic composition)	Very explosive, pyroclastic	Volcanic domes, calderas	Extremely viscous lava, ash-flow deposits, welded tuffs (from pyroclastic flows)	Subduction zones, especially at continental margins, intracontinental rifts, intracontinental hot spots	Yellowstone plateau (Wyoming, Montana), Jemez Mountains (Rio Grande rift, New Mexico), Long Valley Caldera (eastern Sierra Nevada, California)

Explosive volcanism also occurs where felsic continental plates are stretched thin and begin to rift, and at intracontinental hot spots. In both of these cases, large volumes of felsic rock at the base of thick continental crust are melted by heat from the hot rising mafic magma produced by partial melting of the mantle. Eruptions of viscous felsic magmas create the volcanic domes, calderas, and ash-flow deposits characteristic of intracontinental rift zones and hot spots. Rift zones and hot spots may also produce mafic lava, although not necessarily at the same time as felsic. An example of this type of volcanism can be found in the Basin and Range region of southwestern North America, where powerful blasts of felsic ash and pumice have alternated with quiet effusions of basaltic lava throughout the last 20 million years.

At divergent zones and above oceanic intraplate hot spots, effusive eruptions of low-viscosity basaltic lava generally produce near-horizontal lava plateaus and gently sloping shield volcanoes. Table 4-1 shows the relationships between lava types, eruptive styles, types of volcanic cones and products, and plate tectonic settings.

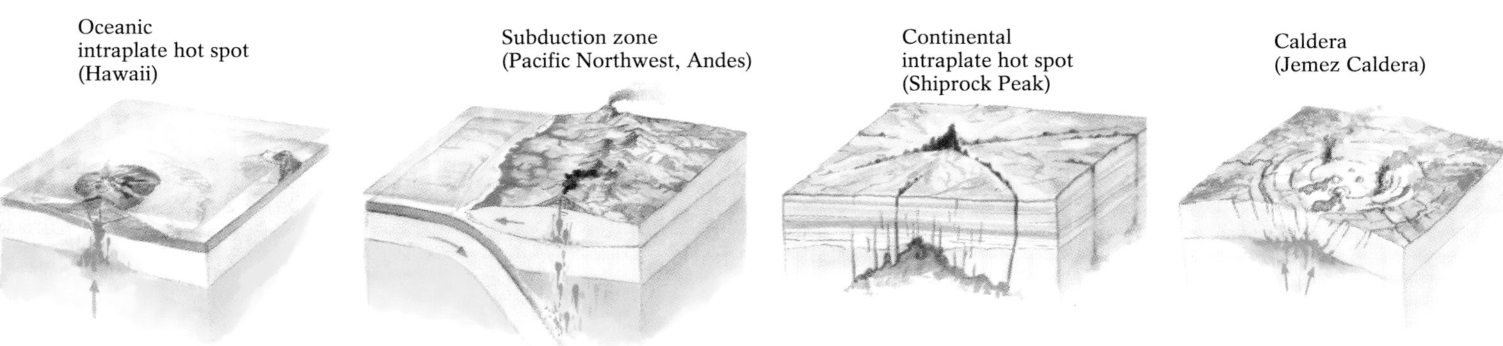

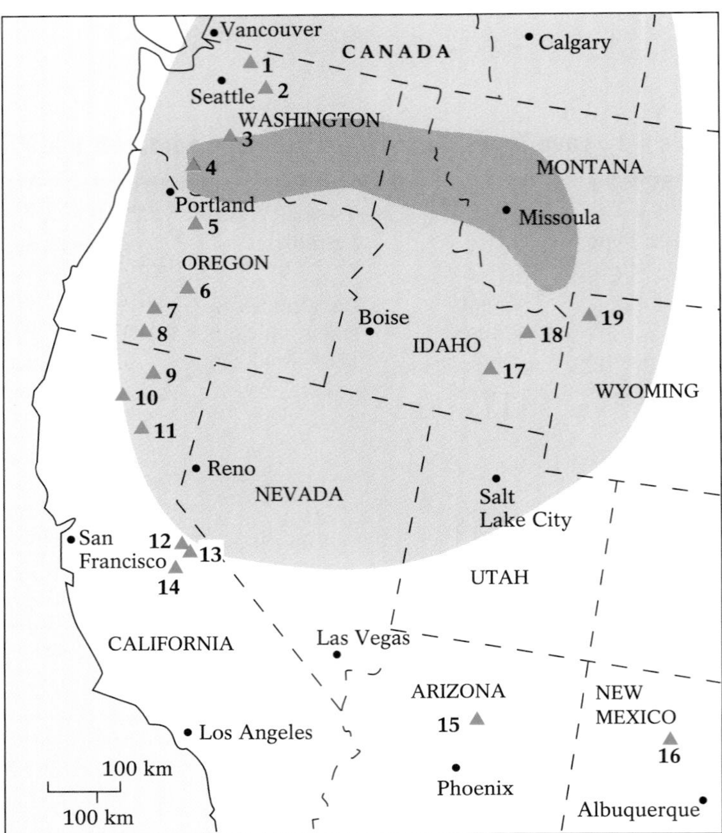

Figure 4-26 A volcanic-hazards map of western North America, showing areas that have experienced recent volcanic activity and those that are the most likely sites of future volcanic events.

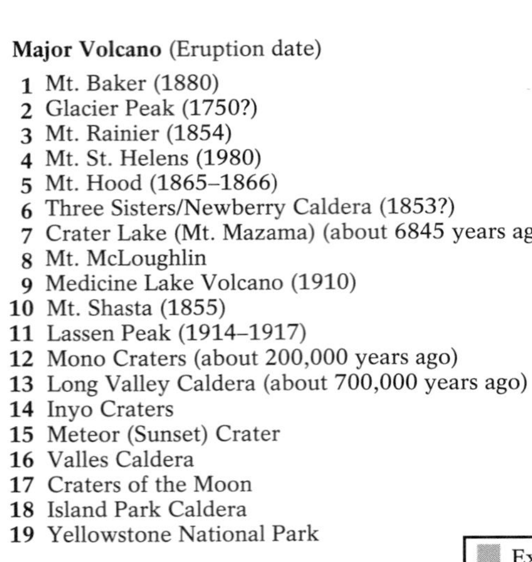

Coping with Volcanic Hazards

Because we can't slow subduction, stop divergence, or chill intraplate hot spots, we will probably never be able to prevent volcanic eruptions. In certain cultures, people rely on faith and prayer to protect them from volcanoes. (Hawaiian legend relates that the fire goddess Pele flies into fits of rage when she is defeated in competition with the islands' young athletic chiefs, stamping her feet in anger to cause earthquakes and summoning rivers of destructive lava in revenge. Some Hawaiians attempt to appease Pele by offering her bushels of sacred ohel berries, flowered leis, and bottles of gin.) For their part, geologists are developing ways to predict volcanic eruptions more accurately, in order to prevent casualties and minimize damage. The first step in such a process is to determine where general volcanic danger zones exist; for purposes of this text, we will focus on North America.

Broad regions of North America are at risk for major damage and destruction from volcanism, especially those western U.S. states and Canadian provinces located on the North American plate where it overrides the subducting Juan de Fuca and Gorda plates and those over scattered intraplate hot spots (Fig. 4-26). Crater Lake in Oregon, Yellowstone National Park in Wyoming, and the Long Valley caldera in California, previously discussed, are examples of recent catastrophic volcanism. The Valles caldera, about 60 kilometers (35 miles) northwest of Sante Fe, New Mexico, probably has considerable eruptive potential as well. Geologists are also wary of the Jemez Mountains in north-central New Mexico, which have experienced numerous explosive ash-flow eruptions in the past few hundred thousand years; this is probably due to a large reservoir of magma about 20 kilometers (12 miles) below the Rio Grande region, believed to be a developing intraplate hot spot.

Many sites in the Cascade Range are immediately threatening. Although most of the range's volcanoes have been relatively quiet in recent decades, Mount Baker and Mount St. Helens in Washington, Mount Hood in Oregon, and Mount Shasta and Lassen Peak in northern California were all actively erupting between 1832 and 1880. Today, Cascade volcanoes threaten major metropolitan areas in southern British Columbia, Washington, Oregon, and northern California. Mount McLoughlin, 50 kilometers (30 miles) from Medford and Klamath Falls in Oregon, and Mount Hood near Portland, which last erupted in 1865, are closely watched for signs that they are awakening. Mount Baker, near Washington's Canadian border, resumed its intermittent rumbling and steam emissions so actively that in 1975 the U.S. Geological Survey predicted, albeit incorrectly, that it would be the next Cascade volcano to erupt. Mount Rainier, the Cascade's grandest peak, last erupted in 1882 with a small whiff of brown ash; its last major eruption occurred some 2000 years ago. In a future eruption, its greatest threat would come from large lahars spawned by steam and tephra emissions onto its glacier-clad slopes. Mount Garibaldi, just north of Vancouver, British Columbia, at the northern end of the Cascade volcanic chain, has eruptive potential as well.

Typical of the Cascades' explosive volcanism was Mount St. Helens' 1980 eruption. Although relatively unspectacular compared to recent volcanic events elsewhere, this was the greatest volcanic disaster in North America's recent history. Highlight 4-2 (following page) documents St. Helens' most recent fury and illustrates some of the hazards facing those who live in the shadow of an active volcano.

Defense Plans

An effective plan to avert volcanic disasters must start with keeping people out of harm's way by zoning against development in the most hazardous regions, building structures that protect life and property, and learning how to predict eruptions accurately to provide ample warning and allow for timely evacuations.

Hawaii, for example, practices *volcanic zoning*, as do other western states. Under such a policy, areas with great potential for danger are set aside as national parks, monuments, and recreation areas, closed to residential and commercial development. As a result, there are no large metropolitan areas near Hawaii Volcanoes National Park; nor are there any on the slopes of Mount Rainier, Mount St. Helens, Crater Lake, Yellowstone National Park, or other hazardous areas.

Battling Lava and Lahars Where preventive zoning is impractical, other means can avert or reduce potential damage, especially from lava flows and lahars. The first recorded attempt to fight a lava flow took place in Sicily during an eruption of Mount Etna on March 25, 1669. About 50 residents of the town of Catania, wearing wet cowskins for protection, used iron bars to poke holes through the hardening crust at the sides of the advancing flow to divert the lava from its path toward their homes. The idea was sound, but the lava's new route was headed directly toward the watchful town of Paterno. With little delay, 500 staunch defenders of Paterno rushed into a somewhat violent confrontation with the Catanians, convincing them to discontinue their innovative lava-control efforts.

The twentieth century has provided more effective methods for diverting lava flows. For example, a carefully positioned explosive device can disperse a flow over a wider area, so that it thins out and cools and solidifies more rapidly. The hardening lava blocks the path of still-flowing lava behind it, forcing it to accumulate upstream and flow along a less damaging route. In 1935, a strategically placed bomb coaxed a flow to detour around the Hawaiian town of Hilo.

Highlight 4-2 *Mount St. Helens*

(a)

(b)

(c)

Figure 4-27 Mount St. Helens' pre-eruption symmetry made it one of the world's most beautiful and photographed volcanoes (**a**). Its eruption on May 18, 1980 (**b**) opened a crater on the volcano's north side, changing its appearance dramatically (**c**).

Mount St. Helens' cone, the youngest in the Cascades at less than 10,000 years of age, came to life with an audible boom on March 30, 1980, after 123 years of silence. Weeks of public anxiety and scientific watchfulness followed, ending on May 18 when an eruption blasted about 400 meters (1300 feet) of rock from the summit, changing what many had considered North America's most beautiful peak to a squat gray crater (Fig. 4-27).

Pre-eruption underground rumblings had signaled the rise of magma into St. Helens' cone. Probably blocked on its way to the central vent by a plug of congealed lava from an earlier eruption, the ascending magma took a sharp turn northward instead. A bulge appeared on the volcano's northern slope and, growing at the rate of 1.5 meters (5 feet) per day, protruded an ominous 122 meters (400 feet) by May 17.

At 8:31 on the morning of May 18, a powerful earthquake beneath the mountain released internal pressure that had been building for weeks. This dislodged the bulge and sent it hurtling down the mountain as an avalanche of debris traveling at 400 kilometers per hour (250 mph). A northward-directed jet of superheated (500°C, or 900°F) ash and gas immediately erupted as a pyroclastic flow, racing downslope with hurricane force (greater than 300 kilometers, or 200 miles, per hour), cutting a swath of complete destruction 30 kilometers (20 miles) wide. The blast and subsequent nuée ardente buried the nearest 12 kilometers (7 miles) of forest land beneath meters of pyroclastics, blew down entire stands of mature trees like matchsticks to a distance of 20 kilometers (12 miles) (Fig. 4-28), and singed the forest beyond for an additional 6 kilometers (4 miles). More than 26 kilometers (16 miles) away, the heat scalded fishermen, who plunged into lakes and streams. On Mount Adams, 50 kilometers (30 miles) away, climbers felt a gust of intense heat just before being bombarded with hot, ash-blasted pine cones.

Meanwhile, a tephra column rose from the summit vent to an altitude of 25 kilometers (15 miles). Swept eastward by the prevailing winds, the dense cloud of gray ash began to fall on the cities and towns of eastern Washington. Yakima, 150 kilometers (100 miles) to the east, received 10 to 15 centimeters (4–6 inches) of ash. In Spokane, farther east, visibility fell to less than 3 meters (10 feet) and automatic street lights switched on at noon. Proceeding across the continent, the ash cloud dusted every state in its path. Hundreds of downwind communities would later spend millions of dollars cleaning up.

Several hours after the start of the eruption, snow and large chunks of glacial ice trapped within the dislodged debris from the mountain's northern slope began to melt. This enormous volume of meltwater mixed with loose material and the eruption's fresh pyroclastics to produce a lahar that rushed 28 kilometers (17 miles) westward down the Toutle River valley at 80 kilometers per hour (50 mph), picking up logging trucks and hundreds of thousands of logs along the way, and buried 123 homes beneath 60 meters (200 feet) of mud.

In all, 60 human lives were lost, along with the lives of 500 blacktail deer, 200 brown bear, 1500 elk, and countless birds and small mammals. The only survivors were burrowers such as frogs and salamanders, which fled into the soft sands of lake shores and stream banks. The human toll would have been much higher if state officials had not heeded the warnings issued in March by the U.S. Geological Survey, after the volcano's initial reawakening, and evacuated the area of most of its year-round residents and closed it to seasonal residents and spring hikers. However, if the eruption had occurred one day later, on Monday, hundreds of loggers at work would have been buried beneath the debris avalanche.

Since 1980, a volcanic dome of highly viscous lava has been rising slowly in St. Helens' crater; it is now 40 stories high. The volcano will probably erupt intermittently, perhaps for decades, as new lavas erupt from the deep scar on its northern slope and eventually restore the mountain's symmetrical cone.

Figure 4-28 The forest north of Mount St. Helens was blown down for miles by the force of the volcano's pyroclastic flow and covered with volcanic ash. The area shown in this photo is 12 kilometers (7 miles) from the crater.

Figure 4-29 A lava wall should be about 3 meters (10 feet) high and have a wide, sturdy base; because the wall is meant to gradually divert lava away from inhabited areas, it should be built at an angle to the expected direction of the lava flow. The wall in this photo, hand-built from lava rocks, has successfully diverted a basaltic flow on the island of Hawaii.

Other solutions have similar effects. In 1973, Icelanders on the coastal island of Heimaey, after contemplating bombing, decided instead to cool a flow with seawater pumped by 47 large barge-mounted pumps anchored in a nearby harbor; the effort was successful. Japanese engineers have designed steel-and-concrete dams that can trap large boulders and slow mudflows to minimize damage and gain time for evacuation. Anywhere, a simple lava wall—a hand-built barrier of boulders or rocks (Fig. 4-29)—can sometimes effectively protect an individual homestead by guiding lava away.

In some parts of Indonesia, a series of ropes are strung across valleys that have a history of lahars; the first movement of mud sets off a siren, warning downstream communities to evacuate. In Japan, closed-circuit video cameras and other sensing devices vigilantly monitor the most vulnerable regions and warn endangered communities. In the wake of the 1985 Armero disaster, such a system is now also in place on Colombia's Nevado del Ruiz.

Predicting Eruptions Geologists have had fair success predicting individual eruptive episodes when they concentrate on a specific volcano *after* an eruptive phase has begun. These monitoring efforts involve measuring changes in a volcano's surface temperature, watching for the slightest expansion in its slope, and keeping track of regional earthquake activity (Fig. 4-30). A laboratory at the University of Washington in Seattle is staffed 24 hours a day to monitor the rumblings of Mount St. Helens—on any given day, it may issue an announcement such as "a nonexplosive, dome-building eruption is predicted within the next 48 hours." [The U.S. Geological Survey missed the call on St. Helens' 1980 blast despite the fact that the mountain was being watched closely by a large team of scientists armed with the latest in prediction technology; however, they did successfully predict the eruption of Mount Pinatubo in the Philippines, evacuating virtually everyone within 25 kilometers (15 miles) before the volcano's powerful blast of May 17, 1991.]

Before a volcano erupts, hot magma rises toward the surface, so any local manifestation of increasing heat may signal an impending event. Ongoing surveys identify new surface hot springs and take the temperature of the water and steam in existing ones. If the escaping steam isn't much hotter than the boiling point of water, surface water is probably seeping into the mountain and being heated by contact with hot subsurface rocks, and all is well for the time being. If, however, the steam is superheated, with temperatures as high as 500°C (900°F), it probably derives from shallow water-rich magma, a sign that an eruption may be brewing. As magma rises, the volcanic cone itself begins to heat up. The overall temperature of a volcanic cone can be monitored from an orbiting satellite equipped with infrared heat sensors to detect the slightest change in surface temperature. This high-altitude technology serves as a simultaneous early-warning system for most of the Earth's 600 or so active volcanoes.

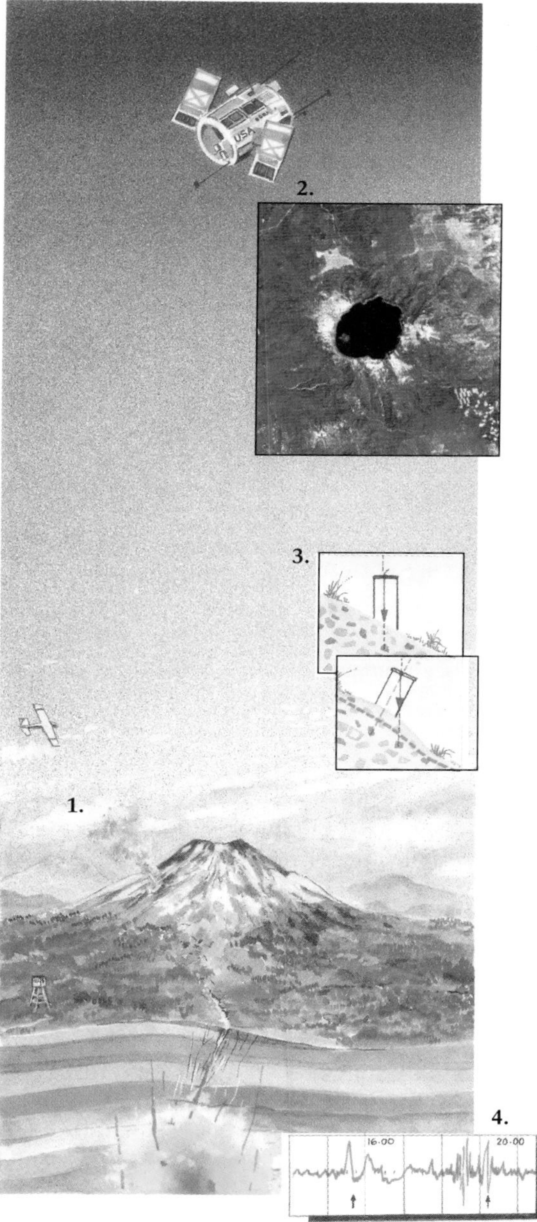

Figure 4-30 Techniques commonly used to predict volcanic eruptions include: watching for escape of superheated steam (1); satellite monitoring of volcanic cone temperature (2); detecting volcanic cone bulges, using tiltmeters (3); and locating increased tremor activity, using seismographs (4).

Active volcanoes expand in volume as they acquire new supplies of magma from below. An increase in the steepness or bulging of a volcano's slope can signal an impending eruption. To detect the inflation of a volcanic cone, a *tiltmeter*, a device like a carpenter's level, is used.

As magma rises, it pushes aside fractured rock and in the process fragments the underground rock a bit more. Because the fracturing causes earthquakes, eruptions are often preceded by a distinctive pattern of earthquake activity called *harmonic tremors*, a continuous rhythmic rumbling. The increased height of rising magma is determined by sensitive equipment that monitors the location of these tremors. The rate at which the magma rises provides an estimate of when an eruption may occur. This is the principal means by which Mount St. Helens' eruptions have been predicted.

Efforts to predict eruptions are thwarted, however, when we are unaware of a region's volcanic potential. Occasionally a new volcano appears suddenly and rather unexpectedly, as was the case in 1943 when the volcano Parícutin developed literally overnight in the Mexican state of Michoacan, 320 kilometers (200 miles) west of Mexico City (Fig. 4-31). The area, though, *was* known to be volcanic, lying just northeast of a subducting segment of the Pacific Ocean.

Extraterrestrial Volcanism

Volcanism occurs (or has occurred in the past) not only on Earth but also on the Moon, Mars, Venus, and one of Jupiter's moons, Io. There are now no active volcanoes on the Moon, but at least one-third of its surface is covered by ancient flood basalts known as lunar *maria* ("seas"). As we saw in Chapter 3, early in its existence, the Moon was struck repeatedly by large meteorites that left deep craters and fractures in its crust. Basaltic lava flowed through those cracks, filling the craters and forming the maria. The smoother-appearing parts of the maria often contain winding trenches, called *rilles*, stretching for hundreds of kilometers; these are probably collapsed lava tubes in the basalt. The Moon's surface also includes a few distinct shield volcanoes, such as those in the Marius Hills.

Figure 4-31 The remains of the town of San Juan Parangaricutiro, which was engulfed during June and July of 1944 with lava from the eruption of Mexico's Parícutin volcano (visible in the background). The eruption, which lasted nine years, began in February of 1943 with the sudden appearance of a small cinder cone in a cornfield about 200 miles west of Mexico City. (Because the flow that buried San Juan Parangaricutiro—the most extensive flow from Parícutin—moved only a few feet per hour, the Mexican Army was able to evacuate the town's 4000 or so residents before any lives were lost.)

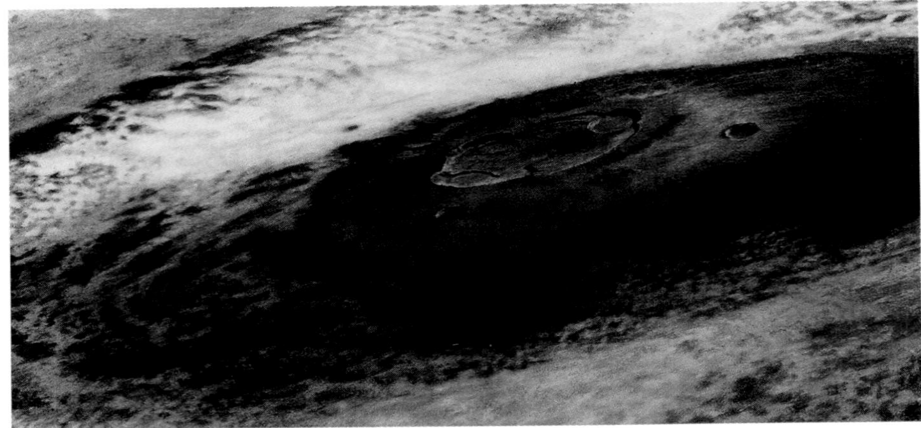

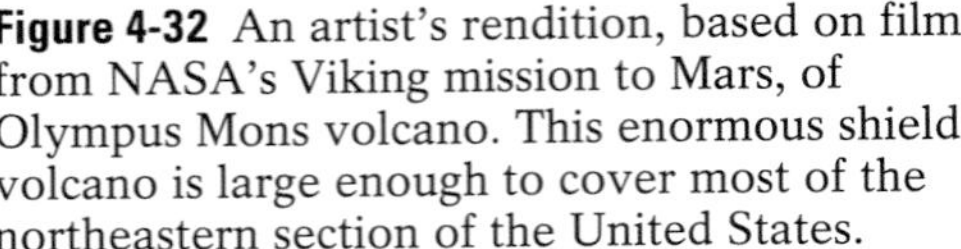

Figure 4-32 An artist's rendition, based on film from NASA's Viking mission to Mars, of Olympus Mons volcano. This enormous shield volcano is large enough to cover most of the northeastern section of the United States.

On Mars, as much as 60% of the surface is covered by volcanic rock derived from approximately 20 centers where volcanic activity was concentrated. Virtually all Martian volcanoes are shields of incredible size. Olympus Mons is approximately 23 kilometers (14 miles) high, more than twice the height of Mount Everest; its diameter is about equal to the width of the state of California (Fig. 4-32). The size of this and other Martian shields suggests that they have remained stationary over their underlying hot spots for a very long time, allowing enormous mountains of volcanic rock to accumulate—either Mars' lithosphere does not move or it moves far more slowly than the Earth's plates. The volume of volcanic rock in Olympus Mons exceeds the total volume of the Hawaiian Islands (a chain of volcanoes that was produced—instead of a single volcanic colossus—because the Earth's Pacific plate *does* move).

Venus also contains large shield volcanoes, some stretching in long chains along great faults in the surface. Thanks to the remarkably sharp radar images from NASA's Magellan satellite, which began orbiting Venus in August 1990, we know that some volcanoes in the Maxwell Montes region are more that 11 kilometers (6.5 miles) tall, higher than Mount Everest. Two shields, Rhea Mons and Theia Mons, are more than 2.5 kilometers (1.5 miles high), 300 kilometers (200 miles) across, and contain 90-kilometer (60-mile)-diameter calderas at their summits. Rhea Mons alone would cover all of New Mexico and much of adjacent parts of Colorado, Texas, and Arizona. Radar also indicates that molten lava lakes may still exist on the Venutian surface. A cluster of seven lava domes discovered in 1991, each more than 15 kilometers (10 miles) in diameter, may be similar in structure and composition to the one growing today in the crater of Mount St. Helens. A series of lava flows, in what appear to be rift valleys similar to those of East Africa, suggest a long sequence of multiple volcanic events on Venus (Fig. 4-33). The ages of these flows and whether they record plate activity on Venus has not yet been determined.

One of Jupiter's moons, Io, is the only other body in our solar system showing direct evidence of volcanic activity. Thus far, eight volcanoes have been detected there by satellite photography, seven of which erupted in one recent 4-month period. Unlike the basaltic lavas produced by the volcanoes of the inner rocky planets, Io erupts molten sulfur and enormous clouds of sulfurous gas (Fig. 4-34). The gas clouds are propelled at speeds approaching 3200 kilometers per hour (2000 mph) to heights 500 kilometers (300 miles) above the surface. Consequently, Io is noted for its yellow-red snowfalls of sulfur, lakes of molten sulfur, and huge multicolored lava flows of black, yellow, orange, red, and brown. Geologists believe that volcanism on Io may result from the frictional heat generated as Io's surface rises and falls in response to the enormous gravitational pull of Jupiter.

Figure 4-33 A computer-generated image of Maat Mons, an 8-kilometer (5 mile)-high volcano on Venus. Lava flows extend for hundreds of kilometers across the fractured plains in the foreground.

Because of their distinctive shapes, volcanic landforms and features are among the most easily identified on satellite images. As satellite technology evolves, future studies of extraterrestrial volcanism will undoubtedly lead to many exciting discoveries that will teach us more about volcanic processes throughout the solar system.

We have now examined igneous activity at relatively shallow depths and the impact this has on the Earth's surface. In later chapters we will explore the more deep-seated processes that generate such near-surface activity. In the next few chapters, however, we will continue to look at the Earth's outermost layer and the variety of rocks that exist there. In Chapter 5, we will see how these rocks are affected by conditions at the planet's surface.

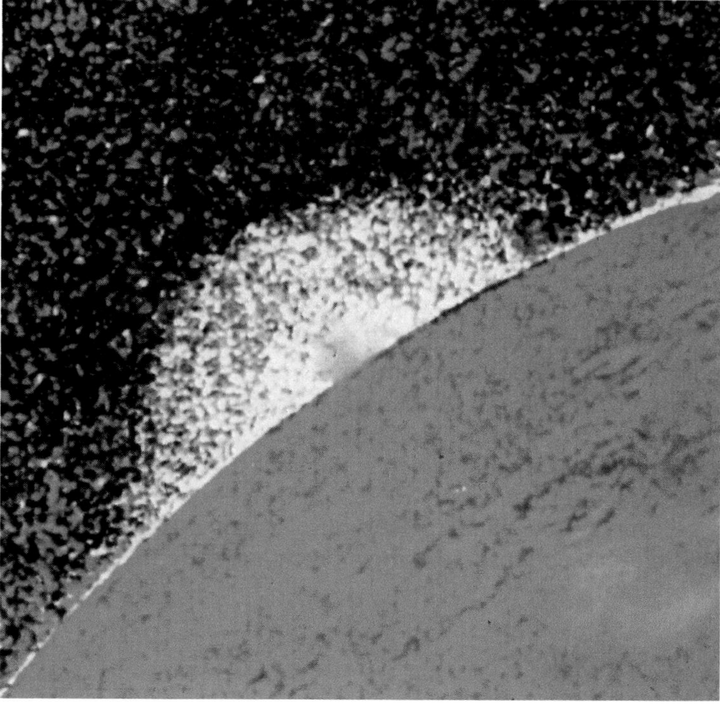

Figure 4-34 A computer-enhanced photo of a volcanic eruption on Jupiter's moon, Io, from data gathered by NASA's Voyager I probe in 1979.

Chapter Summary

Volcanism is the set of geological processes that mark the ascent of magma to the Earth's surface and its expulsion as lava. A **volcano** is the landform that results from the accumulation of lava and rock particles around an opening, or **vent**, in the Earth's surface from which lava is extruded.

Magmas flow and erupt in distinctive ways, depending on their gas content and their **viscosity**, or resistance to flow. Because of its high temperature and relatively low silica content, mafic magma has low viscosity (is highly fluid); it generally erupts (as basaltic lava) relatively quietly, or effusively, because its gases can readily escape and don't build up high pressure. Magmas of intermediate temperature and composition, which erupt as andesitic lavas, are more viscous; these magmas trap their gases, causing pressures to build up until an explosive eruption occurs. Felsic magma, with its high silica content and relatively low temperature, is highly viscous, and generally erupts (as rhyolitic lava) most explosively.

The nonexplosive volcanic eruptions characteristic of basaltic lavas produce lava flows which, when they solidify, are associated with distinctive features such as pahoehoe- and aa-type surface textures, columnar joints, lava tubes, and pillow structures. The explosive volcanic eruptions characteristic of rhyolitic lavas typically eject **pyroclastic** material—fragments of solidified lava and shattered preexisting rock ejected forcefully into the atmosphere. Lava that cools and solidifies as it falls back to the surface forms in a wide range of shapes and sizes; all such particles are collectively called **tephra.** Over time, pressure from overlying materials converts tephra deposits to a rock called **volcanic tuff.**

Explosive ejection of pyroclastic material is usually accompanied by a number of life-threatening primary effects, such as **pyroclastic flows**, or **nuée ardentes** (high-speed, ground-hugging avalanches of hot pyroclastic material), and **lahars** (hot volcanic mudflows). The secondary effects that often follow explosive eruptions include short-term water and air pollution and longer-term, sometimes global, climatic changes.

Lavas of different composition form distinctly different landforms. Despite their different appearances, however, nearly all volcanoes have the same two major components: a mountain, or **volcanic cone**, composed of the volcano's solidified lava and pyroclastics, and a bowl-shaped depression, or **volcanic crater**, within which is the vent from which the lava and pyroclastics emanate. If enough lava erupts and empties a volcano's subterranean reservoir of magma, the cone's summit may collapse, forming a much larger depression, or **caldera**.

Effusive eruptions, usually involving basaltic lava, form gently sloping landforms, largely because the highly fluid lava flows a great distance from the vent. In Hawaii, basaltic magma generally reaches the surface through distinct conduits, or vents; these central-vent eruptions usually produce broad-based cones called **shield volcanoes**. In Iceland, basaltic magma generally reaches the surface through long linear cracks, or fissures, in the Earth's crust, and spreads to produce nearly horizontal lava plateaus.

Explosive **pyroclastic eruptions** involve viscous, gas-rich magmas and so tend to produce great amounts of solid volcanic fragments rather than fluid lavas. Rhyolitic lavas tend to be so viscous that they cannot flow out of a volcano's crater, and so cool and harden within their craters to form **volcanic domes**. Ash-flow eruptions occur in the absence of any volcanic cone at all; they are produced when extremely viscous, gas-rich magma rises to just below the surface bedrock, stretching and collapsing it.

The characteristic landform of pyroclastic eruptions is the **composite cone**, or **stratovolcano**, which is composed of alternating layers of pyroclastic deposits and solidified lava. Pyroclastic eruptions may also produce **pyroclastic cones** or **cinder cones**, composed almost entirely from the accumulation of loose pyroclastic material around a vent. All pyroclastic-type volcanoes produce steep-sided cones, because the materials they eject—solid fragments and highly viscous lavas—do not flow far from the vent.

The various types of volcanic eruptions are associated with different plate tectonic settings. Explosive pyroclastic eruptions of felsic (rhyolitic) lava generally occur within continental areas where plate rifting is taking place, or atop intracontinental hot spots such as the one beneath the Yellowstone plateau of northwestern Wyoming. Most intermediate (andesitic) eruptions occur where oceanic plates are subducting, such as in the Cascades of Washington and Oregon and along the rim of the Pacific Ocean. Effusive eruptions of (mafic) basalt occur generally at divergent-plate margins, such as the mid-Atlantic ridge in Iceland, and above oceanic intraplate hot spots, such as the one beneath the Hawaiian Islands.

The best way to avoid volcanic hazards is to minimize human use of potentially eruptive locations. Structural and strategic defenses include the construction of lava walls, warning systems to facilitate evacuations, and measurements to predict impending eruptions. Prediction techniques include measuring changes in a volcano's slopes, watching for and recording related earthquake activity, and tracking changes in the volcano's external heat flow, both on land (for example, by the appearance of new hot springs) and from space (with the help of infrared satellite imagery).

Volcanism is not restricted to the Earth; it has occurred elsewhere in the solar system in the past, and continues to do so today. Ancient (3-billion-year-old) volcanism is responsible for much of the rock and landform development on the surface of the Earth's Moon. Current or relatively recent volcanic activity has also been detected on Mars, Venus, and Io, one of the moons of Jupiter.

Key Terms

volcanism (p. 95)
volcano (p. 95)
vent (p. 95)
viscosity (p. 98)
pyroclastics (p. 102)
tephra (p. 102)
volcanic tuff (p. 103)
pyroclastic flow (p. 103)
nuée ardente (p. 103)
lahar (p. 104)
volcanic cone (p. 106)
volcanic crater (p. 106)
shield volcano (p. 106)
caldera (p. 107)
pyroclastic eruption (p. 109)
volcanic dome (p. 109)
composite cone (p. 113)
stratovolcano (p. 113)
pyroclastic cone (p. 114)
cinder cone (p. 114)

Questions for Review

1. What criteria do geologists use to designate a volcano as active, dormant, or extinct?
2. Briefly compare basaltic, andesitic, and rhyolitic lava, in terms of their composition, viscosity, temperature, and eruptive behavior.
3. Within a single basaltic lava flow issuing from a Hawaiian volcano, why is pahoehoe lava found closer to the vent than aa lava?
4. Contrast the nature and origin of nuée ardentes and lahars.
5. Describe the three basic types of effusive eruptions (fissure, central-vent, subaqueous) and the volcanic landforms associated with each.
6. How does a composite cone form? What type of lava is associated with a composite cone? How could you distinguish a composite cone from a pyroclastic cone?
7. What types of volcanoes and volcanic landforms are associated with subduction zones? With divergent plate boundaries?
8. Identify three sites in North America that pose a volcanic threat to nearby residents.
9. Describe three techniques that geologists use to predict volcanic eruptions.
10. Compare the volcanism on the Moon to the volcanism on Io, Jupiter's moon.

For Further Thought

1. Look at the photograph above and speculate about the plate tectonic setting where this volcano is found; the composition of the rocks that make up the volcano; and whether eruptions of this volcano tend to be explosive or effusive.
2. The 1980 eruption of Mount St. Helens made a lot of headlines but had no discernible effect on global climate. Conversely, the eruption of the Philippines' Mount Pinatubo in 1991 caused a 1°C drop in global temperature. What differences between these eruptions might explain why one had a sharp effect on climate whereas the other did not?
3. Why do we find andesite throughout the islands of Japan, but not throughout the islands of Hawaii?
4. How would you explain the origin of a volcanic structure composed of 10,000 meters of pillow lava covered by 3000 meters of basalt containing vesicles, columnar jointing, and aa and pahoehoe structure?
5. Under what circumstances might active volcanism resume along the east coast of North America? Within the Great Lakes region of North America?

Weathering Processes

Soils and Soil Formation

Weathering in Extraterrestrial Environments

Figure 5-1 Arches National Park, Utah, shows the results of millions of years of exposure to the Earth's environment. The rocks that were removed to produce these unusual landforms were less resistant to weathering and erosion than are the hardy rocks that remain.

Chapter 5

Weathering: The Breakdown of Rocks

Much of the Earth's most spectacular and unique scenery has been created by quite ordinary environmental factors acting upon rock (Fig. 5-1). These factors, which act constantly and everywhere on Earth, cause individual mineral grains in rock to be removed or altered, producing end products that look much different from the original rocks: The process by which atmospheric agents at or near the Earth's surface cause rocks and minerals to break down is called **weathering**. Weathering is a slow but potent force to which even the hardest rocks are susceptible.

Rocks that have been weakened by weathering are more vulnerable to **erosion**, the process by which moving water, wind, or ice (or, sometimes, sheer force of gravity) incorporates pieces of rock and deposits them elsewhere. The fragments removed from rock by erosion have usually been loosened by weathering, although a high-energy erosive event such as a flood may dislodge even unweathered rock. The effects of erosion on rock tend to be more dramatic than the gradual wearing away of rock by weathering; however, because it generally requires the presence of moving water, wind, or ice, erosion does not occur as universally or constantly as does weathering. Like many geological processes, weathering and erosion are interrelated, and often work in tandem. Together, they produce *sediment*, the loose, fragmented surface material that is the raw material for sedimentary rock (discussed in Chapter 6). However, weathering and erosion are each multifaceted phenomena and must be considered separately. In this chapter we will concentrate on weathering.

Weathering plays a vital role in our daily lives, with both positive and negative outcomes. It frees life-sustaining minerals and elements from solid rock so they can become incorporated into our soils, later to pass into the foods we eat and the water we drink. Indeed, without weathering we would *have* very little food, as weathering produces the very soil in which much of our food is grown. But weathering can also wreak havoc on the structures we build. Countless monuments, from the pyramids of Egypt to ordinary tombstones, undergo drastic deterioration from freezing water, hot sunshine, and other climatic forces. For example, more than half a century after they were sculpted, the seemingly eternal faces of Presidents Washington, Lincoln, Jefferson, and Roosevelt hewn out of the side of South Dakota's Mount Rushmore are in real danger of losing their noses, lips, and mustaches. Engineers and stone masons intervene regularly to secure loosened rocks and

Figure 5-2 A worker repairing weathering damage to Mount Rushmore National Monument, South Dakota.

sharpen the faces' fading features, so that weathering will not forever alter one of North America's most impressive human creations (Fig. 5-2).

On the following pages, we will explore the different ways in which solid rock weathers, the factors that control weathering, and the various practical products that result from it. In later chapters, we will examine erosion and the most common erosional agents: gravity in Chapter 13, streams in Chapter 14, groundwater in Chapters 15 and 16, glaciers in Chapter 17, wind in Chapter 18, and coastal waves in Chapter 19.

Weathering Processes

Rocks can be weathered in two ways. **Mechanical weathering** breaks a mineral or rock into smaller pieces (*disintegrates* it) but does not change its chemical makeup. The only changes induced by mechanical weathering are to such physical characteristics as the size and shape of the weathered structure. **Chemical weathering** actually changes the chemical composition of minerals and rocks that are unstable at the Earth's surface (*decomposes* them), converting them into more stable substances; minerals and rocks that are chemically stable at the Earth's surface are resistant to chemical weathering.

Mechanical and chemical weathering go on constantly and simultaneously in most environments. Mechanical weathering makes rocks more susceptible to chemical weathering by creating more surface area for chemical attack, much as crushing a sugar cube with a spoon causes it to dissolve more rapidly in hot water. By mechanically disintegrating the sugar cube you vastly increase its surface area, so that crystal surfaces previously hidden inside the larger cube are now exposed to the hot water. Similarly, the area of a boulder that is vulnerable to weathering agents consists only of its outer surface until mechanical weathering increases its surface area (Fig. 5-3).

Mechanical Weathering

There are a number of natural processes that reduce rocks to smaller sizes without causing any change in their chemical makeup. In any given location, several of these processes may be working at the same time.

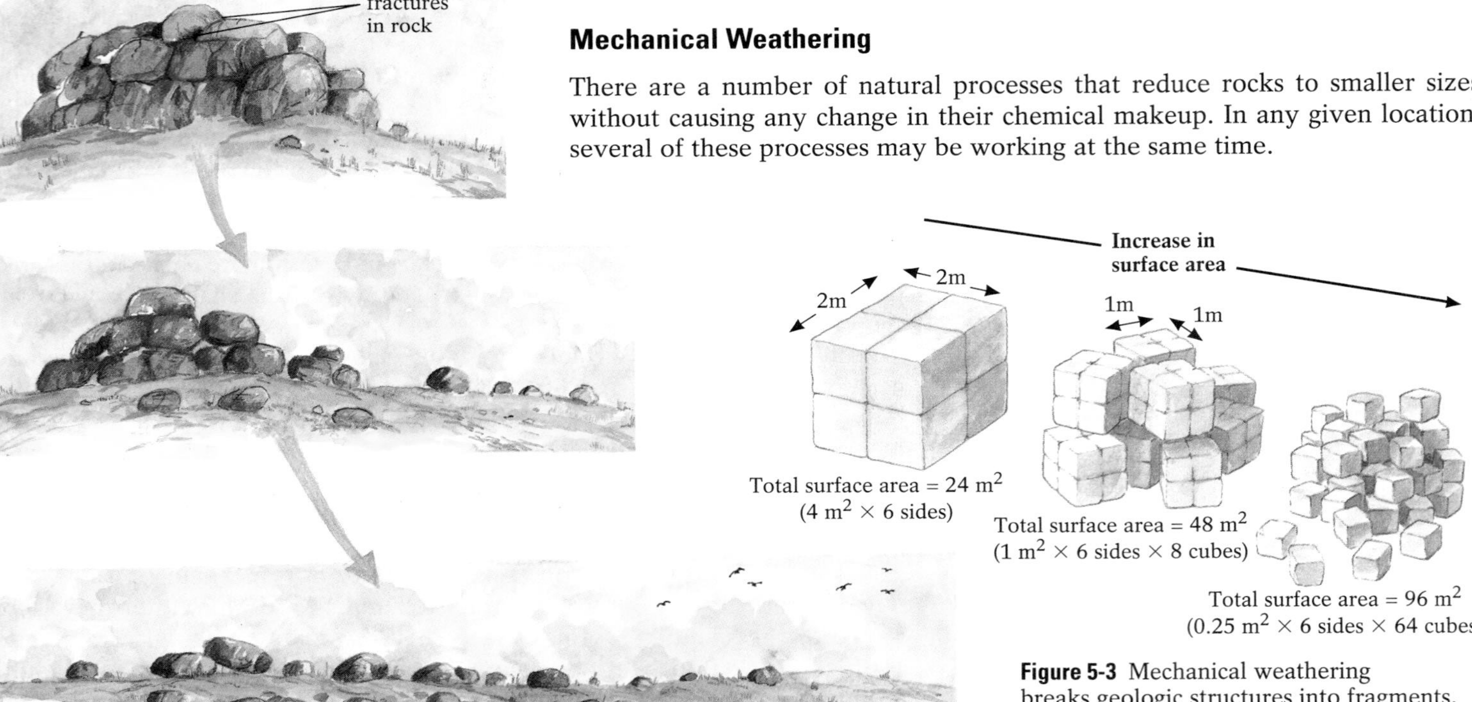

Figure 5-3 Mechanical weathering breaks geologic structures into fragments, increasing the total surface area exposed to the processes of chemical weathering.

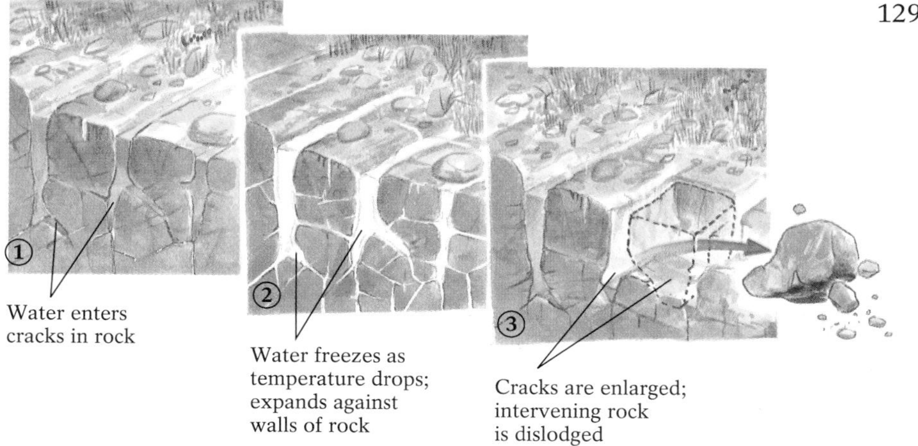

Figure 5-4 Frost wedging occurs when water freezes and expands within cracks in rock, enlarging them. The cracks get bigger with each frost, and may eventually dislodge intervening pieces of rock.

Frost Wedging Water expands in volume by about 9% when it freezes. When water enters pores or cracks in a rock and the air temperature then falls below 0°C (32°F, the freezing point of water), the upper surface of the water freezes first because it is in direct contact with the cold atmosphere. As the cold penetrates downward the rest of the water freezes as well and, because the surface has already frozen, cannot expand upward. This ice thus expands outward, exerting a force far greater than that needed to fracture even solid granite, enlarging the cracks and often loosening or dislodging fragments of rock (Fig. 5-4). In addition, as this wedge of ice forms, it draws more water to itself, which, in turn, freezes and enlarges the wedge, exerting additional outward pressure on the surrounding rocks. This process, called **frost wedging,** is the most effective type of mechanical weathering.

Frost wedging is most active in environments where there is abundant surface water and temperatures often fluctuate around the freezing point of water. When water frozen in a structure's cracks thaws each day in the warm afternoon sun and then refreezes each night, the cracks expand rapidly, causing blocks or slivers of rock to fall with regularity from the structure and collect at its base as **talus** (Fig. 5-5). Frost wedging also occurs when water seeps into roadway cracks and then freezes, causing the notorious pothole obstacle courses found in some northern cities.

Crystal Growth When salt crystals settle to the bottom of a salt-water-filled cavity, because either the water is evaporating or it contains an overabundance of dissolved salt, the growing crystals apply great pressure to the walls of the cavity, prying them farther apart. Rocky shores, where salty sea spray soaks into rocks, evaporates, and leaves behind growing salt crystals, are especially susceptible to weathering by crystal growth. Some well-known stone structures of human design have also been damaged by this process.

Figure 5-5 Frost wedging is particularly evident on mountains in moist temperate regions, where, even in summer, nighttime temperatures often fall below freezing. Here, a cycle of freezing and thawing occurs daily, creating the fresh rock rubble, or talus, that carpets the slopes of many mid-to-high-latitude mountains.

(a)

(b)

Figure 5-6 Cleopatra's Needle as it looked for about 3500 years in Egypt (**a**), and how it looks today, after a century of weathering in New York City's Central Park (**b**).

Among them is Cleopatra's Needle, a granite obelisk that stood for over 3000 years in Egypt without much loss of the finely engraved hieroglyphics on its sides. Before being relocated to Central Park in New York City in 1879, the obelisk was stored at a site where salty groundwater readily penetrated the column. Salt crystallized within small water-filled cracks in the granite as the water evaporated in the hot Middle Eastern air, and this crystal growth caused the initial disintegration of the needle's surface. Upon arrival in New York's humid environment, the same salt crystals absorbed water and expanded, causing further disruption of the surface. Today, little remains of the ancient message inscribed on the obelisk's sides (Fig. 5-6).

Thermal Expansion and Contraction If you have ever sat around a campfire, you may have noticed that thin layers of rock material have fallen off the cracked surfaces of rocks close to the fire. This occurs because each of the different minerals in a rock has its own distinctive physical properties, including the rate and extent to which it expands when heated. **Thermal expansion**, the enlargement of a mineral's crystal structure in response to heat, varies among minerals largely because of the different strengths of their chemical bonds. A grain of quartz, for example, expands about three times as much as a grain of plagioclase feldspar when subjected to the same increase in heat. If a rock that contains both minerals is heated, the expanding quartz grains push against neighboring feldspar grains, loosening and eventually dislodging them. Because rocks are generally poor conductors of heat, the heat does not penetrate very deeply into the rock; instead, the heated outer portion of a rock tends to break away from the cool inner portion. Rock surfaces that are exposed to high daytime and low nighttime temperatures undergo a similar process at a much slower pace. Their component minerals expand with repeated heating and contract upon cooling, eventually causing the rock's outer layer to break apart.

Desert rocks, which are exposed to daily temperature fluctuations of up to 56°C (100°F), should be most susceptible to this process. To test this hypothesis, the effects of thermal expansion and contraction were gauged by heating and then cooling a block of highly polished granite through a range of 38°C (68°F) 89,500 times, the equivalent of 244 years of daily temperature variations. No detectable change occurred in the granite's brightly polished surface as a result. Did this experiment simulate the true environmental conditions of a desert? Is 244 years sufficient to show the effects of thermal expansion and contraction? When the experiment was repeated with a fine mist of water introduced during the cooling cycles to simulate the effects of morning dew, the granite lost its polish, the surfaces of some grains within it became irregular, and cracks appeared near the rock surface. These results suggest that some desert rocks may weather mechanically through the combined agencies of extreme temperature variations and moisture.

Mechanical Exfoliation When erosion of overlying rock or soil exposes a wide area of a large plutonic mass, pressure on the mass is reduced and it expands; since most of the structure remains underground, where surrounding rock continues to exert pressure on it, it expands upward. As the rock in the structure expands, it fractures into sheets parallel to its exposed surface. These sheets may then break loose and fall from the sloping surface of the structure, a weathering process known as **mechanical exfoliation** (Fig. 5-7). Many mountains have a dramatic "stepped" appearance that is produced when large thin slabs of rock, several meters in thickness, exfoliate from underlying rocks (Fig. 5-8).

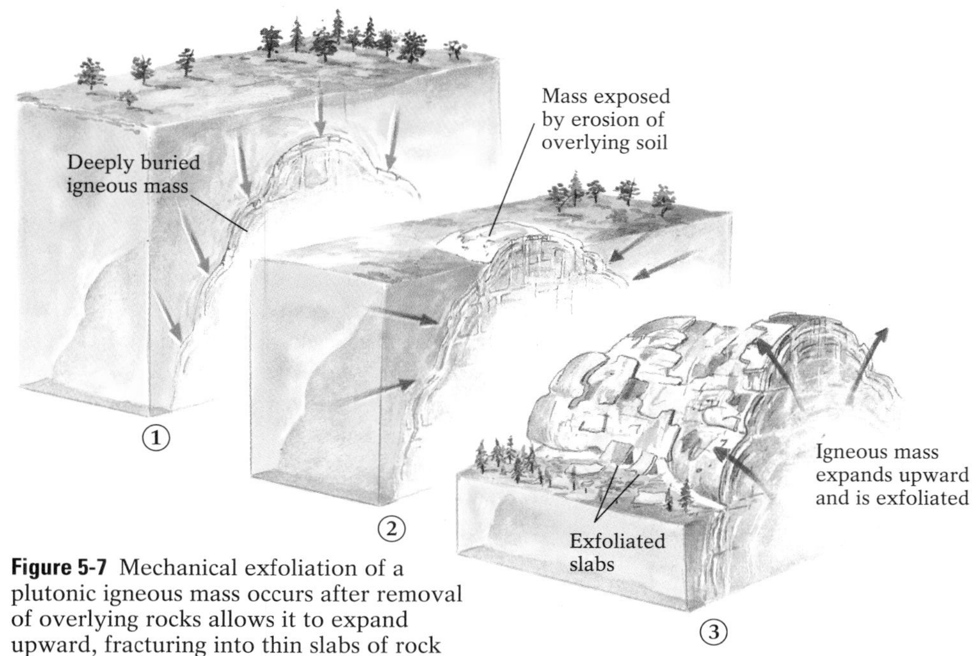

Figure 5-7 Mechanical exfoliation of a plutonic igneous mass occurs after removal of overlying rocks allows it to expand upward, fracturing into thin slabs of rock parallel to its exposed surface.

Figure 5-8 The "steps" of this mountain in California's Yosemite National Park were produced by mechanical exfoliation of the mountain's granitic rock.

Other Mechanical Weathering Processes Almost all rocks contain some cracks and crevices. In the case of surface rocks, plants and trees often take root in these cracks. Although rock would seem strong enough to withstand it, the force applied to a crack by a growing tree root is surprisingly powerful and quite capable of enlarging the crack (Fig. 5-9). The buckled and broken sidewalks of some tree-lined boulevards convincingly demonstrate the weathering ability of such root growth.

Animals also contribute to mechanical weathering, the larger animals by stepping on stones or pebbles and crushing them into smaller particles. Some birds ingest stones, which are used as grinding tools in their digestive processes and eventually released worn down in size. The cumulative activities of even very small creatures such as earthworms and insects can also help break rocks into smaller fragments.

Mechanical weathering by **abrasion** occurs when rocks and minerals collide during transport or when loose transported material scrapes across the exposed surfaces of stationary rock. For example, sediment fragments carried in swirling streams or by gusty winds collide and grind against each other and other rock surfaces, breaking loose fragments into smaller particles and loosening new fragments as well. Similarly, rocks carried along at the base of a glacier are scraped against underlying rocks and are ground to even smaller sizes.

Figure 5-9 Tree roots growing in rock fractures exert an outward force that expands the fractures, mechanically weathering the rocks. This rock has been split by pressure from the growing birch root.

Chemical Weathering

Chemical weathering alters the composition of minerals and rocks, principally through reactions involving water. The water may come from oceans, lakes, rivers, glaciers, the underground water system, or directly from the atmosphere as precipitation (rain or snow).

Water is the single most important factor controlling the rate of chemical weathering because it carries ions to the reaction site, participates in the reaction, and then carries away the products of the reaction. This role is made possible by the charged nature of the two sides of a water molecule (discussed in Chapter 2), and its tendency to break down into its component H^+ and OH^- ions. The three major processes by which rocks weather chemically are dissolution, oxidation, and hydrolysis.

Dissolution In **dissolution**, ions or ion groups from a mineral or rock are removed and carried away by water, often without leaving a visible trace. (Such ions are said to be *in solution*; when the water carrying them eventually evaporates, these substances remain, and are said to have *precipitated* from the solution.) The slightly charged sides of the water molecule attract and remove oppositely charged ions from the surfaces of minerals. Some mineral deposits containing gypsum and halite, for example, are readily dissolved by water molecules (see Fig. 2-9). Dissolution of halite (NaCl) occurs when the positive (H^+) side of a water molecule attracts and dislodges chloride ions (Cl^-) and the negative (OH^-) side of a water molecule attracts and dislodges sodium ions (Na^+).

Often water itself does not remove ions from a mineral or rock, but reacts with another compound in the environment to form a substance—usually an acid—that does. Such is the case when the common sedimentary rock limestone is weathered chemically by dissolution. The process by which limestone is dissolved begins in the atmosphere and soil, where water and carbon dioxide (CO_2) combine to form carbonic acid (H_2CO_3):

$$\underset{\text{Water}}{H_2O} + \underset{\text{Carbon dioxide}}{CO_2} \longrightarrow \underset{\text{Carbonic acid}}{H_2CO_3}$$

Figure 5-10 Weathered and unweathered limestone boulders. In humid environments, such as those in most southeastern states, limestone dissolves extensively along cracks and crevices, leaving behind thousands of cavities that range from minor surface depressions to large underground cave systems.

Although relatively weak, this acid effectively decomposes calcite, the principal mineral in limestone, to form calcium and bicarbonate ions:

$$\underset{\text{Calcite}}{CaCO_3} + \underset{\text{Carbonic acid}}{H_2CO_3} \longrightarrow \underset{\text{Calcium ion}}{Ca^{+2}} + \underset{\text{Bicarbonate ions}}{2\,HCO_3^-}$$

The calcium and bicarbonate ions formed are then carried off in solution by circulating groundwater, often leaving distinct voids in the parent rock (Fig. 5-10).

These two simple reactions have created most of the world's caves (discussed in Chapter 16), where minute fractures in soluble limestone have grown into large underground passages through the gradual dissolution of surrounding rock. Also as a result of these reactions, dissolved calcium and bicarbonate is transferred from the continents to the oceans through the Earth's groundwater and surface-water systems, providing many types of marine creatures with the nutrients that are the raw materials for their shells. Ultimately, the discarded carbonate shells of these sea creatures that now blanket most of the world's sea floors will likely rise above sea level, perhaps uplifted by plate tectonic forces or exposed by receding seas, and become consolidated into carbonate rock (see Chapter 6). In their turn, these newly exposed layers of rock will weather and supply future sea creatures with their calcium and bicarbonate ions, continuing the endless carbonate cycle between land and sea.

Limestone buildings and statues are subject to dissolution just as are natural limestone structures. Because of the formation of carbonic acid in the atmosphere, rainwater is naturally slightly acidic. The more acidic the rain in a given locality, the more rapidly its limestone features, natural or human-made, will be affected. Where rain falls downwind from industrial centers, however, it encounters sulfur dioxide gas (SO_2) that enters the atmosphere as a result of such industrial processes as the burning of sulfurous coal and the smelting of sulfur-rich copper, iron, and nickel ore. Sulfur dioxide combines chemically with atmospheric oxygen and water to form sulfuric acid (H_2SO_4), which is much stronger than naturally occurring carbonic acid and dissolves carbonate rock and building stone quite rapidly.

Oxidation In **oxidation**, a mineral's ions combine with oxygen ions. For example, when iron ions in mafic rocks bond with oxygen ions in the atmosphere, the reaction forms the iron oxide Fe_2O_3 (hematite):

$$4\ Fe^{+2} + 3\ O_2 \longrightarrow 2\ Fe_2O_3$$

In the presence of water, many other iron oxides are formed: Two of the most common are yellow-brown goethite (FeO(OH)) and yellowish limonite ($FeO(OH){\cdot}nH_2O$) (n represents any number). These iron oxides, all commonly called *rust,* form wherever iron-rich minerals and rocks come into contact with water. Rust often stains the surfaces of minerals with high iron contents, such as olivine, pyroxene, and amphibole, and rocks made up of these minerals, such as basalt, gabbro, and peridotite. Like other products of chemical weathering, rust is very stable at the Earth's surface. Thus (much to the chagrin of the owners of aging cars) after rust forms, it does not dissolve.

Other substances, particularly other metals, are also subject to oxidation. Copper, a reddish element with a metallic luster, is more resistant than iron to oxidation, but after long exposure to the atmosphere, it develops a green surface, or *patina*, characteristic of copper carbonates and copper sulfates.

Hydrolysis In **hydrolysis**, H^+ or OH^- ions from water molecules displace other ions from a mineral's structure. Aluminum-rich silicates such as the feldspars, the most abundant minerals in the Earth's crust, are weathered primarily by hydrolysis.

In a typical hydrolysis reaction, potassium feldspar reacts with water's hydrogen ions. The result is a stable clay mineral (a type of silicate similar to the micas), dissolved silica in the form of silicic acid, and potassium ions that have been liberated by the reaction

$$\underset{\text{Potassium feldspar}}{2\ KAlSi_3O_8} + \underset{\text{Hydrogen ions (from water)}}{2\ H^+ + 9\ H_2O} \longrightarrow \underset{\text{Kaolinite clay}}{Al_2Si_2O_5(OH)_4} + \underset{\text{Silicic acid (in solution)}}{4\ H_4SiO_4} + \underset{\text{Potassium ions (in solution)}}{2\ K^{+2}}$$

The clays formed during hydrolysis accumulate within the upper few meters of the Earth's surface, either incorporated into soils or washed out to sea to become oceanic mud. The silicic acid and the potassium ions are generally transported in solution by underground and surface water. Much of the dissolved silica is used either to cement together loose grains of sediment to form sedimentary rocks, or as the raw material from which many marine organisms manufacture their shells and skeletons. The dissolved potassium ions may be extracted from soil water and incorporated into growing plants, or transported to the sea, eventually to be deposited as potassium salt.

Factors that Influence Chemical Weathering

We have seen that water is a key player in all chemical weathering reactions. But water's usefulness as a weathering agent may be enhanced by local temperature and moisture patterns, the activity of living organisms, and the amount of time that the mineral or rock has been exposed to weathering. Finally, of course, the rate at which any mineral or rock will weather depends upon its own chemical stability, or that of its components.

Climate Since water's ions are instrumental in all dissolution, oxidation, and hydrolysis reactions, a climate with ample moisture already provides the

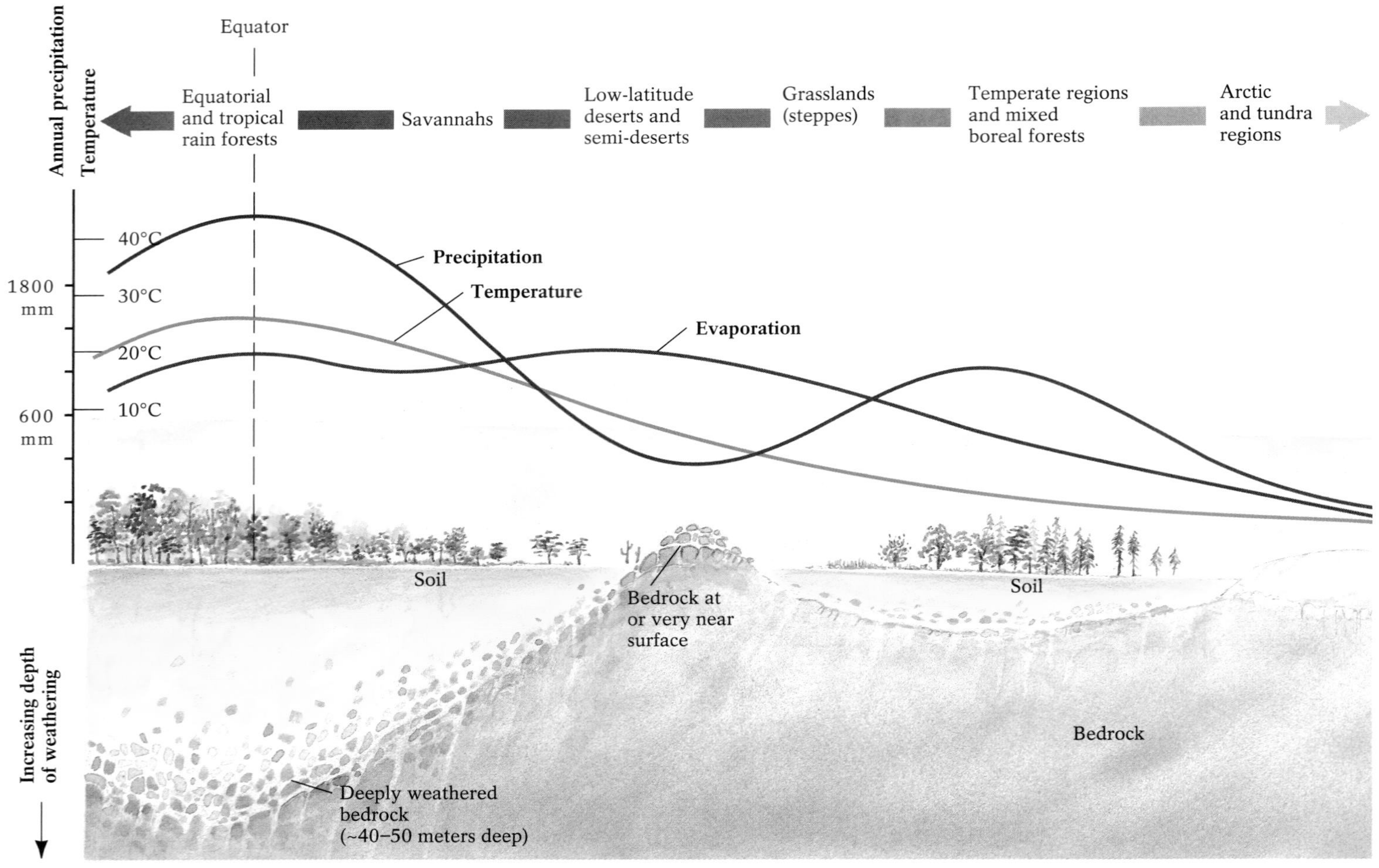

Figure 5-11 The relationship between climate and weathering, shown as a function of the depth to which a region's bedrock (the solid rock immediately underlying the surface soil) is typically weathered. The deepest weathering occurs in the warm, moist tropical zone near the equator, where temperature and precipitation are greatest. Weathering is minimal in the arctic and in the deserts—both regions in which water is in short supply as an agent of chemical weathering.

essential ingredient for chemical weathering. Another key factor that accelerates virtually all chemical reactions is heat. Thermal energy from heat excites the electrons that speed around atomic nuclei, causing them to vibrate more rapidly until they break free of the electrical attraction of the nucleus (discussed in Chapter 2); the greater the heat applied, the more electrons are freed to participate in chemical reactions. In general, the rate of chemical reaction doubles with every 10°C (18°F) increase in temperature.

Because of the combined effects of water and heat, chemical weathering occurs more readily in warm, moist climates than in arid, cold climates. For example, the hot steamy climate of low-latitude tropical areas such as Puerto Rico is optimal for chemical weathering, whereas cold dry places such as Antarctica experience little or no chemical weathering (Fig. 5-11).

Warm, moist climates also promote the growth of lush vegetation; after they die these plants are decayed by microorganisms, producing organic acids that enhance chemical weathering. In addition, photosynthesis by plants produces O_2, promoting oxidation, and animal respiration produces CO_2, which combines with water in soil to form chemically reactive carbonic acid.

Living Organisms Organisms can affect the chemical weathering rate of rocks and minerals by physically causing or increasing their exposure to weathering agents. Burrowing animals, for instance, such as groundhogs, prairie dogs, and even ant colonies, commonly transport unweathered materials from below ground to the surface where they can be weathered. The common earthworm, which churns up underground minerals in the course of consuming the organic matter in soil, is particularly effective at exposing

Figure 5-12 Soil is churned up as earthworms move through it, bringing underground minerals to the surface to be exposed to weathering.

fresh mineral surfaces to the effects of chemical weathering (Fig. 5-12): In humid temperate climates, an average earthworm colony brings 7 to 18 tons of soil per acre to the surface each year; in the tropics, where worms can grow to be 5 feet long, a colony can stir up 100 tons of soil per acre each year. The burrows of such animals are also instrumental in enhancing weathering, by allowing the circulation of weathering agents in air and water.

Time There is a direct relationship between time and weathering: The longer a rock or sediment is exposed to a weathering environment, the more it will decompose. For example, within some of the hills of sediment left behind by the glaciers that once covered much of North America's Great Lakes region, pebbles containing minerals susceptible to weathering have been so thoroughly weathered that they can be rubbed between one's fingers into balls of soft clay. Pebbles of the same composition in other glacially deposited hills in the region, however, are still completely fresh and solid. Assuming the other factors that control weathering affected these hills equally, the hills with the rotted pebbles must have been exposed to the weathering environment for a much longer time (i.e., are much older) than those with the fresh, unweathered pebbles (Fig. 5-13).

Figure 5-13 Hills of glacial deposits of substantially different ages, each produced by different periods of glaciation. Weathered for more than 100,000 years, the older hills in the foreground are less rugged than the younger ones behind them and are marked by thicker soils. Some pebbles within the older hills have weathered to soft clay, whereas similar ones in the younger hills are still fresh and solid.

Figure 5-14 Two gravestone inscriptions from the same cemetery in Williamstown, Massachusetts, showing differential weathering of rock types. The marble gravestone (near right), though exposed to the same climate as the granite gravestone (far right)—and emplaced 50 years *later*—has suffered noticeably more weathering damage. Marble, composed predominantly of the chemically reactive mineral calcite, is much more susceptible to chemical weathering than is granite, a rock composed primarily of the very stable mineral quartz.

Mineral Composition The effect of chemical weathering on rock is determined largely by the *stability* (resistance to chemical change) of the rock's component minerals (Fig. 5-14). A mineral does not tend to change chemically as long as it remains in an environment similar in temperature and pressure to the one in which it originally crystallized from magma. At the Earth's surface, however, a mineral's environment differs from that at which it crystallized. As a result, the mineral becomes relatively unstable, and more likely to change by chemical weathering. As a general rule, the greater the difference between the conditions under which a mineral crystallized and conditions at the Earth's surface, the greater will be the mineral's tendency to change. Minerals that crystallize at high temperatures and pressures are the most unstable at the surface; those that crystallize at lower temperatures and pressures are relatively stable (Fig. 5-15).

When high-temperature minerals such as olivine and pyroxene are exposed at the surface, they quickly begin to oxidize and hydrolyze. The end products of these chemical reactions, such as clay minerals and metallic oxides, are stable at the Earth's surface. Low-temperature minerals such as quartz and mica, which are more stable in the Earth's surface environment, weather chemically at a much slower rate.

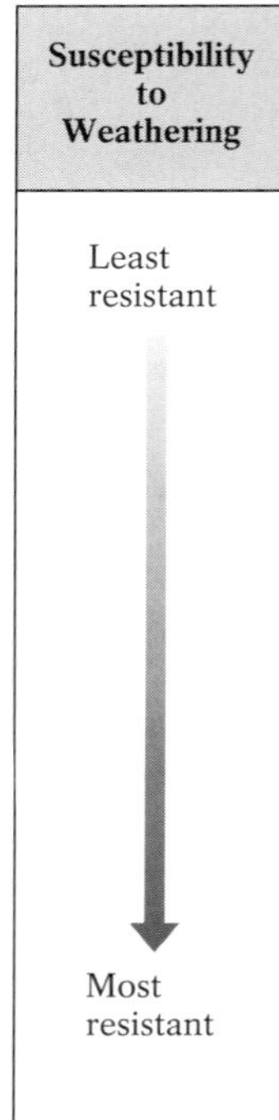

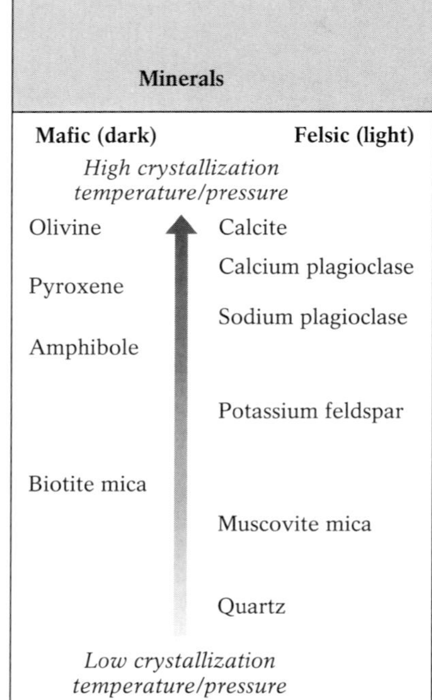

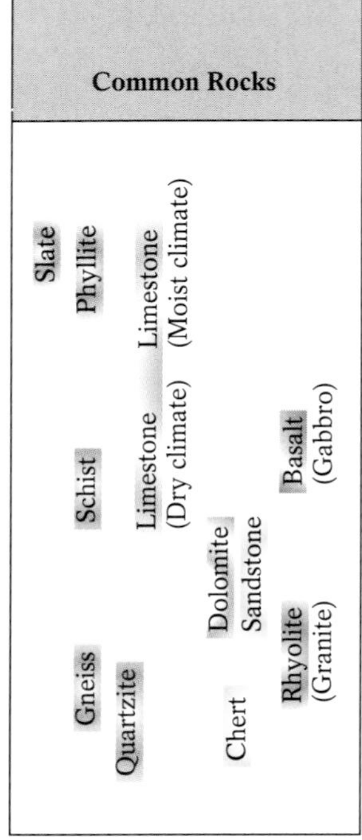

Figure 5-15 Mafic igneous rocks, which crystallize at high temperatures, tend to weather fairly rapidly in the Earth's relatively cooler surface environment. Felsic rocks, which crystallize at lower temperatures—closer to those at the Earth's surface—tend to weather more slowly.

Some Products of Chemical Weathering

After dissolution, oxidation, and hydrolysis remove the soluble ("dissolvable") components in rocks, a combination of relatively insoluble, unreactive, and stable new products of chemical weathering remain. These include the clay minerals and several economically valuable metal ores. Chemically weathered surface rocks are typically found in the form of rounded boulders.

Clay Minerals Clay minerals result primarily from the hydrolysis of feldspars (see Table 2-4). There are several varieties of clay minerals, each with a distinctive chemical composition and set of physical properties that determine its various practical uses; the variety produced from a given rock depends on the climatic conditions under which the hydrolysis occurs. For example, kaolinite ($Al_2Si_2O_5(OH)_4$) is the common product of the hydrolysis of feldspars in a warm, humid climate. To produce kaolinite, the large positive ions (K^+, Na^+,Ca^{2+}) in the parent feldspars must be completely displaced by H^+ ions from water during hydrolysis. The name kaolinite derives from the Kaoling region of northern China, where the mineral is mined extensively and used in the manufacture of fine porcelain. Kaolinite is also the active ingredient in Kaopectate®, because it absorbs the toxins and bacteria that cause intestinal irritation.

In other clay minerals, some large positive ions remain after hydrolysis. For example, smectite, a highly absorbent clay, forms in drier, cooler climates; under these conditions, micas and amphiboles are only partially hydrolyzed, leaving some of their calcium, sodium, and magnesium ions unreplaced. Smectite is used to filter impurities from beer and wine. The negatively charged surfaces of smectite particles attract and bond with the positively charged impurities in those beverages and then settle from the liquids, leaving them clear and pure.

Clay minerals are also used widely in industry as agricultural fertilizers, lubricants on oil-drilling rigs, in the manufacture of bricks and cement, and in the production of paper. Several common household staples owe their effectiveness to some properties of clays. Because of its absorbency and deodorizing properties, for instance, the clay mineral smectite is a key ingredient in cat litter.

Metal Ores Ores are aggregates of minerals that have economic value and can be profitably extracted from their surrounding rocks. Most economically valuable minerals—usually metals—rarely occur in pure form. Aluminum, for example, although quite abundant in the Earth's crust, is generally widely dispersed in clays, feldspars, and micas. (Pure aluminum metal was once so highly prized that Napoleon had banquet cutlery fashioned from it, instead of from silver or gold.)

Intense chemical weathering of feldspar-rich rocks in hot, moist climates produces the aluminum ore bauxite ($Al_2O_3{\cdot}nH_2O$). Bauxite forms because aluminum, being insoluble in water, remains concentrated in soils after most common elements, such as calcium, sodium, and silicon, have dissolved out of them. This important ore is found most often in tropical areas, such as the Caribbean islands and parts of Australia, Africa, and South America. Substantial ancient bauxite deposits can also be found in Georgia, Alabama, and Arkansas, suggesting that the climate in those places was once warmer and more moist than it is today.

Weathering processes also produce or enrich much of the world's mineable iron, copper, manganese, and even silver deposits.

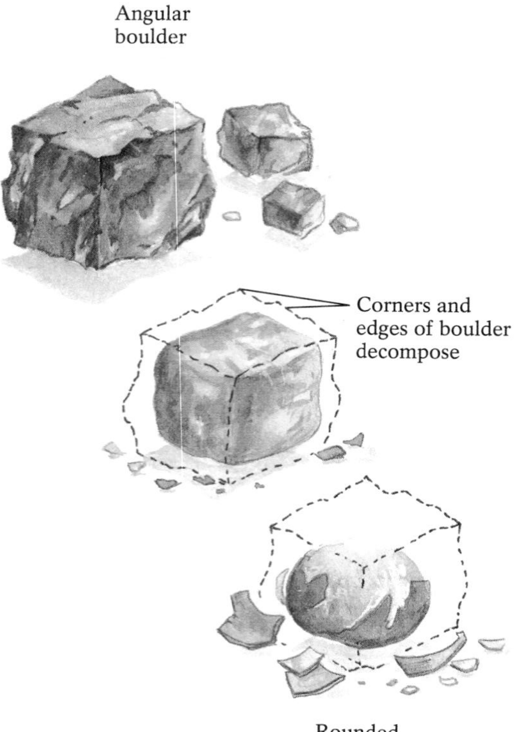

Figure 5-16 Spheroidal weathering. Rounded boulders, such as these basalt boulders in Oregon, evolve from angular ones as the rocks' protruding corners and edges are gradually eroded by chemical weathering.

Rounded Boulders Mechanically weathered rocks are initially quite angular, but in an active chemical-weathering environment, they don't remain so for long. Because the corners and edges of angular rock masses offer more surface area, they are more likely to weather chemically and do so at a faster rate than do the rocks' planar faces. As their corners and edges decompose, angular rocks are eventually converted to smoothly rounded boulders. This process is known as **spheroidal weathering** (Fig. 5-16).

Once a spheroidal shape is attained it remains essentially unchanged, because weathering agents (particularly water) act uniformly across the boulder's entire surface, weathering concentric layers of rock in turn. As a rock layer is weathered it peels away, much like the outer layers of an onion, exposing the unweathered portion beneath it. This process, sometimes called *chemical exfoliation,* occurs because weathering-produced clay minerals absorb water to occupy a greater volume than the unweathered feldspars in rock (Fig. 5-17); as the clay minerals at the surface of a rounded boulder absorb water and increase in volume, they expand outward and separate from the boulder's unweathered interior, leaving the next layer to be acted upon in its turn.

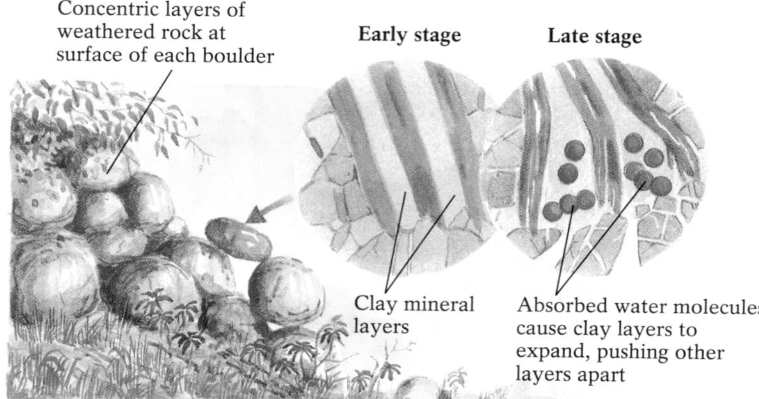

Figure 5-17 After a boulder becomes rounded, its shape is maintained because chemical weathering then occurs uniformly over its surface. The surface of the boulder falls off, in concentric layers, because the clay minerals produced by chemical weathering of feldspars at the boulder's surface occupy a greater volume than the unweathered feldspars in underlying layers.

Figure 5-18 A roadcut in Costa Rica, California, showing layers of regolith (weathered surface sediment and bedrock). Extreme chemical weathering can produce regolith that is more than 60 meters (200 feet) deep.

Soils and Soil Formation

Mechanical and chemical weathering of sediment and the preexisting solid rock, or *bedrock,* underlying it produce **regolith** ("rock blanket"), the fragmented material covering much of the Earth's land surface (Fig. 5-18). Geologists refer to the upper few meters of regolith, which contains both mineral and organic material, as **soil.** Soils, the most common product of weathering on land (oceanic mud is more plentiful), may well be the most valuable of all natural resources. People everywhere depend on fertile soils to support plant growth and provide life-sustaining nutrients in foods.

Soils differ depending on the weathering processes, local climatic conditions, and preexisting materials that produced them. Consequently, they are complex and often difficult to classify. We will first discuss the factors that interact to form soils, and then examine the structure of a typical soil. Finally, we'll take a brief look at the types of soils that help produce the foods we eat.

Influences on Soil Formation

Five factors are particularly important in determining the nature of an area's soil. These factors are: parent material, climate, topography, vegetation, and time.

Parent Material A soil's **parent material** is the bedrock or sediment from which the soil develops. The parent material's mineral content determines both the nutrient richness of the resulting soil and the amount of soil produced (Fig. 5-19). For example, soils forming from basaltic bedrock differ markedly in composition and rate of development from those forming from granitic bedrock.

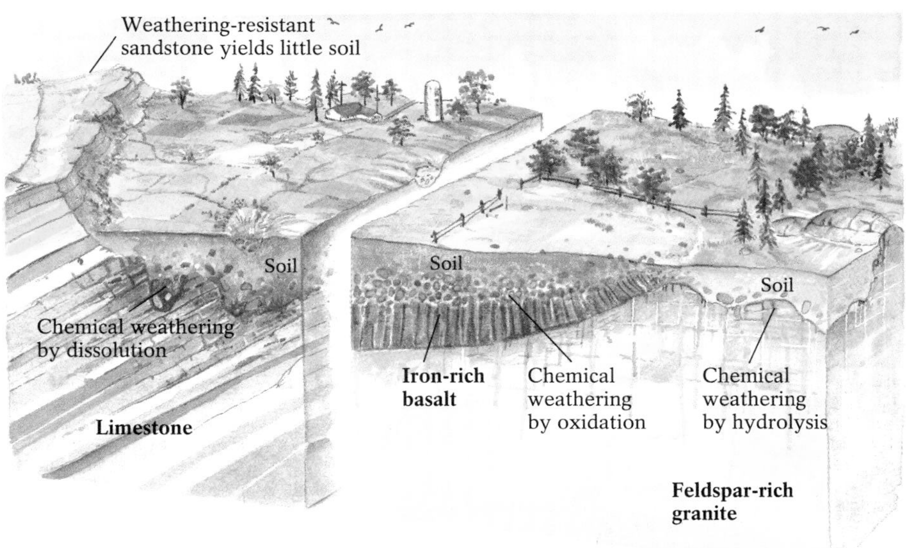

Figure 5-19 The role of bedrock composition on soil development in a moist temperate environment. Advanced chemical weathering of iron-rich basalts (oxidation), fractured limestone (dissolution), and feldspar-rich granite (hydrolysis) produces thick soils on these types of rock. Sandstone, however, tends to remain relatively unweathered because it consists primarily of weathering-resistant quartz grains; as a result, any soil produced on sandstone will be sparse.

The relationship between parent material and soil development is clearly demonstrated on Java and Borneo, two neighboring Indonesian islands with the same climate. The parent materials for Java's soils are largely fresh, nutrient-rich volcanic ash deposits, whereas the parent materials for Borneo's soils consist of numerous granitic batholiths, gabbroic intrusions, and andesite flows. Ongoing volcanism on Java replenishes the positive-ion nutrients—such as potassium, magnesium, and calcium—in its soils, whereas Borneo, lacking fresh ash deposits, has soils depleted of nutrients. The islands' population densities reflect their dramatically different soil fertility and agricultural productivity: Java has approximately 460 inhabitants per square kilometer and Borneo about 2 inhabitants per square kilometer.

Climate An area's climate—the amount of precipitation it experiences and its prevailing temperature—controls the rate of chemical weathering and consequently the rate of soil formation in the area. Climate also regulates the growth of vegetation and the abundance of microorganisms that contribute CO_2 and O_2 to the processes of dissolution, hydrolysis, and oxidation. Chemical weathering and soil formation are most rapid in warm, moist climates, and slowest in cold or dry climates.

Topography **Topography** refers to the physical features of a landscape, such as mountains and valleys, the steepness of slopes, and the shapes of landforms. Topography influences the availability of water and other weathering factors, and the rate of soil accumulation. For example, steep slopes allow rainfall and snowmelt to flow away swiftly; little water remains on them to penetrate the surface, and therefore little or no soil develops there as a result of chemical weathering. Any soil that does form on steep slopes is usually transported downslope before it can accumulate to a significant depth. Conversely, in level, low-lying areas water accumulates and readily infiltrates the ground, enhancing the prospects for chemical weathering and soil development (Fig. 5-20).

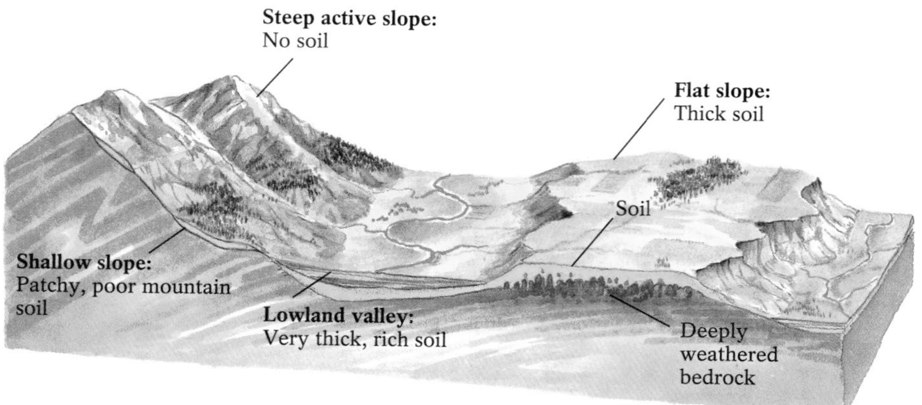

Figure 5-20 The effect of landscape on soil development. Soils are generally thin or nonexistent on steep slopes, because the water required for chemical weathering runs off such slopes (and because any soil that does accumulate would slide downhill). Soils tend to be thickest in lowland valleys, where water and loose material transported from upland come to rest.

Vegetation Vegetation contributes organic matter to soils and produces much of the O_2 and CO_2 involved in chemical weathering reactions. Soils developing on prairie grasslands, for example, receive large quantities of organic matter from surface plant remains and from the decay of extensive subsurface root systems.

Plants also contribute H^+ ions that help weather soils. The H^+ ions, weakly attached to plant roots, replace the large positive ions (such as calcium, potassium, and sodium) in feldspars and other minerals, hydrolyzing them into clay minerals. This exchange helps develop the soil while providing the plants with ions that are nutritionally beneficial to them (and to us) (Fig. 5-21). These ions are returned to the soil when the plants die and decompose, becoming available for the next generation of plants. [If the plants are harvested before they die, however, these ions are permanently removed from the soil. As a result, continuous farming eventually depletes a soil's supply of calcium, sodium, and potassium ions; it then becomes necessary to supplement these with natural or synthetic fertilizers and allow a fallow period (a period during which the soil is not farmed) to let the ions re-accumulate.]

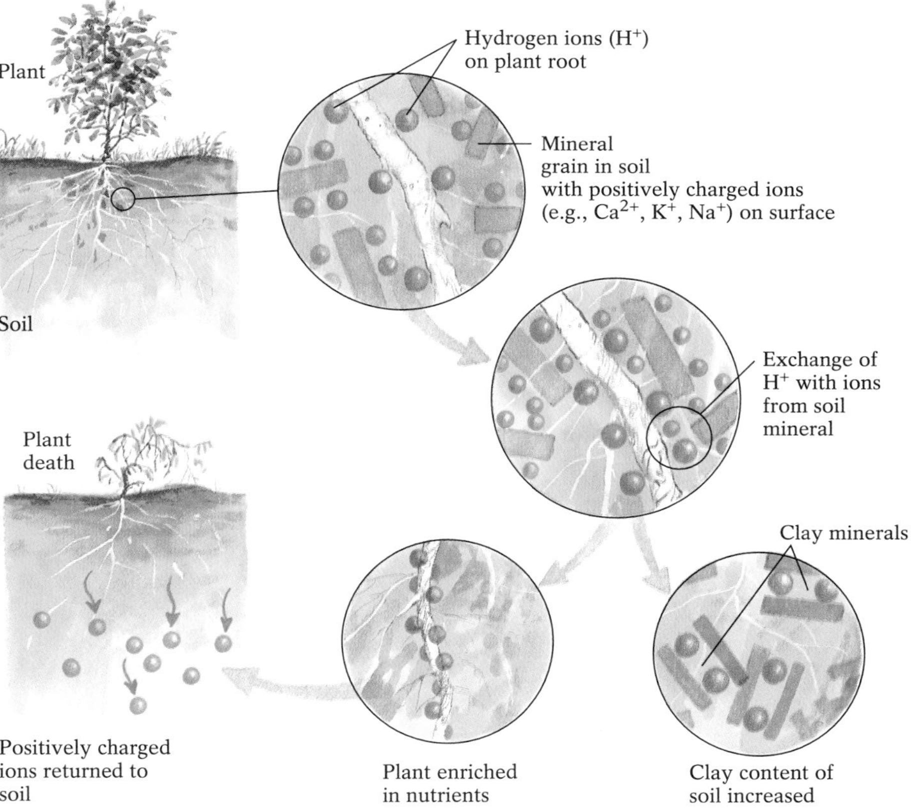

Figure 5-21 Vegetation contributes to soil development through the exchange of H^+ ions, from the surface of plant roots, with positive ions from soil minerals such as feldspars. This process increases the soil's clay content while providing the plant with nutrients, which are returned to the soil when the plant dies.

Time Soils may begin to develop within a few hundred years in warm, wet environments, but may require thousands or hundreds of thousands of years in arid or polar regions. If other factors in soil formation were equal, a landscape exposed to weathering influences over a long period of time would contain a thicker, more well-developed soil than a younger landscape. However, because other factors almost always vary considerably, a thicker soil is not necessarily an older one.

Typical Soil Structure

The effects of weathering are greatest at a soil's surface, where there is direct exposure to the weathering environment. Below the surface, distinct weathering zones develop within a soil as substances are dissolved from the upper layers of the soil and transported to lower levels in solution or as solid fragments suspended in infiltrating water. These transported substances are then commonly precipitated or deposited in the lower soil layers. Thus, the upper part of a developing soil loses some of its original materials, while the lower part gains new components. Each distinct weathering zone is called a **soil horizon.** Soil scientists have identified dozens of horizon types. Among the most common are the O, A, E, B, and C horizons, which are found virtually everywhere within temperate zones; there are other horizons as well, which are found in more specialized environments. The vertical succession of soil horizons in a given location, the location's **soil profile** (Fig. 5-22), is a product of the local soil-forming conditions. Different localities vary in the types of horizons and their depth and degree of development, and thus have different soil profiles.

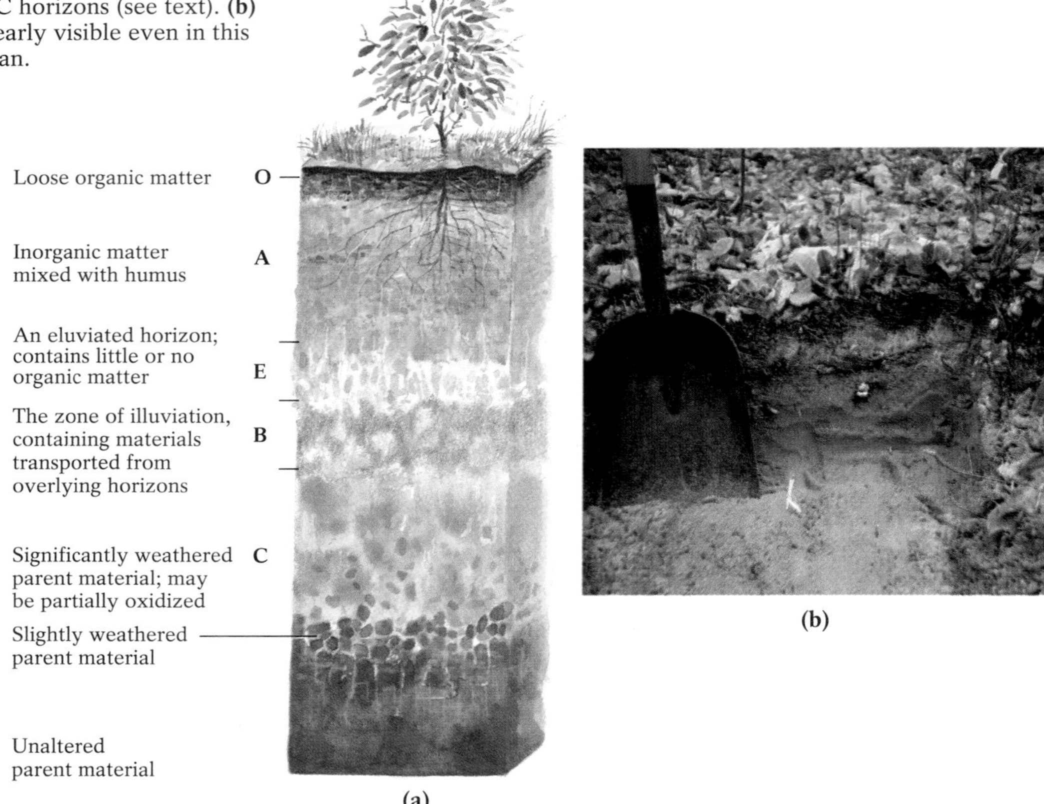

Figure 5-22 Soil profiles. **(a)** The features of a typical temperate-zone soil profile, including the O, A, E, B, and C horizons (see text). **(b)** A vertical succession of soil horizons is clearly visible even in this small section of soil in Margeretta, Michigan.

The upper portion of a soil profile consists of the O, A, and E horizons. The O ("organic") horizon in temperate regions such as North America consists mainly of organic matter, such as recognizable fibers of plant matter. It teems with life, containing 2 trillion bacteria, 400 million fungi, 50 million algae, and thousands of insects in a single kilogram (2.2 pounds). These organisms contribute O_2, CO_2, and organic acids to the developing soil.

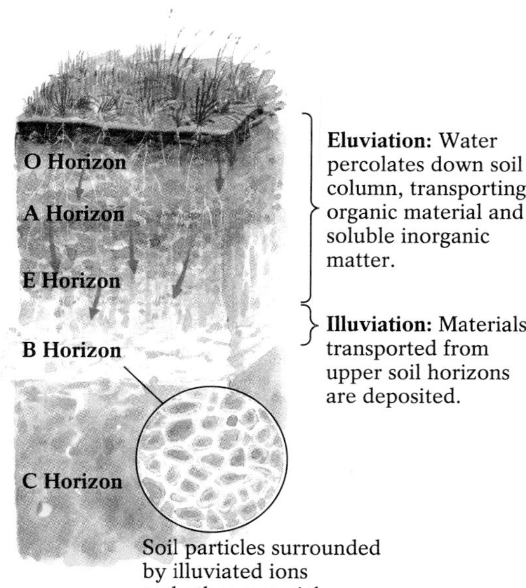

Figure 5-23 Eluviation and illuviation in a soil profile. Material is removed, or eluviated, as water passes through the O, A, and E horizons, and is deposited, or illuviated, as water infiltrates the B horizon.

Figure 5-24 The white layer in this depleted soil in Palo Verde, California, is caliche, carbonate material dissolved from upper soil horizons and precipitated in lower ones. In this case, the overlying soil horizons have since eroded away.

The A horizon consists mainly of inorganic mineral matter mixed with *humus,* a dark-colored, carbon-rich substance derived from decomposed organic material from the O horizon. The thickness of the A horizon depends upon the quantity of decomposed vegetation incorporated into the soil: In a tropical environment with lush vegetation, an obvious A horizon may develop in as little as a few hundred years; in an environment where vegetation is sparse, it may take several thousand years.

The E horizon is a light-colored zone below the A horizon with little or no organic material. Its light color results from dissolution and removal of the iron and aluminum compounds in the upper few meters of regolith. The E stands for *eluviation,* the process by which water removes material from a soil horizon: Fresh water containing organically produced CO_2 percolates downward from the surface, dissolving soluble inorganic soil components and transporting them along with fine soil particles to lower horizons.

The B horizon, which lies below the O, A, and E horizons, contains the materials dissolved or transported mechanically from the upper horizons. This is a zone of *illuviation,* the addition to a lower soil horizon of materials removed and transported by water from upper horizons (Fig. 5-23). There are a variety of B-horizon types, classified according to their predominant components: A Bh horizon, for example, has a high concentration of added humus; a Bo horizon has a high concentration of oxides.

In arid and semi-arid areas, where surface water quickly evaporates, a distinct carbonate-rich horizon is located within or below the B horizon at the depth to which annual rainfall penetrates. This **caliche** layer is formed when brief heavy rains dissolve calcium carbonate in the upper layers of soil and transport it downward; as the water rapidly evaporates, the carbonate precipitates, forming a markedly white layer in the soil (Fig. 5-24). In particularly arid regions, hundreds of thousands of years may be required to precipitate enough calcium carbonate to form a well-developed layer of caliche.

The O, A, E, and B horizons bear little resemblance to the original parent material. The C horizon, however, the lowest zone of significant weathering, consists of parent material that has been partially weathered but still retains most of its original appearance. It may show signs of oxidation from penetrating oxygen-rich groundwater, or it may be completely unoxidized. The C horizon is very thin where there has been little chemical weathering—for example, in a desert—but can be as much as 100 meters (330 feet) thick where chemical weathering is extensive, such as in the warm, wet tropics. Below the C horizon lies unaltered parent material.

Classifying Soils

Until recently, North America was classified by soil scientists as a two-soil continent: The eastern half was covered by *pedalfers* (from the Latin root *ped* for "soil," "al" for aluminum, and "fe" for iron), and the western half by *pedocals* (ped, plus "cal" for the caliche layers typical of the arid West). Pedalfers are relatively fertile, highly organic, iron- and aluminum-rich soils formed in humid temperate environments. Pedocals are relatively infertile, thin, organic-poor soils with high concentrations of calcium carbonate.

As we have seen, however, the complex interaction of parent material, climate, topography, vegetation, and time cause local soil development to vary in significant ways. Modern soil classification (Table 5-1) attempts greater precision than the old "two-soils" system, for very practical reasons: Distinct soil types have specific physical and chemical characteristics that affect our uses for them. Accurate soil classification influences decisions about

Table 5-1 **Classification of Soils**

Soil Order	General Properties	Typical Geologic or Geographic Setting
Entisols	Minimal development of soil horizons; first appearance of O and A horizons; some dissolved salt in subsurface	Young, newly exposed surfaces, such as new flood or landslide deposits, fresh volcanic ash, or recent lava flow; also found in very cold and very dry climates, or wherever bedrock strongly resists weathering
Inceptisols	A horizon well-developed, weak development of B horizon, which still lacks clay enrichment; some evidence of oxidation in B horizon; little evidence of eluviation or illuviation	Relatively young surfaces; cold climates where chemical weathering is minimal, or on very young volcanic ash in tropics, resistant bedrock, and on very steep slopes
Mollisols	Thick, dark, highly organic A horizon; B horizon may be enriched with clays; first appearance of E horizon	Semi-arid regions; generally grass-covered areas having adequate moisture to support grasses but not to cause significant dissolution of soluble materials in upper horizons
Alfisols	Relatively thin A horizon overlying clay-rich B horizon; strongly developed E horizon	Many climates, although most common in forested, moist environments
Spodosols	Much eluviation of A and E horizons, leaving a light-colored, grayish topsoil; aluminum/iron-enriched B horizon stained by dissolved organic material	Moist climates, usually on sandy parent materials (which allow water to infiltrate readily); grasses or trees may provide the organic matter
Aridisols	Thin A horizon with little organic matter overlying thin B horizon with some clay enrichment; caliche layer present in B or C horizons	Arid lands with sparse plant growth
Histosols	Wet, organic-rich soil dominated by thick O and A horizons	Found where production of organic matter exceeds addition of mineral matter, generally where surface is continuously water-saturated; often found in coastal environments
Vertisols	Very high clay content; soil shrinks (upon drying) and swells (upon wetting) with moisture variations	Equatorial and tropical areas with pronounced wet and dry seasons
Oxisols	Shows extensive weathering; highly oxidized B horizon is deep red from layer of oxidized iron	Generally older landscapes in moist climates having tropical rain forests
Ultisols	Shows extensive weathering; highly weathered clay-rich B horizons with high concentrations of aluminum	Very moist, lushly forested climates, often subtropical and tropical

the location of landfills, the design of buildings, and the ways we cultivate soils for food. A modern soils map contains an impressive number of different soil designations. Our current classification scheme names soils according to obvious physical characteristics, describes a soil's clay content, and indicates degree of nutrient depletion. Classification terminology also provides information about moisture content, mean annual air temperature, horizon development, soil chemistry, organic matter content, and even the origin and relative age of the soil. For example, an *entisol* (the root "ent" is derived from *recent*) is a soil that has not yet experienced significant horizon development, and may be a recent flood deposit or fresh volcanic ash. A *vertisol* (which tends to expand *vert*ically) contains clay minerals that swell when moistened and shrink when dried. Vertisols can undermine most structures built upon them.

Figure 5-25 The Angkor Wat temple in Cambodia, showing the varying degrees of durability exhibited by different soil types. The bricks that form the temple's foundation have been fashioned from ultisols, soils that have undergone advanced chemical weathering; these bricks remain remarkably fresh because they are composed of the stable products of this weathering, which resist additional chemical weathering. The general intensity of chemical weathering in this warm, moist climate is obvious from the condition of the temple's columns and statuary, carved from normally resistant sandstone.

Extreme weathering in tropical areas produces *oxisols* (named for their high concentration of insoluble iron oxides) or *ultisols* (for their *ult*imate, or most advanced, degree of soil development). In ultisols, even the ordinarily insoluble quartz has been dissolved away, leaving only the most insoluble elements, such as iron and aluminum. (Oxidized iron makes these soils dark red.) When these soils dry out, they are strong enough to serve as building stones, and are commonly used to build dwellings and other structures in tropical regions (Fig. 5-25); adobe is a fine-grained soil mixture traditionally used as a building material in the American Southwest and Mexico. Oxisols and ultisols are depleted of their potassium, calcium, sodium, and magnesium content and are poor prospects for agriculture. Raising crops with significant nutritional value in soils such as these, composed predominantly of insoluble iron and aluminum, requires advanced agricultural technology and intensive use of fertilizers. The crops of the tropics therefore tend to be the so-called "cash crops" with poor nutritional value, such as coffee, tobacco, sugar cane, palm oil, and cacoa (the prime ingredient in chocolate). While some of us may subsist on that diet, the most agriculturally useful soils are generally those that have weathered less and retain some beneficial nutrients.

Paleosols Sometimes previously buried soils are uncovered that have characteristics different from those of other soils in their regions, suggesting that they formed under different—in particular, ancient—conditions. Buried soils that predate modern soil formation are called **paleosols** ("old soils"). Examples include aluminum-rich bauxite deposits in Georgia, caliche horizons in Connecticut, and deep, highly weathered regolith in southwestern Minnesota. All these presumably formed under climates with very different temperatures and moisture conditions than prevail in those areas today. Similarly, oxisols and ultisols indicative of humid semi-tropical or tropical climates have been found buried beneath the soils forming today in the warm, dry climate of Australia. When we find paleosols, our ability to determine the climates under which they formed enables us to identify climate changes that have occurred in the geologic past.

Figure 5-26 A footprint left on the lunar surface by one of the Apollo II astronauts in July, 1969. Because there is no chemical weathering on the Moon, this footprint will remain for millenia.

Weathering in Extraterrestrial Environments

Weathering as we know it on Earth does not take place on our celestial neighbors. Long suspected, this has been confirmed by recent discoveries about the surface conditions of the Moon, Venus, and Mars. The reasons for the absence of weathering, however, differ in each case.

The Moon has no atmosphere; without atmospheric water, oxygen, or biological activity, there can be no chemical weathering. A mechanical weathering process—the impact of meteorites and micrometeorites—produced the Moon's regolith, which consists primarily of shattered bedrock and glassy fragments expelled from impact craters. The sharp edges of lunar craters, however, even the oldest ones, suggest the absence of Earth-like chemical weathering. Thus, the footprints left in the lunar dust by the Apollo II astronauts—the first to walk on the Moon—are likely to retain their freshness for millions of years (Fig. 5-26).

Venus has a surface temperature of about 475°C (900°F) and an atmosphere composed almost exclusively of CO_2, which traps heat radiating from the planet's surface in a *greenhouse effect.* High temperatures normally promote an increased rate of chemical reaction, but that of Venus is so high that it instantly evaporates every trace of water at the planet's surface. The subsequent absence of water on Venus prevents hydrolysis, carbonation, and oxidation, causing the Venutian landscape to appear remarkably unweathered chemically (Fig. 5-27). Mechanical weathering on Venus, as seen on recent radar images, is probably due to thermal expansion and contraction and exfoliation. High winds at the planet's surface may also be a factor.

Mars, of all planets in our solar system the one whose surface conditions most closely resemble those on Earth, has surface temperatures ranging from –108°C (–225°F) to 18°C (63°F). Mars' thin atmosphere consists largely of CO_2, with small amounts of nitrogen and water vapor. Because of the cool temperatures, most surface water exists as ice, and is generally unavailable for chemical reactions (although it may promote mechanical weathering by frost wedging); the lack of heat also reduces the rate of chemical reactions.

Figure 5-27 Angular rocks on Venus, photographed by the USSR's Venera-13 probe. The sharp, unweathered edges are testimony to the planet's apparent lack of chemical weathering.

Figure 5-28 An unenhanced color photo of the surface of Mars, taken by the United States' Viking 1 lander in July, 1976. The redness of the Martian regolith is most likely due to oxidation of iron-rich rocks and sediments.

There is, however, clear evidence of chemical weathering in Mars' past. The crimson color of the Martian landscape, which we can see easily with a high-quality telescope from a distance of 78,000,000 kilometers (47,000,000 miles), is believed to be due to the reddish iron oxides produced by oxidation of iron-rich bedrock, and soil analyses performed by the versatile Viking lander in 1976 confirmed the high concentration of iron oxides at the planet's surface (Fig. 5-28).

As we have seen, various mechanical and chemical weathering processes can convert solid bedrock to loose, transportable fragments and dissolved ions. Once liberated from their parent bedrock, these fragments and ions are free to move under the influence of such surface-shaping forces as gravity, streams and glaciers, wind, and coastal waves. The processes that transport weathered material, and their subsequent deposition and conversion to sedimentary rock, will be discussed in Chapter 6.

Chapter Summary

Solid rocks can be broken down by weathering, erosion, or, often, a combination of the two. **Weathering** is the slow but constant process whereby rocks are gradually broken down by environmental factors at the Earth's surface. **Erosion** occurs when rock fragments are transported and deposited elsewhere by moving water, wind, or ice; because erosion can only occur in environments in which these factors are present, it does not occur as consistently as does weathering. The products of weathering and erosion contribute to the Earth's sediment, the loose, fragmented geologic material that is the raw material for sedimentary rock.

There are two types of weathering: **mechanical weathering,** which results in the physical disintegration of rock into smaller pieces without changing its chemical composition; and **chemical weathering,** which changes the chemical composition of the weathered rock. Rocks and minerals having structures that are chemically unstable at the Earth's surface are most susceptible to chemical weathering, which changes them into substances that are more stable.

Mechanical weathering may be accomplished by: **frost wedging,** the expansion of cracks in rock as water in the cracks freezes and expands; salt-crystal growth within rock cavities, which forces the cavity walls farther apart; **thermal expansion** and contraction, the alternate enlargement and shrinking of rock as it is repeatedly heated and cooled; **mechanical exfoliation,** the fracturing and removal of successive rock layers that occurs as deep rocks expand upward after overlying rocks have eroded away; penetration of growing plant roots, which expands existing cracks in rock, and the activities of animals; and **abrasion** of transported particles as they collide with each other or with stationary rock surfaces. Rock fragments that fall from a weathered structure and collect at its base are **talus.**

Chemical weathering is largely controlled by climatic factors, such as temperature and, especially, the availability of water. The process of **dissolution,** most effective on soluble rocks such as limestone, occurs when minerals or rocks are decomposed by water or by reaction with naturally occurring acid and the products are carried off by water. **Oxidation,** the reaction of certain chemical compounds with oxygen, works most effectively on iron-rich rocks such as basalt and ultramafic peridotites. **Hydrolysis,** the replacement of major positive ions in minerals (particularly the feldspars) with H^+ ions from water, produces the most common products of chemical weathering, the clay minerals.

The rate at which a given rock or mineral will be chemically weathered depends on a variety of factors: climate (hot, moist regions exhibit more weathering than cold or dry regions); the activity of living organisms; length of time exposed to weathering; and the chemical stability of its components at the Earth's surface. (Minerals that crystallize at very high temperatures, such as olivine and pyroxene, are more readily weathered than cooler-crystallizing minerals such as quartz and mica.) Chemical weathering, the most notable products of which are the clay minerals and several economically valuable metal ores, has the physical effect of rounding previously angular boulders, a phenomenon known as **spheroidal weathering.**

Mechanical and chemical weathering together produce the Earth's **regolith,** the loose, fragmented material that covers much of the Earth's surface, and **soils,** the uppermost, organic-rich portion of the regolith. Soil development is governed by five factors: **parent material** (the bedrock or sediment from which a soil develops), climate, **topography** (the physical features of a landscape), vegetation cover, and time. A developing soil consists of distinct layers having different compositions, called **soil horizons;** the vertical succession of soil horizons in a given location is the location's **soil profile.**

Most temperate-zone soil profiles consist of the typical layers designated (from the surface down) the O, A, E, B, and C soil horizons. **Caliche** is a white carbonate layer, characteristic of desert environments, produced when water carrying dissolved calcium carbonate percolates down from the surface to a lower soil layer and then evaporates.

In recent years, soil classification has become increasingly precise as a guide to land-use decisions; soil scientists now distinguish ten different orders of soil within North America, for instance, whereas they used to distinguish only two. **Paleosols,** or "old soils," are buried layers of ancient soil that may contain evidence of a past climate different from that of today. For example, a paleosol underlying the modern soil in what is currently a moist temperate environment may contain a caliche layer, indicating that a warm arid environment prevailed in that location at some time in the past.

Weathering takes place on other planets in our solar system and on the Moon, although differently than it occurs on Earth. No chemical weathering occurs on the Moon because it has neither atmosphere (thus no O_2 or CO_2) nor surface water; frequent meteorite impacts weather lunar surface rocks mechanically. Because Venus' high temperatures instantly vaporize its surface water, little chemical weathering takes place there, although images of its surface regolith suggest that there is some mechanical weathering on that planet. Although most of Mars' surface water is now trapped in the ground as ice, the characteristic redness of Mars' surface suggests that in the past, climatic conditions did enable water to oxidize its iron-rich bedrock.

Key Terms

weathering (p. 127)
erosion (p. 127)
mechanical weathering (p. 128)
chemical weathering (p. 128)
frost wedging (p. 129)
talus (p. 129)
thermal expansion (p. 130)
mechanical exfoliation (p. 130)
abrasion (p. 131)
dissolution (p. 132)
oxidation (p. 133)
hydrolysis (p. 133)
spheroidal weathering (p. 138)
regolith (p. 139)
soil (p. 139)
parent material (p. 139)
topography (p. 140)
soil horizon (p. 142)
soil profile (p. 142)
caliche (p. 143)
paleosols (p. 145)

Questions for Review

1. Describe the fundamental difference between mechanical and chemical weathering.
2. Discuss three ways that rocks can be weathered mechanically.
3. Explain how the process of mechanical exfoliation works.
4. Of granite, limestone, and basalt, which would be most susceptible to the chemical weathering process of oxidation? To the process of dissolution? Which would weather to produce the most clay?
5. What role does climate play in chemical weathering?
6. Why is quartz a more common mineral in sandstones than plagioclase feldspars? (Hint: See Figure 5-15.)
7. Discuss how soils will vary in a region of irregular topography.

8. Describe the principal characteristics of the major soil horizons, and explain how those characteristics develop.

9. In what ways does the weathering environment of the Moon differ from that on Earth?

For Further Thought

1. Since the Industrial Revolution, we have been burning coal, heating oil, and gasoline at an ever-increasing rate. Combustion of these fuels manufactures carbon dioxide. What would you expect to be the effect of burning these fuels on weathering rates? Explain.

2. Describe what would happen to the physical condition of Cleopatra's Needle (p. 130) if the obelisk was returned to its original home in Egypt.

3. Imagine that the Earth some day becomes devoid of water. How would the nature of chemical weathering change in polar regions? In the arid subtropical deserts? In the equatorial tropics?

4. To the right is a photo of a soil developed on a lava flow in eastern Washington. Judging from the appearance of the soil, what rock type comprises the lava flow? What weathering processes and products are responsible for the color of the soil? Describe the climate that was most likely responsible for this type of weathering.

5. Look around your community, identify the major building stones, and compare their relative states of weathering. Did the local builders make wise choices when selecting their building materials, considering your local climate? Which rock(s) would work best for construction in your area? Which would be poor choices?

The Origins of Sedimentary Rocks

Classifying Sedimentary Rocks

Highlight 6-1 When the Mediterranean Sea Was a Desert

"Reading" Sedimentary Rocks

Figure 6-1 Layers of sedimentary rock (Navajo Sandstone) in Zion National Park, Utah. These rocks are composed of the cemented sand grains of ancient sand dunes. The fascinating "cross-bedded" patterns they exhibit are typical of windblown sands.

Chapter 6

Sedimentation and Sedimentary Rocks

Most of us live where we can't readily view a volcanic eruption or the movement of an active fault, but we probably can walk on a sandy beach or sit by a muddy stream. Even in the heart of New York City, we can watch rivulets of rain collect pebbles and dirt and deposit them into a corner drain. Much of that rainwater will find its way to the Atlantic Ocean, and some of those pebbles to a resting place on the Atlantic sea floor. No matter where you live, you don't have to travel far to find **sediment** (from the Latin *sedimentum,* or "settling"), fragments of solid material that, after being transported some distance by water, wind, or ice, or precipitating out of solution in water, settle down and accumulate, typically in layers (Fig. 6-1).

Sediment, consisting mostly of the products of erosion and weathering (discussed in Chapter 5), can accumulate virtually anywhere on the Earth's surface—from the glaciated summits of the Himalayas, 10 kilometers above sea level, to the deep trenches on the floor of the Pacific Ocean, 10 kilometers below sea level, and everywhere in between. Sediments are continually deposited in lakes, streams, deserts, swamps, beaches, lagoons, and caves, on continental shelves, and at the bases of glaciers throughout the world.

Most sediment is ultimately converted to solid **sedimentary rock.** Sedimentary rocks make up only a thin top layer of the Earth's crust, accounting for barely 5% of the Earth's outer 15 kilometers (10 miles), but they constitute 75% of all the rocks exposed at the Earth's land surface. Sedimentary rocks are our principal source of coal, oil, and natural gas, and much of our iron and aluminum ores; they also store nearly all of our fresh underground water, as well as our cement and other natural building materials.

These rocks also contain clues to the condition of the Earth's surface as it existed in the far-distant past. They record the former presence of great mountains in areas now monotonously flat, and of vast seas that once covered what is now the dry interior of North America. Some sedimentary rocks contain the fossil remains of past life, which tell us much about how the planet has evolved through its history. (Fossils will be discussed in detail in Chapter 8, "Telling Time Geologically.")

This chapter examines the origins of sedimentary rocks, describes how they are classified, and explains the ways in which geologists use them to reconstruct past surface environments. The chapter concludes by showing the relationship of various sedimentary rocks to common plate tectonic settings.

The Origins of Sedimentary Rocks

As we saw in Chapter 5, weathering breaks down rocks both mechanically and chemically. Mechanical weathering breaks down rocks into smaller fragments, without changing their chemical composition, freeing the fragments to be transported to other places. Chemical weathering converts minerals that are unstable in the weathering environment to new, more stable compounds through various chemical reactions; the products of these reactions are often transported away from where they form—either in solution (dissolved in water), or as solids carried by wind, water, or ice—and deposited elsewhere. Our concern in this chapter is what happens to the mobile products of mechanical and chemical weathering—how they are moved from one place to another, deposited as sediment at a new location, and buried by subsequent deposition, and how they eventually become new rocks.

Sediments are classified according to the source of their constituent materials (Fig. 6-2). **Detrital sediment** is composed of transported solid fragments, or *detritus,* of preexisting igneous, sedimentary, or metamorphic rocks. **Chemical sediment** forms from previously dissolved minerals that have either precipitated from solution in water or been extracted from water by living organisms and converted to shells, skeletons, or other organic substances (which are deposited as sediment when the organisms die, or discard their shells). The different types of sediments and the rocks they form are discussed in more detail later in this chapter.

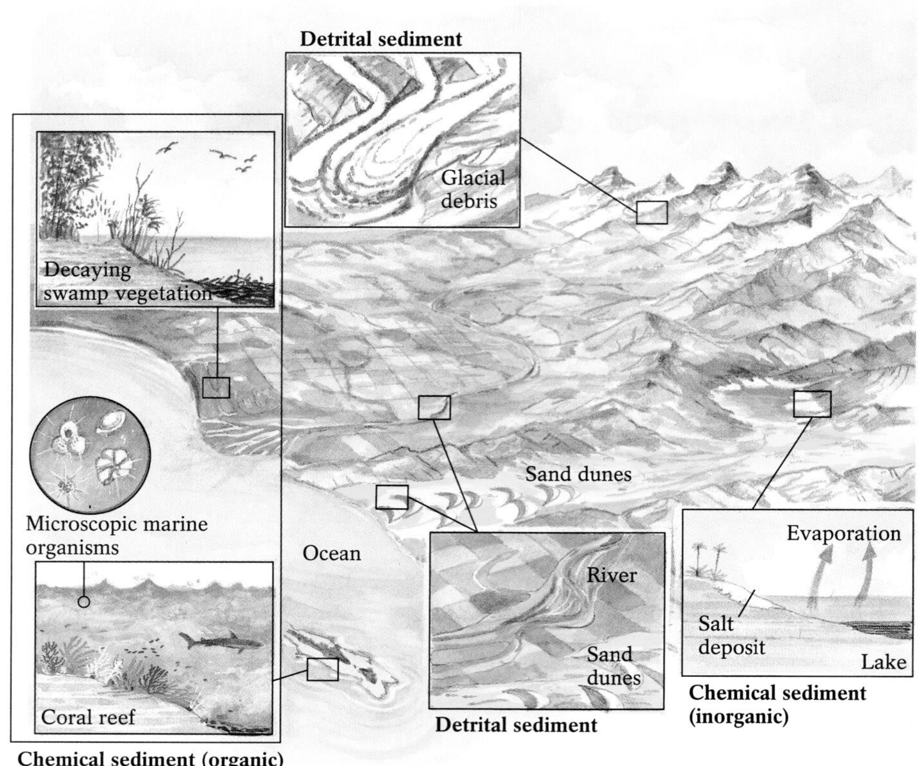

Figure 6-2 Various types of sediments and their origins. Detrital sediments consist of preexisting rock fragments, such as compose glacial debris or river-channel sand. Chemical sediments often consist of minerals precipitated directly from water, such as salt deposits produced by evaporation of small temporary lakes; chemical sediments may also be composed of organic debris, such as partially decayed swamp vegetation or the shells of small marine organisms.

Sediment Transport and Texture

The raw materials from which chemical sediments form—the dissolved products of chemical weathering processes—are transported by the water in which they are dissolved. These materials remain in solution until the temperature, pressure, or chemical composition of the water changes in ways that promote their precipitation, or until a living organism extracts them from solution to manufacture some type of biological structure, such as an outer protective shell or an internal skeleton. In the latter case, the materials are deposited (perhaps after being further transported by the movement of the organism) only after the organism eventually dies or discards its shell.

The vast majority of sediments, however, are detrital, composed primarily of the solid fragments produced by mechanical weathering or erosion of preexisting rocks. (Some detrital sediment may contain the undissolved solid products of chemical weathering.) During transport, detrital sediments are generally carried from high places to low places, largely by the pull of gravity but often with an assist from a transporting medium such as running water, wind, or glacial ice. The loose particles move until the transporting medium loses its capacity to carry them further, such as when a river ceases to flow upon entering relatively still marine water at a coast. At this point, the particles are deposited. Each year, an estimated 10 billion tons of detrital sediment, most of it carried by rivers, are delivered to the world's oceans.

Detrital *sediment texture* is determined by the size, shape, and arrangement of the detrital sediment particles. (Chemical sediments are distinguished by composition more than by texture.) If you've ever held a handful of beach sand or lake-bottom mud, you'll appreciate the variety of sediment textures. A sediment's texture depends on the source of the sediment particles and the medium that transported them.

Grain Size Rock fragments continue to be weathered during transport, so that they are generally reduced in size when they are finally deposited as sediment—rocks are commonly broken, crushed, and abraded while carried by turbulent streams, rasping glaciers, surf crashing against a coast, and violent desert windstorms. The extent to which rock fragments are worn down during transport depends in part on the nature of the parent rocks from which they were removed. Different rocks produce grains of different sizes, shapes, and resistances to weathering. Coarse-grained granite, for example, generally weathers to produce larger grains than those created by the weathering of fine volcanic ash.

Another major determinant of sediment grain size is the nature and energy level of the transport medium. For example, the same pebble that would be pulverized by a creeping glacier might remain unchanged by an oozing mudflow; it might be worn down by a white-water river, but be unaffected by a trickling stream.

Transport media determine the texture of sediment not just by their ability to weather rock fragments, but also by their ability to carry them at all. A raging, flooding river can transport large boulders along with tiny particles and all sizes in between; a gentle wind can carry aloft only minute grains. **Sorting** is the process by which a transport medium "selects" particles of different sizes, shapes, or densities. Wind is the most selective of the transport media; a deposit of windblown silt, with almost all of its particles within a narrow size range, is considered *well-sorted* (Fig. 6-3a). At the other extreme, glacial ice and flooding rivers are unselective, transporting particles with a wide range of sizes. Glacial deposits, which may contain the finest particles

(a)

(b)

Figure 6-3 Differential sediment sorting by transport media. (**a**) Because wind is highly selective of the particles it can transport, this windblown silt near Vicksburg, Mississippi, is limited to very fine sediment particles (i.e., is well-sorted). (**b**) Glaciers are capable of transporting sediment of all sizes. Their deposits, such as this one in Rocky Mountain National Park, Colorado, are typically very poorly sorted.

along with boulders the size of small office buildings, are said to be *poorly sorted* (Fig. 6-3b). Geologists can often determine the medium that transported a sediment by a quick visual estimate of its sorting.

When a moving current (wind or water) eventually loses energy and can no longer carry its suspended sediment load, the sediment particles settle out and drop to the Earth's surface. Because it takes more energy to carry them, the larger, heavier particles are deposited first whenever a current loses energy and slows down; the smallest, lightest particles are carried farthest and deposited last. For example, when a stream flows from a steep mountain slope onto the flat valley plain below, it loses much of its energy and drops its sediment fairly abruptly where the angle of the slope changes sharply; this produces a wedge-shaped body of poorly sorted sediment, called an *alluvial fan,* in which the coarsest grains are deposited first (nearest the foot of the slope) and the smaller grains soon after and slightly farther downstream (Fig. 6-4).

Grain Shape Particles released from rock by mechanical weathering may be jagged and angular, particularly if they originate as irregular grains in a plutonic igneous rock. As we saw in Chapter 5, however, abrading during transport wears off a grain's prominent points. Some transport media are particularly efficient in rounding particles. Swiftly flowing rivers, for instance, bounce pebbles and sand grains around vigorously, so that they collide with other particles and with the river bottom, becoming ever smoother and smaller. Glaciers, on the other hand, embed some of their sediment particles in hundreds of meters of ice, cushioning them from collisions. In general, the more vigorous collisions a particle experiences, and the farther it is carried from its parent rock, the more rounded it becomes.

Grains of the softer minerals, such as gypsum and calcite, become rounded more readily than harder ones, such as quartz. In a recent field study, fragments of soft sedimentary rock became well rounded after only 11 kilometers (6.6 miles) of stream transport. More durable fragments of granite, transported in the same stream, required 85 to 335 kilometers (53–208 miles) of transport to become comparably rounded.

Figure 6-4 An alluvial fan at the south end of Death Valley, in California. This mass of sediment was deposited by a Sierra Nevada mountain stream that, upon reaching the relatively flat Mojave Desert floor, lost the energy it had built up in flowing downslope and could no longer carry its accumulated sediment load.

Sedimentary Structures

Detrital sediments often contain **sedimentary structures,** physical features that reflect the conditions under which the sediments were deposited. For example, the upper surface of a layer of sand may display gentle undulating ripples, which indicates that the sediments were most likely deposited and shaped by flowing water or strong winds. Let's examine some common sedimentary structures and see how they can be used to interpret past environmental conditions.

Bedding (Stratification) **Bedding,** or **stratification,** is the arrangement of sediment particles into distinct layers (*beds,* or *strata*) having different sediment compositions and/or grain sizes. A clear break, or *bedding plane,* is generally visible between adjacent beds. Such bedding planes mark the end of one depositional event and the beginning of the next, usually caused by a change either in the nature of the sediment itself or in the energy with which it was transported. For example, a typical river bed contains distinct layers of different-sized sediments reflecting the river's various sediment-carrying capacities at different times: Interspersed between beds of medium-sized particles reflecting the river's usual sediment load, one could expect to find occasional layers of heavier, coarse-grained particles deposited after high-energy flooding episodes (the only times during which the river would be capable of carrying the heavier sediment). Similarly, a flooding river often deposits a particle load of heavy, coarse-grained sediment on top of finer preexisting sediments in the surrounding area; such a difference in grain size would be visible as a bedding plane in a cross-section of the resulting sediment layers (Fig. 6-5).

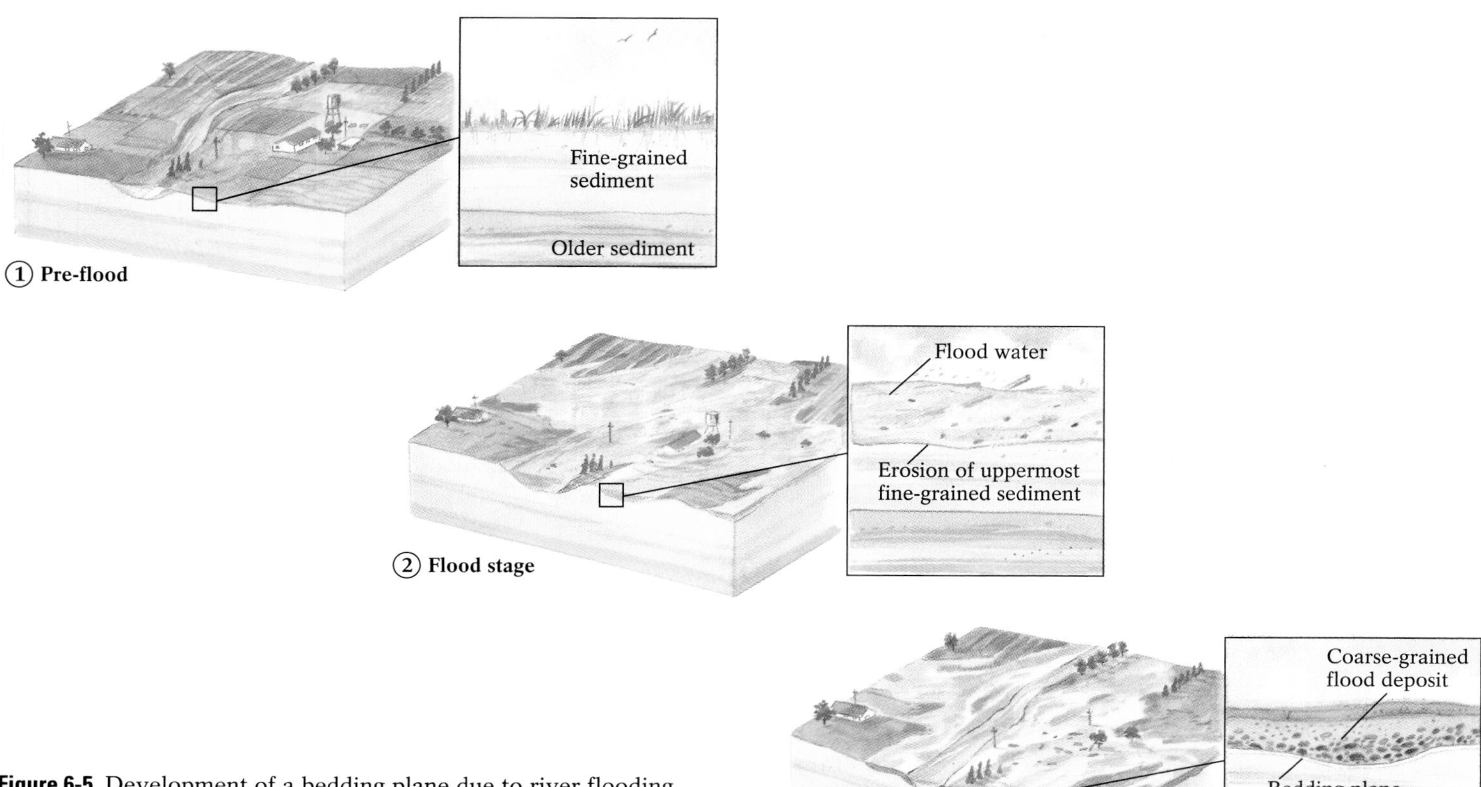

Figure 6-5 Development of a bedding plane due to river flooding. Any depositional event that leaves sediment that differs (either in grain size or in composition) from the preexisting sediment leaves a demarcation, a bedding plane, between the resulting sediment layers.

Graded Bedding When a sediment load containing a variety of sediment sizes is suddenly suspended in still water, its particles will settle at different rates depending on their sizes, densities, and shapes. This produces a **graded bed,** a single sediment layer (formed by a single depositional event), in which particle size varies gradually, with the coarsest particles on the bottom, and the finest at the top (Fig. 6-6a). (The principle of grading can be demonstrated by dropping a handful of unsorted backyard dirt into a tall glass of water—the largest particles will quickly settle to the bottom, while the finest will settle last and land on top of the larger particles.)

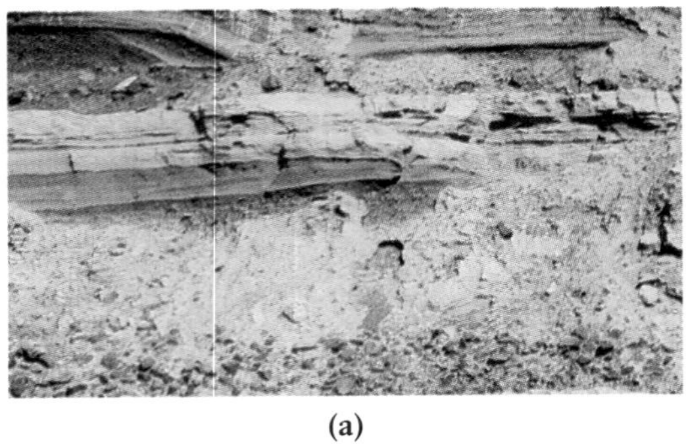

(a)

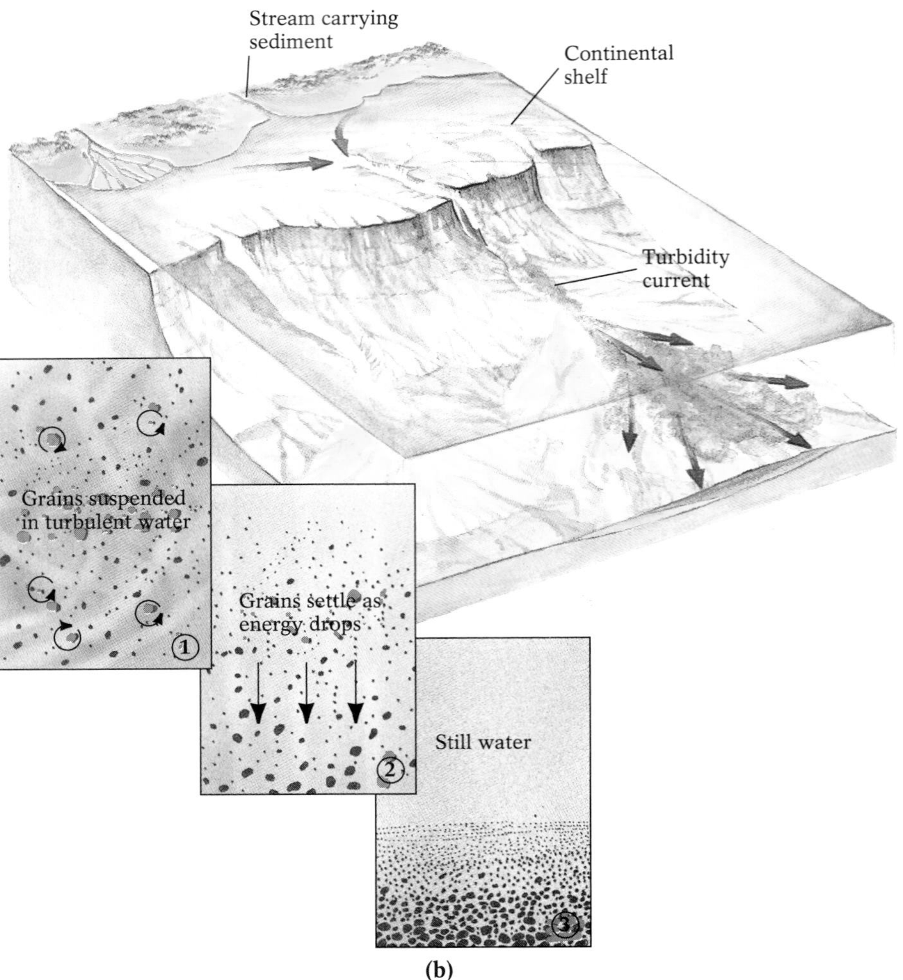

(b)

Figure 6-6 Graded bedding of sediment. (**a**) Graded beds form as particles of different density, size, and shape settle out of a standing body of water into distinct layers. The larger, heavier particles settle to the bottom first and the smaller, lighter particles settle above them. (**b**) Graded sediment is frequently produced by turbidity currents, offshore sediment flows that abruptly lose their energy and drop their particle loads on the ocean floor.

Graded sediments are commonly found on ocean floors, near shores where muddy streams shed their sediment loads onto the continental shelf (the underwater edge of a continent). Offshore sediment accumulates on a continental shelf as an unstable mass that can easily be dislodged (for example, by an earthquake). The result is a dense mixture of sediment and seawater called a *turbidity current* (from the Latin *turbidius,* or "disturbed") that flows rapidly downslope toward the deep-sea floor at 60 kilometers per hour (40 mph) or more (Fig. 6-6b). When a turbidity current reaches the horizontal ocean floor it slows to a virtual standstill, losing its transport energy and dropping its sediment load. The particles from this mixture settle to form a graded bed.

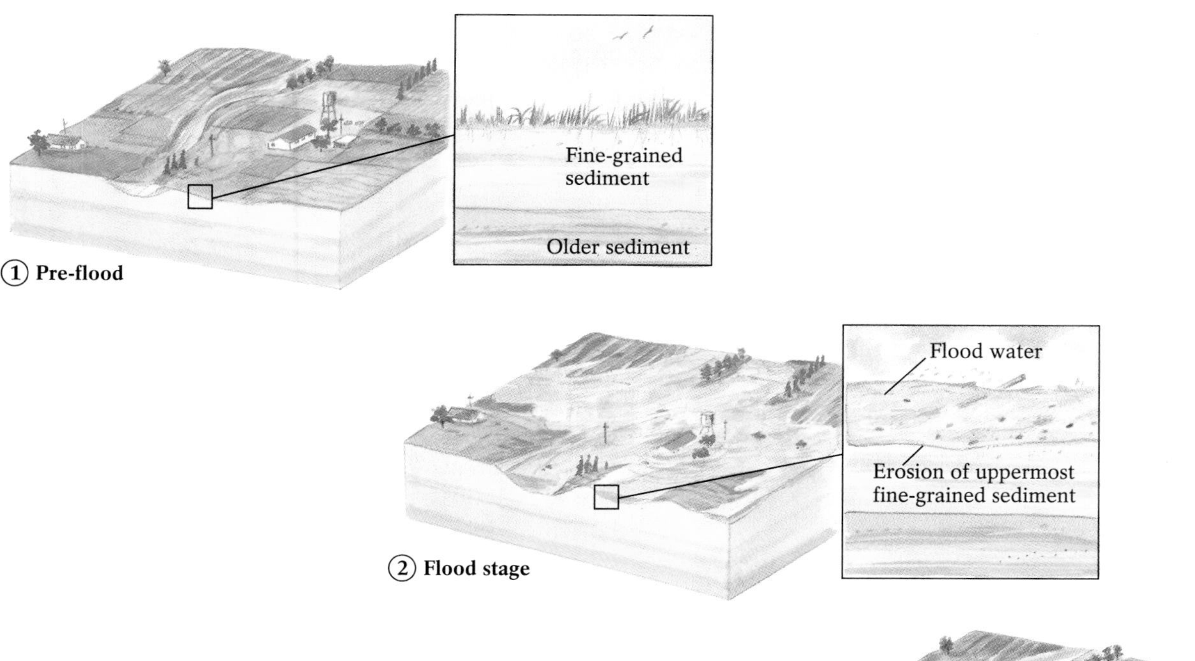

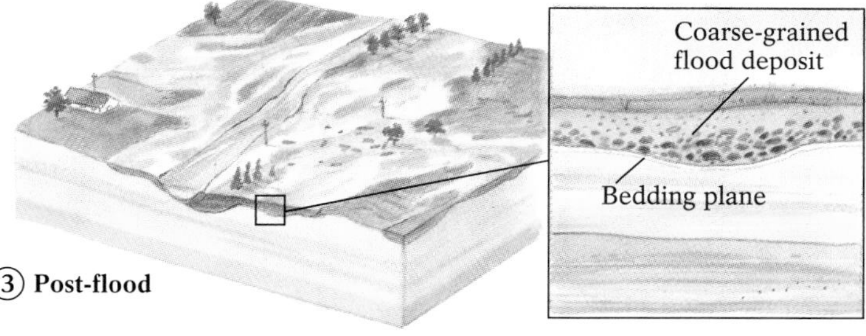

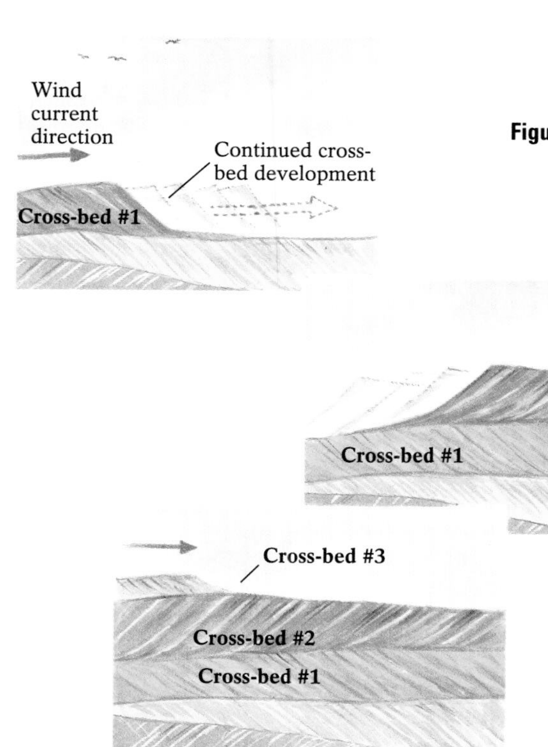

Figure 6-7 The development of cross-bedding in sand dunes.

Cross-Bedding **Cross-beds** are sedimentary layers deposited at an angle to the underlying set of beds. They are made up of particles that have been dropped from a moving current, such as wind or a flowing river, instead of settling through relatively still water or air to form horizontal beds. Cross-beds are often found in wind-deposited sand dunes (Fig. 6-7; see also Fig. 6-1) and in water-deposited ripples at a river's bottom. Since cross-beds always slope toward the downcurrent direction, they record the flow direction of the current that deposited them. The orientation of cross-beds may also help determine if a series of rocks has been overturned by tectonic forces (discussed in Chapter 9).

Surface Sedimentary Features The surface appearance of a layer of detrital sediment often provides clues to the environmental conditions to which it was exposed during or after deposition. For instance, a pattern of wavy lines, or **ripple marks,** preserved on top of a sediment bed indicates that wind or water currents shaped its particles into a series of shallow curving ridges after deposition. The configuration of these ridges, which are often visible on sandy surfaces, reflects the nature of the current that produced them (Fig. 6-8).

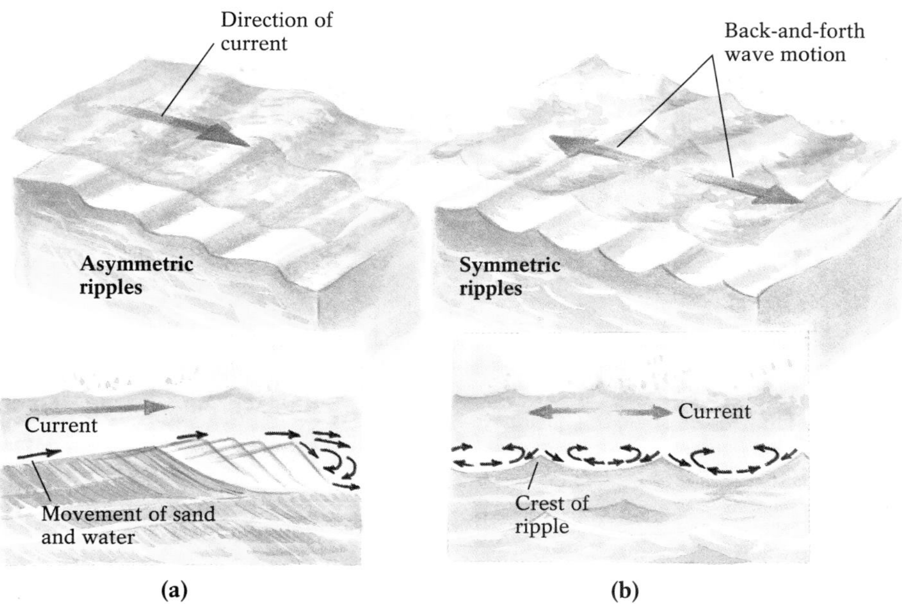

Figure 6-8 Different ripple patterns are produced by different types of currents. (**a**) A current that generally flows in one direction, such as a stream, produces asymmetric ripples: Sand grains roll up the gently sloping upstream side of each ridge and then cascade down the steeper downstream side. (**b**) Symmetric ripples form from the back-and-forth motion of waves in shallow surf zones at the coast, or at the water's edge in a lake. The sharp crest of this typè of ripple always points upward, and thus can be used to determine if a layer of sedimentary rock has been overturned.

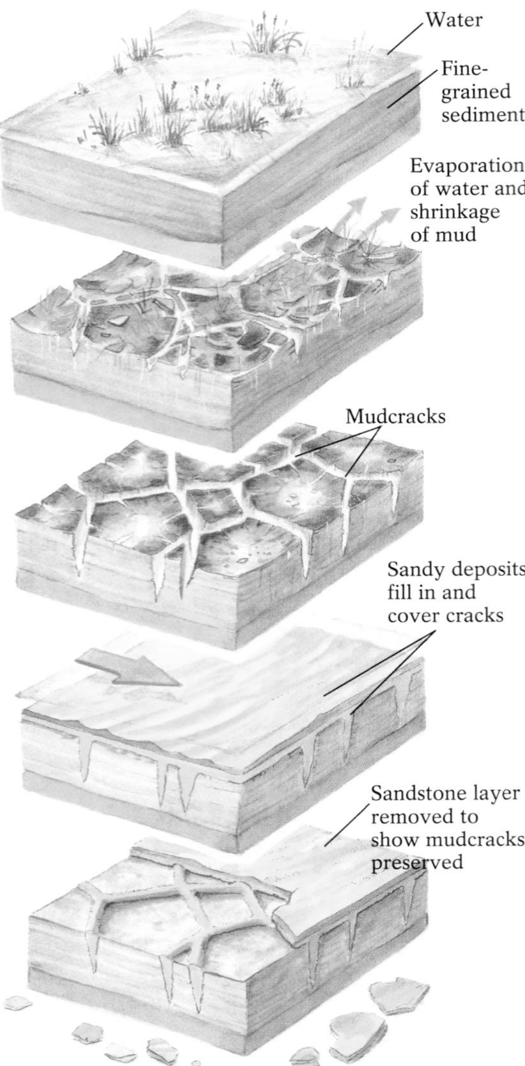

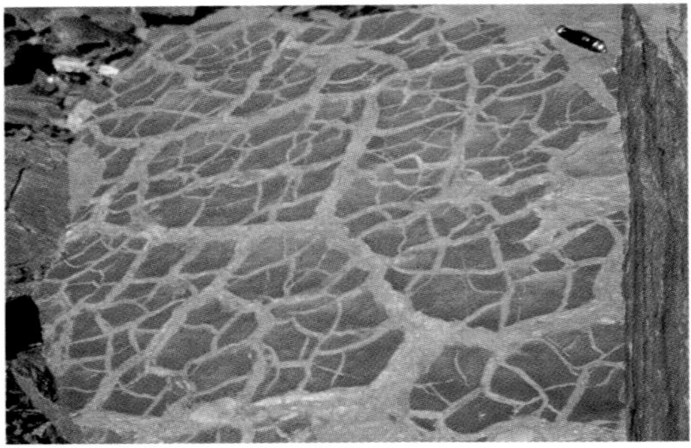

Figure 6-9 The origin of mudcracks. These mudcracks in oxidized red shale in Glacier National Park, Montana, suggest that a body of water once evaporated to dryness where we now find the Montana Rockies.

Mudcracks are fractures that develop when the surface of fine-grained sediment is exposed to the air, dries out, and shrinks (Fig. 6-9). Mudcracks indicate that the watery environment in which the sediment was deposited dried up at some point (as happens, for example, when shallow lakes evaporate). Because they form only at the *top* of a layer of muddy sediment, mudcracks also can be used to determine if a layer of sedimentary rock has been overturned.

Lithification: Turning Sediment into Sedimentary Rock

When a sediment layer is deposited, it buries all previous layers deposited at that location, so that a sedimentary pile may become thousands of meters deep. Sediments buried several kilometers or more beneath the Earth's surface retain heat (produced largely from the decay of radioactive mineral grains) and are compressed by the accumulation of overlying materials. They are also invaded by circulating underground water, which carries dissolved ions. Together, the heat, pressure, and the ions in water change the physical and chemical nature of both detrital and chemical sediments by a set of processes known as **diagenesis;** sometimes this results in the conversion of loose sediment into solid sedimentary rock, or **lithification** (from the Greek *lithos,* or "rock," and Latin *facere,* "to make").

Diagenesis differs from the intense heat- and pressure-related processes that take place in the Earth's interior and cause rocks to melt (Chapter 3) or metamorphose (Chapter 7), as it generally occurs within the upper few kilometers of the Earth's surface, at temperatures less than about 200°C (400°F). During lithification, sediment grains are packed more tightly together, the water between grains is squeezed from the sediment, and, often, chemical cement precipitates and binds the grains together. Diagenesis sometimes also involves the conversion of certain minerals into more stable forms.

If you squeeze a wad of wet clay in your hand, it will get smaller. This illustrates **compaction,** the diagenetic process by which the volume of a sediment is reduced by the application of pressure. As sediments accumulate, pressure from the increased weight of overlying material expels water and air from the spaces between deeply buried sediment grains and packs the grains more closely together. When fine-grained muds (particularly those composed of the clay minerals) are compacted in this way, weak attractive forces between the grains cause them to adhere to each other, converting loose sediment into more cohesive sedimentary rock.

Cementation is the diagenetic process by which sediment grains are bound together by materials originally dissolved during chemical weathering of preexisting rocks. When these dissolved materials eventually precipitate from water circulating through sediment, they cement the sediment grains together (Fig. 6-10). Coarse-grained sediments, such as gravels and sands, are more likely to be cemented than fine-grained sediments, such as silt and clay, because the larger spaces between particles can contain more water and therefore more dissolved materials.

The most common cementing agents include calcium carbonate, silica, and several iron compounds. Calcium carbonate is formed when calcium ions produced by chemical weathering of calcium-rich minerals, such as calcium plagioclase feldspar or calcium-rich pyroxenes and amphiboles, combine with carbon dioxide and water in soil. Silica cements are produced primarily by chemical weathering of the feldspars in igneous rocks. Iron oxides (such as hematite and limonite), iron carbonates (principally siderite), and iron sulfides (such as pyrite) also cement together the grains of coarse-grained sediments.

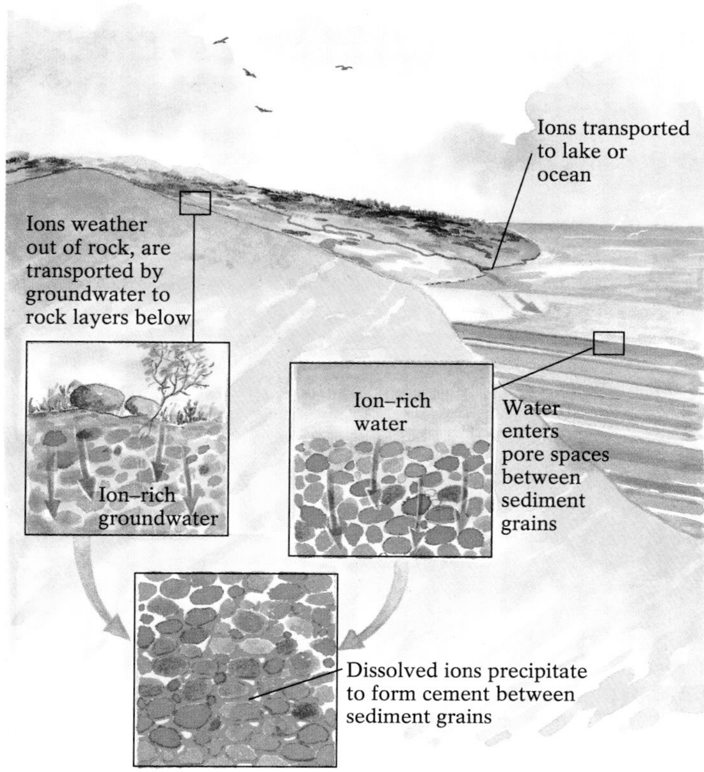

Figure 6-10 Lithification of sediment by cementation. Weathering of source rocks releases ions in solution, which then circulate via groundwater, lake water, or ocean water through coarse-grained sediments. When these dissolved materials precipitate as solid compounds within the spaces around sediment grains (called *pore spaces*), they form a cement that binds the grains together.

The Clinton Formation, an extensive sandstone unit in the southern Appalachian mountains of the southeastern United States, contains so much iron oxide cement that it once served as an important source of iron ore for the steel industry of nearby Birmingham, Alabama.

Compaction and cementation do not affect only rock grains. Because the dissolved products of chemical weathering on land are often ultimately transported to lakes and oceans, the same processes can lithify shells, shell fragments, and other remains of dead organisms that accumulate in these bodies of water. Any rock that consists of preexisting solid particles compacted and cemented together, be they preexisting rock fragments or organic debris, is said to have a **clastic** texture. (Thus, clastic rocks may be detrital or chemical, although they are most often the former.)

The increased heat and pressure associated with sediment burial also promote **recrystallization,** the development of stable minerals from unstable ones. One common mineral that recrystallizes is aragonite ($CaCO_3$), a polymorph of calcite that is secreted by many marine organisms to form their shells, but which in time recrystallizes as stable calcite. This transformation explains the absence of aragonite in ancient carbonate rocks.

Classifying Sedimentary Rocks

As we've learned, sedimentary rocks are generally classified as being either detrital or chemical, depending on their source material. Within each of these broad categories, however, a wide variety of rock types can be found, reflecting the diverse transport, deposition, and lithification processes by which they are formed.

Table 6-1 **Detrital Sediments and Rocks**

Particle Size (mm)	Name of Loose Particle	Name of Loose Sediment	Name of Rock Formed
>256	Boulder	Gravel	
64–256	Cobble	Gravel	Conglomerate (if particles are rounded)
4–64	Pebble	Gravel	Breccia (if particles are angular)
2–4	Granule	Gravel	
0.063–2	Sand	Sand	Sandstone
0.004–0.063	Silt	Silt (Mud)	Siltstone (Mudstone)
<0.004	Clay*	Clay* (Mud)	Shale (Mudstone)

1mm = 0.039 inch

*Note that the term "clay," when used in the context of sediment size, denotes very fine particles of any rock or mineral (as opposed to the term "clay mineral," which refers to a compositionally specific group of minerals); all clay minerals have clay-sized particles, but not all clay-sized particles are composed of clay minerals.

Detrital Sedimentary Rocks

Detrital sedimentary rocks are classified on the basis of their particle sizes (Table 6-1): Shales are the finest-grained; sandstones have grains of intermediate size; and conglomerates and breccias contain the largest grains. Note that all detrital rocks are clastic (i.e., consist of solid particles cemented together).

Conglomerates and breccias are composed of variously sized rock fragments, collectively called gravel; the other detrital sedimentary rocks consist mainly of smaller grains of the abundant rock-forming silicate minerals, such as quartz, feldspar, and the clay minerals. The sand- and silt-sized grains consist mostly of quartz, which, because it is relatively hard and chemically stable (and therefore resistant to chemical weathering), is more likely than other common minerals to survive the journey from its source rocks to its deposition site. Clay minerals, which naturally form fine, flat particles, dominate the fine-sized grains; because they are also chemically stable at the Earth's surface, they, like quartz, can survive the rigors of the Earth's weathering environment. Feldspar minerals, which are less stable at the surface, are present in sediment only if they have not been exposed to significant weathering (as might be the case in environments where chemical weathering is minimal, such as in cold, dry regions, or if they had been buried rapidly enough to be protected from chemical weathering).

Shales More than half of all sedimentary rocks are **shales,** the detrital sedimentary rocks containing the smallest particles (less than 0.004 millimeters in diameter). Because such fine particles only settle out of relatively still waters (in more energetic waters, they would remain suspended), most shales originate in lakes and lagoons, in deep ocean basins, and on river flood plains after floodwaters recede. The extremely fine particles in shales, which are so small that their mineral composition is best analyzed by X-ray diffraction

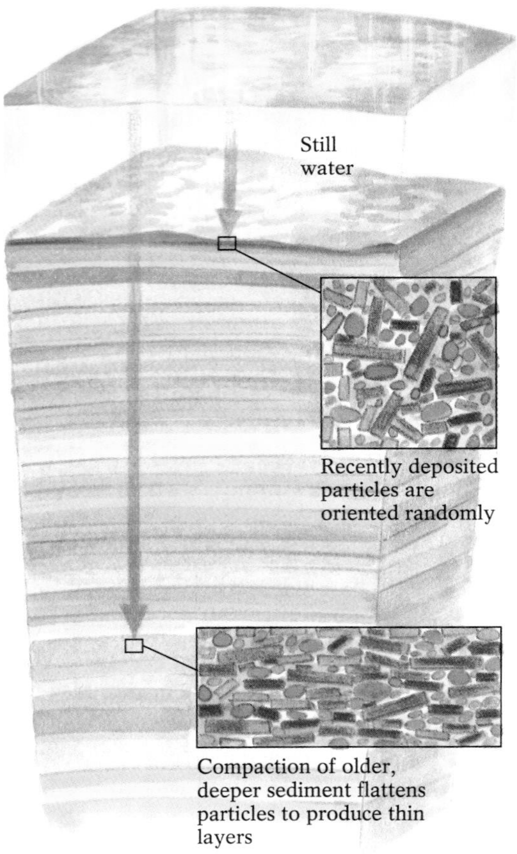

Figure 6-11 When flat or tabular clay and mica grains are initially deposited, they may be oriented randomly. The weight of subsequent deposits, however, causes these grains to "collapse" into a parallel orientation, producing the typical layered appearance of shales.

(a)

(b)

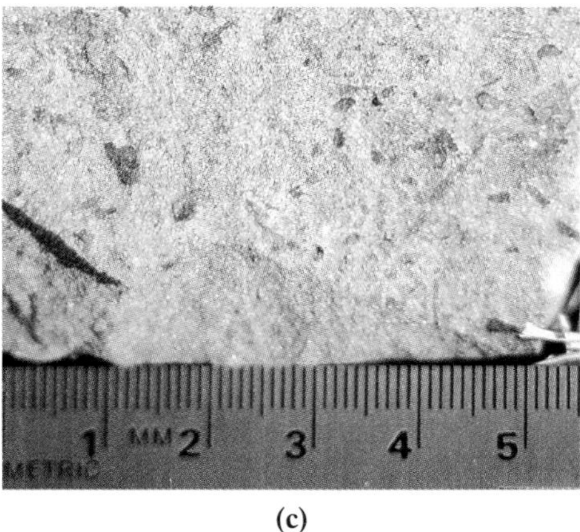

(c)

Figure 6-12 The three major types of sandstone. (**a**) Quartz arenite, composed predominantly of highly rounded quartz grains. One of North America's most celebrated quartz arenites is the St. Peter Sandstone of Minnesota, Iowa, and Wisconsin; it is most prominently exposed in Minneapolis–St. Paul, where the rock is so pure it has been mined, melted, and used to make glass. (**b**) Arkose, containing an abundance of angular feldspar grains. Some of North America's classic arkoses can be found in the Red Rocks area along highway I-70, west of Denver, Colorado. (**c**) Graywacke, distinguished by an abundance of dark volcanic fragments and relatively poor sorting of its particles. Good examples of graywacke can be found in the Ouachita Mountains of Oklahoma and Arkansas, and in the coastal mountains of California, Oregon, and Washington.

(discussed in Chapter 2), consist largely of clay minerals and micas. When these flat or tabular particles are buried beneath hundreds of meters of sediment, compaction flattens them into parallel layers resembling a deck of cards (Fig. 6-11).

A geologist will often place his or her tongue on bits of fine-grained sedimentary rock in order to distinguish the clay-rich shales from the slightly coarser quartz-rich siltstones. When moistened by saliva, the particles of the clay minerals feel smooth, whereas the abrasive quartz grains of siltstones feel noticeably gritty. Some are so dry that they stick to the tongue.

Shales vary considerably in color, depending on their mineral composition. Red shale contains iron oxides that precipitated from water containing both dissolved iron and abundant oxygen. Green shale contains iron oxides that precipitated in an oxygen-poor environment. Black shale forms in water with insufficient oxygen to decompose all of the organic matter in its sediment, leaving a black, carbon-rich residue; such conditions might occur in the still waters of a swamp, lagoon, or deep-marine environment, in which oxygen-rich surface water does not circulate.

Shales have numerous practical uses. They are a source of the clays used for making bricks and ceramics such as pottery, fine china, and tile. Mixing shale with calcium carbonate produces Portland cement, a staple of the construction industry. Oil shale (fine-grained rocks that contain abundant oil) may someday be a key source of energy. The economic value of these and other sedimentary rocks will be discussed in detail in Chapter 20.

Sandstones **Sandstones,** detrital sedimentary rocks whose grains range from 1/16 millimeter to 2 millimeters in diameter, make up about 25% of all sedimentary rocks. The mineral grains in sandstones are generally surrounded by silica or carbonate cement. There are three major types of sandstones, each with its own distinctive composition and appearance. A sandstone composed predominantly of quartz grains (>90%), with very little surrounding *matrix* (the finer material that occupies pore spaces between grains), is a *quartz arenite* (from the Latin *arena,* or "sand"). Quartz arenites are generally light in color, varying from white to tan depending on the color of their cementing agents (Fig. 6-12a). Their grains are rounded and well-sorted, suggesting that they were transported a long distance.

Arkoses (named for a French word that denotes a rock created by consolidation of debris) are distinctive pinkish sandstones containing more than 25% feldspar (Fig. 6-12b). The grains, usually derived from feldspar-rich granitic source rocks, are generally poorly sorted and angular, suggesting short-distance transport, minimal chemical weathering under relatively dry climatic conditions, and rapid deposition and burial.

Graywackes (derived from the German *wacken,* or "waste") are dark, gray-to-green sandstones that contain a mixture of quartz and feldspar grains, abundant dark rock fragments (often of volcanic origin), and a fine-grained clay and mica matrix (Fig. 6-12c). Their poor sorting, angular grains, and the presence of such easily weathered minerals as feldspar suggest that graywackes are deposited rapidly after short-distance transport.

Sandstone's durability has made it a popular building stone, used in the construction of Victorian brownstone houses as well as the gothic-style edifices found on many college campuses. Sandstones also hold much of the world's crude oil, natural gas, and drinkable groundwater, because they often contain a great deal of pore space between sand grains, which is easily saturated with migrating fluids.

(a)

(b)

Figure 6-13 Conglomerates and breccias. The grains in these coarse sedimentary rocks tell us much about their history. (**a**) The roundness of the grains in conglomerates suggests long-distance transport by an energetic transport medium. (**b**) The angularity of the grains in breccias suggests short-distance transport.

Conglomerates and Breccias **Conglomerates** and **breccias**, the coarsest of detrital sedimentary rocks, contain grains larger than 2 millimeters in diameter. In conglomerates, the grains are rounded; in breccias, they are angular (Fig. 6-13). Both contain fine matrix material, and both are typically cemented by silica, calcium carbonate, or iron oxides. The size of conglomerate and breccia grains makes it relatively easy to identify their parent rocks, and the shape of the grains provides clues to their transport path: The rounded particles in conglomerates suggest lengthy transport by vigorous currents; the angular grains in breccias suggest brief transport, as when shattered rock debris accumulates at the base of a cliff.

Chemical Sedimentary Rocks

Whereas detrital sedimentary rocks always consist of distinct fragments of preexisting rocks or minerals compacted and/or cemented together, chemical sedimentary rocks typically consist of an interlocking mosaic of crystals derived from dissolved compounds. There are two kinds of chemical sediments: *inorganic* sediments, which are precipitated directly from solution in water, and *organic,* which are produced by the biological activity of plants and animals. The various types of chemical sedimentary rocks are summarized in Table 6-2.

Table 6-2 **Chemical Sedimentary Rocks**

	Rock Name	Typical Composition
Inorganic	Inorganic limestone	Calcite ($CaCO_3$)
	Evaporites	Halite (NaCl), Gypsum ($CaSO_4 \cdot H_2O$)
	Dolostone	Dolomite ($CaMg(CO_3)_2$)
	Inorganic chert	Chemically precipitated silica (SiO_2)
Organic	Organic limestone	Calcium carbonate remains of marine organisms (e.g., algae, foraminifera)
	Organic chert	Silica-based remains of marine organisms (e.g., radiolaria, diatoms, sponges)
	Coal	Compressed remains of terrestrial plants

Inorganic Chemical Sedimentary Rocks Inorganic chemical sedimentary rocks form when the dissolved products of chemical weathering (Chapter 5) precipitate from solution, which typically occurs when the water in which they are dissolved evaporates or undergoes a significant temperature change. There are four common types of inorganic chemical sedimentary rocks, formed by three distinct processes: inorganic limestones and cherts, which precipitate directly from both seawater and fresh water; evaporites, which precipitate when ion-rich water evaporates; and dolostone, a rock whose origin is still the source of debate.

Limestones, the basic component of which is calcite, or calcium carbonate ($CaCO_3$), account for 10% to 15% of all sedimentary rocks. Most limestones forming today are organic, but under certain conditions limestone also precipitates inorganically, directly from an aqueous solution. Soluble materials usually dissolve at faster rates as water temperature increases—hot coffee, for example, dissolves sugar more quickly than cold. However, the solubility of calcium carbonate is proportional to the amount of carbon dioxide (CO_2) in the water, and warm water typically holds less CO_2 in solution than does cold water. In general, as water warms and the proportion of CO_2 in it decreases, calcium carbonate in the water becomes *less* soluble; thus, more calcium carbonate precipitates and is deposited as *inorganic limestone*. Conversely, as water cools and the proportion of CO_2 in it increases, more calcium carbonate dissolves and less inorganic limestone is produced.

The amount of $CaCO_3$ that remains in solution, however, is affected by factors other than water temperature that also influence the CO_2 content of water. Among these factors are agitation of the water, the presence of photosynthesizing plants, and water depth and pressure. For example, when water is stirred up or agitated, as by wave action, it loses CO_2 to the atmosphere, and therefore tends to precipitate calcium carbonate. Aquatic plants also remove CO_2 from water during photosynthesis, and therefore also promote precipitation of calcium carbonate. A decrease in water pressure allows CO_2 from the water to escape into the atmosphere; thus, because water pressure decreases with decreasing depth, calcium carbonate precipitation is greater in shallow than in deep water.

Inorganic limestone precipitates most readily when several of these factors act together, such as along the Grand Bahama Banks, a shallow submarine platform separated from Florida by the Straits of Florida. Here, relatively pure carbonate muds accumulate as carbonate-rich marine waters wash across broad, shallow continental shelves within 30° of the equator—conditions that combine warm water temperatures, breaking waves and strong currents, and abundant marine plant life to remove CO_2 from solution. As calcium carbonate precipitates in such a setting, it coats sand grains on the sea floor; successive coats of the mineral form concentric layers around the growing grains as tidal currents roll them back and forth along the ocean floor (Fig. 6-14). The result is a bed of spheres called *ooliths* (derived from the Greek *oo*, or "egg," a reference to the sediment's resemblance to fish eggs) about 2 millimeters in diameter. The chemical sedimentary rock formed from this sediment is *oolitic limestone*.

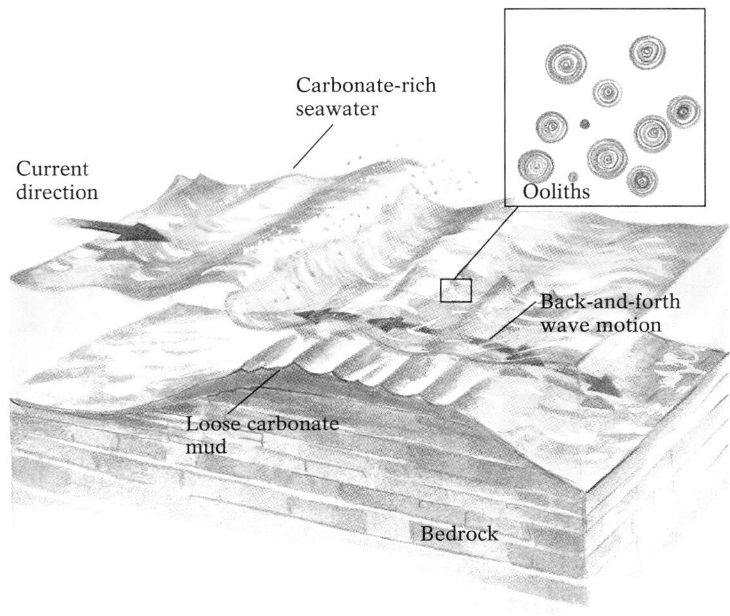

Figure 6-14 The formation of ooliths on a tropical carbonate platform. Calcium carbonate precipitates and coats sand grains as they are rolled along the sea floor by currents, forming spheres called ooliths (shown above). The resulting inorganic chemical sedimentary rock is oolitic limestone.

Inorganic limestone also forms on land, in several geological settings. *Tufa* is a soft, spongy inorganic limestone that forms where underground water emerges at the surface as a natural spring. At the surface, the water encounters a low-pressure environment, warms in the sunshine, and bubbles out in an agitated fashion. All of these factors promote the loss of CO_2, hastening carbonate precipitation. Inorganic limestone in the form of *travertine* forms in caves when droplets of carbonate-rich water on the ceilings, walls, and floors lose CO_2 to the cave atmosphere and precipitate carbonate rock. (See Chapter 16 for a closer look at these cave features.) Both tufa and travertine are used as decorative stones, popular in the design of such public places as banks, office building lobbies, and railroad station waiting rooms.

Evaporites are inorganic chemical sedimentary deposits that accumulate when salty water evaporates (Fig. 6-15). On average, the world's seawater contains, by volume, almost 3.5% dissolved salts. Where marine water is shallow and the climate warm, evaporation increases the concentration of these salts. When evaporation of seawater exceeds the inflow of water, solid crystals precipitate and begin to accumulate at the sea bottom.

Figure 6-15 Evaporite deposits at the Bonneville Salt Flats, west of Salt Lake City, Utah. Modern Great Salt Lake is a small remnant of the much larger Lake Bonneville, a vast lake that existed in Utah about 15,000 years ago when the local climate was cooler, cloudier, and more humid than it is today. Salt deposits such as these formed as most of Lake Bonneville was evaporated by the modern climate, which is warm, clear, and dry.

Evaporites precipitate in an orderly sequence, with those that are the least soluble in water crystallizing first; the most soluble stay in solution until little water remains. Relatively insoluble gypsum ($CaSO_4 \cdot 2H_2O$), for example, begins to precipitate when about two-thirds of a volume of salt water has evaporated. More soluble salts, such as the common table salt halite (NaCl), require about 90% evaporation. The most soluble salts, such as sylvite (KCl) and magnesium chloride ($MgCl_2$), precipitate only after more than 99% of the water has evaporated, and thus are rarely found in evaporites.

Because evaporites dissolve so readily in water, they are rarely evident in moist climates, such as those in the United States' rainy Northwest or humid Southeast. Gypsum is found at the surface in Colorado, Wyoming, and the arid Southwest, where it forms the dunes of White Sands National Monument, New Mexico. *Subsurface* salt deposits, however, underlie much of the rest of the continental United States (in the East, Midwest, Great Plains, Rocky Mountains, and Southwest), wherever inland seas existed in the

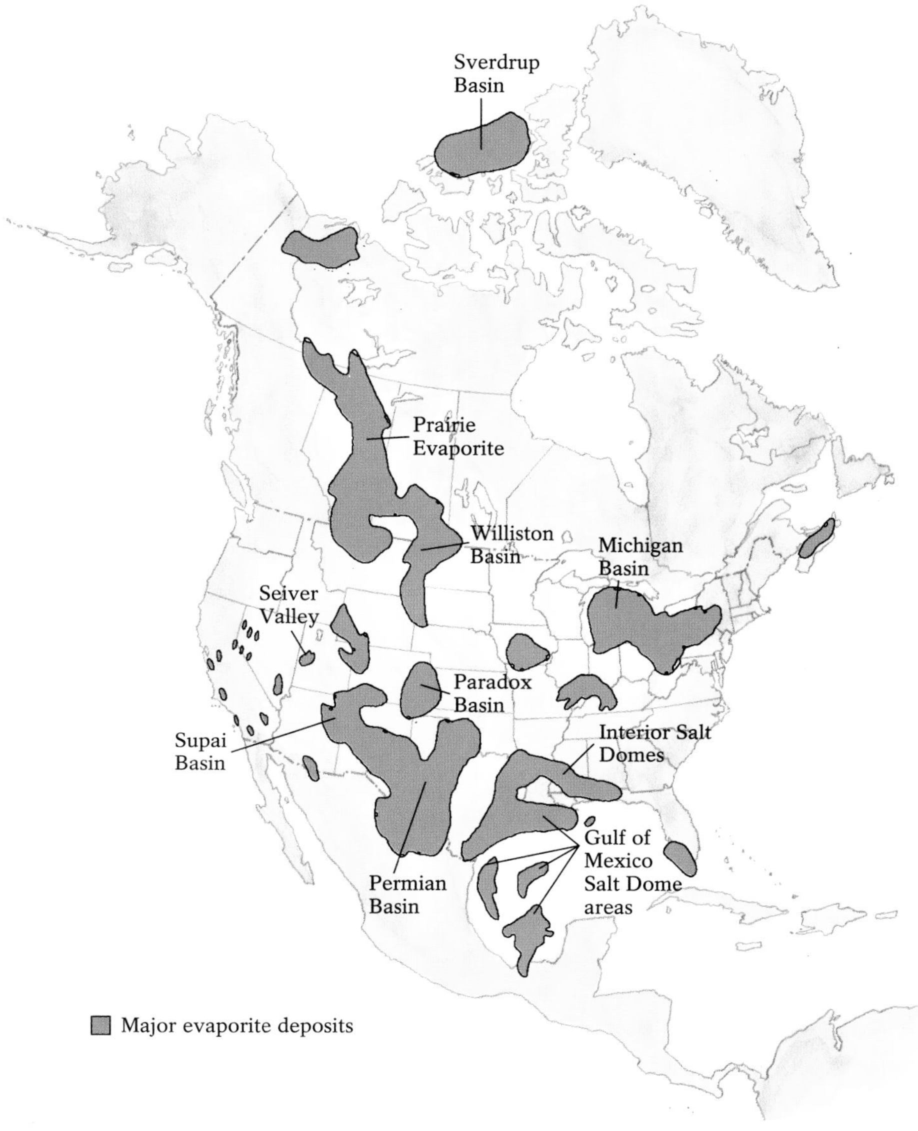

Figure 6-16 The location of major subsurface evaporite deposits in North America. These were produced in the ancient past when, during times of high sea level, salty marine water invaded topographic low spots on the North American continent. The shallow seas thus produced later evaporated during periods of climatic warming, leaving behind thick evaporite deposits. For example, about 400 million years ago, evaporating seas occupying what is now Michigan, Ohio, West Virginia, Pennsylvania, New York, and Ontario produced great deposits of gypsum and halite. Most such deposits were subsequently covered by younger sedimentary rocks.

past (Fig. 6-16). The thickness of salt deposits such as those in central Michigan, which exceed 750 meters (2475 feet), suggests that these seas were vast. If 300 meters (1000 feet) of average seawater evaporated today, only about 5 meters (15 feet) of salt would remain. In order to produce the volume of salt found in Michigan, an ocean 1000 kilometers (600 miles) deep with today's salt content would have had to evaporate completely. It's possible that ancient oceans were saltier than oceans today, but no single ocean could have contained enough salt at one time to account for such a thick deposit. Thus, Michigan's salt probably resulted from long-term precipitation of gypsum and halite, either from a succession of ancient shallow inland seas that periodically evaporated and then were refilled, or from continuous partial evaporation and refilling of a single long-lasting inland sea. One case in which an entire ocean apparently completely evaporated is recounted in Highlight 6-1.

Highlight 6-1 *When the Mediterranean Sea Was a Desert*

The climate in the lands bordering the Mediterranean Sea, such as Morocco, Algeria, Libya, and Greece, can be oppressively hot. Every year this heat evaporates more than 4000 cubic kilometers (960 cubic miles) of the Mediterranean's water, and only about 400 cubic kilometers (96 cubic miles) are replaced by rainfall and inflowing rivers. (If these two were the only means of maintaining its water level, the Mediterranean would be completely dry in about 1000 years.) Today, a massive inflow of Atlantic Ocean water through the Strait of Gibraltar balances the Mediterranean's water deficit, but this has not always been the case.

Deep-sea exploration in the early 1970s revealed a massive evaporite layer more than 2000 meters (6600 feet) thick beneath the floor of the Mediterranean Sea. This is more than 25 times the thickness that would accumulate if the current Mediterranean evaporated to complete dryness. What past conditions could have cause such a great salt accumulation?

One hypothesis proposes that these deposits formed when a barrier restricted water circulation between the Atlantic and the Mediterranean (Fig. 6-17). According to this hypothesis, ongoing convergence between the African and Eurasian plates, which meet beneath the Mediterranean, had by about 8 million years ago gradually raised the sea floor in the area of what is now the Strait of Gibraltar, creating a natural limestone dam that increasingly blocked the inflow of Atlantic waters. The Mediterranean rapidly evaporated as the replenishing water supply diminished. With less water to moderate the semi-tropical heat, temperatures may have risen to 65°C (150°F), further hastening evaporation and producing desert-like conditions. Occasional pulses of Atlantic water over or through the Gibraltar barrier, like a leaky faucet, may have provided the now-dry basin with a periodic supply of salty water which, upon evaporating, added to the thick Mediterranean salt deposits.

Analysis of sediment samples from the Mediterranean's floor indicates that evaporites ceased to be deposited there about 5.5 million years ago, perhaps when a break in the limestone dam that had been letting in just a trickle of Atlantic water grew to become a spectacular waterfall spanning what is now the Strait of Gibraltar—a force that would have removed the barrier completely by about 4 million years ago. A flow 100 times the volume of Niagara Falls would have been required just to balance regional water loss by evaporation and begin to refill the empty Mediterranean basin. To provide enough water to support the sea life of the time (the extent and makeup of which we know from the fossil record), a flow in excess of 1000 Niagaras would have been required, and, even at that flow rate, it would have taken more than a century to refill the basin to its present level. The other oceans of the world, in contributing the water that refilled the Mediterranean, probably fell about 10 meters (35 feet) over that time.

The Rock of Gibraltar is a remnant of the ancient limestone dam, a reminder of the time when an intermittently dry Mediterranean basin separated Europe and Africa. The Mediterranean today never evaporates even to the point of gypsum deposition. But if the African and Eurasian plates continue to converge, a new rocky barrier to Atlantic water inflow may once again rise from the sea floor and the Mediterranean basin may again become dry and salt-filled.

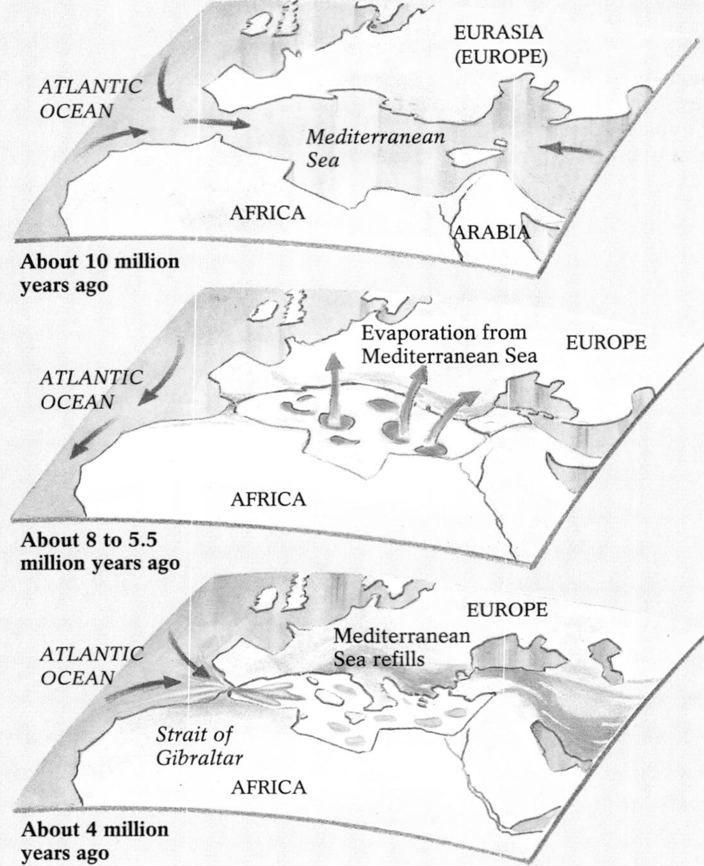

Figure 6-17 The thick evaporite deposits that underlie much of the Mediterranean basin today probably accumulated during a time (8–5.5 million years ago) when a topographic barrier stretched across what is now the Strait of Gibraltar, cutting off the water influx from the Atlantic Ocean. With no water supply to replenish it, the Mediterranean Sea evaporated almost completely, precipitating vast amounts of salt.

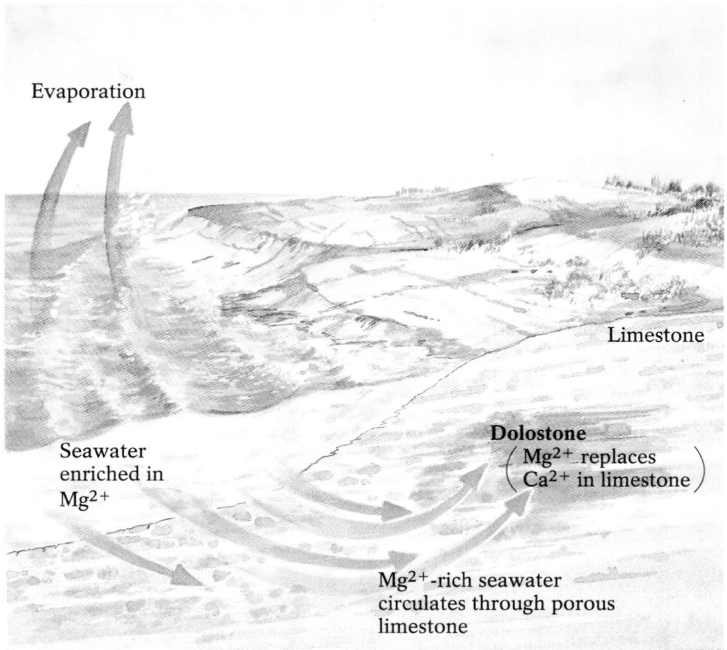

Figure 6-18 One hypothesis for the formation of dolostone: Seawater enriched in magnesium ions (which increase in concentration as the water evaporates) circulates through porous limestone; dolostone forms as the magnesium ions replace calcium ions in the limestone.

Dolostone is rock composed of the mineral dolomite, a calcium-and-magnesium carbonate ($CaMg(CO_3)_2$). Similar to limestone in appearance and chemical structure, dolostone is believed to form when magnesium ions replace some of the calcium ions in a preexisting body of limestone. This can occur when ocean water evaporates to produce a magnesium-rich solution that then circulates through a limestone bed; dolostone production is particularly likely wherever a tropical or semi-tropical climate promotes evaporation of salty marine water near a porous body of limestone (Fig. 6-18).

Although thick dolostone layers are common in ancient sedimentary rock, the mineral dolomite is forming in only a few places today. In the Persian Gulf states, in the Florida Keys, and in the Bahamas, a thin crust of dolomite crystals can be found above the low-tide level on many limestone bodies. Is this the start of the process that produced the thick dolostone beds of ancient sediments? It could be, but the extent of those sediments suggests that evaporative conditions were more widespread in the past than today. Most dolostones apparently formed when worldwide temperatures were warm and sea levels high, and flooding over low-lying areas of continents created shallow inland seas that were susceptible to evaporation. Thick dolostone deposits in La Crosse, Wisconsin, Champaign, Illinois, and Bloomington, Indiana, for instance, suggest that these locales may have once been the shores of tropical inland seas.

Chert is the general name for a group of sedimentary rocks that consist largely of silica (SiO_2) and whose crystals can be seen only through a microscope. Although many cherts have organic origins, the most common are inorganic, forming as a chemical precipitate from silica-rich water. (One type, jasper, owes its reddish color to the presence of an additional small amount of iron oxide.) *Chert nodules,* fist-sized masses of silica commonly found in bodies of limestone and dolostone, are believed to form when portions of these rocks are dissolved by circulating groundwater and replaced by precipitated silica. Because chert is easy to chip and forms sharp edges, it was often shaped by early humans into weapons, cutting blades, and other tools (Fig. 6-19).

Figure 6-19 Stone axes made from chert some 150,000 years ago. Because of the sharp edge that forms when chert is chipped, our ancestors made many of their tools from this hard, silica-rich rock.

(a)

(b)

Figure 6-20 (**a**) Chalk, shallow-marine limestone formed from the carbonate secretions of countless microscopic organisms. (**b**) Accumulation of chalk can eventually produce deposits of impressive size, such as the famed White Cliffs of Dover in England. The cliffs are composed mainly of the skeletons of microscopic marine plants and animals that accumulated about 100 million years ago, when the global sea level was apparently higher and coastal England was under water.

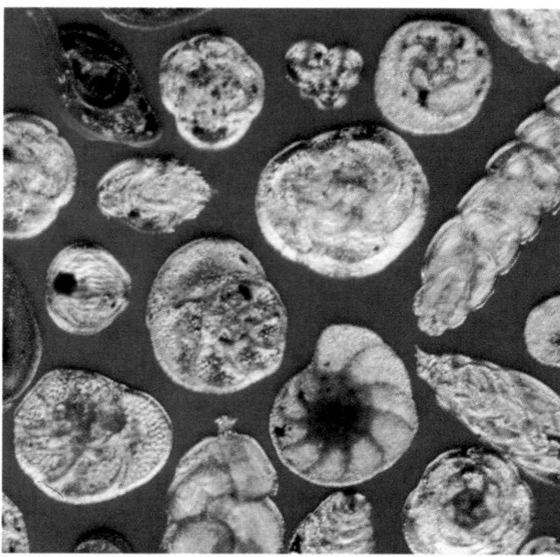

Figure 6-21 Foraminifera, microscopic marine animals whose calcium carbonate shells are an important component of deep-marine limestone. (Magnified 115×.)

Organic Chemical Sedimentary Rocks Organic chemical sedimentary rocks are derived in some way from living organisms. The principal rocks of this type are organic limestones and cherts, composed largely of the skeletal remains of marine animals and plants; and coal, formed from the remains of terrestrial plants. In each case, organic-based chemical sediments are subjected to the same diagenetic processes (compaction, cementation, and recrystallization) that produce the detrital and inorganic chemical rocks.

Nearly all organic **limestones** consist of calcite ($CaCO_3$) that formed in a marine environment. The calcium carbonate shells and internal skeletons of numerous ocean-dwelling organisms form from the $CaCO_3$ in seawater, which is nearly saturated with calcium ions produced by chemical weathering. (The ultimate source of the calcium ions is common calcium-rich minerals such as calcium feldspar, augite, and hornblende, weathered from rocks on land and transported to the sea by rivers and streams.) When these organisms die, their hard parts settle to the sea floor, where they accumulate in great thicknesses and then lithify as clastic organic limestone.

Most organic limestones form in shallow water along the continental shelves of equatorial landmasses, where warm water, plentiful sunlight, and abundant nutrients enable marine life to flourish. In the waters around most Caribbean islands, the Florida Keys, and the east coast of Australia, for instance, live an abundance of microscopic algae that secrete needle-like calcium carbonate casings. As these organisms die, a rain of delicate calcite stalks and branches—so small that 600,000 laid end to end would span this line of type—descends to the shelf floor and accumulates to produce a calcite-rich mud. Other algae and tiny carbonate-secreting animals living within the mud add to the growing volume of carbonate sediment, which appears as a soft white organic deposit called *chalk* (Fig. 6-20).

Organic limestone also forms in deep-marine environments. Although few carbonate-secreting organisms live at the cold, lightless, deep-sea floor, limestone-producing calcium carbonate is supplied by microscopic animals called *foraminifera,* or "forams," that swim and float in the upper 50 meters (160 feet) of the sea (Fig. 6-21). When the forams die, their protective shells accumulate on the sea floor and mix with marine mud to form a deep-sea *ooze.* Once an ooze is buried by subsequent sedimentation, its clays and carbonates become cemented and recrystallized as organic limestone.

Organic **chert** is a chemical sedimentary rock composed mainly of silica from the remains of a variety of marine organisms. Most often, when chert is found in layered beds (rather than the nodules characteristic of inorganic cherts), it is believed to be of organic origin (Fig. 6-22a). One variety of chert, called *flint,* gets its dark gray-to-black color from the presence of carbon-rich matter. Microscopic examination of organic cherts such as flint reveals the remains of silica-based organic debris, such as the shells of single-celled animals called *radiolaria,* the internal structures of single-celled plants called *diatoms,* and the skeletons of larger, more complex animals such as marine sponges (Fig. 6-22b).

Coal is an organic sedimentary rock composed largely of plant remains. Original plant structures, such as fragments of leaves, bark, wood, and pollen, can often be seen in a lump of coal under magnification, and sometimes even with the naked eye. Vegetation generally decomposes quickly at the Earth's surface, but in an environment with little oxygen it may be preserved until it is buried and converted to coal. The stagnant water of a warm, lushly vegetated swamp is ideal for the production of coal because the typically windless conditions produce calm water surfaces, and relatively little atmospheric oxygen circulates down to reach the organic debris accumulating at the swamp's

(a)

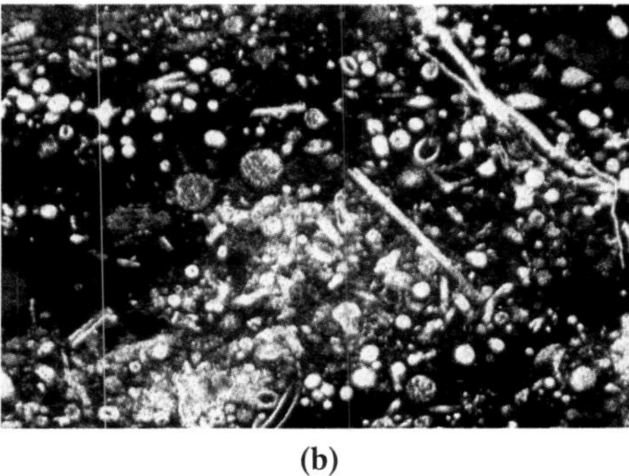

(b)

Figure 6-22 Organic chert, found in the form of layered beds (**a**), forms from accumulation and lithification of the silica-based remains of marine organisms (shown here magnified 20×) (**b**).

bottom; any oxygen that is dissolved in swamp water is quickly depleted by plant decay, leaving the remaining vegetation to accumulate undecayed.

Over millenia, increasing pressure from the weight of overlying sediments expels water, CO_2, and other gases from the accumulating mass of vegetation, and the proportion of carbon in the plant residue is increased. Early in the process, when much of the original plant structure still remains intact, a soft brown material called *peat* is produced; increased heat and pressure create increasingly harder and more compact forms of coal, from soft brown *lignite*, to moderately hard *bituminous* coal, to dense, lustrous *anthracite* (Fig. 6-23).

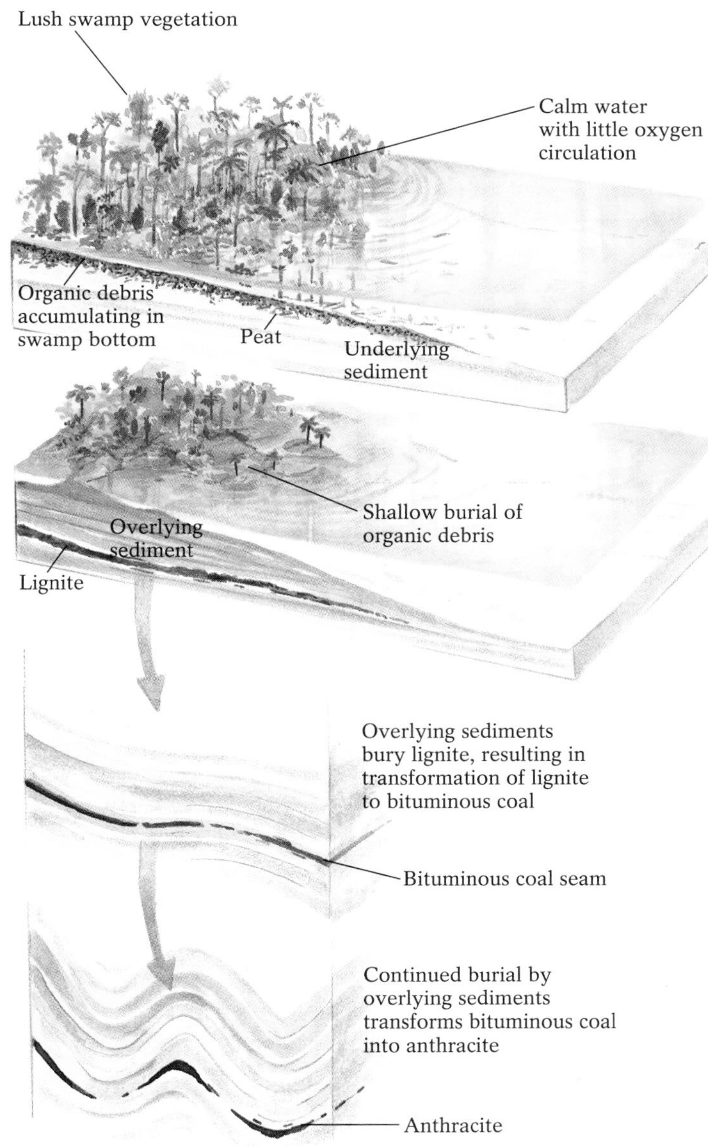

Figure 6-23 The formation of coal from swamp deposits. Abundant organic debris accumulates on the swamp floor and is buried before decaying in the oxygen-poor swamp water; the weight of subsequent deposits and the increased temperatures at greater depths change the debris to progressively harder forms of coal. Above: A body of coal (a "coal seam") exposed along a California–Nevada highway.

Because swamps are created and maintained by water from nearby seas, they are periodically submerged under rising sea levels. When this occurs, plant debris ceases to accumulate and is replaced with sediments typical of marine environments. When the sea level eventually falls again, the swamp is restored and accumulation of plant debris resumes. The cyclical nature of the deposition in such a changeable environment is reflected in the fact that coal beds are generally found alternating with detrital sedimentary rock layers.

Much of North America's coal developed during two principal periods in its geologic past: about 280 million years ago, in what are now the coal-mining regions of the Appalachians in Pennsylvania and West Virginia; and about 75 million years ago, along what are now the bituminous-producing formations of the Rockies and throughout the plains of Montana, Wyoming, North Dakota, and Saskatchewan. Where are North America's future coal deposits? The warm, richly vegetated candidates include the so-called Dismal Swamp of coastal North Carolina, the barrens of the Florida Everglades, and the colorful bayous of Louisiana.

"Reading" Sedimentary Rocks

A sedimentologist can analyze sedimentary rock formations in an area, study their fossils and sedimentary structures, and determine the depositional environment that produced each formation. Ultimately, the region's unique geologic history can be deduced—the sequence of events, such as the rise and fall of sea levels, and possibly even the past plate tectonic settings that determined the present nature of the region.

Sediment Deposition Environments

Sediment can be deposited at virtually any spot on the Earth's surface—from atop the highest peak to the depths of the ocean. Figure 6-24 shows just a sampling of the diversity of depositional environments. These **sedimentary environments** may be *continental* (on a landmass), *marine* (at sea), or *transitional* (in the zone in between). In this section, we will discuss the principal characteristics of these three categories; in later chapters, we will examine the individual environments in greater detail.

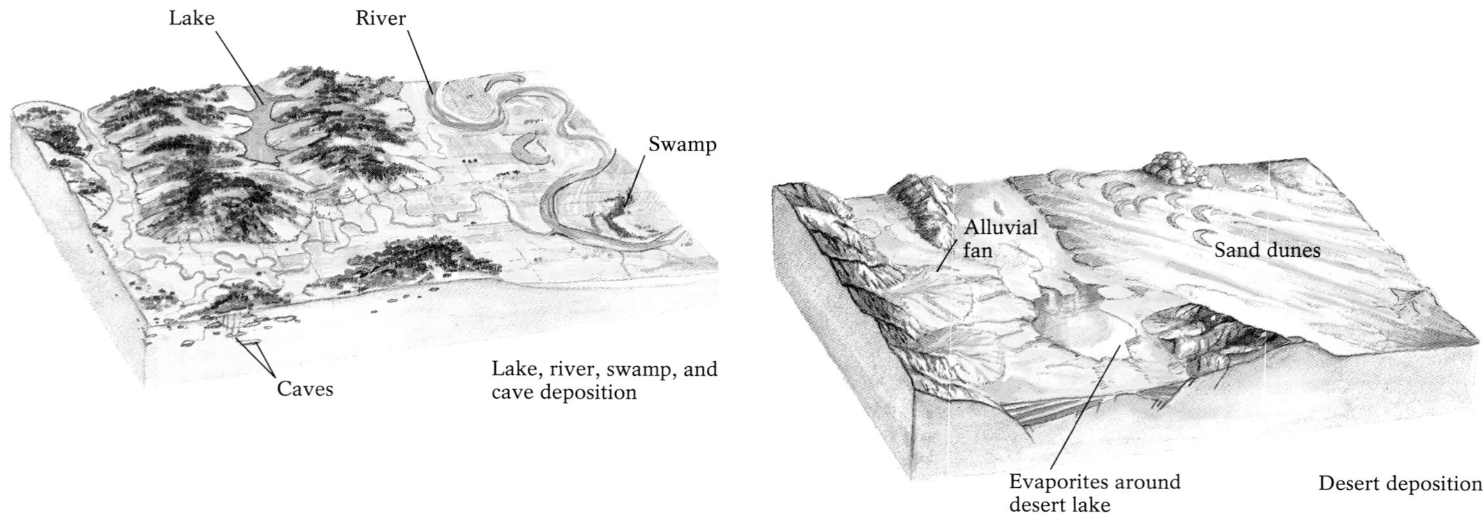

Figure 6-24 Some common geological environments in which sediments accumulate.

At any given time, a sedimentary environment's geological, geographical, biological, and climatic conditions determine the properties of its sediments and leave tell-tale features. These enable geologists to learn about past environments using clues found in ancient rocks.

Continental Environments Sedimentation in continental environments is mostly detrital. Sediment layers contain numerous indicators of past water- and wind-flow directions, and plant and freshwater fossils abound. Rivers, lakes, deserts, glaciers, and caves are all continental depositional environments.

Swift river currents carry and deposit coarse-grained sediments (sand and gravel), forming rippled and cross-bedded structures. Sediments that settle from "standing" floodwaters (i.e., after they've dropped their initial, high-energy load of coarse sediment) tend to be well-sorted, fine-grained, and graded. So are deposits in lake and swamp environments, where sediments also settle from standing bodies of water.

Lake deposits often contain diatoms (siliceous algae) and other organic matter. Because of their lush vegetation, the sediment deposits of swamps may include peat and organic-rich mud.

Continental/Transitional Environment

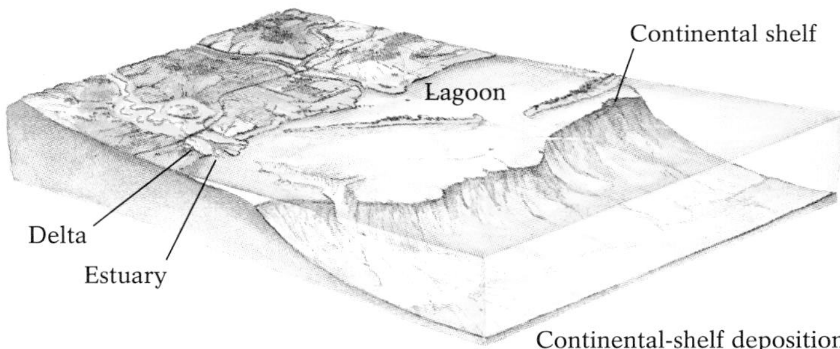

Transitional/Marine Environment

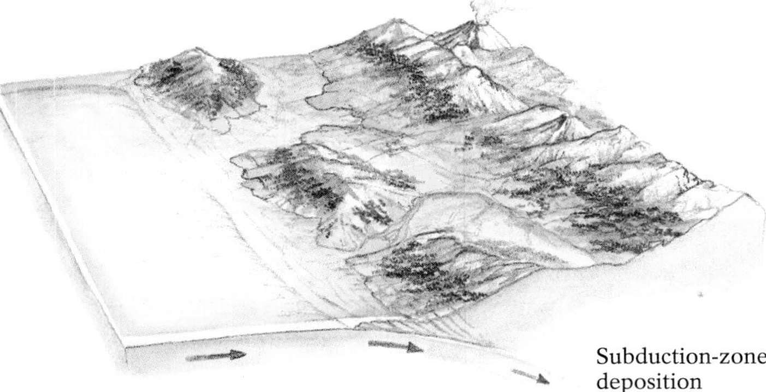

Continental/Transitional/Marine Environment

Transitional/Marine Environment

In caves, calcite precipitated from underground water is deposited as protrusions from the cave's ceilings, walls, and floors. Cave sediments may also include the bones and droppings of cave-dwelling bats, birds, and other animals.

In desert environments, wind is a significant transport medium. Incapable of moving large particles (except in unusual circumstances), winds typically carry aloft only silt- and, occasionally, clay-sized particles; larger sand particles are transported either in a series of short jumps or by rolling along the surface. Thus, desert deposits are generally well-sorted. Some deserts also contain coarse, poorly sorted, poorly stratified alluvial fan deposits, where steep mountain streams have abruptly dropped their loads onto flat desert floors. Evaporites are also found where the extreme heat in desert environments has evaporated temporary water bodies.

In a glaciated landscape, sediments may be deposited from slow-flowing or stagnant melting ice, by swift-flowing meltwater streams, or in the still water of glacial lakes. Thus, glacial deposits vary in composition, texture, and structure. Because flowing ice transports particles of all sizes, ice deposits are typically poorly sorted—as the ice melts, it drops its load in an unstratified heap. Sediments carried by meltwater streams and deposited beyond the glacier's margin are generally coarse and well-rounded, because these streams' swift currents can carry large particles and their turbulent flow promotes the forceful collisions that round sediment grains. Glacial lake deposits are characteristically fine-grained, graded, and well-sorted.

Transitional (Coastal) Environments Along ocean shores, continental and marine sedimentary processes merge. Breaking waves, tides, and ocean currents pulverize soft mineral grains and shells and sweep fine particles out to sea, leaving behind well-sorted, rounded, sand-sized deposits made principally of durable mineral grains, such as quartz.

When a river lets out into an ocean, causing the river's fresh water to mix with the salty seawater, a body of brackish (somewhat salty) water, called an *estuary,* is formed. If the estuary water is not too salty, marine, brackish, and freshwater organisms may live in it, contributing organic debris to its sediment. Estuaries sometimes contain *deltas,* fan-shaped accumulations of well-sorted sediment formed when rivers slow suddenly upon entering the sea (similar to alluvial fans, only formed under water rather than on land); coastal deltas may grow progressively seaward if the sediment is deposited at a faster rate than it is removed by waves, tides, and coastal currents.

Sediment of continental origin may be deposited offshore to form narrow islands that lie parallel to the coastline. A shallow body of water called a *lagoon* is created between the coast and an offshore island. Fine continental sediments may be delivered to lagoons by inflowing streams, and peat and organic mud may also accumulate there.

Continental and marine environments also meet at subduction zones, where explosive volcanism, powerful earthquakes, and mountain building produce a great volume of sediment; some of this sediment is deposited on shore and some is transported offshore by turbidity currents. Transitional deposition also occurs when rising sea levels submerge coastlines, bringing marine environments to continental margins; when this happens, marine sediments are deposited directly on top of continental sediments.

Marine Environments Marine environments vary according to their depth (Fig. 6-25). The shallow-marine environment lies above the continental shelf, 200 meters (about 700 feet) deep or less. Although a continental shelf may extend as many as several hundred kilometers into the sea, most are much narrower, with some being less than 1 kilometer wide. This narrow zone bordering all of the world's continents receives land-derived sediment carried seaward by waves and tides. Because sunlight penetrates about 50 meters (165 feet) below the water's surface, the upper part of the shallow-marine environment abounds in plant and animal life. Thus, its sediments often consist of carbonate-rich sands and mud containing the remains of diverse marine life forms.

The deep-marine environment lies beyond the continental shelf. Here, sediments consist largely of the remains of calcium carbonate- and silica-secreting microorganisms that have died and fallen from the upper 50 meters of the ocean; red and brown clays derived from weathering of ash from continental and submarine volcanoes; land-derived deposits carried to the deep-sea floor by submarine landslides; and a small amount of meteoritic fragments from outer space. These sediments contain few sizable fossils, largely because too little sunlight penetrates this region to support very many bottom-dwelling organisms (at least, not many having hard, fossilizable parts).

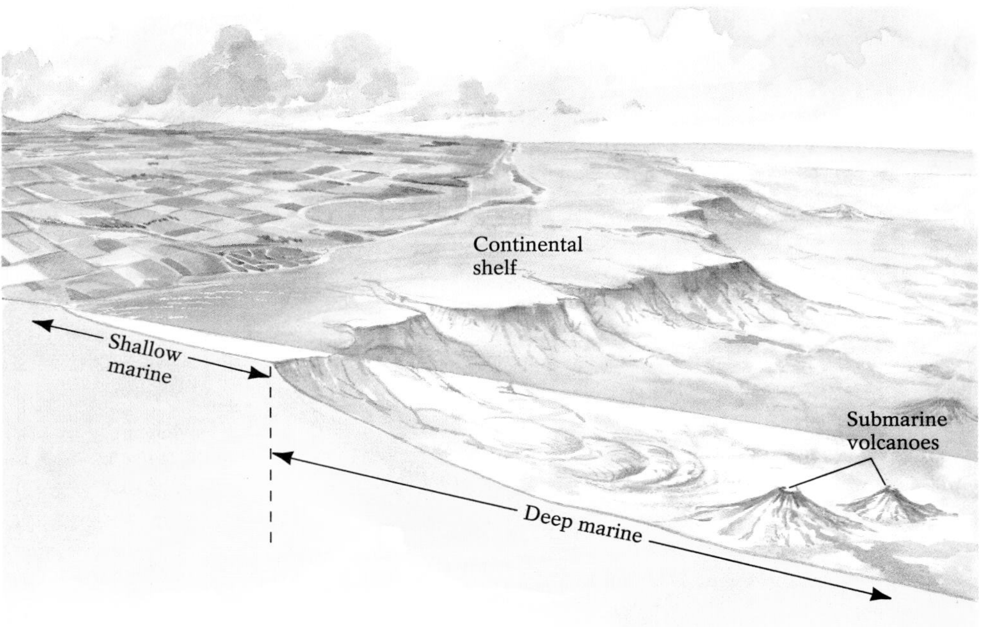

Figure 6-25 Marine sedimentary environments, which are described as being either shallow-marine or deep-marine. Shallow-marine environments lie over continental shelves and the shallow platforms that surround many oceanic islands. Deep-marine environments begin at the foot of the continental shelf.

Sedimentary Facies

Just as sediments deposited in the same place but at different times are recorded as a vertical succession of distinct rock layers, sediments deposited at the same time but in different places are recorded as a *horizontal* continuum of distinct rock types. The set of characteristics (such as mineral content and particle size, shape, and sorting) that distinguish a sedimentary rock deposit from nearby units deposited at the same time is collectively called a **sedimentary facies** (*facies* from the Latin for "form" or "aspect"). (The term "facies" may also refer to the rock itself.)

The nature of a sedimentary facies depends on the particular conditions under which it was deposited; thus, the pattern of facies in a given rock layer reflects the different settings that existed when these rocks were deposited.

For example, if the sediments associated with a modern coastal river were lithified, each of the related depositional settings would appear as a separate facies (Fig. 6-26): Coarse, cross-bedded sandstones and conglomerates would record flowing water in the river's channel; fine shales and organic deposits would mark the river's flood plain; and thinly bedded, graded shales, perhaps containing fossils, would be left behind by floodplain lakes.

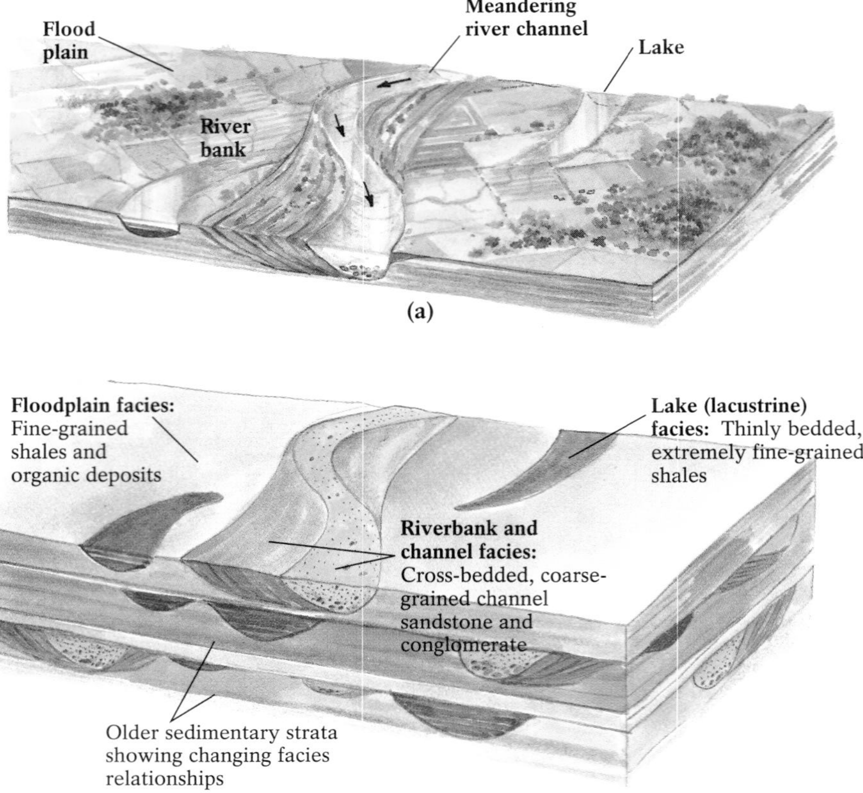

Figure 6-26 An example of sedimentary facies formation. The hypothetical modern river shown in (**a**) is surrounded by a lake-studded flood plain. If we could freeze this scene and convert the sediments to their future sedimentary rocks (**b**), we would find different rocks representing the river channel, the flood plain, and the lakes. Although these rocks are closely spaced geographically, and formed at the same time, they differ in composition and appearance because they were deposited in three different sedimentary settings. (Note that the changing relationships of facies over time, such as when meandering rivers change course, can be seen by examining the vertical succession of rock layers over a broad area.)

We have seen that the vertical succession of rock layers at a specific location shows how the depositional conditions at that spot changed over time. Similarly, the vertical succession of rock layers encompassing various sedimentary settings would show how these settings changed *in relation to one another* over time. Thus, for example, a shift in the course of a river over time would be reflected in the changing relationships of the associated facies in successive rock layers (Fig. 6-26b).

Shifting Sedimentary Facies The sedimentary environment in any given location changes with time, as bodies of water dry up, climate warms or cools, sea levels rise and fall, and so on. These changes alter the particle size and composition of the deposited sediments, which, after lithification, then appear as distinctive facies in the rock record. In Figure 6-27, for instance, note the position of the river, beach, shallow-marine, and deep-marine environments at time A. If the sea level subsequently rose, the shoreline would migrate inland and a marine environment would replace the river environment. Gradually, a coastal beach would replace what was once the river's flood plain, a shallow-marine environment would replace the old beach, and a deep-marine environment would replace the former shallow-marine environment. By time B, these would in turn be replaced by even deeper marine environments. The sediments deposited at any given location at time B would be different from those deposited at the same location at time A, because the environment of deposition would have changed; the sediments deposited at time A would, at time B, be deposited farther inland. Such shifting sedimentary facies, which can provide clues to changes in ancient environmental conditions, are discovered by comparing vertical sequences of rock from neighboring areas.

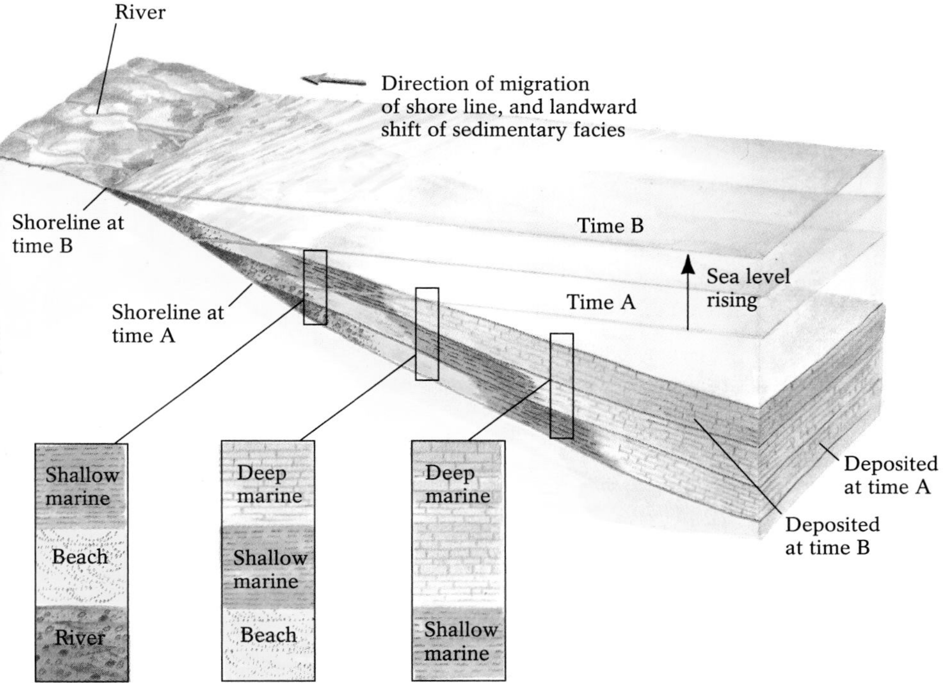

Figure 6-27 Migration of sedimentary environments associated with rising sea level.

Sedimentary Rocks and Plate Tectonics

Some of the Earth's most majestic mountains—the Northern Rockies, the Appalachians, the European Alps, the Urals of Russia, and the Himalayas of China, Nepal, and Tibet—contain sedimentary rock strata that were clearly deposited in marine environments. In them we can see ripple marks that record rising and falling tides and coastal waves, evaporites and mudcracks from past sea-level fluctuations, and thousands of meters of limestone. These rocks, all derived from sediments that accumulated in shallow-marine environments, have been uplifted by plate tectonic forces, as we shall see in Chapter 9.

Sedimentary rocks also provide evidence that lofty mountains once stood where none stand today. The fairly squat Taconic Hills of western Connecticut and Massachusetts and eastern New York are thought to be a remnant of mountains that rose when the plates containing what are now Europe, Africa, and North America collided about 375 million years ago. Rivers carried a great volume of coarse detrital sediment from those now-departed mountains into shallow seas to the west; the cross-bedded sandstones derived from these sediments, themselves subsequently uplifted and then cut through by streams, now form the Catskill Mountains of east central New York.

Specific plate tectonic settings are often associated with certain characteristic sedimentary facies. Recently rifted plate margins, for example, such as the East African rift zone, tend to contain large alluvial fan deposits, volcanic graywackes, and extensive lake deposits—all typical of areas that have been faulted downward during rifting (see Chapter 9). Transform boundaries, such as the San Andreas system in California, are noted for rapid sedimentation of angular, feldspar-rich arkosic sands (formed by the crushing of igneous rock at the plate boundaries). Rapid sedimentation also takes place near sites of continental collisions, such as in the shadow of the still-rising Himalayas, where a 15-kilometer (10-mile)-thick tongue of sediment extends 2500 kilometers (1500 miles) into the Indian Ocean south of Calcutta, India. Volcanoes that rise above a subducting plate supply sediment that tends to be poorly sorted and rich with angular fragments of volcanic rock, testimony to rapid burial and an active sediment source.

Sedimentary Rocks from a Distance

The sedimentary rock types composing a given formation can sometimes be predicted judging only from their appearance from afar and the climate in which they are found. For instance, in moist temperate regions such as the Eastern seaboard of North America, accelerated chemical weathering and erosion remove much soft erodible shale and soluble limestone, leaving prominent ridges of durable, well-cemented sandstone, such as those in the Appalachians of central Pennsylvania. The less resistant rocks usually form the region's long gentle slopes and underlie its broad valley bottoms (Fig. 6-28).

In the arid Southwest, durable sandstones also form the cliffs and soft shales the slopes. In the absence of significant rainfall, however, the Southwest's thick limestone beds experience relatively little weathering and erosion. Thus, they too may form steep cliffs (Fig. 6-28b). From the scenic

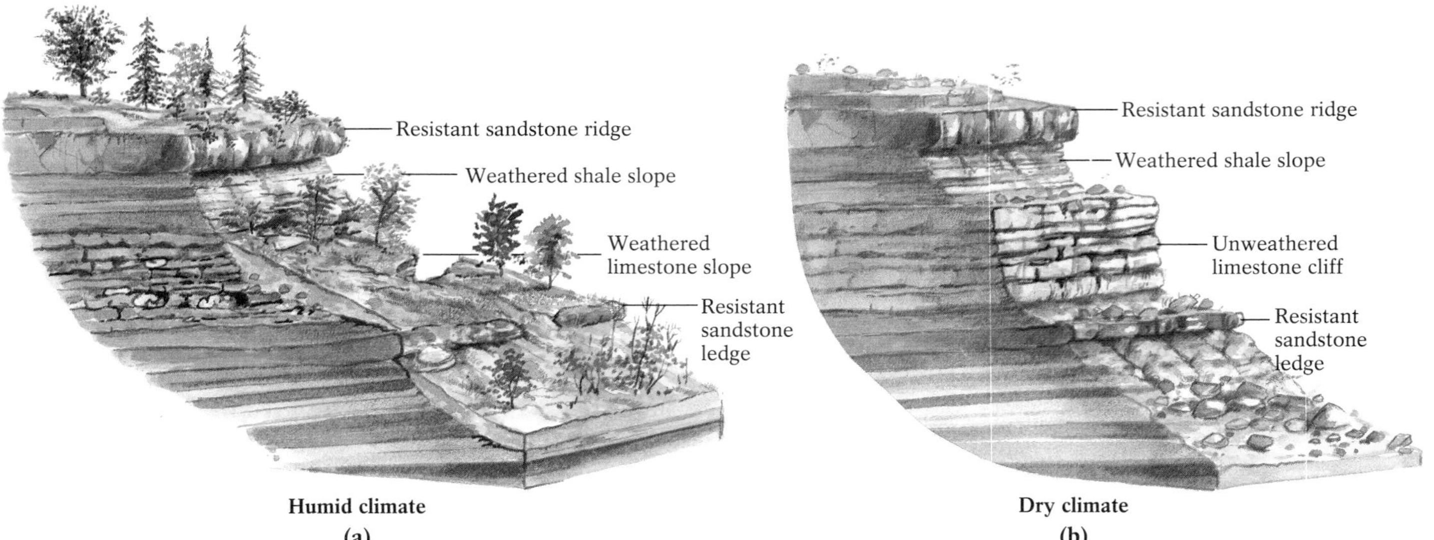

Figure 6-28 Rock types form predictable topographies, often depending on the climate to which they are exposed. In both humid and dry climates, well-cemented, weather-resistant sandstones form prominent ridges and poorly cemented, less-resistant shales form slopes and valleys. Limestones form ridges in dry climates but dissolve away to form slopes in humid climates. Thus, the same sequence of rocks that would form a gently sloping, soil-rich topography in a humid climate (**a**), such as is found in the northeast United States, might form a rugged, soil-poor topography in a dry climate (**b**), such as is found in the southwestern United States.

Figure 6-29 The American Southwest affords numerous vistas of cliffs and canyons in brilliant reds, oranges, pinks, greens, and purples. Here, at Kodachrome Basin in the Painted Desert of Arizona, the colors of the sedimentary rocks are due to their mineral content, the materials that cement their grains, the nature of their iron compounds, and the presence or absence of organic matter.

overlooks along the south-rim road of the Grand Canyon, one can readily distinguish the slope-forming shales from the cliff-forming sandstones and limestones, even from 10 kilometers (about 6 miles).

Sedimentary rocks are more colorful than other types of rocks. Almost every sedimentary rock contains some iron, and even a 0.1% iron content can lend a deep-red hue to an oxidized sandstone. Red, pink, orange, and brown sedimentary rocks are typically continental in origin; their colors developed when they were exposed to air during transport or shortly after deposition, causing their iron to oxidize. In settings where oxygen is in short supply, such as organic-rich lagoons and swamps and in deep-marine settings, the iron in sediment becomes a characteristic green or purple. Low-oxygen, aqueous conditions result in black and dark gray sediments containing undecomposed organic matter. In many sedimentary rock formations, such as some in America's Southwest (Fig. 6-29), color alone tells us much about the environments in which the respective layers formed.

We will be encountering sedimentary processes and rocks throughout the remainder of this book. In Chapter 8, we will see how sedimentary rocks record local, regional, and even global geological events, as well as the evolution of the Earth's animals and plants. In Chapter 9, we will examine how the motion of the Earth's plates causes horizontal beds of sedimentary rock to be uplifted into continent-long mountain ranges such as the Rockies and Appalachians. In Chapters 13 through 19, we will explore how sediment is moved from place to place by gravity, rivers and streams, underground water, glaciers, desert winds, and crashing surf. Finally, in Chapter 20, we will discuss some sedimentary processes that produce many of the Earth's valuable natural resources—from coal, oil, and natural gas, to diamond, iron ore, and the sand and gravel with which we build our cities and highways. But first, in Chapter 7, we will look at the remaining major rock group of the rock cycle (discussed in Chapter 1), the metamorphic rocks.

Chapter Summary

Sediment consists of fragments of solid material derived from preexisting rock, the remains of organisms, or the direct precipitation of minerals out of solution in water. A vast amount of sediment accumulates continuously at the Earth's surface, and most of it is eventually converted to **sedimentary rocks.** Sedimentary rocks make up a thin layer of the Earth's crust that account for about 75% of all the rocks exposed on land.

Sediments are classified according to the source of their constituent minerals. **Detrital sediment** is composed principally of fragments of preexisting igneous, sedimentary, or metamorphic rock. **Chemical sediment** consists of minerals—originally derived from dissolved chemical weathering products—that have either precipitated directly out of solution by inorganic chemical processes or been extracted from solution by organisms and ultimately deposited in the form of shells, skeletons, and other organically derived materials.

All sediments are ultimately deposited at some distance from their point of origin. The nature of detrital sediments in particular is influenced by the transport process, which may be via such agents as flowing surface water, circulating underground water, wind, glaciers, or coastal waves, among others.

Detrital sediment texture—the size, shape, and arrangement of detrital particles in a deposit—is determined largely by transport (in addition to the weathering-resistance of its parent material). During transport, sediment grains undergo **sorting,** a process by which they are carried or deposited selectively, based on the energy of their transport medium and the grain's size, density, and shape. A well-sorted deposit (typical of wind transport) consists of particles of one size; a poorly sorted deposit (typical of glacial transport) contains particles of varying sizes.

Detrital sediments often display **sedimentary structures,** features that develop during or soon after deposition that reflect the conditions under which they were deposited. **Bedding,** or **stratification,** is the arrangement of sediment particles into distinct layers marking separate depositional events. **Graded bedding** is a type of bedding that forms as sediment settles through standing water. In such a case, the coarsest grains settle to the bottom first, and grain size decreases gradually toward the top of a layer. **Cross-bedding** refers to sediment layers that are oriented at an angle to the underlying sets of beds, as is typical of wind-deposited sediments and sediments deposited by moving currents of water. **Ripple marks** are small surface ridges produced by water or wind flowing over sediment after it is deposited. **Mudcracks** occur in the top of a sediment layer when muddy sediment dries and contracts.

After a body of sediment has been buried by subsequent deposits, the increased heat, pressure, and circulating ions to which it is exposed produce a number of changes, collectively known as **diagenesis.** The end result of diagenesis is often **lithification,** the conversion of loose sediment into solid sedimentary rock. **Compaction** is the diagenetic process by which the volume of a sedimentary body is reduced when it is compressed under the weight of overlying materials. **Cementation** of sediment grains occurs when dissolved ions are precipitated in the pore spaces within the sediment. Any rock that is formed by compaction and cementation of separate sediment particles (usually rock or mineral fragments, but sometimes organic debris such as shell fragments) is said to have a **clastic** texture. **Recrystallization** converts certain unstable minerals in sediment into new, stable minerals.

Detrital sedimentary rocks are classified by grain size. They include the fine-grained **shales,** intermediate-grained **sandstones,** and coarse-grained **conglomerates** (when large grains are rounded) and **breccias** (when large grains are angular). Sandstones are subdivided further, on the basis of composition, into quartz-rich arenites, feldspar-rich arkoses, and rock-fragment-rich graywackes.

Classification of chemical sedimentary rocks is based not on grain size, but on the composition of the sediment. They are further classified as being inorganic or organic, depending on how their mineral components were converted from their original dissolved state into solid form. The inorganic chemical sedimentary rocks precipitate directly from water (usually when much of the water has evaporated or it undergoes a significant temperature change). These include: inorganic limestone, consisting primarily of calcium carbonate that precipitates from either seawater or fresh water; **evaporites,** a variety of salts (such as gypsum and halite) that accumulate when seawater evaporates; **dolostone,** a calcium-and-magnesium carbonate believed to form when magnesium ions circulating through limestone replace some of its calcium ions; and inorganic chert, composed largely of silica precipitated from seawater or fresh water.

Organic chemical sedimentary rocks form when organisms extract dissolved compounds from water and convert them into biological hard parts (such as shells and skeletons) that are ultimately deposited as sediment. These include: organic **limestone,** composed of calcium carbonate remains of marine organisms; organic **chert,** composed of silica-based remains of marine organisms; and **coal,** composed of the carbon-rich remains of terrestrial plants.

Sediment accumulates in numerous **sedimentary environments,** which may be continental, transitional (coastal), or marine. Because the properties of any sedimentary rock stem from the specific conditions under which it develops, geologists can distinguish rocks of one depositional setting from rocks of another. The term **sedimentary facies** refers to the set of unique properties that distinguish a given rock from all surrounding rocks deposited in different settings. By noting how sedimentary facies change over a distance as well as over time, we can interpret not only individual sedimentary environments of the past but also their changing relationships to one another.

Key Terms

sediment (p. 151)
sedimentary rock (p. 151)
detrital sediment (p. 152)
chemical sediment (p. 152)
sorting (p. 153)
sedimentary structures (p. 154)
bedding, stratification (p. 155)
graded bedding (p. 156)
cross-bedding (p. 157)
ripple marks (p. 157)
mudcracks (p. 158)
diagenesis (p. 158)
lithification (p. 158)
compaction (p. 158)
cementation (p. 158)
clastic (p. 159)
recrystallization (p. 159)
shales (p. 160)
sandstones (p. 161)
conglomerates (p. 162)
breccias (p. 162)
evaporites (p. 164)
dolostone (p. 167)
limestone (p. 168; *see also* p. 162)
chert (p. 168; *see also* p. 167)
coal (p. 168)
sedimentary environments (p. 173)
sedimentary facies (p. 173)

Questions for Review

1. What are the major types of sedimentary rocks? On what basis are they classified?

2. Briefly describe how sorting and rounding of detrital sediment vary with different transport media, such as wind, rivers, and glaciers.

3. Describe the differences between graded beds and cross-beds. Which of these indicates the flow direction of ancient currents? Which can be used to determine if a sedimentary bed is upside down or right side up? Give an example of a setting in which each forms.

4. Explain the processes of diagenesis, and describe how each affects the physical properties of sediment.

5. Describe the composition and texture of quartz arenites, arkoses, and graywackes.

6. Explain the relationship between carbon dioxide and precipitation of inorganic limestone. Briefly describe the origin of the following types of limestone: oolitic limestone, tufa, and travertine.

7. What are evaporites? Describe two sedimentary environments where evaporites form.

8. Name and describe the origin of three different types of organic chemical sedimentary rocks.

9. Describe how deposition occurs in each of three different *continental* sedimentary environments. How do sediments in deep-marine and shallow-marine environments differ?

10. Name some types of sedimentary facies associated with each of two different plate tectonic boundaries.

For Further Thought

1. Why do we find much more of the evaporites halite and gypsum than the evaporite sylvite? Why do we often find dolostone associated with evaporite deposits?

2. Under what circumstances might we find poorly sorted, angular, arkosic sediments in a coastal environment?

3. Study the photo of a modern salt flat below and speculate about the environmental conditions that existed when these sediments were first deposited. How would the appearance of the present sediments change if the climate in this area became very moist?

4. In the figure below, what can we tell about the route taken by the current that deposited the conglomerate shown?

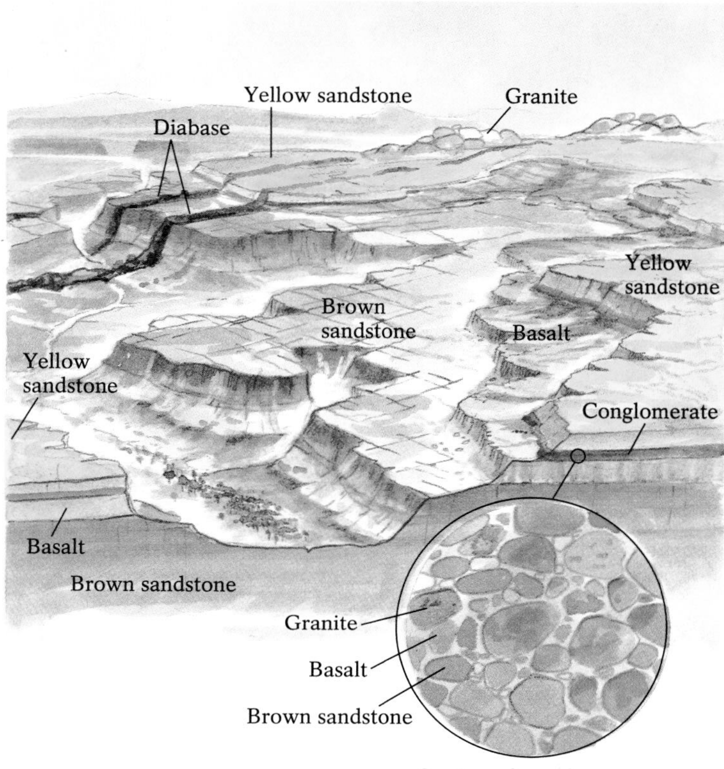

5. Why are oolitic limestones common among the rocks of southern Indiana?

Conditions Promoting Metamorphism

Common Metamorphic Rocks

Types of Metamorphism

Metamorphic Grade and Index Minerals

Plate Tectonics and Metamorphic Rocks

Metamorphic Rocks in Daily Life

The Rock Cycle Revisited

Figure 7-1 Vishnu Schist, deep in the inner gorge of the Grand Canyon. This rock, shown here intruded by more recently created igneous rock, was converted from a soft marine mudstone to a hard, highly contorted metamorphic rock more than 2 billion years ago.

Chapter 7

Metamorphism and Metamorphic Rocks

We saw in Chapter 3 that the magmas that cool to form igneous rocks are created by high temperatures deep in the Earth's interior. We saw in Chapter 6 that the sediments and sedimentary rocks that cover most of the Earth are produced in the relatively low-temperature environment at the Earth's surface. At conditions in between those that produce igneous and sedimentary rocks, the third major type of rocks is produced: **Metamorphic rocks** generally form under the temperature, pressure, and chemical conditions that exist below the zone of diagenesis (discussed in Chapter 6) but above the 50- to 250-kilometer-deep "melting zone" that creates most magmas (see Fig. 3-6). (Although it is believed that metamorphic rock can exist at depths up to 2900 kilometers.) **Metamorphism** (from the Greek *meta,* meaning "change," and *morphe,* meaning "form") is the process by which conditions within the Earth alter the mineral content and structure of solid rock *without melting it.* Any rock—igneous, sedimentary, or a metamorphic rock itself—may be a candidate for metamorphism (Fig. 7-1).

Most of the metamorphic rocks in the United States are buried beneath thousands of meters of sedimentary rock. (Near Topeka, Kansas, for example, the rocks extending from just below the surface to thousands of meters below formed from shallow-marine sediments left behind by inland seas that invaded North America more than 100 million years ago.) The ancient rocks that underlie surface rocks throughout the continent, however, are predominantly metamorphic, and a geologic map of North America reveals exposed or near-surface metamorphic rock throughout much of Canada, the northern portions of the midwestern Great Lakes states, and many of the continent's mountainous regions (Fig. 7-2). Wherever you are on Earth, in fact, there are metamorphic rocks at some depth beneath you.

Metamorphic processes are presumably going on continuously beneath the Earth's surface, but, for obvious reasons, no one has ever seen metamorphism in action. We can observe the destructiveness of lava flowing from Hawaii's Kilauea volcano and watch it congeal into another acre of Hawaiian basalt. We can watch a river carry a load of sand and other loose sediment that may someday form sandstone. But we see metamorphic rocks only after uplift and erosion have stripped away the overlying rocks, and we must be geological detectives to infer, from surface clues, the deep interior processes that produced them. Most of what we know about the causes of metamorphism has come from laboratory experiments and theoretical models that replicate conditions in the Earth's interior.

Figure 7-2 Exposed and near-surface metamorphic rocks of North America. Generally buried beneath kilometers of sedimentary rocks, the continent's metamorphic rocks can be seen only where erosion has removed the overlying rock—principally in very deep river valleys, such as the Grand Canyon, in the cores of mountain ranges, such as the Colorado Rockies and the North Cascades of Washington state, and in the glacially scoured "Canadian Shield" region of Ontario, Quebec, and the adjacent Great Lakes states of Minnesota, Wisconsin, and Michigan.

Conditions Promoting Metamorphism

We saw in Chapter 5 that rocks are more likely to be altered by weathering processes when they are in a new environment than when they remain where they formed. This is a basic principle: Rocks and their constituent minerals are most stable in the environment in which they form, and least stable in markedly different environments. For example, feldspars formed by relatively high-temperature igneous processes rapidly break down chemically at the Earth's surface, an environment with relatively low temperatures, low pressure, and abundant water and atmospheric gases. In their place, clay minerals form, which are more stable than the feldspars under surface conditions. But if the clays became buried beneath tens of kilometers of overlying sediments and heated to 600°C (1110°F), they would in turn be in a new environment of higher temperature and pressure. Under these conditions, the clays would react to form new minerals that would be stable under the new conditions. The resulting materials might even include some feldspars.

Rocks undergoing metamorphism remain in a solid state. Their unstable minerals either recrystallize into new, more stable forms or react with other unstable minerals to produce new minerals with stable atomic arrangements. Such rearrangement is possible because heat and pressure break the bonds between some of the atoms or ions in an unstable mineral, allowing them to migrate to other sites within the mineral, or to another mineral entirely, and rebond. In this way, minute grains in a fine-grained rock are usually replaced by an interlocking mosaic of large, visible grains.

Metamorphic processes never break all the bonds in a rock's minerals—if all its bonds were broken, a rock would melt, becoming a magma (an *igneous* process). Metamorphism occurs when heat and pressure exceed certain threshold levels, destabilizing the minerals in rocks, but do not become high enough to cause melting. The presence of circulating ion-rich fluids also influences metamorphism, as does the composition of the parent rock, which largely determines which metamorphic rocks and minerals can be formed.

Heat

Flour, yeast, water, and a pinch of salt, mixed in the appropriate proportions, are used to make bread. But, without turning on the oven and subjecting these ingredients to a specific temperature, no bread would result from their combination. Similarly, the ingredients for making metamorphic minerals and rocks are present everywhere, but little happens until a source of heat is available.

Heat, which speeds up the pace of all chemical reactions, is perhaps the most important factor contributing to metamorphism. Beneath the Earth's surface, temperature generally increases with depth (see discussion of the geothermal gradient, in Chapter 3); in the Earth's crust and upper mantle, temperature increases at an average rate of 20° to 30°C per kilometer of depth (about 72°F per mile). Temperatures sufficient to metamorphose rocks—greater than about 200°C (about 400°F)—are generally reached at about 10 kilometers (6 miles).

The heat in the Earth's interior has several sources (and, because rocks are generally poor conductors, most heat within the Earth remains there). As we saw in Chapters 1 and 2, radioactive isotopes within the rocky crust and upper mantle emit heat energy as they decay. Some heat is transported by magma intruding from deeper within the Earth. Another source of heat is friction between two bodies of rock grinding past one another along a fault or plate boundary. All of these sources together produce enough heat to promote metamorphism. But heat does not act alone.

Pressure

We measure pressure in *bars*. A bar is equal to the pressure applied to the Earth's surface at sea level by the atmosphere (1 bar = 1.02 kilograms/square centimeter or 14.7 pounds/square inch). Metamorphism requires more than 1 kilobar (1000 bars) of pressure, which is approximately equal to the pressure prevailing about 3 kilometers (2 miles) beneath the Earth's surface. However, since the *temperatures* needed for metamorphism do not normally occur above about 10 kilometers, metamorphism will not occur this shallow unless heat is carried up from greater depths by magma or is induced by friction between tectonic plates.

Like heat, pressure on rocks generally increases with depth as the thickness of the overlying rock increases. The type of pressure associated with deep burial, called **lithostatic** (*litho* meaning "rock," *static* from the Greek *statikos,* which means "causing to stand in place") or **confining pressure,** pushes in on rocks equally from all sides. As a result of lithostatic pressure, a deeply buried rock becomes compressed into a smaller, denser form, but its shape remains the same (Fig. 7-3a).

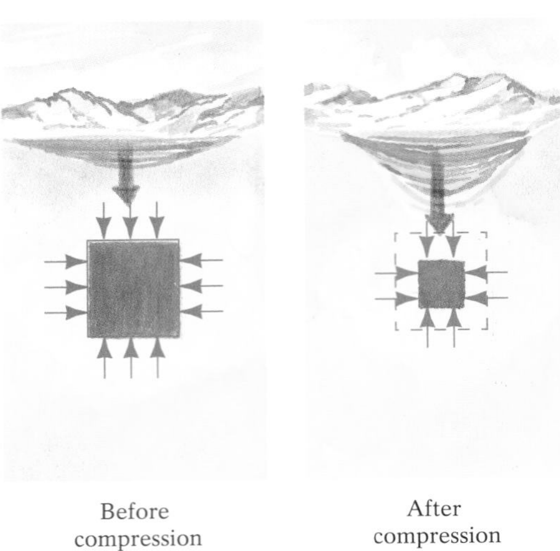

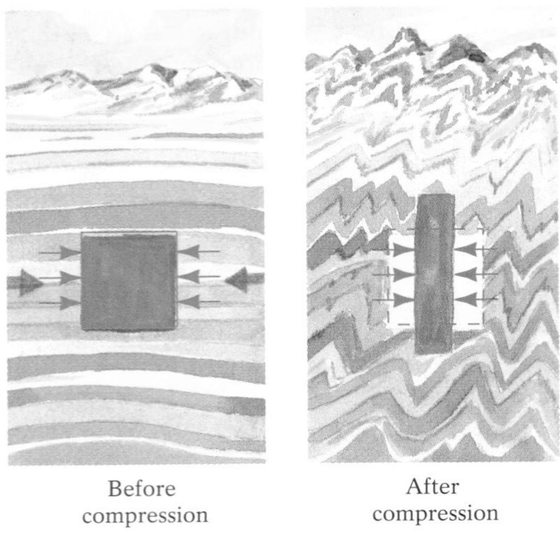

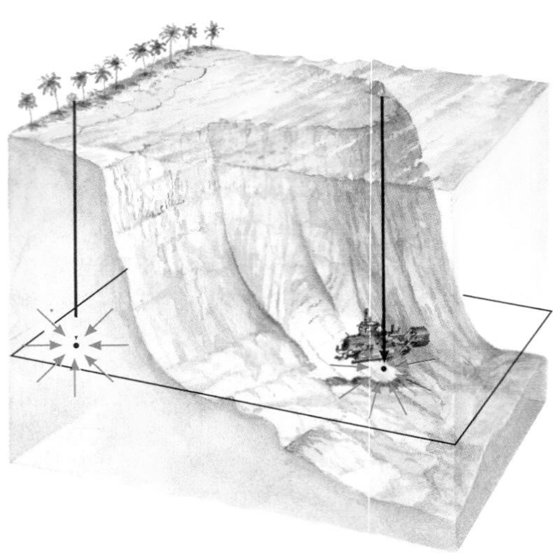

Figure 7-3 (**a**) Deep burial of rocks generally subjects them to lithostatic, or confining, pressure, an inward-pressing force that acts equally from all directions. (Similar pressure is applied to any object immersed in water.) A rock subjected to confining pressure becomes compressed without changing shape. (**b**) A rock subjected to directed pressure has its shape distorted, becoming thinner in the direction of the pushing, or *stress,* and elongated in the direction perpendicular to the stress.

In some geological settings, pressure does not act equally from all directions, but instead acts in one principal plane. Such **directed pressure,** which commonly occurs where tectonic plates collide to form mountains and along various types of faults (discussed in Chapter 9), distorts the shape of rock as shown in Figure 7-3b: The rock becomes flattened in the plane in which the pressure is applied, and lengthened in the plane perpendicular to the pressure. Such pressure may also deform individual components within rocks, such as fossils or bands of minerals, causing them to become stretched and folded (Fig. 7-4).

Rocks subjected to either confining or directed pressure change in a number of ways. Both types of pressure cause pore spaces between mineral grains to close, producing a more compact rock texture, and mineral density to increase. Both can also cause some of a rock's minerals to dissolve, especially when there is water between the grains, and recrystallize. This is because pressure at the contact points between compressed grains causes some bonds there to break. Unbonded ions then migrate to other, lower-pressure sites, where they rebond. The result is a new, more compact mineral structure that is more stable under high pressure (Fig. 7-5).

One change in rocks is produced by directed pressure in particular. When pressure causes some of the bonds between a mineral's ions to break, the unbonded ions migrate from areas of high pressure and rebond at lower-pressure sites. Because low-pressure sites in rocks subjected to directed pressure are usually perpendicular to the direction of the pressure, the recrystallized minerals develop a parallel alignment perpendicular to the directed pressure. If the mineral grains are thin and planar, such as sheet-like mica flakes, their systematic orientation gives the rock a distinctive layered look known as **foliation** (from the Latin *foliatus,* or "leaf-like") (Fig. 7-6).

Figure 7-4 Folded rock in the Blue Ridge Mountains of North Carolina, showing the deformation typical of rocks subjected to directed pressure.

Fluids

In general, if you put two dry chemicals at room temperature into a test tube, nothing will happen. If you heat the test tube with a Bunsen burner, still nothing much will happen. But add one more ingredient—a fluid—and a significant reaction will probably occur. Because particles move about easily in

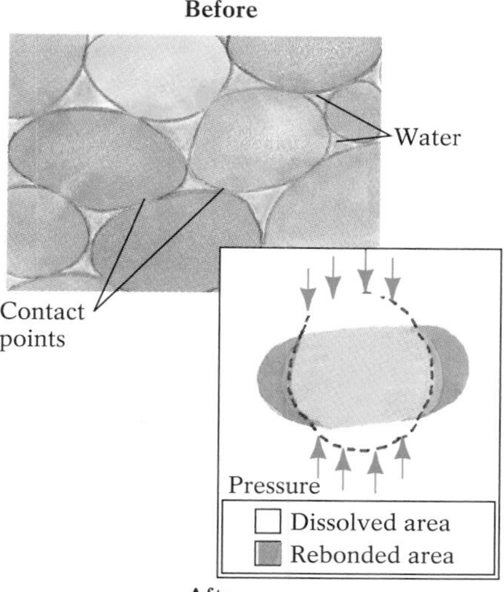

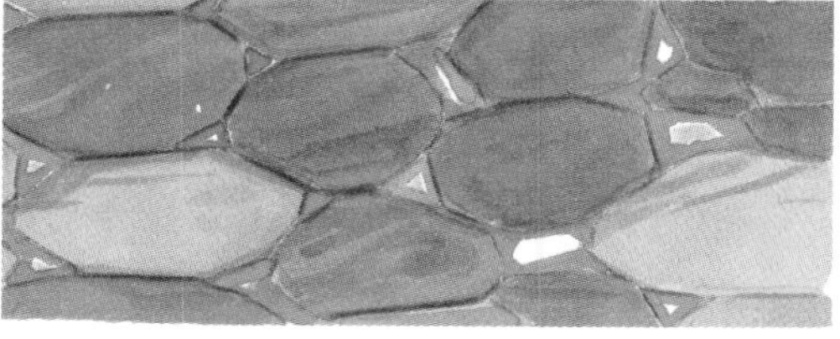

Figure 7-5 Mineral grains may dissolve under the high pressure that exists at the contact points between compressed grains. The pressure strains and then breaks some of the bonds between ions. The unbonded ions then migrate away from the high-pressure contact points to low-pressure areas in the spaces between grains, where they rebond to form more compact, stable structures.

fluids, the presence of a fluid such as a liquid or gas in or around rock under pressure facilitates the migration of unbonded atoms and ions, dramatically increasing the potential for metamorphic activity. For example, water surrounding a mineral grain promotes the exchange of atoms and ions with adjacent grains. Without water, unbonded ions migrate very slowly along grain boundaries or through the minute pathways within a mineral's atomic structure that are expanded by heating.

Water and magmatic gases are the fluids that circulate underground and promote metamorphism. The water may come from various sources. It may have percolated down from the Earth's surface, or been trapped in the spaces within sediment layers or in the cracks of a subducting plate. When magmas cool and crystallize, they may release water and other fluids (such as carbon dioxide or sulfur dioxide gas) to surrounding rocks. Additional water may be released as water-rich minerals, such as clays and amphiboles, are heated. Most underground fluids contain a variety of ions and, as we saw in Chapter 3, acquire more as they migrate through fissures in heated rocks. These hot, ion-rich fluids may add their acquired ions to rocks during metamorphism, changing the chemistry of the rocks and producing a whole new set of minerals. Thus, fluids not only serve as the medium through which a rock's own unbonded ions migrate, but also contribute foreign ions to metamorphic reactions.

Parent Rock

Ultimately, it is the composition of the parent rock that determines which metamorphic rocks and minerals will form under new environmental conditions. No matter what the temperature or pressure, for instance, parent rock devoid of calcium cannot produce a calcium-rich metamorphic rock. In a parent rock that contains only one mineral, metamorphism will produce a rock composed predominantly of that same mineral; the key change will be its recrystallization into a more stable configuration. For example, the calcite grains in a pure limestone will recrystallize into calcite-rich metamorphic limestone (marble). In a parent rock that contains several minerals, their components will combine in different ways to form a number of new minerals. For example, metamorphism of the various clay minerals, quartz grains, mica flakes, and volcanic fragments in graywacke sandstones may release a variety of ions and produce a group of new minerals that were not part of the parent rock.

Common Metamorphic Rocks

Figure 7-6 Directed pressure creates a layered, or foliated, appearance in rock.

The classification of metamorphic rocks is summarized in Table 7-1. Note that some types of metamorphic rocks can be produced from any of a number of different parent rocks; such a metamorphic rock is distinguished according to its appearance and the conditions that formed it rather than its composition. Thus, a gneiss is any coarse-grained, foliated metamorphic rock that formed under high temperature and pressure, regardless of whether it derived from shale, diorite, or granite. Other metamorphic rocks, such as quartzite and marble, are produced by a specific type of parent rock and so are classified by their compositions.

The first criterion that is used to distinguish any metamorphic rock is whether it is foliated or nonfoliated. When the parent rock is a simple single-mineral sedimentary rock such as quartz sandstone, the resulting metamor-

Table 7-1 **Classification and Derivation of Some Common Metamorphic Rocks**

	Rock	Parent Rock(s)	Key Minerals	Metamorphic Conditions
Foliated	Slate	Shale, mudstone	Clay minerals, micas, chlorite	Relatively low temperature and directed pressure
	Phyllite	Shale, mudstone	Mica, chlorite	Low–intermediate temperature and directed pressure
	Schist	Shale, mudstone, basalt, graywacke sandstone, impure limestone	Mica, chlorite, epidote, garnet, talc, hornblende, graphite	Intermediate–high temperature and directed pressure
	Gneiss	Shale, felsic igneous rocks, graywacke sandstone	Quartz, feldspars, garnet, mica, augite, hornblende, staurolite, kyanite	High temperature and directed pressure
Unfoliated	Marble	Pure limestone or dolostone	Calcite, dolomite	Contact with hot magma, or confining pressure from deep burial
	Skarn	Silicate-containing limestone or dolostone	Calcite, dolomite, garnet	Contact with hot magma, or confining pressure from deep burial
	Quartzite	Pure sandstone	Quartz	Contact with hot magma, or confining pressure from deep burial
	Hornfels	Shale, mudstone, basalt	Andalusite, mica, quartz	Contact with hot magma; little pressure

phic rock is usually nonfoliated or so slightly foliated that the foliation is invisible to the naked eye. More dramatic, visible foliation tends to develop when a multimineral rock is subjected to progressively greater heat and directed pressure, as is the case with the sequence of metamorphic rocks that forms from a typical marine shale.

Foliated Rocks Derived from Shale

As temperatures and pressures rise beyond the range of sedimentary diagenesis, clay minerals in shale gradually recrystallize or react to form minute but relatively long, flat mica flakes that become aligned perpendicular to the direction of the applied pressure. This forms the first metamorphic rock to develop from the metamorphism of shale, a subtly foliated rock known as **slate** (from the Old French *esclat*, or "splinter"). Slate tends to break parallel to the mica-rich planes into relatively uniform thin, flat fragments, a pattern known as *slaty cleavage* (Fig. 7-7). (Note the difference between rock cleavage and mineral cleavage, discussed in Chapter 2. Mineral cleavage refers to minerals breaking between planes of atoms or ions in a crystal, whereas rock cleavage refers to rocks breaking between planes of minerals.)

The color of slate depends on the content of its parent shale. Red slates are rich in iron oxide; green slates contain a significant amount of chlorite; purple slates are stained by manganese oxides; and black slates are rich in

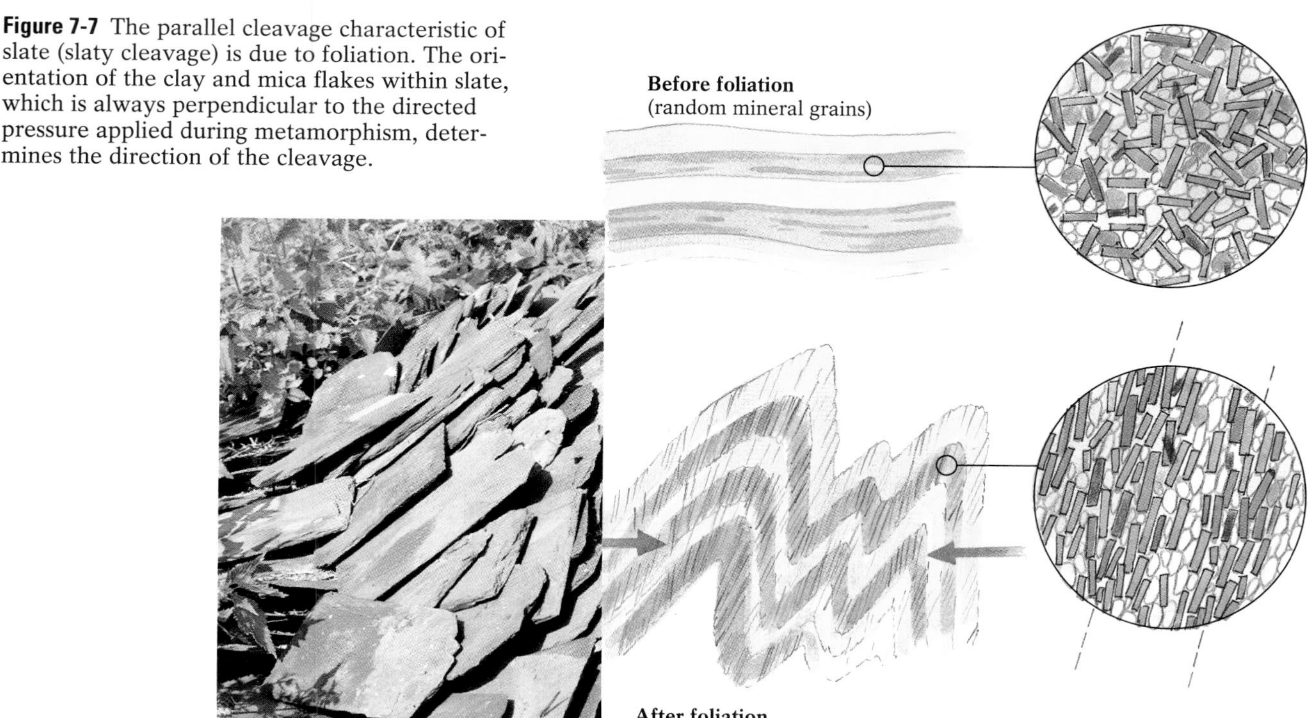

Figure 7-7 The parallel cleavage characteristic of slate (slaty cleavage) is due to foliation. The orientation of the clay and mica flakes within slate, which is always perpendicular to the directed pressure applied during metamorphism, determines the direction of the cleavage.

Figure 7-8 Mica-rich schist showing characteristic schistose foliation and cleavage. The glittery appearance of schistose rocks is due to the presence of their large chlorite and mica grains.

carbon-rich organic material. Whatever their hue, all slates are fine-grained and exhibit slaty cleavage, producing smooth slabs of rock that are commonly used as blackboards, roofing materials, floor tiles, and pool-table tops.

At temperatures of about 300°C (575°F), the microscopic mica and chlorite flakes in slate grow to visible sizes, producing the foliated metamorphic rock **phyllite** (from the Greek *phyllon,* or "leaf-like," for its pronounced thin, wavy foliation). The light reflected off the surfaces of these mica and chlorite grains gives phyllite its characteristic sheen. With increased heat and pressure, the fine mica and chlorite flakes in phyllites continue to grow. When the flakes become as large as about 1 centimeter in diameter, the metamorphic rock is classified as a **schist** (from Greek *schistos,* or "split"), a coarse-grained, strongly foliated rock that splits readily into flat parallel slabs—a distinctive type of rock cleavage known as *schistosity* (Fig. 7-8). (The term schistosity also sometimes refers to the foliation.) Schists have generally metamorphosed to such an extent that any sedimentary structures preserved in the original rock have been obliterated.

As temperatures approach 400° to 500°C (750°–950°F), the minerals in schists become segregated into single-mineral layers. During metamorphic reactions, the liberated ions in unstable felsic minerals such as muscovite mica migrate and recrystallize in distinct light-colored, single-mineral bands containing larger grains of new, stable minerals such as quartz and feldspar. Meanwhile, ions freed from other minerals, such as biotite and the dark-colored amphiboles and pyroxenes, form intervening bands of dark mafic minerals. This process, known as **metamorphic differentiation,** creates the distinctively banded metamorphic rock **gneiss** (from the German *gneisto,* meaning "sparkle," and pronounced "nice").

Figure 7-9 A block of Morton Gneiss, photographed on the Minneapolis campus of the University of Minnesota. The rock's characteristic gneissic banding formed under the high directed pressure and high temperatures typical of mountain building at convergent plate boundaries.

Gneisses are most often found in the cores of ancient mountain ranges that were uplifted during mountain building, and then had their overlying rocks worn away, such as in the Front Range of the Colorado Rockies or in the North Cascades of Washington. Gneisses can also be found in non-mountainous areas where lofty mountains once stood. The 3.7-billion-year-old Morton Gneiss (Fig. 7-9), exposed in the Minnesota River valley of southwestern Minnesota, and the 600- to 700-million-year-old Fordham Gneiss, exposed from mid-Manhattan north in New York City, are remnants of long-departed mountain ranges.

Gneisses exposed to even greater heat will start to melt, marking the beginning of igneous activity and the end of metamorphism. Sometimes a metamorphic rock that has partially melted cools at that stage, resulting in a rock that is part metamorphic and part igneous, known as a **migmatite** (Greek for "mixed rock"). As we saw in Chapter 3, such minerals as quartz and potassium feldspar melt at lower temperatures than biotite mica, amphibole, and pyroxene. When heat in the 600° to 800°C range (1110°–1470°F) is applied to gneiss, its felsic bands begin to melt because their melting points have been reached, but the mafic bands, whose melting points are higher, remain solid. Because the mafic bands are beginning to deform at these temperatures, however, they form intricately contorted patterns under these ultimate metamorphic conditions. The melted felsics cool and crystallize into a granular texture resembling that of plutonic igneous rock (Fig. 7-10).

Figure 7-10 Migmatite from the Chuckwalla Mountains of southeastern California. Migmatites are foliated like metamorphic rocks, but their felsic regions, which cooled from a melted state, contain phaneritic textures like those of plutonic igneous rocks.

Foliated Rocks Derived from Igneous Rocks

Like multimineral sedimentary rocks, igneous rocks exposed to high temperatures and high directed pressure may also react to form schists and differentiate to form gneisses. When basalt, the most common volcanic rock of the Earth's crust, is metamorphosed, its multimineral composition reacts to produce an assemblage of new minerals. Basalt's pyroxenes, olivines, and calcium plagioclases are converted to a group of more stable minerals that includes chlorite (a mica-like iron-and-magnesium silicate) and epidote (an iron-calcium silicate). Because both of these minerals are green, the resulting foliated metamorphic rock is called *greenschist*. When granite and diorite, the most common plutonic rocks of the Earth's continental crust, are metamorphosed under conditions of high temperature and directed pressure, their multimineral compositions react and differentiate to form gneisses. However, because their parent rocks are felsic, these gneisses contain more pronounced quartz and feldspar foliations and fewer mafic foliations than do gneisses derived from shale.

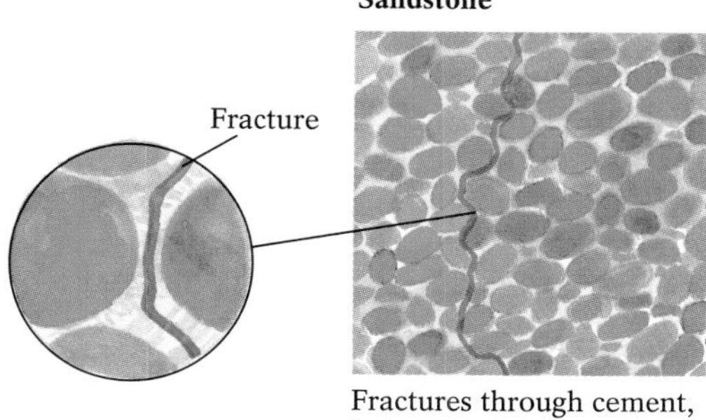

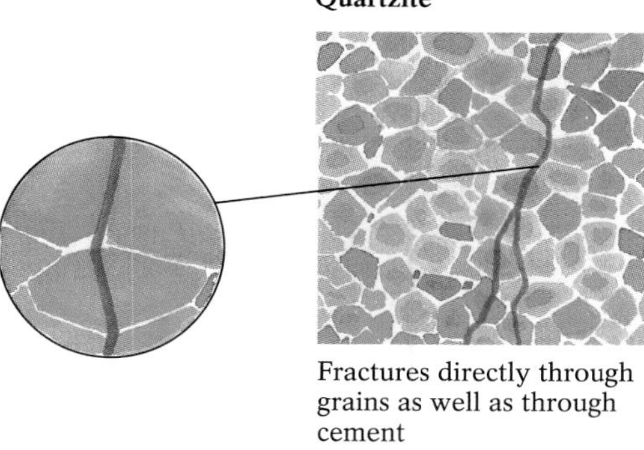

Figure 7-11 When a quartz sandstone is fractured, it tends to break within the cement that surrounds the sand grains. After metamorphism to quartzite, the rock breaks right through the grains.

Nonfoliated Rocks

Nonfoliated metamorphic rocks may be produced by the increased heat and high confining pressure ensuing from deep burial, or they may result from contact with heat from an intruding body of magma.

Sedimentary rocks that consist predominantly of a single mineral typically recrystallize as coarse-grained nonfoliated rocks. When relatively pure limestones and dolostones are metamorphosed, for example, they form a coarsely granular nonfoliated mosaic of calcite and/or dolomite grains called **marble.** In the case of limestones and dolostones containing a significant amount of silicate material, such as clays, feldspars, and micas, metamorphic reactions between the carbonate minerals and the silicates produce a variety of calcium-bearing silicate minerals such as garnet; the coarse-grained, nonfoliated metamorphic rocks containing such calcium-rich silicates are known as **skarn.**

Pure sandstone metamorphoses to **quartzite,** a very durable nonfoliated rock. Although it is sometimes difficult to distinguish a hard, well-cemented sandstone from a quartzite, one need only strike both rocks with a rock hammer and then study their fragments under a binocular microscope. When sandstone is shattered, it tends to break *around* its sand grains because quartz grains are generally stronger than their surrounding cement. During metamorphism, however, the cement recrystallizes, becoming stronger and forming even stronger bonds with the sand grains. Thus, when a quartzite is shattered, the rock tends to break both around the grains *and* directly through them (Fig. 7-11).

When hot magma intrudes a shale or basalt parent rock, the heat from the intrusion drives off mineral-bound water, promoting recrystallization and metamorphic reactions that produce minerals with more compact structures. The resulting rock, **hornfels,** is dark-colored, dense, and hard. Hornfels has a fine-grained texture, because it cools relatively rapidly at the chilled margin of the intrusion.

Types of Metamorphism

Heat, pressure, and chemically active fluids interact in different ways in different geological settings to produce metamorphic rocks. These settings produce several distinctive types of metamorphism. Under some conditions, only a narrow area of rock is affected; in other settings, rocks may be changed over a vast region.

Contact Metamorphism

Any preexisting rock touched by the intense heat of migrating magma immediately begins to metamorphose. This **contact metamorphism** is entirely a result of the heat from the magma and from hot circulating fluids—pressure is not a significant factor in producing contact metamorphic rocks. Because most rocks are poor conductors, the effect of contact metamorphism decreases with increasing distance from the magma. Thus, rocks in direct contact with the intruding magma will be highly metamorphosed whereas those farther away, receiving little magmatic heat, will be only slightly altered. The amount of water in a contact metamorphic rock's mineral structure is sometimes used as an indication of how close to a magma it formed: The closer to the heat source, the more water is driven off; the farther from the heat source, the more water remains.

The effects of contact metamorphism can generally be seen in rock adjacent to any igneous intrusion. Contact metamorphism may extend only a few centimeters or meters around a small igneous dike or sill, or it may extend up to several kilometers around a major batholith. The entire zone of contact metamorphism surrounding an igneous intrusion, which is marked by changes in mineral content and grain texture, is referred to as a metamorphic **aureole** (Fig. 7-12).

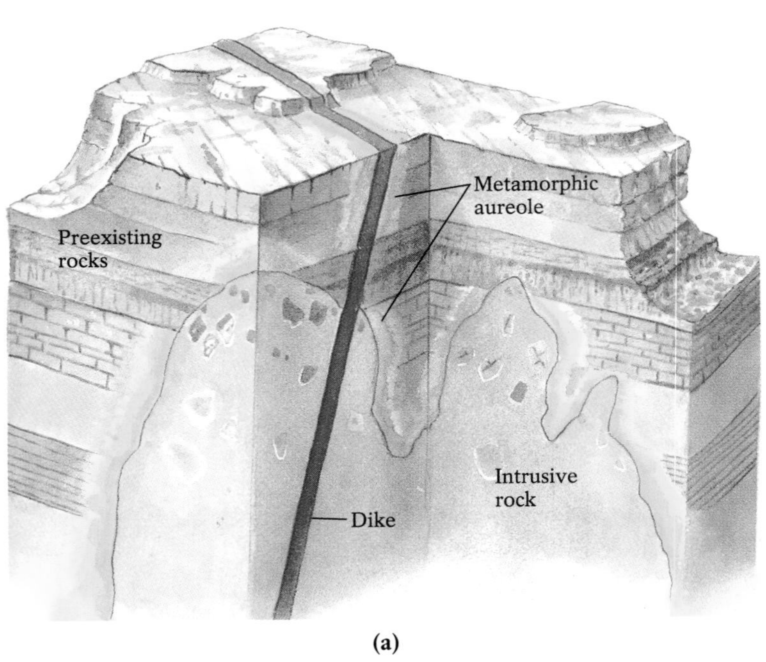

(a)

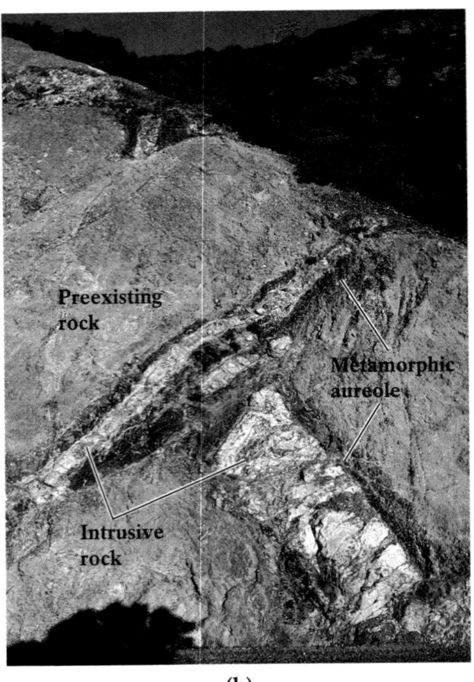

(b)

Figure 7-12 (**a**) Contact metamorphism occurs when magma intrudes preexisting rocks. The rocks adjacent to the magma are strongly metamorphosed; those farther from the region of direct contact are less affected, although they still receive metamorphic heat from circulating ion-rich fluids. (**b**) An igneous dike in preexisting rock, with an obvious metamorphic aureole.

Regional Metamorphism

Unlike contact metamorphism, which has relatively local effects, **regional metamorphism** alters rocks for thousands of square kilometers. It is responsible for the vast regions of exposed metamorphic rock in central Canada and the tracts of metamorphic rock found in the Appalachians of New England, the Rockies, and the Cascades. There are two types of regional metamorphism—burial metamorphism and dynamothermal metamorphism.

Burial Metamorphism **Burial metamorphism** occurs when rocks are overlain by more than about 10 kilometers (6 miles) of rock or sediment, and the confining pressure and geothermal heat at these depths combine to recrystallize their component minerals. Because no directed pressure is involved, burial metamorphic rocks are generally nonfoliated.

Burial metamorphism is occurring today in such locations as the Gulf Coast of Louisiana, where clay minerals lie beneath 12 kilometers (8 miles) of sediment near the bottom of the Mississippi River's deltaic deposits. Samples collected from deep drill holes in the delta indicate that these clay minerals have already begun to metamorphose into minerals more stable at that depth, such as mica and chlorite.

Dynamothermal Metamorphism **Dynamothermal metamorphism** occurs when rocks get caught between two converging plates during mountain building (Fig. 7-13). In such a convergent plate setting, lateral compression—a form of directed pressure—forces some rocks upward and forces some downward to depths of several to tens of kilometers, where these are then additionally subjected to great heat and confining pressure. ("Dynamothermal" refers to both the *dynamic* pressures on the rocks and the *thermal* energy applied by geothermal heat and rising, mantle-derived magma.) Because of the directed pressure that acts on rocks at convergent boundaries, dynamothermal metamorphic rocks are often foliated.

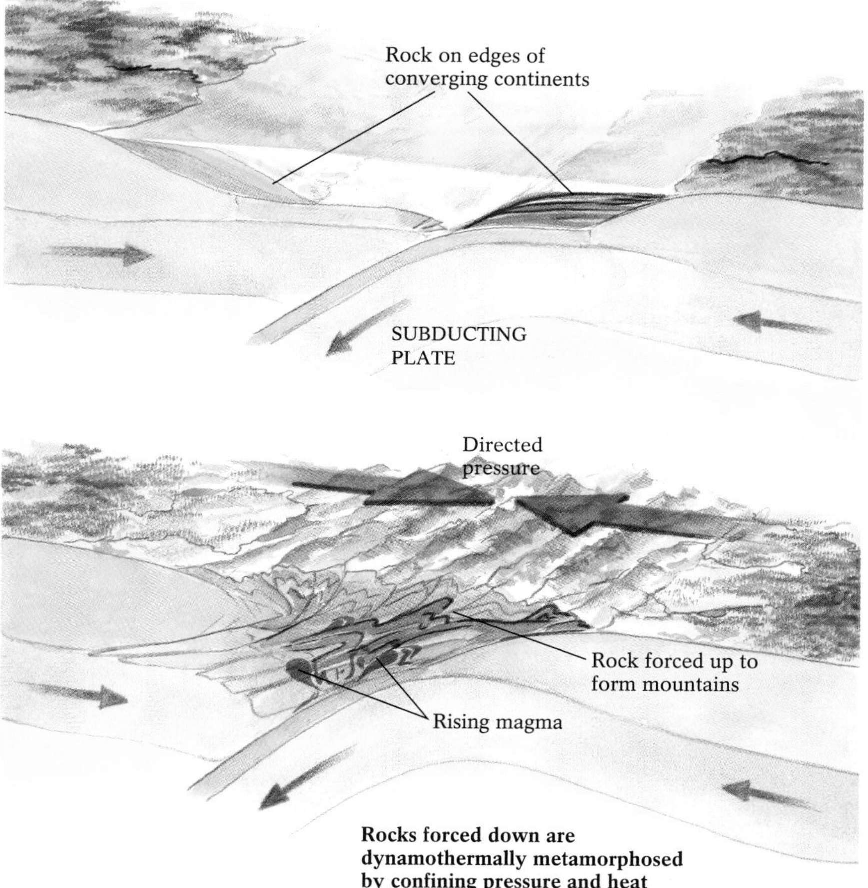

Figure 7-13 Dynamothermal metamorphism is generally associated with the directed pressures and magmatic heat of convergent plate boundaries.

Dynamothermal metamorphism formed the vast regions of metamorphic rock that are the roots of ancient mountains, such as have been exposed by erosion in the northern Great Lakes states of Minnesota, Wisconsin, and Michigan and adjacent parts of southern Ontario and Quebec.

Other Types of Metamorphism

Rocks may be metamorphosed in several ways other than by regional increases in heat and pressure or by contact with intruding hot magmas. Metamorphism can also occur where rocks are invaded by hot water, at fault zones, and at meteorite impact sites.

Hydrothermal Metamorphism **Hydrothermal metamorphism** is the chemical alteration of preexisting rocks by hot water. This water may come from magma, it may derive from the structures of metamorphosed rocks, or it may be groundwater that has percolated down from the surface and been heated by contact with deep subsurface rock. However, most hydrothermal metamorphism occurs within ocean floors, when seawater penetrates cracks near a divergent plate boundary. Descending until it encounters hot basaltic magma, the seawater becomes heated to about 300°C (575°F). Rising then as steam, it passes through overlying mafic rocks and oceanic sediments, dissolving soluble materials and recrystallizing mafic minerals such as olivine and pyroxene into the magnesium silicate serpentine (Fig. 7-14). This process, known as *serpentinization*, is responsible for much of the metamorphism in ocean basins. As the hot, mineral-rich steam ultimately rises to the sea floor, it cools and precipitates its dissolved load, often as valuable concentrations of copper, nickel, iron, and lead.

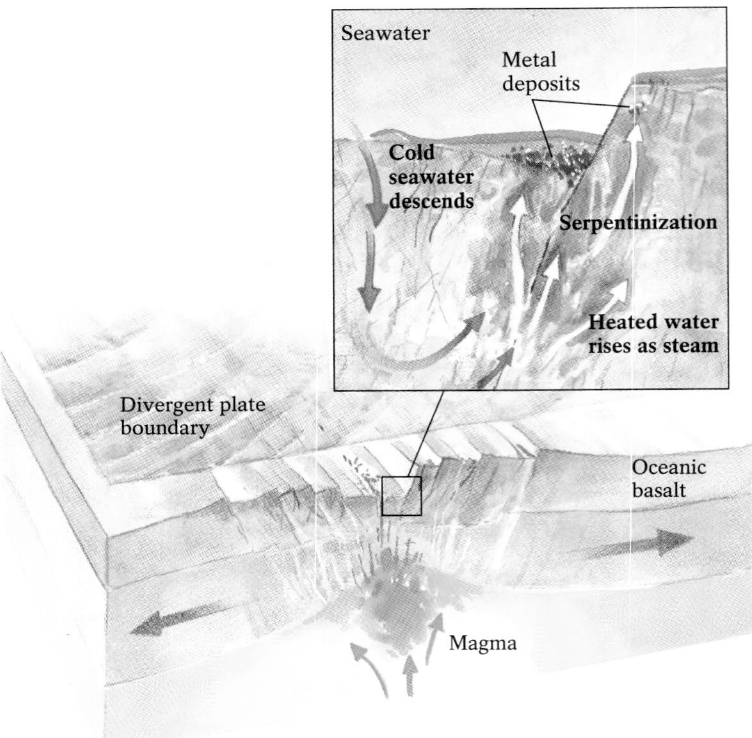

Figure 7-14 Hydrothermal metamorphism occurs most often at mid-ocean ridges, where tectonic plates diverge. Cold seawater comes into contact with hot basaltic magma, heats to its boiling point, rises as steam, and dissolves metals and other ions from basalts and oceanic sediments. This process converts olivine and pyroxene minerals in the basalts to the magnesium silicate serpentine.

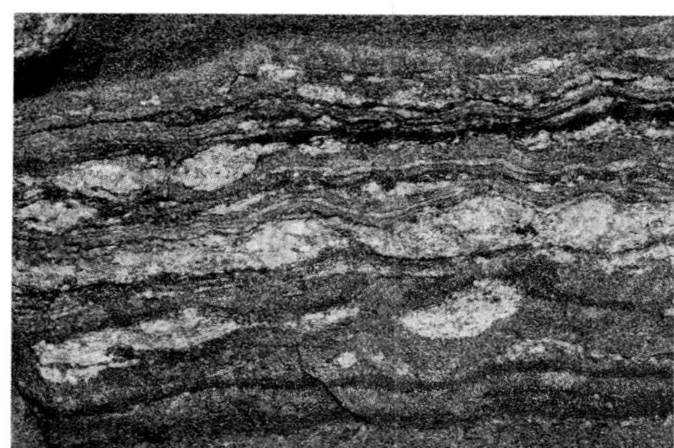

Figure 7-15 Augen gneiss, a common product of fault metamorphism. The augen ("eyes") are the areas of lighter colored rock, which have been flattened and rotated by high pressure.

Fault Metamorphism As rocks grind past one another along a fault, a large amount of directed pressure and considerable frictional heat are generated. The result is **fault metamorphism.** The pressure and heat in the immediate vicinity of an active fault can be high enough to produce foliation and metamorphic recrystallization, although their effects taper off a short distance from the fault. Fault metamorphism may produce unusually large grains (some the size of grapefruits) due to recrystallization, particularly if fluids are available to aid the movement of ions through cracks in the rocks. These large grains are typically flattened and rotated by the grinding motion along the fault into almond-shaped structures known as *augen* (German for "eyes") (Fig. 7-15).

Shock Metamorphism A dramatic form of metamorphism occurs when a meteorite strikes rocks at the Earth's surface. This phenomenon, which is rare because most meteors burn up as they pass through the Earth's atmosphere, is **shock metamorphism.** Shock metamorphism results from the tremendous pressures and temperatures generated at meteorite impact sites, which cause "shocked" minerals in the stricken rocks to shatter and recrystallize. The minerals produced commonly include stishovite and coesite, two extremely high-pressure types of quartz that are found within the curved surfaces of impact craters. Minerals created at meteorite impact sites occur in no other geological setting. The search for impact site metamorphic rocks, both on land and undersea, has produced extensive information on this type of metamorphism. Such study has led to the hypothesis that the Earth's dinosaurs were extinguished by an enormous meteorite impact, perhaps on the Gulf Coast of Mexico, about 65 million years ago. (See Highlight 1-1, "What Caused the Extinction of the Dinosaurs?")

Metamorphic Grade and Index Minerals

Because it is impossible to study directly the environments in which metamorphic reactions naturally occur, geologists simulate these conditions by subjecting minerals to increasing temperatures and pressures in the laboratory. Also in the laboratory, we can combine the chemical compounds that make up common rocks and then heat and stress these mixtures to varying degrees; in this way, geologists can determine which minerals crystallize under various specific conditions of temperature and pressure. From such experiments, the conditions that cause individual rocks to metamorphose are known to within about 10° to 20°C and just fractions of a kilobar of pressure.

Information gained in the laboratory has also allowed geologists to assign to each metamorphic rock a **metamorphic grade,** indicating the degree to which it has been changed from its parent material. *Low-grade* metamorphic rocks retain enough of their original character—such as some bedding and other sedimentary structures, fossils, and many of their original minerals—to enable geologists to readily identify their parent rock. Low-grade metamorphism occurs at relatively low temperatures (200°–400°C, or 400°–750°F) and pressures (about 1–6 kilobars). *High-grade* metamorphic rocks, which have been subjected to much higher temperatures and pressures, lack virtually all of their original structures, fossils, and minerals. High-grade metamorphism occurs at temperatures between 500° and 800°C (950°–1475°F), and pressures of about 12 to 15 kilobars. A variety of intermediate metamorphic grades exist as well.

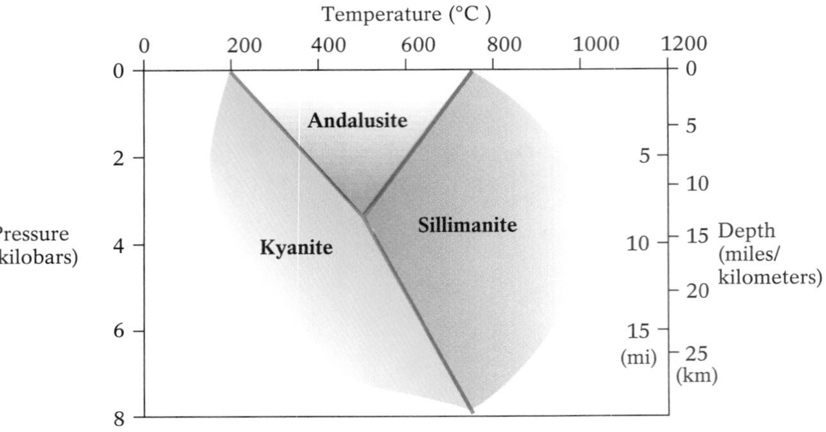

Figure 7-16 The depth, pressure, and temperature relationships that are responsible for the crystallization of the aluminum silicates kyanite, andalusite, and sillimanite. Andalusite is generally found only in rocks that form at relatively low temperatures and pressures.

Mineral Zones

The minerals in metamorphic rocks are clues to the temperature and pressure conditions under which they formed. Some minerals, because they are stable only within a narrow range of temperatures and pressures, serve as indicators of specific metamorphic environments. These are called **metamorphic index minerals,** and areas of rock containing them are designated as specific **mineral zones.**

The presence of a given index mineral throughout a mineral zone indicates that the same specific set of temperature and pressure conditions metamorphosed the entire zone, distinguishing it from other zones that formed under different conditions. For example, the aluminum silicates kyanite, andalusite, and sillimanite—three common index minerals that form only during metamorphism—are often associated with separate mineral zones. Although each of these minerals has the chemical formula Al_2SiO_5, each forms under different temperature and pressure conditions (and therefore has a different crystal structure) (Fig. 7-16). Andalusite forms only at low temperatures and pressures, so metamorphic rocks in an andalusite mineral zone are designated low-grade. Because kyanite and sillimanite form under much higher pressures and temperatures, metamorphic rocks in mineral zones containing these index minerals are considered to be of higher grade than those in an andalusite zone.

Mineral zones are found both in rocks produced by contact metamorphism and in rocks produced by regional dynamothermal metamorphism. In regions of contact metamorphism, the highest-grade metamorphic rocks are found closest to the igneous intrusion, and contain index minerals associated with high temperatures. Lower-grade metamorphic rocks are found at progressively greater distances from the intrusion, and contain index minerals reflecting lower temperatures. A clear example of mineral zones produced by contact metamorphism can be found at Onawa, Maine, where, 365 million years ago, a granitic intrusion invaded and metamorphosed tightly folded sedimentary rocks (Fig. 7-17).

Mineral zones within the rocks of the eastern United States, from eastern Pennsylvania to coastal New England, indicate the variations in metamorphic grade associated with regional dynamothermal metamorphism. About 390 million years ago, the first in a series of collisions occurred between plates that now include Europe, Africa, and North America. These collisions crumpled the eastern edge of North America, initiating the growth of the Appalachian mountains, which extend from Alabama into Atlantic Canada. Compression of the North American plate edge heated and pressed the region's sedimentary parent rocks into a sequence of metamorphic rocks that increase in grade from the largely unmetamorphosed rocks of western Pennsylvania to the intensely metamorphosed zones of Massachusetts and New Hampshire (Fig. 7-18).

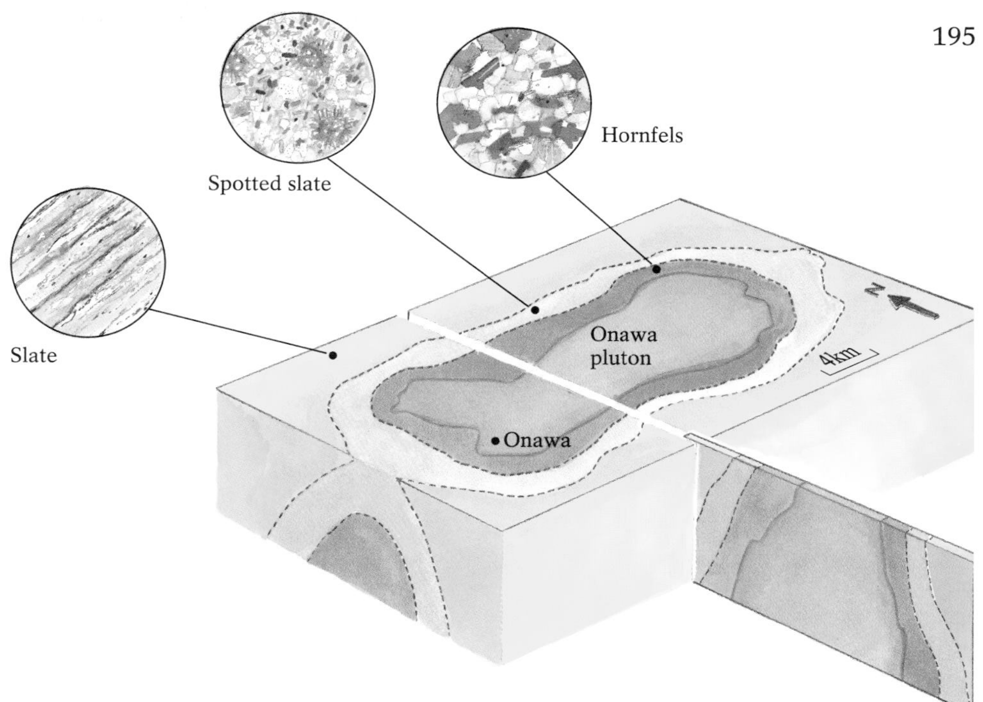

Figure 7-17 Heat from the intrusion of the Onawa pluton in central Maine created a series of metamorphic mineral zones typical of large-scale contact metamorphism. The metamorphic rocks associated with this pluton range from high-grade hornfels, to intermediate-grade "spotted slate" (slate containing a number of large, recrystallized grains), to low-grade slate.

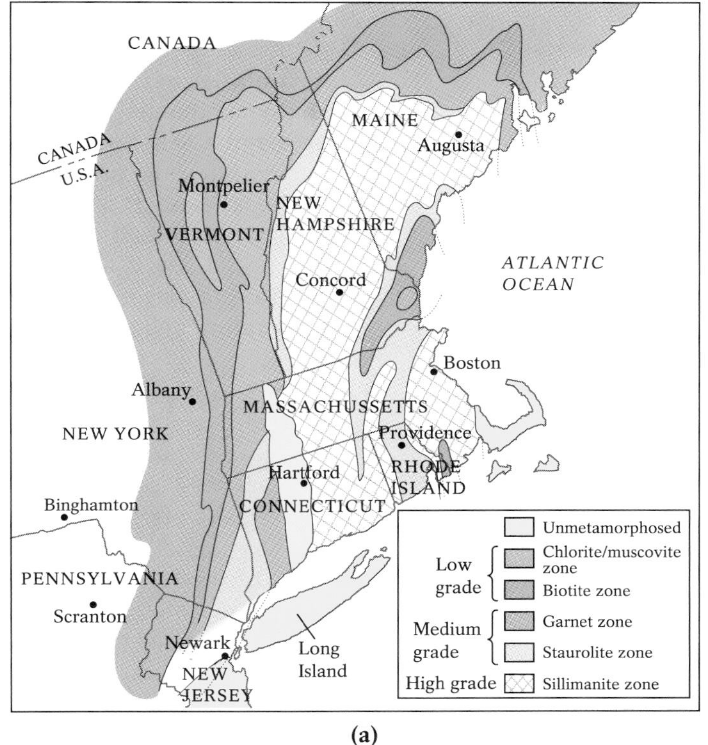

(a)

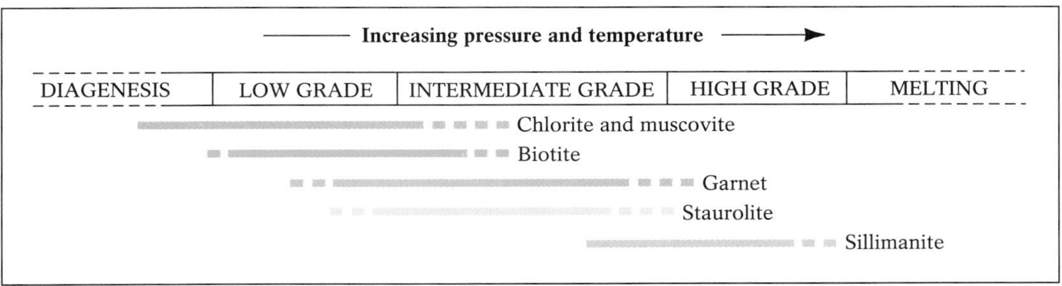

(b)

Figure 7-18 (**a**) The metamorphic mineral zones of the northeastern U.S. These zones resulted from the regional dynamothermal metamorphism associated with several episodes of plate convergence between about 400 and 250 million years ago. (**b**) General metamorphic mineral zones typical of regional dynamothermal metamorphism.

Metamorphic Facies

One major drawback limits the use of index minerals as indicators of metamorphic conditions: If the parent rock did not contain a mineral's component elements, that index mineral will be absent from the metamorphic rocks. For example, don't spend much time looking for the high-grade metamorphic minerals kyanite or sillimanite in most metamorphosed limestones, because there is rarely enough aluminum in limestone to form these minerals. In order to identify past conditions of temperature and pressure, *regardless of the composition of the parent rock,* geologists have formulated the concept of metamorphic facies.

Metamorphic facies are assemblages of different minerals customarily found together in metamorphic rocks. Like an index mineral, a facies indicates the specific temperature and pressure conditions that characterized the development of these metamorphic rocks. Moreover, the presence of such a group of minerals, all associated with a given set of metamorphic conditions, allows us to infer these conditions even if particular individual index minerals are missing from the area. As we saw in our discussion of sedimentary facies in Chapter 6, changes in environmental conditions gradually produce different facies. The facies that we see in metamorphic rocks tell us about changing metamorphic environments.

To understand the concept of metamorphic facies, consider a slab of basaltic oceanic crust and an overlying sediment layer subjected to regional dynamothermal metamorphism within a subduction zone, as well as the nearby continental sediment subjected to contact metamorphism from the rising magma (Fig. 7-19). The basalt and the sediments experience the same temperatures and pressures at each location in the subduction zone. Although the basalt and the sediments will metamorphose to different minerals, because the composition of the parent rocks differ, the new minerals—metamorphosed basalt and metamorphosed sediments—formed at each location constitute a single metamorphic facies because they formed under the same temperature and pressure conditions.

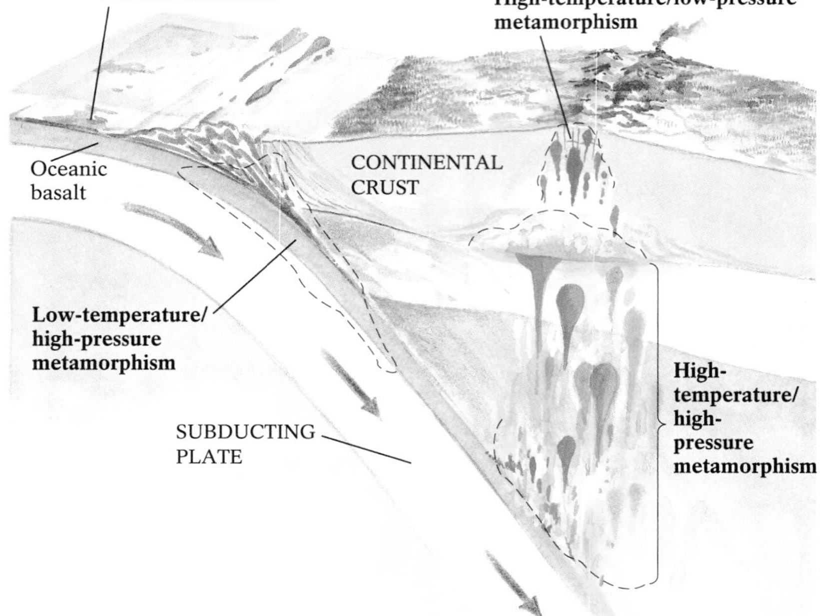

Figure 7-19 The three main metamorphic environments associated with subduction zones: low temperature/high pressure, high temperature/high pressure, and high temperature/low pressure. Each of these produces a separate metamorphic facies, regardless of the composition of the parent rocks.

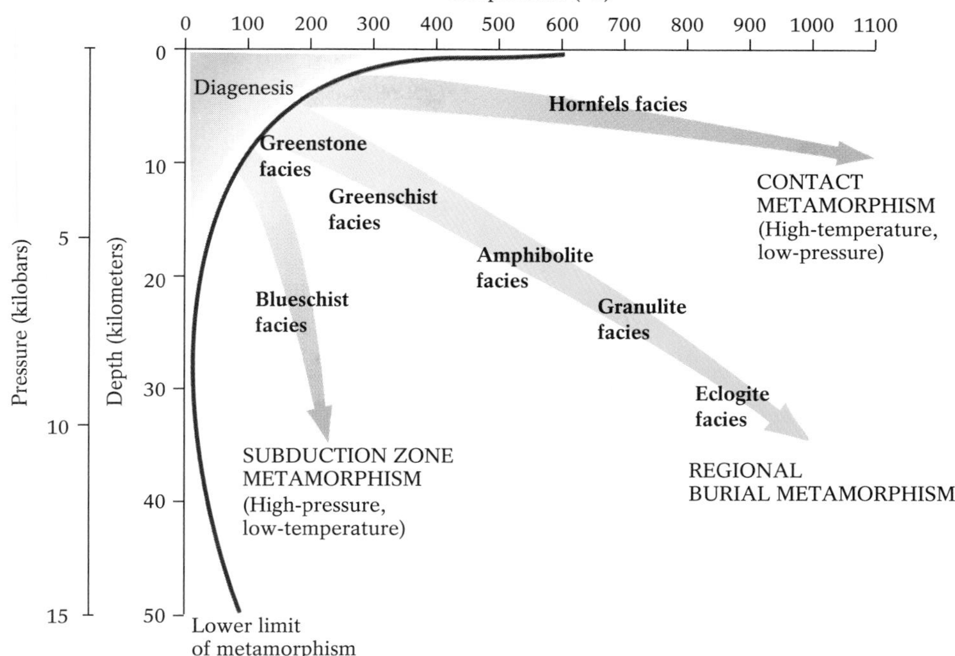

Figure 7-20 The depth, pressure, and temperature relationships that produce the common metamorphic basalts, which are used to identify the various metamorphic facies. Using such a chart, one should be able to extract such information as: What temperature and pressure conditions create the eclogite facies? A rock that is stable at 500°C and 3 kilobars of pressure would belong to which facies?

Because much of the Earth's crust is composed of basalt, there are a number of common metamorphosed basalts that provide a convenient way to name and categorize metamorphic facies (Fig. 7-20). *Hornfels facies* rocks form under the high-temperature/low-pressure conditions generally found around intruding igneous rocks such as batholiths, plutons, and dikes. *Greenstone facies* form under the relatively low temperatures and pressures of regional burial metamorphism, which converts basalts to a greenish, nonfoliated, low-grade metamorphic rock called greenstone. *Greenschist facies* form when the same rocks are subjected to directed pressures, producing chlorite-rich greenschist. *Amphibolite facies* form at intermediate-grade temperatures and pressures, between about 450° and 650°C (850°–1200°F), and at pressures of 4 to 10 kilobars, producing a rock containing a large amount of the amphibole hornblende and some accessory garnet and plagioclase feldspar. If the parent rock contains aluminum-rich micas, clay minerals, and feldspars, amphibolite-facies rocks also contain large grains of intermediate-grade minerals such as staurolite, kyanite, and sillimanite.

The *granulite facies* of high-grade metamorphic rocks develops at higher temperatures and pressures, at which stage virtually all the water in a rock's minerals has been driven out. These ultra-dry conditions inhibit melting, producing minerals that are stable beyond 700° to 800°C (1300°–1475°F); these include quartz and biotite mica, high-temperature pyroxenes, garnets, and intermediate-to-high-grade aluminum silicates such as sillimanite or kyanite.

The *eclogite facies* is a rare assemblage of minerals that forms only in *extremely* high-pressure, high-temperature environments. Found as inclusions in mantle-derived diamond pipes, basalts, and ultramafic intrusions, eclogite-facies rocks consist largely of pyroxenes and iron-rich garnets that crystallize at pressures exceeding 10 kilobars (corresponding to a depth of about 35 kilometers, or 20 miles), and at about 800°C (1475°F). These conditions and the occurrence of eclogites as inclusions in mantle-derived rocks suggest that eclogite-facies rocks probably originate in the upper mantle.

The *blueschist facies* contains minerals that are stable only at high pressures and relatively *low* temperatures. The most prominent is a blue sodium-rich amphibole called glaucophane, which imparts a distinctive bluish tint to the foliated rocks of this facies. Because high pressures exist principally beneath great thicknesses of overlying rock, where high temperatures should also prevail, the conditions that produce blueschist-facies rocks are quite rare. As we shall see in the next section, such conditions typify subduction-zone trenches, where high pressures associated with plate convergence develop at relatively shallow depths and rocks are not subjected to significant amounts of geothermal heat.

Plate Tectonics and Metamorphic Rocks

Plate movements create the heat, pressure, and circulating hot fluids that produce much of the Earth's metamorphic rock. Directed pressure at plate boundaries causes virtually all regional dynamothermal metamorphism, and new magmas generated at subduction and divergent zones are responsible for most contact metamorphism and much of the heat for regional metamorphism as well.

Vast regions of foliated rock suggest that directed pressures played a hand in the rock's metamorphism, with the foliation pattern itself indicating the direction of ancient tectonic compression. Because we know from their high metamorphic grade that some of these rocks formed at great depths, we can conclude that they have experienced considerable uplift and subsequent erosion.

Certain metamorphic rocks indicate specific plate tectonic settings. In subduction zones, oceanic plates converge on a regional scale, and several distinct metamorphic environments can be found (Fig. 7-21). There is a relatively shallow high-pressure/low-temperature zone in the subduction trench,

Figure 7-21 Metamorphic facies characteristic of subduction zones. The high-pressure/low-temperature environment produces blueschist-facies rocks; the high-temperature/high-pressure environment produces granulite-facies rocks; and the low-pressure/high-temperature environment produces greenschist-facies rocks.

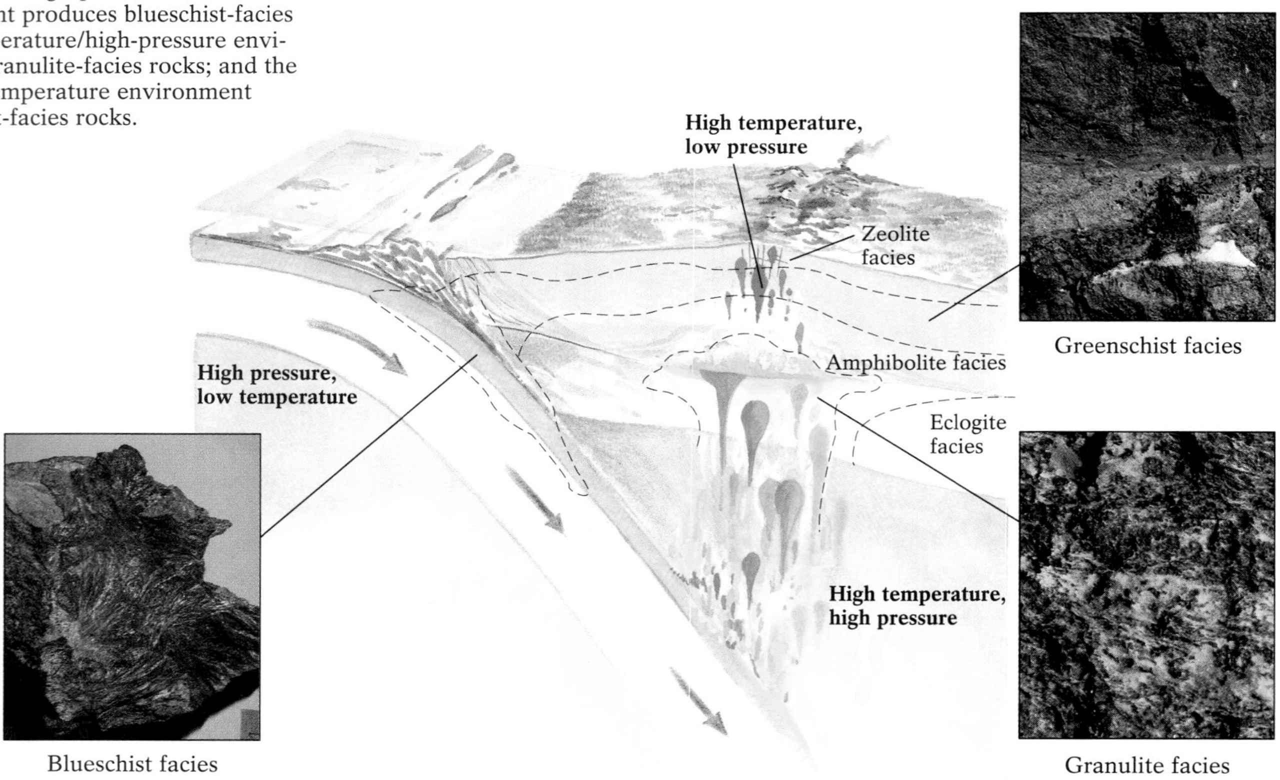

Greenschist facies

Blueschist facies

Granulite facies

Figure 7-22 Three-billion-year-old greenstone from Ely, Minnesota. The pillow structures in this rock are a clue that the parent rock originated as submarine basalts; the fact that the rock is nonfoliated and its structures relatively undeformed indicate that the rock was not subjected to directed pressure.

where the subducting rocks are subjected to directed pressure and undergo blueschist-facies metamorphism. There is a low-pressure/high-temperature zone above the subducting plate but close to the Earth's surface, where heat from rising magmas—coupled with the relatively low burial pressures associated with shallow depths and the directed pressures at convergent plate boundaries—produces greenschist-facies rocks. There is also a high-pressure/high-temperature zone above the subducting plate but at greater depth, where granulite-, amphibolite-, and eclogite-facies rocks form. All metamorphic rocks associated with subduction zones are foliated, because the entire region is subjected to the directed pressures of convergence.

Blueschist-facies rocks are a particularly useful clue to the location of ancient subduction zones and convergent plate boundaries: These rocks form at high pressure and low temperature, conditions that geologists believe occur only when an oceanic plate subducts. The blueschist-facies rocks in western California most likely mark the subduction zone that produced the Sierra Nevada batholiths, some of which can be seen in Yosemite National Park. Blueschist-facies rocks have also been discovered in the Appalachians of northern Vermont, in what may have been a zone of converging plates, volcanism, and great earthquake activity hundreds of millions of years ago.

Ocean basins contain virtually no regionally metamorphosed rocks, although oceanic rocks account for more than 60% of the Earth's crust. Oceanic rocks form principally by volcanic eruptions at mid-ocean ridges where plates diverge; because they are not subjected to directed pressure, they generally do not become foliated (Fig. 7-22). What little metamorphism does occur in ocean basins is either relatively low-grade hydrothermal serpentinization, or contact metamorphism near the crests of mid-ocean ridges due to the heat of rising basaltic magma.

Metamorphic Rocks in Daily Life

Metamorphic rocks are generally strong and durable, for several reasons: Heat and pressure eliminate pore spaces in the rocks, increasing their density; metamorphic reactions replace unstable minerals with stable ones; and recrystallization strengthens the bonds between sediment grains and recrystallized cement. Thus, metamorphic rock is a popular material for the weather-resistant exteriors of office buildings and as foundation stones for such large-scale projects as bridges and dams. Every year in the United States, building and road construction consumes 1.6 trillion kilograms (about 400 million tons) of slate, marble, and quartzite.

Figure 7-23 The white-on-green streaked metamorphic rock serpentinite is a common feature of office building lobbies.

Certain metamorphic rocks are valued highly for their appearance. One of architecture's most prized decorative building stones is serpentinite, the rock produced by hydrothermal metamorphism at divergent plate boundaries (p. 192). This rock, called verd antique ("ancient green") by architects, adorns the interiors of such stately buildings as the United Nations, the National Gallery of Art, and countless office towers, banks, and libraries (Fig. 7-23). When marine limestone and dolostone are deposited in the cracks that often form in serpentinite, the attractively streaked green-serpentine and white-carbonate rock results.

Marble is a metamorphic rock that is highly desirable to sculptors because of its appearance and texture. Because pure marble is snow-white, it shows the details of a sculptor's carving, and, because it is soft, it is responsive to the sculptor's tools. (Calcite, marble's main constituent, measures 3 on the Mohs hardness scale.) The stone from which Michelangelo fashioned his renowned sculptures was a result of metamorphism when the Apennine mountains of Italy, containing a layer of pure marine limestone, were uplifted from the sea (Fig. 7-24). Metamorphosis of impure limestones results in marbles colored by the impurities—pink, red, or brown from iron oxides, gray or black from carbon-rich material, or green from the calcium-magnesium pyroxene diopside or the common amphibole hornblende.

(a)

(b)

Figure 7-24 (a) The Carrara marble quarry in northern Italy. The purity of Carrara marble, which stems from the purity of its parent limestone, has for centuries made it a favored source of sculpting material (**b**).

Steatite, or soapstone, another sculpture material (also often used for laboratory work counters), and many other useful everyday commodities are products of regional dynamothermal metamorphism. Talc, used in talcum powder, and flame-resistant asbestos, used in automobile brake linings and in the safety garb of firefighters, are products of the low-grade regional metamorphism of ultramafic rocks. Garnet, a valuable industrial abrasive because of its hardness (and also January's gem-like birthstone), results from intermediate-grade metamorphism. The high-grade metamorphic minerals kyanite and sillimanite are key components of porcelain casings for spark plugs, because the high temperatures at which they crystallize make them able to withstand intense heat.

Metamorphic rocks also occasionally contain important mineral deposits, which enable the countries in which they're found to prosper in the world's marketplace (see Chapter 20). Zinc (in the mineral sphalerite), lead (in galena), copper (in chalcopyrite and bornite), iron (in magnetite), and gold are but a few of the valuable minerals commonly found in metamorphic rocks.

Potential Hazards from Metamorphic Rocks

While some characteristics of metamorphic rocks have great aesthetic or practical value, others can produce hazardous conditions in natural settings. Foliation, in particular, is potentially dangerous. Because they reduce the strength of metamorphic rocks, slaty cleavage and schistosity can cause slopes to fail and set off landslides, particularly where their cleavage planes are parallel to steep slopes (Fig. 7-25). Building in foliated metamorphic terrains requires careful planning and thorough geological investigation to determine the orientation of the foliation.

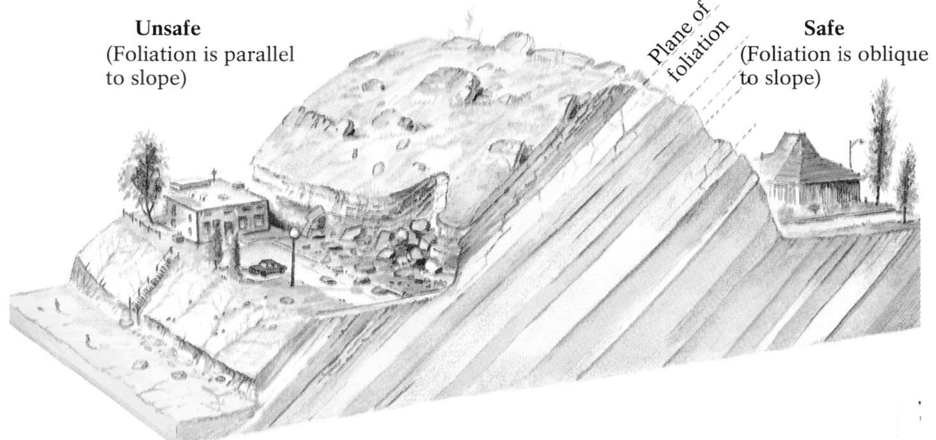

Figure 7-25 Building in areas with metamorphic surface rocks requires consideration of the direction of their foliation planes. The slope on the left side of this figure is unstable because the rock's foliation lies parallel to the slope, and can readily fail, causing a landslide of thick slabs of rock. The slope on the right is stable, because the foliation is oriented oblique to the slope.

The weak planes of metamorphic foliation are also especially susceptible to earthquake damage. A clear indication of the inherent dangers of foliated metamorphic rocks occurred at midnight on August 17, 1959, near Yellowstone Park in Montana, when a powerful earthquake shook the Madison Canyon area. A deeply weathered body of gneiss and schist, the foliations of which paralleled the canyon's south wall (Fig. 7-26), easily slid away from adjacent rock during the shaking. The landslide accumulated debris as it raced downslope at about 100 kilometers per hour (60 mph), covering a campground and its 28 visitors with some 45 meters (150 feet) of rock, soil, and vegetation.

Metamorphic foliation must also be analyzed when designing large-scale engineering projects such as dams. Today, teams of geologists and engineers routinely study local and regional rock formations for evidence of inherent weakness before beginning any such project. In the past, however,

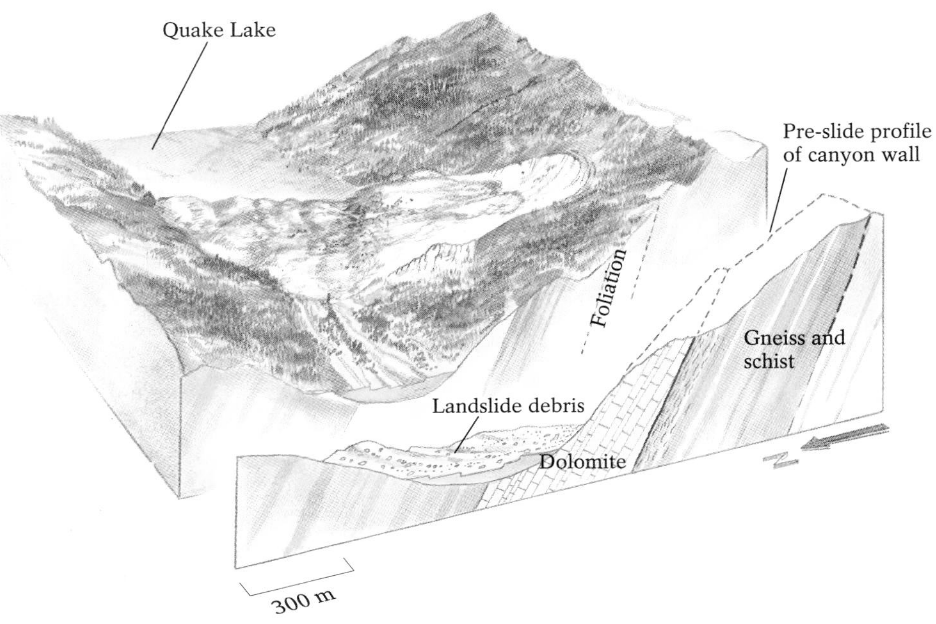

Figure 7-26 The disastrous rock slide of 1959 at Madison Canyon, Wyoming, occurred when an earthquake jolted weak, highly fractured gneiss and schist along their foliation planes parallel to the canyon's south wall. Above: A photo of the landslide aftermath, taken from the northern canyon wall.

this was seldom done. On the night of March 12, 1928, the St. Francis dam in Saugus, California, failed, claiming 500 lives. Its eastern side adjoined loose landslide debris from a weak schist whose foliation planes lay parallel to the canyon wall. This, combined with an undetected fault that traversed the dam site, was largely responsible for the disaster (Fig. 7-27). This disaster improved the way we investigate sites for prospective dams (although community opinions and political pressures may still cause geologists' advice to go unheeded).

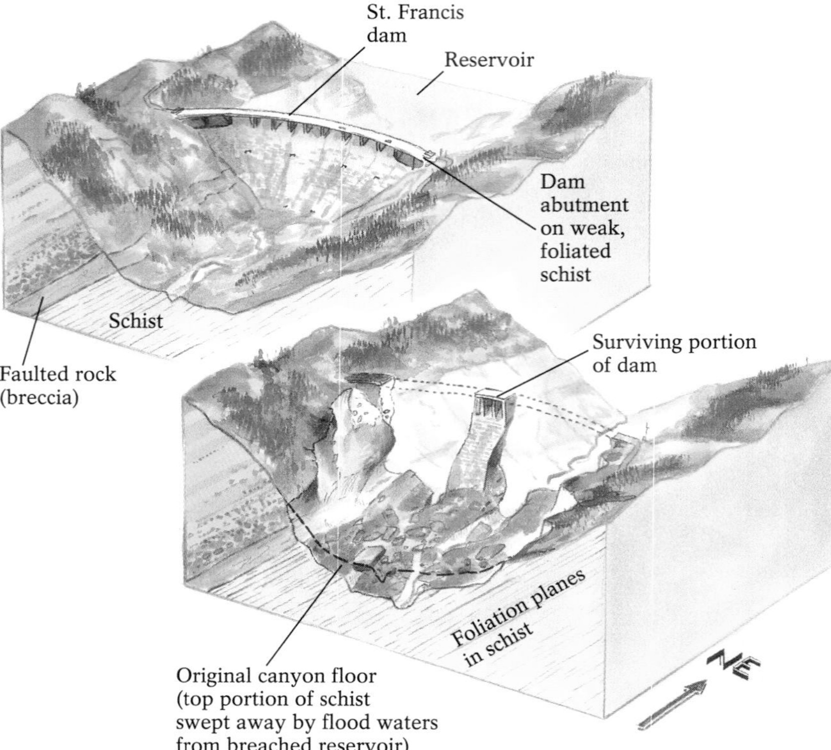

Figure 7-27 The geology of the St. Francis dam, which collapsed in 1928. The weak, foliated rocks that supported one side of the dam and the faulted bedrock below the dam led to the dam's failure.

The Rock Cycle Revisited

When we previewed the rock cycle in Chapter 1 we saw the principal evolutionary paths of the Earth's rocks and the great variety of processes that create them. We have now examined in detail the three basic types of rocks—igneous, sedimentary, and metamorphic—and the major processes (melting, recrystallization, weathering, lithification, and metamorphism) that form and alter them. Now we can look back and appreciate the relationship of these processes to each other and the impact of this relationship on the Earth. Any rock, including an existing igneous rock, can be subducted or remelted by contact with hot magma and form a new igneous rock. Any rock, including an existing sedimentary rock, can be weathered, eroded, transported, deposited, and lithified to form a new sedimentary rock. Any rock, including an existing metamorphic rock, can be heated and pressed to form a new metamorphic rock. There is no end to the cycle. Every body of rock on Earth is constantly changing—although, as we have seen, some processes change rocks more slowly than others.

We have also seen in these first seven chapters how plate tectonic processes contribute to the cycling of the Earth's rocks. We have examined how the igneous processes associated with rifting and divergent zones produce new oceanic lithosphere, and how subduction recycles mafic oceanic lithosphere and its overlying sediments, producing plutons of intermediate and felsic rock. We have seen how rocks uplifted by tectonic forces and exposed to the Earth's surface environment are weathered, eroded, transported, deposited, and lithified to form sedimentary rocks. Finally, in this chapter, we have discussed how the heat and directed pressure at convergent plate boundaries can convert a soft sedimentary rock such as shale to a hard metamorphic rock such as gneiss. Clearly, plate tectonics plays a pivotal role in the evolution of the Earth's rocks.

In the next chapter, we will examine what the Earth's rocks tell us about the past. Then, in the succeeding chapters, we will look more closely at how some of the processes that create rocks also create the Earth's mountains, continents, and ocean floors.

Chapter Summary

Metamorphic rocks, the third major type of rock in the Earth's rock cycle, are formed at temperatures and pressures intermediate between those that form igneous and sedimentary rocks. They are created by **metamorphism** of preexisting rock, during which solid rock is altered in mineral content and structure by heat, pressure, and migrating fluids without melting. During metamorphism many of the original features of a rock may be lost; its mineral grains will generally grow larger and develop an orderly parallel arrangement, and may even recrystallize into new minerals that are more stable under the prevailing conditions.

Most metamorphism occurs at depths greater than 10 kilometers (6 miles) below the Earth's surface, where the geothermal gradient and the weight of the overlying rock contribute the heat and pressure necessary to alter solid rock. Water and magmatic fluids also circulate at these depths, allowing the migration of unbonded ions in rocks undergoing metamorphism. Under any given set of conditions, the composition of the parent rock determines which metamorphic rocks can form.

Rocks at great depths are subjected to **lithostatic,** or **confining, pressure,** which pushes in on rock equally from all sides, making it smaller and denser but maintaining its general shape. Some rocks, particularly those at convergent plate boundaries, are also subjected to **directed pressure,** which distorts the shapes of rocks, flattening them in a single plane along which the pressure is predominantly applied. The minerals in a rock subjected to directed pressure tend to align in parallel planes perpendicular to the plane of the pressure, giving the rock a layered appearance known as **foliation.** All metamorphic rocks are categorized as being either foliated or nonfoliated.

Increased heat and directed pressure are responsible for the sequence of foliated rocks that forms from a typical marine shale, for example. This sequence begins with the fine-grained **slate** and **phyllite,** and progresses to the coarse-grained, more obviously foliated **schist** and **gneiss.** Foliation in slates, phyllites, and schists develops as clay and mica grains become aligned, whereas foliation in gneisses is marked by the formation of distinct, single-mineral layers of feldspars, micas, quartz, and mafic minerals. The mineral bands in gneisses are the result of **metamorphic differentiation,** in which unstable minerals recrystallize as new, more stable minerals in alternating bands of light (felsic) and dark (mafic) minerals. When temperature and pressure increase beyond metamorphic conditions to the threshold of igneous conditions, a "mixed" rock develops with both metamorphic and igneous properties. This rock is called a **migmatite.**

Nonfoliated rocks commonly develop in the absence of directed pressure and when the parent rock is composed largely of a single mineral. **Marble** forms from the metamorphism of relatively pure calcite-rich limestone or dolostone, and **skarn** from the metamorphism of impure (silicate-containing) limestone. **Quartzite** is produced by metamorphism of relatively pure quartz-rich sandstone. **Hornfels** forms where shale or basalt comes in direct contact with hot magma.

Different types of metamorphism occur under different geological conditions. **Contact metamorphism** occurs where rocks are heated by direct or close contact with magma; the zone of metamorphosed rock surrounding a magma is called a metamorphic **aureole. Regional metamorphism** occurs over a broad area of rock, and may be one of two types: **burial metamorphism,** which occurs as a result of the geothermal heat and confining pressure applied to rocks buried beneath more than 10 kilometers of overlying rock; and **dynamothermal metamorphism,** which develops from increased heat and pressure (both directed and confining) in an area of plate convergence. **Hydrothermal metamorphism** involves ion exchange with hot circulating fluids. **Fault metamorphism** occurs where rocks are grinding against one another at active faults. **Shock metamorphism** occurs where rocks are violently heated and compressed by the impact of a meteorite.

The degree to which a metamorphic rock differs from its parent material is described in terms of **metamorphic grade.** Low-grade metamorphism occurs when rocks are heated to temperatures of approximately 200° to 400°C; high-grade metamorphism occurs at temperatures of about 500° to 800°C. The minerals in metamorphic rocks can provide clues to the conditions under which the metamorphism occurred. Some minerals, because they are stable only within a narrow range of temperatures and pressures, indicate specific metamorphic conditions. These are called **metamorphic index minerals,** and areas of rock containing them are designated as specific **mineral zones.** Metamorphic environments may be associated not only with individual index minerals, but may also produce diagnostic assemblages of minerals called **metamorphic facies.** Mineral zones and metamorphic facies delineate areas of rock subjected to the same metamorphic conditions.

Different types of metamorphic rocks and facies form in different plate tectonic settings. A variety of metamorphic rocks develop at different venues within a convergent plate boundary. The high-pressure/low-temperature environment in a subduction trench produces blueschists; the low-pressure/high-temperature environment in the continental crust above the subducting plate produces greenschists; and the high-pressure/high-temperature environment in the mantle directly over the subducting plate produces granulites, amphibolites, and eclogites. Divergent plate boundaries are characterized by relatively low-temperature reactions that involve interaction between warm ocean-ridge basalts and circulating seawater.

Key Terms

metamorphic rocks (p. 181)
metamorphism (p. 181)
lithostatic (confining) pressure (p. 183)
directed pressure (p. 184)
foliation (p. 184)
slate (p. 186)
phyllite (p. 187)
schist (p. 187)
metamorphic differentiation (p. 187)
gneiss (p. 187)
migmatite (p. 188)
marble (p. 189)
skarn (p. 189)
quartzite (p. 189)
hornfels (p. 189)
contact metamorphism (p. 189)
aureole (p. 190)
regional metamorphism (p. 190)
burial metamorphism (p. 190)
dynamothermal metamorphism (p. 191)
hydrothermal metamorphism (p. 192)
fault metamorphism (p. 193)
shock metamorphism (p. 193)
metamorphic grade (p. 193)
metamorphic index minerals (p. 194)
mineral zones (p. 194)
metamorphic facies (p. 196)

Questions for Review

1. What are the four major factors that induce or influence metamorphism?

2. Describe four ways that rocks may change during metamorphism.

3. Define metamorphic foliation, explain how it develops, and list three metamorphic rocks that are foliated.

4. List the sequence of metamorphic rocks that develop during the progression of shale to migmatite. Describe the successive changes in mineralogy and structure that these rocks undergo.

5. Under what conditions do nonfoliated metamorphic rocks generally form? Give examples.

6. List the six common types of metamorphism, and discuss the geological conditions under which they occur.

7. Kyanite, andalusite, and sillimanite are considered metamorphic index minerals. What is a metamorphic index mineral, and why do these three minerals qualify as index minerals?

8. List four different metamorphic facies associated with regional dynamothermal metamorphism, and describe the metamorphic conditions that produce them.

9. What is the plate tectonic significance attached to the presence of the blueschist-facies mineral glaucophane in a rock? What kind of metamorphism generally occurs at mid-ocean ridges?

10. Under what circumstances might the presence of metamorphic rocks lead to landsliding?

For Further Thought

1. Why do quartzites appear to be essentially unfoliated, even when they have been subjected to the directed pressures of regional metamorphism?

2. Why do we rarely find minerals such as chlorite and amphibole directly next to an igneous intrusion?

3. Referring back to Figure 7-12a (p. 190), suggest a reason why some of the rocks crosscut by the intrusive rocks have not undergone contact metamorphism.

4. The Earth's internal temperature was apparently much higher during the first billion years of its history. Speculate about how metamorphism, metamorphic environments, and the distribution of metamorphic rocks would have been different during that early period.

5. From the photo at right, estimate the orientation of the directed pressures that metamorphosed these rocks.

Geologic Time in Perspective

Determining Relative Age

Highlight 8-1 How Fossils Form

Determining Absolute Age

The Geologic Time Scale

Figure 8-1 Niagara Falls, at the border of New York state and Ontario, Canada. Erosion of the rock underlying the falls began about 12,000 years ago, and continues today.

Chapter 8

Telling Time Geologically

In the first seven chapters of this text we explored some basic geological processes, including how plates move, how underground molten rock solidifies into igneous rocks and erupts to produce volcanoes, how solid rocks weather to form loose grains of sand that, in turn, lithify to form sedimentary rocks, and how heat and pressure convert preexisting rocks into hard, durable metamorphic rocks. We've seen that one particular factor is vital to all of these as well as other geological processes—time.

Geological processes have continuously affected the Earth since the planet's birth 4.6 billion years ago. Evidence that they still do so today is commonplace, such as in the eroding of New York's and Ontario's shared Niagara Falls (Fig. 8-1). Most of the major events that have affected the Earth, however—such as the extinction of the dinosaurs 65 million years ago and the breakup of the supercontinent Pangaea some 200 million years ago—culminated long before the first humans evolved. Many of these events, such as the extinction of the dinosaurs, were virtually instantaneous in a geological sense, requiring "only" a few million years or so to occur. Others, such as the breakup of Pangaea, required tens or hundreds of millions of years to unfold.

When thinking about geologic time, we are often overwhelmed by its vastness, and by the amount of time required for such processes as the rise of a mountain range or the creation of an ocean. In terms of Earth history, however, these are relatively rapid events. For example, mountain building by plate convergence (see Chapter 9) generally takes place at the rate of about 1 meter (a little over 3 feet) per 5000 years; at that rate, peaks as high as those of the Rockies or Alps (about 3000 meters, or 9840 feet) would have taken about 15 million years to reach their current altitude. As long a span of time as this seems, it is only about 1/300 of the Earth's existence. Likewise, the Atlantic Ocean, which widens at a rate of about 4 centimeters per year due to plate divergence (see Chapters 1 and 12), will probably take about 1 million years to become 40 kilometers (25 miles) wider; for the Atlantic, which is today about 5000 kilometers (3000 miles) wide, to grow by an additional 2000 kilometers (1200 miles), it would take 50 million years—less than 1/90 of the Earth's current age.

The timing of such events and processes is the subject of **geochronology**, the study of "Earth time" (or the time required for geological processes to occur, relative to the Earth's lifetime). Geochronology helps us comprehend the exceedingly slow pace of certain geological processes and the relatively rapid pace of others. Geochronologists attempt to understand the geological development of every region on Earth, in order to reconstruct the entire sequence of events that makes up the geologic record. This pursuit is an important component of the larger field of **historical geology,** dealing with the origin and evolution of all the Earth's life forms and geologic structures.

To tell time geologically, geologists use a variety of ingenious geologic "clocks." Much of this chapter will explain how trees, radioactive elements, fossils, and other materials function as geologic clocks. But first we must put geologic time into a human perspective, so that we—with our life spans of 80 or 90 years—might begin to grasp the magnitude of the Earth's longevity. Here are three statements that represent different human views of the age of the Earth:

> "High up in the North in the land called Svithjod, there stands a rock. It is a hundred miles high and a hundred miles wide. Once every thousand years a little bird comes to this rock to sharpen its beak. When the rock has been thus worn away, then a single day of eternity will have gone by."

> "Heaven and Earth, Centre and substance were made in the same instant of time and clouds full of water and man were created by the Trinity on the 26th of October 4004 B.C. at 9:00 in the Morning."

> "The Earth is approximately 4.6 billion years old."

The first view is from a Norwegian folktale recounted by Hendrik Van Loon in his *Story of Mankind.* Although imprecise, it does convey some sense of the great extent of Earth history. The second pronouncement was made in 1664 by Archbishop James Ussher of the Irish Protestant Church. Tracing the genealogies of the Old Testament back to the biblical account of Creation, Bishop Ussher calculated an Earth that was precisely 5668 years old. His view reflected the intertwined relationship between Christian theology and the science of his day. It dominated western thought until, in the eighteenth and nineteenth centuries, some naturalists' observations led them to propose that the Earth would have needed much more time to evolve to its current condition. The third view, less poetic than the first and seemingly less precise than the second, is generally accepted by scientists today.

Geologic Time in Perspective

How can we comprehend the magnitude of 4,600,000,000 years? One way is to picture a timeline across the United States, 5000 kilometers (3000 miles) long, on which each kilometer represents about 1 million years (Fig. 8-2). No rocks from the Earth's first 600 million years have ever been identified, so no direct information is available for this period, and the first 600 kilometers of our continent-wide timeline is barren. The oldest rocks known, recently discovered in Canada's Northwest Territories, are dated at 3.96 billion years. Few of our most ancient rocks contain any hint of life, probably because for its first billion years, conditions on Earth were too harsh to support most forms of life. (Although it is possible that very early life forms were extinguished by asteroids and large meteorites colliding with the Earth, which might have generated enough heat to vaporize the planet's first bodies of water and kill everything in them.) The earliest known life forms on Earth were microscopic blue-green algae believed to have lived 3.77 billion years ago; their fossils have been identified in the lithified silt that collected in the bottoms of muddy ponds in what is now northwestern Australia.

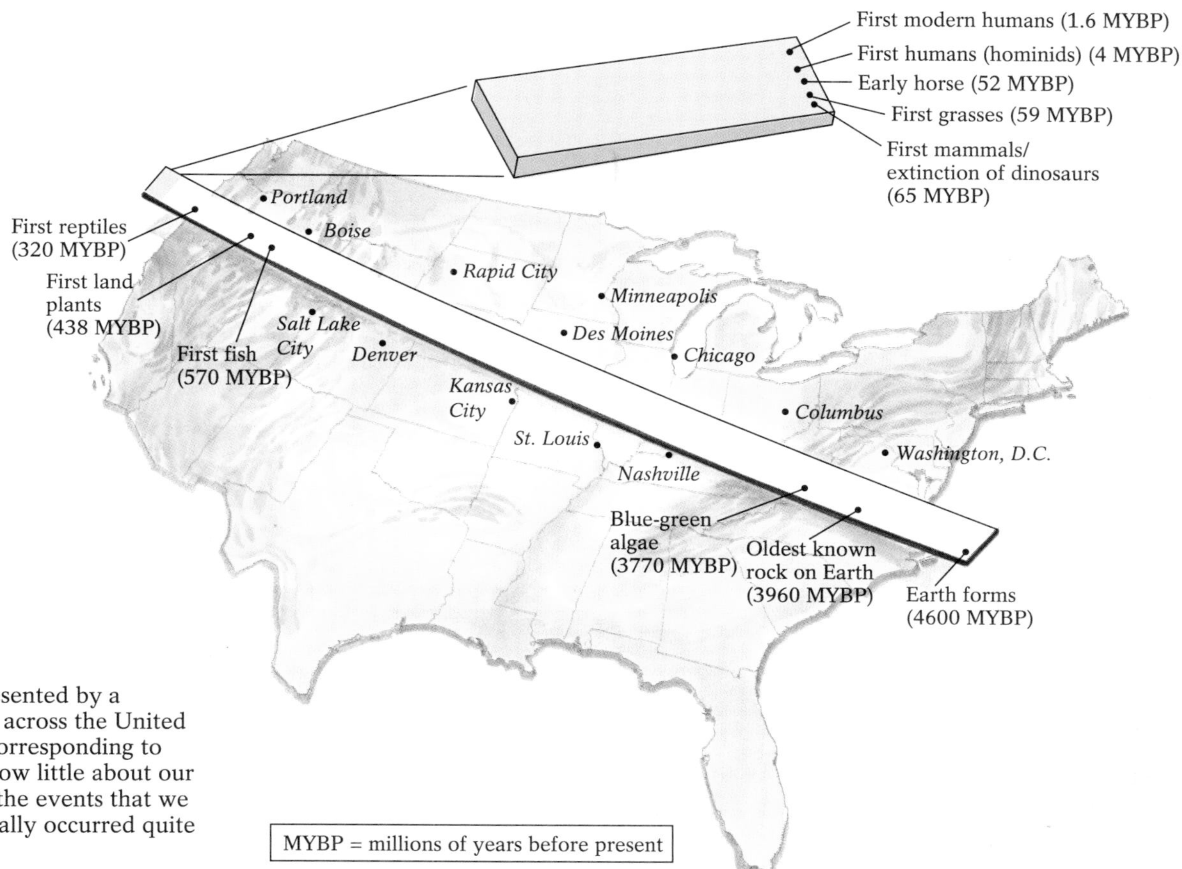

Figure 8-2 Earth history represented by a timeline extending east–west across the United States, with each kilometer corresponding to about 1 million years. We know little about our planet's earliest history, and the events that we know most about have generally occurred quite recently in geologic time.

Over the next 3 billion years, more complex forms of life began to develop and diversify. What allowed these new forms to arise and flourish? Many scientists believe that by about 570 million years ago (at about 2100 kilometers on our timeline), a more hospitable environment had developed on Earth. The atmosphere had evolved sufficiently to be able to deflect a significant portion of the Sun's ultraviolet radiation and burn up most incoming meteorites, and this protective atmosphere enabled an explosion of life to occur in the world's ancestral oceans.

Many of the "great moments" in Earth history are concentrated in the last few hundred kilometers of our timeline. The rise and fall of the dinosaurs, the onset of the Earth's most recent ice ages, the origin of mammals, including human beings, and the arrival of humans on the North American continent are all relatively recent events. Indeed, all of the material taught in a typical history course represents only 0.0000011 percent of Earth history!

How did geologists find out when the Earth's great events actually happened? Why do they believe that the planet is 4.6 billion years old? How did they deduce the ages of the oldest known rock, the oldest fossil, and all the landmark fossils in the evolution of modern species? How do they date the rocks and fossils that continue to be uncovered?

Geologists examine time in two ways. One is **relative dating,** which compares two or more entities to determine which is older and which is younger. Just as you can generally estimate the relative ages of people, because you know the basic characteristics of the bodies and faces of infants, toddlers, children, adolescents, adults, and senior citizens, geologists have learned certain characteristics that help them determine the relative ages of rocks and other geological features. The second way to examine time is **absolute dating,** which establishes approximately how many years ago a specific feature developed or a specific event occurred. To determine the absolute age of any individual, you would have to obtain quantitative information such as that found on a birth certificate or driver's license. But the Earth has no such written record. To determine the absolute age of geological phenomena, geologists use a variety of sources that yield quantitative information, as we shall see shortly. By combining information from both relative and absolute dating, geologists can date with some assurance a wide variety of geological and other materials virtually everywhere on Earth.

Determining Relative Age

Geologists determine the sequence of geological events that has produced an area's rocks using certain basic principles and assumptions, aided by their knowledge of such fundamental processes as sedimentation, volcanism, and erosion.

Principles of Relative Dating

The principles used to determine the relative ages of rocks involve basic scientific logic and understanding of a few spatial relationships, such as where new rock layers form with respect to preexisting layers. Also required is an understanding of the biological concept of evolution and how fossil evidence supports it. With these principles, we can usually answer the questions "which is older?" and "which is younger?", compare the ages of rocks from different geographic regions, and identify gaps in the geologic record.

Uniformitarianism The most basic principle used to interpret Earth history is James Hutton's **principle of uniformitarianism** (discussed in Chapter 1), which states that the geological processes taking place in the present operated similarly in the past. From this, we can assume that ancient earthquakes, volcanoes, floods, and other geological events happened in much the same way as they do today. Our observations of modern geological phenomena, therefore, can help us interpret ancient events (Fig. 8-3). It does not follow from unifor-

Figure 8-3 700,000-year-old ash deposits near Bishop, California (**a**) are similar to those produced by the 1980 eruption of Mount St. Helens (**b**), suggesting that the Bishop area experienced explosive volcanic eruptions comparable to those that occurred at Mount St. Helens more than half a million years later.

(a)

(b)

Figure 8-4 Horizontal beds of sedimentary rock outside of Salt Lake City, Utah. Using the principle of superposition, we can instantly determine that the rocks at the peak of this formation are the youngest and those at the bottom the oldest.

mitarianism, however, that the *rate* at which geological processes take place has necessarily remained the same over time. Some processes, such as mid-ocean ridge divergence, apparently have proceeded at a fairly steady rate. Others, though, such as the erosion of mountain ranges, tend to proceed sporadically, in response to a wide range of variable environmental factors.

Horizontality and Superposition As we saw in Chapter 6, most sediments settle out from bodies of water, and are deposited as horizontal or near-horizontal layers. Lava flows generally also solidify as horizontal layers. This tendency, stated as the **principle of original horizontality,** is fundamental because rock layers are often tilted or deformed by tectonic forces long after they are deposited.

The **principle of superposition** states that rock materials are generally deposited on top of earlier, older deposits; consequently, in any unaltered sequence of rock strata, the youngest stratum will be at the top and the oldest at the bottom. (Much as if each day, after reading the newspaper, you tossed it on top of a growing pile of daily newspapers in the corner of your room—unless the pile were to grow precipitously high and topple over, the recent history of world events would be preserved in the stack, with today's paper on top and progressively older editions below.) The principle of superposition, like the principle of original horizontality, commonly applies to sedimentary rocks and lava flows: In horizontal layers of sedimentary or volcanic rock that have not been disturbed, any given rock bed will be younger than the beds below it, and older than those above it (Fig. 8-4).

When tectonic forces have tilted or even overturned a sequence of rock layers (Fig. 8-5), we must look for readily identifiable sedimentary structures such as ripple marks, mudcracks, graded beds, or cross-bedding (all discussed

(a)

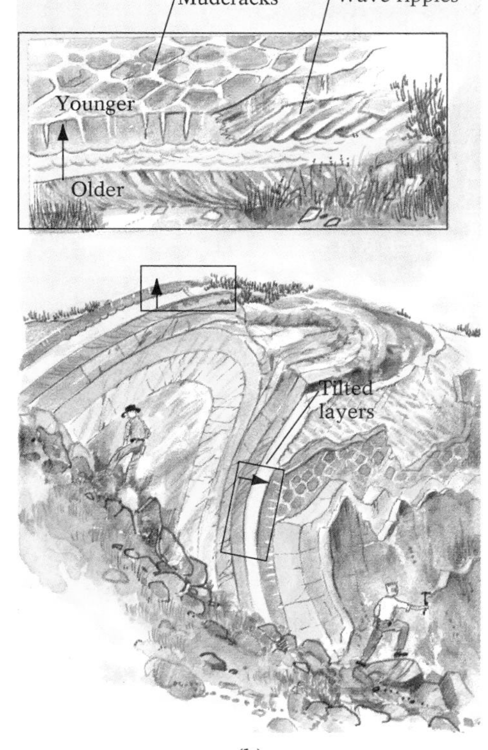

(b)

Figure 8-5 (a) Tilted turbidite beds in Zumaya, Spain. (b) When sedimentary rocks have been displaced from their initial horizontal orientation, we must look for sedimentary structures to identify the top of any individual bed; once the top of a single bed has been determined, the initial orientation of all the beds can be reconstructed using the principle of superposition.

Figure 8-6 Pegmatite dikes (lighter colored rock) cutting across older gneiss.

Figure 8-7 The fault that cuts through this sedimentary rock, in Anza Borrego Desert State Park in California, must be more recent than the deposition of any of the rock layers.

in Chapter 6) to identify the upper surface of any one sedimentary layer. Similarly, vesicles, often found at the tops of lava flows (see Chapter 3), may indicate the upper surfaces of certain volcanic rocks. Once we have identified the top of a rock layer, we can apply the principle of superposition to date relatively the layers below and above it.

Cross-Cutting Relationships Igneous rock often appears within other rock types, indicating that the latter were intruded, or cut across, by molten magma. Because the other rock had to exist first in order to be intruded, it must be older than the rock formed by the intruding magma. The **principle of cross-cutting relationships** states that any intrusive formation, such as a dike or sill (Chapter 3), must be younger than the rock across which it cuts (Fig. 8-6). Cross-cutting relationships also provide relative dates for faults, those fractures in rocks along which there has been displacement (to be discussed in Chapter 9). Faults must be more recent than the rocks they cut through (Fig. 8-7).

Inclusions The **principle of inclusions** states that fragments of other rocks contained within a body of rock must be older than the host rock. Sedimentary conglomerates, for example, contain pieces of preexisting rock, ranging in size from granules to boulders, that were incorporated and deposited along with the finer sediment making up the conglomerate's matrix. Many igneous rocks also contain pieces of preexisting rock that were broken by intruding magma and incorporated when it solidified. Thus, a conglomerate bed must be younger than its included particles, and a body of granite must be younger than its xenoliths (discussed in Chapter 3) (Fig. 8-8).

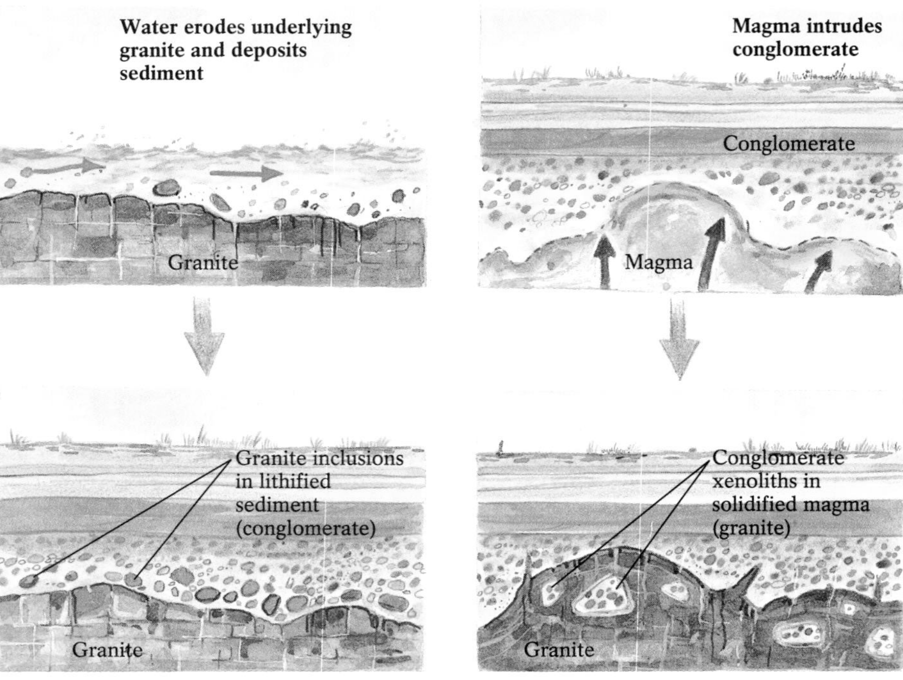

Figure 8-8 Inclusions are always older than the bodies of rock in which they are found. Part (**a**), for example, shows one way in which fragments of preexisting granite can be incorporated into conglomerate sandstone. Conversely, part (**b**) shows how fragments of preexisting conglomerate may become incorporated into granite.

Fossils and Faunal Succession **Fossils** (from the Latin *fossilus* meaning "something dug up") are the remains of ancient organisms, or other evidence of their existence, preserved in geologic material. These are studied by *paleontologists* to reconstruct the evolution of life on Earth. The principles of superposition, cross-cutting relationships, and inclusions can be applied to fossil-bearing rocks to determine the order in which life forms developed and became extinct.

Sedimentary rocks, because they form at the Earth's surface, are more likely than other kinds of rocks to contain the remains of organisms. Even in sedimentary rocks, however, fossils are scarce: Only about 1% of all species that ever existed are believed to have been preserved as fossils. As is shown in Highlight 8-1 (following pages), organisms become fossilized only under very specialized conditions. Many extinct organisms were composed exclusively of soft tissue and thus contained no preservable parts.

Just as geologists' knowledge of the spatial relationships between rock strata has been used to help determine the evolutionary sequence of the fossils contained in them, the reverse is also true. Knowledge of a fossil organism's place on the evolutionary scale can help date a rock unit in which it is found. The **principle of faunal succession** states that the animals on Earth have changed in a definite order through time: The earliest life forms were simpler than those that evolved later, and more recently developed forms are more closely related to contemporary plants and animals than are earlier forms. Thus, rocks containing more complex life forms are generally younger than those containing simpler life forms.

Certain organisms have existed (or did exist) for hundreds of millions of years. Sharks, for example, have inhabited the oceans for more than 400 million years. Other organisms, however, lived only during a relatively brief, specific period of Earth history. When paleontologists find fossils of such short-lived organisms, known as **index fossils,** they know that they have found rocks of a fairly specific age (Fig. 8-9). The species that become the most useful index fossils are those that were geographically widespread during their short time on Earth, as these can eventually be used to date many different—even far distant—rock formations.

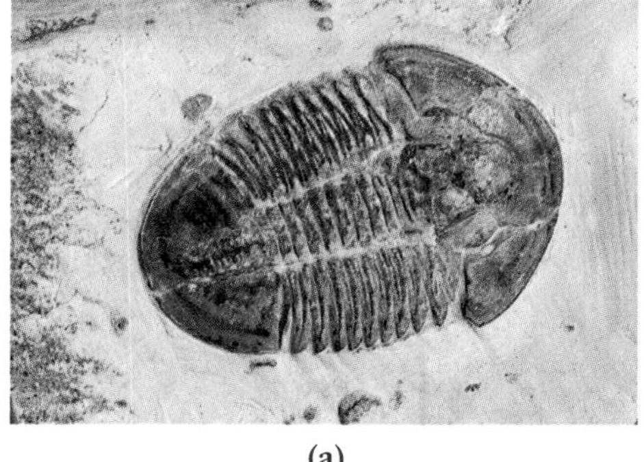

(a)

(b)

(c)

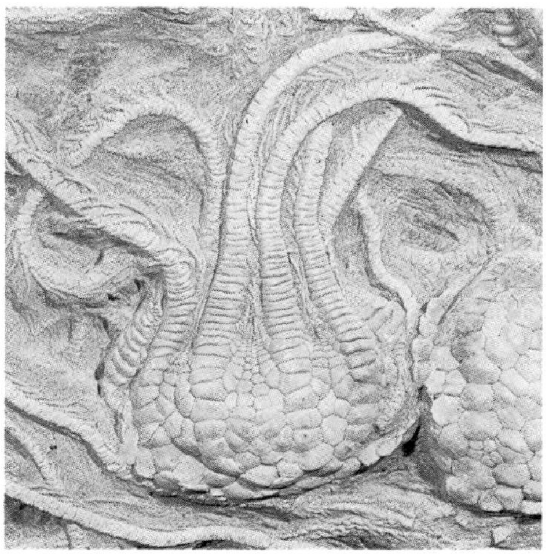

(d)

Figure 8-9 Index fossils. Because each of these organisms lived during a specific period of Earth history, they can be used to date the rocks in which they are found.

(**a**) A 500-million-year-old trilobite; (**b**) 300-million-year-old brachiopods; (**c**) a 200-million-year-old ammonite; (**d**) 100-million-year-old crinoids.

Highlight 8-1 *How Fossils Form*

Most fossils form when the hard parts of organisms become buried in layers of sediment, where they decompose so slowly that their shape is preserved by the rock as it lithifies around them. Geologists have unearthed countless fossils of ancient clam shells, many fish skeletons, and a fair amount of dinosaur bones. They have found few remains of worms, slugs, and jellyfish. This is because organisms with hard parts such as shells and bony skeletons are more likely to remain intact long enough to become preserved in sediment; those composed only of soft tissue would likely decompose rapidly at the surface in most environments. (For the same reason, hard, woody trees are more likely to be fossilized than are plants with no hard parts, such as mosses.) The hard parts of organisms also eventually decompose or dissolve after burial if they come into contact with underground water. For that reason, a fossil is seldom the actual remnant of an organism's original biological substance. After the original bones, shells, and other organic materials have been removed by circulating groundwater, a *mold*—an impression of the organism's form—remains in the enveloping sediment. At a later time, the mold may be filled with other materials, such as occurs when circulating ion-rich groundwater enters a mold and precipitates mineral matter in it. This creates a solid *cast* of the mold, which is what we typically find when we go fossil-hunting. The processes that lead to the creation of molds and casts are depicted in Figure 8-10.

Preservation of an organisms's actual original parts occurs only under unusual environmental conditions. For instance, the skeletons of thousands of mammoths, mastadons, saber-toothed cats, and other late Ice Age mammals were preserved in the viscous tar of the La Brea tar pits, in what is now downtown Los Angeles. Intact woolly mammoths have been dug up from the tundras of Alaska and Siberia, where they had become frozen in Arctic *permafrost* (permanently frozen ground). (Paleontologists and others have even dined on thawed, yet still-edible, mammoth meat.) The original carcasses of entire insects have been recovered from hardened tree sap, or *amber*, in which they became stuck millions of years ago. In each of these cases, the enveloping material did not permit the circulation of water or air, and thus prevented decomposition of the organism's original material.

Even when we don't find fossils showing an organism's physical form, we sometimes find other evidence of its presence.

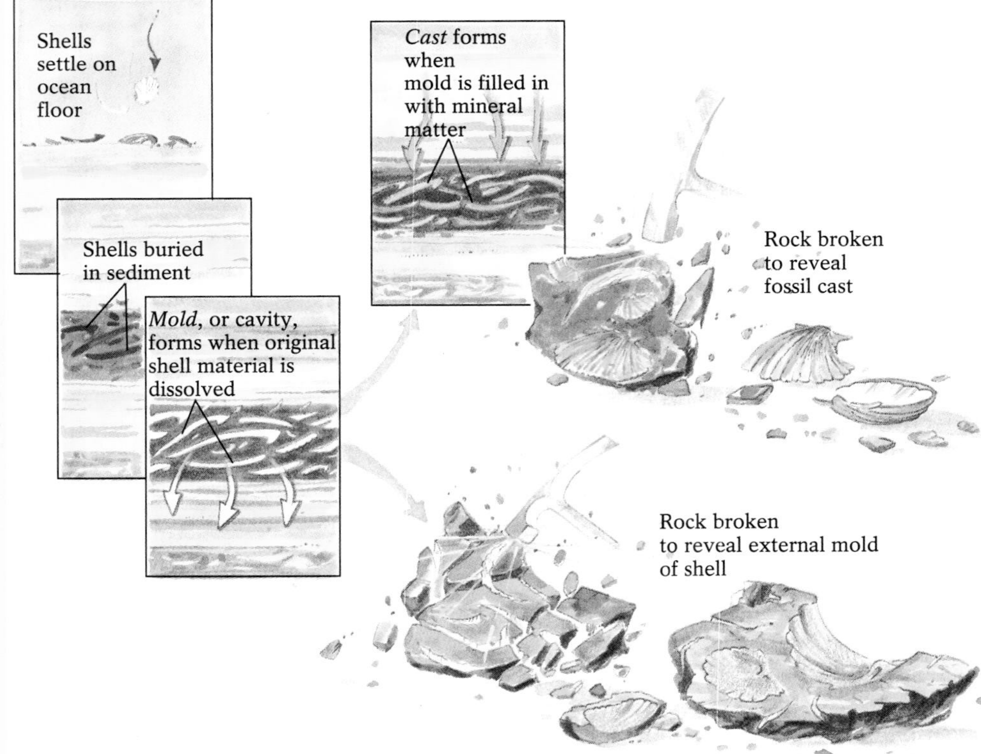

Figure 8-10 An organism that has been buried in sediment may be preserved in the form of a mold or a cast in the resulting rock. A *mold* is an imprint of the organism's external form, left in the surrounding rock long after the organism itself has decayed. After an organism's biological substance has dissolved away, circulating groundwater may precipitate mineral matter into the cavity left in the rock, forming a *cast* of the organism.

Fossils and Faunal Succession **Fossils** (from the Latin *fossilus* meaning "something dug up") are the remains of ancient organisms, or other evidence of their existence, preserved in geologic material. These are studied by *paleontologists* to reconstruct the evolution of life on Earth. The principles of superposition, cross-cutting relationships, and inclusions can be applied to fossil-bearing rocks to determine the order in which life forms developed and became extinct.

Sedimentary rocks, because they form at the Earth's surface, are more likely than other kinds of rocks to contain the remains of organisms. Even in sedimentary rocks, however, fossils are scarce: Only about 1% of all species that ever existed are believed to have been preserved as fossils. As is shown in Highlight 8-1 (following pages), organisms become fossilized only under very specialized conditions. Many extinct organisms were composed exclusively of soft tissue and thus contained no preservable parts.

Just as geologists' knowledge of the spatial relationships between rock strata has been used to help determine the evolutionary sequence of the fossils contained in them, the reverse is also true. Knowledge of a fossil organism's place on the evolutionary scale can help date a rock unit in which it is found. The **principle of faunal succession** states that the animals on Earth have changed in a definite order through time: The earliest life forms were simpler than those that evolved later, and more recently developed forms are more closely related to contemporary plants and animals than are earlier forms. Thus, rocks containing more complex life forms are generally younger than those containing simpler life forms.

Certain organisms have existed (or did exist) for hundreds of millions of years. Sharks, for example, have inhabited the oceans for more than 400 million years. Other organisms, however, lived only during a relatively brief, specific period of Earth history. When paleontologists find fossils of such short-lived organisms, known as **index fossils,** they know that they have found rocks of a fairly specific age (Fig. 8-9). The species that become the most useful index fossils are those that were geographically widespread during their short time on Earth, as these can eventually be used to date many different—even far distant—rock formations.

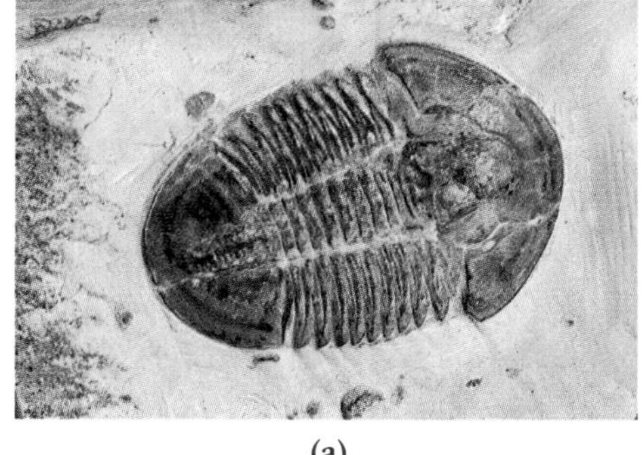

(a)

(b)

(c)

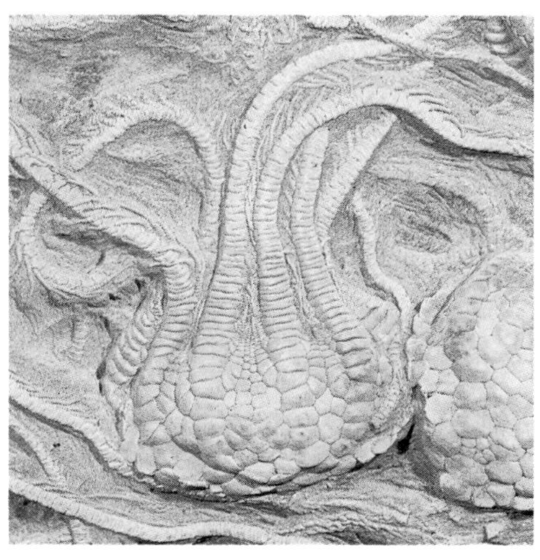

(d)

Figure 8-9 Index fossils. Because each of these organisms lived during a specific period of Earth history, they can be used to date the rocks in which they are found. (a) A 500-million-year-old trilobite; (b) 300-million-year-old brachiopods; (c) a 200-million-year-old ammonite; (d) 100-million-year-old crinoids.

Highlight 8-1 *How Fossils Form*

Most fossils form when the hard parts of organisms become buried in layers of sediment, where they decompose so slowly that their shape is preserved by the rock as it lithifies around them. Geologists have unearthed countless fossils of ancient clam shells, many fish skeletons, and a fair amount of dinosaur bones. They have found few remains of worms, slugs, and jellyfish. This is because organisms with hard parts such as shells and bony skeletons are more likely to remain intact long enough to become preserved in sediment; those composed only of soft tissue would likely decompose rapidly at the surface in most environments. (For the same reason, hard, woody trees are more likely to be fossilized than are plants with no hard parts, such as mosses.) The hard parts of organisms also eventually decompose or dissolve after burial if they come into contact with underground water. For that reason, a fossil is seldom the actual remnant of an organism's original biological substance. After the original bones, shells, and other organic materials have been removed by circulating groundwater, a *mold*—an impression of the organism's form—remains in the enveloping sediment. At a later time, the mold may be filled with other materials, such as occurs when circulating ion-rich groundwater enters a mold and precipitates mineral matter in it. This creates a solid *cast* of the mold, which is what we typically find when we go fossil-hunting. The processes that lead to the creation of molds and casts are depicted in Figure 8-10.

Preservation of an organisms's actual original parts occurs only under unusual environmental conditions. For instance, the skeletons of thousands of mammoths, mastadons, saber-toothed cats, and other late Ice Age mammals were preserved in the viscous tar of the La Brea tar pits, in what is now downtown Los Angeles. Intact woolly mammoths have been dug up from the tundras of Alaska and Siberia, where they had become frozen in Arctic *permafrost* (permanently frozen ground). (Paleontologists and others have even dined on thawed, yet still-edible, mammoth meat.) The original carcasses of entire insects have been recovered from hardened tree sap, or *amber*, in which they became stuck millions of years ago. In each of these cases, the enveloping material did not permit the circulation of water or air, and thus prevented decomposition of the organism's original material.

Even when we don't find fossils showing an organism's physical form, we sometimes find other evidence of its presence.

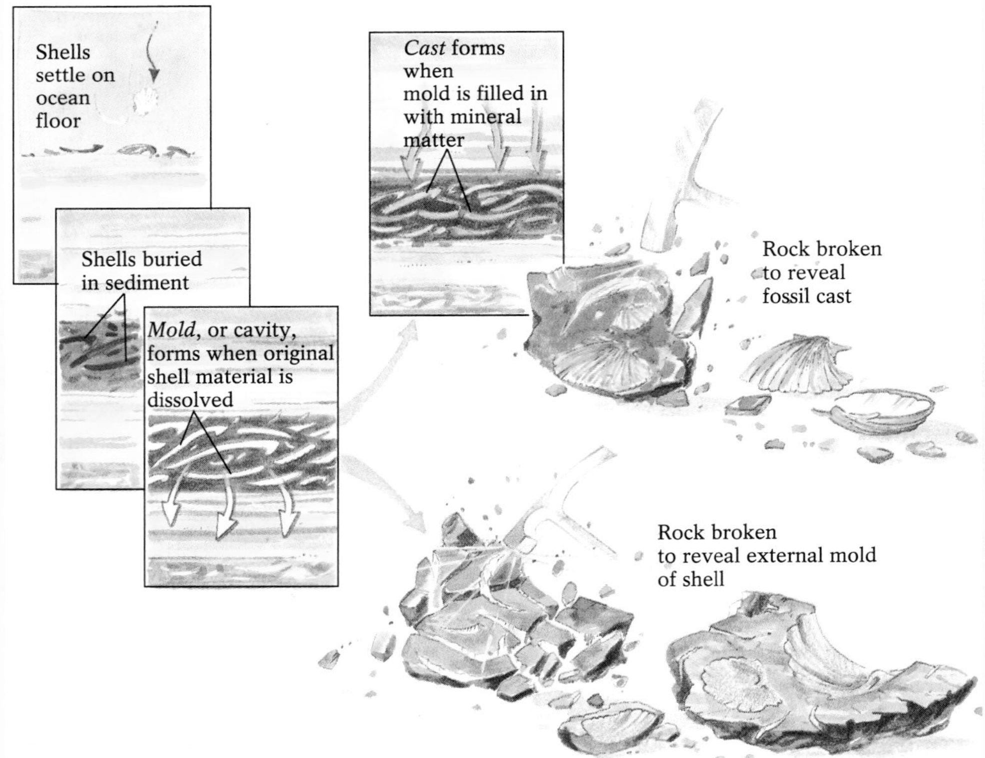

Figure 8-10 An organism that has been buried in sediment may be preserved in the form of a mold or a cast in the resulting rock. A *mold* is an imprint of the organism's external form, left in the surrounding rock long after the organism itself has decayed. After an organism's biological substance has dissolved away, circulating groundwater may precipitate mineral matter into the cavity left in the rock, forming a *cast* of the organism.

There are various forms of such *trace fossils.* If an animal burrowed through sediment, it may have left "tunnels"; if it crawled across soft mud, it may have etched delicate tracks into the sediment. Larger animals that walked around may have left behind footprints. We have even discovered rocks, called *gastroliths,* that dinosaurs swallowed to help grind their coarse food and thus aid digestion, and lithified "droppings," or *coprolites,* from a variety of creatures. Some of the many types of fossils are depicted in Figure 8-11.

(a) (b) (c) (d) (e)

Figure 8-11 Various types of fossils. (**a**) A mold and a cast of a trilobite; (**b**) a cast of a starfish; (**c**) a 3.5-million-year-old fly in amber; (**d**) a saber-toothed cat skeleton from the La Brea tar pits; (**e**) ancient worm burrows. Only the fossils in parts (**c**) and (**d**) contain actual remains of the organisms.

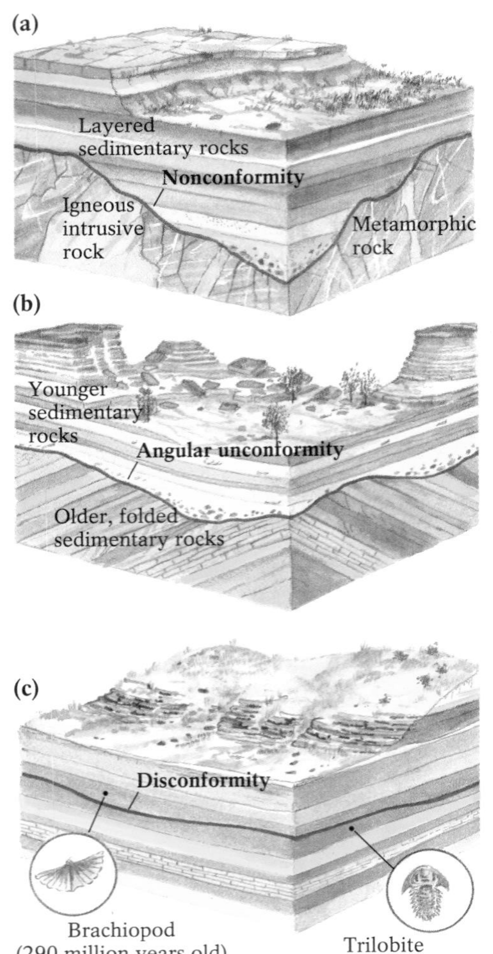

Figure 8-12 The three major types of unconformities between rock layers. (**a**) A nonconformity, between metamorphic or igneous rock and overlying sedimentary rock; (**b**) an angular unconformity, between older, deformed sedimentary rock and younger, undeformed sedimentary rock; and (**c**) a disconformity, between parallel layers of sedimentary rock.

Unconformities

No single place on Earth contains all the rock strata composing the entire geologic record (i.e., all the rock layers ever deposited over the course of Earth history). Gaps in the geologic record are marked by **unconformities,** boundaries separating rocks of markedly different ages. Unconformities occur where erosion has removed rock layers, or where none were deposited during certain geologic periods. They can be quite obvious or they can be subtle. A *nonconformity* is the obvious boundary between an unlayered body of plutonic igneous or metamorphic rock and an overlying layered sequence of sedimentary rock layers (Fig. 8-12a); rock underlying a nonconformity usually shows evidence of having been eroded before the overlying rock was deposited. Somewhat less obvious is an *angular unconformity,* the boundary between a sedimentary rock layer that has been deformed and eroded and a later, horizontal rock deposit that overlies it (Fig. 8-12b). The angular unconformity at Siccar Point in Scotland (Fig. 8-13) has particular historical significance. The most subtle type of unconformity, a *disconformity,* occurs between parallel layers of sedimentary rock (Fig. 8-12c); such a boundary is revealed to be a disconformity if fossil groupings above and below it are found to be of substantially different ages, or if the surface of the lower layer shows evidence of having been significantly eroded.

Virtually every sequence of sedimentary rock layers contains one or more unconformities. In fact, the sedimentary rock record of most regions represents only 1% to 5% of the Earth's existence, so much more of the record is missing than is present. The most complete rock sequences, such as the one exposed in Arizona's Grand Canyon, span only about 20% of the Earth's existence, and that for a limited area (Fig. 8-14).

Figure 8-13 The angular unconformity at Siccar Point, Scotland. James Hutton, the founder of modern geology (Chapter 1), correctly interpreted this formation as showing layers of rock that had been deposited horizontally, lithified, tilted to a nearly vertical position, eroded smooth at the surface, and then been overlain with subsequent horizontally deposited layers—a graphic example of the dynamics that shape the Earth's features, and the amount of time necessary to achieve them.

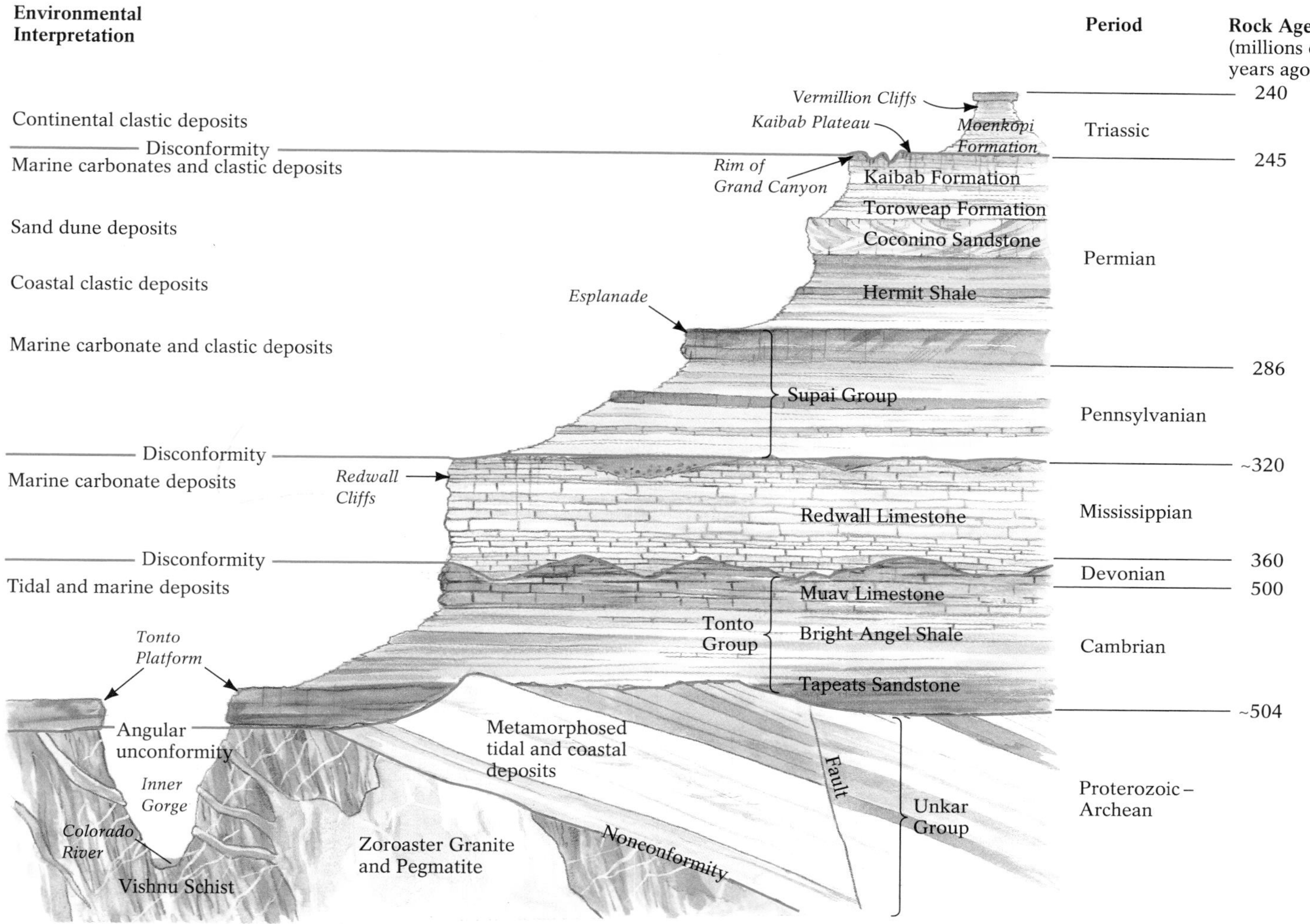

Figure 8-14 The Grand Canyon in northern Arizona affords one of the Earth's most complete exposed rock sequences—almost 2 vertical kilometers (about 1.2 miles) of rock ranging up to 3 billion years in age. Even here, however, there are unconformities representing gaps in the local rock record.

Correlation

The farther apart two rock formations are geographically the less likely it is that they will have identical rock sequences, since different environmental factors would have influenced rock formation and erosion in each area. However, even formations that are physically separated—whether by a few meters or by thousands of kilometers—may contain individual layers that are similar, suggesting that these layers originated at the same time. To determine equivalence in age between geographically distant rock units, geologists look for paleontological or mineralogical similarities between the units, a process known as **correlation** (Fig. 8-15). The presence of markedly similar fossil assemblages in different rock units, for example, suggests that the layers are the same age. Individual index fossils can also be used as a basis for correlation, as all rocks in which such a fossil are found must have formed during the same time period.

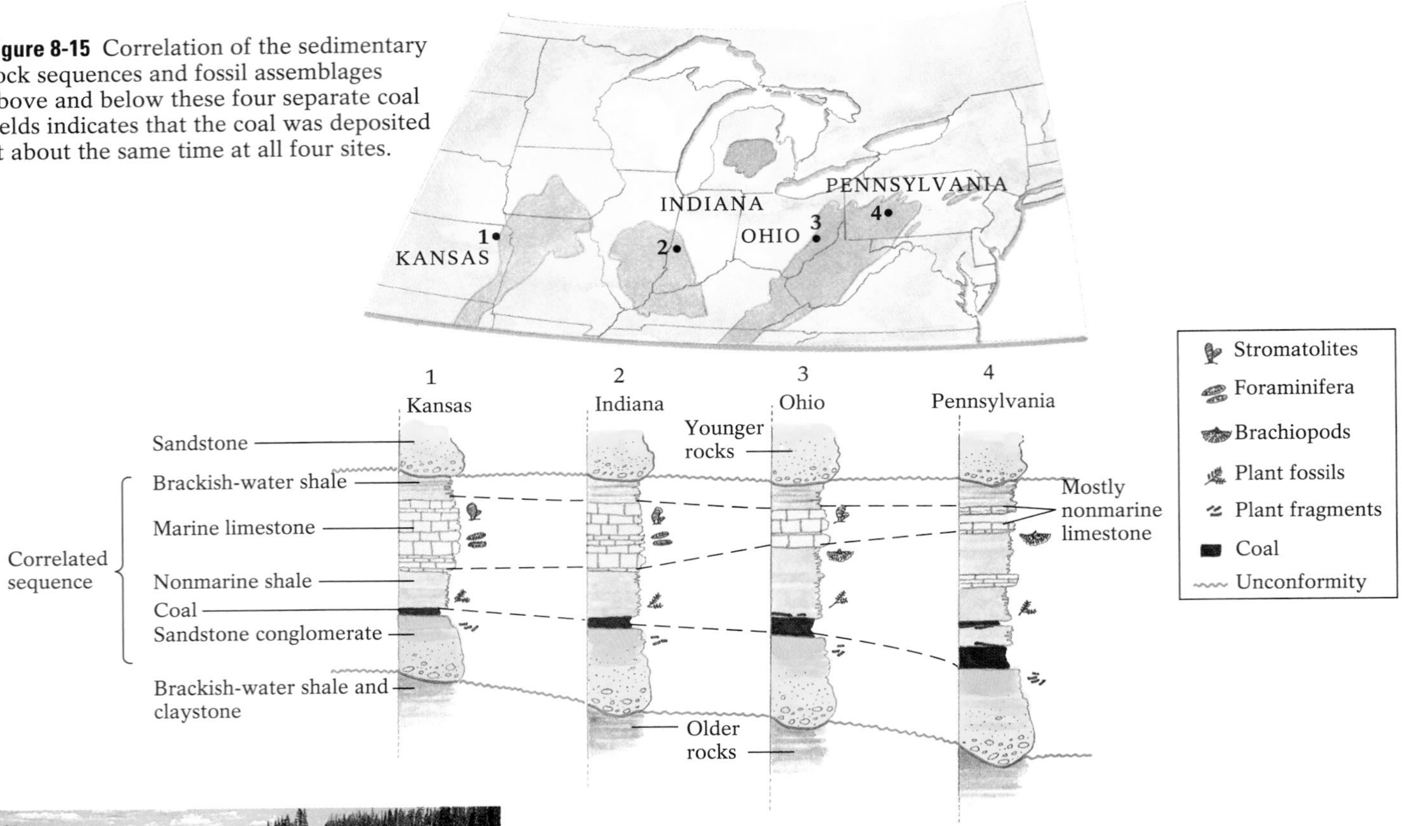

Figure 8-15 Correlation of the sedimentary rock sequences and fossil assemblages above and below these four separate coal fields indicates that the coal was deposited at about the same time at all four sites.

Figure 8-16 Ash from the eruption of Mt. Mazama 6600 years ago, at Crater Lake in Oregon. (The ash from Mazama is the lighter colored layer, underlying a more recently deposited darker layer; the pinnacles were produced by weathering of the ash beds by rain and snow.) The Mazama ash is a key bed found in same-aged rock units throughout the Pacific Northwest.

Two widely separated rock sequences can also be correlated if both contain the same *key bed,* a distinctive stratum that appears at several locations. A key bed records a geological event of short duration that affected a wide area. The eruption of Mount Mazama in the Pacific Northwest about 6600 years ago produced a whitish-tan, highly felsic ash that was scattered so widely that it can be used to correlate the geologic strata of Oregon, Washington, British Columbia, and Alberta (Fig. 8-16).

Distinctive sedimentary facies of the same age occurring over a particularly wide range of formations could indicate global environmental conditions during deposition, such as widespread climate change or a worldwide rise or fall in sea level. The extensive coal beds on both the North American and Eurasian continents, for example, mark a distinctively warm episode that occurred about 300 million years ago; this episode produced substantial vegetation that was later covered by sediments and other deposits and eventually lithified as coal.

Relative Dating by Weathering Features

We saw in Chapter 5 how the processes of mechanical and chemical weathering convert solid rock to loose regolith. Because rock tends to become increasingly weathered over time, comparing different bodies of rock in terms of the extent to which each has been weathered can yield clues to their relative ages. Relative dating by means of weathering characteristics is generally limited to geologic materials that are less than a few million years old.

A fresh break in an old rock often displays a *rind* of weathered material on or near the rock's surface. Such rinds occur when rocks are infiltrated by water, which chemically alters their original composition into some type of weathering product. In general, the longer a rock is exposed to weathering, the thicker its weathering rind will be. Thus, comparative measurement of weathering rinds (among rocks having the same initial composition) can distinguish between glacial or landslide deposits of different ages. It is most accurate for basalt, whose ferromagnesian minerals (olivine, pyroxenes) weather quickly at the Earth's surface.

An *hydration rind* is a similar feature that develops when fresh obsidian, formed from the rapid cooling of felsic magma, is exposed to atmospheric moisture. Through hydration, water from the air is incorporated into obsidian's structure, forming a rind the thickness of which increases over time. The relative ages of obsidian-rich materials can therefore be established by comparative measurements of their hydration rinds. Because obsidian projectile points and cutting blades are often found at ancient sites of human habitation, the hydration rinds of these artifacts can be used to date archaeological finds at such sites.

Knowing the amount of weathering experienced by an area's individual features, from its boulders to its soils, can provide important clues to the relative age of the entire landscape (Fig. 8-17). This can often be judged by sim-

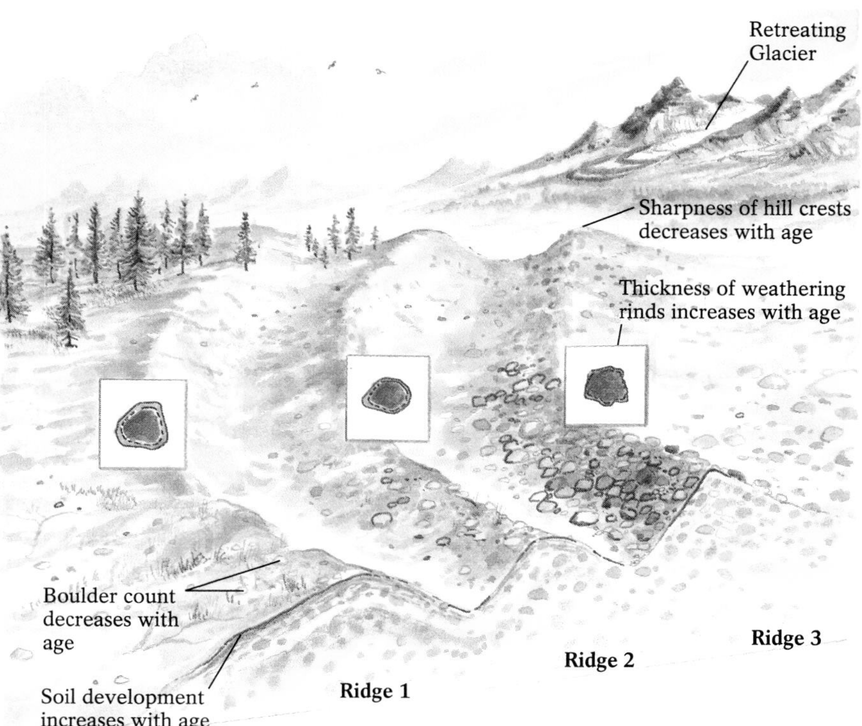

Figure 8-17 This hypothetical landscape shows the variety of weathering phenomena that can be used to determine the relative age of an area and its constituent materials. Ridges 1, 2, and 3 are deposits left behind by the retreating glacier in the background. The sharp hill crests and large number of boulders on ridge 3 indicate relatively little exposure to weathering; the thickness of the weathering rinds on the boulders in ridge 1 indicates that they are fairly old; the soils in ridges 1 through 3 are progressively less developed. These clues all suggest that ridge 1 is the oldest and ridge 3 the most recent of the glacial deposits.

ple observation. In an old landscape, for example, most surface boulders would probably have weathered and broken down, whereas young landscapes would contain a high concentration of intact surface boulders. (An experienced geologist can even estimate the relative age of an intact boulder with a single hearty whack of a geologic hammer: An unweathered boulder's resounding ring is a distinctly different sound from the dull thud of an aged, weathered boulder.) A geologist's trained eye will also note the extent to which weathering and other surface processes have altered the shape of hills. An older landscape will have gentler slopes and rounded, lower hill crests; younger settings will have steeper slopes and sharper hill crests.

Soil, as we saw in Chapter 5, develops over time and reflects the climate, topography, biological activity, and parent material that produced it. The amount of time that was required for a particular soil to develop at the surface can be estimated from the thickness of the overall soil profile, the depth to which its soluble elements have been dissolved away, and the number of horizons that have developed. Thus, it is possible to compare soil-horizon development in topographically and climatically similar landscapes having the same rock types, in order to determine the relative ages of the soils.

Determining Absolute Age

The use of even several relative-dating principles and techniques can tell us only that rock A is older or younger than rock B. Understandably, geologists prefer the more specific "years ago" information yielded by absolute-dating methods. All absolute-dating methods depend on some type of "natural clock"—a process that operates at a constant quantifiable rate over long periods of time. For example, the process of tree growth adds rings to the tree trunk every year. As a result, we can determine a tree's absolute age by counting its rings. Nature has provided a substantial number of such "clocks," which may be used to determine the ages of a range of geologic and paleontological materials.

Radiometric Dating

We saw in Chapter 2 that the atoms of certain chemical elements exist as isotopes with different numbers of neutrons in their nuclei. *Radioactive* isotopes are those whose nuclei spontaneously decay by emitting some neutrons or protons. The decaying radioactive isotope, or **parent isotope,** evolves into a decay product, the **daughter isotope.** The loss or gain of neutrons changes a parent isotope to a daughter isotope of the same element; the loss or gain of protons changes the parent element into an entirely different daughter element with a new set of chemical and physical properties. Eventually, often after a number of intermediate steps, an unstable radioactive parent isotope becomes a stable, nonradioactive daughter isotope.

For example, carbon, with its atomic number of 6, has isotopes with 6, 7, and 8 neutrons. The nuclei of carbon-12 (6 protons, 6 neutrons) and carbon-13 (6 protons, 7 neutrons) atoms are stable. Those of carbon-14 (6 protons, 8 neutrons), however, are radioactive; one neutron spontaneously splits to become one proton and one electron, yielding an atom with 7 protons, 7 neutrons, and 7 electrons—in other words, nitrogen-14 ($^{14}_{7}N$). In this case,carbon-14 is the parent isotope and nitrogen-14 is the daughter.

Radiometric dating uses the continuous decay of the radioactive isotopes, which often become incorporated into the crystal structure of minerals when magma cools, to measure the amount of time elapsed since a rock formation developed. As time passes, a rock will contain less of its initial radioactive parent isotopes and more of their daughter products. (Uranium-238, for example, commonly found in zircon crystals in granitic rocks, decays to form its stable daughter isotope, lead-206. Over time, zircon gains lead-206 as it loses uranium-238.) Radioactive decay rates are constant, unaffected by changes in temperature or pressure or by chemical reactions involving the parent isotope. Thus, by measuring the ratio of parent to daughter isotopes and comparing it with the parent element's known rate of radioactive decay, we can determine the absolute age of a rock.

The time it takes for half the atoms of the parent isotope to decay is the **half-life** of an isotope. For example, if a rock has 10 parent and 10 daughter atoms—a parent-to-daughter ratio of 1:1—the original rock had 20 radioactive parent atoms, and we are seeing the rock after one half-life. After another half-life, the number of remaining parent atoms would halve again, leaving 5 unstable parent atoms and a total of 15 stable daughters—a parent-to-daughter ratio of 1:3. Note that the total number of parent and daughter atoms combined remains the same (in this example, 20) throughout the progression. Regardless of which element is decaying, the proportion of parent to daughter atoms is a predictable ratio at each half-life (Fig. 8-18).

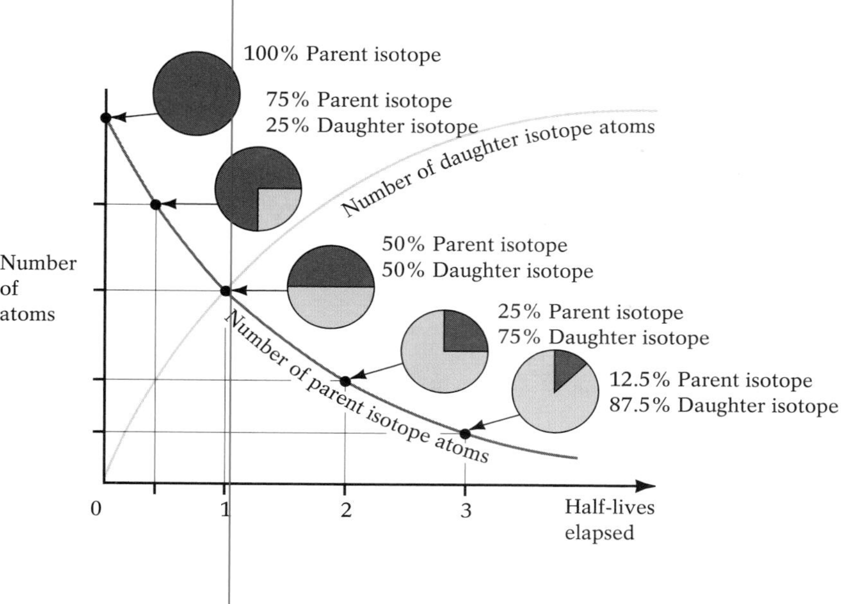

Figure 8-18 Radioactive decay converts a radioactive parent isotope to a stable daughter isotope. With the passage of each half-life, the number of atoms of the radioactive isotope is reduced by half, whereas the number of atoms of the daughter isotope increases by the same quantity. By measuring the ratio of the quantities of the parent to daughter isotopes, geologists can determine the absolute ages of some rocks.

Some decay systems have half-lives that span billions of years. Certainly no one sits in a laboratory waiting for that time to elapse in order to determine the half-life of a given parent–daughter system. Instead, samples containing measured quantities of the radioactive isotope are placed in a device that counts the number of nuclear particles emitted per unit time, and from that we can extrapolate the decay rate in years.

Knowing the specific decay rate experienced by individual isotopes is what enables us to determine the age of a given rock. A *mass spectrometer* measures the precise quantities of parent and daughter isotopes in a substance such as a mineral, rock, bone, tree, or shell, in order to determine their ratio.

From the ratio, we can establish how many half-lives (or fractions of half-lives) have elapsed since the radioactive parent isotope was incorporated into the substance. The number of half-lives can then be multiplied by the known length of one half-life of that isotope to calculate the absolute age of the substance.

Factors Affecting Radiometric Results Radiometric dating is more accurate and useful for igneous than other types of rocks, largely because most radioactive isotopes are concentrated in the Earth's crust, where igneous rocks abound. It is difficult to determine the absolute age of a sedimentary rock radiometrically because sedimentary rocks contain mineral fragments of countless pre-existing rocks of different ages. Dating these fragments yields the age of their source rocks, but not of the sedimentary rock containing them. To determine the time period in which a sedimentary rock formation probably originated, we first radiometrically date the igneous rocks that cross-cut, underlie, or overlie it and then apply the principles of relative dating.

Parent–daughter ratios in metamorphic rocks may be distorted because of the heat, pressure, and circulating liquids and gases associated with intermediate- and high-grade metamorphism. Although these factors do not affect the rate at which isotopes decay, they can affect the relative number of parent and daughter isotopes ultimately found in a rock, because they allow atoms to migrate away from their point of origin. Heating, for instance, may cause the structure of mineral crystals to expand, letting trapped atoms of the daughter isotope escape into the surrounding rock and causing the crystal's age to appear younger than it really is. When a rock that has been heated eventually experiences cooler conditions, such as occurs after mountain building, metamorphism ceases, the mineral structure contracts, and its crystals once again begin to trap the daughter isotope (Fig. 8-19). Although the parent–daughter isotopic ratios of minerals in metamorphic rocks may not reveal the ages of the original minerals or rocks, they can tell us when the metamorphism ended, thus yielding approximate dates for uplift and mountain-building events.

Figure 8-19 Loss of daughter isotopes from rock metamorphosed during mountain building. Heat causes expansion of mineral crystals, allowing daughter atoms to escape and migrate elsewhere. Although the decay of the parent isotope proceeds normally throughout, and daughter isotopes resume accumulating after metamorphism ceases and the crystals contract, the loss of these daughter isotopes during this period skews the resulting parent-to-daughter ratio such that the age of the rock appears younger than it is.

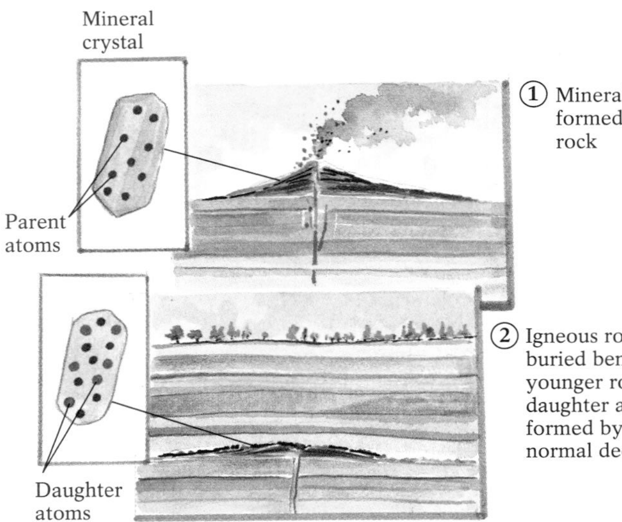

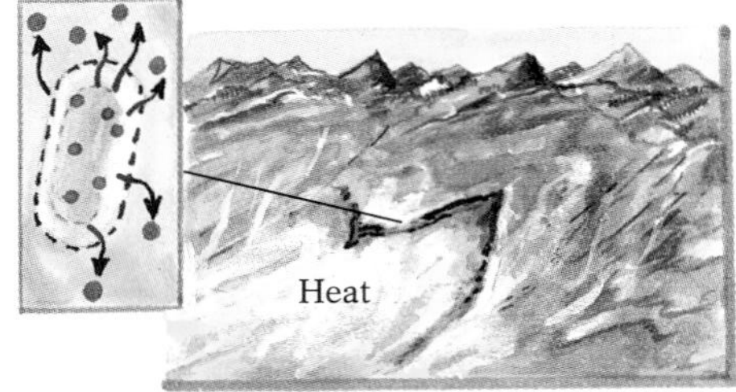

The reliability of radiometric dating can be affected by the condition of the materials dated. In material that is fractured or highly weathered, for example, migrating groundwater may have dissolved away some parent or daughter isotopes, resulting in inaccurate parent–daughter ratios. Fortunately, a recently developed device—the *s*ensitive *h*igh-*r*esolution *i*on *m*icro*p*robe, or SHRIMP—now makes it possible to analyze minute fractions of some crystals and date their unfractured, unweathered regions.

The very age of a substance may limit the effectiveness of dating it by radiometry. Young rocks and minerals, for instance, may not yet have produced a measurable amount of daughter isotopes. For radioactive isotopes with half-lives on the order of hundreds of millions of years, a rock must be 500,000 or more years old to produce a measurable quantity of daughter isotope. Rocks that contain uranium-238, with its half-life of 4.5 billion years, must be at least 10 million years old to yield a measurable amount of lead-206.

Rocks and minerals that are very old may have exhausted virtually all of their parent isotopes. For radioactive isotopes with half-lives of less than 10,000 years, very little parent isotope remains after 100,000 years; after 10 half-lives, there would be only 1/1024 of the parent isotope, and the reliability of radiometric dating would be diminished substantially. The radioactive isotope carbon-14 has a half-life of 5730 years; after about 12 half-lives, or approximately 70,000 years, the amount remaining is too small to be measured accurately.

Uranium, Potassium, and Rubidium Dating Systems There are scores of radioactive isotopes, but only a few are useful for dating purposes. Some decay too rapidly and some too slowly, others migrate into and out of rocks and minerals too readily under ordinary environmental conditions, and still others are simply too seldom found. To be effective for dating purposes, isotope systems must be present in common Earth materials, not be susceptible to gains or losses of parent and daughter isotopes, and have a half-life of at least several thousand years. Sometimes rocks contain more than one radioactive isotope system, allowing us to cross-check the dates yielded by each system.

The most frequently used radioactive isotopes are shown in Table 8-1 (following page), along with their daughter isotopes, half-lives, and the particular minerals or rocks in which they are most often found.

1. **Uranium–lead dating.** Uranium is often found in granites and marine limestones. Virtually all uranium deposits contain three radioactive isotopes—uranium-238, uranium-235, and thorium-232. Each of these decays through a series of steps to form a different isotope of another element—lead. Thus, a rock that contains uranium and thorium provides three radiometric-dating systems. Until recently, the relative scarcity of uranium ore in rocks limited the use of uranium–lead dating, but due to the SHRIMP it is now possible to measure extremely small amounts of uranium and lead within other mineral crystals. Uranium and thorium isotopes have extremely long half-lives, and have been invaluable in dating some of the Earth's oldest rocks (Fig. 8-20).

Figure 8-20 Uranium and lead isotopes, measured using a sensitive high-resolution ion microprobe (SHRIMP), were used to date the oldest rocks yet discovered—this 3.96-billion-year-old Acasta Gneiss from the Yellowknife area of Canada's Northwest Territories.

Table 8-1 **The Major Isotopes Used for Radiometric Dating**

Method	Parent Isotope	Daughter Isotope	Half Life of Parent (years)	Effective Dating Range (years)	Materials Commonly Dated	Comments
Rubidium–strontium	Rb-87	Sr-87	47 billion	10 million–4.6 billion	Potassium-rich minerals such as biotite, potassium muscovite, feldspar, and hornblende; volcanic and metamorphic rocks (whole-rock analysis)	Useful for dating the Earth's oldest metamorphic and plutonic rocks.
Uranium–lead	U-238	Pb-206	4.5 billion	10 million–4.6 billion	Zircons, uraninite, and uranium ore such as pitchblende; igneous and metamorphic rock (whole-rock analysis)	Uranium isotopes usually coexist in minerals such as zircon. Multiple dating schemes enable geologists to cross-check dating results.
Uranium–lead	U-235	Pb-207	713 million	10 million–4.6 billion		
Thorium–lead	Th-232	Pb-208	14.1 billion	10 million–4.6 billion	Zircons, uraninite	Thorium coexists with uranium isotopes in minerals such as zircon.
Potassium–argon	K-40	Ar-40	1.3 billion	100,000–4.6 billion	Potassium-rich minerals such as biotite, muscovite, and potassium feldspar; volcanic rocks (whole-rock analysis)	High-grade metamorphic and plutonic igneous rocks may have been heated sufficientiy to allow Ar-40 gas to escape.
Carbon-14	C-14	N-14	5730	100–100,000	Any carbon-bearing material, such as: bones, wood, shells, charcoal, cloth, paper, animal droppings; also water, ice, cave deposits	Commonly used to date archaeological sites, recent glacial events, evidence of recent climate change, and environmental effects of human activity.

2. Potassium–argon dating. Potassium is one of the Earth's most abundant elements, found in such common rock-forming minerals as biotite and muscovite mica, potassium feldspars, and glauconite (a common mineral in sedimentary rocks). All naturally occurring potassium contains some radioactive potassium-40, which decays to argon-40, giving us the potassium–argon system. A gas that does not readily bond with other elements, argon-40 gas would exist in minerals only as the daughter product of potassium-40 decay. When it is trapped within the crystal structures of certain minerals, it can be measured to yield a reliable

date. However, argon-40 may escape from a rock that is heated during metamorphism or one that is very old or extensively weathered. If this happens, measurement of the remaining argon-40 will falsely suggest that the rock is younger than it actually is. Argon-40 also occurs in small quantities in the atmosphere, and may become trapped in extrusive igneous rock during cooling; measurement in this case could falsely suggest that the rock is older than it actually is.

Because potassium occurs more commonly than uranium and has a similarly long half-life, the potassium–argon system is very useful for dating the Earth's oldest rocks in a wide range of locations. However, even extremely small amounts of argon-40 can also be used to determine the ages of potassium-rich rocks as young as 100,000 years. In east Africa, biotite grains in volcanic deposits above and below sedimentary layers containing fossils of our earliest human ancestors have yielded potassium–argon dates of between 3.6 and 4.0 million years.

3. **Rubidium–strontium dating.** Radioactive rubidium-87, which occurs alongside potassium in numerous minerals and rocks, produces a daughter isotope, strontium-87, that does not escape as a gas. The rubidium–strontium system can therefore reliably date metamorphic as well as igneous rocks. With its 47-billion-year half-life, this system is most effective for rocks at least 10 million years old, and is used primarily to date very ancient rocks, deep-Earth plutonic rocks, and Moon rocks. It also provides a good cross-check of potassium–argon dates.

Carbon-14 Dating The radiometric systems discussed above yield absolute dates for Earth materials that are more than 100,000 years old—the long half-lives of radioactive uranium, potassium, and rubidium render them useless for dating materials younger than this. Radioactive carbon-14, however, has a relatively brief half-life of 5730 years; thus **carbon-14 dating** can be used to date materials from 100 to about 100,000 years old. This time span encompasses the most recent glaciation and climate change, the latest rise and fall of worldwide sea levels, and the rise to dominance of our species, *Homo sapiens.*

Anything that contains carbon—including bones, shells, wood, charcoal, plants, peat, paper, cloth, pollen, and seeds—contains some carbon-14. Carbon-14 is used to date charcoal in ancient campfires, house posts of Native American dwellings, seeds and seafood shells in refuse dumps, and the bones of our predecessors and their prey. It has dated the linen cloth that was wrapped around the Dead Sea scrolls (2000 to 2200 years old), the papyrus on which the history of ancient Egypt was written (about 2100 years old), and even a fragment of the Shroud of Turin (about 700 years old), purported to be the burial cloth of Jesus of Nazareth. All these items have one thing in common—they were all once part of a living organism, or are composed of materials from once-living organisms. It is during life that organisms take up the radioactive carbon that will eventually be used to date their remains.

Carbon-14 atoms occur naturally in the atmosphere, where, along with atoms of carbon's stable isotope, carbon-12, they combine with oxygen to form carbon dioxide (CO_2). Some carbon-14–containing CO_2 dissolves in oceans, lakes, rivers, glaciers, and underground waters, where it is readily taken up by organisms living in or drinking the water. Plants incorporate atmospheric CO_2 during photosynthesis, and use it to produce sugars and starches. When animals eat the plants, then, they also ingest the carbon-14.

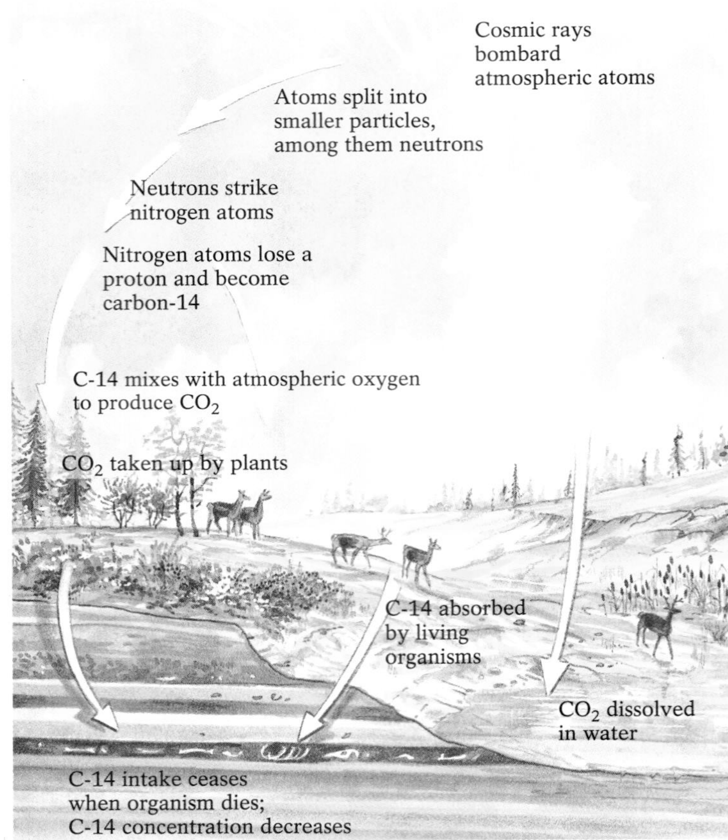

Figure 8-21 The carbon-14 that is eventually used to date long-dead organisms and ancient artifacts is initially created in the Earth's atmosphere. Here, it combines with O_2 to produce CO_2, much of which becomes dissolved in water or is taken up by plants, and is ultimately ingested by animals. While alive, an organism constantly replenishes the carbon-14 in its body. When the organism dies, its intake of carbon-14 ceases, and the carbon-14 it contains decays to its daughter isotope, nitrogen-14. Thus, the less carbon-14 that is left in the remains of an organism, the more time has elapsed since it died.

Thus, carbon-14 finds its way into the cells of most living organisms, where it continues to decay to its daughter isotope, nitrogen-14. As long as an organism is alive and taking in nutrients, it continuously replenishes its supply of carbon-14. As soon as it dies, its supply of carbon-14 immediately begins to diminish. The greater the time elapsed since the death of the organism, the less carbon-14 (and the more nitrogen-14) it will contain (Fig. 8-21).

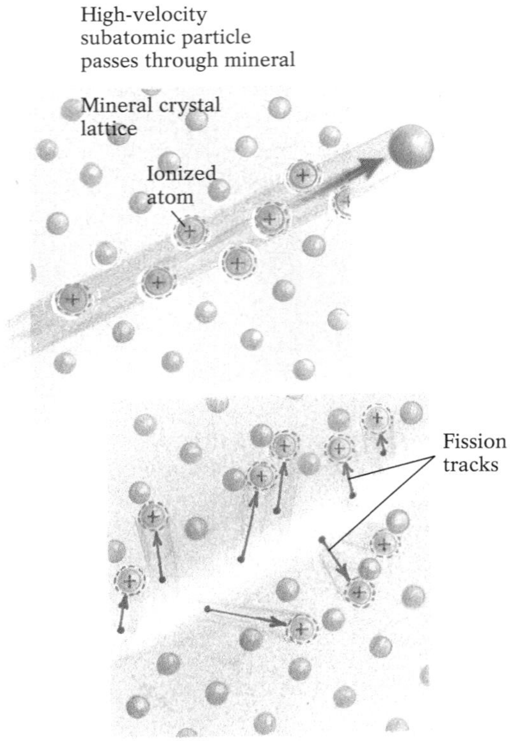

Other Absolute-Dating Techniques

There are some absolute-dating methods that do not rely directly on the decay of a radioactive isotope. Some methods rely on nuclear processes but not on the direct measurement of parent and daughter isotopes. Others derive data from something that produces countable annual layers (such as trees and lakes) or that grows at a fairly constant rate (such as lichen, the simple, crust-like organisms that attach themselves to rocks).

Fission-Track Dating *Fission* is the division of a radioactive atom's nucleus into two fragments of approximately equal size, releasing several subatomic particles. When a radioactive atom that is trapped in a mineral undergoes fission, these subatomic particles move at high speed across the orderly rows of atoms in the mineral's crystal lattice, leaving traces called **fission tracks** within the crystal (Fig. 8-22). Because fission proceeds at a constant rate, emitting the same number of particles per unit time, fission tracks are produced at a constant rate. Thus, the number of fission tracks per unit area of a crystal increases in direct proportion to the age of the sample. Fission tracks can be measured and counted to date minerals that are from 50,000 to billions of years old. This method is particularly useful for dating a wide variety of volcanic glasses and other mineral grains. Fission tracks are usually erased, however, when their host rocks are heated above about 400°C (750°F), and thus cannot be used to date medium- and high-grade metamorphic rocks.

High-velocity subatomic particle passes through mineral
Mineral crystal lattice
Ionized atom
Fission tracks

Figure 8-22 Fission tracks are produced by the decay of radioactive atoms trapped in a mineral's crystal structure. High-velocity decay particles emitted by the radioactive atoms disrupt the orderly arrangement of atoms in the crystal lattice, leaving tracks only a few atoms wide; these are typically flooded with a strong solvent and enlarged so that they may be seen (and counted) with an ordinary microscope.

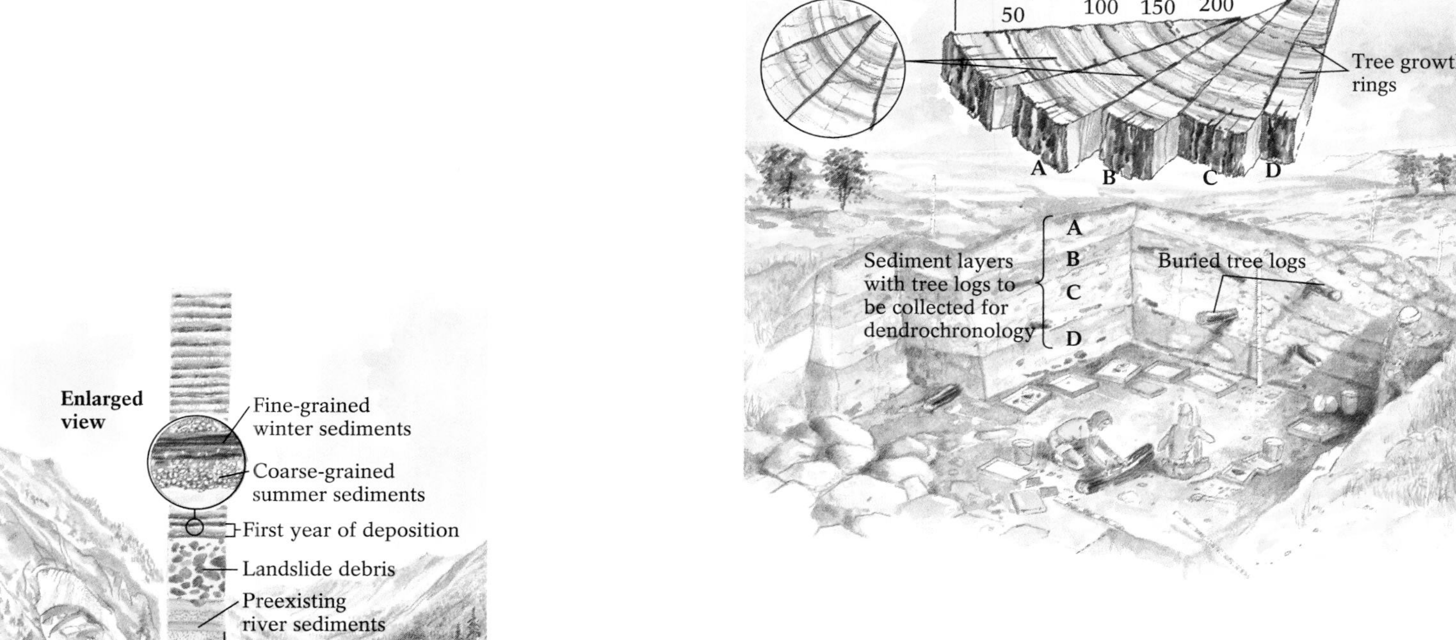

Figure 8-23 Correlation of tree-ring sections in trees or wooden artifacts of overlapping life-spans can establish ages of thousands of years.

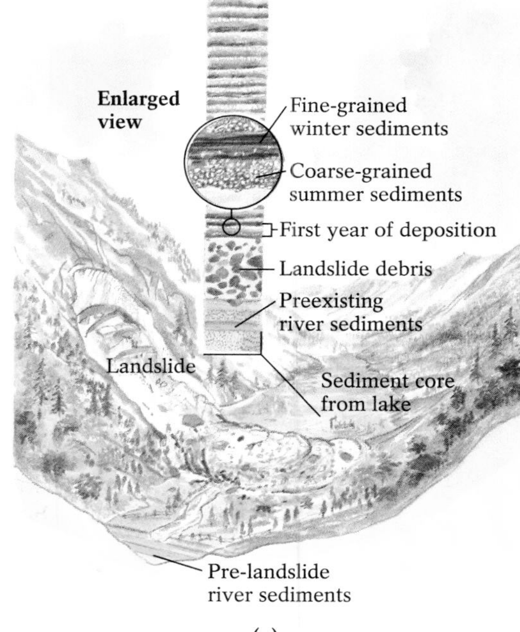

(a)

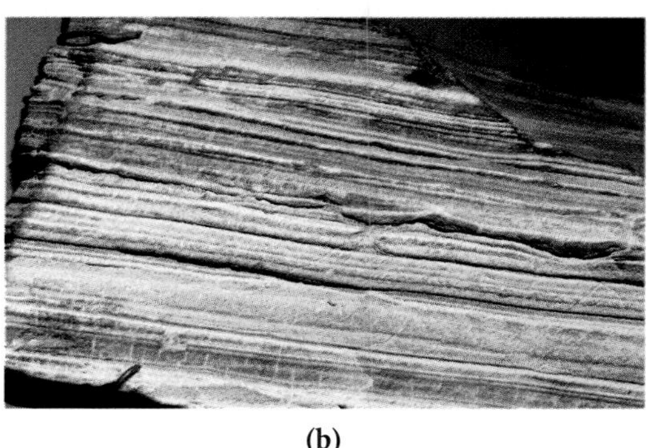

(b)

Figure 8-24 Varves form in lakes that deposit different types of sediments during different times of the year, allowing us to distinguish annual layers. The origin of the lake in part **(a)** can be dated by taking a core sample of its sediments and counting the number of varves deposited since the landslide. **(b)** A photo of varved clay from the Green River Formation in Colorado. The varves were formed in Lake Uinta, which existed about 55 to 45 million years ago.

Dendrochronology (Tree-Ring Dating) In temperate climates, the annual growth of most trees results in concentric sets of dark and light rings in cross-sections of their trunks and branches. By counting the number of rings it has, we can determine a tree's age—an absolute-dating method known as **dendrochronology.** Dating trees in this way can help us date relatively recent geological events such as landslides or mudflows—wherever trees have become established on top of new surfaces. (The older the trees growing on a landslide deposit, the longer ago the landslide occurred.) Because tree rings may be preserved even in long-dead trees, dendrochronology can also be used to date much older events and artifacts. Changes in climatic conditions, such as prolonged droughts, cause the same type of tree-ring variations in all the trees living in the affected area; the ring patterns in these trees can then be compared and correlated, showing that their life spans overlapped. By correlating sections of the ring patterns from progressively older trees in the same area—including fossilized dead trees—dates ranging back about 9000 years have been established (Fig. 8-23). Dendrochronology is also often used by archaeologists to date wooden artifacts, such as house posts and digging sticks.

Varve Chronology Lakes, particularly ones that freeze in the winter, can also produce countable annual layers: **varves** are paired layers of sediment, typically consisting of a thick, coarse, light-colored layer deposited in summer and a thin, fine, dark-colored layer deposited in winter (Fig. 8-24). To count the varves underlying a lake, geologists extract a *core,* a continuous sequence of lake sediment, by drilling through lake mud. The number and nature of the varves in a core reveal how long ago the lake formed and identify events, such as landslides, that affected the lake's existence. For example, a lake produced when a landslide blocks a river valley immediately begins to produce varves. When we later count the varves, we learn when the landslide occurred. Varves can also show a pattern of repeated landslides in an area, information useful for planning the safe development of a community.

Lichenometry *Lichen*—colonies of simple, plant-like organisms that grow directly on exposed rock surfaces—are the basis of a dating method known as **lichenometry.** Given similar rocks and climatic conditions, the larger the lichen colony, the longer the period of time since the growth surface was exposed. The lichen colonies on old tombstones, for example, are larger than those on newer ones. Using tombstones and other stone surfaces of known ages on which lichens grow (such as buildings, bridge supports, and old mine tailings), a lichen-growth curve can be developed to show the rate at which lichen have grown in a given area (Fig. 8-25). Lichen on a surface of unknown age (within the same area) can then be compared to the growth curve to determine the surface's age. Because lichen colonies eventually grow together and can no longer be measured individually, lichenometry is useful only for surfaces less than about 9000 years old. Lichen provide accurate dates for young glacial deposits, rockfalls, and mudflows—all events that expose new rock surfaces on which lichen can grow.

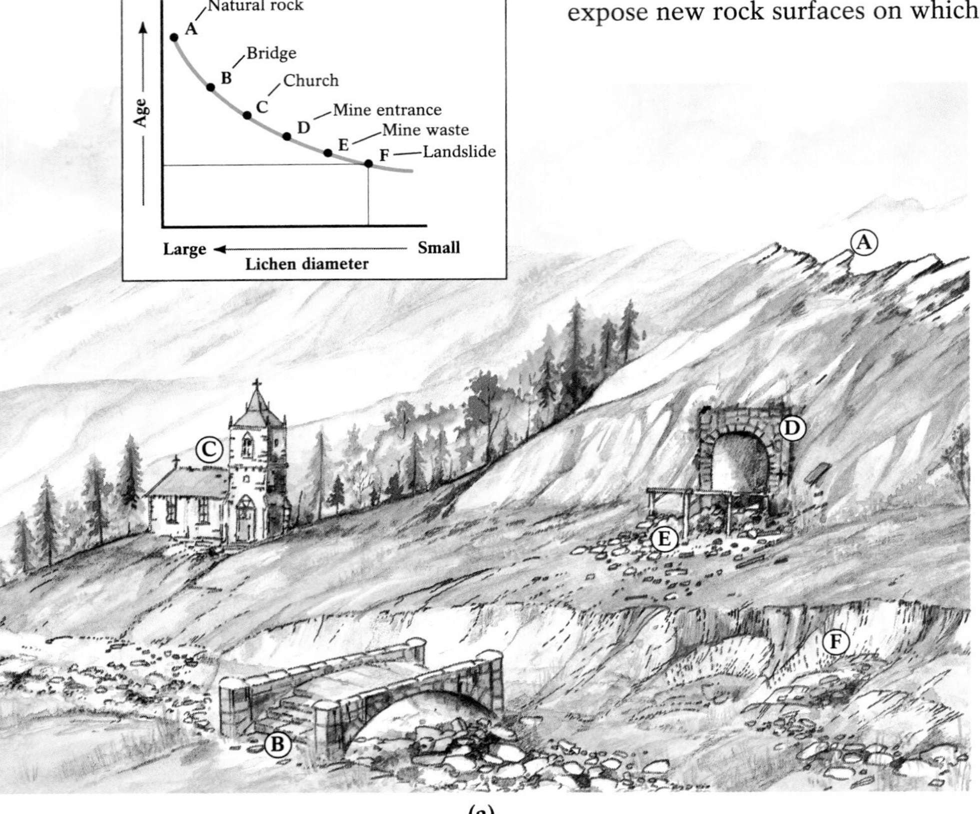

(a)

(b)

Figure 8-25 Estimating the absolute age of a material from the lichen growing on its exposed surface first requires that the rate of lichen growth in the area be determined. By measuring lichen on surfaces of known age, such as the bridge, the church, and the mine entrance in (**a**), geologists plot a growth curve that relates lichen diameters to time. The age of a surface of unknown age can then be estimated by measuring the lichen on it and finding its position on the growth curve. (**b**) Lichen colonies on a granite boulder. (The light-colored areas of rock have been bleached by chemicals in the lichen.)

Combining Relative and Absolute Dating

When geologists and paleontologists study a region for the first time, they use all the dating methods at their disposal—both relative and absolute—to reconstruct its geological history. Some segments of the region's rock record may be more readily dated absolutely because they contain isotopically datable key beds; others, such as fossil-free sedimentary rocks, may only be dated relatively. However, a geological "biography" of any region can generally be written by radiometrically dating every layer that contains radioactive isotopes and then applying all the principles of relative dating to supplement this information. Figure 8-26 shows how the complex history concealed in a region's rocks can be reconstructed, revealing a series of geological events spanning billions of years of Earth history.

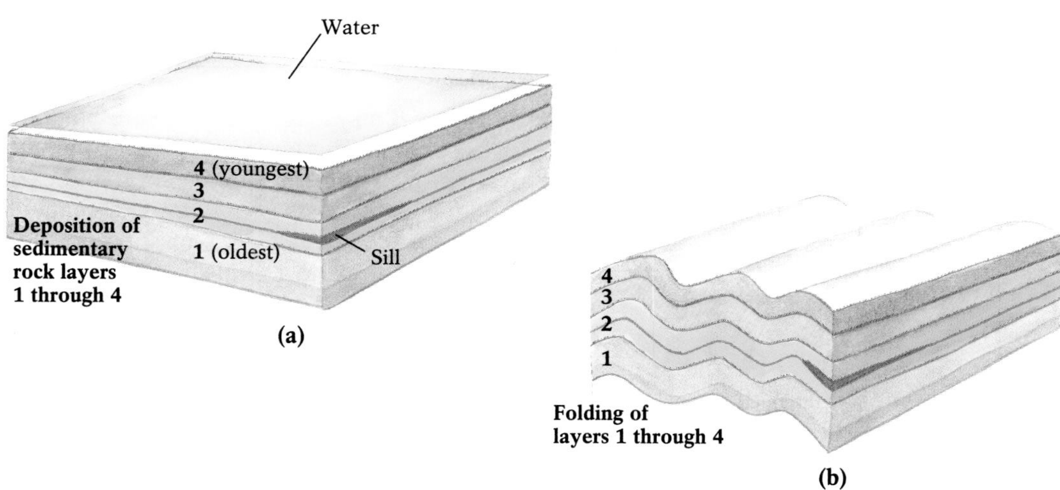

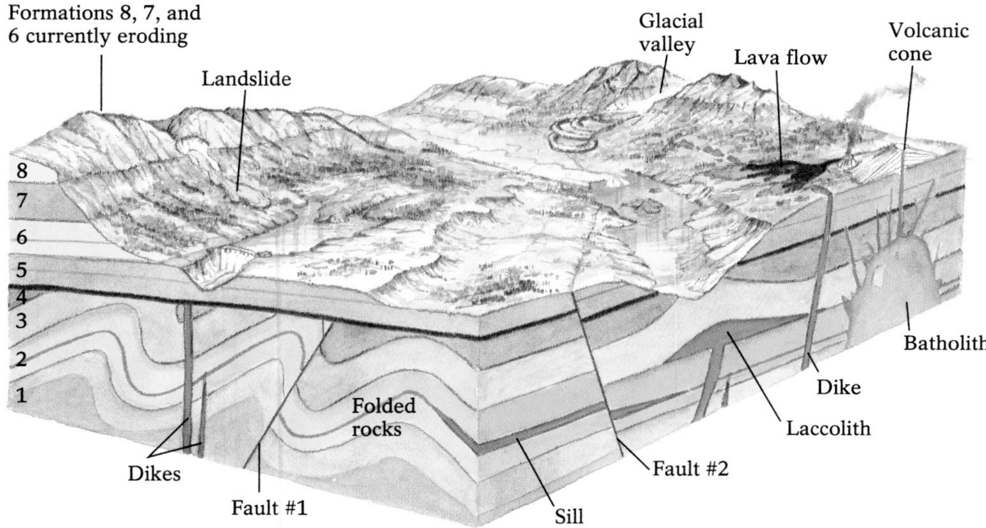

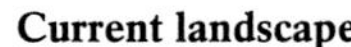
Current landscape

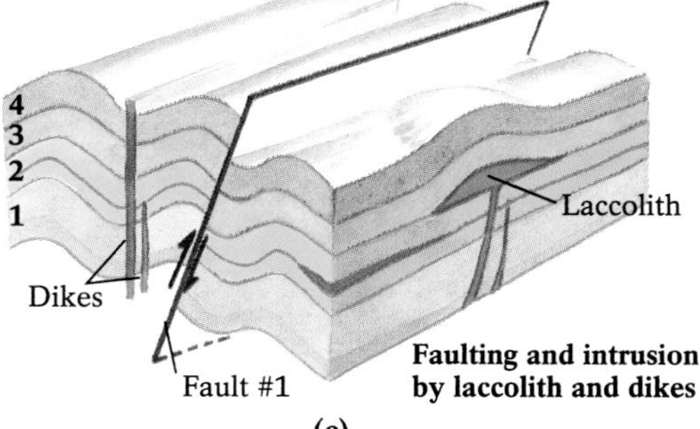

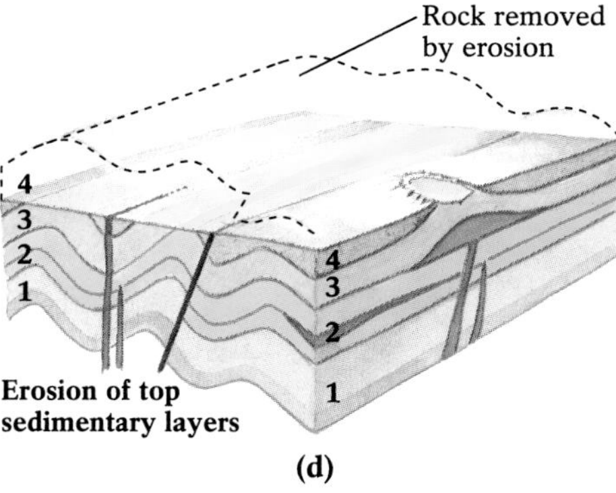

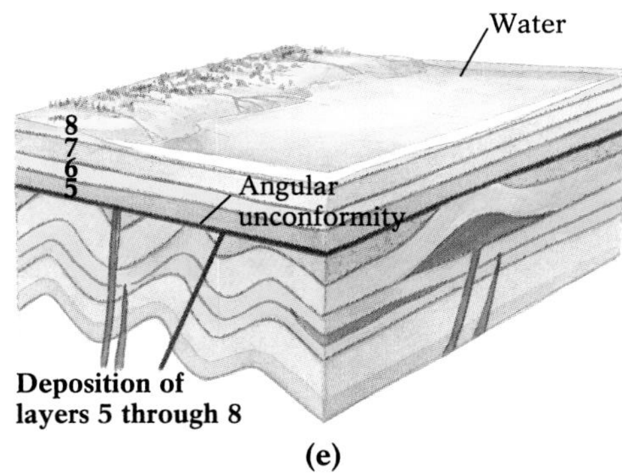

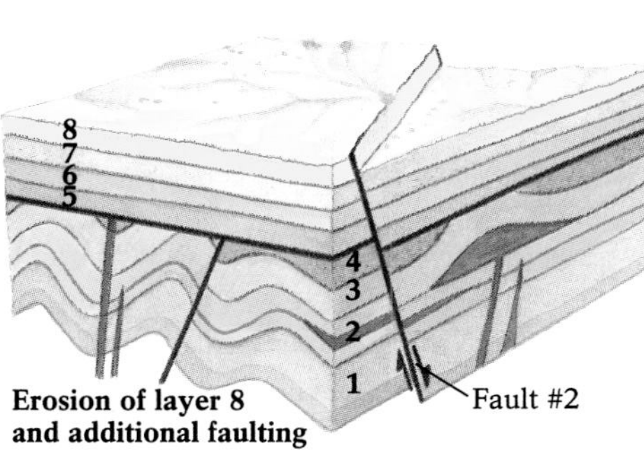

Figure 8-26 The rocks underlying this hypothetical landscape reveal a complex history involving initial horizontal deposition (**a**), folding (**b**), faulting and volcanism (**c**), erosion (**d**), and subsequent deposition, faulting, and erosion (**e** and **f**). (See Chapter 9 for a discussion of folding and faulting.) The geological activity occurring currently at the surface will be similarly recorded in the area's stratigraphy.

The Geologic Time Scale

The **geologic time scale** (Fig. 8-27) organizes all of Earth history into blocks of time during which important events occurred. The basic divisions of the geologic time scale were established during the eighteenth, nineteenth, and early twentieth centuries by geologists and paleontologists who identified changing fossil assemblages in rocks and then applied the principles of superposition, faunal succession, and cross-cutting relationships to establish their sequence. In the last half-century, we have been able to use radiometric dating to assign absolute-age ranges to these divisions.

The largest time spans shown on the geologic time scale are *eons*, which mark major developments such as the first occurrence of life on Earth or the worldwide expansion of multicelled life forms. The earliest is the Hadean ("beneath the Earth") Eon, ranging from the time of the Earth's formation some 4.6 billion years ago to about 3.8 billion years ago, from which time virtually no rocks and no fossils remain. The earliest known life forms appear in

Figure 8-27 The geologic time scale. Absolute ages shown are in millions of years.

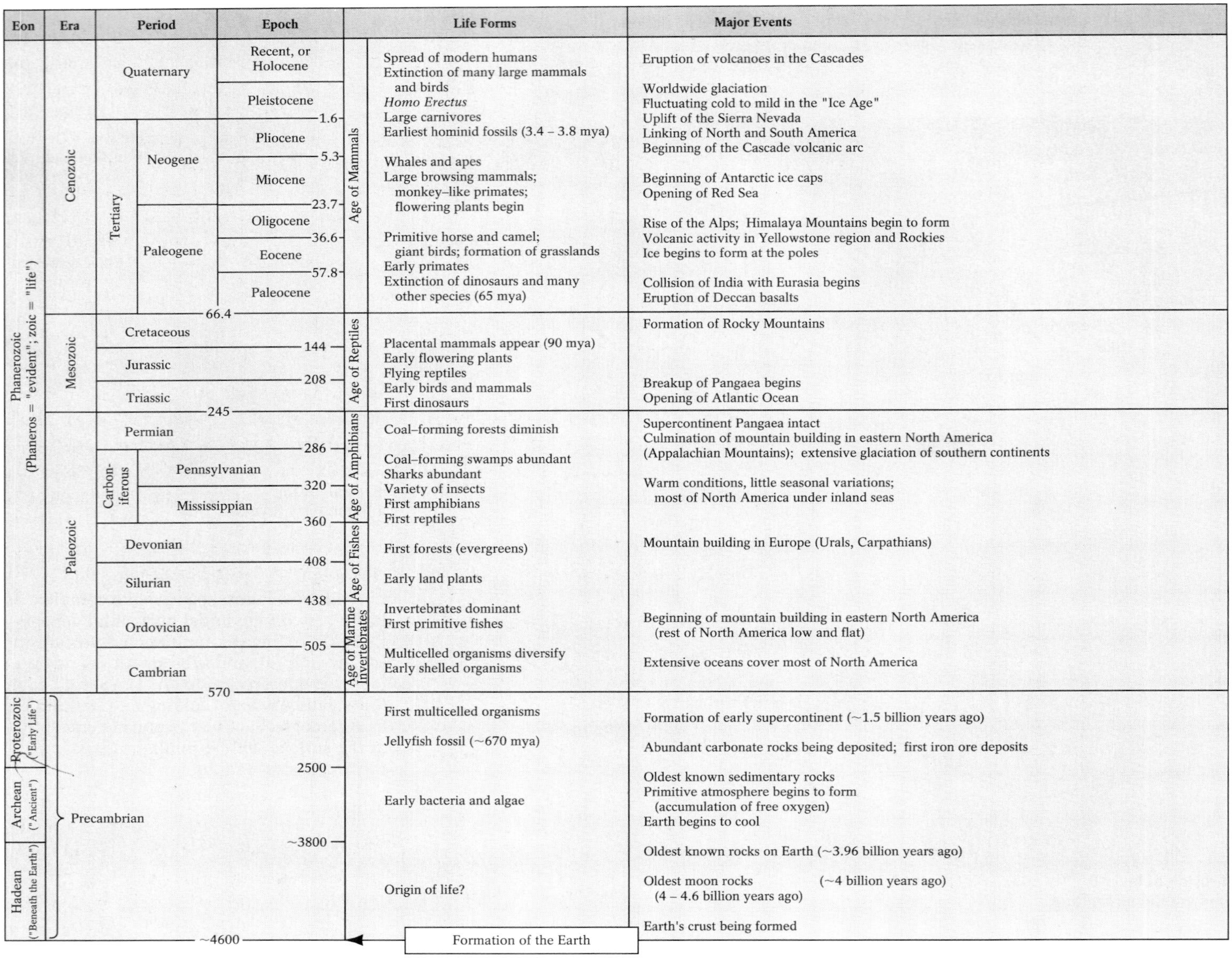

Eon	Era	Period		Epoch	Age at base		Life Forms	Major Events
Phanerozoic (Phaneros = "evident"; zoic = "life")	Cenozoic	Quaternary		Recent, or Holocene		Age of Mammals	Spread of modern humans	Eruption of volcanoes in the Cascades
				Pleistocene	1.6		Extinction of many large mammals and birds; *Homo Erectus*; Large carnivores	Worldwide glaciation; Fluctuating cold to mild in the "Ice Age"
		Tertiary	Neogene	Pliocene	5.3		Earliest hominid fossils (3.4 – 3.8 mya)	Uplift of the Sierra Nevada; Linking of North and South America; Beginning of the Cascade volcanic arc
				Miocene	23.7		Whales and apes; Large browsing mammals; monkey–like primates; flowering plants begin	Beginning of Antarctic ice caps; Opening of Red Sea
			Paleogene	Oligocene	36.6		Primitive horse and camel; giant birds; formation of grasslands	Rise of the Alps; Himalaya Mountains begin to form; Volcanic activity in Yellowstone region and Rockies
				Eocene	57.8		Early primates	Ice begins to form at the poles
				Paleocene	66.4		Extinction of dinosaurs and many other species (65 mya)	Collision of India with Eurasia begins; Eruption of Deccan basalts
	Mesozoic	Cretaceous			144	Age of Reptiles	Placental mammals appear (90 mya); Early flowering plants	Formation of Rocky Mountains
		Jurassic			208		Flying reptiles	
		Triassic			245		Early birds and mammals; First dinosaurs	Breakup of Pangaea begins; Opening of Atlantic Ocean
	Paleozoic	Permian			286	Age of Amphibians	Coal–forming forests diminish	Supercontinent Pangaea intact; Culmination of mountain building in eastern North America (Appalachian Mountains); extensive glaciation of southern continents
		Carboniferous	Pennsylvanian		320		Coal–forming swamps abundant; Sharks abundant; Variety of insects	Warm conditions, little seasonal variations; most of North America under inland seas
			Mississippian		360		First amphibians; First reptiles	
		Devonian			408	Age of Fishes	First forests (evergreens)	Mountain building in Europe (Urals, Carpathians)
		Silurian			438		Early land plants	
		Ordovician			505	Age of Marine Invertebrates	Invertebrates dominant; First primitive fishes	Beginning of mountain building in eastern North America (rest of North America low and flat)
		Cambrian			570		Multicelled organisms diversify; Early shelled organisms	Extensive oceans cover most of North America
Proterozoic ("Early Life")	Precambrian				2500		First multicelled organisms; Jellyfish fossil (~670 mya)	Formation of early supercontinent (~1.5 billion years ago); Abundant carbonate rocks being deposited; first iron ore deposits
Archean ("Ancient")					~3800		Early bacteria and algae	Oldest known sedimentary rocks; Primitive atmosphere begins to form (accumulation of free oxygen); Earth begins to cool
Hadean ("Beneath the Earth")					~4600		Origin of life?	Oldest known rocks on Earth (~3.96 billion years ago); Oldest moon rocks (4 – 4.6 billion years ago) (~4 billion years ago); Earth's crust being formed
							Formation of the Earth	

(a)

(b)

Figure 8-28 (**a**) The fossils in this rock, from the Precambrian Period of the Archean Eon, are *stromatolites,* structures created by a very early form of alga. (**b**) Modern stromatolites in Shark Bay, Western Australia.

rock from the Archean ("ancient") Eon, about 3.8 to 2.5 billion years ago (Fig. 8-28).

Eons are subdivided into *eras,* each of which is defined by its dominant life forms; the Mesozoic Era, for instance, is the "Age of Reptiles" (including dinosaurs). Eras are subdivided into *periods,* which record less dramatic evolutionary changes. Periods, in turn, are subdivided into *epochs,* based on still more subtle biological changes and some nonbiological criteria as well. Each of these successive subdivisions marks progressively smaller units of time. The lengths of time spanned vary, even for subdivisions at the same level, because the key characteristics and geological conditions that define them are not of uniform duration. Thus, the Eocene Epoch of the Tertiary Period of the Cenozoic Era lasted for 21 million years, whereas the Oligocene Epoch of the same period and era lasted only 13 million years.

Life on Earth

Most geologists believe that simple life forms evolved during the Earth's first billion years, but did not survive to give rise to later creatures. Most likely they were eradicated by frequent impacts of the massive meteors that still littered the developing solar system. Impact with a meteor 400 kilometers (250 miles) in diameter and traveling at a speed of 15 kilometers (10 miles) per second could have raised the Earth's surface temperature to as high as 5650°C (3400°F), rapidly vaporizing the early oceans and any life they supported. Only after all the planets and moons of the solar system had finally formed, clearing the surrounding space of most of its remaining debris, could life develop on Earth without hindrance. The first evidence of this earliest life, simple bacteria-like cells, was discovered in 1980 by geologists in northwestern Australia. The cells were found in a layer of sedimentary chert that was dated (by fission-track analysis of the surrounding igneous rocks) about 3.77 billion years—the beginning of the Archean Eon. In South Africa, primitive blue-green algae have since been found in rocks dated at more than 3.3 billion years.

The Archean Eon was followed by the Proterozoic ("early life") Eon 2.5 billion years ago. This was a time of increasing biological diversity, during which the first multicelled organisms arose. Some 670-million-year-old rocks in Australia, for instance, contain fossil evidence of jellyfish, soft marine worms, and a variety of other simple creatures made entirely of soft tissue. These creatures and their descendants flourished until they filled the oceans, 570 million years ago.

The beginning of the Phanerozoic ("visible life") Eon 570 million years ago marks the first point at which marine creatures began to leave behind abundant fossil evidence of hard shells, prominent external spines, and internal skeletons. A number of competing hypotheses have been proposed to account for this development (which, although it distinguishes the Phanerozoic Eon, actually had its start in the late Proterozoic). One claims that the increasing population of algae, which rely on photosynthesis for basic life processes, released sufficient oxygen (a product of photosynthesis) to form the ozone layer in the atmosphere, protecting life forms on Earth from deadly ultraviolet radiation from the Sun. As atmospheric oxygen fueled their metabolism, animals may have begun to grow larger and more complex, requiring a rigid structure to support their larger mass. Another hypothesis is that the oceans—and the creatures in them—contained so much calcium at this time that marine creatures began to excrete calcium, in order to avoid calcium poisoning, and this ultimately became their hard outer shells. A competing hypothesis holds that shells and spiny growths developed as protection against attacks from predators.

The Phanerozoic Eon extends all the way to the present, and has been marked by the development of a remarkable succession of plant and animal life. It is divided into three eras: the Paleozoic, Mesozoic, and Cenozoic. The first part of the **Paleozoic** ("ancient life") **Era** was dominated by marine invertebrates (sea creatures without backbones, such as corals, clams, and other shelled organisms); fish, amphibians, insects, and land plants rose to prominence during the latter part of the Paleozoic. The **Mesozoic** ("middle life") **Era** was dominated by marine and terrestrial reptiles, including dinosaurs; the first birds, mammals, and flowering plants appeared during this era. The **Cenozoic** ("recent life") **Era,** in which we are still living, is distinguished by a rich variety of mammals.

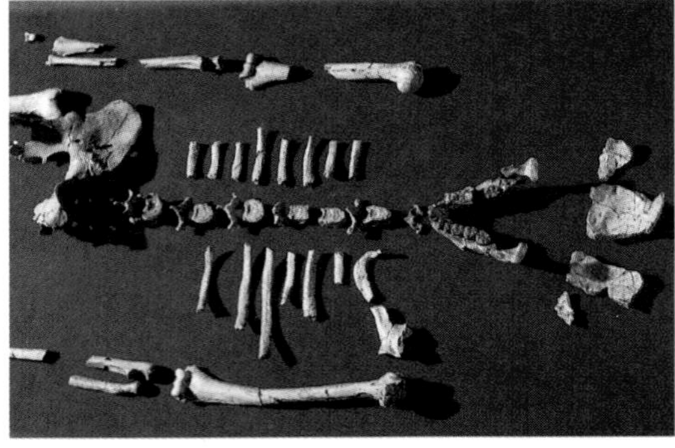

Figure 8-29 The fossil hominid called Lucy, collected in Hadar, Ethiopia, in 1974 by Donald Johanson. Lucy, one of the most complete hominid fossils yet discovered, is believed to be over 3 million years old.

When did human beings arise in the course of Earth history? East Africa may have been the cradle of our species. In what are now Ethiopia and Tanzania, the bones and footprints of the earliest hominids (members of the human family, Hominidae), have been found embedded in sedimentary layers that also contain datable volcanic ash. The oldest dates yielded for these hominid fossils, derived from potassium–argon analysis, are from 3.4 to 3.8 million years ago (Fig. 8-29). Eastern Africa provided not only an hospitable climate and abundant resources for our species to evolve, but also the volcanism and sedimentation that preserved the evidence and allows it to be dated.

The Age of the Earth

Most of the dates cited in this book for rock ages and geological events have been determined by one or more—usually several—of the relative- and absolute-dating techniques just described. The one critical date that we cannot verify by any direct absolute method is the age of the Earth. Many of the Earth's oldest rocks are metamorphic, changed from even older, preexisting rocks. The planet's internal heat, recycling of tectonic plates, catastrophic events such as meteor impacts during the early evolution of our planet, and a host of other geological processes have worked together over these billions of years to change much of the Earth's original material, concealing from us specific evidence of its true age. Yet, although no rock that old has ever been found, most scientists agree that the Earth is about 4.6 billion years old.

Without any of its original material available for radiometric dating, how can we tell how old the Earth is? Because all the components of our solar

Figure 8-30 A lunar boulder, as seen from the Apollo 17 spacecraft in December 1972. (The Earth can be seen in the far distant background.) Fragments of such Moon rocks that have been taken back to Earth for analysis have been found to be about 4.5 billion years old.

Figure 8-31 The Hoba meteorite, in Namibia. Almost every meteorite to have struck the Earth has been radiometrically dated at about 4.6 billion years.

system formed at about the same time, the ages of extraterrestrial rocks provide indirect evidence of the age of our own planet. For example, some exciting clues about the Earth's antiquity have come from Moon rocks collected by American astronauts within just a few hundred meters of their landing craft (Fig. 8-30). These were dated by both uranium–lead and potassium–argon systems to about 4.53 billion years. (There was no Moon-wide search for the oldest rocks—these rocks just happened to be on the ground where the spacecraft landed. Since the Moon's rocks were never recycled by plate tectonic and surface processes, any one of them can represent the Moon's earliest development.) In addition, radiometric dating of virtually every meteorite that has struck the Earth, such as the one in Figure 8-31, has yielded about the same age—4.6 billion years. Most meteorites are remnants of the creation of the solar system, chunks of rock that have not been recycled by planetary tectonics or undergone heating, pressing, or other planetary forces. In all likelihood, they faithfully record the age of the solar system and of the Earth.

Some clever scientific detective work enables us to learn some confirming information about the Earth's date of birth. We have seen that uranium and thorium decay to produce three stable lead isotopes—lead-206, lead-207, and lead-208. Lead has a fourth stable isotope, lead-204, that is *not* a product of radioactive decay. Since it is not created by the decay of other elements, the concentration of lead-204 in the Earth has remained constant, while the concentrations of the other lead isotopes have been increasing at constant rates through radioactive decay since the Earth began. The relative proportions of the four lead isotopes are about the same in every meteorite that does not also contain a radioactive parent, suggesting that these proportions resemble the original mixture of lead in our solar system. By comparing these proportions to those found in Earth rocks, we can determine how much lead of radioactive origin has been added. Scientists have calculated that today's proportions of radioactively produced lead indicate about 4.5 billion years of radioactive decay. So, lunar rocks, meteorites, and the lead isotopes on Earth all convince us that the Earth and the rest of our solar system formed about 4.6 billion years ago.

Since that distant time, geological processes that appear to be excruciatingly slow from a human perspective have accomplished incredible feats. It is humbling to realize, standing before Pikes Peak in Colorado or Mount Katahdin in Maine, that the process of erosion, imperceptibly slow by human time standards, could remove those mountains in a geological instant. The next time you sit by a stream and glimpse the sand grains being carried by the current, consider how little geologic time it would take for the grains to accumulate and lithify to form a hill or a mountain, and how easily geological change is effected when there's "all the time in the world."

This chapter marks the end of Part 1 in this book—the section devoted to building your foundation of geological basics. In this section we described the basic structure of minerals and rocks, and looked at how the major rock groups form: by cooling from a melted state, by lithification of broken-down and dissolved rock fragments into sedimentary rocks, and by changing under the heat and pressure of metamorphic processes. We discussed how plate tectonic movement affects each of these processes. In this last chapter of Part 1, we examined the vast reaches of geologic time and how we measure it. In Part 2, we will examine more closely the dynamic tectonic processes that have combined to create the Earth's most massive features, such as its immense interior layers, its continents and ocean basins, and its lofty mountain ranges. We will also focus on earthquakes, one of the most dramatic effects of these dynamic processes.

Chapter Summary

When geologists determine the ages of rocks and other materials, they are engaged in **geochronology,** the study of time in relation to the Earth's existence. This subject is part of **historical geology,** the branch of the geosciences that involves the study of the Earth's past. To place the origin of an item such as a rock or sediment in the correct historical perspective, geologists apply either **relative dating**—determining how old it is in relation to its surroundings—or **absolute dating**—determining its actual age in years.

Relative dating relies on several key principles. The **principle of uniformitarianism** states that modern geological processes are similar to those that operated in the past. The **principle of original horizontality** states that most sedimentary rocks and lava flows are initially deposited in horizontal layers. The **principle of superposition** states that for tectonically undisturbed sedimentary rocks and lava flows, the uppermost layer in a sequence of rocks is the youngest, with those below it successively older. The **principle of cross-cutting relationships** states that layers cut across by other layers and features, such as igneous dikes and faults, must be older than the features that cut them. Similarly, rocks that are incorporated within other rocks must be older than the rocks that incorporate them. This is called the **principle of inclusions.** Finally, the **principle of faunal succession** states that more complex forms of life succeeded simpler forms through time, and therefore layers of rock containing the remains of more complex organisms are younger than those containing the remains of less complex organisms. The remains of organisms preserved in rock are **fossils;** the most helpful fossils, called **index fossils,** are from species that had wide geographical distribution but lived for only a relatively brief period of time.

When rocks and sediments are being dated relatively, we often find that two physically adjacent layers were actually deposited at distinctly separate periods of time. The boundary between such layers, called an **unconformity,** represents a gap in the rock record (resulting either from erosion of entire rock layers or from periods of nondeposition). To determine the relative histories of geographically distant rock sequences containing different sets of strata and unconformities, geologists establish age equivalence between similar individual rock layers in the different sequences, a process known as **correlation.**

Whenever possible, geologists prefer to date a rock unit absolutely. This can be accomplished when the rock contains a radioactive **parent isotope** that decays to produce a measurable amount of a nonradioactive **daughter isotope** at a constant rate, called the **half-life.** The use of radioactive isotopes to date rocks, some up to 4.6 billion years old, is **radiometric dating.** Most igneous rocks contain some measurable proportion of radioactive isotopes, such as: uranium-238, which evolves by decay into lead-206 **(uranium–lead dating);** potassium-40, which decays to form argon-40 **(potassium–argon dating);** and rubidium-87, which evolves into strontium-87 **(rubidium–strontium dating).** Organic substances up to about 100,000 years old can be dated using **carbon-14 dating.**

Fission tracks are produced in a crystal's structure as high-speed particles emitted during radioactive decay pass through the crystal; some rocks can be dated by measuring the number and length of such fission tracks in their structures. A variety of other absolute-dating techniques have been developed, principally to date rocks and sediments that are less than a few tens of thousands of years old. Some involve counting periodically accumulated layers, such as the annual growth rings in trees—a dating method known as **dendrochronology** (tree-ring dating)—or the seasonally deposited layers of lake-bottom sediment called **varves.** Others involve measuring the accumulation of something that grows at a fairly constant rate, such as the lichen that colonizes exposed rock surfaces; dating using lichen is known as **lichenometry.**

These and other techniques have combined to produce the **geologic time scale,** a chronicle of all of Earth history. They have also enabled us to determine the age of the Earth, how long ago life on Earth originated, and when our early human ancestors lived. The geologic time scale is divided into eons, eras, periods, and epochs. Most of the evidence used to interpret Earth history comes from the Phanerozoic Eon, that segment of time during which evidence of life began to be abundantly preserved as fossils in rocks. The Phanerozoic Eon is divided into the **Paleozoic** ("ancient life") **Era, Mesozoic** ("middle life") **Era,** and **Cenozoic** ("recent life") **Era;** the Paleozoic was dominated by marine invertebrates (such as primitive clams, snails, and corals), and later fish and amphibians, and the Mesozoic and Cenozoic by reptiles (such as dinosaurs) and mammals, respectively.

Key Terms

geochronology (p. 207)

historical geology (p. 208)

relative dating (p. 209)

absolute dating (p. 209)

principle of uniformitarianism (p. 210)

principle of original horizontality (p. 211)

principle of superposition (p. 211)

principle of cross-cutting relationships (p. 212)

principle of inclusions (p. 212)

fossils (p. 213)

principle of faunal succession (p. 213)

index fossils (p. 213)

unconformities (p. 216)

correlation (p. 218)

parent isotope (p. 220)

daughter isotope (p. 220)

radiometric dating (p. 221)

half-life (p. 221)

uranium–lead dating (p. 223)

potassium–argon dating (p. 224)

rubidium–strontium dating (p. 225)

carbon-14 dating (p. 225)

fission tracks (p. 226)

dendrochronology (p. 227)

varves (p. 227)

lichenometry (p. 228)

geologic time scale (p. 230)

Paleozoic Era (p. 232)

Mesozoic Era (p.232)

Cenozoic Era (p. 232)

Questions for Review

1. Briefly explain the difference between relative and absolute dating.

2. Discuss three of the basic principles that are the foundation of relative dating.

3. What qualifies a species to become an index fossil? How are index fossils used in the correlation of sedimentary rock strata?

4. Sketch and label two different types of unconformities.

5. Name three parent–daughter radiometric dating systems, the half-lives of each parent isotope, and the rocks or sediments that are most likely to be dated by each.

6. Briefly discuss two potential problems that may diminish the reliability of an isotopically derived date.

7. Briefly explain how carbon-14 finds its way into the cells of living organisms.

8. Select three absolute-dating methods. Describe their basic principles, and the materials that can be dated by each technique.

9. With reference to Figure 8-27, the geologic time scale, when did each of the following great events in Earth history occur: the origin of the world's iron ores; the first appearance of a protective atmosphere; the origin of flowering plants; the origin of birds; the age of reptiles; the current ice age.

10. If the oldest rocks ever found on Earth are less than 4.0 billion years old, what evidence suggests that the Earth is actually 4.6 billion years old?

For Further Thought

1. Why are obsidian and basalt more susceptible to the development of hydration and weathering rinds than granite and andesites?

2. Using a combination of relative- and absolute-dating methods, derive the history of the hypothetical landscape below. (Go slowly, and don't jump to premature conclusions. Consider all the principles that we've discussed.)

3. Although geologists claim that "the present is the key to the past" (the principle of uniformitarianism), the Earth has most certainly changed throughout its 4.6-billion-year history. Think of two geological processes that operate differently today than they did in the past, and discuss the differences.

4. Suppose you decided not to accept the 4.6-billion-year age of the Earth that geologists propose (primarily from the ages of Moon rocks and meteorites). Devise an alternative strategy for determining the age of the Earth, assuming that you have unlimited funds.

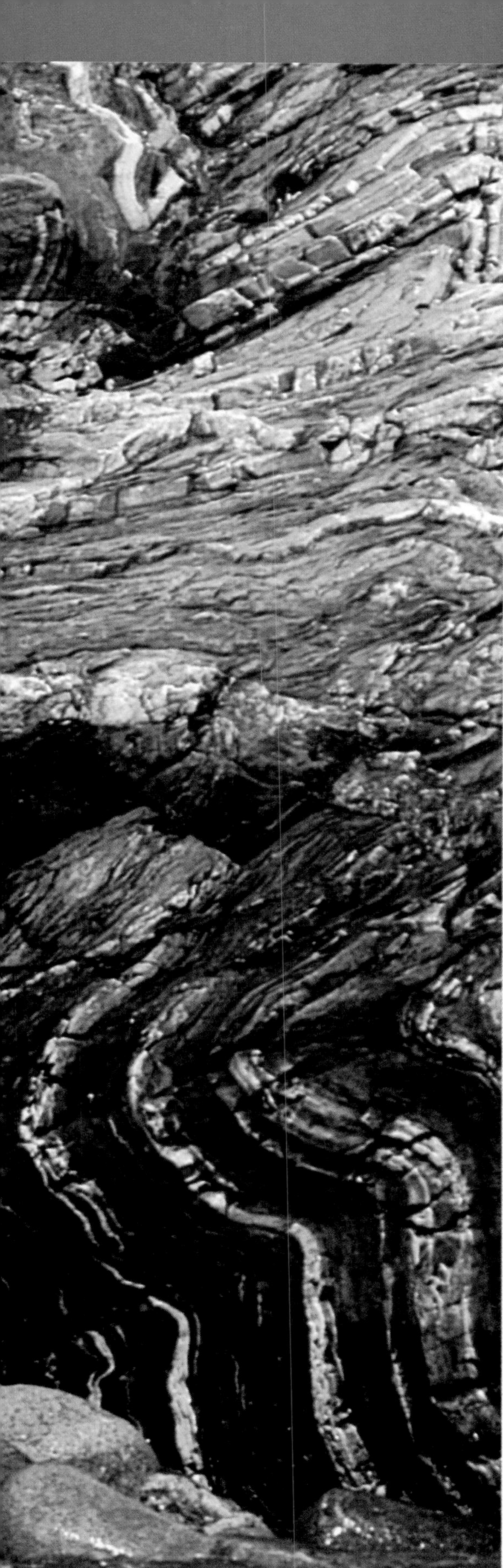

Part 2

Shaping the Earth's Crust

Stressing and Straining Rocks

Folds

Faults

Highlight 9-1 Folds, Faults, and the Search for Fossil Fuels

Building Mountains

Highlight 9-2 The Appalachians: North America's Geologic Jigsaw Puzzle

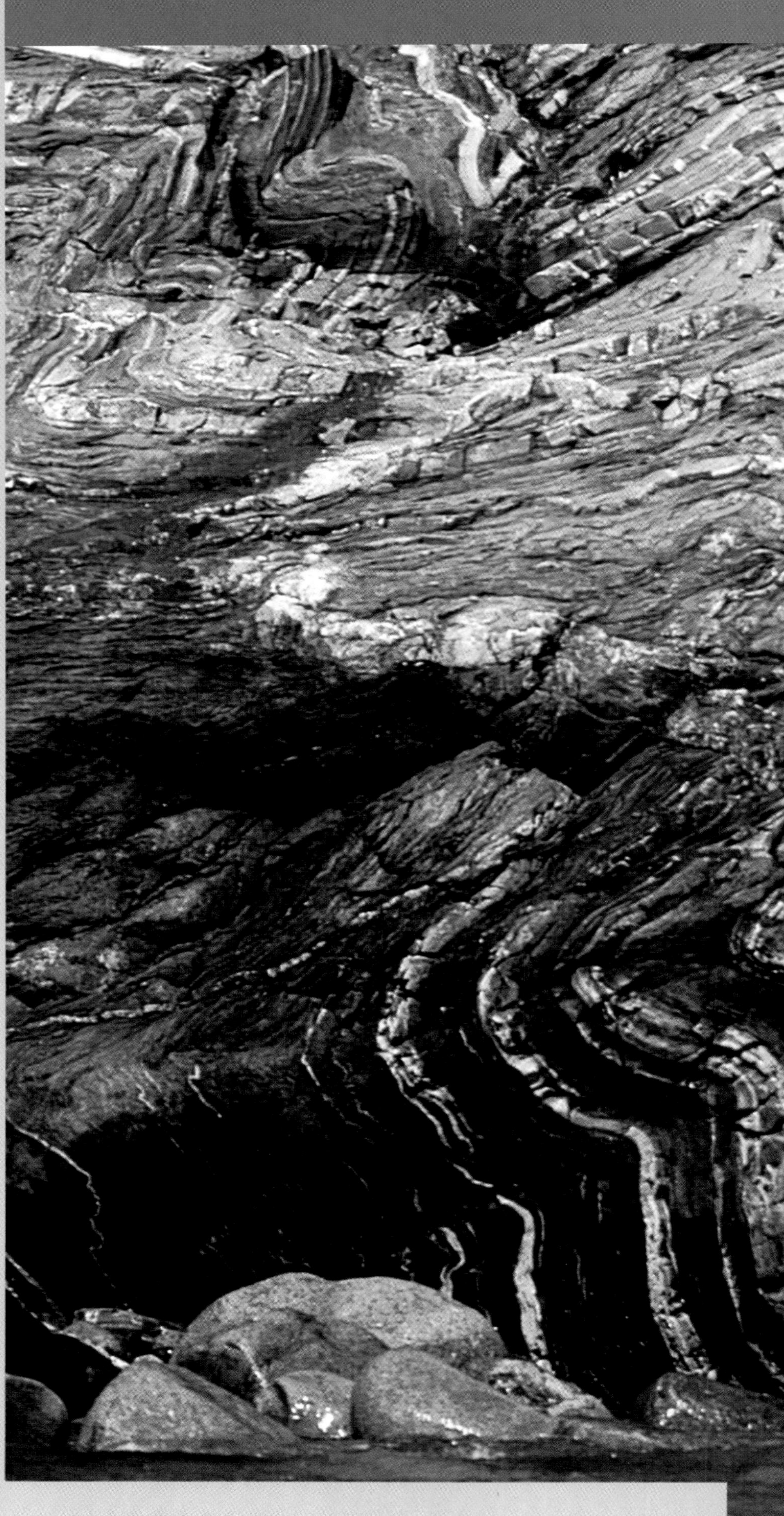

Figure 9-1 Folded strata on Sledge Island, Arkansas, provide dramatic evidence that a force exists powerful enough to distort the shape even of hard rock.

Chapter 9

Folds, Faults, and Mountains

The Rockies, the Alps, the Andes, the Himalayas—something about these peaks draws us to them. Many are inspired by their beauty, and some even risk life and limb to climb them. What enormous forces could have shaped common rocks into massive mountain ranges such as these?

Elsewhere, we can see rocks that look twisted and bent (Fig. 9-1). What powerful forces could so distort such resistant material, contradicting our common belief in the hardness of rock? Throughout Part 1 of this text, we saw how the movement of the Earth's tectonic plates produces most of the large-scale features on the planet's surface, such as its ocean basins and continents. With few exceptions, mountains, too, owe their existence to plate tectonics; the same forces that carry the Earth's plates can tear the edges of those plates apart, rupturing them into huge displaced blocks, or squeeze plate edges together, crumpling and uplifting them into great folds of rock.

Around North America, we can visit young mountains, such as the Cascades of the Pacific Northwest, that continue to rise higher even as you read this book. We can also visit older mountains, such as the Appalachians of eastern North America, that may once have been much loftier but have long since ceased to grow and are now eroding away. We can visit the eroded cores of ancient, now-departed mountains, exposed at mid-continent in the northern Great Lakes states and provinces of south-central Canada.

We can even see evidence of future mountains growing. In the rocks along Sagami Bay near Yokohama, Japan, lives a colony of clams called *Lithophaga,* or "rock eaters." These creatures scoop out small shelters for themselves from the soft rocks at sea level, and wait there for high tide to flood their homes, bringing their meals of marine algae. Moments after Japan's great earthquake of 1923 the land at Sagami Bay shifted upward, leaving rows of *Lithophaga* to starve 5 meters (16 feet) above sea level. There are several even higher rows of abandoned *Lithophaga* dwellings in the cliffs at Sagami Bay, one that correlates with the area's 1703 tremor and one that correlates with its earthquake of 818. The rocks adjacent to the bay have risen roughly 15 meters (about 50 feet) during the past 2000 years.

Several geological phenomena can cause a landscape to move so dramatically. The powerful forces that build up at or near plate edges can crumple and bend rocks, or can break them and shift their position. In this chapter

(a)

(b)

Figure 9-2 (a) These rocks on the west coast of Oregon have been compressed and folded by a past collision of continental plates. (**b**) These rocks, from the Sierra Nevadas of California, have been broken and shifted by powerful rifting-type forces.

and throughout Part 2 of this text, we will focus on the ways that plate movements cause the vast geological processes that deform the Earth's crust, creating many of its structures. The branch of geology devoted to crustal deformation and the creation of mountains is known as **structural geology**.

Stressing and Straining Rocks

Two rock formations that were originally deposited as horizontal marine sediments, but have since been subjected to powerful tectonic forces, are shown in Figure 9-2. The folded sandstone layers of the western Oregon coast were crumpled and folded during a plate collision; the rocks in California's eastern Sierra Nevada mountains were broken and shifted by a plate-rifting motion typifying the tectonics of southwestern North America. Compare these with the photograph of marine sediments near Salt Lake City (Fig. 8-4), which have been relatively undisturbed for hundreds of millions of years.

Where plates interact at converging, diverging, or transform plate boundaries, crustal rocks are subjected to a powerful **stress.** Stress is the force applied to a rock per unit area, usually expressed as kilograms per square centimeter (or pounds per square inch). When a rock is stressed, it becomes deformed, changing in shape and often volume. The change in shape is called **strain.** Rocks may be stressed in three ways, each corresponding to one of the three basic types of plate-boundary movements. Rocks at converging plate margins are pushed together; this type of stress is **compression** (Fig. 9-3a). Compressive stress reduces the volume of a rock. Rocks that have been compressed are generally crumpled, causing them to become thickened vertically and shortened laterally. Rocks at diverging or rifting plate margins are pulled apart; this type of stress is **tension** (Fig. 9-3b). Tensional stress stretches, or extends, rocks so that they become thinner vertically and longer laterally. Rocks at transform plate margins are forced past one another in parallel but opposite directions; this is **shearing stress** (Fig. 9-3c). Shearing stress flattens rocks, alters the angles between their adjacent sides, and slices them into parallel blocks.

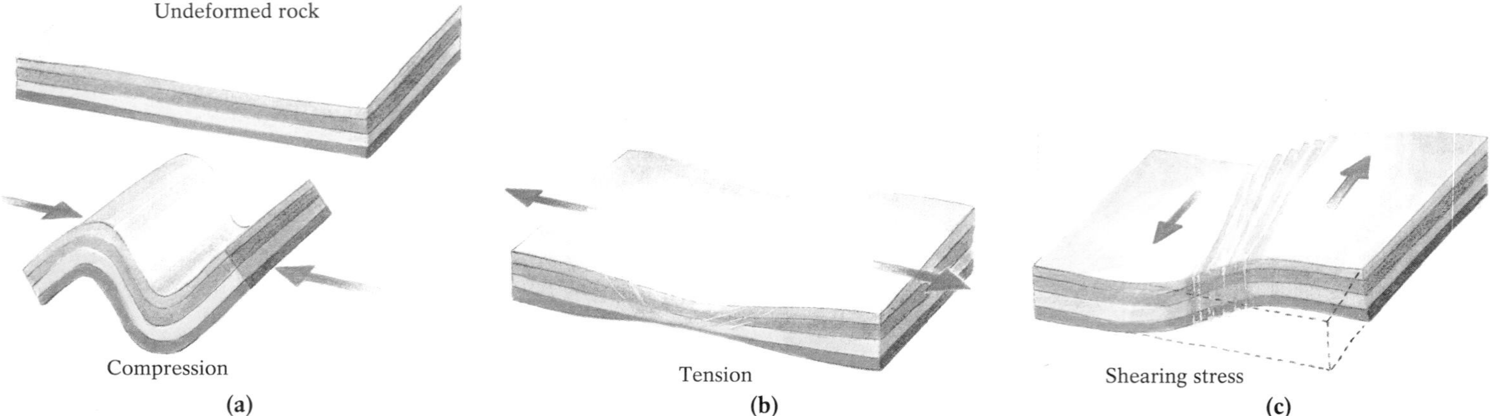

Figure 9-3 The three types of stress applied to rocks, which occur most often at the edges of the Earth's tectonic plates. The edges of converging plates are generally compressed (**a**), becoming thickened vertically and shortened laterally. The edges of diverging plates are generally subjected to tension (**b**), becoming thinned vertically and lengthened laterally. The edges of plates at transform boundaries generally undergo shearing stress (**c**), becoming sliced into parallel blocks of rock.

Types of Deformation

When subjected to stress, rocks strain in different ways (Fig. 9-4). When the stress—whether compression, tension, or shear—is minor, a rock may return to its original shape and volume after the stress is removed, much as a stretched rubber band regains its original shape after use; such a temporary change is described as **elastic deformation**. A rock that is strained elastically is not permanently deformed.

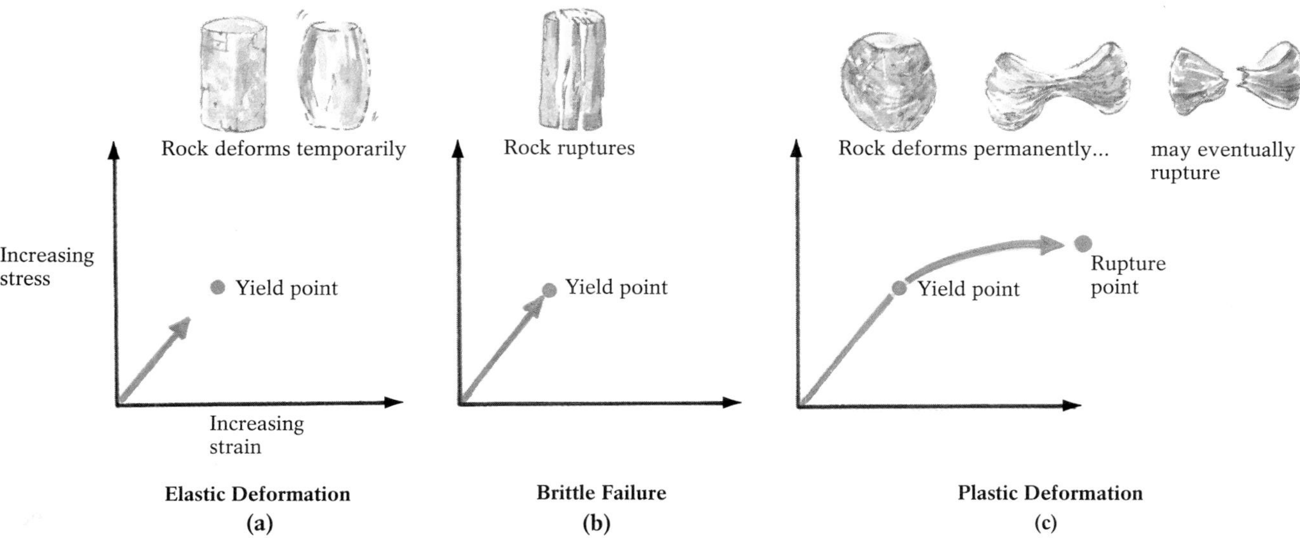

Figure 9-4 The relationship between stress and strain in different types of rock deformation. Within the range of elastic deformation (**a**), application of stress produces a proportionate amount of reversible strain. Stress that exceeds the yield point of a rock causes it to either rupture due to brittle failure (**b**) or deform plastically (**c**). Within the range of plastic deformation, even small amounts of additional stress cause a great deal of irreversible strain.

A greater amount of stress applied to a rock, however, may deform it so it cannot return to its original shape after the stress is removed. The maximum stress a rock can withstand before becoming deformed permanently is its **yield point**, or **elastic limit**. Under some conditions a rock that exceeds its yield point will rupture, a type of permanent deformation known as **brittle failure**; when struck sharply with a rock hammer, for example, even as hard a rock as granite may crack. Under other conditions, a rock that exceeds its yield point will remain intact but will undergo an irreversible change in shape or volume known as **plastic** (or ductile) **deformation.** During plastic deformation atoms in the stressed material move from the areas receiving maximum stress to areas where the stress is lower—much as toothpaste moves upward away from your clenched fist when you squeeze the tube. Sometimes rocks respond to stress by elastic or plastic deformation initially, but then undergo brittle failure if the stress is suddenly increased.

Plastic deformation tends to occur under conditions—such as high temperatures and pressures—that cause atoms to move about and adjust to stress without breaking their bonds. Brittle failure, on the other hand, occurs under conditions in which bonds are likely to break outright rather than adjusting; stress applied under low temperatures and pressures, or applied suddenly, will usually result in brittle failure.

Silly Putty is the perfect medium for demonstrating the different types of rock deformation. Rolled into a ball and bounced gently on a table top, Silly Putty does not change shape appreciably or permanently—it exhibits *elastic behavior* when subjected to a low level of stress. If the ball is slowly pulled in opposite directions until the stress just exceeds Silly Putty's yield point, its shape changes considerably and remains changed even after the stress ceases—it deforms *plastically* when a moderate level of stress is applied gradually. Beyond this point, even small increases in tensional stress cause a great deal of plastic deformation, and the putty is easily pulled into long wispy filaments. Finally, if instead of being stressed gradually the original ball of Silly Putty is pulled firmly and abruptly, it fractures into two sharp-edged pieces, exhibiting *brittle failure*.

Factors Affecting Rock Deformation

The intensity of the applied stress is one factor that determines how rocks will deform, but it is clearly not the only factor. Laboratory experiments that simulate "real-world" conditions have shown that lithostatic pressure, temperature, and the amount of time that a stress is applied all affect the way in which rocks respond to stress.

Lithostatic Pressure Lithostatic pressure from overlying rocks (discussed in Chapter 7) increases with depth. Near the Earth's surface, moderate lithostatic pressure allows the atoms of stressed rock to move freely, causing their bonds to break; as a result, near-surface rocks under sufficient stress are likely to undergo brittle failure. The same rocks, if buried several kilometers beneath the surface, would deform plastically instead, because the greater lithostatic pressure at those depths impedes the movement of atoms, making their bonds less likely to break.

Temperature Subjected to the same amount of stress, rock that would fracture brittlely at low temperatures might soften and become plastically deformed instead when heated. Increased heat can weaken the bonds between atoms without breaking them, allowing the atoms to move more freely. Thus, like a blacksmith's furnace that heats a metal bar so it can be fashioned into a horseshoe, the Earth's internal heat renders brittle rocks more ductile. The amount of heat sufficient to initiate plastic deformation increases with a rock's depth of burial and the associated lithostatic pressure.

Time We can simulate the heat and pressure exerted on rocks deep in the Earth's interior in laboratories, but we cannot simulate the passage of eons of time. Clues in real-world settings, however, hint at the effect of time on deformation—consider, for instance, how sturdy wooden bookshelves gradually sag after decades of sustaining the weight of several heavy volumes. This suggests that a force insufficient to fracture a substance instantaneously may deform it plastically if maintained over a long period of time.

Deformed Rocks in the Field

In the field, visible evidence of strain can be seen at virtually any rock outcrop. Deformation is most readily observed in sedimentary rocks, because they were originally horizontal and laterally continuous. When sedimentary layers deform plastically, their crumpled appearance is clearly visible, and easier to distinguish than similar deformation in a mass of plutonic igneous rock such as granite. Fractures and displacements from brittle failure are also more clearly evident in sedimentary rock, because of the offset of the rock layers.

When geologists find deformed rocks, they want to find out what type of stress (compression, tension, or shearing) was responsible and how much and in what directions the deformed rocks have moved. This information enables them to determine past plate motions and other geological events.

They also want to know the direction in which deformed structures continue underground, because this helps in locating subterranean resources such as oil, coal, natural gas, and numerous valuable metals and other minerals. Geologists estimate the subsurface direction of rocks by determining the orientation of their surface features.

The orientation of a geologic structure or rock layer is determined in relation to the four principal compass directions—north, south, east, and west. The compass direction of the imaginary line forming the intersection between the structure and the Earth's surface is the rock's **strike**. Strike is determined most easily where a rock surface intersects the horizontal plane of a body of water (Fig. 9-5); the direction of the horizontal water line crossing the rock surface, the rock's strike, can be easily determined using a compass.

The **dip** of any structure is the angle at which it is inclined relative to the horizontal. This may be measured with an *inclinometer*, a device much like a carpenter's bubble level contained within a geologic compass. From the angle of dip, it is possible to determine the direction in which the structure is tilted. (Whereas strike is expressed only as a direction, dip requires a measured angle and a direction.) The directions of strike and dip are, by convention, always perpendicular to one another. Measuring the strike, dip angle, and dip direction of a structure establishes its orientation in three-dimensional space. These measurements are shown on geologic maps by a *strike-and-dip symbol.* By making hundreds of strike-and-dip measurements at numerous outcrops and recording the data on a base map or aerial photo, geologists can determine the three-dimensional subsurface form of structures that are barely visible at the surface.

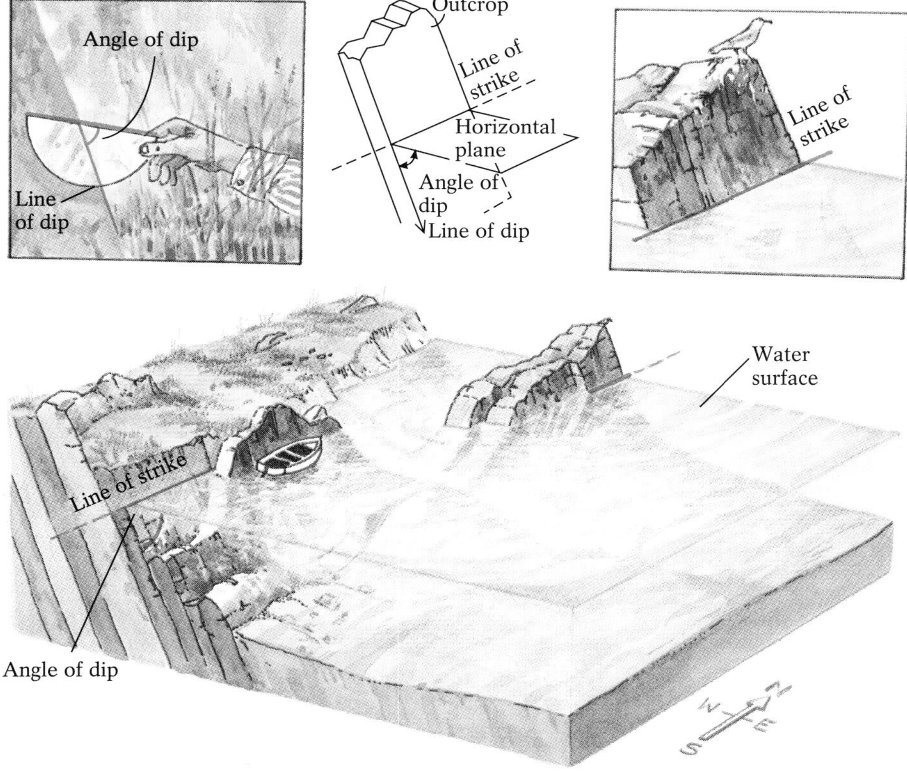

Figure 9-5 Because the surface of a body of water is horizontal, it can be instrumental in determining the orientation of rock outcrops relative to the Earth's surface. The plane of the water traces a strike line across the exposed surface of a partially submerged outcrop, and also provides a horizontal reference against which to measure a structure's angle of dip (determined using an inclinometer). Note that a structure's strike and its dip are always perpendicular.

(a)

(b)

(c)

Figure 9-6 Folded rocks of various sizes. (**a**) A hand specimen of folded gneiss. (**b**) Folded rock strata along Route 23 in Newfoundland, New Jersey. (**c**) A folded mountain in British Columbia, Canada.

Folds

Folds are bends in strata that develop when originally horizontal layers deform plastically. The most commonly observed folded rocks are sedimentary rock sequences. When deeply buried sedimentary rocks are compressed, generally at the edges of converging plates, they fold down and up, forming a series of troughs and arches—much like a rug that bunches up when pushed on a polished hardwood floor. Folds in rock vary in size from the small crinkles visible in some fist-sized hand specimens, through the obvious roadside crenulations seen in rockcuts bordering many major highways (such as those through the Appalachian mountains of eastern North America), to the mountain-sized folds of the Canadian Rockies of British Columbia (Fig. 9-6).

Synclines (from the Greek for "inclined together") are concave, trough-like folds. **Anticlines** (from the Greek for "inclined against") are convex, arch-like folds. A fold's sides are referred to as its *limbs*. Folding produces a series of alternating synclines and anticlines, with those that are adjacent sharing a common limb. Each syncline and anticline has an *axial plane*, an imaginary plane that divides the fold into two approximately equal halves. Where the axial plane intersects the actual bedding planes of the folded rocks is the *axis* of the fold, a line similar in the case of an anticline to the peaked roofline of an A-frame structure, and in the case of a syncline to the keel of a sailboat. In any group of folded rock layers, the innermost rocks of the synclines are the youngest and the innermost rocks of the anticlines are the oldest (Fig. 9-7).

Although in diagrams anticlines may look like topographic hills and synclines like topographic valleys, they are actually structures in the rocks, not surface landforms. Anticlines do not always form hills or ridges, nor do synclines always underlie valleys; the topography in regions of folded rock is largely determined by the different resistances of the folded rocks to erosion.

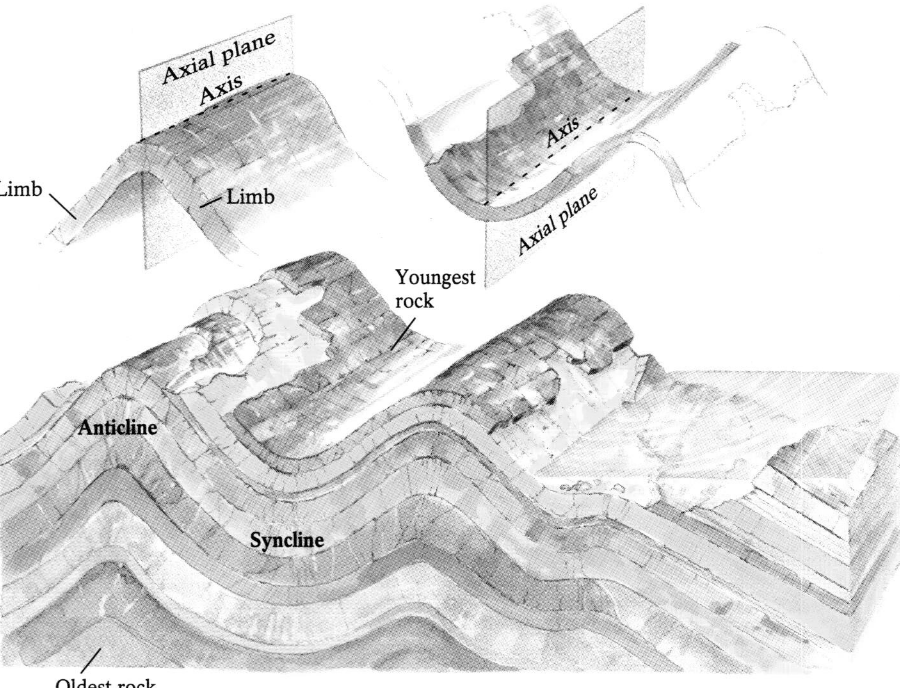

Figure 9-7 The geometry of anticlines and synclines.

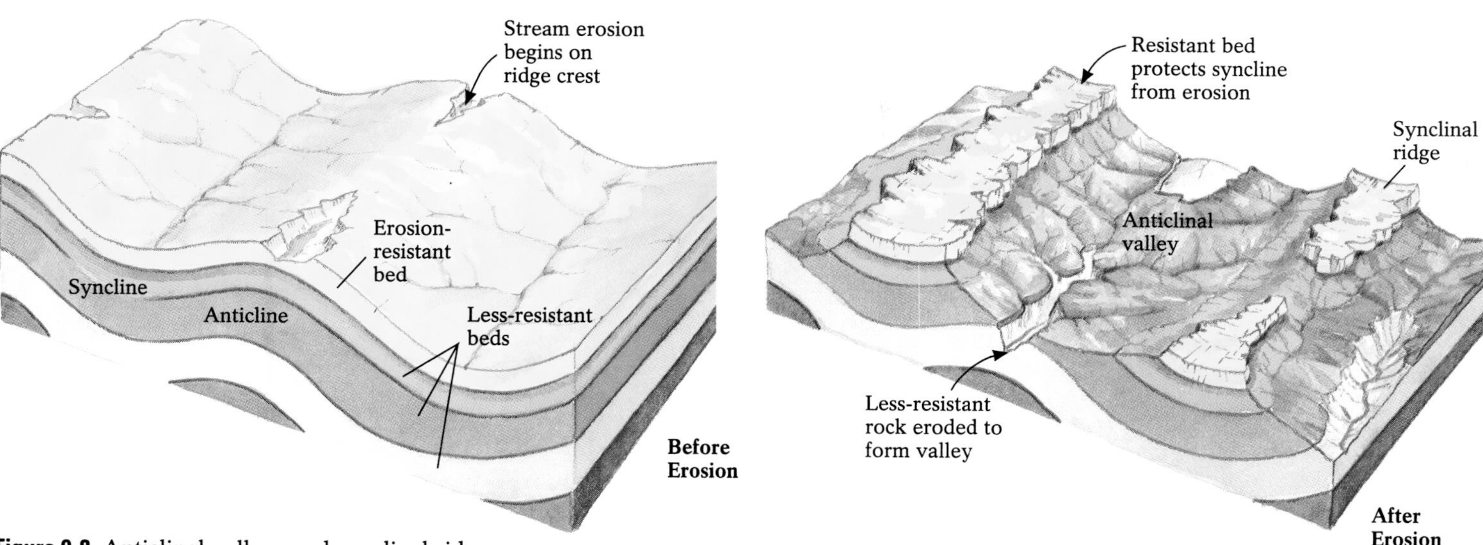

Figure 9-8 Anticlinal valleys and synclinal ridges. The composition of exposed bedrock structures strongly controls their susceptibility to weathering and erosion. Thus, if the axis of an anticline consists of weak erodible rock, its surface expression will be a valley. If the axis of a syncline consists of strong resistant rock, its surface expression will be a ridge.

Thus, if a resistant sandstone were exposed in the trough of a syncline, and readily eroded rocks formed the crests of adjacent anticlines, structural anticlines could form topographic valleys and structural synclines could form topographic ridges (Fig. 9-8).

In the field, regional folding patterns may be largely hidden by vegetation, especially in moist, continuously vegetated regions such as eastern North America. They may also be covered by recent sedimentation, or so weathered and eroded that only scattered small outcrops remain. The orientation of these fragments of sedimentary rock beds, often remnants of the limbs of partially eroded folds, can be established by their strike and dip (see Fig. 9-5), enabling geologists to construct an accurate picture of the regional fold pattern.

Types of Folds

Folds may vary from simple to complex (Fig. 9-9). Under relatively gentle compression, anticlines and synclines appear as broad *symmetrical,* or *open,* structures with near-vertical axial planes and gently dipping limbs inclined at about the same angle. These are most often found in tectonically quiet mid-continent areas. The most common types of folds are the tightly folded anticlines and synclines that occur in belts of folded rock. These are generally found at active convergent plate margins, where compression may range from moderate to intense. Where compression forces one limb to move more than the other, folds may become *asymmetrical.* Prolonged directed pressure may cause an asymmetrical fold to rotate until its axial plane is greatly angled, producing an *overturned* fold, or even until its axial plane is essentially horizontal, virtually parallel to the Earth's surface. Such *recumbent* folds are typically found in highly deformed mountain belts, such as the Northern Rockies, Appalachians, Himalayas, and the European Alps.

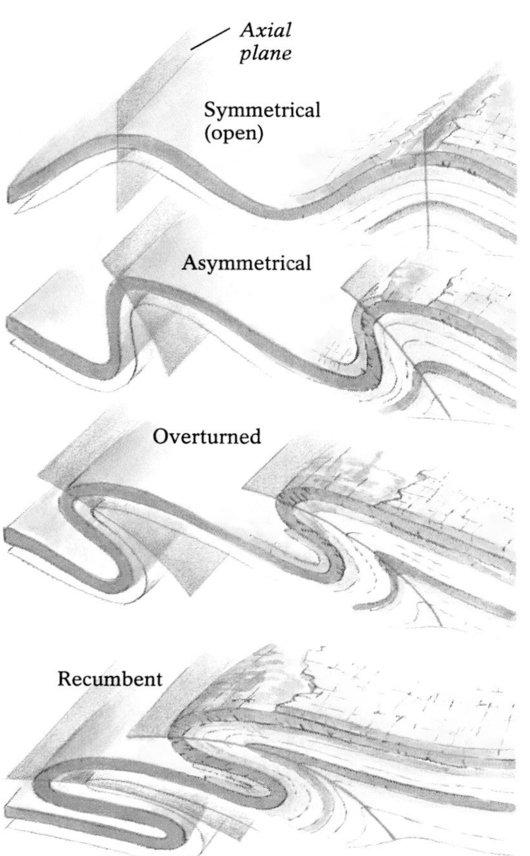

Figure 9-9 Folds vary from broad open structures, with limbs dipping at about the same angles, to overturned and recumbent folds. Such evidence of compression is often found where continental plates have collided.

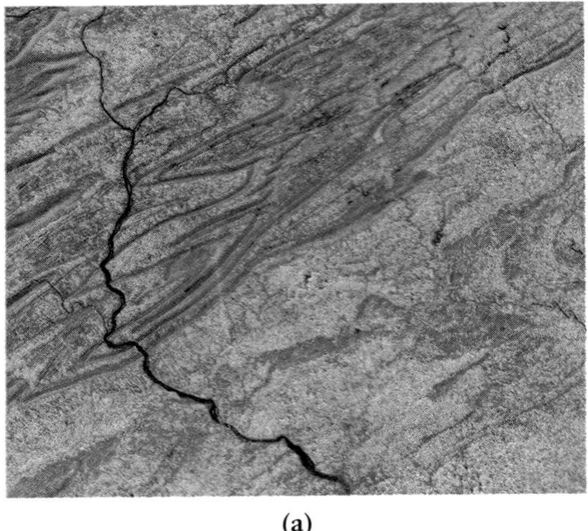

(a)

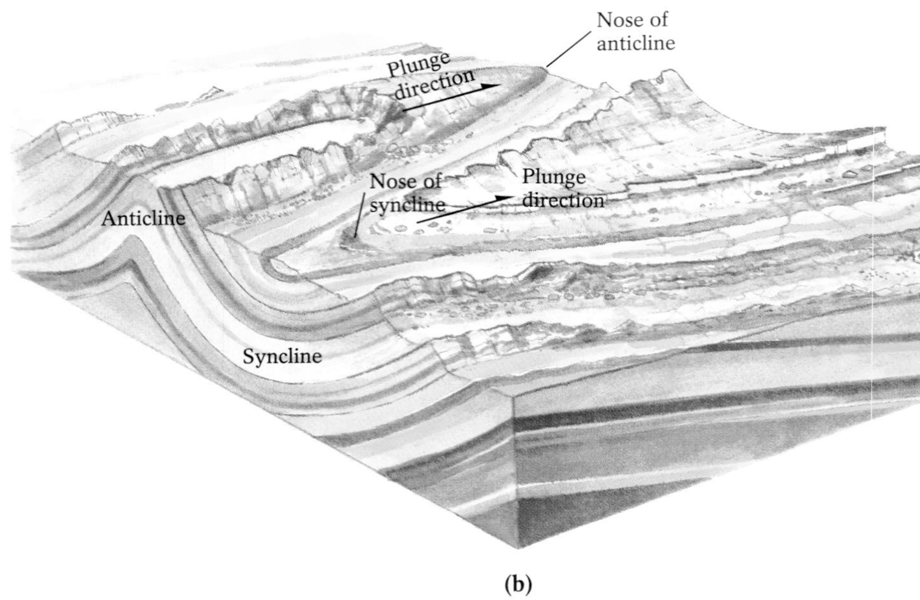

(b)

Figure 9-10 (a) Plunging folds in the Valley and Ridge province of central Pennsylvania, near Harrisburg. These consist of sedimentary beds that were folded and tilted by the tectonic plate collisions that produced the Appalachian mountains. (b) The direction in which the folds are plunging can be determined by identifying the anticlines and synclines, and then noting the direction that the *nose* (the area of maximum curvature of the fold seen on the zigzag pattern of plunging folds) is pointing. The nose of an anticline points toward the plunge direction of a fold, whereas the nose of a syncline points away from its plunge direction.

More complex structures develop when an entire sequence of anticlines and synclines are tilted so that their axes actually intersect the Earth's surface. We identify such *plunging* folds by the characteristic zigzag pattern that emerges as they erode. The Valley and Ridge province of the folded Appalachians of Pennsylvania provides us with some of North America's best examples of plunging folds (Fig. 9-10).

Plate Tectonics and Folding Because folds are generally associated with compression, they occur primarily at subducting and colliding convergent plate boundaries (Fig. 9-11). In subduction zones, loose sediments caught between the converging plates are intensely compressed and often become folded into complex patterns, both within the oceanic trench and in the area between the trench and the growing volcanic arc. Continental plate collisions produce the most extensive types of folding. Too buoyant to subduct, continental lithosphere is pushed up and folded spectacularly. Compressed sediments at the margins of such plates compose such major mountain belts as the Alps, Himalayas, Appalachians, and Canadian Rockies.

Figure 9-11 Folding occurs most often at convergent plate boundaries. (a) At subduction zones, soft marine sediments are compressed within the oceanic trench and between the trench and its associated volcanic arc, creating extensive folding. (b) At continental collision zones, sediments between the plates are intensely folded and metamorphosed.

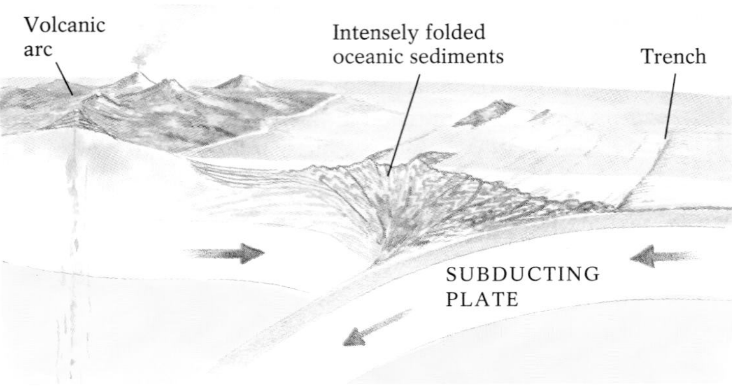

(a)

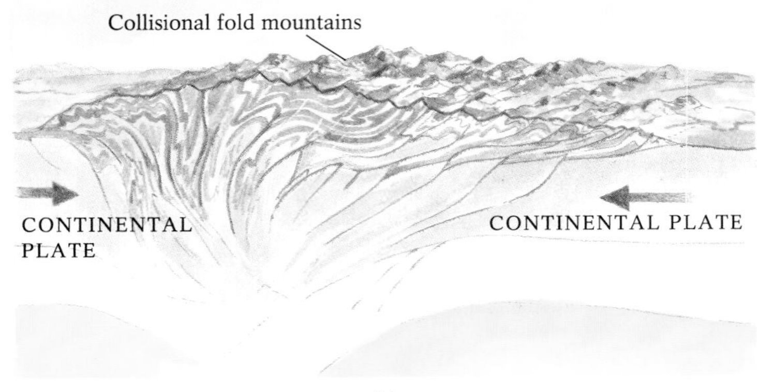

(b)

Domes and Basins A structural **dome** is a subtle oval-shaped bulge on the Earth's surface that, in cross-section, resembles an anticline. As in an anticline, the oldest rocks in a dome are at its center, and all other layers dip away from them (Fig. 9-12). A structural **basin** is a bedrock depression that, in cross-section, resembles a syncline. As in a syncline, the *youngest* rocks in a basin are at its center, and all other layers dip *toward* them. Unlike anticlines and synclines, however, domes and basins sometimes form in mid-continent, mid-plate locations. This leads some geologists to conclude that they cannot be caused only by lateral forces associated with plate movements. It is likely that some of these structures result from vertical forces associated with variations in crustal density. For example, some domes appear to form where low-density materials such as salt, upwelling magma, or warm mantle currents rise toward the surface, pushing flat-lying sedimentary strata upward. Some basins appear to form where materials near and just below the surface are dense enough to depress and deform the materials below them.

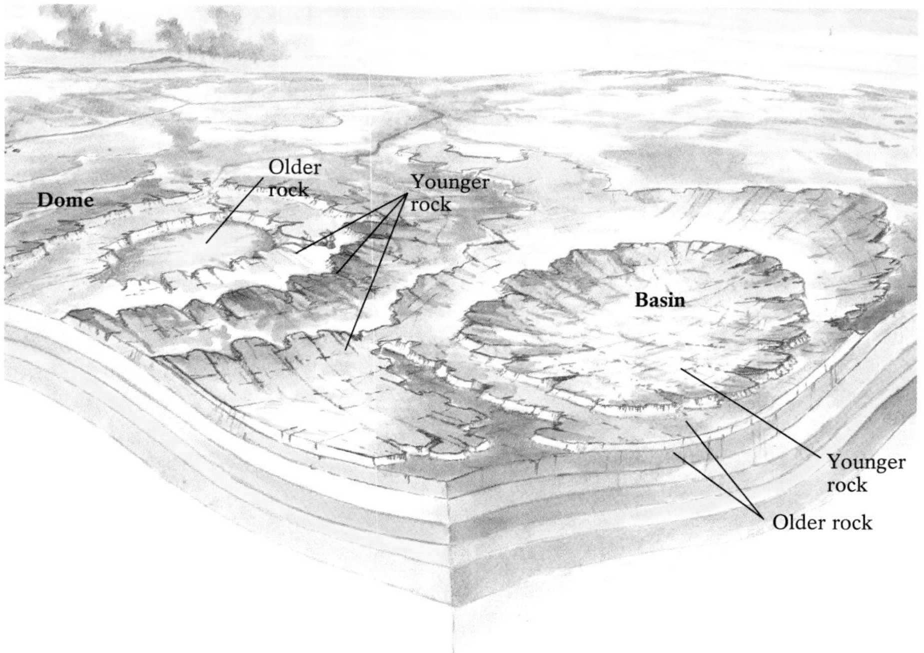

Figure 9-12 Structural domes and basins resemble anticlines and synclines, respectively, in the conformation of their variously aged layers. Because these structures often form far from plate boundaries, however, they may result from vertical motion related to crustal density variations rather than from plate-edge activity.

Domes are more readily seen than basins, standing out prominently at the surface, whereas basins usually become filled with sediment and thus obscured. Noteworthy in North America are the Adirondack dome of northern New York, the Ozark and Nashville domes of the southern Mississippi River valley, and the oil-producing domes of eastern Wyoming. The scenic Black Hills of South Dakota are a large oval dome, whose 2208-meter (7242-foot)-high Harney Peak is the highest point in the state; the region's sedimentary rocks dip away from an exposed core of Precambrian igneous and metamorphic rocks, which are the medium for North America's most prominent sculpture, Mount Rushmore. Noteable structural basins include the oil-and-coal-bearing Williston basin of North Dakota, the Illinois basin of southern Illinois, and the Michigan basin, which forms virtually all of lower Michigan (with the campus of Michigan State University at its center).

(a)

(b)

(c)

Figure 9-13 Products of brittle failure in surface rocks. (**a**) These rocks from Acadia National Park in Maine have been fractured by the application of a variety of stresses, including: the weight of the glaciers that once sat upon them; the cracking that occurs during mechanical exfoliation (Chapter 5); the contraction that accompanies the cooling of plutonic igneous rock; and the stress from past plate-edge interactions. (**b**) The patterned appearance of these joints in Arches National Park, Utah, indicates that these rocks experienced some type of regional stress, perhaps related to plate interactions. (**c**) These faulted rocks have clearly moved relative to each other, as can be seen by the displacement of the various colored rock layers.

Faults

Because most rocks are brittle at low temperature and low lithostatic pressure, virtually every solid rock and nearly all layers of unconsolidated sediment at or near the Earth's surface contain evidence of brittle failure in the form of cracks, or *fractures*. The orientation of fractures in surface rocks provides clues to the stresses that affected the rocks in the past. Some fractures are oriented randomly, and are bordered by rocks that show no evidence of relative movement (Fig. 9-13a). These may be produced by a variety of different, usually fairly localized, stresses, such as those associated with cooling igneous rocks or freezing surface rocks. Fractures that are oriented systematically (for example, at right angles to one another), again with no evidence of relative movement, are called **joints** (Fig. 9-13b). Joints form in response to a specific set of stress conditions, such as those at a plate boundary. The most dramatic examples of cracks in rock are **faults**, fractures along which marked relative movement has taken place (Fig. 9-13c). *Fault blocks* are the rock masses on either side of the *fault plane*, the approximately planar (flat) surface along which the movement occurs. Strike and dip are among the field measurements used to determine the orientation of a fault plane and each fault block's relative direction of movement (Fig. 9-14).

Geologists surveying an area's rock outcrops may find both obvious and subtle clues that help them identify rocks that have been faulted. The most obvious expression of faulting is the visible displacement of rocks, such as

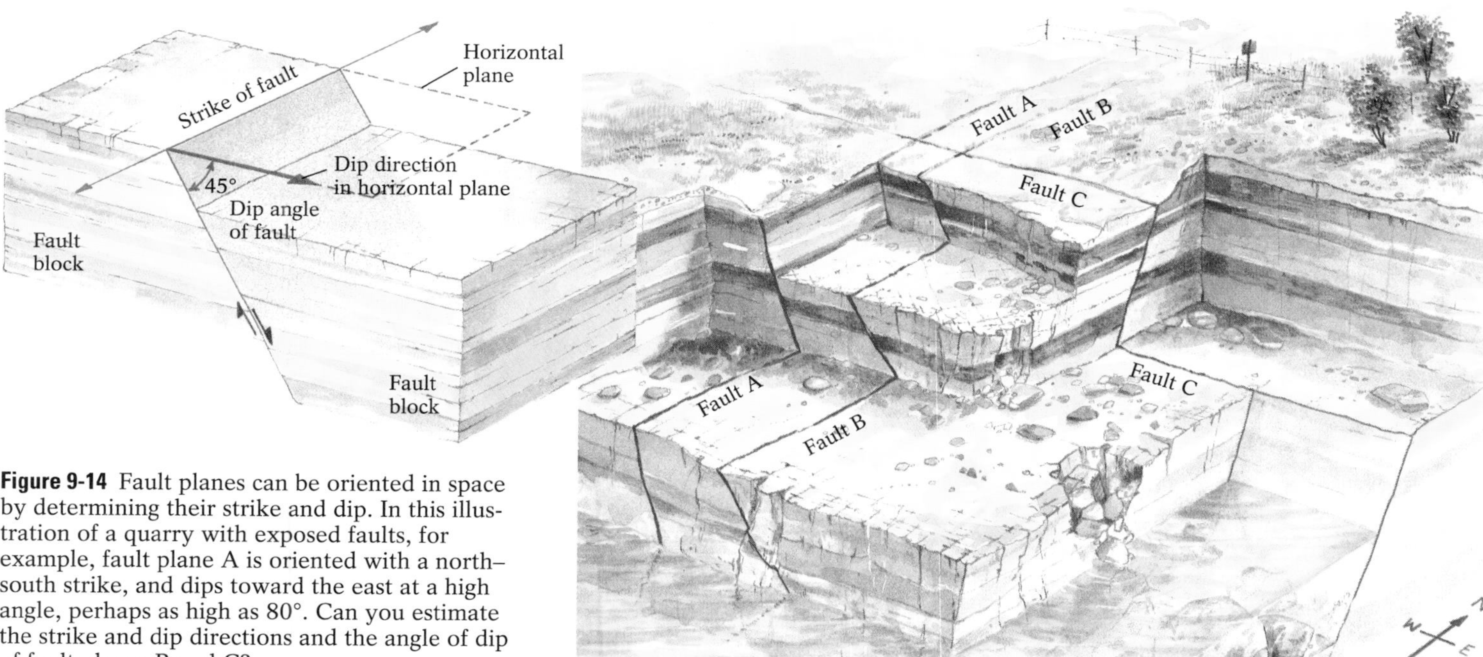

Figure 9-14 Fault planes can be oriented in space by determining their strike and dip. In this illustration of a quarry with exposed faults, for example, fault plane A is oriented with a north–south strike, and dips toward the east at a high angle, perhaps as high as 80°. Can you estimate the strike and dip directions and the angle of dip of fault planes B and C?

that shown in Figure 9-13c. In a rock outcrop that contains only one type or layer of rock, there may be pulverized rock along a fracture, produced by grinding as adjacent rock masses moved. In addition, rocks in contact at a fault plane often have a polished surface called *slickensides*, the result of being abraded and smoothed by pulverized rock acting like jewelers' grit. In more subtle cases, faults may be identified if a local rock sequence is found to be discontinuous with the adjacent stratigraphy. For example, where a distinctive rock bed disappears abruptly, sometimes reappearing at a greater depth some distance away, faulting has very likely disrupted the lateral continuity of the rock layers (Fig. 9-15).

Figure 9-15 By offsetting adjacent blocks of rock, faults disrupt the continuity of individual rock layers. One economic implication of faulting is shown in this illustration, in which the relative movement of the blocks is indicated by arrows. At left, the valuable ore bed that was in the higher fault block has been eroded away, leaving none on that side of the fault. At right, the ore body continues at a slightly greater depth in the lower fault block. If he digs deep enough, the miner at right will find a significant lode of gold, whereas the miner at left will find only a limited amount.

(a)

(b)

Types of Faults

All faults form by shearing motion between adjacent blocks of rock, during which one or both blocks move, or slip, relative to their original position. Such slippage is usually caused by the compressional, tensional, or shearing stresses associated with plate tectonic movement. Faults are generally classified according to the direction of the relative movement between fault blocks—horizontal, vertical, or angled—which is related to the type of stress causing the fault.

In **strike-slip faults**, caused by shearing stress (usually at transform boundaries), fault block movement is largely horizontal, parallel to the strike of the fault plane. North America's most famous strike-slip fault, the San Andreas, underwent one of its greatest recorded displacements on April 18, 1906, the day of the great San Francisco earthquake. If you and a friend had been standing on the shores of Tomales Bay, about 30 kilometers (20 miles) northwest of San Francisco, staring at one another across the San Andreas fault at precisely 5:12 A.M., within seconds you would have been 7 meters (25 feet) to one another's right. The San Andreas fault, an active fracture 1000 kilometers (600 miles) long, extends from well out in the Pacific Ocean beyond Cape Mendocino, northwest of San Francisco, southward beyond California's border into the Gulf of California in western Mexico (Fig. 9-16). Parts of the fault inch along daily, causing relatively little damage (except to the creaking houses directly over the fault and to the sidewalks and curbs, which require frequent repaving), while other segments move less often but in large instantaneous displacements, causing great earthquakes. The extent of movement during the 1906 quake was evident from offset fences and other linear features along more than 400 kilometers (250 miles) of the fault, from Point Arena to San Juan Bautista. By comparison, California's Loma Prieta earthquake of 1989 produced far less displacement—a maximum of 2.2 meters (7 feet) was measured east of Santa Cruz.

Strike-slip faults may stretch for hundreds of kilometers as low linear ridges, which often disrupt the works of both nature and humanity (see Fig. 1-20). As the pulverized rock that characterizes strike-slip faults erodes, several distinctive landforms may be produced: linear valleys, chains of lakes (such as Loch Ness, in Scotland's Great Glen fault zone), sag ponds (depressions) caused by uneven ground settling within the fault zone, and topographic saddles (notches in the skyline of faulted hills) (Fig. 9-17).

Fault movements are more often vertical than horizontal relative to the fault plane. In **dip-slip faults**, the fault blocks move parallel to the dip of the fault plane (Fig. 9-18). Dip-slip faults are often marked by steps in the landscape, formed where the land surface has shifted vertically. Many old mineshafts can be found along dip-slip faults, because precious metals become concentrated there when hot, mineral-rich solutions migrate along faulted rocks. Dip-slip fault blocks are identified by their position relative to the fault plane, using terms derived from mining traditions. The block above the fault plane, from which the miners hung their lanterns, is the *hanging wall*; the block below the fault plane, on which the miners stood, is the *footwall*.

Figure 9-16 (**a**) A map showing the length of the San Andreas fault. (**b**) A section of the fault just south of San Francisco, containing San Andreas Lake and Crystal Spring Reservoir.

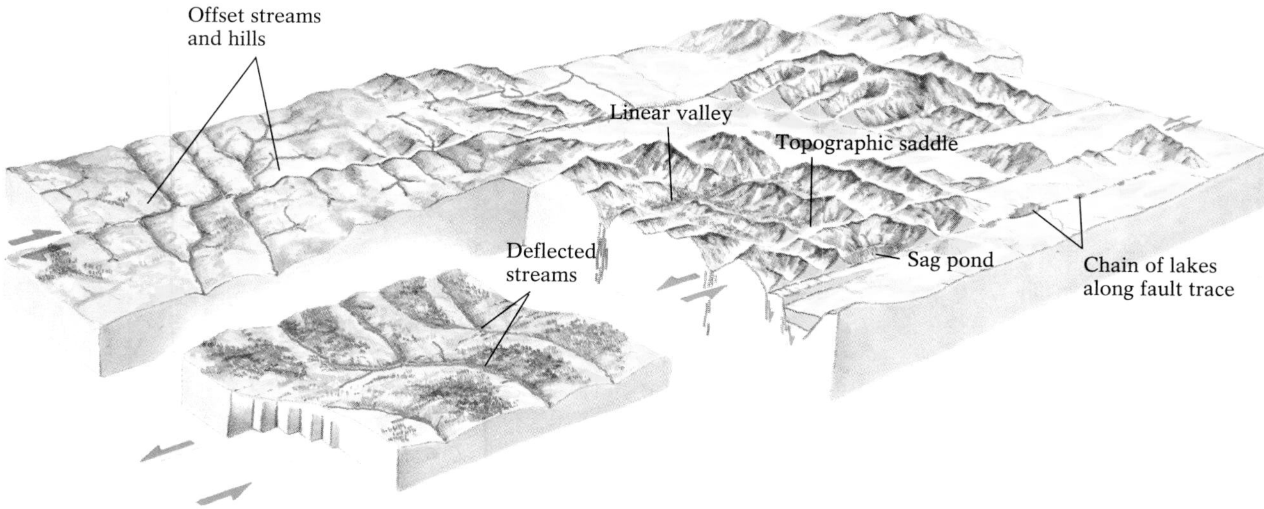

Figure 9-17 Horizontal movement along strike-slip faults causes offset topographic features and distinctive erosional landforms such as linear valleys, lake chains, and sag ponds.

There are various types of dip-slip faults, distinguished by the relative movement of their fault blocks. In a **normal fault**, the hanging wall has moved downward. This type of fault is generally associated with tensional stress, most often where plates are rifting or diverging. At first, tensional stresses stretch and thin the Earth's crust, initially forming a depression and eventually fracturing rocks into faulted blocks. A fault block that drops during normal faulting produces a depression called a **graben** (from the German for "grave"); the blocks that remain standing above and on either side of a

(a)

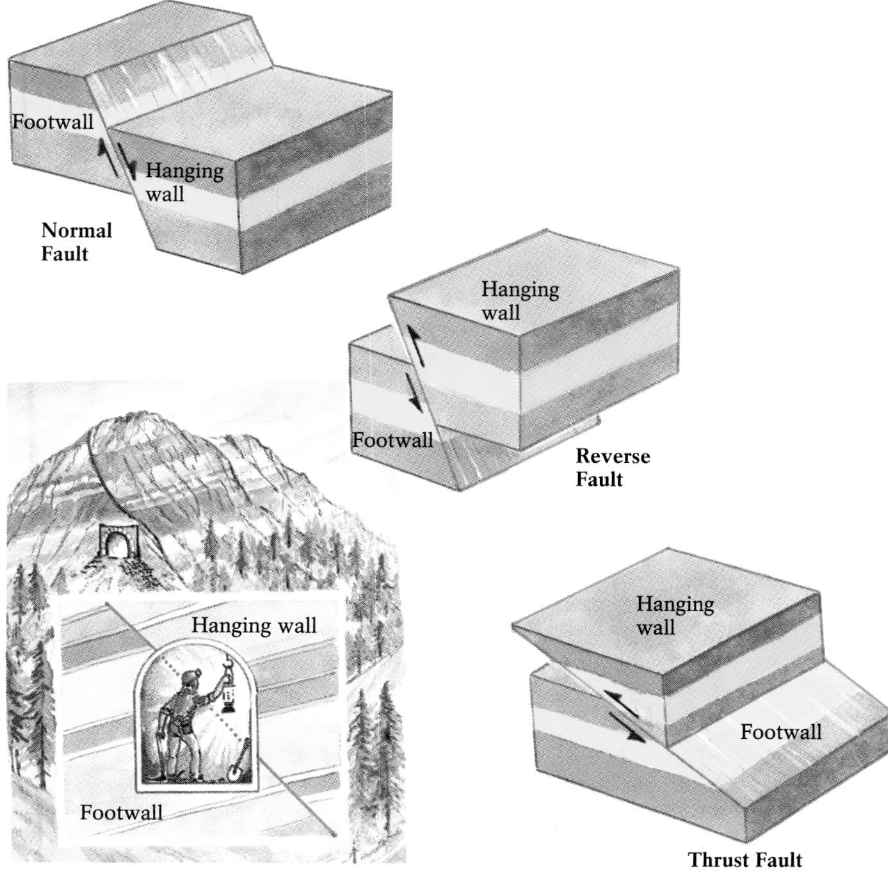

(b)

Figure 9-18 Dip-slip faults. (**a**) The Owens Valley fault, where recent movements have caused major earthquakes. An 1872 earthquake there, which knocked the sleeping residents of Los Angeles from their beds 270 kilometers (180 miles) away, caused 7 meters (23 feet) of instantaneous vertical displacement along the fault. This displacement can now be seen as an abrupt wall across the valley. (**b**) Dip-slip faults are classified by the relative movement of their fault blocks. In normal faults, the hanging wall moves downward relative to the footwall; in reverse faults, the hanging wall moves upward relative to the footwall; thrust faults are low-angle reverse faults.

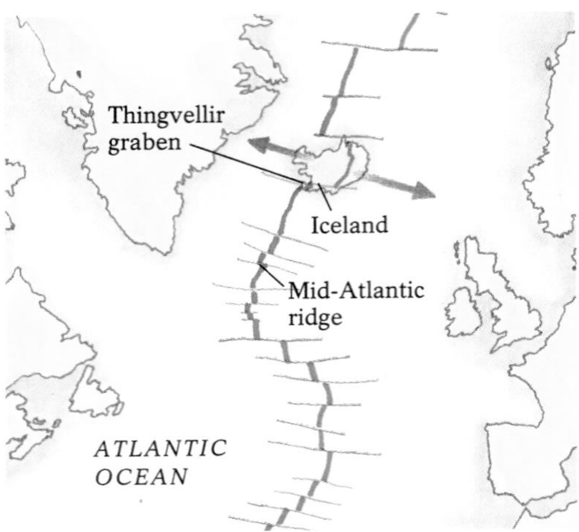

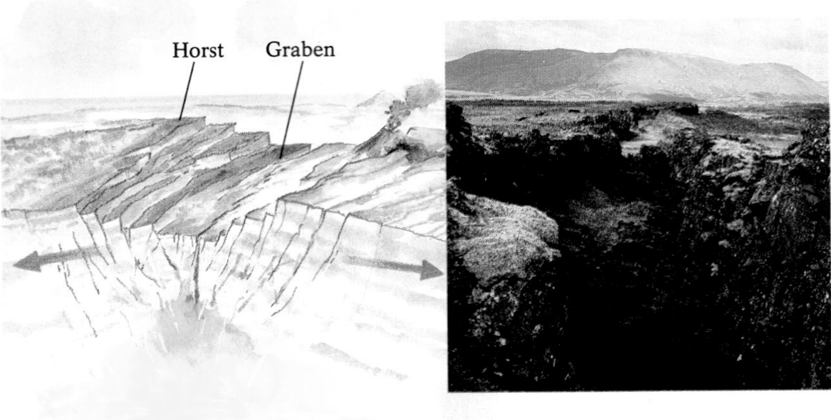

Figure 9-19 Iceland's central valley, the Thingvellir, formed when tensional stress and normal faulting within the diverging mid-Atlantic ridge system created a graben.

graben are **horsts** (from the German for "height"). These features can be clearly seen in the normal faults of East Africa's rift valleys, as well as along the Earth's 60,000 kilometers (40,000 miles) of mid-ocean divergence zones (Fig. 9-19).

In a **reverse fault**, the hanging wall has moved upward relative to the footwall. This type of fault is generally associated with powerful horizontal compression, most often where plates converge. As the hanging wall moves over the footwall, it carries with it deeper, older rocks, transporting them up and over younger rocks to create a rock sequence that violates the principle of superposition (see Chapter 8).

A **thrust fault** is a type of reverse fault in which the blocks move at a relatively low angle (less than 45°). A very low-angle thrust fault (perhaps as low as 5–10°) results in an *overthrust*, in which enormous slabs of rock move horizontally for tens of kilometers. In the Lewis Overthrust in Glacier National Park, Montana, and Waterton-Lakes National Park in Alberta, a 3-kilometer (2 mile)-thick slab of Precambrian marine sedimentary rock was transported more than 50 kilometers (30 miles) eastward, coming to rest atop much younger continental river sediments (Fig. 9-20).

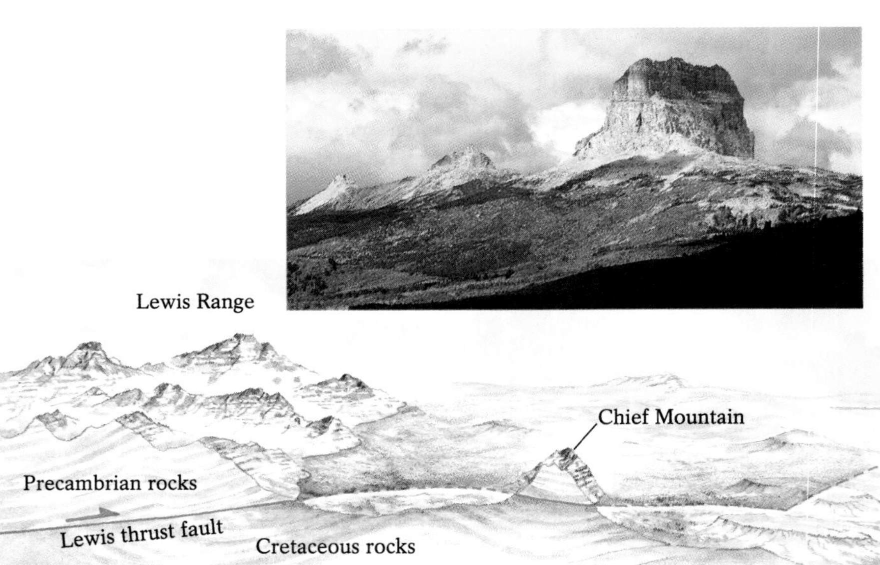

Figure 9-20 The Lewis Overthrust, in Glacier National Park, Montana, formed when compression moved a slab of 800-million–1.1 billion-year-old Precambrian marine sediment on top of a layer of 150-million-year-old continental sediment. Inset: Chief Mountain, formed by subsequent erosion of both layers.

Plate Tectonics and Faulting Each of the three major types of plate margins is associated with a particular type of stress, and thus with a particular type of fault (Fig. 9-21). Normal faults are commonly found where tensional stresses pull apart the Earth's lithosphere, such as along the axes of the world's mid-ocean ridges and where continents are rifting, such as in the north–south Rio Grande valley in New Mexico, where North America may be starting to split apart.

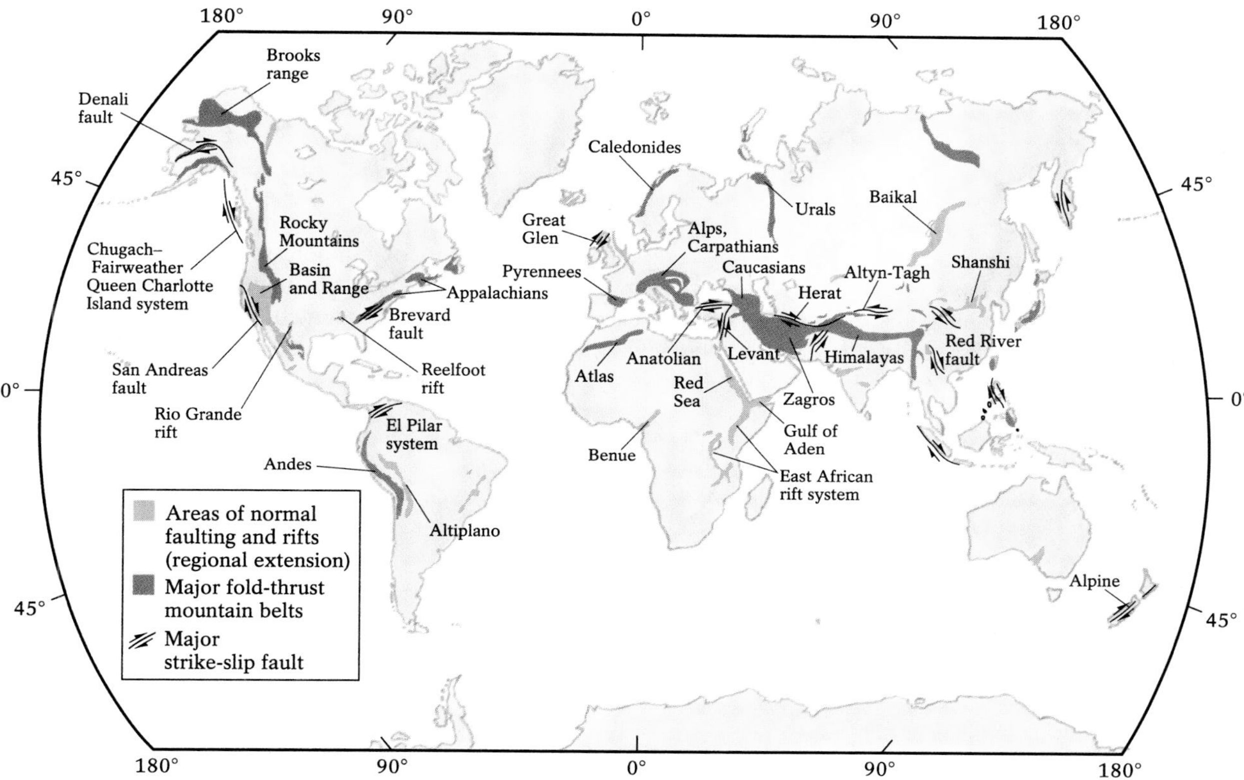

Figure 9-21 The coincidence of different fault types with the Earth's plate boundaries. Normal faults typically occur at rifting and divergent zones. Reverse and thrust faults commonly occur at convergent zones, including subduction zones and zones of continental collisions. Strike-slip faults occur principally at transform boundaries.

Reverse and thrust faults, produced by compressive stresses, are concentrated along convergent plate boundaries, both where oceanic plates subduct and where continental plates collide. The major faults in Japan, the Philippines, and other western Pacific islands are produced largely by convergence of two ocean plates, whereas those in the Andes of South America and Cascades of Washington and Oregon have developed where an ocean plate subducts beneath an adjacent continental plate. Faults in the Alps and Himalayas owe their existence to collisions of two continental plates, those in the Alps from past collisions of Africa and Europe, and those in the Himalayas from the collision of Asia and India. Some of the world's longest continuous faults, such as the San Andreas fault in California and the Anatolian fault in Greece and Turkey, are strike-slip faults that coincide with transform boundaries.

The plate tectonic forces that produce large dip-slip faults in thick sequences of marine sedimentary rocks, as well as those that produce vast folds in marine rocks, sometimes create geologic structures that trap oil and natural gas within rock bodies. Highlight 9-1 describes how knowledge of bedrock structures aids the search for these fuels.

Highlight 9-1 *Folds, Faults, and the Search for Fossil Fuels*

The search for the Earth's dwindling fossil fuel reserves depends on identifying structural features that trap oil and natural gas so that they become concentrated in large recoverable quantities. Oil is produced underground from heating of organic material in marine sediments (the source rocks). The oil then migrates through permeable rocks and fractures until its passage is blocked, and it accumulates in the pores and cracks of other rocks (the reservoir rocks), from which it can be readily collected. Folding may aid the search for oil and natural gas. Less dense than water and the surrounding rocks, these fluids tend to rise toward the surface. In folded terrain, their passage is often blocked at the crests of anticlines, where they then accumulate in the pore spaces of permeable sedimentary rock layers. For this reason, oil searches have often concentrated on known oil-producing marine sediments beneath anticlinal crests. There may even be several oil deposits at different stratigraphic levels within a single anticline (Fig. 9-22).

Unfortunately for late-twentieth-century motorists and residents of cold regions, most large exposed anticlines were drilled by the 1920s, and most of the oil in buried anticlines was detected by geophysical techniques and collected by the end of the 1950s. Although some anticlines continue to produce, geologists today must seek less-obvious oil traps.

Faulting sometimes places an impermeable bed next to oil-bearing permeable ones, creating a barrier to oil flow (Fig. 9-23). Thus, a petroleum geologist looks for faults as a means to identify sites where expensive test drilling is most likely to be productive. Faults and their associated earthquakes may play expensive havoc with property and safety, but they also contribute one important economic benefit: They can help geologists find oil. In Chapter 20, we will consider the future of fossil fuels and evaluate the potential for some alternative energy sources.

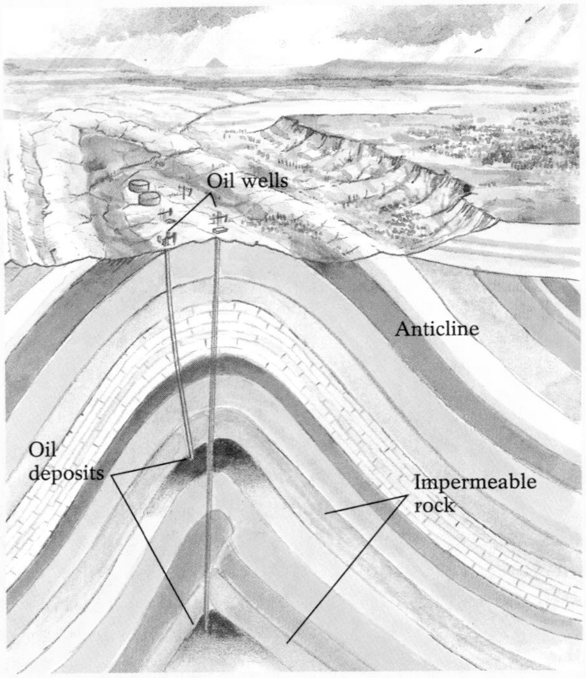

Figure 9-22 The accumulation of oil and natural gas at the crests of anticlines. The low density of these fluids allows them to migrate readily through permeable sedimentary bedrock until they encounter an overlying cap of impermeable bedrock. The fluids become concentrated at the crests of anticlines because these are the highest points within permeable beds.

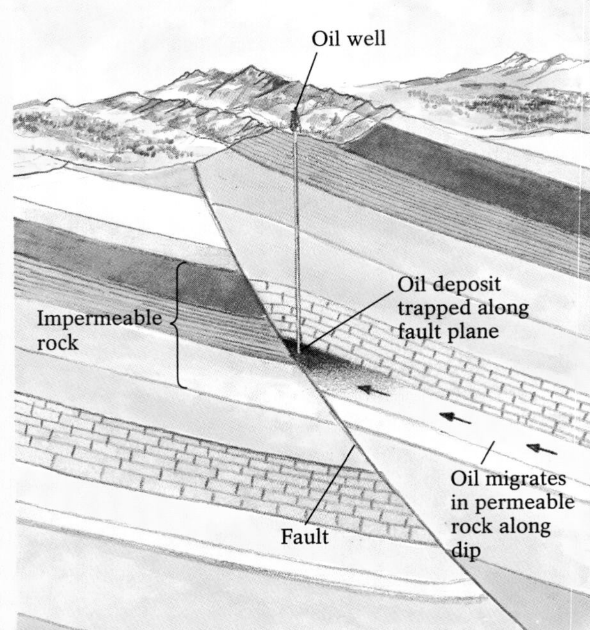

Figure 9-23 The accumulation of oil along fault planes. Oil migrates upward through permeable sedimentary rock because of its low density, but its path to the surface may become blocked by an impermeable bed moved there by faulting.

Building Mountains

A drive down Route 1 along the southern California coast affords an opportunity to see mountain building in action. Near San Clemente, there is a giant stairway of stepped terraces, each of which was cut by waves at sea level and subsequently uplifted (Fig. 9-24). Each remained at sea level during a temporary pause in the uplift of California's coast, long enough for the pounding surf to erode the surface before the next pulse of uplift carried it safely above the wave action. The highest terrace, 400 meters (1300 feet) above present sea level, is the oldest.

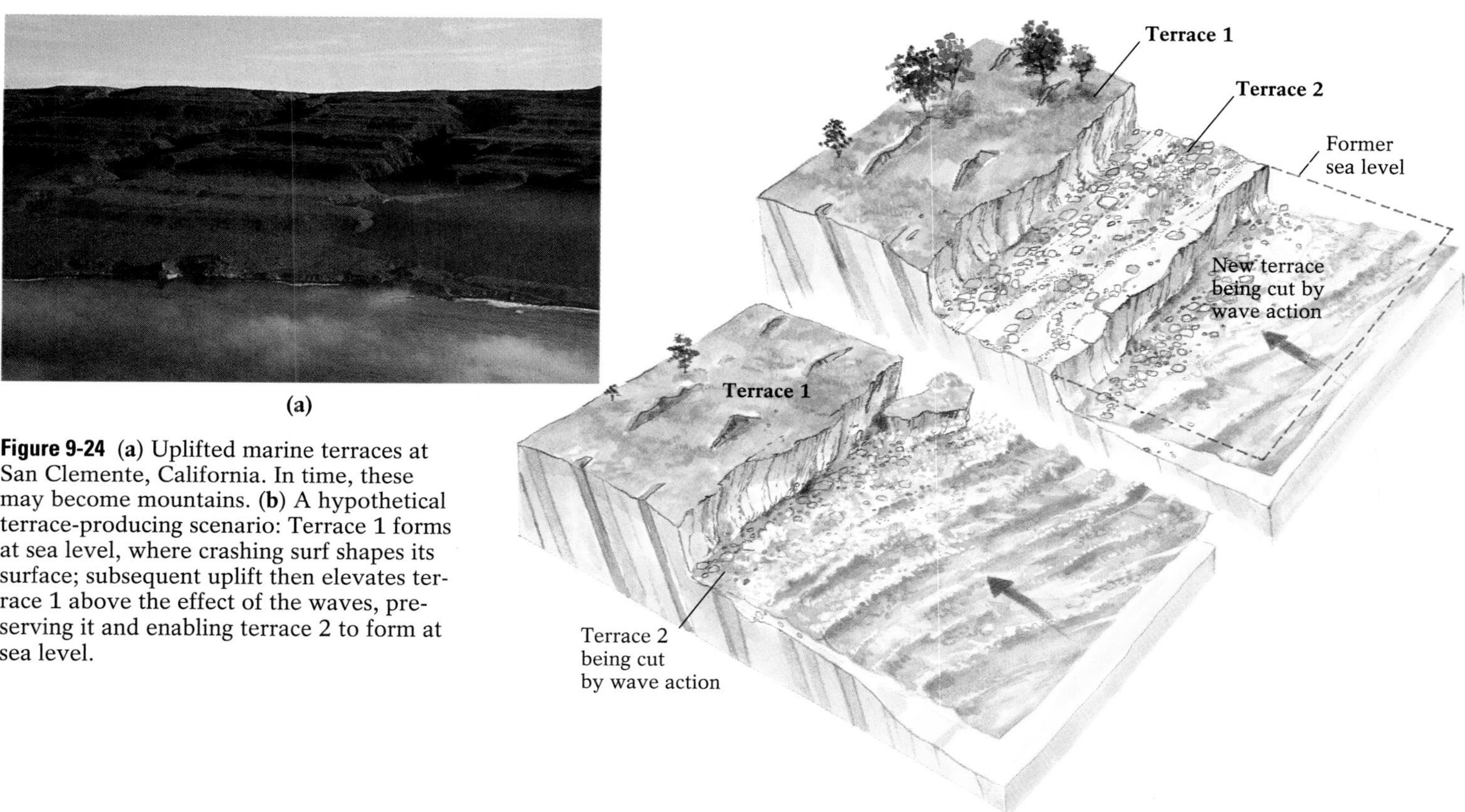

Figure 9-24 (**a**) Uplifted marine terraces at San Clemente, California. In time, these may become mountains. (**b**) A hypothetical terrace-producing scenario: Terrace 1 forms at sea level, where crashing surf shapes its surface; subsequent uplift then elevates terrace 1 above the effect of the waves, preserving it and enabling terrace 2 to form at sea level.

We all know a mountain when we see it, but geologists have a precise way of defining the structures that rise above the Earth's surface. A *mountain* is a part of the Earth's crust that stands more than 300 meters (1000 feet) above the surrounding landscape, has a discernible top or *summit*, and sloping sides. Every continent has mountains, and so does every ocean basin. Some mountains, such as Stone Mountain in Georgia, stand alone, isolated and towering above their surroundings. Some are grouped together in *ranges*, a succession of high peaked structures, such as the volcanic Cascades of the Pacific Northwest. Some ranges occur in continent-long mountain groups, or *systems*, such as the Appalachian mountain system, which comprises the Great Smokies of Tennessee and North Carolina, the Blue Ridge of Virginia, the Catoctin Mountains of Maryland, the Poconos of Pennsylvania, the Catskills of New York, the Taconics of Connecticut, the Berkshires of Massachusetts, and the Green and White Mountains of Vermont and New

Hampshire. Mountains vary in composition and shape, even within a system, as a consequence of their diverse origins. Indeed, there are so many ways to make mountains that even adjacent ones may have been formed by completely different processes at different times (Fig. 9-25).

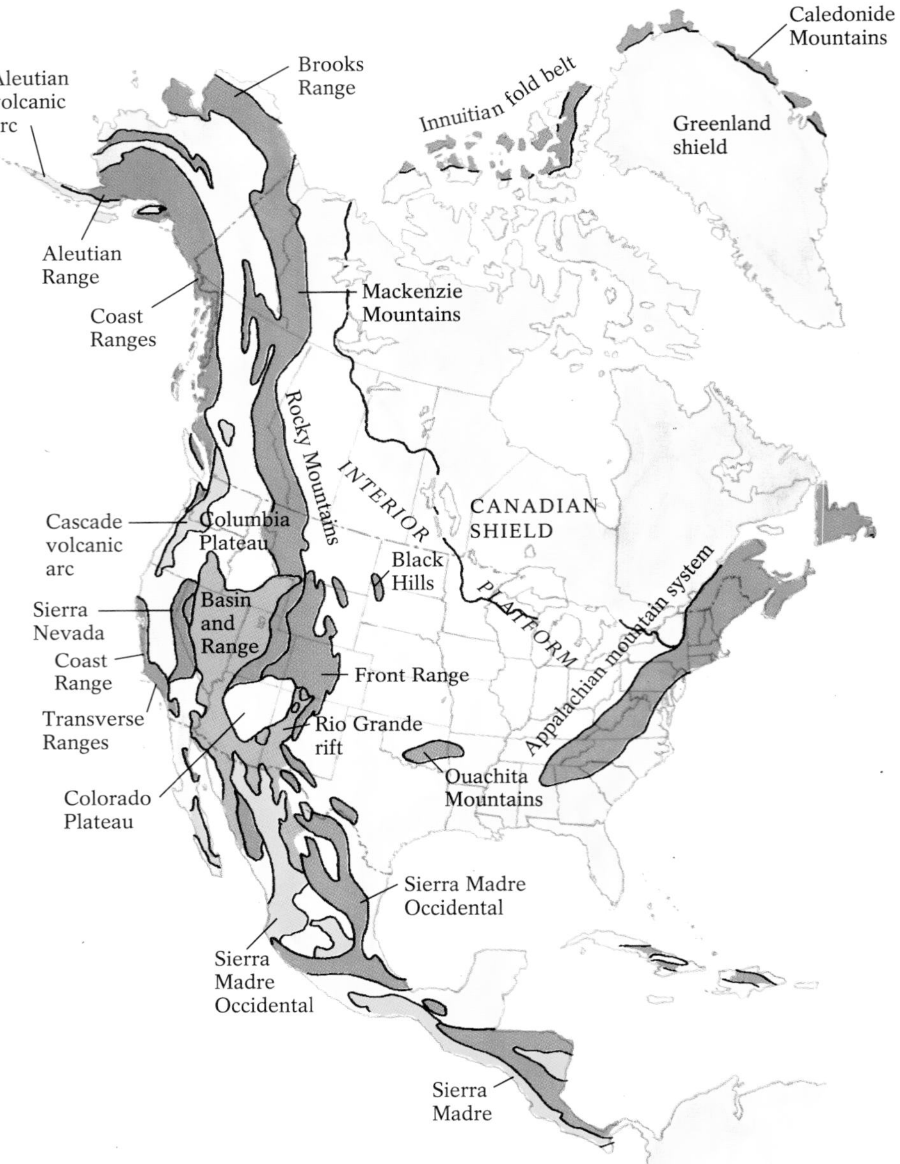

Figure 9-25 The mountain ranges of North America. Some of these have formed from volcanism, some from folding and thrusting at the convergent plate boundaries associated with past continental collisions, some from normal faulting due to recent tensional stresses in western North America, and some from uplift within the interior of the North American plate.

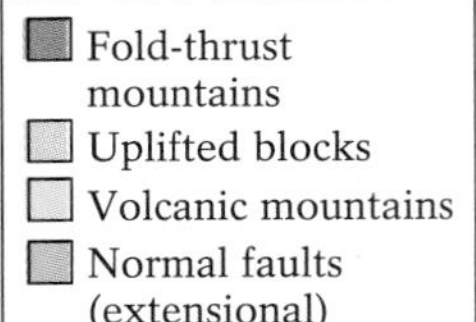

Some mountains, such as the Catskill Mountains of upstate New York, are actually just the uplands that remain after streams have cut deep valleys into a plateau. Some, such as Mauna Loa and Mauna Kea of Hawaii, are undeformed accumulations of basalt that erupted from sea-floor hot spots. Others, especially complex systems such as the Appalachians of eastern North America and the Alps of southern Europe, formed by multiple episodes of sedimentation, intense folding, thrust faulting, plutonism, volcanism, and metamorphism during collisions of broad plate edges.

When we look at mountains, we are truly seeing eons of the planet's past. A mountain and its rocks represent several periods of geologic and tectonic activity. Except in the case of volcanic mountains, the rocks within a mountain are usually much older than the mountain itself, having existed for a long time before being uplifted. Thus, today's mountains may have formed from one tectonic process, but their component rocks generally formed much earlier and from entirely different processes.

Types and Processes of Mountain Building

Mountains have been formed during most of the Earth's past, and are forming today. Some, such as the Himalayas of India, China, and Tibet, are young and still rising. Others, such as the Appalachians and the Urals of central Europe, are so old that the forces that created them ceased to operate hundreds of millions of years ago, and the only processes affecting them today are weathering and erosion. Some mountains are so old that they have been completely worn away, and are now flat regions in continental interiors. These remnants, called *continental shields*, are exposed as highly metamorphosed rocks; they are the deeply eroded cores of some of the Earth's earliest mountains, in some cases uplifted more than 3 billion years ago.

Geologists refer to the processes of mountain building as **orogenesis** (from the Greek *oro*, "mountain," and *genesis*, "birth"). We have already examined one of the Earth's principal orogenic processes—volcanism—in Chapter 4. Volcanic mountains form around volcanic vents from the accumulation of lava flows and pyroclastic materials. Some, such as Hawaii's shield volcanoes, form atop intraplate hot spots. Others, such as the submarine peaks of the mid-Atlantic ridge, erupt at divergent plate boundaries. Still others occur in subduction zones: South America's Andes and North America's Cascades were built where oceanic plates subducted beneath continental plates, and such volcanic island peaks as Alaska's Aleutians and the Pacific's Philippines formed where one oceanic plate subducted beneath another. Our concern in this chapter is mountains formed from processes that uplift materials in the Earth's crust. There are three principal nonvolcanic types of mountains, each originating in a distinctive tectonic setting and each identified by its geological structure: fault-block mountains, fold-and-thrust mountains, and upwarped mountains.

Fault-Block Mountains **Fault-block mountains** are bounded on at least one side by high-angle normal faults. They generally form where the Earth's crust has been stretched, thinned, and ultimately fractured by tensional stresses. The crust is broken into a number of high, tilted horsts interspersed with much lower grabens. Fault-block formations can be seen across the western United States (see Fig. 9-25), eastward from the Sierra Nevada of California, through Nevada, Utah, Arizona, to Colorado and New Mexico, and from Mexico north to the Tetons of northwestern Wyoming. This is the Basin and Range province, a region of normally faulted grabens and horsts. Picturesque Jackson Hole is a graben at the foot of a fault block that rises 2000 meters (about 7000 feet) to the Grand Tetons.

The Basin and Range province, like most mountain systems, has experienced multiple mountain-building periods. Most rocks in this region probably originated as sediments and oceanic crust that accumulated off the coast of North America hundreds of millions of years ago. During this time, the Pacific and North American plates periodically converged, compressed, folded, and thrust-faulted these rocks against what was then North America's western coast, producing the Sierra Nevada and Rocky Mountain ranges. Geologists believe that compression of the continental margin ended about 30 million years ago, when not only the eastward-moving oceanic plate but also the diverging oceanic ridge farther west that had produced it were completely subducted. The southwestern landscape between the Sierra Nevada and Rocky Mountain ranges then became stretched, and was broken by steep normal faults into scores of individual fault-block mountain ranges and intervening sediment-laden basins (Fig. 9-26). Extensive volcanism accompanied the faulting, as plumes of basaltic magma rose to the surface through deep-seated fractures.

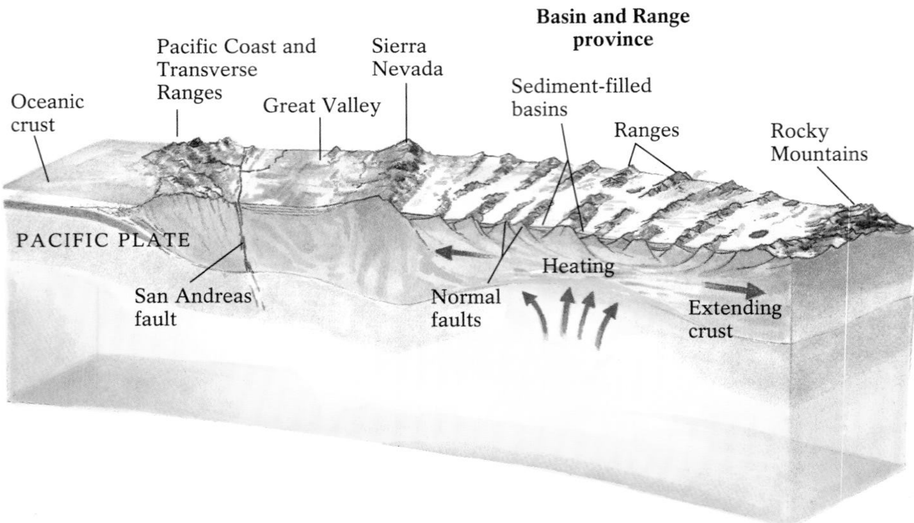

Figure 9-26 A proposed tectonic model for the origin of the Basin and Range province of southwestern North America. (See also Figure 9-25.) Although we know that the Basin and Range province was created by normal faulting due to crustal stretching, it is not known for certain what caused this process.

A number of hypotheses have been offered to explain how the normal faults of the Basin and Range province originated. One proposes that the faulting occurred when magma, generated by melting of the subducted oceanic plate, spread laterally beneath the overlying lithosphere, thinning and stretching it until it fractured. Another hypothesis suggests that the region's volcanism, faulting, and crustal thinning arose when a new independent hot spot developed within the southwest portion of the North American plate. The origin of the Basin and Range province is clearly a result of crustal stretching, thinning, and faulting in the American Southwest, but a geological consensus on the precise mechanism has not been established.

Fold-and-Thrust Mountains **Fold-and-thrust mountains** develop where continental plates collide and produce exceptionally high mountain systems (3 kilometers, or 2 miles, or taller). Typically, these rocks consist predominantly of marine sediments that have been intensely folded, thrust faulted, and, in places, metamorphosed and intruded by large plutons. Most have been deformed more than once, due to a succession of mountain-building events. Fold-and-thrust systems include the European Alps, the Himalayas of India, Tibet, and China, the Urals of central Europe, the Northern and Canadian Rockies, and the Appalachians of eastern North America (see Highlight 9-2, p. 260).

Upwarped Mountains **Upwarped mountains** are those that form when a large area of the Earth's crust is gently bent into broad regional uplifts without much apparent deformation of the rocks. After erosion removes the overlying sedimentary strata from such a feature, a rugged core of durable older igneous and metamorphic rocks often stands prominently above the surrounding sedimentary terrain.

Because many upwarped mountains are located far from plate boundaries, our current understanding of plate tectonics does not explain their origin. Geologists have not yet reached a consensus on the forces that drive upwarping. Some believe upwarps are due to local vertical forces at plate interiors, in contrast to the horizontal forces that dominate at plate edges. For example, crustal upwarping seems to involve the ascent of low-density material. This may be caused by localized increases in the Earth's interior heat (heating expands material and reduces its density), or, alternatively, lower-density mantle materials may be forced upward as heavier materials sink toward the Earth's core. Either way, older buoyant rock probably rises, arching overlying layers of sedimentary rock.

In many places, upwarped mountains consist of ancient Precambrian rocks that became exposed after younger sediments had eroded. We can see this in the Adirondack Mountains of northern New York, which are believed to have formed from a combination of upwarping and plate tectonic forces. About 1 billion years ago, the rocks now at the core of the Adirondacks were compressed, folded, thrusted, and metamorphosed at a collisional plate boundary, and then intruded by igneous plutons. By the beginning of the Paleozoic Era 600 million years ago, these complex proto-Adirondacks had been eroded down to a fairly flat, or *low-relief*, surface on which younger sedimentary rocks were deposited. Then, for unknown reasons, upwarping occurred during the Cenozoic Era (the last 65 million years) after hundreds of millions of years of relative stability. During the last 2 to 3 million years, streams and glaciers eroded the uplifted Paleozoic and Precambrian rocks into deep valleys and high-standing mountains, creating the 1000 meters (3500 feet) of relief (the difference between an area's highest and lowest points) that now typifies the region (Fig. 9-27). Local earthquake activity as recent as 1989 suggests that uplift continues in the Adirondacks today, although the area is thousands of kilometers from a plate boundary.

Figure 9-27 The Adirondack Mountains of northern New York. The center peak in the background is Mount Marcy, the Adirondack's tallest at 1630 meters (5344 feet).

Highlight 9-2 *The Appalachians: North America's Geologic Jigsaw Puzzle*

The Appalachians of eastern North America are a mosaic of folded, thrusted, and metamorphosed provinces that evolved over nearly a billion years of Earth history. The system extends 3000 kilometers (2000 miles), with a north–south orientation from Lewis Hill (810 meters, or 2672 feet, high) in eastern Newfoundland to Cheaha Mountain (730 meters, or 2407 feet, high), about 75 kilometers (40 miles) east of Birmingham, the highest point in Alabama. To its west is the Appalachian plateau, a region of relatively unfolded, unmetamorphosed, coal-bearing rocks; during Paleozoic time, these were lushly vegetated wetlands adjacent to the rising Appalachians. To its east lie the more recent deposits of the Atlantic coastal plain.

The Appalachians' width of 600 kilometers (about 400 miles) comprises three provinces, separated by major thrust faults. Each province developed at a different time and has distinct rock formations, but all were affected by the same series of mountain-building episodes. West to east, the Valley and Ridge province consists of a thick sequence of relatively unmetamorphosed Paleozoic sediments that were folded and thrust to the northwest by compression from the southeast; the Blue Ridge province consists of highly metamorphosed Precambrian and Cambrian crystalline rocks; the Piedmont province consists of metamorphosed Precambrian and Paleozoic sediments and volcanic rocks that were intruded by granitic plutons (Fig. 9-28).

A legion of geologists have been studying these mountains for a century in an effort to decipher their complex history. The most likely hypothesis for the evolution of the Appalachians begins with Late Precambrian plate collisions (about 1.1 billion years ago) that assembled a pre-Pangaea supercontinent. About 800 to 700 million years ago this Precambrian supercontinent began to break up, with North America rifting from Eurasia and Africa. This gave rise to an ancestral Atlantic Ocean and an additional continental fragment that was separated from the North American plate by a marginal sea (Fig. 9-29a). Then, about 700 to 600 million years ago, subduction began within the ancestral Atlantic Ocean, and an island arc developed above this subduction zone (Fig. 9-29b). The oceanic crust in the marginal sea also began to subduct, between 600 and 500 million years ago, and a volcanic arc developed in the continental fragment (Fig. 9-29c); meanwhile subduction continued beneath the island arc, and the ancestral Atlantic Ocean began to shrink. By about 500 million years ago the marginal sea had entirely subducted, resulting in the collision of the continental fragment with the eastern edge of the North American plate (Fig. 9-29d). In this collision, the "foreign" rocks of the continental fragment were thrust-faulted northwesterly over younger continental and marine sediments along the continental shelf of the North American plate. This first mountain-building episode in the long geologic evolution of the Appalachian system is called the Taconic orogeny. It resulted in the formation of what are now called the Blue Ridge Mountains of Maryland and Virginia. Preserved in the Blue Ridge Mountains are highly deformed masses of gabbro, serpentinite, and pillow basalts from the oceanic crust that once underlay the marginal sea, as well as metamorphosed Precambrian-through-Cambrian-age marine sediments from deeper parts of the marginal sea. These rocks belong to what is now the western or inner Piedmont province in eastern North America. Excellent exposures of these rocks may be seen in Newfoundland, metropolitan Baltimore and Washington D.C., and east of the Blue Ridge Parkway in Virginia and North Carolina.

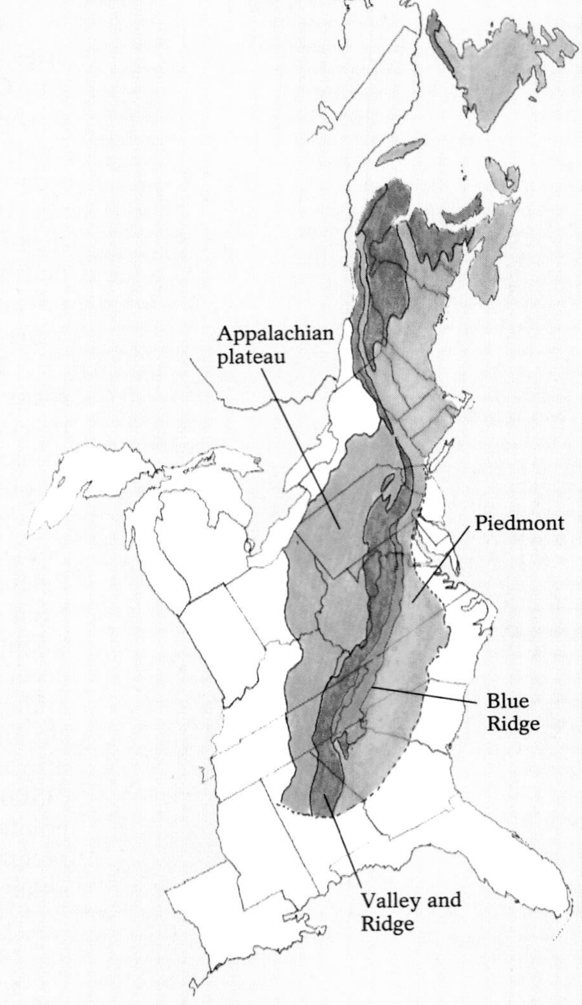

Figure 9-28 The provinces of the Appalachian mountain system, in eastern North America.

The second mountain-building episode, the Acadian orogeny, occurred 400 to 350 million years ago, at the end of the Devonian Period (Fig. 9-29e). The ancestral Atlantic Ocean continued to close as Africa began its northwestward convergence. The island arc developed earlier (see Fig. 9-28b) now collided with the eastern margin of the North American plate, pushing the Blue Ridge and western Piedmont on top of the plate and farther to the northwest. These rocks form the eastern Piedmont province. Much of New England's widespread metamorphism (discussed in Chapter 7) and many of the East Coast's granite plutons (such as those in Acadia National Park in Maine) resulted from this collision and the subduction of the ancestral Atlantic Ocean crust beneath the eastern edge of the North American plate.

The third and final mountain-building collision, the Allegheny orogeny, occurred between 350 and 270 million years ago (Fig. 9-29f). The continental edge of Africa approached North America as the ancestral Atlantic Ocean continued to subduct, an episode recorded in the plutonic intrusive rocks seen today along the eastern edge of the Piedmont province. The orogeny culminated when the continental crust of Africa collided with North America, finally closing the ancestral Atlantic Ocean and forming the supercontinent Pangaea. This collision produced the fold-and-thrust mountains of the Valley and Ridge province, and thrust the Blue Ridge and Piedmont provinces farther inland. In Africa, a corresponding orogenic belt is marked by the Mauritanide Mountains.

The last great East Coast tectonic event was the opening of the modern Atlantic, about 200 million years ago (Fig. 9-29g). At this time the African plate was once again separated from the North American plate, but a healthy chunk of the African plate was left attached to eastern North America from New York City to Florida. Thus, the three plate collisions that produced the Appalachians added to the East Coast an ancient (Precambrian) fragment of North America that had earlier been rifted away, an island arc that originated on the African-European side of the ancestral Atlantic basin, and a massive slice of the African plate. The Appalachians' complex geology and their long history of multiple orogenic events typify all fold-and-thrust mountain systems.

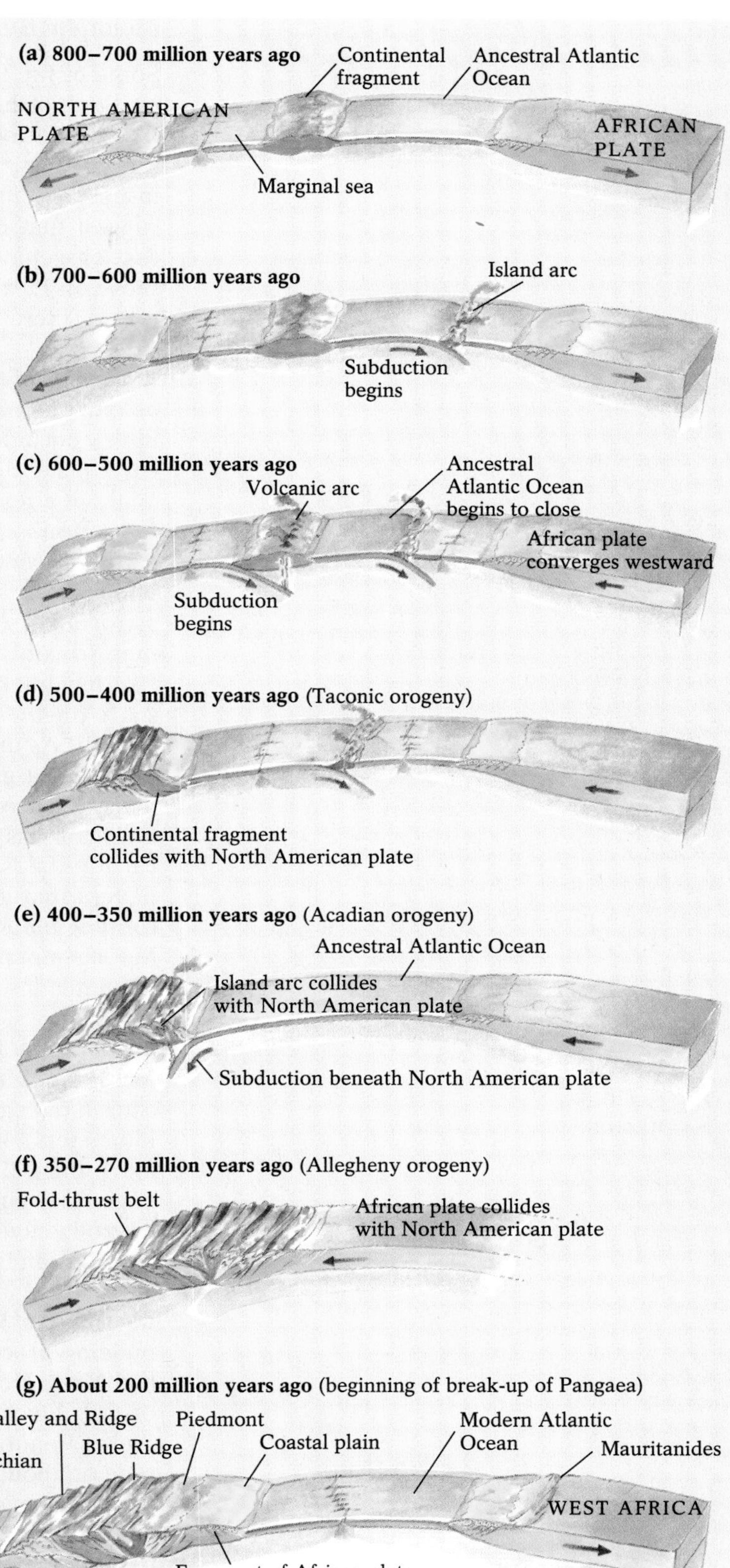

Figure 9-29 A model for the evolution of the Appalachians.

Mountain Building on Our Planetary Neighbors

Some of the Earth's planetary neighbors have mountains that resemble those on Earth; others have mountains that appear to have formed from processes that do not occur on our planet. The three bodies nearest to the Earth—the Moon, Mars, and Venus—have been studied by unstaffed satellites, telescopic inspection, and, in the case of the Moon, by human visits. The data collected suggest that Earth-like plate tectonics is absent on the Moon and Mars, and that orogenic processes on these two bodies must therefore be different from those on Earth. Plate tectonics may, however, operate on Venus—if so, orogenesis there may be more Earth-like.

Mountains of the Moon The Moon does have major highlands, some as lofty as our Mount Everest, that call for an orogenic explanation. Yet, geologists have concluded that the Moon is tectonically dead because highly sensitive earthquake-detecting devices have documented relatively few quakes there. Without stress and strain, and faulting and folding, how did the lunar mountains form?

The Moon's highlands consist of rocks that apparently formed about 4.5 billion years ago (the date consistently yielded by rocks collected by Apollo astronauts). At that earliest period of its existence, the Moon was just beginning to cool down after being heated by meteoroid bombardment over its first few hundred million years. The impact and accretion of these high-speed space fragments caused the Moon's outer 100 to 150 kilometers (60–90 miles) to melt. As the meteoroid impacts gradually became less frequent, the Moon's magma sea gradually cooled. The first components to crystallize were low-density calcium feldspars, which rose to the surface. This primordial lunar crust consists of anorthosite, a rock that is almost completely composed of the calcium feldspar anorthite. Meteoroid impacts over the last 4 billion years have apparently dislodged chunks of anorthosite and hurled them about, forming mountains of rock debris. Thus, in the absence of plate tectonics, the Moon's mountains developed through a surface process of fragmentation and accumulation.

Mountains of Mars In the apparent absence of Earth-like plate tectonics on Mars, the principal mountain-building process on this planet seems to be volcanism on a grand scale. Olympus Mons, in Mars' Tharsis region, is about 600 kilometers (400 miles) in diameter (about the size of the state of Ohio) and more than 23 kilometers (14 miles) high. As we learned in Chapter 4, this huge size suggests that Olympus Mons is a shield volcano that has continued to grow, having remained indefinitely atop its magma source due to a stationary Martian lithosphere.

Mountains of Venus Venus may well be tectonically active; if so, its mountain-building processes may be more like those on Earth than those of the Moon and Mars. Until recently, however, Venus' 25-kilometer (15 mile)-thick cloud of CO_2 blocked the view of conventional telescopes and prevented us from seeing Venutian orogenesis in action. This CO_2 cover, which begins at an altitude of about 70 kilometers (45 miles) above the planet's surface, creates the solar system's ultimate greenhouse effect by trapping heat radiated from the planet's surface. Even when Earth probes penetrated the cloud, surface temperatures of 500°C (900°F) and the overwhelming atmospheric pressure (90 times our own) caused the spacecrafts' cameras to malfunction. A Soviet Venera probe in 1975 survived for about an hour, sending back a few fuzzy photographs of the surface. In 1990, the U.S.-launched Magellan probes used radar, which can pierce clouds, and finally provided scientists with exciting images of the mountainous topography of Venus.

The Magellan images show a pattern of lowlands and highlands that have led some geologists to speculate about ongoing plate tectonic activity on Venus. The highlands may be analogous to the Earth's prominent continents, whereas the plains may be analogous to the Earth's low-lying oceanic crust. The highlands, which constitute only about 8% of Venus' surface, are mostly plateau-like structures, although there are also a few discrete mountain ranges (Fig. 9-30). Maxwell Montes on the Ishtar highlands, for example, rises to a height of 11 kilometers (about the elevation of Mount Everest) above the surrounding lowlands. Venus also contains structures that have been tentatively identified as 5-kilometer (3-mile)-high shield volcanoes and enlarged tectonic valleys, possibly grabens. Of course, we are unable to confirm these speculations until our ability to study the Venutian surface improves.

This chapter has often referred to the connection between tectonic stress, faults, mountain building, and earthquakes. The next chapter focuses specifically on earthquakes: their causes, their often-catastrophic effects, and the ways humans have attempted to cope with them.

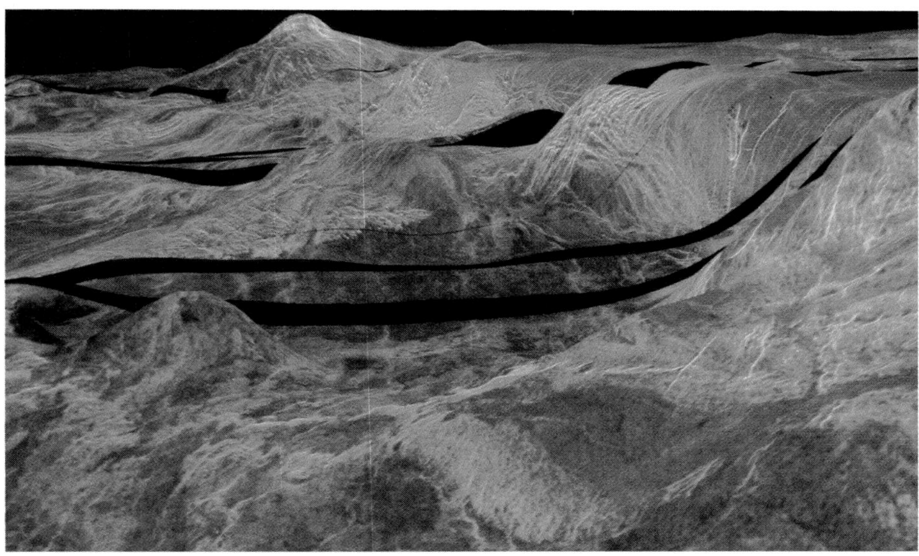

Figure 9-30 Venus' Ishtar mountains and adjacent lowlands, from the high-resolution radar images provided by the Magellan probe in 1990.

Chapter Summary

The rocks that make up the Earth's lithosphere are often subjected to great forces—particularly at the edges of tectonic plates—and respond by bending or breaking. **Structural geology** is the study of such deformed rocks. When a sufficient force, or **stress** (force per unit area of rock), is applied to rocks, they may **strain**, changing in shape or volume. There are three principal types of stress: **compression**, which squeezes rocks together; **tension**, which stretches and pulls rocks apart; and **shearing stress**, which grinds rocks by forcing them past one another in opposite but parallel movement. When only a minor amount of stress is applied, rocks undergo **elastic** (temporary) **deformation**. A greater amount of stress may cause a rock to exceed its **yield point**, or **elastic limit**, and undergo permanent deformation. Under certain conditions, particularly when a great amount of stress is applied rapidly to relatively cool and shallow rocks, the rocks break, or undergo **brittle failure**. Under other conditions, particularly when stress is applied gradually to relatively warm, deep rocks, the rocks undergo **plastic** (permanent) **deformation** without breaking. When geologists find deformed rocks at the surface, they determine the rocks' orientation in space in order to estimate their subsurface structure. To do this, they measure the rocks' **strike**, the compass direction of the line forming the intersection between the rock and the Earth's surface, and its **dip**, the angle at which it is inclined relative to the horizontal.

When rock layers deform plastically, they form a series of **folds**, including trough-like **synclines** and arch-like **anticlines**. Folds occur most often at convergent plate boundaries, such as within the trenches of subduction zones and at the collision zones between masses of converging continental lithosphere. Plastic deformation of rocks also produces oval-shaped bulges, or **domes**, in the Earth's surface as well as bedrock depressions, or **basins**.

When rocks deform by brittle failure, they may exhibit randomly distributed fractures, systematically distributed **joints**, or **faults**, cracks along which marked movement has taken place. Geologists classify faults according to the relative direction in which the affected rocks, or fault blocks, have slipped. **Strike-slip faults** are marked by horizontal movement, whereas **dip-slip faults** are marked by vertical movement. The most common types of dip-slip faults are **normal faults**, which develop under tensional stress. In a normal fault, the fault block that drops produces a depression called a **graben**; the blocks that remain standing above and on either side of a graben are **horsts**. **Reverse faults** are dip-slip faults that develop under compressional stress. **Thrust faults** are low-angle reverse faults. In normal faults, the fault block above the fault surface (the hanging wall) has moved downward relative to the fault block below the fault surface (the footwall). In reverse faults, the hanging wall has moved upward relative to the footwall. Strike-slip faults are generally found at transform plate boundaries, normal faults at divergent plate boundaries, and reverse faults at convergent plate boundaries.

The most dramatic effect of stress-induced rock deformation is the creation of mountains, or **orogenesis**. Tensional forces may cause regional normal faulting, which produces uplifted ranges, or **fault-block mountains**, separated by downdropped basins. Compressional forces produce combination **fold-and-thrust mountains**. **Upwarped mountains** are broad regional uplifts that may form by the rise of low-density material below the Earth's surface.

Mountain-building also occurs on some of our planetary neighbors. Because the Moon and Mars appear to be tectonically inactive, their mountains are believed to have formed by entirely different orogenic processes. Lunar mountains were probably produced mainly by meteoroid impacts, whereas Martian mountains seem to be largely of volcanic origin. The mountains on Venus, however, may have risen in response to Earth-like tectonic forces.

Key Terms

structural geology (p. 240)
stress (p. 240)
strain (p. 240)
compression (p. 240)
tension (p. 240)
shearing stress (p. 240)
elastic deformation (p. 241)
yield point (p. 241)
elastic limit (p. 241)
brittle failure (p. 241)
plastic deformation (p. 241)
strike (p. 243)
dip (p. 243)
folds (p. 244)
synclines (p. 244)
anticlines (p. 244)
domes (p. 247)
basins (p. 247)
joints (p. 248)
faults (p. 248)
strike-slip faults (p. 250)
dip-slip faults (p. 250)
normal faults (p. 251)
graben (p. 251)
horsts (p. 252)
reverse fault (p. 252)
thrust fault (p. 252)
orogenesis (p. 257)
fault-block mountains (p. 257)
fold-and-thrust mountains (p. 258)
upwarped mountains (p. 259)

Questions for Review

1. What are the three principal types of tectonic stress, and at which type of plate boundary does each develop?
2. Compare brittle failure and plastic deformation of rocks. Where does each type of deformation commonly occur within the Earth?
3. List three factors that affect how rocks deform and describe the effects of each.
4. Determine which way is north and then hold this book so that its direction of strike is 45° west of north, and it dips 45° and toward the northeast.
5. Draw an anticline and syncline, and label the limbs, axial plane, and axis of each. Why does the core of an anticline contain the formation's oldest rocks, and the core of a syncline contain its youngest rocks?
6. Distinguish between fractures, joints, and faults.
7. What are four ways to identify faults in the field?
8. Draw two simple sketches illustrating the difference between normal faults and reverse faults. Label the hanging wall and footwall; use arrows to show their relative direction of movement.
9. Where within deformed geological terrain might you look for a subsurface accumulation of oil?

10. Briefly discuss two North American examples of different styles of orogenesis.

For Further Thought

1. In what types of rocks might you expect to find fractures but no faults?

2. Identify the type of fold and type of fractures that appear in the photo below.

3. Speculate about how the sedimentary layers in the diagram below became arranged in their present stratigraphic order.

4. Suggest future plate tectonic scenarios that would result in development of normally faulted grabens in Minnesota and the creation of Appalachian-type fold-and-thrust mountains along the Louisiana coast.

5. What would the Moon's surface look like if plate tectonics had been active there for its entire history? How might the Earth's surface look if plate tectonics had never developed here?

The Causes of Earthquakes

Highlight 10-1 Alaska's Good Friday Earthquake, 1964

The Effects of Earthquakes

The World's Principal Earthquake Zones

Coping with the Threat of Earthquakes

The Moon: A Distant Seismic Outpost

Figure 10-1 Flames erupting from a ruptured gas main in Granada, California—a result of the Northridge earthquake of 1994.

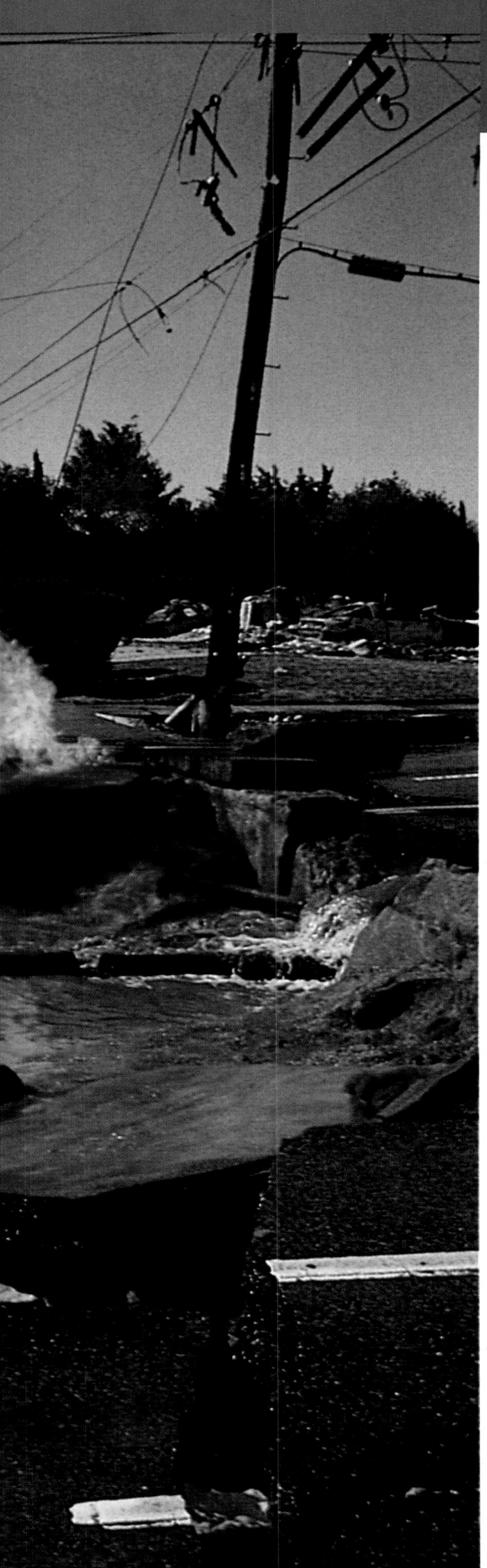

Chapter 10

Earthquakes

In recent years, Los Angeles, San Francisco, Tokyo, Mexico City, and many other cities have experienced cataclysmic earthquakes. In such heavily populated and industrialized areas, the damage caused when vast tracts of land are shaken by powerful tremors can be devastating. Skyscrapers, highways, hospitals, and homes collapse, and thousands of people die in the wreckage, are dislocated, or lose their most treasured possessions. Governments shift abruptly into emergency mode as water supplies are cut off, electric lines snap, and fires erupt from severed gas lines (Fig. 10-1). In short, life is disrupted for entire communities until clean-up and rebuilding is complete. The forces that cause such devastation are, in most cases, the very same forces that created our planet's most impressive mountain ranges.

During the Loma Prieta earthquake of 1989 (named for a mountain peak near the quake's point of origin), tremors sped through the rocks and soils of northern California at 11,000 kilometers per hour (7000 mph), toppling buildings, freeways, and bridges, rupturing gas mains and starting fires, and setting off landslides throughout the San Francisco Bay area. The quake emanated from the 35-kilometer (20-mile)-long South Santa Cruz mountain segment of the San Andreas fault, about 80 kilometers (50 miles) southeast of San Francisco and 16 kilometers (10 miles) northeast of Santa Cruz. Although the public attention was riveted on fires in the Marina district along San Francisco's waterfront and rescue efforts on Oakland's collapsed Nimitz Freeway, the most extensive damage occurred closer to the quake's point of origin, in the towns of Santa Cruz, Los Gatos, and Watsonville. In all, the Loma Prieta earthquake caused billions of dollars in damage and claimed 65 lives.

Earthquakes, among the greatest catastrophes that geological forces can produce, have devastated numerous civilizations throughout history. Virtually every culture that has experienced them has sought to explain their cause, often by invoking images of giant creatures that carry the Earth through the sky, occasionally stumbling and jostling the planet. In Japanese tradition, the source of all tremors is a giant catfish; in many native North American traditions, a giant tortoise; and for the farmers of Mongolia, an enormous frog. The Wanyamwasi of western Africa believe that the Earth is supported on one side by mountains and on the other by a giant, who causes the Earth to tremble whenever he relinquishes his grip to embrace his wife.

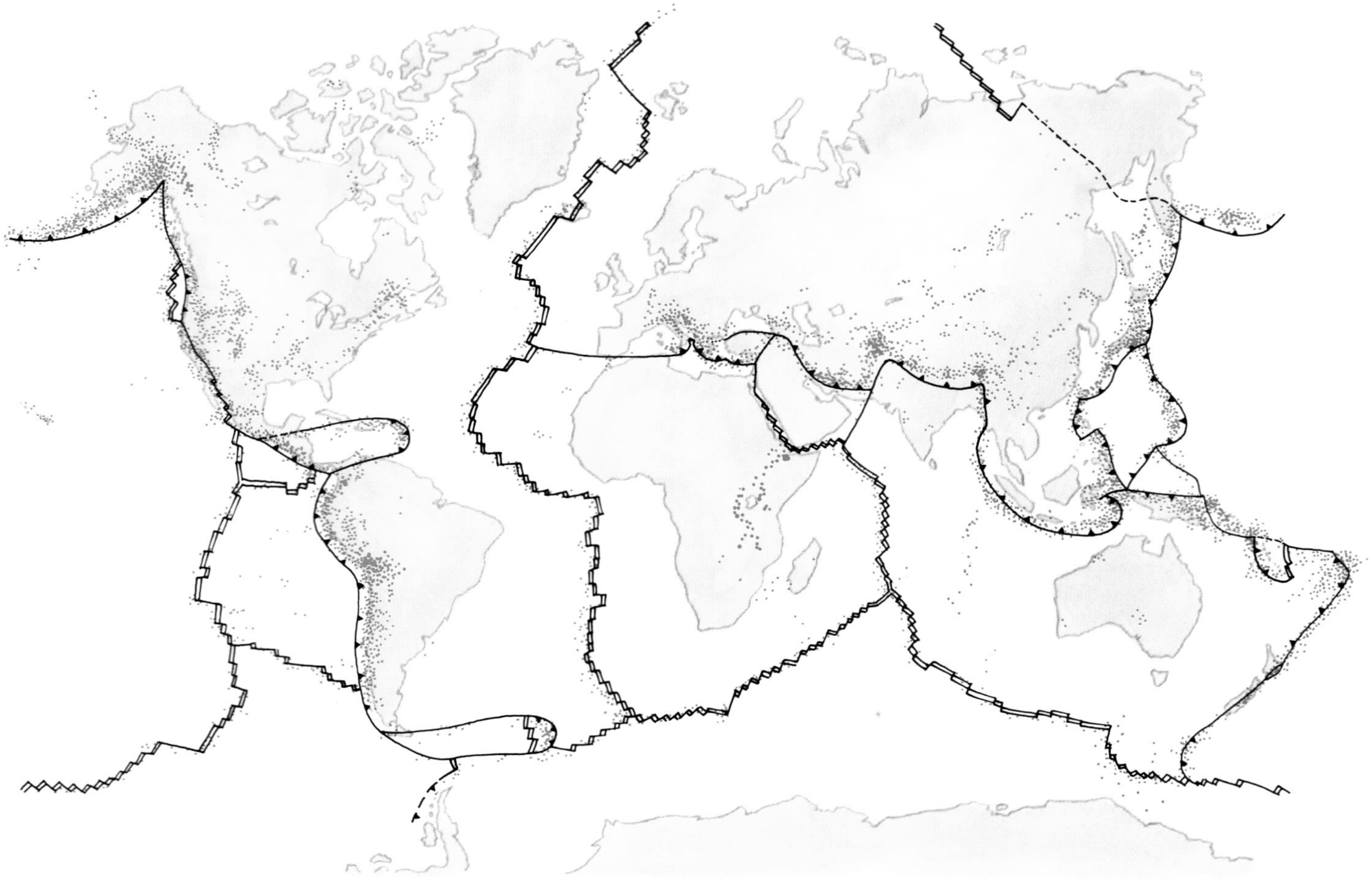

Figure 10-2 The worldwide distribution of earthquakes occurring during the last 100 years. Notice that earthquakes are concentrated at or near plate boundaries.

In this chapter, we will discuss the causes of earthquakes and how geologists study and evaluate them. We will also examine how we have learned to cope with their effects, and how we are trying to predict and prepare for them.

The Causes of Earthquakes

The trembling of the ground during an **earthquake** is most often caused by the sudden release of energy accumulated in deformed rocks. In Chapter 9 we saw that rock becomes deformed, or strained, when subjected to stress, usually at or near plate boundaries (Fig. 10-2). Deep within the Earth, when rock is exposed to stresses beyond its yield point, it deforms plastically to form folds. At shallower underground levels, however, where rock is relatively cool and subjected to less lithostatic pressure, it deforms elastically until—if the stress is not relieved—it ruptures. At these depths, stressed rocks accumulate *strain energy*, which builds until the rocks are either allowed to return to their original shape or they rupture (much as a stretched rubber band, if it is not allowed to snap back to its original shape, eventually breaks). Rocks that rupture create new faults; rocks at preexisting faults, although hampered by friction between the fault blocks, eventually shift under stress. Both cases result in sudden rock movement and the release of pent-up strain energy.

The precise subterranean spot at which rocks begin to rupture or shift is an earthquake's **focus** (Fig. 10-3). The quake's **epicenter** is the point on the Earth's surface that is directly above the focus, and thus generally feels the greatest impact from the quake. (A quake's surface effect generally decreases

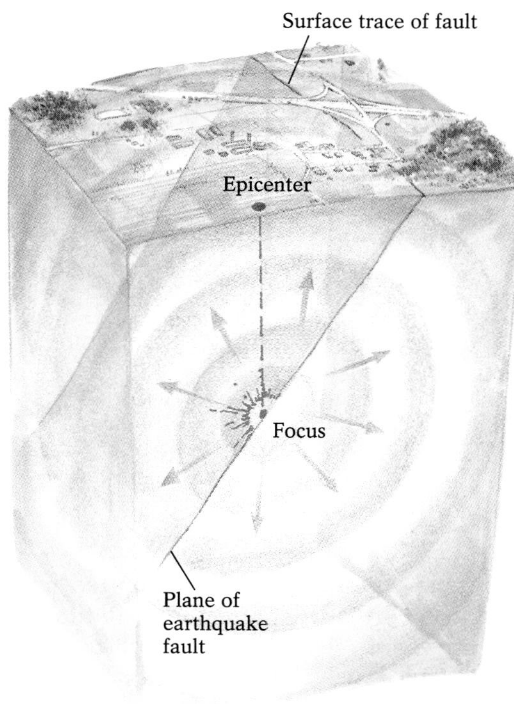

Figure 10-3 At the moment an earthquake occurs, pent-up energy is released at the quake's focus and transmitted through the Earth. The point on the Earth's surface directly above a quake's focus is its epicenter.

with increasing distance from its epicenter.) After a major earthquake, the rocks in the vicinity of the quake's focus continue to reverberate as they adjust to their new positions, causing numerous, generally smaller, earthquakes, or **aftershocks**. Aftershocks, which may continue to occur for as long as one to two years after the initial quake, can produce significant further damage by shaking already weakened structures. In the aftermath of the main shock from southern California's 1994 Northridge earthquake, thousands of aftershocks occurred, causing new damage and adding significantly to the quake's $15-billion-plus clean-up and rebuilding cost.

After their stored strain energy has been released during an earthquake, the rocks along a fault cease to move and the fault blocks become temporarily locked in place by the friction between them. As the rocks undergo subsequent stress, and strain energy again accumulates, the friction holding the fault blocks is eventually overcome and they lurch, releasing newly accumulated energy in the form of another earthquake. This explains why in most cases movement along faults is generally sporadic, and earthquakes occur periodically rather than continuously.

Unfortunately for the citizens of California, seismologists agree that neither the Loma Prieta nor the Northridge quake was the long-awaited "Big One." California is long overdue for an earthquake of even greater magnitude than either event—one powerful enough to relieve the energy that has been building along the San Andreas and nearby faults for as much as 130 years.

Seismic Waves

When rocks break or a locked fault lurches free, the released energy is transmitted at great speed through the surrounding rocks in all directions. Earthquake energy, like all types of energy, is transmitted from one place to another in the form of waves. When a large stone is tossed into a placid lake, the energy from its fall is transferred to the water on impact, ripples through the water in all directions, and eventually dies out at some distance. When someone speaks, sound energy produced by the vibration of the person's vocal cords is transmitted through the air in the form of sound waves. At the instant an earthquake occurs, the released energy is transmitted through the Earth as **seismic waves**. (The term *seismic*, from the Greek for "shaking," refers to anything involving earthquakes.)

Seismology, the study of earthquakes and the Earth's interior, is based on information yielded by two main types of seismic waves: **body waves**, which transmit energy through the Earth's interior in all directions from an earthquake's focus, and **surface waves**, which transmit energy along the Earth's surface, moving outward from a quake's epicenter.

Body Waves There are two types of body waves, distinguished by the speed and type of motion with which they travel through the Earth. Primary, or **P waves**, are the fastest seismic waves—passing through the Earth's crust at a velocity of 6 to 7 kilometers (about 4 miles) per second—and are the first to arrive at an earthquake-recording station following an earthquake. P waves are initiated when rocks near an earthquake's focus are compressed (reducing their volume) by the effects of breakage or sudden slippage at the focus. These rocks expand elastically past their original volume after the wave continues on to compress adjacent rocks, only to be compressed again as the next wave of energy passes. In this way, rocks in the path of P waves in turn compress and expand in the direction in which the waves are traveling much as the coils in a Slinky® toy vibrate when stretched and then struck (Fig. 10-4).

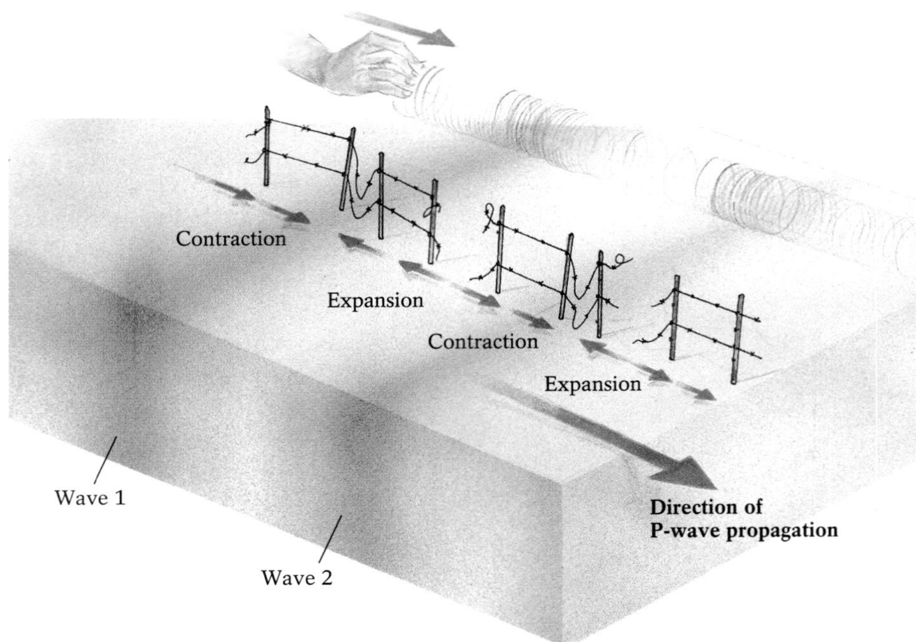

Figure 10-4 Primary waves, or P waves, are compressional, resulting in the alternate contraction and expansion of rocks in a direction parallel to the direction of wave propagation. The back-and-forth motion that displaces rocks as P waves pass (like the vibrating coils of a Slinky toy) can cause powerlines to snap.

The energy imparted by striking one end of the Slinky is transmitted by a wave of compression, like a P wave, that moves from one end of the toy to the other. During an earthquake, the arrival of P waves at the Earth's surface is marked by a series of sharp jolts as surface rocks alternately compress and expand with passing waves of energy.

Secondary, or **S waves**, are slower by half than P waves—passing through the Earth's crust at a velocity of about 3.5 kilometers per second (about 2 miles per second)—and therefore take longer to arrive at earthquake-recording stations. The motion of rock in an S wave is similar to that of a rope fixed at one end and flicked sharply up and down at the other: The rocks move up and down, *perpendicular* to the direction in which the wave travels (Fig. 10-5). (An S wave also resembles the stadium "wave" that has become popular at sporting events, in which successive individuals stand and then sit: The motion of the components of the wave—the fans—is perpendicular to the direction of the wave as it travels around the stadium.) As an area of rock moves up or down in an S wave, it creates a drag on neighboring

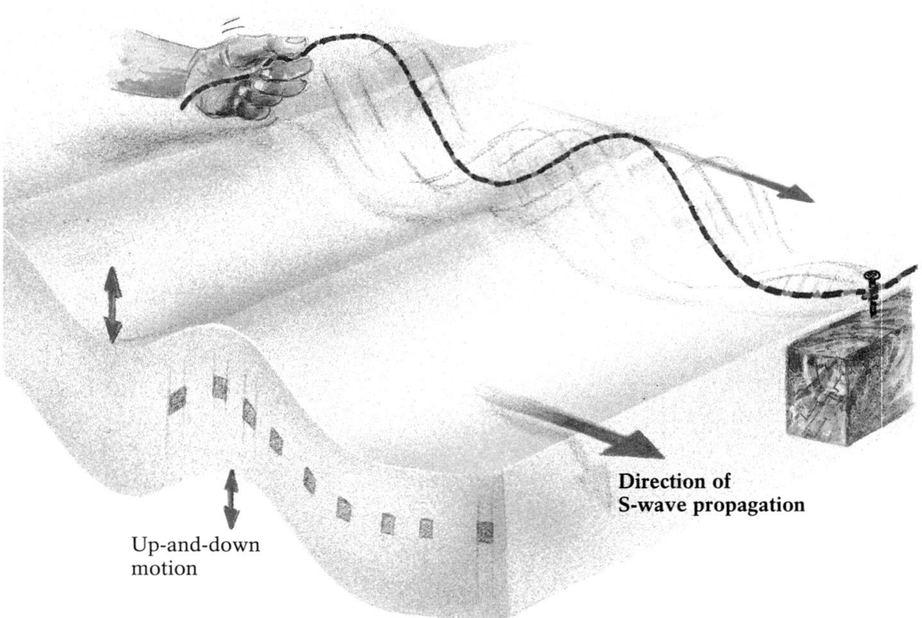

Figure 10-5 S waves, or secondary waves, are shearing waves. They generate an up-and-down motion that displaces adjacent rocks in a direction perpendicular to the direction of wave propagation.

rocks, pulling them along as well. This movement has a shearing effect on the rocks (see Chapter 9), changing their shape but not their volume. During an earthquake, P waves strike with a succession of compressional jolts, while S waves impart a wriggling motion that in large quakes can snap off chimneys and jostle objects from shelves.

Surface Waves When an earthquake occurs, some of the body waves that move outward from the focus spread up toward the epicenter, where they cause the surface to vibrate. This vibration generates surface waves, which travel within the upper few kilometers of the Earth's crust. These are the slowest seismic waves and arrive last at earthquake-recording stations, traveling through the crust at a velocity of about 2.5 kilometers per second (about 1.5 miles per second). There are two main types of surface waves, one having a side-to-side whipping motion like a writhing snake, the other a rolling motion that resembles ocean swells (Fig. 10-6). People who have experienced this second type of surface wave have likened it to a brisk walk across a waterbed, and some have become "seasick." Both types of surface waves occurring together cause objects to rise and fall while being whipped from side to side; this combination is responsible for most earthquake damage to rigid structures.

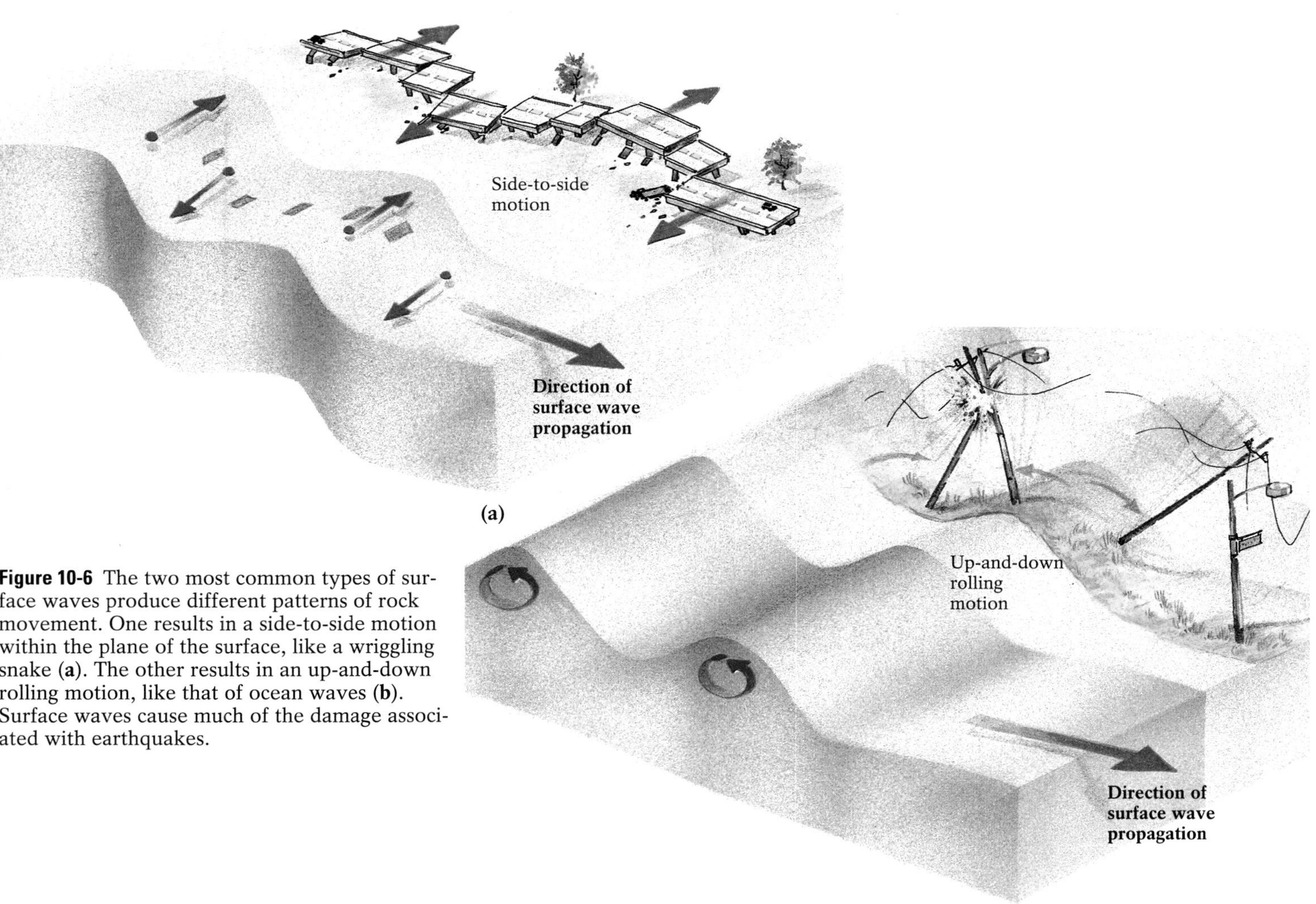

Figure 10-6 The two most common types of surface waves produce different patterns of rock movement. One results in a side-to-side motion within the plane of the surface, like a wriggling snake (**a**). The other results in an up-and-down rolling motion, like that of ocean waves (**b**). Surface waves cause much of the damage associated with earthquakes.

Measuring Earthquakes

An earthquake's strength depends on how much of the energy stored in the rocks is released. Traditional efforts to measure earthquakes compared their *intensity*, judged by the extent of the destruction they caused. People's descriptions were the basis for comparison, and the variable, subjective nature of human recollections often led to imprecise determinations. In 1902, the Italian seismologist Giuseppe Mercalli developed a standard list of increasing levels of damage to human-made structures, including descriptions of how the ground shakes and the landscape responds at each level (Table 10-1).

Table 10-1 **Abridged Mercalli Intensity Scale**

Intensity Value	Description
I	Not felt except by a very few under especially favorable circumstances.
II	Felt only by a few persons at rest, especially on upper floors of buildings. Delicately suspended objects may swing.
III	Felt quite noticeably indoors, especially on upper floors of buildings, but many people do not recognize it as an earthquake. Standing automobiles may rock slightly. Vibration like passing of truck. Duration estimated.
IV	During the day felt indoors by many, outdoors by few. At night some awakened. Dishes, windows, doors disturbed; walls make creaking sound. Sensation like heavy truck striking building. Standing automobiles rocked noticeably.
V	Felt by nearly everyone, many awakened. Some dishes, windows, and so on broken; cracked plaster in a few places; unstable objects overturned. Disturbance of trees, poles, and other tall objects sometimes noticed. Pendulum clocks may stop.
VI	Felt by all, many frightened and run outdoors. Some heavy furniture moved; a few instances of fallen plaster and damaged chimneys. Damage slight.
VII	Everybody runs outdoors. Damage negligible in buildings of good design and construction; slight to moderate in well-built ordinary structures; considerable in poorly built or badly designed structures; some chimneys broken. Noticed by persons driving cars.
VIII	Damage slight in specially designed structures; considerable in ordinary substantial buildings, with partial collapse; great in poorly built structures. Panel walls thrown out of frame structures. Fall of chimneys, factory stacks, columns, monuments, walls. Heavy furniture overturned. Sand and mud ejected in small amounts. Changes in well water. Persons driving cars disturbed.
IX	Damage considerable in specially designed structures; well-designed frame structures thrown out of plumb; great in substantial buildings, with partial collapse. Buildings shifted off foundations. Ground cracked conspicuously. Underground pipes broken.
X	Some well-built wooden structures destroyed; most masonry and frame structures destroyed with foundations; ground badly cracked. Rails bent. Landslides considerable from river banks and steep slopes. Shifted sand and mud. Water splashed, slopped over banks.
XI	Few, if any, (masonry) structures remain standing. Bridges destroyed. Broad fissures in ground. Underground pipelines completely out of service. Earth slumps and land slips in soft ground. Rails bent greatly.
XII	Damage total. Waves seen on ground surface. Lines of sight and level distorted. Objects thrown into the air.

Source: U.S. Federal Emergency Management Agency (FEMA)

The **Mercalli intensity scale** made it possible to rank earthquakes by intensity, but it could not locate an earthquake's epicenter accurately; nor could it distinguish an extremely strong earthquake that occurred far away from a separate, less severe event that occurred close by at about the same time. The Mercalli scale could err also because damage to structures is due not only to the actual intensity of ground motion, but also to the quality of construction and the nature and stability of the local soil. Moreover, this scale is totally ineffective in uninhabited areas on land or under the sea, where there are no human-made structures to sustain damage. A more accurate scale, based on quantitative measurement of the amount of energy released by a quake, was soon devised.

The Modern Seismograph A **seismograph** is a machine that measures the intensity of earthquakes by sensing and recording the seismic waves they generate; it produces a **seismogram**, a visual record of the arrival times of the different waves and the magnitude of the shaking associated with them (Fig. 10-17). Seismographs are located at more than 1000 stations around the world. Comparisons of recordings made of a single earthquake event at numerous stations enable seismologists to locate the quake's epicenter and determine how much energy was released.

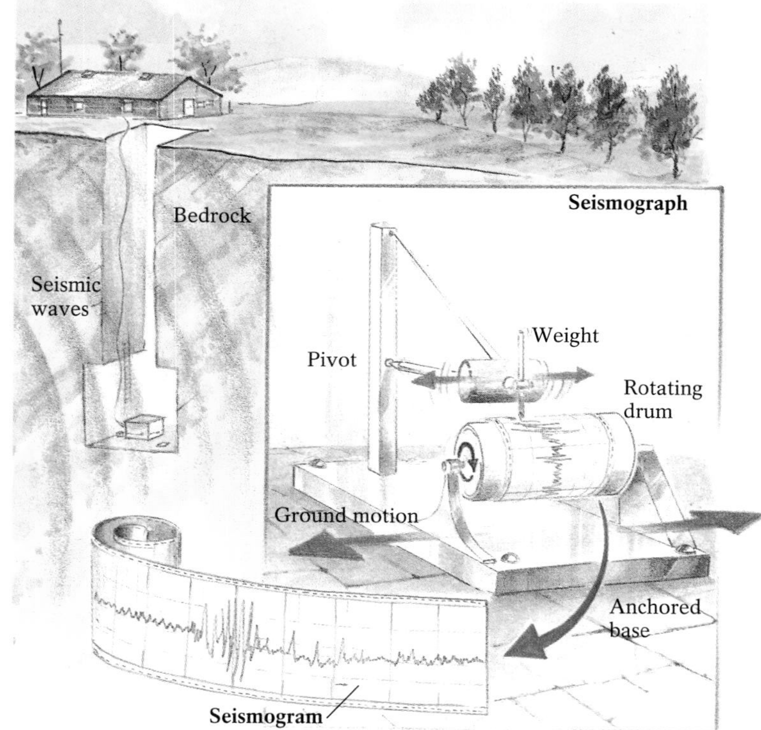

Figure 10-7 The functioning of a traditional seismograph. At its most basic, a seismograph consists of a mass suspended by a spring or wire from a base that is firmly anchored. (The entire seismograph is usually encased in a protective box and bolted to the Earth's crust meters below the surface.) The base moves with the Earth during an earthquake, while the suspended mass remains motionless. When a seismic wave jostles the seismograph beneath the suspended stylus, the stylus records the shaking on a roll of paper that turns at a steady rate on a rotating drum anchored to the base, producing a tracing called a seismogram. The amplitude of the seismogram's squiggles is proportional to the amount of energy released by the earthquake. (Note that seismographs can also be oriented vertically, to record vertical ground movement.)

A modern seismograph may receive signals from a *seismometer*, a device that amplifies wave motion electronically so that even weak or distant disturbances are detectable. Seismometers are sensitive enough to detect passing traffic, high winds, and nearby crashing surf. They can detect earthquakes that occur on the opposite side of the globe. Seismometers can also detect underground nuclear explosions, and are used to monitor nuclear test-ban treaties.

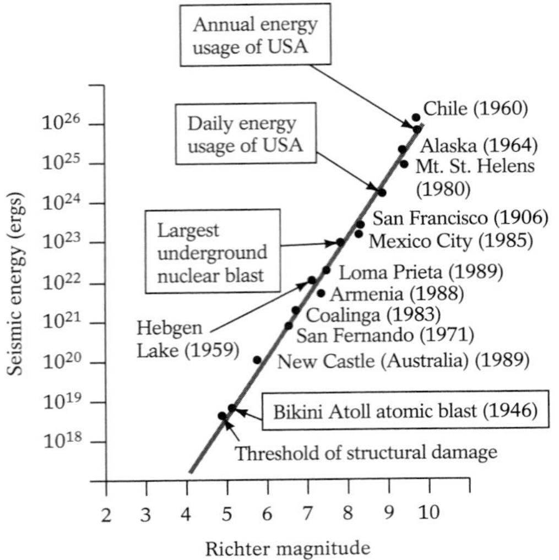

Figure 10-8 The Richter scale indicates the relative magnitude of earthquakes based on the energy they release (which is in turn determined by the amplitude of seismogram tracings from the quakes). It is logarithmic in nature, with each successive unit of magnitude representing a 30-fold increase in energy released.

The Richter Scale In 1935, Charles Richter of the California Institute of Technology proposed a way to use seismograms to determine an earthquake's *magnitude*, or the amount of energy released during the quake. The **Richter scale** correlates the *amplitude* of the largest peak traced on a seismogram during a quake (half the distance from the crest to the trough of the traced peak) to the amount of energy released by the quake. (The measured amplitude must be adjusted for the seismograph's distance from the earthquake's focus, to take into account the weakening of seismic waves over distance.)

The Richter scale is logarithmic, with each successive unit corresponding to a 10-fold increase in the amplitude of the seismogram tracings. Hence, the amplitude of the largest peak on a seismogram marking an earthquake of magnitude 5 is 10 times greater than that produced by a quake of magnitude 4, and 100 times that of a magnitude-3 event. The actual amount of energy released during tremors of different Richter magnitudes varies even more. For each unit increase in the Richter scale, there is about a 30-fold increase in energy released; thus, the energy released by a magnitude-6 earthquake is about 30 times greater than the energy released by a magnitude-5 quake and 900 times greater than the energy released by a magnitude-4 event (Fig. 10-8).

To understand the logarithmic nature of the Richter scale, consider metropolitan Los Angeles, which seismologists believe is struck by an earthquake with a magnitude greater than 8 every 160 years or so. The last quake of this magnitude occurred there in 1857. One might think that an occasional magnitude-6 quake would relieve a significant amount of accumulated strain energy and forestall or prevent the next "Big One." But the energy released by a magnitude-8 earthquake is 900 times that released by a magnitude-6 quake. Nine hundred earthquakes of magnitude 6 would therefore be needed to dissipate the accumulated strain energy and avert the next great quake (Richter magnitude greater than 8.0). Since southern California experiences only one quake of that magnitude every 5 years on average, the region can still expect a seismic event 900 times as violent as any in recent memory.

The Richter scale has no upper limit, but virtually all rock types would fail and release their stored strain energy before accumulating enough energy to produce an earthquake of 9.0 on the Richter scale. Thus, 9.0 to 9.5 appears to be the effective upper limit of the scale.

Monitoring Seismic Waves Most active seismograph stations are in locations that experience many earthquakes. In recent years there has been increased international cooperation in the field of seismology to standardize equipment and procedures and to share data freely. Since the early 1960s, stations from many nations have fed their data directly to the Earthquake Information Center of the U.S. Geological Survey in Golden, Colorado. The Center's high-speed computers continually analyze incoming signals, quickly locating earthquake epicenters and determining quake magnitudes.

Earthquakes that measure 1 to 2 on the Richter scale occur hundreds of thousands of times each year, but are known only to seismological laboratories; quakes must be of magnitude 3.5 or higher to be felt by humans. A 4.5 quake can cause minor damage. A quake of magnitude 6 is considered major and does extensive damage. Every 5 to 10 years, a great earthquake with a magnitude of 8 or greater strikes the Earth. The Alaskan quake of 1964, North America's largest since the development of the modern seismograph and the Richter scale, measured 8.4. The energy released by that quake, the most intensively studied in history, equaled that of about 1 billion tons of TNT. This powerful quake is discussed further in Highlight 10-1.

Highlight 10-1 *Alaska's Good Friday Earthquake, 1964*

Figure 10-9 The damage to downtown Anchorage, Alaska, during its 1964 Good Friday earthquake illustrates the effects of a magnitude-8.4 quake.

At 5:36 P.M. on Good Friday, March 27, 1964, Carol Tucker was shopping in the new J.C. Penney store in downtown Anchorage. Schools had been closed that day for the holiday, workers had left their offices early, and downtown shops were largely deserted. Most of the citizens of Anchorage were at home, preparing their evening meal.

As she walked through the third-floor chinaware department, Ms. Tucker, a longtime Alaska resident, felt the familiar jiggle of an earth tremor. But this time the shaking did not subside, but rather grew more pronounced, shattering china and exploding display cases in a shower of glass. When Ms. Tucker finally reached an exit, she paused for a moment for no apparent reason before emerging into the chaos of downtown Anchorage. This moment of hesitation probably saved her life. At that instant, the massive concrete facade of the store plunged to the street below, crushing a young man already on the street. The initial quake, measuring 8.4 on the Richter scale, lasted for nearly 4 minutes. Buildings were damaged throughout a 100,000-square-kilometer (40,000-square-mile) area (Fig. 10-9). Ice on arctic rivers and lakes shattered throughout a 250,000-square-kilometer (100,000-square-mile) area. Buildings swayed perceptibly 2500 kilometers (1500 miles) to the south in Seattle, and the ground rose and fell 2 centimeters (1 inch) as far away as Houston, Texas, and Orlando, Florida.

One hundred and fifteen Alaskans died in the Good Friday earthquake. The toll was remarkably low for an event of this magnitude, largely because of the sparse population in the Anchorage area and the timing of the tremor. Had the quake struck when offices and schools were filled, this city of 150,000 would have suffered fatalities in the thousands.

For 3 days, aftershocks rumbled through buildings that had remained standing during the main quake, shaking some to the ground. During the next 18 months, 10,000 aftershocks, some as strong as 7 on the Richter scale, rattled the stricken area. In all, the damage totaled $300 million, 40 times the original purchase price of the entire Alaska territory.

Locating an Earthquake's Epicenter

The different velocities of the various types of seismic waves can be used to determine the location of a quake's epicenter. Just as two sprinters who set out together and run at different speeds will reach the finish line at different times, so do the P and S waves from an earthquake reach a seismograph station at different times. For example, a seismograph located 600 kilometers (400 miles) from an earthquake's epicenter might begin to record P waves about 100 seconds after the quake begins, and S waves about 70 seconds later. Because we know the velocity at which each type of wave travels through various rock types, we can use the difference between their arrival times and our knowledge of the rocks through which they traveled to calculate the distance from seismograph stations to the earthquake's epicenter.

By comparing the arrival times of seismic waves from earthquakes of known origin, recorded at stations around the world, geologists have developed travel-time charts that convert the lag between P- and S-wave arrival times to the distance to a quake epicenter. For quakes of unknown origin, the lag between the arrival times of the two types of waves is used to calculate the distance from the seismograph stations to the earthquake's epicenter (Fig. 10-10a).

The linear distance from an earthquake to a single seismic station will not by itself identify the earthquake's location, since the quake could have occurred at that distance from the station *in any direction*. However, this distance may be represented on a map as the radius of a circle centered on the station, and similar circles drawn from three different stations would intersect at the quake's epicenter (Fig. 10-10b). To pinpoint the precise location of an earthquake, at least three seismic stations must cooperate.

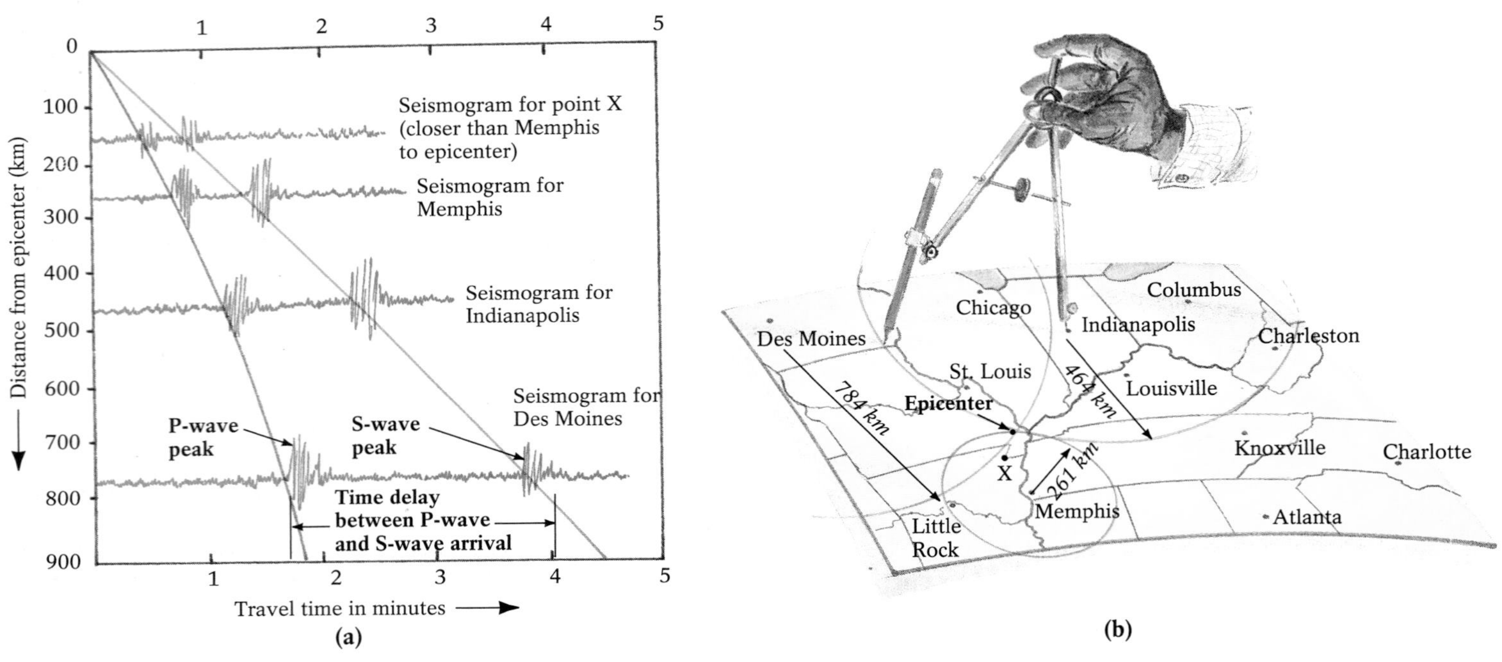

Figure 10-10 (a) Seismograms from various seismograph stations, recording the arrival times of P and S waves at each from an earthquake in the New Madrid, Missouri, area. The closeness of the two wave types at the Memphis, Tennessee, recording station indicates that the earthquake occurred nearby. Using our knowledge of the velocities of P and S waves through the Earth's crust, and the 120-second and 186-second delays at Indianapolis, Indiana, and Des Moines, Iowa, respectively, we can determine each station's distance from the earthquake epicenter. (b) By drawing a circle around each of these stations on a map, with each circle's radius proportional to the station's distance from the quake's epicenter, we can determine the location of the epicenter—it is marked by the single point at which the three circles intersect on the Earth's surface.

The value of being able to locate an earthquake's epicenter was demonstrated on May 23, 1989, when an earthquake occurred that registered 8.3 on the Richter scale. Moments after the quake, a worldwide network of more than 100 state-of-the-art seismological facilities transmitted data to high-speed computers at the U.S. Geological Survey installation at Golden, Colorado. Within minutes it was determined that the earthquake had occurred along the Macquarie ridge, a submarine mountain chain 800 kilometers (500 miles) southwest of New Zealand. As we will discuss shortly, a powerful quake that strikes the ocean floor often produces enormous sea waves that can devastate coastal regions. A sea-wave alert was issued for coastal communities around the Pacific Ocean basin, in adequate time to successfully evacuate all the people in these areas.

Earthquake Depth and Magnitude

When an earthquake occurs, news accounts generally identify its location with reference to the nearest town or city. In truth, an earthquake occurs at some depth below such a site, at the quake's focus. We describe an earthquake as shallow if its focus is less than 70 kilometers (45 miles) below the surface, intermediate if its focus is from 70 to 300 kilometers (45 to 180 miles) below the surface, and deep if its focus is more than 300 kilometers (180 miles) below the surface. Earthquakes have been known to occur up to a maximum depth of about 700 kilometers (450 miles), but more than 90% of all earthquakes emanate from depths of less than 100 kilometers (60 miles) and virtually all catastrophic quakes originate within 60 kilometers (40 miles) of the surface. Large earthquakes seldom occur at greater depth because rocks in that environment have been softened by heat and have lost some of their ability to store strain energy. Only within a few tens of kilometers of the surface are rocks brittle and elastic enough to accumulate a vast amount of energy. The depth of the focus of the 1964 Good Friday earthquake in Alaska, for instance, was 33 kilometers (20 miles); the focus of the 1994 Northridge earthquake was 21 kilometers (13 miles) deep, that of the 1989 Loma Prieta earthquake was about 18 kilometers (12 miles) deep, and that of the San Francisco earthquake of 1906 was only 15 kilometers (10 miles) deep.

There are limits to the magnitudes of quakes at different depths. Shallow earthquakes have been recorded with Richter magnitudes as high as 8.6, and intermediate ones have magnitudes up to 7.5. Deep quakes, however, are rarely recorded with Richter scale magnitudes greater than 6.9. A magnitude-8.2 earthquake that struck Bolivia in June 1994 did occur at a depth of almost 700 kilometers (400 miles); seismologists are analyzing the data from this quake to determine how such deep, presumably warm plastic rocks could store enough energy to produce such a powerful tremor.

The Effects of Earthquakes

During an earthquake, many geological effects occur simultaneously: Large areas of the ground shift position as displacements occur along faults; landslides and mudflows are set off by the passage of surface waves; in places, the seemingly solid earth starts to flow as if liquid; surfaces of bodies of water rise and fall; groundwater levels fluctuate as the Earth compresses and expands with the passage of P waves; and enormous sea waves, generated principally by submarine fault displacements, strike coastlines.

Figure 10-11 Wallace Creek, on the Carrizo Plain in southern California, offset by the San Andreas fault.

Ground Shifts

One of the most obvious geological effects of an earthquake is large-scale shifting of the landscape. After fault blocks have unlocked and moved and pent-up energy is released, geological features and geographical landmarks are often markedly displaced (Fig. 10-11). During the 1906 San Francisco quake, a 432-kilometer (270-mile)-long rupture opened along the strike-slip San Andreas fault and the west side of the fault lurched laterally northward 6.4 meters (21 feet), as determined by measuring displacements of fences and other linear features near the epicenter. By comparison, the lateral displacement produced by the 1989 Loma Prieta quake was only 2.2 meters (7 feet). During the 30 million years of motion along the San Andreas fault, the rocks and landforms on the west side have shifted an estimated 564 kilometers (350 miles) northward.

Vertical ground shifts occur during movement along dip-slip faults. In southern California's 1994 earthquake, L.A.'s Santa Susana Mountains rose about 30 centimeters (1 foot). During the Good Friday earthquake of 1964, 500,000 square kilometers (200,000 square miles) of Alaska and the adjacent sea floor dropped downward by normal faulting in some places, and thrust upward by reverse faulting in others. The maximum upward land shift totaled 12 meters (39 feet). In places, the sea floor dropped as much as 16 meters (52.5 feet). While coastal locations that had dropped below sea level suffered extensive flooding, uplift at the coast left harbors, fisheries, and docks high and dry, requiring those facilities to be relocated.

Landslides and Liquefaction

A large earthquake's violent shaking often jostles and dislodges large masses of unstable rocks and soils from hillsides, causing them to rush downslope. Such *landslides* may also occur as rock fragments become detached from bedrock, thick layers of sedimentary rock slide along weak bedding planes, foliated metamorphic rock slides along foliation planes, or loose sediment cascades downslope. On June 7, 1692, a relatively moderate quake sent the entire city of Port Royal, on the Caribbean island of Jamaica, sliding into the sea, where it finally came to rest beneath 15 meters (50 feet) of water. The city, picturesquely infamous as a base for numerous pirates and other scoundrels, was built on loose, steeply sloping sediment in an area prone to periodic quaking. When marine archaeologists discovered the city in 1959, it lay protected and virtually intact beneath 3 meters (10 feet) of marine silt; in one kitchen, a copper kettle still contained the evening's meal of turtle soup.

Earthquakes may also cause **liquefaction**, the conversion of unconsolidated sediment with some initial cohesiveness to a mass of water-saturated sediment that flows like a liquid. When fine, moist sediment is jostled during an earthquake, the pressure on the water between sediment grains increases, forcing them apart. Without frictional contact between adjacent grains, the sediment loses its cohesiveness, and what was once firm ground becomes a slurry of mud. During Alaska's 1964 quake, the sediment beneath the Turnagain Heights neighborhood of Anchorage liquefied (Fig. 10-12). The devastated area, worthless for future development, was bulldozed into Earthquake Park. During the Loma Prieta quake, liquefaction of the bay muds and landfill soils beneath the Nimitz Freeway, the Oakland area, and the Marina district of San Francisco was responsible for the collapse of structures and much of that quake's death toll.

Seiches

Seismic waves cause the water in an enclosed or partially enclosed body of water, such as a lake or bay, to move back and forth across its basin, rising and falling as it sloshes about. This phenomenon is called a *seiche* (pronounced "saysh"). In 1964, the water at one end of Kenai Lake, south of Anchorage, rose 9 meters (about 30 feet), overflowed its banks, and flooded inland before reversing its direction. The soils around the lake were stripped down to bare bedrock by the back-and-forth motion of the water. During Montana's Madison Canyon quake of 1959, the Hebgen Lake dam overflowed due to a seiche that oscillated back and forth across the lake every 17 minutes for 11 hours.

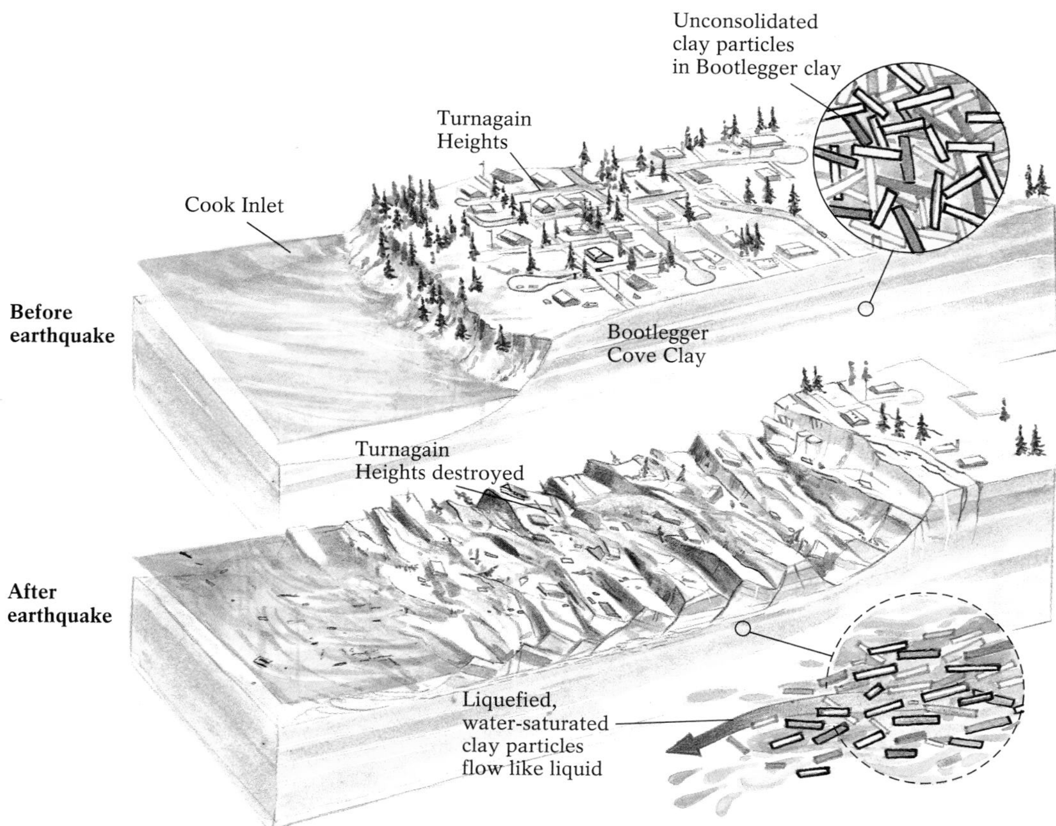

Figure 10-12 Liquefaction occurs when the grains in a layer of wet, fine-grained sediment are shaken during an earthquake. In 1964, the clay layer beneath the entire Turnagain Heights area of Anchorage, Alaska, liquefied, resulting in a chaotic landscape of jumbled homes and asphalt blocks.

Seiches may develop at great distances from powerful earthquakes, sometimes so far from the epicenter that no other effects of the quake are felt. The Alaska earthquake set off seiches in reservoirs as far away as Michigan, Arkansas, Texas, and Louisiana. Water rose and fell 2 meters (6.5 feet) in bays along the Texas Gulf Coast, damaging boats that were buffeted by the waves. Even swimming pools in Texas, 6000 kilometers (4000 miles) from the quake's epicenter, developed seiches.

Tsunami

A Japanese folktale tells of a venerable grandfather who owned a rice field at the top of a hill. One day, just before the harvest, the old man felt the sharp jolt of an Earth tremor, and from his hilltop vantage point saw the sea pull back from the shore. Curious villagers rushed out to explore the exposed tidal flats and collect shellfish. From experience, the old man knew of the grave danger to his neighbors. With his grandson by his side, he dashed about his fields, setting fire to his crop. The villagers saw the smoke, and hurried up the hill to aid their neighbor. As they beat out the flames, they saw the old man scurrying ahead, setting new fires, near the hill's crest. Hoping to prevent him from destroying all that he owned, they rushed uphill to stop him. Moments later, they saw a tremendous wall of water surging onshore, flooding the flats where they had been standing, and they understood that the old man had sacrificed his livelihood to save their lives.

Figure 10-13 The famous print *Beneath the Waves Off Kanagawa*, by Japanese artist Katsushika Hokusai (1760–1849), depicts the helplessness of a group of boatmen in the face of tsunami-sized waves.

More than any other country, Japan, with its thousands of kilometers of island coastlines nestled precariously within one of the Earth's principal seismic zones, has suffered from **tsunami**, Japanese for "harbor wave" (Fig. 10-13). Tsunami ("tsunami" is both singular and plural) are frequently more than 30 meters (100 feet) high and can move faster than 800 kilometers per hour (500 mph), making them far higher and faster than ordinary waves. At Cape Lopatka on Kamchatka Island, in 1739, one reached a height of 65 meters (210 feet), the equivalent of a 20-story building.

Most tsunami originate from a large instantaneous displacement of the sea floor during submarine faulting, either when the sea floor drops from normal faulting or rises from reverse faulting. When a section of the sea floor drops downward, a trough in the ocean surface results, temporarily withdrawing water from the coast. A wave is created as water rushes in to fill the trough, overcompensates, and travels to land as a potentially catastrophic crest—a tsunami (Fig. 10-14a). When a section of the sea floor is thrust upward, water is displaced upward, pulling along water from the shore, and arrives at the coast as a tsunami (Fig. 10-14b).

Tsunami usually consist of several waves that may arrive at irregular intervals of several hours. Moreover, tsunami can travel great distances from a quake epicenter and are generally undetected before striking a coast. In 1960, a tsunami generated by a catastrophic quake in southern Chile struck the Hawaiian shore at Hilo 7 hours later, where it took 61 lives; 22 hours after the quake the tsunami had traveled 17,000 kilometers (11,000 miles) to reach Honshu and Hokkaido in Japan, where it took 180 lives.

Tsunami that strike a coast at high tide often cause more deaths than the earthquakes that generate them: Of the 131 deaths caused by the 1964 Alaskan quake, 109 resulted from tsunami that buffeted the southern Alaskan coast. Tsunami from that quake also hit the coasts of Oregon and California, causing 16 deaths there. At Crescent City, California, the first wave arrived at low tide. It crested at about 4 meters (14 feet) above sea level, destroying several lumber boats and releasing a morass of giant redwood logs. After three additional small waves residents returned to the coast, believing the worst was over. A fifth wave struck at high tide, however, cresting at more than 10 meters (33 feet) and washing 12 people out to sea.

At sea, tsunami are barely perceptible. With crests of successive waves separated by as much as 160 kilometers (100 miles), these waves are mere bumps on the sea surface, usually less than 1 meter (3 feet) high. (Contrary to the Hollywood myth perpetuated in the film *The Poseidon Adventure*, in which a ship is overwhelmed in the open sea by a tsunami 30 meters, or 100

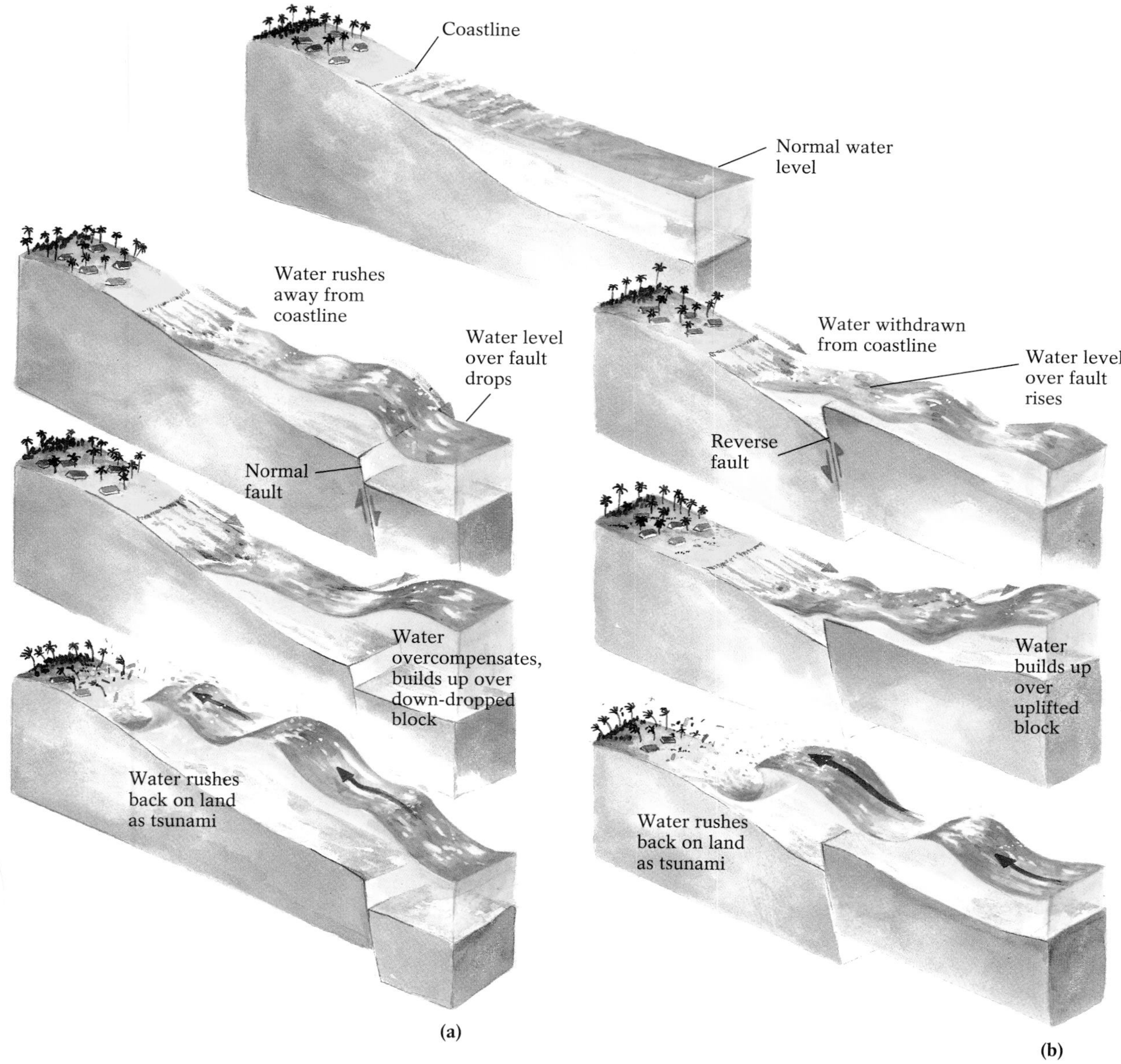

feet, high.) Only in the shallow waters of enclosed bays and harbors, where drag against the ocean floor causes the fast-moving waves to bunch up, are tsunami visible as devastating walls of water.

Fires

Indirectly, earthquakes also cause fires, the severity of which depends on such factors as the type of building materials in use, whether gas lines are ruptured, and whether water mains are broken. On September 1, 1923, when a powerful earthquake struck the Tokyo–Yokohama area of Japan at midday, thousands of open cooking fires were overturned. Fanned by high winds, the fires coalesced and spread until, by evening, the glow from the blaze was bright enough to read by 15 kilometers (10 miles) from the city. Before the flames were extinguished, 500,000 buildings in the mostly wood-frame city were destroyed and 120,000 people had perished. Each year, Japan observes a national day of earthquake preparedness in commemoration of this tragedy.

After the earthquake, Tokyo was rebuilt with heightened awareness—with broader avenues to ease passage of fire-fighting equipment and concrete instead of wooden buildings. Millions of liters of water are stored in underground cisterns and quake-resistant structures, and the city's 30,000 taxis are equipped with fire extinguishers.

Earthquake damage to gas mains, oil tanks, and electrical power lines typically cause fires in inhabited regions, and, because water mains are frequently ruptured as well, fire-fighting capability is often impeded. During its earthquake of 1906, San Francisco's wooden buildings were tinder for a blaze that caused 80% to 90% of the city's damage. The fire raged uncontrolled for three days, largely because water lines had broken into hundreds of segments that subsequently lost pressure. Finally, it became necessary to create a firebreak by dynamiting rows of buildings. Sadly, it is believed that the main blaze was kindled several hours after the quake by a woman on Hayes Street who, shaken by the events of the day, decided to restore normalcy to life by preparing a ham-and-eggs breakfast. Unfortunately, the flue of her stove had been damaged during the shaking; the resulting fire consumed more than 500 square blocks of the city's central business district.

The World's Principal Earthquake Zones

As we have seen, the vast majority of earthquakes are concentrated at plate boundaries, whereas plate interiors are generally seismically inactive. Nevertheless, major earthquakes do occasionally strike within a plate interior: Charleston, South Carolina, Boston, Massachusetts, and southern Missouri, all located far from fault-prone plate boundaries, have each experienced a major quake within the past 250 years.

Earthquake Zones at Plate Boundaries

The depth of earthquakes varies significantly among the different types of plate boundaries (Fig. 10-15). At oceanic divergent zones, most earthquakes are generally extremely shallow, occurring at less than 20 kilometers (12 miles) in depth, because strain energy can build up only within the uppermost, brittle portion of an oceanic plate. Deep earthquakes are virtually nonexistent in these zones because the asthenosphere, located about 100 kilometers (60 miles) beneath oceanic plates, consists of partially melted, softer material that deforms plastically and therefore produces no earthquakes. This is true as well of continental rift zones, continental collision zones, and along continental transform boundaries. Earthquakes in these zones cannot occur at depths greater than about 50 to 80 kilometers (30 to 65 miles) —the maximum thickness of the brittle portion of continental plates.

At subduction zones, however, shallow quakes occur where the subducting plate begins its descent at an ocean trench, but progressively deeper quakes are initiated within the descending slab to a depth of about 700 kilometers (450 miles). The region within which earthquakes occur at subducting oceanic boundaries is called the **Benioff zone**, named for the American seismologist Hugo Benioff, who first observed that earthquake foci are progressively deeper inland from ocean trenches.

From the onset of subduction at an ocean trench to a depth of about 300 kilometers (200 miles), intense friction between the descending slab and the overriding plate causes ruptures along the slab's brittle upper surface, resulting in shallow and intermediate-depth earthquakes. These are generally the most powerful subduction zone earthquakes. As the slab descends to greater

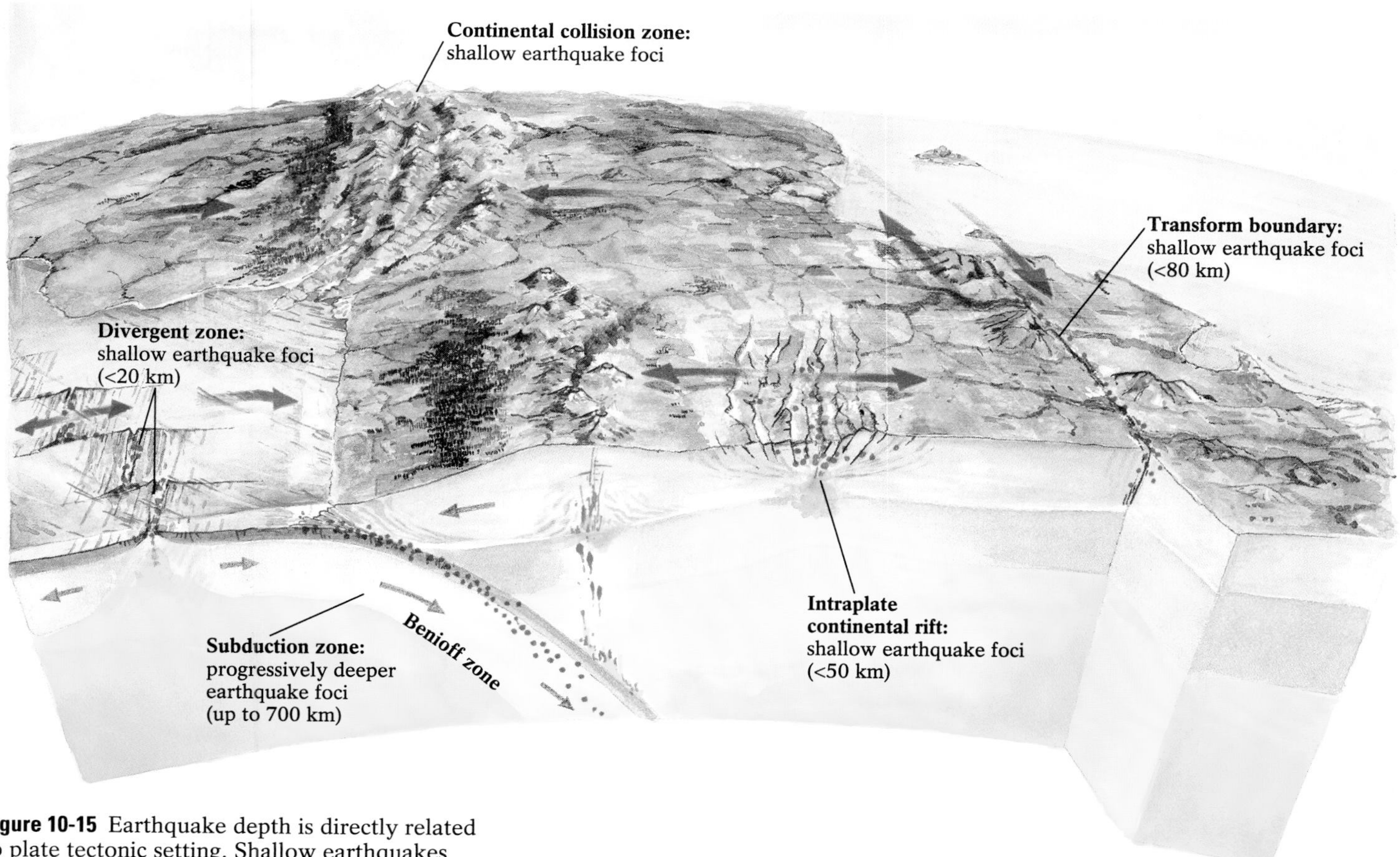

Figure 10-15 Earthquake depth is directly related to plate tectonic setting. Shallow earthquakes occur where plates are rifting, diverging, or undergoing transform motion. Deep-focus earthquakes occur exclusively at subduction zones, within the brittle portion of the descending slab. The slab remains largely intact and brittle (except for some minor melting that may take place at its very top) as it passes through the partial-melting zone. Thus, it continues to generate earthquakes until it descends to about 700 kilometers, at which point it has warmed sufficiently to deform plastically.

depths, the center of quake activity shifts to the still-brittle interior of the slab, which has not yet been in the Earth's interior long enough to warm up and soften. By the time the slab descends to approximately 700 kilometers (450 miles) below the surface, it has been heated sufficiently to soften throughout, and thus deforms plastically instead of undergoing brittle failure. Hence, earthquakes rarely occur at depths below 700 kilometers.

The magnitude of earthquakes varies considerably among the different types of plate boundaries, with almost all major earthquakes occurring along either convergent or transform boundaries. (Earthquakes at divergent plate boundaries are typically of relatively low magnitude, because rocks being stretched tend to break before accumulating too much strain energy.) Every year, about 80% of the world's earthquake energy is released in the Pacific Rim region, where the oceanic plates of the Pacific Ocean basin subduct beneath adjacent plates along most of the ocean's perimeter (see Figure 10-2). This region includes the earthquake zones of Japan, the Philippines, the Aleutians, and the west coasts of North, Central, and South America. Most of the remaining 20% of the Earth's earthquake energy is released annually along the collision zone stretching through Turkey, Greece, Iran, India, and Pakistan to the Himalayas, southern China, and Myanamar (Burma).

Japan The western edge of the Pacific plate descends steeply beneath the eastern edge of the Eurasian plate at the Japan trench. Japan alone is shaken by 15% of the planet's released seismic energy; each year, its residents experience more than 1000 earthquakes of magnitude 3.5 or greater. Since 1900, Japan has suffered 25 earthquakes as powerful as San Francisco's great quake of 1906. Landslides, tsunami, fires—Japan has experienced the worst of each.

Yet, Japan's next quake may be its worst. As at other quake-prone plate boundaries, Japan experiences periodic earthquakes whenever one of its faults accumulates enough energy to overcome the frictional resistance between fault blocks and breaks free. The more time that elapses since the last major earthquake, the more energy accumulates, and the more likely it is for the next one to be quite powerful. Historical records from A.D. 818 to the present indicate that cataclysmic quakes have struck Tokyo once every 69 years on average, and of Japan's nine major quakes since the Tokyo quake in 1923, none has occurred in metropolitan Tokyo, where more than 20 million people now live. The stress on Tokyo's rocks has been building for more than 70 years. Another great quake is also probably overdue in the densely populated Tokai region to the south, where the last major quake released accumulated energy on December 24, 1854.

Figure 10-16 Effects of the Mexico City earthquake of September 19, 1985 (magnitude 8.1). Due to the violent shaking of Mexico City's geological foundation of unconsolidated lake clay, damage to the city—500 kilometers from the quake epicenter—was actually greater than that in areas closer to the epicenter.

Mexico and South and Central America Earthquakes have claimed hundreds of thousands of lives in Chile and Peru, along the west coast of South America, where the Nazca plate (Fig. 1-12) flexes downward to form the Peru–Chile trench and the Andes mountains. The Peruvian quake of May 31, 1970, leveled virtually every dwelling along a 70-kilometer (45-mile) strip of the coastline, taking 120,000 lives, demolishing 200,000 buildings, and leaving 800,000 people homeless. Similarly, subduction of the Cocos plate causes the tremors that have periodically devastated such cities as Mexico City, Mexico, and Managua, Nicaragua (Fig. 10-16).

Western North America Western North America from southern California to Alaska owes much of its extensive earthquake history to interaction between the Pacific plate and the North American plate. The southern end of the transform boundary between these plates extends for 1000 kilometers (600 miles), from Baja California to Cape Mendocino northwest of San Francisco. Between Cape Mendocino and Vancouver, British Columbia, the Pacific–North American transform boundary is interrupted by two small plates (the Gorda plate off the coast of northern California and Oregon, and the Juan de Fuca plate off the coast of Washington and southern British Columbia) that are being subducted beneath westward-drifting North America. The Cascade volcanoes, from Lassen Peak and Mount Shasta in northern California to Mount Garibaldi north of Vancouver, are products of this subduction zone. The transform boundary between the Pacific and North American plates resumes north of Vancouver and extends to Alaska, where the Pacific plate subducts beneath the North American and Eurasian plates, producing the volcanic Aleutians. Thus, the transform boundaries and subduction zones of the western states and provinces form an almost continuous length of earthquake-prone country.

Intraplate Earthquakes

On June 10, 1987, a rare mid-plate earthquake unnerved the citizens of Lawrenceville, Illinois, and 16 surrounding states and Canadian provinces, registering 5.0 on the Richter scale. It triggered alarms at a nuclear power plant, cut phone service, and shook hospital patients from their beds in Iowa and West Virginia.

Though generally shallow (less than about 50 kilometers, or 30 miles), mid-plate earthquakes tend to be of lower magnitude than those at active plate margins, apparently because less strain energy builds at mid-plate. However, the older, colder (and therefore more brittle), and less fractured rocks at mid-plate transmit seismic waves more efficiently than the young, warm rocks at active plate margins. Thus, mid-plate earthquakes, such as those in eastern North America, are felt over a significantly larger area than are plate-boundary quakes such as those in the west.

Locating mid-plate faults to determine their history and predict their future seismic activity can be difficult, particularly in eastern North America where heavy vegetation conceals rocks and their faults. Because tectonic activity there is infrequent and sporadic, any surface evidence of it is usually obscured or eradicated with the passage of time. Seismic wave studies, however, show the crust of eastern North America to be riddled with a deep irregular network of old, near-vertical faults. Those along the East Coast may date from the rifting of North America from Europe and Africa during the initial break-up of the supercontinent Pangaea, 200 million years ago. Those in the Midwest, south of Lake Superior, may date from an older rifting event believed to have begun about 1.1 billion years ago.

We don't know exactly what causes intraplate earthquakes. Some hypotheses cite processes that are unrelated to plate tectonic motion. One, for example, proposes that after long-term erosion has removed vast quantities of surface materials, the unloaded crust becomes buoyant, rising and generating enough stress to arouse old faults. Another suggests that the load of mid-continent sediment deposition weighs down the crust, the pressure reactivating its faults. Still another hypothesis proposes that these earthquakes may be triggered by excessive rainfall, which infiltrates and saturates the network of subsurface fractures, reducing friction between adjacent blocks of rocks and reactivating long-dormant faults. (In support of this hypothesis: The great Charleston, South Carolina, quake of 1886 followed two years of unusually high rainfall.)

Other hypotheses view intraplate seismic activity as a precursor to future plate configurations, perhaps an early warning of rifting. According to this hypothesis, plate rifting is initiated by rising masses of heat-softened mantle material. Such thermal plumes are alleged to have set off quake activity near Boston and Charleston, and extremely young plutons that have been identified by seismic studies in those areas are considered evidence of incipient rifting and future plate-edge locations.

A simpler model holds that plate motion creates stress on the already weakened and faulted crust at some intraplate sites, causing infrequent but sometimes powerful quakes. The lateral motion of the westward-drifting North American plate, for example, may create frictional stress within the plate as it rides atop the underlying asthenosphere. The strain energy from these stresses may accumulate over long periods, until it exceeds the strength of the rocks or the resistance along their faults.

Earthquakes in Eastern North America Because there have been relatively few earthquakes to release pent-up seismic energy in the interior of the North American plate, large quakes may be looming in this region. Seismologists working with earthquake records that date back to 1727 predict that there is a better-than-even chance of a magnitude-6 earthquake striking east of the Rocky Mountains before the year 2020.

Past eastern earthquakes have included a sharp tremor, with an estimated magnitude of 5, that struck New York City on August 10, 1884—felt from Washington, D.C., to Maine, the quake hurled horsecars from their tracks on Fifth Avenue and sent a two-ton safe sliding across the lobby of the Manhattan Beach Hotel in Brooklyn. In 1929 a powerful tremor toppled 250 chimneys in Attica, New York, and in 1983 the Adirondack Mountains of New York were rocked by a quake of magnitude 5.2. In Massachusetts, earthquakes occurred near Plymouth in 1638, at Cape Ann 80 kilometers (50 miles) northeast of Boston in 1755, and in Cambridge in 1775. The 1755 quake, felt from Nova Scotia to South Carolina, shattered roof gables on colonial homes, destroyed 1500 chimneys, and snapped the gilded-cricket weather vane off the roof of Faneuil Hall in Boston. In Canada, an earthquake emanated from fractured rocks 20 kilometers (12 miles) beneath the evergreen forests of the small lumber town of Chicoutimi, 150 kilometers (100 miles) north of the city of Quebec, on November 25, 1988; measuring a surprising 6.0 on the Richter scale, it could be felt as far away as Washington, D.C. One of the most powerful eastern earthquakes, estimated at magnitude 7, razed much of Charleston, South Carolina, in August of 1886 and took the lives of 60 of its citizens.

The most powerful recorded seismic events of eastern North America were the New Madrid, Missouri, quakes, which began on December 16, 1811, and lasted for 53 days, ending February 7, 1812. There were three main shocks and more than 1500 powerful aftershocks, of which the most violent was the last—it shook a 2.5-million-square-kilometer (1-million-square-mile) area from Quebec to New Orleans and from the Rocky Mountains to the eastern seaboard. Church bells tolled in Boston, 1600 kilometers (1000 miles) away, windows rattled violently in Washington, D.C., and pendulum clocks stopped in Charleston. Judging from observations by fur trappers and an eyewitness account from naturalist John Audubon, it is believed that the three principal tremors may have exceeded Richter magnitudes of 8.5.

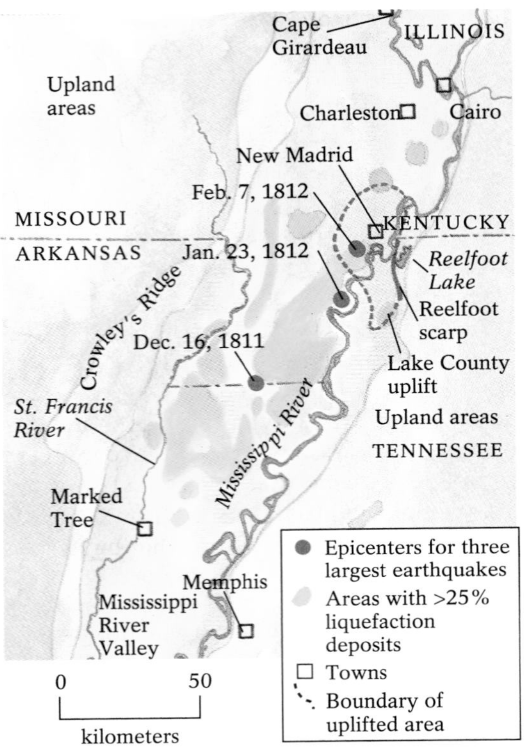

Figure 10-17 The effects of the New Madrid, Missouri, earthquakes of 1811–1812 included liquefaction of sediments, uplift of some areas, and subsidence of other areas to form several new lakes.

Near the estimated epicenter of the New Madrid quake, entire forests were flattened when trees snapped from the violent shaking. Large fissures opened, some too wide to cross on horseback. Geysers of sand, water, and sulfurous gas erupted as the shaking ground compacted. Thousands of square kilometers of prairie land subsided, flooding to form St. Francis and Reelfoot Lakes in northwestern Tennessee as well as swamps that today stretch 300 kilometers (190 miles) from Cape Girardeau, Missouri, to northern Arkansas (Fig. 10-17). Other areas were uplifted, exposing lake beds that have since dried out. The Mississippi River, dammed in places by uplifts and landslides, flowed wildly elsewhere, with waves that overwhelmed many riverboats. The flow of the river even reversed direction where uplifts changed lowlands into highlands. Whole river-channel islands disappeared, and offsets along the river produced new cliffs and waterfalls.

Because they occurred at a mid-plate location and produced a great variety of landscape changes, the New Madrid events are of considerable academic interest to geologists. The study of these events is especially urgent because of the possibility that the area could again experience seismicity of the same magnitude. In the nineteenth century, the Mississippi valley was sparsely populated. The same area today, from St. Louis to Memphis, is home to more than 12 million unprepared citizens.

Coping with the Threat of Earthquakes

Can anything be done to prevent damage from earthquakes, and protect the millions of people in North America and elsewhere who live in vulnerable regions? The U.S. government has prepared a checklist of helpful, logical "do's and don'ts" (Table 10-2), but there is even more that can be done.

Table 10-2 **Earthquake Do's and Don'ts**

What to Do Before an Earthquake

Check for potential fire risks, such as defective wiring and leaky gas connections. Bolt down water heaters and gas appliances.

Know where and how to shut off electricity, gas, and water at main switches and valves.

Place large and heavy objects on lower shelves. Securely fasten shelves to walls. Brace or anchor top-heavy objects.

Do not store bottled goods, glass, china, and other breakables in high places.

Securely anchor all overhead lighting fixtures.

15 Survival Items to Keep on Hand

Portable radio with extra batteries

Portable fire escape ladder for buildings with multiple floors

Flashlight with extra batteries

First aid kit containing any specific medicines needed by members of household

First aid book

Fire extinguisher

Bottled water

Adjustable wrench for turning off gas and water

Smoke detector

Matches

Canned and dried foods for one week for each member of household

Nonelectric can opener

Telephone numbers of police, fire, doctor

Portable stove with propane or charcoal

Cash (banks and automatic teller machines may be closed or inoperable)

What to Do During an Earthquake

If you are outdoors, stay outdoors; if indoors, stay indoors. During earthquakes, most injuries occur as people enter or leave buildings.

If indoors, take cover under a heavy desk, table, bench, or in doorways, halls, or against inside walls. Stay away from glass. Don't use candles, matches, or other open flames either during or after tremor. Extinguish all fires.

If in a high-rise building, don't dash for exits; stairways may be broken or jammed with people. Never use an elevator.

If outdoors, move away from buildings, utility wires, and trees. Once in the open, stay there until shaking stops.

If in a moving car, drive away from underpasses and overpasses. Stop as quickly as safety permits, but stay in vehicle. A car may shake violently on its springs, but it is a good place to stay until tremors stop. When you drive on, watch for fallen objects, downed wires, and broken or undermined roadways.

The Most Common Causes of Earthquake Injuries

Building collapse or damage

Flying glass from broken windows

Falling pieces of furniture, such as bookcases

Fires from broken gas lines, electrical shorts, and other causes, aggravated by lack of water caused by broken water mains

Fallen power lines

What to Do After an Earthquake

Be prepared for aftershocks.

Check for injuries; do not move seriously injured persons unless they are in danger of sustaining additional injury.

Listen to the radio for latest emergency bulletins and instructions from local authorities.

Check utilities. If you smell gas, open windows and shut off main gas valve. Leave building and report gas leakage to authorities. If electrical wiring is shorting out, shut off current at main meter box.

If water pipes are damaged, shut off supply at main valve. Emergency water may be obtained from hot water tanks, toilet tanks, and melted ice cubes.

Check sewage lines before flushing toilets

Check chimney for cracks and damage. Undetected damage could lead to fire. Approach chimneys with great caution. Initial check should be from a distance.

Do not touch downed power lines and objects touched by downed lines.

Immediately clean up spilled medicines, drugs, and other potentially harmful materials.

If power is off, check your freezer and plan meals to use foods that will spoil quickly.

Stay out of severely damaged buildings. Aftershocks may shake them down.

If you live along a coast, do not stay in low-lying coastal areas. Do not return to such areas until local authorities tell you that the danger of a tsunami has passed.

Source: U.S. Federal Emergency Management Agency (FEMA)

Limiting Earthquakes Caused by Humans

Any earthquake-defense plan should start with policies to eliminate quakes caused by humans, such as those that have been initiated by building dams to create large reservoirs. The additional water accumulated in these reservoirs applies considerable weight to underlying rocks and eventually seeps into deep faulted rocks, reducing the friction between adjacent rock masses and activating old, previously stable faults. Dam building may thus stimulate seismic activity in areas that have been inactive for centuries.

The Hoover Dam, impounding Lake Mead, was erected across the Colorado River at the Arizona–Nevada border in an area that had been relatively free of earthquakes. In 1936, shortly after the dam was completed, a series of 600 tremors began that lasted for 10 years. The largest, measuring about 5 on the Richter scale, was felt throughout the region. It is believed that the 42 cubic kilometers (10 cubic miles) of water contained in young Lake Mead had depressed fractured crustal blocks and lubricated subsurface fractures in the bedrock, causing the blocks to slip.

Another way humans may cause earthquakes was revealed during the early 1960s in Denver, Colorado. This community is built over fractured billion-year-old granite that had been quake-free for the 80 years preceding 1962. From 1962 until 1965, however, it experienced 700 tremors measuring from 3 to 4.3 on the Richter scale; the largest caused significant damage. In 1962, engineers at the U.S. Army's Rocky Mountain arsenal had drilled a 3660-meter (12,000-foot)-deep well into faulted bedrock beneath the Denver area, and every month as much as 16 million liters (4 million gallons) of liquid waste were injected into it under high pressure. The high fluid pressure widened bedrock joints and the liquid waste lubricated potential slip surfaces along major faults. When the arsenal was informed of its responsibility for the tremors, it discontinued the project and the level of quake activity was dramatically reduced (Fig. 10-18).

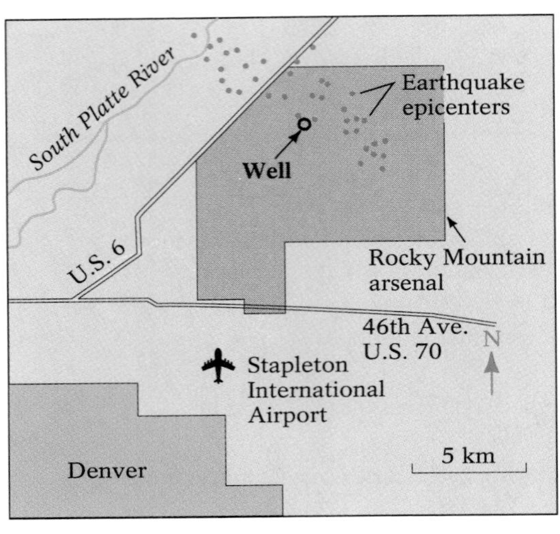

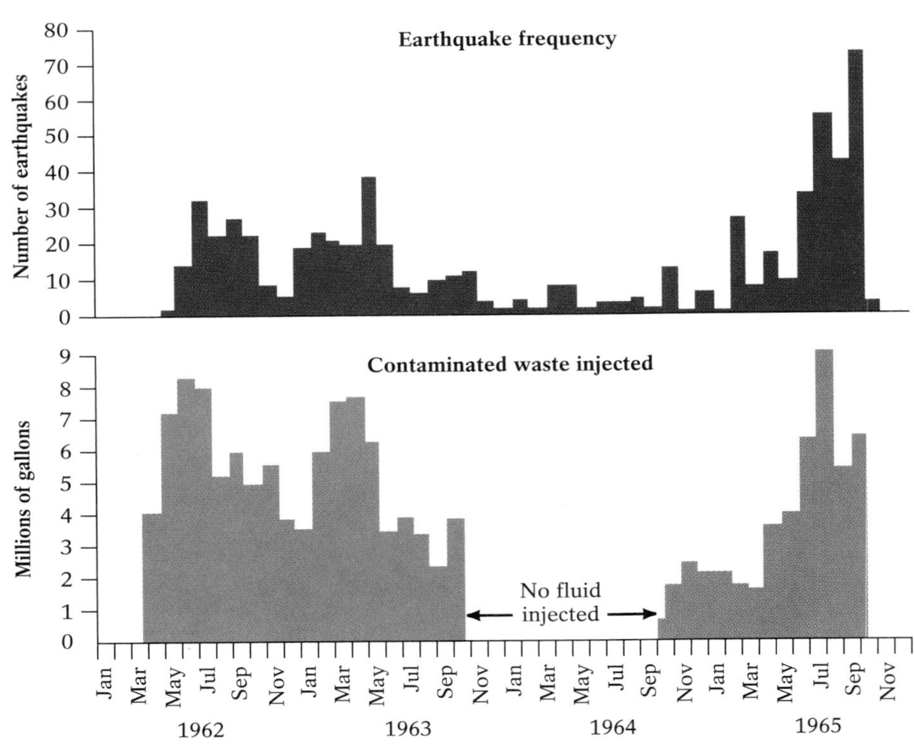

Figure 10-18 Graphs correlating a series of earthquakes that occurred in Denver during the early 1960s with the high-pressure disposal of liquid waste deep in the region's underlying fractured bedrock during those years. Note how the frequency of seismic events coincided with the timing and quantity of waste disposal.

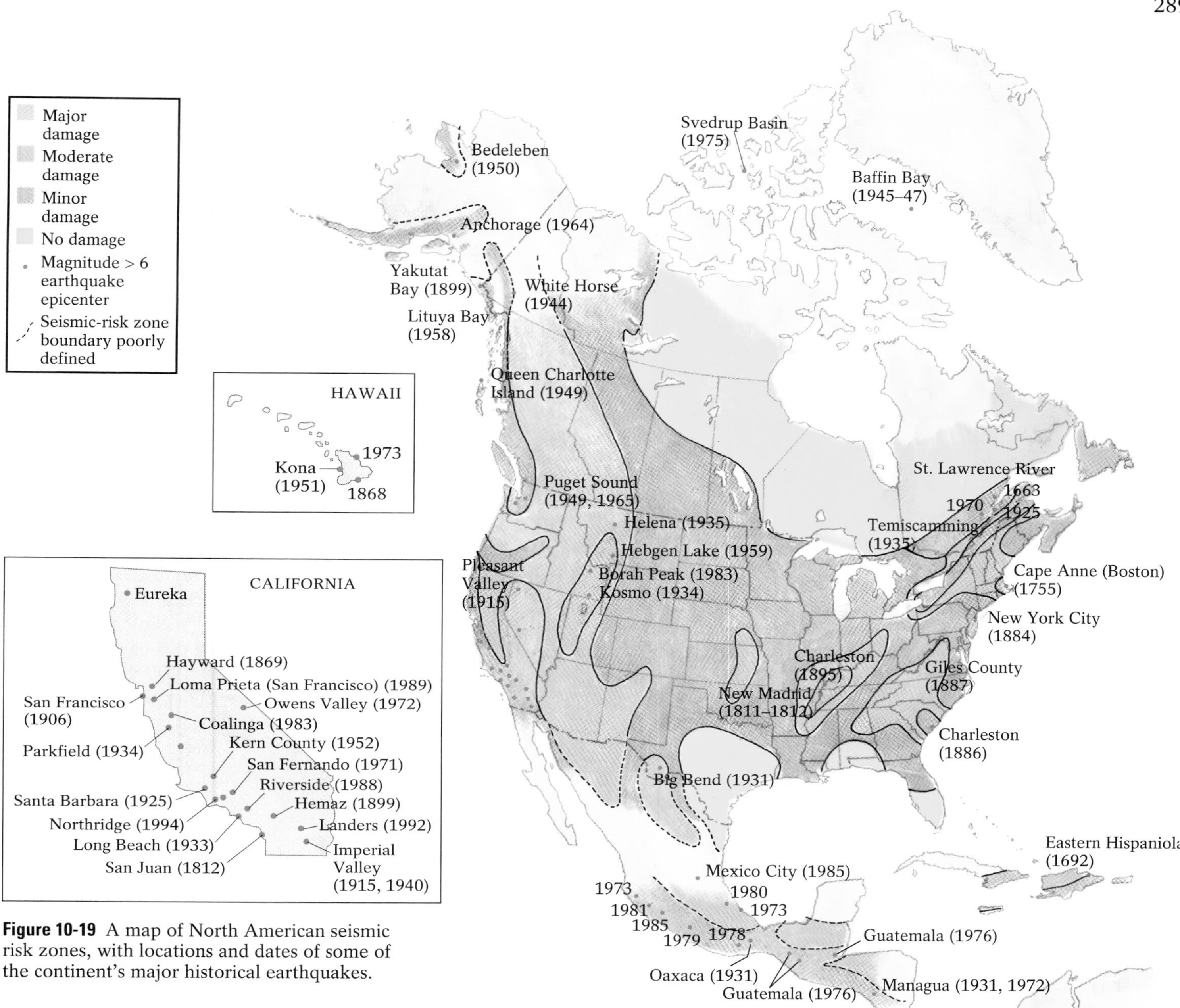

Figure 10-19 A map of North American seismic risk zones, with locations and dates of some of the continent's major historical earthquakes.

Determining Seismic Frequency

Seismologists study the earthquake history of an area to determine the statistical probability of future earthquakes there. If you hear that there is a 40% chance of a magnitude-7.5 earthquake striking the Palm Springs section of the San Andreas fault within the next 30 years, for instance, that "prediction" is based on the frequency and magnitude of past events in that area. Maps such as that in Figure 10-19, though common, have limited usefulness in making predictions. They identify areas that have experienced quakes in the past and that should therefore take earthquake precautions, but they do not indicate the frequency of destructive tremors; for example, Charleston, South Carolina, with its one large 1886 event, is represented in the same way as southern California, with its frequent tremors. Moreover, because such a map relies on historically recorded events, its information is too recent to tell us much about the pattern of seismic events over thousands of years. To develop a long-term history for North America, we must attempt to date fault and quake activity that preceded the 400 to 500 years of recorded North American history.

In California, for example, substantial written records exist for only the past 150 years, and constitute too brief a set of earthquake observations to determine past quake frequency. To date earlier seismic events in California, we first look for geological evidence of a fault, such as topographic breaks in the landscape, linear valleys that may follow a fault line, areas of crushed rock, small lakes that may have formed where impermeable crushed rock prevents surface water from seeping underground, or any linear features that have been offset by fault movements (such as the orange grove in Figure 1-20 and the creek in Figure 10-11).

When a fault is found, geologists look for evidence of all past seismic events related to the fault and try to date them to determine the fault's *recurrence interval*, the average amount of time between its associated quakes. One way of doing this is to examine the geologic strata composing the area. Trenches excavated across the San Andreas fault, for example, reveal a complex stratigraphy that includes vertically offset organic deposits. Carbon-14 dating and the principle of cross-cutting relationships determine when the offsets occurred, enabling us to estimate when their related seismic events may have taken place (Fig. 10-20). There is evidence northeast of Los Angeles that nine major events have taken place there during the past 1400 years, with an average recurrence interval of 160 years. This fact is unnerving to southern Californians, given that the last great earthquake in that area occurred in 1857.

Figure 10-20 Trenches excavated across a fault expose the stratigraphy of the fault zone. According to the principle of cross-cutting relationships, any movement along a fault must be more recent than the age of the youngest rock or sediment that it displaced, and older than the oldest rock or sediment that it did not displace. Thus, each of the seismic events shown here can be dated relative to one another, and its absolute age estimated in relation to the organic deposit. Photo: Vertically offset deposits near Palmdale, California, showing evidence of a previous earthquake there.

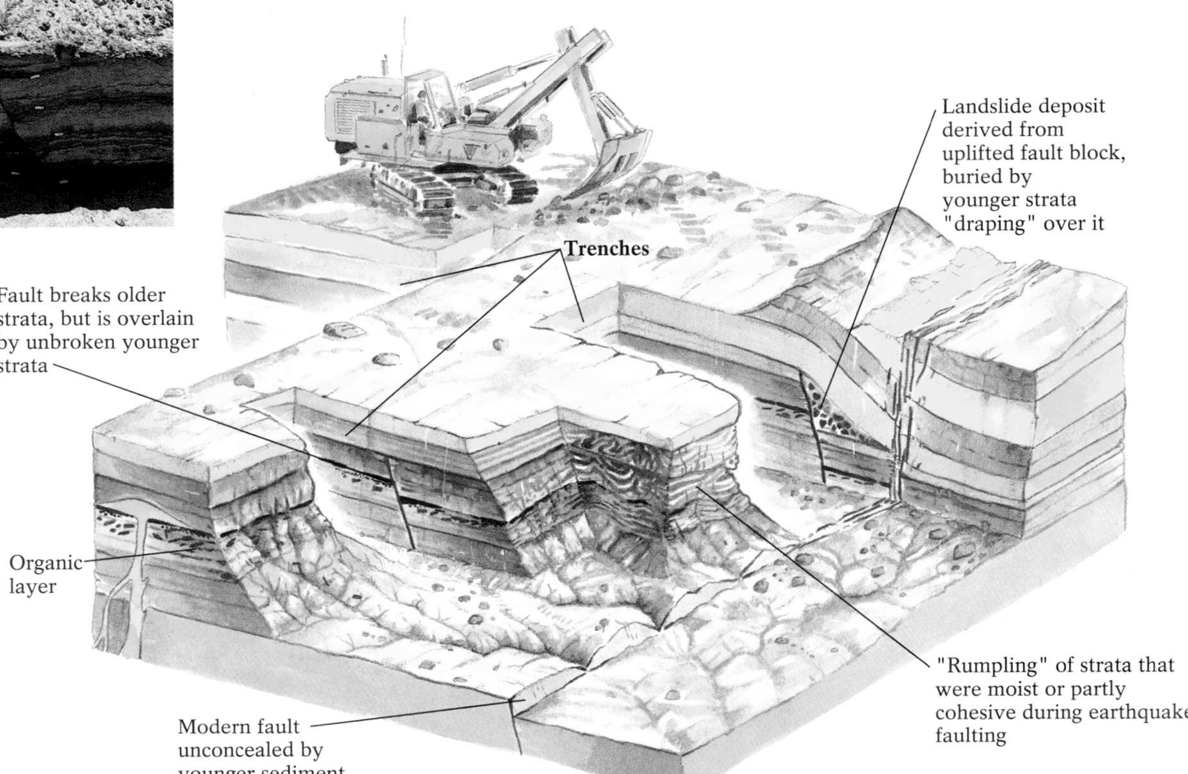

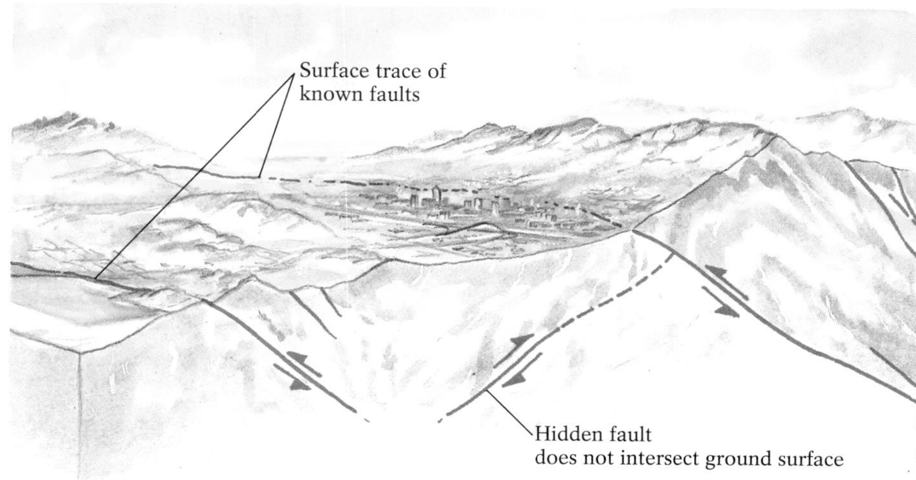

Figure 10-21 Hidden faults, because they are not evident at the surface, often go undetected until an earthquake occurs. One such hidden fault is believed to have caused the Whittier earthquake of 1987, which claimed eight lives and caused $360 million dollars in damage in Los Angeles' Pasadena area. We know now that Los Angeles is underlain by numerous hidden faults, such as those recently discovered at Echo, MacArthur, and Elysian Parks.

Seismologists can estimate the frequency of past earthquakes only for faults they can see. Even with lasers that can measure fault displacements of hundredths of a millimeter and ultrasensitive seismometers that can detect the faintest tremors on the opposite side of the Earth, undiscovered faults may still lie beneath downtown Los Angeles. After an earthquake shook Whittier, California, on October 1, 1987, local geologists were able to identify numerous hidden faults in the Los Angeles area (Fig. 10-21). These had little or no surface expression until the quake exposed their location through compressed pavement, uplift, and fractures at the surface. One such hidden fault, part of the Oak Ridge fault system, was responsible for the quake that devastated the northern San Fernando Valley in January 1994. Although its existence had been suspected for at least half a century, its extent and depth were unknown until surface damage and aftershocks gave clues to its location. Another recently discovered fault, the Elysian Park fault, runs along Wilshire Boulevard through Hollywood toward Santa Monica, threatening such well-known landmarks as Dodger Stadium, Beverly Hills, and the world-famous "HOLLYWOOD" sign. The discovery of two additional hidden faults in Los Angeles, one running through Echo Park and the other through MacArthur Park, has further heightened the city's awareness. Although the San Andreas is more likely than these recently discovered faults to produce a powerful earthquake of, say, 7.5 to 8.0 in magnitude, it is located 50 kilometers (30 miles) east of downtown Los Angeles; consequently, its earthquakes may be less damaging than lower-magnitude tremors generated on hidden faults directly beneath the city.

Building in an Earthquake Zone

Safe use of earthquake-prone regions requires planning where to locate various structures, and engineering those structures to survive the regions' expected quake magnitudes. Certain structures, such as nuclear power plants and large dams, are considered *critical facilities* because their destruction by an earthquake would take a heavy toll in human lives. Critical facilities must be sited on stable ground, and as far as possible both from active faults and from major population centers and heavily traveled highways. Every effort must be made to locate hidden faults. Extensive geological surveying must also precede the siting of public facilities that would be essential for coping with a quake disaster, such as the fire and police departments, hospitals, other civil-defense installations, and the local communications network.

Ground Stability When seismic waves pass through adjacent areas of solid bedrock and soft, wet sediment, the shaking is generally more pronounced in the sediment. While the bedrock shakes in short sharp jolts, the loose sediment develops a continuous rolling motion that can magnify the quake's intensity. Structures built on unconsolidated materials therefore typically suffer far more damage than comparable ones located on solid bedrock (Fig. 10-22). During the 1906 San Francisco quake, for instance, the buildings on wet bay mud and landfill sustained four times the damage of those on nearby bedrock; during the 1989 Loma Prieta earthquake, the waterfront areas again suffered the greatest damage. The destruction seemed to hopscotch around the Bay Area, finding those structures sited on the most unstable materials.

Figure 10-22 A photo showing differential earthquake damage in Varto, Turkey, in 1953. Variations in the destructiveness of earthquake shaking are often controlled by the underlying geologic materials: The collapsed houses shown here were built on the unconsolidated sands of an old river channel, while the surviving houses rest on a solid bedrock bench.

The epicenter of the 1985 Mexico earthquake was near Acapulco, yet the damage was greater and the death toll far higher in more-distant Mexico City. Acapulco is built on solid volcanic bedrock, which shook far less violently than the ancient lake clays on which Mexico City is situated. The California and Mexico City events demonstrate that in quake-prone areas, geological surveys are essential to identify solid bedrock upon which communities can be safely built. Potential home buyers should inquire about the ground beneath their prospective property.

Structural Stability The amount of damage sustained by a building during an earthquake is related to the magnitude and duration of the quake, the nature of the ground on which the building stands, and the building's design. The earthquake that destroyed several cities in Armenia in 1988, taking more than 50,000 lives, was of the same magnitude as California's 1989 Loma Prieta quake, which took 65 lives. India's catastrophic earthquake of 1993 was of lower magnitude than either of these, but took almost 18,000 lives. The greater damage and loss of life in Armenia and India was largely due to the poor construction of the buildings there. San Francisco's structures were built to meet strict standards, and with the expectation of having to survive high-magnitude quakes; constructed using superior designs, methods, and materials, they were better able to withstand the shaking when it occurred.

Figure 10-23 Many of the structures that collapsed during India's earthquake of 1993 were constructed of stones and other inflexible materials.

Figure 10-24 The Transamerica Building, in San Francisco, California. The pyramidal shape of this structure makes it less likely than more traditional buildings to suffer extensive earthquake damage.

The most secure structures are built of strong and flexible but relatively light materials. Wood, steel, and reinforced concrete (containing steel bars) tend to move *with* the shaking ground. Unreinforced concrete, heavy masonry, stucco, and adobe (mud brick) are generally inflexible. Their components move independently and in opposition to the shaking, battering one another until the structure collapses (Fig. 10-23).

Well-designed and engineered skyscrapers may be the safest buildings. A building with structural elements (floors, walls, ceilings) that are joined firmly may sway in a quake like a tree in the wind, but it is not likely to collapse. Distributing a building's weight with the greater proportion at the bottom also enhances safety. The momentum that develops when a top-heavy building swings from side to side may topple it. The pyramidal Transamerica Building in San Francisco (Fig. 10-24), which tapers progressively toward the top and therefore has most of its weight concentrated in its lower floors or in its below-ground foundation, is able to sway with the shaking Earth without developing momentum near its top. Buildings that suffer most during large tremors are those with unsupported roofs that extend across a broad span, such as bowling alleys, supermarkets, and shopping malls.

Another general rule about building in an earthquake zone is the simpler, the better. Projections or decorative elements that are not securely attached may loosen or fall during a tremor. Many nineteenth-century buildings feature projecting decorations secured only by dried-out mortar. Los Angeles and San Francisco have recently begun requiring owners to secure such features with special braces.

Earthquakes also play havoc with highways, endangering motorists and severing vital transportation lifelines. We know now that elevated roadways must be strongly connected to their support columns, that the columns themselves must be of equal height so that they flex at the same rates, and that, wherever possible, these roadways should be sited on solid bedrock. (Unfortunately, North America's greatest period of highway building was during the 1950s and 1960s, when less was known about earthquake engineering.) The collapse of Oakland's Nimitz Freeway during the 1989 Loma Prieta earthquake demonstrates that multilevel roadways are particularly inappropriate in quake-prone areas, although California has many such structures. Costly efforts to increase the strength and stability of California roadways and bridges after the Loma Prieta earthquake did pay off during the 1994 Northridge earthquake, however; among the dozens of L.A.'s bridges and overpasses that had been upgraded, only one collapsed in that quake.

Earthquake Prediction: The Best Defense

In March of 1966, two powerful tremors rocked a broad area 200 kilometers (120 miles) south of Beijing, the capital of the People's Republic of China. The death toll and property damage were so appalling that Premier Chou En Lai declared the "People's War on Earthquakes." For his army, Chou recruited 10,000 professional seismologists and 100,000 amateurs from the ranks of China's college students, telephone operators, and factory workers. Two hundred fifty regional seismic stations were set up to coordinate the huge volume of data collected by these observers. The effort paid off handsomely on February 4, 1975.

In late 1974 and early 1975, residents of Liaoning province in southern Manchuria reported drawing well water that bubbled with high levels of radon gas; among other precursor signals of an impending earthquake, thousands of snakes crept out of winter hibernation and froze to death on icy roads (see p. 297). Chou's army focused their attention on Haicheng, the region's population center with 90,000 inhabitants. On February 1, a series of seismic spasms struck the city; the largest, on the morning of February 4, measured 4.8 on the Richter scale. Early that afternoon Chinese seismologists issued an earthquake advisory, ordering residents to extinguish their cooking fires and evacuate to the city's parks and fields. At 7:36 P.M., a powerful tremor, 7.6 on the Richter scale, destroyed almost 90% of the tile-roofed dwellings of Haicheng. A disaster of enormous proportions was averted, demonstrating the effectiveness of accurate earthquake prediction.

Figure 10-25 Citizens of Beijing living in tent cities on the streets, after the earthquake of 1976.

One year later, however, in May of 1976, the same network of professional and amateur seismologists failed to predict the time and location of the greatest natural disaster of this century. With a Richter magnitude of 8.0, this earthquake struck the coal-mining and industrial center of Tangshan, located about 150 kilometers (90 miles) east of Beijing, taking 650,000 lives and injuring 780,000. Fearing aftershocks, Beijing's residents lived on the streets, in fields, and in parks for three months (Fig. 10-25). These two events—one glorious hit, one tragic miss—characterize the current state of earthquake prediction.

To justify an evacuation such as the one in Haicheng, seismologists monitor and analyze certain pre-quake signals to make short-term predictions on the order of months, weeks, or days. Long-term prediction, on the order of decades or years, is often used to establish building codes and construction standards and to site critical facilities such as nuclear power plants. Recent efforts at long-term prediction have focused on areas where earthquake activity has been unexpectedly absent. Recent efforts at short-term prediction have involved monitoring the buildup of stress in fault zone rocks.

Long-Term Prediction from Seismic Gaps When two adjacent plates move, strain energy accumulates in the rocks along their boundaries. Although this energy probably accumulates uniformly throughout the seismically active zone, it may be released irregularly, with earthquakes occurring in different places and at different times. Some fault segments release their strain energy continuously, as fault blocks move without locking up. This nearly constant motion, called **tectonic creep**, prevents the large buildup of strain energy that causes great earthquakes, but it can gradually deform or break any features subjected to it (Fig. 10-26). The creeping Hayward fault in Berkeley, California, runs the length of the football field at Memorial Stadium on the University of California campus, bending the seats in the bleachers and damaging a large drainage culvert beneath the field.

Figure 10-26 A low wall bent by tectonic creep along the Calaveras fault in California. The creeping of the fault continuously confounds road-construction engineers, who can't seem to keep the curbs straight.

Locked fault segments that have not experienced seismic activity for a long time are prime candidates for major earthquakes. These locked segments are known as **seismic gaps**. Once a seismic gap is identified, geologists estimate the rate of plate movement and determine the amount of strain energy built up in the fault segments—the longer the period of seismic "silence," the more strain energy is being accumulated, and the greater the eventual earthquake is likely to be. Figure 10-27 shows the areas in the Western Hemisphere that have not suffered their expected share of earthquakes in recent times.

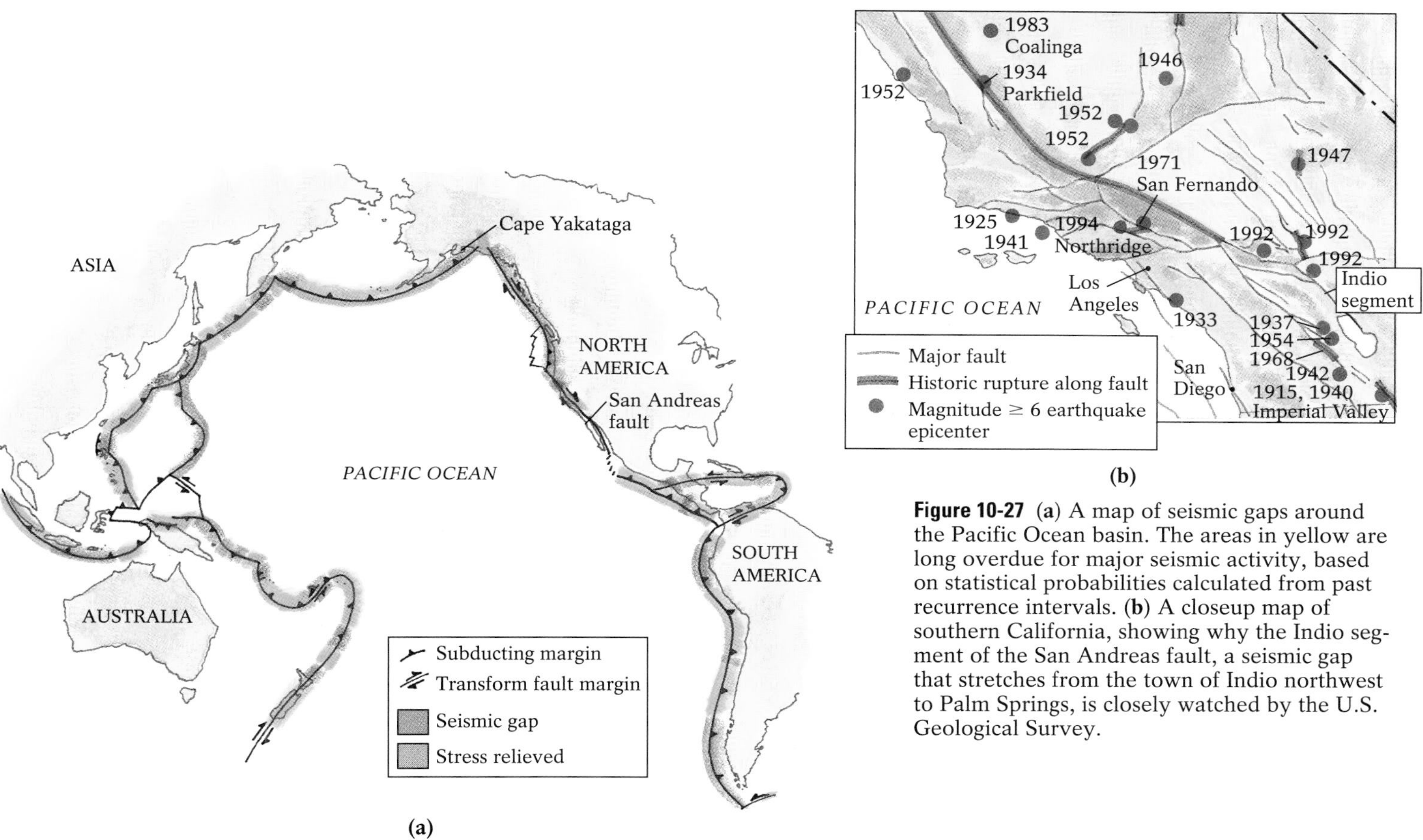

Figure 10-27 (a) A map of seismic gaps around the Pacific Ocean basin. The areas in yellow are long overdue for major seismic activity, based on statistical probabilities calculated from past recurrence intervals. (**b**) A closeup map of southern California, showing why the Indio segment of the San Andreas fault, a seismic gap that stretches from the town of Indio northwest to Palm Springs, is closely watched by the U.S. Geological Survey.

In southern Alaska, the area around Cape Yakataga has been unnervingly quiet seismically for more than a century. Based on the amount of accumulated strain in the rocks, geologists predict that a magnitude-8 earthquake is due there soon. Fortunately, the area is sparsely populated. The entire stretch of the San Andreas fault in densely populated southern California has also been seismically quiet for too long. Statistically, it is considered a strong possibility that a magnitude-8 earthquake will occur in southern California within the next 30 years. (Offset streams and other strain indicators along the San Andreas suggest that the long-overdue "Big One," when it occurs, may generate as much as 11 meters (35 feet) of instantaneous displacement.) In 1992, a series of magnitude-6 quakes around the Big Bear Lake area east of Los Angeles aroused concern that the rocks there are beginning to yield to their accumulated strain energy.

Short-Term Prediction by Measuring Dilatancy When rocks are stressed, they develop countless minute cracks long before they rupture completely. These cracks begin to appear when accumulated strain energy approaches about one-half the amount needed to produce rock failure. As cracking continues, rocks expand in volume, or *dilate*. **Dilatancy** produces a number of side effects that geologists can monitor to make short-term earthquake forecasts (Fig. 10-28):

1. *Swarms of micro-earthquakes.* As hairline cracks open and rocks dilate, countless micro-earthquakes (magnitude less than 1) occur that are detectable only with the most sensitive seismometers. A swarm, or cluster, of these tremors sometimes precedes a major quake. Micro-earthquakes that are followed by a more powerful quake are called **foreshocks**.

2. *Tilt or bulge in rocks.* The surface of a dilated rock may bulge upward. Sensitive tiltmeters measure changes in surface slope, and frequent land surveys detect changes in ground elevation. In 1964, a major earthquake in Niigata, Japan, was preceded by 10 years of surface bulging near the quake's epicenter.

3. *Changes in seismic wave velocity.* When rocks dilate and cracks develop, the velocity of any seismic waves passing through them slows. Later, if groundwater fills the cracks, the seismic waves speed up again because water transmits seismic waves more efficiently than does air in the cracks. When rocks begin to fracture, seismologists determine the extent and rate of fracturing and the presence of groundwater by periodically setting off small explosions nearby and monitoring any changes in seismic wave velocities.

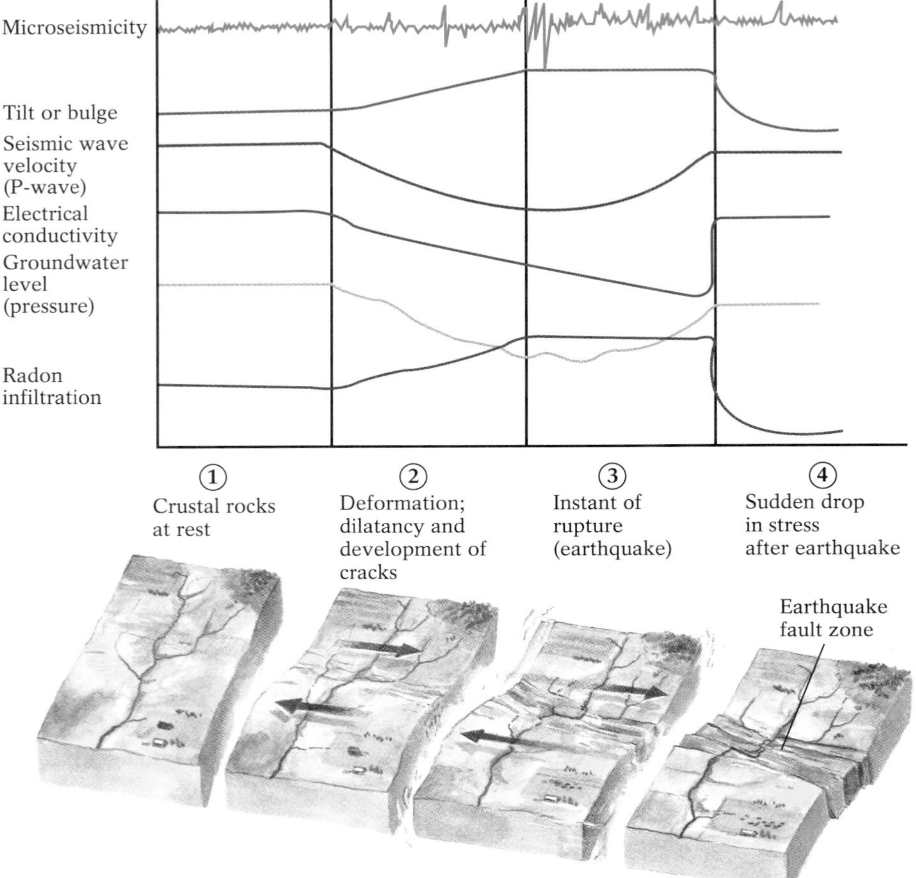

Figure 10-28 Some phenomena caused by the dilatancy of highly stressed rocks, and their relationship to seismic activity.

4. *Variations in electrical conductivity.* Rock is a relatively poor conductor of electricity, and becomes an even poorer one when dilated by air-filled cracks. But when water, a good conductor, enters a network of connected fractures in a stressed rock, electrical conductivity increases, sometimes preceding an earthquake by days or hours. Thus, improved electrical conductivity may indicate that water is filling new fractures in highly stressed rocks that are about to fail and generate an earthquake.

5. *Anomalous radiowave signals.* Geologists have sometimes noticed that there are fewer radio waves in the atmosphere several days before an earthquake. They hypothesize that as underground electrical conductivity increases before an earthquake, it draws radio waves from the atmosphere into the ground. In May 1984, a marked reduction in radio waves preceded by 6 days a magnitude-6.2 earthquake near Hollister, California. Proponents of radiowave monitoring claim that it will someday provide short-term warning—as little as 3 to 5 days—of impending earthquakes.

6. *Changes in groundwater level and chemistry.* As rocks dilate and groundwater seeps into the fresh cracks, the water level in nearby wells may drop perceptibly, and some wells may even run dry. Groundwater chemistry is monitored as well, to detect increased proportions of elements that normally reside tens of kilometers beneath the surface but which rise and infiltrate the groundwater in surface wells after stressed rocks dilate and crack. Prior to recent powerful quakes in China and Armenia, geologists observed a sharp increase in radioactive radon gas dissolved in well water near the epicenters. The radon, produced by the radioactive decay of uranium in deep granitic rocks, is believed to have migrated through fractures in the dilated rocks into the water of near-surface wells. Constant monitoring of the radioactivity of groundwater may therefore provide pre-quake warnings by signaling increased dilatancy in area rocks.

Unusual Animal Behavior When dilatancy occurs, the cracking rocks evidently emit high-pitched sounds and minute vibrations imperceptible to humans but noticeable by many animals. Unusual animal behavior has often been observed by scientists and farmers shortly before earthquakes. Dogs have been known to howl incessantly, as they did in the San Francisco streets the night before the great 1906 earthquake. Cattle, horses, and sheep refuse to enter their corrals, and ordinarily timid rats leave their hideouts and make their way fearlessly through crowded rooms. Shrimp crawl up onto dry land, fish jump from water, ants pick up their eggs and migrate, and ground-dwelling birds, such as chickens, roost up in trees. In cold weather, snakes leave warm subsurface hibernation dens only to freeze on icy ground. Such episodes of extraordinary animal behavior may serve as pre-quake warnings.

Predicting Parkfield Earthquakes Consider the following series of numbers: 1881 . . . 1901 . . . 1922 . . . 1934 . . . 1966. Those are the years in which Parkfield, California, was struck by a major earthquake on its segment of the San Andreas fault. The sequence would have looked better if the 1934 was 1944, but the U.S. Geological Survey, overlooking the anomaly, concluded in 1988 that Parkfield was in a seismic cycle of 20 to 22 years. The Parkfield earthquakes had been remarkably similar in nature and magnitude. Each began with foreshocks that struck within 2 kilometers (1.25 miles) of the epicenter of the main shock, and all had Richter magnitudes in the 5-to-6 range. For these reasons, Parkfield was the subject of the first long-term earthquake prediction made by the U.S. Geological Survey in 1988: that there was a 95% statistical probability that a quake of magnitude 5.5 to 6.0 would occur there before the end of 1993.

Parkfield, a tiny hamlet of 34 citizens, has been overrun by geophysicists on 24-hour call since the mid-1980s. A large concentration of instruments has been installed to detect foreshocks and to document all of the possible precursor signals that may accompany a major quake. On November 13, 1993, an earthquake that measured 4.8 on the Richter scale struck Parkfield, prompting the U.S. Geological Survey to issue a high-level alert on November 15 that predicted a magnitude-6 earthquake within 72 hours. After three days passed and the quake did not materialize, the alert was canceled, signaling that seismologists had yet to develop the ability to accurately predict even the most "predictable" of earthquakes.

Can Earthquakes Be Controlled?

Although we cannot slow or stop the motion of the Earth's tectonic plates, we may someday be able to control or prevent earthquakes by reducing the energy buildup along segments of individual faults. The U.S. Geological Survey has drawn up a plan to prevent the San Andreas fault from locking by converting dangerous seismic gaps to zones of less-threatening tectonic creep (Fig. 10-29). Hypothetically, a stressed segment could be temporarily locked in place by drilling deep wells at each end of the segment and withdrawing the lubricating pore water from each end. Then, ever so cautiously, water would be injected into a well in the center of the segment. In theory, this would lubricate the fault so that its energy would be released gradually by tectonic creep, and only between the locked ends of the controlled segment. By proceeding in small segments along the entire fault, pumping water into locked areas in turn, it may be possible to dissipate much of the fault's energy before it accumulates to deadly levels.

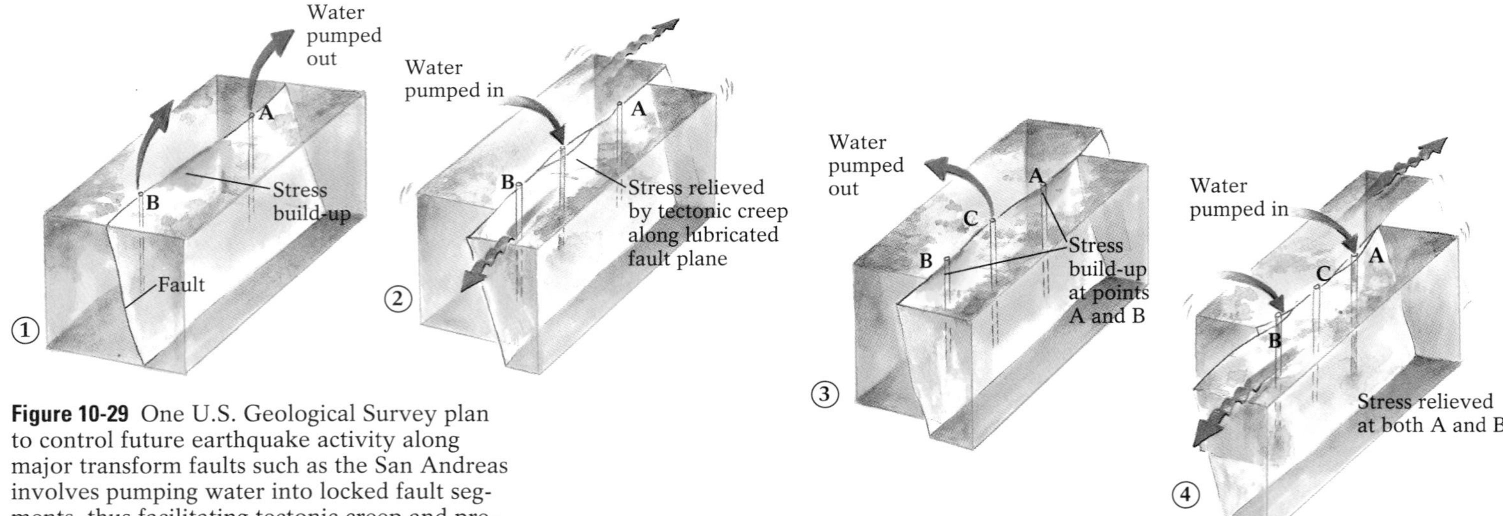

Figure 10-29 One U.S. Geological Survey plan to control future earthquake activity along major transform faults such as the San Andreas involves pumping water into locked fault segments, thus facilitating tectonic creep and preventing a major buildup of strain energy.

Is such an earthquake-prevention strategy really plausible? No one knows yet if injection of water would cause only small displacements. What if the proposed procedure instead unleashed the accumulated energy of more than 130 years and caused the next great quake in the Los Angeles area? The plan's supporters believe it deserves serious consideration *after* the next San Andreas earthquake, so the injected water could unleash only a modest accumulation of energy.

The Moon: A Distant Seismic Outpost

In addition to our burgeoning knowledge of the pattern of terrestrial earthquakes, geologists are compiling data on extraterrestrial quakes. America's Apollo astronauts installed highly sensitive seismometers on the Moon that have been sending data to earthbound scientists for more than 20 years. The lunar seismometers are so sensitive that they can detect the impact of a softball-sized meteorite that strikes anywhere on the Moon's surface, and many of the recorded "moonquakes" are actually meteorite impacts.

Even so, only about 3000 weak moonquakes (Richter magnitude 0.5 to 1.5) are recorded annually, in comparison to the million or more earthquakes that occur each year. The low magnitude and relative scarcity of moonquakes suggest that plate tectonic processes are largely inactive on the Moon.

Those moonquakes that are not attributable to meteorite impacts appear to be concentrated in the lunar interior, from 600 to 800 kilometers (400 to 500 miles) deep (Fig. 10-30). Most occur when the Moon is closest to the Earth, suggesting that the gravitational attraction of the Earth may induce slight adjustments in deep fractured lunar rocks, causing them to release small quantities of strain energy.

This chapter has presented many of the principles underlying the science of seismology. In Chapter 11, we will explore variations in the behavior of seismic body waves, and what they have taught us about the structure and composition of our planet's otherwise inaccessible interior.

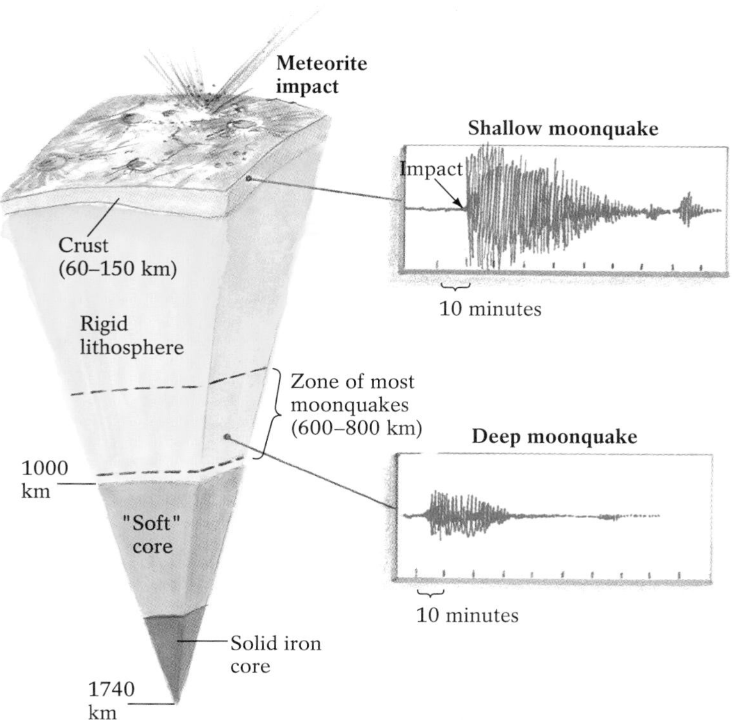

Figure 10-30 The Moon has two centers of seismic activity: one at the surface, where meteorite impacts break rocks and generate small tremors, and one in the Moon's mantle, where fractured rocks may be stressed in response to the Earth's gravitational attraction.

Chapter Summary

Earthquakes are vibrations of the ground that occur when strain energy in rocks is suddenly released, causing rocks to shift along preexisting faults or fracture, creating new faults. The precise subterranean spot where rocks begin to fracture or shift is a quake's **focus**, and the point at the surface directly above the quake's focus is its **epicenter**. A major quake is generally followed by **aftershocks**, which occur as rocks adjust to their new positions.

Earthquake energy is transmitted through the Earth as **seismic waves**. The study of earthquakes and the Earth's interior is known as **seismology**. The most common seismic waves include **body waves**, which transmit energy through the Earth's interior, and **surface waves**, which transmit energy along the Earth's surface. There are two types of body waves: **P waves**, which cause rocks in their path to be alternately compressed and expanded in a direction parallel to the wave's movement; and **S waves**, which cause rocks to shear and move perpendicularly to the wave's direction.

The intensity of earthquakes, quantified as a function of the damage caused in human terms, is estimated with the **Mercalli intensity scale**. A **seismograph** records the arrival times of seismic waves at a specific location by tracing lines on a **seismogram**. The **Richter scale**, which quantifies earthquake magnitude (a measure of the energy released by an earthquake), is based on the amplitudes of those traced lines. Because seismic waves travel at different velocities, the lag in arrival times between the faster-moving P waves and slower-moving S waves at different seismograph locations (at least three) can be used to pinpoint the epicenter of an earthquake. Each year hundreds of thousands of earthquakes are measured that have Richter magnitudes of 1 to 2. Quakes exceeding 3.5 on the Richter scale can be felt by humans, magnitudes of 6 denote major quakes, and once every 5 to 10 years, exceptionally powerful earthquakes occur that exceed 8 on the Richter scale.

The effects of high-magnitude earthquakes include large ground shifts along faults, landslides, **liquefaction** (the conversion of formerly stable fine-grained sediment into a fluid mass), seiches (the back-and-forth sloshing of water in enclosed bays), large fast-moving sea waves called **tsunami**, and fires.

The world's principal earthquake zones are located at or near plate boundaries, with the deepest quakes generally occurring at subduction zones. The foci of earthquakes at subduction zones have a distinctive pattern—they occur at progressively greater depths inland from ocean trenches. The area in which such a pattern occurs is called a **Benioff zone**. The subduction zones surrounding the Pacific Ocean basin, particularly near Japan, Alaska, and the west coasts of North, Central, and South America, and the broad collision zone stretching from the Arabian peninsula to the Himalayas of Asia are the world's most seismically active regions. Powerful historic earthquakes at such localities as New Madrid, Missouri, and Charleston, South Carolina, indicate that as-yet unexplained tectonic forces are also at work within plate interiors.

Human attempts to cope with the threat of earthquakes involve limiting quakes caused by humans, assessing local seismic history and future risk, land-use planning, building quake-resistant structures, and, most importantly, developing ways to predict earthquakes. Long-term predictions, on the order of tens to hundreds of years, focus on the frequency of past earthquake events. Such predictions are based on the premise that while some areas along an active fault are constantly releasing their energy as they inch along by **tectonic creep**, other areas within seismic zones are locked and accumulating a large amount of energy. Such zones, called **seismic gaps**, are considered overdue for an earthquake. Short-term predictions, on the order of months, weeks, or days, center around the phenomenon of **dilatancy**, the volumetric expansion that occurs as minute cracks begin to open in highly stressed rocks. Dilatancy may produce swarms of micro-earthquakes—called **foreshocks** if they precede a larger quake—as well as tilting or bulging of surface rocks; changes in seismic wave velocities, electrical conductivity, and groundwater levels and chemistry; anomalies in radiowave signals; and unusual behavior in animals. These are some of the most important signals used to predict earthquakes.

Lunar seismic activity takes the form of "moonquakes," tremors that are much less frequent and smaller in magnitude than our earthquakes. Moonquakes are believed to be produced primarily by meteorite impacts and by lunar rocks adjusting to the Earth's gravitational pull.

Key Terms

earthquake (p. 268)
focus (p. 268)
epicenter (p. 268)
aftershocks (p. 269)
seismic waves (p. 269)
seismology (p. 269)
body waves (p. 269)
surface waves (p. 269)
P waves (p. 269)
S waves (p. 270)
Mercalli intensity scale (p. 273)
seismograph (p. 273)
seismogram (p. 273)
Richter scale (p. 274)
liquefaction (p. 278)
tsunami (p. 280)
Benioff zone (p. 282)
tectonic creep (p. 294)
seismic gap (p. 295)
dilatancy (p. 296)
foreshocks (p. 296)

Questions for Review

1. Draw a simple diagram showing the difference between the focus and the epicenter of an earthquake.
2. Describe the differences between P and S waves.
3. What is the fundamental difference between the Mercalli intensity scale and the Richter scale? How much more powerful than a magnitude-5 earthquake is a magnitude-7 earthquake?
4. Draw a simple sketch to illustrate how geologists use P- and S-wave arrival times to locate the epicenter of an earthquake.
5. Describe how earthquakes cause tsunami and liquefaction.
6. Why are the foci of divergent-zone earthquakes limited to depths of less than 100 kilometers, whereas those of subduction zones extend to 700 kilometers? Why are there no earthquakes below 700 kilometers? If you heard that an earthquake occurred at a depth of 300 kilometers, what could you conclude about the quake's geological setting?
7. Name three geographical areas that suffer significantly from earthquakes, and specify the plate tectonic setting in which each is located.
8. Discuss how human activity was responsible for a swarm of earthquakes that struck Denver, Colorado, during the early 1960s.
9. How would you determine whether a given location had experienced earthquakes in the past? On what basis could you predict the probability of an earthquake's occurring there in the future?
10. Before moving into a community, what should you look for to determine if it is especially vulnerable to earthquake damage?

For Further Thought

1. Speculate about the nature of the geologic materials (solid bedrock or loose sediment) within the map area in the figure below. Where do you think the major faults lie?

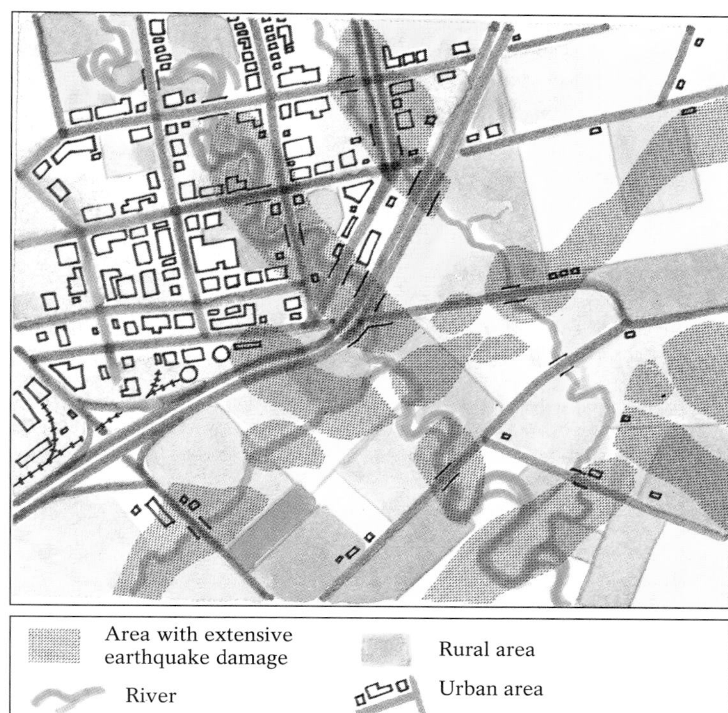

2. If you lived in an area that had suffered no earthquakes during recorded history, but which then began to experience occasional tremors, how would you explain the apparently sudden onset of seismicity? Propose at least three hypotheses.
3. What evidence would you need to estimate the Richter magnitude of an earthquake that occurred before the advent of the seismograph?
4. If you were a city planner in an earthquake-prone region, responsible for coordinating plans for an urban center that will include schools, hospitals, fire and police stations, residential areas, parks, commercial areas, a mass-transit system, power plants, dams, etc., how would you proceed in determining the optimal location of all these components of the community?
5. Suppose all earthquake activity on Earth ceased. What set of geological circumstances could explain such a change?

Investigating the Earth's Interior

Highlight 11-1 Seismic Tomography: A Planetary CAT Scan

Taking the Earth's Temperature

Highlight 11-2 Simulating the Earth's Core in the Laboratory

The Earth's Gravity

Magnetism

Figure 11-1 Because they are carried up from deep in the Earth's mantle, diamonds such as this one provide clues to the nature of the planet's interior.

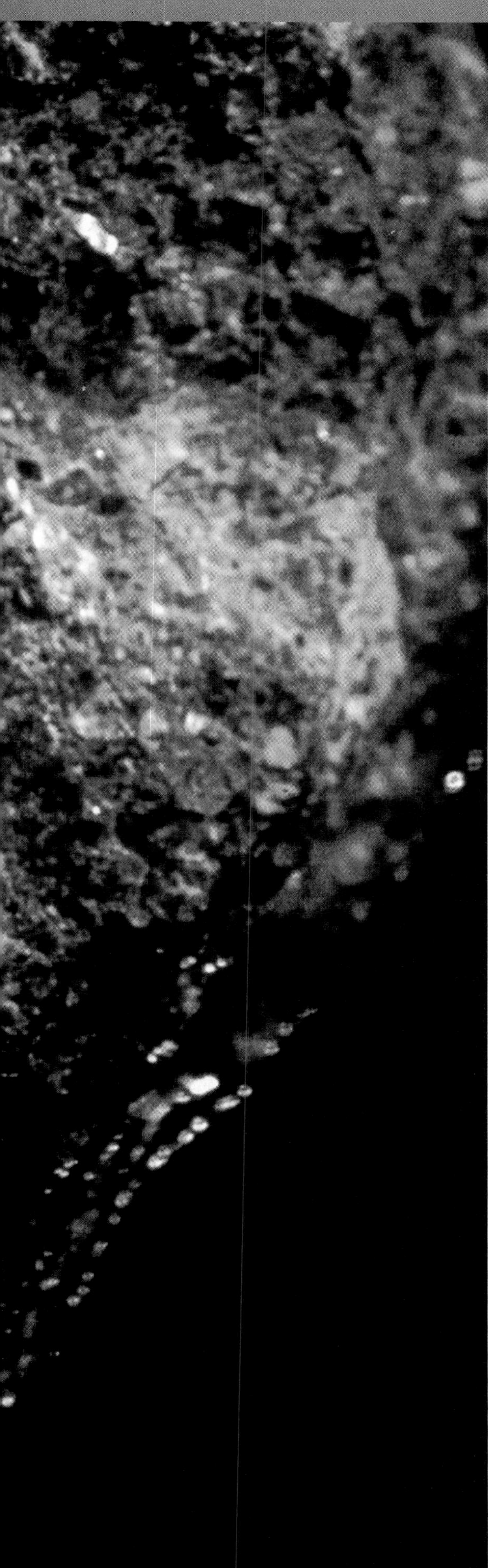

Chapter 11

Geophysical Properties of Planet Earth

What is the Earth's interior like? In 1864, author Jules Verne, in *Journey to the Center of the Earth*, imagined the subterranean world as a yawning abyss filled with giant sea serpents and other grotesque creatures. Today we know what's really in the Earth's interior: layers of rocks and metals, subjected to increasing heat and pressure and becoming increasingly dense at progressively deeper levels. The story we can now tell of the origin and structure of the Earth's interior is at least as exciting, if less fanciful, than Verne's fantastic imaginings.

Human beings have lived in space for months at a time, using telescopic probes to study both our immediate galactic neighborhood and the distant cosmos. They have walked on the Moon, collecting its rocks and conducting numerous experiments on the lunar surface. They have also dived in submarines to the floors of the Earth's oceans to study the secrets held there. The Earth's deep interior, however, remains the one frontier that no one is likely ever to visit in person. The deepest hole ever drilled (begun in northern Russia in 1970, and still in progress) extends only 0.2% of the distance to the planet's center, piercing 13 kilometers (8 miles) into the Earth.

Only in rare circumstances do rocks from deep in the Earth appear at or near the surface. Volcanic eruptions that carry unmelted subterranean rocks within rising deep-origin magmas allow us a few indirect glimpses of the makeup of the Earth's interior (Fig. 11-1). Rocks from as deep as 200 kilometers (125 miles) have been found within diamond-bearing kimberlite pipes, which are products of deep explosive volcanism. Where continental plates have collided in ages past, rocks from tens of kilometers deep—even ultramafic rocks from the upper mantle—have sometimes been transported to the surface along reverse faults. These relatively few exposures of deep-origin rocks confirm that the composition of the Earth's interior varies with depth, but still leave unanswered many questions regarding our planet's internal structure.

This chapter explains how geologists explore the Earth's interior. We will discuss how the behavior of seismic waves is used to probe the deep interior, and how geologists use the information gained to speculate about the Earth's internal layering. We will discuss the Earth's internal heat and its gravitational and magnetic fields, and see how magnetism in rocks is used to trace past movements of the Earth's plates. Because study of the fundamental properties of the Earth's interior depends on the physical principles underlying seismic waves, heat flow, gravity, and magnetism, this branch of the geosciences is called **geophysics.**

Investigating the Earth's Interior

Most of what we know about the Earth's interior comes from analysis of variations in the speed of seismic waves. Like all waves, seismic waves tend to travel in a straight line and at an unchanging velocity as long as they are passing through a homogeneous medium of constant temperature and pressure. But comparison of data collected at seismograph stations around the world shows that seismic waves occasionally slow down or speed up. Their varied speeds indicate that they are passing through materials of varied composition, structure, temperature, and pressure. From this we conclude that the Earth's interior is not homogeneous, nor are its temperature and pressure the same at all depths.

The Behavior of Seismic Waves

In Chapter 10 we learned that P waves compress the materials through which they pass. These may be any medium—solid, liquid, or gas. However, different media conduct P waves at different rates. After being compressed by a passing P wave, an elastic medium such as solid granite returns to its original volume more swiftly than does an inelastic medium such as a liquid or gas. Hence, P waves accelerate when they pass through solids of greater elasticity and slow down when they enter liquids or gases (fluids), which are inelastic.

The inelasticity of fluids also explains why S waves disappear altogether upon entering them. S waves move through material by shearing, a form of deformation that temporarily displaces adjoining material, which returns to its original shape after the wave has passed. Fluids cannot deform and then return to their original shape—they simply flow away from the shearing stress. Consequently, S waves can only pass through solid media, and die out completely when they encounter an inelastic fluid medium within the Earth's interior. The changing velocities of P and S waves as they pass through media of different elasticities determine the paths they take through the Earth's interior, and are the basis of many seismological studies.

Because a seismic body wave emanates from an earthquake's focus in all directions, its effect on the surrounding rock is like an ever-widening sphere traveling out from the focus (see Fig. 10-3). This effect is known as a *wavefront*. Of course, earthquakes emit not just one but a series of waves, so seismic behavior is often illustrated using a concentric series of wavefronts. The direction of wavefront movement is represented by arrows, or *rays*, drawn perpendicular to the wavefront's surface. Visualizing the paths of seismic waves in this way helps us understand many of the phenomena studied by seismologists.

When seismic waves move from one medium to another of different elasticity—such as between the layers of the Earth—their paths suddenly change. If they approach the boundary between the adjoining media at a relatively small angle, they may bounce, or *reflect*, off the surface, much as a flat stone can be made to skip off the surface of a lake. If they approach at a somewhat greater angle, they may enter the new medium and undergo a change of velocity, causing them to bend, or *refract* (Fig. 11-2a).

Figure 11-2 The velocity of seismic waves changes as they pass through materials of different elasticities. (**a**) Seismic waves speed up or slow down as they pass between two layers, causing the waves to bend, or refract. (**b**) Because density and rigidity increase with increasing depth, seismic waves speed up as they descend within individual layers, causing their paths to curve back toward the surface.

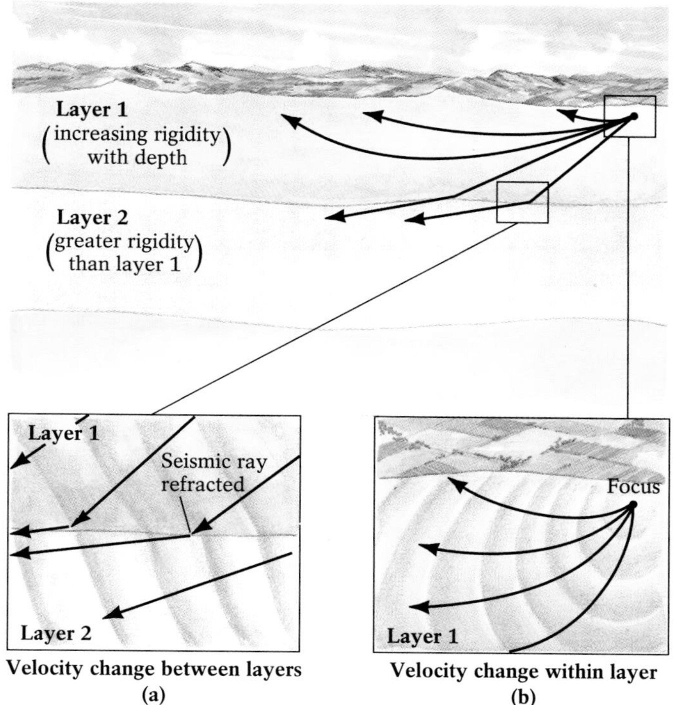

Seismic wave velocities also change as waves travel within individual layers of the Earth's interior. Seismologists have determined that as seismic waves pass through a layer of the Earth's interior their velocity increases with depth, so that the deeper segment of a wavefront travels faster than segments that are less deep, and much of the wavefront curves back toward the Earth's surface (Fig. 11-2b). Given our knowledge that the speed of seismic waves increases with the rigidity and density of the materials through which they pass, we can therefore infer that any given layer of the Earth becomes more rigid or dense with increasing depth.

Using the great volume of data from seismographic readings around the world, seismologists have been able to estimate the density, thickness, composition, structure, and physical state of each portion of the Earth's interior. From this, we have learned that the Earth is composed of the three principal concentric layers examined briefly in Chapter 1—a thin outer crust, a large underlying mantle, and a central core—and have come to better understand the complex nature of these layers (Fig. 11-3).

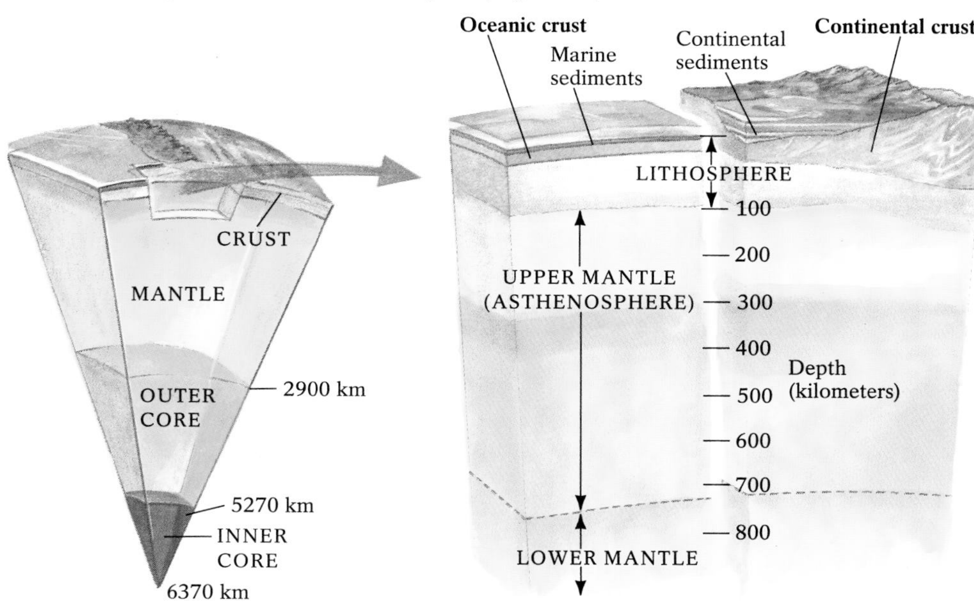

Figure 11-3 As well as delineating the three major components (crust, mantle, and core) of the Earth, seismological studies provide details about the different natures of the layers—such as that continental crust is significantly thicker and less dense than oceanic crust (see text).

The Crust

The crust, the rocky outer segment of the Earth, is composed principally of silicate-rich igneous rocks. The crust is the most accessible of the Earth's layers: We can sample it directly from both the continents and the sea floor by drilling. As a result, we know that continental crust is typically granitic and relatively thick, whereas oceanic crust is typically basaltic and relatively thin.

The crust has also been studied extensively by seismological surveying. In the case of earthquakes occurring 50 kilometers (30 miles) or less below the Earth's surface, seismic waves recorded within about 1100 kilometers (680 miles) of the focus have passed exclusively through the crust. If we know the precise time and location of such an earthquake, the exact distance from the focus to each seismograph, and the exact time when a wave is first recorded on each seismogram, we can determine the wave's velocity through the crust. By comparing velocities derived in this way to those of seismic waves passed through different rock types in laboratory experiments, geophysicists have been able to estimate the composition of deep, otherwise inaccessible regions of the crust. (Seismological measurement of the shock waves generated by underground nuclear testing during the 1960s also provided a great deal of data that has been used to determine the composition and structure of the Earth's crust, thus deriving scientific benefit from the Cold War's arms race.)

When an earthquake occurs in San Francisco, for example, the seismic waves that are recorded at Fresno State University in Fresno, California (about 290 kilometers, or 180 miles, away) have traveled only through continental crust. Because we have known the precise timing of some of San Francisco's past earthquakes and the precise arrival times of the P waves recorded in Fresno, we have determined that P waves move through the granitic continental crust between these cities at a velocity of about 6 kilometers (3.7 miles) per second.

Because basaltic oceanic crust is more rigid than continental crust and therefore transmits seismic waves more efficiently, P waves travel faster through oceanic crust. Seismological studies have shown that P waves travel through basalt and its underlying plutonic equivalent, gabbro, at about 7 kilometers (4.4 miles) per second.

Continental Crust At its thinnest, where plates are being stretched or rifted, continental crust is generally less than 20 kilometers (12.5 miles) thick; where continental plates have collided to form mountains, the crust may be as much as 70 kilometers (45 miles) thick. In North America, the thinnest crust—in the basins between mountain ranges in Nevada, Arizona, and Utah—is only 20 to 30 kilometers (12.5–19 miles) thick. The thickest crust, in the Rocky Mountains of Montana and Alberta and the Sierra Nevada of California, is more than 50 kilometers (30 miles) thick.

Slight variations in seismic wave velocities indicate that the upper portion of continental crust is different in composition from the lower portion. In upper continental crust, P-wave velocity is about 6 kilometers (3.7 miles) per second; this region is a complex jumble of basalt and andesite, numerous granitic intrusions, vast subsurface regions of high-grade metamorphic rocks, and a nearly continuous blanket of sediment and sedimentary rock. When P waves pass through the deepest parts of continental crust, they speed up to about 7 kilometers (4.4 miles) per second. The increased velocity is believed to result from either a pressure-induced increase in the rigidity of the lower crust or a change in composition to metamorphosed gabbro. The average composition of continental crust, with its near-surface felsic rocks and deeper mafic rocks, is probably equivalent to the composition of andesite or diorite. The density of continental rock varies between 2.7 grams per cubic centimeter (170 pounds per cubic foot) (2.7 times as dense as water) near the surface and 3.0 grams per cubic centimeter (190 pounds per cubic foot) near its mafic base.

Oceanic Crust Oceanic crust is more difficult to sample than continental crust because it lies below about 4 kilometers (2.5 miles) of seawater on average. However, new methods of drilling enable geologists to collect ocean-floor rocks, and seismic studies make it possible to infer the structure and composition of deeper segments of oceanic crust. Together, these two types of investigation give us an accurate view of the structure and composition of ocean crust. Whereas a variety of processes have formed continental crust, only eruptions of basalt and intrusions of gabbro at oceanic divergent plate boundaries have formed oceanic crust. As a result, its composition is about the same everywhere. Oceanic crust from top to bottom typically consists of a 200-meter (700-foot) deposit of marine sediment, a 2-kilometer (1.2-mile) layer of submarine pillow basalts, and a 6-kilometer (3.7-mile) layer of gabbro. The average density of oceanic crust is about 3.0 grams per cubic centimeter (190 pounds per cubic foot), which is the approximate density of both basalt and gabbro.

The Crust–Mantle Boundary Often, two sets of seismic waves of the same type and emanating from the same source at the same time arrive at a single seismograph station at different times. Such occurrences led geologists to conclude long ago that such waves traveled at different velocities along different paths. One set took a shorter, more direct route through the Earth's crust; the other took a longer route through the mantle, but actually arrived earlier because its velocity increased as it entered and passed through rigid mantle material (Fig. 11-4).

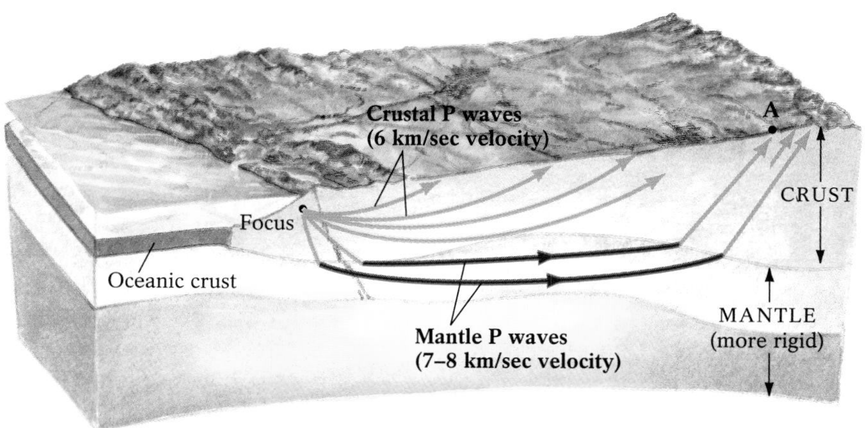

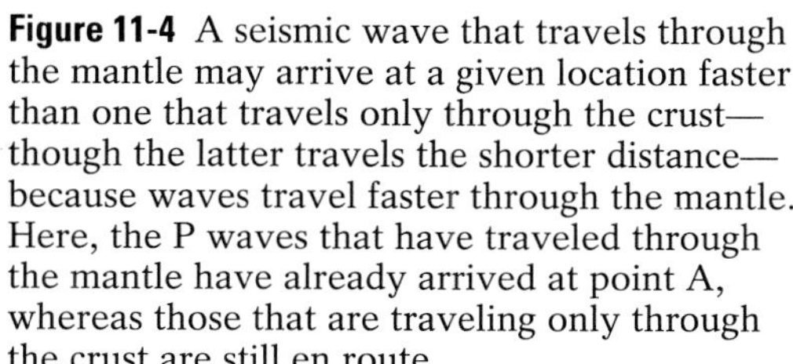
Figure 11-4 A seismic wave that travels through the mantle may arrive at a given location faster than one that travels only through the crust—though the latter travels the shorter distance—because waves travel faster through the mantle. Here, the P waves that have traveled through the mantle have already arrived at point A, whereas those that are traveling only through the crust are still en route.

To understand how a seismic wave that has traveled a longer distance can arrive at a seismograph earlier than one that has traveled a shorter distance, think of the motorist who drives a longer distance on a freeway rather than a shorter distance on a city street with traffic lights, stop signs, and slower speed limits. Unless snarled by rush-hour traffic, a longer freeway trip at higher speeds will still dispatch the motorist to his or her destination sooner than a shorter, slower journey through downtown traffic.

In 1910, Croatian seismologist Andrija Mohorovičić attributed the differing arrival times of seismic waves to a **seismic discontinuity,** an abrupt change in seismic wave velocity that occurs because of a marked change in the composition of materials within the Earth's interior. The boundary between the base of the crust and the top of the more rigid mantle is therefore named for its discoverer—it is known as the Mohorovičić discontinuity, or simply the **Moho**.

The Mantle

The mantle, the layer below the crust, is the largest segment of the Earth's structure. Accounting for more than 80% of the planet's volume, it extends to a depth of about 2900 kilometers (1800 miles). At greater depth, as pressure increases, the density of mantle rock increases as well, from about 3.3 grams per cubic centimeter (210 pounds per cubic foot) at the Moho to 5.5 grams per cubic centimeter (340 pounds per cubic foot) at the mantle–core boundary. Because of the high temperatures and pressures to which they are subjected, mantle rocks deform and actually flow, albeit quite slowly, in convection currents and in response to changes in the weight of the overlying crust. In the short term, however, such as the time it takes for seismic waves to pass through them, mantle rocks behave as elastic solids, allowing the passage of both compressional P waves and shearing S waves. Hence, the mantle displays characteristics of both solid and not-so-solid materials.

The minerals in mantle rocks are dense silicates that contain iron and magnesium, such as olivine and pyroxene (discussed in Chapter 2). The specific minerals and their internal structures vary as pressure and temperature increase with greater depth. We know these mineralogical variations in the mantle exist because P- and S-wave velocities, which generally increase steadily as the waves penetrate more deeply, change abruptly at certain depths within the mantle. These abrupt changes in seismic wave velocity indicate distinct mineralogical changes and mark the boundaries between the sublayers of the mantle: the upper mantle, the transition zone, and the lower mantle.

The Upper Mantle Some of what we know about the upper 400 kilometers (250 miles) of the mantle derives from the chunks of it incorporated into rising magma or the small amounts exposed in mountain ranges. The composition of the upper mantle is dominated by peridotite, a coarse-grained plutonic rock composed largely of olivine, pyroxene, calcium-rich and magnesium-rich garnet, and other metamorphic minerals that form under high-pressure conditions.

At the Moho, the velocity of descending seismic P waves increases to about 8 kilometers (5.0 miles) per second as they pass from the crust to the mantle, and continues to increase within the mantle because its rigidity increases with depth. The rate of increase in velocity is not constant, however: P waves accelerate to 8.3 kilometers (5.2 miles) per second by about 100 kilometers (62 miles) beneath the Earth's surface, but then slow to less than 8 kilometers per second and remain at this lower velocity until they reach a depth of about 350 kilometers (217 miles). This 250-kilometer (155-mile)-thick region—another seismic discontinuity—is known as the **low-velocity zone** (Fig. 11-5). S-wave velocities are also markedly slower in this region.

Seismic waves slow in the low-velocity zone because it is partially molten (10% or less) and thus less rigid than the rest of the mantle. That this is the only portion of the mantle containing any liquid is a result of the unique interplay of temperature and pressure in the low-velocity zone.

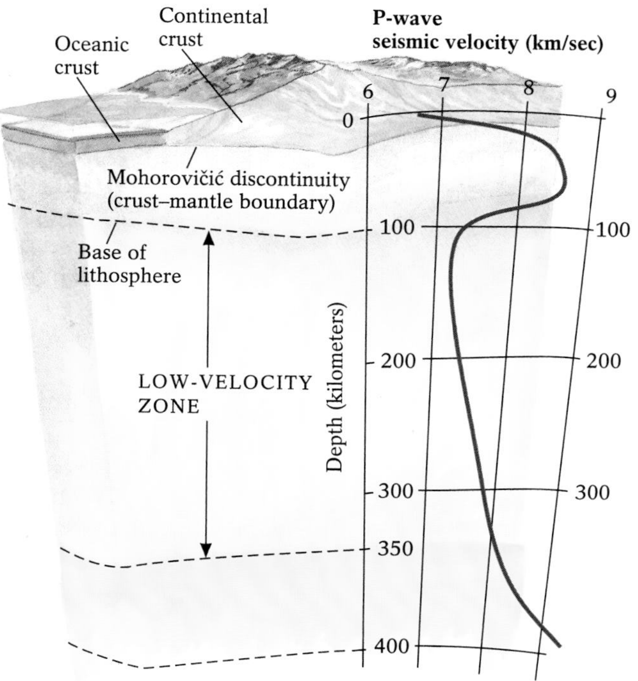

Figure 11-5 The velocities of seismic waves vary across the boundaries of the Earth's interior layers. These seismic discontinuities, such as occur at the Mohorovičić boundary (or Moho) and the low-velocity zone within the upper mantle, result from the changing density, rigidity, and mineralogical composition of the Earth at various depths.

Both temperature and pressure increase with depth in the Earth's interior, although temperature (which melts rocks) initially increases at a faster rate than pressure (which tends to keep rocks solid). Within the low-velocity zone, the greater temperature increase predominates; above and below this zone, however, the effects of pressure overcome the effects of temperature to maintain the solid state of the mantle (Fig. 11-6). As we discussed briefly in Chapter 1, the crust and the uppermost segment of the mantle, which make up the Earth's plates, constitute the lithosphere. The partially melted, low-velocity zone beneath the plates constitutes the asthenosphere. The most significant consequence of the existence of the low-velocity asthenosphere is plate tectonic motion. The slow flowing of this partially melted zone is believed to be what transports the Earth's tectonic plates. (This and other hypotheses proposed to explain plate motion are discussed in Chapter 12.)

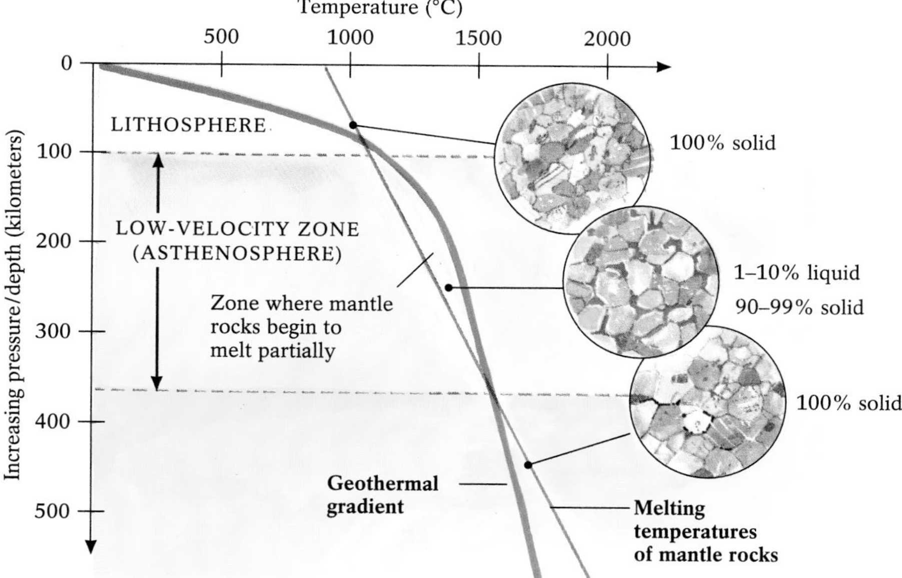

Figure 11-6 The partially melted condition of the low-velocity zone (asthenosphere) of the upper mantle is caused by the temporary dominance of rising temperature over increasing pressure. Above and below this 250-kilometer (155-mile) thick layer, mantle rock is 100% solid.

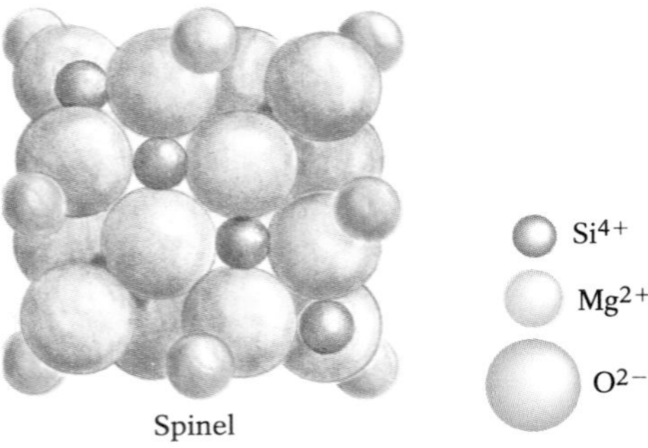

Figure 11-7 Within the mantle's transition zone, increasing pressure causes ultramafic mantle minerals to collapse into denser, more compact structures. In one such phase change, the structure of olivine gives way to that of spinel, a mineral with the same composition as olivine but a more compressed structure.

The Transition Zone Seismic wave velocities increase in the zone below the asthenosphere, due to mineralogical changes that produce an increase in the rigidity of mantle rocks there. No rocks have ever been collected from below the asthenosphere, but laboratory experiments have simulated the pressures and temperatures of this region and tested their effects on ultramafic mantle-type rocks. These experiments suggest that increasing pressure causes changes in the structure of mantle minerals at two points. At about 400 kilometers (250 miles) and again at about 700 kilometers (440 miles) below the Earth's surface, mineral structures are compressed without changing composition. In the first *phase change,* magnesium olivine is compressed to form the structure of the mineral spinel (Fig. 11-7). In the second phase change, spinel and other mantle minerals, such as garnet, are compressed further to form simple metallic oxides such as MgO, FeO, and SiO_2, causing about a 10% increase in density and a corresponding increase in rigidity. These effects define the mantle's **transition zone.**

The Lower Mantle The lower mantle extends from 700 to 2900 kilometers (440–1800 miles) below the Earth's surface. Despite extremely high temperatures, pressure from the weight of the overlying crust and upper mantle is great enough to keep lower-mantle materials solid. This pressure further compresses minerals and crystal structures and, with increasing depth, produces a steady increase in the density and rigidity of lower-mantle rocks. As a result, the velocity of P waves reaches about 13.6 kilometers (8.5 miles) per second at the base of the mantle. Laboratory studies simulating the pressure and temperature conditions of the lower mantle suggest that it probably consists of dense magnesium silicates and oxides, such as ilmenite ($MgSiO_3$) and periclase (MgO).

The Mantle–Core Boundary The boundary between the mantle and the core is about 2900 kilometers (1800 miles) below the surface, almost halfway to the center of the Earth (or about the distance from Boston to Houston). Of all the Earth's internal horizons, the mantle–core boundary—another prominent seismic discontinuity—has the most dramatic effect on seismic waves. This is where the greatest change in the composition and rigidity of Earth materials takes place. At the boundary, the velocity of P waves suddenly decreases from 13.6 kilometers (8.5 miles) per second to 8.1 kilometers (5 miles) per second, and S waves vanish altogether. Because we know that P waves are slowed substantially by materials of low rigidity, and that shearing S waves are not transmitted at all by a liquid medium, we can infer that the outer portion of the core is a liquid.

At the mantle–core boundary, highly compressed ultramafic silicate- and oxide-rich rock gives way to a molten iron–nickel mix with densities ranging from 10 to 13 grams per cubic centimeter (620–810 pounds per cubic foot). Until recently, this boundary was thought to be a distinct, relatively featureless surface. New technology, however, has revealed quite a different picture. An innovative approach to deep-Earth investigation, derived from a sophisticated medical diagnostic tool, is described in Highlight 11-1.

The Core

The Earth's core, slightly larger than the entire planet Mars, accounts for 3486 kilometers (2167 miles) of the planet's total radius of 6370 kilometers (3959 miles). Although it constitutes only one-sixth of the Earth's total volume, the core is so dense that it makes up more than one-third of the planet's

Highlight 11-1 *Seismic Tomography: A Planetary CAT Scan*

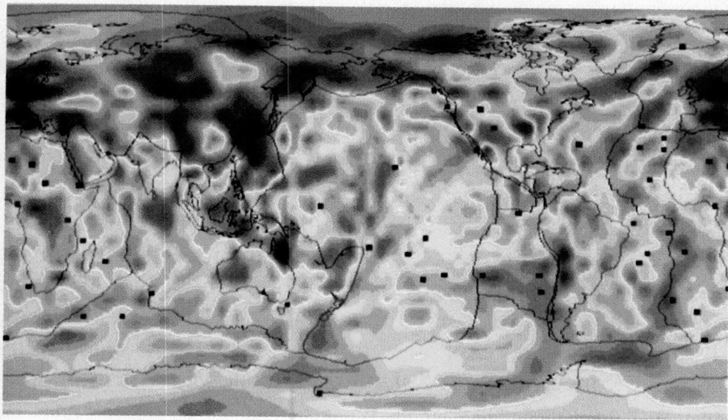

Figure 11-8 A seismic tomography image of the upper mantle at a depth of 410 kilometers (254 miles), showing hot spots (black dots), hot regions (increasingly dark shades of orange) and cool matter (blue). The continents are outlined in black. The Pacific Ocean (center) appears as a very warm region with several hot spots. Geophysicists are analyzing tomographic data to determine relationships among interior heating, tectonic activity and other processes, and surface features.

How do we study the Earth's interior? To examine the mantle, the mantle–core boundary, and even the core, geophysicists have adapted the CAT scan (computerized *a*xial *t*omography), developed for medical purposes. In medical scanning, a patient swallows or is injected with a radioactive (but safe) substance. X-ray-like images are made from numerous angles along successive planes of a portion of the patient's body, and a computer calculates the amount of radiation absorbed by the internal organs at every point along each plane to produce a series of three-dimensional representations. The sequence provides a view of every part of every organ within the scanned part of the body.

Inspired by the CAT scan, geophysicists have developed **seismic tomography**, a process that uses seismic waves to provide three-dimensional views of the Earth's interior. Powerful computers at the California Institute of Technology have synthesized data from 25,000 recent earthquakes from around the world to generate a series of cross-sectional views of the core. A sequence of such cross-sections provides a three-dimensional image of the core. This technique has recently shown, for example, that the outer surface of the Earth's core is not smooth and featureless; rather, it projects up into the mantle in some places, whereas elsewhere the mantle extends downward into the liquid outer core. Some of the core's peaks extend higher into the mantle than Mount Everest, the highest mountain at the Earth's surface, and some of its valleys are six times deeper than the Grand Canyon.

So far, there is no consistent evidence showing a relationship between the features of the mantle–core boundary and the Earth's surface geology. Seismic tomography scans show that mantle–core boundary peaks of 10 kilometers (6 miles) lie below such varied surface features as the Gulf of Alaska, the deserts of eastern Australia, the volcanic islands of the central North Atlantic, and the vast plains of south-central Asia. Conversely, 10-kilometer-deep valleys at the boundary lie below such varied geologic structures as the Philippine Sea, the highlands of Mexico, and the limestone reefs of the East Indies.

The cause of these mantle–core boundary irregularities is unknown. Is the comparatively "cold" mantle sinking into the hotter outer core, or do rising mantle convection currents draw the fluid core surface upward into peaks? Additional tomographic studies may someday provide the answer.

Seismic tomography is also probing other features that lie at unseen depths. It is being used to construct three-dimensional images of subducted plates in order to identify former plate boundaries now located at plate interiors. It is also helping geologists identify mantle hot spots and other characteristics, which may provide clues to the location of warm convecting mantle that may drive plate movements (Fig. 11-8).

total mass. Because of its great depth, pressures within the core exceed 3 million times atmospheric pressure. Core temperatures are believed to exceed 7600°C (13,700°F).

The core most likely consists of a mixture of ultra-dense metals such as iron and nickel, along with a few other elements (perhaps sulfur and silicon).

An iron–nickel core is hypothesized for several reasons: 1) It is consistent with seismic wave data; 2) the composition of meteorites, which are thought to be similar to the materials that formed the Earth, is predominantly iron; and 3) mathematical analysis of the densities of the whole Earth and of its interior layers predicts a core density equal to that of an iron–nickel mixture at core pressures. The average density of the core is estimated to be 10.7 grams per cubic centimeter (670 pounds per cubic foot).

Seismic wave data suggest that the core has two layers: a liquid outer core 2270 kilometers (1411 miles) thick, surrounding a solid inner core with a diameter of 1216 kilometers (756 miles), or slightly larger than the Moon. Both probably consist of iron and nickel mixtures, but the interplay of temperature and pressure within them differs. Heat dominates in the molten outer core, while pressure dominates in the solid inner core. Differentiation may also be at work: The entire core may originally have been liquid, but as it cooled during the last 4 billion years, its denser elements (primarily iron) sank to form an inner core with a high concentration of high-melting-point elements. At the same time, lighter components, perhaps sulfur and silicon, migrated upward to mix with the remaining molten iron and nickel, forming an outer core where the concentration of elements with lower melting points made a molten state more likely.

The Liquid Outer Core Geologists monitoring seismic wave arrival times after earthquakes have discovered that no S waves appear within an arc of 154° lying directly across the globe from an earthquake's epicenter. This region, analogous to the shadow cast by an object that intercepts light waves, is called the **S-wave shadow zone.** This seismic shadow must be due to the S waves' inability to penetrate the intervening materials. Since S waves do not travel through a liquid medium, geologists have concluded that a liquid region exists at the Earth's center, and that its outer boundary is indicated by the border of the S-wave shadow zone (Fig. 11-9).

Figure 11-9 Because shear waves cannot pass through an inelastic medium, S waves stop when they encounter the Earth's liquid outer core. This creates the S-wave shadow zone, a broad area on the opposite side of the Earth from an earthquake's epicenter where no S waves appear.

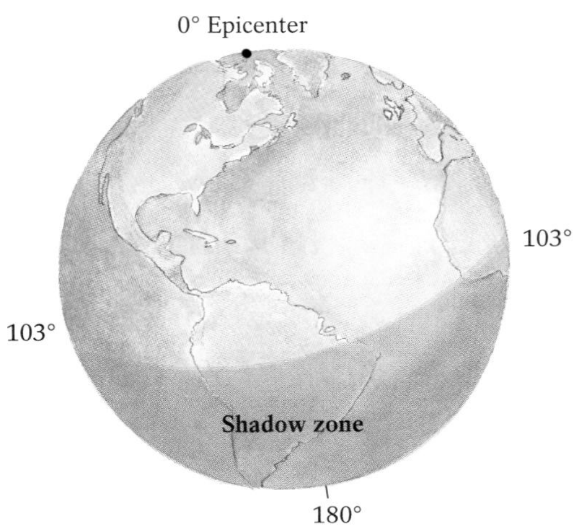

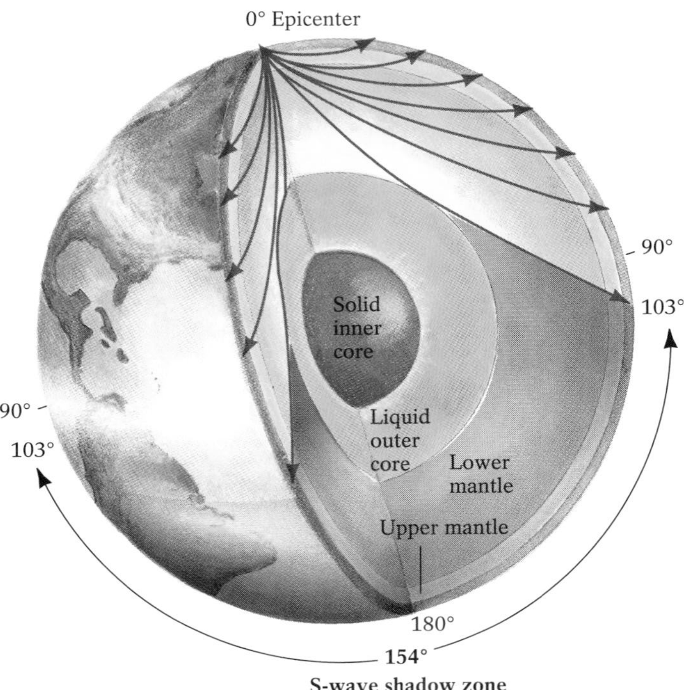

P waves, which do travel through liquids, cast a different kind of shadow. P waves reach the side of the planet opposite the epicenter of an earthquake, but arrive there somewhat later than would be expected judging from their initial speed, and are missing altogether in the zone between 103° and 143° from the epicenter. These phenomena can be explained by known P-wave behavior at solid–liquid boundaries. When a P wave grazes the mantle–core boundary, it reflects and continues on a curved path to the surface, where a seismograph records it at about 103° from the epicenter. When a P wave enters the liquid outer core, it is refracted, so that it always emerges 143° or more from the epicenter (Fig. 11-10). No P waves arrive in the region between 103° and 143° of an epicenter, forming the **P-wave shadow zone.**

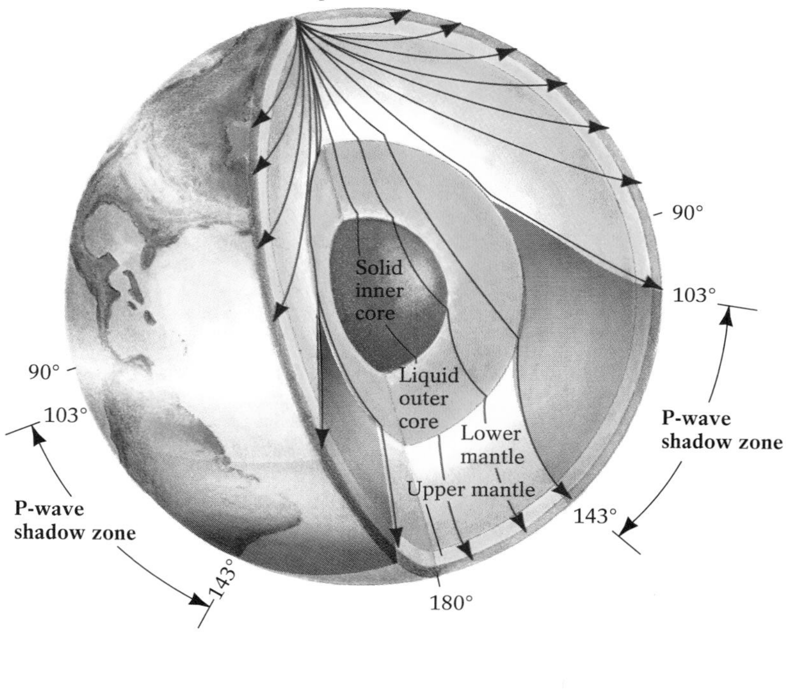

Figure 11-10 Depending on the angle at which they approach the Earth's liquid outer core, P waves either pass through it and refract or they reflect off its surface. This creates a band on the opposite side of the Earth from an earthquake's epicenter in which no P waves appear, or the P-wave shadow zone.

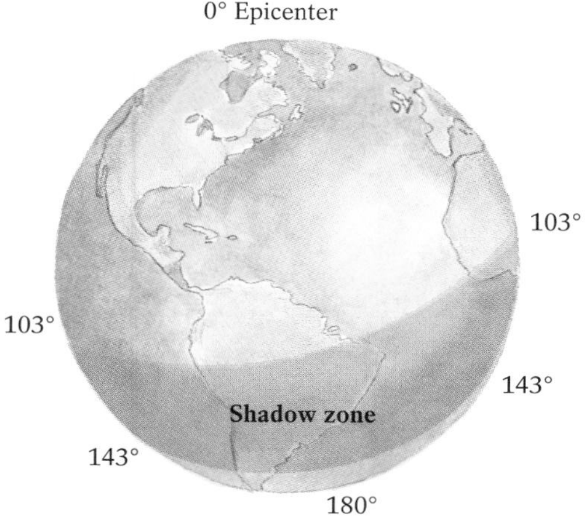

The Solid Inner Core Once the existence of the liquid outer core was known, geophysicists puzzled over another enigma: Why did the P waves that passed directly through the entire planet arrive on the other side *earlier* than expected? Because P waves would travel faster in a solid region, they hypothesized that the waves' early arrival indicated that some part of the core was solid.

To test this hypothesis, seismologists analyzed the arrival times of P waves generated by underground nuclear test blasts, for which time and place of origin were known precisely. These studies showed conclusively that the velocity of these waves increased abruptly as they entered the inner core, supporting the hypothesis that the inner core is indeed solid. Knowing the velocities of P waves as they pass through the crust, mantle and outer core made it possible to determine their velocity increase through the inner core. The change in velocity, together with the reflection and refraction patterns of P waves, establish that the depth of the outer core–inner core boundary is 5100 kilometers (3170 miles). Figure 11-11 summarizes how P and S waves behave as they pass through the Earth's interior.

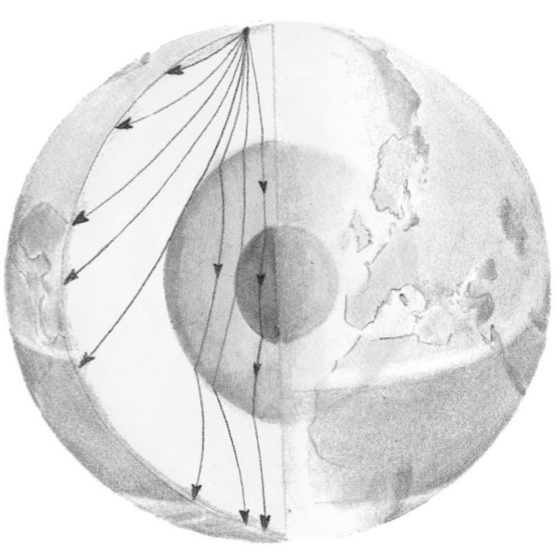

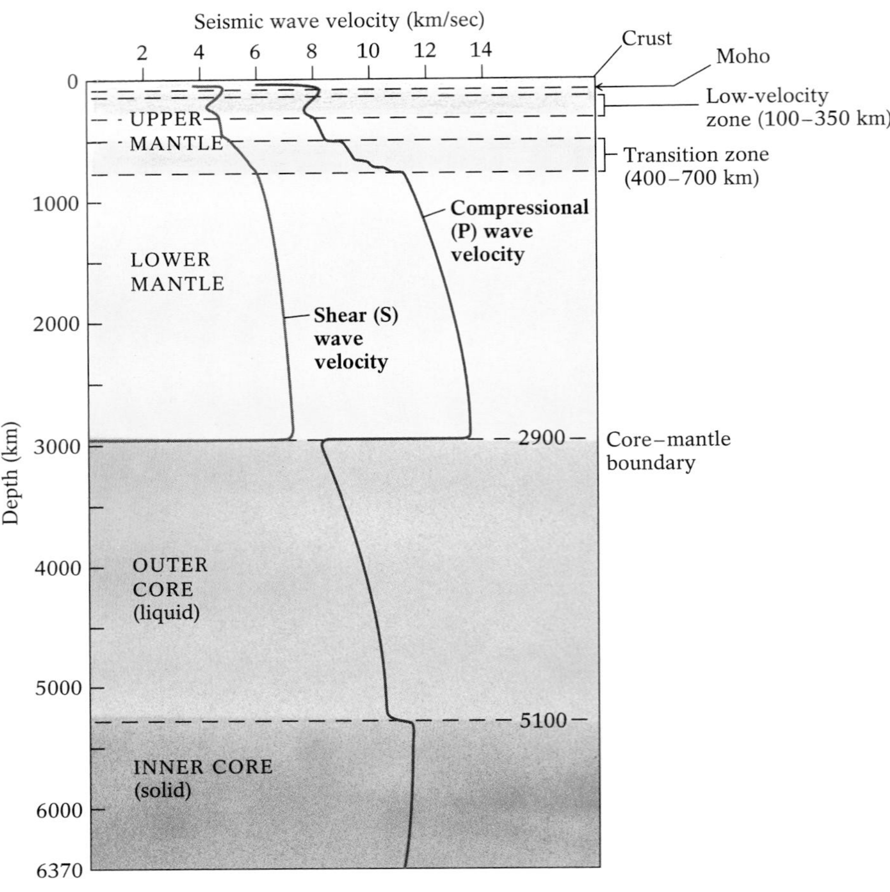

Figure 11-11 The behavior of P and S waves as they pass through the Earth's interior layers. P waves accelerate as they cross the crust–mantle boundary (Moho), slow down as they enter the less rigid asthenosphere (or low-velocity zone), speed up again as they approach and pass through the lower mantle, slow precipitously as they enter the less rigid liquid outer core, and speed up again as they pass through the solid inner core. Like P waves, S waves accelerate as they cross the crust–mantle boundary and slow down dramatically as they enter the asthenosphere; these waves speed up again as they enter the lower mantle, but stop altogether when they reach the liquid outer core.

Taking the Earth's Temperature

The pressure at any depth in the Earth's interior can be estimated by totaling the pressures applied by each of the overlying layers. There is no similar correlation between depth and temperature. To estimate interior temperatures, we must apply scientific logic to correlate the few direct observations we can make with experimentally derived data and inferences.

Direct evidence of the Earth's interior temperatures is quite limited. Miners who search for gold in South Africa's Western Deep Levels near Johannesburg report that it gets hotter as they descend. Air conditioning is required to offset temperatures of 40°C (104°F) at the bottom of the mine. Recall from Chapter 3 that the geothermal gradient is the rate at which the Earth's internal temperature increases with depth. Near the surface, the geothermal gradient is about 30°C per kilometer, meaning that the temperature at 1 kilometer beneath your geology classroom would be about 30°C (86°F) warmer than at the base of the building. If the temperature continued to increase at that rate, mantle rock would melt at depths below 100 kilometers (60 miles). From seismological studies, however, we know that the mantle at that depth is still solid. This is partially because the pressure at such depth inhibits melting (see Chapter 3); in addition, the high concentration of radioactive isotopes in crustal rocks adds heat at and near the surface but not elsewhere in the interior, affecting the geothermal gradient. (Although the deep inner Earth may lack radioactive decay as a heat source, it does maintain a high temperature through other means—perhaps from the residual heat produced in the Earth's earliest stages of formation and heat released as the liquid outer core solidifies.)

Determining Interior Temperatures

To remain solid at any given depth, a region's temperature must be below the melting points of its constituent materials at the prevailing pressure. From our knowledge of the composition of a solid portion of the Earth's interior and the pressure at that depth, we can determine the melting temperature of each of its components and thus derive the maximum temperature possible at that depth. For example, we know that the crust and upper mantle down to the low-velocity zone are solid, and we have an idea of the mineral components of these regions; therefore, we can determine the maximum temperature of each of these layers at the pressures we would expect at these depths.

The low-velocity zone is partially melted, so it must be hotter than the melting points of some of its ultramafic rocks. For 2500 kilometers (1500 miles) below this zone, down to the mantle–core boundary, the mantle is solid again, suggesting that rising pressures in this zone result in high melting points, exceeding the range of its slowly rising temperatures. The same logic suggests that in the liquid outer core, where temperatures must be higher than the melting points of its materials, the effect of heat once again dominates that of pressure; from inferences of the outer core's composition, geologists estimate that its temperature is 4800°C (8650°F) or higher. In the solid inner core the effect of pressure again exceeds that of heat, keeping the melting points of its materials higher than the prevailing temperatures there; from inferences of the inner core's composition and estimates of its components' melting points at the extremely high prevailing pressures, we believe its temperature to be 7600°C (13,700°F) or lower at the inner core–outer core boundary. The center of the core may be somewhat hotter at the higher pressures of that depth.

Geologists have been able to estimate the extreme temperature and pressure conditions of both core regions ever more accurately through experimentation, due to the recent technological advances described in Highlight 11-2 (following page).

Heat at the Earth's Surface

The amount of internal heat near the Earth's surface varies significantly from one place to another. Because uranium isotopes, the primary source of radioactivity on Earth, are generally associated with granite rocks, there is little heat produced by radioactive decay where there is no granite. Because oceanic crust is mafic and contains virtually no granitic rocks, geologists long assumed that oceanic rocks were cooler than their granitic continental counterparts. However, thousands of temperature measurements taken during recent deep-sea probes show that in some places even more heat rises from oceanic than from continental crust: Along divergent plate boundaries and at intraoceanic hot spots, rising thermal plumes convect internal heat from great

Highlight 11-2 *Simulating the Earth's Core in the Laboratory*

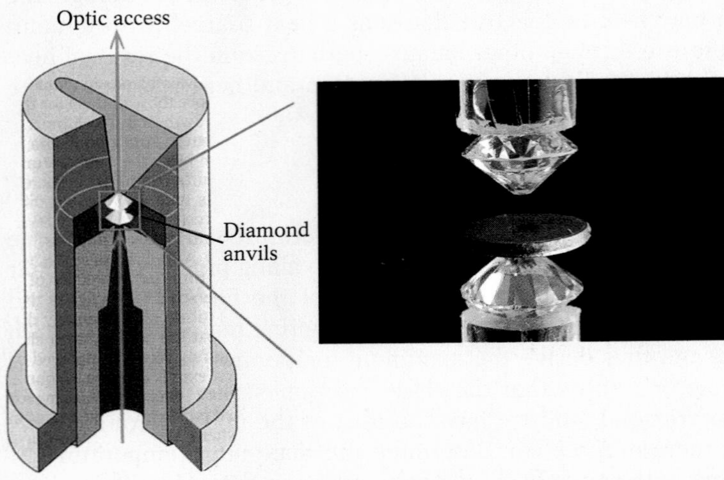

Figure 11-12 A diamond-anvil high-pressure cell, a device that simulates conditions in the Earth's lower mantle by subjecting samples to extreme pressure and heat.

Until the last few decades, laboratory simulations of the Earth's internal heat and pressure could recreate conditions only up to depths of 300 kilometers (200 miles). The jaws of a hydraulic press could produce pressures about 100,000 times atmospheric pressure, but this is only about 1⁄30 of the estimated pressure at the core. A great advance came in the 1960s with the introduction of the diamond-anvil high-pressure cell, a device that squeezes a sample substance between two diamonds and heats it with lasers (Fig. 11-12). This produces a pressure of 1.7 million times atmospheric pressure and a temperature of 3000°C (5400°F). Although this corresponds to conditions in the lower mantle, it still falls far short of the estimated conditions at the core.

In 1987, using further technological advances and considerable ingenuity, several laboratories found a way to simulate estimated inner core pressures of 3 to 4 million times atmospheric pressure. By detonating chemical explosives wrapped around rocks that had been fitted with internal electrical sensors, scientists at the University of California at Berkeley, the California Institute of Technology, and the University of Illinois generated shock waves that compressed the rocks to within the range of core pressures. Sensors recorded various properties of the compressed rocks in the few millionths of a second before they were destroyed by the shock waves.

In another recent experiment, metal and plastic "bullets" were shot into a thin film of iron at a velocity of 27,000 kilometers per hour (17,000 mph). This simulated the application of core-level pressures to the materials that most likely make up the core. The results of these pioneering experiments yielded new data about the behavior of Earth materials at the core's extremely high pressures, and melting-point estimates of these materials were revised on the basis of the new information. Previous estimates of melting points at the mantle–outer core and outer core–inner core boundaries, and hence of temperatures in those regions, had been significantly lower.

depths beneath the oceans, perhaps from as deep as the mantle–core boundary. Most of the heat of the oceanic crust, therefore, is residual heat from the formation of oceanic lithosphere at divergent plate boundaries, and not the result of near-surface radioactive decay, which predominates in continental crust.

Is the Earth Still Cooling?

Heat always flows from warmer to cooler areas. In the Earth's interior this is accomplished by convection, which continually transports heat from deep in the Earth up to the surface, where it dissipates in the atmosphere, thus contributing to the planet's cooling. (Although, because of the Earth's great size and the limited speed at which convection occurs in rock, this cooling process progresses very slowly.) The overall pattern of terrestrial heat flow is complicated by other factors, however, including several sources of ongoing heating.

The decay of radioactive isotopes has been giving off heat since the Earth's formation. Although their volume is continually decreasing and the amount of heat they produce is decreasing accordingly, radioactive isotopes are still a significant source of heat in the Earth's continental crust. Friction produced by the motion of the Earth's plates, particularly at subduction zones, is also a heat source. In addition, solidification of the core releases heat deep within the planet. The net effect of the processes contributing to heating and cooling is unknown; it is not clear whether the Earth is indeed cooling down.

The Earth's Gravity

The Earth hurtles through space at approximately 107,000 kilometers (66,500 miles) per hour on its annual journey around the Sun, and a point on the equator travels around the spinning Earth's axis at a velocity of 1700 kilometers (1060 miles) per hour. Why then don't you fly off that chair you're sitting on? **Gravity**, the force of attraction that all objects exert on one another, pulls you and everything around you toward the Earth's center. This force is what accounts for every object's *weight* at the Earth's surface.

The force of gravity acts in direct proportion to the product of the masses of the attracted objects, and is inversely proportional to the square of the distance from the center of one mass to the center of the other, according to the following (the symbol ∞ means "is proportional to"):

$$\text{Force of gravity} \propto \frac{\text{mass of A} \times \text{mass of B}}{\text{distance}^2}$$

The Earth's enormous mass provides the strong gravitational attraction that affects everything on or near it, and greatly overpowers the attraction between smaller objects in its vicinity (Fig. 11-13). Because the force of gravity decreases sharply as the distance between objects increases, a meteor hurtling through space close to the Earth may be drawn to our planet by a greater gravitational force than that exerted by the much larger, but more distant, Sun.

Figure 11-13 Because the mass of the Earth is so much greater than any object on its surface, all objects, such as these fallen rocks, are attracted by gravity toward the Earth's center.

Measuring Gravity

Gravity measurements help us to explore the Earth's interior by giving us clues as to what may lie below various features on the Earth's surface. A *gravimeter* measures variations in the Earth's gravity. It contains a mass suspended from a spring so sensitive that it registers even such minute differences in gravitational attraction to the Earth's center as would be produced by placing the gravimeter on a desk instead of on the floor. Thousands of gravimetric measurements taken everywhere from mountain tops to ocean floors have shown that the force of the Earth's gravity varies significantly. The larger variations are due to altitude and latitude. The pull of gravity is somewhat lower at the top of a high mountain than at sea level because of the greater distance to the Earth's center. The pull of gravity is stronger at the poles than at the equator because the rotation of the Earth makes it bulge at the equator and flatten out at the poles; as a result of the bulge, the equator is 21 kilometers (13 miles) farther from the center of the Earth than are the

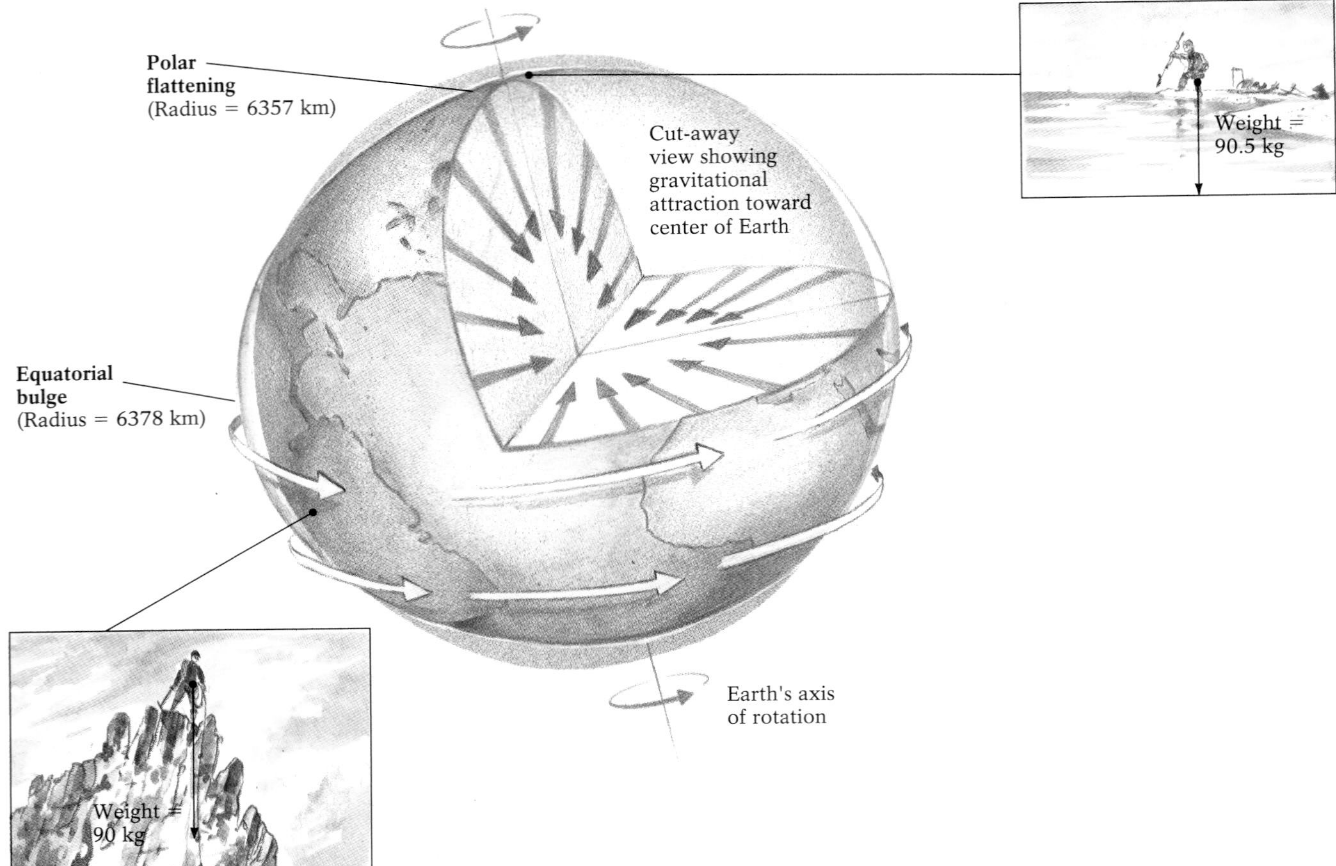

Figure 11-14 The gravitational attraction of the Earth varies from place to place on its surface. Because the planet is not spherical, and its surface is irregular, latitude and topography affect an object's distance from the Earth's center and consequently its weight. A person who weighs 90.5 kilograms (199 pounds) at sea level at the North Pole, for example, would weigh only 90 kilograms (198 pounds) in the equatorial mountains of South America, where the gravitational pull is weaker.

poles. This explains why a person would weigh a bit less standing on top of a high peak near the equator than standing at sea level near the North or South Pole (Fig. 11-14).

Gravity Anomalies

Even after adjusting for the effects of altitude and latitude, gravity readings are still not the same everywhere on Earth. The remaining differences between actual gravimetric measurements and the theoretical values we would expect are called **gravity anomalies**. Gravity anomalies are due to variations in density within the Earth's interior, which affect the Earth's mass at such locations and consequently its gravitational pull.

Negative gravity anomalies, attractions lower than the theoretical value, are caused by relatively low-density rocks or sediments below the surface in the midst of denser country rock. They are often found at continental mountain ranges, which generally have deep roots of low-density felsic rock, or where subducted basaltic lithosphere is surrounded by denser mantle rocks. Negative anomalies also occur along oceanic trenches, which are filled with water and low-density sediment, and therefore register a lower gravitational attraction than the denser oceanic crust that surrounds them. Analysis of negative gravity anomalies can have practical applications: Along the Gulf Coast of Louisiana and Texas, anomalously low gravity signals the presence of subsurface low-density salt deposits, produced by evaporation of ancient oceans and now surrounded by denser sedimentary rocks (Fig. 11-15). As we shall see in Chapter 20, impermeable salt deposits often trap migrating crude oil. Thus, gravity studies are sometimes used to search for crude oil.

Positive gravity anomalies, attractions higher than the theoretical value, result from the presence of high-density rocks below the surface, often where the Earth's crust is thinnest. Positive anomalies often occur over ocean basins, for example, where high-density mantle rocks lie relatively close to the surface below thin oceanic crust, and are also found where continental crust has

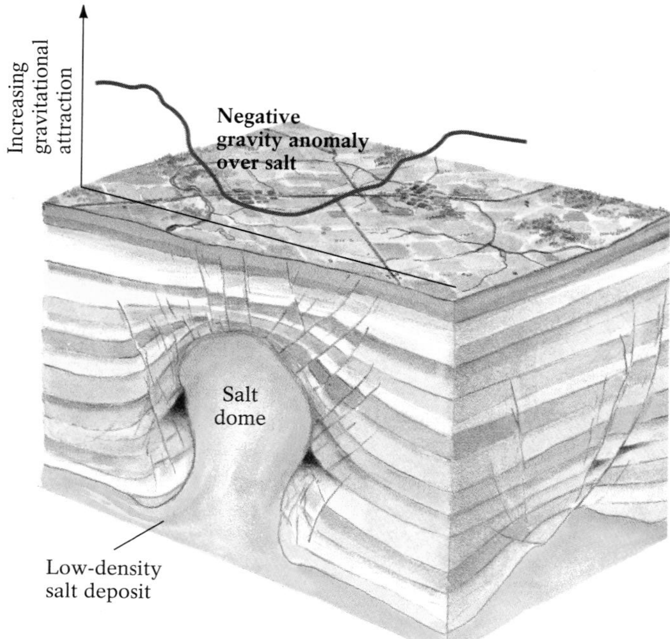

Figure 11-15 A negative gravity anomaly may be caused by a concentration of low-density salt below the surface.

thinned, perhaps as a prelude to plate rifting. One positive anomaly stretches southward along the Minnesota–Wisconsin border from Lake Superior through Iowa and Kansas to Oklahoma. Known as the mid-continent gravity high, this anomaly marks the location of an ancient rift in the North American plate into which dense mantle material rose 1.1 billion years ago (Fig. 11-16a). The rifting stopped, however, before the continent divided, leaving mid-continent basalt flows at the surface and high gravity readings marking the presence of a shallow, 50-kilometer (30-mile)-wide zone of dense subsurface rock.

Not all positive anomalies are associated with thin crust. They can also occur where masses of ultramafic mantle rocks have been caught at or near the surface between colliding plates to become part of such continental mountains as the Alps and Appalachians. Higher-than-expected gravity readings may also show us where large bodies of dense metallic ore are concentrated in the crust (Fig. 11-16b). (See Chapter 20.)

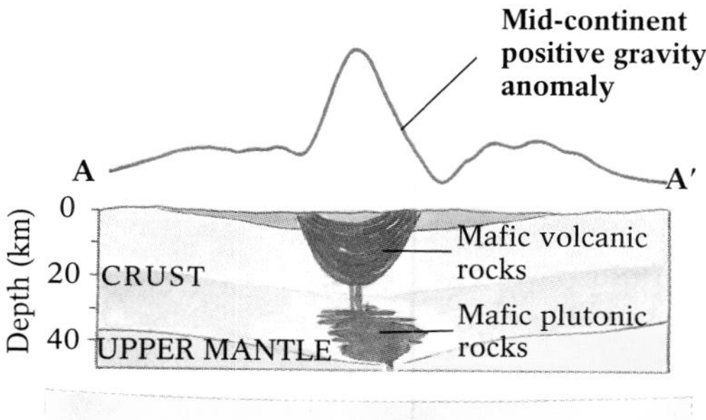

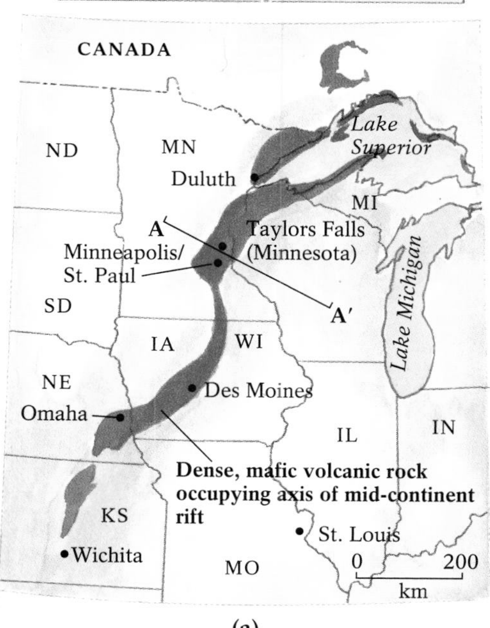

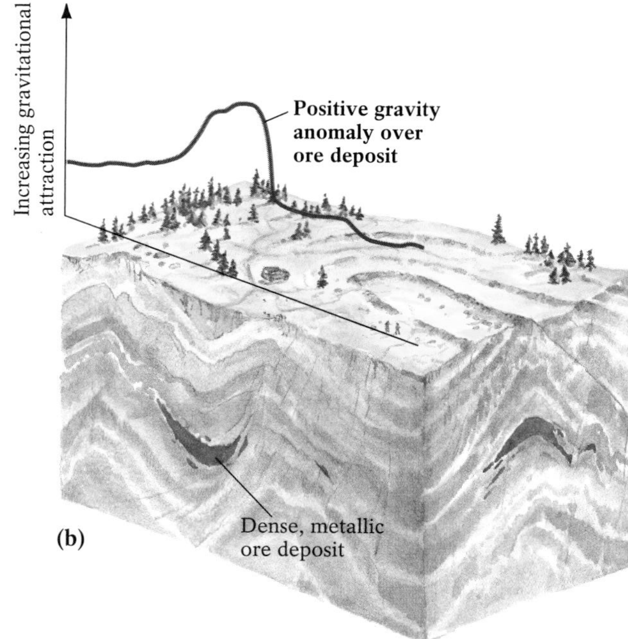

Figure 11-16 Positive gravity anomalies are caused by the presence of high-density rocks near the surface. (**a**) The mid-continental gravity high, a nearly continent-long positive gravity anomaly in North America, probably formed when the North American plate began to rift 1.1 billion years ago and a body of mantle-derived mafic rock was introduced into the continental rocks of the Midwest. (**b**) A positive gravity anomaly may also be caused by a near-surface concentration of dense metallic ore.

Isostasy

Each segment of the Earth's lithosphere "floats" on the underlying denser, partially molten asthenosphere of the upper mantle. However, because areas of the lithosphere vary significantly in density and thickness, gravity pulls some of its segments (those with more mass) more deeply into the asthenosphere. In addition, because the weight and dimensions of any given segment change over time, its depth relative to the mantle also changes over time. This equilibrium between lithospheric segments and the asthenosphere beneath them is called **isostasy.**

Isostasy explains why an iceberg that breaks off a coastal glacier may initially plunge below the surface of an adjacent bay, but immediately bobs back up. Because ice is less dense than water, the iceberg is buoyed upward and then adjusts its position until it displaces a volume of water equal to its own total weight. Water is 10% more dense than ice, so a volume of water equal to the weight of the iceberg will amount to 90% of the iceberg's volume. This explains why every iceberg, like every ice cube floating in a water glass, has 10% of its volume above the surface of the water and 90% below (Fig. 11-17a). If some of the ice melts, the iceberg immediately adjusts its position to maintain the same 10:90 proportion of ice above and below the water.

Like a floating iceberg, each segment of the Earth's lithosphere, floating on denser underlying asthenosphere, rises or sinks to achieve its own isostatic equilibrium (Fig. 11-17b). Because continental crust is less dense than oceanic crust, it is more buoyant—a larger proportion of it floats above the asthenosphere. Because continental crust is also much thicker than oceanic crust, however, it also extends far deeper into the asthenosphere. If we could look below the surface of the mantle, we would see that tall continental mountain ranges have proportionately deep "roots" extending far into the asthenosphere; segments of oceanic crust, which are much less thick than continental crust, have much smaller subsurface portions.

Figure 11-17 According to the principle of isostasy, the depth to which a floating object sinks into underlying material depends on the object's density and thickness. (**a**) All floating blocks of ice, being of the same density, sink in water so that the same proportion of their volume (90%) is submerged; the thicker the block of ice, the greater this volume will be. (**b**) Continental crust, because it is of lower density than oceanic crust, is more buoyant and has a larger proportion of its volume above the mantle; because it is so thick in comparison to oceanic crust, however, it extends farther into the mantle (has a deep "root").

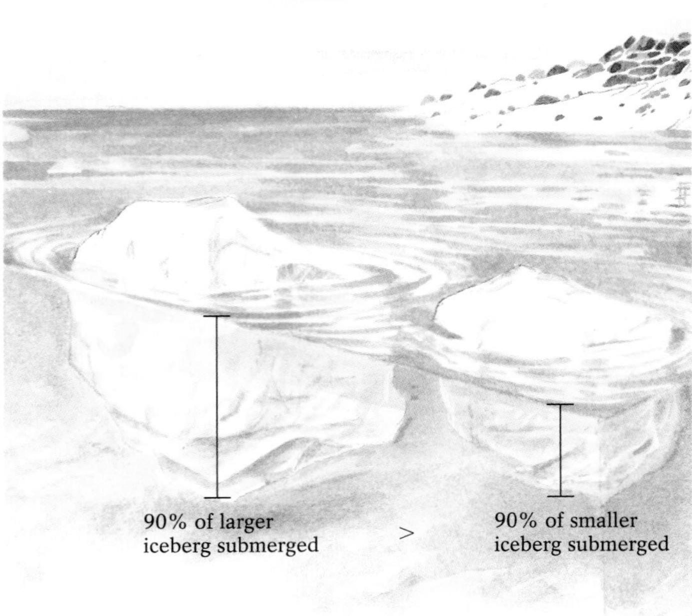

(a)

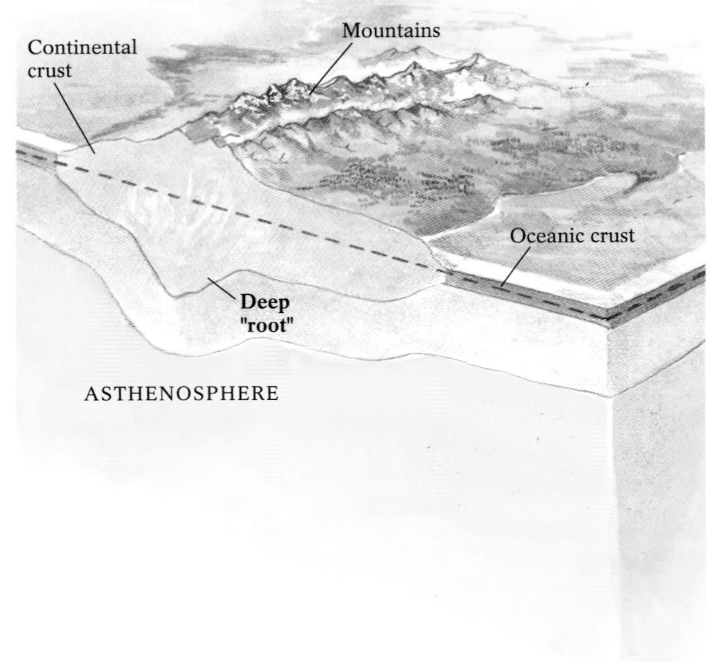

(b)

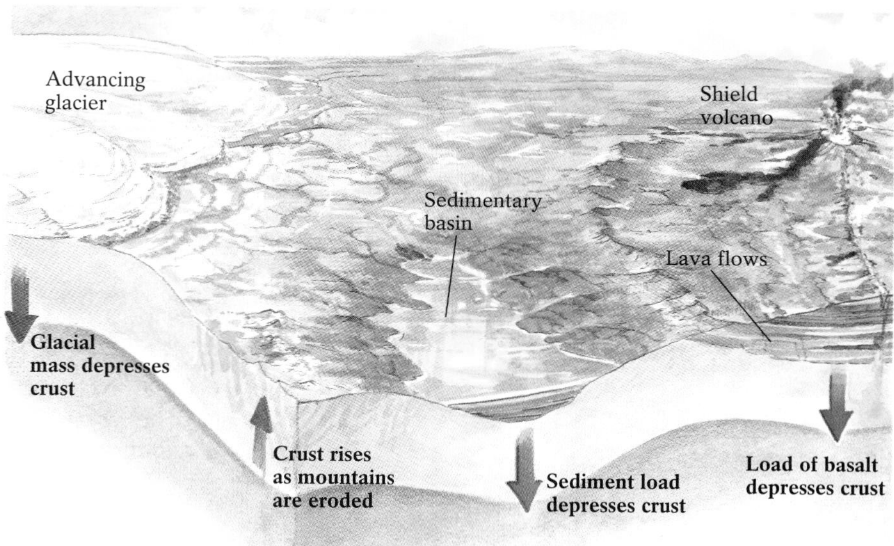

Figure 11-18 Isostatic adjustments in the Earth's crust. As mass is added to a segment of the lithosphere, it tends to subside. As mass is subtracted from a segment of the lithosphere, it tends to rise.

Each segment of the lithosphere is constantly undergoing isostatic adjustment, as mass is added to its surface in some places and removed in others. Isostatic response to variations in crustal loads is similar to the effect of cargo on a ship: The more cargo, the deeper in the water the ship will ride. As lava flows, fresh sediment, or advancing glaciers are added to a given segment of the Earth's crust, that segment gradually sinks deeper into the asthenosphere. Conversely, when materials are removed from a given segment of the crust, as a hill erodes or a glacier melts, that segment slowly rises (Fig. 11-18).

Magnetism

About 2500 years ago, the Greeks noticed that some dark rocks near the Asia Minor city of Magnesia had a strange property: Objects made of iron became attached to them. For centuries magnetic rocks were thought to have wonderful abilities—to comfort the sick, stop hemorrhages, cure toothaches, increase a person's gracefulness, and even reconcile estranged spouses. A truly practical purpose for magnetic rock was discovered 1500 years ago by the Chinese, who observed that a suspended sliver of such rock would turn freely and always come to rest in the same position. Thus was invented the magnetic compass. And in 1600, England's Sir William Gilbert proposed that the Earth itself behaves like a giant magnet, setting the stage for our modern understanding of magnetism.

Magnetism is the force, associated with moving electricity, that enables certain substances to attract or repel similar materials. In our discussion of minerals in Chapter 2, we examined the role of electrical charge in attracting charged particles. Magnetism arises not from the charges on the particles, but from the *motion* of negatively charged particles. As we saw in Chapter 2, electrons are negatively charged particles that are in constant motion; certain patterns of electron motion create magnetic properties in some substances.

Electrons move not only by orbiting the nucleus of an atom, but also by spinning about their own axes. Each orbiting, spinning electron creates and is surrounded by its own minute region of magnetic influence, or **magnetic field.**

Every substance contains innumerable spinning electrons, but only certain substances are magnetic. When electrons spin randomly, their magnetic fields cancel each other and the substance containing them is not magnetic. In certain substances, however, electrons align their magnetic fields with those of other electrons and tend to spin in one direction more than in another. The magnetic fields of these electrons enhance each other, giving the substance itself an overall magnetic field. When magnetic substances are subjected to a strong external magnetic field, their internal fields become further aligned and they become more strongly and permanently magnetized. The intensity of a magnetic field can be measured with a sensitive device known as a *magnetometer.*

Although magnetic fields are invisible, the influence of their lines of force can be seen by shaking iron filings onto a sheet of paper placed over a bar magnet. The filings form a pattern that shows numerous closed loops of magnetic force emerging from the magnet's north pole and entering its south pole (Fig. 11-19).

Figure 11-19 The magnetic field of a simple dipolar bar magnet includes a north pole, from which the lines of magnetic force emerge, and a south pole, at which the lines of magnetic force reenter the magnet. In the same way that electrically charged particles attract opposite and repel like charges, magnetic north and south poles attract one another while like poles repel each other.

Certain compounds of iron, such as magnetite (Fe_3O_4), are strong natural magnets; the rocks of Magnesia contained magnetite. Mafic rocks, such as basalt and gabbro, usually contain some magnetite. Other forms of iron, such as hematite (Fe_2O_3), become temporary magnets after exposure to a strong external magnetic field, and lose their magnetism soon after the external field is removed. A rock can be *demagnetized* by chemical weathering, lightning strikes, and heat, all of which can alter the orientation of electrons. Heat, for example, causes electrons to vibrate and shift from their aligned positions into random orientations. Magnetite loses its magnetism completely when heated to 580°C (1075°F).

The Earth's Magnetic Field

The Earth has a magnetic field that penetrates and surrounds the planet. It extends into space more than 60,000 kilometers (37,000 miles) beyond the Earth and its atmosphere, and every cubic centimeter of the planet is permeated with invisible lines of magnetic force. Because the lines of force of the Earth's magnetic field curve, the angle at which they intersect the Earth's surface varies from place to place: They are perpendicular to the surface at the planet's magnetic poles, and parallel to the surface at the equator. At all points in between, the lines of magnetic force intersect the surface at an

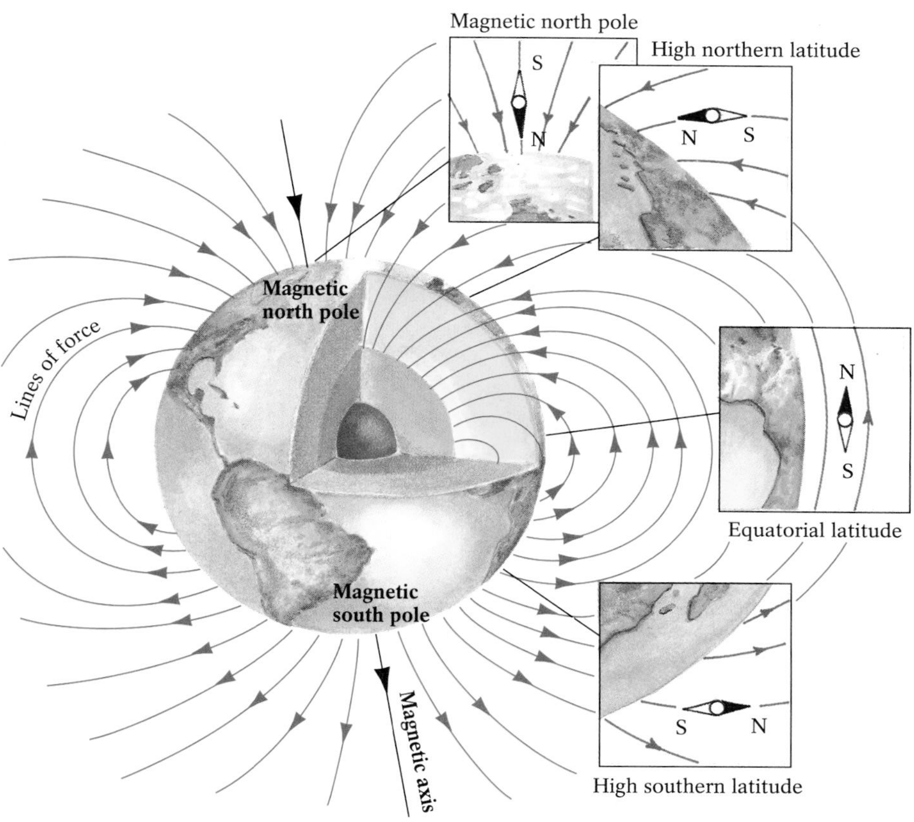

Figure 11-20 The Earth's prevailing magnetic field. Unlike those from the simple bar magnet in Figure 11-19, lines of magnetic force currently emerge from the Earth at the planet's magnetic south pole, located near McMurdo Sound in Antarctica, and reenter at the planet's magnetic north pole, near Prince of Wales Island in the Canadian Arctic. (Note that the planet's magnetic poles do not correspond exactly to its geographic North and South Poles.) A freely suspended magnetic needle at the Earth's surface, aligning itself along the planet's magnetic force lines, would settle perpendicular to the surface at the Earth's magnetic poles, parallel to the surface at the equator, and at various angles to the surface at points in between.

angle that increases toward the poles, and decreases toward the equator (Fig. 11-20). As we will see, the Earth's magnetic north and south poles periodically move and even switch places, a phenomenon that can help geologists decipher the history of its rocks.

The Origin of the Earth's Magnetic Field We still know relatively little about what causes the Earth's magnetic field. Rapid, frequent changes in the Earth's magnetic field argue against a permanent magnet in the Earth's interior. Seismological studies that have clarified our understanding of the Earth's interior show no evidence of a huge concentration of magnetite ore acting as a permanent magnet in the Earth's interior. Besides, most naturally occurring magnetic minerals can exist only within the Earth's upper 30 kilometers (20 miles); below that depth they would be demagnetized by the Earth's internal heat. Thus, the Earth's global magnetic field cannot be a product of simple rock magnetism. What could generate an Earth-sized magnetic field?

Our present understanding of the origin of the Earth's magnetic field derives from what we know about the movement of electrons. Physicists in the nineteenth century discovered that a magnetic field can be created by an electrical current, and an electrical current can be generated spontaneously by moving a substance capable of conducting electricity through an existing magnetic field. Such a system is known as a *self-exciting dynamo.* We use this knowledge today to generate electricity in power plants by rotating an electrical conductor in a magnetic field. A self-exciting dynamo stimulates itself to produce more electricity, which in turn produces a stronger magnetic field, which then produces more electricity, and so on.

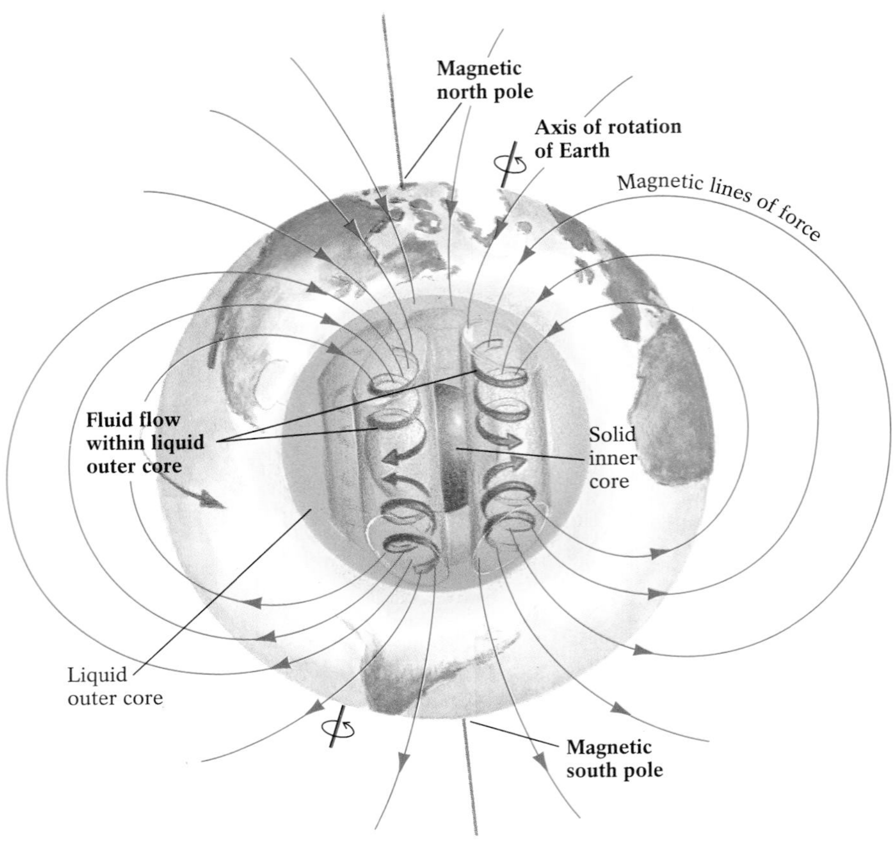

Figure 11-21 The Earth's magnetic field is generated in part by the flow of electrically conductive fluid in its outer core (see text).

The Earth functions as a gigantic self-exciting dynamo, with its magnetic field generated by movement of electrons in the molten iron in its liquid outer core. The electrical current in the outer core is in turn generated by movement of the Earth through the Sun's powerful magnetic field, and is enhanced by convection of the outer core material (promoted by heat from the inner core) and the rotation of the Earth, which sets the liquid in motion and determines the direction of convection (Fig. 11-21). These moving currents generate the Earth's magnetic field, which generates more electrical currents, and the system continues to stimulate itself.

Magnetic Reversals The Earth's magnetic field changes, perhaps more than any global feature other than the weather. For instance, the location of the magnetic poles changes over time with respect to the geographic poles. At present, the magnetic north pole is drifting westward at a rate of about 0.2° per year. In addition, at intervals averaging a half million years, the Earth's magnetic field actually weakens and gradually vanishes. (Since 1830, the strength of the Earth's field has decreased by about 6%. If it continues to decrease at that rate, the current field will disappear in about 2000 years.) After the field disappears, it starts to develop again, often with *reversed polarity*—magnetic north and magnetic south have exchanged places hundreds and perhaps even thousands of times during Earth's history. The effects of such **magnetic reversals** were first discovered in France at the turn of the century, when scientists noticed that some layers of volcanic rock were magnetized in the opposite direction from other layers.

Magnetic reversals are believed to be a result of variations in outer-core convection. As the amount of heat produced by internal sources varies, so does the outer core's convective flow. Turbulence arising from this variable flow may strengthen or weaken the magnetic field. The times during which the Earth effectively has no magnetic field may last for several centuries. As a new field gradually develops, its strength may fluctuate erratically, but it eventually builds until it is well established. A magnetic reversal—from one stable field to a stable field of reversed polarity—is believed to take between 1000 and 5000 years. Today's polarity, which is characterized by lines of magnetic force that emerge from the magnetic south pole and reenter the magnetic north pole (see Fig. 11-20), began about 780,000 years ago; this polarity is referred to as *normal*. When the Earth's magnetic field reverses, the lines of force emerge from the magnetic north pole and reenter the magnetic south pole; this polarity is referred to as *reversed*. Today, the needle in a hand-held compass points toward the north; during a period of reversed polarity, the needle would point toward the south.

Paleomagnetism Evidence of past changes in the Earth's magnetic field is often found in the geologic record, a phenomenon known as **paleomagnetism**. Such changes are most clearly detected in mafic igneous rocks, such as basalt, and in lake and marine sediments. As mafic lava cools, small crystals of magnetite are formed that are, for a while, free to move and orient themselves with the Earth's magnetic field (Fig. 11-22a). By the time the entire lava body has solidified, the magnetite crystals are frozen in place, recording the alignment of the Earth's magnetic field at that time. Similarly, when magnetite grains settle through a relatively still body of water, such as a lake or protected bay, they are free to rotate like compass needles and become aligned with the Earth's field as they fall (Fig. 11-22b). After they are buried by subsequent sediments, they can no longer rotate freely. Magnetite-rich sediments and lava thus create a permanent record of the Earth's past magnetic fields.

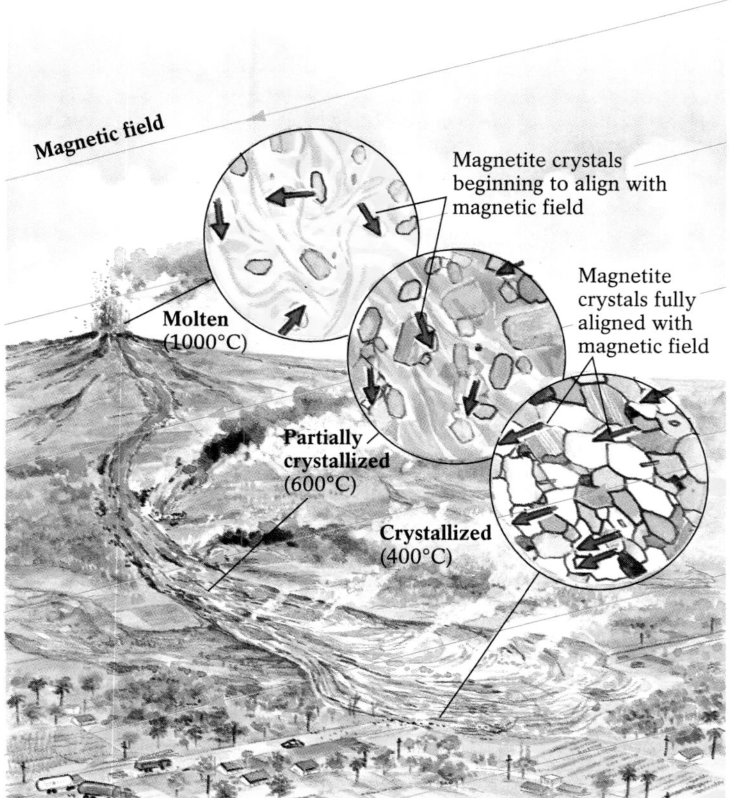

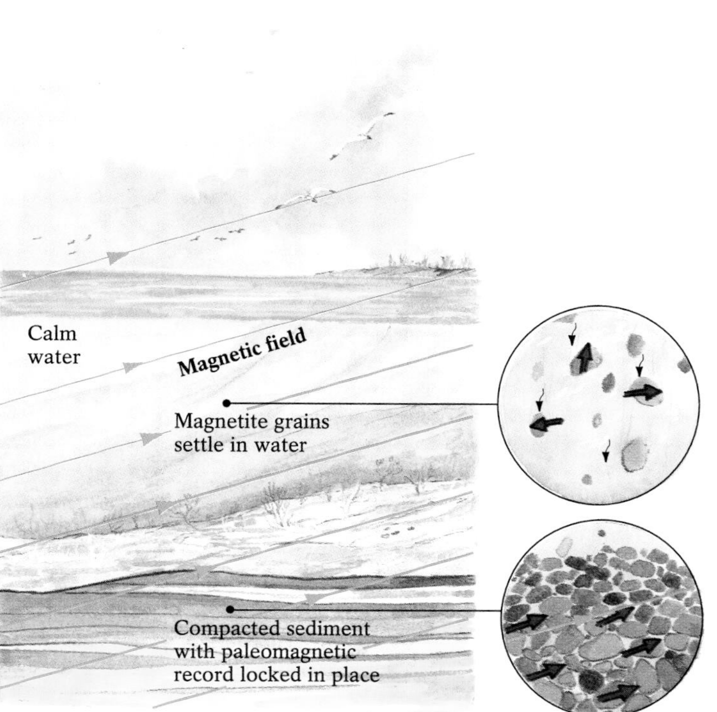

Figure 11-22 (**a**) Magnetite crystals in molten basalt flows are free to become aligned with the Earth's prevailing magnetic field. Once the lava solidifies, the magnetite crystals are fixed in place, recording the nature of the Earth's magnetic field at the particular time and place. (**b**) Magnetic grains of sediment settling through relatively still bodies of water are free to become aligned with the Earth's field as they fall. Once they are buried by subsequent deposits they are locked in place, leaving a paleomagnetic record of the field.

The terrestrial ("on land") record of the Earth's magnetic field reversals is highly susceptible to loss due to weathering and erosion; consequently, it is at best incomplete. The most complete record is found at the ocean floor, where there is uninterrupted marine sedimentation and little or no erosion (Fig. 11-23). In addition, sea-floor spreading provides a unique opportunity to study past magnetic field reversals—they are recorded in the pattern of basaltic lavas that have continuously erupted and cooled at mid-ocean ridges (see Chapter 12). The ocean-floor record of fluctuations in the Earth's field shows that during the past 75 million years alone, the Earth's field has reversed itself as many as 171 times. Periods of stable polarity, either normal or reversed, have lasted from 25,000 to several million years. There is no definite pattern to the length of time between magnetic field reversals, and geophysicists do not know for certain when the next reversal will occur or what effects it will have on Earth's inhabitants.

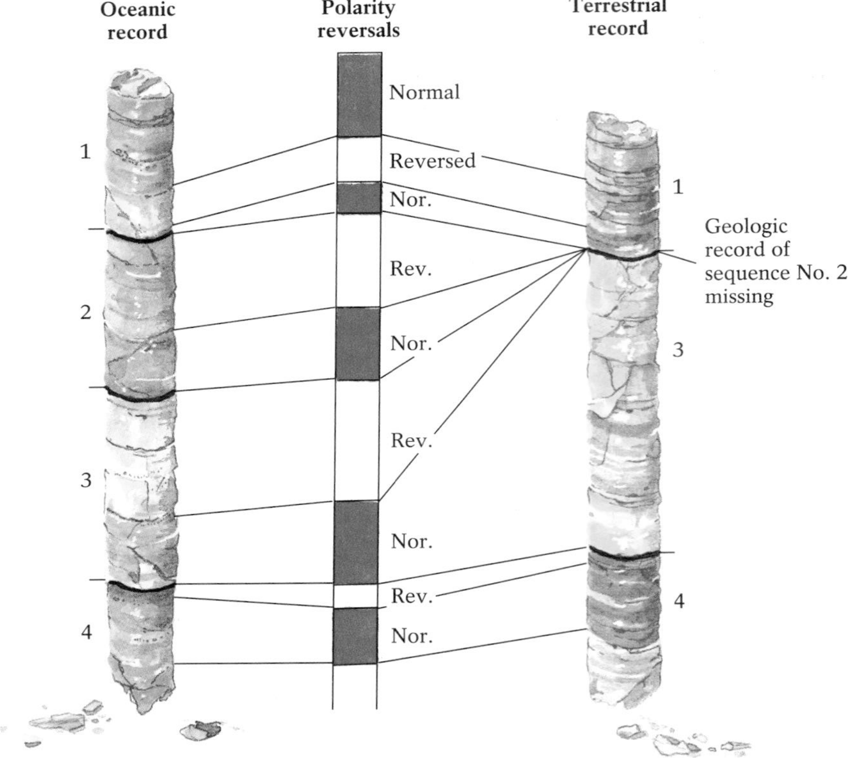

Figure 11-23 The terrestrial record of magnetic reversals is interrupted by erosion and the intermittent nature of continental volcanism and sedimentation. The oceanic record, generated by uninterrupted marine sedimentation, is more complete.

Paleomagnetism can even help us reconstruct past geographic arrangements of the Earth's landmasses. Because a rock's magnetite crystals align with the Earth's magnetic field at the time of the rock's formation, the crystals function as a kind of "paleocompass" needle, pointing to their original magnetic poles. We can determine the rock's original geographic latitude by comparing the angle the crystals form with the Earth's surface to the angle at which the lines of force of the Earth's magnetic field intersect the surface. (The angle increases with increasing latitude; see Figure 11-20.) The extent to

Figure 11-24 The Wrangell Mountains of Alaska. Igneous rocks in these mountains contain magnetite grains that indicate they formed near the equator, suggesting that they have been moved a great distance by plate motion. Sedimentary rocks in the mountains contain fossils that corroborate the paleogeographic origin of the igneous rocks.

which any magnetite-bearing rock has been moved from its original position is presumed to be due to movement of the tectonic plate on which it is located. Paleomagnetism makes it possible to estimate how far and in which direction a rock (and therefore its plate) has moved (Fig. 11-24).

Other Magnetic Fields in Our Solar System

To generate a magnetic field, a planet must: 1) be differentiated in such a way as to concentrate conductive materials in its interior; 2) maintain a flowing liquid layer; and 3) spin on its axis to set the liquid in motion. All of the planets and moons of our solar system are spinning; many may have differentiated interiors. But only some appear to contain liquid interior layers, without which a self-exciting magnetic field cannot be produced.

The Moon does not appear to contain a liquid interior layer, and so we would not expect it to generate a magnetic field. Indeed, it does not. The Moon is partially differentiated, however, and its elements could not have migrated into layers according to their different densities if it had not at one time been at least partly melted. Magnetometer-toting astronauts and Earth-bound lunar scientists have found aligned magnetite crystals in lunar rocks, confirming that the Moon's small iron-rich core was once molten and generated a magnetic field. These rocks, magnetized billions of years ago when they formed, have not been reheated and demagnetized since. This suggests that the Moon has been tectonically inactive for quite some time.

Mars' notable red appearance suggests a high concentration of iron-rich rocks at the surface. The red planet, therefore, may not be differentiated enough for the iron at its surface to have settled to its core. Without a conductive core, Mars would not have a magnetic field. This was confirmed in September 1976, when the sensitive magnetometer on the U.S. Viking probe detected no evidence of a magnetic field on Mars.

Mercury, like Earth, appears to have undergone extensive planetary differentiation. Because of Mercury's unusually high density, planetary geologists believe it has an extremely large, iron-rich core, and that all or most of Mercury was once molten. Although Mercury rotates so slowly on its axis (its day lasts 59 hours) that it probably could not generate the necessary currents in its liquid core to form a self-exciting dynamo, the magnetometer on the 1973 Mariner 10 Mercury probe did detect a weak magnetic field (about 1% the strength of the Earth's). Its pattern of lines of force is much like that of the Earth's field. The existence of a magnetic field on Mercury supports the hypothesis that the planet's core is at least partly molten.

Jupiter and Saturn, two large outer planets in our solar system, are composed principally of hydrogen in liquid form. At extremely high pressures liquid hydrogen exhibits the properties of a metal, providing these planets with a conductive inner layer. Both planets spin quite rapidly, with Jupiter's day lasting 9 hours 50 minutes, and Saturn's about 10.5 hours. Their rapid spin and their conductive liquid interiors give these planets the strongest magnetic fields in the solar system. The strength of Jupiter's field was measured by Pioneer and Voyager probes as at least 10 times that of the Earth; Saturn's field is nearly as strong as Jupiter's.

Our knowledge of the geophysical properties of planet Earth has grown considerably during recent years, thanks in part to rapid advances in space-satellite technology. In the next chapter, we will see how these and other technological advances are enabling Earth scientists to track the movements of the continents, study the details of the ocean floor, and reconstruct past positions of the Earth's plates.

Chapter Summary

Geophysics is the study of the fundamental properties of the Earth's interior, especially heat flow, gravity, and magnetism. Much of the data that contribute to this study come from the seismic body waves (compressional P and shearing S waves) that emanate from the focus of an earthquake and are recorded by the worldwide network of seismological stations.

By analyzing the waves' arrival times as recorded by seismographs, geophysicists have identified the thickness, density, composition, structure, and physical state of the layers of the Earth's interior. We now believe that the Earth comprises the following: continental and oceanic crust, each of differing composition, thickness, and structure; a **seismic discontinuity**—a boundary where seismic waves are refracted as they speed up or slow down because of a marked change in rigidity—at the base of the crust and the top of the mantle, known as the **Moho;** a thick multilayered mantle whose upper portion contains a partially melted region, the **low-velocity zone** (another seismic discontinuity), in which seismic waves slow down temporarily; a distinct **transition zone** between upper- and lower-mantle zones; a seismic discontinuity at the mantle–core boundary; and a two-part core consisting of a liquid outer core and a solid inner core. **Seismic tomography**, a research technique adapted in principle from medical CAT scans, uses seismic waves to generate three-dimensional images of the Earth's interior.

The **S-wave shadow zone** and **P-wave shadow zone** are segments of the Earth opposite an earthquake's focus in which S and P waves are not recorded. Because shearing S waves do not travel through a liquid, and P waves are refracted when they enter a region of lower rigidity such as a liquid, their absence in the S- and P-wave shadow zones confirms that the Earth's outer core is liquid.

Because there are few direct means of measuring the Earth's deep interior temperatures, estimates are derived using scientific logic and laboratory experimentation. If the Earth is solid at a particular depth, the temperature there must be lower than the melting point of the material at that pressure. Recent experiments suggest that the temperature at the mantle–outer core boundary exceeds 4800°C (8650°F) and that the temperature of the outer core–inner core boundary is about 7600°C (13,700°F).

Gravity is the force of attraction that any object exerts on another; it is proportional to the product of the objects' masses, and inversely proportional to the square of the distance between their centers. The Earth's gravitational attraction is far greater than that of any object on or near it. Gravity at the Earth's surface is affected by altitude (there is a greater attraction in lowlands than on mountain tops), latitude (there is a greater gravitational attraction at the poles than at the equator), and the composition of the underlying interior layers (there is a greater attraction where high-density materials are concentrated). **Gravity anomalies** are local deviations in the Earth's gravity caused by variations in density in the planet's interior. Negative gravity anomalies, where the force of the attraction is lower than expected, occur in mountainous areas where low-density continental crust is particularly thick, above sediment-filled oceanic trenches, and where there are rising subterranean bodies of salt. Positive gravity anomalies, where the force of attraction is greater than expected, are commonly found where high-density rocks lie at shallow depths. They occur where dense mantle material rises beneath rifting or diverging plate boundaries, where mantle rocks are found at or near the surface within collisional mountains, and above high concentrations of metallic ores.

Isostasy is the constant depth-adjustment of the Earth's lithospheric segments as they float on the denser underlying asthenosphere. Surface processes that add mass to the segments, such as volcanism, glaciation, sedimentation, and mountain building, cause them to sink; processes that remove Earth materials from the segments, such as erosion and glacial melting, cause them to rise.

Magnetism is a property of some materials, associated with the movement of electrons, that causes them to attract or repel other materials having this property. Geophysicists have determined that the Earth has a region of magnetic influence, or **magnetic field**, similar to that of a simple bar magnet. This field is probably due to the flow of charged iron particles within the Earth's liquid outer core. The invisible lines of force generated by the Earth's magnetic field emerge from the magnetic south pole and reenter the magnetic north pole. The Earth's magnetic field periodically weakens and changes the direction of its polarity, a phenomenon known as **magnetic reversal. Paleomagnetism** is the preservation in the geologic record of past changes in the Earth's magnetic field. It is used primarily as a tool to reconstruct past geographic arrangements of the Earth's landmasses.

Other planets in our solar system have a magnetic field if they are differentiated, have a conductive liquid layer, and spin, keeping the liquid in motion.

Key Terms

geophysics (p. 304)
seismic discontinuity (p. 308)
Moho (p. 308)
low-velocity zone (p. 308)
transition zone (p. 310)
seismic tomography (p. 311)
S-wave shadow zone (p. 312)
P-wave shadow zone (p. 313)
gravity (p. 317)
gravity anomalies (p. 318)
isostasy (p. 320)
magnetism (p. 321)
magnetic field (p. 321)
magnetic reversals (p. 324)
paleomagnetism (p. 325)

Questions for Review

1. Describe the behavior of a seismic wave as it enters a more rigid medium.

2. Describe how oceanic and continental crust differ in composition and thickness.

3. What is the Moho, and how was it discovered?

4. What is the major difference between the upper mantle and the lower mantle?

5. What is the S-wave shadow zone, and what causes it?

6. What are three possible sources of the Earth's internal heat?

7. Discuss how and why the Earth's gravitational pull varies with topography and latitude.

8. Describe two places where you might find positive gravity anomalies, and two places where you might find negative gravity anomalies.

9. Briefly describe two geological settings where land surfaces rise isostatically, and two settings where surfaces subside isostatically.

10. Briefly discuss our current understanding of how the Earth's magnetic field originated.

For Further Thought

1. The outer core of the Earth's interior is in a liquid state. Why doesn't the liquid rise to the surface and erupt as lava?

2. In films showing the first humans to walk on the Moon, the astronauts seemed to be jumping and bounding due to the weak force of gravity. Why is the gravitational attraction of the Moon so much weaker than that of the Earth? How much gravitational attraction would be likely on the surface of a planet that is much larger and composed of much denser material than the Earth?

3. What kind of gravimetric measurement would you expect to make if you were standing on an enormous body of dense platinum ore at the bottom of a canyon in Antarctica?

4. If the Earth's internal heat supply were to be completely exhausted, what would happen to the asthenosphere and how would isostatic adjustments at the Earth's surface be affected? How would the Earth's magnetic field be affected?

5. Has the Earth always had a magnetic field? What evidence supports your answer?

Determining Plate Velocity

The Nature and Origin of the Ocean Floor

Highlight 12-1 **The Unseen World of Divergent Zones**

Highlight 12-2 **Convergence and the Birth of the Himalayas**

The Origin and Shaping of Continents

The Driving Forces of Plate Motion

Figure 12-1 A satellite image of the Northern Hemisphere, centering on the North Pole.

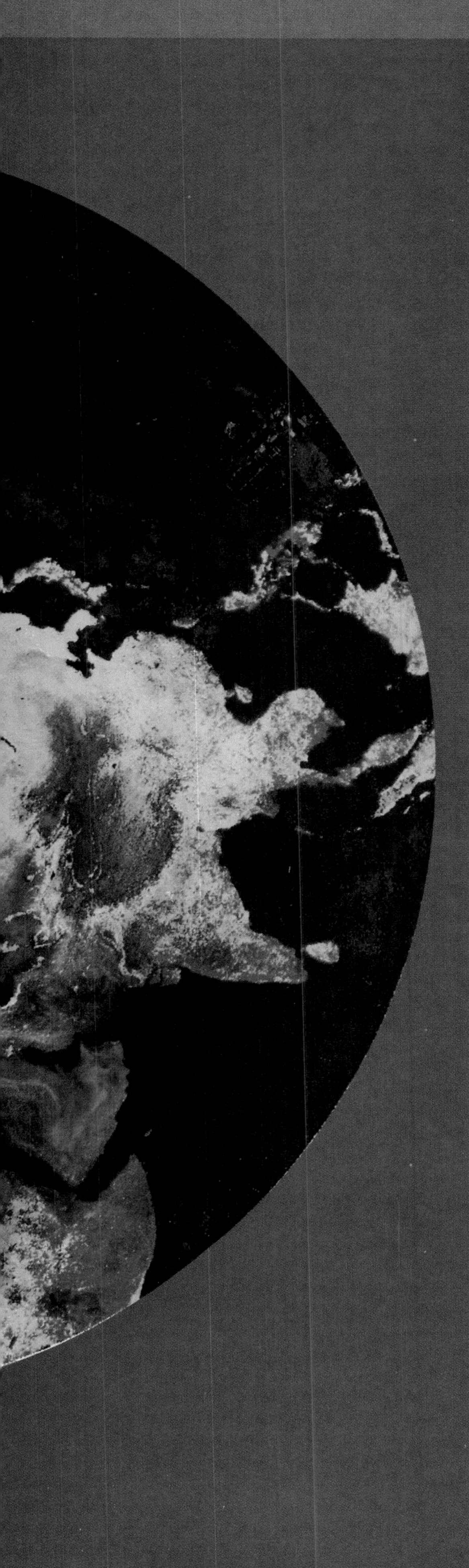

Chapter 12

Plate Tectonics: Creating Oceans and Continents

Since 1957, when the U.S.S.R. launched the first space satellite (Sputnik 1) into orbit around the Earth, satellite technology has vastly improved our knowledge of our planet. In addition to providing dramatic still pictures of the Earth's oceans and continents (Fig. 12-1), data from satellites can show us how these features change and move over time.

Every year, ground-based lasers bounce concentrated beams of light off the Laser Geodynamics Satellite (LAGEOS), recording the amount of time required for the beams to make a round-trip journey. With that information, the precise distance between the satellite and the ground station can be determined. Because LAGEOS speeds around the Earth in the same direction and at exactly the same velocity as the Earth rotates on its axis, its position over the planet is fixed. Any change in the laser beam's travel time therefore represents a change in the geographic position of the landmass containing the ground station. Such changes in position have been detected by instruments at ground stations in widely scattered locations—but the extent of change varies among them. Thus, space-age technology confirms that the Earth's plates do indeed move, and suggests that they do not all move at the same rate.

We have discussed plate movement throughout the first eleven chapters of this text because virtually every aspect of geology—from the origin of earthquakes and volcanoes to the rise and fall of sea level—is affected by plate motion. We examined the development of the theory of plate tectonics and the evidence for it in Chapter 1, and in later chapters detailed some of the basic assumptions of that theory:

- The Earth's lithosphere consists of rigid plates averaging 100 kilometers (60 miles) in thickness.
- The plates move relative to one another by divergence, convergence, or transform motion.
- Plates form at divergent plate boundaries and are consumed by subduction at certain convergent plate boundaries.
- There are three basic types of convergent plate boundaries: those that occur where ocean plates subduct beneath other ocean plates; those that occur where ocean plates subduct beneath adjacent continents; and those that occur where two continental plates collide.
- Most earthquake activity, volcanism, faulting, and mountain building takes place at plate boundaries.

- Plates generally do not deform internally; that is, the centers of plates tend to be geologically stable.

The theory of plate tectonics addresses the origin of both the continents and the oceans. Its predecessor, Alfred Wegener's continental drift hypothesis, generally ignored the ocean basins—it was primarily concerned with past movements of the continents. When Wegener first proposed his hypothesis in 1915 the ocean basins were almost entirely unexplored, and Wegener, like virtually all geologists of his time, believed that ocean floors were very old, featureless plains. (He thought the oceans passively surrounded the drifting continents, which plowed through them on their way to distant points on the globe.) This lack of knowledge about the geology of the sea floor caused most of the conceptual problems that consigned the continental drift hypothesis to decades of rejection. Intense deep-sea exploration, which began in the 1950s, led to the discovery of the spreading ocean ridges and revealed the true nature of the ocean's role in plate motion. Only then could continental drift become a useful concept, underlying today's better understanding of plate tectonics.

In this chapter we will take a large-scale view of plate motion and its effects. We will examine the rates of plate motion, the types of landforms created by moving plates, and the driving mechanisms for plate motion. We will see that plate movements are largely responsible for the creation of the Earth's ocean basins and the geology of the continents.

Determining Plate Velocity

A plate's velocity is a measure of its speed and direction of movement. We can estimate a plate's speed in *absolute* terms only when there is a fixed point outside of the plate to use as a reference (so that we can monitor changes in distance from the fixed point to any feature on the moving plate). Given two or more moving plates and no fixed reference point, we can only determine each plate's velocity *relative* to the others (Fig. 12-2).

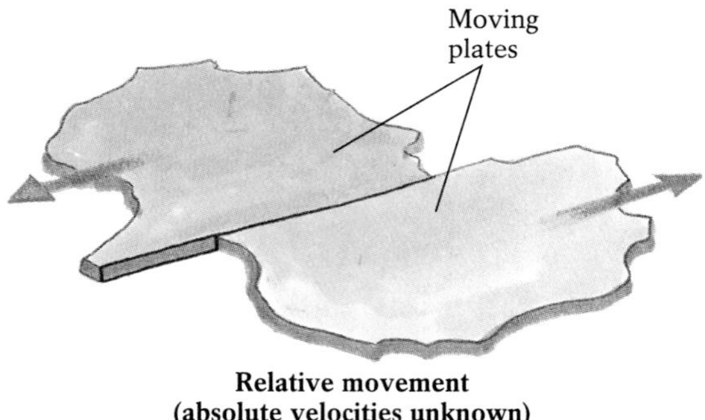

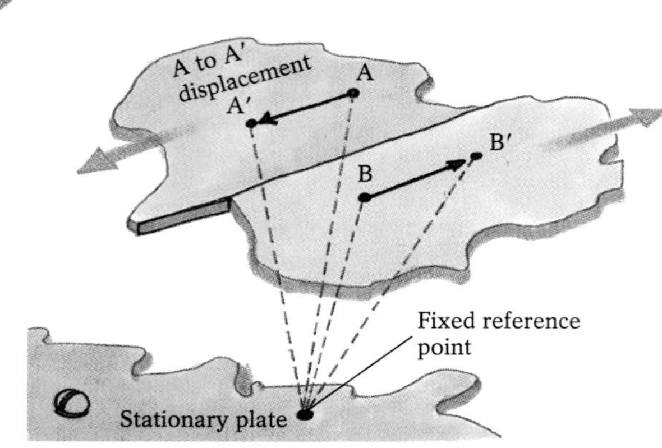

Figure 12-2 Determining relative and absolute plate speeds. If two plates are both moving, and there is no fixed reference point by which their displacement can be measured, we cannot establish the absolute speed of either one; the changing distance between the moving plates reflects only their motion *relative* to one another. A plate's *absolute* speed is determined by measuring its changing distance from a fixed point, such as a feature on an adjacent stationary plate.

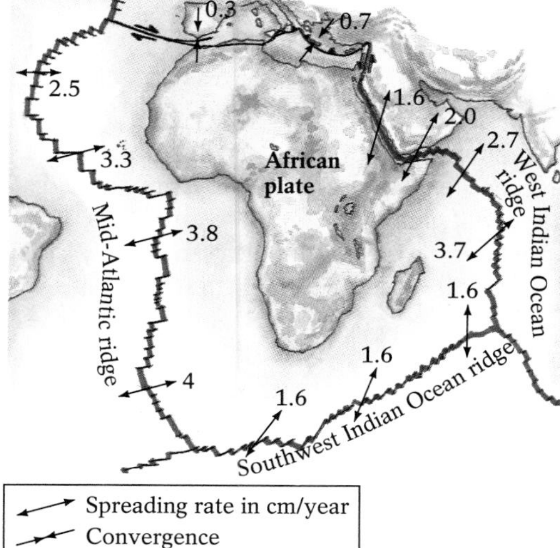

Figure 12-3 The African plate, surrounded by divergent plate boundaries, has probably been stationary for millions of years. The absolute rate at which surrounding plates move can be estimated using Africa as a fixed reference point.

Although all plates can move, they are not all always in motion. At present, some seem to be immobilized. The African plate, for example, is nearly surrounded by plate boundaries that are diverging at similar rates. Equal forces from all directions may be holding Africa in place. If the African plate is in fact stationary, it can be used as a reference point to measure the velocity of the surrounding moving plates (Fig. 12-3).

Hot spots, localized regions where plumes of hot mantle material rise below the base of the lithosphere, provide fixed reference points to determine both plate speed and direction. These pockets of heat in the upper mantle may be moving extremely slowly as the deep mantle convects, but they are effectively stationary relative to the faster-moving lithospheric plates above them. About three dozen hot spots, from which magma rises to form volcanoes, are active today (Fig. 12-4), and perhaps 100 or so have been active in the past 10 million years. Many are at or near divergent plate boundaries, but some (such as the one beneath Yellowstone National Park in Wyoming) are in plate interiors, where they fuel intraplate volcanism.

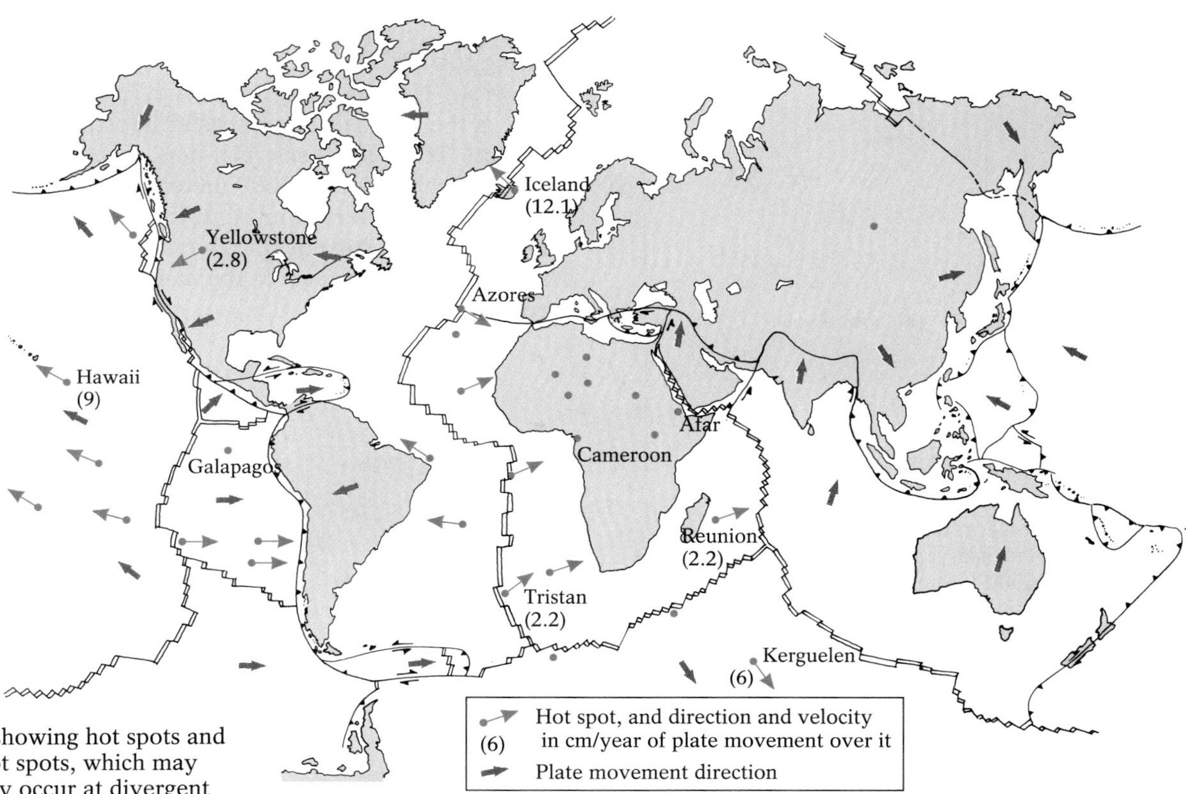

Figure 12-4 A world map showing hot spots and absolute plate motion. Hot spots, which may produce volcanoes, usually occur at divergent plate boundaries, but can also be found under plate interiors. Because hot spots are essentially fixed relative to the faster-moving plates above them, they can be used as references to determine the absolute velocities of the world's tectonic plates.

Plate Movement over Hot Spots

Hot spots beneath oceanic plates produce volcanoes on the sea floor directly above them. As a plate moves across a hot spot, a chain of submarine volcanoes is formed, each of which may grow above sea level and become an island. Eventually, active volcanoes become extinct as they pass beyond the hot spot, and new volcanoes form at the end of the trail of volcanic islands.

The direction of the plate's movement can be deduced from the locations of the extinct volcanoes, and its speed can be determined from the ages of the volcanoes and the distances between each and the hot spot.

Volcanoes that have not grown above sea level form conical submarine mountains called **seamounts.** Some island volcanoes that originally formed over oceanic hot spots have since been worn flat by weathering, wave action, and stream erosion. Called **guyots** (pronounced "ghee-owes"), these are now also found below sea level, as the lithosphere that carried them cooled, became more dense, and subsided deeper into the mantle.

All oceans contain numerous hot spot islands, seamounts, and guyots, generally arranged in long chains that form as drifting plates trail away from hot spots. Most prominent is the Hawaiian Island–Emperor Seamount chain, which extends first to the northwest then north, through more than 6000 kilometers (4000 miles) of the central Pacific, ending at the Aleutian trench (Fig. 12-5). Potassium–argon dating of basalt from the entire chain indicates that these volcanic features developed over a span of more than 75 million years, indicating that this hot spot is long-lived.

The Hawaiian Islands are relatively recent features formed by the central Pacific hot spot. Kauai, whose rocks date from 5.6 to 3.8 million years ago, is the oldest, northwesternmost island in the chain. (The island chain is growing toward the southeast, so the Pacific plate must be moving toward the northwest.) Hawaii, the "Big Island," which has nearly continuous eruptions today, is the youngest, southeasternmost, and only actively volcanic island in the chain. To its southeast is a new seamount, the future island of Loihi, the peak of which is nearly 5000 meters (16,500 feet) above the sea floor. Loihi's summit is still several thousand meters below sea level; if Loihi's eruptions continue at their current rate, however, it will replace the island of Hawaii as the volcanic centerpiece of the central Pacific in another few hundred thousand years.

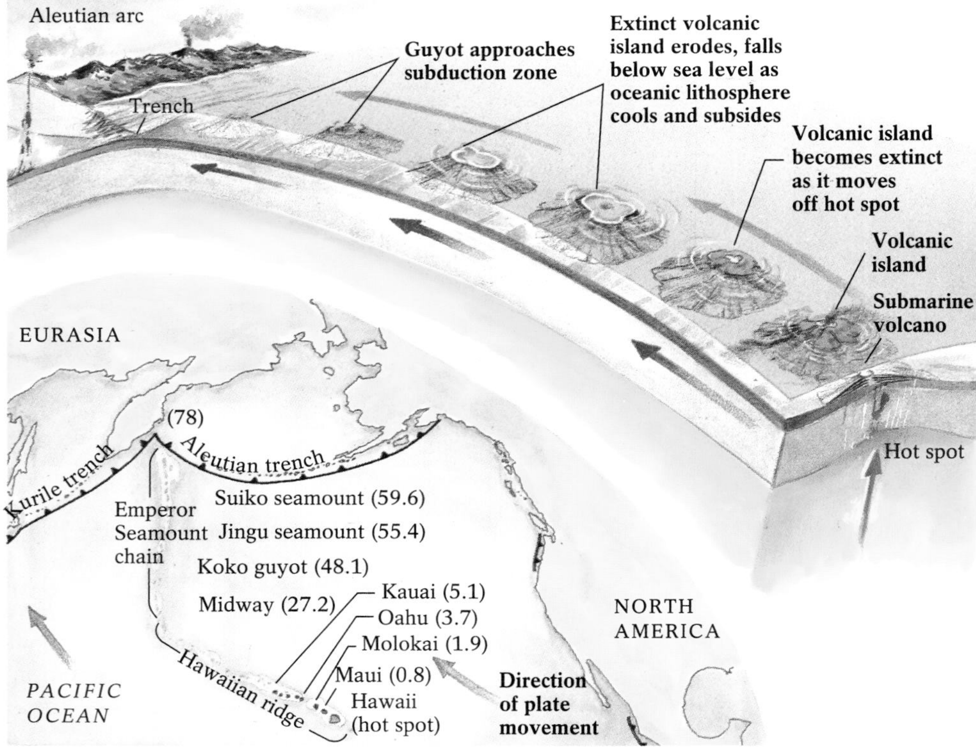

Figure 12-5 The same mid-Pacific hot spot has fueled the eruptions that produced every volcanic island and submarine mountain in the 6000-kilometer (4000-mile)-long Hawaiian Island–Emperor Seamount chain. As the moving plate carries each island away from the hot spot, the volcanoes lose their heat source and become extinct. Weathering and erosion gradually reduce their height above sea level and ultimately they become submerged, as the lithosphere cools and subsides isostatically. (Volcano ages, in millions of years, are shown in parentheses.)

There are several other central-Pacific seamount chains. All are marked by a noticeable bend, presumably formed after the Pacific plate changed direction from due north to its present northwesterly trend. Basalt from volcanoes at these bends dates to about 40 million years ago; thus, we know that's when the Pacific plate changed direction. The rate of Pacific plate movement can be estimated using the Hawaiian hot spot as a fixed reference point. For example, Midway Island is located about 2700 kilometers (1700 miles) northwest of the hot spot, and its basaltic rocks are 25 to 27 million years old. Dividing the distance from the hot spot by the age of the rocks, we conclude that the Pacific plate has moved at an average speed of about 10 centimeters (4 inches) per year.

Hot spots beneath continental plates do not produce such obvious volcanic chains, probably because much of their magma cools and solidifies within the thick continental lithosphere and never reaches the surface. Nevertheless, some chains of igneous formations may be the eroded remains of volcanoes formed over continental hot spots long ago. One such sequence on the North American plate stretches from southwestern Idaho to northwestern Wyoming. Its older features, the basaltic plains of the Snake River, formed about 5 million years ago; the younger geysers, volcanic cliffs, and uplifted plateaus of Yellowstone National Park formed in the last million years or so.

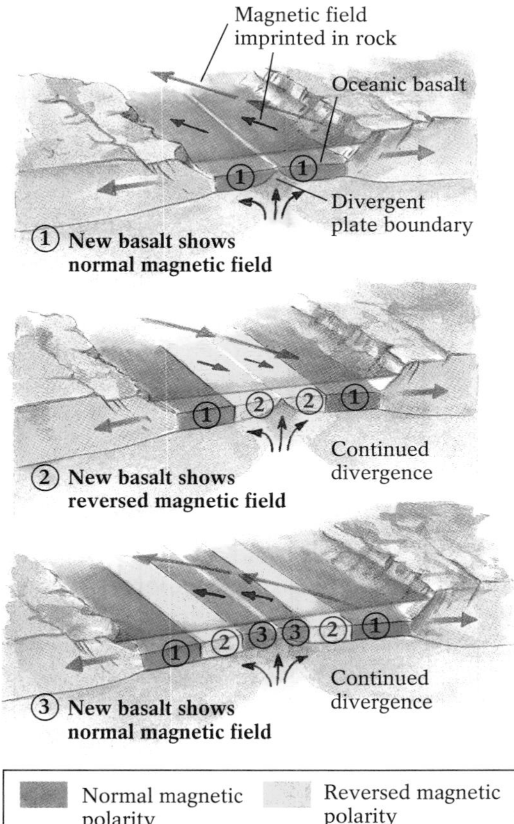

Figure 12-6 Marine magnetic anomalies reflect reversals in the Earth's magnetic field. As basaltic lavas cool and solidify at mid-ocean ridges, their magnetite crystals become aligned with the prevailing direction of the field. Each resulting stripe of basalt has either normal (like today's field) or reversed magnetism (opposite to today's field).

Tracking Magnetic Field Reversals

Rates of oceanic plate motion can also be determined from our growing knowledge of the periodic reversals of the Earth's magnetic field (discussed in Chapter 11). After World War II, military vessels bearing magnetometers traveled across the Pacific and Atlantic Oceans, seeking metal from wrecks of submarines and other military hardware that may have sunk to the ocean floor. The magnetometers recorded a puzzling pattern of magnetic field variations: Alternating stripes of slightly stronger and weaker magnetism along the ocean floor paralleled the mid-ocean ridges.

At first, technicians were perplexed by these bands of variable magnetism, and assumed their equipment was malfunctioning. In the early 1960s, however, Fred Vine, a young Cambridge University student, named the magnetic stripes **marine magnetic anomalies** and proposed that they were evidence of sea-floor spreading at divergent plate boundaries. Vine and his associate Drummond Matthews explained that as basaltic lava cools at spreading mid-ocean ridges, it becomes magnetized in the direction of the Earth's magnetic field at the time of eruption; basalts that erupted at times of reversed polarity would consequently display the magnetic reversal (Fig. 12-6). Magnetometers towed above a normally magnetized stripe record a slight strengthening (about 1%) of the magnetic field; towed above a stripe formed under reversed magnetism, they record a slight weakening of the field. (This is because the Earth's magnetic field is enhanced in the vicinity of rocks of normal polarity, and weakened near rocks of reversed polarity.)

Dating any rock along a marine magnetic reversal tells us the age of the basalt along the entire length of the reversal. Because oceanic plates grow continuously at divergent zones and therefore all contain an uninterrupted record of the oscillations of the Earth's magnetic field, the sequence of magnetic anomalies in oceanic basalt can be used to date any area of sea floor. In addition, marine magnetic anomalies have been used to estimate rates of plate motion. By measuring an anomaly's distance from the spreading ridge, the rate of spreading and therefore of plate motion can be determined: For example, if an anomaly that is known to be 4.5 million years old is located 45 kilometers from the ridge crest, the spreading rate is 1 centimeter per year.

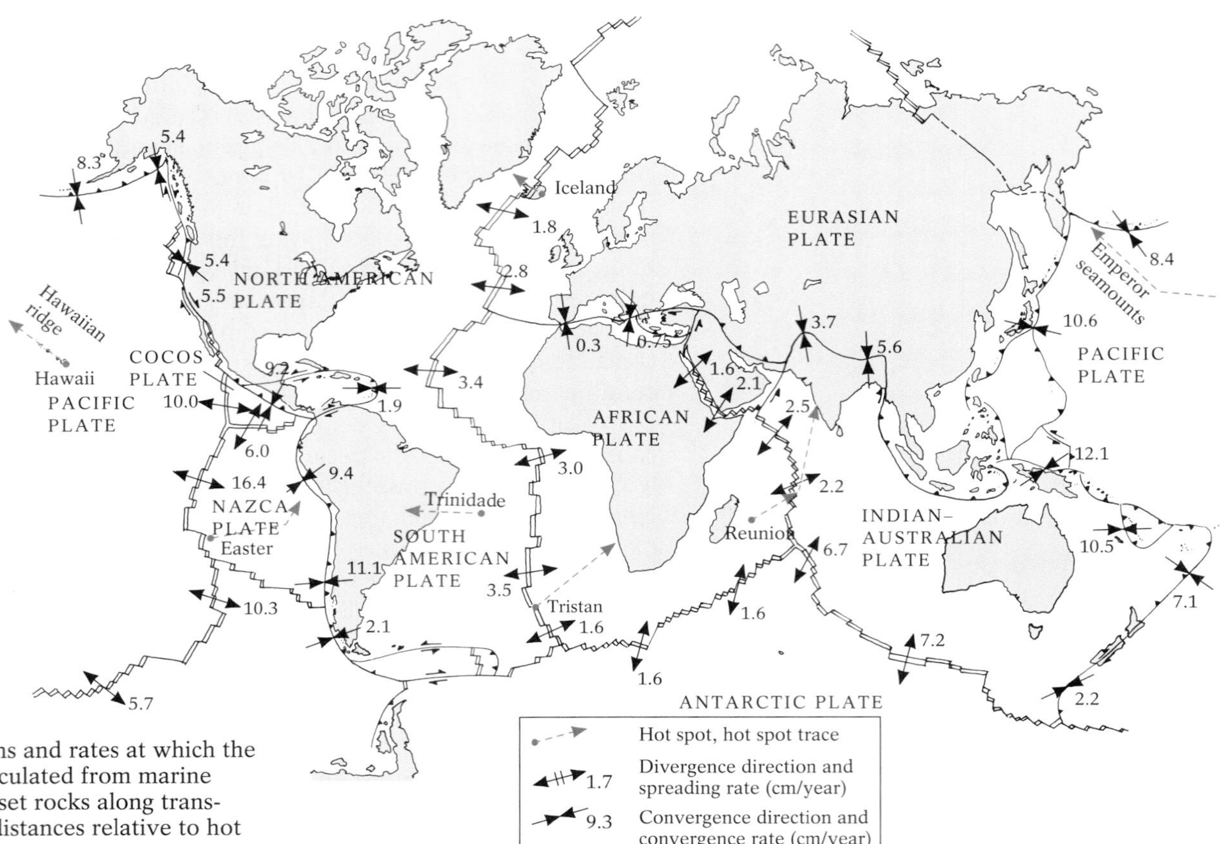

Figure 12-7 The directions and rates at which the Earth's plates move, calculated from marine magnetic anomalies, offset rocks along transform faults, and island distances relative to hot spots.

How fast, then, do plates move? Using orbiting satellites, quasi-stationary continents, oceanic hot spots, and magnetic anomalies, we can measure rates of plate motion and verify their accuracy. The Pacific, Nazca, Cocos, and Australian-Indian plates, among the Earth's fastest, move more than 10 centimeters (4 inches) per year (Fig. 12-7). The North and South American, Eurasian, and Antarctic plates, among the slowest, move from 1 to 3 centimeters (0.4–1.2 inches) per year.

The Nature and Origin of the Ocean Floor

If the ocean basins were drained of their water, we would see chasms as deep as 11,000 meters (36,000 feet), volcanic ranges thousands of kilometers long, chains of flat-topped mountains, long linear fractures oozing lava, and faulted cliffs stretching like walls for great distances. Most of these features were discovered and mapped in the early 1940s, as a by-product of the U.S. Navy's World War II search for enemy submarines and safe passages. The principal method of mapping at that time used **echo-sounding sonar**. A ship using this method emits a sharp pinging noise. The resulting sound waves travel at a speed of 1500 meters (5000 feet) per second, the speed of sound in seawater. The time that elapses as the sound waves bounce off the sea floor and return to a listening device is used to calculate the depth of the ocean bottom and map its topographic features. After World War II, increasingly sophisticated technology became available. Today, we can map subsurface ocean-floor rocks by means of **seismic profiling**, which uses more powerful energy waves. Some of these waves reflect off the sea floor, but others penetrate the surface and reflect off layers of underlying sediment and rock.

Deep-sea drilling and submersible vessels take samples from and photographs of the sea floor. Between 1965 and 1980, a group of universities, research laboratories, and government agencies jointly funded and staffed the Deep Sea Drilling Project (DSDP). The voyages of its *Glomar Challenger* research vessel produced a total of more than 96 kilometers (60 miles) of deep-sea cores to be studied for clues to the origin of ocean basins. In addition, oceanographers on ALVIN, the best-known deep submersible, do field work at the sea floor; they have discovered new lifeforms and geological processes. Even more recently, remote-controlled video cameras have continued and expanded deep-sea exploration.

Exciting new findings have come from space satellites that can indirectly photograph the entire ocean floor by bouncing a pencil-thin beam of microwaves between the spacecraft and the sea surface. NASA's SEASAT, a satellite dedicated to studying the ocean floor, has confirmed that the ocean's surface contains bulges and depressions that reflect variations in sea-floor topography. The added mass of a chain of submarine mountains produces a positive gravity anomaly (see Chapter 11 for a discussion of gravity anomalies) that attracts water, resulting in a sea-surface bulge as high as 30 meters (100 feet). The lesser mass of a deep-sea trench produces a negative gravity anomaly and a corresponding sea-surface depression as much as 60 meters (200 feet) deep. In this way, vast unexplored areas of the world's ocean floors have been charted by satellite (Fig. 12-8).

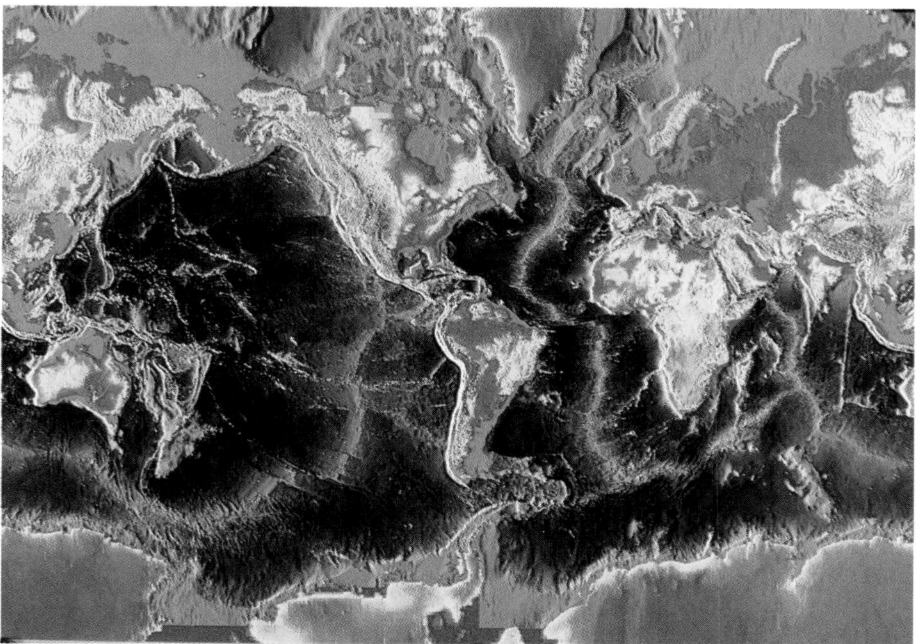

Figure 12-8 This map, which shows the significant variations in the morphology of the sea floor, was created using satellite information on the elevation of the seawater surface. Positive gravity anomalies produced by masses such as submarine mountain ranges attract large quantities of seawater, causing an upward expansion of the ocean surface (light blue areas); conversely, negative gravity anomalies produced by bedrock depressions cause the ocean surface to be lowered (dark blue areas).

Rifting and the Origin of Ocean Basins

Studies of the topography and bedrock composition of the ocean floor have enabled us to speculate about the origin and evolution of ocean basins. As we've learned, oceans form where an existing continent rifts into two or more smaller continents that then diverge. Rifting begins when a warm current of mantle material, rising at a hot spot, stretches and thins the continental crust above, causing its surface to break into a three-branched fracture (Fig. 12-9).

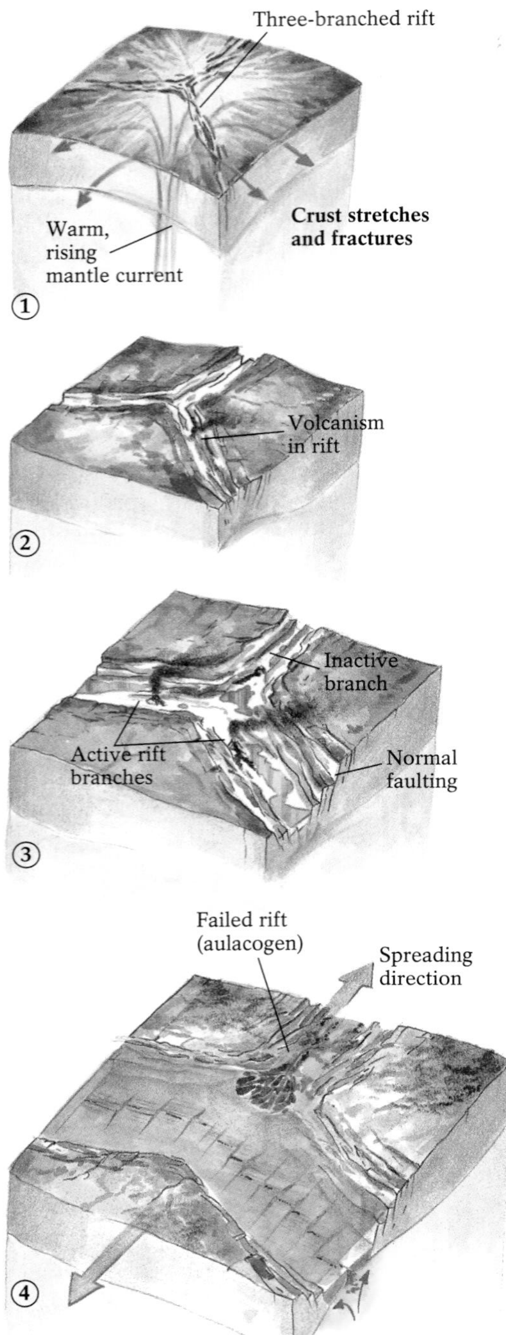

Figure 12-9 Rifting of a continental plate. Rising warm currents from the Earth's interior flow beneath the continental lithosphere, stretching and then tearing it. The initial branched rift typically consists of three fractures, two of which continue to diverge. The third, the aulacogen, becomes inactive.

As the mantle current flows beneath the rifting lithosphere, two of the fractures separate further, while the third becomes inactive. This inactive fracture forms a linear depression known as an *aulacogen* (Greek for "furrow"), or *failed arm*. Many of the world's great river valleys, such as North America's Mississippi, South America's Amazon, and Africa's Niger may be the failed arms of ancient rifts. Intraplate earthquakes in the upper Mississippi valley, such as those in the New Madrid area of Missouri and Tennessee, may be due to the release of stresses built up along such an ancient fracture.

The two actively rifting fractures are marked by high heat flow, normal faulting, frequent shallow earthquakes, and widespread basaltic volcanism. Normal faulting at rift fractures produces steep-walled rift valleys in which sedimentation is particularly vigorous, because of the sharp topographic contrast between the valley floors (graben) and the high-standing horsts marking the edges of the rifted plate (Fig. 12-10).

As rifting progresses, the rift valley widens, possibly expanding until it reaches an adjacent ocean, which floods into the low-lying graben. If the new seaway is later isolated from the adjacent ocean, perhaps due to a drop in sea level, the water in it is likely to evaporate, particularly in hot, arid climates. At this stage in the rifting process, rising and falling sea levels may result in thick layers of evaporites alternating with shallow-marine clastic sediments in rift valleys.

Further rifting deepens the valley and the thick sediment load sinks; more seawater floods in, enough to prevent complete evaporation and evaporite deposition. Sediments accumulating in the seaway are deposited directly on rifted continental rock at first, and then on the new oceanic crust that forms from the basalts rising and cooling in the rift. Sediments from the raised margins of the rift accumulate offshore. The high-standing cliffs above a rifted valley erode, and may eventually be reduced to sea level. As the rocks at the rift margins cool, the land surface subsides and the one-time rift edges become the foundations for continental shelves. After rifting is complete, the original rift edges are no longer plate margins, nor are they volcanically or seismically active. They are now **passive continental margins.** The actual plate margins are half an ocean away, at the still-active spreading center.

There are currently a number of places where the various stages of rifting can be observed. Uplift, high heat flow, and stretching and thinning of continental crust appear to be taking place in the Rio Grande area of Colorado and New Mexico, and the Basin and Range province of Arizona, Nevada, and Utah. In the future these may become rifts that will cleave North America into two or more smaller continents, perhaps separated by a major seaway. More advanced rifting can be seen in the Afar Triangle in East Africa, where a three-branched fracture has already separated formerly contiguous Africa and Arabia. The rift's two active arms are the Red Sea and the Gulf of Aden (Fig. 12-11), where divergence continues. The East African branch of the three-part rift, today occupied by the large lakes near which much fossil evidence of early humans has been found, is probably an aulacogen. The Great Rift valley of East Africa has not grown wider for several million years and it appears that rifting may have halted there altogether, or be very slow. It may never evolve into an ocean, but instead remain a sediment-filled topographic depression that will someday accommodate Africa's next great river system. The Red Sea, on the other hand, is on the threshold of becoming a true ocean. As soon as continued rifting fully breaches the Sinai and Somali peninsulas, the Red Sea will connect the Mediterranean and the Indian Ocean. It has already begun to develop an area of true oceanic basalt crust with marine magnetic anomalies.

The final stage in the evolution of rifted continental margins can be found today on the west (North and South American) and east (European and

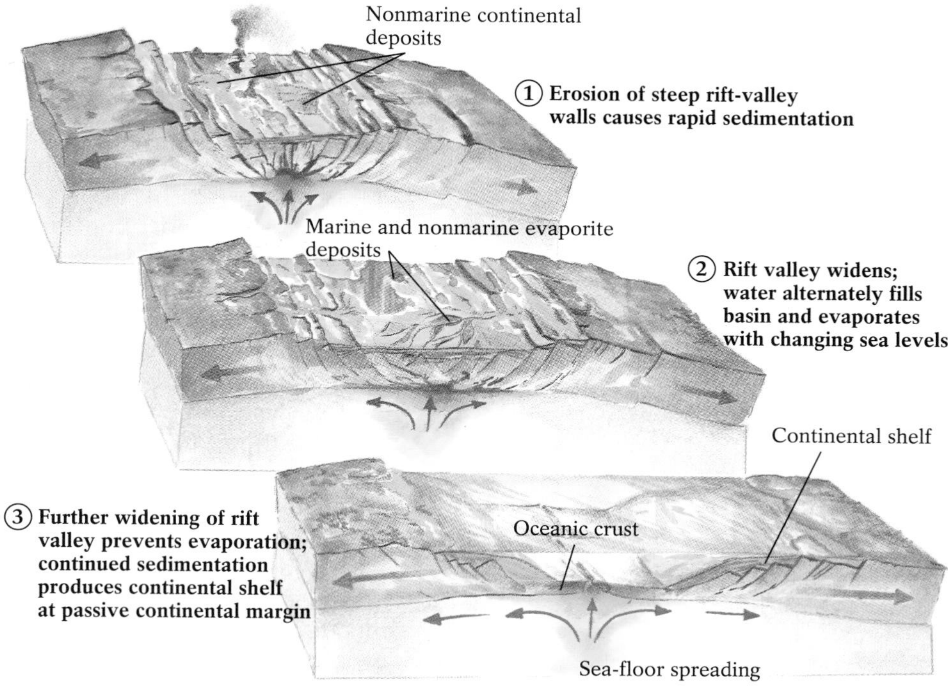

Figure 12-10 The growth of an oceanic basin in a rift valley, and the types of sedimentation that occur during its various stages.

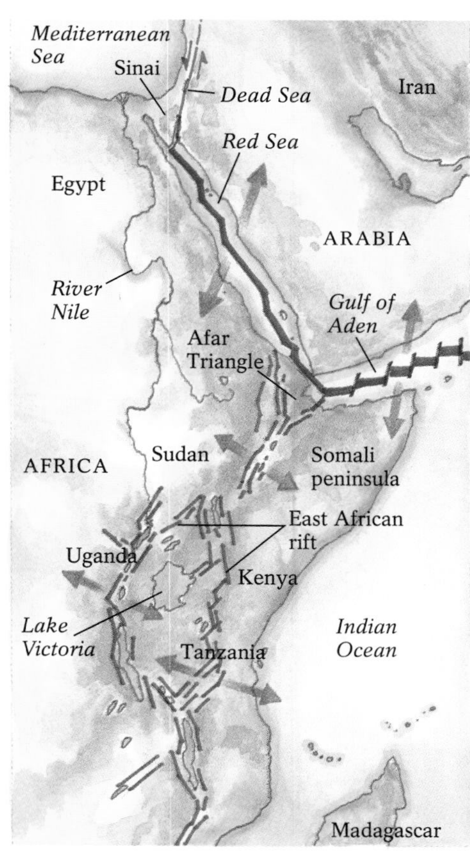

Figure 12-11 The East Africa rift zone has two actively expanding fractures—the Red Sea and the Gulf of Aden. The valleys and gorges of Kenya and Tanzania are the failed arm, or aulacogen, of the rift.

African) coasts of the Atlantic Ocean. Some evidence of the Atlantic's initial rifting lies beneath the sediment of these continental shelves. On land, much of the rifting record can be seen in the fault-bound, 200-million-year-old sediments of the northeastern United States and the numerous outcrops of basalt found along the Eastern seaboard from New England to North Carolina. Faulted basins and rocks of the same age and composition occur throughout the British Isles and northwestern Africa. We now know that 180 million years ago, these British and African rocks coexisted with today's North American rocks on the supercontinent of Pangaea. Their journey to their present locations began with the rift that separated the Americas from Europe and Africa. It may have taken as many as 20 distinct hot spots to provide the thermal energy to rift Pangaea into today's major continents, most of which had taken shape by about 65 million years ago.

Rifting sometimes ceases following an initial period of faulting and volcanism, but we do not yet understand why. One such aborted rift is believed to have begun about 1.1 billion years ago, as an outpouring of mafic lava reached the surface in the middle of the North American continent. Basalts and gabbros of this age are found along the north and south shores of Lake Superior in Minnesota and Wisconsin. Farther south, similar rocks form a linear zone of anomalously high gravity buried beneath younger Cambrian rocks. This mid-continent gravity high, tens of kilometers wide, stretches through Iowa and Missouri to the Oklahoma panhandle (see Fig. 11-16). Had this rifting continued, midwesterners today might be taking trans-oceanic flights between Milwaukee and Minneapolis.

Divergent Plate Boundaries and the Development of the Ocean Floor

Once rifting is complete, oceanic crust begins to form on either side of the spreading center, which lies between the two new plate margins. As the two plates grow, an underwater mountain range, or **mid-ocean ridge**, develops between them, formed by the continuing underwater eruption and accumulation of basaltic lava. Today, mid-ocean ridges stretch continuously for about 65,000 kilometers (40,000 miles) across all the major basins (see Fig. 1-15).

They can be as wide as 1500 kilometers (900 miles), and in places their peaks rise more than 3 kilometers (about 2 miles) from the ocean bottom. Mid-ocean ridges are the largest raised topographic features on the Earth's surface, and their length and breadth account for approximately 23% of the Earth's total surface area.

Whereas most continental mountain systems consist principally of folded metamorphic and sedimentary rocks, and massive batholiths of felsic igneous rock, the ocean-ridge chain contains only relatively undeformed basalt. Most ridges are split down the middle by an *axial rift valley*. These valleys form as a large central block drops downward by normal faulting along the *axial ridge crest*. Some axial rift valleys are deeper than the Grand Canyon and three times as wide. Recent dives by submersibles directly into these valleys have brought back evidence of remarkable, previously unknown, biological and geological processes, which are described in Highlight 12-1.

Oceanic lithosphere everywhere has a similar structure. Its upper surface to about 200 meters (700 feet) below the sea floor consists of unconsolidated sediment of siliceous or carbonate ooze from the remains of microscopic marine organisms, fine reddish-brown clay from weathering of iron-rich marine lavas, or both. Below this is a 2-kilometer (1.2-mile)-thick layer of oceanic basalt, its top characteristically pillowed from its eruption as lava under water. Under the basalt lie 5 to 6 kilometers (3-4 miles) of gabbro, formed from slow plutonic crystallization and crystal settling. A layer of the ultramafic mantle rock peridotite is at the base of a typical ocean plate.

This entire sequence of ocean-floor rock may be altered chemically as seawater penetrates its dikes, faults, and fissures. The water reacts with the pyroxene in the basalt and gabbro to form green chlorite, and with the magnesium olivine in the ultramafic peridotite to form the magnesium silicate mineral serpentine. These reactions may eventually produce *serpentinite*, a soft, greenish rock with a contorted, snake-like appearance. The geological name for the group of rocks that make up oceanic lithosphere is the **ophiolite suite** (from the Greek *ophis*, or "serpent", and *lithos*, or "rock") (Fig. 12-14).

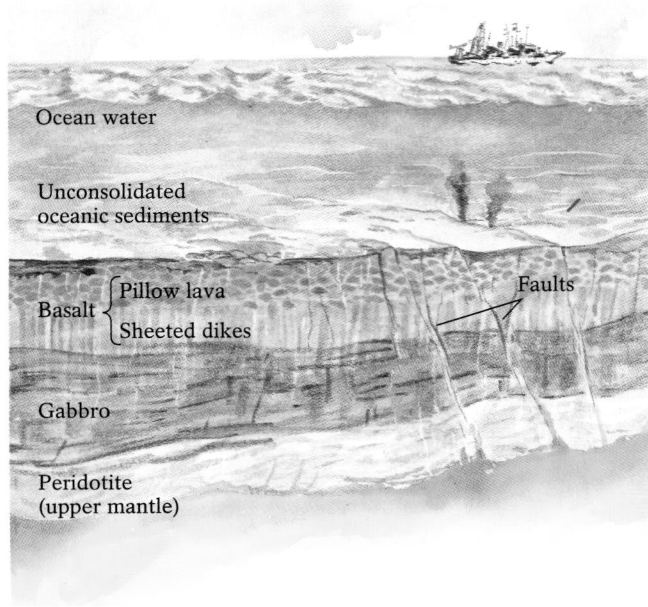

Figure 12-12 The layers of the ophiolite suite, which make up oceanic lithosphere.

Highlight 12-1 *The Unseen World of Divergent Zones*

Since the early 1970s, oceanographers using deep-sea submersibles have studied the rift valleys of the mid-Atlantic ridge, the Galápagos ridge off the coast of Ecuador and Peru, and the East Pacific rise south of Baja California. They have photographed the eruption of pillow lavas and the chemical interaction of cold seawater and warm basalt. Recent dives into the valleys of the East Pacific rise and the Juan de Fuca ridge (off the coast of Washington state) returned with videos showing plumes of black sulfurous "clouds" of mineral-laden water rising from vertical chimney-like structures. These hydrothermal vents emit extremely hot water carrying various gases such as sulfur dioxide and water vapor and numerous metallic sulfides that nourish a complex community of hitherto unknown lifeforms. Giant clams and exotic tube worms feed on bacteria that thrive in high temperatures, in a world untouched by sunlight (Fig. 12-13).

Vast quantities of valuable minerals are accumulating within axial rift valleys. Seawater seeps down into newly formed oceanic crust, becomes superheated by the underlying magma reservoir, and is then able to dissolve copper, iron, zinc, cobalt, silver, and cadmium out of the warm mafic rocks. These mineral-rich waters, when heated to about 400°C (750°F), rise and erupt at the sea floor, creating dark plumes called *black smokers* (Fig. 12-14). Contact with cold seawater causes minerals to precipitate from the plumes, encrusting the basalt flows around each plume vent to form a chimney-like structure. Similar ore combinations are found within many of the Earth's folded mountains; it is likely that they are ancient slabs of oceanic crust that originated in this manner.

Figure 12-13 Bacteria around volcanic vents at a mid-ocean ridge derive energy from heat-generated chemical reactions involving compounds such as hydrogen sulfide (H_2S). More complex creatures, such as these giant tube worms, subsist on these bacteria.

Figure 12-14 Black smokers, hot plumes of mineral-rich water vented at volcanically active regions of the sea floor. Accumulations of precipitated minerals form chimney-like structures around the plumes. The "smoke" here consists of hot water and grains of iron, copper, and other sulfide minerals.

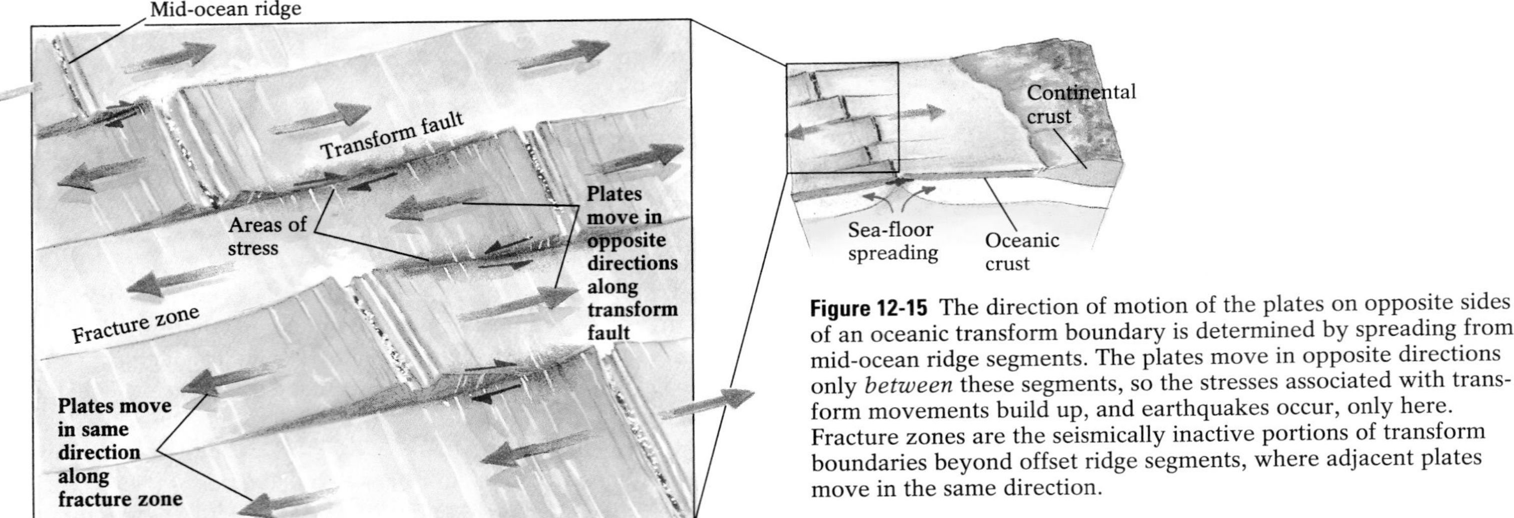

Figure 12-15 The direction of motion of the plates on opposite sides of an oceanic transform boundary is determined by spreading from mid-ocean ridge segments. The plates move in opposite directions only *between* these segments, so the stresses associated with transform movements build up, and earthquakes occur, only here. Fracture zones are the seismically inactive portions of transform boundaries beyond offset ridge segments, where adjacent plates move in the same direction.

Transform Boundaries and Offset Mid-Ocean Ridges

We said earlier that the Earth's ocean-ridge system is essentially continuous, but, as Figure 1-15 showed, it is not represented by a smooth continuous line. Rather, the ridge consists of countless segments interrupted by abrupt perpendicular offsets. These offsets occur when plates slid past one another in opposite directions at transform boundaries.

As we saw in Chapter 9, transform boundaries contain faulted plate blocks that move in opposite directions, producing stresses that may cause earthquakes. Earthquakes associated with oceanic transform boundaries almost always occur between the offset ridge segments, where the opposing motion of the two spreading ridges generates enough shearing stress to break rocks (Fig. 12-15). Because beyond this inter-ridge area plate movement is in the same direction on both sides of the transform boundary, little shearing stress is produced and earthquake activity is minimal.

California's San Andreas fault zone is a partially oceanic transform boundary, extending from the northern end of the East Pacific rise in the Gulf of California to the southern end of the Gorda ridge off the coast of southern Oregon and northern California. It was created by a complicated sequence of events involving both oceanic transform movement and subduction (Fig. 12-16).

Residual northern Farallon (Juan de Fuca) plate
NORTH AMERICAN PLATE
San Andreas fault
Gulf of California
East Pacific rise
Residual southern Farallon (Cocos) plate
Gorda ridge
PACIFIC PLATE
Present
Transform boundary moves on land
Baja California begins to separate
3 million years ago
Transform boundary lengthened
10 million years ago
NORTH AMERICAN PLATE
Baja California
Southern Farallon plate
Northern Farallon plate
Transform boundary
PACIFIC PLATE
Spreading center
About 30 million years ago

Figure 12-16 The process that created the San Andreas fault began about 30 million years ago, as the westward-drifting North American plate made contact with a segment of the spreading center separating the Farallon and Pacific plates. The North American plate had not yet reached a more northern segment of the spreading center, which was offset to the west by a transform boundary; subduction of the southern portion of the spreading center under North America effectively split the Farallon plate into separate northern (now called the Juan de Fuca) and the southern (now called Cocos) plates. As the southern Farallon subducted, the transform boundary between the northern and southern plates became longer; about 3 million years ago it "moved" onshore, becoming the San Andreas fault. Also about this time, the Cocos plate spreading center (the East Pacific rise) met Baja California, separating it from the rest of the continent and forming the Gulf of California. Today, the North American plate continues to override what remains of both the northern and southern Farallon plates; in a few million years, the San Andreas fault will most likely extend the full length of the west coast.

Convergence and Subducting Plate Margins

Oceanic plates diverging from mid-ocean ridges eventually encounter less-dense plates and are subducted under them, becoming reabsorbed into the Earth's mantle. Large segments of the ocean floor have been swallowed in this way. The Pacific, for instance, is a shrinking ocean, with subduction occurring along most of its margins. The spreading ridge of the East Pacific rise is no longer in the ocean's center because the eastern part of the Pacific has been consumed by subduction along the west coasts of the Americas. As oceanic plates descend, the overriding continental plates become elevated; the large-scale destruction of Pacific oceanic lithosphere has produced the great western mountain ranges of the Andes, Sierra Nevada, Cascades, and Rockies.

Pacific plate subduction also explains why there is an asymmetrical magnetic anomaly pattern off the coast of Washington and British Columbia, while a beautifully symmetrical pattern surrounds the mid-Atlantic ridge at Iceland (Fig. 12-17). Most of the oceanic plate east of the Gorda and Juan de Fuca spreading ridges has been consumed by subduction, thus eliminating most of the magnetic anomaly pattern east of the ridges. Neither of the oceanic plates forming at the mid-Atlantic ridge is subducting; therefore their anomaly patterns are symmetrical.

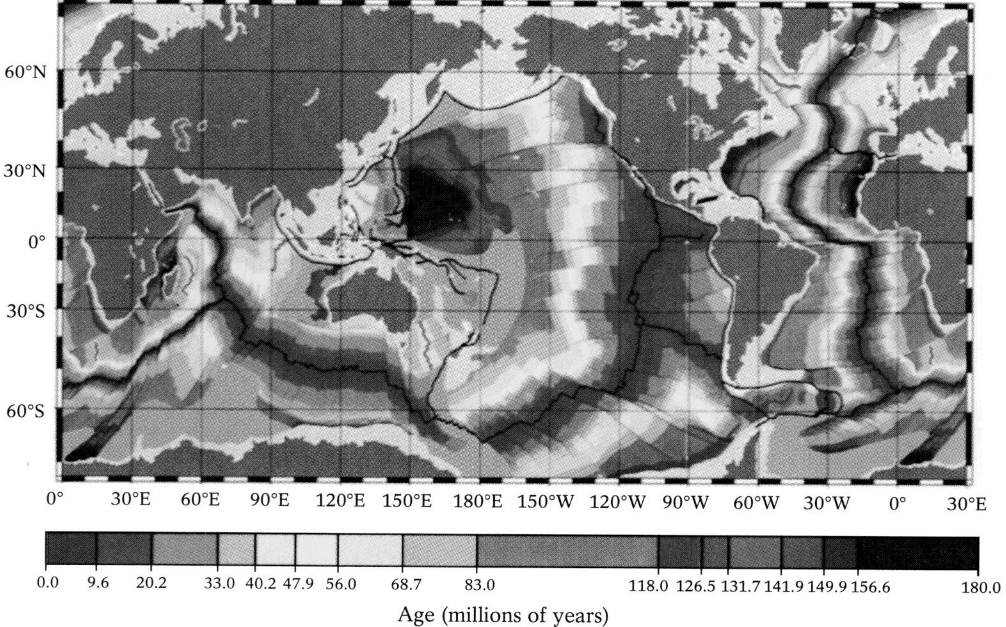

Figure 12-17 A world map showing the ages of oceanic lithosphere segments, as determined by dating marine magnetic anomalies. The colored stripes represent oceanic lithosphere of the ages indicated in the key. The width of each stripe is proportional to the spreading rate at mid-ocean divergent plate boundaries. A symmetrical anomaly pattern (upper right) characterizes a mid-ocean spreading center. An asymmetrical pattern (center) shows that subduction has consumed part of an oceanic plate.

Ocean Trenches and Sedimentation **Ocean trenches** are deep, linear, relatively narrow depressions in the Earth's surface that develop where oceanic plates subduct. Some Pacific trenches, particularly in the tectonically active western region, are between 40 and 120 kilometers (25–75 miles) wide and can be thousands of kilometers long. The deepest is the 11,022-meter (36,372-foot)-deep Marianas trench near the island of Guam. Sediment scraped from the subducting plate and eroded from the overriding plate accumulates in these trenches, which may become completely filled where sedimentation rates are high, such as next to rapidly rising mountain belts. The trench off Oregon and Washington, for example, has become filled during the last 3 million years by sediments eroding from the rising Cascade Mountains.

Figure 12-18 A mélange along the Sonoma County coast in California. These rocks, a jumbled mixture of sea-floor and land-derived materials that have undergone high-pressure/low-temperature metamorphism, formed in the ocean trench associated with the subduction of the Farallon plate under North America.

As a cold oceanic plate subducts, a mixture of fine-grained deep-sea sediments, coarser land-derived sediments, siliceous and carbonate oozes, submarine basalts, and serpentinized gabbros and peridotites is packed against the inner wall of the trench. Caught in the high-pressure zone between converging plates, the resulting mass solidifies to become rocky material that is sliced, crushed, and thrust into a chaotic jumble called a **mélange**. Relatively cold slices of oceanic lithosphere and associated sediments are subducted rapidly to depths of 30 kilometers (20 miles) or more, where they metamorphose under the unique combination of low-temperature and high-pressure conditions. The blueschist minerals that commonly form within a mélange indicate metamorphism within a subduction zone. (See Chapter 7 to review subduction-zone metamorphism.)

In North America, mélanges that have been uplifted subsequent to their formation are found within the Franciscan rocks of the coast ranges of California, a product of the consumption of the Farallon plate (Fig. 12-18). They are also in the Klamath Mountains of southern Oregon, in the Kootenay Mountains of eastern British Columbia, in the coastal mountains of south-central Alaska, in the Blue and Wallowa Mountains of eastern Oregon, and in the Appalachians of New England and the Canadian maritime provinces. Each of these mélanges strongly indicates past subduction.

Features of Subduction Zones The process of subduction produces a variety of characteristic structures, illustrated in Figure 12-19. Mélanges and other thrust-faulted rocks pile up within an ocean trench to form an **accretionary wedge**, a mass of sediments and oceanic lithosphere scraped from the subducting plate and plastered onto the edge of the overriding plate. As rocks continue to accumulate, the wedge thickens and is lifted isostatically, causing a linear range of mountains to form just inland of the trench; the coast ranges of California, such as those near Big Sur (Fig. 12-20), originated in this way. Farther inland there may be a **volcanic arc**, a chain of volcanoes fueled by magmas melting and rising from the subducting plate.

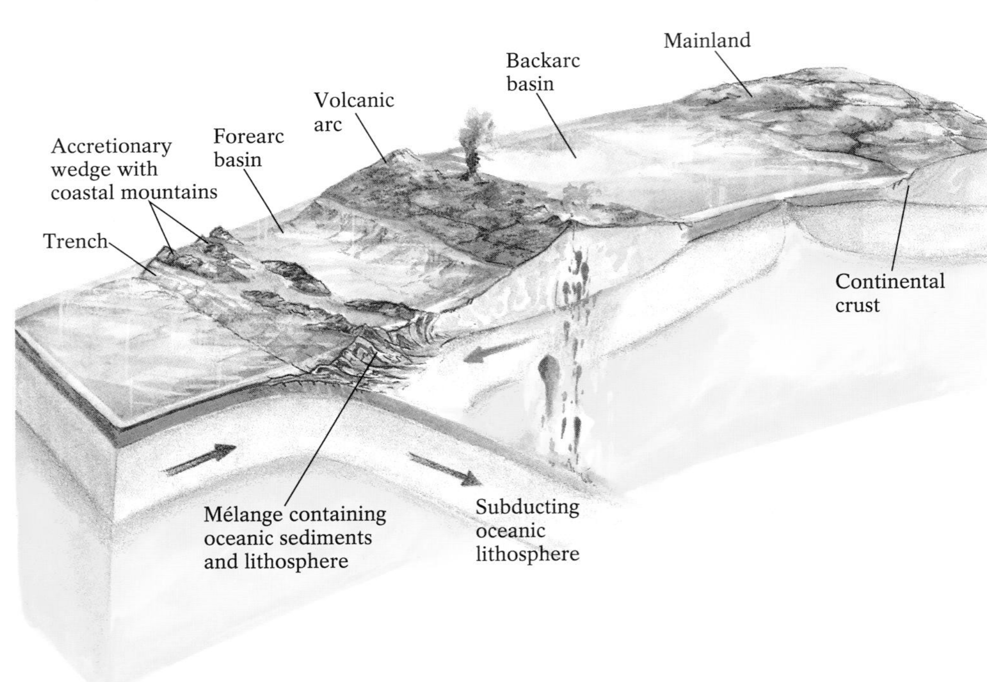

Figure 12-19 Subduction-zone features. (See text for discussion.) These features can be found at the subduction zones between converging oceanic and continental plates or between two plates of oceanic lithosphere.

Figure 12-20 Coastal mountains typically begin as an accretionary wedge at a subduction zone. These coastal mountains near Big Sur, California, originated as offshore turbidites of continental origin which became packed in the Farallon trench and were dragged downward tens of kilometers before being uplifted isostatically.

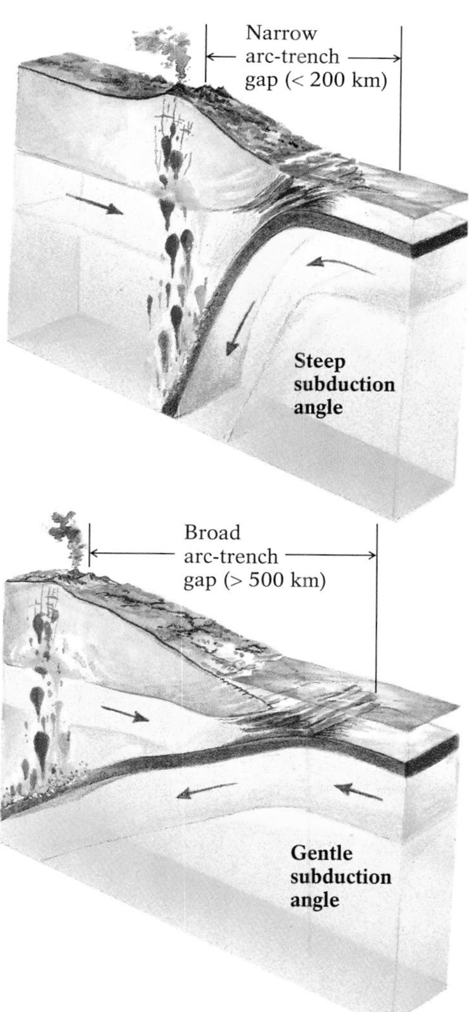

Figure 12-21 The breadth of an arc-trench gap is proportional to the size of the angle of subduction: A steep angle produces a narrow arc-trench gap, and a gentle angle produces a broad arc-trench gap.

Between the accretionary wedge and volcanic arc is a sediment-trapping depression called the **forearc basin**. Sediment eroded from both the uplifted accretionary wedge and the volcanic arc accumulates in the forearc basin. Thus, this sediment may be of either marine or terrestrial origin.

The distance between the ocean trench and the volcanic arc is called the *arc-trench gap*. The breadth of the arc-trench gap is determined by the angle at which the subducting plate descends. Steep subduction transports a descending plate relatively quickly to greater depths and higher temperatures, generally producing narrow arc-trench gaps and narrow forearc basins. Where the angle of subduction is relatively gentle and therefore requires more lateral distance to reach greater depths, wide arc-trench gaps and forearc basins form (Fig. 12-21).

On the inland side of a volcanic arc, another sediment-trapping depression, or **backarc basin**, forms. Sediments deposited here are derived from the eroding volcanic arc as well as from continental streams flowing toward the sea. Backarc basins form when, during subduction at an ocean trench, the overriding plate becomes stretched and thinned on the landward side of the volcanic arc. This may be due either to local mantle currents created by the subduction or to tensional stresses resulting from gravitational pull on the subducting plate; in either case, the crust of the overriding plate may stretch so much that it eventually rifts, enabling currents of basaltic magma to rise and solidify into new oceanic crust, a process called **backarc spreading** (Fig. 12-22). Backarc spreading beneath the Sea of Japan, for instance, is causing the volcanic island arc of Japan to move eastward, as the backarc region expands from the creation of new sea-floor crust between Japan and the Chinese mainland. Backarc spreading is marked by thinning plates, high heat flow, normal faulting, frequent earthquakes, and basaltic volcanism.

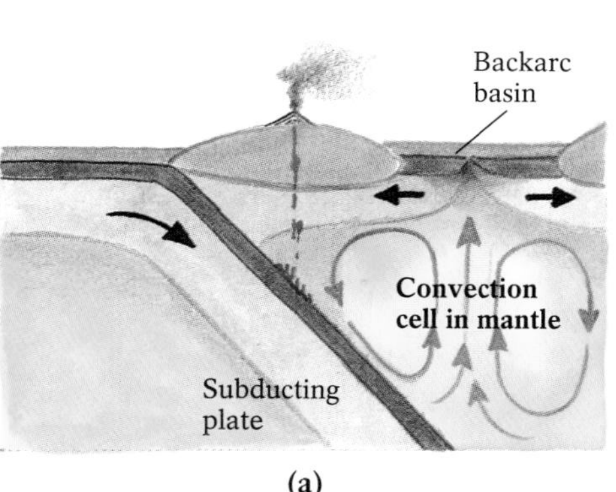

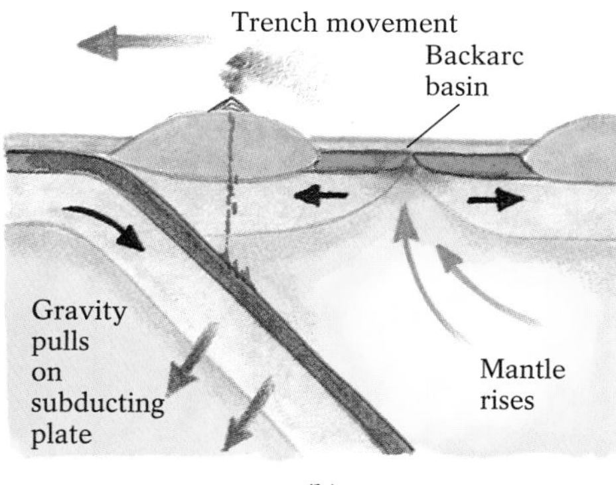

Figure 12-22 Two proposed mechanisms of backarc spreading. **(a)** Subduction-induced mantle currents. **(b)** Tension from gravitational pull on the subducting plate.

Convergence and Continental Collisions

Given enough time, the entire oceanic portion of any subducting plate is reabsorbed into the Earth's mantle. At this point, the plate's continental portion (if it has one) reaches the subduction zone and may encounter another continental plate there. This results in the third principal type of plate convergence—continental collision. As we have seen, continental plates have relatively low density and thus are buoyant. When the forward edges of two continental plates converge, neither can subduct. With nowhere to go but up, the lightweight continents collide and compress each other's rocky boundaries, becoming welded together into a single larger block of continental rock. A lofty mountain range marks the site of the collision. The world's highest mountains, the Himalayas of southern Asia, were formed and continue to grow because of the ongoing continental collision between China and India (Highlight 12-2).

The boundary between collided continents is a **suture zone**. During a collision, continental crust within a suture zone thickens because one plate is thrust slightly beneath the other. The crust also thickens there because slices faulted from the colliding edges of both plates are swept into fold-and-thrust mountains. Because fragments of an oceanic plate are often trapped between colliding plates and become attached to the continents, mountains high above sea level sometimes contain large masses of ophiolite rocks. The complex folded and thrust-faulted mountains in suture zones may also contain anything else that lies in the path of the colliding plates. This is why remnants of stratovolcanoes and felsic batholiths from collided volcanic arcs, or metamorphosed mélanges from the subduction trenches of departed oceanic plates, can sometimes be identified within a mountain's rocky structure.

Collisions such as these produced many of the Earth's mountain ranges, including the Alps in southern Europe, the Appalachians in eastern North America, and the north–south Urals that formed during collision of Europe and Asia. Today, continental plates are actively colliding in only one place: from the northern shore of the eastern Mediterranean, where Africa impinges on southern Europe, across to the Himalayas, where India continues to butt up against southern Asia. This collision produces frequent earthquakes, including Armenia's devastating tremors in the late 1980s.

Based on evidence from the past, ancient suture zones are likely sites for rifting in the future. One convincing piece of evidence is the rift that currently exists along the ancient African–European suture zone, in which the Mediterranean Sea has formed and re-formed several times; the sea lies between the African and Eurasian plates, the multiple collisions of which have in the past intermittently uplifted the European Alps. After a lengthy period of rifting, divergence, and ocean building, convergence and subduction may resume, reuniting formerly contiguous landmasses. Along the northern shore of the Mediterranean today, subduction is again closing the gap between Africa and Europe, which will eventually form a brand-new set of Alps. Italian volcanism, devastating Greek and Turkish earthquakes, and uplift of the island of Cyprus from the sea floor are some hints of the impending collision.

Similarly, the modern Atlantic Ocean opened 180 million years ago along a suture zone that had formed 20 million years earlier, when North America, Europe, and Africa collided to form the supercontinent Pangaea, creating the Appalachian mountain system (see Chapter 9). But rifting does not always precisely follow the line of an old suture zone: When Pangaea rifted apart, a large block of Africa, hundreds of kilometers wide, remained attached to North America east of the Appalachians, stretching from New York to Florida.

Highlight 12-2 *Convergence and the Birth of the Himalayas*

The creation of the Himalaya Mountains, formed from the convergence, collision, and suturing of India to Asia over millions of years, represents a significant event in the geologic history of the Earth (Fig. 12-23). Among its other effects, the uplift of this mountain chain forever changed the climate of Asia by isolating it from the southern oceans.

The process leading to the Himalayan orogeny began during the Mesozoic Era, when India separated from the Pangaean assembly of continents (see Chapter 1). About 180 million years ago, India together with Madagascar broke away from Pangaea and began to drift rapidly northward. (Soon after, Madagascar rifted away from India.) The Tethys Ocean, which separated the Indian continent from Asia, was being consumed along a subduction zone south of Asia. The drifting India passed over the stationary Reunion hot spot about 65 million years ago, a period marked by the eruption of the extensive Deccan basalt flows. (Today, the Reunion hot spot is about 5000 kilometers, or 3000 miles, southwest of India in the Indian Ocean.) For the next 30 million years India continued to move rapidly northward as subduction along the southern edge of Asia continued. The collision of India with Asia began about 35 million years ago, when the northern margin of India collided with the Tibetan microcontinent. By about 10 million years ago, the Tethys Ocean was entirely closed and the Himalayan orogeny entered its next stage, which continues to the present day. Sediments on the leading edge of the Indian plate and accretionary wedges, volcanic arcs, and batholiths along the southern edge of the Asian plate were folded and thrust-faulted up onto the continents to form the modern Himalayan mountain belt.

Continued convergence of India into the Asian continent results in repeated thrust faulting of the leading edge of the Indian plate, increasing the thickness of the continental crust beneath the emerging Himalayas. This increased thickness is seen today as the highest plateau on Earth, the Tibetan Plateau. Several suture zones mark the boundary between the colliding continents and continental fragments trapped within the collisional zone. India is still moving into Asia today, at the rate of about 5 centimeters per year.

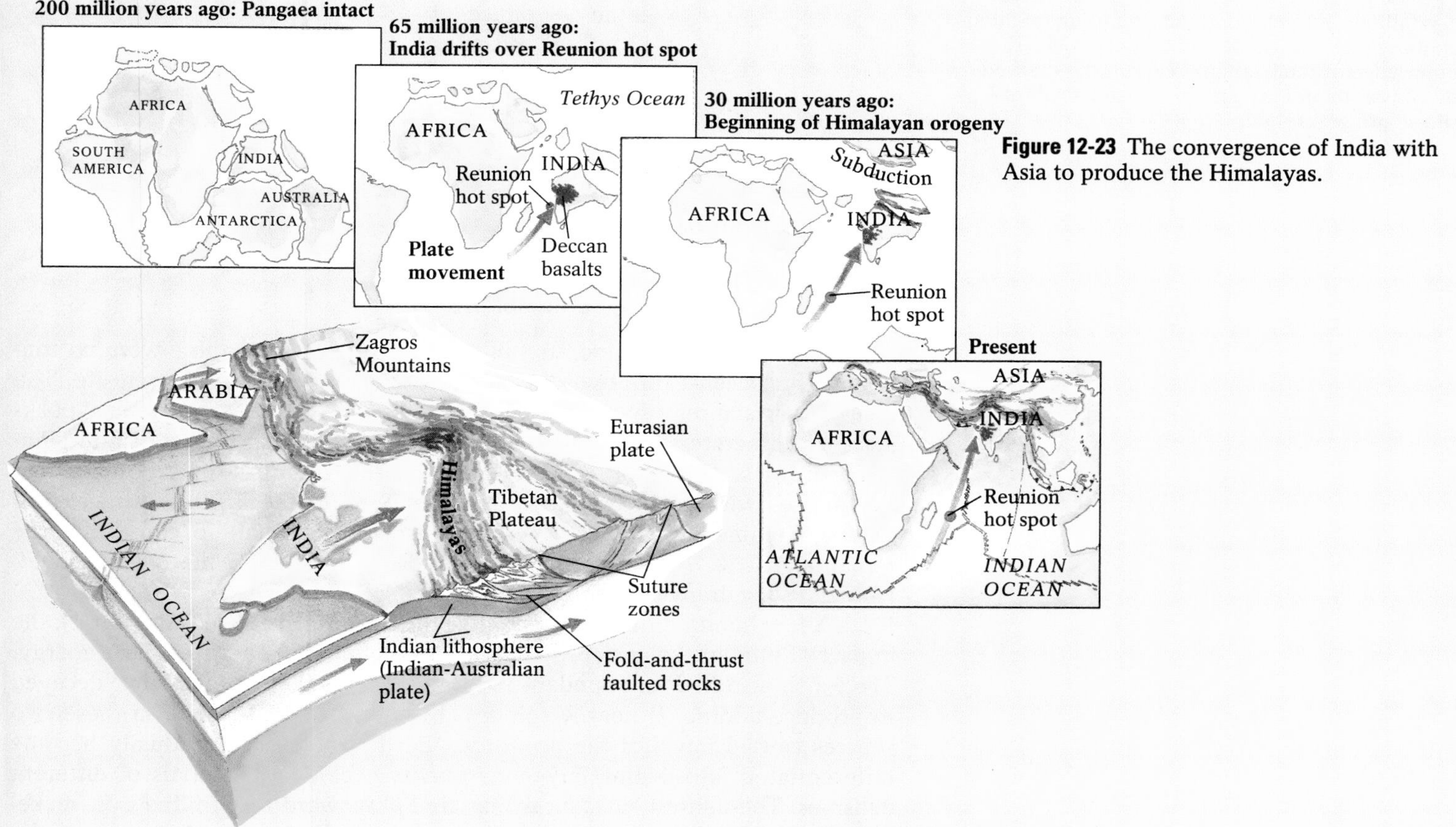

Figure 12-23 The convergence of India with Asia to produce the Himalayas.

The Origin and Shaping of Continents

Because continents do not subduct, continental lithosphere, once formed, becomes a permanent part of the Earth's surface. As a result, there are rocks up to 4 billion years old on the continents. For clues to the earliest stages in Earth history we must look here, not to the younger ocean floors.

Every continent has the same basic components (Fig. 12-24). The oldest parts are the **continental shields**, broad areas of exposed crystalline rock in continental interiors that have not changed appreciably for more than a billion years. Every continent contains at least one large shield area. North America's Canadian Shield extends across much of Canada from Manitoba to the Atlantic coast, dipping into the northern United States from northern Minnesota and Wisconsin to upstate New York's Adirondack Mountains.

Surrounding the continental shield is the **continental platform**, where the continental shield is covered by a veneer of younger sedimentary rock. Together, the continental shield and platform constitute the **craton**, a continental region that has been tectonically stable for a vast period of time. At the edges of the craton, near the borders of continents, are coastal mountains, coastal plains, and continental shelves.

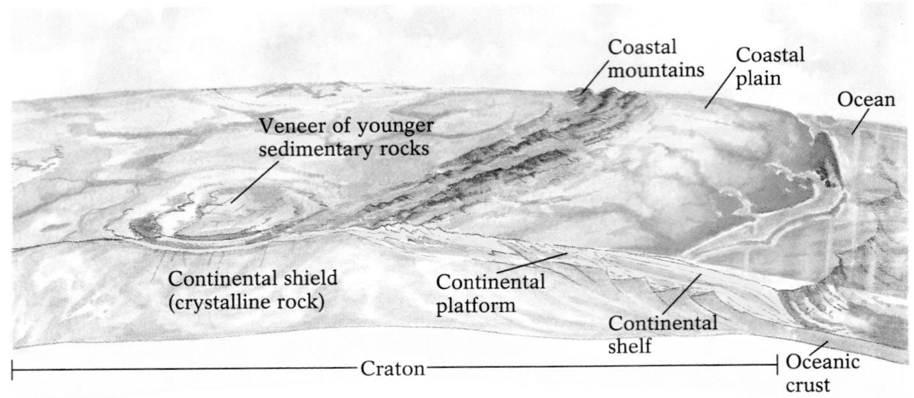

Figure 12-24 The anatomy of a continent. Every continent contains a tectonically stable nucleus, or craton, that consists of one or more ancient, crystalline continental shields and a surrounding continental platform of sedimentary rock. The outer edge of the continent is typically marked by coastal mountains and plains and a submarine continental shelf.

The Origin of Continental Lithosphere

In the Earth's first few hundred million years of existence, there were no continents, no oceans, no atmosphere, and, obviously, no land or sea life. The planet's surface may have looked similar to that of the Moon today, pockmarked with craters caused by the impact of countless fragments of interplanetary debris. By about 600 million years ago, however, continental landmasses with primitive plants covered 30% of the Earth's surface; vast oceans, teeming with a great diversity of plant and animal life, constituted the remaining 70%. A thick protective atmosphere covered the entire planet. How did such dramatic changes come about?

A few hundred million years after the Earth formed (see Chapter 1), the impact of infalling solar debris converted gravitational energy to heat energy. The Earth also had an abundant supply of radioactive isotopes that decayed to produce additional heat. Geologists hypothesize that heat from these two sources caused much of the primordial Earth to melt and gradually become differentiated into distinct layers as gravity separated materials of different densities. The densest substances migrated downward to form the iron–nickel core, and the lightest substances floated upward to form the rocks of the Earth's initial crust (see Chapter 11).

Four billion years ago, the newly formed mantle was hotter and less viscous than it is today, and probably flowed more rapidly. Vigorous mantle convection currents brought great volumes of magma to the surface, and basaltic lava spewed from numerous hot spots across the Earth's surface. Enormous clouds of steam and other gases erupted as well, forming the Earth's first atmosphere and condensing to form the first global ocean. Evidence from recently discovered rocks in Australia, Greenland, and North America supports these hypotheses. Now metamorphic, these ancient rocks (3.8 to 4.1 billion years old) were originally sedimentary, suggesting that some type of atmosphere must have existed to weather sediment from preexisting rocks. Some even appear to have been transported by and deposited in bodies of water. We can therefore conclude that as early as 4 billion years ago, the Earth had both an atmosphere and bodies of water.

The first felsic continental lithosphere formed as a result of plate tectonic processes somewhat different from those of today. Before there were continents, early landmasses probably consisted of huge volcanic islands enlarged by great volumes of mafic (basaltic) and ultramafic lava erupting from the mantle (Fig. 12-25). This primitive continental lithosphere was probably quite hot and thus too buoyant to subduct. As eruptions continued, the islands grew and thickened, and early sedimentary basins developed around them. Some of the islands may have collided and formed larger landmasses. Older sections cooled and became more dense, until some parts sagged downward into the hotter interior, folding and dragging the sedimentary basins along with them. The basalts and ultramafics became partially melted, and differentiated to produce the first intermediate and felsic magmas. Continued partial melting and differentiation of these rock materials eventually created new sets of volcanic islands that consisted largely of metamorphic basalts called greenstones and felsic plutons of varying compositions. This is the combination of rock types found in most ancient continental shields.

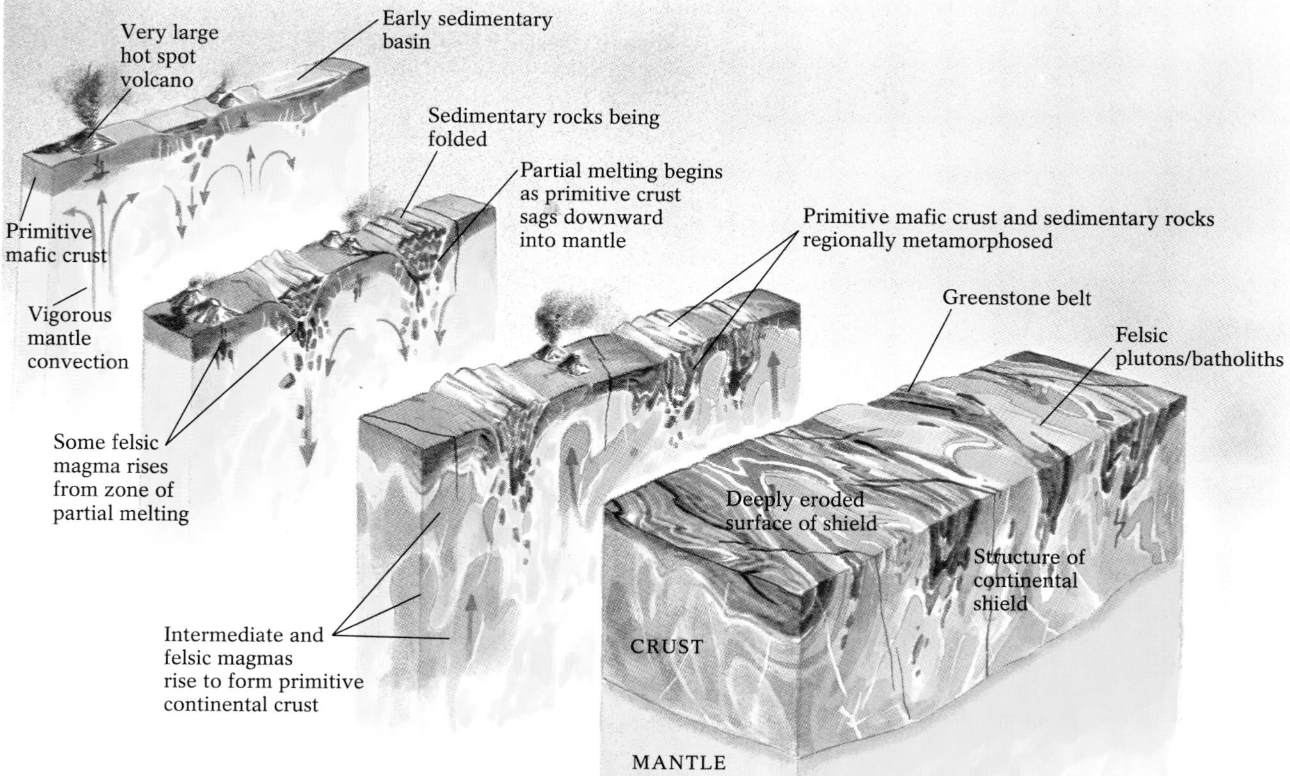

Figure 12-25 Continental shields formed when the Earth's interior was considerably hotter than it is today. Vigorous convection brought warm mantle material to the surface, where it erupted as mafic and ultramafic lavas. As the flows cooled and became more dense, they subsided into deeper regions, where they partially melted and differentiated, eventually solidifying as intermediate and felsic plutons. These ancient low-density plutons are the oldest rocks on Earth and, together with metamorphic greenstone belts, make up continental shields.

After small, lower-density felsic landmasses developed, subduction began to occur where dense oceanic lithosphere and the light continental lithosphere converged. About 2.5 billion years ago, recycling of primitive oceanic lithosphere produced some of the world's first andesites and their characteristic stratovolcanoes (Chapter 4). In time, the continental nuclei became enlarged as drifting continental plates collided, fold-and-thrust mountains were uplifted, and forearc mélanges accumulated. These masses of ancient continental lithosphere have been weathered and eroded over the ages, removing overlying sediments to expose their deep plutonic and metamorphic roots. The landmasses subsequently rifted, creating numerous centers of continental growth around which additional subduction and accretion took place, further increasing the extent of continental lithosphere. Today, these centers of growth are the Earth's stable shields.

Nearly all the continental shields were created during the Precambrian Era. The ages of shield rocks tell us that most continental growth occurred 3.5 to 3.2 billion years ago, a particularly intense period of continent formation. By about 2.0 billion years ago, as much as 75% of today's area of continental lithosphere may already have formed. Most of the remainder formed during two other brief periods: 1.9 to 1.7 and 1.2 to 0.9 billion years ago. In the last 900 million years, plate tectonic activity appears to have slowed considerably, probably because the Earth's internal heat supply has diminished. Little continental material has been added in the past 600 million years, even though the geographic configuration of the continental lithosphere has changed radically. While some continental lithosphere is always returning to the mantle as part of the cycle of continental erosion–marine deposition–subduction, an approximately equal amount has presumably been produced by the cycle of subduction–partial melting–volcanic arc eruption. Hence, the Earth's overall volume of continental material does not appear to be changing at present.

The core of the Canadian Shield, exposed in Ontario, western Quebec, and the northern Great Lakes states, began growing between 3.8 and 2.5 billion years ago. Progressively younger regions, added on during subsequent collisions between landmasses, surround these ancient rocks (Fig. 12-26). Like those of other continents, the North American nucleus is composed of large granitic intrusions, complexly deformed metamorphic rocks, and greenstone belts containing relatively undeformed metamorphosed basalts.

Greenstone belts, named for their abundance of the green metamorphic mineral chlorite, are linear masses of metamorphosed pillow basalts, tens to hundreds of kilometers long, surrounded by metamorphosed oceanic mudstones. These are probably ancient ocean-floor rocks that were incorporated into a continental shield, perhaps during early episodes of plate subduction and collision. Almost all greenstone belts are found in rocks about 2.5 billion years or older.

The Changing Shape of the Earth's Continents

Bordering North America on both the Atlantic and Pacific coasts are rocks that are no more than 600 million years old. Recent evidence indicates that many of these younger rocks are geological immigrants, fragments of distant plates that have been transported here by convergent plate motion and attached to our continent by collisions. Such fault-bounded rock bodies, which originated elsewhere, are **displaced terranes**. The hunt for such terranes is an exciting new aspect of continental geology.

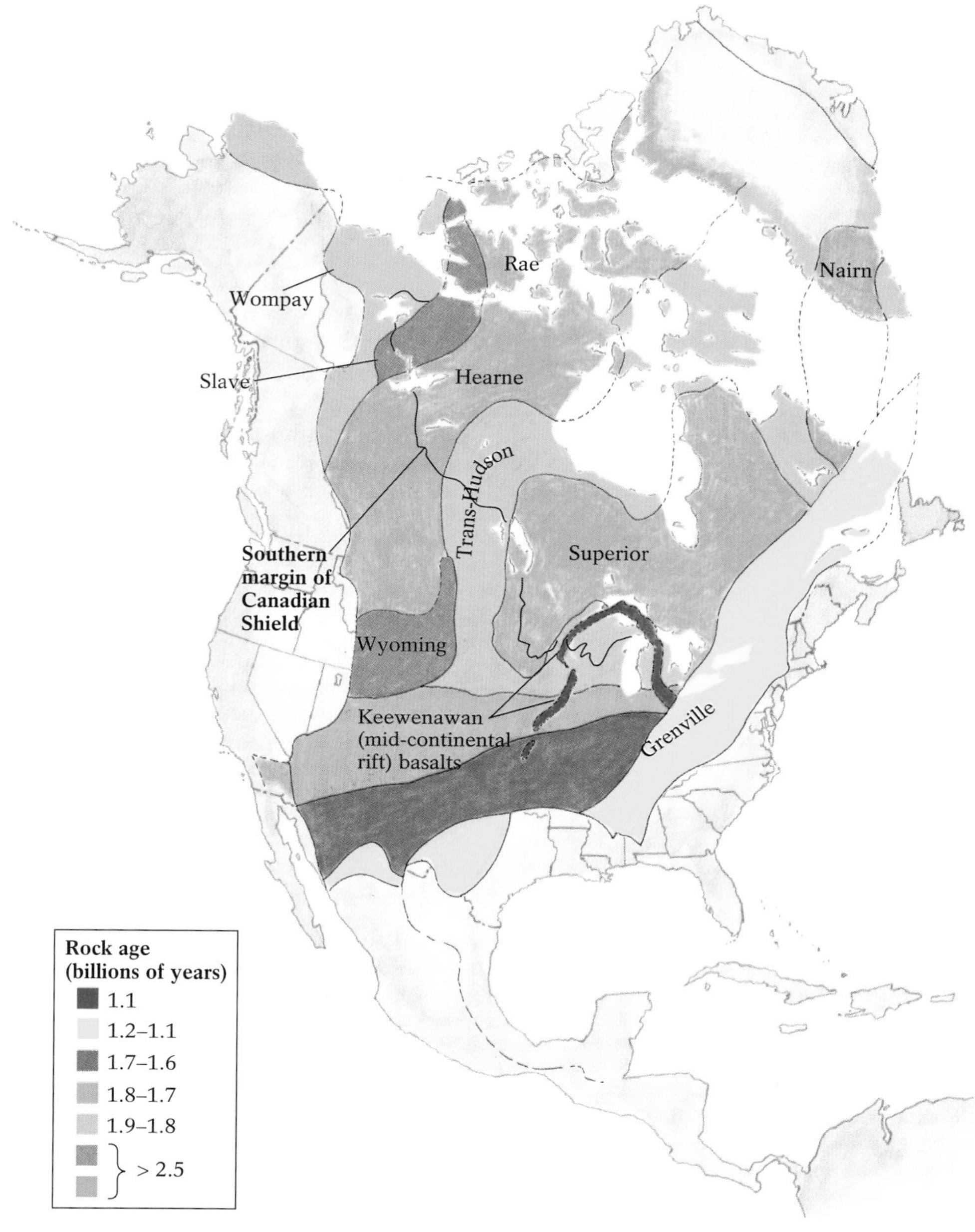

Figure 12-26 The geological provinces of North America. Rocks of the Canadian Shield, North America's ancient crystalline nucleus, are exposed throughout central and southern Ontario and Quebec as well as northern Michigan, Wisconsin, and Minnesota. In all other places, however, the provinces are covered with younger rock.

Displaced terranes can be distinguished from their surrounding rocks by their different ages, geological structures, stratigraphies, fossil assemblages, and/or magnetic properties. Some may have been island arcs, formed in an ocean basin and then towed to a continental margin by subduction of the intervening ocean plate. Others are **microcontinents**, pieces of continental lithosphere broken from larger distant continents by rifting or transform faulting. Displaced terranes are located and mapped by searching for geological evidence of former plate margins. As we have seen, former convergent boundaries may contain evidence of subduction and suturing, such as mélanges, blueschist minerals, and slices of ophiolite rocks.

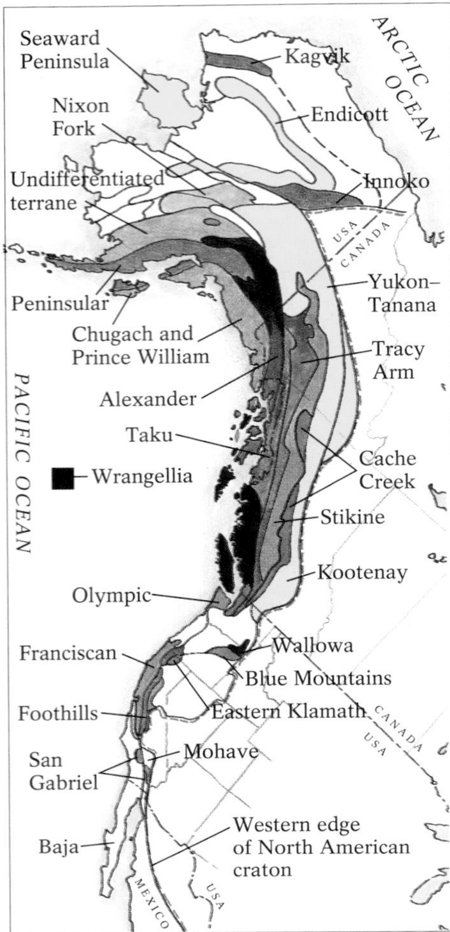

Figure 12-27 The displaced terranes attached to western North America. These are believed to be the remnants of assorted microcontinents, islands, and other unsubductable landmasses attached to the continent during subduction of past Pacific plates.

More than 100 fault-bounded regions of various sizes, accreted to western North America from Alaska to California, have been identified as displaced terranes (Fig. 12-27). Most were swept against the western boundary of the westward-drifting North American plate after about 200 million years ago, when Pangaea rifted apart and the mid-Atlantic ridge began to develop. With an ancient plate of the Pacific basin gradually subducting beneath the continent's western edge, dozens of buoyant fragments were accreted to North America's western shore. The continent's "original" west coast probably extended from what is now central British Columbia south through Idaho to Utah or Arizona. Virtually all lands farther west (except for some younger volcanics and sediments) are probably displaced terranes.

One of the most extensive displaced terranes of western North America is Wrangellia, stretching more than 3000 kilometers (1900 miles) from the Wrangell Mountains east of Anchorage, Alaska, through Vancouver Island, British Columbia, to Hells Canyon, Idaho. Wrangellia's rocks and fossils are distinctive: Throughout, there are thick sequences of basalts, shallow marine limestones, and the fossil clam *Dionella*, a species found in rocks native to Asia but nowhere else in North America. Wrangellian basalts display a magnetic record suggesting that they formed near the equator. This microcontinent may have drifted northeast from equatorial Asia for 7000 kilometers (4400 miles) or more before colliding with North America about 100 million years ago.

The displaced terranes of eastern North America (Fig. 12-28) have a longer and more complex tectonic history. The Appalachian mountain system, for example, is believed to contain foreign island arcs and pieces of Africa added to North America during the several stages of the closing of an ancient Atlantic Ocean between 500 and 250 million years ago (see Chapter 9). In the eastern Canadian maritime provinces of New Brunswick, Newfoundland, Prince Edward Island, and Nova Scotia and adjacent parts of eastern Maine, Massachusetts, and Rhode Island, there are displaced terranes that record closing oceans, accreted island arcs, continental collisions, transform faulting, and rifting, all from about 600 to 150 million years ago.

The Earth's Plates before Pangaea

The geologic record becomes progressively more difficult to read as we go back in time; most of what we know about the Earth's plate tectonic history occurred within the last 200 million years, after the breakup of the supercontinent Pangaea. However, recently developed techniques using paleomagnetism are yielding more precise knowledge of where ancient rocks formed geographically, and continuing efforts to identify displaced terranes have given us a clearer picture of ancient landmasses. As we try, cautiously, to reconstruct plate configurations that predate the breakup of Pangaea, we must always remember that even though rifting often takes place at suture zones, the continental blocks that collided to assemble the supercontinent were probably not the same shape as the ones we know today.

Five hundred million years ago, well before Pangaea formed, there was an earlier supercontinent, Gondwana, somewhere near the South Pole (Fig. 12-29). Gondwana, which contained all of the southern-hemisphere landmasses, would eventually become the southern part of Pangaea. At that time, there were also three northern landmasses, each probably separated from the others by a sizable ocean. These independent continents—which would become Laurasia, the northern half of Pangaea—included most of what is now North America, what is now northern Europe, and a combination of what is now southern Europe and parts of Africa and Siberia.

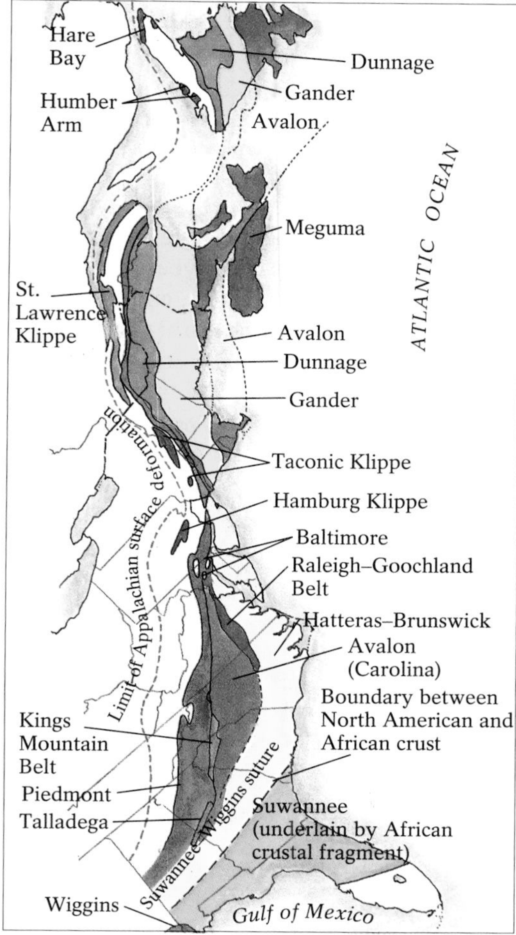

Figure 12-28 The displaced terranes of eastern North America, many dating from the assembly of the supercontinent of Pangaea. They remained attached to North America after Pangaea rifted apart to form the modern Atlantic Ocean.

The pre-Laurasian landmasses and Gondwana began converging about 420 million years ago. The partially subducted sea floor from this event remains today as ophiolite rocks in Nova Scotia and Newfoundland. This collision produced the northern Appalachians and corresponding ranges in the British Isles and Norway, and contributed some displaced terranes to northeastern North America. The Avalon terrane of Nova Scotia, for example, may have been an island arc trapped between the colliding pre-Laurasian continents. Meanwhile, Siberia collided with northern Europe, completing the formation of Laurasia and producing the Ural Mountains of central Russia.

Over the next 100 million years, Laurasia and Gondwana collided and formed Pangaea. The final collision in the assembly of the supercontinent was between Africa, a separate continent that is now southeastern North America, and North America. This last stage in the formation of Pangaea created the southern Appalachians. By about 225 million years ago, at the close of the Paleozoic Era, Pangaea was one vast landmass that stretched from pole to pole. Virtually all of the Earth's continental lithosphere remained joined together for the next 50 million years or more, until Pangaea began to rift and water flowed in to form the modern oceans.

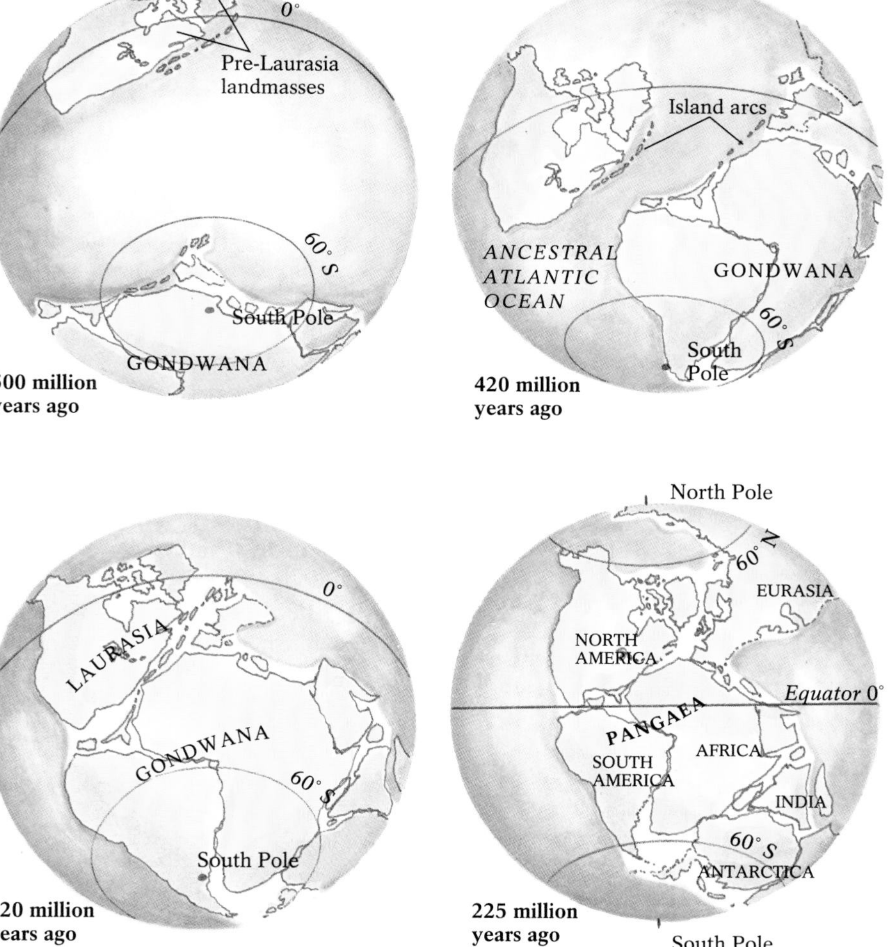

Figure 12-29 The origin of the supercontinent Pangaea.

Looking into the Plate Tectonic Future

By projecting current plate motions into the future and speculating about the location of the future rifts and subduction zones, geologists have developed a probable view of the plate tectonic world 50 to 100 million years from now (see Fig. 1-29). Australia will likely collide with southeast Asia, and transform motion along the San Andreas fault will cleave western California completely from the North American continent, leaving it free to move northward as a separate microcontinent. The glamorous beaches of the Mediterranean's Riviera may become rugged landlocked deserts, wedged between the colliding edges of the African and Eurasian plates. There may be continued rifting in East Africa and future rifting in the American Southwest. Subduction may resume someday along the east coast of North America, bringing explosive volcanism and powerful earthquakes to such places as New York City, Boston, and Philadelphia. Computer models predict that within the next few hundred million years, all of the Earth's landmasses will reunite as another Pangaea-like supercontinent.

The Driving Forces of Plate Motion

We have documented conclusively that plates move, and we now know a great deal about the plate tectonic history of the Earth, but we still cannot explain with certainty what drives plate motion. This is akin to knowing what a car looks like and how fast it goes, but having no idea what makes it run. Fortunately, because there is excellent evidence for plate motion and we can identify its effects, we can accept the general theory of plate tectonics even though we do not yet understand all of the forces behind it.

As we say in Chapters 1 and 11, heat from the Earth's interior probably sets plates in motion. Heat rises through the mantle by convection, the flow of currents in a fluid due to variations in their temperature. (Hot materials are less dense and therefore rise; cold ones are more dense and therefore sink.) Slow convection currents in the warm asthenosphere produce movement of the cold, brittle lithosphere directly above it. When plate tectonics was first proposed three decades ago, the plates were viewed as passive hitchhikers on the flowing asthenosphere. Recent findings, however, suggest that plates may contribute actively to their own mobility.

Ridge Push, Slab Pull, or Plate Sliding?

There are a number of factors that may play a role in the movement of the Earth's plates; three of these are illustrated in Figure 12-30. When an ocean ridge grows at a divergent plate boundary, does the rising magma wedge itself between adjoining oceanic plates and actively push them apart? Would this push be powerful enough to move a 5000-kilometer (3000-mile)-wide plate and force its distant edge down into the mantle at a subduction zone? If plates are being pushed, we might expect them to fold up accordion-style, which doesn't happen. In fact, ocean-exploring submersibles diving deep into axial rift valleys have recently identified thousands of tears and fissures within oceanic lithosphere. This strongly suggests that plates are being *stretched* at divergent boundaries, not pushed and compressed. Does this stretching and tearing indicate that oceanic plates are being actively pulled? This seems theoretically possible. Old, cold oceanic lithosphere is denser than the warm asthenosphere and therefore sinks. The lower portion of a descending slab also becomes more dense as some of its light felsic components melt and rise,

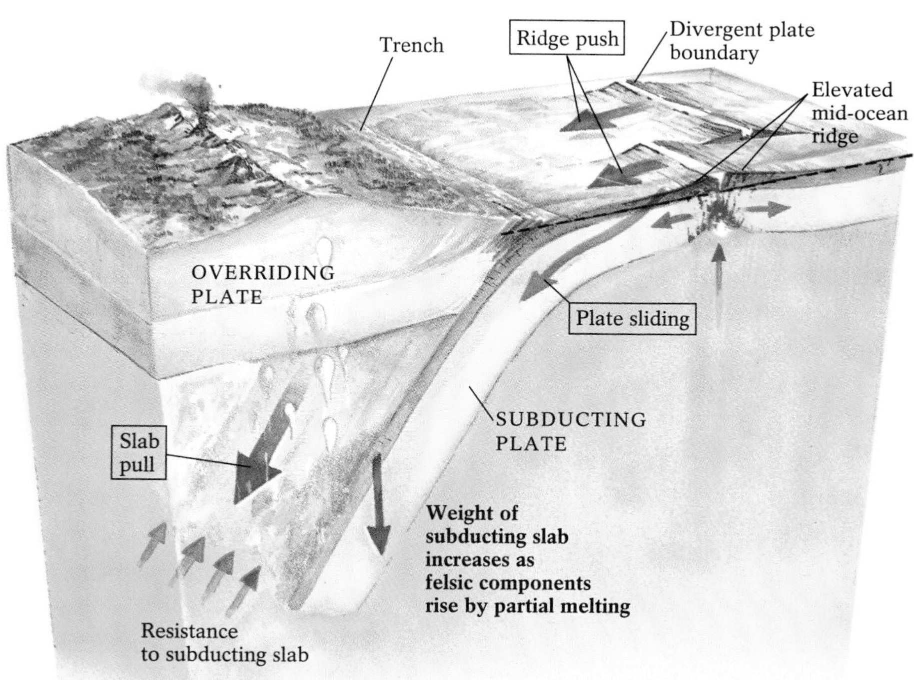

Figure 12-30 Three factors that may help drive plate tectonics: *ridge push* at divergent plate boundaries; *slab pull* as old, dense plates subduct into the mantle; and gravity-induced *plate sliding* at ocean ridges.

leaving its dense mafic components to become more concentrated. The increased density may actively tow oceanic plates into their subduction-zone graves.

Is there any evidence that this process does actually occur? In some places, where two continents have collided and are now joined, the subducting oceanic plate that once lay between them continues to descend deeper into the mantle. Recent studies also show that rates of plate motion are proportionate to the amount of a plate's margin undergoing subduction: The fastest-moving plates subduct along a significant proportion of their margins, whereas the slowest-moving plates generally lack subducting margins entirely. This supports the hypothesis that gravitational pull on a descending plate contributes to general plate motion. But it seems unlikely that this force is powerful enough to influence divergence at the opposite side of the plate, thousands of kilometers away. Moreover, lithospheric plates are so brittle they would most likely break if such large-scale pulling forces were transmitted through them for thousands of kilometers. Furthermore, the Atlantic Ocean's floor diverges from the mid-Atlantic ridge at about 2 centimeters (nearly an inch) per year, though no slabs sink along any of its margins. Some force other than slab pull must therefore be responsible for the motion of the Atlantic Ocean segments of the North and South American, Eurasian, and African plates.

Another hypothesis suggests that gravity causes oceanic lithosphere to move laterally away from elevated mid-ocean ridges. The continuing eruption of basaltic lava at ocean ridges produces high mountain ranges of still-warm, low-density rocks. Newly formed oceanic lithosphere may be literally sliding down the slope of the uplifted asthenosphere beneath ridge crests. Mathematical calculations indicate that oceanic plates could slide down even gentler slopes at a rate of several centimeters per year, about the observed rate for the mid-Atlantic ridge. In the initial stages of rifting, however, plates begin to diverge before an oceanic ridge is created. Hence, gravity alone cannot account for plate motion.

Geologists have not yet reached a consensus regarding the relative importance of these possible tectonic influences. Are oceanic plates driven by different forces or combinations of forces than drive continental plates? Do these forces, perhaps acting together, play a more important role in driving plates than the convection cells discussed in Chapter 1?

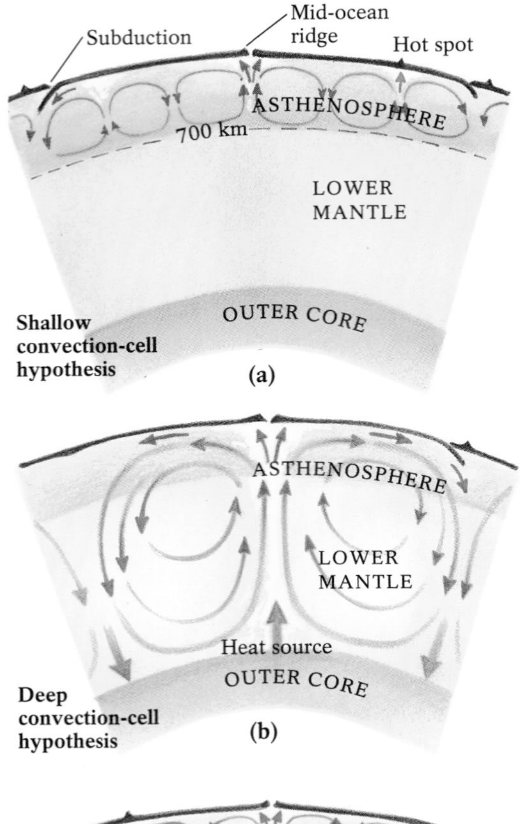

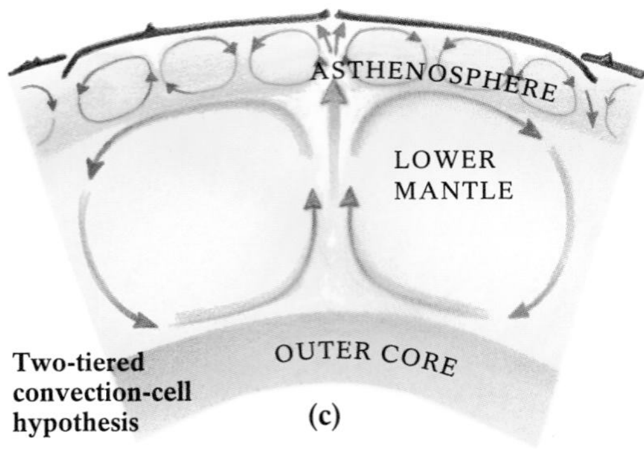

Figure 12-31 Several hypotheses have been proposed for the configuration of the Earth's convection cells. (**a**) The cells may be shallow, confined to the asthenosphere. (**b**) The cells may extend through the entire mantle to the outer core. (**c**) Some hypotheses propose two sets of convection cells that meet at a depth of 670 kilometers (420 miles), the zone within the mantle where seismic waves are known to accelerate in response to changes in the chemistry and structure of mantle materials (see Chapter 11).

Convection Cells, Revised

Although many geologists accept that convective flow in the mantle is an important mechanism of plate movement, some question whether convection cells can exist within the solid lower mantle. If they do, does the mantle flow rapidly enough to drive the Earth's plates at their observed velocities? Advocates of mantle convection still have questions about the dimensions of convection cells (Fig. 12-31). Are they shallow, largely confined to the soft, mobile asthenosphere? Do they extend down to 700 kilometers (450 miles), the depth to which subducting plates are known to penetrate? Do they fill the entire mantle, energized by heat rising from the Earth's liquid outer core? One model proposes that convection cells are stacked in two tiers, with large, deep, slow-moving cells transmitting heat from below to power small, shallow, fast-moving cells above them, which carry the Earth's lithospheric plates.

Seismic tomography (discussed in Chapter 11) can generate three-dimensional images delineating warm and cold regions of the mantle. (Recall that warmer, less rigid rocks slow the passage of seismic waves, whereas colder, more rigid rocks transmit seismic waves more efficiently.) In one seismic tomography image, cold continental rocks appear to extend beneath the North American and Eurasian plates to depths of at least 400 to 600 kilometers (250–370 miles), uninterrupted by the warm asthenosphere that rises up at ocean ridges and hot spots. This temperature pattern casts doubt upon the simple convection-cell model, proposed during the early years of the plate tectonic revolution, that requires cold continental lithosphere to move as a passenger on warm currents in a shallow asthenosphere.

Thermal Plumes—A Possible Alternative to Convection Cells

Some geophysicists have proposed that deep-Earth heat rises to the asthenosphere as scattered vertical columns of warm upwelling mantle material called **thermal plumes**, rather than as distinct convection cells. These narrow plumes, 100 to 250 kilometers (60–160 miles) in diameter, rise beneath both continents and oceans and at plate boundaries as well as plate interiors (Fig. 12-32). They are manifested at the Earth's surface as hot spots and may carry enough energy to move plates. Thermal plumes may originate within the asthenosphere or at the mantle–core boundary, a depth of 2900 kilometers (1800 miles). Thermal plumes—like the hot spots they generate—may be considered stationary relative to the faster-moving lithospheric plates that override them. But as it isn't clear where they originate, they may actually be carried along with the slow convection of the lower mantle.

Thermal plumes are generally believed to lift up the overlying lithosphere, which becomes domed; they then spread laterally beneath the lithosphere and in doing so apply a dragging force to the base of the lithosphere that causes it to rift. The fact that many thermal plumes are located at or near diverging plate boundaries is viewed as more than coincidental. The thermal plume beneath Iceland, for example, is considered by some to be the cause of the island's high altitude above sea level.

The thermal-plume hypothesis separates divergence from subduction completely. In this view, cooling mantle material descends slowly through the entire mantle—not, as in the convection-cell hypothesis, at individual plate boundaries above the descending arcs of convection cells. This might explain why there is no subduction at the margins of the Atlantic Ocean.

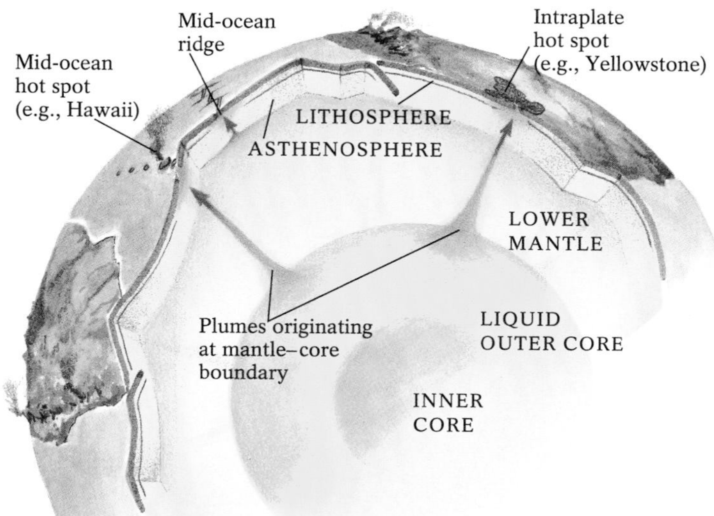

Figure 12-32 Thermal plumes are an alternative to convection cells as an explanation of the force that drives plate motion. They are also believed to fuel surface hot spots such as those that created the Hawaiian Islands and Yellowstone's volcanoes and hot springs. The source of the heat that creates plumes remains uncertain, although some geologists believe that the plumes rise from the mantle–core boundary, heated by the liquid outer core.

Plate-Driving Mechanisms: A Combined Model

It seems unlikely that a single unified model can explain all plate movements. More likely, the driving mechanisms of plate tectonics are a combination of ridge push, slab pull, gravity-induced sliding, convection cells, and thermal plumes. Some form of convection beneath the lithosphere apparently transports mantle heat upward to drive the mid-ocean plate divergence of Atlantic-style tectonics. Once plate motion is initiated by mantle convection and oceanic ridge crests have developed, gravity may induce the lateral motion of oceanic plates as they slide from topographically high crests over the underlying soft, weak asthenosphere. Gravity also independently enhances plate subduction by returning cold, dense lithosphere to the Earth's interior; this mechanism may be driving Pacific-style tectonics. In the future, of course, we may learn about other driving forces of which we are now unaware.

In the mid-1960s, plate tectonics was an exciting new hypothesis. Since then, it has graduated to full-fledged theory as Earth scientists have found evidence confirming most of its precepts. Although most now believe that tectonic plates exist and move, the oceans and continents continue to be explored intensively in an effort to answer the numerous questions that remain.

Chapter Summary

New technology such as satellite-based studies of plate motion and geomagnetic research enables Earth scientists to study the mechanisms of plate tectonics. The actual velocity of plates can now be measured directly, by aiming ground-based lasers at reflectors on satellites in fixed orbits. **Hot spots,** areas of particularly high heat flow in the mantle that may fuel intraplate volcanism, can serve as fixed reference points to measure absolute plate velocities. The pattern of structures that forms as a plate moves over a hot spot can reveal both plate speed and direction. Such structures include volcanic islands such as the Hawaiian Islands; submarine mountains, or **seamounts;** and older, eroded **guyots,** which initially grew above sea level but subsided below sea level over time. **Marine magnetic anomalies**, the stripes of alternating magnetic polarity in oceanic lithosphere, are also used to estimate rates of divergence and, hence, of plate motion.

Our understanding of the Earth's tectonic past comes largely from our growing knowledge of the evolution of the world's ocean floors. In the last four decades, marine geologists and oceanographers have used **echo-sounding sonar** to study the topography of the sea floor and **seismic profiling** to study its underlying layers.

The accumulation of geophysical data has helped to clarify the sequence of events that produces ocean basins. Rifting begins when a warm current of mantle material rises under a continental plate, causing it to stretch and thin until it tears. Three radiating rift valleys typically form simultaneously; two of these eventually diverge, while the third—the failed arm, or aulacogen—commonly becomes inactive and fills with sediment or a stream network. After rifting between two new plates is complete, the rift edges become inactive tectonically, forming **passive continental margins.**

As divergence continues, a full-blown seaway develops; new oceanic lithosphere forms at a **mid-ocean ridge,** where mantle-derived basalts erupt to produce a linear chain of volcanic mountains. The typical rock sequence within oceanic lithosphere, called the **ophiolite suite,** consists of a veneer of deep-marine sediment, underlying layers of pillow basalts and related intrusive gabbros, and a bottom layer of ultramafic mantle peridotite; the entire suite may be chemically altered by reaction with infiltrating seawater. The mid-ocean ridge generally becomes divided into short, offset segments where it is cut by oceanic transform boundaries.

A diverging oceanic plate gradually cools and becomes more dense, and may eventually sink back into the Earth's interior by subducting under a less dense plate far from the mid-ocean ridge. An **ocean trench** develops where a subducting plate flexes downward into the mantle, forming a depression in the Earth's surface. Subduction produces several rock types and geologic structures: Chaotic mixtures of oceanic sediments and ophiolite rocks called **mélanges** pile up within the trench, forming masses of rock called **accretionary wedges** that become attached to the edge of the overriding plate; as the subducting plate sinks into the warm mantle its rocks partially melt and erupt, forming a chain of volcanoes called a **volcanic arc;** and sediments accumulate in topographic depressions called **forearc** and **backarc basins. Backarc spreading**, divergence within the backarc basin, sometimes separates these coastal features from the mainland.

Once an oceanic plate has been completely subducted, the continental blocks on either side undergo continental collision and become connected to form a larger continent. The boundary between them, the **suture zone,** is often marked by a collisional fold-and-thrust mountain range such as the Himalayas. Suture zones, because they are weak spots in the Earth's lithosphere, are frequently the sites of subsequent rifts.

The oldest oceanic rocks are only 200 million years old, so we rely on the older, more complex geology of the continents to tell us of the Earth's earlier tectonic history. The nucleus of a continent consists of ancient bodies of rock called **continental shields;** these are generally surrounded by younger sedimentary rocks, which form the **continental platform.** Together the shield and platform constitute the continental **craton,** that portion of a continent that is tectonically stable. The rocks exposed at the outer edges of the continents are often **displaced terranes,** whose fossils, stratigraphic sequences, rock types, and magnetic signatures suggest that they originated elsewhere. **Microcontinents** are pieces of lithosphere broken from larger continents and may eventually become displaced terranes. Information from all these rocks has allowed geologists to speculate cautiously about the past and future configurations of the Earth's landmasses.

Although geologists have learned much about the origin of both oceanic and continental rocks, they remain uncertain about the mechanisms that drive plate motion. New data have cast some doubt on a simple convection-cell model for plate motion. Plate motion may instead be driven by the combined effects of **thermal plumes,** rising currents of hot mantle rock, and gravity, which pulls newly formed plates down the slopes of mid-ocean ridges and older plates back into the mantle at subduction zones.

Key Terms

hot spots (p. 333)
seamounts (p. 334)
guyots (p. 334)
marine magnetic anomalies (p. 335)
echo-sounding sonar (p. 336)
seismic profiling (p. 336)
passive continental margins (p. 338)
mid-ocean ridge (p. 339)
ophiolite suite (p. 340)
ocean trench (p. 343)
mélange (p. 344)
accretionary wedge (p. 344)
volcanic arc (p. 344)
forearc basin (p. 345)
backarc basin (p. 345)
backarc spreading (p. 345)
suture zone (p. 346)
continental shields (p. 348)
continental platforms (p. 348)
craton (p. 348)
displaced terranes (p. 350)
microcontinents (p. 351)
thermal plumes (p. 356)

Questions for Review

1. Describe three methods of determining the velocity of plate motion.

2. Describe three methods of mapping deep-sea topography.

3. What is an aulacogen? How does it form?

4. List four phenomena that accompany continental rifting.

5. Describe what happens at a rift margin, from the onset of rifting to the formation of an ocean.

6. Draw a simple sketch of an ophiolite suite (include all of the rock types and structures).

7. Why do the earthquakes associated with oceanic transform boundaries generally occur only between offset oceanic ridge segments?

8. List the three common types of convergent plate boundaries, and name a geographic example of at least two types.

9. Draw a sketch of a convergent boundary between an oceanic plate and a continental plate. Be sure to include all of the important associated landforms.

10. Briefly describe the fundamental differences between the convection-cell hypothesis and the thermal-plume hypothesis of plate tectonics.

For Further Thought

1. Of the two patterns of marine magnetic anomalies below, which shows a faster rate of sea-floor spreading? Explain.

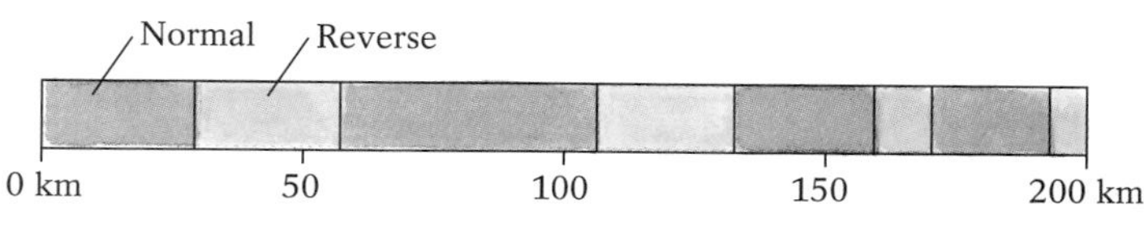

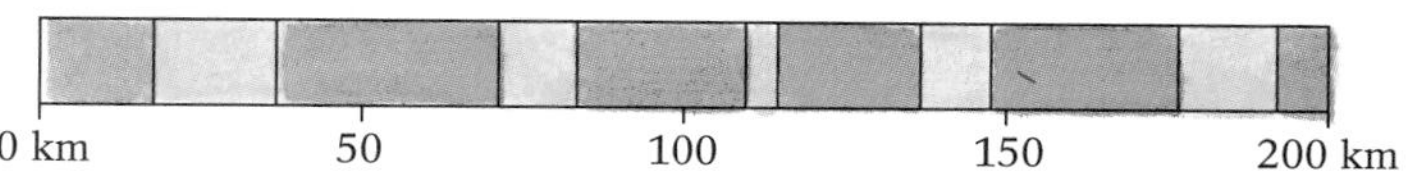

2. What would be some of the major geologic repercussions if a new rift opened between Ohio and Indiana?

3. How would the east coast of North America change if the oceanic segments of the plates that make up the Atlantic Ocean basin began to subduct? How would it change if the Atlantic Ocean lithosphere subducted completely?

4. How might the development of a new subduction zone along the East Coast affect plate interactions on the West Coast?

5. If you had unlimited financial resources, how would you determine the true driving mechanisms of plate tectonic motion?

Part 3

Sculpting the Earth's Surface

What Causes Mass Movement?

Types of Mass Movement

Highlight 13-1 The Gros Ventre Slide

Avoiding and Preventing Mass-Movement Disasters

Highlight 13-2 How to Choose a Stable Homesite

Extraterrestrial Mass Movement

Figure 13-1 A rockslide in Zion National Park, Utah.

Chapter 13

Mass Movement

At 11:37 P.M. on August 17, 1959, an earthquake registering 7.1 on the Richter scale shook thousands of square kilometers in the vicinity of West Yellowstone, Montana. The soil surface rolled like sea waves, and lakes and rivers sloshed back and forth. A wall of water rushed across Hebgen Lake and over the Hebgen dam, sweeping away a number of campgrounds on its way downstream. The quake also triggered a landslide of more than 80 million tons of rock and weathered regolith, including boulders up to 9 meters (30 feet) in diameter, which hurtled downslope at speeds reaching 150 kilometers (90 miles) per hour. As the slide tumbled into Madison Canyon, it compressed and forcefully expelled the air in its path, producing hurricane-force winds that battered the valley. Two-ton cars were blown into the air; one flew more than 10 meters (33 feet) before being dashed against a tree. By the time the slide mass finally came to rest, it had covered the valley floor with 45 meters (150 feet) of bouldery pulverized rubble, buried U.S. Highway 287, and taken the lives of 28 campers.

The Madison Canyon disaster is an extreme example of **mass movement**, the process that transports quantities of Earth materials (such as bedrock, loose sediment, and soil) down slopes by the pull of gravity (Fig. 13-1). Every slope is susceptible to mass movement. Sometimes, as in Madison Canyon, the movement is as fast as several hundred miles per hour and involves a great mass of material, dislodged from a steep canyon wall, that may travel more than 80 km (50 miles) from its source. More often the movement is quite slow, even imperceptible, and involves just the upper few centimeters of loose soils on a gentle hillside.

Of the 20,000 lives lost as a result of all natural disasters in the United States during the years 1925 to 1975, only about 500 were lost as a direct result of mass movement. Damage associated with mass movement, however, accounted for a staggering $75 billion, compared to only $20 billion for all other natural catastrophes. Thus, while hardly any of us are likely to perish in a mass-movement event, there is a strong likelihood that we will incur some costs—perhaps because of a cracked home foundation or a living room filled with flowing mud—as a result of this geological process. In this chapter, we will explore the nature of various mass-movement processes. We will examine the underlying causes of mass movement, the geological factors that promote and trigger it, and some of the ways its dangers can be prevented.

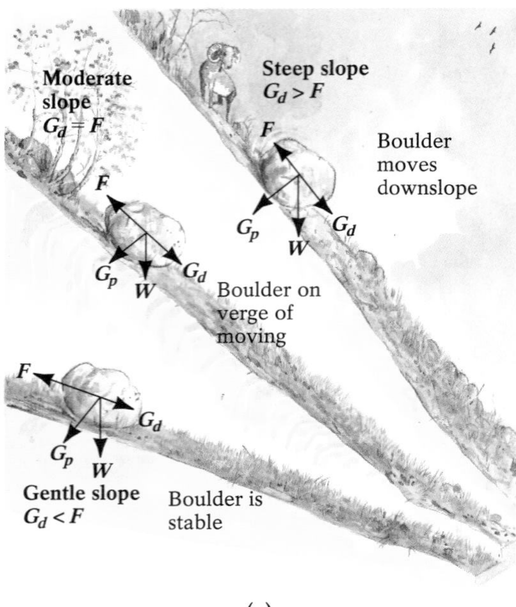

(a)

(b)

Figure 13-2 (a) The principal force tending to hold materials in place on a slope is friction (F); the principal force tending to drive materials downslope is the parallel component of gravity (G_d). These forces are affected by several factors, including the steepness and slipperiness of the slope and the weight (W) and water content of the material on the slope. (**b**) This boulder sits on a steep hillside in Acadia National Park, Maine, without sliding off because the friction holding it in place exceeds the gravitational force that would drive it downslope.

What Causes Mass Movement?

The underlying cause of all mass movement is the same: Materials on a slope become loosened and, when pulled by gravity, move downslope. Various factors loosen materials and promote downslope movement; others hinder such movement. Mass movement occurs when the factors that drive materials downslope overcome the factors that resist downslope movement.

The principal factor driving mass movement is gravity, which constantly coaxes materials downslope. Two principal factors provide resistance to mass movement: the friction between a slope and the loose material at its surface; and the strength and cohesiveness of the material composing the slope, which prevents it from breaking apart and slipping at its surface. The steepness of the slope, the water content of the materials, the amount of vegetative cover, and the slope's history of human and other animal disturbances are other factors that affect a slope's mass-movement potential.

Gravity, Friction, and Slope Angle

Two main factors determine whether loose material, such as a boulder, on a slope will stay put or slide: gravity and friction. Gravity promotes downslope movement; friction resists it. Gravity has a component parallel to the slope (G_d, for $\text{Gravity}_{\text{downslope}}$) that works to pull surface materials downslope, and one perpendicular to it (G_p, for $\text{Gravity}_{\text{perpendicular}}$) that contributes to the effects of friction. Friction (F) is the force that opposes motion between two bodies in contact. The amount of friction between a boulder and a slope depends on the magnitude of G_p and the slipperiness of the surfaces in contact.

The boulder in Figure 13-2a is likely to slide if G_d is relatively high or if F is relatively low. The steeper the slope, the greater the force G_d exerts, and the more likely it is that material will slide down it. In addition, if G_d is high then G_p is low, and friction, which is partially dependent on G_p, decreases. A rounded boulder on a smooth slope can roll, however, in which case it is not held back by friction. Such a boulder can roll downslope even if G_d is relatively low.

Several natural and artificial processes can create steeper slopes. Faulting, folding and tilting of strata, river cutting, glacial erosion, and coastal wave cutting are all natural processes; quarrying, road cutting, and waste dumping are some of the human activities that can steepen and destabilize slopes (Fig. 13-3). Likewise, a reduction in friction, such as when water infiltrates the contact zone between loosened materials and the slope surface, can initiate downslope movement.

Slope Composition

The likelihood that mass movement will occur is also influenced by the materials that make up the slope. A slope may be composed of any combination of solid bedrock, weathered bedrock, soil, vegetation, and a variable amount of water. Under certain conditions, some of these promote slope stability; in other situations, they promote instability.

Solid Bedrock Solid rock tends to be completely stable even when it constitutes a vertical cliff. Its degree of cementation (if clastic sedimentary rock) or the interlocking pattern of its crystals (if igneous, metamorphic, or chemical sedimentary rock) imparts strength to the rock that usually exceeds the

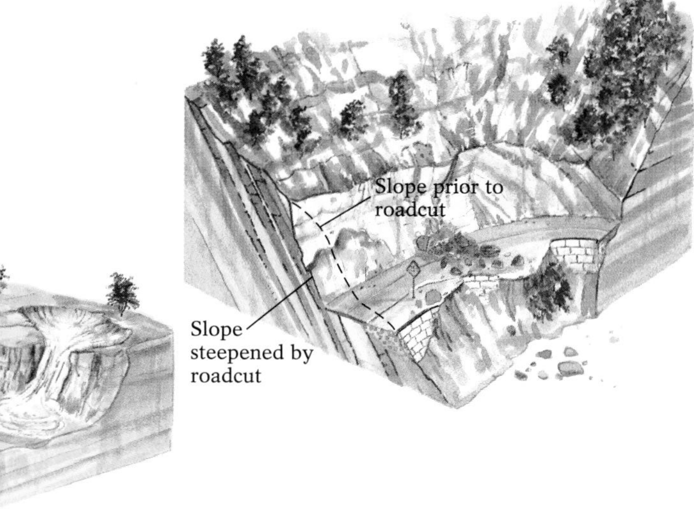

Figure 13-3 Some common processes that oversteepen slopes.

downslope force trying to pull it apart. The stability of rock is reduced, however, if any of the following geological events or circumstances leave it vulnerable to breaking apart under gravity:

- tectonic deformation shatters the rock into a network of joints, fractures, or faults;
- the mechanical weathering processes of freeze–thaw, thermal expansion and contraction, exfoliation, or root penetration open significant cracks in the rock;
- the rock is sedimentary (such as sandstone), and developed bedding planes during deposition;
- the rock is soluble (such as limestone) and large cavities form within it from dissolution;
- the rock is igneous and developed a joint pattern during cooling (such as columnar basalt);
- the rock is a foliated metamorphic rock with marked cleavage or schistosity (such as slate or schist).

When a plane of weakness, whether a fault, crack, joint, or bedding plane, lies roughly parallel to the surface of a steep slope, the rock is even more likely to break along that plane and slide downslope, and the slope is thus even more unstable (Fig. 13-4).

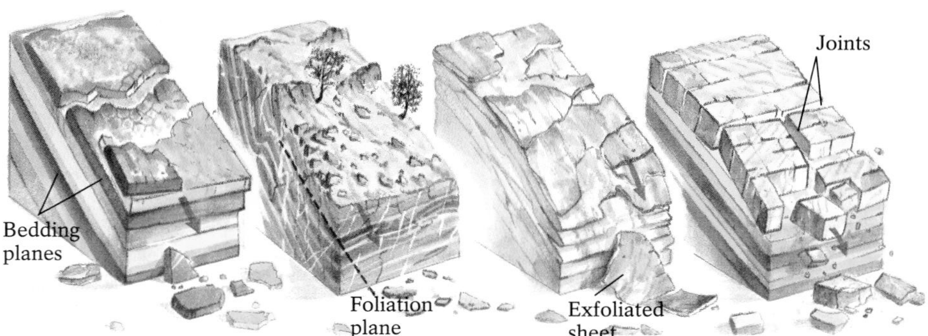

Figure 13-4 Slopes such as these, with planes of weakness parallel to their surface, are especially susceptible to mass movement.

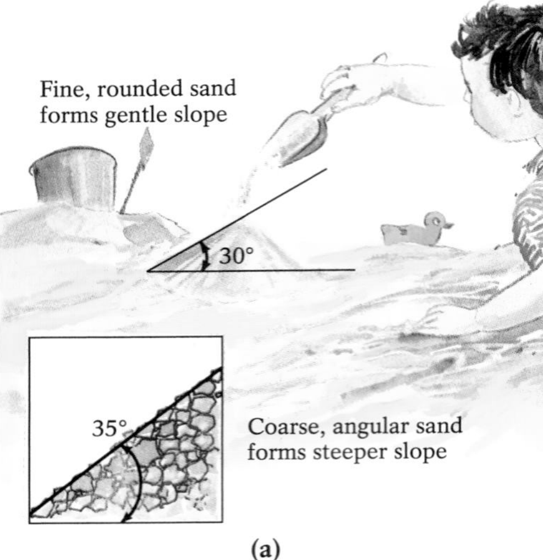

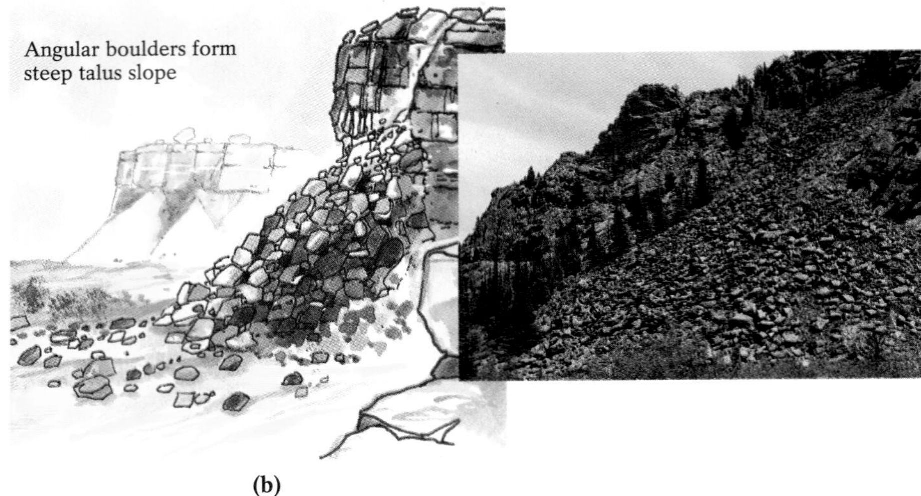

Figure 13-5 The angle of repose of unconsolidated materials depends largely on particle size and shape. (**a**) Coarse, angular sand forms a steeper slope than does fine, rounded sand. (**b**) A talus slope, composed of large and irregularly shaped boulders, can form slopes greater than 40°.

Unconsolidated Materials A slope composed of loose dry material remains stable only if the friction between its components and underlying materials is greater than the downslope component of gravity (G_d), which increases with the steepness of the slope. Different materials form stable slopes at different slope angles. If you pile dry sand up, for example, it forms a small conical hill; when the hill attains a certain slope, the slope angle remains constant, even when you add more sand. The extra sand simply cascades down the slope, gradually piling up around its base. The maximum angle at which an unconsolidated material is stable is that material's **angle of repose**. For dry sand, as in sand dunes, the angle of repose generally ranges between 30° and 35° (Fig. 13-5a).

The angle of repose of any loose material is partly determined by the size and shape of its particles. Large, flat, angular grains with rough, textured surfaces generally have steeper angles of repose than smaller grains that are rounded and smooth. Imagine stacking a pile of highly polished marbles, and another of a batch of cookies. The flatter shape and bumpiness of each cookie's surface enables us to stack them more steeply than the marbles. In nature, the steepest slopes, greater than 40°, are maintained by large, highly angular boulders. These large boulder piles, called **talus slopes**, often form at the feet of cliffs that have been weathered by frost wedging (Chapter 5) (Fig. 13-5b).

The tendency of fine-grained loose material to move also depends on how its particles are arranged. Fine particles that are deposited rapidly are generally arranged somewhat chaotically, in a structure that resembles a house of cards (Fig. 13-6a). Although this structure appears unstable, the friction between the particles at their contact points can impart enough strength to resist mass movement for centuries. Loose particles that are deposited slowly, however, are generally arranged in a more organized fashion, in a structure that resembles a stack of pancakes (Fig. 13-6b). In this configuration, the particles tend to slide past one another, promoting mass movement.

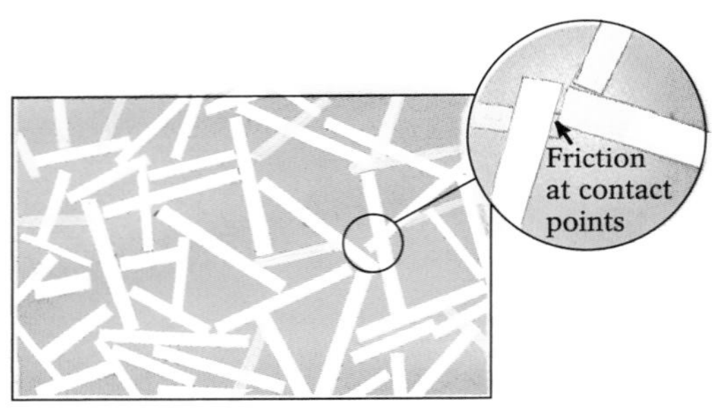

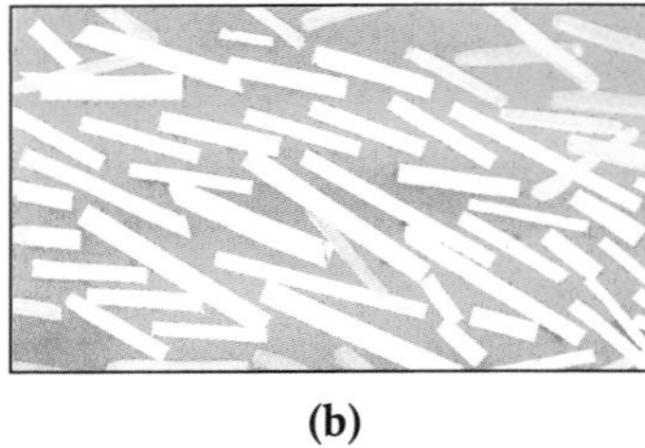

Figure 13-6 The effect of particle arrangement on stability. (**a**) When loose material is deposited rapidly, its particles are oriented at angles to one another, producing friction between the grains and imparting strength to the material. (**b**) When particles are deposited slowly and in parallel planes, friction is reduced and so is the strength of the material.

Figure 13-7 The cohesiveness of partially saturated sediment is demonstrated by the intricate sand sculptures into which it can be molded, such as this one at Cannon Beach, Oregon.

Vegetation Vegetation, especially the extensive and deep root networks of large shrubs and trees, binds and stabilizes loose unconsolidated material. Removal of vegetation by forest fire or clear-cutting for timber or farming allows loose material to move downslope, especially shortly after a rainstorm. Several decades ago in the town of Menton, France, farmers decided to remove olive trees with their deep, stabilizing roots from the area's steep slopes in order to plant more profitable shallow-rooted carnations. This horticultural miscalculation contributed to landslides that took 11 lives. Similar widespread tree-cutting in the forests of the Philippines was a primary cause of the tragic landslides that claimed 3400 lives in November 1991.

Water Water, more than any other factor, is likely to cause previously stable slopes to fail and slide. Initially, a small amount of water actually increases the cohesiveness of loose material by binding adjacent particles by *surface tension*, a weak electrical force on the surface of a drop of water that attracts and holds other particles, such as sand grains (Fig. 13-7). Water also fosters the growth of stabilizing vegetation. But excessive water can promote slope failure, by reducing the friction between surface materials and underlying rocks or between adjacent grains of unconsolidated sediment. The role water plays in reducing friction between an object and a slope can be seen using the flat, plastic water slides found in many backyards during the summertime. The thin film of water between the slide and slider reduces friction dramatically, allowing the slider to slide a long distance and reach a high velocity. Even solid bedrock can be moved with the aid of water: Infiltrating water reduces the friction between adjacent rock masses that are separated by a plane of weakness (such as a bedding plane, fault, or joint) (Fig. 13-8).

Figure 13-8 The effects of water on surface friction and the movement of slope material. (**a**) Water can cause failure in slopes of solid bedrock by lubricating planes of weakness in the rock. (**b**) Water reduces friction between a rock mass and even the most gentle underlying slope. This boulder in Death Valley, California, has reached its current position by sliding on the wet surface of the underlying clay.

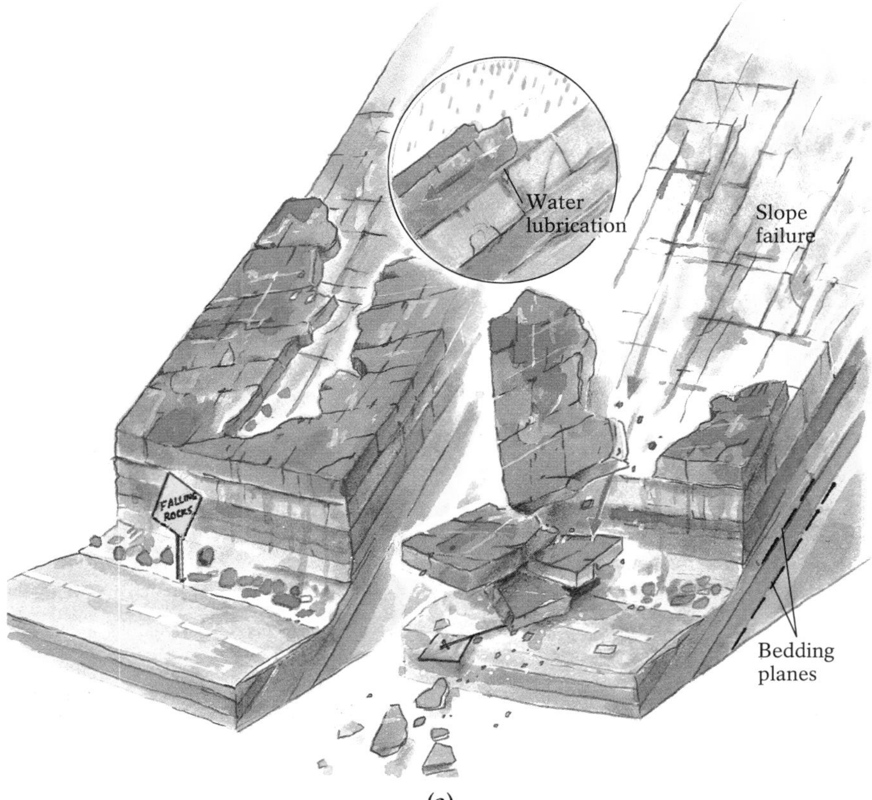

(a)

(b)

Figure 13-9 Saturation with water promotes mass movement in unconsolidated slope materials by decreasing the internal friction and electrostatic attraction between particles.

Saturation with water also reduces cohesiveness between individual sediment grains. In general, an unconsolidated sediment's cohesiveness lies in the internal friction between grains in contact and the weak attractive forces (electrostatic attraction) between grains. When water completely surrounds sediment grains, it isolates them from adjacent grains and effectively eliminates the friction and the electrostatic attraction between them. Thus, although damp soil may be more cohesive than dry soil, wet, saturated soil becomes a formless mass of flowing mud as individual grains are forced apart (Fig. 13-9). Water also promotes slope failure by adding to the weight of slope materials, thereby increasing G_d. A stable mass of dry sand, with slopes at or below the angle of repose, may have as much as 35 to 40% of its volume occupied by dry pore spaces. If a prolonged rainstorm fills those pores with water, a considerable weight is added, and the once-stable slope can fail.

Setting Off a Mass-Movement Event

Before a stable slope becomes unstable and fails, it develops a fragile balance, or equilibrium, between the forces that tend to drive movement downslope and the forces that tend to resist movement. Some event may then tip the delicate balance in favor of the driving forces, setting off downslope movement. The trigger mechanisms, or immediate causes, of mass movement can be either natural or human-induced.

Natural Triggers Mass-movement events can be set off by a number of other climatic and geological events. Torrential rains, earthquakes, and volcanic eruptions can all send loose materials moving downslope. Heavy rainfalls, for example, can quickly saturate thick clayey regolith that may have taken millenia to accumulate. At some point, the slope's material can no longer withstand the downslope force of gravity, and it slides downhill. In 1967, a three-hour electrical storm in central Brazil triggered hundreds of slope failures, taking more than 1700 lives.

An earthquake can dislodge massive blocks of bedrock and enormous volumes of unconsolidated material. The series of powerful earthquakes that struck New Madrid, Missouri, between December 1811 and February 1812 initiated hundreds of mass-movement events along a 13,000-square-kilometer (5000-square-mile) area of the Mississippi Valley. In the most tragic event of its kind in recent memory, the 1970 earthquake in Peru, which measured 7.7 on the Richter scale, shook the Andes so violently that more than 150 million cubic meters (5 billion cubic feet) of ice, rock, and soil buried several mountainside towns and their 30,000 inhabitants.

Volcanic eruptions, particularly those that propel large quantities of hot ash onto snow- or glacier-covered slopes, produce enormous volumes of meltwater. The water mixes with fresh ash and other surface debris to form a muddy slurry that flows rapidly down steep volcanic slopes, burying landscapes and communities below. Much of the loss of life and property during the 1980 eruption of Mount St. Helens resulted from such flows.

Human-Induced Triggers Mass-movement events can also be set off by a variety of human activities. Mismanagement of water and vegetation, oversteepening and overloading of slopes, mining miscalculations, and even the vibrations of loud sounds can trigger slope failure.

When we over-irrigate slopes for farming, install septic fields that leak sewage, divert surface water onto sensitive slopes at construction sites, or water our sloping lawns profusely, we introduce liquids that destabilize slopes by reducing friction. In one case of inadvertent water mismanagement, a Los

Angeles family went on vacation and left their lawn sprinklers on. They returned to find their hillside lawn and home sitting on the highway in the valley below.

When we clear-cut forested slopes or accidentally set forest fires, we eliminate the deep, extensive root networks that bind loose materials together. When we cut into the bases of sensitive slopes to clear land for homes or roads—especially if we dump the removed or other material at the top of the slopes—we oversteepen the slopes and jeopardize their equilibrium. Building housing developments on hillsides or cliffsides, as is common in California, can cause mass movement when slopes bearing the extra weight of roads and buildings are later undermined by heavy rainfall.

The rumble of a passing train, the crack of an aircraft's sonic boom, and the blasting that accompanies mining, quarrying, and road construction can each trigger mass movement. Vibrations from these activities can separate grains of loose sediment, eliminating the friction between them.

Sometimes human activities such as mining can combine with natural factors to increase the probability of mass movement along a slope. Such was the situation that led to the Turtle Mountain landslide of 1903, near the Canadian Rockies town of Frank, Alberta, which claimed 70 lives (Fig. 13-10). It is believed that removal of a large volume of coal near the mountain weakened an already unstable structure containing numerous joints and fractures. The month preceding the slide had been a wet one in the southern Alberta Rockies, and water from copious snowmelt probably entered and lubricated the fractures in Turtle Mountain, further increasing its potential for mass movement.

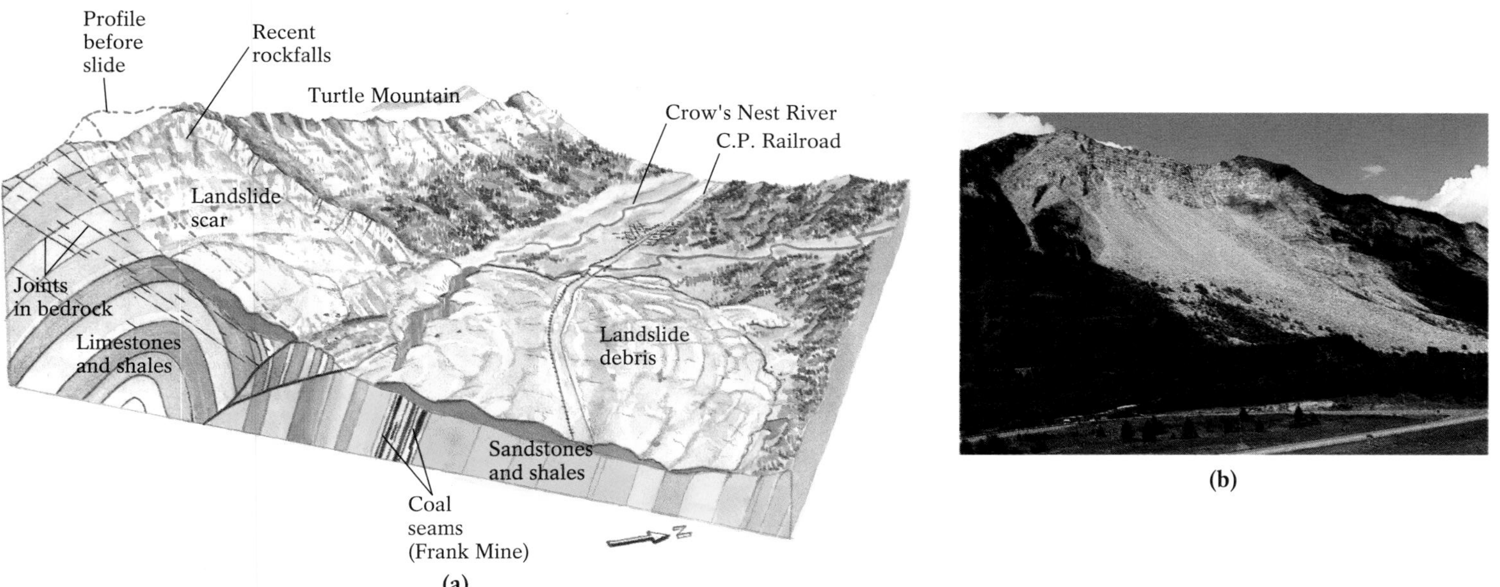

Figure 13-10 The 1903 Turtle Mountain landslide, in Frank, Alberta. (**a**) The disaster began when an enormous mass of limestone broke free and moved rapidly downslope to a protruding rock ledge where it was launched airborne toward the valley below. After its 900-meter (3000-foot) drop, the rock struck the weak shales and coal seams in the valley and shattered into a great avalanche of crushed rock that spread at speeds as high as 100 kilometers (60 miles) per hour to a distance of 3 kilometers (2 miles) across the valley. The mass of crushed rock had such momentum that it actually ascended 120 meters (400 feet) up the opposite side of the valley. Ironically, 16 men working in the mines survived by digging their way out through a soft coal seam. (**b**) Turtle Mountain as it looks today, showing evidence of more recent rockfalls within the scar from the 1903 slide.

Types of Mass Movement

There is no universally accepted classification scheme for mass movements. Ask a soil scientist, a geologist, and a civil engineer to classify a mudflow (a rapidly flowing slurry of mud and water) and you'll likely get three different answers. Geologists, however, generally classify mass movements based on the speed and the manner in which the materials move downslope.

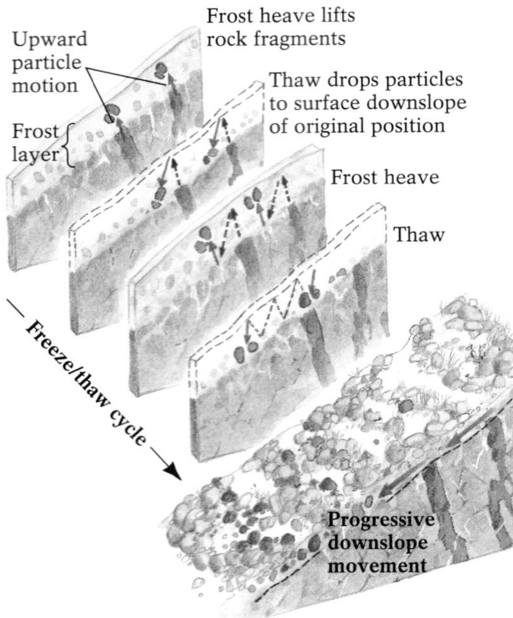

Figure 13-11 Frost-induced creep. Soil particles are displaced when ice forms beneath them, pushing them toward the surface. The downslope component of gravity carries the displaced particles a short distance downslope as the frost thaws, and the particles sink back down.

Slow Mass Movement

Creep, the slowest mass movement, is measured in millimeters or centimeters per year. It occurs virtually everywhere, even on the gentlest slopes. Creep generally affects unconsolidated materials, such as soil or regolith, that rarely exceed a few meters in depth. Because of friction with underlying slope material, a mass of material undergoing creep tends to move faster at its surface and more slowly at deeper layers, causing implanted objects such as lamp posts and telephone poles to slant.

Loose material experiencing creep is continuously rearranged as each individual particle responds to the influence of gravity. A particle can be dislodged and set in motion in any number of ways, such as by the burrowing of an insect or larger animal, surface trampling of passing animals, or the splash of raindrops. Even the swaying of plants and trees in the breeze can cause their roots to move enough to jostle loose soil cover.

In sufficiently cold environments, where soil water freezes periodically and ice develops in the ground, the volume of the near-surface materials expands upon freezing and displaces soil particles at the slope surface. As thawing occurs and the near-surface materials contract, the displaced surface particles are pulled slightly downslope by gravity before settling back into a stable position (Fig. 13-11). In clay-rich soils, periodic wetting and drying swells and shrinks materials and moves soil particles in a similar manner. Each of the countless particles in such soils or regolith is gradually but constantly being nudged downhill, even on a gentle slope.

A special variety of creep occurring principally in cold regions is **solifluction** ("soil flow"). This comparatively fast form of creep develops where the warm sun of the brief summer season thaws only the upper meter or two of soil or regolith. Because the underlying *permafrost* (permanently frozen ground) is impermeable to water, the thawed soil becomes waterlogged and flows downslope at rates of 5 to 15 centimeters (2–6 inches) per year (Fig. 13-12).

A good indication that creep has occurred exists when structures that were originally upright are no longer vertical. Fences usually appear tilted on creeping slopes. Older gravestones are generally more slanted than younger ones in the same cemetery because they have been creeping for a longer period. Telephone poles inserted vertically in a creeping slope near Yellowstone National Park developed 8° inclinations after 10 years. Trees on creeping slopes, because they continually adjust their shape to remain vertical and absorb maximum sunlight, often have contorted trunks (Fig. 13-13).

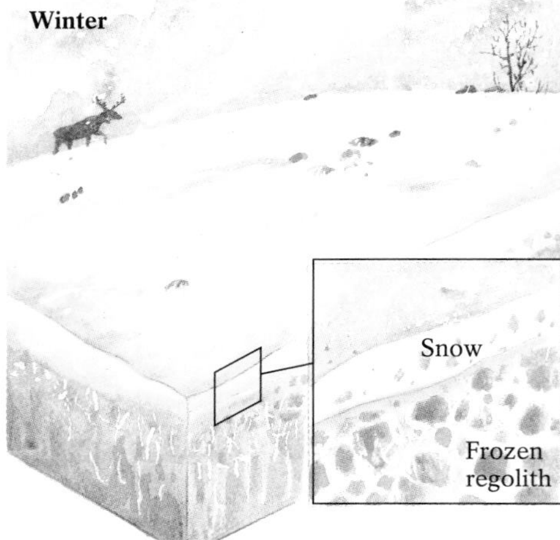

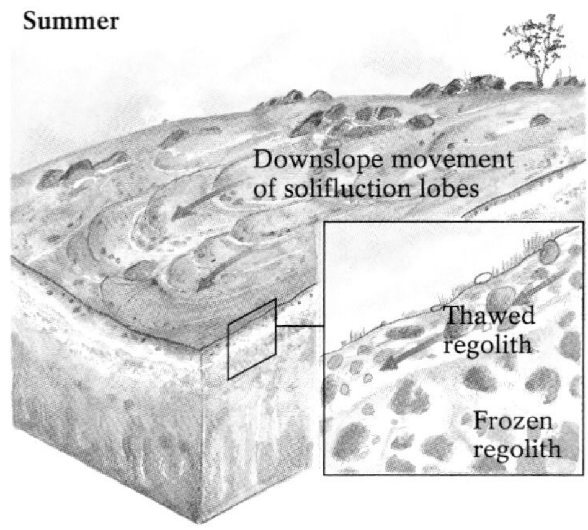

Figure 13-12 Solifluction occurs where the soil and regolith is frozen to depths of hundreds of meters. During the brief summer season, the surface materials thaw out to a depth of several meters and flow downslope with the aid of the water released by melting. Solifluction typically creates lobe-shaped masses of slowly moving sediment.

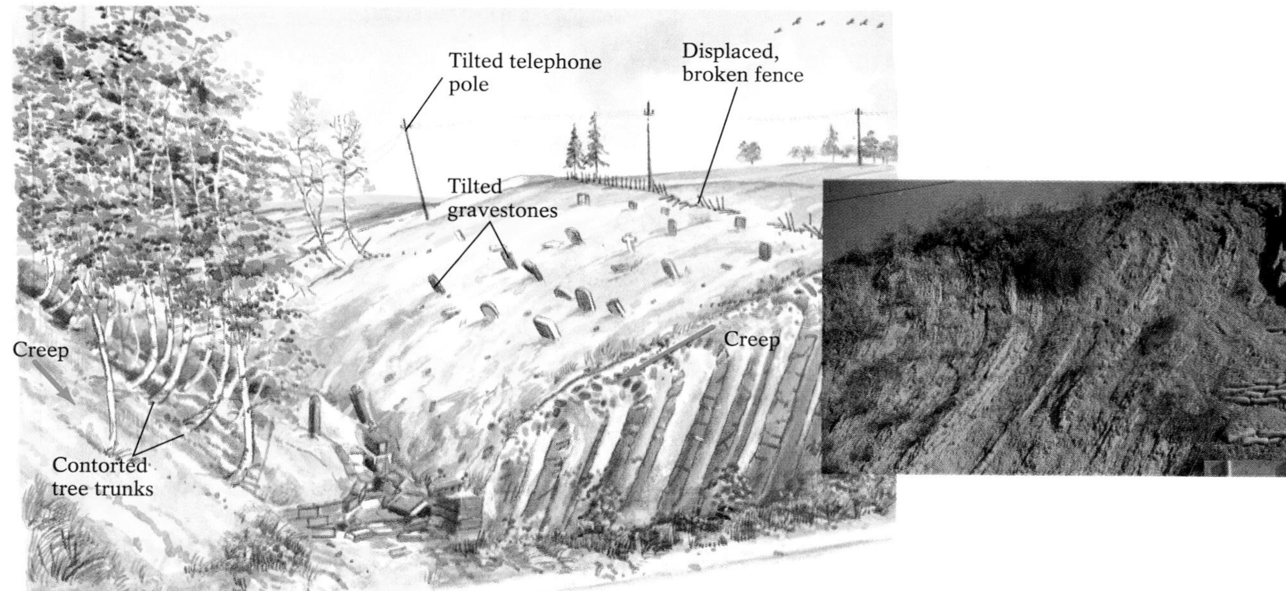

Figure 13-13 Creep affects every near-surface feature, including soil, rock, vegetation, and a variety of shallow-rooted human-made items such as fences, gravestones, and utility poles.

Rapid Mass Movement

Rapid mass movements, rather than occurring at the rate of centimeters per year like creep, can proceed at kilometers per hour or even meters per second. Rapid mass movements are further classified according to their different types of motion (Fig. 13-14). A **fall** occurs when loose rock or sediment is dislodged and drops from very steep or vertical slopes. A **slide** occurs when a mass of rock or sediment is dislodged and moves along a plane of weakness, such as a fault, fracture, or bedding plane. A slide that separates along a *concave* surface is a **slump**. A **flow** occurs when a mass of rock fragments or sediment moves downslope as a highly viscous fluid. Each type of movement can occur at varying speeds and include a variety of materials. Often one type of rapid mass movement evolves into another. When a fall's material strikes the ground, for example, it may continue to rush downslope as a high-speed flow containing a mixture of shattered rock, soil, and vegetation.

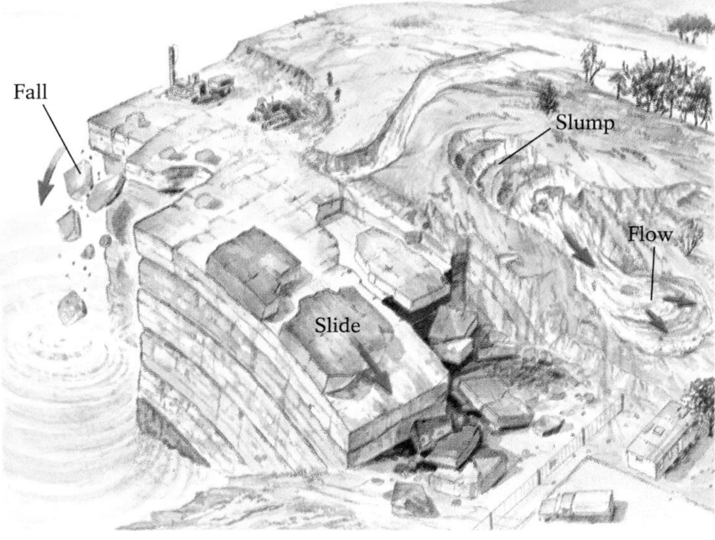

Figure 13-14 Rapid mass movements include falls, slides, slumps, and flows. (See text.)

Figure 13-15 A rockfall at the base of a coastal cliff, Great Australian Bight, South Australia.

Falls A fall is the fastest type of rapid mass movement, and can be very dangerous. It occurs when rock or sediment breaks free from a steep or vertical slope, often at a plane of weakness, and plummets through the air to the ground or water below at an acceleration of 9.8 meters (32 feet) per second squared (Fig. 13-15). A small, 150-cubic-meter (5300-cubic-foot) rockfall that landed directly on a bus in Beaver County, Pennsylvania, in 1942 took 22 lives. A fall that occurs at the coast may create a large sea wave. A spectacular coastal rockfall, set off by an earthquake, sent 31 million cubic meters (1 billion cubic feet) of bedrock plummeting 900 meters (3000 feet) into Lituya Bay near Anchorage, Alaska, in 1958. An enormous wave, hundreds of meters high, raced across the bay, lifting boats anchored immediately inside the bay mouth and transporting them out to sea.

Slides and Slumps A slide consists of a single intact mass of rock, soil, or unconsolidated material. It may involve a small displacement of soil over solid bedrock, or a huge displacement of an entire mountainside. Large slides may be preceded by days, months, or even years of detectable accelerated creep. In the largest slides, a great mass of material detaches along an identifiable plane of weakness, or **slip plane**.

The nature of a slip plane varies with the local geology. Unconsolidated sediments often develop slip planes where there is a change in sedimentary material—along the boundary between a layer of sand and a layer of clay, for example. In solid bedrock, slip planes develop where planes of weakness exist within the rock itself (see Fig. 13-4). In sedimentary rocks, the slip plane is often a bedding plane within a weak rock layer, such as thinly bedded shale that is inclined toward an open space, such as a river valley. In plutonic igneous rocks, the slip plane is usually a large joint produced by exfoliation. In metamorphic rocks, particularly those of high grade, the slip plane is likely to occur along foliations such as schistose bands. Highlight 13-1 describes a very large slide that occurred along a boundary between sandstone and shale.

A slide's slip plane is generally flat, or planar. A slide that separates along a concave slip plane, moving in a downward and outward motion, is a slump. Slumps are distinguished by the crescent-shaped scar that forms in the landscape where the slump has detached. The steep, exposed cliff face that forms where a slump mass has pulled away from the rest of the slope is a **scarp.** Slumps do not usually travel far from their point of origin, nor do they move at high velocities. The slump block itself, because of the nature of its motion, usually remains intact (Fig. 13-16).

Figure 13-16 Slumping on cliffs in Dorset, England.

Flows A flow is any mixture of solid, mostly unconsolidated particles that moves downslope like a viscous fluid. It may be wet or dry. Like other mass movements, flows are classified by their velocity and composition. They typically move more swiftly and are more dangerous when their water content is high. In February 1969, for example, a drenching rainstorm moved into southern California from the Pacific, dumping 25 centimeters (10 inches) of rain in nine days. In the next six weeks, the Los Angeles area received an additional 60 centimeters (25 inches) of rain, more than twice the normal amount for what is typically the "rainy season." The area's precipitation patterns are notoriously erratic: The previous summer had seen a prolonged drought, in which brush fires devegetated hundreds of thousands of acres of the Santa Monica and San Gabriel Mountains, which consist largely of clay-rich marine shales. The torrential rains triggered numerous flows, having saturated a clayey soil cover, denuded of vegetation by fire and drought, over impermeable bedrock. One hundred lives were lost, 15,000 people were displaced from mud-covered homes, and a billion dollars in property damage was incurred as a result of these flows.

Highlight 13-1 *The Gros Ventre Slide*

Figure 13-17 Sheep Mountain, along the Gros Ventre River in Wyoming, was the site of one of North America's greatest mass-movement events. (a) Landslide debris dams the Gros Ventre River. (b) The geology of the Sheep Mountain slide.

On June 23, 1925, just east of the small town of Kelly, Wyoming, near Jackson Hole, the largest known slide in United States history rumbled down the south slope of Sheep Mountain. More than 38 million cubic meters (1.3 billion cubic feet) of sandstone blocks, shale slabs, and loose soil and regolith were transported more than 600 meters (2000 feet) into the valley of the Gros Ventre River (Fig. 13-17a). One enormous sandstone mass, traveling intact, brought with it the entire overlying pine forest. The momentum of the slide mass was so great that it hurtled upslope on the opposite side of the valley more than 100 meters (330 feet) before it slid back down, finally coming to rest on the valley floor. There the slide mass formed a natural dam, 75 meters (250 feet) high, blocking the Gros Ventre River to create a lake 8 kilometers (5 miles) long and 70 meters (240 feet) deep. The lake filled so quickly that only 18 hours after the slide, a house 18 meters (60 feet) above the river was floated off its foundation by rising lake water.

Why did this particular mountainside slide? In Sheep Mountain, layers of clay-rich shales alternate with massive beds of limestone and sandstone (Fig. 13-17b). The beds are inclined toward the valley at an angle greater than 20°. Water from that spring's abundant snowmelt and heavy rainfall could not drain into an impermeable shale layer several meters below the surface, causing all the overlying rock to become saturated. The added weight of the water, combined with the moistening of the slip plane above the shale, caused an entire layer of near-surface sandstone to slide.

The area was sparsely populated, and fortunately no lives were lost *during* this monumental event. But after heavy May rains two years later, the new lake breached the natural dam, cutting a channel 15 meters (50 feet) wide through the slide mass. A 5-meter (16-foot)-high wall of water rushed downhill and washed out the town of Kelly in a furious flood that took the lives of six residents.

Figure 13-18 The Slumgullion mudflow in San Cristobal, Colorado. This mudflow formed when rock near the summits of these mountains became saturated with water after a high spring melt. The saturated mass, intermixed with glacial debris, flowed 11,500 feet into the valley below.

Flows may contain a wide variety of materials, such as loose rocks, soil, trees and other vegetation, water, snow, and ice. The velocity of a flow is determined by its water content, the nature of its materials, the nature of the underlying materials, the angle of the underlying slope, and the vertical distance it has traveled. **Earthflows** are relatively dry masses of clayey or silty regolith which, because of their high viscosity, typically move as slowly as a meter or two per hour (though they sometimes move as fast as several meters per minute). Because they flow so slowly, they are rarely life-threatening, although they can do a good deal of damage to structures in their paths. A typical earthflow moves at a rate faster than creep but more slowly than a **mudflow**, which is a swift-flowing slurry of regolith mixed with water. Mudflows generally form in humid areas marked by a thick blanket of regolith and are often triggered by a heavy rainstorm.

Mudflows are typically fine-grained, consisting 80% or more of sand-sized and finer grains of soil and regolith. The consistency of mudflows can vary from that of wet concrete to that of muddy water. Because they are typically saturated with water, mudflows tend to flow through topographic low spots, such as valleys and canyons, just as streams do, and can move considerable distances on slopes as gentle as 1 to 2° (Fig. 13-18).

Mudflows are most likely to develop after heavy rainfalls on sparsely vegetated slopes with abundant loose regolith. These conditions are commonly found in the canyons and gullies of semi-arid mountains, where loose sediment accumulates between infrequent storms. With few roots to bind loose materials, the occasional cloudburst washes large quantities of sediment into arroyos (dry stream channels) and canyons. The saturated sediment then rushes rapidly down the channel.

Occasionally, a partly waterlogged solid clayey sediment becomes a highly fluid mudflow in an instant, when water separates its sediment particles and reduces the friction between them. **Quick clays** are sediments that are solid at one moment and liquid the next (Fig. 13-19). The trigger mechanism for a quick-clay flow may be an earthquake, an explosion, crashing thunder, or even the vibrations from heavy vehicles and passing trains.

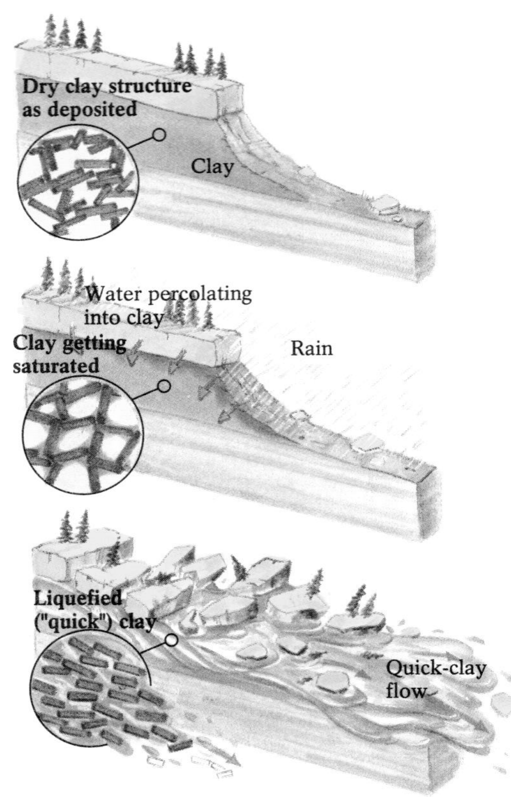

Figure 13-19 The development of quick-clay flows.

Certain regions of North America are particularly vulnerable to quick-clay flows. Fifteen thousand years ago, much of eastern Canada and New England was covered by an ice sheet several kilometers thick, the weight of which compressed the land beneath it. When the ice finally melted, the land was below sea level. The seawater that flowed inland left behind marine-clay deposits—a loose, disorganized framework of platy clay particles. On November 12, 1955, near the town of Nicolet, Quebec, along the Gulf of St. Lawrence, 165,000 cubic meters (5.8 million cubic feet) of this marine clay liquefied into quick clay that flowed toward the Nicolet River valley. The flow stopped just meters before engulfing the crowded local cathedral, and there were only three fatalities. A similar, but much larger and more catastrophic quick-clay flow occurred on May 4, 1971, in the Quebec village of Saint-Jean-Vianney, where a layer of marine clay liquefied instantaneously, and 6.9 million cubic meters (240 million cubic feet) of clay buried 40 homes and took 31 lives (Fig. 13-20).

Volcanic eruptions can produce catastrophic mudflows known as *lahars* (see Chapter 4), when hot volcanic ash melts snow or glacial ice or when the rainfall that often accompanies an eruption saturates fresh volcanic ash and the accumulated soil and ash layers from previous eruptions. A lahar may rush downslope at very high speeds. The 1980 eruption of Mount St. Helens set off a lahar that traveled between 29 and 55 kilometers (18–35 miles) per hour, sweeping away homes, bridges, and everything else in its path.

Figure 13-20 Saint-Jean-Vianney, Quebec, after the quick-clay flow of 1971.

Lahars can also occur during a volcano's noneruptive periods. Mount Rainier has generated more than 60 lahars during the last 10,000 years. A major one took place about 5000 years ago, when the summit of the mountain became unstable after millenia of steam emissions had converted solid volcanic andesite to loose clay. More than 450 meters (1500 feet) of the mountain, estimated to have stood at 4850 meters (16,000 feet) at the time, flowed to the lowland below. We can infer it was a noneruptive lahar because no ash layers exist to record an eruption. Mount Rainier continues to vent steam, and thus alter its andesitic summit to clay, raising great concern about the possibility of another noneruptive lahar in the future. For this reason, the U.S. Geological Survey monitors Mount Rainier vigilantly and has drafted an advisory evacuation plan for nearby communities, where more than 50,000 people live today.

Figure 13-21 A debris flow caused by a one-night flood in a small stream, in Britannia Beach, British Columbia.

Debris flows, like mudflows, tend to follow topographic low spots, are common in sparsely vegetated mountains in semi-arid climates, and are triggered by the sudden introduction of large amounts of water. Debris flows consist of particles that are generally coarser than sand-size, and often contain boulders one meter or more in diameter (Fig. 13-21). Hence, they require a steeper slope than other types of flow to set them off. A mass movement that begins as a fall or slide can eventually break into finer fragments and continue downslope as a debris flow. Debris flow velocities range between 2 and 40 kilometers (1–25 miles) per hour.

Swifter and more dangerous than debris flows are **debris avalanches,** common in the Appalachians, the Green and White Mountains of New England, and the Cascades and Olympic Mountains of the Pacific Northwest. They occur on very steep slopes, are also triggered during and after heavy rains, and are enhanced when vegetative cover is removed by fire or logging. During an avalanche, the entire thickness of soil and regolith may become detached from the underlying bedrock and rush downslope through narrow valleys. In 1973, when Hurricane Camille brought 60 centimeters (25 inches) of rain to Virginia's Blue Ridge Mountains, debris avalanches claimed 150 lives.

(a)

(b)

Figure 13-22 Triggered by a major earthquake, the Yungay debris avalanche of May 1970 buried several towns, taking 30,000 lives. (**a**) Before the avalanche. (**b**) After the avalanche.

The high velocity of debris avalanches, which under certain topographic conditions can even propel debris through the air, accounts for the great damage they can cause. In Yungay, Peru, a powerful earthquake on May 31, 1970, dislodged an avalanche of debris that hurtled downslope at speeds exceeding 200 kilometers (120 miles) per hour (Fig. 13-22). On patches of the slope in the avalanche's path, grass and flowers—including shrubs more than 1 meter (3.3 feet) high—were undisturbed, suggesting that the debris flew through the air *above* them. Compressed air, trapped below rapidly falling debris, forms a supporting air cushion that reduces friction between the moving debris mass and the underlying surface. Eventually, as gravity pulls the debris down to the slope's surface, this air cushion is expelled with a powerful wind gust. A tangle of downed trees that appear to have been uprooted by a great blast of air is often seen at the edges of large debris avalanches.

Avoiding and Preventing Mass-Movement Disasters

Efforts to prevent mass-movement disasters must begin with a prediction of the likelihood of slope failure in a given area. This involves analysis of both the local geology and the historical records of past mass-movement events. The next step is avoidance of human activities that could contribute to slope failure. Finally, a mass-movement defense plan must prevent failure of existing slopes by stabilizing those at risk.

Predicting Mass Movements

Mass movements may well be the most readily predicted of all geological hazards. Simple eyewitness accounts of past and ongoing mass movements are the best clues to their future behavior. In the spring of 1935, when the autobahn was being laid through some clay deposits between Munich, Germany, and Salzburg, Austria, a series of slides caught the German engineers by surprise. Had they heeded the construction crews' observation that the slope "wird lebendig" (was "becoming alive"), they might not have been so shocked.

The next step in prediction is thorough terrain analysis. Geologists study the composition, layering, and structure of slope materials, determine their water content and drainage properties, measure slope angles, search for field evidence of past mass-movement events, and bury instruments in boreholes on slopes to monitor the movement and deformation of slope materials. The geological surveys of the United States and Canada have issued slide-hazard maps based on such terrain analyses (Fig. 13-23). The United States Office of Emergency Preparedness also keeps an inventory of slide-prone areas.

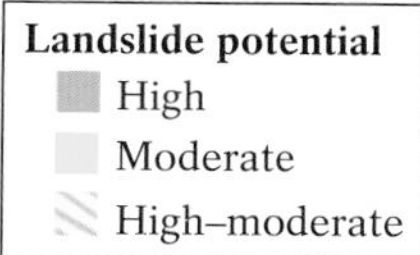

Figure 13-23 A mass-movement hazard map for the United States and Canada. Mass-movement hazards have been concentrated where slopes are steep, such as in western mountains and the Appalachians; where annual rainfall is high, such as in the humid Southeast and misty Northwest; where droughts and human activities destabilize sensitive slopes, such as in southern California; and in the permafrost regions of Canada, where repeated freezing and thawing contribute to slope instability.

Figure 13-24 Lithified mudflow deposits from the Pliocene Epoch, in Borrego, California.

Evidence of earlier slope failures includes talus slopes, landslide scars, and jumbles of slide and flow debris. Unlike most other sediments, mass-movement deposits are usually quite coarse and bouldery, and very poorly sorted. Only a few mass-movement deposits (such as those from mudflows) are crudely stratified, a consequence of large, weighty boulders settling toward the bottom of the flow (Fig. 13-24). Often transported within a protective muddy matrix, the boulders in mass-movement deposits tend to be quite angular. Because they were transported only short distances, nearly all of the rock fragments in such deposits are locally derived.

Earlier slope failures are not always easy to detect, however. Many very large slumps and slides are difficult to identify after they have been modified by creep and revegetated. Because they are often so large that up close they cannot be recognized, aerial photographs are often required to detect the crescent-shaped scarp at the head of a slump or the hilly chaotic terrain that marks a past slide or flow.

We can establish a pattern of recurrence for mass-movement events in a particular area if we can date prior mass-movement events and determine their trigger mechanisms. Such information would be very useful to municipal planners and the local populace. Vegetation is usually the key to dating. A flow on an arid treeless slope provides little evidence as to when it occurred. A flow on a vegetated slope, however, often incorporates organic material that can be dated with carbon-14. Dendrochronology and lichenometry (Chapter 8) can also help date a mass-movement event. A tree engulfed by a flow or slide may be bent by the flowing mass and reveal a datable disruption in its tree-ring pattern; the extent of lichen growth on a shattered rock can reveal how long ago the rock was broken (Fig. 13-25).

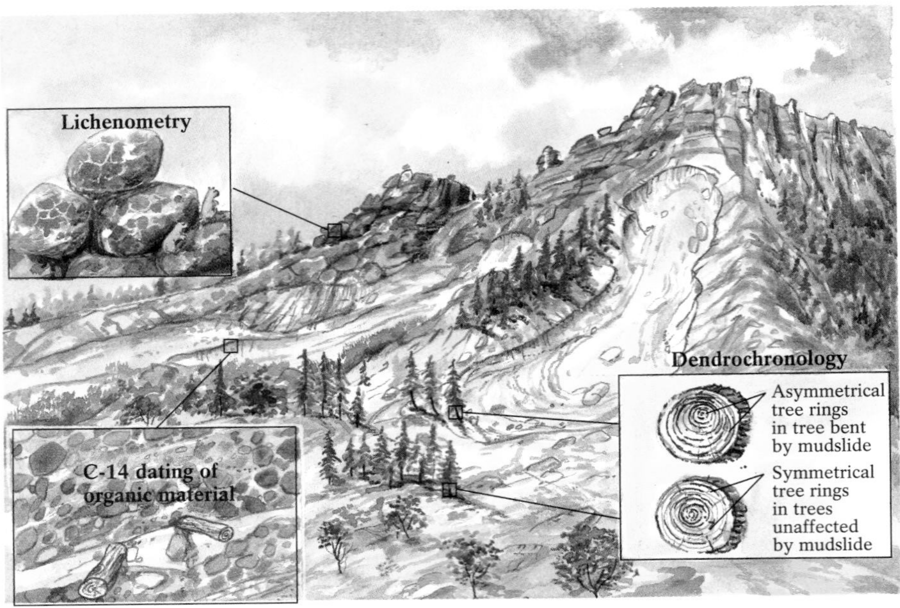

Figure 13-25 Methods of dating mass-movement events. Carbon-14 dating can be used if the moving mass incorporated trees or other types of vegetation. Lichenometry can be used if bedrock exposed or boulders shattered by the event have since become colonized by lichens. Dendrochronology can be used if the moving mass encroached upon living trees that survived the event: The rings of an affected tree grow asymmetrically from the moment it is bent by the mudflow, so we can determine how long ago the mudflow occurred by counting the asymmetrical rings.

After the dates of mass-movement events are established, they can be compared to the dates of other events that are considered possible trigger mechanisms. If heavy rain is a suspected trigger, mass-movement dates are compared to climatic records; if the records confirm a correspondence, it becomes possible to predict how an area's slopes will respond to various amounts of rainfall in the future. We know, for example, that most slides in the San Francisco Bay area have occurred after storms that dropped more than 15 centimeters (6 inches) of rain on steep, already water-saturated slopes.

We can never predict exactly when a mass-movement event will occur, but we can be alert to the potential dangers of specific slopes. Where property is valuable, or where mass movement poses a particular threat to safety, expensive landslide-warning devices that signal a slide or flow in progress can be installed. In the slide-prone southern Rockies, the Denver and Rio Grande Railroad has installed electric slide detectors—wired to respond to the high pressure of an encroaching slide mass—upslope from certain bridges. Contact with a moving mass breaks an electrical circuit, sending a "stop" signal to approaching trains.

Changes in animal behavior may also signal an impending mass-movement event, much as they do for earthquakes. Shortly before the entire Swiss village of Goldau was destroyed in 1806, the town's livestock acted nervously and all the beekeeper's bees abandoned their hives. Within hours, a block of rock 2 kilometers (1.2 miles) long and 300 meters (1000 feet) wide broke loose from a steep valley wall and buried the town and its 457 inhabitants in a massive rockfall.

Avoiding Mass Movements

Given our knowledge of the factors that cause mass movements and our ability to identify potential hazards, it is prudent to avoid building on sensitive slopes, especially since few insurance companies cover homes and property for mass-movement losses. But even after a risk map has been drafted and the potential for slides and flows identified, people in susceptible areas are reluctant to abandon ancestral homelands, commercially or agriculturally valuable properties, and scenic hillsides. Thus, avoidance of mass movement is rarely practiced, as the overdevelopment of southern California's steep canyon slopes and unstable ocean-front cliffs attests. Highlight 13-2 examines the importance of mass-movement avoidance to prospective home buyers.

Highlight 13-2 *How to Choose a Stable Homesite*

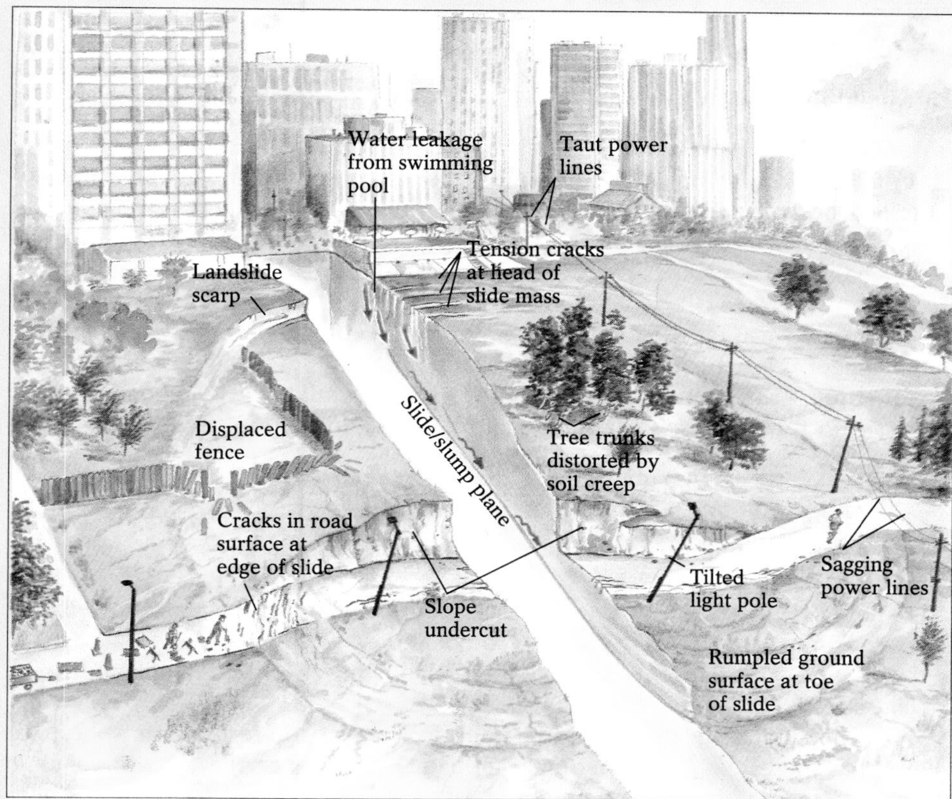

Figure 13-26 Various signs of past, current and potential future mass movement in an urban area. If, for example, the power lines in a neighborhood are very loose, it suggests that the poles that hold them have moved closer together, as one might expect at the toe of a slide where the slope is bunching up like a rumpled carpet. At the head of the slide, because the poles on the slide mass are moving away from those upslope, the lines are often very taut. (The poles in the middle of a slide mass may keep their original spacing because the mass may not be deforming much internally.

If you do not already do so, at some point in the future you may wish to own your own home. Having studied physical geology, you will want to ensure that your dream house doesn't fall victim to a geological nightmare.

Suppose you're exploring southern California's scenic beachfront locales. You happen upon the mosaic sign for the town of Portuguese Bend and notice that the beautiful ceramic signpost is cracked in two, and your newly grown geologic antenna is raised. You check the local real-estate listings and find a house priced at $50,000 that should be worth $500,000. Your intuition warns you that something is amiss. What else should you look for? How can you tell—in southern California or anywhere else in the world—if you're in mass-movement country?

First examine the property itself. Are there any signs of an old mud or debris flow? Look for exposed rocks or fresh breaks in the landscape that reveal the underlying geologic material. Then investigate the entire neighborhood. Is there evidence of a slump scar upslope where a block may have broken away? Is your dream house site in the middle of a large potential slide mass? Evidence of current or past activity may be hundreds of meters or even several kilometers away. As you drive through the community and its environs, look for fences that are out of alignment, and for power and telephone lines that seem too slack in some places and too taut in others (Fig. 13-26).

Look carefully at the house itself, and if possible at neighboring ones too. Look for large cracks in the foundation (small cracks may be due to initial drying and settling of the concrete). Doors and windows that stick may indicate that once-linear structural features are now out of line, although poor craftsmanship or high moisture content may also be responsible. A cracked pool lining might explain why a swimming pool doesn't retain water. Finding only one such problem may not indicate danger, but the presence of several problems should send a strong warning. If the geology, topography, and hydrology of a homesite all raise questions of slope stability, the site may well be subject to progressive slope failure. Your suspicions might be confirmed by checking newspaper accounts and the records of the local housing authority, by contacting the state geological survey, and by interviewing the property's neighbors.

People in developing nations must often overcultivate unstable slopes to feed the hungry, overgraze already sparse vegetation to maintain their livestock, and deforest slopes to provide wood for dwellings and heating. All of these actions promote mass movement. As long as we continue to use and inhabit unstable slopes in these ways, *prevention* will remain the most practical way to deal with the threat of mass movement.

Preventing Mass Movements

It is far cheaper and safer to stabilize a slope before it fails than to clean up afterwards. Unfortunately, the dollars lost to mass movements continue to exceed by a factor of 50 the money spent to prevent them. Civil engineers estimate that more than 95% of all slope failures could be prevented with regional, local, and site surveys and aggressive remedial action.

The first step in developing a prevention plan for any given site is to acquire a clear understanding of its subsurface geology and identify its potential trigger mechanisms. Unfortunately, detailed studies require costly drilling and sophisticated analytical techniques to determine the composition and physical properties of subsurface materials. In addition, in some settings the relevant geological features are highly variable, and tests must be conducted at a number of locations. Elsewhere, however, the local stratigraphy is similar virtually everywhere within the region. The entire city of Seattle, for example, which is one of the urban landslide capitals, is susceptible to slope failure by virtue of regional geology. The Puget Sound region of Washington state, including Seattle, typically contains a layer of permeable sand overlying a layer of impermeable clay, which promotes the buildup of excess water near the surface and consequently frequent damaging landslides and mudflows (Fig. 13-27).

The final step in preventing slope failure is to determine how best to enhance the forces that resist mass movement or reduce the forces that promote it. Several methods can be used, often simultaneously, to increase slope stability. *Nonstructural methods,* which do not require large, costly engineering efforts, involve management of vegetation and introduction of soil-strengthening agents. *Structural methods,* which are very costly, involve building structures such as retaining walls or actively modifying the terrain.

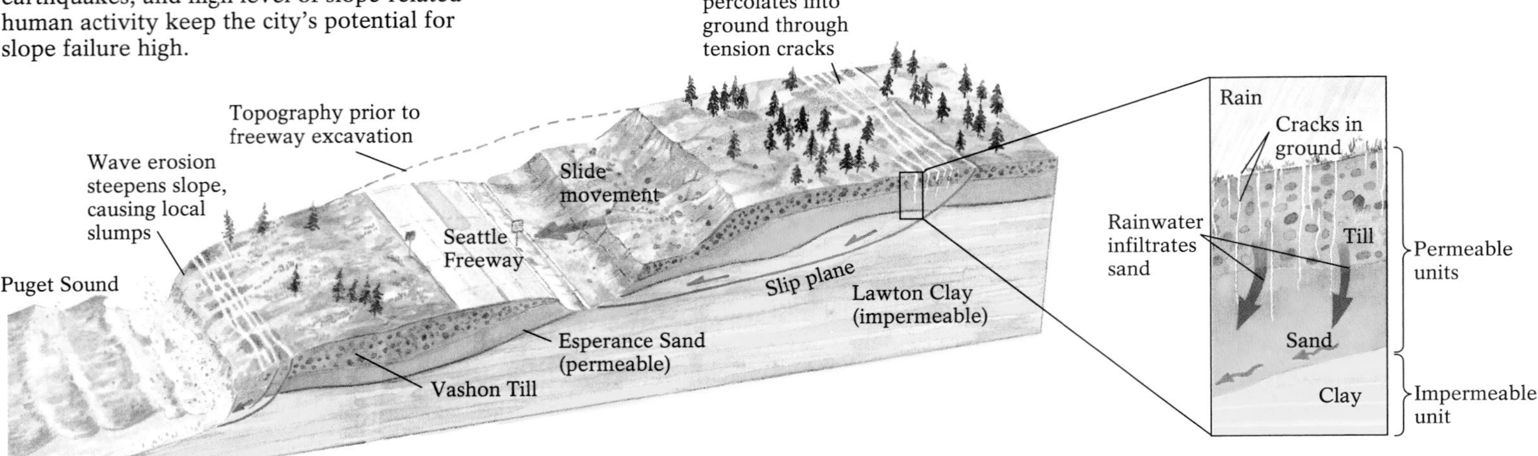

Figure 13-27 The slide-prone geology of Seattle, Washington. Virtually all of Seattle is built on 15,000-year-old glacial deposits consisting, in part, of a permeable layer of sand overlying an impermeable layer of clay. Water from Seattle's well-known rainfall infiltrates the sand and filters down until it encounters the clay. There the water is trapped, high water pressures build up, and the overlying sand layer is buoyed upward. The sand layer is then likely to slip—especially during the rainy months of January, February, and March—taking along with it some of the underlying clay. In addition to slip-plane lubrication, Seattle's oversteepened hills, occasional earthquakes, and high level of slope-related human activity keep the city's potential for slope failure high.

Nonstructural Approaches One nonstructural method of preventing mass movement is to plant fast-growing trees and plants that will develop extensive, deep root networks that bind up near-surface soils. In addition, broad-leaved species such as elms, maples, and oaks protect slopes with their large leafy canopies; some falling rain lands on the leaves and evaporates, reducing the likelihood of soil saturation.

Another nonstructural approach utilizes chemical solutions that bind clay-mineral particles, which can be added to soil or regolith to increase the strength of loose slope materials. This can even be done after a flow has begun: In Norway, a quick-clay flow was halted by bulldozing sodium- and magnesium-chloride salts into the moving mass. Cement can also be injected into a slide or flow mass to lend strength and cohesiveness to unconsolidated sediment.

Structural Approaches Structural mass-movement prevention efforts can be aimed at modifying a slope, building supports for it, or reducing its water content. One way of modifying a slope for safety is to *unload* it, by removing excess weight such as buildings, other structures, and loose soil. An overly steep slope can also be *graded* by moving material from the top and spreading it at the toe to reduce the slope angle (Fig. 13-28a). More aggressive approaches involve removing all material overlying a potential slip plane (Fig. 13-28b) or cutting flat terraces into a slope to reduce weight and create areas where small slide masses can come to rest (Fig. 13-28c).

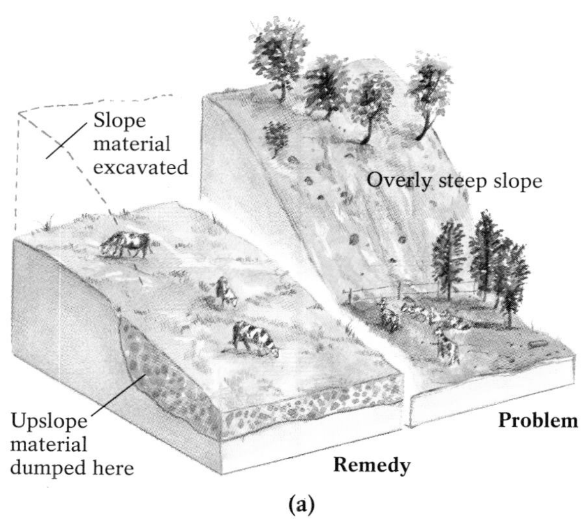

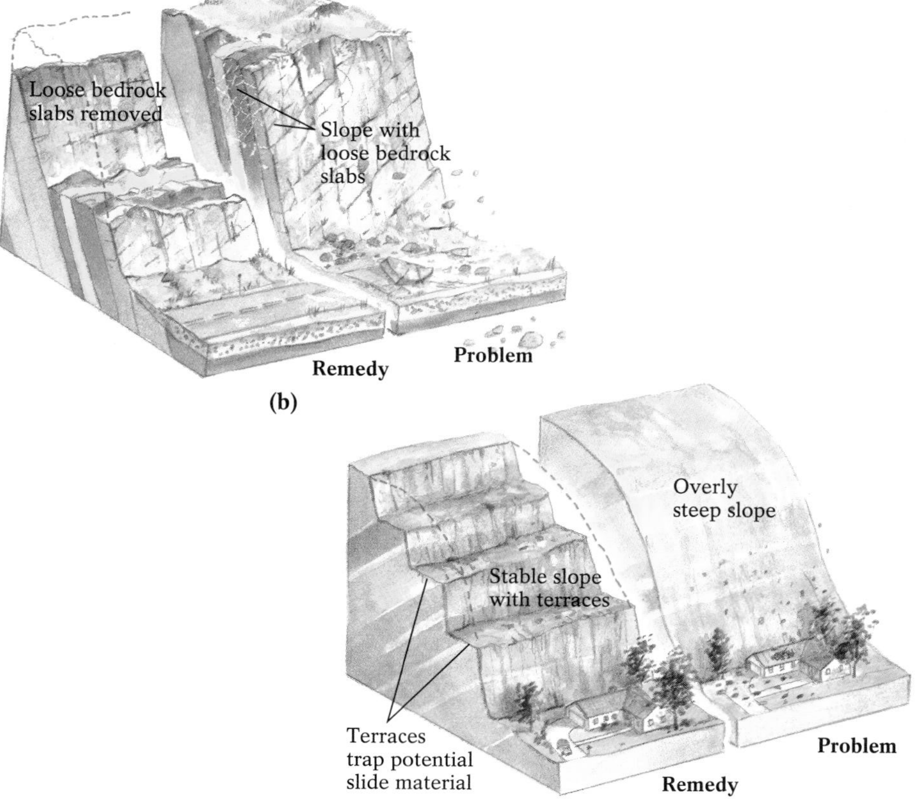

Figure 13-28 Structural methods of slide prevention involve manipulating a slope so that the forces driving downslope movement are diminished or the forces resisting movement are enhanced. (**a**) Grading a slope requires moving material from its top to its base to decrease the slope angle. (**b**) Loose slope material may be removed to prevent its falling unexpectedly. (**c**) Terraces may be cut into a slope to reduce the slope angle and to catch any falling debris.

Figure 13-29 Structural supports to prevent slope failure include rock bolts and retaining walls. (**a**) Rock bolts stabilize loose jointed granite on Storm King Mountain, overlooking Storm King Highway in Westchester, New York. (**b**) A retaining wall is built to support the ground underlying Route 659, near Blacksburg, Virginia.

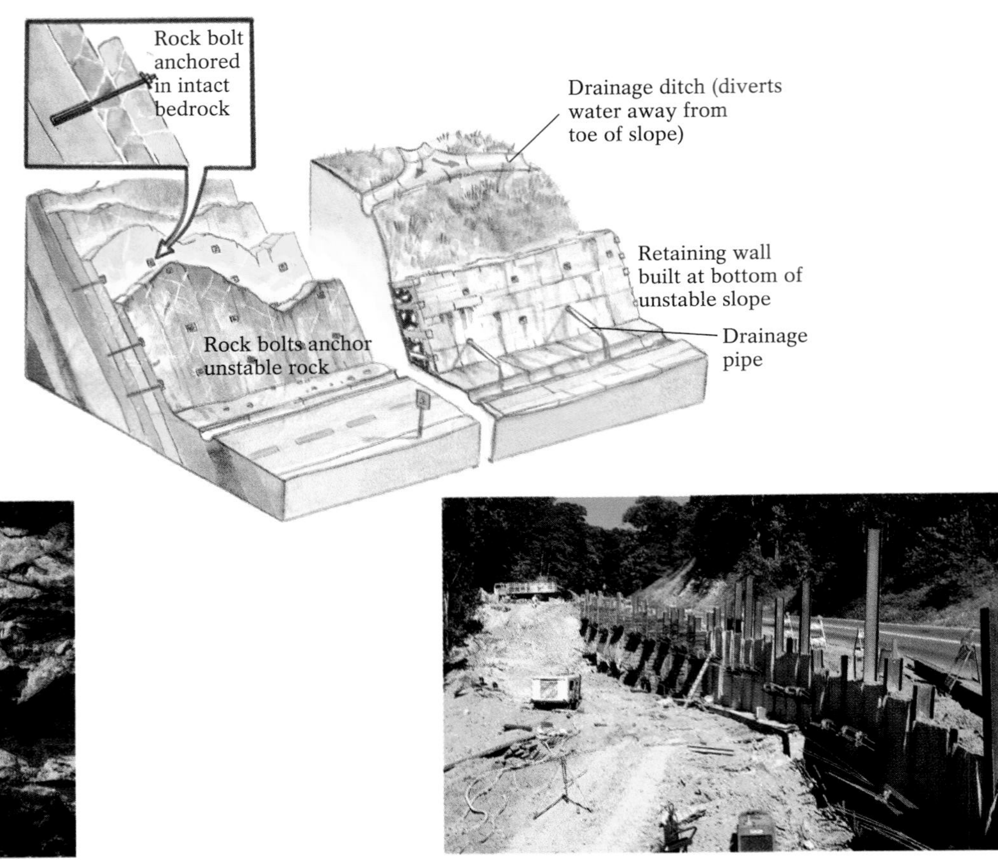

(a)

(b)

Installation of bolts or pins can hold a potential rockslide in place (Fig. 13-29a), but it is a costly procedure used only in particularly dangerous situations or to protect valuable property. Retaining walls, such as are used to shore up excavation walls at construction sites, can be built to provide support to the base of a slope; to do this, steel or wooden piles are driven through loose surface materials into underlying stable materials (Fig. 13-29b).

Perhaps the best way to reduce the likelihood of mass movement is to reduce the water content of a slope, either by preventing water from entering or by removing it after it has entered. To prevent infiltration, small earthen barriers can be erected to divert water away from a sensitive slope. Covering a slope with thin sheets of plastic or a layer of concrete can also prevent infiltration. Where a slope is already saturated, drains, such as perforated pipes or a layer of coarse permeable gravel, can be installed to collect water and channel it away. Water can be pumped from a slope to reduce the weight overlying potential slip planes, increase friction along the planes, and decrease swelling and weakening of soil clay. In Pacific Palisades, California, tunnels were excavated through an unstable slope and hot air circulated through them to dry subterranean pore water.

Slopes can also be frozen to stabilize them temporarily. During construction of the Grand Coulee Dam in Washington, liquid nitrogen was circulated to freeze 377 points in the dam site's slope. Seventy-three thousand kilograms (80 tons) of soil and regolith, making up a section of slope 12 meters (40 feet) high, 30 meters (100 feet) long, and 6 meters (20 feet) thick, were frozen every day.

Extraterrestrial Mass Movement

Images from NASA's Apollo, Viking, and other missions reveal that mass movement also occurs on two of Earth's nearest neighbors—the Moon and Mars. Although the Moon's gravity is only one-sixth that of Earth, mass movement is one of the few ways in which its surface has been modified. Because of the absence of water on the Moon, mass movement is limited to dry processes. Lunar mass movement is triggered primarily by meteorites; numerous daily impacts set off avalanches of rock and regolith. Once an impact crater has formed, slumps occur on the crater's inner walls, which are generally steeper than the angle of repose of loose lunar regolith. Marks left by rolling and sliding boulders may remain unchanged for millions of years on the Moon's virtually weathering-free surface (Fig. 13-30).

The surface of Mars exhibits a wider range of mass-movement features. The trigger mechanisms of Martian slides and flows include "marsquakes" and related faulting. Periodic melting of the planet's subsurface permafrost may have released large amounts of water in the past that oversteepened slopes and promoted development of massive slides, slumps, and debris flows. Such melting may have formed rapid mudflows that eroded all or parts of the Valles Marineris, one of Mars' largest canyons (Fig. 13-31).

In this chapter we have explored the many ways that gravity drives erosion of the Earth's surface. In Chapters 14 through 17, we will examine the role that water plays—surface stream water, unseen groundwater, and frozen glacial water—in transporting weathered materials and modifying the Earth's landscapes.

Figure 13-30 A lunar impact crater showing landslide scarps and slumps, as photographed by NASA's Apollo 16 mission.

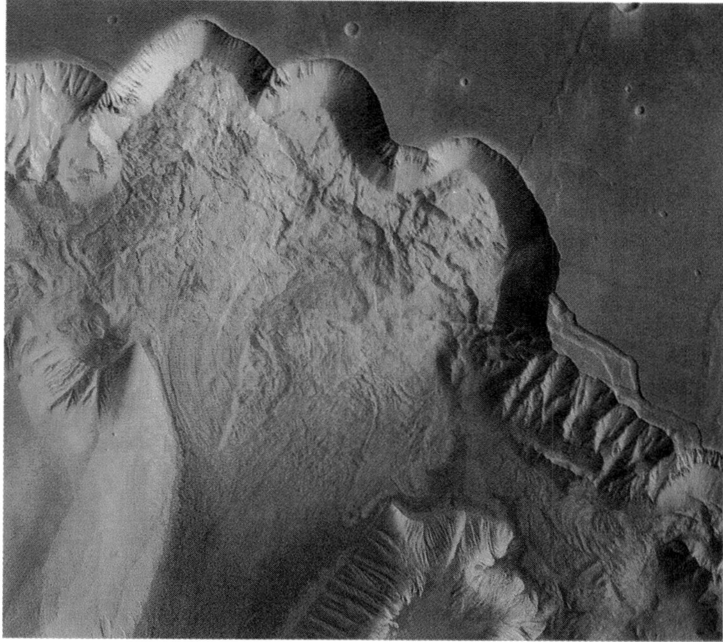

Figure 13-31 Mass movement on the Martian surface can be seen in the Valles Marineris, a 2-kilometer (1.2-mile)-deep canyon. Satellite images show that some of the flows are similar in form to those on Earth, and may have traveled as far as 30 kilometers (20 miles) across the canyon floor. Slumps that look quite similar to their counterparts on Earth can also be seen along the margin of the canyon.

Chapter Summary

Mass movement is the process that transports masses of Earth materials down slopes by the pull of gravity. It occurs when the factors that drive material downslope overcome the factors that tend to resist downslope movement. The force of gravity is the principal driving force, and the strength and cohesiveness of slope materials and the friction between them and the underlying slope are the principal factors resisting mass movement.

The steepness of the slope, the nature of its materials (for example, unconsolidated sediment or solid bedrock), the amount of water within the materials, the slope's vegetative cover, and the slope's history of human and other disturbance are factors that determine whether mass movement will occur. Of these, slope steepness (and its effect on the downslope component of gravity) and the slope's water content (and its effect on friction) are the primary determinants of mass-movement potential. If a slope is composed of solid bedrock, mass movement is enhanced when the bedrock is jointed, faulted, foliated, bedded, or contains cavities. The effect of slope steepness is particularly pronounced for loose unconsolidated material, which is stable only until it reaches its **angle of repose**, the maximum angle loose material can maintain without downslope particle motion. In general, large, irregularly shaped particles have a higher angle of repose. This can be seen on **talus slopes**, the large boulder piles that accumulate at the base of mountains, which often maintain slopes of 40° or more. Water contributes to mass movement by increasing the weight of slope materials, reducing friction between planes of weakness, and reducing internal friction between loose grains in unconsolidated sediment.

Mass movement can be triggered by a variety of natural and artificial causes. Earthquakes, volcanic eruptions, heavy rains, and snowmelt are common natural triggers. Ill-advised excavation of unstable slopes and overloading of unstable slopes by dumping or overbuilding are just a few of the ways that human activity triggers mass movement.

Mass-movement processes are classified by their velocity and their composition. In general, slow mass movement, called **creep**, occurs on virtually all slopes composed of unconsolidated soil or regolith. Creep may be enhanced by burrowing organisms, frost action, the impact of raindrops, and trampling by passing animals. A special variety of creep occurring principally in cold regions is **solifluction**, the movement of waterlogged soil over frozen ground.

There are several types of rapid mass movement. In a **fall**, materials are dislodged and fall from a steep or vertical slope without contacting the slope. In a **slide**, rock or regolith detaches along a plane of weakness, or **slip plane**, and moves as a single intact mass along the slope. A slide that separates along a concave slip plate is a **slump**, which leaves behind a crescent-shaped **scarp**, the steep, exposed cliff face that remains where a slide mass has pulled away. A **flow** is a slurry of loose material and varying amounts of water.

Flows may contain a wide variety of materials, such as loose rocks, soil, trees and other vegetation, water, and ice. **Earthflows** are relatively dry and quite slow-moving; **mudflows** are faster-moving wet slurries of regolith mixed with water. Occasionally, a partly waterlogged mass of solid clayey sediment becomes a highly fluid flow in an instant, when water separates its sediment particles and reduces the friction between them. Such sediments, called **quick clays**, can be produced by an earthquake, explosion, or any other source of vibration. **Debris flows** consist of particles coarser than sand-size, and may contain boulders one meter or more in diameter. The fastest variety of flow is a **debris avalanche**; these occur on very steep slopes, and are generally set off by heavy rains.

A slope's potential for mass movement can be predicted from analysis of its stability, evidence of past events, historical records, the use of landslide-warning devices, and changes in animal behavior. Prevention of mass movement may follow a nonstructural strategy, such as forest fire prevention, increased planting of vegetation with extensive root systems, and avoidance of excavation and overloading of sensitive slopes. Structural approaches include unloading slopes, grading slopes, building retaining walls, sealing slopes from further water infiltration, and draining water from slopes.

Mass movement also occurs on the Moon and Mars. Lunar mass movement is dry and is triggered by meteorites; Martian mass movement may be caused by periodic melting of its permafrost, and is triggered by "marsquakes" and related faulting.

Key Terms

mass movement (p. 363)
angle of repose (p. 366)
talus slopes (p. 366)
creep (p. 370)
solifluction (p. 370)
fall (p. 371)
slide (p. 371)
slump (p. 371)
flow (p. 371)
slip plane (p. 372)
scarp (p. 372)
earthflow (p. 374)
mudflow (p. 374)
quick clay (p. 374)
debris flow (p. 375)
debris avalanche (p. 375)

Questions for Review

1. Draw a simple sketch that illustrates the components of the force of gravity that act on an object located on a slope. Include the force of friction.

2. List at least one type of weakness plane that can be found within each of the three common types of rock (igneous, metamorphic, and sedimentary).

3. What does the "angle of repose" of an unconsolidated sediment describe? What is the general particle size and shape of a sediment that would lie at a steep angle of repose?

4. Describe at least three ways that water contributes to mass movement.

5. List three different mass-movement trigger mechanisms.

6. What is the fundamental difference between a flow and a slide?

7. Describe how quick clays change instantaneously from solids to liquids.

8. Describe two ways that you might date a landslide.

9. If you were thinking of moving into a slide-prone region, what would you look for to determine if your prospective home was located on a stable slope?

10. Draw a simple sketch illustrating how slopes are graded.

For Further Thought

1. How would prospects for mass movement change if a completely clear-cut slope was replanted and became densely reforested? How do you suppose the recent forest fires in Yellowstone National Park have affected the stability of the park's slopes?

2. In which North American states or provinces would you expect to find the most mass movement taking place? Explain your reasoning.

3. Discuss the possible ways that plate tectonic activity could increase the likelihood of mass movement in a given area.

4. What type of mass movement is shown in the photo below? What steps would you take to shore up this material?

5. Why is there a greater quantity and wider variety of mass-movement events on the Earth than on other planets in our solar system?

Streams

The Geological Work of Streams

Highlight 14-1 **The Mississippi River and Its Delta**

Controlling Floods

Highlight 14-2 **The Great Midwestern Floods of 1993**

Stream Evolution and Plate Tectonics

Extraterrestrial Stream Activity: Evidence from Mars

Figure 14-1 Kaskaskian Island, Illinois, during flooding of the Mississippi River in August 1993.

Chapter 14

Streams and Floods

The view from an airplane on a clear day suggests that rivers are among the most common geological features on the Earth's surface. They flow in virtually every geological and geographical setting, with the exception of large segments of the Arctic and the entire Antarctic, streaming down mountainsides, crossing farmlands, and bisecting or bordering cities and towns. Most major cities are located on the banks of rivers, having been founded at convenient river crossings or by waterfalls or rapids that prevented boats from proceeding upstream. Historically, the success and prosperity of most major cities have been proportional to the size of their rivers. A river provides many benefits: a steady supply of water for home and industrial use, transportation, recreation, and irrigation. In modern times, rivers are the source of clean electric power. Crops grown in the fertile soils near rivers provide nourishment for a third of the Earth's human population.

Rivers also bring floods, the most universal of natural disasters (Fig. 14-1). Most streams flood every 2 or 3 years, causing injuries, drowning deaths, and starvation due to loss of crops and livestock, damaging property, disrupting transportation networks, and spreading epidemic diseases caused by insects and microorganisms in contaminated floodwaters.

The amount of water in, on, and above the Earth—an estimated 1.36 billion cubic kilometers (326 million cubic miles)—has remained fairly constant for more than a billion years. About 97.2% of Earth's water is in its oceans; 2.15% is frozen in ice caps and glaciers; and the remaining 0.65% is in lakes, streams, groundwater, and atmosphere (Fig. 14-2). Most of the Earth's water originated in its mantle and surfaced as **juvenile water,** the steam that accompanies volcanic eruptions; some water is returned to the interior when water-rich oceanic plates subduct.

Water heated by solar energy evaporates from the Earth's surface and enters the atmosphere. Because air temperature decreases with altitude, water vapor cools as it rises and condenses into microscopic water droplets that coalesce to form clouds. If a cloud becomes dense enough to form droplets that are too heavy to remain suspended, the water precipitates as rain, snow, sleet, or hail, some of which falls back into the ocean and the rest onto land. This precipitation is referred to as **meteoric water.**

97.2% Oceans
500-gallon water tank
Two 5-gallon buckets
2.15% Ice caps and glaciers
Three 1-gallon milk bottles
0.65% Lakes, streams, groundwater, and atmosphere

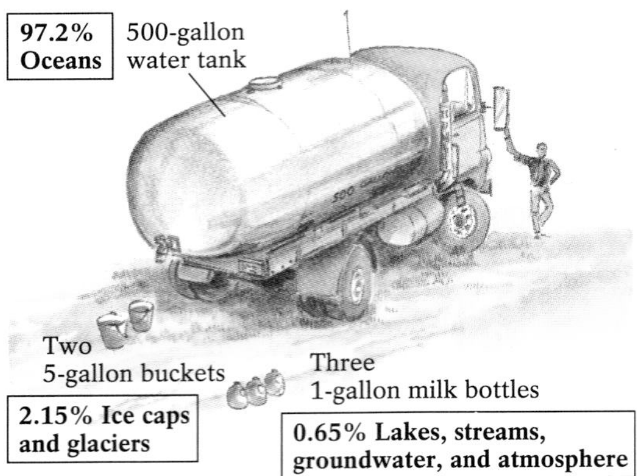

Figure 14-2 Distribution of the Earth's water, by relative volume.

When water falls on land, several things may happen to it. Snow on a high peak may be stored in the snowpack or in a glacier for tens, hundreds, or even thousands of years. Rain that falls into a lake may also remain stored for many years. Water that falls on plants and soil may evaporate into the atmosphere or be absorbed into the ground, where it may be picked up by plant roots, absorbed by soil particles, or pass into the underground water system. From there it may enter a stream, from the surface or underground, and run to the sea. The Earth's water moves perpetually among its oceans, land, and atmosphere and, via plate subduction and volcanism, between its mantle and its surface. This phenomenon is referred to as the **hydrologic cycle** (Fig. 14-3).

In this chapter, we will explore the surface-water segment of the hydrologic cycle and learn how rivers flow and how they create distinctive landforms by erosion and deposition. We will also see why they flood and discuss efforts to alleviate flooding.

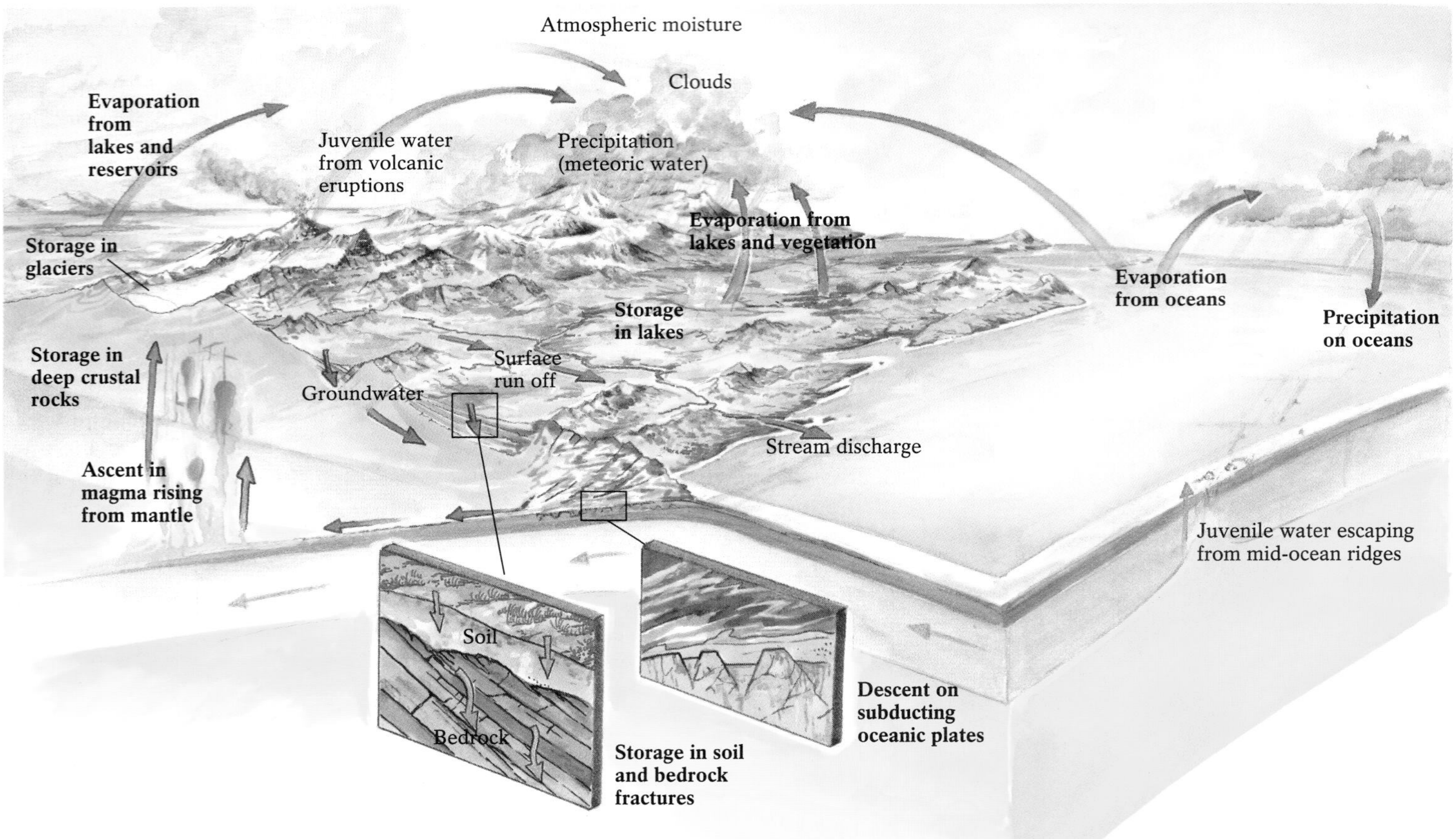

Atmospheric moisture
Clouds
Evaporation from lakes and reservoirs
Juvenile water from volcanic eruptions
Precipitation (meteoric water)
Evaporation from lakes and vegetation
Storage in glaciers
Evaporation from oceans
Precipitation on oceans
Storage in lakes
Storage in deep crustal rocks
Surface run off
Groundwater
Stream discharge
Ascent in magma rising from mantle
Juvenile water escaping from mid-ocean ridges
Soil
Bedrock
Descent on subducting oceanic plates
Storage in soil and bedrock fractures

Figure 14-3 The hydrologic cycle. All the water that falls on the Earth's land surfaces from the atmosphere eventually enters the vast oceanic reservoir and then returns to the atmosphere through one or more of the pathways of the cycle.

Figure 14-4 North America's drainage divides and major drainage basins. The drainage basin of the Mississippi River encompasses roughly 40% of the lower 48 states in the United States.

Streams

From the smallest creek to the mighty Mississippi River, any surface water whose flow is confined to a narrow topographic depression, or channel, is a **stream,** whether it is known locally as a river, creek, brook, or run. (Geologists use the term *stream* and *river* interchangeably.) There are an estimated 2 million streams in the United States. The flat land immediately surrounding a stream channel, which would be submerged if the stream were to overflow its banks, is called its **flood plain.**

Every stream has a water-collecting area known as a **watershed** or **drainage basin;** this is the total area from which precipitation (rain and snowmelt) reaches a stream. Some streams receive water from a network of smaller streams, called **tributaries,** in their drainage basin. In North America, watersheds can be less than a square kilometer (the area drained by a small tributary) or as much as 3.2 million square kilometers (1.25 million square miles), the area drained by the Mississippi River system. Every drainage basin is bounded by an area of higher topography, a **drainage divide,** that separates it from an adjacent drainage basin. Divides may be low ridges between two small tributaries or continent-spanning mountain ranges (Fig. 14-4). The Rocky Mountain portion of North America's continental divide, for example, stretches from the Canadian Yukon to New Mexico; rain falling anywhere along this divide flows either westward toward the Pacific Ocean or eastward toward the Atlantic Ocean (via the Gulf of Mexico).

At a drainage divide, rainfall first flows downslope as *overland flow,* which consists of broad sheets of unconfined water only a fraction of a centimeter thick. These are a stream's *headwaters.* Wherever headwater encounters slight surface depressions or changes in surface composition, it erodes narrow depressions called *rills.* This is the start of a stream's *transport area,* in which flowing water shapes and travels through progressively larger channels. Downslope, where the rills carry more water, they merge into larger, branching channels; these are the tributaries, which in turn carry water and sediment to the main stream, or **trunk stream.** The trunk stream traverses the greatest part of the transport area and carries the most water and sediment. At the downstream segment of the trunk stream, its *mouth,* its water and sediment load begins to disperse. Here, numerous small channels, or **distributaries,** may branch off, carrying water and sediment across lowlands to an ocean (Fig. 14-5).

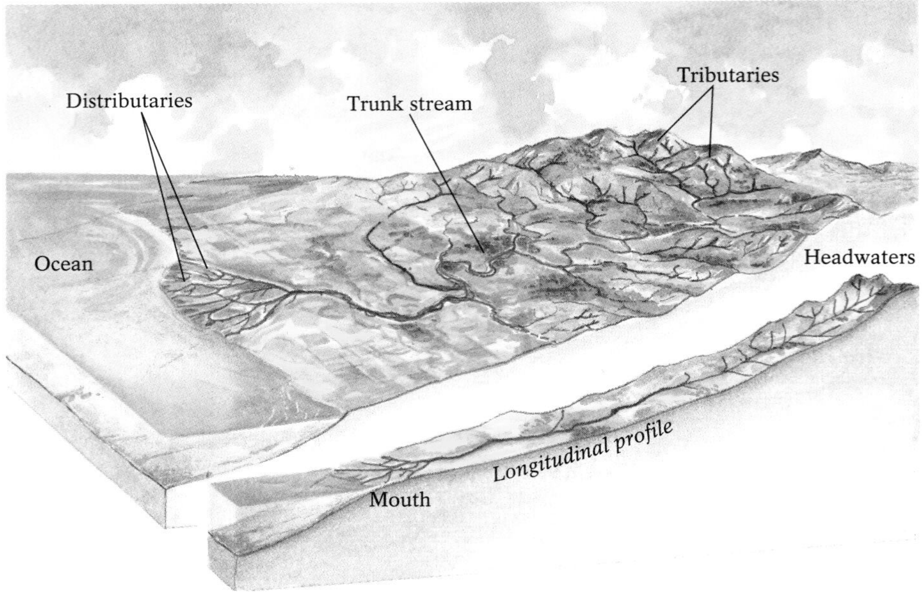

Figure 14-5 A stream system network. All major stream systems consist of tributaries that coalesce to form a trunk stream. The trunk stream transports water and sediment downstream until it splits into a network of smaller distributaries which deliver the system's water and sediment to the sea. The longitudinal profile of a stream (from its headwaters to its mouth) is characteristically concave.

Stream Flow and Discharge

The flow of a stream is driven by its **gradient,** or slope—its vertical drop in elevation over a given horizontal distance. The steeper its gradient, the faster a stream flows. A stream's gradient, which is expressed in terms of meters per kilometer (or feet per mile), depends upon the topography over which it flows, and generally decreases from its headwaters to its mouth. Gradients can be 50 meters per kilometer (265 feet per mile) or more in the mountains, and 5 meters per kilometer (25 feet per mile) or less on lowland plains. In North America, gradients range from 66 meters per kilometer (350 feet per mile) for the upper 6.5 kilometers (4 miles) of the Uncompahgre River in the Colorado Rockies to 0.1 meters per kilometer (0.5 feet per mile) for the lower Mississippi River downstream of Cairo, Illinois. All streams, even those that appear calm, flow *turbulently* (from the Latin for "turmoil"), with their water moving in swirls and eddies. Although the overall direction of flow is downstream, some water may swirl upward, like dry leaves caught in autumn gusts, or descend violently, like the vortex produced as the last of your bathwater disappears down the drain.

Stream velocity, expressed in meters (or feet) per second, is a measure of the distance a stream's water travels in a given amount of time. Slow-moving streams have velocities of 0.27 meters per second (less than 1.0 feet per second); swift-flowing streams can have velocities exceeding 10 meters per second (about 33 feet per second). The water in a stream does not have uniform velocity; disregarding rocks and other barriers, its *local* velocity depends partly on the gradient of the channel, and partly on where in the channel the water is flowing. Water velocity is slowest at the sides and bottom because of friction between the water and the channel. Water velocity is greatest in the center of a stream in a straight segment of the channel, equidistant from the banks and just below the surface, where the water feels no friction from either the bed or the atmosphere. Where a stream curves, velocity is greatest at the outside of the curve and least at the inside.

The texture of a stream bed acts along with the gradient to determine velocity. Upstream in the headwaters, the bouldery roughness of the channel bed creates significant frictional drag and sends chaotic streamflow swirling upward, downward, and sideways. Thus, most of a stream's energy is expended against its bed, and its actual *downstream* velocity is relatively low. Downstream, velocity actually increases despite a decrease in gradient, partly because the stream flows over a smoother bed of sand, silt, and clay, which lowers frictional resistance to flow.

Related to stream velocity is **stream discharge,** the volume of water passing a given point on the stream bank per unit of time. Discharge, usually expressed in cubic meters of water per second, can be determined if the stream's width, depth, and velocity are known. The formula is

$$\text{discharge} = \text{width} \times \text{depth} \times \text{velocity}$$

A stream's long-term discharge is dictated principally by the size of its drainage basin and the amount of precipitation into it. The discharge of minor tributaries can be as small as 5 to 10 cubic meters (180 to 350 cubic feet) per second. The discharge of the largest river in North America, the Mississippi, is about 18,000 cubic meters (600,000 cubic feet) per second. The discharge of the largest river on Earth, the Amazon of South America—which drains an area equal to about 75% of the continental United States—is about 200,000 cubic meters (7 million cubic feet) per second, about one-fifth of the Earth's entire freshwater stream flow. To appreciate this amount of water, consider that one day's discharge of the Amazon would supply New York City's freshwater needs for more than 5 years. A stream's daily discharge is also influenced by such climatic conditions as the amount and timing of precipitation and the quantity of snowmelt, and by the ability of the local soils to absorb water.

The Geological Work of Streams

Streams erode, carry, and deposit sediment. Although streams carry only about one-millionth of the planet's water, they are its most important geological agents of surface change. A stream cuts down through uplifted land toward its **base level,** the lowest level to which it can erode its channel. The *ultimate* base level of all streams is sea level. However, streams may encounter *temporary* base levels, such as lakes or extremely durable layers of bedrock, that halt their downcutting for considerable lengths of time. When global sea levels fall, as happens during periods of glaciation when a great volume of seawater is converted to ice, coastal streams respond quickly by cutting downward to the lowered base level. Conversely, during periods of warming when sea level rises, streams deposit their sediments further inland, where their discharge encounters the rising sea.

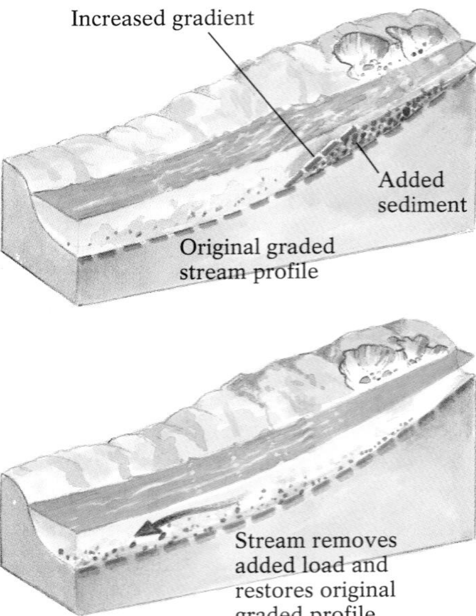

Aggradation

(a)

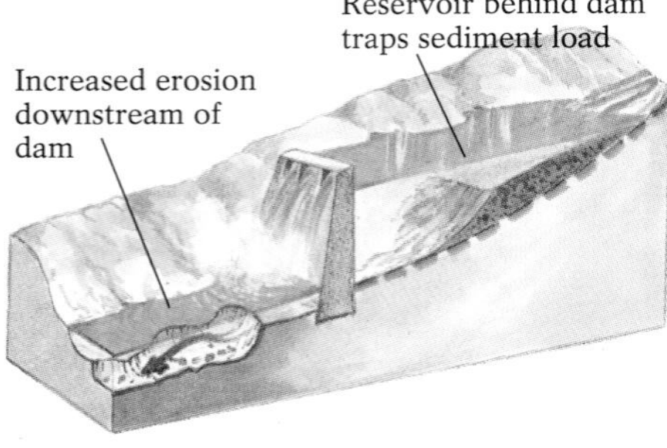

Degradation

(b)

Figure 14-6 Aggradation and degradation of graded streams. (**a**) When a new load of sediment is added to a graded stream, the stream becomes aggraded; that is, its gradient is instantaneously increased, causing the stream to flow faster and begin to remove the added sediment load. In time, the stream removes the added load and returns to its original gradient. (**b**) When a stream's sediment load is reduced or eliminated (for example, by settling in a new reservoir), the stream becomes degraded downstream; that is, it erodes more of its bed, thereby decreasing its gradient.

Graded Streams

Stream behavior is one of the most dynamic geological processes. Streams respond immediately to changes in their environment. When a torrential storm dumps 25 centimeters (10 inches) of rain on a drainage basin, the stream quickly rises, flows more rapidly, erodes more sediment, and deposits that sediment downstream wherever its flow is blocked or slowed. If normal faulting occurs across its channel and the downstream fault block drops down, the stream begins immediately to erode the fault blocks with the greater energy that it derives from its raised position.

In an unchanging environment, a stream's gradient would be maintained at such an angle that the stream would flow just swiftly enough to transport all the sediment supplied to it from the drainage basin, with little net erosion or deposition. In reality, however, a stream's environment is always changing, and its gradient is constantly adjusting to achieve this balance between erosion and deposition. A stream in such a state of equilibrium is called a **graded stream.** The equilibrium of any graded stream is only temporary, lasting just until the next change in its environment: A sudden increase in sediment load, such as from a volcanic mudflow or the collapse of a stream bank, or a rapid increase in discharge, such as from a heavy rainstorm, will disturb a stream's equilibrium. A graded stream responds *instantaneously* to any change in sediment load, discharge, or base level in such a way as to absorb the effects of the change and to establish a new graded state.

If the sediment load entering a graded stream increases or decreases, the stream's gradient compensates to maintain sediment transport and flow velocity. For example, if several loads of coarse gravel are dumped into a graded stream that previously transported only fine sands and silt, the stream initially cannot transport this new material. Instead, the coarse sediment settles onto the stream bed, causing the channel gradient to steepen instantaneously, a process known as **aggradation** (Fig. 14-6a). The new steeper gradient causes the stream to flow more swiftly, eroding the added sediment until the graded equilibrium is restored. The streams of central California aggraded in this fashion during the Gold Rush days of the last century, when mining operations dumped millions of tons of debris into them.

Conversely, a graded stream that suddenly has its normal sediment load decreased has a steeper gradient and a swifter flow than it needs to carry the reduced load. Such a stream scours its bed, eroding more sediment and thereby reducing its gradient, a process called **degradation** (Fig. 14-6b). Degradation typically occurs downstream from a dam. Construction of Hoover Dam, for example, caused the Colorado River to degrade its slope downstream as far south as Yuma, Arizona, a distance of 560 kilometers (350 miles). Responsible dam engineers consider the possibility that scoured slopes might eventually undermine their dam, and build protective structures to prevent this.

Stream Erosion

The principal result of streams' erosive power is the creation and deepening of valleys, the Earth's most common continental landforms. The composition of a valley's rocks and sediments strongly affects the rate at which it erodes. It could take thousands or millions of years for a stream to cut a valley through a granite batholith, whereas a sizeable valley can be eroded in unconsolidated sands by a single powerful flood.

A stream left to its own processes would cut vertically through uplifted terrain like a saw cutting through a board, forming a chasm with near-vertical

Figure 14-7 The Yellowstone River in Wyoming cuts down rapidly and vertically through the uplifted rocks of its canyon. Broadening of the upper portion of the canyon, caused largely by mass movement, produces the stream valley's characteristic V shape.

walls. Why then do most river valleys have a distinctive V shape? The reason is that as the valley is cut, overland flow and mass movement remove loosened material from its slopes (Fig. 14-7). For example, although downcutting by the Colorado River is largely responsible for the 1800 meters (5940 feet) of *vertical* relief of the Grand Canyon (the difference in elevation between its highest and lowest points), overland flow and mass movement eroded the 21-kilometer (13-mile) width *across* the canyon.

In temperate and humid regions, chemical weathering contributes to valley development by breaking down exposed bedrock, so that it is more easily removed by overland flow and mass movement. Thus, stream valleys in such climates have gradually sloping walls. In arid regions, however, chemical weathering is negligible; with little loose sediment to remove, valleys here are steep-walled canyons.

A stream's erosive power increases with its velocity; the increase in rate of erosion is approximately equal to the square of the increase in velocity. Thus, if a stream's velocity doubles, its erosive power increases by a factor of four. For this reason, a stream's erosive power increases significantly during a flood, when its velocity and discharge increase dramatically.

Processes of Stream Erosion Streams erode their channels by abrasion, hydraulic lifting, and dissolution. **Abrasion** is the scouring of a stream bed by transported particles. Fine particles suspended in the stream constantly burnish the bed's surface. Some large pebbles rotate in swirling eddies, carving circular depressions called *potholes* into the bedrock (Fig. 14-8), while others are bounced against the underlying bedrock. Larger rocks are rolled along the bed. Erosion by abrasion is most efficient in swift-flowing, sediment-laden floodwaters.

Hydraulic lifting, erosion by water pressure, occurs when turbulent streamflow through fractures in the bedrock dislodges sediment grains and loosens large chunks of rock. Like abrasion, hydraulic lifting is most active during high-velocity floods; a 1923 flood along the Wasatch Mountain front of central Utah lifted 90-ton boulders and transported them more than 8 kilometers (5 miles) downstream.

When a stream flows across and dissolves soluble bedrock such as limestone, dolostones, and evaporites, dissolution also contributes to stream erosion. Every year, an estimated 3.5 million metric tons of rock are dissolved from the continents and carried out to sea. The Niagara River, which flows between Lake Erie and Lake Ontario on the border between Ontario, Canada and New York state, carries 60 tons of dissolved rock over Niagara Falls every minute.

Figure 14-8 A stream and potholes in Jasper National Park, Canada. The potholes were abraded in the solid bedrock by stones swirling in the stream. Potholes can be as deep as 5 meters (17 feet) and as large as 2 meters (7 feet) in diameter. The stone that ground a pothole is often found lying at its bottom.

Drainage Patterns Streams erode their networks of tributary valleys in distinctive **drainage patterns.** These patterns are determined largely by the way in which erosion is affected by the topography, composition, and structure of the terrain (Fig. 14-9). A geologist can estimate the composition and structure of a landscape simply by looking at its stream drainage pattern from an airplane, or by examining the web of fine blue stream lines that indicate the pattern on a topographic map.

When the rocks or sediments that underlie a drainage system are uniform in composition and undeformed, they tend to erode into a branching, *dendritic* drainage pattern (from the Greek *dendron,* or "tree"). In general, dendritic patterns develop on relatively flat sedimentary rocks, newly exposed coastal muds, and newly exposed massive plutonic igneous or metamorphic rocks (Fig. 14-9a). The drainage pattern of a landscape with topographic peaks, such as young volcanic cones, lava domes, and structural domes (discussed in Chapter 9), consists of a network of streams draining from a central high area like the spokes of a wheel. This is *radial* drainage (Fig. 14-9b).

A *rectangular* drainage pattern, which looks like a grid of city streets (Fig. 14-9c), forms when stream erosion enlarges joints, fractures, and faults in bedrock. Because these features usually develop in perpendicular sets, the stream pattern shows many right-angle bends. A *trellis* drainage pattern, in which narrow valleys are separated by parallel ridges, develops where easily eroded rocks alternate with erosionally resistant rocks. Short, steep tributaries drain down the ridge slopes perpendicular to the main stream valleys, creating a resemblance to the parallel slats of a garden trellis (Fig. 14-9d).

Figure 14-9 Four types of drainage patterns, produced by stream erosion of various bedrock types and under various structural conditions. (**a**) Dendritic drainage patterns often develop on undeformed sedimentary rock of uniform composition. (**b**) Radial drainage patterns usually form on volcanic cones of homogeneous composition. (**c**) Rectangular drainage is often found where bedrock is cut by perpendicular joints. (**d**) The best place to look for trellis drainage is in an area with long parallel folds of sedimentary rock, such as the Ridge and Valley province of the Appalachians.

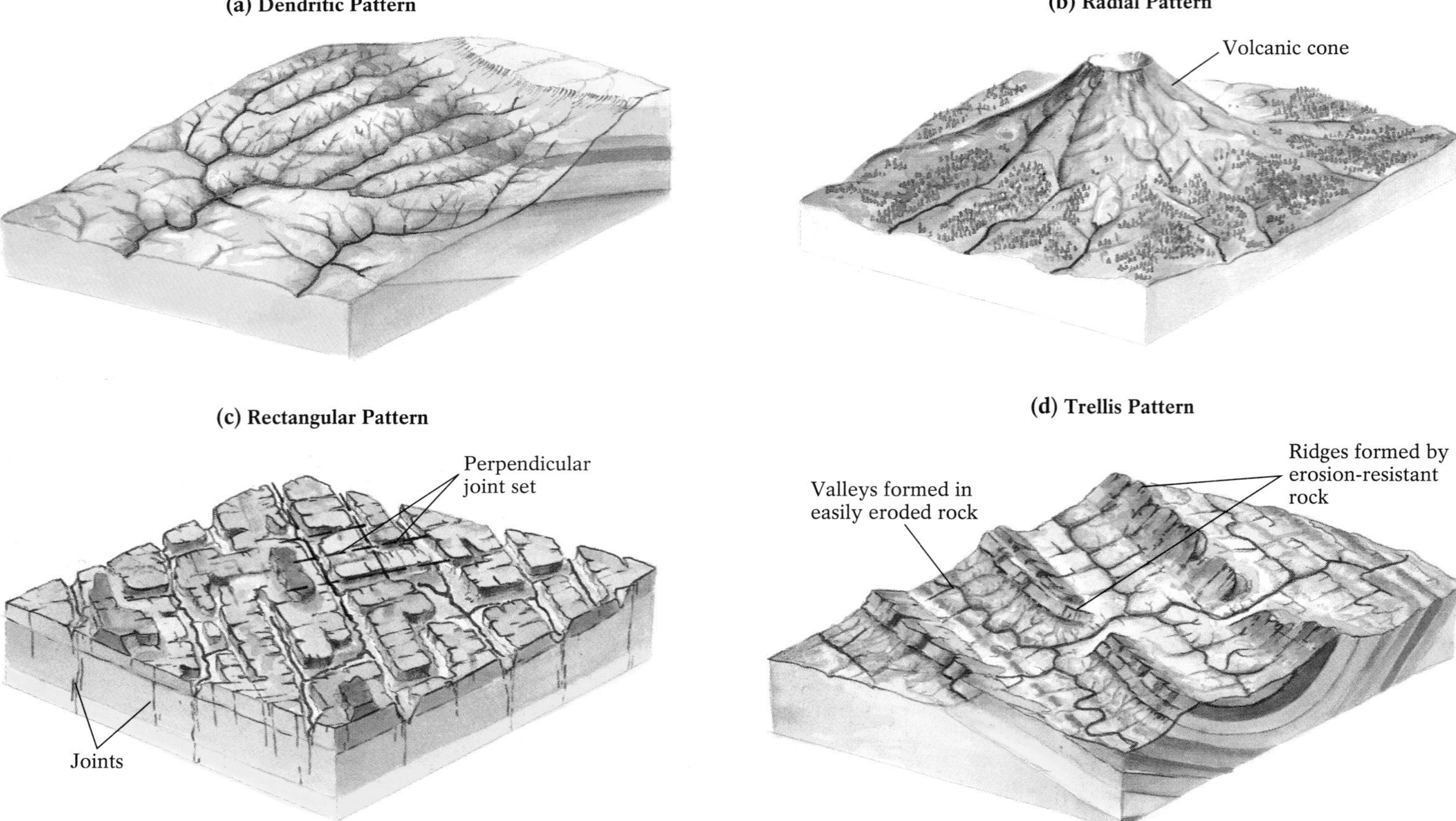

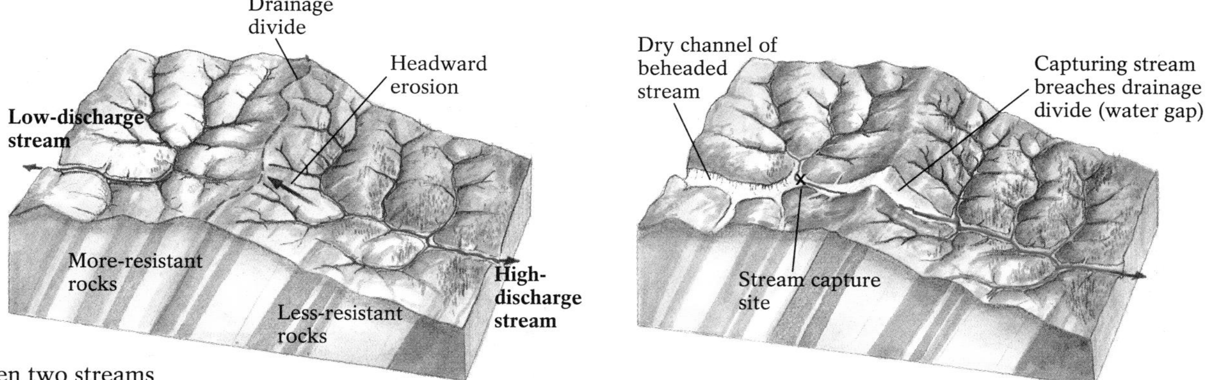

Figure 14-10 Stream piracy. When two streams erode headward toward a drainage divide, the stream that erodes more vigorously may breach the divide and capture, or "pirate," the headwaters of the other. The stream that loses its headwaters is said to be "beheaded."

Stream Piracy The opposite sides of a ridge usually differ in slope, and the stream on the steeper slope of a divide will tend to flow more swiftly and erosively. One side of a divide will also be eroded more effectively if it has less-resistant surface rocks, sediments, or soils. Similarly, if one side of a divide receives more rain than the other side, its streams are usually more vigorous and have greater, more erosive discharges. When two streams erode headward toward a drainage divide, the stream that erodes more vigorously may erode a channel through the divide itself, capturing the headwaters of streams on the other side; this process is colorfully named *stream piracy* (Fig. 14-10). A stream that has lost its headwaters in this fashion, described as *beheaded,* simply stops flowing. A *water gap* is a steep-walled rocky gorge cut through a ridge or mountain range by a pirating stream. (If a pirating stream is later diverted from its course and its gorge is abandoned, the water gap becomes a dry notch in the ridge known as a *wind gap.*)

Superposed and Antecedent Streams A *superposed stream* is one that has cut downward through a number of rock layers of different compositions and structures, while maintaining the same drainage pattern that is established at the original surface. This situation is common when a stream becomes established on flat, undeformed sedimentary rock overlying folded subsurface rocks or masses of plutonic igneous rock; as the overlying sedimentary rock erodes away, its drainage pattern is imposed upon the lower layers regardless of their differing natures and topographies (Fig. 14-11).

Figure 14-11 Superposed streams. Whenever we see a drainage pattern that seems inconsistent with the surrounding topography—such as a dendritic pattern cutting across a folded landscape—it may mean the stream was established on a different, overlying layer of rock that has since eroded away. Photo: The Delaware River, a superposed stream that forms the border between Pennsylvania and New Jersey. The Delaware Water Gap (near top of photo) was created as the river cut down through a formerly buried ridge, carving out a steep-walled gorge.

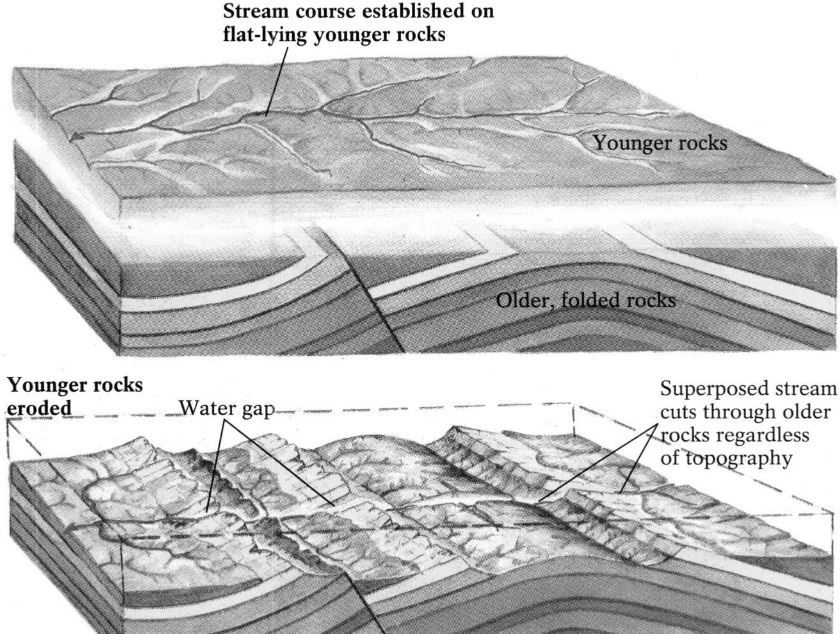

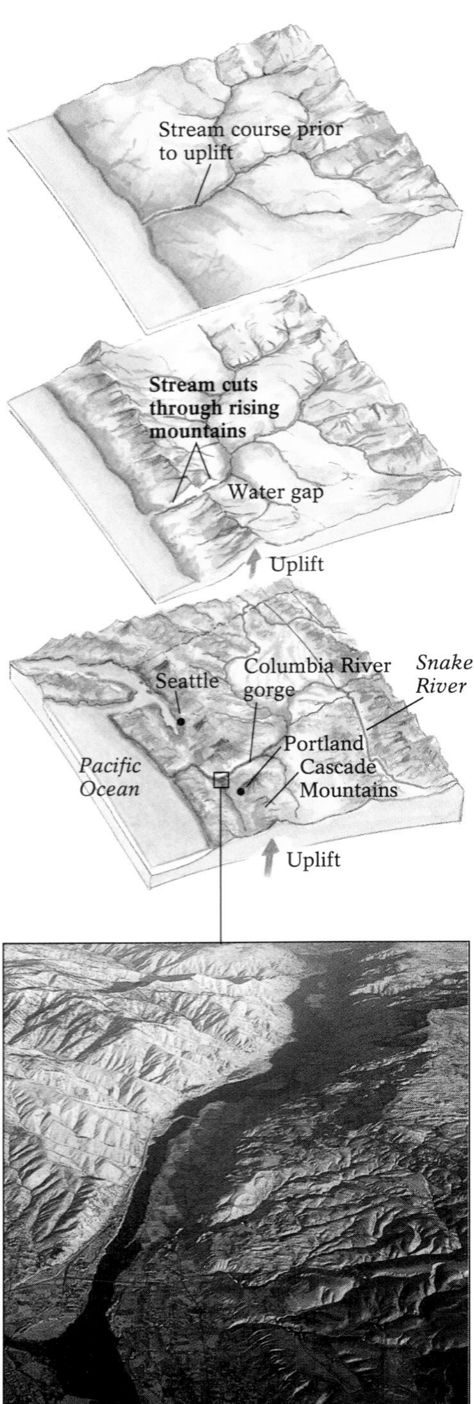

Figure 14-12 Antecedent streams. The Columbia River gorge, which cuts across the rising Cascades mountain range in Washington state, is a typical antecedent stream valley.

Antecedent streams are older streams that cut through recently uplifted landscapes. If uplift is more rapid than a stream's ability to cut down through the heightened rock, a topographic barrier results and the stream is simply diverted; if uplift is slow and steady, however, stream-cutting keeps pace and the stream maintains its course, cutting a water gap through the rising ridge or mountain range (Fig. 14-12).

Channel Patterns Stream channels can be straight, braided, or meandering, and can change from one form to another as a stream crosses different landscapes; thus, a single stream may contain segments of various shapes between its headwaters and its mouth. Straight channels are rare, but can occur where linear fractures, joints, and faults in resistant bedrock direct a stream's path. Straight segments also occur in areas of active uplift, where steep slopes compel the headwaters of a stream to take the most direct route downslope.

Networks of converging and diverging streams separated by narrow sand and gravel bars are known as **braided streams** (Fig. 14-13). They develop where the sediment load is unusually high, forcing a stream to deposit its excess sediment on its channel bed; in places, these deposits build up until they break through the water surface, forming islands around which streamflow diverges. Braided streams often occur where a stream's banks consist of loose, highly erodible sediment. Braided streams also form where the slope of a stream bed decreases abruptly (such as at the foot of a mountain), reducing the stream's ability to carry its sediment, or where the stream's discharge drops sharply (as in deserts, where much of a stream's water evaporates or seeps into the sandy desert floor).

Meandering streams (from the Latin *maendere,* "to wander") wind their way across relatively flat landscapes in fairly evenly spaced loops. The loops, or *meanders,* form as a stream erodes its banks, and erosion and deposition work together to exaggerate its curves (Fig. 14-14). Because flow velocity increases toward the outside bank of a curve, the water is most turbulent and erodes the most sediment there; the sediment eroded from the outside bank is then carried downstream and deposited on the *inside* bank of the next curve, where velocity and turbulence are minimal. In this way, a meandering stream transfers sediment from its erosional outer banks to its depositional inner banks, with very little net erosion or deposition for the stream as a whole.

Figure 14-13 A braided channel of the Toklat River, in Denali National Park, Alaska.

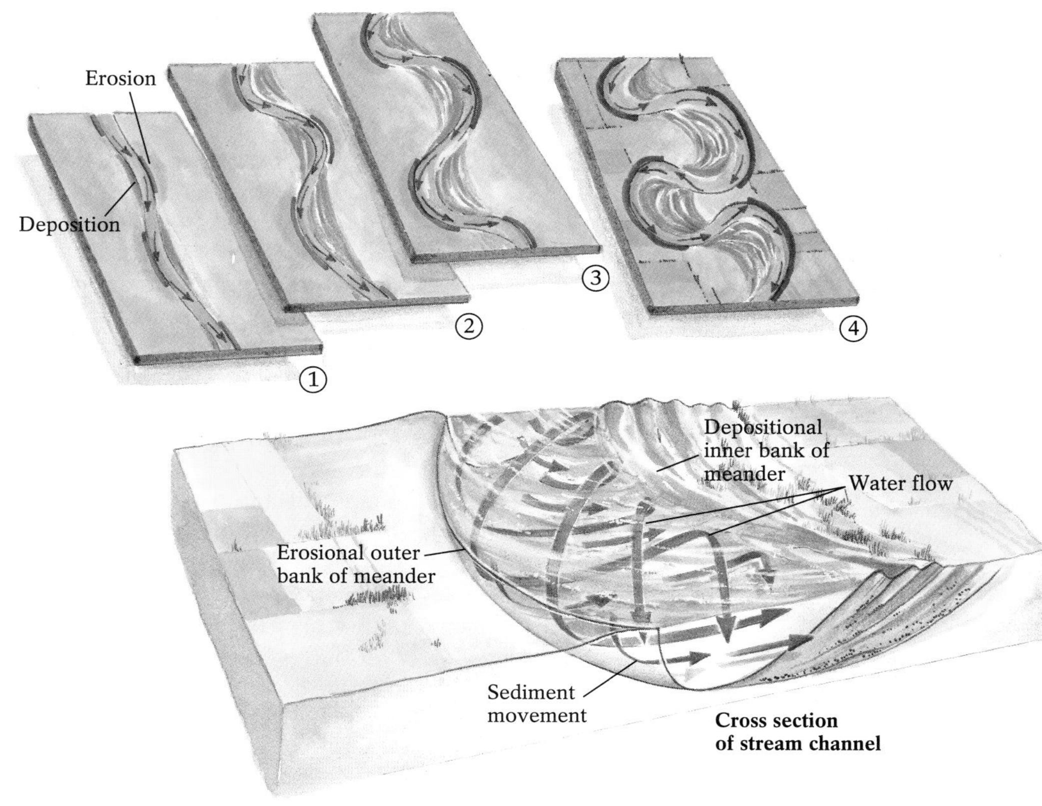

Figure 14-14 The evolution of meandering streams. As a somewhat curving stream flows, the high-velocity water at the outside of each bend produces a cross-channel corkscrew-like component of flow that erodes the outside bank of the bend. The material eroded from the outside of the bend is then deposited downstream at the inner bank of the next bend. Meander bends grow increasingly pronounced as progressively more sediment is removed from their outside banks and added to their inside banks. Above: The meandering Little Bear River, in Utah.

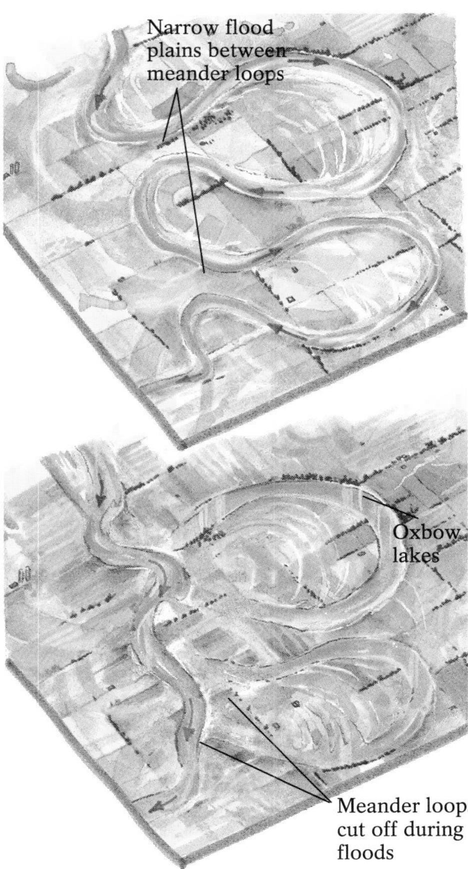

As erosion and deposition along a meandering stream progress, meander curves typically become more pronounced and closer together, until they form a series of loops separated only by thin strips of flood plain. During a flood, a stream may cut through a separating strip, bypassing an entire loop (Fig. 14-15). The cut-off meander loop then becomes a crescent-shaped body of standing water called an **oxbow lake,** so named because its shape resembles that of the curved collars worn by farmers' oxen. Subsequent flood deposits may fill oxbow lakes, forming *meander scars.*

From 1765 to 1932, the Mississippi River cut off about nineteen meanders between Cairo, Illinois, and Baton Rouge, Louisiana, equal to about 400 kilometers (250 miles) of its course. During the Civil War, General Ulysses S. Grant, commander of the Union Army, tried unsuccessfully to create a cutoff so that Union forces traveling by river could elude the Confederate guns at Vicksburg, Mississippi; nature finally achieved the cutoff (the "Centennial" cutoff) in 1876, when Vicksburg was separated from the river's active channel. Despite all its cutoffs, the Mississippi has not been shortened. Existing meander curves have expanded and once-straight cutoff sections have themselves begun to meander.

Figure 14-15 Creation of oxbow lakes. When a stream's meander bends are so pronounced that the land between successive loops is narrowed to thin strips, bends may be completely bypassed or "cut off" (especially during floods), forming bodies of standing water.

Waterfalls and Rapids Waterfalls and rapids occur at sudden drops in topography along a stream's course, usually where erosion has removed softer sections of bedrock, leaving more resistant rock as a "step" in the stream's profile, or where faulting has lowered or raised a portion of the stream's profile. Whitewater rapids are often relics of waterfalls whose "steps" have been largely eroded, so that water rushes turbulently over irregular rocky beds. A waterfall lasts only as long as the conditions that create it. The energetic plunging water of a waterfall erodes a deep pool at its base, called a *plunge pool,* that undermines the step from which the water falls. As the step is eroded progressively back, the waterfall migrates upstream; the height from which the water plunges decreases until the fall no longer exists and the stream reestablishes a graded profile (Fig. 14-16). One waterfall, created by the Madison Canyon earthquake of August 1959 in the Yellowstone area of Montana, lasted barely a year. As a result of the quake, a tributary to the Madison River—Cabin Creek—was forced to flow over a fresh 3-meter-high fault scarp. By June 1960, the scarp was almost completely eroded, and the waterfall was replaced by a small set of rapids. Five years later, all evidence of both the rapids and the waterfall were gone.

Figure 14-16 The evolution of waterfalls and rapids. When the graded profile of a stream is interrupted by a protruding resistant ledge or by faulting (or relative uplift of any kind), the stream at first cascades over the "step" in the landscape. Eventually, the force of the free-falling water excavates a plunge pool at the foot of the falls, undermining the cliff face and causing the upstream retreat of the falls. The falls are gradually eroded down to rapids, which are in turn eventually completely eroded, returning the stream to its graded state. (**a**) Whitewater falls in Nantahala National Forest, North Carolina; this water drops 411 feet in a distance of 500 feet. (**b**) A significantly smaller waterfall in Franconia Notch State Park, White Mountain National Forest, New Hampshire. (**c**) Rapids along the Ellis River in White Mountain National Forest.

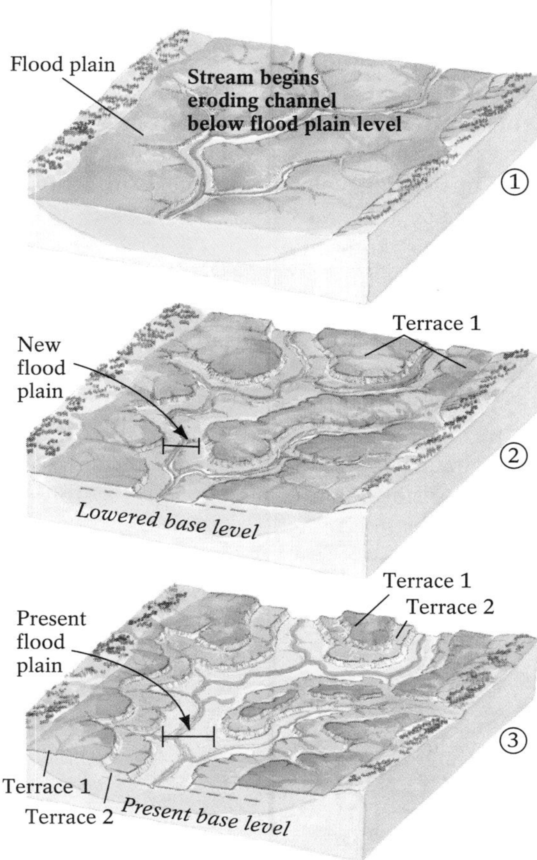

Figure 14-17 The creation of stream terraces. A terrace forms when a stream erodes its channel below the level of its flood plain, either because its discharge has increased or its base level has been lowered. When a stream has undergone a series of such changes, it may form a series of tiered terraces in a single valley. Photo: Terraces above the Cave River at Arthur's Pass, New Zealand.

Thousands of waterfalls and rapids exist in North America, from picture-postcard falls such as Yosemite Falls in California's Sierra Nevada to the countless small falls and rapids that dot the springtime streams of the East (found especially where streams flow from the durable igneous and metamorphic rocks of the Appalachians into the soft sediments of the Atlantic coastal plain). Best known, of course, is Niagara Falls, which has become North America's best-known falls, although Yosemite Falls is nine times as high.

Niagara Falls, which is now about 55 meters (176 feet) high and 670 meters (2200 feet) wide, has been wearing itself away for 12,000 years (see Fig. 8-1). It formed when the last great North American ice sheet retreated north of the Great Lakes region, uncovering a ridge of resistant dolostone. Since then, the falls has retreated southward 11 kilometers (7 miles) from its point of origin at Lake Ontario. Billions of liters of water thunder annually through the 50-kilometer (30-mile)-long gorge of the Niagara River between Lake Erie and Lake Ontario, plunge over the falls, and erode the shale bed below the resistant cap of dolostone, undercutting the falls.

The future of Niagara Falls is uncertain. At its average past rate of retreat (about 1 meter, or 3 feet, per year), Niagara Falls would have eroded the remaining 30 kilometers (20 miles) of dolostone cap rock all the way to Lake Erie in about 30,000 years. However, the United States and Canada currently divert about 75% of the river's discharge into four large tunnels to generate thousands of megawatts of hydroelectric power (enough to supply more than 2 million homes). The diversion has slowed the retreat of the falls to less than 0.5 meter (2 feet) per year, possibly doubling its remaining life to more than 60,000 years.

Stream Terraces When a stream's discharge increases over a long period of time (perhaps because of climate change) or when its base level is lowered (perhaps because of a global sea-level change or local tectonic movement), it usually erodes its channel to a lower level. This leaves its flood plain high and dry, forming a gently sloping topographic bench known as a **stream terrace.** A terrace usually stands high enough above a stream so that it is untouched by even the most extreme flood. Terraces are often found in sets within a single stream valley, remnants of a once-continuous floodplain surface that spanned the entire valley before being cut down to progressively lower levels (Fig. 14-17).

Stream Transport

The world's streams carry about 45 trillion cubic meters of water to the sea every year, along with 9 to 10 billion tons of sediment. The Mississippi River alone delivers more than 1 million tons of sediment *daily* to the Gulf of Mexico. A stream's velocity, the composition and texture of its sediment, and the characteristics of the bedrock through which it flows determine the amount and type of sediment that it transports. High velocity is required to erode and transport large boulders. High velocity is also required to pick up small, flat clay particles and mica flakes, because electrical forces on their surfaces cause them to cling to the stream bed; once in the streamflow, however, fine clay particles are transported long distances, often remaining suspended in the streamflow all the way to the sea.

The maximum load of sediment that a stream can transport is its **capacity.** Capacity is expressed as the volume of sediment passing a given point on the channel bank in a given amount of time. Capacity is proportionate to discharge: The more water flowing in the channel per second, the greater the volume of sediment that is transported in that time.

The diameter of the largest particle that a stream can transport is the measure of a stream's **competence.** Competence is proportionate to the square of a stream's velocity: The greater a stream's velocity, the larger the particles the stream can transport. Thus, when velocity doubles—in a flood, for example—competence increases by a factor of four. This relationship explains why a stream that ordinarily carries only fine gravel can sweep boulders downstream during a flood. During the 1933 Tehachapi flood in California, steam locomotives of the Santa Fe railroad were carried hundreds of meters downstream and then buried by tons of transported stream gravel.

Sediment Load Streams transport sediment in different ways, depending on particle size (Fig. 14-18). Very fine solid particles are usually distributed within stream water as a **suspended load.** Coarse particles that move along the stream bottom form the **bed load.** Other sediment is carried invisibly as dissolved ions in the water, forming the **dissolved load.**

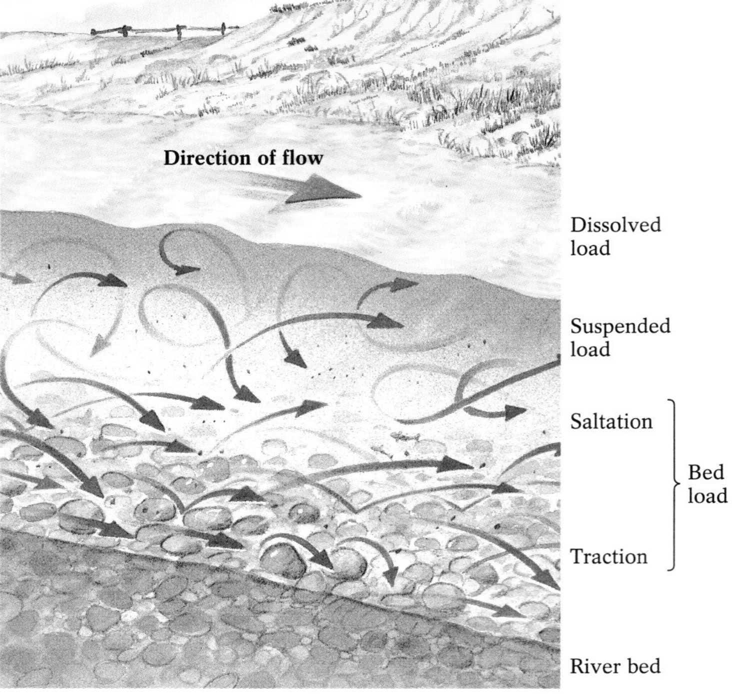

Figure 14-18 Sediment distribution and movement within a stream. (See text.)

Most of the world's stream-borne sediment, about 7 billion tons annually, travels as suspended load. These fine particles generally remain suspended in a stream, like confetti caught in swirling air currents, because a stream's turbulence and velocity prevent the particles from settling to the bed. About 70% of the Mississippi River's annual load of 500 million tons of sediment is suspended. The characteristic red color of the Colorado River is due to suspended particles of red silt and sand eroded from the bedrock of the Grand Canyon region. Because the Colorado flows through an arid region with little vegetation to hold soils in place, it carries about the same total volume of suspended sediment as the Mississippi, although its drainage area is only one-fifth the size and its discharge is much less.

More than 10% of the world's stream-borne sediment is transported by the remarkable Huang Ho (Yellow River) of central China. Only about 50% by volume of the Huang Ho flow is water, the other 50% being sediment. During floods, as much as 90% of the river may be sediment, reducing its flow rate to that of molasses. The sediment comes from the 200-meter (650-foot)-thick blanket of yellow windblown silt through which the river flows. There is little vegetation to anchor the loose silt in this extremely arid region, and periodic storms wash a huge volume of sediment from the adjacent slopes. The Huang Ho's annual sediment load would be enough to build a wall 25 meters (80 feet) high and 25 meters wide that would encircle the planet.

The volume of a stream's suspended load is largely determined by the stream's velocity and discharge, but particle shape and density also play a part. Flat particles and low-density particles tend to remain suspended. A flake of low-density muscovite mica, which is also flat (has a large surface area) and therefore resists settling through the water column, can travel thousands of kilometers without dropping onto the stream bed. Denser particles settle at a faster rate than less-dense particles. A gold particle, whose density is nineteen times that of water (specific gravity = 19.3), settles more rapidly than a grain of quartz sand (specific gravity = 2.65).

Unlike its suspended load, which moves constantly, a stream's bed load moves only when the water's velocity becomes great enough to lift or roll larger particles. The coarse material that makes up bed load has a range of particle sizes. The sandy portion moves along the bed with a bouncing motion known as *saltation* (from the Latin *saltare,* "to jump"). Saltating sand grains are lifted from the bed by hydraulic lifting or by the impact of other saltating grains. Once lifted into the stream, these grains are carried upward and forward by the stream's turbulence and velocity until the pull of gravity brings them down to the bed. The larger, heavier cobbles and boulders usually move by a rolling motion called *traction.* The largest boulders can be moved only by major floods. You can often see them settled into the bottom of a stream, covered with algae; no algae would grow if the boulders were often in motion.

Natural waters are never pure H_2O, but contain ions dissolved from surrounding bedrock. These are carried by streams as their dissolved load. (The coating that adheres to tea kettles is precipitated from the dissolved load in water supplies.) The most common dissolved ions are bicarbonate, calcium, sodium, and magnesium, with smaller amounts of chlorides, sulfates, nitrates, silica, and organic acids. The amounts and kinds of dissolved ions in stream water depend on the factors that have caused chemical weathering—such as the climate, vegetation, and bedrock composition—and the acidity of the water. Velocity has no effect on a stream's ability to carry dissolved load, although it might affect the rate at which material enters solution. In general, dissolved load is especially great in streams that flow through warm, moist regions of low relief, lush vegetation, and outcrops of soluble bedrock such as limestones and evaporites.

The volume of a stream's dissolved load also depends on the topography of the drainage basin, which controls whether precipitation runs off rapidly as sheet flow and stream flow or instead infiltrates the groundwater system. Dissolved ions are most abundant in local groundwater, whose slow movement through soluble bedrock promotes dissolution more effectively than contact of a flowing stream with its bed.

Stream Deposition

When the effect of gravity on stream particles overcomes the effect of stream velocity, the particles are deposited, with the heaviest particles settling out first. Deposited stream sediments are described collectively as **alluvium** (a word of French and Latin derivation meaning "washed over"). Up to 75% of stream alluvium is dropped into a stream's channel, onto its flood plain, onto flat valley floors at the foot of steep gradients, and into standing bodies of water along its course; the remainder settles where the stream enters the sea.

In-Channel Deposition During normal (nonflooding) flow, a stream with large amounts of coarse bed-load sediment often deposits some of its load in its channel as mid-channel bars. As sand and gravel accumulate, the stream is diverted around the channel bars and becomes braided (see Fig. 14-13). The bars become increasingly resistant to erosion as vegetation begins to grow on them.

In-channel deposits may also be in the form of **point bars,** where sediments scoured from the outside banks of meandering streams are deposited on the streams' inner banks (see Fig. 14-14). As a point bar grows, the water flowing over the bar becomes shallower, increasing friction between the water and the surface of the bar and decreasing stream velocity even further, thus promoting even more deposition (Fig. 14-19). Because point bars are where a meandering stream drops its heaviest particles, gold, platinum, and silver are often concentrated there. The gold deposits in the rivers of the Yukon and Alaska beckoned thousands during the gold rushes of the nineteenth century (discussed in Chapter 20).

Figure 14-19 Growing point bars on the Madison River, Montana.

Floodplain Deposition When a stream's channel is unable to contain the water coursing through it, the water flows over its banks and onto the surrounding flood plain. Because of their increased velocity, channeled floodwaters carry a large volume of sediment; as this water spreads over the relatively flat flood plain, its velocity decreases and it drops its sediment load. Thus, as its name suggests, a flood plain is a major repository of flood-borne stream sediment. Some typical features associated with flood plains are illustrated in Figure 14-20.

Natural levees (from the French *lever* "to raise") are ridges of sediment deposited on both banks of a stream during successive floods, as the decrease in velocity, turbulence, and competence at the banks causes floodwater to drop its sediment load. The coarser sediment accumulates adjacent to the channel to form the levees, and the finer sediment settles farther out on the flood plain, enriching agriculture for miles around. Natural levees, which grow higher with each flood, tend to be the highest points on a flood plain, and can sometimes prevent lower-volume flows from overflowing the channel banks. Between floods, trees that prefer well-drained growth sites may take root on the coarse-grained levees, enhancing their usefulness in flood prevention as well as their appearance (Fig. 14-21). Roads, towns, and farms tend to prosper along the gentle, relatively dry, landward slopes of levees.

Beyond the levee slopes lies the **backswamp,** that portion of the flood plain, usually near the river, where deposits of fine silts and clays settle from standing waters following a flood. After flooding subsides, the water left standing on the flood plain usually evaporates or slowly infiltrates the groundwater system. If a flood plain is marked by surface depressions, however, water may remain at the surface as environmentally valuable **wetlands**—lakes, marshes, and swamps that serve as habitats for migrating

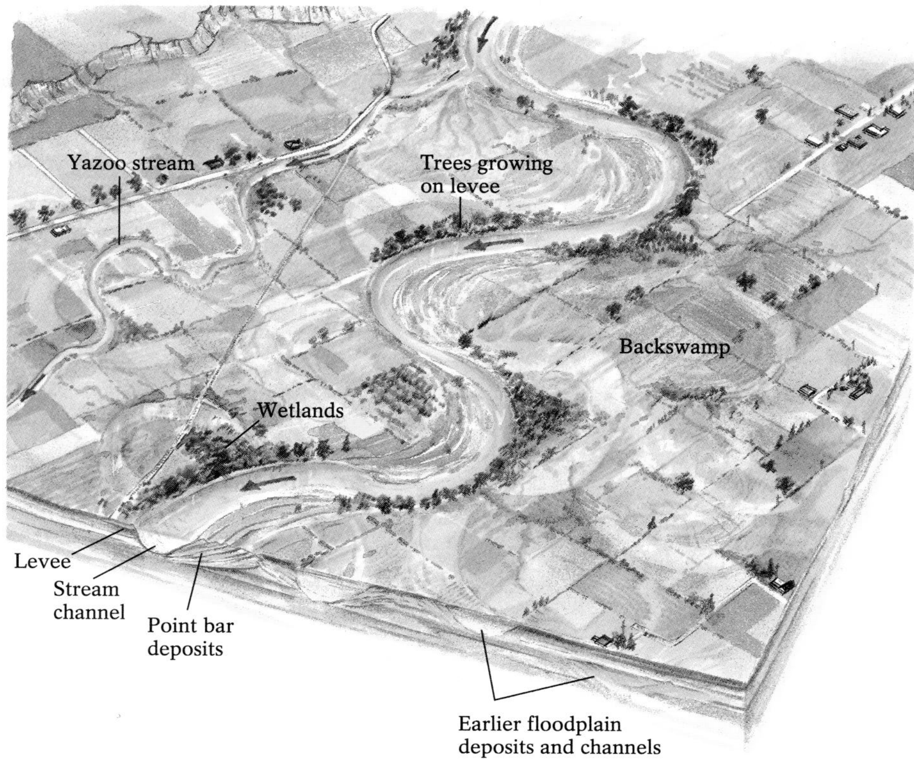

Figure 14-20 Floodplain features, produced when streams overflow their banks and deposit their sediment on the surrounding land. (See text.)

birds and other wildlife and from which the groundwater system is replenished. Sometimes wetlands are drained artificially to allow farming or other land uses, such as industry or housing.

Floodplain landscapes are often reshaped by phenomena unrelated or only indirectly related to flooding. The point bars deposited within a stream's channel, for example, become major parts of a flood plain's geological structure. As a meandering stream erodes the outside curves and deposits sediment on the inside curves of its meanders, its channel migrates snake-like through its own previous floodplain deposits of sand and gravel. A mature flood plain may thus overlie a succession of ancient, migrated, partially eroded point bars (see Fig. 14-20). Meanders migrating over a flood plain may threaten the land on which a community is located. The town of New Harmony, in southwestern Indiana, waited anxiously for years as the meandering Wabash River approached it. The river would have reached the town in 1994 if civil engineers had not halted it, less than a kilometer to the north, by erecting a concrete-block barrier that stabilized its eroding banks and prevented further erosion.

Figure 14-21 A vegetated levee on the bank of the Mississippi River. Natural levees, composed of coarse sediment, are generally the best-drained parts of the flood plain and consequently are the most capable of supporting trees that cannot tolerate saturated soil conditions.

Growing levees can also affect neighboring floodplain features, by diverting the course of tributaries from the main stream channel, either temporarily or permanently. Diverted tributaries may be forced to flow parallel to the main channel for tens or hundreds of kilometers before they are able to cross a low spot in the levee and join the trunk stream. Such parallel streams, isolated at the margin of a flood plain, are called *yazoo streams;* they are named for the Yazoo River, which shadows the Mississippi River for 320 kilometers (200 miles) before finally joining it near Vicksburg, Mississippi.

Figure 14-22 An alluvial fan in Death Valley, California. Alluvial fans form when streams flowing from steep mountain slopes encounter a sharp reduction in slope at the foot of the mountain. Here they flow more slowly and less energetically, dropping their sediment loads as fan-shaped deposits.

Deposition at Breaks in Slope When a stream carrying coarse sediment leaves a narrow, high-gradient mountain valley and flows out onto a broad, relatively flat valley floor, its gradient is sharply reduced, causing a substantial loss of transport energy. Free from the confinement of the narrow valley, the stream spreads out in a characteristic shape across the lowland plain as it deposits its sediment load, forming a triangular deposit known as an **alluvial fan** (Fig. 14-22).

Deposition into Standing Water When a stream enters a standing body of water such as a lake or ocean, its flow decreases abruptly and then stops, and it deposits its suspended load of fine sand, silt, and clay. This sediment usually accumulates in the standing water to form a **delta,** a roughly triangle-shaped alluvial deposit that fans outward from its apex at the mouth of a stream (Fig. 14-23). (This formation was named by the Greek historian Herodotus, who noted in the fifth century B.C. that the sediment at the mouth of the Nile River resembled the uppercase Greek letter delta, Δ.) Because it is heavier, the coarser stream sediment is deposited first, close to the mouth of the stream; this produces the delta's angled *foreset beds.* Finer sediments are carried farther from the mouth, where they are deposited horizontally in layers known as *bottomset beds.* The greater volume of the nearshore sediment gradually expands the foreset beds outward, extending the delta and burying previously deposited bottomset beds. Flooding at the surface of the expanding delta spreads thin, horizontally layered sediments, known as *topset beds,* on top of the inclined foresets.

A delta grows outward as long as its stream deposits more sediment than is eroded from the delta by waves and shoreline currents. The Earth's great deltas can be found where major rivers—such as the Mississippi, the

Figure 14-23 (**a**) The anatomy of a delta. (**b**) The Tengarito River delta, New Zealand. (**c**) A cross-bedded deltaic deposit from the Cretaceous Period, in Colorado.

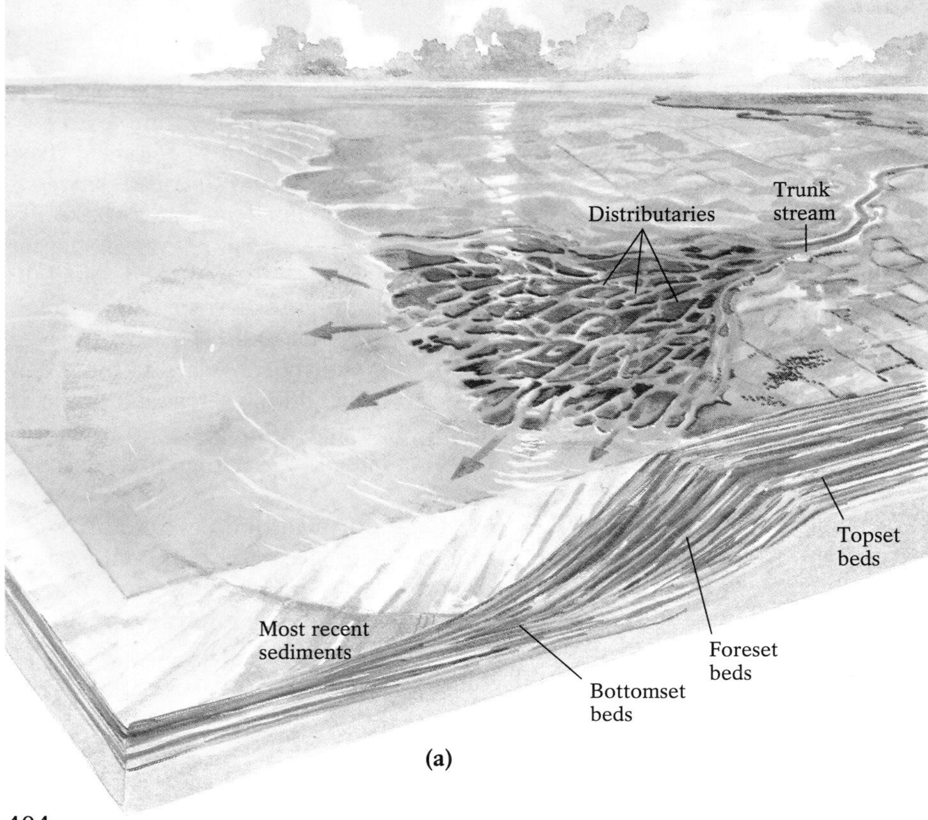

(a)

(b)

(c)

Ganges in India, the Indus in Pakistan, and the Nile in northern Africa—deliver large sediment loads to relatively quiet coastal waters. At the mouths of other large rivers, such as the Amazon in South America and the Niger in Africa, vigorous waves and currents immediately sweep away discharged sediments, and deltas do not form. In North America's Pacific Northwest, powerful winter storm waves remove virtually all the sediment carried by the Columbia River, which consequently has no delta. The St. Lawrence River, which rises in Lake Ontario and flows past Montreal and Quebec, has no delta for two reasons: because much of the region's sediment settles to the bottom of Lake Ontario and the other Great Lakes and never enters the river, and because the river cannot accumulate a delta-building load on its relatively short 600-kilometer (380-mile) journey to the Gulf of St. Lawrence and the northern Atlantic.

As a stream's delta grows, its gradient decreases; consequently, it flows more slowly and gradually loses the capacity to carry its suspended load. As more sediment is dropped in a stream channel it becomes clogged, and the pent-up flow branches into a new network of distributaries. The channel also becomes shallower, making the stream more likely to flood and break through its levees. Freed in these ways from its original channel, the stream may discover a more direct route to the sea (with a steeper gradient) and begin to build a new delta lobe at its new outlet (Fig. 14-24). Soon the younger distributaries carry most of the sediment load, and the new delta grows. At the same time the abandoned delta segments, without a replenishing sediment supply, begin to erode away. Thus, the shapes and locations of a stream's active and abandoned deltas change continuously.

With their fertile soils of fresh floodplain silt and their intricate networks of navigable waterways, deltas have served for thousands of years as coastal centers of agriculture and commerce. Ancient deltas are valuable sources of oil and natural gas (deriving from the decomposition of marine fauna) and coal (from accumulation of plant remains in stagnant deltaic swamps). Highlight 14-1 focuses on the history and economic significance of North America's greatest delta.

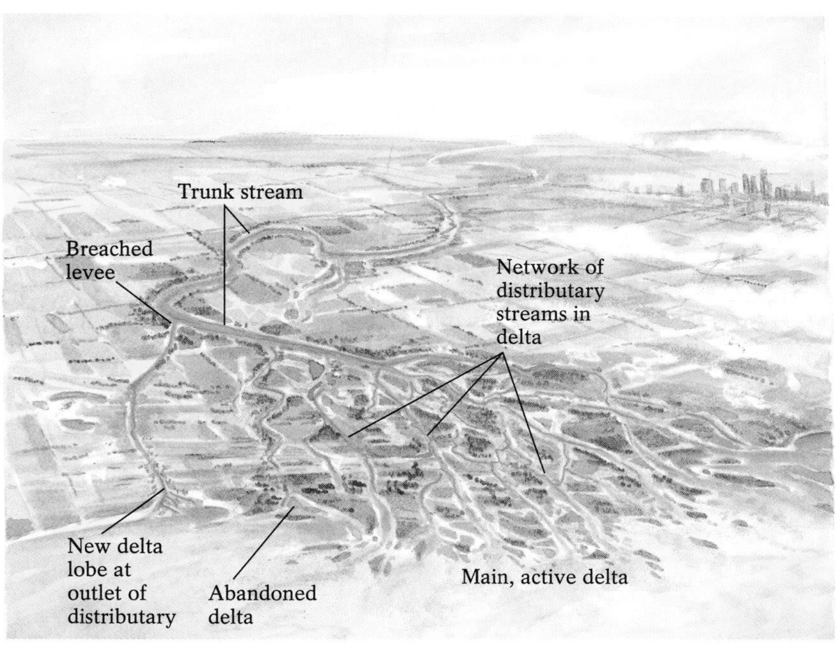

Figure 14-24 The large sediment loads and gentle gradients of deltaic streams promote in-channel deposition. The streams are then forced to diverge around these deposits or overflow their channel walls, forming new distributaries and deltas.

Highlight 14-1 *The Mississippi River and Its Delta*

The Mississippi River begins its 3750-kilometer (2350-mile) journey as a creek a few meters wide and 10 centimeters (4 inches) deep that trickles from Lake Itasca in the pine forests of northern Minnesota. In its first 2100 kilometers (1300 miles), the river cascades over bedrock falls and rapids wherever it is not dammed or restrained. South of Cape Girardeau, Missouri, the Ohio joins the Mississippi and they form a majestic alluvial valley—the much-tamer lower Mississippi valley—which stretches for more than 1600 kilometers (1000 miles) and reaches widths of 48 to 200 kilometers (30–125 miles). By the time the Mississippi empties into the Gulf of Mexico 200 kilometers (120 miles) south of New Orleans, it carries water and sediment from the Missouri, the Ohio, the Arkansas, and more than 100,000 other streams, and has deposited fertile silt over more than 770,000 square kilometers (300,000 square miles) of floodplain farmland (and, occasionally, over one or more cities).

Much of the lower Mississippi valley was once a shallow marine bay located in an aulacogen (see Chapter 12), a failed rift arm that formed about 200 million years ago when North America split from Europe and Africa. The Mississippi's first delta grew southward into this bay, filling the aulacogen with a succession of southward-growing deltas. As successive ice ages caused the Earth's water to accumulate in glaciers during the last 3 million years (discussed in Chapter 17), global sea levels fell and the Mississippi's marine bay receded. During this period, the Mississippi transported the great supply of sediment produced by the vast ice sheets that covered Canada and the Great Lakes region of the United States, and its delta grew steadily. Beginning 12,000 years ago, following the last period of worldwide glaciation, the Earth's climate warmed and sea levels rose. About 9000 years ago, the Mississippi began to construct its present-day delta (Fig. 14-25).

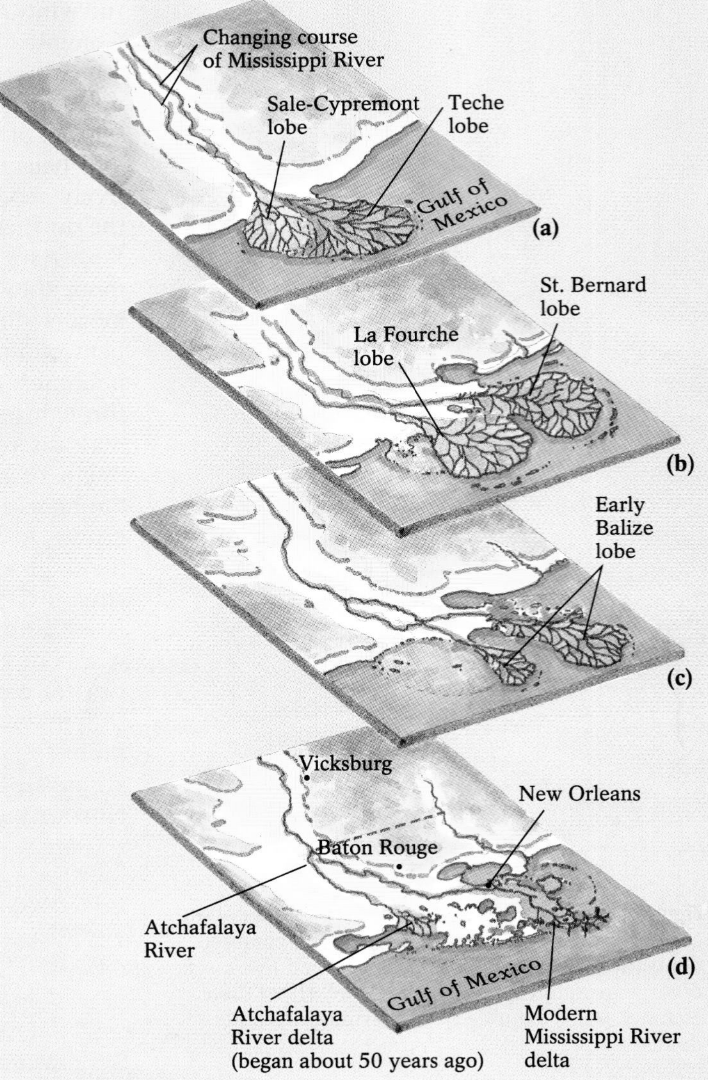

Figure 14-25 The evolution of the Mississippi River delta plain. (**a**) The Sale-Cypremont delta (active 7500 through 5000 years ago), one of the earliest known deltas of the Mississippi, was abandoned when the river diverted its sediment to the Teche lobe (active 5500–3800 years ago). (**b**) Delta growth shifted farther east when the St. Bernard lobe (active 4000–2000 years ago) was deposited. Delta building moved west again, forming the La Fourche lobe (2500–1500 years ago). (**c**) By about 1000 years ago, the modern birdfoot (Balize) lobe began to be constructed. (**d**) In the last 100 years or so, the Atchafalaya River, a distributary, began to carry more of the Mississippi's flow. Were it not for flood-control structures built by the U.S. Army Corps of Engineers, the Atchafalaya would probably become the main channel of the Mississippi, as it presents a shorter route to the Gulf of Mexico.

Today the Mississippi delta consists of at least seven distinct lobes, each of which was once the river's active delta. Together, they constitute about 40,000 square kilometers (15,000 square miles) of real estate added to Louisiana (Fig. 14-26). The active portion of today's delta, known as the Balize delta, is 1000 years old. Its bird's-foot shape has been built by three major distributaries whose gradients are now too gentle to transport sediment effectively. The inactive lobes were each abandoned as the river discovered shorter, steeper routes to the gulf. Many more abandoned lobes have eroded away completely, and overlying sediments have depressed others below sea level. Most of the very large Balize delta is under water. The Mississippi River *bayous,* the colorful network of lakes and minor streams in southern Louisiana, are abandoned channels crisscrossing inactive lobes.

For several decades, the Mississippi River has been trying to cut a steeper, far shorter (225-kilometer/150-mile) course along the Atchafalaya River, 100 kilometers (60 miles) to the west of its present channel. If and when it succeeds, the lower 500 kilometers (300 miles) of the present river would be cut off, imperiling the livelihood of Baton Rouge and New Orleans and rendering worthless the great investments made in navigational improvements along its course. Human intervention has thus far preserved the river's course. In 1973 a winter of unusually high precipitation produced raging floods in the lower Mississippi valley, which might have opened the Atchafalaya channel and caused the river to bypass New Orleans, leaving it a sleepy bayou town instead of a center of river commerce. However, floodgates were opened to divert water from the main channel to Lake Ponchartrain, thereby preventing the river from breaking through its levee and taking the Atchafalaya course. Continued dredging to deepen the lower Mississippi channel enough to accommodate the river's flow has thus far saved Baton Rouge and New Orleans. But the river may yet have its way.

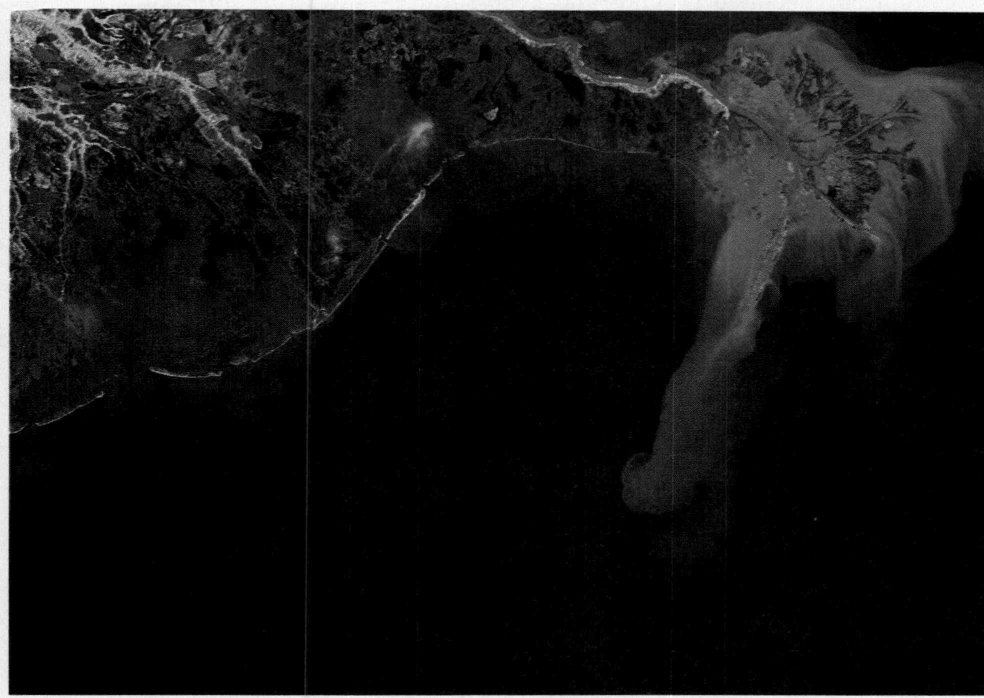

Figure 14-26 The modern delta of the Mississippi River. Today's delta, shaped like a bird's foot, consists of the modern segment centered at Head of Passes, Louisiana, and a number of inactive, abandoned lobes that no longer receive sediment. During the past century, the river has attempted to shift away from this lobe and establish a new course along the Atchafalaya River. Unless prevented by human intervention, it will eventually do so.

Controlling Floods

Most streams overflow their banks every 2 or 3 years, causing flooding. This is usually the result of local weather—simply too much rain falls or snow melts, producing too much water for a channel to carry. Floods can be caused by a succession of storms as well as a single storm. If a storm falls on unsaturated ground, the soil may soak up the water and delay its entry into streams. After the ground becomes saturated, however, it is unable to absorb the next downpour, and the excess water runs off the surface directly into local streams, causing them to rise rapidly.

The geology of the surface and the topography of a terrain help determine whether water runs rapidly off the surface into streams or infiltrates the groundwater system, where it could remain long enough to avert flooding. Some of the factors that affect runoff are whether the bedrock near the surface is permeable, like sandstone, or impermeable, like granite; whether the rock is highly fractured; whether the surface soils are clay-rich types that swell when wet and seal the surface against infiltration; and whether the local topography is steep or gently sloping (Fig. 14-27).

Figure 14-27 The Big Thompson Creek flood. Big Thompson Creek, in the Colorado Rockies northwest of Denver, received more than 30 centimeters (12 inches) of rain on the evening of July 31, 1976. Because this volume of precipitation, which nearly equaled the average annual precipitation for the area, could not be absorbed by the steep exposed bedrock of the canyon, discharge in the creek rose rapidly to more than four times its previous record. The surface of the creek rose from two-thirds of a meter (about 2 feet) to more than 6 meters (20 feet). In all, the flood took 144 lives and caused $35 million in damage. (Many victims would have survived had they climbed upslope instead of trying to outrun the wall of water that thundered down the narrow gorge.)

Flooding is also a likely result when heavy rains fall on an area whose vegetation has been reduced by extensive forest fires, droughts, or widespread clear-cutting. Trees and their root systems tend to keep soils open and enhance water infiltration, thus reducing the likelihood of floods. As communities expand into undeveloped areas, new roads, buildings, and parking lots replace vegetated grounds and keep surface water from infiltrating into the groundwater system, increasing the potential for flooding.

Flood Prediction

Floods are usually seasonal, occurring during heavy spring rains and snowmelt in temperate climates and during rainy seasons elsewhere. Most streams have an annual pattern of flooding, to which a region's residents may be compelled to adjust—by keeping rowboats in basements, building structures on stilts or cinderblocks, maintaining supplies of sandbags, and so forth. It is the *unusual* flooding event, which produces much more water than stream systems can handle, that causes the greatest damage. Can such unusual floods be predicted? How can prediction help allay their destructive power?

Flood prediction is necessarily imprecise; to a large extent we are subject to the irregularities of weather. The best we can do is analyze the frequency of past floods and use the data to determine the statistical probability of a major flood occurring within a given time. Stream discharge is monitored over long periods of time by plotting *hydrographs,* which show how discharge varies day-to-day and year-to-year and how discharge is affected by human activities such as urbanization and forest clear-cutting (Fig. 14-28). Hydrograph data for a stream are then used to plot its flood-frequency curve (Fig. 14-29), from which we can estimate the probability of a particular flooding discharge being equaled or exceeded in that stream in any one year.

For example, after recording a stream's discharge over a 100-year period, we determine that it has a peak discharge as great as 2000 cubic feet per second (cfs) once every 10 years; thus, there is a 1-in-10 chance (10% probability) of attaining such a discharge in any particular year. When that discharge does occur, we call the event a 10-year flood. Similarly, recorded data may show that the discharge of the same river reaches 3500 cfs once every hundred years; thus, there is a 1-in-100 chance (1% probability) of

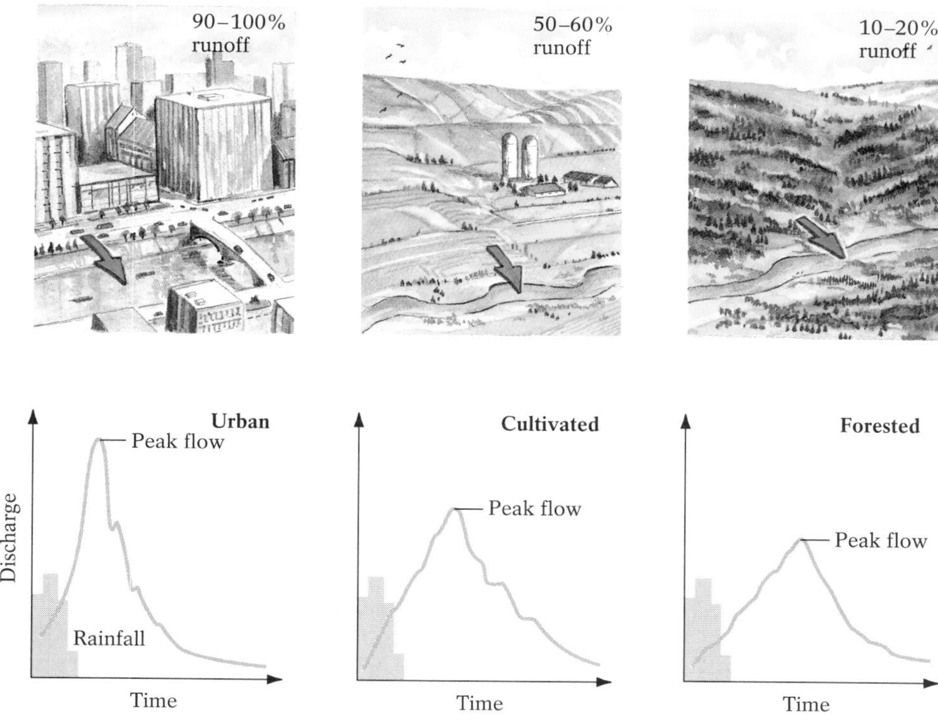

Figure 14-28 Hydrographs record how a stream's discharge varies with time. A peaked hydrograph curve indicates that a stream's discharge has risen dramatically in a short space of time, suggesting that precipitation and other surface water have flowed rapidly into the stream with little delay in the groundwater system. Because urban areas are largely paved over with asphalt and cement, little surface water infiltrates the groundwater system and much runs off into streams; thus, urban streams tend to have peaked hydrographs. By comparison, the hydrographs of streams in undeveloped areas are generally less peaked (changes in discharge are more gradual).

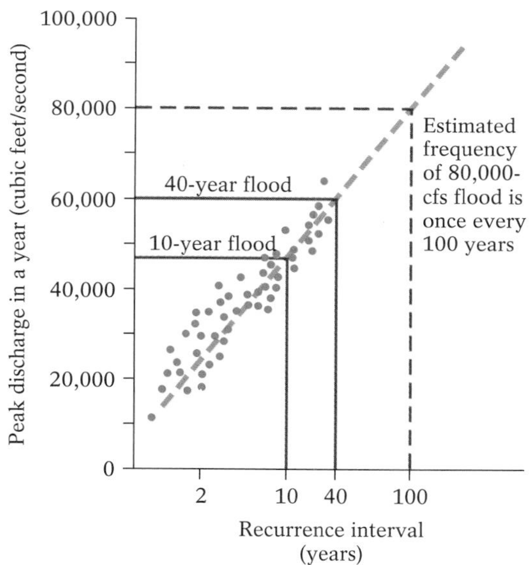

Figure 14-29 A flood-frequency curve shows how often large discharges and floods of various magnitudes have occurred throughout a stream's recorded history.

attaining such a discharge in any particular year. Such an event is called a 100-year flood. These probabilities are updated as the discharge record for a stream lengthens with time.

Note that the discharge associated with the 100-year flood of one stream is not necessarily the same as that of any other; every stream has its own unique flood-frequency curve with specific discharges associated with floods of particular magnitudes.

Nature, however, seldom conforms precisely to our expectations. After an area endures one 100-year flood, there will not necessarily be a 99-year respite before the next one—portions of the Mississippi River have had to cope with 100-year floods in 1943, 1944, 1947, and 1951. Moreover, because of inadequate early record-keeping, flood frequency is sometimes hard to define. The Mississippi's flood of 1993 has been classified by some agencies as a 200-year flood and by others as a 500-year event.

Flood Prevention

A number of defensive strategies—structural, nonstructural, and combinations of both—have been used to try to curb the catastrophic effects of floods. The structural approach involves some form of construction to contain the stream within its banks, divert some of its water to temporary storage facilities, or accelerate flow through the stream system.

The most common stream-containment structures are *artificial levees,* earthen mounds built on the banks of a stream's channel to increase the volume of water that the channel can hold. Artificial levees have been built along the Mississippi River to protect croplands since the eighteenth century. Spillways are cut through the artificial levees to allow rising water to escape into human-made subsidiary channels, thus reducing the likelihood of a flood

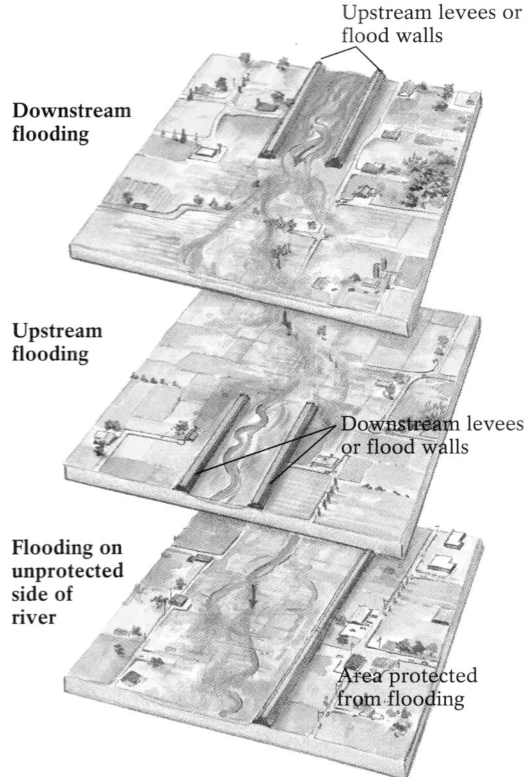

Figure 14-30 Artificial levees and flood walls must cover the full length of a river and be built on both of its banks; otherwise, flooding is merely diverted.

downstream. Concrete *flood walls,* which are more expensive, have been constructed by the U.S. Army Corps of Engineers to prevent overflow from the channel at strategic locations, such as along the commercial centers of riverside cities such as St. Louis.

To be effective, artificial levees and flood walls must be built on both sides of a stream and must extend for its full length (Fig. 14-30). If they are built only upstream, the flooding stream will "let loose" downstream after being denied access to its flood plain along the controlled section. If they are built only downstream, the upstream floodwater will wash over the flood plain behind them. If artificial levees and flood walls are constructed on only one side of a stream, the flood plain on the uncontrolled side receives a double share of floodwater.

Flood-control dams are often built upstream from densely populated or economically important areas. These earthen or concrete structures shunt excess water to temporary storage facilities that can contain enough water to prevent overflow. Here the excess water either evaporates or seeps into and replenishes the groundwater system. Such structures have non-flood-related uses as well; dams along the Tennessee, Mississippi, Missouri, Colorado, and Columbia Rivers generate hydroelectricity, provide irrigation water for agriculture, supply municipal water to numerous communities, and create regional recreational facilities.

Channelization is a structural alteration made to a stream's channel in order to speed the flow of water and thus prevent it from reaching flood height. Channelization may involve clearing obstructions such as fallen trees that slow flow or widening or deepening a channel by dredging to increase its capacity. More radical channelization is used to increase a stream's gradient and therefore its velocity. This usually involves cutting off meanders to straighten a stream. The shorter, straight channel will have a steeper gradient than before, and its increased velocity will transport more water—perhaps rapidly enough to prevent flooding.

Channelization can also have other, more inconvenient, effects on streams. This method has been used since 1910 to control the Blackwater River, a tributary of the Missouri River, southeast of Kansas City. The Blackwater, 54 kilometers (34 miles) long with a gradient of 1.7 meters per kilometer (9.3 feet per mile), had flooded regularly because it flowed too slowly to accommodate large storm discharges. Engineers excavated a straight channel that increased both its gradient and velocity, but also caused the more active river to scour its bed more energetically. The river subsequently grew from an initial depth of about 4 meters (14 feet) to an average depth of more than 12 meters (39 feet), and from 9 meters (30 feet) wide to more than 71 meters (233 feet). As the river widened, bridges were undermined and had to be replaced. Only constant maintenance now keeps the Blackwater River straight.

Structural solutions to flood problems are very expensive. The greater the potential flood, the more costly the structural defense and the more resistant communities are to undertaking it. Structural defenses can also give local residents a false sense of security; the more protected they feel, the closer to a stream they build their homes. Consider what might happen if a structure is designed to impound a 50-year flood and the region is then struck by a 100-year event. For these reasons a nonstructural strategy, if feasible, is always preferable.

Nonstructural defenses against flooding depend on identifying high-risk areas, instituting zoning regulations to minimize development in them, and managing resources to minimize the amount of water entering a stream chan-

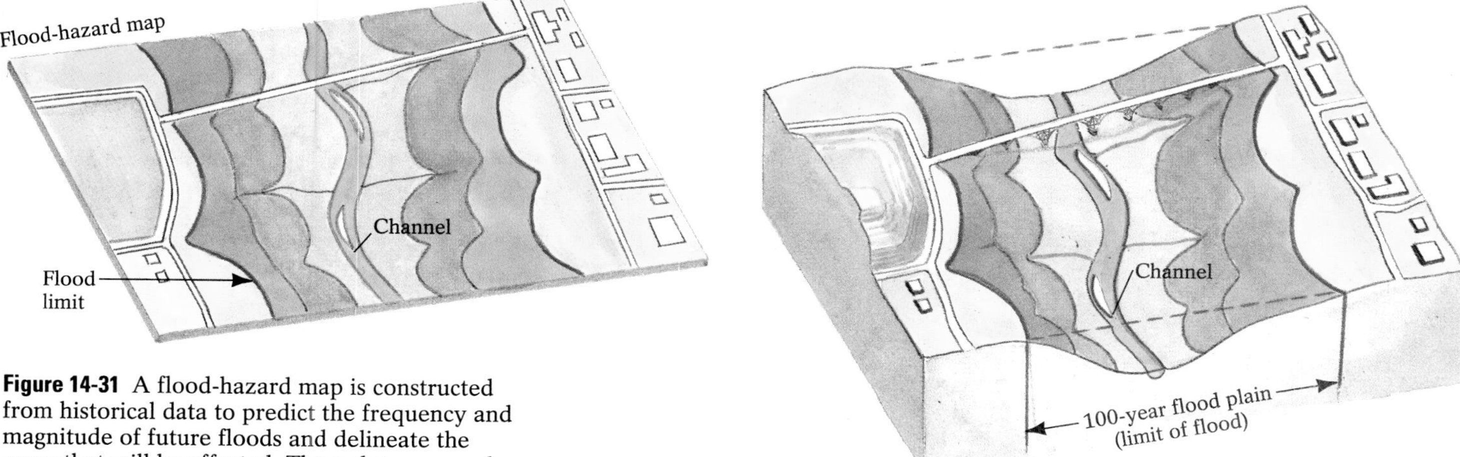

Figure 14-31 A flood-hazard map is constructed from historical data to predict the frequency and magnitude of future floods and delineate the areas that will be affected. These data are used to establish floodways—the areas of a river valley that would be covered by floods of various magnitudes—and institute regulations restricting development in flood-prone areas.

Floodplain Zones

"Floodway district": regulated floodway; kept open at all times; no flood-control measures, no building

"Floodway fringe district": flood-prone, protected by flood-control measures; building allowed if protected by "flood-proofing"

"Floodway limit": area subject to flooding only by very large discharge (100-year flood)

nel at any one time. These approaches first require a thorough geological and botanical survey to identify flood-risk factors. For example, the natural distribution of water-loving plants helps identify areas that have flooded regularly in the past. Geologists map and date ancient flood deposits and analyze their thicknesses and textures to estimate the timing and magnitude of prehistoric floods. These data are used to create a flood-hazard map showing *floodplain zones*, or *floodways*—the corridors of land adjacent to the stream that would be covered by floods of particular magnitudes (Fig. 14-31). Such graphic information guides communities in drafting zoning ordinances to minimize development in high-risk areas.

For example, a 12-kilometer (7-mile)-wide floodway was identified after the disastrous Rapid City, South Dakota, flood of June 10, 1972, and many buildings were moved from it. Today little new construction takes place in the floodway, which is used mostly for ballfields, golf courses, and other sporadically used spaces. Structures in the floodway that could not be moved or abandoned are protected by artificial levees and flood walls.

An important part of nonstructural regional planning for flood-hazard reduction is the prevention of excessive forest clearing and the prompt suppression of forest fires. This helps maintain the vegetation cover, which as we know promotes the absorption of water into the groundwater system and reduces stream discharge.

How have we fared in our efforts to prevent floods? During June and July of 1993, a succession of storms dumped an extraordinary amount of rain on the drainage basin of the Mississippi River in Minnesota, Iowa, Illinois, and Missouri, triggering the river's worst flood on record and causing billions of dollars in damage to croplands and other property. Highlight 14-2 focuses on the tragic flooding of 1993 and illustrates how nature can confound our flood-prevention strategies.

Highlight 14-2 *The Great Midwestern Floods of 1993*

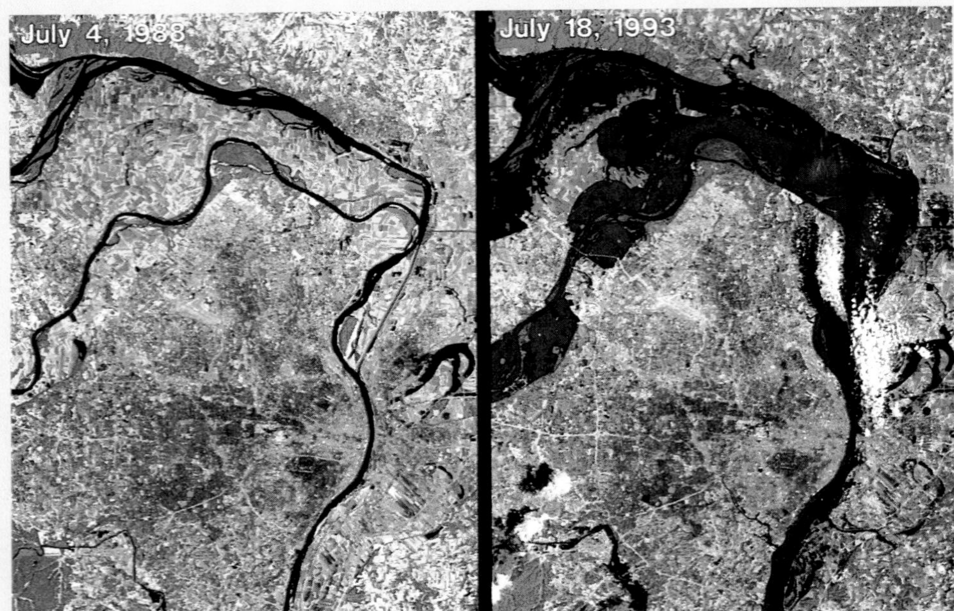

Figure 14-32 Satellite photos taken before and during the flood of July 1993 show the extent of the flooding at the confluence of the Mississippi and Missouri Rivers.

On June 11, 1993, a foot of rain fell in southern Minnesota. Four days later, over 11 more inches fell in about the same area. Thus began the wettest June, and North America's worst flood, since weather records began to be compiled in 1878. Before the wet weather ended over two months later, the upper Mississippi River had surged over its banks along an 800-kilometer (500-mile) stretch from St. Paul, Minnesota, to St. Louis, Missouri. It swept away numerous artificial barriers and in places spread as far as 13.5 kilometers (8.5 miles) from its channel (Fig. 14-32).

The swirling floodwaters undermined interstate bridges, washed away roads, and brought to a halt all barge traffic along the upper Mississippi, the economic artery of the Midwest. Rushing, muddy floodwaters, carrying uprooted trees and junked cars, swept away thousands of homes and businesses, displaced more than 50,000 people, took dozens of lives, and caused more than 10 billion dollars in property damage and agricultural losses. For the first time in anyone's memory the waters of the Mississippi and Missouri overflowed their levees and washed together, 30 kilometers (20 miles) from their confluence north of St. Louis, closing the bridge at West Quincy, Missouri—the only connection between Missouri and Illinois for more than 300 kilometers (200 miles).

Life throughout a 12-state region was disrupted in countless ways. In Hannibal, Missouri, Mark Twain's boyhood home, the Mississippi crested (reached its peak) at 8.6 meters (28.6 feet), which was 5 meters (16.5 feet) above flood stage; children caught catfish on streets where they had ridden bicycles only days before. Fortunately, a recently built, 11-meter (33-foot)-high flood wall saved the historic downtown area and the Twain homestead. In Davenport, Iowa, where residents had preferred preservation of their river views to building levees and flood walls, ducks swam lazily in the 4 meters (14 feet) of water covering the outfield of Davenport Stadium, the home of the Quad City River Bandits, the town's minor league baseball team. Two-hundred and fifty thousand Iowans in Des Moines went without drinking water for days after floodwaters contaminated the municipal water supply with raw sewage and chemical fertilizers.

The residents of the lower Mississippi valley, south of Cairo, Illinois, were spared the flooding, which affected only the upper Mississippi valley. Why was the lower Mississippi spared while the upper was overwhelmed? One reason is that the channel of the upper Mississippi is relatively narrow and shallow, and so cannot accommodate a great volume of water. At St. Louis, for example, the channel is only 18 meters (60 feet) deep and 575 meters (1900 feet) wide. By comparison, the channel of the lower Mississippi is wide and deep. At Vicksburg, Mississippi, the channel is 37 meters (123 feet) deep and 970 meters (3200 feet) wide, and thus is able to carry a much larger discharge. More importantly, the states that comprise much of the watershed of the upper Mississippi (Minnesota, Wisconsin, Iowa, Missouri, and northern Illinois) and that of the Missouri, its major tributary (Nebraska and the Dakotas), received months of soil-saturating storms, whereas the lower Mississippi watershed experienced below-normal precipitation. (The lower Mississippi valley receives 65% of its water from the Ohio River and its tributaries, which flow from western Pennsylvania near Pittsburgh through Ohio, West Virginia, Indiana, Kentucky, and southern Illinois.) As a result, there was no flooding in the lower Mississippi valley states of Arkansas, Tennessee, Mississippi, and Louisiana.

Did human activity—65 years of "managing" the Mississippi River to prevent floods—actually contribute to the magnitude and tragedy of the flood of 1993? Before the last 100 years or so, the Mississippi and its tributaries determined their own boundaries. During periods of high water, the rivers broke through or overflowed natural levees, flooding tens of thousands of square kilometers of the surrounding, largely uninhabited, lands. In the twentieth century, millions of people have migrated to the region, building cities, towns, and large farms along the rivers' banks. When a flood in 1927 took 214 lives, Congress enacted the first Mississippi River Flood Control Act, assigning the U.S. Army Corps of Engineers the daunting task of confining the river to its channel.

The Corps of Engineers' efforts consist of about 300 dams and reservoirs and thousands of kilometers of artificial levees and concrete flood walls, all designed to prevent the river from spilling onto its natural flood plain. The system also contains numerous pumping stations, spillways, and diversion channels designed to divert water for storage in temporary holding basins. But the events of 1993 showed that the Corps' efforts have largely failed, perhaps because such flood-control structures simply cannot contain an *extraordinary* flood. By confining such great discharges to a channel, the retention structures actually caused the swollen rivers to flow more rapidly and violently, thus damaging the very structures designed to restrain them. And by denying the river access to its natural flood plains, the structures caused the streams to rise higher than they would have, insuring that once they did breach the levees, their floods would be more damaging. Furthermore, the existence of artificial levees and flood walls bred a false sense of security that encouraged the growth of cities, towns, and farms closer to the river banks than was really safe.

What does the future hold for the residents of the upper Mississippi valley? Certainly more flooding, but perhaps less human interference with the river's natural behavior. Some communities have proposed that all flood-retention systems be eliminated and that zoning limit future development close to the region's rivers, at least within their 100-year floodways. Others look longingly at St. Louis' 16-meter (52-foot)-high concrete flood wall, which saved that city's downtown business district when the Mississippi reached its record crest at 14.2 meters (47 feet). The debate continues between those who believe we can tame the mighty Mississippi and those who know we cannot.

Stream Evolution and Plate Tectonics

Our understanding of the evolution of stream systems is, like other geological issues, being reassessed in light of our growing knowledge of plate tectonics. Plate-margin stresses produce much of the local uplift that is then eroded by downcutting streams to create spectacular waterfalls, distinctive terraces, and deeply incised river channels (Fig. 14-33). Tectonic uplift also produces the rising mountains, such as the Cascades at the convergent boundary between the North American and Pacific plates, through which such antecedent streams as the Columbia and Snake Rivers of the Pacific Northwest flow.

Figure 14-33 The Goosenecks of the San Juan River, southeastern Utah. These meanders have been deeply incised into the Colorado plateau, which has been tectonically uplifted due to interactions between the North American and Pacific plates.

Tectonic activity often determines the character of the regional stream drainage pattern. For example, tectonic stresses cause the faults and fractures that erode to produce a rectangular drainage pattern, whereas folded mountain belts at convergent plate boundaries tend to display trellis drainage. Radial drainage patterns evolve as streams dissect the young volcanoes that are generally found at active plate boundaries.

Many large stream systems originate in the folded mountains at convergent margins; they cross stable mid-plate interiors and then build distributary systems and deltas at passive plate margins. For example, the Amazon River rises in the Andes of western South America, where the South Pacific's Nazca plate subducts beneath the South American plate. The river then flows eastward across the continent for thousands of kilometers before emptying into the Atlantic Ocean. Similar tectonically produced drainage divides can be found in the Alps of southern Europe, which rose from the convergence of the African and Eurasian plates; the Himalayas, produced by convergence of the Indian and Eurasian plates; and the North American Rockies, originating from convergence of the North American and Pacific plates.

Extraterrestrial Stream Activity: Evidence from Mars

Braided channels, stream-modified islands, dendritic drainage patterns, and catastrophic-flood topography on the surface of Mars were clearly photographed by the 1971 Mariner 9 spacecraft and the 1976 Viking probe. On Earth, such features are characteristically formed by flowing water. Yet there is virtually no free water on Mars' surface; all the liquid water at its surface and in its atmosphere would barely fill a small swimming hole. Atmospheric pressure on Mars is so low the atoms on a liquid's surface can easily escape, so water vaporizes instantaneously. Moreover, the planet's atmosphere is so thin that it does not trap heat reradiated from the surface. Thus, the Martian surface is too cold for water to exist in liquid form. How then can we explain the spectacular fluvial (river-related) landforms photographed on Mars?

Some geologists propose that Mars' water is trapped as ice below ground, in the pores of the regolith, and that this water may from time to time be released catastrophically when the ice is melted by intrusions of magma. Most of what appear to be major flood channels are located in Mars' southern volcanic highlands, whose large angular rocks may have been formed after magma melted a vast volume of underground ice, causing the rock to collapse and flooding the surface, carving deep canyons. When the floodwaters evaporated soon afterward, braided channels punctuated by teardrop-shaped islands of sediment were left. We cannot determine how long ago Mars' catastrophic floods took place, but the numerous meteorite-impact craters superimposed on the channels suggest that they may be at least hundreds of millions of years old.

Although absolutely no rain falls on Mars today, in the early life of the solar system water may have been created there from volcanic outgassing of the planet's interior, as it was on Earth. Mars likely had a more substantial atmosphere at that time, and therefore some water could have survived at the surface. With a more significant atmosphere (and a prevailing greenhouse effect) and greater atmospheric pressure, Mars was almost certainly warmer than it is today. Some of Mars' fluvial-looking features, such as what appear to be meandering channels (Fig. 14-34) and dendritic drainage patterns, suggest that 3 to 4 billion years ago Mars' surface may have been dotted with lakes and streams nourished by frequent rainstorms. This early period coincides with the solar system's period of maximum meteorite activity. By breaking through the Martian crust and allowing subsurface magmas to rise to the surface, meteorites may have been responsible for Mars' early volcanism and thicker atmosphere. More recently there have been fewer meteorite impacts and less volcanism, and therefore Mars' present atmosphere is cool, thin, and dry. Today the planet's surface is frigid, arid, and devoid of fluvial activity.

Figure 14-34 A 3-mile-wide channel north of the Martian equator, photographed in 1972 by Mariner 9. This and other apparently fluvial features on Mars are believed to have formed billions of years ago, when the Martian climate and atmosphere allowed free water to exist and flow at the planet's surface.

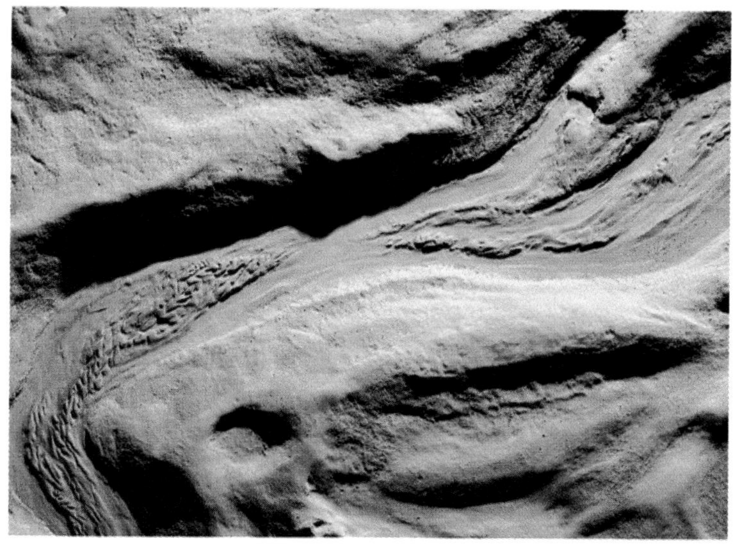

For most of us, rivers are the most obvious component of the Earth's hydrologic cycle, and have the greatest impact on our daily lives. In the future, however, as surface waters suffer increasing contamination from the burgeoning human population, the largely unseen groundwater component may take on greater importance as perhaps the best source of clean fresh water. In the following chapter, we will examine how groundwater accumulates and flows, and how we can tap the reservoir beneath our feet while preserving its quality.

Chapter Summary

Virtually all of the Earth's water initially derives from the planet's interior, principally as **juvenile water,** the steam that accompanies volcanic eruptions. Virtually all of the Earth's surface water derives from the atmosphere as **meteoric water,** the moisture that falls as precipitation. All water on the planet is eventually cycled among its oceans, land, and atmosphere and between its mantle and its surface, a process called the **hydrologic cycle.**

Any surface water whose flow is confined to a narrow topographic depression, or channel, is a **stream.** The flat land bordering a stream's channel is its **flood plain.** All streams are surrounded by **watersheds,** or **drainage basins,** the areas of land from which they acquire their water. The topographic highland that separates two adjacent drainage basins is a **drainage divide.** Rainfall initially flows from a drainage divide across the basin's slopes as broad shallow sheets of water that are not confined to channels. This flow erodes increasingly large channels in the surface, finally creating a network of small **tributaries** that feed water into a main or **trunk stream.** The trunk stream, in turn, may eventually split into another network of small channels, or **distributaries,** which usually empty into an ocean.

Stream velocity is governed principally by the slope, or **gradient,** of the stream bed and the bed's roughness, which may cause friction and impede flow. Velocity is directly related to **stream discharge,** the volume of water passing a point on the stream bank during a given unit of time.

Streams tend to cut downward into bedrock until they reach their **base level,** the lowest point to which a stream can erode (usually sea level). After some streams reach base level, they enter a temporary state of equilibrium to which the stream's gradient has been delicately adjusted so that the stream transports all the sediment supplied to it with very little net erosion or deposition. A stream in this state is called a **graded stream.** If more sediment is introduced to a graded stream, it responds by instantaneously increasing its gradient, a process called **aggradation.** If sediment is removed from a graded stream (such as in a reservoir behind a dam), it responds by instantaneously lowering its gradient, a process called **degradation.**

Erosion, combined with mass movement, is responsible for the development of a stream's characteristic V-shaped valley. The rate of erosion is directly proportional to the stream's velocity. A stream erodes its bed by **abrasion, hydraulic lifting** of loose particles, and dissolution. Streams develop characteristic **drainage patterns,** networks of interconnecting tributaries, trunk streams, and distributaries that reflect the underlying geology. The most common drainage network is the dendritic pattern, which develops on rocks or sediments of uniform composition and topography.

In response to the local topography, climate, and geology, streams or segments of streams may be straight, **braided,** or **meandering.** When two consecutive bends in a meandering stream are close to one another, the stream can cut through the thin strip of flood plain that separates them, abandoning the cut-off segment and leaving it as an **oxbow lake.** When a stream erodes downward to a new base level, it often leaves a fairly flat-lying surface, or **stream terrace,** that marks the stream's former position.

A stream's ability to transport sediment depends largely on its discharge and velocity. The volume of sediment transported by a stream, controlled principally by its discharge, is called its **capacity.** The maximum size of its transported particles, largely a function of stream velocity, defines a stream's **competence.** The sediment in a stream can travel as a **suspended load** in the water column, bounce or roll along the stream bed as a part of its **bed load,** or travel in solution as a **dissolved load.**

Ultimately, when the pull of gravity overcomes the velocity and turbulence of streamflow, transported particles settle from a stream. These particles, called **alluvium,** may be deposited within the channel (forming mid-channel bars and **point bars**), on the stream's flood plain (building **natural levees** and **backswamp** deposits that mark a flood plain's **wetlands**), at sharp breaks in slope (building **alluvial fans**), and in bodies of standing water (constructing **deltas**).

Floods occur when precipitation cannot infiltrate an impervious or saturated ground surface and therefore flows into a stream so fast that the channel capacity is exceeded. Such conditions are quite common in extensively paved urban areas. In their efforts to predict floods, geologists document the historical pattern of floods in a region and develop a predictive model based on statistical probability. To cope with the inevitability of floods, communities build flood-control structures, such as artificial levees and flood walls, alter channels to allow water to pass through more swiftly (a strategy called channelization), and designate floodway zones in which structural development is restricted.

The evolution of streams, like most other geological systems, has recently been reevaluated in light of our growing understanding of plate tectonics. Ultimately, plate motions may control, among other things, a stream's base level, disruptions to its profile (by faulting and uplift), and its supply of sediment (particularly in volcanic areas).

Past stream activity seems to have been instrumental in shaping the Martian landscape. Geologists suspect this from a host of photographed fluvial features, such as meandering channels, streamlined mid-channel islands, and dendritic drainage patterns, which suggest that the Martian surface experienced a prolonged fluvial period early in its history, before the planet's atmosphere thinned and its surface cooled.

Key Terms

juvenile water (p. 387)
meteoric water (p. 387)
hydrologic cycle (p. 388)
stream (p. 389)
flood plain (p. 389)
watershed (p. 389)
drainage basin (p. 389)
tributary (p. 389)
drainage divide (p. 389)
trunk stream (p. 390)
distributary (p. 390)
gradient (p. 390)
stream discharge (p. 391)
base level (p. 391)
graded stream (p. 392)
aggradation (p. 392)
degradation (p. 392)
abrasion (p. 393)
hydraulic lifting (p. 393)
drainage patterns (p. 394)
braided streams (p. 396)
meandering streams (p. 396)
oxbow lake (p. 397)
stream terrace (p. 399)
capacity (p. 400)
competence (p. 400)
suspended load (p. 400)
bed load (p. 400)
dissolved load (p. 400)
alluvium (p. 402)
point bar (p. 402)
natural levee (p. 402)
backswamp (p. 402)
wetlands (p. 402)
alluvial fan (p. 404)
delta (p. 404)

Questions for Review

1. Briefly describe the main elements of the hydrologic cycle, including the processes responsible for water's changes in state. List three ways water can be delayed on land before eventually reaching an ocean.

2. How is stream discharge measured? How do changes in stream velocity, width, and depth affect stream discharge? Calculate the discharge of a river that is 70 meters wide and 20 meters deep and that flows at a velocity of 10 meters per second.

3. Discuss two potential causes of stream aggradation.

4. Sketch three different drainage patterns.

5. Briefly explain how one stream might pirate the water of another.

6. Draw a sketch illustrating the origin of an oxbow lake. How does an oxbow lake become a meander scar?

7. How do the sediment particles that are most likely to be transported in suspension differ from those transported by traction? What is the difference between a stream's competence and its capacity?

8. What are the principal causes of flooding?

9. Describe the differences between structural and nonstructural approaches to flood control.

10. List three ways in which plate tectonics may affect the evolution of a stream profile.

For Further Thought

1. Where would you expect to find a stream that exhibits high competence but low capacity? Where would you expect to find one that exhibits low competence but high capacity?

2. What would be the effect on the Earth's fluvial landscapes if plate tectonic activity suddenly ceased?

3. How would the world's streams respond if climatic warming melted the Antarctic ice sheet?

4. How would the potential for flooding change over a long period of time if all of the dams along the Mississippi River were removed?

5. What kind of deposits do you think underlie the farmland shown in the photo below?

Groundwater Recharge and Flow

The Water Table

Finding and Managing Groundwater

Highlight 15-1 **Should You Hire a Water Witch?**

Highlight 15-2 **Who Owns Groundwater?**

Maintaining Groundwater Quality

Figure 15-1 Siloam Pool in Jerusalem. For centuries the city's sole water source, the pool holds water from the only nearby spring, which was first channeled to a pool within the city walls in the eighth century B.C.

Chapter 15

Groundwater

For thousands of years, wells and natural springs have supplied clean, abundant groundwater to human communities throughout the world. Located in the centers of many villages, wells have been focal points for meeting, bartering, and sharing the news of the day; in other places, water from nearby surface springs has been rerouted to pools within village walls (Fig. 15-1). Even today, the presence of an adequate supply of uncontaminated groundwater can determine whether a region or community will grow and prosper. Pure water has even become a commercially valuable commodity, and supermarkets and gourmet restaurants are able to sell bottles of ordinary groundwater at exorbitant prices.

Humans can survive for weeks and even months without food, but can live without water for only a few days because it is the medium in which virtually all biological processes take place. Our bodies require about 4 liters (1 gallon) of water per day—1.2 billion liters (300 million gallons) of water per day for North America's 300 million people; this water is drawn from the continent's freshwater reserves. We use more than 1.8 trillion liters (450 billion gallons) of water per day for domestic purposes (cooking, bathing and other sanitary needs, and lawn watering), for agriculture (livestock and irrigation), and for industry (processing natural resources and manufacturing the goods that maintain our way of life).

As we saw in Chapter 14, the Earth's hydrosphere extends from the top of the atmosphere to approximately 10 kilometers (about 6 miles) below the Earth's surface and comprises oceans, glaciers, rivers and lakes, atmospheric water vapor, and groundwater. This water is constantly moving through the hydrologic cycle (see Fig. 14-3). The cycle's largely unseen groundwater, only 0.6% of the world's fresh water, provides more than 50% of our drinking water, 40% of our irrigation water, and 25% of the water used in industry.

Groundwater use doubled between 1955 and 1985 as the U.S. population grew. Throughout North America, we are withdrawing groundwater reserves that took thousands of years to accumulate, and supplies (particularly in the Southwest) are running low. Moreover, groundwater in many places has been contaminated by human, animal, and industrial waste. Some of the most urgent issues facing the world's citizens have to do with groundwater: where to find it, how to keep it clean, who owns it. In most areas, climate and local geology are capable of providing a continuing supply of cool, refreshing, healthy groundwater for present and future generations if we learn to manage and preserve it.

Groundwater Recharge and Flow

Groundwater *recharge* is the infiltration of water, mostly from precipitation, into the Earth's surface. After infiltration, groundwater flows through soils, rocks, and other Earth materials until some emerges as groundwater *discharge,* or outflow, returning to the surface as streams or springs or remaining temporarily in lakes, ponds, or wetlands.

The amount of water that infiltrates groundwater systems depends on the condition and type of local surface materials, the nature and abundance of vegetation in the area, the area's topographical features, and the amount of precipitation (Fig. 15-2).

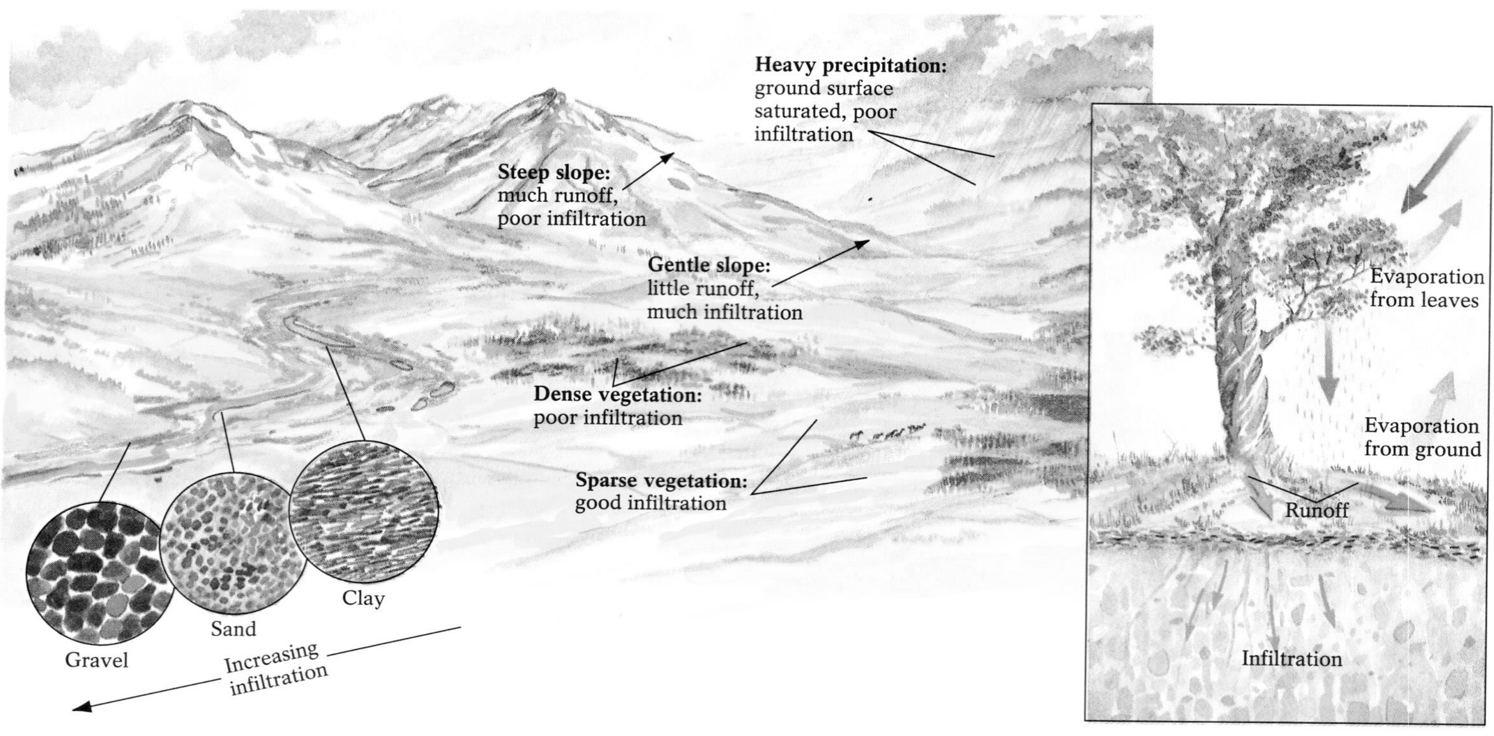

Figure 15-2 Factors affecting infiltration. The amount of water that enters the groundwater system is influenced by bedrock composition, vegetation patterns, surface slope, and climate. Gentle, lightly vegetated slopes composed of permeable materials are most conducive to infiltration.

Condition and type of surface materials. Extensive infiltration is promoted by the abundant pore spaces in loose soils and unconsolidated sands and gravels. Exposed bedrock that is fractured or inherently porous, such as coarse sandstone, also allows substantial infiltration by surface water. On the other hand, unconsolidated clay, which consists of closely packed flat particles with minute pore space, impedes infiltration, as does unfractured crystalline bedrock, such as that composed of plutonic igneous rocks (for example, granite and diorite) or high-grade metamorphic rocks (such as gneiss).

Vegetation. A third or more of the precipitation in heavily vegetated areas falls on the leafy canopies of trees and other vegetation and evaporates before reaching the ground. About a fourth lands on the ground but either evaporates from the ground surface, runs off as stream flow, or is temporarily stored at the surface in lakes or glaciers. The remaining precipitation seeps into the ground if surface conditions permit. Infiltration of the water that does enter the ground is enhanced by the plant and tree roots, which open pathways in soil into which the water can seep.

Topography. Surface water runs slowly down gently sloping terrains, so it has ample time to seep into the ground; but on steep slopes and cliffs, surface water runs rapidly downslope and into nearby streams before any large percentage of it infiltrates a groundwater system.

Precipitation. The amount of precipitation experienced over long- and short-term periods affects the amount of groundwater recharge in any kind of terrain. Extended droughts, for example, may curtail groundwater recharge significantly for several years or more. Shorter-term recharge variations, such as those from spring rains and snowmelt, generally replenish the groundwater supply seasonally. The type of precipitation also affects the degree of infiltration. A driving rainstorm packs surface soils and washes fine clayey particles into soil pores, clogging them and impeding infiltration. A series of drenching storms in close succession increasingly favors runoff because the first storm saturates the surface, and water from subsequent storms is unable to find pathways into the soils. Such was the case during the midwestern floods of 1993 (see Chapter 14). Snowmelt infiltrates the groundwater system to some extent, but it may also enter streams or evaporate.

Movement and Distribution of Groundwater

Water is drawn into the ground by gravity and moves to various depths depending on the properties of the soils, sediments, and rocks that it encounters (Fig. 15-3). As water passes through the region of weathered regolith, some of its molecules are attracted and bound to clay minerals in the soil and are used by plants. The attraction between water and the charged surfaces of clay minerals reduces both evaporation and infiltration from this near-surface region. Water that is not bound by clay minerals or lost to evaporation moves down through a **zone of aeration,** or unsaturated zone, where the pore spaces in rocks and soils contain both water and air. Some water may descend farther, into the **zone of saturation,** where every available pore space is filled with water. The **water table** is the boundary between the zones of aeration and saturation. The lower part of the aeration zone, which can range from a few tens of centimeters to several meters above the water table, is the **capillary fringe;** in this region, the effect of gravity is countered by the attraction of the water molecules to mineral surfaces and to other water molecules (which causes water in narrow channels to rise by *capillary action*) and pressure from the saturated zone below, each of which causes the water to move back *upward* from the water table into the unsaturated zone. The saturation zones in most places do not extend to great depths because the pressure of overlying rock forces deep fractures to close and some rock to recrystallize into impermeable metamorphic rocks. For these reasons most groundwater stays within the upper 1000 meters (0.6 mile) or so of continental crust.

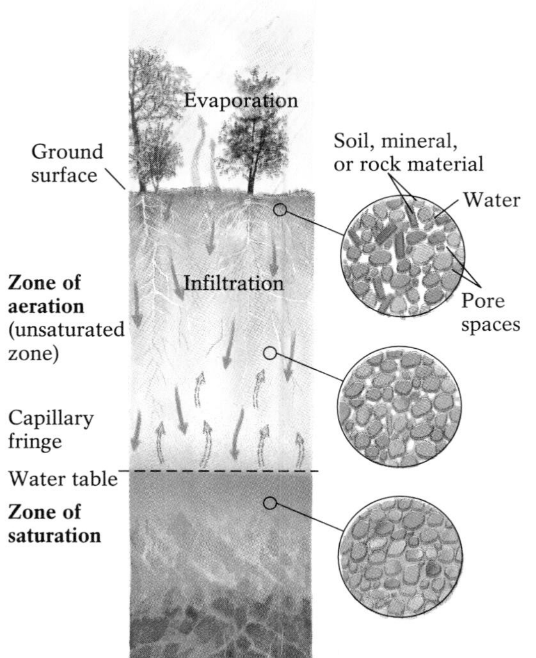

Figure 15-3 The subsurface distribution of groundwater. (See text.)

Porosity **Porosity** is the volume of pore space compared with the total volume of a soil, rock, or sediment. Porosity (expressed in percent) determines how much water the material can hold. *Primary* porosity is the porosity that develops when rocks form; an example is the vesicles or spaces that remain near the top of basalt after lava cools. *Secondary* porosity is that which develops after a rock has formed, usually as a result of faulting, fracturing, or dissolution.

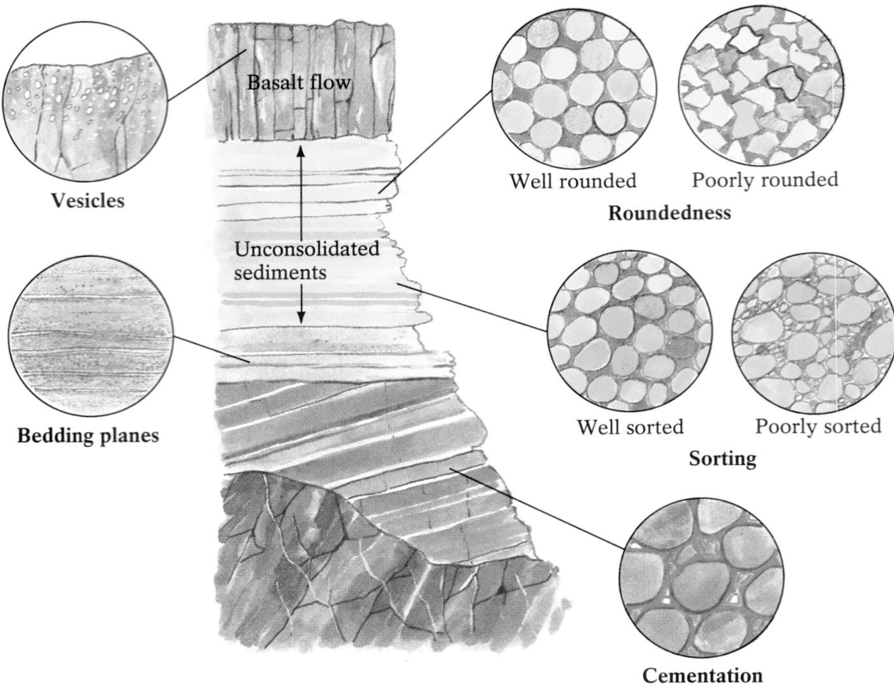

Figure 15-4 The primary porosity of sedimentary rocks is affected by the grain roundedness, sorting, and cementation of its particles. The presence of vesicles and bedding planes also increases a rock's primary porosity.

Primary porosity in sedimentary rocks is determined by such factors as the variation in grain size and shape and the degree to which the grains are cemented (Fig. 15-4). Well-rounded grains tend to have larger pore spaces and therefore hold more water. When a sediment contains grains of various sizes (is poorly sorted), the finer particles tend to move into the voids among the coarser particles, clogging the pores and reducing porosity. When loose sediments are converted to sedimentary rock by cementation, the cement fills the pore spaces and further reduces porosity. Fine clayey muds (though they impede groundwater recharge) hold much more water when saturated than coarse sediment because they contain a high percentage of minute pores. Fresh mud from the Mississippi River's delta south of New Orleans can consist of 80% water.

Permeability The crucial factor (other than the amount of infiltration) that determines the availability of groundwater is not how much water the ground holds, but whether the water can flow through the pores. **Permeability** is the capability of a substance to allow the passage of a fluid. Permeability of rock or sediment is determined by the size of its pore spaces and the extent to which they are connected. If pores are very small, as in clayey sediment, water molecules adhere as a fine film to adjacent particles and effectively prevent other water molecules from passing through. Only when pores are relatively large can water flow.

Even in a porous material the passage of groundwater is impeded if the pores are sealed off from one another. For example, the pores in a very porous limestone may not be connected, and so water does not flow through it. In other cases, a material may be essentially nonporous but still be permeable because it has a network of interconnected fractures that can provide a modest flow of water (Fig. 15-5). When such fractures are not joined throughout the rock body, the flow can never be large; a rock such as basalt, however, which often develops a highly interconnected fracture system as it cools, can usually serve as a reliable source of groundwater.

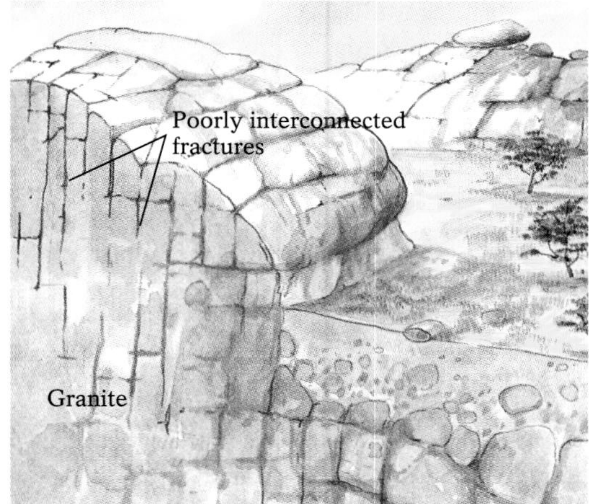

Figure 15-5 Pore connection and permeability. Although basalt is a largely nonporous igneous rock, this basaltic lava flow is highly permeable because of its connected network of columnar joints. Although the granitic bedrock is also fractured, its fractures are unconnected; thus this granitic outcrop is relatively impermeable.

Groundwater Potential and Flow Rate Surface water in a stream flows down its channel's slope under the influence of gravity, from higher to lower elevation. Groundwater also flows by gravity from high to low areas, but its flow is also driven by differences in pressure on the water from the weight of overlying water and the surrounding rocks. Geologists describe the **potential** of groundwater flow at any depth below the water table as a combination of the influence of gravity and the pressure on the water at that depth.

Groundwater generally flows from areas of high potential, such as high spots in water tables under hills, to areas of low potential, such as low spots in water tables under valleys. In some cases, groundwater compressed beneath the great weight of overlying rocks and water beneath a hill may even rise *against* gravity into low-potential zones, such as valley-floor lakes or river bottoms, which are under atmospheric pressure only.

Groundwater flows between any two points that differ in elevation and/or pressure. The flow rate is partially controlled by the **hydraulic gradient,** the difference in potential between two points divided by the lateral distance between the points (Fig. 15-6). Groundwater tends to flow most swiftly between two points when one has a much greater potential than the other and the distance between them is small.

The flow rate of groundwater is also affected by the nature of the material through which it flows. Every material has an intrinsic **hydraulic conductivity** that reflects, among other things, the sizes, shapes, and degree of sorting of its grains. Material such as coarse, well-rounded, well-sorted gravel has high hydraulic conductivity, whereas fine-grained, angular, poorly sorted sediment has low hydraulic conductivity.

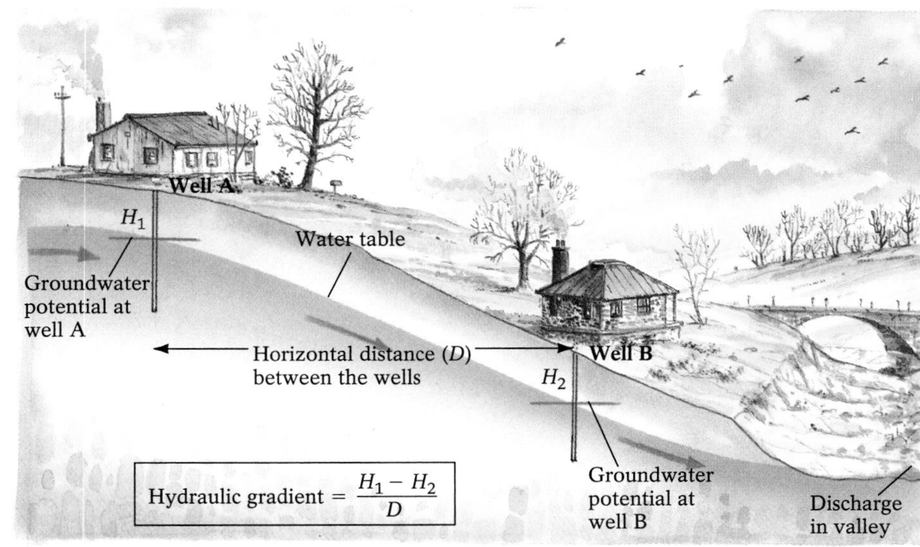

Figure 15-6 The hydraulic gradient is the difference in potential between two areas ($H_1 - H_2$) divided by the distance between them (D). Groundwater flows from areas of high potential (beneath hills) to areas of low potential (beneath valleys); flow is generally faster when the distance between the two areas is shorter.

The flow rate of groundwater was first calculated in the nineteenth century by Henry Darcy, a French hydraulic engineer. Darcy reasoned that the rate at which groundwater flows between two wells is directly proportional to the difference in potential between the wells and to the hydraulic conductivity of the materials through which the water flows. Thus, where the hydraulic gradient and the hydraulic conductivity are greater, groundwater flows more swiftly than where the hydraulic gradient and hydraulic conductivity are lower.

Geologists use two simple methods to measure the rate of groundwater flow at any location. When the groundwater flow path is known, they inject a harmless dye into a well and note the time of injection and the time the dye arrives at a nearby well downslope. Dividing the distance between the wells by the time elapsed yields the flow rate. Groundwater flow rates over longer distances can be determined by radiocarbon dating. (Carbon-14 dating is discussed in Chapter 8, "Telling Time Geologically.") Rainwater incorporates carbon-14 as it passes through the atmosphere and soils. As the groundwater moves slowly for a long distance underground, some of its carbon-14 undergoes radioactive decay. The distance between the groundwater recharge site and the well from which the water is withdrawn divided by the carbon-14 age of the water yields the flow rate.

Groundwater flow is extremely slow, averaging between 0.5 and 1.5 centimeters (0.2–0.6 inches) per day through moderately permeable material such as poorly sorted sand. Flow through unfractured crystalline rocks such as granites and gneisses may be only a few tens of centimeters per year. The swiftest flow, through well-sorted, coarse, uncemented gravels or highly fractured basalts, can reach 100 meters (330 feet) per day. In comparison, even sluggish streams generally flow at least 2 meters (6.6 feet) per second.

It is the slow movement of groundwater that accounts for its availability for human use. If groundwater flowed as rapidly as rivers, it would not be stored for long in the ground. However, because of its slow flow rate, the water we draw today from a deep well may have fallen as rain thousands of years ago. In the hot, dry parts of Arizona, radiocarbon dating indicates that the deep groundwater is more than 10,000 years old; it was precipitated during the last worldwide glaciation when Arizona's climate was cool and cloudy enough for rainwater to infiltrate the groundwater system instead of evaporating, as much of it does today.

The Water Table

Lakes, swamps, and year-round streams are places where the water table intersects the Earth's surface. Where there is no such surface water, the depth of an area's water table can be determined by digging a hole progressively deeper into the ground; the depth at which groundwater begins to seep into the hole—indicating that the surrounding material is saturated with water—marks the height of the local water table. In an area that receives significant rainfall every year, water may appear in a hole less than 1 meter (3.3 feet) deep. A hole dug on a beach usually begins to accumulate water at considerably less than a meter below the surface, whereas in an arid locale, water does not appear until a hole reaches tens of meters below the surface (Fig. 15-7).

The water table varies in depth according to local topography and prevailing climate. Its configuration often parallels topography, although its highs and lows are less pronounced than those of the landscape (Fig. 15-8). The

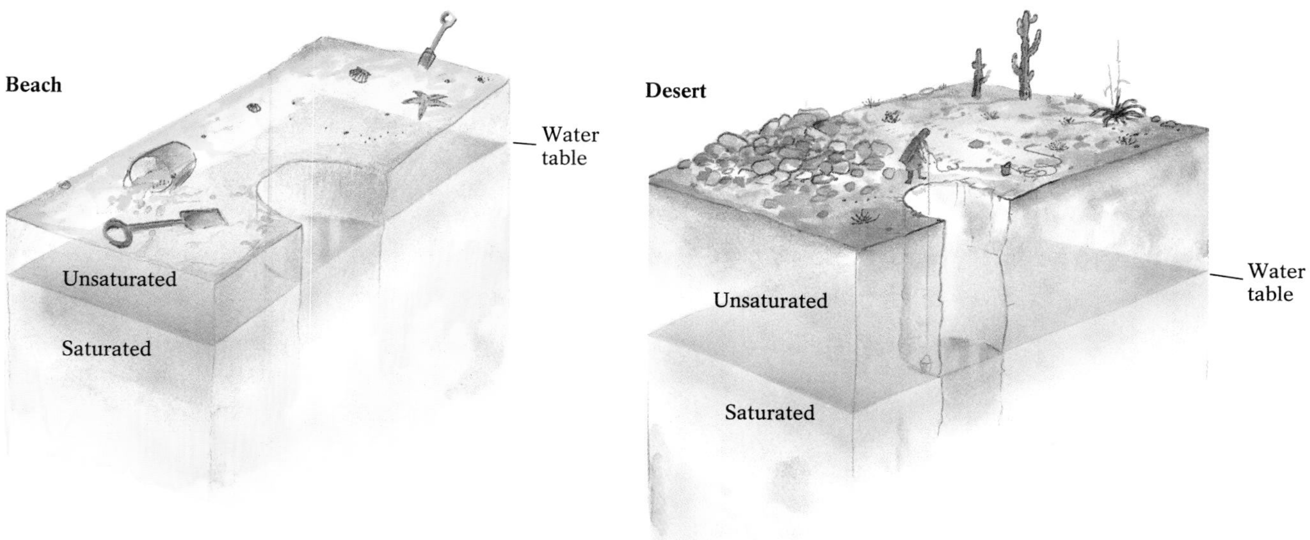

Figure 15-7 Variations in water table depth. Groundwater pools at the bottom of a shallow hole at the beach, where the water table is close to the surface. In the desert, where the water table lies far beneath the surface, the hole must be much deeper.

depth of the water table is generally established by a long-term balance between recharge and discharge, despite seasonal climatic fluctuations. (Water tables tend to be higher in winter and lower in summer.) Major climatic events, such as storms and droughts, can raise or lower the water table temporarily, as can variations in the amount of water we extract from the ground.

A prolonged dry spell, such as the one that struck the northeastern United States during the early 1960s, can lower the water table dramatically. During that drought, the region experienced a 2-meter (6.6-foot) lowering of the water table, and unscrupulous contractors constructed many homes in areas that normally have high groundwater levels. When the drought ended in the mid-1970s, recharge caused the water table to rise to normal levels, and many Long Island and New Jersey homeowners discovered that their basements were in the zone of saturation.

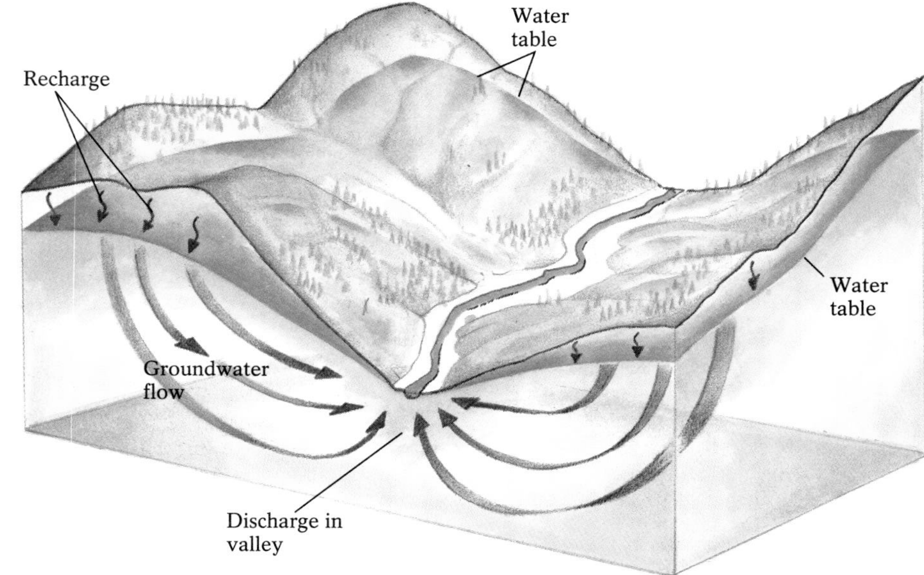

Figure 15-8 Water-table configuration and topography. When precipitation falls on an irregular landscape, the slopes receive the most rainfall because they constitute the largest surface area. The water that infiltrates the hills must then travel through a considerable expanse of soil and rock to reach a stream; it flows slowly, causing it to mound up below the surface of the hills. If groundwater recharge were to cease for a long time—for instance, during a drought—the water-table mounds would flatten out in the hilly areas as groundwater flow continued; in most humid areas, however, rainfall is frequent enough to replenish the supply beneath the hills and maintain the undulations of the water table.

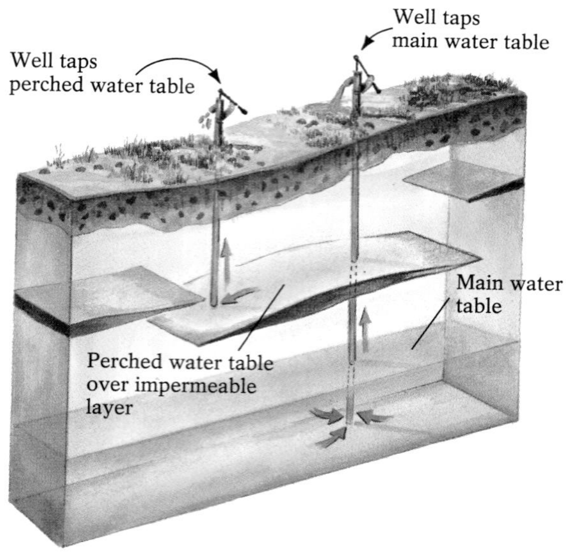

Figure 15-9 Perched water occurs in the zone of aeration when a local impermeable layer intercepts descending water. To insure a steady year-round flow of groundwater, wells must be drilled to below the main—not the perched—water table.

Sometimes a local impermeable layer (such as a lens of clay) obstructs downward flow of water through a region's zone of aeration and water accumulates above it, saturating part of the overlying zone. In this case, what appears to be the regional water table is actually a **perched water table,** a locally saturated area within the zone of aeration (Fig. 15-9).

Aquifers: Water-bearing Rock Units

An **aquifer** (from the Latin "to bear water") is a permeable body of earth material that transmits groundwater and also stores significant amounts. Aquifers, which occur in a variety of geological settings, are the sources of the groundwater we withdraw for our domestic, agricultural, and industrial needs. The most productive aquifers are composed of unconsolidated sand and gravel, well-sorted poorly cemented sandstones, or highly jointed limestones and basalts.

Unconfined aquifers lie very near the water table, with little or no overlying rock or sediment, and their water is usually at atmospheric pressure. Most local groundwater comes from unconfined aquifers made of loose slope material, sands, gravels, and floodplain deposits left by streams and rivers, sands and gravels transported by recent glaciers, and young, jointed lava flows such as those in Hawaii and the Pacific Northwest (Fig. 15-10). The Ogallala formation, a major unconfined aquifer, is a poorly sorted, generally uncemented mixture of clay, silt, sand, and gravel produced by a combination of these sources; a component of the High Plains regional aquifer system, it underlies much of Nebraska and parts of South Dakota, Wyoming, Colorado, New Mexico, Kansas, Oklahoma, and Texas. The Ogallala supplies 1.2 trillion liters (317 billion gallons) of water per year to irrigate 14 million naturally arid acres, yielding 25% of the U.S. feed-grain exports and 40% of flour and cotton exports.

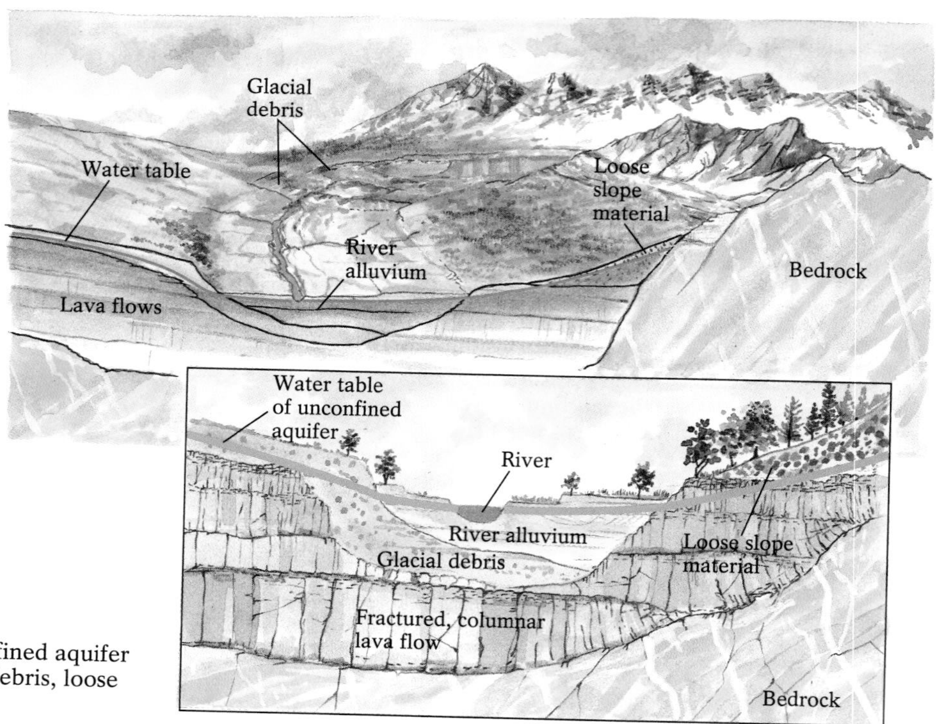

Figure 15-10 This composite landscape shows an unconfined aquifer composed of four different types of materials: glacial debris, loose slope material, jointed lava flows, and river alluvium.

Confined aquifers are sandwiched between rock layers that are either effectively impermeable or have very low permeability. Impermeable layers are called **aquicludes;** low-permeability layers are called **aquitards.** Unlike unconfined aquifers, which are usually local in range, confined aquifers exist over broad regional areas and typically lie at greater depths. They can even be found at great depths under deserts—such as in Nevada, eastern Utah, southern Arizona, and parts of southern Montana—where precipitation has entered the groundwater system by infiltrating exposed permeable rocks in mountain ranges, nearby or distant. The Dakota Sandstone, a confined aquifer whose water supply originates in the Black Hills of South Dakota, distributes water to the Great Plains of the eastern Dakotas and Nebraska.

Artesian Aquifers Under certain geological conditions, the water in a confined aquifer can rise against the downward pull of gravity, and may even gush from the ground. This occurs when the confined aquifer is tilted at an angle to the Earth's surface, so that it is exposed at the surface high in the mountains. Because the surrounding aquicludes prevent water from escaping, the weight of the water higher in the inclined water column presses on the water below it. When the aquifer is tapped, either naturally by a fault or artificially by a well, this pressure drives the water upward. This type of aquifer is referred to as **artesian** (Fig. 15-11)—named from the town of Artois, near Calais, France, where water has been flowing unaided from the ground for centuries. Artesian water tends to rise toward the elevation of the point of recharge at the surface, though friction between the water and the surrounding rocks and sediments constrains the tendency somewhat. The level to which the pressurized water would rise in the absence of friction is called the **potentiometric surface** of the aquifer. When the potentiometric surface is high enough above the Earth's surface, groundwater flows freely from the ground.

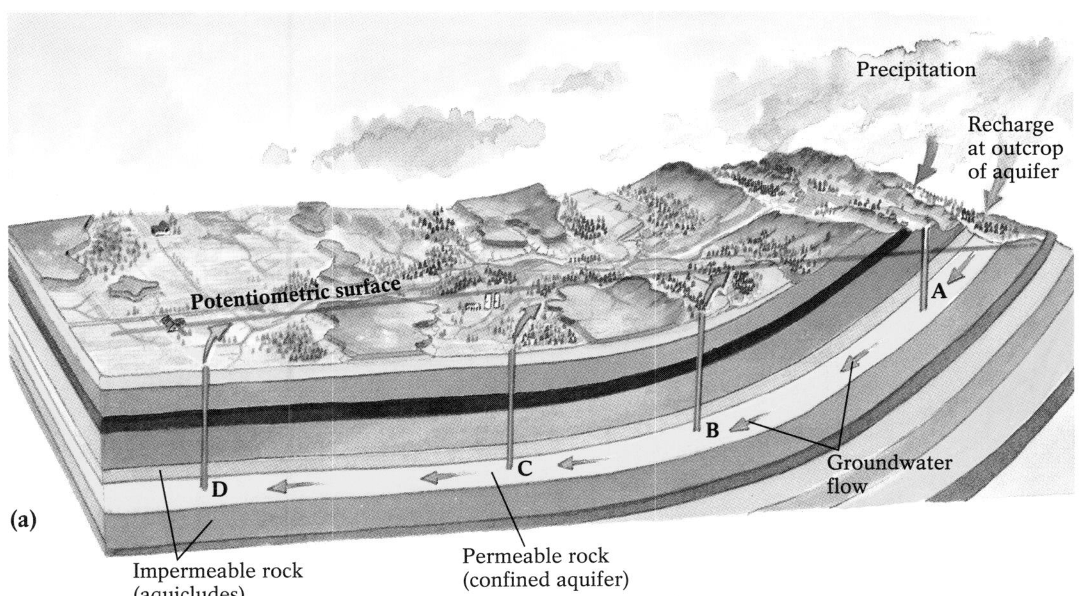

Figure 15-11 Artesian aquifers. (**a**) High pressure causes groundwater to rise in the wells at points B, C, and D, where the potentiometric surface is significantly higher than the well sites (see text). Water does not flow unaided from well A because it lies at the same level as the potentiometric surface. (**b**) Water gushing from a natural artesian well in the Dakota Sandstone aquifer (circa 1910). When this aquifer was first discovered and tapped, water gushed freely from wells under unusually high artesian pressures. During the last 75 years, more than 10,000 wells have been drilled to tap the Dakota Sandstone aquifer. Now, however, extensive withdrawal of water has reduced the pressure within the aquifer to the point that the potentiometric surface is no longer higher than the land surface of the eastern Dakotas. Today, water must be actively pumped where it once flowed forcefully and unassisted.

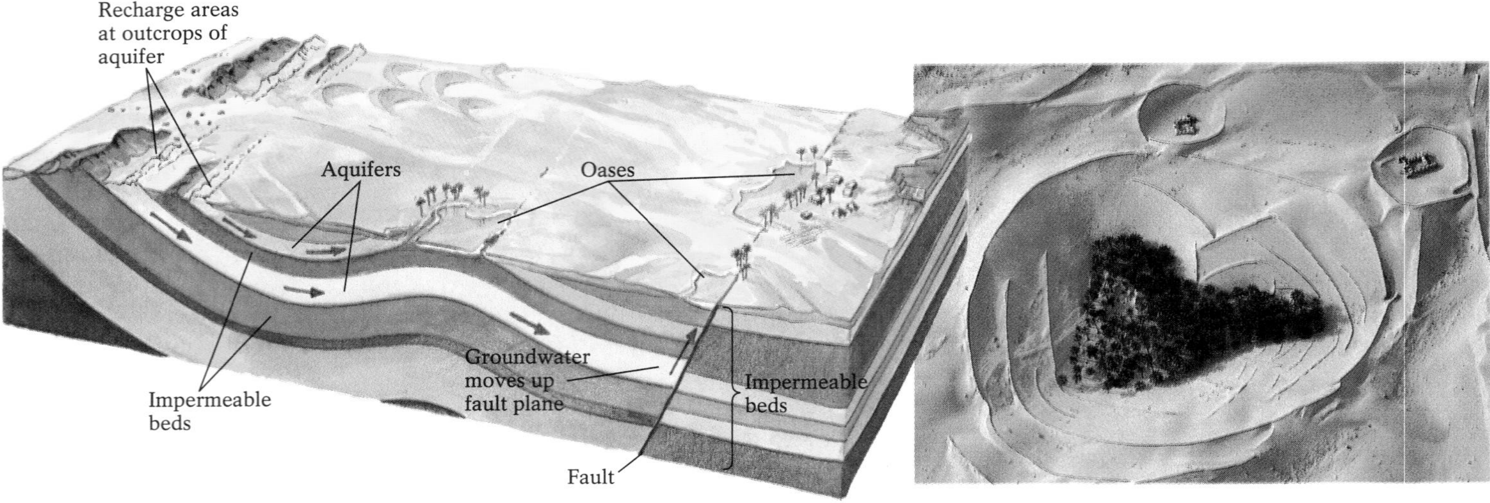

Figure 15-12 Desert oases. Oases occur in arid climates where artesian water rises to the surface, such as along a fault or anticline.

Water under artesian pressure can surface in unexpected places. Oases, for example, are pools of artesian water (Fig. 15-12). An aquifer that receives its recharge from mountains surrounding a desert may extend hundreds of meters beneath the arid surface. Water flows to the surface where the rocks of the aquifer crop out at the surface or where a fault extends down to the aquifer.

Most modern water supply systems are designed to simulate artesian conditions. Have you ever wondered why when you turn on a faucet in your home, the water flows out at high pressure? It is because the tall water towers in your city or town act like elevated recharge areas, and the municipal water pipes act like confined aquifers. The difference in elevation between the tower and your home causes the water to flow from your faucets under pressure (Fig. 15-13).

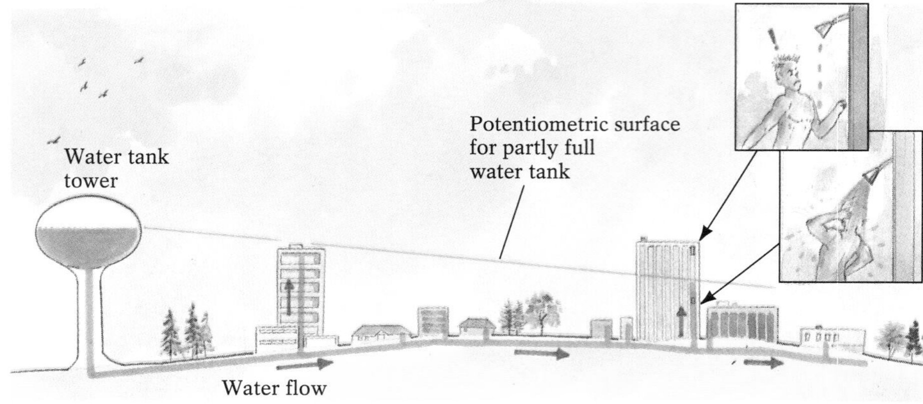

Figure 15-13 A municipal water tower functions in much the same way as the recharge area of an artesian aquifer. Water on the lower floor of a building rises under high pressure because its outlet is located below the potentiometric surface. On the top floor, water pressure is low because its outlet is located at a higher elevation than the potentiometric surface.

Natural Springs and Geysers

Natural springs are sites at which groundwater flows to the surface, usually laterally, and issues freely from the ground. These may develop where groundwater encounters an impermeable bed and is forced to flow around it, or where erosion or other processes have lowered the surface so that it intersects the water table (Fig. 15-14).

Groundwater is insulated from fluctuations in air temperature and therefore tends to approach the mean annual air temperature for the region.

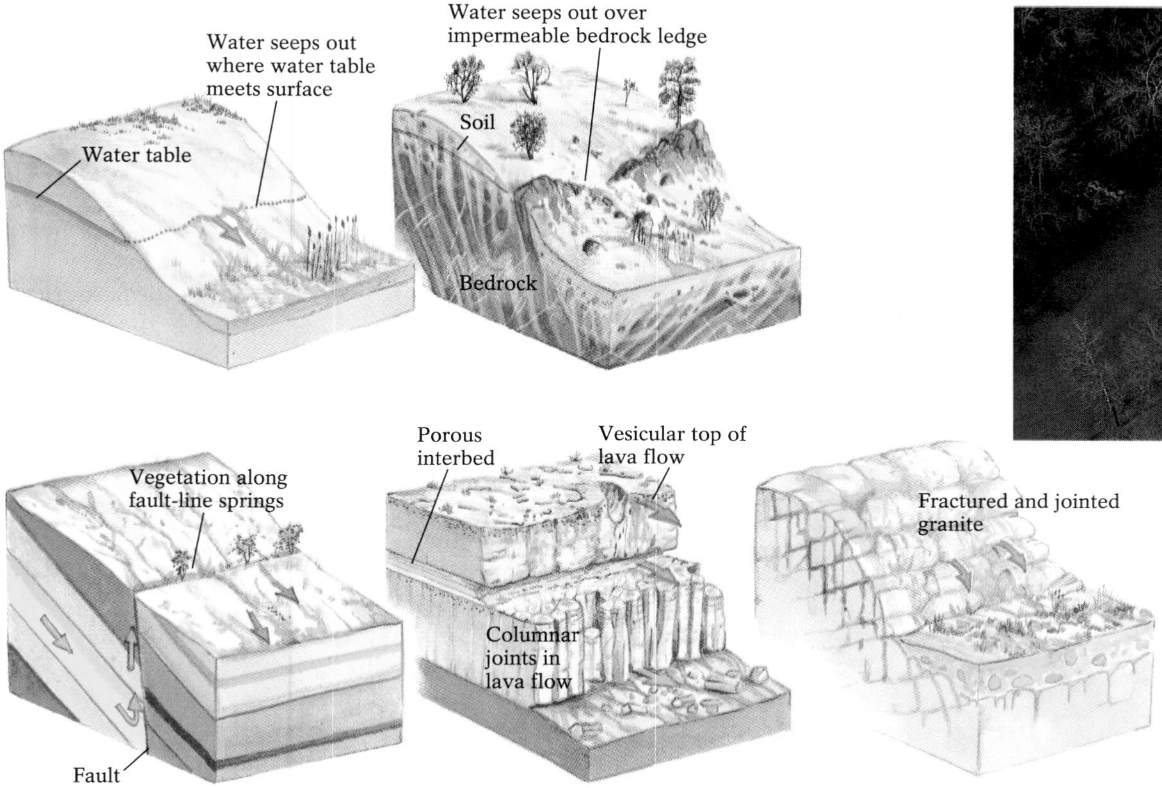

Water seeps out where water table meets surface

Water table

Water seeps out over impermeable bedrock ledge

Soil

Bedrock

Vegetation along fault-line springs

Fault

Porous interbed

Vesicular top of lava flow

Columnar joints in lava flow

Fractured and jointed granite

(a)

(b)

Figure 15-14 Natural springs. (**a**) Various geological settings in which natural springs occur. (**b**) Big Spring in the Missouri Ozarks, which may reach a flow rate of 2 billion liters (530 million gallons) of water per day, is a result of the local water table intersecting a valley wall.

A *thermal* (or hot) spring is a spring whose flow is at least 6°C (about 11°F) higher than this mean temperature. Most of the more than 1000 thermal springs in North America occur in the West, near the Rocky, Cascade, Olympic, and Sierra Nevada Mountains. The warmth of these springs' water is due primarily to recent encounters with magmas. After percolating downward for hundreds of meters, the groundwater reaches and is heated by warm igneous rocks, in some places to the boiling point. As the hot water rises, it usually dissolves large quantities of soluble rock. After it reaches the surface, the water evaporates, leaving behind the minerals as a deposit called sinter (Fig. 15-15).

(a)

(b)

Figure 15-15 Sinter deposited at Mammoth Hot Springs in Yellowstone National Park, Wyoming. (**a**) Here, hot groundwater has infiltrated and dissolved carbonate deposits and then precipitated them at the surface to create white cliffs of travertine. Sometimes the hot water incorporates sulfur on its way to the surface; when this happens, the spring emits the distinctive rotten-egg odor of sulfur dioxide, and the deposits have a yellowish tinge. (**b**) Steam vents, or *fumaroles*, in the sinter.

Most eastern hot springs, such as those of the Appalachians and Ouachitas, are not associated with hot igneous rocks. There, deeply circulating groundwater passes through rocks that are heated by the Earth's geothermal heat flow (see Chapter 11 for a discussion of the Earth's internal heat), hundreds of meters below the Earth's surface. For example, the water that flows from Warm Springs, about 90 kilometers (about 60 miles) south of Atlanta, Georgia, enters the Hollis formation at Pine Mountain at an average temperature of 16.5°C (about 62°F). After a subterranean journey of 3 kilometers (about 2 miles) along a curving path that descends to a depth of 1100 meters (3300 feet) under Pine Mountain, it emerges at Warm Springs under artesian conditions with a temperature of 36.5°C (about 98°F).

An intermittent surface emission of hot water and steam is called a **geyser,** from the Icelandic word *geysir,* "to rush forth." Geysers occur where groundwater that descends through extensive underground chambers or fractures is heated by an underlying source (a shallow magma chamber or a body of young warm igneous rock) and is then pushed up by steam under great pressure (Fig. 15-16). Pressure of the overlying water column inhibits molecular vibration in the deeper water, raising the boiling point above 100°C (212°F). The superheated water expands, pushing some water out of the geyser opening, reducing the pressure on the deep water, and lowering its boiling point. The deep water instantly flashes to steam, forcefully expelling all the water from the column. A period of recharging and reheating then occurs, followed by another eruption; the amount of time between eruptions varies with the groundwater supply, the permeability of the rock, and the complexity of the network of passages that delivers the water to the heat source.

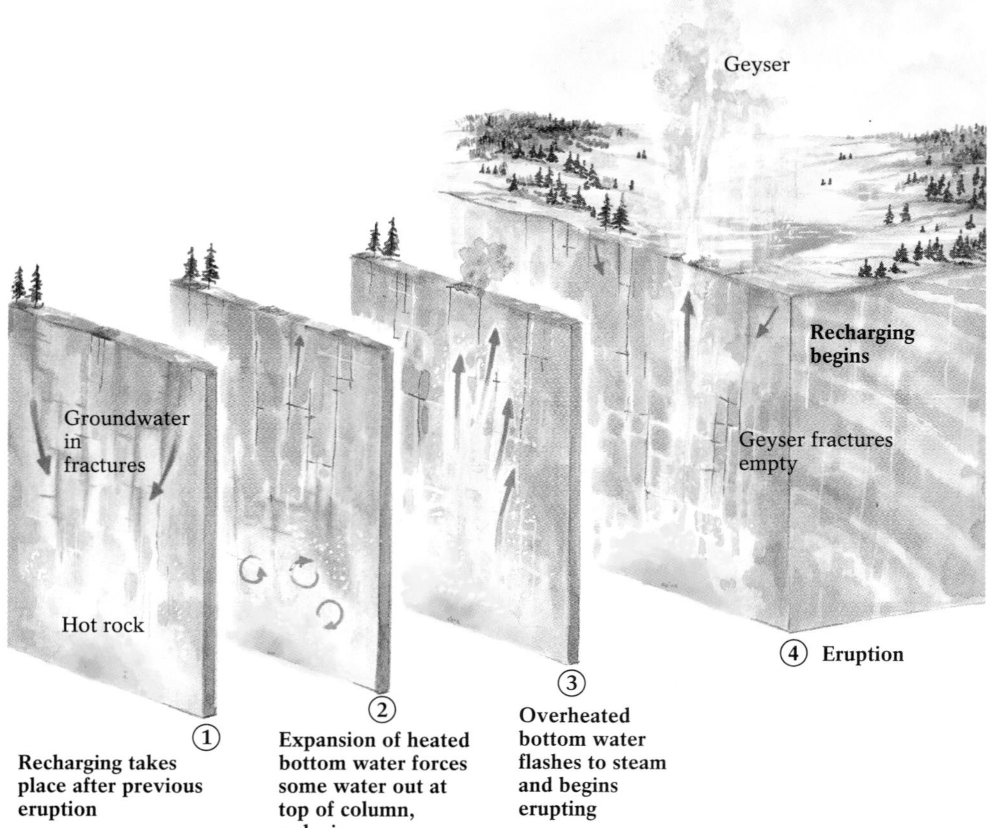

Figure 15-16 Geysers develop when groundwater encounters a shallow heat source and erupts at the surface as boiling water and steam. Water pressure at the bottom of the water column inhibits boiling initially, but as water near the top of the column is pushed out by expansion of the heated bottom water, the pressure on the column is reduced and the water instantaneously turns to steam. Photo: Strokkur geyser in Iceland.

The best-known geyser fields are in areas of current or recent volcanism. Volcanism at the divergent zone that bisects Iceland and at the subduction zone under New Zealand produces the geysers of those regions. In North America, geysers in northern California derive from the subduction that fuels the volcanism of Mount Shasta and Lassen Peak. The hot-spot volcanism of Yellowstone National Park powers the park's numerous geysers. Yellowstone's Old Faithful geyser, so named because of its remarkable punctuality, used to shoot a jet of steam and boiling water to a height of 50 meters (170 feet) every 65 minutes. Recent earthquake activity in the region, however, has altered the geyser's plumbing, affecting its recharge time and making it somewhat less faithful.

Many countries with hot springs and geysers are developing ways to use the geothermal energy to help meet human energy needs; we will discuss some of these in Chapter 20.

Finding and Managing Groundwater

To acquire groundwater in most areas, we dig wells that intersect the water table. But how do we decide where to dig? The best strategy for locating groundwater and digging productive wells begins with detailed knowledge of the configuration of the local water table (Fig. 15-17). Because every body of surface water represents an intersection with the water table, we should locate all lakes, year-round streams, natural springs, and swamps. A topographic map showing their locations provides a key to the underlying water table.

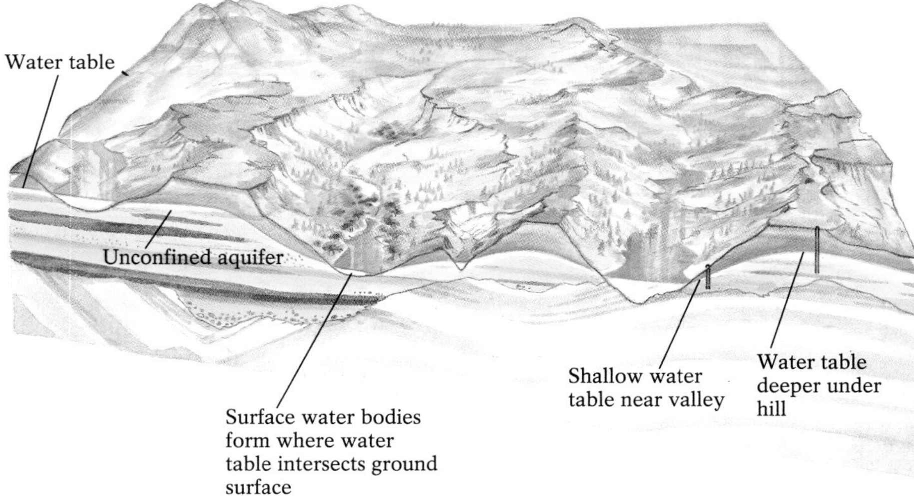

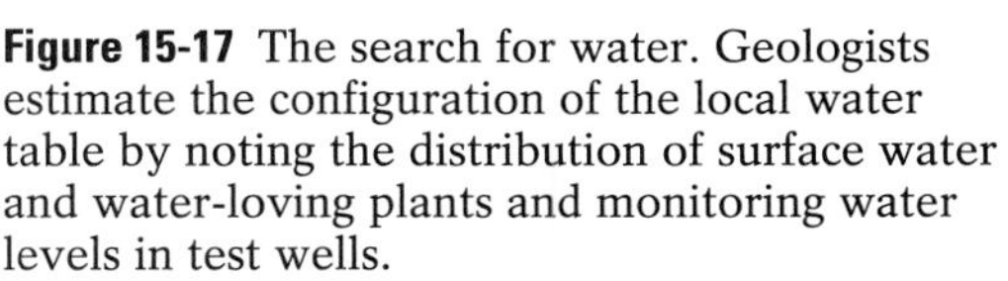
Figure 15-17 The search for water. Geologists estimate the configuration of the local water table by noting the distribution of surface water and water-loving plants and monitoring water levels in test wells.

An area's vegetation also provides clues to the depth of its water table. Some plant species have extensive, deep root systems for capturing water far below the surface, while others have shallow root systems and grow only where abundant water is close to the surface. The location of such plants yields a useful estimate of the depth of the water table.

Even with these clues, it may still be necessary to drill a series of small test wells to locate the optimal place to drill a water-producing well. Because the water in a well dug in an unconfined aquifer rises to the level of the water

Highlight 15-1 *Should You Hire a Water Witch?*

There are more than 25,000 water witches in the United States today—and a multitude of people who swear by them. What may be an early description of the art of water witching appears in the Judeo-Christian bible: "And Moses lifted up his hand, and with his rod he smote the rock twice; and water came forth abundantly, and the congregation drank and their beasts also."

Water witches walk an area with a forked stick (or a switch, rod, or wire) in hand until the stick seems to twist, dip, or jerk uncontrollably downward toward the place where underground water will be found. They claim that the presence of water initiates the behavior of the stick. Are these people frauds? Probably most are not. Many dowsers don't even charge for their services. But do they really find water? Skeptics suggest that the water witch manipulates the rod by holding it in delicate balance so that the slightest small-muscle movement turns it downward. Controlled studies have compared the success rate of water witches to chance and found no significant difference. One study in Australia compared geologists' efforts to find water with those of water witches and concluded that the witches caused twice as many dry holes to be dug as the scientists. At Iowa State University, water witches invited to find water along a prescribed course across the campus could not locate the water mains under their feet.

But water witches do often locate water. Geologists believe that most of these successes stem from the fact that in most humid climates water is virtually everywhere. Other successes may be due to years of experience at finding water in a specific region and to some geological common sense: Over the years, the witches, like most observers, would learn that wells dug in valleys produce water more often than those dug on hilltops, and that certain plants flourish where water lies near the surface. They may even acquire an intuition about groundwater flow and the relation between groundwater and permeable rocks.

Thus, as experienced water witches walk the landscape, they probably process the real information necessary to ascertain the presence of water. Perhaps when they notice the optimal conditions are met their muscles flex subconsciously, and the rod responds.

table, test wells show the depth of the water table at particular places. With these data and with the knowledge that the water table lies deeper under hills than under valleys, a reasonable estimate of the water table's configuration can be made.

Some people hire water witches to search for water. Also known as dowsers, these people claim to possess the ability to locate water with forked willow sticks or metal rods. Highlight 15-1 discusses whether it is possible that a water witch can really find water using these devices.

Threats to the Groundwater Supply

As soon as a well is drilled and groundwater withdrawn, issues arise regarding how to manage the water supply and deal with problems related to its use. Rural villages have modest groundwater needs and usually have been able to meet them by withdrawing water from shallow wells in unconfined aquifers or by tapping perched water. However, many cities have had to find new sources of uncontaminated groundwater, and residents in the suburbs have had to drill deeper wells to reach higher-yield confined aquifers. As urban areas continue to sprawl, incorporating both suburbs and rural areas, population growth, the arrival of new industries, and other changes in the local economy, including agriculture, have so increased the need for water that groundwater supplies in many areas are being severely overused.

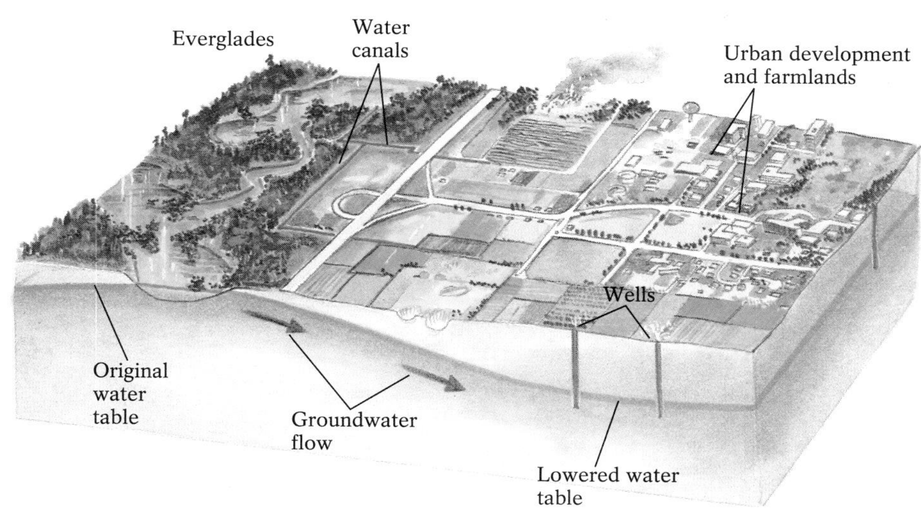

Figure 15-18 Overuse of groundwater. The water table beneath southern Florida used to intersect the land surface across a broad area, creating the Everglades. But recently, withdrawal of water from Florida's groundwater reservoir has lowered the water table and caused much of the Everglades to dry out. The lesson is clear: When the water table sits at the surface, it takes only a slight alteration in its height to cause far-reaching ecological disruption.

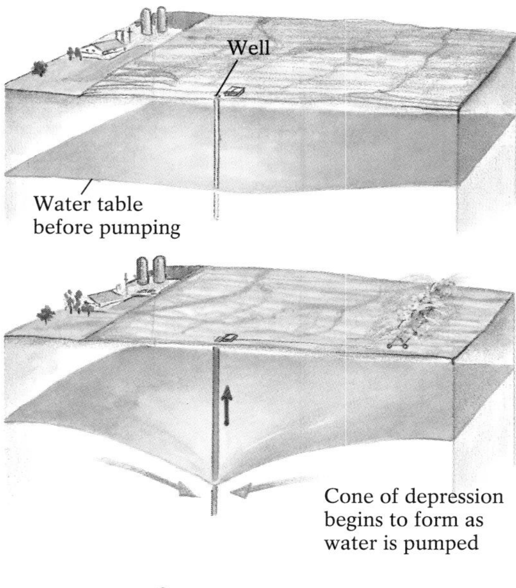

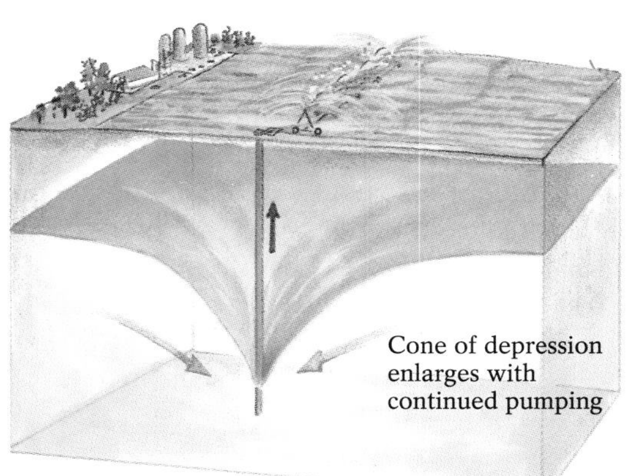

An example of the growing demand on groundwater systems can be seen in the imperiled Everglades of Florida. Although the Everglades appear to be broad areas of standing surface water, they are actually fed by groundwater that is flowing imperceptibly toward Florida's southern coast (Fig. 15-18). Increased pumping of groundwater due to rapid population growth and exportation of water from the Everglades to Florida's farmlands and coastal cities have greatly lowered the regional water table. Consequently, much of what was once a continuous wet, marshy surface is now dry much of the year, with forest fires smoldering for months in the dried-out vegetation and organic muck.

When groundwater withdrawal exceeds recharge, several major problems may result. The water table may become lowered over a growing geographic range as groundwater is depleted, land above the depleted water source may sink, or salty marine water may infiltrate the freshwater supplies.

Groundwater Depletion When water is withdrawn from a well, the water table in the vicinity is drawn down around the well, forming a **cone of depression** (Fig. 15-19). You can simulate this depression by inserting a drinking straw a few millimeters into a glass of water. When you begin to drink, the surface of the water is depressed around the straw; when you stop drinking, the water flows back immediately into the depression. Because groundwater flow is so slow, a cone of depression in a water table does not refill immediately. It can even become a permanent feature around a well as long as withdrawal continues.

Figure 15-19 The water table around a well is drawn down when water is pumped out, creating a cone of depression.

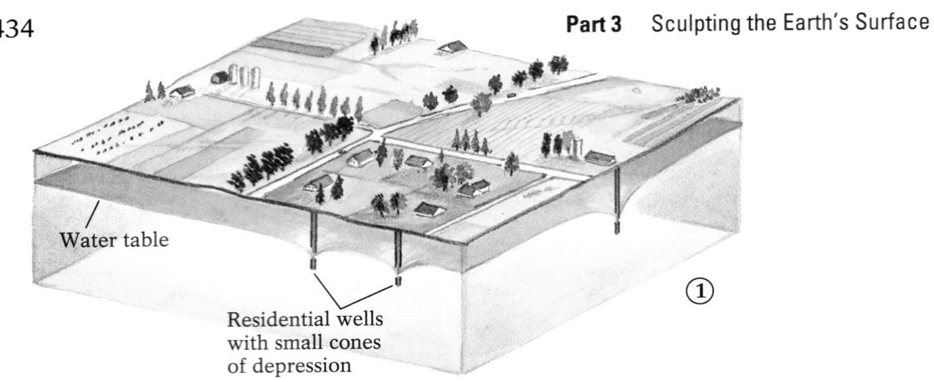

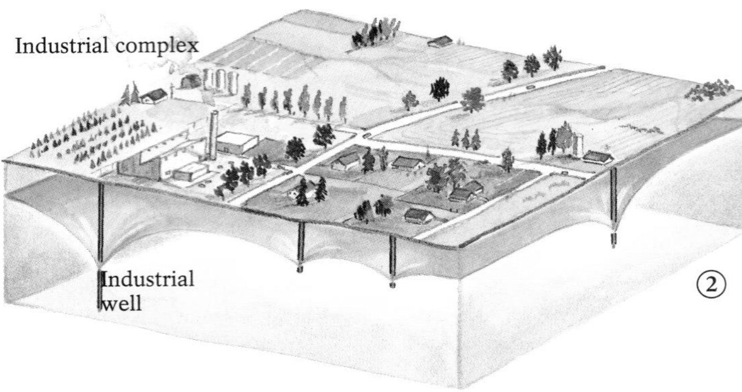

Figure 15-20 The effect of development on a water table. Stage 1: Community consists of farms and some suburban homes having small wells. Stage 2: Industrial development begins, replacing some of the farmland, and groundwater withdrawal increases. Stage 3: A large industrial complex covers much of the farmland, lowering the water table to below the depth reached by the residential wells.

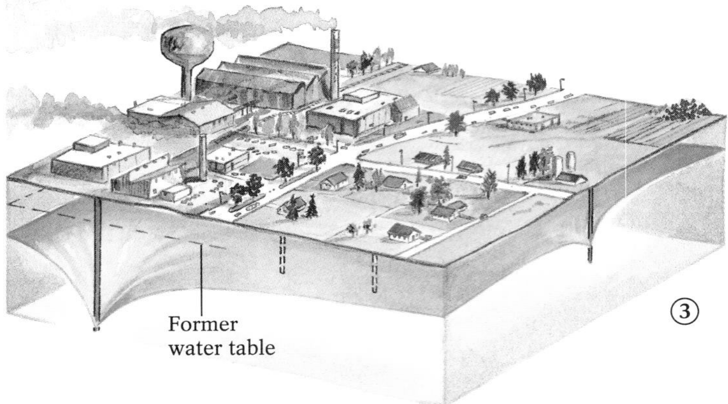

In moist temperate regions, cones of depression are insignificant when domestic use is moderate because recharge generally exceeds the modest rate of groundwater withdrawal. Where there is much irrigation or great industrial activity, however, large overlapping cones of depression can develop as withdrawal from one or more wells removes water on which adjacent wells depend. For example, the arrival of a water-intensive industry can create a cone of depression that lowers the water table over a sizeable area, leaving neighboring wells dry (Fig. 15-20). As a result, neighboring water users must drill deeper (and more expensive) wells to water below the cone of depression.

A reality of North America today is that, in many areas, the rate of groundwater recharge is too slow to keep pace with our increased need for groundwater. When groundwater is used at a rate that exceeds natural recharge, it is essentially being "mined"; eventually the water table falls below the bases of local unconfined aquifers. At that point, the aquifers become largely depleted. Groundwater mining is particularly serious in arid regions

Highlight 15-2 *Who Owns Groundwater?*

The United States has no uniform code for groundwater rights. Legal ownership of water is determined by state and local statutes and a few broad underlying principles. In most of the arid western states, a doctrine known as *prior appropriation* gives priority to the earliest users of groundwater, with latecomers having subordinate rights. With an accurate chronology of first use, there can be little dispute regarding rights to groundwater under this doctrine. During a shortage, the last to use groundwater is the first to lose it. In most of the eastern United States, the *riparian doctrine* gives all users equal (and virtually unrestricted) rights regardless of when their ownership began.

Can property owners legally exploit their groundwater to the detriment of their neighbors? In colonial America, water disputes were decided by the traditional English Rule of Capture, or Absolute Ownership, which stated that the water under the land is part of the land. This rule allowed any property owner to use as much water as she or he saw fit, even to the detriment of other people. Only if it could be shown that damage to others resulted from malicious intent could the excessive user be held liable.

In most states today, landowners can use water *only on the land from which it was taken,* and they are liable for any sale or transfer of water from their land that harms a neighbor's water supply. Also fundamental to the American rules of water use is the principle of "reasonable and beneficial use." Domestic use (within a household) has always been considered reasonable and beneficial, as has watering livestock in cattle-raising states, and as has irrigation in the arid West (but not where there is no compelling need to irrigate). As the nation became increasingly industrialized, manufacturing joined the list of reasonable uses. In California, the use of water to mine gold is still classified as reasonable (a relic of bygone days), as is the use of water for oil exploration in Texas and Louisiana.

In recent years, a few states, including California, have established a doctrine of correlative rights, by which each user's share of groundwater is proportional to her or his share of the overlying land. Again, the principle of reasonable and beneficial use is applied. Thus, the answer to the question "Who owns groundwater?" depends in some cases on the state in which the question is asked.

such as Phoenix, Arizona, where extensive irrigation and enormous losses to evaporation place an impossible demand on the groundwater reservoir.

The effect of a single overdrawn well on neighboring wells raises the important legal issue of who owns groundwater. Unlike other resources that stay put, such as metals and gems, groundwater flows from under one property to another. Our recent ability to determine the direction of groundwater flow enables us to estimate groundwater supplies. But even when we know how much there is, the question remains: Who owns it? Highlight 15-2 provides some answers.

Land Subsidence When more groundwater is withdrawn from an aquifer than is replenished by recharge, less water is present to help support the overlying load of rock, sediment, and soil. The weight of the overlying material pushes down, compressing the aquifer, and the land surface undergoes **subsidence.** Excessive withdrawal of groundwater has led to subsidence in New Orleans (more than 2 meters, or 7 feet, of subsidence), Las Vegas (more than 1 meter, or 3 feet), Mexico City (more than 7 meters, or 23 feet), and central California (more than 8 meters, or 26 feet). By displacing segments of the surface and subsurface, subsidence damages structures such as roads, buildings, and water and sewer lines.

Subsidence in coastal areas can cause flooding during high tides and storms. For example, Houston, Texas, and Venice, Italy, are both coastal cities built on unconsolidated deltaic deposits. Over years of domestic and industrial withdrawal of groundwater, these deposits and their aquifers have become compressed, causing the land surface to subside (Fig. 15-21). Venice,

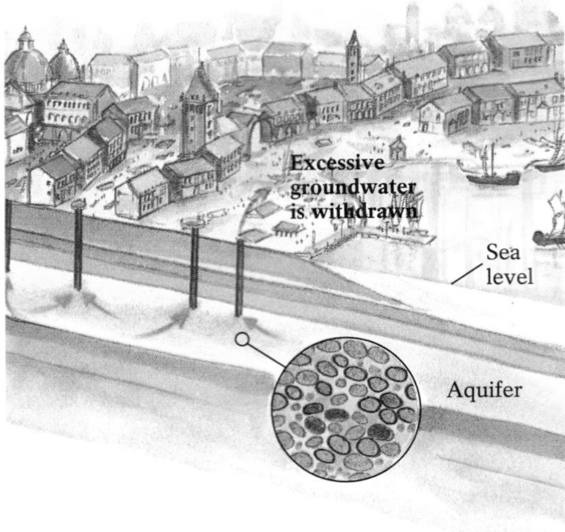

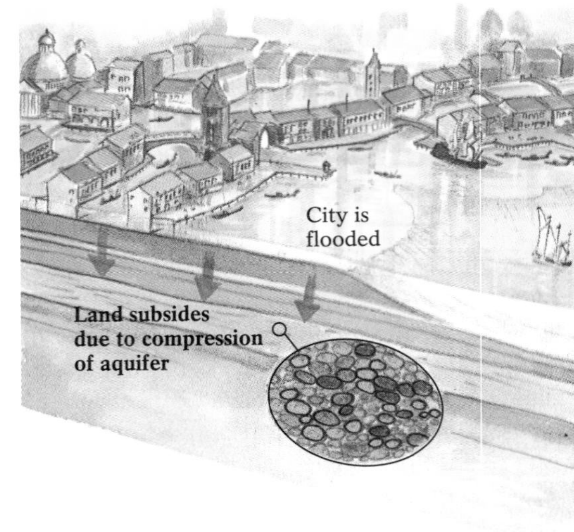

(a)

(b)

Figure 15-21 Subsidence in coastal areas. (**a**) Excessive groundwater withdrawal in a coastal city such as Venice, Italy, causes its aquifer to become compressed and the land to sink below sea level. (**b**) In Venice today, canals serve as streets. Recent efforts to repressurize Venice's aquifer through artificial recharge (pumping water into the aquifer) and to regulate the withdrawal of groundwater have slowed its subsidence rate somewhat, but at considerable cost.

which has subsided more than 3 meters (10 feet) during the last 1500 years, is plagued by frequent flooding from the Adriatic Sea, which damages the city's historic and architectural treasures. The Houston–Galveston area, one of North America's fastest-growing metropolises, has subsided 1 to 2 meters (3.3–6.6 feet) since 1906; this compounds flooding problems in a city that is battered by frequent hurricanes.

Saltwater Intrusion Aquifers in coastal communities, particularly those on islands and peninsulas, are threatened by the intrusion of salt water. As a coastal aquifer is infiltrated by precipitation on a nearby watershed, it may at the same time be infiltrated by salty marine water from the seaward side of the landmass. Because fresh water is less dense than salt water, and therefore floats above it, the freshwater–saltwater boundary is usually well defined. As long as groundwater withdrawal is not excessive, natural recharge provides a constant pool of fresh water above the salt water. However, any increase in withdrawal that disturbs the recharge–discharge balance allows more salt water to enter the aquifers, where it may eventually completely replace the fresh water (Fig. 15-22).

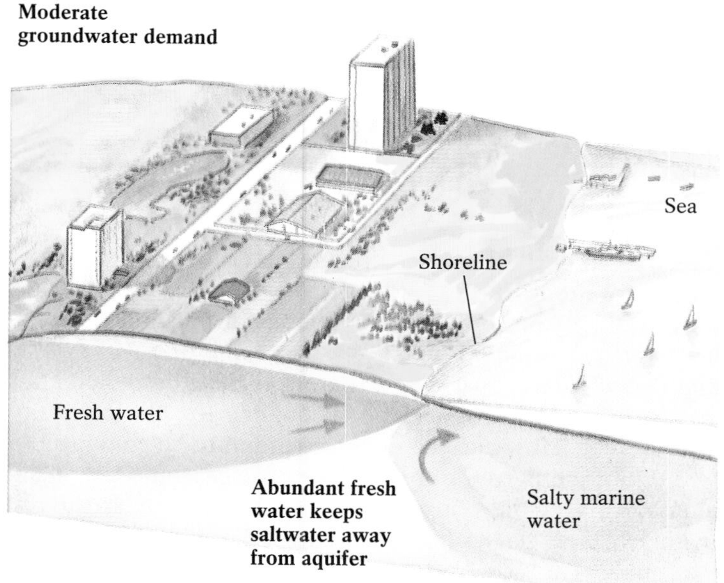

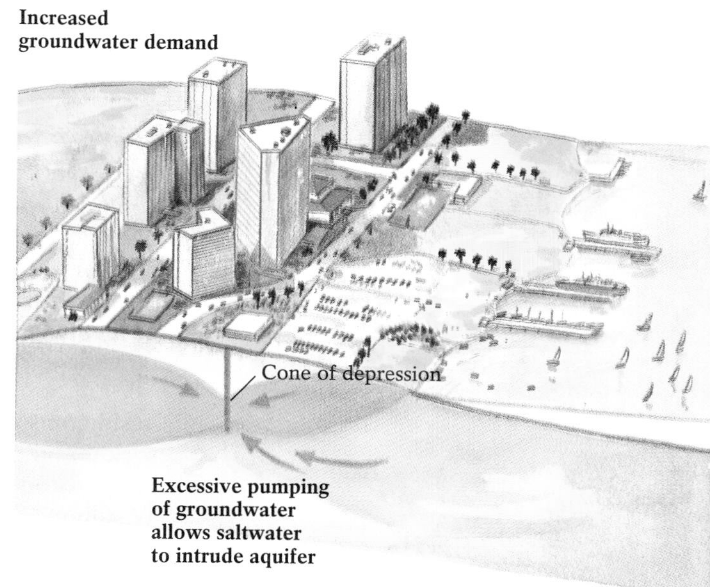

Figure 15-22 Saltwater intrusion of an aquifer in a coastal setting. Excessive withdrawal of groundwater allows marine water to penetrate aquifers at such sites, and saltwater sometimes flows from the faucets in coastal residences as a result.

During the 1930s, the groundwater demands of the 3 million residents of Brooklyn, New York, caused its freshwater aquifer to become an Atlantic Ocean saltwater aquifer. Since then Brooklyn has had to import its water from upstate New York, at considerable expense. In neighboring Nassau County, Long Island, New York, where there was rapid suburbanization in the 1950s and 1960s, many wells have been intruded by salt water; residents must now pay for their fresh water and for the new pipelines that supply it from the mainland. Coastal communities along the Gulf of Mexico, the Carolinas, Georgia, Florida, and California are similarly threatened.

Dealing with Groundwater Problems

The problems of groundwater depletion, subsidence, and saltwater invasion are being confronted throughout the world. Groundwater geology, or *hydrogeology,* which deals with these and other subsurface water problems, is one of the most rapidly growing fields in the earth sciences. Some remedies have been proposed, and a few have been successfully implemented. Most solutions enhance local recharge, reduce withdrawal and discharge, or develop alternative water sources.

Figure 15-23 A recharge basin/Little League field in Long Island, New York. To counteract high groundwater withdrawal rates, basins such as this one are excavated in order to collect precipitation and release it into the groundwater system gradually.

Enhanced Recharge Human intervention can supplement inadequate natural recharge in a community by means of projects that increase the local supply of water. For example, a steep slope can be graded (see Chapter 13) so that more water infiltrates the ground. In areas of moderate to heavy rainfall, open recharge basins can be constructed to hold rainwater for gradual release into the groundwater system (Fig. 15-23). In drier regions and where withdrawal consistently exceeds precipitation, water can be imported and pumped into the ground.

Fresh water is now commonly transported to replenish dwindling groundwater supplies. In and around densely populated Chicago, where natural recharge is considerable but still inadequate, fresh water from Lake Michigan is pumped into the ground. Surface water from the Sierras is transported by the Owens Valley aqueduct to the arid San Fernando Valley of Los Angeles, where it is injected into the groundwater system to maintain the water table. In Long Island, New York, waste water from industrial air conditioners is recaptured and returned to the ground.

Some communities that have exhausted their groundwater supplies borrow or buy groundwater from areas of abundance, an arrangement known as an *interbasin transfer*. But how can water be transported long distances without substantial losses through seepage and evaporation? Rapidly growing Denver receives much of its water through pipelines and tunnels that stretch from the western slopes of the Rockies. Similarly, New York City draws most of its municipal water supply from rural upstate regions through an extensive pipeline and reservoir system.

New York and Denver are primarily intrastate transfers; but when interbasin transfers cross state lines and international boundaries, the costs skyrocket and politics intervene. How can the donor community be convinced to relinquish its water? When southern California covets water from Oregon and Washington's Columbia River, the outcry in Portland can be heard in San Diego. The Great Lakes states have already declared the lakes off limits to thirsty southwestern states. And it will take remarkable international cooperation—and money—to convince the underpopulated, water-rich areas of Canada to provide water to high-demand areas in the United States and Mexico.

Conservation During dry spells, local authorities declare water emergencies. Typical emergency measures include prohibiting auto washing; limiting lawn watering to early morning or evening hours (to reduce evaporation) and to two or three times each week; serving water in restaurants only upon request; and turning off faucets while brushing teeth and shaving. Industrial conservation is also required, with stiff penalties for excessive use. Table 15-1, which shows how much water is consumed by some common household activities, illustrates how everyone can conserve water.

Table 15-1 **Water Use in an Average Household**

Activity	Volume of Water Use
Dishwashing	40 liters (10 gallons)
Toilet flushing	12 liters (3 gallons)
Showering	80–120 liters (20–30 gallons)
Bathing	120–160 liters (30–40 gallons)
Washing-machine load	80–120 liters (20–30 gallons)
Drinking and food preparation	4–8 liters (1–2 gallons)

Other Water Sources Another solution to water shortages may be *desalinization*, the removal of salt from seawater. Seawater can be distilled, which involves solar evaporation or boiling, followed by condensing and collecting the salt-free vapor, or it can be passed through a filter that removes the salts. Desalinization is costly, but in arid lands with extensive seacoasts, it can be cost-effective. Desalinization plants have operated for some time in Israel and Egypt, and there is hope that the process may become economically feasible in many other nations in the future (Fig. 15-24).

Recently, another means of providing a new water source was the subject of an international symposium at Iowa State University. Would it be feasible to tow massive icebergs from Antarctica to water-desperate countries in the arid Middle East? Because 90% of the ice would melt or evaporate in

Figure 15-24 A desalinization plant on the Netherlands Antilles island of Curaçao, off the northwest coast of Venezuela.

transit, researchers proposed that only icebergs at least 10 kilometers (6 miles) across should be considered. The cost and possible environmental repercussions of such a venture make it unlikely that the iceberg solution will be practical anytime in the foreseeable future.

Maintaining Groundwater Quality

Groundwater for drinking and some other human uses must be relatively pure. As we discussed in Chapter 5, because most groundwater is slightly acidic it dissolves the soluble components of rock and sediments through which it passes. Thus, groundwater naturally contains numerous dissolved ions—some benign, such as those that result from the dissolution of various carbonates and chlorides, and some harmful, such as arsenic, mercury, and selenium.

Groundwater, however, can also contain manufactured contaminants that have been introduced into the groundwater system in various ways. On a small, local scale, virtually every household periodically disposes of toxic substances into the municipal refuse system or storm drains. A check of typical household trash finds half-used cans of paint, cleansers, and solvents. Remember that can of bug spray you threw out, the burned-out light bulbs, and that broken thermometer? Such poisons and heavy metals leach from city dumps into the groundwater system, posing significant threat to groundwater quality and community health. The amount of these toxins, however, is minor compared with agricultural and industrial wastes or the by-products of medical research. Among the most threatening, because large quantities are often released, are insecticides and fertilizers, salt and other chemical ice-retardants washed from roadways, carcinogenic industrial by-products from lumber mills and factories (such as PCBs), biological wastes from cattle feedlots and slaughterhouses, and sewage from overworked septic fields and broken sewer lines.

Some human-made contaminants require extreme caution in handling and disposal. Nuclear waste at certain levels causes illnesses, birth defects, and death, and may remain dangerously radioactive for thousands of years. For that reason, when a disposal site for nuclear waste is finally selected, it must be absolutely impermeable so that there is absolutely no possibility of leakage to the groundwater system. We will examine this issue in greater detail in Chapter 20.

Pollutants enter the groundwater system when wastes are disposed of improperly, such as in receptacles that may rust and leak or when they are tossed haphazardly into poorly designed landfills or are dumped illegally on unmonitored land and allowed to seep directly into an aquifer. Groundwater contaminants are often detected only after they reach a well, by which time an entire aquifer may be affected. In western Minnesota in 1972, for example, a number of employees of a company became seriously ill from arsenic poisoning, even though arsenic had nothing to do with their occupations. Environmental detective work traced the arsenic to the company's new well. During the Dust Bowl years of the 1930s the western plains were struck by an infestation of locusts. To protect crops, farmers laced bran with arsenic and scattered it over their fields to kill the insects. In 1934, after the locusts were gone, the unused bran was buried. Forty years later a new well received recharge from the disposal site.

Once contaminants reach the groundwater reservoir, it may take many years to flush them out because groundwater flows so slowly. In the last few decades, local, state, and federal governments have enacted laws for monitoring groundwater quality, and environmental protection agencies regulate all varieties of waste disposal. The state of Montana, for example, strictly governs the sizes, materials, and construction methods used for septic tanks and drain fields.

Natural Groundwater Purification

Some aquifers are self-cleansing. For example, most deep groundwater passes slowly through rocks of low permeability and emerges quite pure. Natural purification involves three processes that combine to eliminate such toxins as sewage bacteria, larger viruses, and other suspended solids.

- Some contaminants adhere to clay particles as the water percolates through the soil.
- Other contaminants are completely decomposed by oxidation in the soil.
- Many different organic solids are consumed by various microorganisms.

Raw sewage that passes through soil, even if only about 30 meters (100 feet) of moderately permeable materials, is partially purified naturally. (This is actually the only treatment of sewage discharge from a septic tank or drain field.)

The cleansing ability of a soil, sediment, or rock depends primarily on whether its permeability is such that impurities in the recharge adhere to mineral surfaces. If an aquifer is too permeable, water passes through too quickly for contaminants to be removed, and if the pores are too large, contaminants pass right through them. The permeability of an unconfined aquifer can be estimated with a percolation test: digging a hole, filling it with water, and noting how long it takes for the water to seep from the hole. If the water vanishes quickly—for example, within minutes or even hours—the soil may be too permeable to cleanse contaminated water. In rural areas, building permits

Figure 15-25 Landfill sites. When geologists consider the selection of a geologically sensible site for a municipal landfill, especially in a humid climate with a shallow water table, they try to minimize the potential for groundwater contamination by avoiding areas where coarse, highly permeable gravels overlie highly permeable bedrock. Basalt flows with unseen lava tubes, limestone terrains with extensive cavern systems, and granites with complex and connected fracture systems would all promote the swift passage of polluted groundwater directly to your well. The ideal solution is to build the landfill on a thick, relatively impermeable layer of soil and sediment overlying impermeable bedrock.

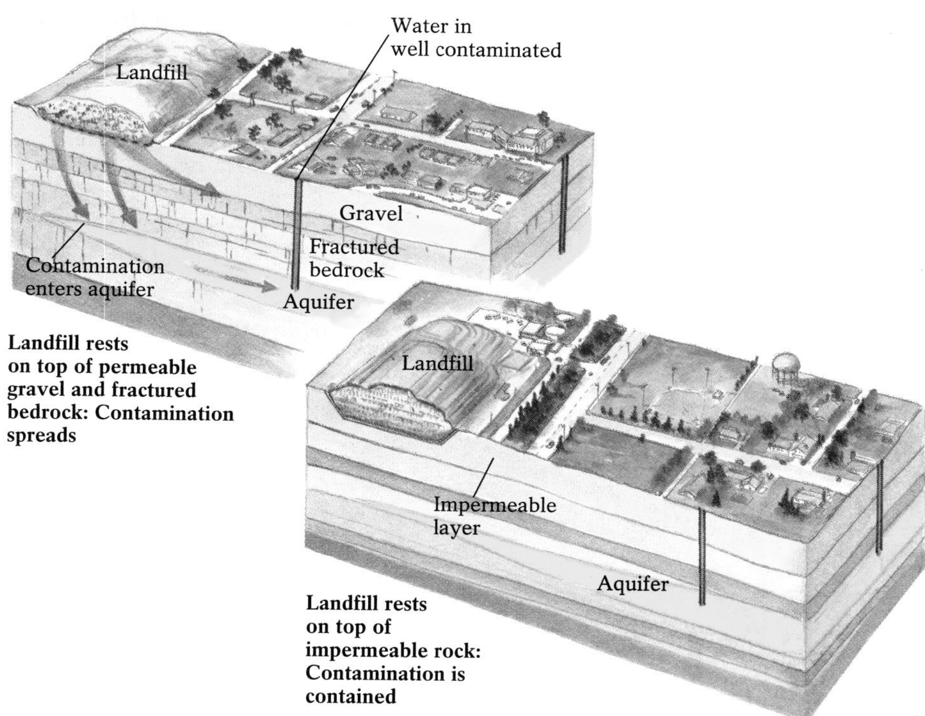

are not issued unless the municipal water-quality agency has done percolation tests and determined the soils to be suitable for household septic systems.

Some rock types remove contaminants less effectively than others. Certain bedrock aquifers, such as those containing large joints, cavities, and faults, allow water to pass through too swiftly to be cleansed. Consequently, major pollution sources such as sewage plants, landfills, and cattle feedlots should be located where fairly fine-grained impermeable soils overlie unfractured, low-permeability bedrock (Fig. 15-25).

Cleaning Contaminated Aquifers

Unfortunately, dissolved chemical pollutants in groundwater are generally unaffected by natural cleansing. When an aquifer is found to be polluted, the first step is to identify and, if possible, close down the pollution source. This can be done by taking samples from local streams and investigating the waste-disposal policies of local industries; the point at which contaminants enter the stream system and (perhaps) the culprit can be identified in this way. But if the source of the chemical pollutants was buried years earlier, or if the contaminants have entered a complex regional-groundwater network by illicit dumping some distance away, detection may be more difficult and time-consuming.

The second step is to discontinue groundwater use while the aquifer is being cleansed, assuming that another water source is available. Cleansing can take decades, given the average rate of groundwater flow. Groundwater systems can be cleansed in several ways. The tainted water can be pumped out, or fresh water can be pumped in until the contaminants are sufficiently diluted. Chemicals that either remove or neutralize the contaminants can be introduced into the system; however, this treatment is seldom feasible because groundwater flow is usually too slow for the agents to percolate throughout the system. Instead, contaminated groundwater is often pumped to a surface holding pond for treatment and then returned to the groundwater system. As this discussion has suggested, restoration of a polluted aquifer to good health is costly and technically demanding.

Groundwater is not only essential to the well-being of human communities, but it also accomplishes a considerable amount of geological "work" as it passes slowly through certain types of subsurface rock. In the next chapter, we will examine the principal geological product of groundwater flow—the caves that develop when naturally acidic groundwater percolates through soluble carbonate bedrock.

Chapter Summary

Surface water—flowing in rivers, standing in lakes, or falling as precipitation—enters the ground wherever topography, geological composition, and vegetation cover permit infiltration. The optimal conditions for development of a large reservoir of groundwater include a well-vegetated, gently sloping or near-level landscape composed of fractured bedrock or coarse well-sorted sediment.

Water enters the ground under the influence of gravity, which carries it downward in soil and rock until it fills every available connected pore space. A **zone of aeration** near the surface may not be water-filled because of surface evaporation; this zone contains many air-filled pores. Below it lies the **zone of saturation,** in which all available pore spaces are water-filled. The upper surface of the saturated zone is the **water table.** Immediately above the water table, within the zone of aeration, is the **capillary fringe,** into which water rises from the water table due to the molecular attraction of water molecules and the high pressure in the zone of saturation (compared with the zone of aeration).

Water flows underground when the geological materials are sufficiently porous and permeable. **Porosity,** the percentage of pore space relative to the total volume of soil, rock, or sediment, is a measure of how much water can be held by earth materials. **Permeability** is a measure of the ability of rock or sediment to transmit a fluid; the presence of connected fractures in solid bedrock and the coarse, well-sorted texture of unconsolidated sediment and soil increase permeability.

Groundwater flows between two points because of differences in **potential,** the energy that arises from the difference in elevation and water pressure between the points. The rate at which the potential changes for a lateral distance is called the **hydraulic gradient.** Because of their varying permeabilities, every material differs in its ability to transmit groundwater, a property known as **hydraulic conductivity.**

Regional water tables vary in depth according to local topography and prevailing climate. If a layer of impermeable material lies within the zone of aeration, infiltrated water may be retained locally, saturating part of the zone of aeration and forming a **perched water table.**

Aquifers are permeable, water-bearing bodies of geologic materials. Those found at the surface, such as within floodplain deposits and glacial gravels, are not overlain by impermeable cap rock and are called *unconfined* aquifers. *Confined* aquifers, found at greater depth, are sandwiched between impermeable rock layers called **aquicludes** or rock layers of low permeability called **aquitards.** When an aquifer's water is under high pressure from large elevation differences between recharge and discharge sites, the water rises above the level of the aquifer and gushes from the ground. Geologists describe such self-pumping aquifers as **artesian.** The level to which such pressurized water would theoretically rise, in the absence of friction, is the aquifer's **potentiometric surface.**

The regional water table can often be identified from the location of surface-water features such as rivers, lakes, and **natural springs,** places where the Earth's surface intersects the water table and groundwater flows out without human assistance. By drilling test wells, and with the knowledge that the water level in wells approximates the local groundwater table, geologists can estimate the location of the regional water table. Groundwater, particularly where it is converted to steam by contact with shallow magmas or warm, recently crystallized igneous rocks, may erupt from the ground as **geysers.**

Human activity can disturb the groundwater system through overwithdrawal and contamination. In areas experiencing rapid growth and where formerly rural land has become suburban or urban, increased population and industrial development cause groundwater demand to rise sharply, and the water table may drop significantly. Large **cones of depression,** local depressions in the water table, develop around wells. Lowered water tables make it necessary to dig deeper wells at higher costs. They also promote **subsidence** of the land surface as depleted aquifers become compressed.

In coastal regions, excessive groundwater withdrawal may cause salt water to infiltrate subsurface aquifers.

To compensate for critical groundwater shortages, local communities may transfer water from distant drainage basins and institute rigorous conservation plans. Concern about groundwater quality is also growing, as evidence of various pollutants, including hazardous and toxic wastes, appears in our water supply. Government agencies are paying increased attention to preventing aquifer contamination and eliminating existing sources of pollution.

Key Terms

zone of aeration (p. 421)

zone of saturation (p. 421)

water table (p. 421)

capillary fringe (p. 421)

porosity (p. 421)

permeability (p. 422)

potential (p. 423)

hydraulic gradient (p. 423)

hydraulic conductivity (p. 423)

perched water table (p. 426)

aquifer (p. 426)

aquiclude (p. 427)

aquitard (p. 427)

artesian (p. 427)

potentiometric surface (p. 427)

natural spring (p. 428)

geyser (p. 430)

cone of depression (p. 433)

subsidence (p. 435)

Questions for Review

1. Discuss three factors that affect groundwater recharge.

2. What is the fundamental difference between the zone of aeration and the zone of saturation?

3. What is the difference between primary and secondary porosity? List three factors that control a rock's primary porosity. Describe a rock that would have high permeability, but low porosity.

4. What is a water table? What is a perched water table?

5. What is an aquifer? Where would you look for an unconfined aquifer? A confined aquifer?

6. Draw a simple sketch showing the conditions necessary to create an artesian groundwater system.

7. What are some geological circumstances that might create a natural spring?

8. Explain why certain geysers erupt at regular time intervals.

9. Describe two negative consequences of unwise groundwater management in coastal settings.

10. Briefly describe two ways to enhance dwindling groundwater supplies. Briefly describe two ways in which groundwater might be purified.

For Further Thought

1. How might the Earth's groundwater be affected if the greenhouse effect raised global temperatures significantly? What might happen to the Earth's groundwater if there is continued reduction or loss of tropical rain forests?

2. Speculate about the role that groundwater might have played during the Earth's first billion years, when plate tectonic activity was probably more intense.

3. In the landscape illustrated in Figure 15-17 (p. 431), what are some features—other than those labeled—that might provide clues to the configuration of the water table? Where else would you decide to drill test wells?

4. How would you solve the groundwater problems facing the Florida Everglades?

5. Suppose you were placed in charge of insuring the groundwater needs of Los Angeles, California, for the twenty-first century. Name five problems you would face and the solutions you would propose.

How Limestone Is Dissolved

Caves

Karst Topography

***Highlight 16-1* Karst Landscapes in History**

Protecting Karst Environments

Figure 16-1 Cave deposits in Carlsbad Caverns, New Mexico.

Chapter 16

Caves and Karst

Caves are unique geological structures that form when groundwater, flowing unseen beneath the Earth's surface, slowly dissolves and gradually enlarges minute openings in soluble bedrock. The result is sometimes spectacular enough to attract tourists, artists, photographers, and scientists alike (Fig. 16-1). Why would anyone choose to spend time in a wet, dark, 50-centimeter (20-inch)-high crawlway hundreds of meters beneath the Earth's surface? Add to that the pungent aroma of ammonia from decaying bat droppings and the appeal of this popular recreational activity becomes even harder to understand. For most speleologists—the scientists who study caves—as well as for nonscientist "spelunkers," this compulsion is fueled by a sense of adventure and an eagerness to explore the unknown. The depths of the oceans and the probably countless numbers of undiscovered caves are the remaining major frontiers of geological exploration on Earth, and both have drawn increased attention in recent decades. As a result, our knowledge of these hitherto uncharted regions has expanded considerably.

The features that are created when groundwater dissolves soluble rocks underground or when surface water dissolves exposed soluble rocks are known collectively as **karst.** (The term *karst* comes from the Slavic *kars,* which means "a bleak, waterless place.") Beneath the surface, karst may be a single cavern or complex networks of caves, some with spectacular deposits of precipitated limestone. At the surface, karst may appear as towering monoliths or other novel landforms or as numerous circular depressions in the ground; sometimes streams flow briefly before plunging into one of these depressions and disappearing from the surface. In this chapter, we focus on the unique geology associated with soluble rocks, and pay special attention to the environmental sensitivity of this geological setting.

How Limestone Is Dissolved

Although almost all rocks can be dissolved under certain conditions, very few rocks are dissolved easily and rapidly. Gypsum and halite, which are deposited when bodies of water evaporate, are minerals that do dissolve readily in water (see Chapters 2 and 6); the caves near Las Vegas, Nevada, formed in gypsum deposits. But because of evaporites' limited distribution, and because in moist environments they dissolve completely, they seldom

Figure 16-2 This deposit of terra rosa in Indiana consists of the oxidized iron-rich clay that remains after the calcite fraction of limestone bedrock has dissolved.

produce karst. Limestone is by far the most abundant soluble rock, and it is the material that forms nearly all caves and karst. The mineral calcite is the most soluble component of limestone. A limestone, however, may consist of as little as 50% calcite, if it incorporated other components while it formed. In marine environments, for example, a common secondary component of limestone is fine-grained iron-rich clay from weathering of marine basalts. When the calcite in such impure limestone is completely dissolved, the clay becomes concentrated at the surface as a deep red soil. This *terra rosa* ("red earth") often overlies fresh, undissolved limestone (Fig. 16-2).

Limestone, insoluble in pure water, is dissolved readily in natural rainwater, which acquires carbon dioxide (CO_2) as it passes through the atmosphere and soil. (CO_2 in soils is produced by respiration of soil organisms and the decomposition of the remains of plants and other organisms.) Water and CO_2 combine to form carbonic acid (H_2CO_3), which attacks and dissolves limestone by the chemical-weathering process of carbonation (see Chapter 5).

Carbonic acid reacts with the calcite ($CaCO_3$) in limestone to produce calcium bicarbonate ($Ca(HCO_3)_2$). As the calcium bicarbonate is carried away in solution, voids develop in the bedrock. As groundwater containing calcium bicarbonate flows through unconsolidated sediment, some calcite may precipitate in the pore spaces between grains and act as cement to consolidate them, forming such common detrital sedimentary rocks as sandstone. Water containing dissolved calcium bicarbonate ("hard water") makes soap difficult to lather and leaves an unappealing residue on anything it washes. It also produces hard, white calcium carbonate deposits in pipes, water heaters, and tea kettles. Water softeners that exchange sodium ions for calcium ions can be used to solve hard-water problems in household water supplies. Most of the calcium bicarbonate produced by limestone dissolution, however, is carried out to sea, where marine organisms use it to form their shells and skeletons or it is eventually precipitated as inorganic limestone.

Limestone dissolution occurs most rapidly at and immediately below the Earth's surface. Carbonic acid is usually more concentrated here; moreover, farther underground, water is more likely to be saturated with dissolved calcium bicarbonate and therefore less able to dissolve additional calcium carbonate. For these reasons, most limestone dissolution occurs within a few hundred meters of the surface.

Another factor determining the rate of limestone dissolution is the extent of fracturing, jointing, and bedding. Bedrock dissolves more rapidly when it has extensive fractures or joints, which enable more water to infiltrate and produce greater surface area for dissolution; similarly, thin beds form caves more rapidly than thick beds because they tend to have more numerous and more closely spaced joints and fractures. In time, as water continues to infiltrate joint, fracture, or bedding-plane networks, what begins as a barely perceptible pattern of minute cracks can become an intricately branched network of caves and connecting passageways.

Climate and topography also affect dissolution rates. Climate determines precipitation and temperature, and thus the amount of unfrozen water available to dissolve rock. Climate also promotes or retards some secondary factors that affect dissolution rates; abundant rainfall and warm temperatures, for example, support the growth of lush vegetation, which, as it decays, adds substantial amounts of acid-forming CO_2 to soil. The effect of topography can be seen on moderately sloped terrains where regional groundwater flows readily from hills to valleys, enhancing the rate of calcium carbonate dissolution. On steep slopes, water runs off too rapidly to infiltrate and limestone dissolution is inhibited.

(a)

(b)

Figure 16-3 Differential dissolution. (**a**) Less-soluble fossils project above a partially dissolved limestone surface. (**b**) The height of the limestone pedestals beneath their protective sandstone boulders is a measure of the local dissolution rates in Yorkshire, England.

Karst is unlikely to develop in polar regions, where little unfrozen water is available to dissolve soluble rock and the thick permafrost layer is a barrier to water infiltration. In addition, the thin soil and regolith of polar regions generally contain little CO_2 because of sparse vegetation and the resulting lack of plant decay. Karst is also unlikely to develop in hot desert regions, where the lack of water and sparse vegetation preserve rather than dissolve limestone. Caves found in these environments are usually relics of ancient times, indicating a past period of greater precipitation and more temperate climate.

Because the many factors that control calcium carbonate dissolution vary enormously, dissolution rates are most accurately estimated on a very small scale. For example, in a churchyard cemetery in cool, damp England, gravestones cut from fossil-rich limestone dissolve at a rate of about 5 millimeters (0.2 inches) per 100 years. Fossil sea lily stems in these rocks are more resistant to dissolution than the surrounding fine-grained limestone, and so the rate of dissolution can be determined by the extent to which the fossils project above the gravestone's surface (Fig. 16-3). By knowing the year a particular gravestone was cut, and measuring the difference in relief of the fossils and the stone's surface, we can estimate the rate at which the limestone is dissolving (Fig. 16-3a).

On a somewhat larger scale, a similar approach can be used to determine some regional rates of dissolution. Yorkshire, England, was covered by advancing ice sheets until about 12,000 years ago. As the glaciers melted, they left behind large boulders of relatively insoluble sandstone that protected the limestone surface immediately beneath them. The unprotected surface dissolved and the protected areas became 50-centimeter (20-inch)-high pedestals on which the boulders now rest (Fig. 16-3b). The dissolution rate of this limestone is approximately 4.2 millimeters (0.17 inches) per 100 years (50 centimeters divided by 12,000 years), almost the rate determined from the gravestone fossils about 100 kilometers (60 miles) away.

Another means of determining limestone dissolution rates relies on a micro-erosion meter that can directly measure surface lowering to within 0.0005 millimeters. Measurements in arid regions of Australia indicate a dissolution rate between 1.6 and 2.9 millimeters (0.63–0.11 inches) per 100 years. In one of Australia's temperate regions, with high precipitation and extensive vegetation, measured dissolution rates have approached 10 millimeters (0.39 inches) per 100 years—a rate as fast as or faster than those of some other surface processes, such as landsliding and stream cutting.

Dissolution rates can even be estimated by measuring the volume of solutes in discharging groundwater. In many areas today, limestone is steadily dissolving in minute cracks below the water table. One such site is Big Spring in the Missouri Ozarks near Poplar Bluff (see Fig. 15-14), where millions of liters of groundwater emerge at the surface every day, carrying 190 metric tons of dissolved limestone into the nearby Current River. This large quantity of dissolved material clearly indicates that large underground voids must be forming.

Caves

Caves are natural underground cavities, the below-surface expression of karst. The most common geological products of limestone dissolution, caves have always held a special fascination for humankind. They were widely used for shelter or ceremonial purposes throughout southern Europe tens of thou-

Figure 16-4 This famous panorama of bulls, horses, and other creatures was painted on a cave wall in Lascaux, France, some 15,000 years ago. The cave in which it was found may have been a ceremonial site for performing rituals designed to ensure successful hunts. The prehistoric painters crushed ochre and hematite to produce yellow, red, and brown powdered pigments; crushed manganese yielded black, dark brown, and violet. The pigments were applied directly to the damp limestone cave walls and ceilings.

sands of years ago (Fig. 16-4). Greek mythology tells us that Zeus was born in a cave. The Japanese sun goddess Amaterasu sought refuge in a cave, plunging the world into darkness. In Anglican lore, King Arthur and his knights and their hounds are believed to be still slumbering in a Welsh cave, waiting to be summoned into battle once again.

Information yielded by caves is invaluable to anthropologists, biologists, archaeologists, geologists, and other scientists. Caves contain records of ice-age climate changes and human physical and cultural development. They harbor the remains of extinct animals such as the cave bear, the woolly mammoth, and other organisms that evolved as cave dwellers, enabling us to study how species evolve and adapt to unique environmental conditions.

Almost every state and province in North America contains some type of cave, and geologists believe that more than half the caves in North America have not yet been discovered. There are major systems in the Black Hills of South Dakota; in northern Florida; in the mountains of Montana and Wyoming; in the Guadalupe Mountains of New Mexico and Texas; in the Appalachians of Pennsylvania, New York, Maryland, Virginia, West Virginia, Tennessee, and Alabama; in the Ozarks of Arkansas, Missouri, and Oklahoma; and in the Great Lakes region of Indiana, Illinois, Iowa, and Minnesota. In central and eastern Canada and much of New England, caves tend to be relatively smaller and less extensive than elsewhere in North America because the bedrock in these regions, the continent's ancient crystalline craton (see Chapter 12), contains little limestone. Moreover, these northerly regions were glaciated for much of the past 2 million years, and have had only about 10,000 years for cave development since the ice melted.

Some of the continent's most spectacular caves have been designated as national parks and monuments to be preserved for future generations. Others, however, have been treated in mercenary, and often destructive, ways. Large quantities of delicate and rare cave crystals have been hacked out by amateur collectors. Caves have been dynamited to create more commercially accessible entrances. In Kentucky, gypsum crystals from caves have been mined for use as paint pigments. However, most of North America's caves have survived, largely because their relative inaccessibility protects them from human disturbance.

Cave Formation

The early stages of cave development progress slowly, because minute cracks in limestone are initially too small to allow significant water infiltration. Thousands of years may pass before small amounts of percolating water dissolve enough bedrock to enlarge the cracks significantly. As the cracks grow larger, more water infiltrates and flows more vigorously, vastly accelerating the rate of dissolution. The enlarging underground rivulets capture water from narrower cracks; eventually a few primary passages dominate the underground drainage, becoming the main caverns and connecting passageways of a growing cave system.

When caves first begin to form, dissolution occurs primarily in the zone of saturation, probably at or just below the water table. There, groundwater combines with inflowing CO_2-rich water that is not yet saturated with solutes and can readily dissolve limestone. This period of intense dissolution along joints, fractures, and bedding planes below the water table forms the intricate honeycomb pattern of many major cave systems (Fig. 16-5a).

The second stage of cave formation occurs if the water table drops below its original level, leaving caves high and relatively dry above it. At this point, the cave environment becomes one of open air (Fig. 16-5b). With the

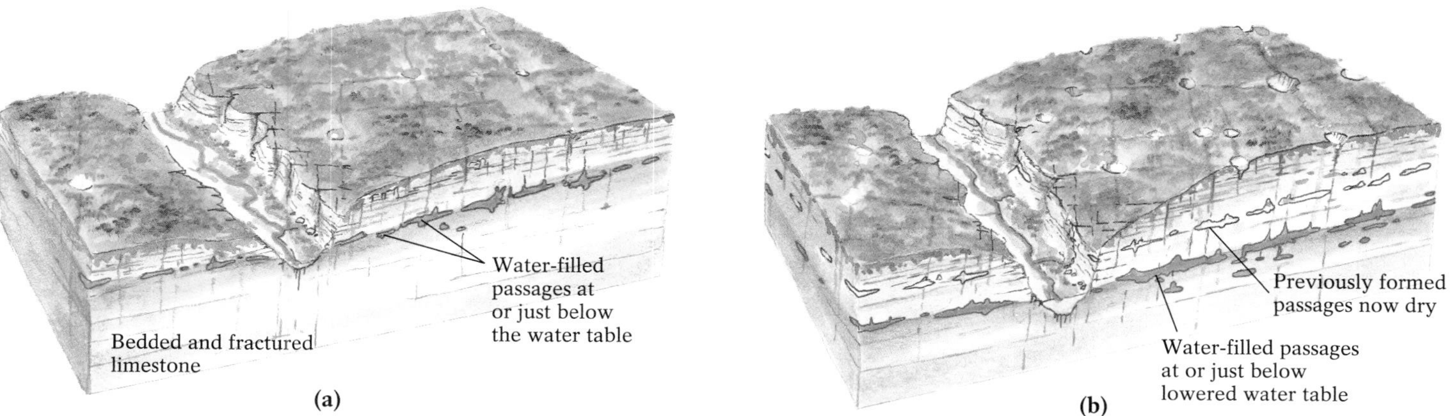

Figure 16-5 The two-stage process of cave formation. (**a**) Stage one: Acidified groundwater that fills all joints, fractures, and bedding planes below the water table dissolves limestone and forms large caverns and connecting passageways. (**b**) Stage two: The water table drops below the caverns and passageways, leaving them in an open-air environment.

Figure 16-6 An underground river carved this deep channel by dissolution and abrasion.

water table lowered, erosion by fast-flowing subterranean streams may begin to carve potholes and canyon-like slots into a cave's bedrock floor (Fig. 16-6). Because the roof of such a cave is no longer supported by water, it may collapse. A cave-roof breakdown in Indiana's large Wyandotte Cave left a 40-meter (130-foot)-high pile of rubble on the cave floor. Cave structures can also be damaged by earthquakes. During the New Madrid earthquake of 1811 in Missouri, Mammoth Cave—300 kilometers (200 miles) away in Kentucky—experienced not only swaying cave walls but also broken roofs that rained down on terrified workers in its saltpeter mines.

Cave Deposits

Once a network of cave chambers and connecting passageways has formed and the water table has dropped below the base of the system, water percolating downward from the surface enters its open-air spaces. Because this water is generally rich in calcium bicarbonate, two things can happen: 1) in the open-air environment of the cave, some water can evaporate, increasing the concentration of calcium bicarbonate in the remaining water until it becomes saturated; and 2) much of the CO_2 dissolved in the water can escape into the air in the cave, reversing the carbonation process and causing carbonic acid to be converted into CO_2 and water (see Chapter 5). The high $Ca(HCO_3)_2$ concentrations and the CO_2 loss cause $CaCO_3$ to precipitate from solution, becoming deposited on cave surfaces and building up a variety of formations known as **speleothems.** Speleothems consist largely of the rock **travertine**, the name applied to calcium carbonate when it forms cave deposits.

Stalactites and Stalagmites **Stalactites** are stony travertine structures, resembling icicles, that hang from cave ceilings. Their name is derived from the Greek *stalaktos,* which means "oozing out in drops." Stalactites form as successive drops of water, one at a time, fall from a crack or joint and enter a cave. As the water evaporates or loses carbon dioxide, it leaves behind a trace of calcium carbonate. Initially, the center of each growing stalactite is a hollow tube, like a soda straw, through which the next drop enters the cave. Eventually precipitated travertine clogs the tube, and the drops of water are forced to find an alternate route into the cave, usually at the top or side of the stalactite. As the water drips down the stalactite's outer surface, it deposits more travertine and creates the irregular shape of a typical stalactite.

Stalagmites are the travertine deposits that accumulate on the floors of caves from water that drips from their ceilings. Their name is derived from the Greek *stalagmos,* or "dripping." Some of the water that drips from joints and cracks falls onto the cave floor, where it evaporates and releases CO_2,

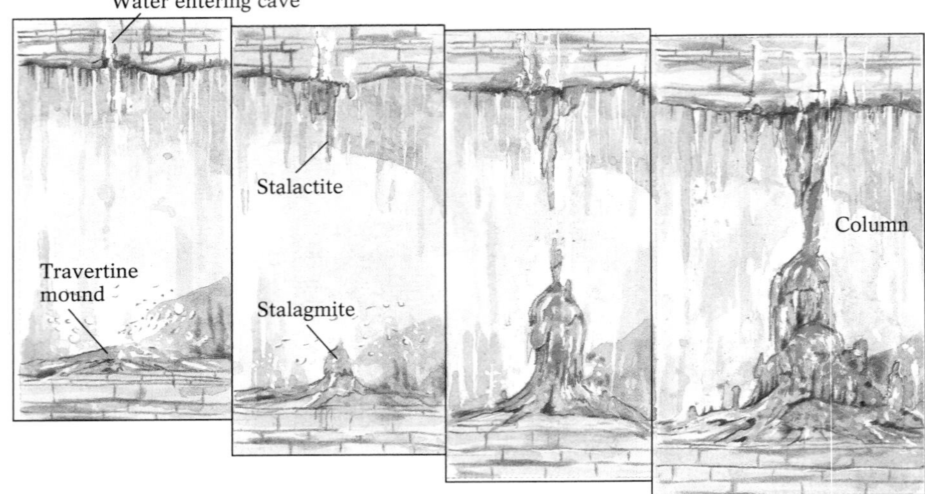

Figure 16-7 The creation of stalactites, stalagmites, and columns from precipitated travertine.

precipitating travertine. Stalagmites may grow as high as 30 meters (100 feet), and sometimes merge with stalactites to form single speleothems, called *columns,* that reach from floor to ceiling (Fig. 16-7).

Other Cave Deposits A variety of other delicate and beautiful shapes are assumed by speleothems (Fig. 16-8). Encrustations that form where water drips from the ceiling are called *dripstones*. Those made by moving water are called *flowstones*. Speleothems that form below a joint or crack in the ceiling of a cave may produce *banded draperies*, or *drip curtains*. *Rimstone dams* form when travertine precipitates over obstructions to water outflow from a cave pool. *Helictites* are twig-like structures whose central tubes are so narrow that water passes through them by capillary action (see Chapter 15). Because capillary action is not dependent on gravity, helictites can grow from any point on a cave ceiling, wall, or floor, and in any direction.

Another type of unusual speleothem, *cave pearls,* form as calcium carbonate precipitates in concentric layers around a sand grain or other sand-sized particle. This may occur when water drips from the cave ceiling into a bicarbonate-rich pool of water. The constant drip keeps the pool's water perpetually agitated, which releases carbon dioxide, precipitating travertine onto small fragments at the pool's bottom. The constant agitation keeps the growing pearl in motion so that it does not adhere to any surface, enabling travertine to be added to it uniformly in layers. *Cave popcorn* grows on a larger speleothem as water within the speleothem moves to its surface and evaporates, leaving small clumps of precipitated travertine around surface openings.

Figure 16-8 Various types of speleothems (see text). **(a)** Stalactites in their early stages of formation resemble hanging soda straws. **(b)** Stalactites hanging from the cave ceiling and stalagmites growing upward from the floor merge to form a *column.* **(c)** *Helictites* form by capillary action and can grow in any direction. **(d)** *Cave pearls* form when layers of travertine precipitate around a sand grain or similar particle. **(e)** Giant *cave popcorn* forms by the precipitation of travertine at the surface openings of large speleothems. **(f)** A speleothem resembling a fried egg.

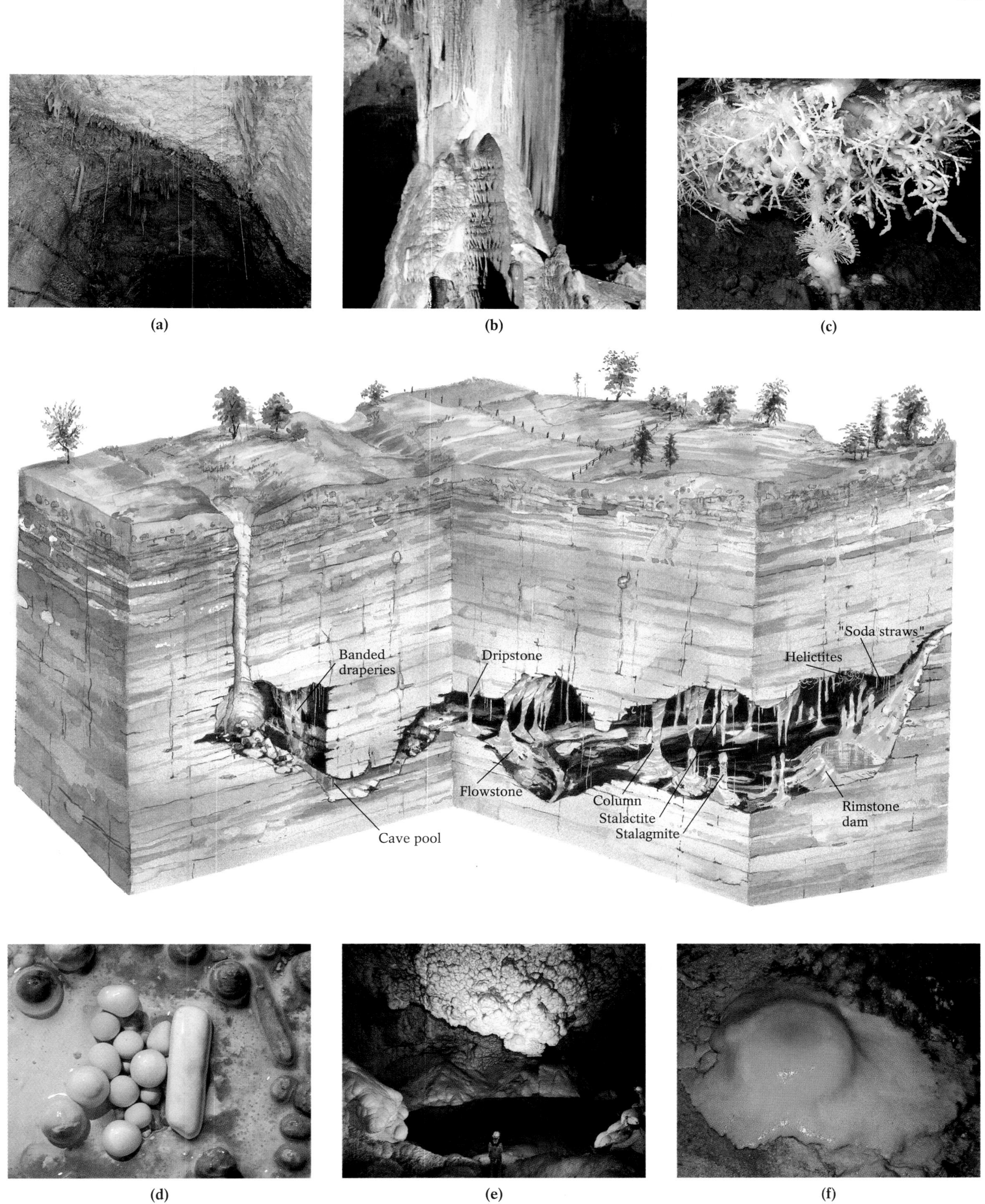
(a)
(b)
(c)
Banded draperies
Dripstone
"Soda straws"
Helictites
Flowstone
Column
Stalactite
Stalagmite
Rimstone dam
Cave pool
(d)
(e)
(f)

Speleothem Growth Speleothem growth is promoted by the same factors—solubility and porosity of surface materials, climate, topography, and vegetation—that influence the infiltration of groundwater and contribute to its acidification. Conditions favor the formation of speleothems in caves in Virginia, Kentucky, and Florida, where they grow as rapidly as 1 centimeter per year. In regions with relatively impermeable, less soluble rocks, cold or arid climates, and limited vegetation, such as northern Alberta and British Columbia, speleothems grow at a rate of less than 1 centimeter per 100 years. The 20-meter (65-foot)-long stalactites in Carlsbad Caverns in arid New Mexico have also grown at less than 1 centimeter per 100 years, and therefore record hundreds of thousands of years of growth.

We can observe the growth record of a speleothem by cutting a cross section through it. For example, a horizontal slice of a stalactite displays concentric layers of travertine. Although not always annual like tree rings, these growth layers do record the history of the speleothem. In caves under landscapes that were glaciated during the most recent ice age, speleothems display travertine layers that have cracked. These cracked layers indicate interruptions in speleothem growth that occurred because the ground froze or became covered with ice, preventing water from infiltrating. As a result, caves became dry and speleothems became coated with dust blown about by cave winds. As the climate warmed, the ice departed, the ground thawed, groundwater again infiltrated, and speleothem growth resumed. By dating the travertine on either side of a cracked, dusty layer within a speleothem, the time and duration of a glacial period can be estimated (Fig. 16-9). Studies in France and England indicate that speleothem growth in these areas ceased about 90,000 years ago and resumed about 15,000 years ago, spanning the period of the last worldwide glacial expansion (see Chapter 17).

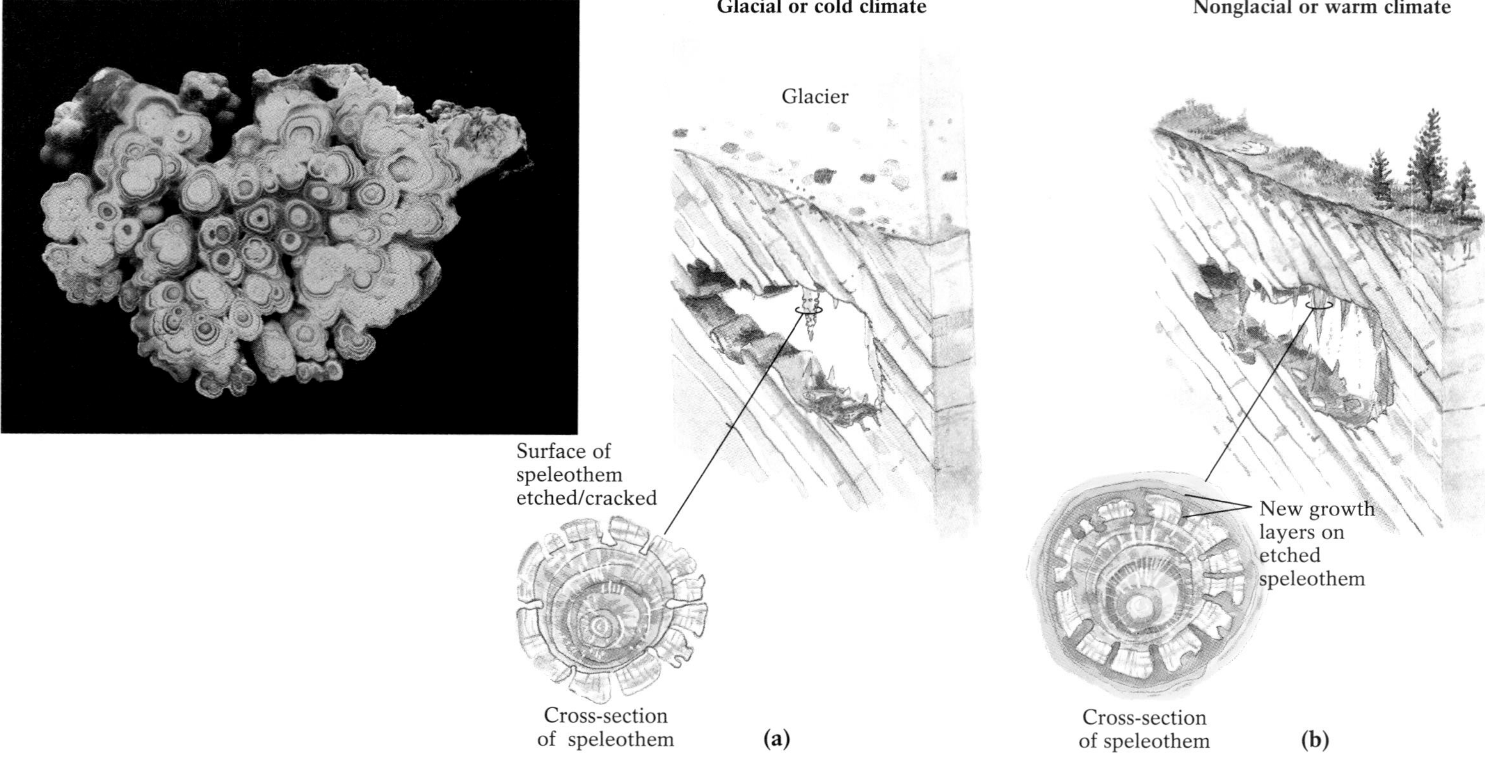

Figure 16-9 Speleothem growth is affected by climate. (**a**) During cold periods, when the ground is frozen or when ice covers the land above a cave, water cannot infiltrate the limestone bedrock and speleothem growth ceases. Without water, speleothems dry out, their surfaces crack, and cold winds from the surface spread a layer of dust on them. (**b**) During warm periods, water is free to infiltrate and speleothem growth resumes. The time and duration of glacial and nonglacial periods can therefore be estimated by dating the layers within speleothems. Below: A sliced speleothem from Kokaweaf Cavern, California, reveals concentric layers that can be dated by uranium-isotope dating.

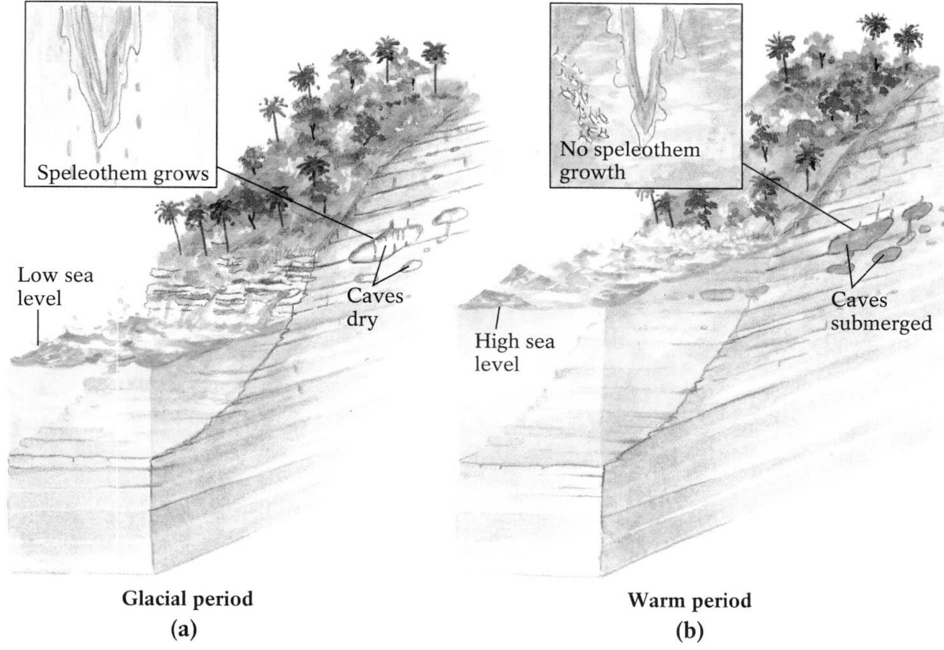

Figure 16-10 Speleothem growth in low-latitude and equatorial coastal caves is affected by the rise and fall of sea levels. (**a**) During worldwide glacial periods, global sea level is lower, coastal caves dry out, and speleothems tend to grow. (**b**) During warm periods, glaciers melt, sea level rises, and speleothem growth ceases because caves are filled with water.

Even speleothems in tropical coastal regions never covered by glaciers can be used to date past ice ages. Here, travertine deposits record the rise and fall of sea level instead of the comings and goings of ice. When glaciers around the world expand during an ice age, marine water is converted to snow and ice on land and sea levels fall worldwide. The water level in coastal caves drops, producing the air-filled cave environment required for speleothem growth (Fig. 16-10). When glaciers melt during warmer periods, sea levels rise and caves located at or below the new sea level fill with seawater. As a result, there is no speleothem growth because there is no air-filled cavity into which water can drip.

Two Famous Caves

Among the best-known caves in North America are Howe Caverns in Cobleskill, New York, Wyandotte Cave in southern Indiana, Luray Caverns in Virginia, and the Sonora Caverns in Texas. All are popular tourist attractions, and many of these and other caves have been preserved as parks so that they can be administered by state and federal agencies and protected by law. Two of North America's most spectacular cave systems, Mammoth Cave in Kentucky and Carlsbad Caverns in New Mexico, have been included in the U.S. national parks system because of their unique character and grand scale.

Mammoth Cave Mammoth Cave is located in moist, temperate west-central Kentucky between Louisville and Nashville, Tennessee. Its vast labyrinth of connected galleries, some 80 meters (250 feet) high, contains underground lakes, rushing rivers, and spectacular waterfalls. Mammoth Cave was explored intensively centuries ago by native North Americans, and more recently by westward-migrating pioneers. In the nineteenth century the cave, known for its exceedingly long dry tunnels, was considered a natural wonder and rivaled Niagara Falls as the main tourist attraction of eastern North America. In 1972, a connection was found linking the Mammoth Cave system to the adjacent Flint Ridge system, making the entire complex, with over 400 kilometers (250 miles) of interconnected passages, the world's longest surveyed system (Fig. 16-11).

Mammoth Cave's geological history began about 350 million years ago, when the St. Genevieve and St. Louis limestones were deposited on the floor of a shallow sea. Cave formation began in the limestone perhaps 30 million years ago, but its most active phase began 2 million years ago. At that time, a massive ice sheet originating in central Canada near Hudson Bay advanced southward toward what is now Kentucky. Although the ice never covered the Mammoth Cave region, it blocked large westward-draining rivers, diverting them into the local stream systems. These enlarged streams cut so deeply through the regional limestone plateau that they drew water down from the area's zone of saturation, lowering the water table and draining the cave system.

Largely because the internal open-air environment of Mammoth Cave is extremely dry, there has been no significant speleothem development. Most water within the Mammoth–Flint system flows as subterranean streams that have entered the caverns through local surface depressions where bedrock joints intersect. Most of the area is overlain by moderately impermeable cap rock that inhibits infiltration of water. Only in a few places, where this rock has been eroded, does enough water enter to produce speleothems.

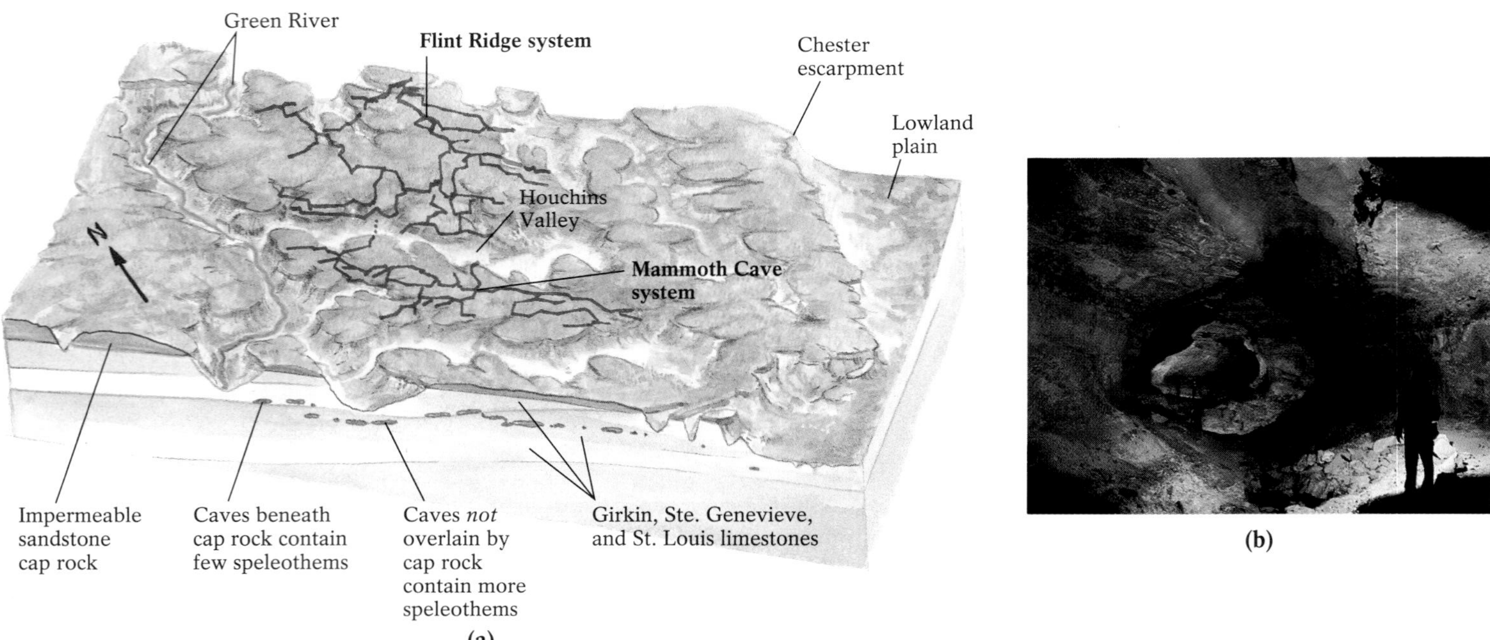

Figure 16-11 The Mammoth–Flint Ridge cave system in Kentucky (see text). (**a**) Because most of it is overlain by relatively impermeable sandstone "cap rock," Mammoth Cave is fairly dry and undergoes limited speleothem development. (**b**) A smooth-walled passageway in Mammoth Cave.

Carlsbad Caverns At sunset one hot June day in 1901, a young cowboy named Jim White noticed an odd black cloud above an area of parched ground in southeastern New Mexico. The cloud, swirling and lurching like an erratic tornado, was darker near the ground and grew diffuse at greater altitude. When the curious cowboy drew closer, he discovered that the cloud was an enormous swarm of millions of bats, exiting from what appeared to be a bottomless pit. White tossed a flaming cactus into the abyss and estimated that it fell more than 200 feet before it struck bottom. He had discovered Bat Cave, part of Carlsbad Caverns, North America's most spectacular cave system (Fig. 16-12).

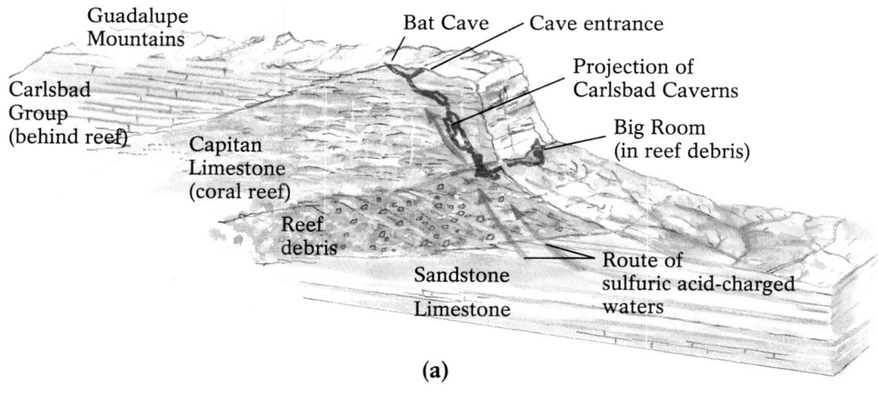

(a)

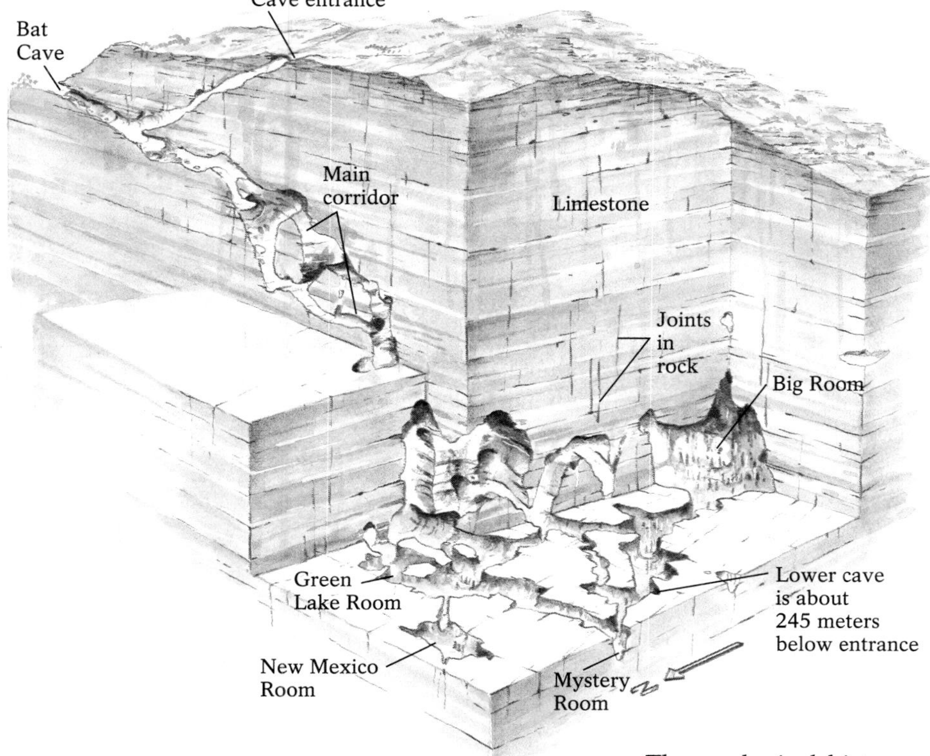

(b)

(c)

Figure 16-12 Carlsbad Caverns, New Mexico (see text). (**a**) A geologic cross section of the Guadalupe Mountains, showing the location of the cave system. (**b**) Most of the large cave "rooms" of Carlsbad developed at a depth of 245 meters (800 feet) below the ground surface. (**c**) Well-developed speleothems in one of Carlsbad's rooms (see also Figure 16-1).

The geological history of the Carlsbad region began about 250 million years ago, when what is now arid eastern New Mexico was periodically covered by a shallow inland sea in a semi-tropical environment. The region's rocks from that period are largely reef deposits composed of a framework of calcite-secreting algae, sponges, and the remains of moss-like sea creatures called bryozoans. Periodic evaporation of the shallow sea left behind gypsum and other evaporites and vast quantities of sulfur-rich salty water.

About 60 million years ago, when the southern Rockies uplifted to form the Guadalupe Mountains, the ancient reef was folded and faulted. The joints and fractures resulting from this tectonic upheaval enlarged to become the enormous galleries of the cavern system, but by an unusual process. Within the reef deposits, hydrogen sulfide (H_2S) compounds from the salty marine water combined with oxygen to produce sulfuric acid (H_2SO_4); Carlsbad's caverns were created by sulfuric acid–charged waters that migrated *upward* by capillary action through the jointed reef limestone, instead of by the typical cave-forming descent of infiltrating water charged with carbonic acid.

In the last 2 to 3 million years, additional uplift promoted further cave development. A spectacular forest of speleothems grew during the most recent ice age. During periods of worldwide ice expansion, New Mexico was cooler and wetter; in this moist climate, the water that fell as rain evaporated more slowly than it does today. Hence there was plenty of water to infiltrate the area's limestone and dissolve and precipitate it as the massive speleothems of Carlsbad Caverns. The oldest stalactites in the caverns are about 1 million years old. Virtually every type of speleothem can be found here, from delicate helictites to cave popcorn.

Nonlimestone Caves

Most caves form when acidified water dissolves limestone, but some form in other ways—within solidifying lava flows, by the force of crashing surf against coastal bedrock, and within the bases of melting glaciers. Caves form in lava when basaltic flows cool quickly, forming a thin black crust at the surface while the interior remains molten and continues to flow (discussed in Chapter 4). Eventually the molten lava drains from the crust, leaving behind a lava tube, examples of which can be seen in Oregon, Washington, northern California, Idaho, and Hawaii (Fig. 16-13). Lava dripping from the roof of a lava-tube cave ("lavacicles") may accumulate on the cave floor to form deposits that resemble the stalagmites of limestone caves.

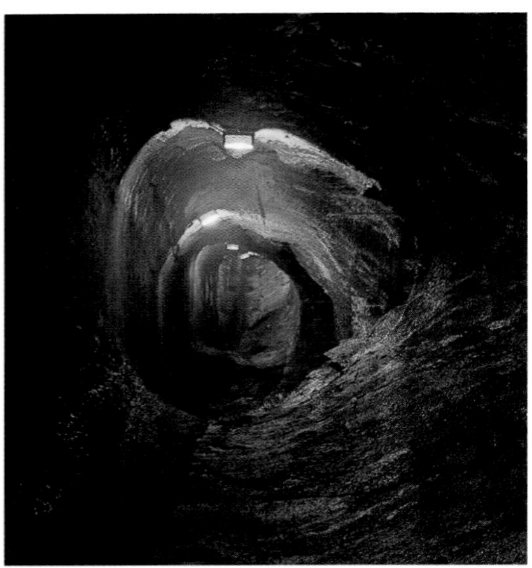

Figure 16-13 Thurston lava tube, in Hawaii, is a cave that developed within a basaltic lava flow.

Exposed coastal bedrock eroded by pounding surf can form *sea caves* (discussed further in Chapter 19). Fractured, jointed, faulted, or otherwise less resistant rocks are most likely to erode in this manner. Oregon's Sea Lion Cave, home to hundreds of seals, and Maine's Anemone Cave in Acadia National Park were both sculpted by surf.

Ice caves are formed within glaciers by melting caused by the heat of meltwater streams, which flow at the bottom of many glaciers and large perennial snow fields during the summer months. The melting that creates ice caves may lead, however, to their eventual disappearance (Fig. 16-14).

Figure 16-14 An ice cave in Muir Glacier, Glacier Bay National Park, Alaska. This ¼-mile-long cave was part of a glacial remnant that separated from the main glacier about 25 years ago. The photo was taken in 1984—the cave was gone by 1990.

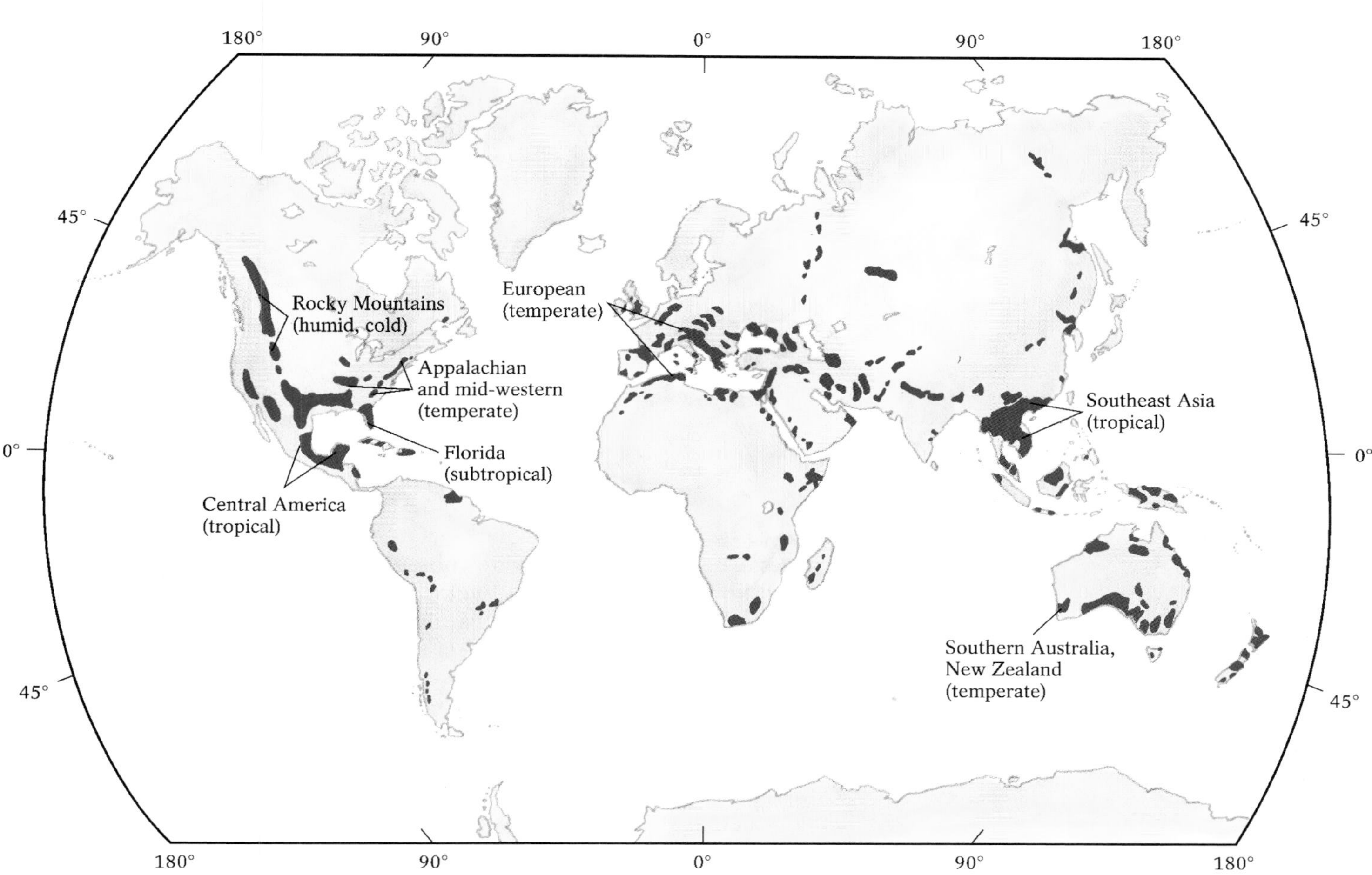

Figure 16-15 The worldwide distribution of known karst landforms. Little or no karst is found in arctic regions, where water is often frozen and thus unavailable for dissolution, soils contain little decaying organic matter, and permafrost impedes infiltration of water. In temperate regions and in cool, humid, mid-latitude regions, water falls as rain or snow, and the ground is unfrozen most of the year and readily absorbs water; here, dissolution reactions can proceed at a relatively rapid pace. In subhumid and arid regions near the equator, low precipitation levels and high rates of evaporation combine to protect limestone from dissolution. These regions contain virtually no karst. The prodigious rainfall and lush vegetation in the humid tropics provide the optimal environment for limestone dissolution, and karst landscapes are highly developed there.

Karst Topography

Karst topography, the *surface* expression of karst, is caused by dissolution of exposed soluble bedrock such as limestone, dolostone, and gypsum. Any of the millions of square kilometers of soluble bedrock that lie at or near the Earth's surface are likely to become or are already karst. Extensive karst can be found in China, Southeast Asia, Australia, Puerto Rico, Cuba and elsewhere in the Caribbean, the Yucatán peninsula of Mexico, and much of southern Europe; about 15% of the area of the contiguous 48 states is karst land, including substantial parts of Virginia, Alabama, Florida, Kentucky, Indiana, Iowa, and Missouri (Fig. 16-15).

Karst Landforms

Karst landforms range from broad plains studded with small, almost imperceptible, surface depressions to such unusual features as stream valleys containing no streams (or streams that flow at the surface and then suddenly disappear), spectacular natural rock bridges, and huge monoliths composed of insoluble rock (Fig. 16-16). Some of these landforms occur virtually everywhere there is extensive surface carbonate bedrock; others, particularly those caused by rapid dissolution acting over long periods, occur only in tropical environments.

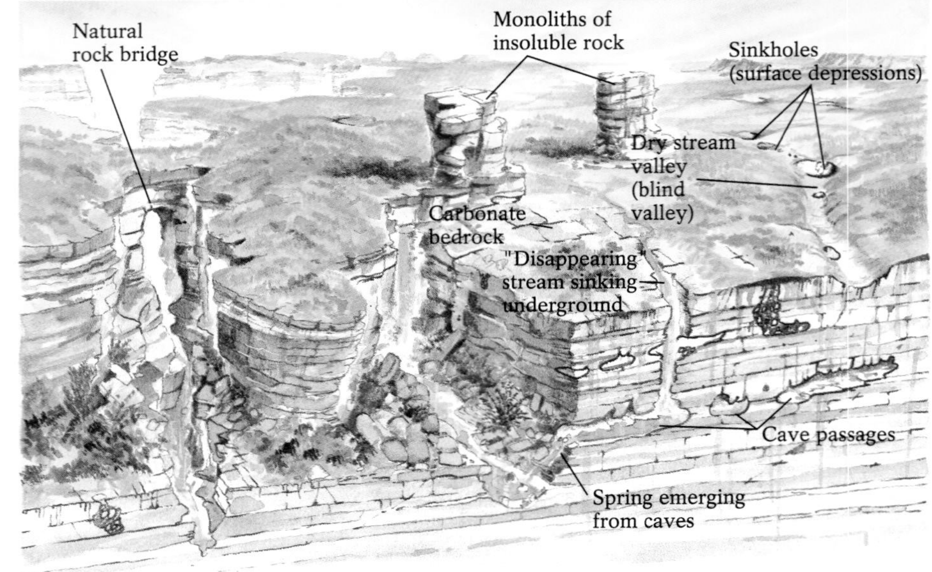

Figure 16-16 Typical landforms associated with karst topography (see text).

Sinkholes **Sinkholes** are circular surface depressions that appear in most limestone terrains; they provide subterranean drainage for surface water. Streams flowing across a well-developed karst plain generally drain through sinkholes, creating passageways that descend well underground, often to caverns below. Sinkholes usually occur together in great numbers over broad expanses of limestone bedrock containing numerous joints and fractures. Central Kentucky alone has 60,000 sinkholes; southern Indiana has more than 300,000.

Solution sinkholes form when limestone at or just below the surface is dissolved by groundwater rich in carbonic acid. The diameters of these sinkholes widen at the surface over time, and are progressively narrower below ground, forming their characteristic funnel shape (Fig. 16-17). Solution sinkholes generally develop on flat or gently sloping landscapes where water tends to remain before draining, promoting prolonged contact and extensive dissolution. On steeper terrains, water is drained from the surface too rapidly to allow for significant dissolution of near-surface rock.

Figure 16-17 The origin of solution sinkholes. Photo: A sinkhole in Minnesota, viewed from ground level.

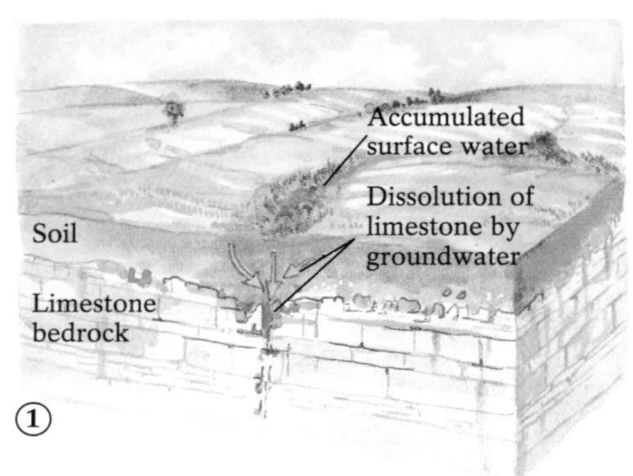

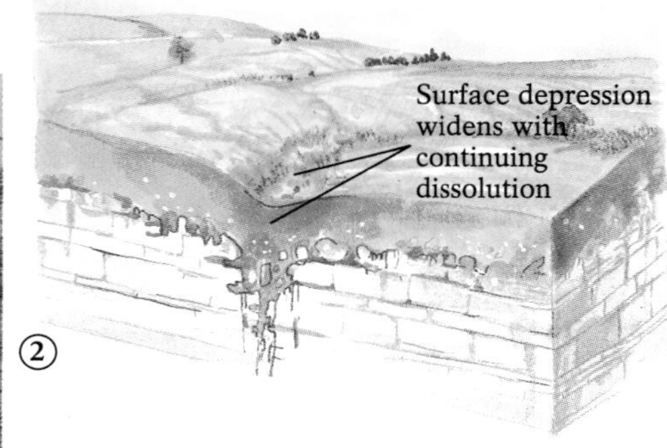

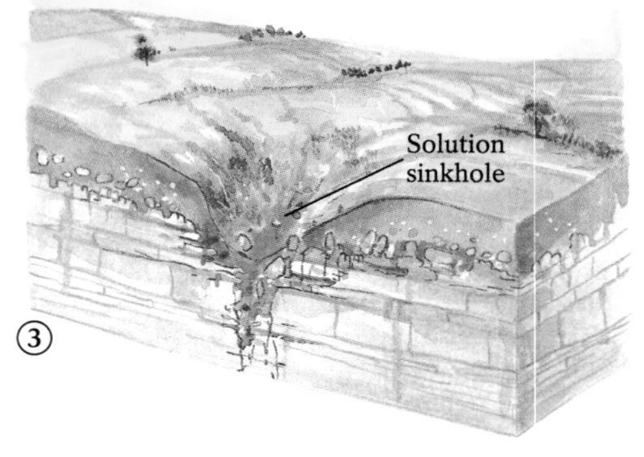

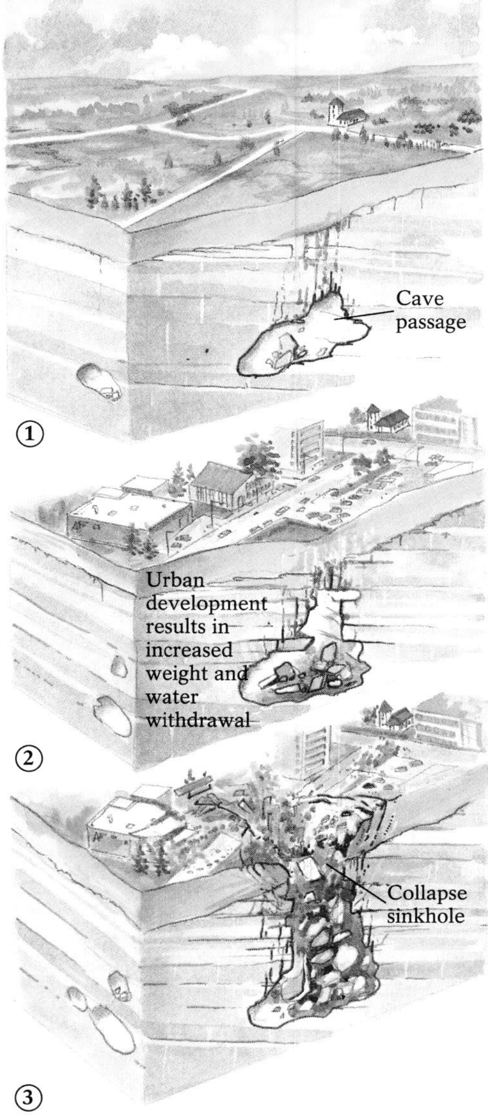

Figure 16-18 The origin of collapse sinkholes.

Collapse sinkholes form when the roofs of caves collapse under the weight of overlying rocks and soils (Fig. 16-18). This often occurs suddenly and unpredictably. They are usually deeper than solution sinkholes, and typically have steep, sometimes vertical, sides, and rocky and irregular floors covered with rubble. Most collapse sinkholes form following the lowering of the regional water table due to a lengthy drought (Fig. 16-19a), during which water that had helped support the overlying surface would have been removed (see Chapter 15). A soaking rain following a drought can add considerable weight to such a newly unsupported surface, causing it to collapse into the caverns below. Collapse sinkholes can also form when communities lower their water tables by excessive withdrawal of groundwater (Fig. 16-19b). In recent years, as the population of the southeastern United States has exploded and a great volume of groundwater has been extracted to meet rapidly rising demand, the regional water table has dropped and the surface in many areas has been left unsupported. As a result, thousands of new collapse sinkholes have appeared throughout the Southeast.

A sinkhole that is deep enough to intersect the local water table rapidly fills with water to form a lake. The Mayan civilization that flourished in the Yucatán peninsula in Mexico between A.D. 600 and 900 depended exclusively on water from *cenotes,* the region's deep, steep-sided sinkhole lakes. Archaeologists have excavated now-dry cenotes and retrieved treasures of jade, gold, and copper artifacts cast into them to please the rain god, Chacmul, to insure a steady supply of water.

Sinkhole lakes that lie above the local water table often last only as long as rubble or sediment clogs the sinkhole's lower outlet. When the outlet reopens, the water rushes through it into caverns. In the nineteenth century, a sinkhole at what is now Alachua Lake in the Lake District of north-central Florida was receiving the surface drainage from the surrounding plain, and the water was regularly disappearing into the local cave system. By 1871, enough organic and mineral debris had washed in to clog the sinkhole's outlet and form a lake 13 kilometers (8 miles) long and 6 kilometers (4 miles) wide. The lake remained until, 20 years later, the clogging debris was washed into the underlying caverns by descending water; with the outlet opened again, all the lake's water drained out. Even today, expensive Lake District property often becomes devalued overnight as sinkhole lakes drain like unstoppered bathtubs, and a lakefront view becomes one of foul-smelling mudflats.

(a)

(b)

Figure 16-19 A drop in water-table level is a common cause of collapse sinkholes. (**a**) A drought-induced sinkhole in Winter Park, Florida. The sinkhole formed in 1981, when the roof of a large subterranean cavity caved in without warning, swallowing a three-bedroom bungalow, a portion of a Porsche dealership, and half of a municipal swimming pool. A drop in the local water table brought on by the lengthy drought had removed the support supplied by groundwater to the roof of the cavity. The crater that resulted is about 122 meters (400 feet) wide and 38 meters (125 feet) deep. (**b**) The December Giant, a collapse sinkhole in an isolated section of Shelby County, Alabama. The crater, 140 meters (450 feet) across and 50 meters (165 feet) deep, formed with an earth-shaking rumble in December 1962, after heavy November rains soaked a surface left unsupported by heavy groundwater pumping nearby.

Figure 16-20 Streams in karst often disappear suddenly down sinkholes that open into caverns and passageways below the surface. In southern Indiana, streams with suggestive names such as Sinking Creek and the Lost River flow underground for more than 10 kilometers (6 miles).

Sinkholes can also be a boon to groundwater management, as they can be used to control the water level and flood potential of local streams. In Springfield, Missouri, for example, an area prone to periodic flooding, excess surface water is channeled toward a few selected sinkholes for subsurface disposal. In Kentucky, at the Bowling Green airport, engineers devised a system of drainage ditches that terminate at specific sinkholes, where their water is safely transferred underground.

Disappearing Streams and Blind Valleys Water soaks rapidly into an absorbent karst plain: Surface streams often drain completely into sinkholes in a matter of minutes or hours, becoming **disappearing streams** (Fig. 16-20). Because disappearing streams exist at the surface so briefly, they rarely meander and seldom contribute to flooding. The climate of central and western Pennsylvania, for example, would normally promote flooding, but few floods occur there because most surface water is quickly transferred underground through the region's sinkholes.

Some disappearing streams eventually return to the surface as natural springs. How can we tell if the water gushing from a natural spring is the same water that disappeared down a sinkhole some kilometers away? Geological researchers sometimes trace the water of a disappearing stream by dyeing it with a harmless chemical or biological material and stationing volunteers at springs throughout the area to await the emergence of the dyed water (Fig. 16-21). If a subterranean stream system's passageways are well-connected, the water may take only a few hours to travel distances of several kilometers. Another way to trace the flow path of a disappearing stream is to put pollen grains from plants not native to the region in the water and, using pollen traps installed at the spring outlets, note if and when the grains arrive. (Karst studies in Minnesota, for example, have used eucalyptus-tree pollen imported from Australia.)

(a)

(b)

Figure 16-21 Tracing disappearing stream water. (**a**) Dye is poured into a sinkhole. (**b**) Colored water emerges from a natural spring several kilometers away.

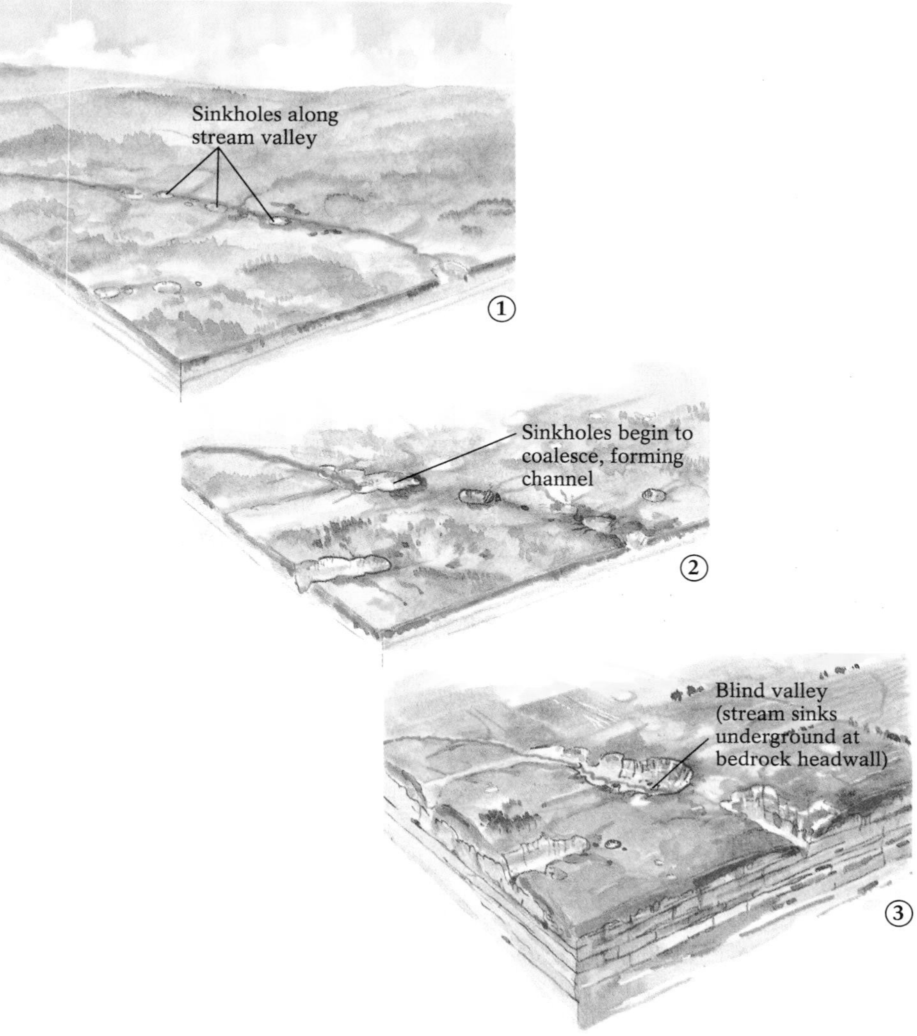

Figure 16-22 The origin of blind valleys (see text). Photo: The Tarn Gorge, a blind valley in southern France, shown during high discharge after a storm.

Blind valleys are channels that form when adjacent sinkholes, their diameters increased by continued dissolution and erosion, coalesce to form larger sinkholes (Fig. 16-22). Blind valleys contain water only during periods of heavy rain, when the water cannot drain into the ground completely. When a stream flows through a blind valley, it tends to lower the valley floor much faster upstream, where streamflow is greater; at the downstream end of the valley, where streamflow has dissipated, a bedrock *headwall* develops at which the stream seems to disappear, becoming diverted underground.

Natural Bridges As a series of neighboring sinkholes expand and coalesce, the surface overlying broad sections of underground stream channels may collapse. Segments of surface material that do not collapse form **natural bridges** over the exposed channel. Perhaps the most famous natural bridge in North America is in the Blue Ridge Mountains of Virginia, where a remnant of massive dolomitic limestone forms a bridge about 30 meters (100 feet) long and 40 meters (130 feet) high (Fig. 16-23). The bridge was purchased by Thomas Jefferson in 1774 (from King George III, for the equivalent of $2.50) in order to preserve it as a natural wonder and resource. More recently, it has been incorporated into U.S. Highway 11, saving millions in bridge-construction dollars.

Figure 16-23 Natural Bridge, in the Blue Ridge Mountains 20 kilometers (12 miles) southwest of Lexington, Virginia, spans Cedar Gorge Creek.

Highlight 16-1 *Karst Landscapes in History*

Figure 16-24 French paratroopers prepare to blow up a Viet Minh arms depot in a cave near Langson, Indochina border, in July 1957.

For centuries, partisans, guerrilla warriors, and bandits have enjoyed the protection of natural hideouts provided by karst landscapes. In the former Yugoslavia during World War II, outnumbered and ill-equipped partisans tormented their German invaders from karst refuges. In 1954, the guerrilla armies of the Viet Minh confounded French troops by hiding in and attacking from the maze of karst cockpits, caves, and blind valleys in what was then North Vietnam (Fig. 16-24). In 1957, Fidel Castro and his supporters planned their insurgency against Cuba's dictatorial regime from the shelter of the karst terrain of the Sierra Maestra on the island's southeastern coast. In the nineteenth century, Jesse James and his gang hid in the karst valleys and caves of the Missouri Ozarks to avoid capture and punishment for their bank and train robberies. And, since Prohibition began in 1919 (continuing even after its repeal in 1933), Appalachian karst land in the eastern United States has sheltered the illegal distilleries that manufacture moonshine whiskey.

Figure 16-25 Karst towers in the Kwangsi region of southern China. The limestone that once surrounded the towers has been carried away in solution.

Tropical Karst Landforms Limestone dissolution is rapid in tropical climates, where it is enhanced by abundant rainfall and lush vegetation. Because it is rapid, and has been relatively undisturbed by ice ages, it has produced karst formations that differ from those found in temperate climates. **Cockpit karst** consists of numerous closely spaced, irregular depressions and steep-sided conical hills. The depressions, or *cockpits,* result from enhanced dissolution of soluble rock. The hills, or *towers,* which can protrude more than 200 meters (660 feet) above the rapidly dissolving rock that surrounds them, consist of less soluble, less fractured rock (Fig. 16-25). Excavation of a cockpit generally ceases when groundwater reaches an insoluble layer and can no longer cut vertically by dissolution. Thus, the depth of a cockpit and consequently the relative heights of its neighboring karst towers are limited by the thickness of the soluble carbonate bedrock.

Because of their surface hollows and dry valleys, karst landscapes have helped determine the outcome of some significant political and social events in recent history, several notable instances of which are explored in Highlight 16-1.

Protecting Karst Environments

Well-developed cave and karst environments are typically marked by numerous sinkholes that are connected to a vast underground network of large caverns and passageways. In such places, water readily enters the ground and moves through it so rapidly that there is little opportunity for contaminants to be filtered out by soils or impermeable bedrock. Thus, cave and karst landscapes are extremely sensitive to careless handling of waste. For example, when Hidden River Cave in Kentucky opened to visitors in 1916, it boasted elegant galleries and memorable subterranean boat rides. But for decades, nearby towns disposed of their sewage into several large sinkholes, hopelessly contaminating the regional groundwater system. By the mid-1930s, Hidden River Cave had developed such vile odors that tourism there came to an end.

Garbage dumped into a sinkhole will quickly contaminate a local water supply (Fig. 16-26). During the 1960s, the community of Alton, Missouri, in beautiful Ozark country, disposed of much of its refuse into a 15-meter (50-foot)-deep dry sinkhole near the town. In 1969, after an unusually rainy winter and spring, 9 meters (30 feet) of water accumulated in the sinkhole. Concerned that Alton's sinkhole disposal was affecting regional groundwater quality, the U.S. Forest Service introduced 230,000 liters (61,000 gallons) of dyed water into the sinkhole's standing water and surveyed groundwater sources for miles around. Three months later, dyed water containing a high concentration of undesirable material from Alton's refuse was detected 25 kilometers (16 miles) away at the neighboring town of Morgan Springs, indicating that Alton's careless dumping was adversely affecting the entire region's groundwater quality.

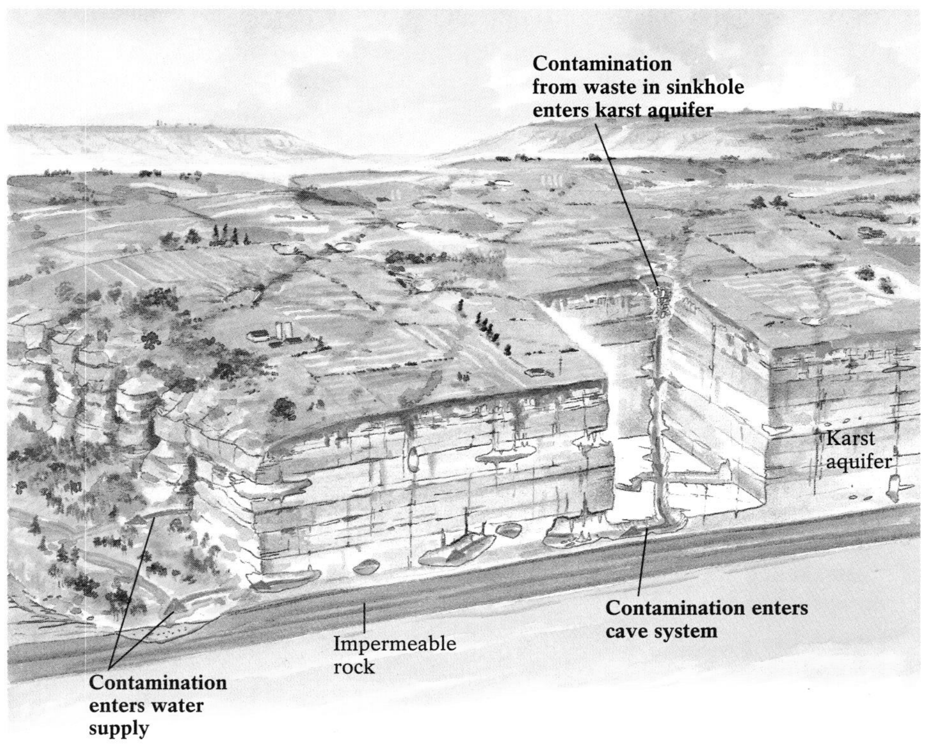

Figure 16-26 Disposal of refuse in sinkholes quickly contaminates karst aquifers. Above: A refuse-filled sinkhole.

As we saw in Chapter 15, some groundwater contamination is removed by natural processes; groundwater usually moves so slowly that many contaminants are gradually eliminated by filtration and oxidation. Water in karst, however, drains so swiftly through open underground passages that there is not enough time for its pollutants to be filtered out or consumed by microorganisms. In karst regions, polluted recharge moves into the groundwater through numerous entry points—a large cave system may have hundreds or even thousands of sinkholes. Moreover, such an extensive, complex system typically holds a huge volume of water. For these reasons, chemical treatment of contaminated karst aquifers is not feasible. Because they are so open to contamination and impossible to treat effectively, aquifers in karst regions are unreliable as sources of drinkable water, and should not be used for waste disposal of any kind.

In Chapters 14, 15, and 16, we have examined the surface and subsurface geological work accomplished as flowing liquid water passes through various segments of the hydrologic cycle. In the next chapter, we will look at how flowing frozen water—as glaciers—has substantially changed the regions that are now, or have in the past, been cold enough to freeze the Earth's surface and subsurface water.

Chapter Summary

Karst terrains develop from the dissolution of soluble bedrock. Nearly all karst is the result of the dissolution of calcite in limestone by carbonic acid. Dissolution rates are largely controlled by the composition and structure of the bedrock, the amount of rainfall and vegetation in the area, and local topography. Karst terrains contain both subsurface and surface features not found in any other geological setting.

Extensive cave systems are often found below the surface of a karst landscape. **Caves** are natural underground cavities that generally form as carbonate bedrock dissolves along preexisting joints, fractures, faults, and bedding planes. Most caves develop by a two-stage process: In the first stage, the local water table is high; cave chambers and connecting passageways form from a system of water-filled bedrock fissures. In the second stage, the water table has dropped, and because the cave is now located above the water table, its rooms and passageways exist in an open-air environment. As water that percolates through overlying bedrock enters the cave, it can evaporate or release its carbon dioxide, both of which result in dissolved limestone being precipitated. Cave deposits, called **speleothems,** include **stalactites,** which grow downward from the cave ceiling, and **stalagmites,** which grow upward from the cave floor. The limestone that makes up speleothems is called **travertine.** Nonlimestone caves can also form, as lava tubes in solidified basalt flows or as voids in eroded coastal bedrock or partially melted glaciers.

Karst topography, the surface expression of karst, is dominated by conical depressions called **sinkholes** that are often found by the thousands in areas of exposed flat-lying limestone bedrock. Solution sinkholes form by dissolution of limestone by carbonic acid in groundwater; collapse sinkholes form when the roofs of caves collapse from the weight of overlying materials. Because sinkholes usually occur in large numbers and close together, surface water rarely traverses a karst plain completely. Most often, streams literally drain from the surface as they encounter sinkholes. Such streams are called **disappearing streams.** The diverted water may reappear elsewhere as a natural spring. Coalescing sinkholes create **blind valleys,** which seem to terminate downstream at bedrock walls, where their streams sink underground. Blind valleys contain water only after heavy rains. When a network of sinkholes coalesces, the surface overlying underground stream channels can collapse. Segments of the surface that do not collapse form **natural bridges.** The development of karst topography is most pronounced in the moist tropics. There, limestone dissolution creates **cockpit karst,** a landscape marked by closely spaced cockpits surrounded by towers of residual less-soluble bedrock.

Because surface water is transmitted rapidly into and through the groundwater system before contaminants can be filtered out or consumed by microorganisms, karst regions are particularly sensitive to groundwater contamination.

Key Terms

karst (p. 445)
cave (p. 447)
speleothem (p. 449)
travertine (p. 449)
stalactite (p. 449)
stalagmite (p. 449)
sinkhole (p. 458)
disappearing stream (p. 460)
blind valley (p. 461)
natural bridge (p. 461)
cockpit karst (p. 462)

Questions for Review

1. Name three different soluble rocks. Which is most likely to form caves, and why?
2. List four factors that affect the rate of limestone dissolution.
3. Briefly explain the role that vegetation plays in the development of karst.
4. Briefly describe how caves form. Where is the water table located during the two stages of cave formation?
5. How do stalactites differ from stalagmites?
6. Explain how speleothems can be used to determine past fluctuations in global sea levels.
7. Discuss two fundamental geological differences between Mammoth Cave and Carlsbad Caverns.
8. Briefly describe two types of caves formed by processes other than dissolution.
9. Describe the two ways that sinkholes form.
10. Why is a karst landscape particularly susceptible to groundwater pollution?

For Further Thought

1. Describe what may happen to the calcium carbonate that enters solution as caves are created. How might this calcium carbonate eventually reenter the rock cycle?
2. How would you determine the local limestone dissolution rate in your town or city?
3. Discuss three reasons why you would not expect to find caves in polar regions.
4. What geological features do you think are shown in the aerial photos below, taken over the same area of the Midwest 43 years apart? Why might these features have grown in number and size between 1937 and 1980?

5. Speculate about how plate tectonics might affect the origin of a cave and its deposits.

Figure 17-1 Briksdalsbreen, a part of Norway's Jostedalsbreen glacier, which is the largest glacier in Europe.

Chapter 17

Glaciers and Ice Ages

Glaciers provide some of the most spectacular scenery on Earth (Fig. 17-1), drawing millions to Alaska and other high-latitude regions in which they are common. Glaciers also help shape the very landscapes on which they sit—the Earth's surface has probably been changed more by glaciation than by any other process. Without glaciers, the fjords of Norway would not exist. North America would not have its Great Lakes, Niagara Falls, Hudson Bay, Puget Sound, or the 15,000 lakes of Minnesota. There would be no Cape Cod in Massachusetts and no fertile rolling hills in the Midwest and southern Canada. The peaks of the Rockies and Cascades would be less impressive, and California's Yosemite Valley would lack its sheer-faced cliffs. Rivers such as the Missouri and Ohio would drain north to the Arctic and Atlantic Oceans rather than south to the Mississippi River and the Gulf of Mexico. If the Earth had no glaciers today, the shapes of the continents themselves would be substantially different because sea level would be about 70 meters (230 feet) higher than it is. Landlocked cities such as Memphis, Tennessee, and Sacramento, California, would be seaports, and San Francisco, New York, and many other coastal cities would be mostly under water.

But where are the glaciers that have changed so much of our continents? And how can so small a fraction of the hydrologic cycle (see Fig. 14-2) accomplish so much change? The answers lie in the occurrence of **ice ages,** the dozen or so periods—each lasting tens of millions of years—during which the planet's climate was substantially cooler than usual and glaciers covered a significant portion of the Earth's land surface. During an ice age, climatic fluctuations cause glaciers to alternately grow and advance, during *glacial periods,* and thaw and retreat, during *interglacial periods.*

The Earth is currently in the midst of an ice age—the **Quaternary ice age,** named for the Quaternary Period of the Cenozoic Era (see Fig. 8-27, the geologic time scale)—that has spanned the last 1.6 million years of Earth history. This ice age has been marked by many episodes of glacial advance and retreat, primarily during the **Pleistocene Epoch** of the Quaternary, which ended about 10,000 years ago. At the height of glacial expansion in the late Pleistocene, about 18,000 years ago, ice covered about 30% of the planet's land surface, including most of northern Europe, northwestern Asia, Canada, and the northern United States (Fig. 17-2); in places the ice was 4 kilometers (2.5 miles) thick. We are now living in an interglacial perioa of significant warming that

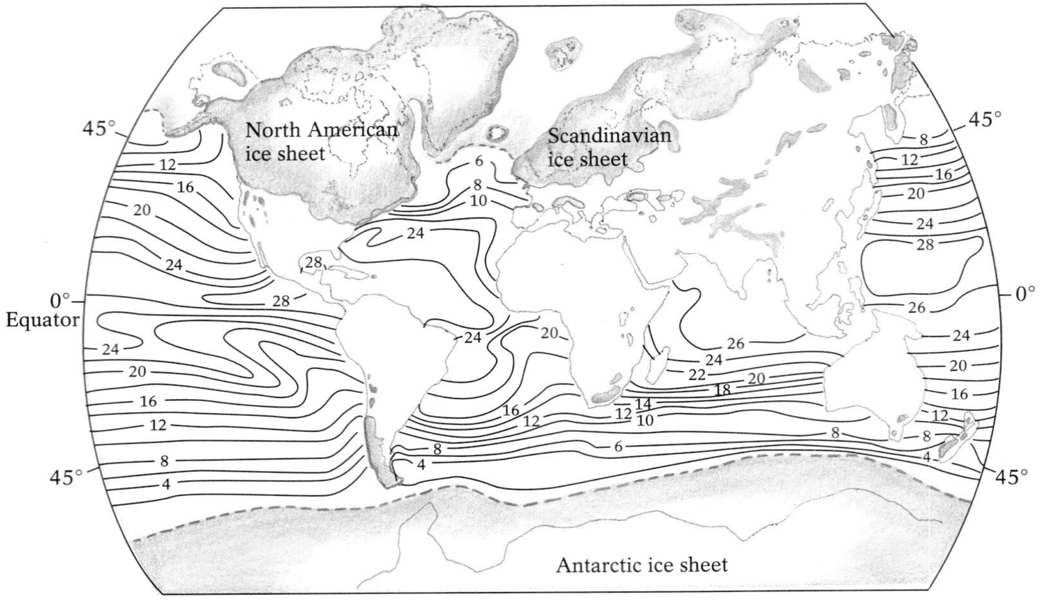

Figure 17-2 A "climap" showing glacier distribution during the last major period of worldwide glacial expansion. During this expansion, which reached its maximum about 18,000 years ago, ice covered about 30% of the Earth's land surface. The contour lines indicate ocean temperatures in degrees Celsius; note how the temperatures drop with proximity to the glacial margins.

began about 12,000 years ago. Today, glaciers cover only about 10% of the world's land surface. At very high latitudes, such as in the Arctic, Antarctica, and Greenland, glaciers are fairly common and can exist at any elevation, even at sea level; in mid-latitude regions, however, glaciers exist only at high elevations (at about 2500 to 3000 meters, or 8000 to 10,000 feet), such as in the Northern Rockies of Alberta, British Columbia, and Montana (Fig. 17-3).

In this chapter, we discuss what glaciers are, where they are today and where they were in the past, and how they have shaped the landscape. We also consider why certain periods of Earth history have been marked by worldwide glacial expansions whereas others have not, and speculate about when glaciers may again cover vast areas of North America.

Figure 17-3 The relationship between glaciers and geographic latitude, shown along a line from Alaska to the tip of South America. In low-latitude, warmer regions (0–30°) glaciers are found only at very high elevations (more than 5000 meters, or 17,000 feet). In middle latitudes (30–50°) they can occur at somewhat lesser heights (2500–3500 meters, or 8000–11,000 feet). But most glaciers today are found in high-latitude regions (over 60°) such as Alaska and Antarctica, where they can exist even at sea level.

Glacier Formation and Growth

A **glacier** is a moving body of ice that forms from the accumulation and compaction of snow. Glaciers flow downslope or outward under the influence of gravity and the pressure of their own weight.

Glacier formation begins with a snowfall and the accumulation of snowflakes, delicate hexagonal (six-pointed) crystals of frozen water (Fig. 17-4). The initial snowpack tends to be fluffy, because at first the shape of the snowflake crystals keeps the flakes separated; about 90% of a pile of newly fallen snow consists of space between crystals. The density of fresh snow is only 0.1 grams per cubic centimeter (that of liquid water is 1.0 grams per cubic centimeter).

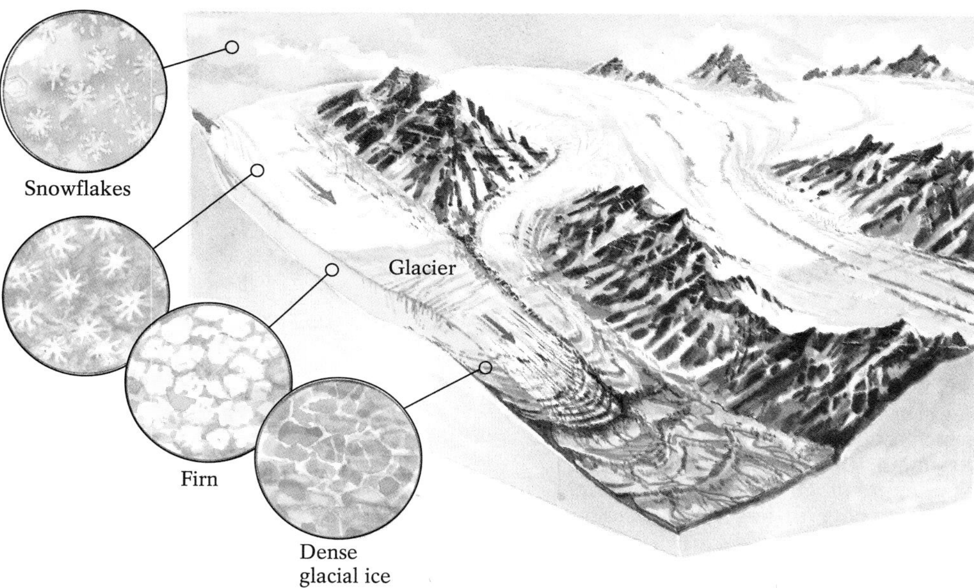

Figure 17-4 The formation of glacial ice from snow (see text). Hexagonal snowflakes that fall on a glacier are gradually changed into rounded crystals. As overlying snow buries the crystals and exerts pressure on them, they become packed into increasingly dense *firn* and finally glacial ice.

Soon, the air circulating through the spaces in the snowpack causes the fragile points of some snowflakes to sublimate (a process in which a solid changes directly into a gas). Some snow is thus directly converted to water vapor, which then recrystallizes in the open spaces between flakes. As more snow falls, the weight of the overlying snow melts some of the contact points between crystals; the water produced by this *pressure melting* migrates to open, pressure-free spaces in the snowpack and refreezes, binding the snow crystals together. (Pressure melting also occurs under the blades of an ice skater. The skater's weight, pressing downward on the narrow blades at each moment, melts a small amount of ice which refreezes the instant the weight is removed. The result is that the skater is always gliding on a thin film of water.)

Eventually, the processes of sublimation and recrystallization and pressure melting and refreezing convert large, ornate snowflakes to small, nearly spherical crystals that are more tightly packed. **Firn** (from the German for "last year") is the well-packed snow that survives a summer melting season; the density of firn is about 0.4 grams per cubic centimeter. Dense firn is made up of interlocking crystals of glacial ice after nearly all the air has been squeezed out by repeated pressure melting and refreezing. Ultimately, the density of glacial ice reaches 0.85 to 0.90 grams per cubic centimeter, about the same as an ice cube.

Glaciers form and grow when snow accumulation is greater than losses due to summer melting and sublimation. Where summer warmth completely removes even heavy winter accumulations of snow, such as in Minnesota, glaciers do not occur. Cool, cloudy, brief summers that minimize melting probably contribute more to glacier growth than very low winter temperatures or thick accumulations of snow.

In mountainous regions, glacier formation and growth are further influenced by the elevation at which the snow lands and the steepness of a mountain's slope (Fig. 17-5). Mountainside glaciers usually form at or above the **snowline**, the lowest topographic limit of year-round snow cover; snow always exists above the snowline, but snow that falls below the snowline, where temperatures are warmer, usually melts in summer. Snow also tends to accumulate more on gentle slopes, where it is unlikely to slide downslope. Glacier formation often also depends on the orientation of a mountain's slope and the direction of snow-drifting winds. In the Northern Hemisphere, snow is more likely to survive the summer on north-facing than on south-facing slopes, because the former receive less direct sunlight. More snow accumulates, and glacier formation is more likely, on the leeward slopes of mountains, where drifts from the windward side generally settle. In the mountains of western North America, where the prevailing winds blow from west to east, most glaciers form on northeastern slopes.

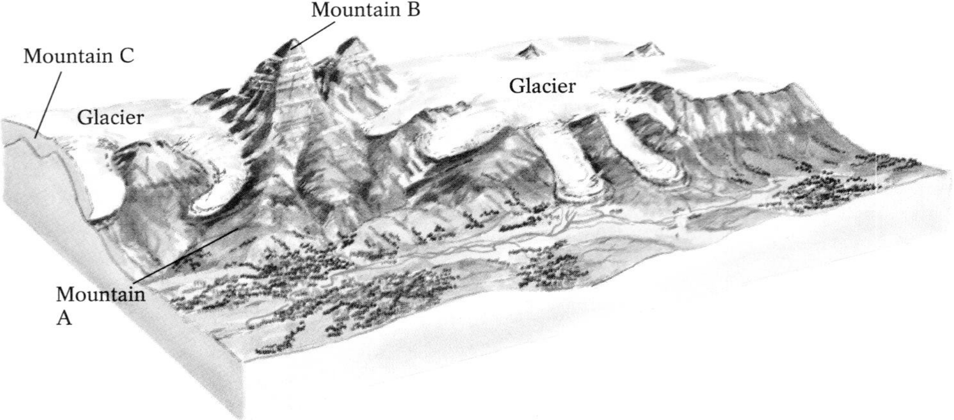

Figure 17-5 Elevation and slope steepness are two factors influencing glacier formation. Mountain A does not support the development of a glacier because its summit is lower than the regional snowline, below which summer temperatures are too warm to support glaciers (see text). Mountain B, the summit of which is well above the snowline, nevertheless does not support glacier development because it is too steep and snow flows downhill rather than accumulating. Mountain C has a gently sloped shape and much of its summit is above the snowline, making it an optimal site for glacier development.

The time required for fresh snow to become glacial ice varies with the rate of snow accumulation and the local climate. In snowy climates with average annual air temperatures close to the melting point of ice (0°C), snow may be converted to ice in only a few decades. In extremely cold polar settings, where snowfall is typically low and little melting takes place, glacial ice formation may take thousands of years.

Classifying Glaciers

Glaciers are classified broadly by whether the local topography confines them or allows them to move freely (Fig. 17-6). **Alpine glaciers** are confined by surrounding bedrock highlands. Because they are confined, they are relatively small. There are three types of alpine glaciers: **Cirque glaciers** create and occupy semicircular basins on mountainsides, usually near the heads of valleys; **valley glaciers** flow in preexisting stream valleys; **ice caps** form at the tops of mountains.

Figure 17-6 Types of glaciers (see text). (**a**) Angel Glacier, a cirque glacier on Mount Edith Cavell, Jasper National Park, Canada. (**b**) A valley glacier in the North Cascade Mountains of Washington. (**c**) An ice cap in the Sentinel Range, part of the Antarctic continental glacier. (**d**) A tidewater glacier at Kenai Fjords National Park, Alaska.

Another type of glacier originates as a confined alpine glacier but flows onto an adjacent lowland where, unconfined, it can spread radially. These are *piedmont* ("foot of the mountain") glaciers. Piedmont glaciers that flow to coastlines and into seawater are *tidewater* glaciers. The only completely unconfined form of glacier is a **continental ice sheet,** an ice mass so large that it blankets much or all of a continent. Modern continental ice sheets cover Greenland and Antarctica. The Antarctic ice sheet actually consists of two ice sheets separated by the Transantarctic Mountains. It is up to 4.3 kilometers (nearly 3 miles) thick and occupies an area about 1.5 times that of the continental United States. Eighteen thousand years ago, such vast ice sheets covered North America to south of the Great Lakes, and western Europe south to Germany and Poland.

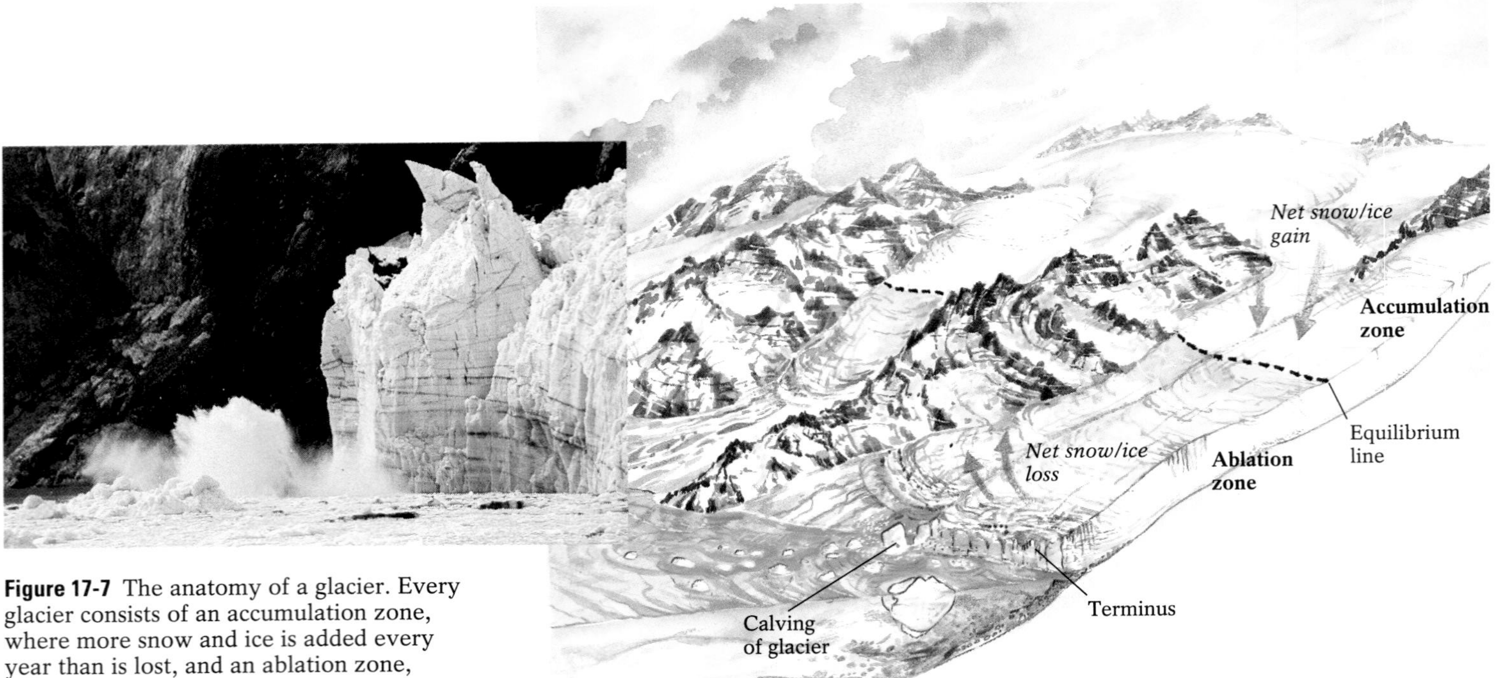

Figure 17-7 The anatomy of a glacier. Every glacier consists of an accumulation zone, where more snow and ice is added every year than is lost, and an ablation zone, where more snow and ice is lost than is added. An equilibrium line, where the amount of snow and ice added approximately equals the amount lost, forms the boundary between the two zones. In addition to melting and sublimation, a glacier may lose ice by calving if it terminates in a body of water. Inset: Calving of Margerie Glacier in Glacier Bay National Park, Alaska.

The Budget of a Glacier

The *budget* of a glacier is the difference between the glacier's annual gain of snow and ice and its annual loss, or *ablation*, of snow and ice due to melting and sublimation. If accumulation exceeds ablation for several consecutive years, the budget is positive, and the glacier increases in thickness and area. As a glacier expands, its outer margin, or **terminus,** advances downslope in confined glaciers and outward in unconfined glaciers. If ablation exceeds accumulation for several years, the budget is negative, and the glacier decreases in size and its terminus retreats. If accumulation is about the same as ablation, a glacier's budget is balanced and its terminus remains stationary. Thus, the position of the terminus is largely determined by the glacier's long-term budget.

All glaciers, from the smallest cirque glacier in Montana to the Antarctic ice sheet, have a **zone of accumulation,** a **zone of ablation,** and an **equilibrium line** separating the two zones (Fig. 17-7). The position of the equilibrium line can change every year, depending on the gain or loss of glacial volume. The zone of accumulation, identifiable by a blanket of snow that survives summer melting, is nourished principally by snowfall, sometimes augmented by avalanches from surrounding slopes. Ablation in middle and low latitudes is due mainly to summertime melting; in high latitudes, where summer temperatures may be below freezing, it is due to sublimation. The zone of ablation is recognizable in summer by its expanse of bare ice. Glaciers that terminate in bodies of water, such as tidewater glaciers, can also lose ice by *calving,* a process in which chunks of ice are broken off by wave or tidal action. Calving is the source of icebergs.

Glacial Flow

Whether a glacier's terminus is advancing due to accumulation, receding due to ablation, or stationary, ice within the glacier tends to flow forward from the accumulation zone toward the ablation zone. In the accumulation zone, snow becomes compacted into ice and flows down toward the glacier's bed; in the ablation zone, ice flows upward toward the surface and outward toward the glacier's edges. The extent of the flow depends largely on the temperature of the ice, which may vary from place to place within a glacier. The closer the temperature of ice is to its melting point, the more rapidly it flows.

Glaciers flow by a combination of two mechanisms: internal deformation and basal sliding. In **internal deformation,** a glacier's ice crystals deform under pressure from overlying ice and snow and slip past one another (particularly when the glacier is relatively warm and water exists at crystal boundaries), or fracture and move along planes of weakness. All glaciers, even those that remain frozen to their beds, move to some extent by internal deformation. In **basal sliding,** warmer glaciers such as those in mid-latitude mountain ranges thaw at their bases, producing a film of water that enables the glacier to slide along its bed. Glaciers in warm climates are more likely than those in extremely cold climates to move by both basal sliding and internal deformation, and thus they usually flow faster. All of the factors involved in glacial flow are illustrated in Figure 17-8.

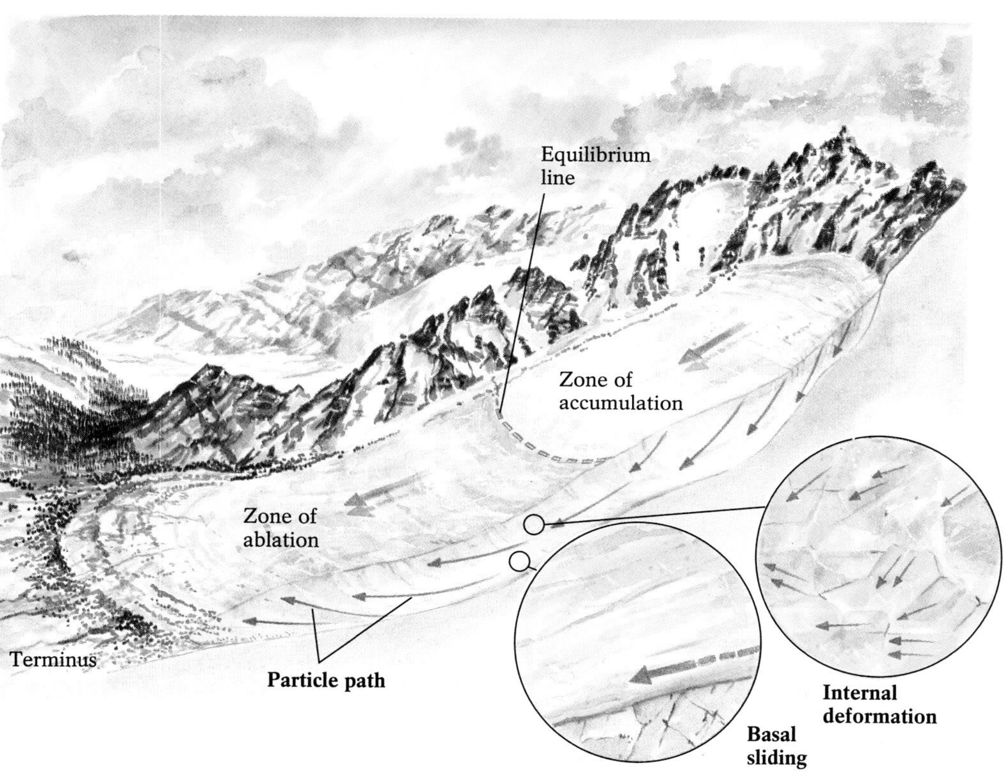

Figure 17-8 The mechanics of glacial flow. In general, particles within a glacier move downward in the accumulation zone, parallel to the ground at the equilibrium line, and upward in the ablation zone. Movement of the glacier as a whole results from a combination of internal deformation of its ice crystals and basal sliding.

Velocity of Glacial Flow A glacier's velocity depends on factors inherent in both the ice and its location. The temperature of the ice controls the availability of water for basal sliding and enhanced ice-crystal slippage; its thickness provides the pressure that causes ice-crystal deformation. The steepness of the bedrock slope (under alpine glaciers) or of the ice-surface slope (of continental ice sheets) increases the effect of gravity on basal sliding. Cold, nonsliding ice sheets flow slowly, usually only a few meters per year; warmer alpine glaciers on steep slopes can flow 300 meters (1000 feet) or more per year.

Figure 17-9 A satellite photo of the Hubbard Glacier surging. Basal melting caused the Hubbard Glacier to surge in the summer of 1986, clogging the mouth of Russell Fjord with ice and turning a large marine bay into a lake.

Some glaciers periodically **surge,** or accelerate, usually in brief episodes lasting several months to a few years that are separated by longer periods (10–100 years) of normal flow. Surging glaciers can move up to a hundred times faster than their normal velocity. Surges apparently occur when a large volume of water accumulates at the base of a glacier, reducing friction and facilitating basal sliding. The ice is apparently lifted from its bed by high water pressure, much as when a car hydroplanes on wet highway pavement. Although there is no way to observe the action of the water directly, the large quantity of water that flows from a glacier's terminus as a surge comes to an end supports this hypothesis.

The most rapid surge known, on the Kutiah Glacier in Northern Pakistan, reached a peak velocity of 110 meters (about 350 feet) per day and lasted for several months. One newsworthy glacier surge took place during the summer of 1986 on the Hubbard Glacier in the St. Elias Mountains, near Glacier Bay National Park in southeastern Alaska (Fig. 17-9). Its velocity increased from its usual 30 to 100 meters (100–300 feet) per year to about 5 kilometers (3 miles) per year. Advancing across the mouth of Russell Fjord, the surging glacier blocked the outlet of a large marine bay, turning it into a rapidly filling freshwater lake and separating thousands of marine mammals from their accustomed saltwater habitat. As the lake rose to about 20 meters (65 feet), it swamped trees along the shore and threatened to overflow the Situk River. Fortunately, the surging ceased and the ice in the fjord broke up, restoring the bay.

The Work of Glaciers

We mentioned at the start of this chapter that many striking and familiar landforms are the work of past glaciers. The processes of glacial erosion and deposition have been more effective than almost any other geological process in shaping the features of North America, northern and central Europe, and Asia, especially the alpine highlands on all seven continents.

Glacial Erosion

Glaciers remove materials from the landscape by erosion. Rapidly flowing glaciers that move largely by basal sliding cause the most erosion, particularly within the accumulation zone where ice moves down to the glacier's bed. Glaciers erode by abrasion and by quarrying. **Glacial abrasion** occurs when rock fragments embedded in the base of a glacier scrape the surface of underlying bedrock like sandpaper on wood. A glacier whose basal ice contains fine, gritty rock fragments may eventually polish the rock surface below it to a high shine. Ice carrying coarser rock fragments can cut long **striations,** or scratches, into the bedrock surface, ranging in scale from small bedrock scratches to sizeable grooves many meters wide and deep (Fig. 17-10). Striations are usually oriented in the same direction as the ice flow, and so can indicate the direction taken by ancient glaciers.

(a)

(b)

Figure 17-10 Bedrock striations produced by glacial abrasion. **(a)** Glacially polished and striated rock along the shore of St. John's Bay, Newfoundland. **(b)** Large glacial grooves in bedrock on Slate Island in Lake Superior, Ontario, Canada.

Glacial abrasion is enhanced when: 1) there is a steady supply of fragments to sustain abrasion; 2) the glacier's base contains fragments that are harder than the underlying bedrock; 3) the sliding velocity of the ice is rapid; and 4) the underlying bed consists of readily eroded materials. Deep striations occur when a large supply of durable pebbles or cobbles (such as granites or gneisses) is carried by a rapidly sliding glacier over soft sedimentary bedrock (such as shale or limestone). The abrading stones themselves become abraded and may eventually be pulverized into a silty rock powder called *glacial flour,* which imparts a distinctive teal-green color to rivers that flow through glaciated regions.

Glacial quarrying occurs when a glacier lifts masses of bedrock from its bed. This process occurs most often on bedrock that is jointed or fractured, whether by preglacial tectonic activity or by frost wedging. (Frost wedging, discussed in Chapter 5, is common in the frigid climates associated with advancing glaciers, and when meltwater at a glacier's base seeps into bedrock cracks and refreezes.)

Abrasion and quarrying together can shape a rock mass into a distinctive asymmetrical form, a *roche moutonnée* (French for "sheep-like rock") (Fig. 17-11). A glacier usually gently abrades the side of the rock mass from which it advances (the upglacier direction) but deeply quarries the side toward which it flows (the downglacier direction). A roche moutonnée is aligned lengthwise in the direction of glacial flow. Its upglacier side is a gentle humpbacked slope, whereas the downglacier side is an abrupt steep drop. A roche moutonnée's asymmetry provides another indicator of the direction of glacier flow.

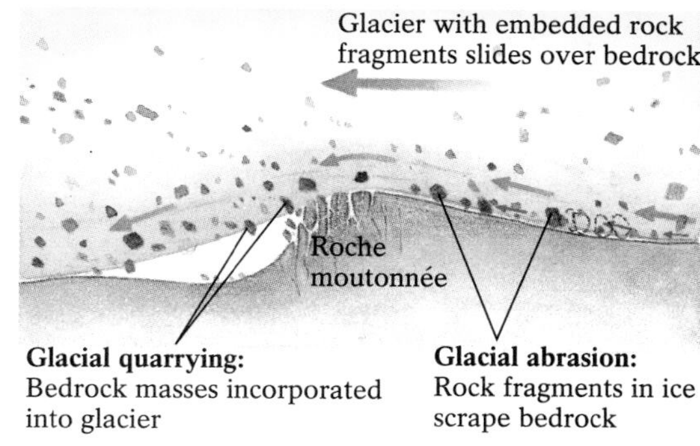

Figure 17-11 Glacial abrasion and quarrying combine to shape a roche moutonnée, like this one in Yosemite National Park, California. This distinctive erosional feature is produced when a glacier overrides a bedrock hill, abrading its upglacier side and quarrying its downglacier side.

Erosional Features of Alpine Glaciation Alpine glacial erosion can sculpt smooth mountain slopes into spectacular rough-hewn peaks and precipitous gorges. The process begins with the creation of **cirques,** deep horseshoe-shaped basins that develop in shady depressions at or above the local snowline. Here, lack of sunlight and accumulation of wind-drifted snow support the growth of perennial snowfields. Meltwater from these snowfields seeps into the surrounding bedrock during periods of partial thaw. Then, when the weather gets colder, the water refreezes, loosening rock fragments by frost wedging. Mass movement and surface runoff subsequently remove some of the loosened debris, enlarging the basin. As snow accumulates in the basin, it forms a cirque glacier. Quarrying and abrasion by the glacier's flow continue to erode the cirque, making it progressively deeper and longer.

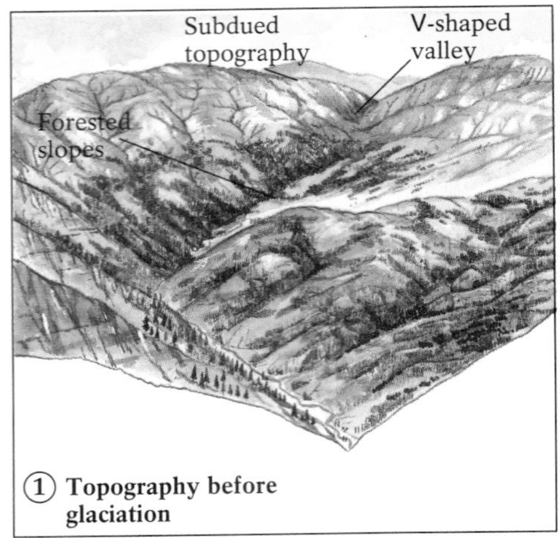

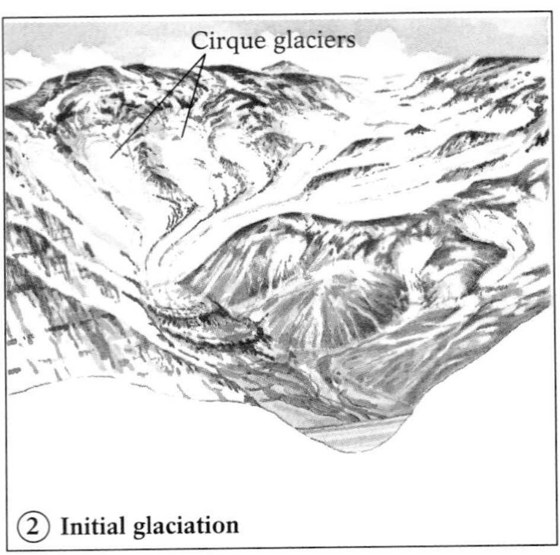

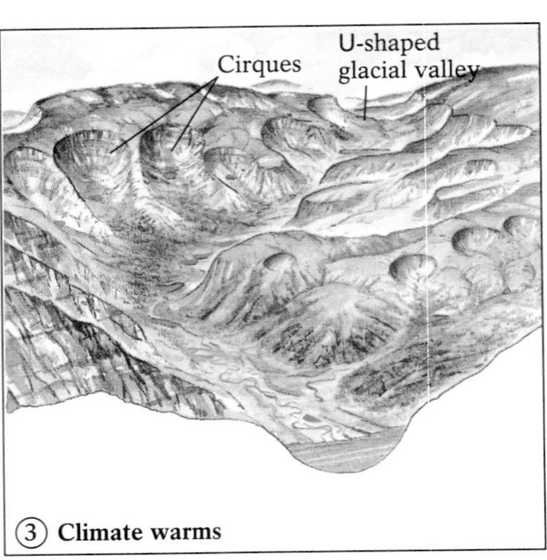

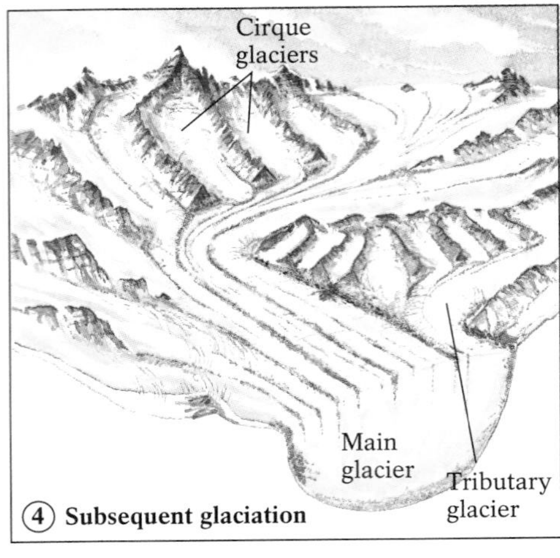

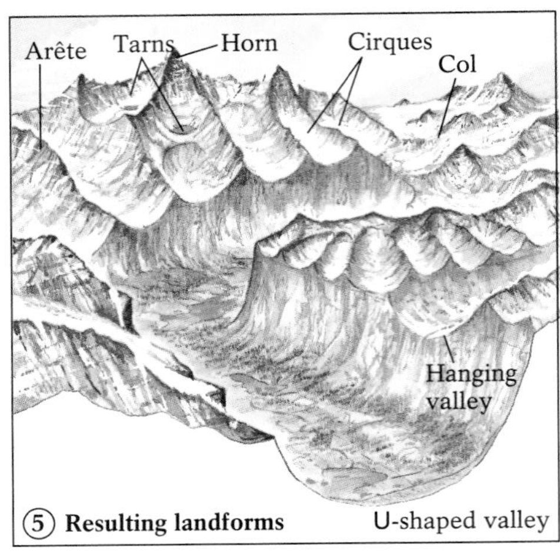

(a)

(b)

(c)

Figure 17-12 Effects of alpine glacial erosion. (**a**) Growth and proliferation of cirques on mountainsides eventually produces horns, arêtes, cols, hanging valleys, and fjords. Another common indicator of glacial erosion is the conversion of V-shaped stream valleys into U-shaped glaciated valleys. (**b**) A U-shaped valley in Glacier National Park, Montana. (**c**) U-shaped valleys that become flooded with seawater produce fjords; this one is Bela Bela Fjord in British Columbia, Canada.

A number of cirque glaciers can form on the same mountain at the same time, producing a variety of unique alpine landforms (Fig. 17-12). When two cirque glaciers on opposite sides of a mountain erode headward and converge, they form a sharp ridge of erosion-resistant rock at their top and sides called an **arête** (from the French for "fishbone," because a series of arêtes resembles a fish skeleton when viewed from the air). Continued headward erosion of a cirque can erode part of an arête, producing a mountain pass, or **col** (from the Latin *collum,* or "neck"). Cols sometimes provide natural routes through imposing mountain ranges; Colorado's State Route 40, for example, traverses Berthoud Pass, a col through the Rockies near Winter Park. When three or more cirque glaciers erode a mountain, a steep peak, or **horn,** develops. The Matterhorn in the Swiss Alps is perhaps the best known of these forms, but horns also dot the skyline of Banff and Jasper National Parks in the Canadian Rockies.

If the climate warms over time, a cirque glacier may melt until all that remains is water. This forms a cirque lake, or **tarn,** in its basin. If the climate cools and the glacier's budget is consistently positive, a cirque glacier may spread beyond its basin and flow downslope along a preexisting stream valley. As such a valley glacier moves it erodes the valley floor and walls, ultimately transforming the narrow V-shaped stream valley into a broad-floored U-shaped glaciated valley (Fig. 17-12b).

If cooling continues, small glaciers may develop in tributary stream valleys and eventually join to form a main valley glacier. The main glacier is thicker and flows faster than its tributaries, eroding its underlying bedrock more rapidly. Eventually, it undercuts the tributary valleys, leaving them perched above the main glacier as hanging valleys. If a warm period causes melting, waterfalls will cascade from the hanging valleys; Yosemite Falls in Yosemite National Park, with a vertical drop of 735 meters (2425 feet), is an example.

If seawater submerges a U-shaped glaciated valley, a deep, saltwater-filled **fjord** is formed (Fig. 17-12c). Some fjords were once thought to be stream valleys drowned by rising sea levels. The depth of most fjords, however, is greater than the postglacial rise in sea level. Tidewater glaciers apparently continue to erode their basins below the sea, thus carving them to increasingly great depth.

Erosional Features of Continental Ice Sheets Erosion by continental ice sheets produces striations and roches moutonnées as well as much larger features. An ice sheet can abrade and quarry entire mountains, producing large-scale asymmetrical landforms called *whalebacks,* similar to but bigger than roche moutonnées. As a continental ice sheet flows over preglacial stream valleys, it carves monumental U-shaped troughs. The Finger Lakes of western New York—eleven long, deep troughs—were excavated by ice-sheet erosion of preglacial valleys (Fig. 17-13a), as were North America's five Great Lakes (Fig. 17-13b), the northern section of Puget Sound in Washington, and Scotland's Loch Ness.

(a)

(b)

Figure 17-13 Continental ice sheets erode preglacial lowlands to leave deep basins and valleys. (**a**) During the Pleistocene Epoch, ice accumulating in the Lake Ontario basin (in what is now upstate New York) overflowed and moved south over the landscape, enlarging preexisting stream and river valleys to form deep linear troughs as well as countless smaller depressions, which eventually became filled with glacial meltwater and are today known collectively as the Finger Lakes. Lake Ontario, itself a glacially carved basin, is north of the Finger Lakes in this satellite photo. (**b**) A satellite photo showing the five Great Lakes of North America (from left to right, Lakes Superior, Michigan, Huron, Erie, and Ontario). The much smaller Finger Lakes are shown at lower right.

Continental ice sheets also erode on a continental scale. Because most of a glacier's erosion affects the bedrock under its accumulation zone, the ice sheet that covered much of northern North America 18,000 years ago scoured its accumulation zone bare of loose rock and soil. The most intensive erosion occurred in the southern portion of the Canadian Shield and in the region of soft sedimentary rocks to the south, in southern Ontario and Quebec, Minnesota, Wisconsin, Michigan, Illinois, and New York. This pronounced erosion produced the Great Lakes, which are more than 500 meters (1600 feet) deep in places.

Glacial Transport and Deposition

The load of debris eroded by a glacier is eventually deposited, usually when the glacier's ice begins to melt. All glacial deposits are called **glacial drift.** (The name originated in the nineteenth century, when people thought that such deposits had been carried to their resting places by icebergs floating on the biblical floodwaters. We continue to use the term today, but with more understanding of the components of these deposits.)

Figure 17-14 Large granite erratics on a lake shore in Superior National Forest, Minnesota. The erratics were carried by the Laurentide continental ice sheet about 15,000 years ago, and probably came from due northeast, passing over southwestern Ontario.

The size of the particles that a given flow of water or air can carry is limited to a narrow range, depending on flow velocity and volume. Thus, deposits left by water or air are typically well-sorted (see Chapter 6). Because of the high viscosity of ice, however, even extremely large masses of embedded rock cannot settle through the ice, and are transported. As a result glacial deposits are typically poorly sorted, often ranging in size from fine clays to house-sized boulders. Most glacial drift is deposited after being carried only a few kilometers, but some materials can be transported for remarkably long distances. A rock that differs from most others in a glacial deposit can sometimes be traced back to its source outcrop and its transport distance determined. For example, chunks of copper have been found in glacial deposits in southern Illinois; this copper is not native to Illinois but comes from Michigan's Upper Peninsula, 960 kilometers (580 miles) north. An exceptionally large, glacially transported rock that has been eroded from one type of bedrock and deposited on another is a **glacial erratic** (Fig. 17-14).

Some glacial debris is deposited directly by the ice that carried it; other debris is picked up and later deposited by meltwater streams that flow on, within, beneath, and in front of the ice. Meltwater streams deposit glacial sediment wherever their paths take them—on land, into the lakes that form in front of the ice, or into the sea.

Till Drift that is deposited directly from glacial ice forms a distinctive sediment called **glacial till** (from the Scottish word for "obstinate," an allusion to Scotland's plow-resisting rock-and-clay soils). Till is deposited almost exclusively in the ablation zone, either by being plastered onto the underlying glacial bed by flowing ice or by sloughing off the glacier's surface as it melts (Fig. 17-15a). Tills are characteristically unstratified and unsorted, and generally consist of large rock fragments surrounded by a finer-grained matrix of sand, silt, or clay. Because all the particles in a till were eroded from the bedrock over which the glacier passed, geologists can judge the rock types that were present along the glacier's route; they can then use this information to determine the glacier's flow direction.

Sometimes the stones in till are economically valuable. Tills throughout the Midwest, for example, have occasionally contained diamonds. In 1876, in a southeastern Wisconsin town, a well digger unearthed a pale yellow stone and gave it to his employer, who thought it was common quartz or topaz and later sold it for one dollar. It turned out to be a 15-carat diamond and was

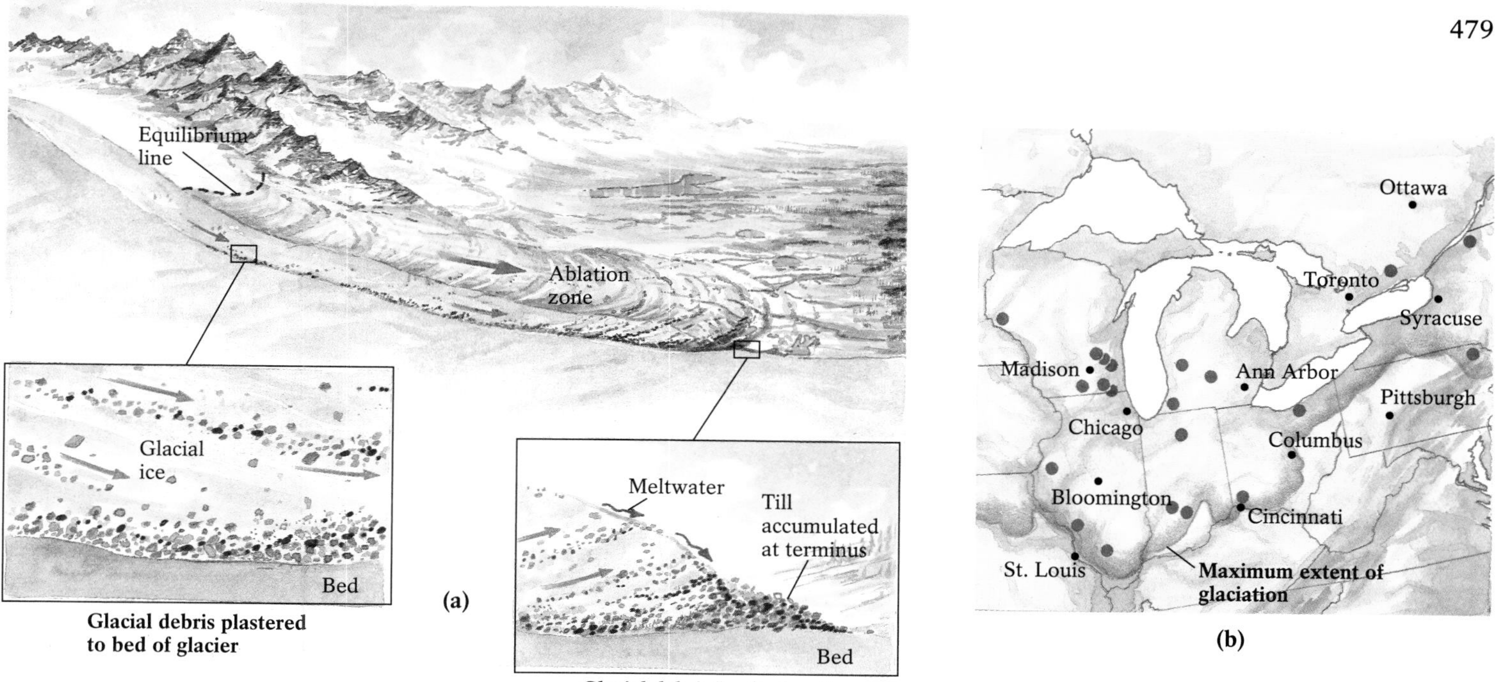

Figure 17-15 Deposition of glacial till. (**a**) Tills are deposited downglacier from the equilibrium line, either by being plastered to the glacial bed by the weight of the glacier or by "melting out" at the glacier's surface. (**b**) Known locations of diamond-bearing tills around the Great Lakes.

eventually sold to the American Museum of Natural History in New York City (from which it was stolen in 1964 and never recovered). Similar stones have been found throughout the Great Lakes region (Fig. 17-15b).

Glacial till often forms a **moraine,** a mass of drift that accumulates at the terminus of a glacier. (The name comes from *morena,* a word used by farmers to describe ridges of rocky debris in the French Alps.) Moraines occur as bands of hills marking the various advances and retreats of a glacier (Fig. 17-16). Even when the terminus of a glacier remains stationary, ice continues to flow from the accumulation zone to the ablation zone; as it does so, it erodes and transports sediment to the terminus where the load piles up. The longer the terminus remains stationary, the more till the glacier will deposit in one place and the larger the resulting moraine. If the climate warms and ablation begins to exceed accumulation, the glacier recedes, leaving a *terminal moraine* marking the farthest advance of the ice.

Figure 17-16 Glacial advance and retreat and the deposition of moraines (see text). At the end of the sequence shown, moraine 2 is the glacier's terminal moraine—that which marks the maximum extent of the glacier. Moraines 3 and 4 are recessional moraines, marking the positions of the glacier's terminus during periods when the glacier's retreat was halted by temporary cooling trends. Moraine 1 has been overridden and eroded. Inset: A moraine deposited by Exit Glacier at Kenai Fjord in Alaska.

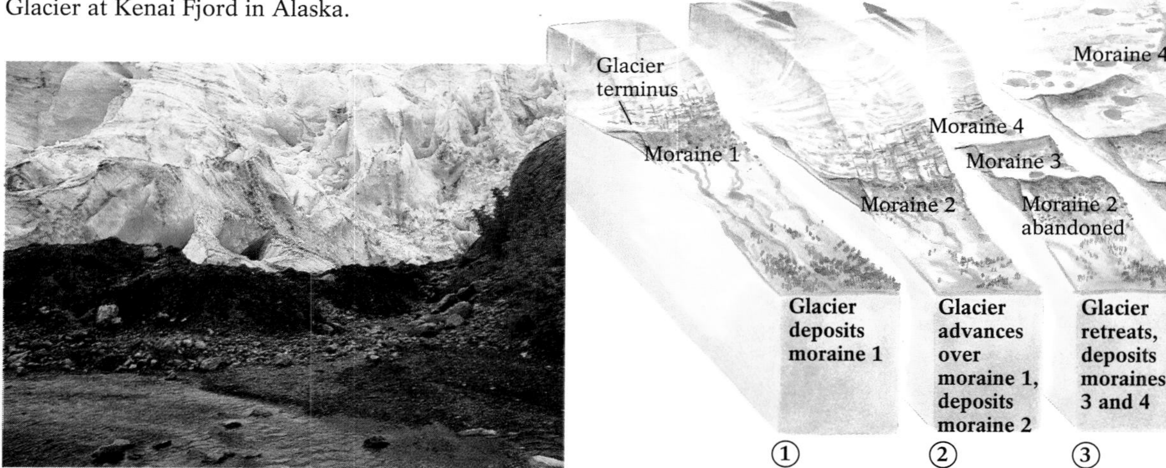

If warming is interrupted by a return to glacial conditions, halting glacial recession and causing the terminus of a receding glacier to remain stationary for a few hundred years, the glacier may deposit a *recessional moraine*, upglacier from and usually parallel to its terminal moraine. A glacier that recedes intermittently may deposit several recessional moraines. Continental ice sheets also produce terminal and recessional moraines (Fig. 17-17).

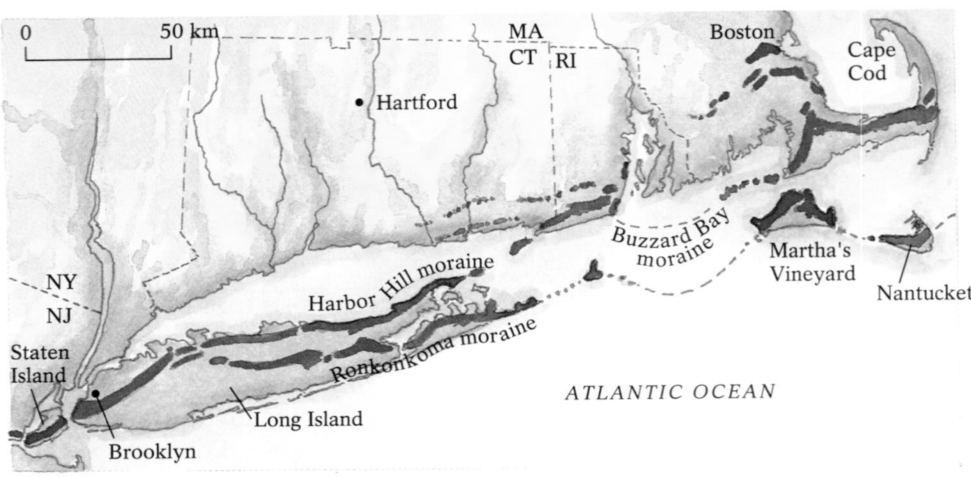

Figure 17-17 Terminal and recessional moraines were deposited over a large area of the northeastern United States coast by the North American ice sheet. The recessional Harbor Hill moraine, deposited about 14,000 years ago, passes through Prospect Park in Brooklyn, New York, continues across the north shore of Long Island, and then appears farther north, where it forms the mid-island hills of Cape Cod. The terminal Ronkonkoma moraine, deposited about 20,000 years ago, forms central Long Island and continues east and then north to Martha's Vineyard and Nantucket Island in Massachusetts Bay. The map indicates local names for the moraines, such as "Buzzard Bay Moraine" and "Vineyard Moraine."

Two other types of moraines—lateral and medial—are formed only by alpine glaciers. Because they are topographically confined by their valleys, these glaciers have distinct sides adjacent to the valley walls where till is deposited, forming *side* or *lateral moraines*. Lateral moraines also contain any material that has fallen from the valley walls into the crevices between the ice and the walls. When two valley glaciers merge, their sediment loads are carried downvalley along the sides of the two distinct ice streams (unlike merging river tributaries, the sediment loads of which would be turbulently mixed). When these loads are deposited, they remain adjacent to one another, forming a *medial moraine* (Fig. 17-18).

Figure 17-18 Medial moraines form when adjacent glaciers converge, causing the lateral moraines at their edges to run together. Here several medial moraines are developing along the branches of the lower Kaska Wulsh Glacier in the St. Elias Mountains, Yukon Territories, Canada.

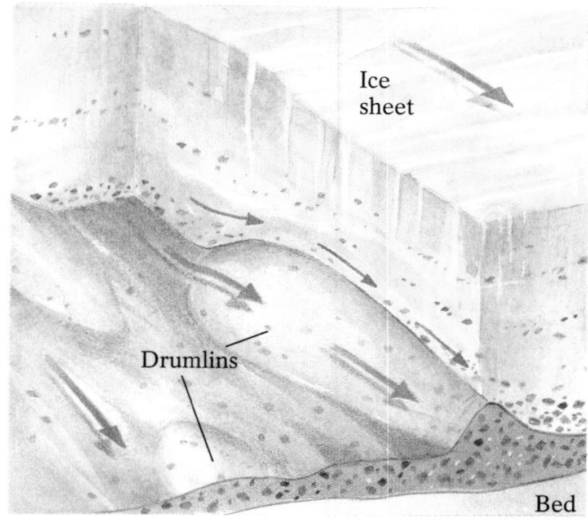

(a)

(b)

Figure 17-19 (**a**) Ice sheets passing over preexisting moraines can exert enough pressure to reshape them, forming low oval hills called drumlins. Drumlins are usually 25 to 30 meters (80–100 feet) high, 0.5 to 1 kilometer (0.3–0.6 miles) long, and 400 to 600 meters (1300–2000 feet) wide. (**b**) A drumlin field east of Rochester, New York.

If climate cools significantly after glaciers have remained stationary for a long period of moraine building, the glaciers may advance over previously constructed moraines, incorporating their sediments and redepositing them at "new" terminal moraines. When ice sheets, thicker and more massive than alpine glaciers, override moraines, they apply enough pressure to reshape the moraines into gently rounded, elongated hills called **drumlins** (from the Gaelic word *druim,* for "ridge" or "hill"). A drumlin has an asymmetrical profile similar to that of the inverted bowl of a kitchen spoon; it is streamlined, being blunt on its upglacier end and gently sloping on its downglacier end, due to having been formed beneath actively flowing ice (Fig. 17-19).

Drumlins, which range in height from 5 to 50 meters (17–170 feet), are typically aligned parallel to one another and oriented in the direction of ice flow. They generally occur in clusters of hundreds or thousands in areas once covered by continental ice sheets. One drumlin field, east of Rochester in west-central New York, contains more than 10,000 drumlins. Other notable drumlin locations include western Nova Scotia, southern Manitoba and Saskatchewan, central Minnesota, eastern Wisconsin, and northern Michigan. Perhaps the continent's most famous drumlins are north of Boston where the Battle of Bunker Hill (itself a drumlin) was fought on Breed's Hill, an adjacent drumlin.

Deposits from Glacial Meltwater Streams Melting may be continual near the end of a period of glaciation, or a glacier may melt seasonally, especially during the summers in temperate climates. Meltwater streams flowing from a glacier sort and stratify the drift they deposit, creating glacial sediments that are very different from till.

Glacial meltwater may flow on top of, in front of, and beneath a glacier. Sediment deposited on top of a glacier fills in depressions in its surface; sediment deposited immediately in front of a glacier may become deltas in glacial lakes. The most common sediment from glacial meltwater, a mixture of sand and gravel particles, is deposited by braided streams downstream of the terminus as **outwash.** Outwash plains, characterized by broad, gently sloping surfaces, accumulate beyond the front of a glacier.

Erosion of outwash produces a mass of fine silt, which is often later transported by powerful glacial-age winds. The fertile plains along the Mississippi River near Vicksburg, Mississippi, though far removed from any direct effects of glaciation, were buried about 18,000 years ago under a thick layer of silt eroded from the outwash of the North American ice sheet.

Figure 17-20 Fertile loess farmland in Washington state.

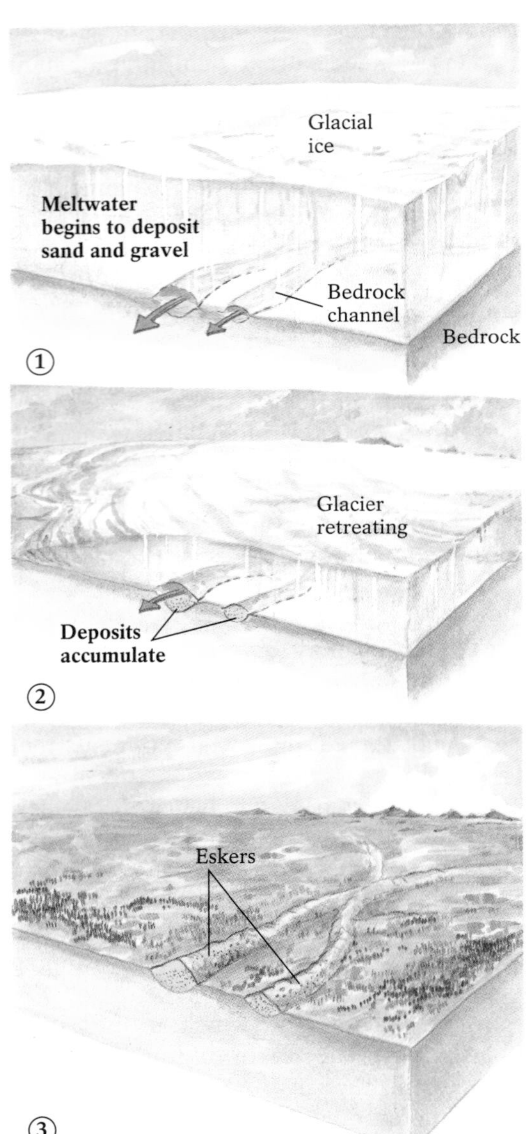

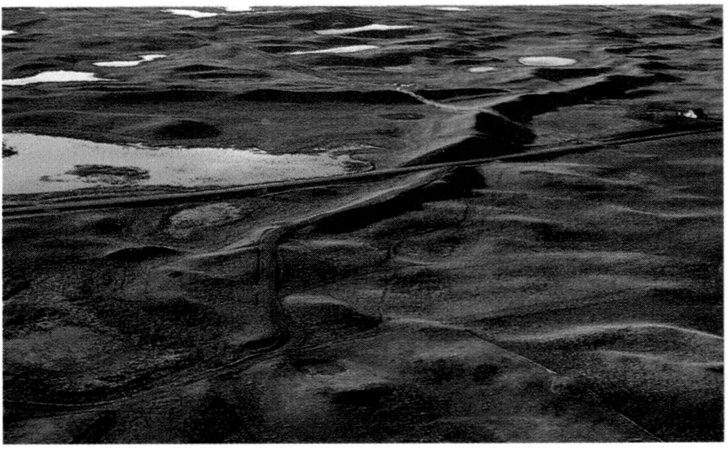

Figure 17-21 The origin of eskers. Eskers typically form under the ablation zone of a glacier, where meltwater erodes curved channels in the bedrock and deposits its load en route. When the glacier retreats, the meltwater flows away or evaporates, leaving meandering ridges of stratified, cross-bedded sand and gravel. Photo: Eskers in Coteau des Prairies, South Dakota.

Although the ice sheet extended only as far as northern Iowa, its outwash was transported down the Mississippi River to the Gulf of Mexico and deposited along its banks; winds blowing across the drying outwash eroded it and transported the fine particles to the surrounding land. Such wind-blown silt deposits are known as **loess** (from the German for "loose"), and are commonly found downwind from exposed, drying outwash. Throughout the Mississippi River valley (and beyond the fronts of European ice sheets as well) are buff-colored, near-vertical bluffs composed of loess. (We will discuss loess in greater detail in Chapter 18.) The loess eroded from outwash plains provides some of North America's finest farmland (Fig. 17-20).

Meltwater deposits of sand and gravel can also accumulate underneath glacial ice, forming sinuous ridges known as **eskers**. Eskers form beneath the ablation zone, as meltwater flowing along a glacier's bed carves S-shaped bedrock channels and deposits sand and gravel in them (Fig. 17-21). Eskers can be seen throughout southern Canada and in much of the northern United States, including Maine, Michigan, Wisconsin, Minnesota, the Dakotas, and eastern Washington. They may be less than a kilometer long or more than 150 kilometers (100 miles) long, and can be 30 meters (100 feet) high. If you're interested in seeing an esker you may have to move quickly, because eskers are being mined extensively for their commercially valuable sand and gravel and are rapidly disappearing from the landscape.

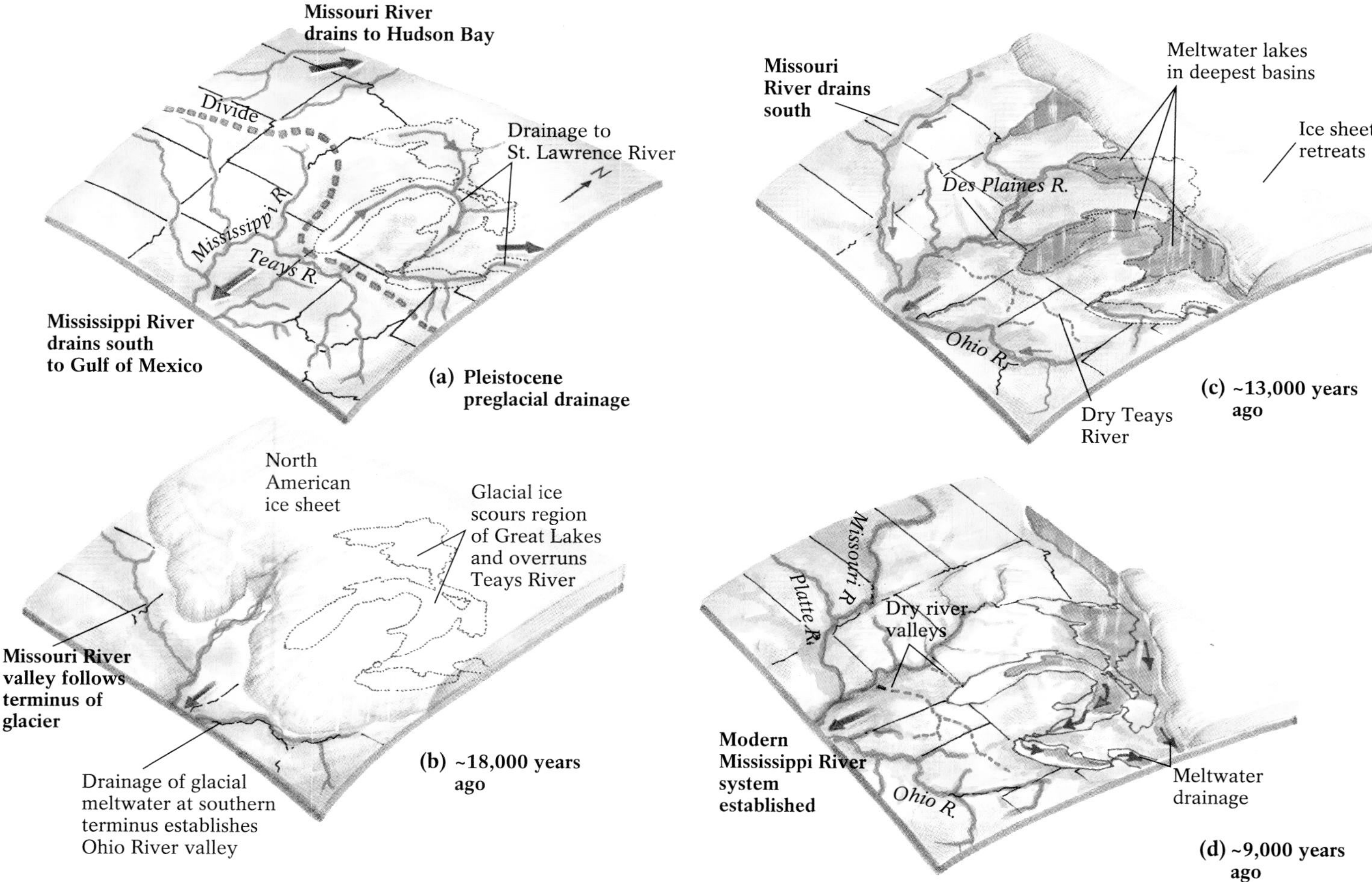

Figure 17-22 The effect of the North American ice sheet on the Great Lakes region. (**a**) Before the period of glacial expansion in the Pleistocene, streams flowed through the broad valleys that today are the Great Lakes. The major east–west tributary of the Mississippi River was the Teays River, which flowed through what are now Illinois, Indiana, and Ohio. North of the Teays was an elevation that served as a divide, north of which rivers such as the Missouri flowed northward. (**b**) At the maximum extent of Pleistocene glaciation about 18,000 years ago, the North American ice sheet completely covered the Great Lakes region, deepening the lakes' basins by erosion. It also blocked the flow of the Missouri River, diverting it southward. (**c**) With warming, the ice sheet melted and retreated northward, leaving accumulations of meltwater in the deepest basins. By about 13,000 years ago, the Great Lakes and the Mississippi and its major tributaries were filled. Water in shallower basins, such as the Teays River, flowed off or evaporated, leaving dry valleys. (**d**) By 9000 years ago, the ice sheet had completely retreated from what is now the United States, and the current drainage system of this region was well established.

Other Effects of Glaciation

In addition to the direct effects of glacial erosion, transport, and deposition, glaciers and glacial periods also produce other direct and indirect changes on land, at sea, and in the atmosphere. They may even contribute to the evolution, migration, and extinction of various plants and animals. The following are some of the many signs of past worldwide glacial expansion.

Glacial Effects on the Landscape Glaciers and the conditions associated with glacial periods can change landscapes in a number of ways other than by erosion and deposition. The movement of the ice itself can disrupt preexisting features, and its weight can produce depressed and uplifted areas. Changes in climate can create vast areas of frozen ground and large landlocked bodies of water.

Expanding ice sheets can act like continent-wide dams, altering the courses of preexisting streams. The Missouri River and its tributaries, for example, once flowed north to Hudson Bay, until they were blocked and diverted by the North American ice sheet. Now the Missouri River follows the ice sheet's southern terminus and then joins the Mississippi River on its way south to the Gulf of Mexico (Fig 17-22). A similar drainage disruption, which caused a great series of floods, is discussed in Highlight 17-1.

Highlight 17-1 *The Channeled Scablands of Eastern Washington*

Today the Clark Fork River of northwestern Montana is a freely flowing mountain stream, but 13,000 years ago it became blocked by a broad lobe of ice protruding from a retreating Cordilleran ice sheet, which covered western Canada to the Canadian Rockies (Fig. 17-23a). The lobe of ice, known as the Purcell Lobe, flowed through a wide U-shaped valley extending from southern British Columbia to northern Idaho; in the process it dammed the Clark Fork's westward flow, creating a lake that became swollen by meltwater from the wasting ice. Known today as Glacial Lake Missoula, it was more than 300 meters (1000 feet) deep and occupied an area about the size of Lake Michigan.

As meltwater continued to flow into it, Glacial Lake Missoula rose until it overflowed the ice dam, carving deep channels into the ice until eventually the ice dam ruptured, releasing the accumulated waters of Glacial Lake Missoula in a torrent (Fig. 17-23b). More water flooded eastern Washington than the combined flow of all rivers in the world today. The flood's velocity has been estimated, from the sizes of deposited boulders, at from 50 to 75 kilometers per hour (30–50 miles per hour).

The effects of this episode are still evident today in the Pacific Northwest (Fig. 17-24). The flood produced numerous rivers up to 75 kilometers (50 miles) wide and 250 meters (900 feet) deep, the courses of which are now marked by giant dry waterfalls and massive streambed ripples. It also excavated enormous channels by tearing hundreds of meters of basaltic lava from the Columbia River plateau, leaving thousands of remnant lava *scabs*—masses of basaltic rock that survived the flood intact. The region is thus aptly named the Channeled Scablands.

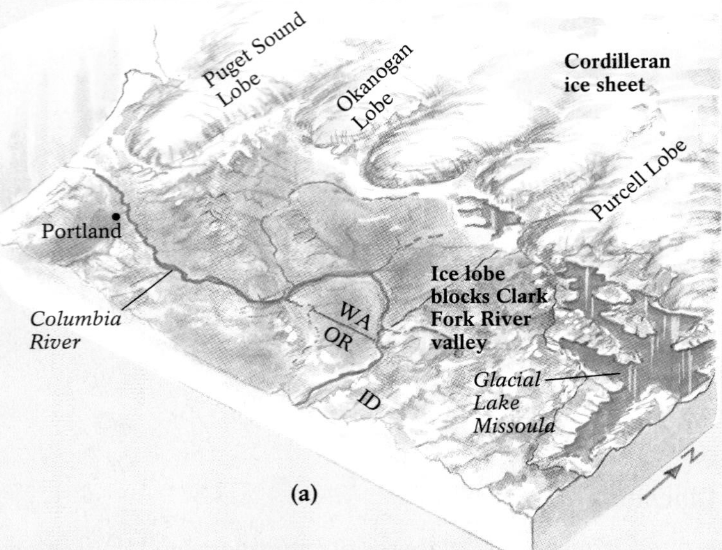

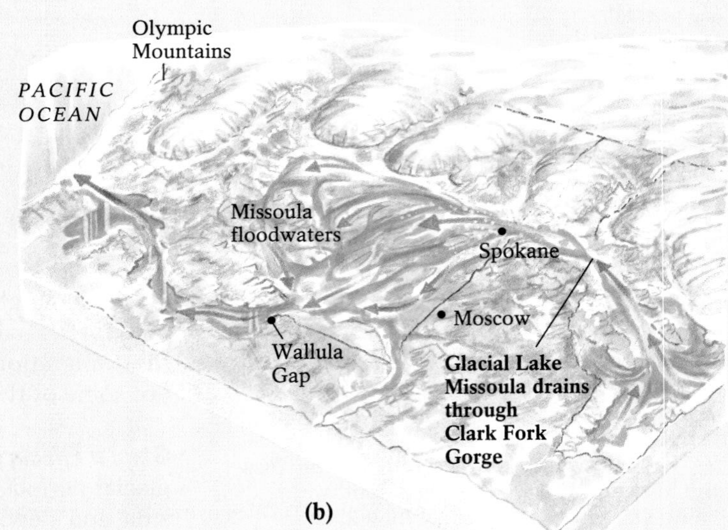

Figure 17-23 The creation and draining of Glacial Lake Missoula. (**a**) About 13,000 years ago, the Purcell Lobe of the retreating Cordilleran ice sheet blocked the Clark Fork River valley, creating Glacial Lake Missoula. (**b**) As the glacier retreated, its meltwater swelled the lake until it finally overflowed its dam, releasing more than 20 million cubic meters (more than 750 million cubic feet) of water per second.

Figure 17-24 Effects of the Missoula flooding. (**a**) Massive rivers created by the flooding produced giant bedrock ripples, such as these in northwestern Montana. (**b**) This basalt bedrock near Coulee City, Washington, is etched by dry channels and falls where floodwaters once streamed. (**c**) As Glacial Lake Missoula grew, its rising shoreline carved wave-cut benches that can now be seen on the side of Sentinel Mountain at the University of Montana campus.

Geologists speculate that such great floods occurred repeatedly in eastern Washington during the Pleistocene ice age, as glaciers periodically advanced and retreated. We know that the last flood took place about 13,000 years ago because Mount St. Helens erupted at the same time—ash layers from the eruption are interlayered with the last of the flood deposits and provide geologists with a dating marker. Massive ice sheets and unimaginable floods, plus a fiery volcanic cataclysm, make Washington state of 13,000 years ago a geologist's dream: The evidence left behind gives geologists a storehouse of information about the region's recent past.

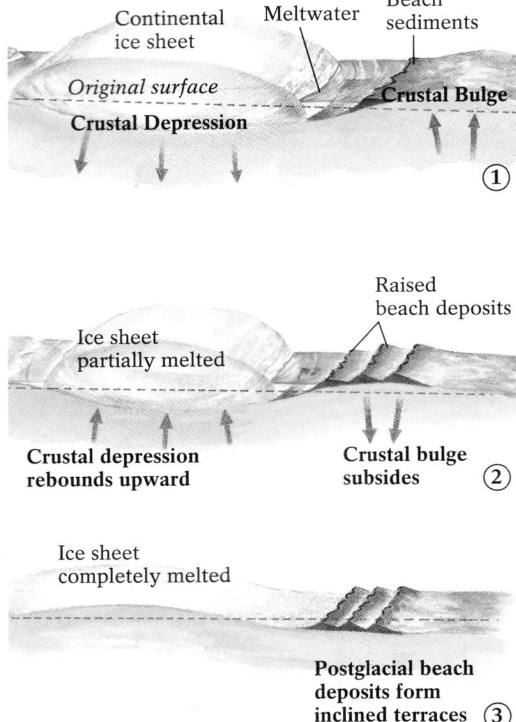

Figure 17-25 Formation of terraces from crustal rebound. Meltwater accumulates in the space between a retreating ice sheet and the crustal bulge at its margin, leaving sediments. As the ice sheet retreats further and the crust slowly returns to its original position, the bulge subsides, leaving the sediment deposits as slightly inclined terraces.

A continental ice sheet weighs so heavily on the Earth's crust that the asthenosphere (in the upper mantle) is pushed away from the center of the ice sheet toward its margins, creating a crustal depression under the glacier and topographic bulges along its borders. When the climate warms and the ice sheet melts, the displaced asthenosphere gradually returns to its original position, the depressed crust slowly rebounds, and the marginal bulges are eliminated. (This is analogous to what happens to a water bed when you lie on it. The weight of your body pushes the water from beneath you, forming a depression in the bed's surface, and toward the edges of the bed, forming a bulge. When you get up, the water at the edges flows back to its original position, and the depression is eliminated.) The effects of crustal rebound can be seen throughout Scandinavia and around the Hudson Bay region of east-central Canada, in the form of uplifted terraces (Fig. 17-25). The process of crustal rebound, which continues today in those areas, takes thousands of years because of the extremely high viscosity of the flowing asthenosphere.

When ice sheet margins bulge due to crustal depression and later subside, flooding may become a serious problem. The southwestern Netherlands, located near the margin of the former European ice sheet, once stood well above sea level atop just such a bulge. Since the departure of the ice sheet about 12,000 years ago, the depressed crust has been rebounding and the related bulge has diminished, causing the southwestern Netherlands to subside—in places to a position below sea level. Since the thirteenth century, only Holland's well-known dikes have kept the North Atlantic from flooding its valuable farmland.

The nonglaciated areas adjacent to glaciers and ice sheets have extremely cold climates, and the pervasive frost action in such *periglacial* (near-glacial) environments produces a distinctive set of landscape features. (These features are also found in some high-latitude regions—such as northern Alaska, arctic Canada, and Siberia—that are also extremely cold, despite being far from glaciers.) Perennially frozen ground, or **permafrost,** is characteristic of high-latitude periglacial environments. Today, permafrost underlies about 25% of the Earth's land surface, mostly in remote polar regions. Permafrost in Siberia extends to a depth of 1500 meters (5000 feet), and sections in the Canadian Arctic and Alaska reach depths of 1000 meters (3300 feet) and 600 meters (2000 feet), respectively. Most of the world's permafrost is thought to have developed during the last worldwide glacial period, between about 90,000 and 12,000 years ago. The presence of frozen woolly mammoths in the permafrost of Siberia and Alaska has made carbon-14 dating of those deposits possible.

A permafrost layer consists of soil, sediment, and bedrock that is continually at or below 0°C. During the summer—even brief high-latitude summers—the thin upper zone of permafrost, or active layer, thaws and becomes extremely unstable, and is apt to trigger mass movements. Buildings and other structures in permafrost areas must be designed to minimize the flow of heat from the structures to the ground. Oil pipelines (carrying warm crude oil) and heated buildings are usually elevated above the ground surface, allowing cold air to circulate beneath them and keep the surface frozen.

The ground surface in periglacial areas often has a variety of unique features (Fig. 17-26). For example, as periglacial regolith alternately freezes and thaws, the frozen subsurface materials expand upward, slowly pushing stones up until they surface, often in surprising arrangements: *Sorted circles,* for instance, form as stones are pushed laterally into circular patterns. When soil temperatures plunge well below 0°C (32°F), the ground in periglacial areas freezes, contracts, and cracks to form large polygonal shapes, or *patterned ground.* When the surface water that flows during brief warm periods drains

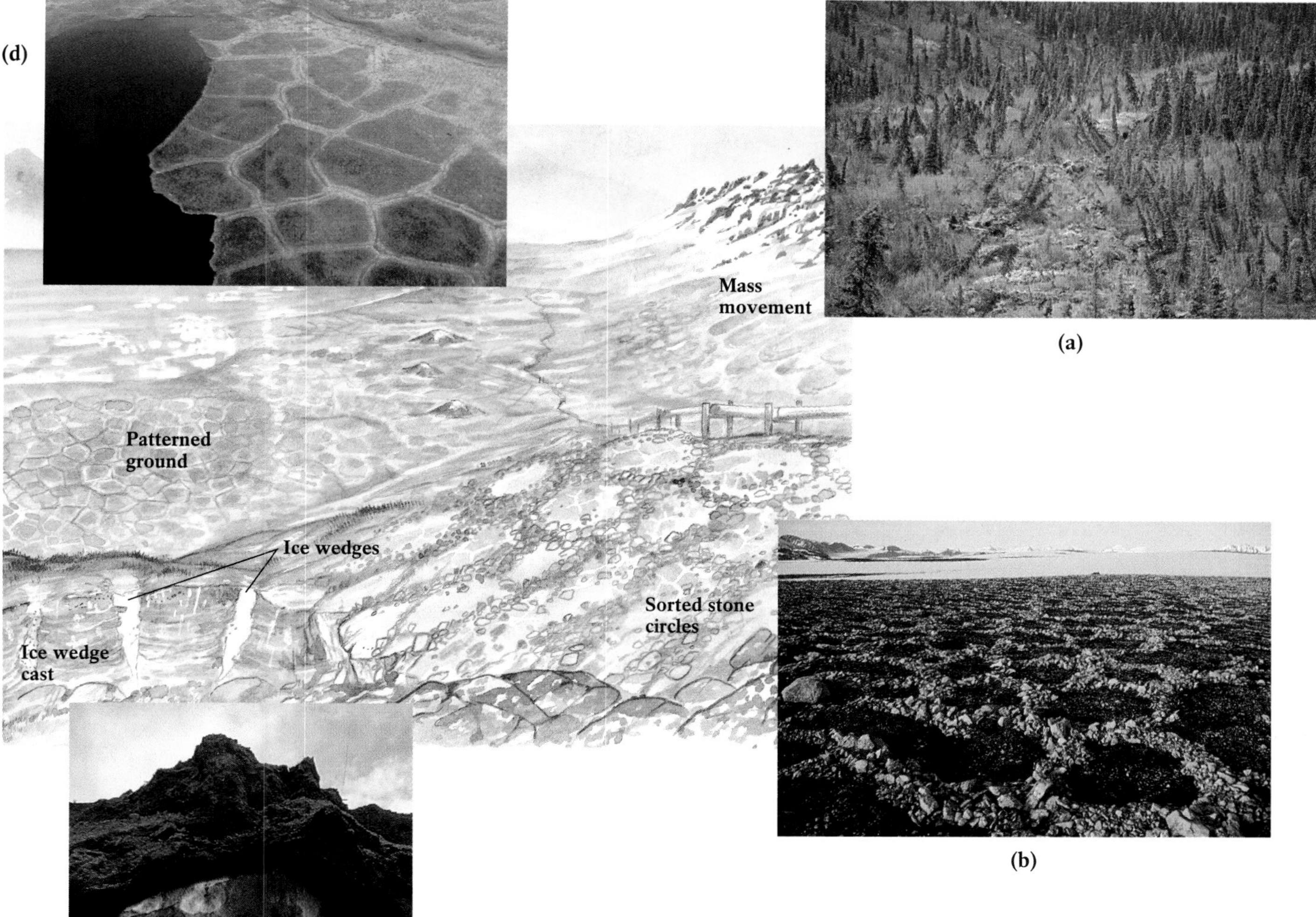

Figure 17-26 The features of a periglacial environment include: (**a**) "slipped" areas resulting from permafrost thawing and subsequent mass movement, such as this forest in Mount McKinley Park, Alaska; (**b**) sorted stone circles, such as these in western Spitsbergen, Norway; (**c**) ice wedges, such as this one along the Yukon River in Galena, Alaska; and (**d**) patterned ground, as shown here in the Arctic.

into the cracks at the edges of the polygons, it freezes and forms *ice wedges.* When ice wedges thaw in spring and summer, the water drains out, leaving cavities that may be filled by inwashing sediment, which subsequently becomes compacted to form *ice-wedge casts.*

The presence of *relict* ("fossilized") periglacial features indicates that periglacial environments once existed along the front of the North American ice sheet from New Jersey and Pennsylvania west to the Dakotas, and at higher elevations south along the crest of the Appalachians. Pleistocene-age ice-wedge casts in New Jersey and Connecticut, more than 1500 kilometers (1000 miles) south of the current permafrost limit, suggest that this region was once the periglacial frontier of the North American ice sheet.

The indirect effects of ice sheets may also extend well beyond periglacial regions. When air from warmer regions encounters air chilled by ice sheets, the result is cloudy, cool, rainy weather beyond the ice sheet's terminus. During the last glacial period, such a humid and cool climate prevailed in North America. This climate produced more precipitation than falls today, and less of it evaporated. The remaining water accumulated in landlocked basins, forming **pluvial lakes** (from the Latin *pluvia,* "rain"). This indirect effect of glaciation even reached as far south as Death Valley, California, and the deserts of the American Southwest, covering vast areas of those now-arid lands with shallow lakes. The largest pluvial lake, Lake Bonneville, covered much of Utah, eastern Nevada, and southern Idaho; Utah's Great Salt Lake is a small remnant of it (see Fig. 6-15).

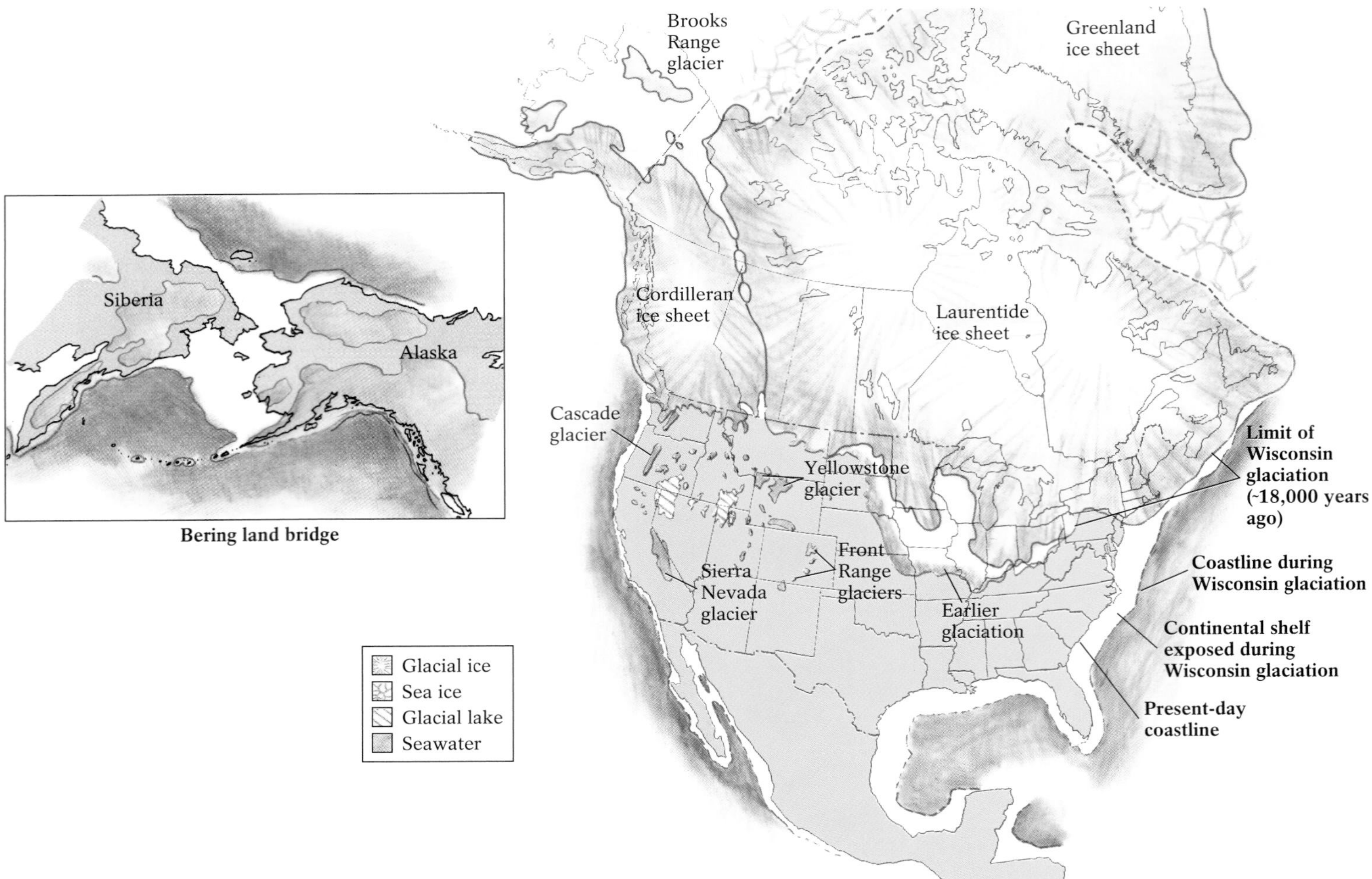

Figure 17-27 During the Wisconsin glaciation, falling sea levels exposed vast areas of the continental shelves. These included strips of land that served as bridges between landmasses. One exposed land strip in particular altered the course of human history in the Western Hemisphere —the strip of land now covered by the shallow Bering Strait, separating Alaska and Siberia. This was, until about 12,000 years ago, a land bridge that permitted human and animal traffic to cross in both directions. Most American archeologists agree that this land bridge provided the route by which the first human inhabitants of North America arrived from Asia.

Glacial Effects on Sea Levels Because the water in glacial ice ultimately comes from the oceans, global sea levels drop when glaciers expand. As a result, during glaciations much land that was under water becomes exposed, and the coastlines of the continents change. For example, during the so-called Wisconsin glaciation 18,000 years ago, when the global sea level was about 140 meters (about 500 feet) lower than it is today, North America's Atlantic coastline extended more than 150 kilometers (100 miles) east of what is now New York City onto the continental shelf (Fig. 17-27). From fossilized tree stumps and mastodon bones recovered from North America's continental shelves, we know that the exposed continental shelves were covered by spruce and pine forests and traversed by herds of migrating ice-age mammals.

Lowered sea levels also expose "land bridges," connections between landmasses that are separated by narrow seaways in nonglacial times. These strips of land are actually exposed areas of continental shelf. For example, between about 100,000 and 40,000 years ago, during a period of worldwide glacial expansion, Britain and France were connected by a land bridge that today is covered by the shallow English Channel. Over it, humans migrated from the European mainland to the British Isles. Similarly, during the Wisconsin glaciation, until about 12,000 years ago, a land bridge linked Siberia in Asia to Alaska in North America; it is now the Bering Strait. Over this land bridge, humans and giant mammals such as mammoths and mastodons migrated into North America, and other animals, such as the camel and the horse, migrated from North America into Asia and Europe.

Glacial Effects on Flora and Fauna When climate changes and glaciers alter their habitats, animals may be unable to find sufficient food and nesting sites, and the life cycles of plants may be disrupted. During periods of glacial expansion, ice sheets overrun vast tracts of land, temperatures become colder, skies become cloudier, and sunlight decreases. Plants and animals may migrate to more suitable environments, evolve in ways that enhance their ability to survive the changed conditions, or become extinct.

As ice advanced to its last major maximum about 18,000 years ago, many of the plants and animals of North America and Europe migrated as much as 3000 kilometers (2000 miles) southward and competed for space and food with the flora and fauna already there. As the climate warmed and the ice sheets finally melted, environments changed again, leading to another series of migrations and further adaptations. The formerly habitable continental shelves were inundated by the rising sea, grasslands gave way to returning forests, and plants and animals began to populate newly deglaciated but still-barren lands. Organisms that could adapt to environmental changes survived both the Pleistocene's cold glacial and warm interglacial periods. Such ice-age creatures as the ground sloth, mammoth, and saber-toothed cat could not adapt to the warmer postglacial climate and became extinct.

Humankind was affected as well. With the warmer climate that followed the end of the last glacial period, migrating humans could find sustenance in newly habitable territories and human settlements expanded. The warmer climate and availability of new animal and plant species that could be exploited for food may have spurred the development of new tool technologies and hastened the discovery of agriculture. As surpluses accumulated and less time was spent hunting and gathering food, trade developed and complex societies emerged in South and North America, Africa, Europe, and the Near East.

The Causes of Glaciation

What causes glaciation? What makes ice advance and retreat during an ice age? The answers to these questions are complex, and there is no consensus among those who study them. One frustrated student concluded that there are at least as many hypotheses to explain the causes of glaciation as there are glacial geologists (scientists who study the geological effects of glaciers).

Many geologists have concluded that ice ages require two things: sizeable landmasses at or near the poles, and land surfaces whose average elevation is relatively high. These conditions enhance the prospects for an ice age by putting landmasses at latitudes and elevations where the climate is colder and glaciers are most likely to grow. The one phenomenon (which we have discussed throughout this text) that can both move landmasses to polar regions and raise them to higher elevations is, of course, plate tectonics.

The slow global cooling that started about 50 million years ago and led to the onset of the most recent ice age began after the landmass we now call Antarctica had moved to the South Pole; other major landmasses had meanwhile moved north of the Arctic Circle. As plates converged, several mountain ranges were rising steadily at their edges, producing elevations that rose well above preexisting regional snowlines. The rising Andes and Rockies were oriented north–south and were near oceans; therefore, they intercepted moist marine air carried by westerly winds, forcing it to rise above their peaks into the cold, high-altitude air, increasing precipitation and enhancing snow and ice accumulation.

But plate motion cannot move lands to and from the poles or raise and lower lands swiftly enough to account for the cycles of repeated glacial advances and retreats, which have lasted for periods of approximately 100,000 years during past ice ages. These short-term fluctuations in climate and glaciation may have been due to one or more of a variety of causes, including volcanism, meteorite impacts, and variations in the amount of solar energy reaching the Earth.

Volcanism could not have caused all the expansions and retreats of Pleistocene ice sheets because it does not occur at regular intervals and its effects are too brief. Large volcanic eruptions do lower world temperatures by as much as 1 to 2°C (2–4°F) for up to several years because volcanic ash and gases released into the atmosphere reflect and absorb solar radiation, causing less sunlight to reach Earth. However, no direct relationship between major eruptions and ensuing ice expansions has been proven, perhaps because the lower temperatures are so short-lived. Worldwide glacial expansion, for example, did not follow such cataclysmic eruptions as those of Mount Mazama and Krakatoa (see Chapter 4).

Large meteorite impacts also hurl veils of dust into the atmosphere and screen out significant solar radiation, lowering global temperatures. But again, they are not periodic and their effects are short-lived, so they are unlikely to be a principal cause of glacial expansion. It is possible, however, that a major strike may hasten climatic cooling already developing in response to other factors.

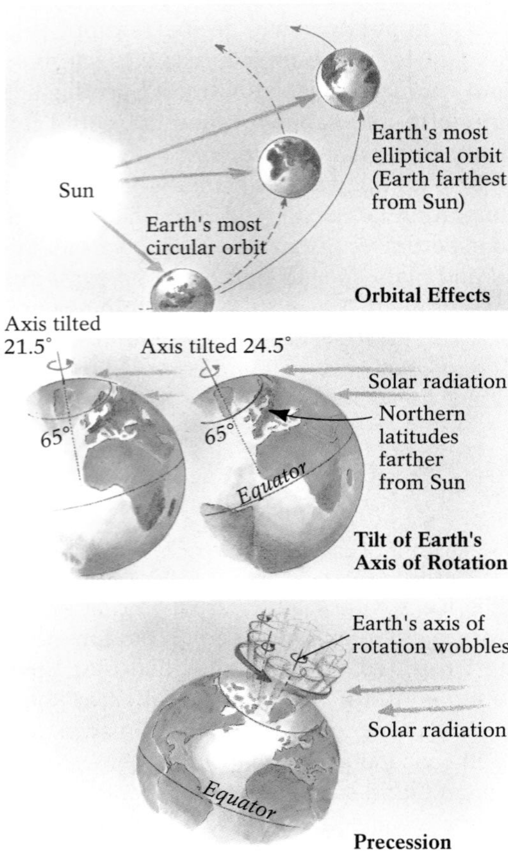

Figure 17-28 Milutin Milankovitch proposed that there are three astronomical factors that may interact to affect the amount of solar radiation striking the Earth at any given time, thereby influencing climate and, possibly, the periodic advance and retreat of glaciers in the region of 65°N latitude. The climate at 65°N latitude cools when the Earth's orbit around the Sun is at its most elliptical, when the tilt of the Earth's axis moves this region farther from the Sun, and when the Earth is not completely steady on its axis, but wobbles slightly. Each of these three factors has a different period of recurrence, but when they have coincided in the past glaciation has tended to occur.

Variations in the Earth's Orbital Mechanics

Many glacial geologists believe that glacial fluctuations during an ice age are caused primarily by variations in the Earth's position and orientation relative to the Sun. This hypothesis, first proposed by Yugoslavian astronomer Milutin Milankovitch around 1930, explains that whereas the total amount of solar radiation the Earth receives may vary but little, periodic changes in where and when that radiation is received have profound effects on global climate. For example, the Milankovitch hypothesis suggests that ice sheets have developed repeatedly in the vicinity of 65°N latitude, which today is largely ice-free. But when this region receives less solar radiation and becomes sufficiently cold, and when oceans provide ample moisture to large landmasses such as North America, Scandinavia, and Asia, then winter snows falling on these lands are able to survive the summer melting season, accumulating from year to year until glaciers form. The region in the vicinity of 65°S latitude has experienced significantly less glacial expansion because there is far less land in the Southern Hemisphere on which glaciers can grow. Once a global glacial expansion begins and the world's climate cools, glaciers do expand in Southern Hemisphere highlands such as the Andes.

Milankovitch calculated the effects of three periodic astronomical factors to show how they combine to lower the radiation levels at 65°N, and why they happen to do so at 100,000-year intervals. These three factors are the shape of the Earth's orbit around the Sun, the tilt of the Earth's axis toward or away from the Sun, and the wobble, or precession, of Earth's axis and equatorial plane as the Earth spins (Fig. 17-28). Each factor varies in cycles of different length, but according to Milankovitch the three cycles coincide at times, producing centuries or even millenia of low summertime radiation at 65°N. When this happens, summers are cool and brief, and alpine glaciers and continental ice sheets expand. Additional factors, however, must also contribute to the growth of glaciers, or else ice ages would occur with more precise periodicity than the geologic record indicates.

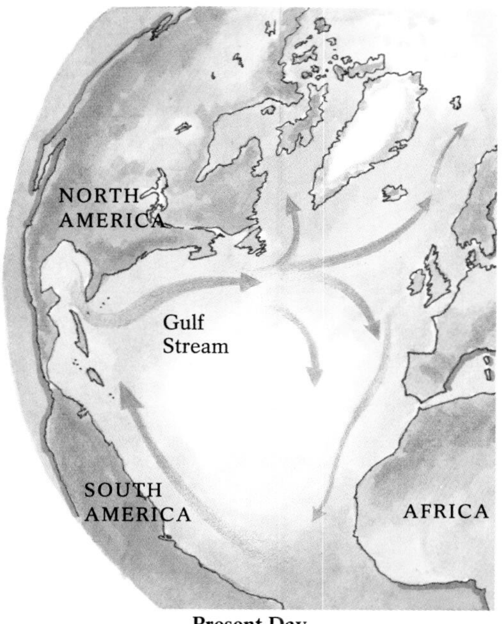

Present Day

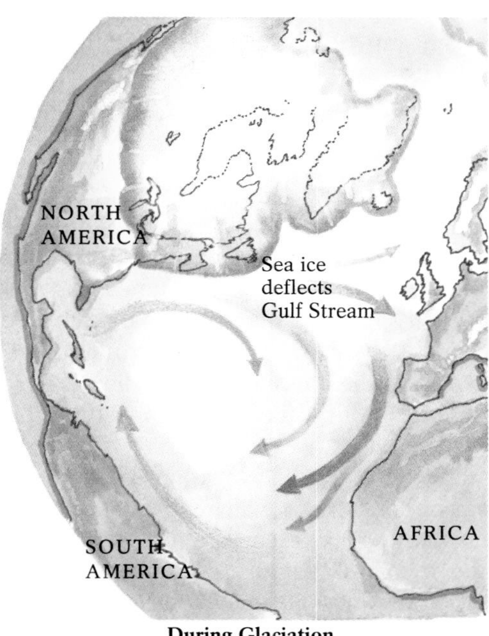

During Glaciation

Figure 17-29 During periods of glaciation, sea ice deflects the warm ocean currents of the Gulf Stream away from northern latitudes, causing these areas to become colder.

When astronomical factors initiate a period of glaciation, they set off other effects that can reinforce global cooling. The most important of these is *albedo,* the percentage of incoming solar radiation reflected from surfaces. As more incoming radiation is reflected, less is available to warm the Earth's surface, and thus the climate cools. Because light-colored surfaces, such as snow, are more reflective than dark-colored surfaces, such as exposed bedrock or soil, the more snow cover there is on Earth the faster additional cooling takes place, causing glaciers to grow and advance.

During periods of glaciation sea ice also tends to increase, deflecting climate-moderating ocean currents. For example, as sea ice expands in the North Atlantic, it deflects the tropical Gulf Stream away from the poles and toward the west coast of Africa (Fig. 17-29). Deprived of the Gulf Stream's warmth, the polar region becomes colder and its glaciers grow and advance.

When the astronomical cycles combine to produce lengthy periods of high summertime radiation at 65°N, melting is enhanced, glaciers begin to shrink and recede, and the albedo effect is reversed as the newly exposed dark bedrock absorbs solar energy. As the climate warms, sea ice melts as well, enabling warm ocean currents to return to arctic waters, warming the region. Eventually, all factors coincide to produce an interglacial period of maximum warming, and the glaciers at 65°N melt away altogether.

Reconstructing Ice-Age Climates

Until 65 million years ago, average global temperatures were as much as 10°C (18°F) warmer than today and there were few, if any, ice masses on the Earth, even in the planet's polar regions. Over the last 65 million years the Earth's climate has gradually cooled, as landmasses have moved toward the poles; by about 1.6 million years ago, the planet had entered an ice age from which it has yet to emerge.

During the last worldwide glacial expansion, some 18,000 years ago, temperatures in New England averaged about 5 to 7°C (9–12°F) colder than today, making what is now Boston as cold as Anchorage, Alaska. Temperatures in continental interiors, such as in the American Midwest, may have been as much as 10°C (18°F) colder because interiors lack moderating factors such as proximity to oceans and their warming currents.

Our estimates of past temperatures come largely from studies of the remains of temperature-sensitive plants and animals recovered from sedimentary deposits of known age. The terrestrial record of climate change, however, is incomplete. One geologist has likened this record to a blackboard that has been partially erased and written over repeatedly, so that most earlier markings are illegible. The record in various places has been erased by glacial erosion, or is inaccessible under layers of younger glacial deposits. Nevertheless, a variety of different types of evidence make us certain that, during ice ages, cold periods of worldwide glacial expansion alternate with warm periods marked by glacial retreats. For example, as we saw in Chapter 16, speleothems in caves under formerly glaciated areas contain a record of alternating glacial and nonglacial conditions (see Fig. 16-10).

The distribution of particular species of plants depends on the temperatures and precipitation characteristics of their environments; for example, plants that prosper in cold, dry south-central Canada could not survive in warm, wet Louisiana. Our knowledge of the conditions that support modern vegetation makes it possible to estimate the climate that supported similar or related species in the past. Pollen from ancient plants that became mixed into lake-bottom mud can be dated, usually by carbon-14 (see Chapter 8). A pollen diagram constructed from these data shows how a region's pollen, and

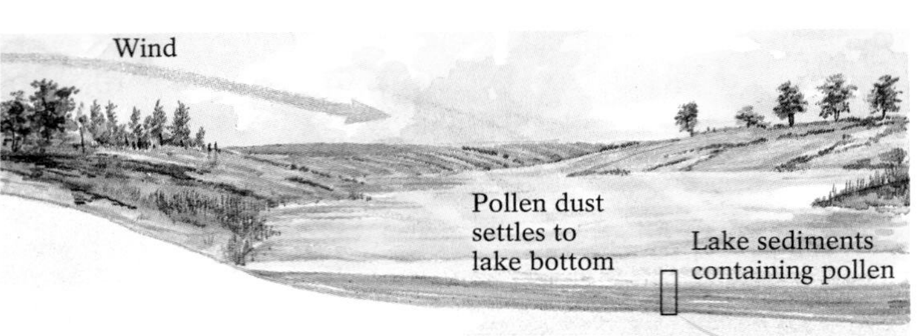

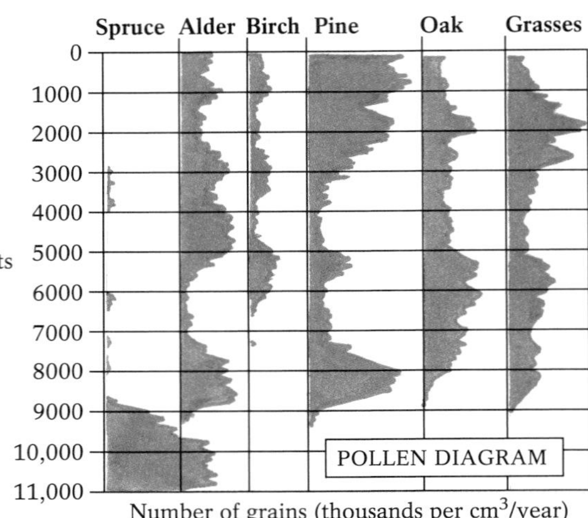

Figure 17-30 A pollen diagram, derived from the number of pollen grains preserved in datable sediment layers, can determine which plant species flourished in a particular location and when. In this way we can estimate past climatic conditions. For example, pollen from numerous lake beds throughout the temperate, treeless prairies of North America's Central Plains indicates that during the last major glaciation this part of the continent was covered with lush forests of spruce, tamarack, alder, and birch. Because these trees grow today farther north, in southern Canada, we can conclude that the mid-continent region was significantly colder 15,000 to 20,000 years ago than it is today.

thus its vegetation, has varied through time (Fig. 17-30). Past climates can then be reconstructed by comparing ancient plant assemblages with related modern plants and their associated climates.

Surviving glaciers also preserve a terrestrial record of climate change. Glaciologists (scientists who study the snow and ice in glaciers) extract cores of ice from glaciers and ice sheets that are some kilometers thick by boring through the ice down to the underlying bedrock. Examination of these ice cores has shown that the ice layers that accumulated during periods of glacial expansion contain twelve times as much airborne dust as those from interglacial times. This suggests that the atmosphere during glacial times was more turbulent than it is today. Winds during glacial periods may have been stronger than normal as a result of frequent clashes of warm and cold air masses; such winds might have swept across the continents, including exposed areas of the continental shelves, eroding dried-out mud and depositing it on glaciers.

If the terrestrial record is like a partially erased blackboard, the marine record is like a thick paint chip from an apartment wall whose many colors record how many times the wall was painted. Whereas the terrestrial record may show interruptions, the marine record is usually continuous, largely because there is little erosion underwater. Thus, in many places, the entire record of global glaciation has been preserved. To yield absolute time frames for these events, a combination of methods is used: carbon-14 dating of organic remains found in the more recent upper sections of cores of deep-sea muds, isotope and fission-track dating of ash layers, and geomagnetic-stratigraphy dating of the older lower sections (see Chapter 8).

Another way to identify past climates is by comparing fossil and living foraminifera (*forams* for short; see Fig. 6-21), microscopic marine organisms that produce coiled shells. We know that the direction of coiling of modern foram shells is controlled by water temperature; thus, the direction of coiling of fossil foram shells in marine sediment can be used to derive the water temperature at the time of deposition. By dating the sediments and rocks in which such fossil forams are found, we can tell what temperatures prevailed at different periods of time (Fig. 17-31).

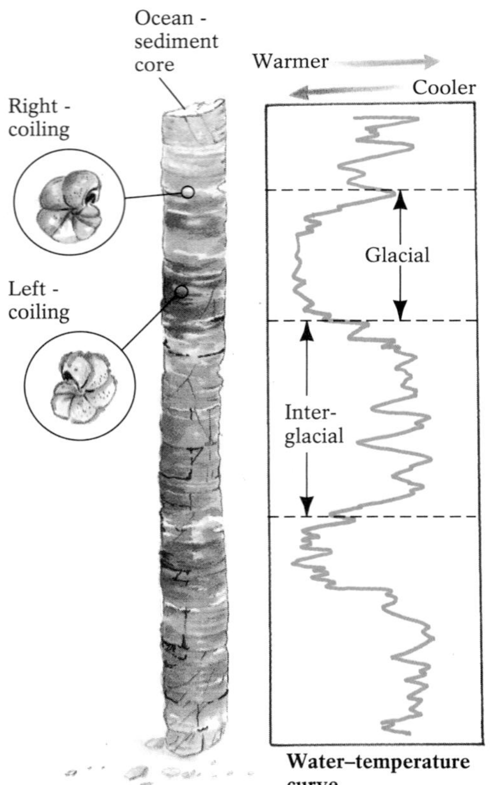

Figure 17-31 The coiling patterns of fossil forams can be used to determine ancient climate changes.

Figure 17-32 Glacial grooves and tillite in the arid landscape of South Africa suggest that this area was once near the South Pole, and was cold enough to be glaciated.

The Earth's Glacial Past

The Earth has experienced numerous ice ages throughout its 4.6-billion-year history. Our ability to study the earliest ice ages is limited, because erosion of the ancient geologic record and its burial beneath more recent rocks and sediments have left only scattered and fragmentary evidence from the ancient ice ages. Nevertheless, the geologic record yields clear evidence of at least three ice ages in the Precambrian Era. The oldest Precambrian ice age, which occurred about 2.2 billion years ago, is known from layers of *tillite,* a sedimentary rock produced by lithification of till.

The geologic record of global glaciation after the Precambrian is more widespread and better preserved. Layers of tillites and "fossilized" moraines and eskers show that during the Paleozoic Era, about 500 million years ago, an ice sheet occupied what is now one of the world's hottest and driest lands, the Sahara of northern Africa. Before plate tectonics moved all the continents to their present locations, the Sahara was located at the South Pole. Ironically, one may now swelter in the merciless African sun while standing within reach of an ancient esker.

The most complete record of any pre-Pleistocene glaciation was left by an extensive glaciation that occurred during the Permian Period, about 250 to 300 million years ago. This ice sheet striated the bedrock and deposited extensive tillites in regions that are now Australia, South America, India, and South Africa (Fig. 17-32)—at the time, these constituted the supercontinent of Gondwana, located in the vicinity of the South Pole.

There is no evidence of glaciation from the Mesozoic Era, which lasted from 225 to 65 million years ago. Apparently the Mesozoic was a time of global warming, when the Earth's landmasses had moved from the poles to subtropical and tropical latitudes. In northern Alaska, there are Mesozoic rocks that contain remnants of coral reefs, tropical vegetation, and dinosaur fossils.

Although the Earth's most recent ice age is associated with the start of the Pleistocene Epoch 1.6 million years ago, recent evidence from forams and other marine species suggests that surface waters began to cool at least 50 million years ago (Fig. 17-33), during the Eocene Epoch. Cooling continued worldwide, and by about 35 million years ago world temperatures had dropped 8 to 10°C (15–18°F). At about this time, large domes of ice may have begun to grow on Antarctica; striated stones, believed to have been carried northward by floating icebergs, are found today in the south Pacific in the muddy oceanic sediments from that period. Gradual global cooling continued, and by 12 to 10 million years ago ice had formed in the mountains of Alaska.

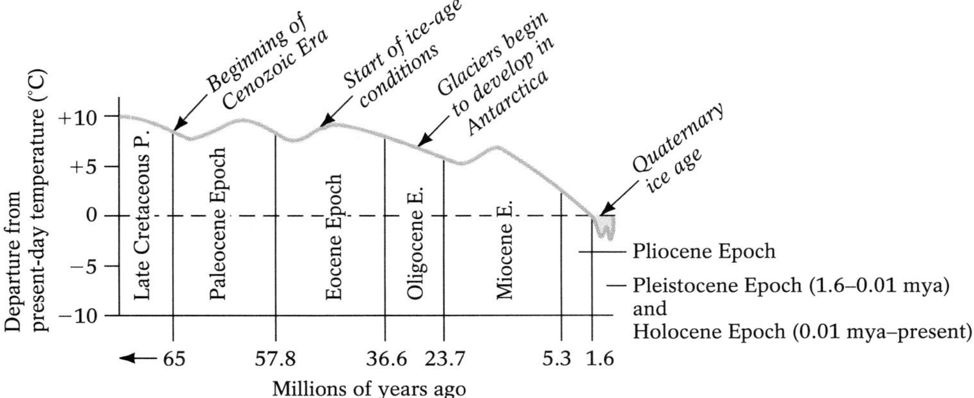

Figure 17-33 Changes in global climate during the Cenozoic Era. The most recent (Quaternary) ice age is associated with the start of the Pleistocene 1.6 million years ago, but the Earth had been cooling for about 50 million years prior to that.

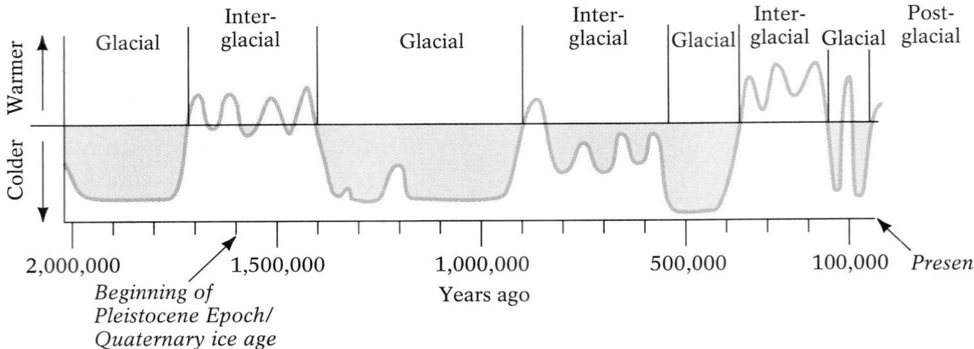

Figure 17-34 Changes in climate during the Quaternary ice age. The Pleistocene Epoch was marked by long periods of glaciation interspersed with brief warming interglacials. (The Holocene portion—about 10,000 years ago to the present—of the Quaternary has seen a more even alternation of glaciation and thawing.)

By 3 million years ago, an ice sheet had buried the island of Greenland, and 500,000 years later, ice was accumulating on the high plateaus and mountains of North America and Europe. The Quaternary ice age, with its alternating mid-latitude ice-sheet advances and retreats, was about to begin.

Glaciation in the Pleistocene

During the past 1.6 million years the Earth has experienced a succession of alternating glacial and interglacial periods (Fig. 17-34). During the glacials, the Earth's climate became colder by about 5 to 10°C (9–18°F), and its ice cover expanded from the polar regions to the middle latitudes. During the interglacials, the Earth's climate warmed to its present temperatures or several degrees warmer, and its ice receded from the middle latitudes but remained in polar regions. At least four times during the Quaternary, ice sheets developed in Canada and advanced southward into the northern United States, then retreated. Tills and outwash deposits exposed in Kansas and Nebraska have been strongly modified by weathering and erosion over their long existence since the early Pleistocene; the slopes of these old moraines are gently rounded and subdued by mass movements (discussed in Chapter 13), and their tills are deeply oxidized. (See Chapter 5 for a discussion of the chemical-weathering process of oxidation.) Deposits from more recent Pleistocene glacial events, especially of the last 100,000 years, are relatively unweathered and unmodified by mass movement.

The location of North America's youngest glacial deposits indicates that the continent's last ice sheet extended from the eastern Canadian Arctic south to New York City and west to the foothills of the Rocky Mountains in Montana, an area of more than 15 million square kilometers (6 million square miles). This ice sheet, the *Laurentide*, probably began to grow in Canada west of Hudson Bay and in Labrador and Newfoundland, where the relatively high topography and cold climates promoted snow and ice accumulation. Nourished by moisture from the Atlantic Coast and the Gulf of Mexico, the Laurentide ice sheet advanced east and south. Another glacial center, the *Cordilleran* ice sheet, formed farther west, between the Canadian Rockies and the Coast Range of British Columbia. It probably began as a network of alpine glaciers that expanded and merged to form an ice cap that eventually covered the adjacent valleys and plains.

At the time of maximum glacial expansion, a continuous wall of ice stretched 6400 kilometers (4000 miles) across the full breadth of southern Canada and the northern United States. This ice sheet, called the North American ice sheet, stripped an average of 15 to 25 meters (50–80 feet) of regolith and bedrock from central and southern Canada and the northern Great Lakes region, depositing it from Cape Cod to northern Washington state as drift averaging about 15 meters (50 feet) deep (Fig. 17-35).

Figure 17-35 The effect of the North American ice sheet, which reached its greatest extent about 13,000 years ago. In the zone of maximum glacial erosion, flowing ice and meltwater from the Laurentide ice sheet excavated some soft preglacial bedrock to about 30 meters (100 feet) below sea level.

Between about 13,000 and 12,000 years ago, the world's climate warmed, ice sheets began to melt, and the North American ice sheet largely retreated from the United States and southern Canada. Brief periods of cooling temporarily interrupted the warming trend, and some glaciers even advanced slightly before resuming their retreat. During one such brief cooling period, the portion of the ice sheet that flowed through the Lake Michigan basin moved south of Green Bay, Wisconsin. There it overran trees that had begun to grow in the warmer climate, incorporating them into a thin layer of till (Fig. 17-36). Although the ice sheet had retreated well into Canada by about 10,000 years ago, it took another 3000 to 4000 years for northern Canada's ice caps to disappear.

Figure 17-36 A log in glacial till, in Two Creeks, Wisconsin. The log was buried about 12,000 years ago, when a temporary cooling period caused the retreating North American ice sheet to re-advance over newly established forests.

Recent Glacial Events

Alternating warm and cool periods also characterized the **Holocene Epoch**—the last 10,000 years. The Holocene began with the postglacial warming that melted most of the Pleistocene's mid-latitude ice. From about 8000 to 6000 years ago, the Earth's climate was about 2°C (3.6°F) warmer on average than today. Moderate cooling since then has produced many new glaciers in alpine settings, but it has not been sufficiently cold, or cold long enough, to reconstitute mid-latitude ice sheets. During the most recent cold snap—the "Little Ice Age" from about 700 to 150 years ago—winters were colder and snowier than average and summers relatively cool and wet. Some winters were so cold that eighteenth-century New Yorkers could sometimes walk across the frozen Hudson River to New Jersey. Snowlines in alpine areas descended sufficiently to cause significant glacial expansion in the world's major mid-latitude mountain ranges. Occasionally whole alpine communities were overrun by the ice.

Sustained warming began by 1850 and continues today, although vestiges of pack ice from the Little Ice Age lingered long enough in the North Atlantic to doom the maiden voyage of the *Titanic* on April 15, 1912. By 1920 there was virtually no floating ice south of the Arctic circle, and today most glaciers in North America and Europe are still retreating (Fig. 17-37), although more slowly than earlier in the century. We can compare modern glacial positions to those shown in oil paintings and lithographs of the early 1800s to establish the amount and rate of glacial retreat. A recent field and photographic survey of 200 glaciers in Alaska found 7% advancing, 63% retreating, and 30% virtually standing still. One that is retreating, the Columbia Glacier in Alaska, flows from a deep coastal fjord into Prince William Sound, where it calves icebergs into the Sound's shipping lanes, posing a direct threat to the economy of the state.

Figure 17-37 The retreating Athabasca Glacier in the Columbia Icefield of western Canada. The signposts indicate past positions of the glacier's terminus.

The Effects of Human Activity on Glaciers Human activities may be contributing to global warming and hastening the melting of the Earth's remaining ice masses. In recent years, our use of refrigeration, air-conditioning, fire extinguishers, and aerosol cans has introduced into the atmosphere a vast quantity of gases called chlorofluorocarbons (CFCs). These gases—together with the increased atmospheric carbon dioxide from our burning of coal, oil, and natural gas—are believed to be causing both a reduction in the atmosphere's ozone layer (which shields the Earth from some solar radiation) and an increase in the *greenhouse effect* (which may accelerate melting of glacial ice).

The greenhouse effect promotes global warming in roughly the way a glass-walled botanical greenhouse creates a warm environment—by allowing solar radiation in and not allowing heat out. Carbon dioxide in the Earth's atmosphere allows solar radiation to pass through it and warm rocks, soils, water, and vegetation on the Earth's surface, and absorbs the heat that rises from the surface (Fig. 17-38). As a result the Earth's heat, trapped in the lower atmosphere and unable to radiate back into space, remains to warm the Earth below. Any mechanism that increases the concentration of atmospheric CO_2 and CFCs promotes climatic warming and accelerates the melting of glaciers.

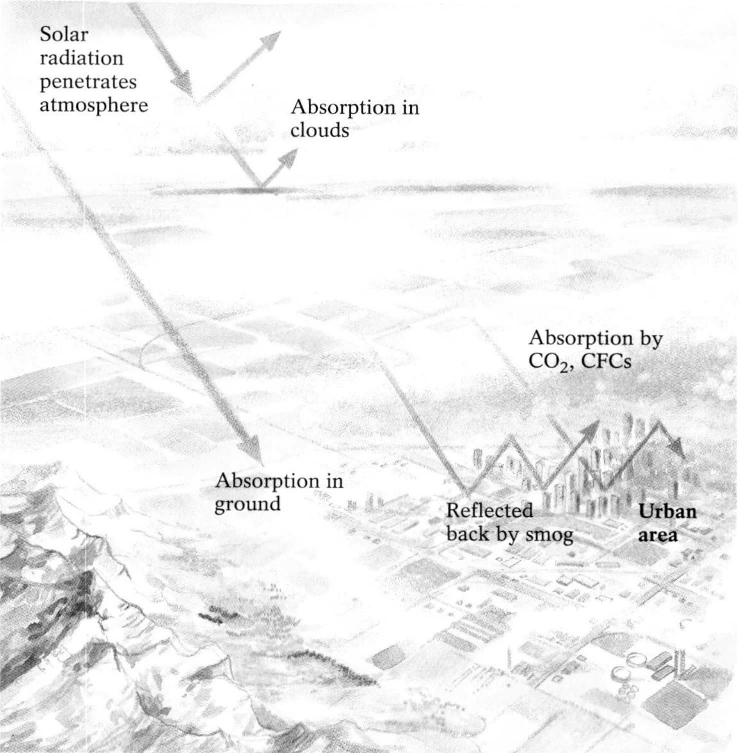

Figure 17-38 Human activity can contribute to global warming—and thereby glacial melting—in various ways, and inadvertently increase glacial melting. The greenhouse effect develops when CO_2 and other lower-atmosphere gases absorb heat radiating from the Earth's surface. If too much heat enters the atmosphere, it is unable to dissipate; the lower atmosphere becomes increasingly warm as the heat accumulates, causing temperatures on Earth to rise as well.

Human activity has been increasing atmospheric CO_2 levels recently in two principal ways: by introducing vast amounts of CO_2 from the burning of carbon-rich fossil fuels and by destroying forests, especially the vast tropical rainforests, which remove great quantities of CO_2 from the air during photosynthesis. Since 1850, as the use of fossil fuels has surged worldwide, atmospheric CO_2 concentration has risen by 25%. If this rate of CO_2 production continues, people in the next century may have to live with CO_2 levels double that in the pre-industrial era; such levels could cause global temperatures to rise by 4 to 8°C (7–14°F).

This global temperature increase would be enough to melt a significant amount of the Earth's polar ice and permafrost and to raise global sea levels by as much as 50 to 100 centimeters (1–3 feet) by the year 2030. Sea level has already risen 20 to 25 centimeters (about 10 inches) since 1900 because of the melting of glacial ice and the thermal expansion of sea water. The predicted additional rise in sea level of 25 to 75 centimeters (10–30 inches) would surely flood such low-lying cities as New Orleans, Louisiana, and Miami, Florida, during large storms.

Figure 17-39 The theoretical effect on North America if all the ice on Earth today were to melt completely. The global sea level would rise about 70 meters (230 feet), submerging most coastal areas and even some inland areas.

If human activities caused *all* of the world's ice to melt within a few thousand years, global sea level would rise about 70 meters (230 feet) (Fig. 17-39). The water in New York Harbor would reach to the armpits of the Statue of Liberty, and most of the Atlantic Coast from New Jersey to the Carolinas would be submerged. The entire state of Florida would be covered by the Atlantic, much of Alabama, Mississippi, and Louisiana by the Gulf of Mexico, and much of interior California by the Pacific. The rising sea would also completely inundate such low-lying places as the Netherlands, Bangladesh, Bermuda, and numerous Pacific-island nations, effectively removing them from the world map.

Human-induced global warming could also result in expansion of the world's deserts, necessitating major shifts in agriculture and forestry. Of course, there would be national winners and losers. Canada would do well: Cold regions now unsuitable for agriculture would become productive as the climate that now supports the Midwestern corn belt of Iowa and Minnesota would migrate north to Ontario and Manitoba. But the Great Plains wheat belt of the United States might become too warm and dry for productive agriculture. Recent Midwestern droughts and the resulting food shortages and price hikes may be an early glimpse into the Earth's balmy future.

The Future of Our Current Ice Age How can we predict when the next period of worldwide glacial expansion will occur, when we can't even predict with certainty whether tomorrow's ballgame will be rained out? The Pleistocene record suggests that interglacials have an average length of 8000 to 12,000 years and are followed by 10,000 to 20,000 years of slow intermittent cooling before glacial conditions return. We are now in an interglacial that began about 10,000 years ago, so it may be drawing to a close.

If human activities are producing a greenhouse effect, it is likely to prove temporary (especially if we succeed in reversing CO_2 and CFC buildup). But if we let CO_2 production go unchecked for the next 2000 years, the very high level of CO_2 buildup will coincide with a warming trend predicted by Milankovitch's Earth–Sun orbital factors. The result will be an unusually intense warming and a period of *superinterglaciation*. If, however, human-induced warming can be curtailed, the Milankovitch factors predict that ice sheets will again cover vast areas of North America and Europe about 23,000 years from today.

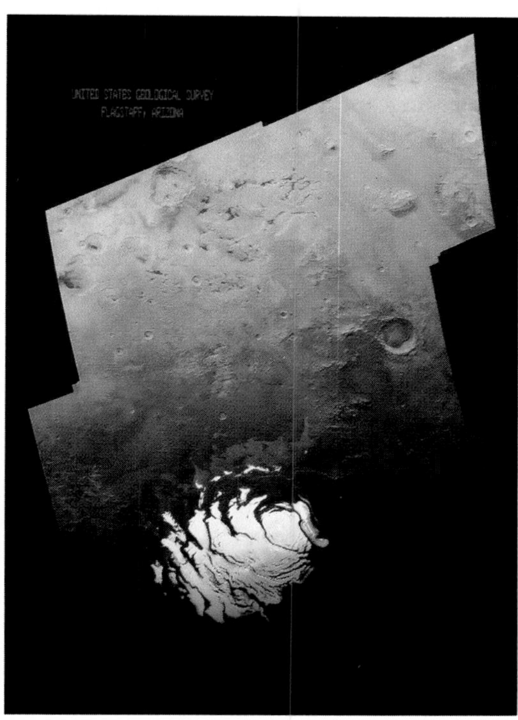

Figure 17-40 An ice cap near the south pole of Mars, photographed by the Viking Orbiter 2 during Mars' southern summer in 1972.

Glaciation on Mars

Other planets in our solar system also experience glaciation. Recent satellite surveys of the surfaces of the large outer planets—Jupiter, Saturn, Uranus, and Neptune—and some of their moons have revealed polar ice caps and extensive ice cover. Closer to home, on Mars, there are white patches near the poles that appear to grow and shrink seasonally, much as the Earth's polar ice does (Fig. 17-40). Mars' polar ice is probably only a few meters thick, and most of it may be frozen CO_2 (dry ice), the main component of its atmosphere. In Mars' summer, its northern hemisphere ice cap recedes rapidly at first, and then more slowly. The rapid shrinkage may be caused by sublimation of frozen CO_2 at the edges of the ice cap, where it is thinnest. Whatever ice lasts through the summer is probably frozen water, which evaporates imperceptibly slowly in Mars' atmosphere.

Satellite photographs of Mars reveal troughs and ridges in the polar regions that resemble the glaciated valleys and large moraines on Earth. The terrain around the ice caps consists of alternating horizontal layers of light and dark materials, which may be composed of stratified outwash or loess. As many as 50 layers, each of which is 15 to 35 meters in thickness (50–120 feet), have been counted at a single site. Confirmation of a complex history of multiple glaciations will probably have to await a visit to Mars by a glacial geologist.

Chapter Summary

Glaciers are bodies of ice, composed of recrystallized snow, that flow downslope or spread radially under the influence of gravity. **Ice ages** are the dozen or so periods of Earth history when the planet's climate was substantially cooler and glaciers covered a significant portion of the land surface. Climate fluctuations during an ice age cause the Earth's glaciers to alternately grow and advance, during glacial periods, and thaw and retreat, during interglacial periods. The Earth's current ice age, the **Quaternary ice age,** spans the last 1.6 million years of Earth history. During the **Pleistocene Epoch** of the Quaternary Period (1.6 million–10,000 years ago), ice sheets expanded to southern Canada and the northern United States, and to northern and central Europe.

Glaciers form as snow that has survived a summer melting season first becomes well-packed **firn** and then refreezes to ice. Glaciers tend to form in a climate marked by moist snowy winters and, especially, cool cloudy summers that minimize melting. Glaciers are classified on the basis of their topographic setting. Relatively small **alpine glaciers,** such as **cirque glaciers** (which erode basins into a mountainside), **valley glaciers** (which occupy existing stream valleys), and **ice caps,** are confined by surrounding mountains. They typically form at or above the regional **snowline,** the lowest topographic limit of year-round snow cover. **Continental ice sheets,** which are unconfined, cover vast topographic lowlands; they exist today only at very high latitudes such as in Antarctica and Greenland.

The **terminus** of a glacier is its outer margin. In a glacier's **zone of accumulation,** the addition of snow exceeds the volume lost; in the **zone of ablation,** more snow melts than is added. An **equilibrium line** between the two zones marks the location where accumulation and ablation are equal. The ice in a glacier flows downward toward its underlying bed in its accumulation zone and upward from its bed in the ablation zone; thus, a glacier flows internally even when its terminus is stationary. Overall glacial flow occurs either by **internal deformation,** where individual ice crystals fracture or slip past one another, by **basal sliding** across a thin film of meltwater that accumulates at the glacier's base, or some combination of both. Some glaciers with a good deal of water at their base accelerate periodically, a process known as **surging.**

Small-scale glacial erosion, or **glacial abrasion,** is caused by small particles embedded in a glacier's bed; they make linear scratches, or **striations,** in the underlying bedrock surface. On a larger scale, **glacial quarrying** breaks off and removes large masses of bedrock. Glacial erosion generally occurs in the accumulation zone, as downward-flowing ice impinges on the glacier's bed. Alpine glaciers carve smooth, unglaciated slopes into sharp, jagged peaks that may display horseshoe-shaped depressions, or **cirques,** sharp pointy peaks, or **horns,** long serrated ridges, or **arêtes,** and distinct breaks in the ridgeline, or **cols.** When cirque glaciers melt, a cirque lake, or **tarn,** forms in the basin. Together, glacial abrasion and quarrying remove a large quantity of bedrock and sediment, producing the asymmetry of glacially eroded hills, converting V-shaped stream valleys into U-shaped glaciated valleys, and eroding coastal valleys to depths well below sea level, producing water-filled valleys called **fjords.**

Glaciers tend to deposit their load in the ablation zone, where melting releases the debris that was contained in the ice. Deposition can occur directly from glacial ice or after transportation by meltwater. All glacial deposits are collectively known as **glacial drift.** Drift consisting of unsorted, unstratified sediments deposited from ice is **glacial till.** Drift consisting of sorted, stratified sediments deposited from meltwater is **outwash.** A till can contain **glacial erratics,** large blocks of bedrock deposited on bedrock of different composition, generally after being transported a long distance. Outwash is generally deposited beyond a glacier's terminus, forming an outwash plain. It is also a primary source of **loess,** the wind-blown dust commonly found downwind from exposed, drying outwash. Meltwater sediments deposited in channels at the glacier's base form a sinuous ridge called an **esker.** Tills and outwash deposited at a glacier's margin combine to form a series of hills called **moraines.** An advancing glacier may override an earlier moraine or any other preexisting hill, producing an asymmetrical hill called a **drumlin.**

Glaciers depress the crust beneath them and disrupt preglacial drainage systems. Other effects associated with glacial periods, which can be seen even beyond the actual glaciated areas, include: freezing of soils and regolith to form **permafrost** and other periglacial features; creation of **pluvial lakes** in what are now arid regions; fluctuations in global sea levels; and the evolution, migration, and extinction of various flora and fauna, including our human ancestors.

Geologists try to predict future climate trends by reconstructing patterns of ice-age climate change. From the terrestrial record, they study fossil periglacial features, cave deposits, fossil pollen from cores of lake mud, and the chemical composition of old glacial ice. From the marine record, they study the distribution of temperature-sensitive microfauna such as forams.

Plate tectonics has moved the Earth's continents toward the poles and raised mountains above snowlines, thus preparing the Earth for ice ages. Milankovitch hypothesized that variable astronomical factors—such as the shape of the Earth's orbit around the Sun, the tilt of its axis, and the wobble of its equatorial plane as it rotates on its axis—moderate the amount of solar radiation that reaches the Earth. When these factors coincide to cause worldwide cooling, glaciers advance. The high albedo of ice-covered glacial regions causes them to reflect heat away from the Earth, thus enhancing cooling, whereas the low albedo of dark, ice-free regions causes them to absorb incoming solar radiation, thus enhancing warming. Thus, the more ice-covered land there is the cooler the climate will be, and the more dark, soil-covered land that is exposed the warmer the climate will be—that is, until Milankovitch or other factors coincide to reverse the trend.

Ice ages have occurred in the Precambrian and Paleozoic Eras and during the Pleistocene Epoch of the Cenozoic Era. Human activity may affect the course of glaciation in the future. Activities that release radiation-absorbing gases into the atmosphere can cause a warming greenhouse effect. For the last 10,000 years, the **Holocene Epoch,** Earth has been in an interglacial period of generally moderate warming, with a few cooling periods of only a few hundred years each. Some geologists suggest that we are now experiencing an unusually warm human-induced interglacial. The gradually changing position of the continents and the ongoing variations in solar radiation will combine to produce another worldwide glacial expansion in about 20,000 to 25,000 years.

Key Terms

ice ages (p. 467)
Quaternary ice age (p. 467)
Pleistocene Epoch (p. 467)
glacier (p. 469)
firn (p. 469)
snowline (p. 470)
alpine glacier (p. 470)
cirque glacier (p. 470)
valley glacier (p. 470)
ice cap (p. 470)
continental ice sheet (p. 471)
terminus (p. 472)
zone of accumulation (p. 472)
zone of ablation (p. 472)
equilibrium line (p. 472)
internal deformation (p. 473)
basal sliding (p. 473)
surge (p. 474)
glacial abrasion (p. 474)
striation (p. 474)
glacial quarrying (p. 475)
cirque (p. 475)
arête (p. 476)
col (p. 476)
horn (p. 476)
tarn (p. 477)
fjord (p. 477)
glacial drift (p. 478)
glacial erratic (p. 478)
glacial till (p. 478)
moraine (p. 479)
drumlin (p. 481)
outwash (p. 481)
loess (p. 482)
esker (p. 482)
permafrost (p. 486)
pluvial lake (p. 487)
Holocene Epoch (p. 495)

Questions for Review

1. What are three types of topographically confined glaciers?

2. Draw a simple diagram showing the basic components of all glaciers: the accumulation zone, ablation zone, and equilibrium line. Indicate where glacial erosion and deposition are most likely to occur.

3. Explain how the two glacial flow mechanisms, basal sliding and internal deformation, work.

4. Distinguish glacial abrasion from glacial quarrying. Describe several distinctive features or landforms produced by each of these two erosional processes.

5. List five prominent erosional features produced by alpine glaciers.

6. Describe the origin and location of terminal, recessional, lateral, and medial moraines.

7. Describe the difference in texture and appearance of till and outwash.

8. What three major effects of glaciers occur during ice ages in areas far removed from glaciers?

9. What factors are believed to cause ice ages and their fluctuations in ice volume?

10. Discuss three methods that geologists use to reconstruct past climates.

For Further Thought

1. In the Northern Hemisphere, the lateral moraines of alpine glaciers are often larger on the south-facing sides of east–west oriented valleys. Why?

2. Some eskers actually "climb" up and over topographic ridges, in seeming defiance of gravity. How can the water that deposits eskers flow uphill?

3. If latitude 65°N is crucial in terms of the Milankovitch theory, why isn't latitude 65°S equally crucial? (Hint: Look at a globe or a world map.)

4. Speculate about what might happen to the world's climate and glaciers if the Antarctic ice sheet surged in the surrounding oceans, and vast areas of the ocean were displaced by floating ice.

5. Identify the "mystery glacial landform" in the photograph below.

Identifying Deserts

Highlight 18-1 How Hot Is Hot?

The Work of Water in Deserts

The Work of Winds in Deserts

Highlight 18-2 America's Dust Bowl

Wind Action on Mars

Desertification

Figure 18-1 Steep-sloped, angular rock formations in Monument Valley, Utah.

Chapter 18

Deserts and Wind Action

Monument Valley of southeastern Utah, which displays the sharply angular slopes and austere barrenness characteristic of an arid landscape (Fig. 18-1), is unlike any landscape seen earlier in this text. Monument Valley's unique appearance, like that of the Earth's other deserts, is caused primarily by its extreme lack of water. Without appreciable water, there can be little chemical weathering—soils are thin, dry, and crumbly, and winds readily sweep loose particles into dunes or sandblast exposed rock surfaces with them. These characteristics are typical of **deserts,** regions that receive very little annual rainfall and are generally sparsely vegetated.

Every major continent, although surrounded by water, contains at least one extensive dry region. Deserts account for as much as one-third of the Earth's land surface—more area than is occupied by any other geographical environment. Relatively few deserts, however, resemble the popular image of endless tracts of drifting sand. In North Africa's Sahara, the world's largest desert (*sahara* means "desert" in Arabic), only 10% of the surface is sand-covered; and even the Arabian Desert, Earth's sandiest, is only 30% sand-covered. Desert climates are *arid,* or characterized by dryness, and are found in cold as well as torrid regions. Such diverse landscapes as ice-bound Antarctica, the fog-shrouded coasts of Peru and Chile, and the near-continuous 8000-kilometer (5000-mile) stretch of land across northern Africa and the Arabian peninsula to southern Iran are all considered deserts, based on their extreme lack of surface water.

The term *desert* is misleading in its implication that the land is literally deserted, devoid of life. Hot deserts are home to some of the Earth's hardiest plants and animals (Fig. 18-2), which have developed through evolution specific adaptations to extremely dry conditions. Some hot-desert plants produce seeds that can endure 50 years of drought, some have small waxy leaves that minimize water loss to evaporation, most have thick, spongy stems that store water from the occasional cloudburst, and most produce deep root systems to tap groundwater supplies. Some of these plants resemble dead twigs for months or even years on end, until the occasional downpour arouses them into a brief but memorable bloom.

Animals in hot deserts include insects, reptiles, birds, and mammals. Some birds fly hundreds of kilometers to find a source of water, which they carry back to their young in absorbent abdominal feathers. Some desert rodents may live their entire lives without a single drink of water, absorbing

(a)

(b)

(c)

Figure 18-2 Some desert life forms. (**a**) Desert shrubs and flowers in southwestern Colorado. (**b**) A roadrunner in the Sonoran Desert. (**c**) A desert snake eating a lizard.

the moisture they need from the plants they eat. Most desert animals are nocturnal, avoiding activity during the hottest and driest times of day, and venturing out only after temperatures have dropped considerably.

In this chapter, we will examine the processes that shape a desert's unique landscape. We will see how water's brief, intermittent appearances actually contribute significantly to the evolution of desert landforms. We will especially discuss the role of wind, which is another major agent of surface change in deserts. We will also look at how human activity contributes to the expansion of deserts, and at efforts to use desert lands for agriculture and human habitation.

Identifying Deserts

How do we determine if an area is dry enough to be called a desert? One common guideline defines a desert as any region that receives less than 250 millimeters (10 inches) of precipitation annually, an amount that in most areas is insufficient to sustain crops without irrigation. The world's driest deserts, however, are well under the defining 250-millimeter mark. A very arid region may receive all its annual rainfall in one sudden cloudburst, the water from which may evaporate or be absorbed by the desert regolith in minutes, leaving virtually no trace. After a cloudburst, there may be no precipitation for months or even years. The world's driest region is the Sahara, which on average receives only 0.4 millimeters (0.016 inches) of precipitation annually. The Sahara may actually have no measurable precipitation for years; its annual average may be accounted for by a few storms over a decade. By the early 1990s, the northern Sahara had been virtually rainless for more than 20 years.

A more useful method of identifying deserts is by means of an **aridity index,** a ratio of a region's potential annual evaporation (a function of its yearly receipt of solar radiation) to its recorded average annual precipitation. In an area with an aridity index of 1.0, the amount of annual precipitation equals the area's potential for evaporation. Such an area would have a humid climate. An area with an aridity index greater than 4.0—where the potential for evaporation is at least four times greater than the annual precipitation—is classified as a desert. In two of the driest spots on Earth, the eastern Sahara of Africa and the Atacama Desert of Peru, the aridity index is 200; such areas are described as *hyper-arid.* An area with an aridity index between 1.5 and 4.0 is classified as *semi-arid;* semi-arid regions, such as the Great Plains east of the Colorado Rockies, are capable of supporting a greater diversity of life than deserts.

Highlight 18-1 *How Hot Is Hot?*

Although deserts may occur in any climatic zone, most—and the best known—are in hot places. Where are the hottest places on Earth? At El Azizia, Egypt, a desolate outpost in the Libyan Desert of northeastern Africa, the air temperature rose to a blistering 58°C (136°F) in the shade on September 13, 1922. The North American record, 56°C (133°F), occurred in Death Valley, California, in July 1913.

In such oppressive temperatures, the human body can lose a gallon of perspiration between sunrise and sunset. If water intake is not maintained, severe dehydration ensues as the body rapidly draws on the water reserves stored in body fat and blood. Sweat glands become overworked, the body cannot maintain its normal temperature, and a high fever results. In a few hours, blood circulation slows and ultimately death occurs.

Where there is no surface moisture to evaporate, humidity is low and there are few clouds to filter daytime sunlight or retain the heat that radiates at night from sun-heated rock surfaces. Consequently, days become even hotter, and nighttime temperatures plummet by as much as 40°C (72°F) from daytime highs. At night, heat dissipates rapidly to the atmosphere and temperatures often drop below freezing.

In 1974, the young British journalist Richard Trench vividly described life in the thermally fickle Sahara of Algeria:

> The writer will not soon forget arising at 3:00 A.M.—in the Algerian Sahara in early September. The bucket of water for washing presented a thin film of ice, and a heavy wool sweater and leather jacket were comfortable while riding in an open jeep. By 9:30 A.M., the air temperature was nearing 85°F; by 11:30 A.M., a pocket thermometer registered over 105°F and at 2:00 P.M. the same thermometer registered 127°F. What passed for a local pub in a nearby oasis cooled its beer by wrapping the bottles in wet sacking and laying them in the sun. Evaporation occurs at an almost unbelievable rate under such circumstances and effective ground moisture levels are fantastically low. The foregoing account amounts to a record of a daily temperature variation of at least 95°F. The beer was delicious.

Types of Deserts

Deserts may be cold, temperate, or hot. Cool deserts may even retain much of their moisture and use it effectively. In a cold desert, such as in northern Scandinavia, the 250-millimeter (10-inch) annual precipitation can support a dense forest because the region's low temperatures retard evaporation, making moisture available for tree growth. But there are no forests in warm southern Nevada, which has the same amount of annual precipitation, because the warm climate hastens evaporation. The best-known deserts are far hotter and drier than most. The deserts of northern Africa, among the world's hottest, driest places, are described in Highlight 18-1.

Subtropical Deserts Deserts can be found at all latitudes, but most are located in the subtropical belts between 20° and 30° of latitude on either side of the equator. How can these extremely dry subtropical deserts exist adjacent to equatorial areas drenched in tropical rain? The equatorial zone receives direct solar radiation virtually year-round, and the warm air there evaporates large amounts of moisture from equatorial oceans; as the air warms, it expands, becoming less dense, and, forced up as cooler air displaces it, it begins to rise, carrying water vapor upward like steam rising from a pot of boiling water. As the air rises, it immediately begins to cool. By the time it reaches an altitude of about 10 kilometers (6 miles), the air has cooled enough that its capacity to hold water vapor is sharply reduced, and it releases its moisture as the torrential rains that fall upon equatorial lands.

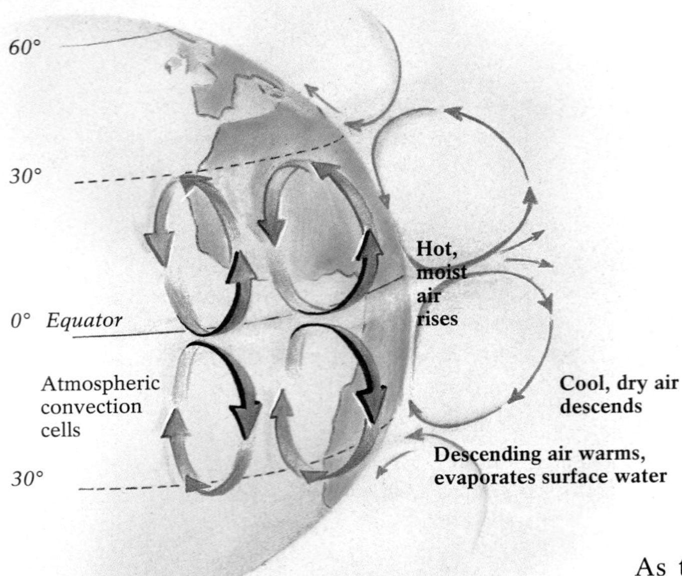

Figure 18-3 Atmospheric convection cells form as warm, moist equatorial air rises, cools at about 10 kilometers (6 miles) above the surface, releases its moisture as tropical rain, spreads northward and southward from the equator, and then descends in the subtropical latitudes 20° to 30° north and south of the equator. The now-dry air warms and is able to evaporate ground moisture rapidly, producing desert conditions on land.

As the now-dry tropical air continues to rise and cool, it becomes denser; no longer capable of rising, it begins to spread laterally. By the time it reaches a latitude of 20° to 30° north or south of the equator, it is dense enough to descend to the surface; there the air warms and gains increased capacity to hold water vapor, enabling it to evaporate surface water. Under the pressure of overlying descending air, some of the warm air approaching the surface spreads toward the equator where it is heated further, evaporates a great deal of water, and again enters the cycle of rising warm air and sinking cool air. This cycle of heating, rising, cooling, descending, and reheating is an atmospheric *convection cell* (Fig. 18-3), similar to the kind in the mantle that helps drive plate motion. The air that spreads toward the poles creates an important set of prevailing winds, discussed later in this chapter.

It is these air-circulation patterns that produce the two subtropical belts of desert lands shown in Figure 18-4. Just north of the Tropic of Cancer in the Northern Hemisphere are the Sahara, the Arabian Desert of the Middle East, and the parched landscapes of Mexico and the American Southwest. Straddling the Tropic of Capricorn in the Southern Hemisphere are the Kalahari and Namib Deserts of southwestern Africa, the Atacama Desert of Peru and Chile, and the interior deserts of the Australian outback.

Figure 18-4 The worldwide distribution of the Earth's deserts. Most deserts occur in the two subtropical belts between 20° and 30° north and south of the equator.

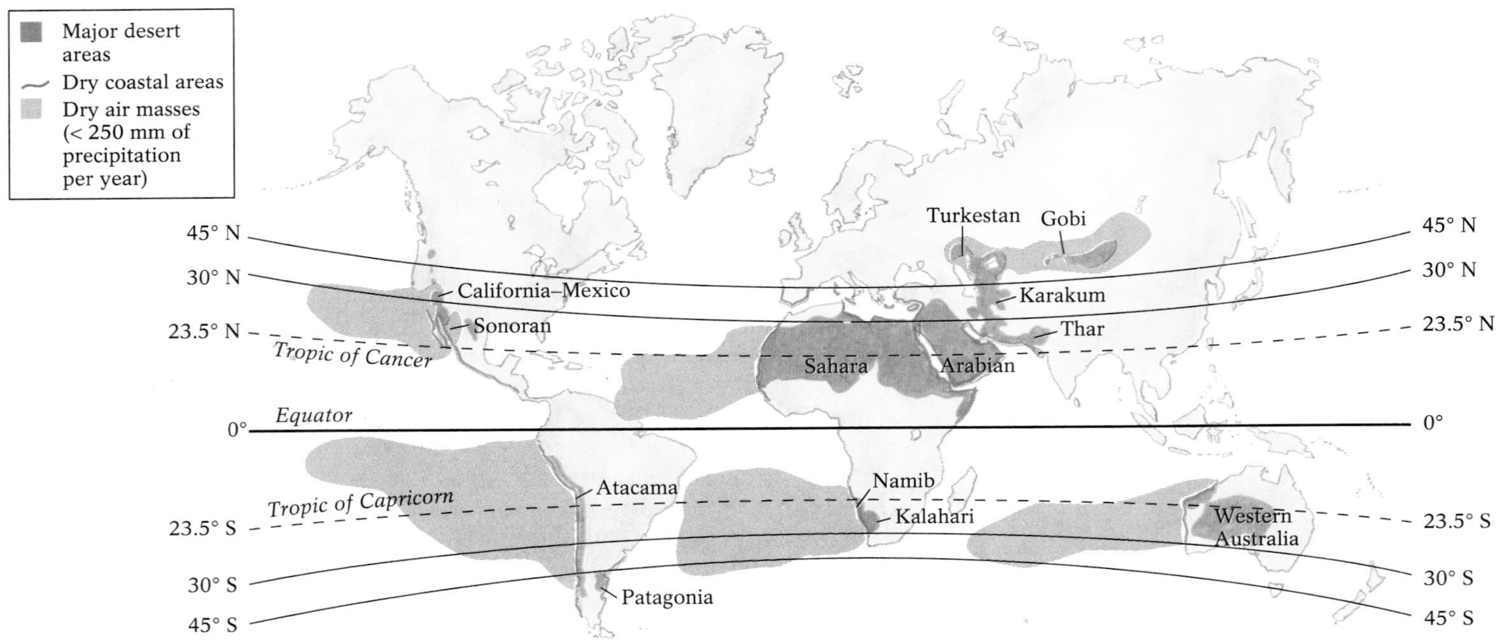

Rain Shadow Deserts Many other, smaller deserts form on the leeward side of mountain ranges. As moist air on the windward side rises and cools, it loses much of its capacity to hold water, which precipitates on the windward slopes and leaves the leeward slopes and the regions beyond them dry. This process, by which moist air is "wrung dry" by mountains, creates the **rain shadow effect.** The desert lands of Nevada, Arizona, and eastern California, for example, are leeward of the Sierra Nevada mountain range and therefore in its rain shadow; Pacific air masses are dry by the time they reach these deserts (Fig. 18-5).

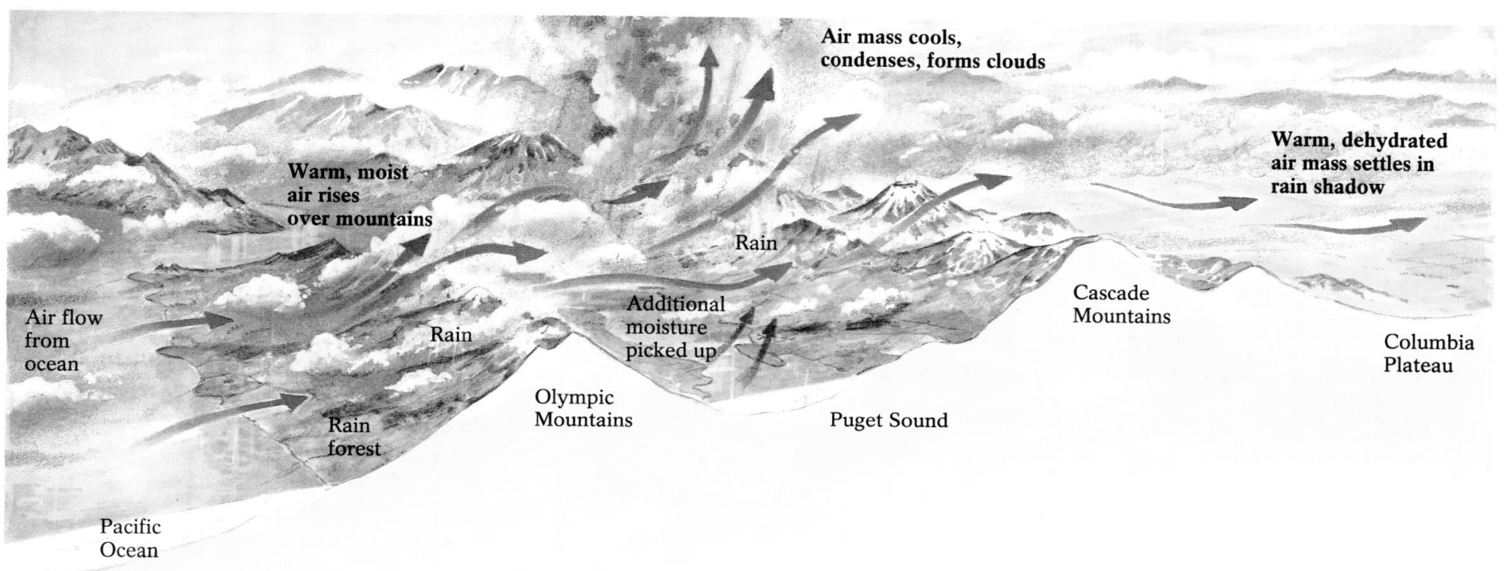

Figure 18-5 The rain shadow effect results when moist air masses lose their moisture as they pass over a large topographic barrier, such as a mountain range. This causes the region in the lee of such a barrier to be dry. In western North America Pacific Ocean moisture rises west of the Olympic Mountains, cools, and precipitates over the Olympic rain forest of coastal Washington, where annual precipitation is more than 2500 millimeters (100 inches). After picking up additional moisture over Puget Sound, the air mass condenses and rain falls on the western side of the Cascade Mountains. The now-dehydrated air descends on the mountains' lee side, where it becomes warm and able once again to hold moisture, producing rapid evaporation and the treeless landscape of the arid Columbia Plateau, where annual precipitation is less than 250 millimeters (10 inches).

Continental Interior Deserts Deserts also form from an absence of moisture in the interiors of continents, far from oceans. In central Asia, the Gobi Desert and the Taklamakan ("the place from which there is no return") exist principally because of their great distance from water. Air masses that reach them have already lost virtually all their moisture after crossing thousands of kilometers of land. The location of these Asian deserts is a direct consequence of plate tectonics: They developed about 40 million years ago after the Indian subcontinent became attached to Asia, enlarging the continent and leaving central Asia several thousand kilometers away from the nearest ocean. (See Highlight 12-2, "Continental Convergence and the Birth of the Himalayas.") The aridity of the area is enhanced because it lies leeward of the mighty Himalayas.

Deserts Near Cold Ocean Currents Some deserts, however, are quite close to bodies of water. The Atacama Desert of Peru and Chile owes its extreme aridity to its warm subtropical climate and to the Pacific Ocean's cold Humboldt current, which flows along the coast of western South America. The current originates in the icy Antarctic and flows northward toward the equator, cooling the air parcels above it so they hold little water. Consequently, the Peruvian coast seldom experiences more than a drizzle, and annual precipitation is less than 1 millimeter; the cold ocean air loses its remaining moisture to condensation, shrouding the Peruvian coast in a persistent veil of fog.

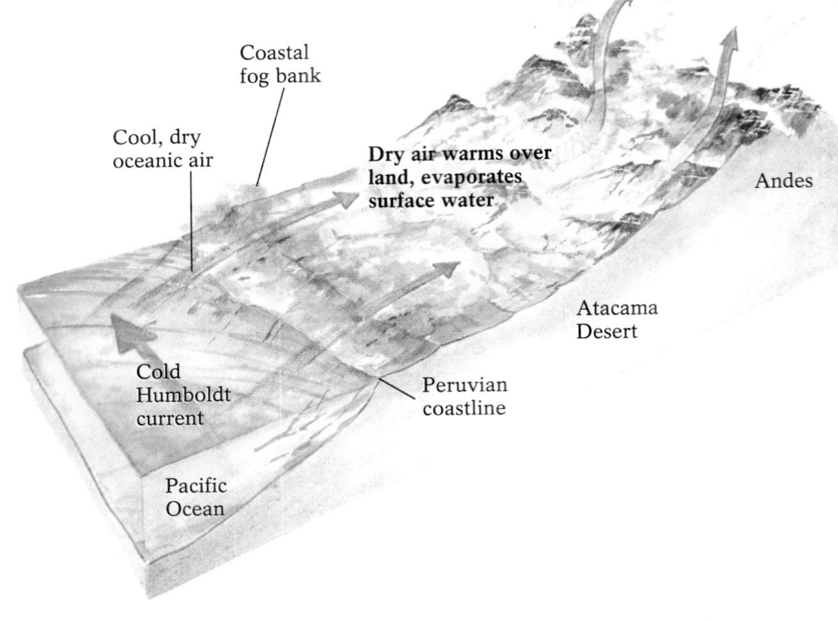

Figure 18-6 Cool dry air from above the cold Humboldt current blows onshore along the Peruvian coast. The air warms over the land surface, and, because its capacity to hold water increases, it evaporates virtually all surface water from coastal Peru. The result is one of the Earth's driest places—the Atacama Desert, just west of the Andes.

Onshore breezes move the dehydrated coastal air inland, where it warms in the subtropical sun. Now the air can hold more water, and so it evaporates nearly all surface water from the region. Thus, a combination of subtropical warmth, a cold ocean current, and onshore winds remove nearly every drop of moisture from western Peru's air (Fig. 18-6). A similar situation occurs along the southwest coast of Africa, where cold Atlantic currents and hot African land surfaces combine to create the hyper-arid Namib and Kalahari Deserts.

Polar Deserts Some places are both extremely arid and numbingly cold. Because they receive little solar radiation, such high-latitude locations as northern Greenland, Arctic Canada, northern Alaska, and Antarctica have temperatures that remain below freezing year-round, even during their brief summers when the sun never sets. But global atmospheric circulation brings only cold dry air to these lands; thus they qualify as deserts as well.

Figure 18-7 Although oxidation, like other forms of chemical weathering, proceeds slowly in deserts, it has created colorful rocks such as these of the Painted Desert in northeastern Arizona. Oxidation of iron-bearing silicates in the area's sandstones produces red-orange iron oxide, which paints these rocks their characteristic hues.

Weathering in Deserts

Such desert features as steep angular cliffs, sharp-edged stones, and relatively thin soils give the impression that little or no weathering takes place there. Mechanical-weathering processes are at work, however, and it is these that produce most of a desert's loose sandy regolith. As discussed in Chapter 5, extreme daily temperature fluctuations cause rock surfaces to expand and contract repeatedly. Eventually individual grains flake off the surface and join the desert's supply of shifting sands. At the same time, ongoing evaporation of any available water precipitates dissolved salts within rock fractures as salt crystals. As salt crystals grow, they can exert enormous pressure on adjacent mineral grains, acting like minute crowbars to force them apart.

Because water is necessary for chemical reactions, it was long believed that the lack of water in deserts meant that chemical weathering did not operate there. Cold desert nights, however, promote condensation of dew on rocks, and some deserts enjoy a brief rainy season. What moisture there is, along with organic acids produced by desert vegetation, supports a small amount of chemical weathering. Even hyper-arid regions may contain enough moisture to oxidize iron-rich silicates, resulting in the colorful hues of many desert rocks (Fig. 18-7). The extremely slow chemical weathering in deserts also produces distinctive loose thin soils, *aridisols,* which have low organic content and are completely dry for at least six months of the year.

Figure 18-8 Petroglyphs carved into desert-varnished rocks by native North Americans 2000 years ago. Because petroglyphs such as these have not been "re-varnished" since they were carved, we can deduce that desert varnish forms quite slowly.

Many desert rock surfaces are covered with **desert varnish,** a thin, shiny red-brown or black layer of manganese and iron oxides. When this coating appears even on pure white quartzite rocks that initially contained no manganese or iron, we must ponder its origin. One hypothesis proposes that high winds transport manganese- and iron-rich clays from nearby sources and forcefully plaster them onto dew-dampened rocks; the clays adhere and are oxidized by oxygenated dew. Another hypothesis proposes that microbes in the desert soil concentrate manganese on rock surfaces. Whatever the process, it is slow and apparently unique to the desert environment (Fig. 18-8).

The Work of Water in Deserts

Although desert landscapes appear vastly different from those in other environments, they are shaped by many of the same processes that act on virtually all land surfaces. Despite conditions of extreme aridity, water is the primary sculptor of desert landforms. Brief occasional cloudbursts cause the rapid erosion and subsequent deposition that create the desert landscape (Fig. 18-9).

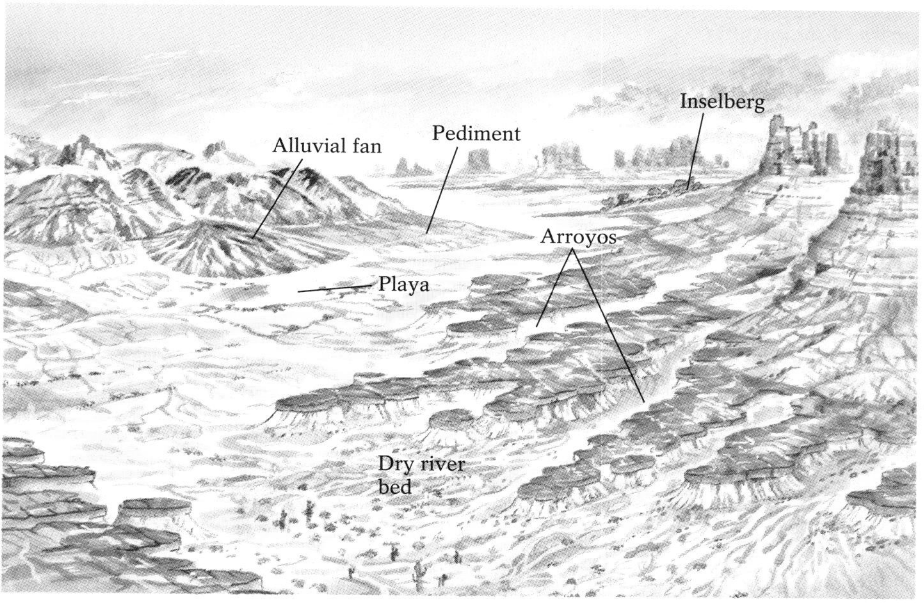

Figure 18-9 Desert landforms produced by water (see text). Intermittent surface-water flow in deserts produces arroyos, pediments, and inselbergs. Although the arroyo pictured here is dry, it fills rapidly with rushing water during a desert cloudburst. Deposition from desert streams creates alluvial fans and playas.

Stream Erosion in Deserts

In arid regions, there is only a minimal network of soil-binding roots to anchor sediment fragments, which are easily removed by flowing water. Moreover, infrequent desert storms produce immediate and intense surface runoff that is not slowed by vegetation as it would be in grasslands or forests. The rapidly moving water of short-lived desert streams can, over thousands of years, erode numerous **arroyos,** stream channels that are dry most of the year. Desert streams can also erode entire mountains, depositing thousands of meters of sediment into channels and basins to construct great alluvial fans that slope gently from the remnant mountains to the desert floor.

A violent desert thunderstorm can drop 5 centimeters (2 inches) of rain in minutes, and set off a flash flood. Because water isn't readily absorbed by compacted, sun-baked ground that contains few roots, it rapidly overflows existing stream channels. One storm can send a 3-meter (10-foot)-high wall of water and sediment sweeping through an arroyo with virtually no warning. During a brief, intense storm, water velocity in an arroyo becomes so rapid that there is little meandering or lateral migration. These occasional rushing streams excavate their channels vertically, eventually forming long, narrow, steep-sided canyons. (Anyone wandering in an arroyo when a flash flood occurs would have difficulty escaping because of the near-vertical canyon walls.) As a storm abates and the velocity of floodwaters decreases, the water infiltrates or evaporates, leaving behind a high volume of coarse sediment covering the canyon floors. Little happens between floods to change the structure or appearance of these deposits.

When desert streams remove sediment from a mountain, the resulting erosional surface is a **pediment** (from the Latin for "foot"), which lies at an angle of 5° or less and may extend for many kilometers from the base of an eroding mountain. Pediments enlarge slowly as mountains recede by the repeated action of running water, perhaps until nearly all of the mountain's mass has been eroded. Pediments are usually covered thinly with loose debris that is being transported downslope. Very large pediments may have begun to form under the moist, pluvial conditions of the Pleistocene (discussed in Chapter 17), when there was more moisture, runoff, and weathering than today; it is unlikely that the arid environment in which pediments are currently found could have provided sufficient water to erode that much sediment, even over several thousand years. The reduced desert waterflow of today may be moving only materials already loosened by weathering in a past climate, but not significant additional material.

After advanced pediment development, part of the mountain may remain as a resistant, steep-sided bedrock knob. Such a residual erosional feature is called an **inselberg** ("island mountain" in German), and is typically composed of a durable rock such as granite, gneiss, or well-cemented sandstone. A **bornhardt,** named for the German geologist Wilhelm Bornhardt (who first studied them, in eastern Africa), is a large rounded or dome-shaped inselberg that forms slowly as erodible rock surrounding a mass of more resistant rock is removed (Fig. 18-10). Much of the weathering that produces bornhardts takes place underground, where moisture is more plentiful than at the desert surface. Because a bornhardt stands prominently above the surrounding plain, its smooth steep slopes shed water quickly, minimizing its prospects for further weathering, and accelerating the weathering rate of the surrounding rocks and sediments.

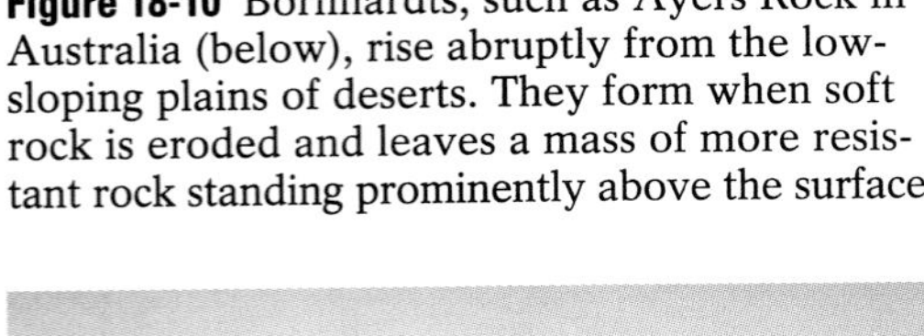

Figure 18-10 Bornhardts, such as Ayers Rock in Australia (below), rise abruptly from the low-sloping plains of deserts. They form when soft rock is eroded and leaves a mass of more resistant rock standing prominently above the surface.

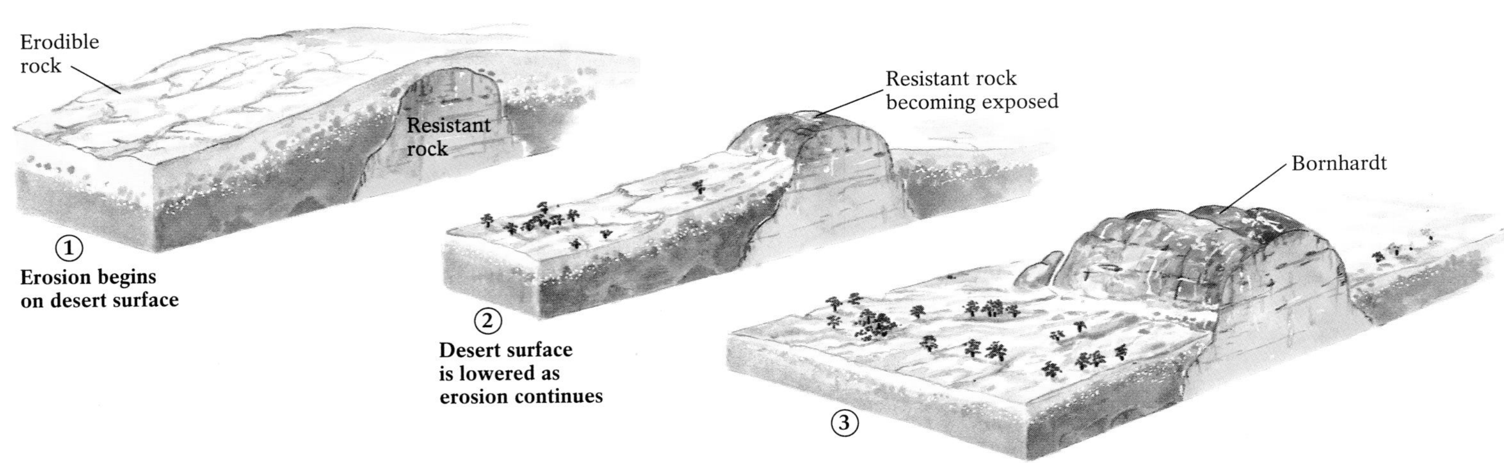

Stream Deposition in Deserts

Whenever erosion takes place, deposition is sure to follow. In moist, temperate regions, most products of erosion are carried to and deposited in the nearest ocean. But in arid regions there are few streams long enough to reach the nearest ocean, and deposition usually occurs close to the erosion site. Wherever surface water evaporates or infiltrates, it drops the sediment load washed from the surrounding steep mountain slopes, forming alluvial fans and dry lake beds, the characteristic features of arid-region deposition.

The water from occasional desert torrents flows rapidly at first, typically confined to a narrow arroyo with steep walls. At the outlet of the arroyo where the canyon walls end, the flow slows appreciably as the unconfined water spreads out. As the water's velocity slows, it loses its ability to transport sediment, and deposits its load in the form of an alluvial fan (see Figure 6-4). Each mountain stream that reaches the desert plain produces its own alluvial fan, proportional in size to the drainage area of the stream. The slope of the fan surface, about 10° at the top, decreases gradually until it merges with the desert floor. The steeper slope near the top of the fan contains the coarsest sediment, dropped first as the stream slows, which maintains a relatively high angle of repose (see Chapter 13). The sediment generally becomes increasingly finer toward the bottom of the fan.

Water that infiltrates the coarse upper fan sediments may reemerge at the foot of the fan as a spring. Such springs are often marked by groves of mesquite trees in the American Southwest (something to remember if you ever find yourself thirsty in that part of the world). Alluvial fan sediments may contain enough water to irrigate the adjacent desert area for cultivation—as Native Americans in the Southwest have done for more than 1500 years—or contribute to the water needs of a city. San Bernardino, California, for example, draws some of its water from the alluvial fan deposits on which it is located.

During brief periods of higher-than-average precipitation in a desert's interior, water drains toward topographic lows, where it may collect as a temporary lake. Such a lake may evaporate in a few days or weeks, however, leaving behind a **playa** (Spanish for "beach"), a dry lake bed on the desert floor (Fig. 18-11a).

A playa's dry bed typically consists of a combination of precipitated salts initially dissolved from soluble rock, and fine-grained clastic sediment (Fig. 18-11b). The salts, commonly a mixture of sodium and potassium carbonates, borates, chlorides, and sulfates, may form a blinding white residue

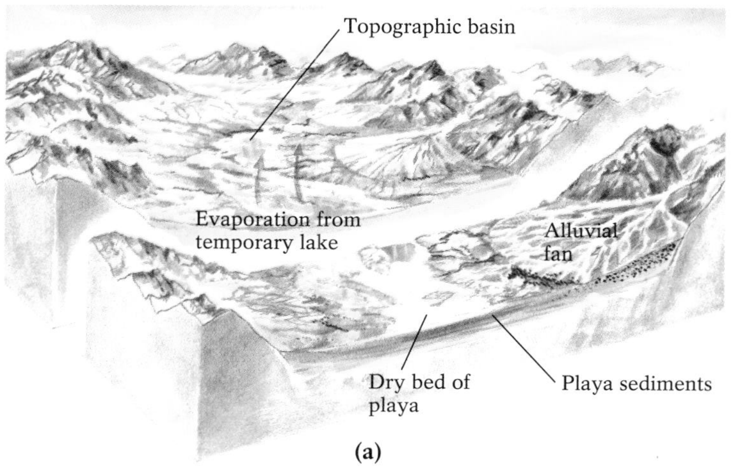

Figure 18-11 (**a**) A playa is a dry basin in the desert floor from which a temporary lake has evaporated. (**b**) Evaporite deposits in Devil's Golf Course, Death Valley, California. Ninety meters (295 feet) below sea level, Devil's Golf Course is a playa that was a significant obstacle to westward migration in the late nineteenth century.

or a bizarre landscape of jagged pinnacles. Industrial chemicals have been mined from the playas of the Southwest for more than one hundred years; Death Valley has long been a source of sodium borate, or borax ($Na_2B_4O_7 \cdot 10\ H_2O$), used in pottery glazes, household cleansers, and as a hardening agent for alloys and shielding material against nuclear radiation. (Borax, originally carted away by "twenty-mule teams," became the trade name for a cleanser made famous in commercials narrated by former president Ronald Reagan during the 1950s.)

If its inflowing water does not encounter and dissolve soluble rock, a playa may contain only fine-grained clay-rich sediment. Such a surface can be baked so hard by the desert sun that it can become a natural landing strip for aircraft. This is why Rogers Dry Lake, on Edwards Air Force Base in the Mojave Desert, north of Los Angeles, is the optimal landing site for NASA's space shuttles.

When precipitation and runoff from surrounding mountains exceed evaporation and infiltration on the desert floor over an extended time, a large lake may form that can last for years or centuries, even in the ultra-dry desert air. With no outlets from which water can drain, most desert lakes become extremely saline (salt-rich) as evaporation continually concentrates dissolved salts. Large saline water bodies of this type include the Salton Sea in southern California, the Great Salt Lake in Utah, the Dead Sea, a basin that descends to 392 meters (1286 feet) below sea level (Fig. 18-12), and Lake Chad, a 22,000-square-kilometer (8500-square-mile) body of water in the southern Sahara.

Figure 18-12 The Middle East's Dead Sea, which is essentially a desert lake, is so saline that swimmers float noticeably higher in it than they do in the ocean.

The Work of Winds in Deserts

Winds, a major force in shaping deserts, are air currents set in motion by heat-induced changes in air pressure. (High-pressure air is cool, and generally flows as wind into regions of lower-pressure air that is warmer.) Near the equator, a zone of low pressure prevails as sun-warmed air becomes less dense and rises. As we saw earlier in our discussion of the origin of subtropical deserts, this air cools and becomes more dense as it ascends, eventually losing its buoyancy and, still at high altitudes, spreading north and south of the equator. At about 20° to 30° of latitude in both directions, the cooler, denser air sinks back toward the surface. The result is two permanent subtropical high-pressure zones, the *horse latitudes*, which are noted for their calm, clear, warm weather. (The name originated when sailing ships, without a driving breeze, stalled; when food and drinking water were exhausted, a ship's cargo of horses was the sailors' only source of sustenance.) The high-pressure air of the horse latitudes ultimately spreads, close to the Earth's surface, back toward the equator, filling the void created by rising low-pressure equatorial air. Some of the high-pressure air goes the other way, toward the poles, drawn there by relatively low air pressure near 60° of latitude.

Global wind patterns do not follow a strictly north–south or south–north path. The Earth's rotation about its axis imparts an east–west component to the flow of air above its surface. In the Northern Hemisphere, descending air currents are deflected toward the right of the direction in which they are moving; in the Southern Hemisphere, they are deflected toward their left (Fig. 18-13). The apparent deflection of any freely moving body (such as air, water, or a missile) by the Earth's rotation is called the *Coriolis effect.*

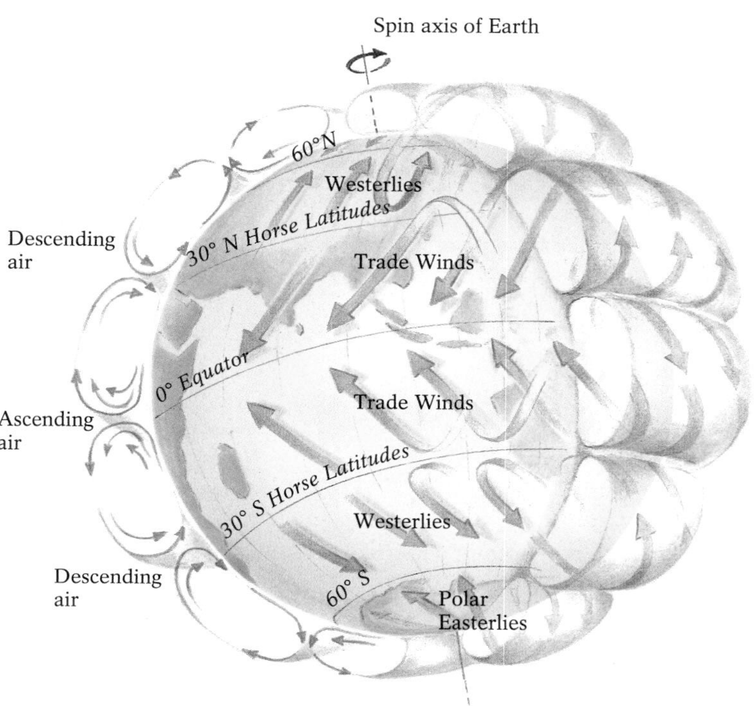

Figure 18-13 The Earth's global wind patterns. The Earth's rotation causes the deflection of moving currents of air (and water as well), so that in the Northern Hemisphere descending air masses are deflected to their right as they approach the Earth, and in the Southern Hemisphere they are deflected to their left. This deflection is known as the Coriolis effect.

In the Northern Hemisphere, the descending air of the subtropical high that returns south toward the equator is deflected to its right, or from east to west. This deflection produces northeasterly or *trade winds,* which prevail along the warm, low-latitude trade (an archaic meaning of "trade" is "track" or "course") routes taken in the eighteenth and nineteenth centuries by merchant sailing vessels. (Winds are designated by the direction from which they are blowing; hence a north wind blows from the north.) The descending air of the subtropical high that turns toward the North Pole is also deflected to the right, or from west to east, by the Earth's rotation, producing the mid-latitude winds, or westerlies. The westerlies prevail between about 30° and 60° of latitude in both hemispheres; in the Northern Hemisphere, they dominate the wind pattern for most of North America, Europe, and Asia.

Global wind patterns have probably remained relatively constant throughout the Earth's history, because the orientation of the Earth's axis and its distance from the Sun have been relatively constant. The consistent difference in the amount of heat received by the equator and poles has probably always produced high- and low-pressure zones that generate winds. The direction of the Earth's rotation has always been the same, so the Coriolis effect has always deflected winds in the same direction as it does today.

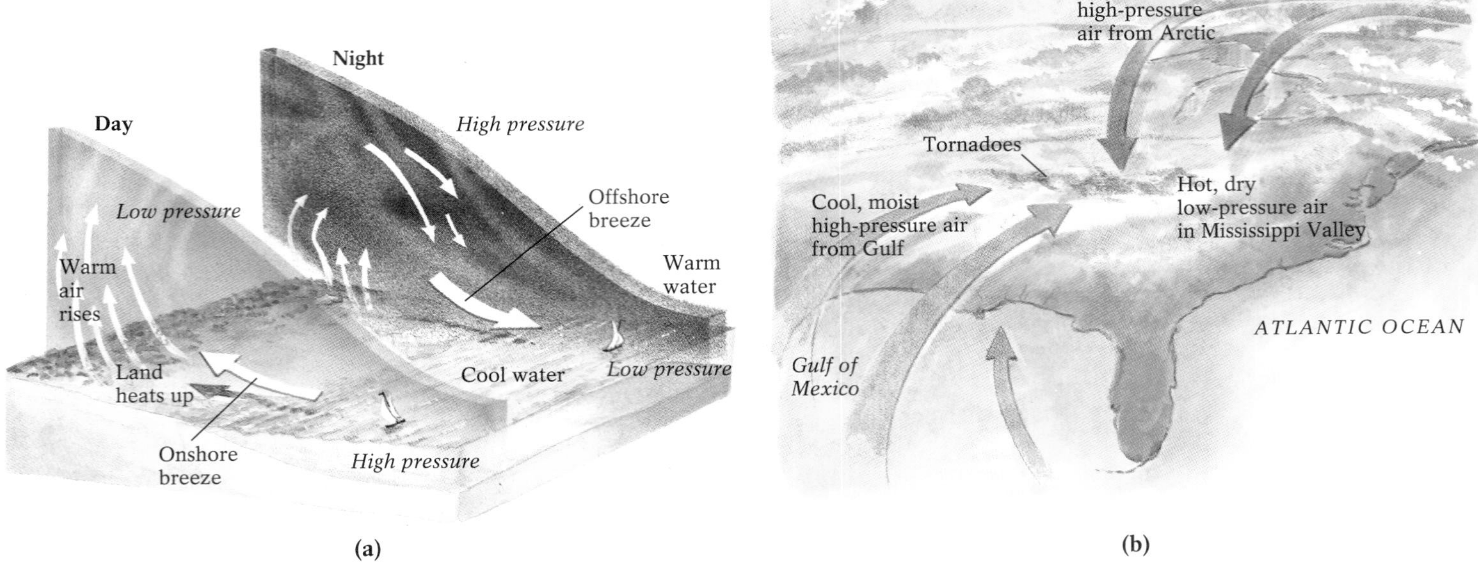

Figure 18-14 On both local and continental scales, onshore and offshore winds develop because solid ground warms and cools quickly, whereas water warms and cools more gradually. (**a**) On a warm afternoon land heats up more rapidly than the ocean, and a zone of low pressure forms over the land; cooler high-pressure air then blows from the sea to the land as an onshore breeze. At night, the sea cools slowly, retaining more daytime heat than the land. As the land cools, higher air pressure develops over it, sending an offshore breeze out to the low-pressure zone over the ocean. (**b**) During the summer months, the entire North American continent heats up, producing a large temporary zone of low pressure overhead; higher pressure develops over the relatively cooler surrounding ocean. (This differs from the winter pattern of cold land/high pressure and relatively warm ocean/lower pressure.) This pressure differential produces southerly winds that funnel moist air from the Gulf of Mexico up the Mississippi Valley, where it may encounter cold air descending from the Arctic. The resulting clash of cold, dry arctic air and warmer, moist gulf air produces tornadoes in the middle of the continent.

Low-altitude winds, such as the pleasant onshore and offshore breezes that moderate coastal climates, can also be caused by local differential heating of the atmosphere (Fig. 18-14a). When this occurs on a continental scale, it sometimes has catastrophic results. For example, differential heating of land and ocean draws moist air from the cooler, higher-pressure zone of the Gulf of Mexico into the warmer, lower-pressure zone of the Mississippi Valley during the warm spring and summer months. This produces warm, moist southerly winds that, if they encounter cold, dry air descending from the Arctic, can cause tornadoes (Fig. 18-14b). Differential heating similarly causes the gusty winds of deserts: During the day, heated surface air rises to create a low-pressure zone; in the evening, relatively cooler and higher-pressure surface air rushes in, often at a velocity of 100 kilometers (60 miles) per hour or more, to replace it.

Erosion by Wind

When you walk along a beach on a blustery day and your legs feel the sting of blowing sand, you are experiencing erosion by wind. The erosive power of wind was evident after a fierce 24-hour windstorm swept through California's San Joaquin Valley on December 12, 1977. Three-hundred-kilometer (190-mile)-per-hour winds removed nearly 100 million tons of topsoil from a 2000-square-kilometer (775-square-mile) area. This produced a blinding dust and sand storm that severely damaged crops, buildings, and automobiles and, during a period of zero visibility, killed five motorists and injured many more.

Wind is, after water, the second most effective agent of surface change in deserts. Wind cannot erode solid rock in the way that flowing water or ice can, by dislodging cemented grains from sedimentary rock or quarrying crystalline bedrock from outcrops of igneous and metamorphic rock. Desert landscapes, however, are more vulnerable to wind erosion than most other geographical settings because they are covered with loose materials, with little vegetation to anchor the soil and hold its particles down. In deserts, winds remove large quantities of loose, unconsolidated material.

Deflation Wind gusts blowing across a dry treeless desert lift sand- and silt-sized particles from the surface, but leave behind larger and heavier pebbles and cobbles. Clay-sized particles also remain because they adhere to the surface. The removal of large quantities of loose material is **deflation.** Deflation excavates distinct depressions in a desert floor and leaves a coarse pavement of large particles.

Figure 18-15 A blowout caused by deflation in Sand Hills State Park, Texas.

Deflation usually lowers the landscape slowly, up to a few tens of centimeters per thousand years. In extraordinary cases, however, such as occurred during North America's Dust Bowl of the 1930s, as much as one meter of fine-grained topsoil can be deflated in just a few years. Land surfaces may deflate over broad expanses or in local areas only. Small lowered regions, known as *blowouts,* form where surface vegetation has been disturbed by animal trampling, overgrazing, range fires, drought, or human activity. Blowouts are often little more than 3 meters (10 feet) across and 1 meter (3.3 feet) deep. Because water helps make loose deflatable material more cohesive, blowouts are seldom deeper than the local groundwater table. Thousands of blowouts dimple the semi-arid Great Plains of North America from Texas to Saskatchewan (Fig. 18-15).

Deflation basins are large blowouts; they occur where local bedrock is particularly soft or where faulting has produced a broad area of crushed bedrock. Sizeable deflation basins are shaped by the winds that blow across the western Great Plains. Near Laramie in southeastern Wyoming, soft fine-grained bedrock has been deflated to produce Big Hollow, a depression 15 kilometers (9 miles) long, 5 kilometers (3 miles) wide, and 30 to 50 meters (100–160 feet) deep.

A **desert pavement** is a surface layer of closely packed stones left behind by deflation (Fig. 18-16). It develops rapidly at first because sand and silt are readily carried off, but more slowly after surface particles are removed and the remaining coarse material prevents further erosion of the underlying fine-grained materials. A mature desert pavement, one or two layers of pebbles thick, and with virtually no exposed fine regolith, may take hundreds or thousands of years to develop. When a mature desert pavement is disturbed, for example by a dirt bike, fine regolith is once more exposed and deflation resumes until the protective pavement is restored.

Figure 18-16 Desert pavements form from the progressive deflation of fine particles, concentrating coarse materials at the surface. Photo: A desert pavement in the Sahara.

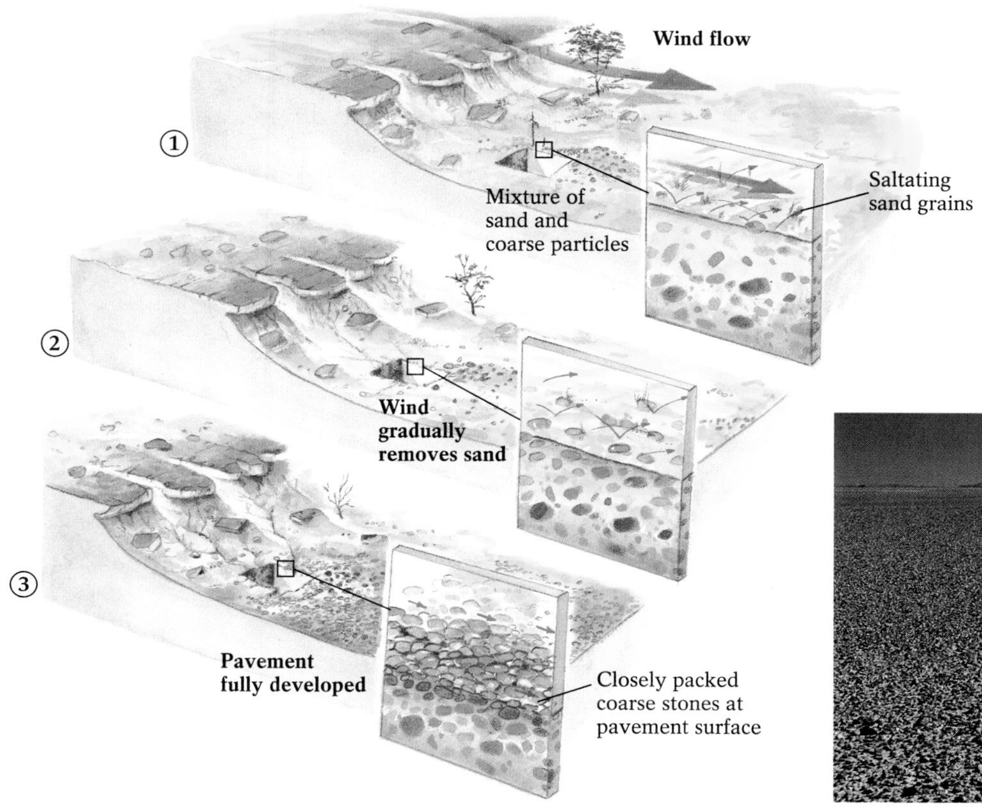

Figure 18-17 In desert areas, such as this one in Goblin Valley, Utah, it is common to see balanced rocks perched precariously on narrow pedestals that have been cut by wind abrasion.

Abrasion **Wind abrasion** occurs when a wind-borne supply of eroded particles is hurled against a surface and sandblasts it. In a Saharan sandstorm, several hours of wind abrasion can strip the finish from a jeep and pit the windshield until it is no longer transparent. Wind-eroded sand consists largely of quartz, one grain of which is 2500 times as dense as air. The collision of quartz grains and other grain or bedrock surfaces can weaken the cement between sedimentary grains, fracture grains in igneous and metamorphic rocks, and dislodge particles previously loosened by weathering.

Wind abrasion produces several distinctive desert features, such as the rock pedestal shown in Figure 18-17. Wind-shaped stones with flat sharp-edged faces, called **ventifacts** (Latin for "wind-made"), may be strewn across a desert floor. Ventifacts' characteristic polished facets with pitted "orange-peel" textures are produced by prolonged wind abrasion on the side of a ventifact that faces the prevailing wind direction; the orientation of the sandblasted facets on a group of ventifacts indicates the dominant wind direction. Ventifacts with more than one faceted side may have been shaped by prevailing winds that changed direction or blew from more than one direction, but it is most likely that their position changed, exposing a new surface to wind abrasion (Fig. 18-18). A stone's position can readily be changed by the actions of animals, powerful storm winds, floodwaters, a cycle of frost wedging, or undercutting at the base of the stone by wind abrasion.

Figure 18-18 Ventifacts form when wind blowing predominantly from one direction abrades desert-floor stones, creating flat surfaces and sharp edges. As the wind changes direction or the stones shift position, exposing other surfaces to wind abrasion, more facets are produced on the newly exposed surfaces. Photo: Ventifacts in Death Valley, California.

Figure 18-19 Field of yardangs, Egypt.

Wind abrasion also produces a large-scale topographic feature, the streamlined desert ridge called a *yardang* (from the Turkish *yar*, meaning "steep bank"). Yardangs most commonly form where strong unidirectional winds abrade soft sedimentary bedrock layers, leaving behind more resistant layers. The resulting landform, which somewhat resembles the inverted hull of a sailboat (Fig. 18-19), is surrounded by wind-abraded depressions. Yardangs are usually wider on the windward side and taper leeward. They may be only a few centimeters long or as much as tens of kilometers in length and tens of meters in height.

Transportation by Wind

In deserts, wind-transported particles roll or saltate along the surface like the sand grains near the bottom of a stream (see Chapter 14), or travel meters above the surface, suspended by the wind's turbulence. Two interacting and opposing factors are involved in the movement of even the smallest particle of dust: the driving force of the wind and the particle's inertia, or tendency to stay in place.

Other modes of transport, such as rivers and glaciers, lift and move a particle whenever their driving force exceeds the particle's resistance to movement. Wind, however, must achieve a much greater relative velocity because it cannot lift a loose particle directly from the surface: Vegetation and coarse sediment on the ground interrupt its passage just above the surface, creating a layer of virtually "dead air." To move any particle in this still layer, wind must reach a velocity able to overcome the inertia of the largest grains. As these grains begin to move, rolling at first, they collide with other particles, propelling them upward. Once airborne, particles are carried forward by the wind in parabolic (curved) paths until gravity pulls them down. Large saltating grains are overcome by gravity fairly soon, and fall back to the surface, striking other grains and in turn setting them in motion. Saltating larger grains mobilize the smaller particles that winds can then carry in suspension (Fig. 18-20). If wind velocity drops below the level required to move large grains, transport of mid-size grains continues until the wind dies down further and even the smallest grains no longer saltate, ending the chain reaction that set the particles in motion.

Bed Load Like the water-borne sediment that moves along the bottom of a stream, the portion of a desert's wind-transported sediment that moves on or close to the ground is called *bed load*. Air is not dense enough to lift and

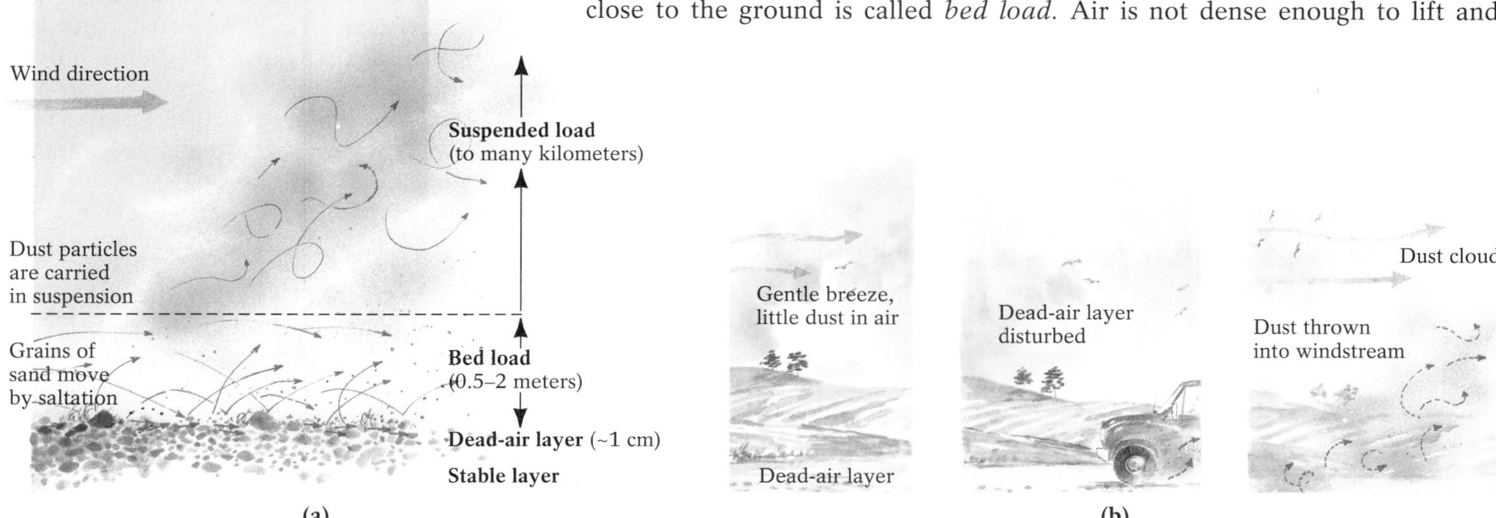

Figure 18-20 Transport of wind-borne sediment. (**a**) Like stream-borne particles, wind-borne particles may travel by rolling or saltating along the surface or by being carried in suspension above the surface. More velocity is needed to transport particles in wind, however, because obstructions at ground level slow air movement, creating a dead-air layer. (**b**) On a relatively windy day, little dust rises from a dry country road because the particles do not extend above the dead-air layer, and thus are unaffected even by strong wind gusts. When a passing truck or a tractor disturbs the dead-air layer, it throws dust up above the zone of motionless air and into the windstream, and a thick cloud of dust rises behind its tires.

carry a windstream's bed load, however; its particles move by saltation. The nature of the surface, the shape and density of the particle, and the force of the collision that sets it in motion determine the height to which a grain will saltate. A grain propelled from a soft bed of sand can attain a height of only 50 centimeters (20 inches) or less, whereas a grain may saltate to a height of 2 meters (6.6 feet) from a surface covered by a continuous pavement of stones. (Imagine tossing unpopped popcorn on a shag carpet, then on a polished hardwood floor.)

A violent wind can churn up a rock-strewn desert and produce a bed load that extends to a height of 1 to 2 meters (3.3–6.6 feet); the air above this dense, surface-hugging cloud of particles is typically quite clear. A person caught in such a sandstorm can be submerged in what appears to be a swirling pond of sand, as illustrated by an eyewitness account by British journalist Richard Trench, who survived a Saharan sandstorm in 1974:

> Suddenly the wind began to rise, blowing in short and powerful gusts. The surface of the desert, normally so still, was growing restless. As the wind rose, so the desert rose too. It was dancing about my feet. It hurled itself against my calves, whirled around my body and beat at my bare arms. Nor did it stop there. It grabbed my neck, stung my face, encrusted itself in my throat, blocked my nostrils, and blinded my eyes. I felt alone and feared death by drowning.

Suspended Load Like the water-borne sediment carried above a stream bed, the portion of a desert's wind-transported sediment carried long distances downwind without settling back to the surface is called *suspended load.* Most of the suspended load in an airstream consists of *dust,* particles about 0.15 millimeters or less in diameter. Dust may include silts and clays from soils, volcanic ash, pollen grains, airborne bacteria, minute plant fragments, charcoal from forest fires, fly ash from the burning of coal, small salt crystals from evaporation of sea spray, and finely crushed glacial flour deflated from exposed dry outwash. Many dust particles are relatively flat, with a large surface area relative to their weight; this facilitates uplift, like the wings on an airplane.

Because their shape and low density effectively counter the pull of gravity, dust particles lifted above the dead-air layer by turbulent winds can be swept into the upper atmosphere, thousands of meters high, where they can remain for several years and travel thousands of kilometers. Distinctive red dust from the Sahara has been found 4500 kilometers (3000 miles) away on the Caribbean island of Barbados, 3000 kilometers (2000 miles) away on Parisian roofs, and even further north in Sweden, where it occasionally discolors the winter snows. In North America in the 1930s, dust from the drought-stricken fields of the Great Plains of Texas and Oklahoma fell on fresh snow in New England. In March 1935, dust from eastern Colorado was carried by sustained 80-kilometer (50-mile)-per-hour winds to upstate New York, 3000 kilometers (2000 miles) away, where it darkened the midday skies. Highlight 18-2 further describes America's Dust Bowl.

Deposition by Wind

If wind can excavate blowouts and deflation basins and darken the sky with a veil of dust, what happens when it stops blowing? Airborne particles begin to fall, with the largest, heaviest saltating grains dropping closest to their source, and the flattest, least dense particles of dust dropping farthest from their source. Wind thus sorts its deposits. The appearance and location of wind's depositional landforms are determined by the size and amount of sediment, whether it is carried as bed load or in suspension, the constancy and direction of the wind, and the presence of stabilizing vegetation.

Highlight 18-2 *America's Dust Bowl*

Figure 18-21 Ominous clouds of dust obscured the midday sun during the Dust Bowl years of the 1930s. This photo was taken in April 1935, in Mills, New Mexico.

In the early 1930s, the plowed wheat fields from the Texas Panhandle to the prairie provinces of Canada (Alberta, Saskatchewan, and Manitoba) were stripped of their rich topsoil by strong winds that blew across the drought-parched landscape. Southeastern Colorado, western Kansas and Nebraska, and the Texas and Oklahoma panhandles were particularly hard-hit. Towering columns of swirling dust, called *black rollers,* darkened the sky at noon (Fig. 18-21). In May 1934, powerful winds lasting for 36 hours swept dust from the Great Plains into a dense black cloud that cast a 2000-kilometer (1200-mile)-long shadow across the eastern half of the continent. In autumn, prairie dust fell in upstate New York as "black rain," and in winter in the mountains of Vermont as "black snow."

The dust in the Great Plains was sometimes so dense that it could penetrate the fabric of any garment and bury entire fields of crops. Many people and farm animals died from suffocation or "dust pneumonia," a malady akin to miner's silicosis (a condition brought on by the inhalation of mine dust). As one eyewitness reported:

> These storms were like rolling black smoke. We had to keep the lights on all day. We went to school with the headlights on and with dust masks on. I saw a woman who thought the world was coming to an end. She dropped down to her knees in the middle of Main Street in Amarillo and prayed out loud: Dear Lord! Please give them a second chance.

A combination of natural and human factors produced the Dust Bowl. Much of the Great Plains had been overcultivated. Settlers had plowed the thick tough prairie sod to plant great fields of wheat until, by 1929, more than 100 million acres were under cultivation. For years, rainfall was abundant and harvests were plentiful. But removal of the protective grass cover left the land vulnerable to wind deflation when drought came. In the early 1930s, precipitation decreased to less than 50 centimeters (20 inches) per year and winds gusted regularly at speeds greater than 15 kilometers (9 miles) per hour. Crops failed everywhere. With no roots from trees or crops to anchor it, hundreds of millions of tons of rich topsoil were blown away. Thousands of farmers, their crops blighted and their soils depleted, abandoned their lands and moved on. The westward migration of "Okies" to California was immortalized in John Steinbeck's epic novel, *The Grapes of Wrath.* Songwriter Woody Guthrie captured the bleak outlook of the times in "So Long, It's Been Good To Know You," written on April 14, 1935, the date of one of the decade's worst dust storms.

By 1939, a few wet years, along with state and federal attempts to improve farming practices, had ended the Dust Bowl tragedy. The prairie states and provinces became habitable and cultivatable again. Damaging droughts occurred between 1950 and 1980, but with extensive irrigation and crop rotation, there was less threat of a recurrent major problem. Residents of the plains have learned to adjust to naturally recurring dry periods.

Dunes: Bed-Load Deposition **Dunes** are wind-built mounds or ridges of sand. Sometimes they are covered by vegetation. A field of many sand dunes generally resembles a sea with whitecaps; the surfaces of individual dunes, like those of ocean waves, are covered with ripples (see Chapter 6 for a discussion of ripple marks). Sand dunes form in both arid and humid climates, wherever there is a sufficient supply of sand that is initially not stabilized by vegetation (vegetation often, however, gains a foothold later), and where strong winds blow constantly. Dunes are found along the sandy shores of oceans, seas, and large lakes (such as Lake Michigan), and near large, dry, sandy flood plains.

In North America, dunes are found in such scenic places as Florence, Oregon; Alamosa, Colorado; Padre Island, Texas; Burns Harbor, Indiana; Long Island, New York; Cape Cod, Massachusetts; and Cape Breton, Nova Scotia.

Dunes typically form where an obstacle, such as a hill or picket fence, interrupts the flow of saltating sand, or the wind slows to a speed at which it can no longer transport sediment. Dunes also form where a narrow obstacle, such as a clump of vegetation or a large rock, forces the windstream to diverge and go around it. An obstacle creates a *wind shadow* of calmer air leeward (downwind) of it, and a smaller one upwind of it as well. The calm leeward air cannot keep saltating grains aloft, so they settle out to start or add to a growing dune (Fig. 18-22).

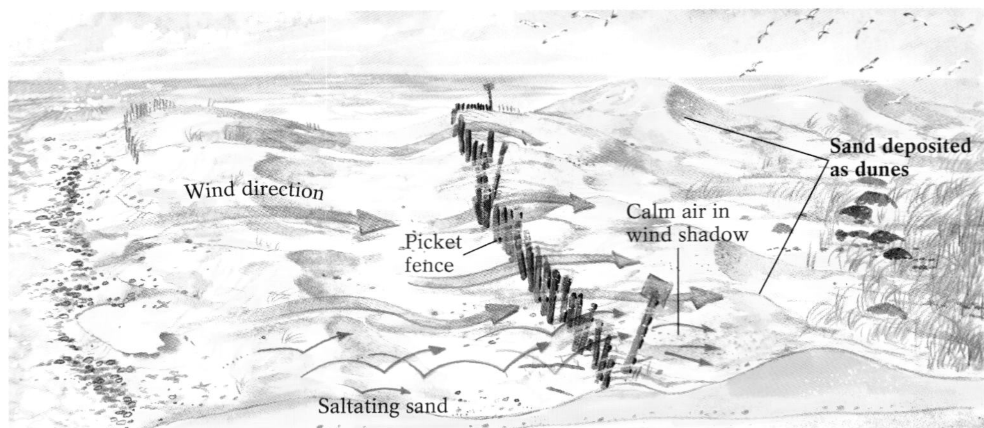

Figure 18-22 Sand dunes often form when a surface obstacle such as a boulder, hill, or picket fence interrupts the flow of migrating, saltating sand. A wind shadow is created leeward of such an object, and sand settles and accumulates on the ground.

As a dune grows, it itself becomes an obstacle, trapping sand in its own lee. Sand dunes can also develop windward of an obstacle. For example, westerly winds pick up large quantities of sand as they blow across the barren sandstone cliffs and deserts of the American Southwest. When these winds reach the lofty Sangre de Cristo Mountains of southern Colorado, they slow down and drop their load onto the Great Sand Dunes National Monument (Fig. 18-23).

Dunes accumulate more rapidly on sand-covered surfaces than on desert pavement. They continue to grow as long as sand-carrying winds blow from the same general direction, or until they reach the height limit imposed by the local wind pattern. Dunes reach their height limit when the wind speed that would be required to carry a load of sand over the dune and drop it on the lee side becomes too great to deposit any additional sand, and the wind's load is carried past the dune. Most dunes are from 10 to 25 meters (35–80 feet) high. Those of Saudi Arabia and China exceed 200 meters (660 feet), but they are complexes of dunes deposited one atop another, not individual dunes (Fig. 18-24). The sand-sized particles that make up dunes are most often weathering-resistant quartz grains. Gypsum grains, eroded from evaporite deposits on the slopes of the San Andres Mountains to the west, compose the dunes at White Sands, New Mexico; the dunes on the island of Bermuda comprise the broken shells of marine organisms.

Most dunes are asymmetrical, sloping gently (about 10°) on the windward side and more steeply (about 34°) on the leeward, at the angle of repose of dry sand (discussed in Chapter 13). As sand grains saltate up the gentle windward slope and reach the dune crest, they spill over to the leeward wind shadow. Here the decreased wind velocity quickly deposits the particles on the steep leeward slope, or **slip face.**

Figure 18-23 The dune field of the Great Sand Dunes National Monument continues to grow as loose material eroded from the arid lands to the west and carried by westerly winds is dropped against the Sangre de Cristo Mountains of southern Colorado.

Figure 18-24 Complexes of dunes in the Gobi Desert, western China.

As sand saltates up the exposed windward side and is deposited on the sheltered leeward side, dunes migrate downwind (Fig. 18-25). Small dunes tend to migrate faster and farther, because there is less sand to be moved. The rate of migration ranges from a few meters per year for large dunes in vast sandy deserts with gentle variable winds, to hundreds of meters per year for small dunes on bare, rocky desert floors with strong unidirectional winds. Coastal dunes may even migrate seaward and extend a coastline offshore by several kilometers (Fig. 18-26).

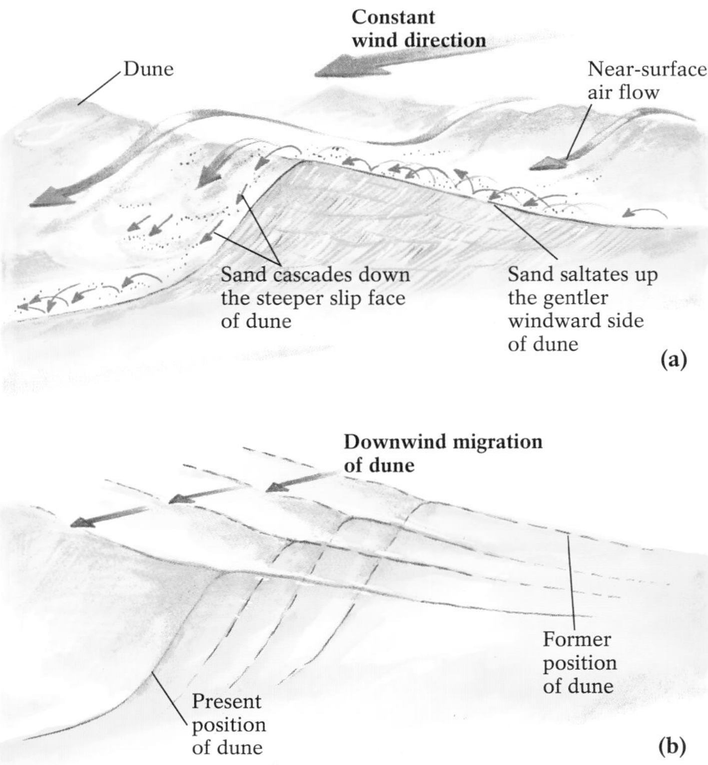

Figure 18-25 (**a**) Sand saltates up a dune's windward side, then cascades down its slip face. (**b**) This progressive transport of sand from the windward to the leeward side of dunes causes downwind migration of the dunes.

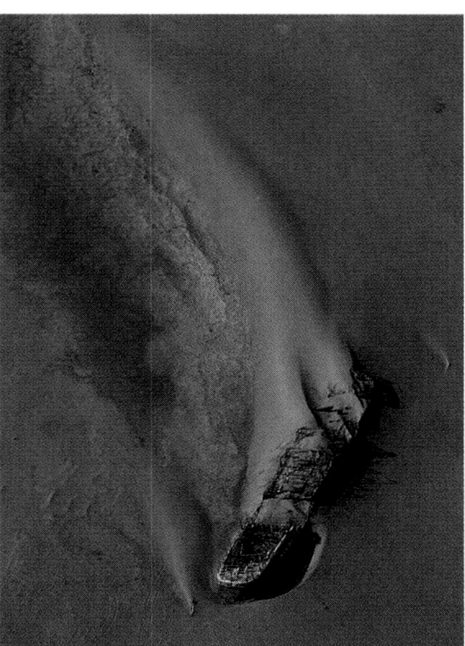

Figure 18-26 The stranded German freighter *Eduard Bohlen*, which ran aground off Africa's southwestern coast in 1912, now lies sand-locked more than a kilometer inland. The migrating and drifting sands of the Namib Desert have extended the coast seaward.

Figure 18-27 Sand encroaching on Interstate 84, Columbia River Gorge, Oregon.

Migrating dunes can menace anything downwind of them—buildings, forests, roads, or railroads. Dune migration across major interstate highways requires frequent and costly sand removal to keep the roads open (Fig. 18-27). Sand movement can be halted by planting a continuous grass cover on a dune, where climate permits. If vegetation is removed, perhaps by motorcycles, dune buggies, or all-terrain vehicles, wind gusts erode new blowouts and dunes resume their downwind march.

Dune Shapes Dune shapes are determined by local conditions, type of sand, degree of aridity, the nature of the prevailing winds, and the type and amount of vegetation present. **Transverse dunes** are a series of parallel ridges that typically occur in arid and semi-arid regions where sand is plentiful, wind direction constant, and vegetation scarce (Fig. 18-28a). These dunes form perpendicular to the prevailing wind direction and have a gentle windward slope and steep leeward slip face. In the Sahara they can be as large as 100 kilometers (60 miles) long, 100 to 200 meters (330–660 feet) high, and 1 to 3 kilometers (0.6–2 miles) wide. Transverse dunes can also develop along the shores of oceans and large lakes, where abundant sand is shaped by strong onshore winds; such dunes dot the southeastern shore of Lake Michigan at the Indiana Dunes National Lakeshore in Indiana, and at Warren Dunes State Park in Michigan.

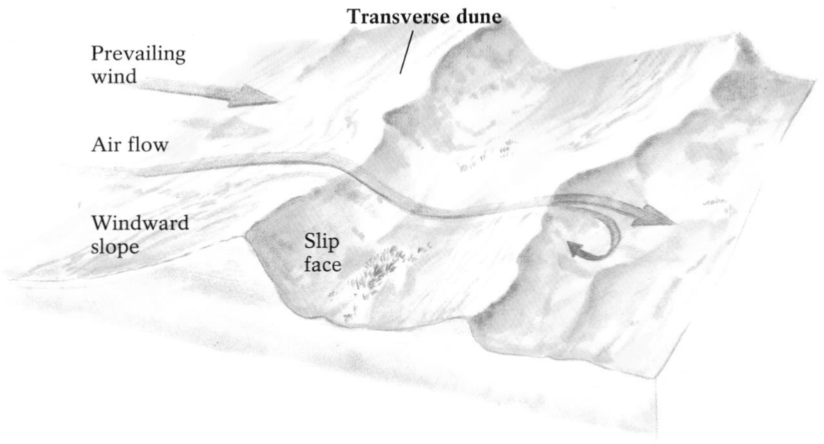

(a)

Figure 18-28 Common dune types. (**a**) Transverse dunes on Mesquite Flat, Death Valley, California.

Longitudinal dunes are also parallel ridges. They form when sand supply is moderate and wind direction varies within a narrow range (Fig. 18-28b). Longitudinal dunes are oriented parallel to the prevailing wind direction. Small ones may be only 60 meters (200 feet) long and 3 to 5 meters (10–16 feet) high. In the Libyan and Arabian Deserts, where strong winds blowing from several directions converge over the dune crest to create one sinuous form, they can be as large as 100 kilometers (60 miles) long and 100 meters (330 feet) high.

Barchan dunes (pronounced *bar´-kane*) are crescent-shaped ridges that form perpendicular to the prevailing wind as sand begins to accumulate

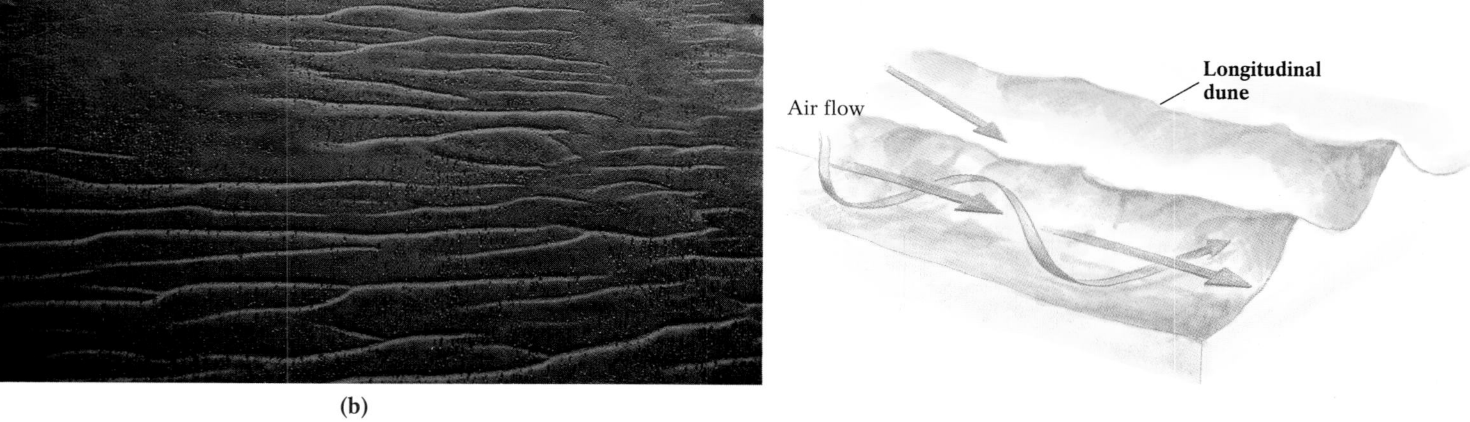

Figure 18-28 (**b**) Longitudinal dunes in the Simpson Desert, central Australia.

around small patches of desert vegetation (Fig. 18-28c). Barchans develop in arid regions on flat, hard ground where there is little available sand and wind direction is fixed. As they grow, barchans become thicker and higher in their centers where air flow is impeded most and more sand is deposited. Because their horns, the points of the crescents, are thinner than their centers, the horns migrate downwind more rapidly, thus extending the barchan with its characteristic sharply pointed horns in the downwind direction.

Parabolic dunes are horseshoe-shaped, differing from barchans in that their horns point upwind (Fig. 18-28d). They commonly form along sandy ocean and lake shores, the only appreciable dune areas outside of deserts. Parabolic dunes develop from transverse dunes that are exposed to accelerated wind deflation, especially after removal of some vegetation. A small deflation hollow forms on the transverse dune's windward side, and the wind-excavated sand piles up downwind. As the hollow grows, the wind becomes concentrated at its center, speeding the migration of that portion of the dune. The horns, a remnant of the original transverse dune, are usually still covered by vegetation and remain anchored in place; the rest of the parabola continues to migrate downwind, forming a horseshoe shape that can become quite elongated.

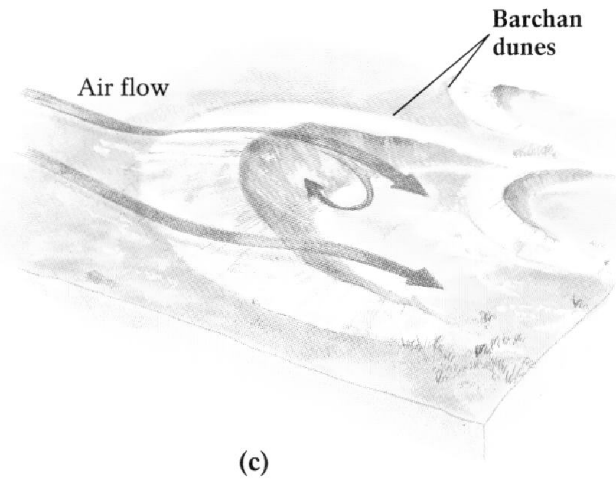

Figure 18-28 (**c**) Barchan dunes in the Baja Desert of Baja California.

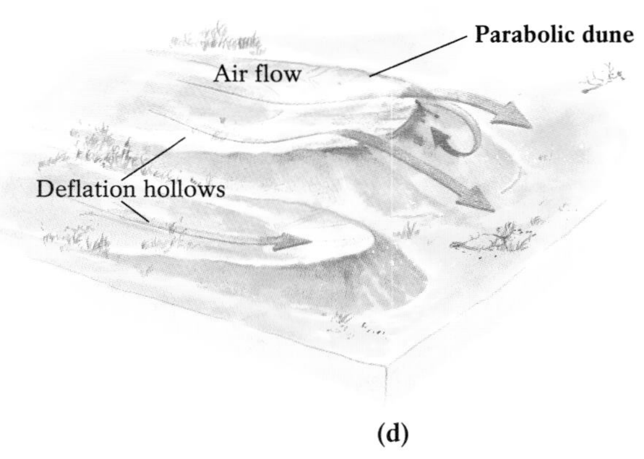

Figure 18-28 (**d**) Parabolic dunes.

(e)

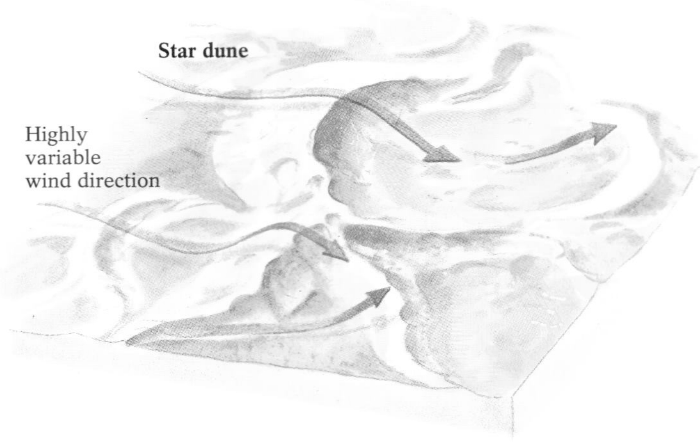

Figure 18-28 (e) Star dunes in the Namib Desert.

Star dunes, the most complex of the dune types, form when winds blow from three or more principal directions, or when wind direction is constantly shifting (Fig. 18-28e). They tend to grow vertically to a high central point and may have three or four arms radiating from the center. Continued variability of wind direction causes star dunes to remain relatively fixed in position.

Loess: Suspended-Load Deposition Loess (discussed in Chapter 17), is a wind-borne silt deposit that resembles fine-grained lake mud. Unlike water-borne sediments, which are typically deposited at topographic low spots, loess deposits cover hills, slopes, and valleys evenly, in a manner that suggests that they literally fell from the sky. Most loess grains are quartz, feldspar, mica, or calcite. Slight oxidation of accessory iron minerals gives loess its characteristic yellow-brown or ocher color. The terrestrial origin of loess is also indicated by the presence of shells from such terrestrial creatures as air-breathing snails, as well as by the scattered mammal bones, worm burrows, and plant-root tubes that loess commonly contains.

Loess is always deposited downwind of a plentiful supply of silt. Much of the loess in America's Midwest is a yellow-brown deposit that apparently originated when coarse-to-medium silt particles (0.01–0.06 millimeters in diameter) were eroded from drying glacial outwash and deposited downwind. The streams draining from the melting ice carried large volumes of silt-sized glacial flour. As the outwash plains dried during the warm interglacial periods, the prevailing winds from the west swept up the fine silt and deposited it downwind to the east. More than 500,000 square kilometers (200,000 square miles) of the land east of the Missouri and Mississippi Rivers is loess-covered, including much of the farmland in Iowa, Illinois, Indiana, and Missouri. Similar deposits are found east of the Rhine River in Germany, in other central European countries, and in central Asia.

Some loess, for example that west of the Mississippi River in eastern Kansas and Nebraska, originated in nonglacial environments. These deposits apparently were eroded by wind from the ancient dust-storm deposits that make up the sand hills of northwestern Nebraska. Loess now lying on the Palouse Hills of southeastern Washington was lifted by westerly winds from the drying flood plain of the mighty Columbia River to the west. The world's largest loess deposits are in northern China. Their phenomenal 300-meter (1000-foot) thickness derives from the nearly endless supply of silt in the vast Gobi Desert to the west; this airborne dust settles and washes into the Huang Ho (Yellow River) and Yellow Sea, giving them their distinctive color.

Because loess originates in both glacial environments and arid deserts, places where little chemical weathering occurs, the particles are remarkably fresh. Consequently, loess retains even its more soluble mineral constituents, making loess-covered lands, such as those in Iowa, extremely fertile. Loess particles are typically angular because, being airborne, they experience few of the collisions that tend to round particle edges. Their texture thus promotes the relatively high porosity of loess deposits, which adds to their moisture-holding capacity and high agricultural potential.

Reconstructing Paleowind Directions

Winds of the past, or *paleowinds,* have left traces of their intensity and prevailing direction in dunes and loess deposits that have since turned to stone. These traces can help us learn more about the geologic past. When geologists study a sandstone formation, they search for clues that determine whether it was deposited by wind or water. Excellent sorting, cross-bedded sands displaying the angle of repose of dry sand, the occasional discovery of sharply faceted ventifacts, the presence of sand grains polished and pitted by mid-air collisions, and the absence of ultra-light mica flakes (which are easily swept away in suspension) suggest a wind-deposited sandstone. After a wind-related origin is confirmed, geologists can use any of several wind-generated features to reconstruct ancient wind directions. These include the shape of yardangs, the position of ventifact facets, the asymmetry of dunes, and the thickness and texture of loess deposits.

Sand dunes rarely survive intact in sedimentary rocks; they are usually buried by later dune migration. But although the gentle windward slope and the steep leeward slip face of a dune form may vanish, the inclined bedding of its original slip face may be preserved under the sands of advancing dunes, creating dune-type cross-bedding. The leeward slip-face beds are inclined at the angle of repose of dry sand (30°–34°), and indicate the paleowind direction that prevailed when the bedding layer was formed (Fig. 18-29).

Certain properties of loess, like those of other wind-generated features, can also be used to estimate past wind directions. A loess layer is generally thicker near its source and thins downwind, where its particles are finer. The 10-meter (33-foot)-thick layers of loess that occupy the eastern banks of southward-flowing rivers in the Midwest thin to less than a few centimeters in the East. Hence, we know that even 12,000 years ago, the prevailing winds were westerly, as they are today (Fig. 18-30).

Figure 18-29 Numerous ancient, lithified sand dunes are found in national parks in the arid Southwest. The Navajo Sandstone, for example, crops out throughout Zion National Park in southwestern Utah. It is a memento from 180 million years ago, when what is now Utah was on the continent's west coast. (This was before the lithosphere on which California is located was added to the continent.) We can be sure that the Navajo Sandstone originated on land, and therefore was deposited by wind, because dinosaur footprints have been found in the formation. From the orientation of the cross-beds, we can reconstruct the northerly and westerly paleowinds that formed the migrating coastal sand dunes that later became Checkerboard Mesa.

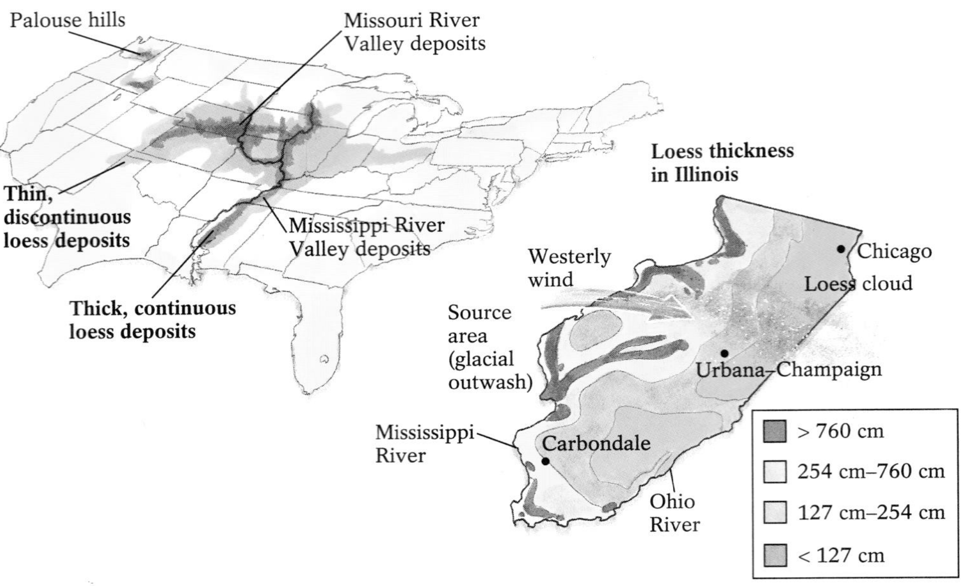

Figure 18-30 The loess that covers the rolling hills in the Mississippi River valley becomes thinner and finer-grained with greater distance east of the Mississippi River. This suggests that it was deposited by westerly winds.

Reconstruction of paleowind directions enables geologists to deduce ancient plate tectonic events. Winds blow from west to east over the British Isles today, a wind pattern that has probably varied little through time for that latitude. But British geologists discovered that the winds that formed the dune-type cross-bedding of the New Red Sandstone apparently blew from the east and northeast. Applying their understanding of plate tectonics, they now believe that the New Red Sandstone was probably deposited about 200 million years ago. At that time England was somewhere to the south in a belt of easterly winds, and has since drifted northward into the belt of westerlies as a consequence of plate motion. The southern origin of these rocks has been confirmed by recent studies of the paleomagnetism of British rocks.

Wind Action on Mars

The Earth is not the only windswept planet in the solar system. All of the planets are heated differently by the Sun's rays; those that have atmospheres also exhibit evidence of wind action. Mars is an especially windy planet.

The Mariner 9 space probe went into orbit around Mars on November 13, 1971, after a five-month journey from Earth. Its first photographs were, disappointingly, of a planet-wide dust storm. For three months, dust that rose as high as 6 kilometers (4 miles) obscured the entire Martian surface. On Earth, where the air is more than 100 times as dense, it would have taken winds in excess of 160 kilometers (100 miles) per hour to raise dust to that height. On Mars, however, the weak gravitational force enables dust particles to ascend four times as high as those carried by comparable wind gusts on Earth. Because Mars' atmosphere is so thin, however, very little dust actually moves. Although dust storms obscure surface features when viewed from orbit, visibility is not significantly impaired for cameras located within the storms themselves. Fine oxidized particles that are perpetually suspended in the Martian atmosphere contribute to the planet's pink-red appearance. Shifting deposits are the cause of the frequent surface changes observed on Mars, where patches of light-colored dust are often succeeded by darker ones; strong local winds apparently lift and transport light-colored fine particles, exposing the dark bedrock below. Leeward streaks of light-colored dust covering dark rocks are especially common downwind of the rims of impact craters.

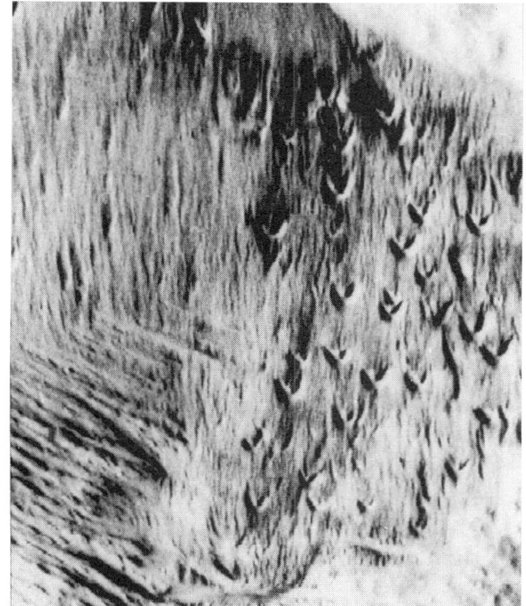

Figure 18-31 These crescent-shaped ridges on Mars are remnants of barchan dunes produced by the planet's strong surface winds.

Mars' global dust storms apparently start when summer comes to its southern hemisphere. As the edges of the Martian frozen carbon-dioxide polar cap rapidly sublimate (see Chapter 17), large temperature differences arise between the still-solid polar cap remnants and the warming surrounding landscape. These temperature differences produce differential air pressures, which in turn lead to high winds, producing dust storms that spread northward until they engulf the planet's entire surface for months.

After the dust finally settled in 1972, Mariner 9 was able to give Earth-bound scientists exciting photographic images of the Martian surface. A dune field, 60 by 30 kilometers (40 by 20 miles) was sighted on the floor of a large crater. Viking probes landed on the Martian surface a few years later, on July 20 and September 3, 1976. The landers descended onto loess-covered plains, where they photographed interspersed ventifacts, barchan and longitudinal dunes, yardangs, and other wind-generated landforms (Fig. 18-31).

Desertification

Desertification is the invasion of desert conditions into formerly nondesert areas. The common symptoms of desertification are a significant lowering of the water table, a marked reduction in surface-water supply, increased salinity in natural waters and soils, progressive destruction of native vegetation, and an accelerated rate of erosion.

Northern and western Africa have been experiencing rapid desertification for 2000 years. After the end of the last major worldwide glaciation, about 8000 years ago, the Namib Desert of southwestern Africa was a lush savannah that supported advanced Stone Age societies (Fig. 18-32). It remained a fertile grassland, teeming with wildlife, for the next 6000 years. We can reconstruct the land features of that ancient world using radar imaging. Radar waves cannot penetrate moist soil. In humid areas, radar is either absorbed by soil moisture or reflected, creating an image of the surface. The regolith of hyper-arid deserts, however, is virtually transparent to radar; the reflected image shows the contours of the underlying bedrock. In November 1981, the space shuttle Columbia cruised about 200 kilometers (120 miles) above the eastern Sahara, using radar to map the configuration of the bedrock beneath the extremely dry Selima sand sheet, a vast, flat, featureless sea of sand. The radar waves penetrated 5 kilometers (3 miles) of sand, and were reflected to reveal an ancient surface of rocky hills and deep, wide river valleys (Fig. 18-33).

Figure 18-32 This cave painting depicts falling rain. It is found today in the parched Namib Desert of southwestern Africa, but was painted thousands of years ago in the moist, temperate climate characteristic of the region at that time.

Figure 18-33 A satellite image of the Selima sand sheet, enhanced by space shuttle radar imagery in 1982. The topography of the bedrock under the sand shows an integrated network of stream valleys.

Two factors were involved in turning the northern African savannah and grassland into the modern Sahara. One—drought—is a natural and unavoidable factor; the other—overpopulation and land mismanagement—is human-induced. Drought or overpopulation, or both, can start the process of desertification. When a land's inhabitants cannot produce enough to provide for their needs, they tend to overgraze their cattle, plant crops without replenishing the soil, cut down trees for shelter and fuel, and draw more water from springs and other sources than is naturally resupplied. Gradually soils become depleted of their nutrients, and the removed trees and their roots do not grow back. Without vegetation, soils cannot hold water or prevent wind erosion. Water from occasional cloudbursts washes away the unbound topsoil and, in hot climates, evaporates before it can enter the groundwater supply. Eventually the parched land is left with virtually no productive capacity. The process of desertification, once begun, tends to be self-perpetuating. Left with no source of sustenance, the land's inhabitants must migrate in search of food, water, and shelter.

As much as 35% of the world's land, now barely sustaining a population of 850 million, consists of the semi-arid margins of already arid lands. Overpopulation and drought may convert them in the next few decades from dry but habitable grasslands to deserts unable to sustain significant human populations. As many as 70 nations are affected; about half of these are in Africa, where the Sahara is advancing southward by as much as 50 kilometers (30 miles) per year, and choking dust storms and newly formed dune fields already threaten some northern and western cities.

The Sahel, the region of northern Africa immediately south of the Sahara, has been particularly hard-hit by desertification in recent decades. The worst drought of the century struck in the early 1970s, displacing 20 million people and their herds of cattle, sheep, goats, and camels. Animals that were not led away died. Although the drought was the immediate cause of the disaster, inadvertent land mismanagement had set the stage for it decades before. From 1935 to 1970, the Sahel's population doubled and progressively larger herds grazed on the marginally productive land. A few abnormally wet years in the early 1960s increased the region's productivity temporarily, encouraging farmers to plant more crops and expand their herds and grazing lands. Nomadic herdsmen from the south rushed to the Sahel, because it had become an area of scattered groves of trees (for fuel and shelter), abundant seasonal grasses (for grazing), and agricultural fertility (the primary crops were cotton, beans, millet, sorghum, and maize).

But in 1970, no rain fell at all. The large herds quickly demolished the existing grasses; the rich topsoils, without binding roots, were easily eroded by wind. By the next year's rain, the ground was baked so hard as to be impermeable, and surface runoff only accelerated soil erosion. Without crops, starving people were forced to eat their remaining grain seeds, eliminating all hope of new planting. Trees were destroyed by foraging animals and people, eroding the soil further and depriving the region of fuel and building materials. Animals died by the millions. The starving population migrated to the region's larger cities, which were tripled in size by the refugee camps that sprang up around them. Worldwide relief efforts were too little and too late to prevent the deaths of hundreds of thousands of people from starvation, malnutrition, and associated diseases. The drought continued into the 1980s, ending only in the mid-1990s (Fig. 18-34).

Figure 18-34 Skeletons of a cattle herd in the sun-baked lands of the Sahel, which underwent rapid desertification from 1970 through the early 1990s.

In the United States, overpopulation and urban growth, overgrazing, and excessive groundwater withdrawal, particularly in the arid and semi-arid Sunbelt states, have already increased soil salinity and accelerated erosion. Today, about 27% of U.S. lands that are not already deserts face encroaching desertification.

Reversing Desertification

To reverse desertification, available water must be delivered where it is needed. To supplement the scant rainfall typical of arid regions, copious volumes of water can often be pumped from deep aquifers, channeled from distant lakes or rivers, or drawn from the ocean and desalinized.

Some desert communities, such as Palm Springs, California, are naturally supplied by deep groundwater that originated as rain or snow on moist highlands hundreds or thousands of kilometers away. This water can be pumped for use in irrigation. In Africa, rain falling in the eastern highlands near the equator enters the Nubian Sandstone, an aquifer that extends more than 3000 kilometers (2000 miles) north, where it lies deep beneath the Sahara; faulting and folding has brought the aquifer's waters to the surface in the form of springs, where oases flourish (see Fig. 15-12). The quantity and quality of artesian water at oases is determined by the rate of flow through the aquifer. Swift flow brings water that is relatively pure and supports date palms and other vegetation, but when water flows slowly through rocks, it dissolves a considerable amount of salt and may be undrinkably saline, even toxic, by the time it surfaces at an oasis. Some desert groundwater may be a relic of infiltration and deep storage from moister climatic conditions thousands of years ago.

In some extremely parched places, inventive technologies have been developed to reclaim recently desertified lands and to make portions of ancient deserts more habitable for human populations. Systems of wells, channels, and collecting pools are in use in the Middle East to collect and store infrequent storm runoff underground, to protect it from evaporation. Throughout the region, farmers apply a specially designed plastic mulch to their farmlands or cover their fields with plastic sheeting, punctured for plant stems, to retain irrigation moisture. Computers monitor and regulate water flow through pipes and canals, delivering water where it is needed at times when evaporation is minimal. In Israel, "drip" agriculture—using perforated garden hoses that snake amid plantings—delivers water constantly to individual plants, literally drop by drop. Water is even drawn from the Mediterranean Sea and desalinized for agricultural use. In the American Southwest, billions of liters of Colorado River water have been channeled into the Sonoran Desert of southern Arizona, providing sufficient irrigation to produce food for hundreds of thousands of people (Fig. 18-35). These are but a few of the many techniques that can enable arid lands to support a human population.

Figure 18-35 A circle of irrigated land in the Yuma Valley of Arizona's Sonoran Desert. Irrigation is largely responsible for the agricultural productivity and rapid population growth of this region.

Efforts to reverse desertification can also focus on controlling wind erosion. Trees serve as an effective windbreak to halt shifting sands, and help to retain both soil cover and groundwater. Migrating dunes can be stabilized by planting fast-growing trees such as poplars, which have deep, sand-binding roots. In northeastern China, an entire forest, 500 by 800 kilometers (300 by 500 miles), was planted upwind of 90,000 acres of prime farmland. Dunes in China have even been leveled by hand and covered with topsoil to create productive new farmland.

Many ecology-minded individuals, however, reject the notion that deserts are lifeless environmental wastelands that could and should be modified by humankind for its own purposes. Rather, they view deserts as vibrant natural environments—valuable ecosystems teeming with distinctive flora and fauna—that should be protected and preserved in their natural state, in much the same way we have sought to save the Earth's tropical rain forests. Perhaps the best approach we can take to desert lands is to appreciate the beauty and ecological uniqueness of the ancient, naturally occurring ones, while learning to manage their marginal lands more wisely to prevent desertification of now-habitable semi-arid regions.

Chapter Summary

"**Desert**" is a somewhat relative term used to describe dry regions where vegetation is typically sparse. It can be defined as an area that receives less than 250 millimeters (10 inches) of precipitation annually, or as a region with an **aridity index,** or ratio of potential annual evaporation to average annual precipitation, greater than 4.0. There are five principal types of desert: 1) the subtropical deserts that girdle the Earth between about 20° to 30° latitude both north and south of the equator, where dry air descends toward the surface as part of the global wind-circulation system; 2) deserts on the dry leeward side of major mountain ranges, created by the **rain shadow effect**; 3) interior deserts, in the centers of continents; 4) coastal deserts that develop where prevailing onshore winds have been cooled by cold oceanic currents; 5) polar deserts, which are both extremely cold and dry.

In the dry desert environment, such processes as weathering and sediment transport operate in different ways and at different rates than in moist temperate climates. Weathering in deserts is mostly mechanical. What little chemical weathering there is consists of manganese and iron oxidation; the resulting oxide stains on rocks are **desert varnish.**

Although one seldom sees surface water in deserts, many of their characteristic landforms have been etched by rapidly flowing, short-lived streams produced by occasional storms. Streams move through, and deepen, the numerous channels that cross a desert; a channel that carries a stream during periods of high discharge but is dry most of the year is an **arroyo.** The mountains that border many deserts recede by the action of surface-water erosion to produce large-scale, gently inclined surfaces called **pediments.** An **inselberg** is a steep-sided knob of durable rock; it may be what remains of a mountain following advanced pediment development. **Bornhardts** are large, dome-shaped inselbergs that form when soft rock surrounding a mass of resistant rock erodes.

In deserts, the transporting ability of surface runoff diminishes rapidly due to evaporation or infiltration. But while surface runoff lasts, it carries sediments that it deposits to form distinctive features: Alluvial fans accumulate where a slope ends and the desert floor begins. A **playa** is a dry lake bed that develops when a desert-floor lake evaporates.

Winds are initiated by the pressure differences caused by differential heating of the Earth's atmosphere. The resulting convection cells set north–south winds in motion, which are deflected to the east or the west by the Earth's rotation, depending on latitude. Because soil-binding moisture and vegetation are scarce in deserts, the land surface is particularly susceptible to modification by wind action. Wind erodes by **deflation,** removing finer particles from the surface. The layer of pebbles left behind is known as a **desert pavement.** Wind-carried particles are hurled against rock surfaces, effectively sandblasting the rock; this is **wind abrasion.** The windward surfaces of individual rocks and boulders are beveled by wind abrasion to form asymmetrical **ventifacts.**

Particle motion begins when wind reaches a velocity that is high enough to overcome the inertia of the largest grains in the dead-air layer at ground level. The grains begin rolling and collide with other grains, initiating saltation. Saltating grains make up a wind stream's coarse bed load. Finer particles are carried aloft in suspended load. The bed load is deposited when the wind's transport energy decreases below a critical level; wind often slows in the wind shadow leeward of a surface obstacle. **Dunes,** hills of loose wind-borne sand, are the result. The size, shape, and orientation of a sand dune is determined by the amount of sand available, local vegetational cover, the intensity of winds, and the constancy of wind direction. Most dunes are asymmetrical, with a gradual windward slope and steep leeward **slip face.** There are five principal types of dunes: **transverse dunes** are parallel ridges that develop perpendicular to the prevailing wind direction; **longitudinal dunes** are parallel ridges that develop parallel to the prevailing wind direction; crescent-shaped **barchan dunes,** whose horns point downwind, lie perpendicular to prevailing winds; horseshoe-shaped **parabolic dunes,** whose horns point upwind, also lie perpendicular to prevailing winds; and **star dunes** grow vertically to a high central point with three or four radiating arms.

The suspended load is deposited as loess, which consists of silt-sized particles. With adequate moisture, loess produces an extremely fertile soil. Because their grain size and bed thickness diminish downwind, loess sheets deposited in the past provide clues to paleowind directions, as do lithified cross-bedded dunes.

Mars experiences planet-wide seasonal dust storms. The planet's weaker gravitational attraction enables its dust to ascend higher than it would in comparable winds on Earth. Viking probes have photographed loess-covered plains, ventifacts, dunes, and other wind-generated landforms.

In the late twentieth century, conditions that promote the growth of deserts are overtaking many formerly semi-arid regions. Drought and overpopulation can cause **desertification,** the invasion of desert conditions into formerly nondesert areas; this is happening on every continent except Antarctica. The symptoms of desertification include significant lowering of the water table, marked reduction in surface-water supply, increased salinity in natural waters and soils, progressive destruction of native vegetation, and accelerated rates of erosion. Newly formed desert lands can be reclaimed through irrigation, controlled planting, diversion of surface waters from moist regions, and use of deep groundwater resources where available.

Key Terms

desert (p. 503)
aridity index (p. 504)
rain shadow effect (p. 507)
desert varnish (p. 509)
arroyo (p. 509)
pediment (p. 510)
inselberg (p. 510)
bornhardt (p. 510)
playa (p. 511)
deflation (p. 514)
desert pavement (p. 515)
wind abrasion (p. 516)
ventifact (p. 516)
dune (p. 519)
slip face (p. 520)
transverse dune (p. 522)
longitudinal dune (p. 522)
barchan dune (p. 522)
parabolic dune (p. 523)
star dune (p. 524)
desertification (p. 527)

Questions for Review

1. Briefly describe three different types of deserts and the conditions that contribute to their aridity.

2. Draw a simple sketch to illustrate the rain shadow effect.

3. Cite evidence that both chemical and mechanical weathering occur in arid regions.

4. List three landforms that are formed by the work of water in arid regions.

5. Why do coastal breezes tend to blow onshore during the day and offshore at night?

6. Briefly describe how desert pavement forms.

7. Draw a simple sketch to show how ventifacts form.

8. Sketch the basic shape of a transverse dune, viewed from the side. What is the essential difference between a barchan dune and a parabolic dune?

9. Describe three ways geologists can reconstruct paleowind directions.

10. What are two causes of desertification? How do they produce their effects?

For Further Thought

1. Speculate about how changes in the configuration of the Earth's continents through plate tectonic activity might increase or decrease the total area of arid regions.

2. What would happen to the distribution of deserts if global warming were to have a significant effect on the Earth's climate?

3. Determine the prevailing wind direction of the landscape in the photograph below.

4. Describe how global wind patterns may have changed through Earth's history. (Hint: Bear in mind that the Earth's rate of rotation has been slowing throughout its history.) Speculate about the future of the Earth's global wind patterns.

5. What actions would you favor in combating the global trend toward increased desertification?

Waves, Currents, and Tides

Processes That Shape Coasts

Highlight 19-1 **Lake Michigan's Vanishing Shoreline**

Types of Coasts

Highlight 19-2 **North Carolina's Outer Banks**

Sea-Level Fluctuations and Coastal Evolution

Figure 19-1 A space shuttle photograph of Cape Cod, Massachusetts, showing its kilometers of shoreline and offshore islands.

Chapter 19

Shores and Coastal Processes

Shores and coasts, at once scenic and educational, are wonderful places to observe natural processes, particularly the action of waves, tides, and nearshore currents. North Americans have a passion for vacationing near shores and coasts and more than half of us live within 80 kilometers (50 miles) of the Atlantic or Pacific Ocean or near one of the Great Lakes.

All shores change constantly through natural processes. Sometimes those processes act rapidly and dramatically. On January 2, 1987, for example, a powerful winter Nor'easter gouged more than 20 meters (65 feet) of dunes and beaches from Nauset Beach, at the bend in the "elbow" of Cape Cod along the Atlantic coast of Massachusetts (Fig. 19-1). The beach, a 20-kilometer (12-mile)-long pile of sand that had been scraped from the sea floor by waves and tides over the past 4000 years, had sheltered the bayside town of Chatham and its fishing fleets for centuries. But on that night, Nauset was breached by 6-meter (20-foot)-high waves that also swept away nearly a kilometer of the cape's offshore islands, leaving no barrier to protect Chatham from the Atlantic's waves and storms.

The breach of Nauset Beach had some benefits. Chatham's fishermen acquired a shortcut to their Atlantic fishing grounds. More importantly, increased wave and tidal activity flushed nitrates and phosphates out of accumulated septic fields adjacent to the bay, which had contaminated or driven off most of the edible sea life in the bay; that left cleaner water and restored the bay's aquatic life. Now bay scallops flourish again in an environment with lower nitrate levels, and striped bass and bluefish have returned after years of pollution-impaired fertility.

A **shoreline** is the boundary where a body of water meets the adjacent dry land. The entire region bordering a body of water is a **coast** (Fig. 19-2); coasts extend inland until they encounter a different geographical setting, such as a mountain range or a high plateau. A *shore,* the strip of coast closest to a sea or lake, is often a sandy strip of land, or a *beach.* In this chapter we consider the variety of processes that shape and change our coasts—the waves, currents, and tides that erode and deposit coastal materials and the plate movements that raise or lower coasts above or below sea level. We also look at how human activity affects the evolution of coasts, and at people's attempts to manipulate coastal environments to reduce damage to coastal lands and their distinctive character.

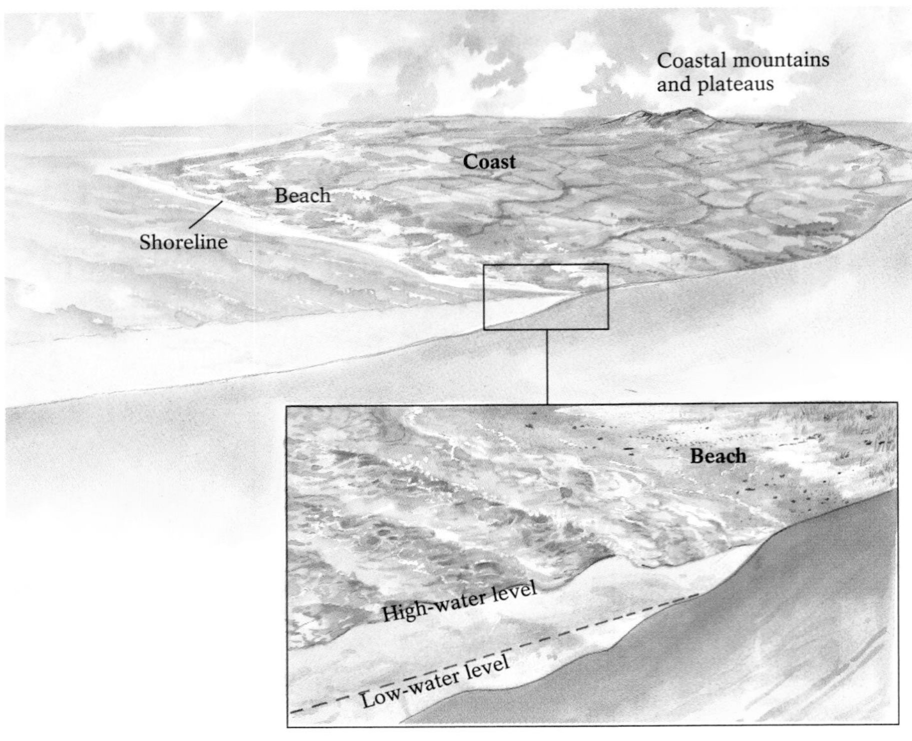

Figure 19-2 A typical coast consists of a shoreline, where the ocean meets the land, and a beach, a narrow strip of the coast washed by the water from breaking waves.

Waves, Currents, and Tides

Moving water is the great agent of geologic change at the Earth's coasts. Water can be set in motion by the wind, which produces most waves and currents, and by the combined effects of the gravitational pull of the Moon and Sun and the rotation of the Earth, which alternately raise and lower water surfaces, producing tides.

Waves and Currents

All waves—be they earthquake, sound, water, or any other kind—transport energy. The principal source of the energy transported by water waves is solar radiation, which (as we saw in Chapter 18) heats the atmosphere more in some regions (such as near the equator) than in other regions (such as near the poles). This creates zones of low atmospheric pressure (more heated) and high atmospheric pressure (less heated). When air flows from high-pressure to low-pressure zones as wind, it drags across the surface of any water body in its path, reshaping the surface for the moment.

Wind drag causes the surface water of an ocean or lake to alternately rise to form a *crest* and fall to form a *trough* (Fig. 19-3). A wave's *height* is the vertical distance between crest and trough. In the open ocean, waves commonly measure 2 to 5 meters (7–18 feet) high; during hurricanes, wave heights can exceed 30 meters (100 feet). Wave *length* is the distance between two adjacent waves, measured from crest to crest or trough to trough (Fig. 19-3). Ocean waves can be 40 to 600 meters (135–2000 feet) apart. The time required for one wave length to pass a stationary point is a wave's *period.* Ocean waves commonly have a period of a few seconds. Wave *velocity,* the speed at which an individual wave travels, is typically 30 to 90 kilometers per hour (20–60 miles per hour) in mid-ocean.

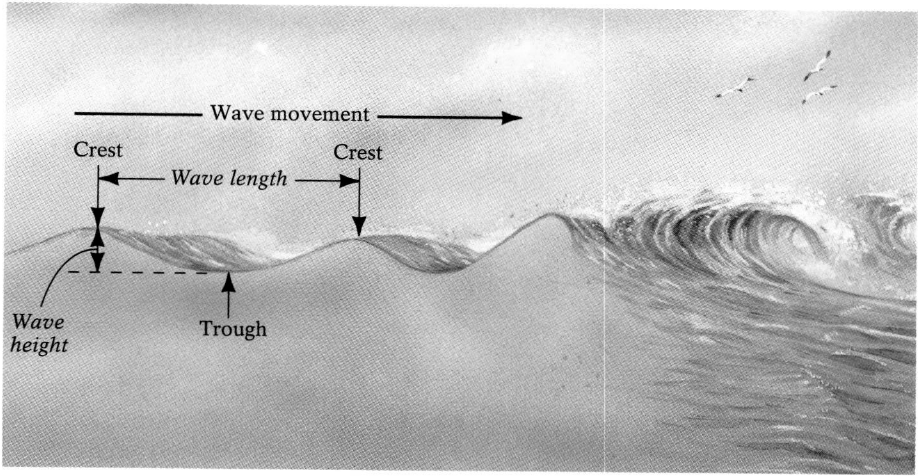

Figure 19-3 The components of a wave.

The Movement of Waves Wave heights, lengths, periods, and velocities are determined by the speed and duration of the wind, the constancy of wind direction, and the distance the wind travels across the water surface. The distance the wind travels is its *fetch.* A brisk wind that blows for a long time from the same direction across a large body of water produces a closely spaced series of large, fast-moving waves. Fetch is greatest, and waves largest (in height), where waves travel without being interrupted by landmasses. Pacific Ocean waves tend to be very large, in part because the Pacific is so huge and its landmasses so widely spaced that the fetch of its winds is great. Similarly, the band of the global ocean system that is south of the southern tip of South America is virtually uninterrupted by any landmasses. This globe-circling fetch also generates some of the world's largest waves, commonly exceeding 15 meters (50 feet) in height. The largest wind-generated wave ever observed, sighted in the northern Pacific in 1933 by the captain of the USS *Ramapo,* was 34 meters (115 feet) high. This wave was generated by persistent gale-force winds.

A breeze of only 3 kilometers per hour (2 mph) is the minimum that can set a perfectly calm water surface rippling in small waves. If you blow forcefully across the surface of a bowl of water, you can create sizeable waves; as you gradually reduce the force of your breath, the height of the waves diminishes.

When waves from several storms blowing from different directions and distances converge in one area of an ocean, complex sets of waves of various heights and periods arrive at the shore at irregular intervals. Regardless of where they originated, waves with similar velocities and periods can become synchronized (be "in synch"), reinforcing one another to produce large waves; this is called *constructive interference.* Waves with different velocities and periods may be unsynchronized ("out of synch"), thereby partly canceling one another and producing small waves; this is *destructive interference.* Some of the big waves awaited by surfers at Malibu Beach, California, may originate in Antarctic storms, growing larger from constructive interference by waves from Hawaiian rain squalls. When you watch waves, look for irregular arrival intervals and sharply variable heights—you will be seeing the cumulative effect of storms blowing throughout the ocean basin.

Although mid-ocean waves move outward from a wind source, only the wave *form* moves significantly outward; the water within the wave has a rolling circular path, rising and falling as the wave passes but moving only a short distance from its original position. This **oscillatory motion** dies out as ocean depth increases, and is virtually absent below a depth equal to about one-half the wave length (defined earlier as the distance between two successive crests). This depth is called the *wave base*; water below the wave base is undisturbed by waves passing above. In deep water surface waves have no effect on the sea floor, but in shallow water a wave base may intersect the sea floor, disturbing its loose sediments.

After a journey of perhaps thousands of kilometers, waves cross a continental shelf and approach a coast. As the shelf becomes progressively shallower landward, the sea-floor depth decreases to the wave base, and the water's oscillatory motion is interrupted by the sea floor. A set of waves whose length is 100 meters (330 feet), for example, begins to drag against the sea floor at a depth of 50 meters (165 feet).

When a wave touches bottom, its velocity decreases because some of its energy is spent moving loose sediment on the sea floor. The seaward waves, which are still in deep water, follow the shallow-water waves, which are slower. As a result, the length and period of the following waves decrease and their heights increase. That is, the waves bunch up and steepen, and appear at shorter intervals. As a wave continues into the near-shore shallows, its crest moves faster than its bottom; eventually, the crest overruns the rest of the wave and falls over and *breaks*, dispersing the wave's energy. The offshore shallow area where waves break is the *breaker zone*. The water in a breaking wave moves landward as *surf*, low foamy waves. The motion of surf is **translatory,** signifying that the oscillatory motion of the water has changed to motion in which the water itself actually moves forward (Fig. 19-4).

The water that hurtles up the beach after a wave breaks is known as *swash*. The water that returns to the sea is called *backwash*. Swash and backwash are common sources of the sand that rolls back and forth along the shore, eroding some from coastal rocks and bringing some inland from farther out at sea.

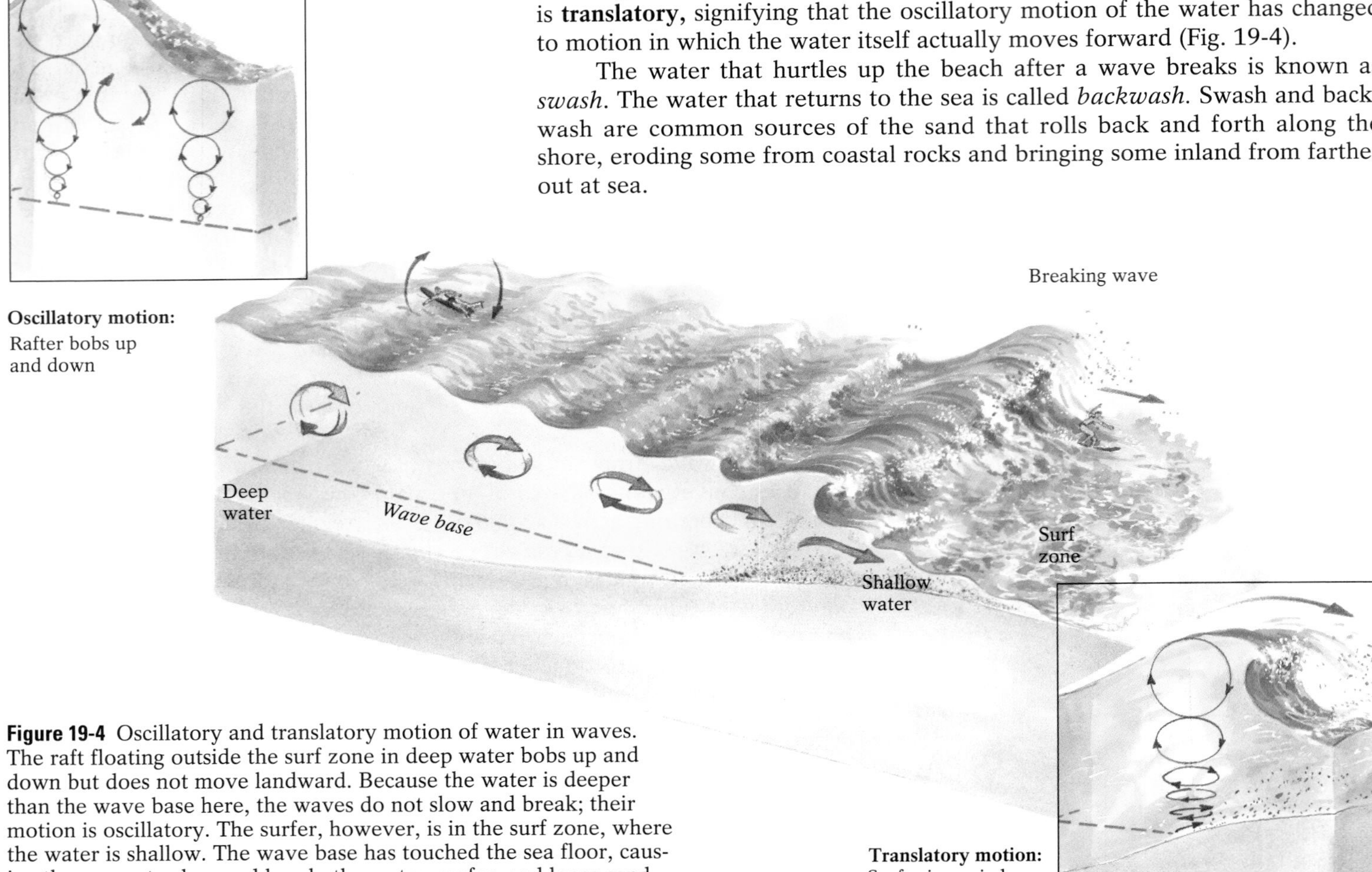

Figure 19-4 Oscillatory and translatory motion of water in waves. The raft floating outside the surf zone in deep water bobs up and down but does not move landward. Because the water is deeper than the wave base here, the waves do not slow and break; their motion is oscillatory. The surfer, however, is in the surf zone, where the water is shallow. The wave base has touched the sea floor, causing the waves to slow and break; the water, surfer, and loose sand particles are carried shoreward by translatory motion.

Wave Refraction and Coastal Currents Waves generally strike a coast at an angle, with part of a wave entering shallower water sooner than the rest. When the first-arriving part touches bottom it loses energy and slows down, and its crest breaks. The rest of the wave, still in deeper water, continues to move at a high speed and pivots around the slow, shallow-water segment, much as a marching band turns a corner (by having the marchers closest to the corner take small slow steps while their comrades farthest from the corner take large quick steps). This **wave refraction,** or bending, causes the last-arriving portion of the wave to be nearly parallel to the coast before breaking (Fig. 19-5). Waves that first approach the beach at a 50° to 60° angle are typically refracted to less than a 5° angle.

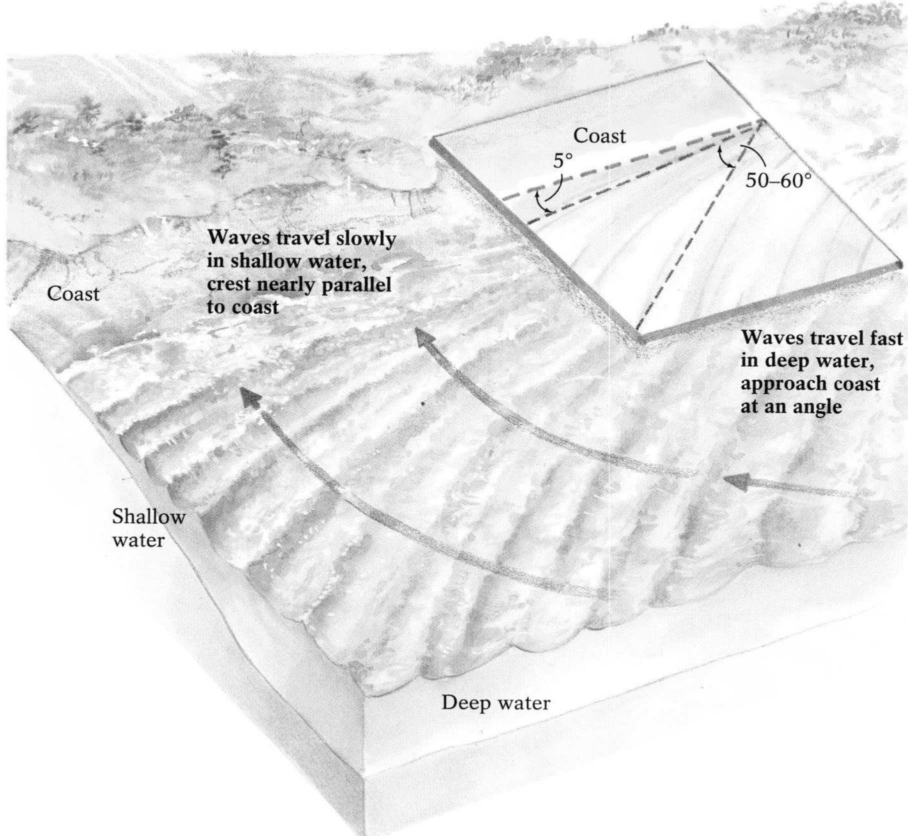

Figure 19-5 Wave refraction. Virtually all waves approach the coast at an angle, but the slower velocity of the part of the wave front that first reaches shallow water bends the incoming wave front until it is almost parallel to the coastline.

As each refracted wave breaks and strikes the coast, its surf pushes the swash ahead of it up the beach at a small angle. Backwash then returns to the sea perpendicular to the shoreline. The combined swash and backwash of the waves creates a turbulent **longshore current** that flows nearly parallel to the shoreline in the surf zone, carrying sediment—and sometimes swimmers—meters down the coast (Fig. 19-6).

Geologists and oceanographers have monitored longshore currents—using iridescent dye or radioactive grains of sand—and have learned that their velocity is usually between 0.25 and 1.0 meters per second. It can, however, be several meters per second and powerful enough to lift loose bottom sediment into suspension and transport it down the coast for great distances. Some sand on the Outer Banks of North Carolina, for instance, originated on the rocky coast of Maine, 1500 kilometers (1000 miles) to the north.

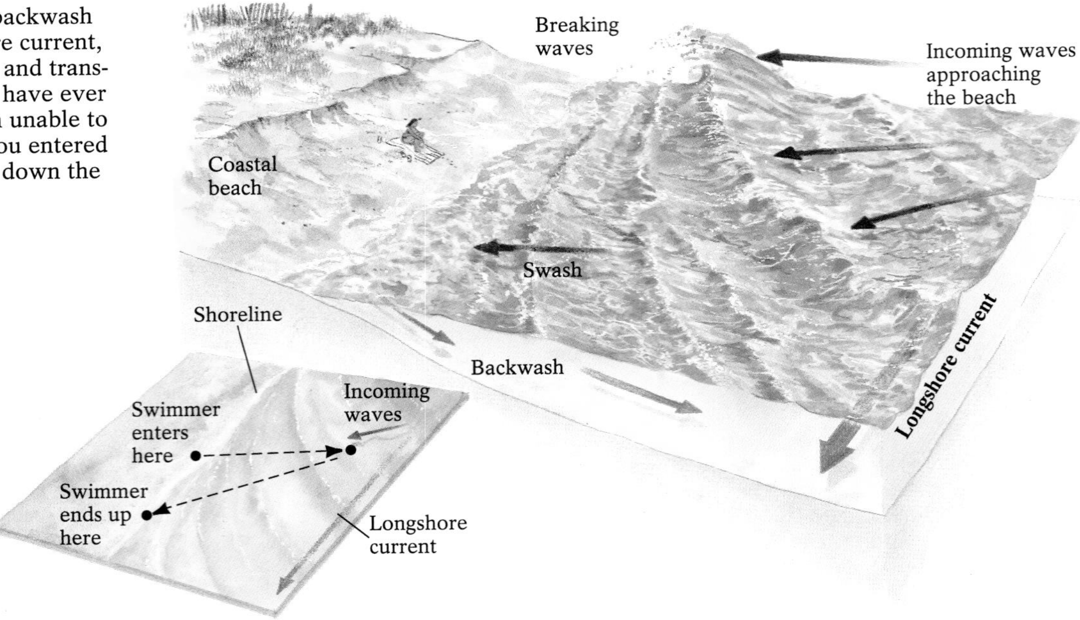

Figure 19-6 The combined swash and backwash of breaking waves produce a longshore current, which travels parallel to the shoreline and transports sediment along the coast. If you have ever gone swimming in the ocean and been unable to immediately locate the place where you entered the water, you were probably carried down the beach by the longshore current.

Rip currents, known to swimmers as undertow, flow straight out to sea, moving water and sediment *perpendicular* to the shoreline (Fig. 19-7). Rip currents occur when water that accumulates in the surf zone by translatory motion moves seaward. They usually interfere with incoming waves, causing them to break before reaching the beach or even preventing them from breaking altogether. (Swimmers caught in powerful rip currents can bypass the current by swimming parallel to the shore for a few tens of meters; surfers, however, should ride rip currents seaward for a swift, effortless return to the breaker zone.)

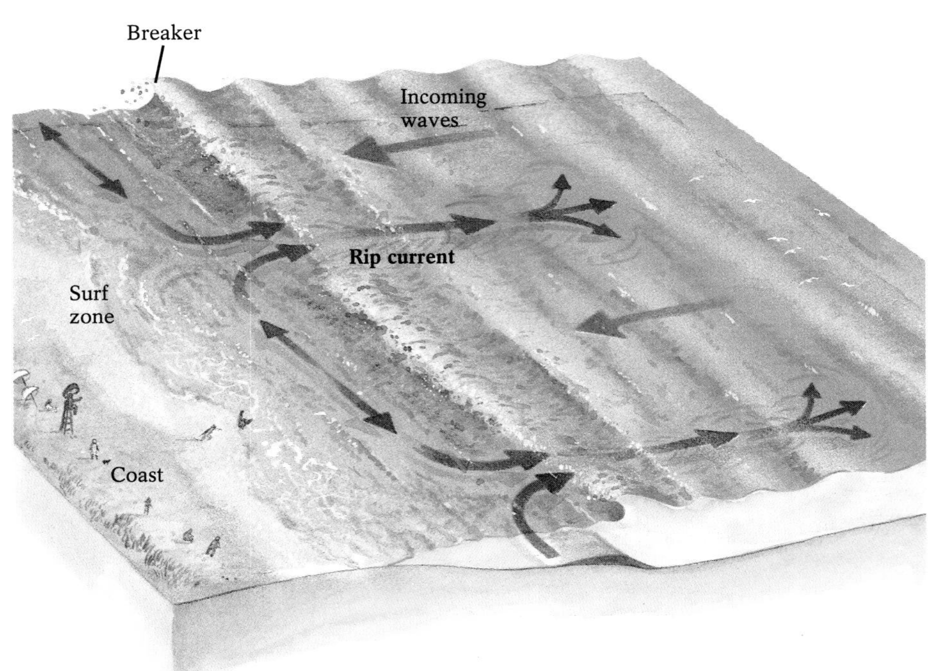

Figure 19-7 A rip current occurs when backwash builds up in the surf zone and then flows seaward through the incoming waves. An ocean swimmer who is unable to return to shore because of a strong seaward rip current should swim parallel to the coast for a short distance to escape it.

Tides

Tides are the twice-daily rise and fall of the surfaces of oceans and large lakes, which move shorelines alternately landward and seaward. They result from the gravitational pull of the Moon and the Sun and from the rotation of the Earth. As we saw in Chapter 11, the force of gravity increases as the masses of the involved bodies increase. But the force of gravity is also inversely proportional to the square of the distance between the bodies; that is, it decreases as the distance between the bodies increases. The mass of the Sun, and therefore its gravitational force, is about 180 times as strong as that of the Moon. But the Sun is 390 times farther away from us, and therefore exerts only 40% as much gravitational pull on the Earth as the Moon does. Because the Moon exerts the dominant gravitational force, the portion of the Earth's oceans that is facing the Moon at any given time is pulled moonward into a bulge, or *high tide*.

As the Earth rotates on its axis, its coastal regions experience two high tides each day, one when they are on the side of the Earth facing the Moon, and the other when they are on the opposite side, away from the Moon. The opposite-side tide occurs because the Earth is a spinning sphere, largely covered by oceans, whose water would escape from the surface if not for the Earth's gravitational pull. On the side of the Earth away from the Moon, the Moon pulls on the oceans less strongly because it is farther away, by the added 12,000 kilometers (nearly 8000 miles) of the Earth's diameter. Thus, the Moon's gravitational attraction is at a minimum on the more distant side of the Earth, and the spinning force that pulls the oceans away from the Earth is at its maximum, causing the second tidal bulge. At any given moment, the two tidal bulges are occurring at the places on Earth that are closest to and farthest from the Moon. As the Earth rotates, the tidal bulges "move" around the Earth's oceans (Fig. 19-8).

Also at every moment, two *low tides* are occurring on opposite sides of the Earth. These are located halfway between the Earth's closest and its farthest points from the Moon, where the gravitational pull of the Moon and the effect of the Earth's rotation are minimal.

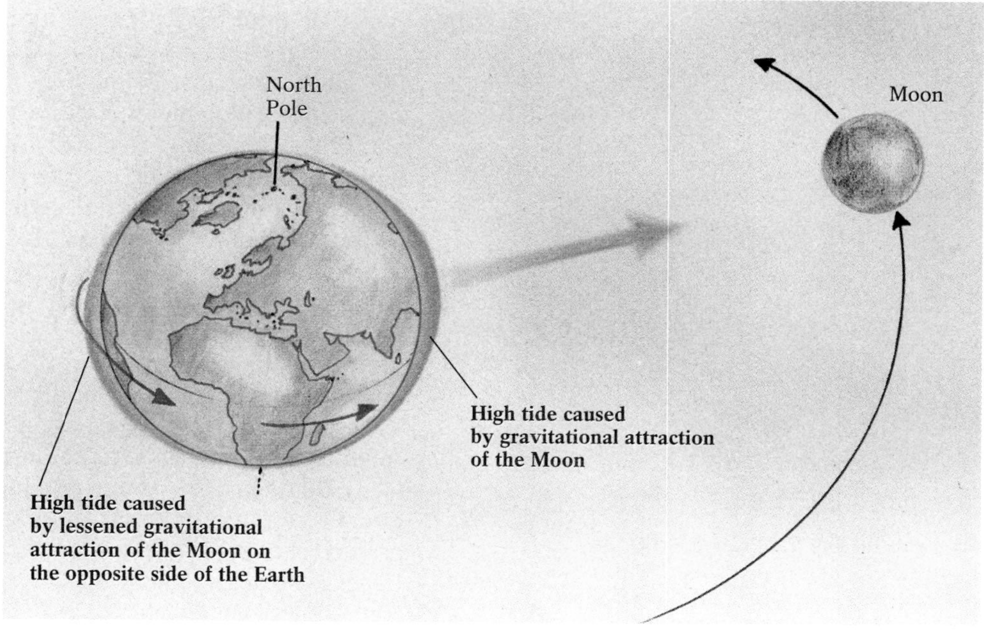

Figure 19-8 Tidal bulges. The Earth's high tides occur where the gravitational attraction of the Moon is strongest (where the Moon is closest) and where it is weakest (where the Moon is farthest away), and so are influenced by the rotation of the Earth.

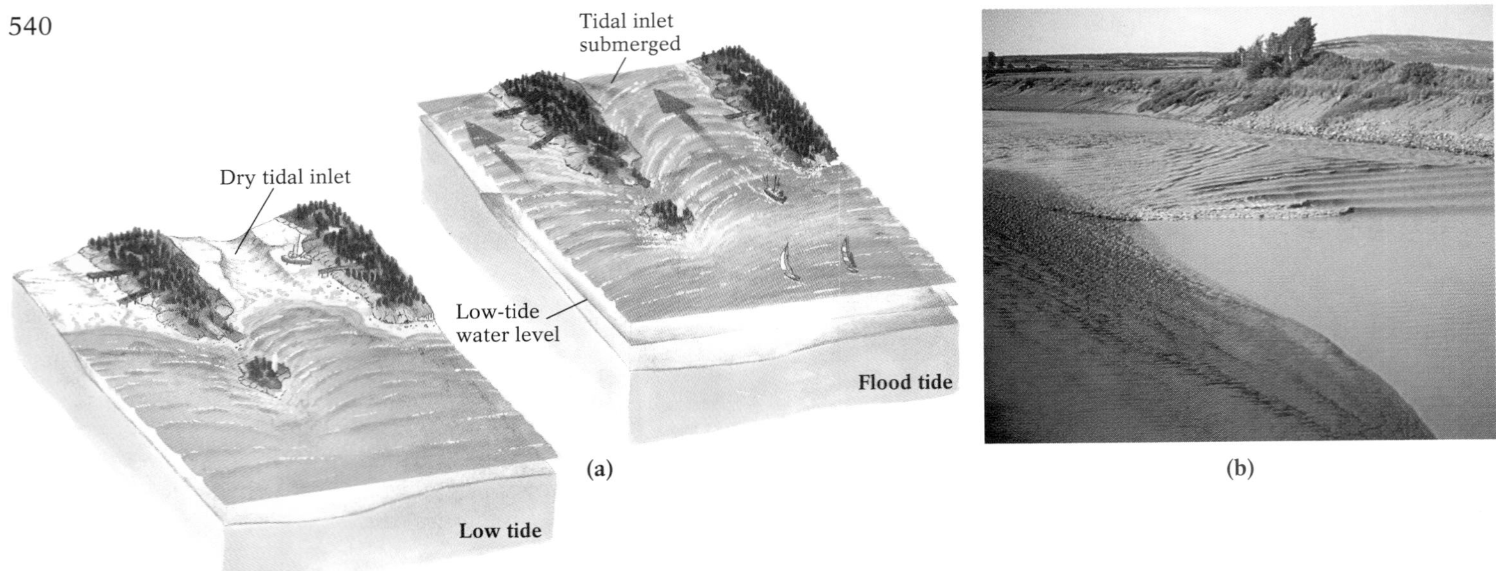

Figure 19-9 The effect of tides on a tidal inlet. (**a**) During low tides, the inlet is dry; during flood tides, the rising water carries boats and sediment inland. (**b**) A tidal bulge may migrate up rivers as a turbulent *tidal bore*, such as this one in the River Hebert in Nova Scotia.

(a)

(b)

Figure 19-10 The tidal range at the Bay of Fundy, between the Canadian Maritime provinces of New Brunswick and Nova Scotia. (**a**) Eroded columns of rock at high tide. (**b**) The same columns at low tide.

A rising, or **flood tide,** elevates the water surface of an ocean, advancing the shoreline landward (Fig. 19-9a). On a gently sloping coast, high tide advances farther inland than on a steeply sloping coast. A rising tide can migrate up a river: After each low tide, a tidal bulge moves into New York City's harbor and more than 200 kilometers (120 miles) up the Hudson River to Troy and Albany. Flood-tide currents can reach 25 kilometers per hour (15 mph) and can even reverse the downstream surface flow of rivers. A flood tide whose upstream movement is turbulent is called a **tidal bore** (Fig. 19-9b). The tidal bore of the St. John River, which empties into the Bay of Fundy, a narrow arm of the Atlantic Ocean between the Canadian Maritime provinces of New Brunswick and Nova Scotia, migrates upstream at 10 to 15 kilometers per hour (6–9 mph). As flood tides move upstream, they transport fresh, oxygenated water into stagnant coastal areas.

A falling, or **ebb tide,** lowers the water surface of an ocean, and the shoreline recedes seaward. Ebb-tide currents transport sediment and organic nutrients seaward from coastal marshes and lagoons, nourishing marine life offshore.

The difference between local low and high tides is an area's *tidal range.* Tidal ranges vary according to the irregularity of the coastline and the size of the body of water. Flood tides funneled into restricted bays and estuaries cause a large volume of additional water to pile up in a small area; in this situation the tidal range is usually quite large. (As we shall see in Chapter 20, a large tidal range can be used to spin a turbine and generate electrical power.) The tidal range at Seattle, in the relatively narrow constriction of Puget Sound, is from 3 to 4 meters (10–13 feet); but in the Bay of Fundy it is even larger, up to 20 meters (75 feet) (Fig. 19-10). When the coastline is relatively straight, the additional water of a flood tide is spread over a large area, moderating its effect; as a result, the tidal range is usually small, as little as 0.5 meters (about 1.5 feet), as in the open Pacific along Hawaiian beaches. Lakes and other inland water bodies—especially large ones such as the Great Lakes—also experience flood and ebb tides, but these bodies usually are too small to have a sizeable tidal range.

Tidal currents do not produce major geological effects. They can, however, scour the bottoms of shallow tidal inlets and transport some fine-grained sediment. When tidal currents coincide with large storm waves, their effect is strongly enhanced. When a storm strikes at high tide, waves penetrate much farther inland than at low tide, perhaps flooding beachfront communities and eroding beach cliffs ordinarily protected by wide beaches.

Processes That Shape Coasts

The most significant factor shaping shorelines and coasts is the constant battering of waves. Wherever wave energy is concentrated, rock and sediment are eroded; wherever wave energy is dissipated, the eroded material is deposited. For example, paired erosion and deposition removes rock and sediment from exposed coastal cliffs and deposits them in quiet-water bays; the overall effect is to straighten an irregular coastline. Other processes that modify coasts are the twice-daily rise and fall of tides, the biological processes that produce coastal swamps and forests and organic reefs, the long-term rise and fall of sea level, and the geological processes of tectonic uplift and subsidence.

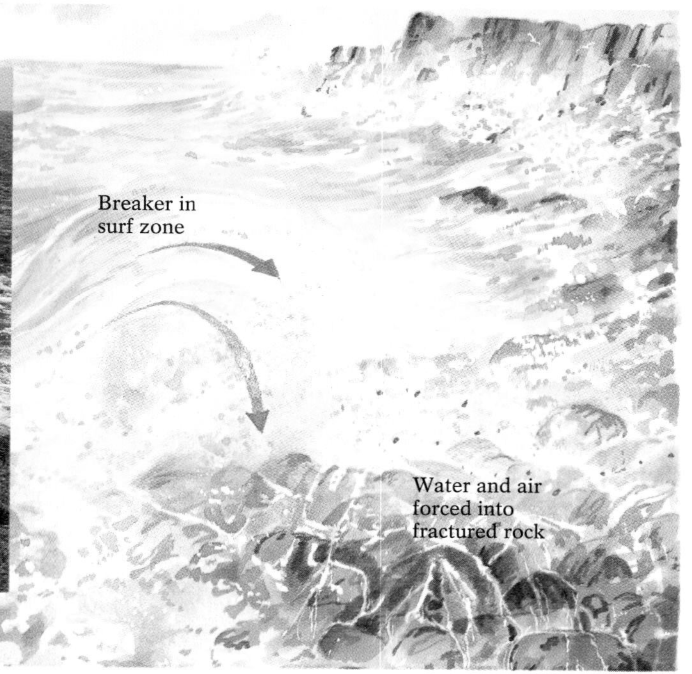

Figure 19-11 Crashing surf erodes solid bedrock by forcing water and air into fractures in the rock. Photo: Waves crashing against bedrock at Schoodic Point, Acadia National Park, Maine.

Coastal Erosion

When winds set the sea in motion, the moving water applies stress against its container—the shoreline. Breaking waves can hurl thousands of tons of water against coastal rock and sediment with a force powerful enough to be recorded on nearby seismographs. An average 14,000 waves strike the exposed rocks and beaches at a given coast every day. Waves erode principally by forcing water and air under high pressure into rock crevices (Fig. 19-11). A small, 2-meter (6.5-foot)-high wave, for example, produces about 15 metric tons of pressure per square meter of rock or sediment surface (more than 2000 pounds per square foot). Although this force is applied for only a fraction of a second, it is repeated every six or so seconds and is sufficiently powerful to dislodge large masses of bedrock or sediment. Storm waves can also hurl large rocks through the air to impressive heights. At the lighthouse at Tillamook, Oregon, a 61-kilogram boulder (approximately 130 pounds) was tossed over the light and crashed through the keeper's roof, 40 meters (130 feet) above sea level. After waves loosen and remove rocks, they keep hurling them against the coastline.

Wave-induced erosion is affected not only by wave size and energy but also by the erodibility of local rocks or sediments; erodibility depends on the strength of the exposed rock or sediment and the extent to which it is jointed or fractured. Softer or more fractured bedrock is usually more easily eroded than harder, less fractured rocks, whether the erosion is due to streams, winds, or waves.

The slope of the local sea floor helps determine how much wave energy will strike a given coast. Waves tend to break offshore when the slope is gradual and the water remains shallow for a considerable distance, as is true of the continental shelf. In such locations, wave bases intersect the sea floor well before the shore, and wave energy is dissipated in churning up the sea-floor sediments. But when the slope is steep, incoming waves may never touch bottom—instead they strike the coast with full force. This is particularly common where a coastline is irregular and contains **headlands,** cliffs that project into deep water.

The orientation of a coastline relative to prevailing storm winds also influences coastal erosion. A coast that is oriented perpendicular to the most common wind direction will often receive the full force of incoming waves. For example, any segment of the west coast of North America that is oriented north-to-south is directly in the path of waves driven by Pacific westerlies. Any segment oriented east-to-west is more likely to take Pacific storm waves at an angle.

Erosion along about 85% of California's 1100 kilometers (700 miles) of Pacific coastline claims an average of 15 to 75 centimeters (0.5–2.5 feet) annually (Fig. 19-12). Erosion is also claiming vast tracts of land along the Gulf Coast of Texas and Louisiana. In a recent 9-month period, more than 3 meters (10 feet) of the coastline of Chambers County, Texas, has been lost to Galveston Bay southeast of Houston. Along the East Coast, the beaches of North Carolina have been worn back by as much as 20 meters (70 feet) during the past decade, largely by storms. The landmark 63-meter (208-foot)-high lighthouse at Cape Hatteras, which was built more than 1500 meters (5000 feet) inland in 1879, now stands, imperiled, at the shore. A similar light on Morris Island near Charleston, South Carolina, which once stood on dry land, is now about 500 meters (1700 feet) offshore and surrounded by water.

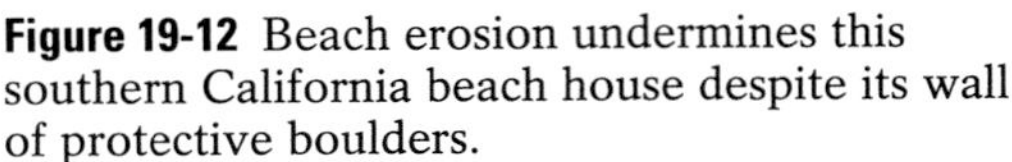
Figure 19-12 Beach erosion undermines this southern California beach house despite its wall of protective boulders.

Highlight 19-1 *Lake Michigan's Vanishing Shoreline*

Figure 19-13 Waves generated by westerly winds easily erode the loose glacial sediments of this coastal bluff at St. Joseph, Michigan, on the eastern shore of Lake Michigan.

Thousands of Michiganders live in towns from Benton Harbor to Ludington on the eastern shore of Lake Michigan, and thousands more from nearby Chicago vacation there. For the past two decades, residents and visitors have watched as erosion has removed lakefront beaches and caused the lake's eastern bluffs to recede (Fig. 19-13). This beach and cliff erosion is unusually rapid because the area is composed largely of readily eroded unconsolidated glacial deposits, and because powerful winter storm winds push large waves against the lake's eastern shore. To add to the problem, recent balmy winters have reduced the lake ice that protects the bluffs from winter storm waves. Also, several years of unusually high precipitation have produced record high lake levels, so that water now covers part of the shore. In 1964, when the water level of the Great Lakes was 1.5 to 2.0 meters (5–7.5 feet) lower than today, wide beaches received and dissipated much of the wave energy that now strikes the bluffs directly. The few relatively dry years that have occurred did not lower lake levels sufficiently to reduce the rate of erosion, and the bluffs continue to recede, about 0.4 meters (1.5 feet) annually. It would take about 5 years of drought to restore the lakes to their pre-1964 levels.

Human activity has hastened erosion of the lake's beaches and cliffs. For instance, overdevelopment—building too many homes and other structures along the lakeshore—has removed much of the protective vegetation from bluff tops as well as vegetation that stabilizes the sands of energy-absorbing dunes. Meanwhile, disposal of wastewater and sewage in septic fields and on bluff-top farmlands has saturated cliff slopes, reducing the strength of their sediments and promoting mass movement.

What then is the future of Lake Michigan's scenic bluffs? The warmer, wetter climate of the past two decades may continue; in fact, the drier period of the past century may have been an anomaly, in which case the lake has been returning to its normal level. If so, accelerated bluff erosion and shore loss may be long-term phenomena along Lake Michigan's shore.

Large inland lakes are also susceptible to coastal erosion. In North America, the eastern shores of the Great Lakes, which are often buffeted by powerful late-autumn storm waves, have suffered from rapid coastal erosion in recent years. Highlight 19-1 focuses on recent erosion problems along the eastern shores of Lake Michigan.

Landforms Produced by Coastal Erosion The great energy of waves that strike headlands rapidly wears down their cliffs into a series of distinctive erosional features. Initially, a *wave-cut notch* forms at the base of the cliff. Continued wave action enlarges the notch until it undercuts the cliff, removing the foundation of the overlying rock masses so that they fall into the surf. After enough rock has fallen, the remaining cliff base becomes a **wave-cut bench.** Meanwhile the surf, laden with rock debris, further abrades the cliff face and enlarges the bench. Eventually, the cliff retreats so far that it is no longer in the breaker zone. The wide bench then protects the cliff from further erosion to some extent, except during large storms. But the bench can also accelerate headland erosion, by enhancing the refraction of incoming waves against the headland's flanks.

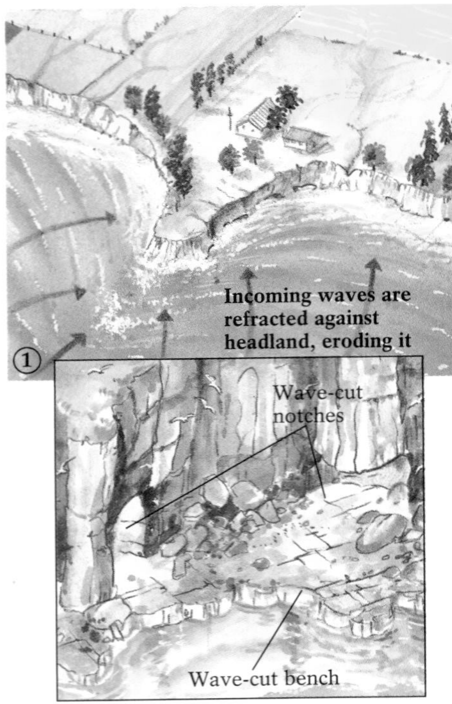

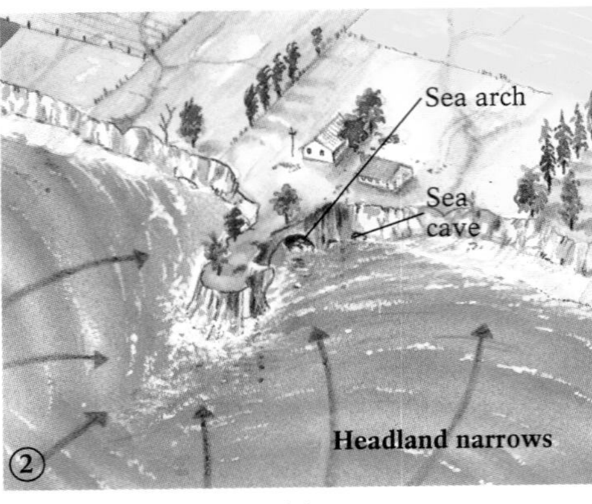

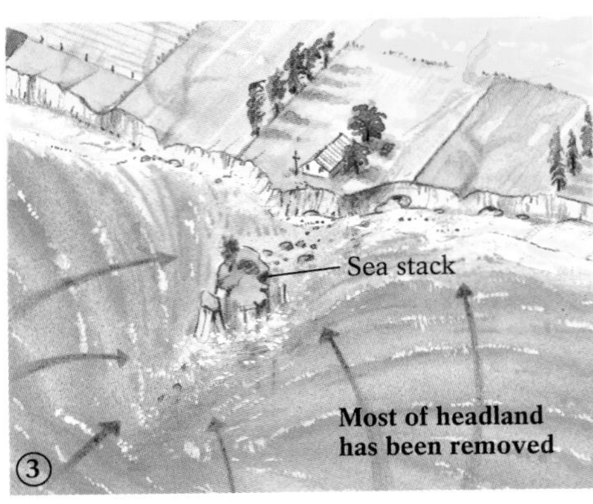

(a)

(b)

(c)

Figure 19-14 The evolution of erosional coastal landforms. (**a**) Wave fronts are refracted against the flanks of headlands, forming wave-cut notches and benches. The notches erode to form sea caves as the headland narrows. Two sea caves, eroding at both sides of a headland, form a sea arch. Finally, the arch collapses, isolating a sea stack. (**b**) Sea caves on Cape Kildare, Prince Edward Island, Canada. (**c**) A sea arch at Land's End, England.

As waves approach a wave-cut bench, the wave front encounters the shallow bottom and slows down, breaking over the bench. The waves in the adjacent deeper-water bays pass the headland without touching bottom. Thus they can travel faster, refracting toward the flanks of the exposed headland and concentrating their energy there. This produces a series of notable coastal landforms, which are especially dramatic when the headland cliffs contain rocks of differing resistance to wave erosion. At first, battering waves erode the cliff rock to form **sea caves.** Further wave action can excavate caves on both sides of the headland until they join, forming a **sea arch.** Continued wave erosion can cause the supporting foundation of the arch to dwindle and finally collapse. Only an isolated remnant of the original headland, a **sea stack,** may remain (Fig. 19-14).

Protecting Against Coastal Erosion Human efforts to prevent the effects of coastal erosion usually involve building structures that deflect waves so they do not strike the coast in force. In Texas, desperate ranchers have constructed barriers of abandoned cars to prevent waves from the Gulf of Mexico from washing away their land. Sand dunes can provide a first line of defense against the encroaching sea by absorbing the energy of waves, but they tend to migrate with the wind unless they are stabilized in some way. On the Outer Banks of North Carolina, hardy vegetation has been planted so its roots can hold soil against wind erosion, and sand has been added to create a higher and continuous dune line. On the southern shore of New York's Long Island, driftwood and old tires are piled up in some places to fortify the area's natural dunes.

Larger protective structures reduce coastal erosion by absorbing the brunt of wave energy. **Riprap** is a heap of large angular boulders piled seaward of the shoreline (Fig. 19-15a). Eventually breaking waves remove the riprap, but not for several years or even decades, and riprap is relatively easy and inexpensive to rebuild. **Seawalls** are sturdy, longer-lasting structures, generally built parallel to the shore, that are designed to withstand the pounding of the highest recorded waves in an area (Fig. 19-15b). Seawalls are large and solid enough to repel waves seaward to divert some of their energy. They are, however, expensive to build and maintain. A $12-million, 6.4-meter (20-foot)-high seawall was completed in 1905 at Galveston, Texas, after the hurricane of 1900, the worst natural disaster in North American history (which took 6000 lives and destroyed two-thirds of Galveston's build-

(a)

(b)

Figure 19-15 (**a**) Riprap at Puerto Penasco, Mexico, off the Gulf of California. (**b**) A seawall fends off incoming waves in the Sandbridge area of Virginia Beach, Virginia.

ings). The sandy beaches directly in front of the smooth-faced wall were quickly eroded by the wave energy that it diverted. Without the beach, the waves struck the wall directly, breaching it in several places. The wall is still in service, but damaged portions must be rebuilt periodically to maintain its effectiveness.

Coastal Transport and Deposition

Deposition occurs in coastal settings for much the same reasons as deposition by wind, rivers, and glaciers occurs: The supply of sediment exceeds the ability of the current to transport it. When the energy needed for transport drops below a critical level, the sediment can no longer be carried. Wave energy may be lost or interrupted for a variety of reasons: in response to seasonal variations in wind velocity and wave force; when water depth increases abruptly; when waves, refracted from headlands, are channeled into bays; and when any barrier, natural or artificial, prevents waves from reaching the shore, interrupting the longshore currents.

When more sediment is delivered to a shore than is removed or redistributed by nearshore currents, the excess sediment is deposited, most often as a beach. A **beach** is defined as the dynamic, relatively narrow segment of a coast that is washed by waves or tides and covered with sediments of various sizes and compositions. It may consist only of sand, or it may also contain coarse gravel or even cobbles. Beach sediment is typically white (from quartz grains) or beige (from shell fragments), but may even be black (from mafic volcanic fragments); it often accumulates as windblown sand *dunes* (see Chapter 18) on the landward side of the beach. A beach's boundaries stretch from the low-tide line landward, ending where topography changes—for example, at a sea cliff or sand-dune field, or where permanent vegetation begins. A typical beach, illustrated in Figure 19-16, is composed of a **foreshore,** which extends from the low- to the high-tide line, and a **backshore,** which extends from the high-tide line to the sea cliff or vegetation line. A beach's profile consists of several components with variable slopes. The steepest part of the foreshore is its seaward margin, called the **beach face,** which receives the swash of breaking waves. The backshore, which receives swash only during major storms, may contain a **berm,** a horizontal bench or

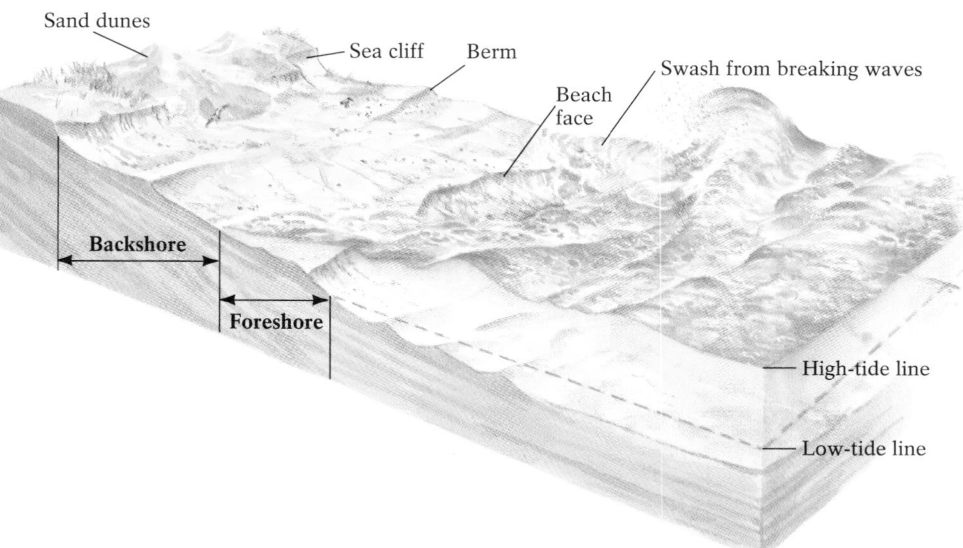

Figure 19-16 The components of a typical beach (see text).

landward-sloping mound of sediment deposited by storm waves. Some beaches contain several parallel berms from different storms, others none, depending on the extent of storm activity.

More than 90% of beach sediment originates at inland and upland sources and is delivered by coastal streams. An estimated 15 billion cubic meters of river sediment is deposited at the world's shorelines every year. Some is derived from erosion of headlands and beach cliffs and a smaller amount from the offshore surf zone. After sediment arrives at the shoreline, it begins its longshore journey when it first encounters a breaking wave. Ordinarily, far more sediment is carried away by the longshore current than is transported offshore by rip currents and storm waves or to backshore dunes by sea breezes. A beach is naturally maintained when the amount of sediment received by the shore balances the amount exported by longshore currents.

Longshore Transport Energy for transport of coastal sediment comes primarily from waves and nearshore currents. When waves break at some oblique angle to the shore, they create a longshore current that transports sediment in the same direction as the wind and waves. The current moves sediment either on the beach face, as **beach drift,** or within the surf zone, as **longshore drift.** In either case, waves lift the grains and move them by swash at an oblique angle to the shoreline; they then return the grains seaward by backwash straight down the slope of the beach face (in the case of beach drift) or the surf zone (in the case of longshore drift). Normally, sand is carried offshore more effectively in the surf zone than on the beach face, but in a storm, a grain on a beach can travel downcurrent as much as 1000 meters (3300 feet). Longshore drift can transport enormous quantities of sand; for example, more than 1.5 million tons of sand a year pass the California coast at Oxnard, and 750,000 tons of sand a year, some eroded from granitic cliffs on the Maine coast, pass by Sandy Hook, New Jersey.

Landforms Produced by Coastal Deposition If the longshore current that drives longshore drift suddenly encounters deeper water, such as at the entrance to a bay, the waves that drive the current no longer touch bottom. They do not break, and the current is interrupted. All the sediment being carried is deposited at that point as a **spit,** a finger-like ridge of sand that extends from land into open water. As more sediment is added, the spit can grow by tens of meters per year; the growth depends on the supply of sand and the intensity of wave energy. Where waves or currents into a bay are particularly strong, a growing spit becomes curved, forming a **hook.** A spit becomes a **baymouth bar** if it grows completely across a bay entrance (Fig. 19-17).

Coastal deposition also occurs when the waves that drive the longshore current find their path to the shore interrupted, perhaps by the shallow continental shelf or some other natural offshore landform, or by a human-made structure. Where the continental shelf has a gentle slope and waves dissipate much of their energy by breaking farther offshore, longshore currents are reduced. In such a case the sediment loads delivered by streams to the beaches cannot be carried off, and the beaches tend to be wide.

One natural structure that intercepts incoming waves and thus interrupts longshore currents is a sea stack. Waves break on the seaward surface of the stack, but quiet water prevails on its landward side. Wave-borne sand is deposited in the wave-free zone on the landward side of the stack, because there is no wave energy to carry it along the coast. It accumulates as a *tombolo,* a sandy landform that grows from the mainland to a stack (Fig. 19-18).

(a)

(b)

Figure 19-17 Deposition of spits, hooks, and baymouth bars. These features typically develop when the sediment being transported by longshore transport is deposited where water velocity is reduced, such as at the mouth of a bay. (**a**) A spit at Cape Hencopen, Delaware. (**b**) Baymouth bars on the south shore of Martha's Vineyard, Massachusetts.

Figure 19-18 Deposition of a tombolo landward of a sea stack. The sand that makes up the tombolo is deposited because the sea stack absorbs the wave energy, interrupting the longshore current. Photo: Tombolos connecting offshore islands to the mainland at Duxburg Bay, Massachusetts.

(a)

(b)

Figure 19-19 (**a**) Santa Barbara Harbor in 1931. (**b**) Santa Barbara Harbor in 1977. The breakwater at Santa Barbara was built in the 1920s to provide safe harborage for pleasure boats. However, it interrupted the longshore current, causing sand to be deposited behind it and on the seaward edge of the beach. This extended the beach but closed off the passage from the marina to the sea. In addition, wave action immediately down the coast from the breakwater caused beach erosion there. To solve the dual problems of excessive deposition behind the breakwater and beach erosion down the coast, sand is regularly dredged from behind the breakwater and pumped downcurrent to the depleted beach area—a costly substitution of human energy and money for the interrupted wave energy.

Human-Induced Coastal Deposition Some human-made structures disrupt the natural balance between the amount of sediment delivered to the shore and the amount removed by the longshore current, causing beaches to grow in some places and shrink in others. **Breakwaters**, walls designed to intercept incoming waves, are built to create quiet, wave-free zones that curtail coastal erosion and protect boats in harborages. Breakwaters are often oriented parallel to the coasts. About 70 years ago, a breakwater was constructed offshore at Santa Barbara, California, to protect private boats from being battered by Pacific waves. But the structure interrupted the longshore current, causing sediment to be deposited seaward of the beach (Fig. 19-19). After 30 years, the beach had widened by hundreds of meters, clogging the harborage.

Groins are shore-protection structures that jut out perpendicular to the shoreline in order to interrupt longshore drift and trap sand, and thus restore an eroding beach. They are typically built where a wide sandy beach, such as Miami Beach, is vital to a community's economic life. **Jetties** are structures, typically built in pairs, that extend the banks of a stream channel or tidal outlet beyond the coastline. They direct and confine channel flow to keep channel sediment moving, thus preventing sediment from filling the channel. Both groins and jetties broaden up-current beaches by trapping sand there (Fig. 19-20); however, because they remove sand from the longshore transport system, they can cause erosion and narrowing of down-current beaches.

Beach Nourishment: Restoring Shrinking Beaches Human-caused beach shrinkage may occur when inland dams on rivers intercept beach-bound sediment. Flood-control dams built in the last 35 to 40 years in the Missouri–Mississippi River system, for example, have trapped half the sediment bound for the Gulf of Mexico. This has prevented natural replenishment of the sediment removed by longshore currents from the Louisiana coast and delivered to Gulf Coast islands off of Texas, thus jeopardizing beaches in both states. Similar beach losses occur along the California coast wherever dams intercept and retain the sediments of coast-bound rivers.

To compensate for sediment loss from a shrinking beach, sand can be imported. The sand for such *beach nourishment* is typically dredged from nearby lagoons, inland sand dunes, and offshore sand bars. In one successful beach-nourishment project, along Mississippi's Gulf Coast at Biloxi and

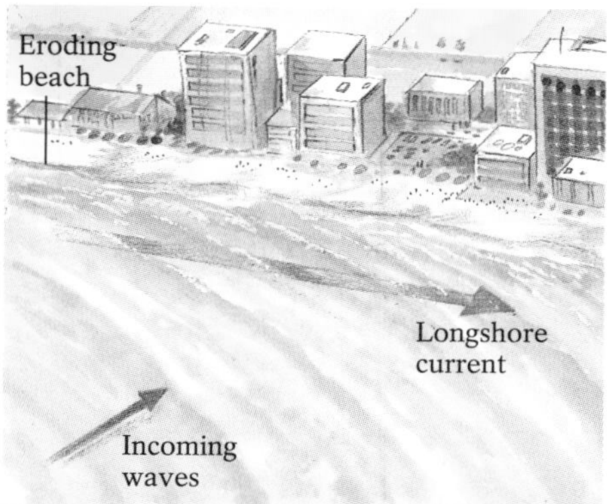

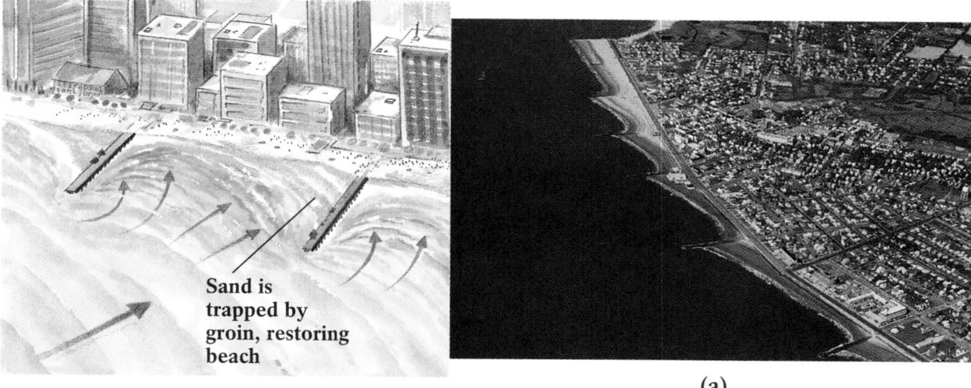

(a)

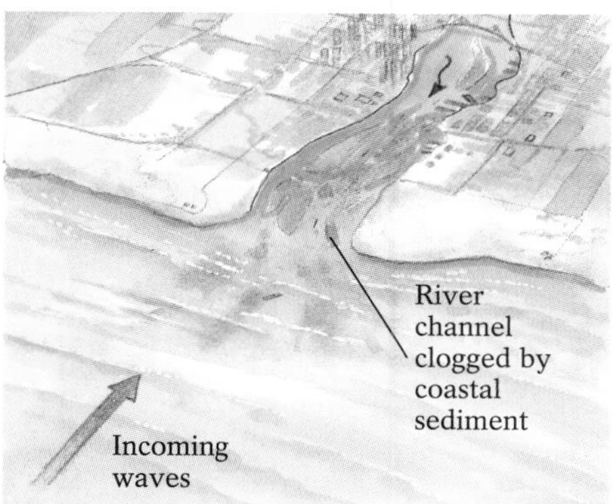

(b)

Figure 19-20 By interrupting longshore currents, groins and jetties cause longshore drift to accumulate on their upcurrent side, broadening beaches there. (**a**) Groins off Cape May, New Jersey. (**b**) A cruise ship in a jetty in Miami Beach, Florida.

Gulfport, a mixture of sand and mud was pumped from offshore sites to create the world's longest artificial beach (30 kilometers, or 20 miles, long). The gulf's gentle waves remove the fine mud particles, leaving behind a sandy beach that has made the area into a recreation site.

But beachfront resort communities often find that the costs of beach nourishment far outweigh the benefits. A beach-nourishment program in Miami Beach, Florida, halted decades of beach erosion and widened the beach to 100 meters (330 feet). The sand was imported from the Everglades at a cost of $64 million over a 10-year period. This sand consisted of quartz and fine clay particles rather than the coarse shell fragments of Miami's natural beach. When wave erosion carried off the clay, the offshore water became turbid, damaging coral reefs (which flourish only in clear, sunlit water). The same problem plagues the reef at artificially nourished Waikiki Beach in Hawaii.

Successful beach nourishment requires expert knowledge of the local longshore currents and their normal seasonal variations. At Long Branch, New Jersey, 459,000 cubic meters of sand intended for beach nourishment was dumped too far offshore, beyond the reach of the longshore current that was supposed to carry it to the beach. At Ocean City, Maryland, where a $5-million nourishment project was completed just before the winter-storm season, 60% of new beach growth was lost to powerful Atlantic waves. Beach nourishment is a difficult and costly strategy for maintaining beaches.

Types of Coasts

Every coast is shaped by a combination of erosional and depositional processes that is distinctive to its geological setting (Fig. 19-21). So far, we have looked at such processes as wave-driven longshore and rip currents, rising and falling tides, and sediment delivery by coastal streams. However, there are other erosional and depositional processes that also produce and modify coasts. It is on the basis of these formative processes that coasts are classified.

(a)

(b)

Figure 19-21 Cape Flattery, Washington (**a**), is an erosional coast with scores of sea stacks and arches and a rugged, bouldery shoreline. The processes that have shaped it differ from those at Cape Hatteras, North Carolina (**b**), which is a depositional coast characterized by sandy beaches, dunes, and barrier islands.

Primary coasts are shaped by nonmarine processes, many of which are illustrated in Figure 19-22. Glacial erosion, for example, produced such primary coasts as the fjords of Alaska and British Columbia; glacial deposition of terminal and recessional moraines shaped the northeast Atlantic from Cape Cod, Massachusetts, to the southwestern edge of Long Island at Brooklyn, New York. Stream deposition produces primary coasts such as the Gulf Coast of Louisiana, where stream deltas extend into the marine environment. Long-term rising sea levels produce bays and estuaries by flooding river systems; the postglacial sea-level rise that began about 13,000 years ago created the primary coasts of Chesapeake Bay and Delaware Bay along the mid-Atlantic coast of North America. Primary coasts can also be shaped by the interaction of biological processes with geological processes, such as at carbonate-reef coasts.

A **secondary coast** is shaped predominantly by ongoing marine erosion or deposition, such as the processes described in the previous section. Secondary erosional coasts contain cliffed headlands, wave-cut terraces, and an assortment of sea caves, stacks, and arches. Secondary depositional coasts include beaches, spits, hooks, and tombolos. A secondary coast can be both erosional and depositional, as may be the case with offshore barrier islands.

Barrier Islands

Barrier islands are nearly continuous ridges of sand, parallel to the main coast and separated from it by a bay or lagoon. Barrier islands can rise as much as 6 meters (20 feet) above sea level, and are 10 to 100 kilometers (6–60 miles) long and 2 to 5 kilometers (1.5–3 miles) wide. They are usually composed of a relatively narrow beach, perhaps 50 meters (165 yards) wide, and a broader zone of inland dunes that makes up most of the island.

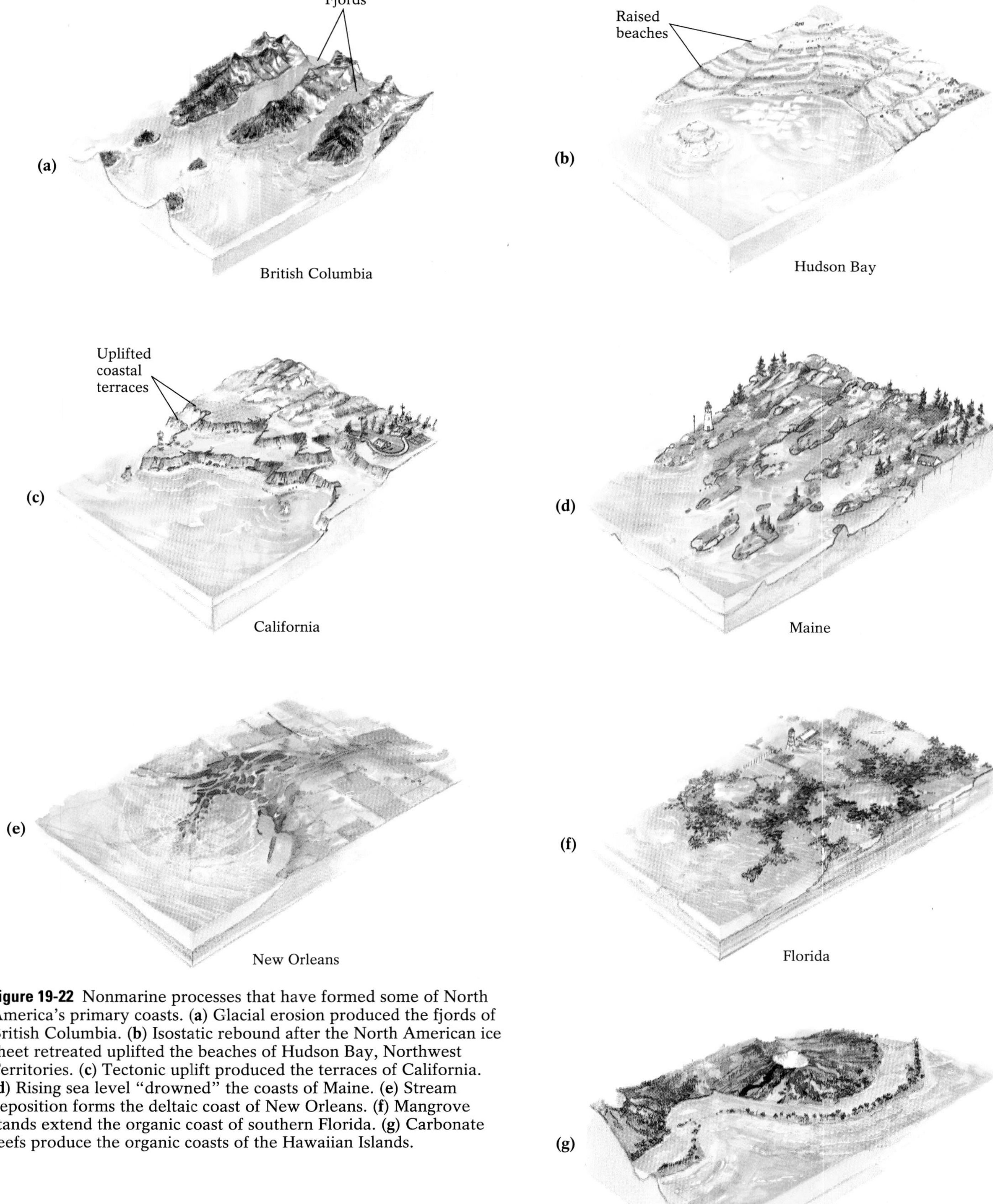

Figure 19-22 Nonmarine processes that have formed some of North America's primary coasts. (**a**) Glacial erosion produced the fjords of British Columbia. (**b**) Isostatic rebound after the North American ice sheet retreated uplifted the beaches of Hudson Bay, Northwest Territories. (**c**) Tectonic uplift produced the terraces of California. (**d**) Rising sea level "drowned" the coasts of Maine. (**e**) Stream deposition forms the deltaic coast of New Orleans. (**f**) Mangrove stands extend the organic coast of southern Florida. (**g**) Carbonate reefs produce the organic coasts of the Hawaiian Islands.

Figure 19-23 A satellite image showing Padre Island, off the southeast coast of Texas. Padre Island, North America's longest barrier island, stretches from Port Isabel to Corpus Christi.

Barrier islands are the most common North American coastal feature, lining the East Coast for 1300 kilometers (850 miles) from eastern Long Island, New York, south to Florida, and then continuing for another 1300 kilometers along the Gulf Coast to eastern Texas. This string of 295 islands is interrupted only by occasional tidal inlets or by the flow of major streams such as the Hudson River at New York Harbor or the Delaware River at Delaware Bay. Padre Island, North America's longest beach, is a classic barrier island (Fig. 19-23).

There is some debate about whether barrier islands are secondary coasts consisting of depositional landforms or whether they are secondary coasts consisting of combined erosional–depositional landforms. In the depositional model, barrier islands are considered to be elongated spits, projecting from bends in the coastline, which have been breached by tidal currents or powerful storm waves to form what appear to be separate islands. In the erosional–depositional model, waves breaking offshore above the broad gently sloping continental shelf move sand from the bottom and deposit it as long sand bars that continue to grow higher until they rise above sea level; these elongated bars are subsequently breached by storm waves to form the islands. In this model, the islands trend parallel to the coast because the refracted waves that lift and deposit sand to form them trend parallel to the coast.

Still other geologists hold that barrier islands are primary coasts, remnants of a sand-dune system that bordered the continent during the last period of worldwide glacial expansion. Because sea level at that time was more than 100 meters (330 feet) lower than today, coastal dunes accumulated some distance seaward of what is now the coastline. As the glaciers melted, the dunes were surrounded and isolated by the rising sea.

Whatever the origin of barrier islands, once they begin to form, erosional and depositional processes combine to shape them. Continued deposition causes the islands to migrate landward. Onshore winds, inflowing tidal currents, and storm waves transfer sand from the seaward side of these islands to the lagoon side, forming tracts of dunes and fan-shaped deltaic deposits (Fig. 19-24). The landward migration rate for barrier islands along the Atlantic Coast ranges from 0.5 to 2 meters (1.5–6.5 feet) per year. This rate increases considerably during extraordinary storms. Ultimately, the islands may merge with the mainland, extending continental land seaward.

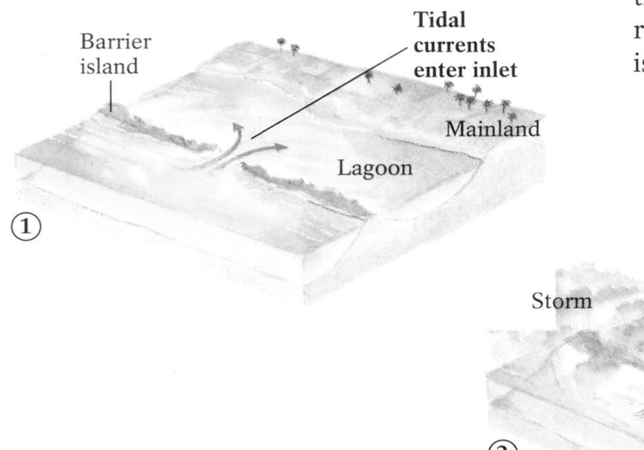

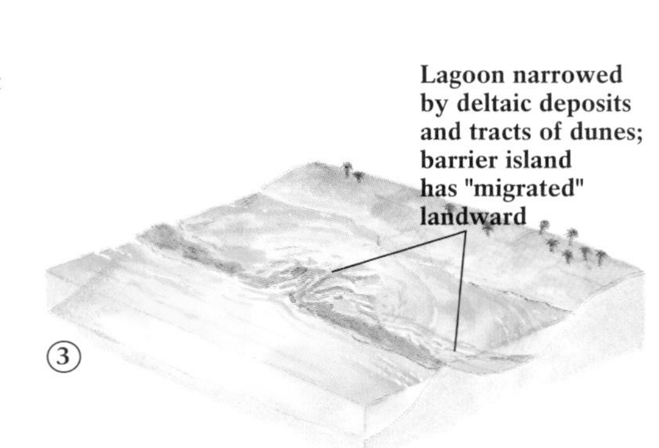

Figure 19-24 Migration of barrier islands. Barrier islands such as those along the Atlantic coast of North America tend to grow landward as sand from their seaward side is transferred by incoming storm waves, producing interior dunes and deltas within the lagoons separating the islands from the mainland.

Human Impact on Barrier Islands North America's long chain of barrier islands is immeasurably valuable as the first line of defense against storm waves in the hurricane-prone Southeast and the Gulf Coast. But these islands are also coveted by nearby city-dwellers as prime destinations for weekend escapes. Although many barrier islands have been set aside for recreation and as wildlife preserves, and others are privately owned and undeveloped, more than one-quarter are vacation resorts, overbuilt with condominiums, hotels, and casinos. Rehoboth Beach, Delaware, Ocean City, Maryland, Virginia Beach, Virginia, Hilton Head, South Carolina, Miami Beach, Florida, Gulf Shores, Alabama, and Galveston, Texas, are but a few of the well-known barrier-island communities developed in this century. Miami Beach, which had a population of 644 permanent residents in 1920, had a population of 96,298 in 1989.

Developers at Ocean City, Maryland, have raised so many luxury hotels and condominium apartment buildings that the recently wild natural seashore is now worth more than $500 million per mile. But their ill-conceived alterations to the beach are jeopardizing their investment. To give every room an ocean view, developers leveled much of the island's protective frontal dunes. Erosion followed, as did a costly beach-nourishment program that may never restore the beaches to their pre-disturbance dimensions. Only 3 kilometers (2 miles) south of Ocean City, Assateague Island, home to herds of rare wild ponies, has been set aside as a national seashore and wildlife habitat. Because its frontal-dune line is preserved by law, Assateague Island's beaches have suffered little erosion. The protective role of barrier islands, the processes that shape them, and the costs of human interference with natural coastal processes are also seen on North Carolina's Outer Banks (Highlight 19-2 on following page).

Organic Coasts

Some coasts, particularly in mangrove swamps and along reefs, are formed by erosion and deposition in conjunction with biological processes. Mangroves and some other trees live in standing tidal water in tropical climates. Mangroves grow from an extensive web of long roots that rise above the water. This root system dissipates much of the energy of waves entering the swampy mangrove forest and also traps fine sediment, creating a quiet-water environment and expanding an island's area. Hardy mangrove seedlings take root in the growing tidal mud flats, enabling the forest to grow seaward. In North America, mangrove swamps can be seen expanding along the southern tip of the Florida mainland (Fig. 19-25); mangrove coasts can also be found on some Caribbean islands and in the tropics of southeast Asia.

Figure 19-25 A typical mangrove coast in the Florida Everglades. Mangrove coasts form as the protruding roots of mangrove trees trap sediment.

Reefs (as we saw in Chapter 6) grow near continental or island shores from the carbonate remains of corals, algae, and sponges. The warm near-surface water in which reefs flourish is limited to a geographic band from about latitude 30°N to 25°S. Clear water is necessary because most reef organisms are filter-feeders, which suck in a large quantity of water and filter out microscopic plankton to eat; they could not survive if they ingested the suspended particles in turbid water. Finally, reef organisms require moderately salty water, so reefs are unlikely to form where a large influx of fresh water dilutes the local seawater or where evaporation concentrates the seawater, making it too salty.

The active portion of a reef is generally located near the sea surface, where sunlight and the algae on which reef organisms feed are plentiful. The active zone can extend downward only as far as sunlight penetrates, to about 75 meters (250 feet). When sea level rises even slightly, as it is doing today, every reef grows vertically to meet the sunlight needs of its inhabitants.

Highlight 19-2 *North Carolina's Outer Banks*

The Outer Banks, along the east coast of North Carolina, is a wonderful outdoor laboratory for assessing the effects of human interference with an efficient natural system. This string of barrier islands includes both a developed coast, the Cape Hatteras National Seashore, and an undeveloped coast, the pristine Cape Lookout National Seashore (Fig. 19-26).

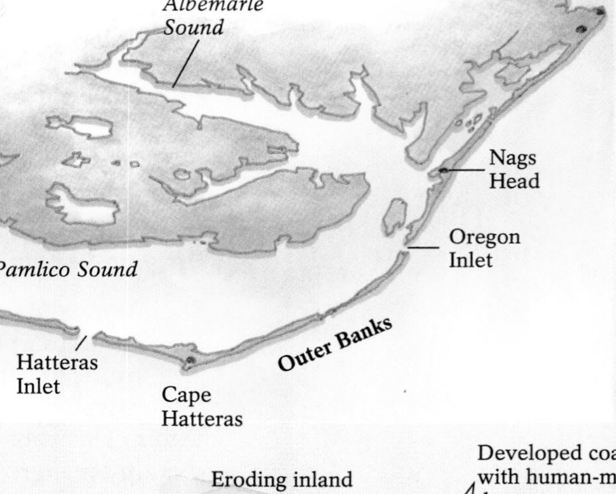

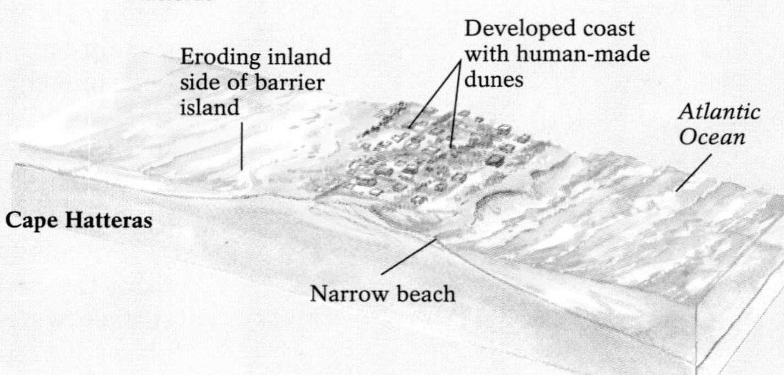

Figure 19-26 Various sections of the Outer Banks of North Carolina have evolved differently, partly in response to the role played by the National Park Service. Human-made frontal dunes protect the narrow beaches of Cape Hatteras (see also Fig. 19-21b); the undisturbed Cape Lookout area, on the other hand, has wide beaches.

The Cape Hatteras coast has been stabilized for more than 50 years by a 10-meter (35-foot)-high human-made line of dunes that protects State Highway 12 from flooding by winter storms and periodic hurricanes. Built beyond the reach of damaging salt spray and wind- and water-borne sand, the artificial dunes are lushly vegetated. But during storms the dunes deflect wave energy much as a seawall does, redirecting it toward the seaward beaches and enhancing erosion there. Since the dunes were built, erosion has narrowed the island's 200-meter (about 650-foot)-wide beaches to only 30 meters (about 100 feet). The inland side of the Outer Banks along Pamlico Sound is also being eroded, largely because the high artificial dunes keep away the sand that would otherwise wash over lower natural dunes; this interrupts the natural tendency of a barrier island to migrate landward. Disruption of the natural barrier-island system along the Hatteras section has forced North Carolina to compensate with an expensive beach-nourishment program that has so far failed to halt erosion. If erosion of that segment of the Outer Banks continues at its present rate, little of Cape Hatteras may be left in the future.

At Cape Lookout, on the other hand, the banks are naturally adjusted to the area's periodic storms. Broad beaches and low dunes absorb erosive energy from storm waves. In the natural system, the vegetation growing along salt marshes has evolved to be somewhat resistant to salt spray and flooding. It traps landward-migrating sand, replenishing the natural interior dunes. When a powerful storm hits, the island grows and migrates landward, instead of eroding.

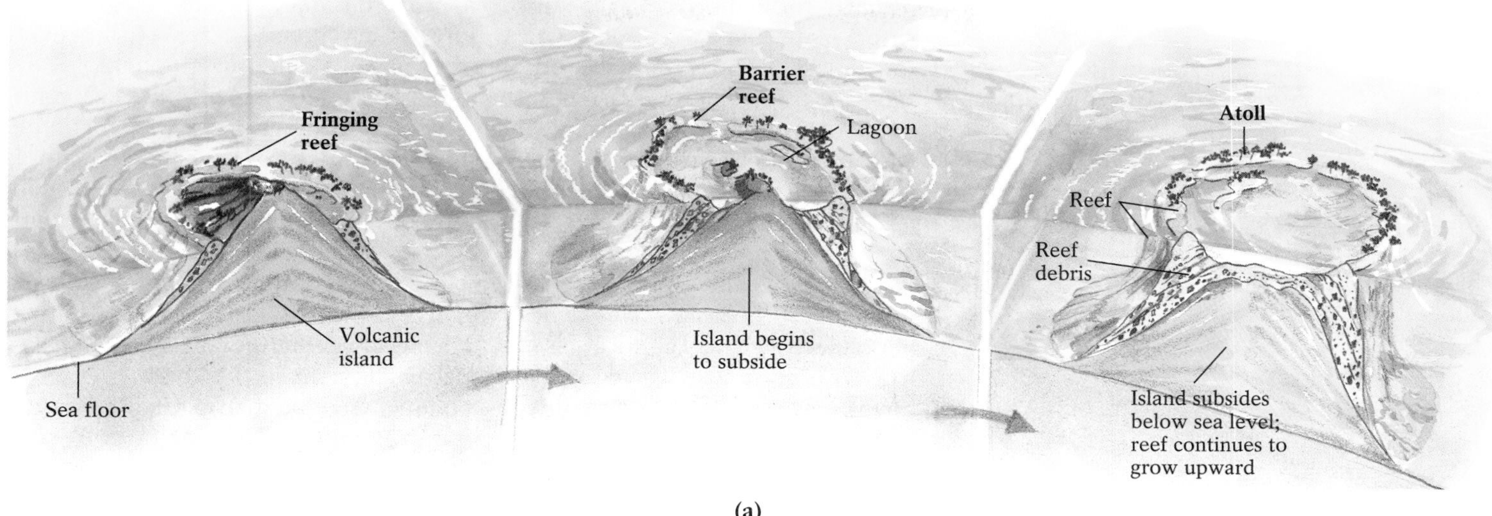

(b)

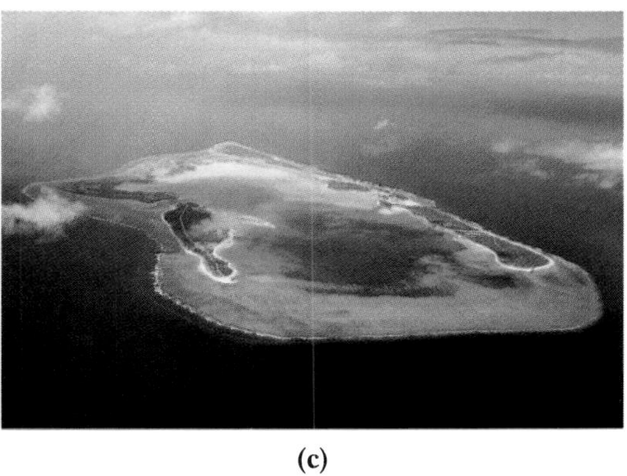
(c)

Figure 19-27 The evolution of carbonate reefs. (**a**) A fringing reef forms on the side of a subsiding volcanic island. As the island subsides, the fringing reef becomes a barrier reef, separated from the volcano by a circular lagoon. After the island sinks completely below sea level, only the reef remains visible—as an atoll—because it continues to grow toward the water's surface. (**b**) Great Barrier Reef, Australia. (**c**) Wake Island, a coral atoll in the Pacific Ocean.

Carbonate reefs are of three basic types: fringing, barrier, and atoll (Fig. 19-27). **Fringing reefs** are built directly against the coast of a landmass, such as along the margins of volcanic islands in the Caribbean and the South Pacific. They are usually from 0.5 to 1.0 kilometers (0.3–0.6 miles) wide and grow seaward toward their organisms' food supply.

Barrier reefs are built on the local continental shelf and are separated from the mainland by a wide lagoon. Incoming waves break against the reef, which protects the mainland coast from wave erosion. Some barrier reefs begin to form as a fringing reef around a subsiding landmass, such as a volcanic island. As the island subsides and much of its area descends below sea level, the reef grows vertically so that the surface of its active zone is always within a few meters of sea level.

Atolls, the most common reefs, are circular structures that extend from very great depth to the sea surface and enclose relatively shallow lagoons. Atolls are barrier reefs that once surrounded oceanic islands that subsequently subsided completely below sea level. This explanation of the origin of atolls was first proposed by Charles Darwin in 1859, and was confirmed in the 1940s by deep drilling of the Bikini and Eniwetok atolls in the South Pacific prior to their selection as atomic bomb test sites. Bikini was found to consist of more than 700 meters (2300 feet) of coralline limestone atop a volcanic island. Surf crashing against an atoll piles the reef debris above sea level.

The Plate Tectonic Settings of Coasts

The different characteristics of North America's East, West, and Gulf coasts are largely due to their distinctive plate tectonic settings. Plate setting strongly determines the steepness of the local continental shelf, the behavior of waves, and consequently the prospects for coastal erosion and deposition. Every major type of plate boundary develops a distinct coastal style.

Divergent plate boundaries that have recently rifted—such as exist in the Gulf of California between Baja California and the Mexican mainland and in the Red Sea between the African and Arabian plates—are bounded by high normal-faulted scarps. Their coasts have steep continental shelves, are vulnerable to the head-on attack of wave action, and consequently feature such recently formed erosional landforms as wave-cut benches and sea stacks. As rifted plate margins move away from a warm spreading center, however, they subside and eventually become tectonically quiet and topographically subdued (see Chapter 12). These plate margins, referred to as passive continental margins, have broad, gently sloped continental shelves. Incoming waves break offshore on the shallow shelf, creating such depositional features as spits, beaches, and barrier islands. The East Coast of North America, with its depositional landforms, clearly shows coastal evolution at a passive continental margin.

Coasts that are shielded by an offshore island arc from the direct onslaught of oceanic waves typically evolve principally by marine deposition, particularly where the local climate enhances weathering and enables rivers to transport and deliver a large volume of sediment to the coast. In such an environment large rivers can deposit enough sediment to extend coastlines by producing large deltas; for example, the Mekong River delta in tropical southeast Asia can survive in the South China Sea, relatively protected in the lee of the island arcs of the Philippines and Malaysia.

Along convergent plate boundaries, coasts generally have narrow, steep continental shelves, and tectonic uplift continuously produces sea cliffs beyond the shoreline. Without offshore shallows to dissipate wave energy, incoming waves strike the coast directly, forming such erosional features as sea stacks and arches and wave-cut benches. The northwest coast of North America, with its assortment of erosional landforms, exhibits coastal evolution at a convergent plate margin.

The East Coast of North America The East Coast is bounded by a broad, tectonically quiet continental shelf that is slowly subsiding under the weight of continent-derived, river-transported sediment. The entire coast is also slowly being submerged as a result of the global sea-level rise of the past century. Variations in the East's coastal topography are primarily due to its local glacial history and its varied bedrock. Glacial erosion carved deep fjords and impressive U-shaped valleys into the soft sedimentary rocks along the northeastern coast of Canada and on the eastern coasts of Baffin Island and the Labrador Peninsula. Glacial erosion of exposed granite plutons produced the low, rocky coasts of the Canadian Maritime provinces and of Maine and New Hampshire to the south. As the local ice sheets melted, isostatic rebound of the land surface lifted (and continues to lift) their coasts and continental shelves, so that oncoming waves now shape these coasts by erosion.

Figure 19-28 East Coast shorelines, such as this one at Curritick Sound, Virginia, are characterized by wide sandy beaches and offshore islands.

The coast from Boston to Long Island, New York, a zone of glacial deposition, consists primarily of unconsolidated sediments. Vulnerable to attack by crashing surf, these deposits have been eroded and then carried by longshore transport to be redeposited as spits, baymouth bars, barrier islands, and beaches (Fig. 19-28). South of the glacial terminus in New York City, to the Florida Keys, the East Coast features large bays such as Delaware Bay, Chesapeake Bay, and Pamlico Sound in the middle-Atlantic states, the barrier islands and inland lagoons and marshes that parallel the coastline for more than 1000 kilometers (about 600 miles), and the reefs of tropical southern Florida.

The West Coast of North America The West Coast from Alaska through California is completely different from the East Coast. For much of its length, converging plates and tectonic uplift have produced a steep narrow continental shelf and rising coastal mountains. The stretch of coast from Alaska to northern Washington was glaciated during recent ice expansions, and consequently contains primary coasts with major westward-draining fjords and U-shaped valleys. South of the glacial terminus, from southern Washington

Figure 19-29 The stepped terraces at Palos Verdes Hills in southern California are typical of a West Coast uplifted by plate tectonics. Each terrace is a wave-cut bench formed when that terrace was at sea level. Southern California's coast is being lifted tectonically from the interaction of the Pacific and North American plates, but this tectonic activity occurs sporadically, separated by periods of relative tectonic stability during which the wave-cut benches have time to develop.

through Oregon and into southern California, wave erosion has cut into the uplifted terraces, shaping headlands into a near-continuous chain of cliffs and rugged offshore islands; eroded sediment is deposited as narrow beaches in protected bays (Fig. 19-29).

Southern California's beaches are narrow, in part because flood-control and irrigation dams on coastward-draining rivers trap stream sediment so it cannot reach and replenish the coast. The narrow beaches, in turn, expose sea cliffs to greater storm-wave erosion than wide beaches, which would absorb and dissipate some of the wave energy. In recent years, the West Coast has also sustained accelerated erosion attributed to El Niño, the warm ocean current that shifts periodically to the eastern Pacific. In the early 1980s, El Niño caused a 10 to 15 centimeter (4–6 inch) rise in the Pacific sea level, unusually high tides, severe winter storm waves, and accelerated erosion.

Sea-Level Fluctuations and Coastal Evolution

During the last period of maximum glacial expansion, sea level was approximately 130 meters (425 feet) lower than it is today, as water that evaporated from the ocean basins became incorporated into the vast Pleistocene ice sheets. The melting of continental ice sheets and the return of meltwater to the world's oceans, which started at the end of the Pleistocene glacial period about 10,000 years ago, has since caused global sea levels to rise, rapidly at first and then gradually, stabilizing near modern levels between 4000 and 5000 years ago.

This global trend of rising sea levels may be moderated by a number of local factors, including tectonic movement at plate edges; isostatic movements due to postglacial growth or shrinkage of ice masses, or sediment accumulation or erosion; and local withdrawal of groundwater or oil.

Tectonic activity at convergent plate boundaries elevates the land by reverse faulting, causing a relative drop in sea level. Tectonic motion at divergent plate boundaries lowers the land surface by normal faulting, causing a relative rise in sea level. Extensive glaciation depresses the Earth's crust under the weight of the ice; when the ice melts, the crust rebounds isostatically (see Chapter 17), causing the previously glaciated regions to rise and relative sea level to drop. For example, coastal Viking villages in Scandinavia are now far inland because the land on which they sit—once depressed under the weight of a glacier—has gradually been uplifted by postglacial isostatic rebound.

Human activity can indirectly cause a relative rise in sea level. For example, global sea levels have begun to rise recently as average temperatures have increased, which is thought to be due to our widespread use of fossil fuels and the resulting greenhouse effect (discussed in Chapter 17). The warmer atmosphere causes the upper few meters of the oceans to become warmer and expand, thus raising sea levels. Higher temperatures may also be melting the world's remaining glaciers, causing some of their waters to return to the oceans.

Humanity's pressing needs for fresh groundwater and a steady flow of oil and gas also contribute to local sea-level rises. When too much of any of these resources is extracted from source rocks, the subsurface rocks become compressed (discussed in Chapter 15), the land surface subsides, and relative sea level rises. Local subsidence of the land near coastal Galveston, Texas, is tied to overpumping of water and oil from the local bedrock, leading to the encroachment of the Gulf of Mexico onto Galveston's shore (Fig. 19-30).

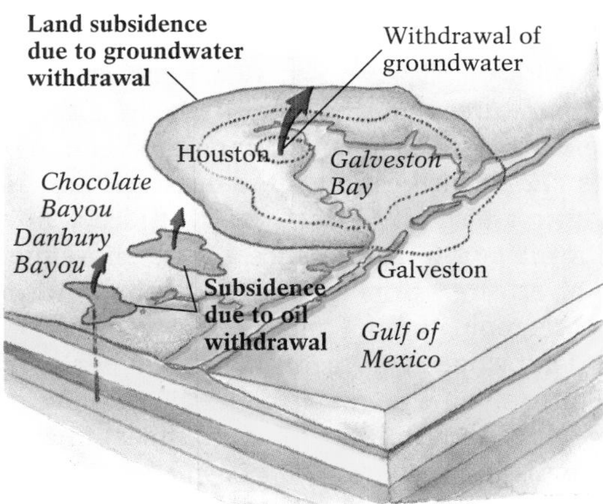

Figure 19-30 Subsidence of the Galveston, Texas, area due to excessive groundwater and oil withdrawal has caused the local sea level to rise relative to the coast.

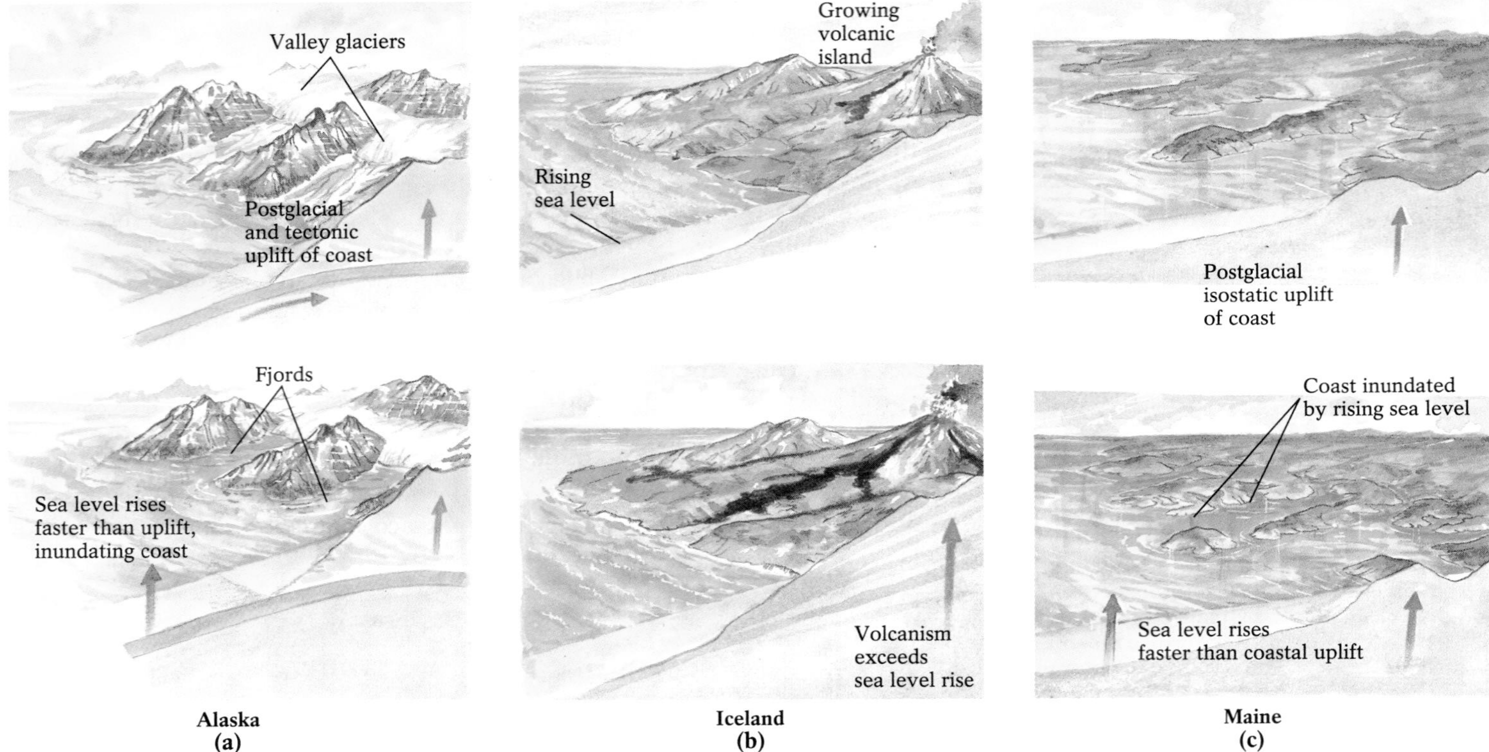

Figure 19-31 The different types of sea-level changes occurring in (**a**) Alaska, (**b**) Iceland, and (**c**) Acadia National Park along the Maine coast illustrate how tectonics, recent glacial history, and global sea-level change combine to determine local sea levels.

The factors that combine to drive local sea-level fluctuations are illustrated along the coasts of Alaska, Iceland, and Acadia National Park in southern Maine (Fig. 19-31). Coastal Alaska is currently rising, due to the ongoing convergence of the Pacific and North American plates and to the minor crustal rebound that occurs as Alaska's glaciers shrink. Worldwide sea-level, however, is rising faster than Alaska, and the Alaskan coast is being inundated, producing its incredible fjord scenery. Iceland, on the other hand, is rising rapidly as a consequence of its active volcanism and its position atop the growing mid-Atlantic divergent ridge. Because Iceland's increase in height exceeds the pace of rising worldwide sea level, it is emerging from the North Atlantic. Acadia National Park in Maine is located at a passive continental margin, and is not rising much tectonically. It is, however, rising isostatically because it was heavily glaciated during the Pleistocene. Nevertheless, like Alaska, Acadia's uplift rate falls short of the postglacial rise in worldwide sea level. Thus, its coast is also slowly being covered by the rising Atlantic.

Several studies have shown that sea level has risen by about 10 centimeters (4 inches) during the past century. With a possible increase of about 0.6 to 1.0°C (1–1.8°F) in global temperatures over the next 40 years, the rate at which sea level is rising will accelerate. Some climate experts predict that it will rise from 30 to 70 centimeters (12–28 inches) over the next century. In places, such as the low-lying Atlantic and Gulf Coasts, this will accelerate coastal erosion and cause a significant landward advance of the shoreline. Higher sea levels accelerate coastal erosion, principally by enabling storm waves to penetrate farther inland, threatening homes and other structures.

Because the Atlantic Coastal Plain slopes very gently, a sea level rise of as little as 30 centimeters (12 inches) along the coast of Pamlico Sound in North Carolina would shift the shoreline about 3 kilometers (2 miles) inland. River mouths would be inundated by rising seas, causing coastal estuaries to extend farther upstream. For every 10-centimeter (4-inch) rise in sea level, it is estimated that the freshwater–saltwater interface in Atlantic estuaries would migrate an additional 1 kilometer (0.6 mile) upstream. If world climate warmed sufficiently to melt *all* of the Earth's residual glacial ice, there would be a 60 to 70 meter (200–250 foot) rise in worldwide sea level—enough to dramatically affect geography everywhere. In North America, the entire Florida peninsula and New York, Boston, and Philadelphia would find themselves completely submerged; huge bays would penetrate the continent on both coasts, and new coastal cities would arise in such now-inland places as St. Louis, Missouri, Memphis, Tennessee, San Antonio, Texas, Albany, New York, and Fresno, California.

Protecting Coasts and Coastal Environments

As we have seen, while coastal erosion is removing prominent headlands, deposition of beaches in adjacent bays and construction of baymouth bars that trap continental sediments are filling in coastal bays; the overall effect of these processes should be the straightening of the world's coasts (Fig. 19-32). Why then are many of the world's coasts still irregular? As with other geological systems, the processes that shape coasts rarely proceed to completion; in the long term, they are almost always interrupted. Thus, the only predictable feature of coasts is change.

Individual humans have little power to control long-term geological change. Our governments, however, can often act to protect us from the inevitability of coastal change. A few states offer low-interest loans to property owners willing to relocate ocean-front homes to less-vulnerable sites. Twenty-nine of the 30 coastal and Great Lake states have coastal-zone management programs. Many states prohibit construction of permanent coastal structures that would be vulnerable to swamping by rising sea levels. They also prohibit construction of barriers that interfere with natural longshore currents. Environmental organizations can also affect coastal evolution, sometimes by purchasing environmentally sensitive coastal property to preserve it and protect it. For example, the Nature Conservancy of Washington, D.C., has recently bought long parcels of the barrier islands off Virginia to prevent future overdevelopment of these fragile coastal environments.

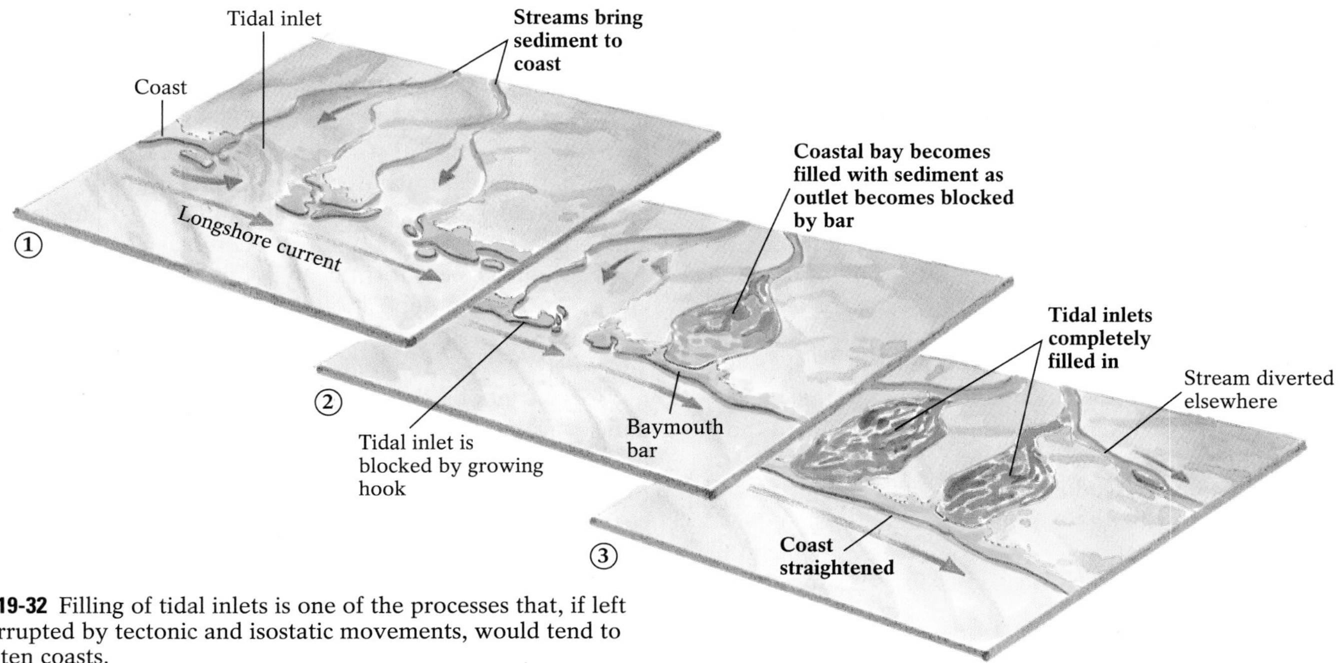

Figure 19-32 Filling of tidal inlets is one of the processes that, if left uninterrupted by tectonic and isostatic movements, would tend to straighten coasts.

Chapter Summary

Shorelines are the places where bodies of water meet dry land. Landward of ocean shorelines are **coasts,** the strips of land bordering an ocean that extend inland to a different geographical setting. Coasts are shaped by the sediments delivered to them, primarily by streams, and the waves that erode and redeposit those sediments.

Waves are caused by wind dragging across the sea surface. Wind speed, duration, and direction, and the extent of uninterrupted water surface across which the wind is blowing, determines the height, velocity, and wavelength of waves. A wave moving across deep, open ocean exhibits **oscillatory motion,** in which water particles move in circular orbits. As a wave moves toward the coast, it usually first encounters the continental shelf, where the ocean floor rises. In the shallow water zone above the continental shelf, friction between the wave and the ocean floor slows the lower part of the wave while the upper part of the wave steepens until it overruns its base and breaks. After a wave breaks, its water no longer moves in a circular path but instead is sent hurtling forward by **translatory motion.**

When a wave approaches the coast from an angle, the first part of the wave front to encounter shallow water slows down, causing the rest of the wave front to swing around until it is nearly parallel to the coastline; this is **wave refraction.** Virtually all waves strike the coast at some angle; their water moves up the beach at an angle but takes a perpendicular path back to the ocean. This zigzag motion generates a **longshore current,** which moves parallel to the coast. Longshore currents transport most coastal sediment. Longshore currents may be crossed by **rip currents,** which develop when water piled up in the surf zone flows seaward, perpendicular to the coast.

Tides, the twice-daily rise and fall of the ocean surface, are primarily due to the gravitational pull on the ocean surface by the Moon and from the force created as the Earth spins on its axis. **Flood tides** elevate the sea surface and cause the shoreline to move inland, sometimes migrating up rivers. A flood tide whose upstream movement is turbulent is called a **tidal bore.** Falling or **ebb tides** lower the sea surface and cause the shoreline to move seaward, carrying land-derived sediment out to sea.

Most coastal erosion occurs when waves strike a coast with sufficient energy to remove loose materials. When the water immediately offshore is deep, the waves crash with their full force against coastal **headlands,** cliffs that jut seaward. Erosion of a headland eventually undercuts the cliff, removing the support from overlying rock masses, which fall into the surf. The remaining cliff base becomes a **wave-cut bench,** a terrace washed by breaking waves. Wave refraction at headlands bends incoming waves against the sides of the headlands, eroding them to produce **sea caves.** A **sea arch** forms when two sea caves have eroded completely through the headland. When a sea arch collapses, an isolated offshore bedrock knob, or **sea stack,** forms. Coasts can be protected from erosion by such structures as a barrier of loose boulders, or **riprap,** or a concrete **seawall** built parallel to the coast.

Coastal deposition also produces a variety of coastal landforms, the most common being a **beach**—a dynamic, relatively narrow segment of a coast washed by waves or tides and covered with sediment of various sizes and compositions. Beaches consist of a **foreshore** area, which extends from the low-tide to the high-tide line, and a **backshore** area, which extends from the high-tide line to the sea cliff or inland vegetation line. The steepest part of the foreshore is its seaward margin, or **beach face.** Most beaches also contain one or more **berms,** horizontal benches or landward-sloping mounds of storm-deposited sediment. Sediment transported along the beach face by swash and backwash moves by **beach drift;** sediment transported offshore within the surf zone moves by **longshore drift.**

When a longshore current suddenly encounters deeper water, such as at the entrance to a bay, the sediment it transports is deposited as a **spit,** a finger-like ridge of sand that extends from land into open water. Where waves or currents into a bay are particularly strong, growing spits can become curved, forming a **hook.** A spit that grows completely across a bay entrance is a **baymouth bar.** Coastal deposition occurs when wave energy decreases below a critical level. Wave energy can be dissipated by wind variation or sudden increases in water depth, or when the waves are intercepted by natural or artificial structures offshore. Human-built structures that intercept waves or currents, causing deposition, include **breakwaters,** built parallel to the coast, and **groins** and **jetties,** built obliquely or perpendicular to the coast.

Primary coasts form from nonmarine processes such as glaciation, stream deposition, tectonic activity, flooded river systems, and biological activities. **Secondary coasts** are products of coastal erosion and deposition. **Barrier islands** are nearly continuous ridges of sand, parallel to the main coast and separated from it by a bay or lagoon. They may be secondary coasts, consisting of depositional or combined erosional and depositional landforms, or they may be primary coasts consisting of the remnants of a drowned ancient coastal sand-dune system.

Organic coasts include those formed by mangrove tree roots in swamps and those formed by carbonate reefs. Carbonate reefs are commonly found along the margins of volcanic islands in the Caribbean and the South Pacific. **Fringing reefs** initially surround the island and grow seaward toward their organisms' food supply. **Barrier reefs,** separated from the island's coast by a wide lagoon, form and grow vertically as the island subsides. After a volcanic island completely subsides, its barrier reef becomes an **atoll,** a circular structure that extends from very great depth to the sea surface and encloses a relatively shallow lagoon.

Plate tectonic settings influence the nature of coastlines. Recently diverged plate margins and convergent plate boundaries have steep continental shelves, and their coastlines include erosional landforms such as wave-cut benches and sea stacks; passive continental margins have broad continental shelves, and their coastlines typically contain depositional features such as beaches and spits. Sea-level fluctuations also affect coastlines: Rising sea levels cause shorelines to migrate inland, and falling sea levels cause them to migrate seaward.

Key Terms

shoreline (p. 533)
coast (p. 533)
oscillatory motion (p. 536)
translatory motion (p. 536)
wave refraction (p. 537)
longshore current (p. 537)
rip current (p. 538)
tide (p. 539)
flood tide (p. 540)
tidal bore (p. 540)
ebb tide (p. 540)
headlands (p. 542)
wave-cut bench (p. 543)
sea cave (p. 544)
sea arch (p. 544)
sea stack (p. 544)
riprap (p. 544)
seawall (p. 544)
beach (p. 545)
foreshore (p. 545)
backshore (p. 545)
beach face (p. 545)
berm (p. 545)
beach drift (p. 546)
longshore drift (p. 546)
spit (p. 546)
hook (p. 546)
baymouth bar (p. 546)
breakwater (p. 548)
groin (p. 548)
jetty (p. 548)
primary coast (p. 550)
secondary coast (p. 550)
barrier island (p. 550)
fringing reef (p. 555)
barrier reef (p. 555)
atoll (p. 555)

Questions for Review

1. Draw a simple diagram showing wave crests, wave troughs, wave height and amplitude, and wave length. Define wave period and fetch.

2. Briefly describe the difference between the oscillatory and translatory motion of waves.

3. Why are most wave fronts nearly parallel to the shoreline on arrival?

4. How does a longshore current develop?

5. Explain why the Earth's coasts experience two high tides and two low tides each day.

6. What four principal factors control the rate of coastal erosion?

7. Draw a sketch showing the main components of a beach. Using arrows, indicate the source of most beach sediment and how it is most commonly removed.

8. Describe how the following depositional coastal landforms develop: spit, hook, baymouth bar, tombolo.

9. Discuss how breakwaters, groins, and jetties interrupt longshore transport and affect coastal erosion and deposition.

10. Discuss the three proposed models for the origin of barrier islands.

For Further Thought

1. Why do you suppose surfing is more popular on the western than on the eastern coast of North America? Speculate about the prospects for surfing on the western coast of Europe or Africa.

2. Suppose you acquired some valuable beachfront property in Oregon. How would you protect this land from coastal erosion? (Note: You have unlimited financial resources.) In your plan, try to minimize the negative secondary effects that follow most cases of human interference with natural coastal systems.

3. What would be the effect on California's beaches if all the state's dams were removed?

4. How would the eastern coast of North America be affected if widespread subduction resumed there?

5. How might the coasts of North America change if the Earth entered another period of worldwide glacial expansion?

Fossil Fuels

Alternative Energy Resources

Mineral Resources

Future Use of Natural Resources

Figure 20-1 The need for the Earth's dwindling resources necessitates mining in remote areas. This gold mine was excavated out of a jungle in northeastern Brazil.

Chapter 20

Human Use of the Earth's Resources

This text has often referred to the many natural resources that are useful or essential in people's lives today, whether for industrial or personal use. The Earth's resources were believed until recently to be unlimited; today we face serious shortages of numerous essential materials (Fig. 20-1). For example, scientists believe that the world's recoverable supply of crude oil, from which we get gasoline, will last only another 50 years at the current rate of use. How have we managed to exhaust our stores so quickly? Can we compensate for these losses?

The answers lie in the growth rate of the world's population, the quantity of natural resources each individual uses, and the search for alternative resources. In the United States alone, each person directly or indirectly uses about 10,000 kilograms (22,000 pounds) of raw materials each year, most of which is stone and cement for the construction of roads and buildings, but which also includes about 500 kilograms (1100 pounds) of steel, 25 kilograms (55 pounds) of aluminum, and 200 kilograms (440 pounds) of industrial salt (mostly for cold-weather road maintenance). Each American also uses nearly 3800 liters (1000 gallons) of oil per year. Collectively, Americans account for about 30% of world oil consumption. The United States, with only about 6% of the world's population, uses about 30% of its minerals, metals, and energy. One American uses up to 30 times as much material and energy as a person in an emerging nation.

Natural resources are distributed unevenly among nations and continents. Valuable materials are abundant in some geological settings; elsewhere they are in short supply. Some nations, such as Canada, the United States, and Russia, have a wealth of varied natural resources (Fig. 20-2); others, such as Japan, have few. Some former resource producers have virtually exhausted their domestic supplies and must now import from other nations. Great Britain, once a great mining nation that exported tin, copper, lead, and iron, must now import those commodities. Meanwhile, some smaller, developing nations have vast supplies of a few important materials. Guyana and Surinam, on the northeast coast of South America, and Jamaica, in the Caribbean, possess some of the world's richest supplies of aluminum. But no nation is self-sufficient in all essential resources.

Resource consumption worldwide is rising at an accelerating rate as world population increases (now at 5.6 billion—three times what it was in 1920—and expected to double by 2040) and people everywhere strive for

Figure 20-2 Giant dredges are used to search for gold in Siberian mines.

the benefits of technological development. Unless new supplies of depleted resources are identified, or substitutes for them are found, and industrial development is managed in ways that limit resource depletion, people everywhere will be forced to change their ways of life.

Reserves are natural resources that have been discovered and can be exploited for profit with existing technology. We know where reserves are, and we have the means to extract them; their economic value in the marketplace exceeds the cost of their extraction. **Resources** are deposits that we know or believe to exist, but that are not exploitable today, whether for technological, economic, or political reasons. Estimates of resources hidden beneath the surface are arrived at through exploratory drilling, geophysical modeling, and extrapolation from known reserves. World oil reserves, for example, are estimated at 700 billion barrels (a *barrel* is a volume equaling 159 liters or 42 U.S. gallons), whereas world oil resources are thought to be about 2 trillion barrels.

Resources may become reserves if their profitability increases. In the 1970s, for example, when the price of gold surged to $800 per ounce, deposits that had not been worth developing at lower prices immediately became highly profitable ores. (An **ore** is a mineral deposit that can be mined for a profit; this is an economic, not a geological, term.) Conversely, reductions in the price of a material on world markets can lead some countries to import rather than mine and develop it themselves, changing a profitable reserve into an unprofitable resource. This was illustrated by the recent decline of the U.S. steel industry, which languished when it became cheaper to import steel from Korea than to mine the iron and other materials used to produce steel domestically.

A low-value deposit may become an ore body if economies improve, political events enhance access, or new technology develops. But if economies languish, the political climate prevents access to deposits, or new technologies render the resource obsolete, a once-profitable reserve may become a relatively valueless resource.

Access is often controlled by politics: The Persian Gulf War of 1991 caused an increase in oil prices as some Middle Eastern oil reserves became unavailable resources (Fig. 20-3). Other deposits that might be extracted profitably today are located in national parks or wilderness areas; these would be reserves if they could be exploited, but remain resources because they are protected by legislation in order to avoid environmental disruption. Political events have been affecting resource development for millenia, as has changing technology. Five thousand years ago, humans made tools of stone, especially chert and obsidian; when copper, iron, and bronze came into use, and especially after smelting developed, stone for tool-making lost value.

Figure 20-3 During the 1991 Persian Gulf War, the Iraqi military ignited thousands of Kuwaiti oil wells. Political conditions in certain oil-producing nations have prompted oil-poor industrial nations to seek alternative energy sources.

Some natural resources are *renewable,* meaning they are naturally replenished over short time spans (such as trees) or available continuously (such as sunlight). *Nonrenewable* natural resources form so slowly that they are typically consumed much more quickly than nature can replenish them; these include fossil fuels such as coal, oil, and natural gas, and metals such as iron, aluminum, gold, silver, and copper. These materials are like crops that can only be harvested once. We cannot "grow" another crop of nonrenewable resources, although some, such as copper and aluminum, can be recycled for reuse. Some resources may be either renewable or nonrenewable, depending on how we use them. Soil, for example, is a renewable resource if sound agricultural practices are followed. It is nonrenewable when depleted of its nutrients by overplanting, or when it is allowed to erode after being overgrazed, overplanted, or deforested.

As we seek both to maintain our standard of living and, at the same time, protect the global environment into the twenty-first century, we must all become more knowledgeable about the Earth's energy and mineral resources. To use dwindling resources wisely, we need to understand how they formed, where they are abundant, the environmental consequences of their use, and how long the known supplies are likely to last. This chapter addresses these and other crucial issues in regard to fossil fuels, alternative energy sources, and both metallic and nonmetallic minerals. (We looked at other environmental resource issues in previous chapters, especially water supply and pollution in Chapters 15 and 16.)

Fossil Fuels

Worldwide, we get relatively little of our energy from renewable sources such as solar energy, the energy produced by the rise and fall of tides, wind power, and power from streams. We get most of our energy from nonrenewable **fossil fuels,** derived from the organic remains of past life. The principal fossil fuels are oil, natural gas, and coal. Oil and natural gas are by now well past their peak production period. Extraction of the world's reserves of crude oil peaked around 1975, and at the current rate of use these reserves will be virtually exhausted by about 2040. Nevertheless, for economic and technological reasons, the nations of the world continue to draw nearly 95% of their total energy from a dwindling supply of fossil fuels that are, in practical terms, nonrenewable (at least on a human time scale).

Figure 20-4 The first commercial oil well was developed in 1859 in Titusville, Pennsylvania, by Edwin Drake (at right). Drake's well, which was 21.2 meters (70 feet) deep, yielded 35 barrels of oil per day.

Petroleum

The most common and versatile of the fossil fuels is **petroleum,** a group of gaseous, liquid, and semi-solid substances composed chiefly of **hydrocarbons,** molecules consisting entirely of hydrogen and carbon. Typically pumped from the ground as dark, viscous crude oil, petroleum is refined to produce propane for camp stoves, motor oil and gasoline for cars, tar and asphalt for roads, and the natural gas and heating oil that warm our homes and workplaces. It is also the major ingredient in plastics, synthetic fibers, dyes, cosmetics, explosives, certain medicines, certain fertilizers, and records, tapes, and compact discs.

Humans have used petroleum for thousands of years. Forty-five hundred years ago, Babylonians collected crude oil bubbling from natural pools to make glue for attaching metal projectile points to spears. In what is today Iraq, oil seeping from rocks in the valleys of the Tigris and Euphrates Rivers was used in mortar for setting bricks, in grout that set tiles in ancient mosaics, and to waterproof boats.

Modern use of petroleum began in 1816, when gas extracted from coal was first used in Baltimore's gaslights. Combustible gas was discovered in 1821, when a water well in Fredonia, New York, was accidentally ignited, producing a spectacular flame. Wooden pipes were installed to carry the gas to 66 gaslights in downtown Fredonia. Commercial use of oil began in 1847, when a merchant in Pittsburgh started bottling and selling natural "rock oil" as a lubricant for machines in the home and workplace. In 1852 Canadian chemists, using oil from what is now Oil Springs, Ontario, first refined kerosene from rock oil for use in home lamps, which soon eliminated much of the candle and whale-oil industries. The oil-well industry was born on Sunday, August 27, 1859, when Edwin Drake of Titusville, Pennsylvania, pumped oil from the first true oil well (Fig. 20-4).

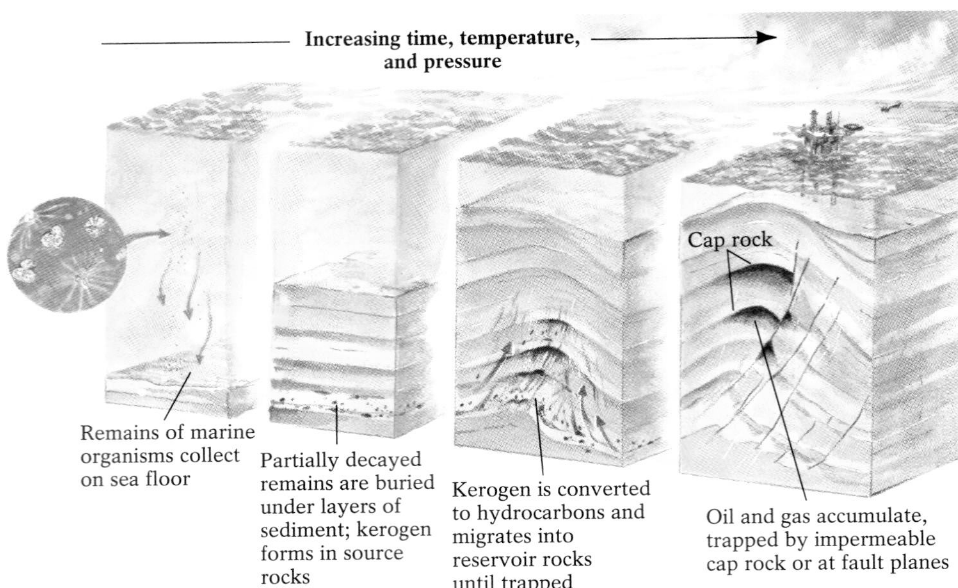

Figure 20-5 Petroleum begins as a large accumulation of partially decayed microorganisms in marine mud. The remains of the organisms are eventually converted to kerogen as the marine mud lithifies to become source rocks. Geothermal heat "cooks" the kerogen, which becomes petroleum as its organic molecules break down into a variety of hydrocarbons. Petroleum is expelled from its source rocks and migrates through permeable reservoir rocks, accumulating when further movement is blocked by some structural or stratigraphic feature, such as an impermeable cap rock that serves as a trap. Erosion and faulting may disrupt oil traps, enabling oil to escape to the surface (see Highlight 9-1).

The Origin of Petroleum Petroleum production generally begins in marine basins in tropical environments, where a rich diversity of microscopic plants and animals exists (Fig. 20-5). When the organisms die they start to decay by oxidation, but eventually the oxygen in bottom waters is depleted, and decay ceases. The organic remains may then become buried under layers of sediment and additional organic material, preventing subsequent decay. As sediments accumulate, pressure and geothermal heat convert the organic molecules to a substance called **kerogen,** a solid waxy organic material. Kerogen is converted to various liquid and gaseous hydrocarbons at temperatures between 50 and 100°C (100–200°F) and at a depth of 2 to 4 kilometers (1.2–2.5 miles).

At the start of this process, kerogen's large, complex organic molecules form highly viscous hydrocarbons such as tar, petroleum jelly, and paraffin wax. With increasing heat, these molecules break down to form smaller, simpler, less viscous ones, such as those in diesel oil, kerosene, and gasoline. Above about 100°C (200°F), liquid petroleum is converted to various natural gases, ranging from those with relatively complex molecules, such as octane and butane, to the simplest, lightest natural gases—propane, ethane, and methane. At temperatures of about 200°C (400°F) and at depths of 7 kilometers (4 miles) or more, the lightest gas, methane, escapes upward to areas of lower pressure, and rocks no longer contain hydrocarbons.

The rocks in which hydrocarbons form are called **source rocks;** they are typically shales and siltstones lithified from fine-grained organic-rich muds. Oil and gas are rarely found in source rocks because most liquid and gaseous hydrocarbons are readily expelled from their compacting source muds. They tend to migrate upward into adjacent permeable **reservoir rocks,** such as well-sorted sandstones and highly fractured or porous limestones, and continue to migrate until they are trapped by an impermeable cap rock or faults and folds in rocks.

Geological activity can destroy oil traps as well as create them. Uplift and erosion can remove the trapping cap rock, allowing oil or gas to escape at the surface. A new fault, or the extension of an old one, can breach an oil trap and allow its oil to seep out. Due to these processes, much of the oil and gas formed before about 65 million years ago has long since escaped from its traps and evaporated at the surface. More than 60% of today's oil-producing wells are in rocks that are 65 million years old or younger.

Geologists estimate that less than 0.1% of all marine organic matter buried at the sea floor eventually gets trapped as usable petroleum. In some settings, there may not be enough geothermal heat to convert organic matter to kerogen and petroleum. In others, deposits may have experienced high temperatures but not at great enough depth, enabling shallow-forming hydrocarbons to escape without being trapped. The conditions required to produce, trap, and retain hydrocarbons are rarely found together, which explains why most marine rocks are petroleum-free. We do not know how long it takes for oil and gas to form. No known petroleum sources are less than 1 to 2 million years old, so the process must take at least that much time.

Oil Shale and Oil Sand **Oil shale**, a black-to-brown clastic sedimentary rock consisting of a mixture of waxy kerogen and fine mineral grains, is a common source rock. However, because it was never buried deeply enough to raise its temperature to that required to convert kerogen to oil, oil shale has retained its hydrocarbons. Based on current and anticipated increased rates of usage, the estimated 2 to 5 trillion barrels of shale oil underlying the United States (which has about two-thirds of the world's oil shale) could supply America's petroleum needs for more than 500 years. According to conservative estimates, this country's largely untapped oil shales (Fig. 20-6) contain more than ten times the known oil reserves of the Middle East. Economics, technology, and the need to protect the environment, however, keep us from using this oil to heat our homes and fuel our cars. Current world prices for crude oil are not high enough to justify the cost of developing this resource.

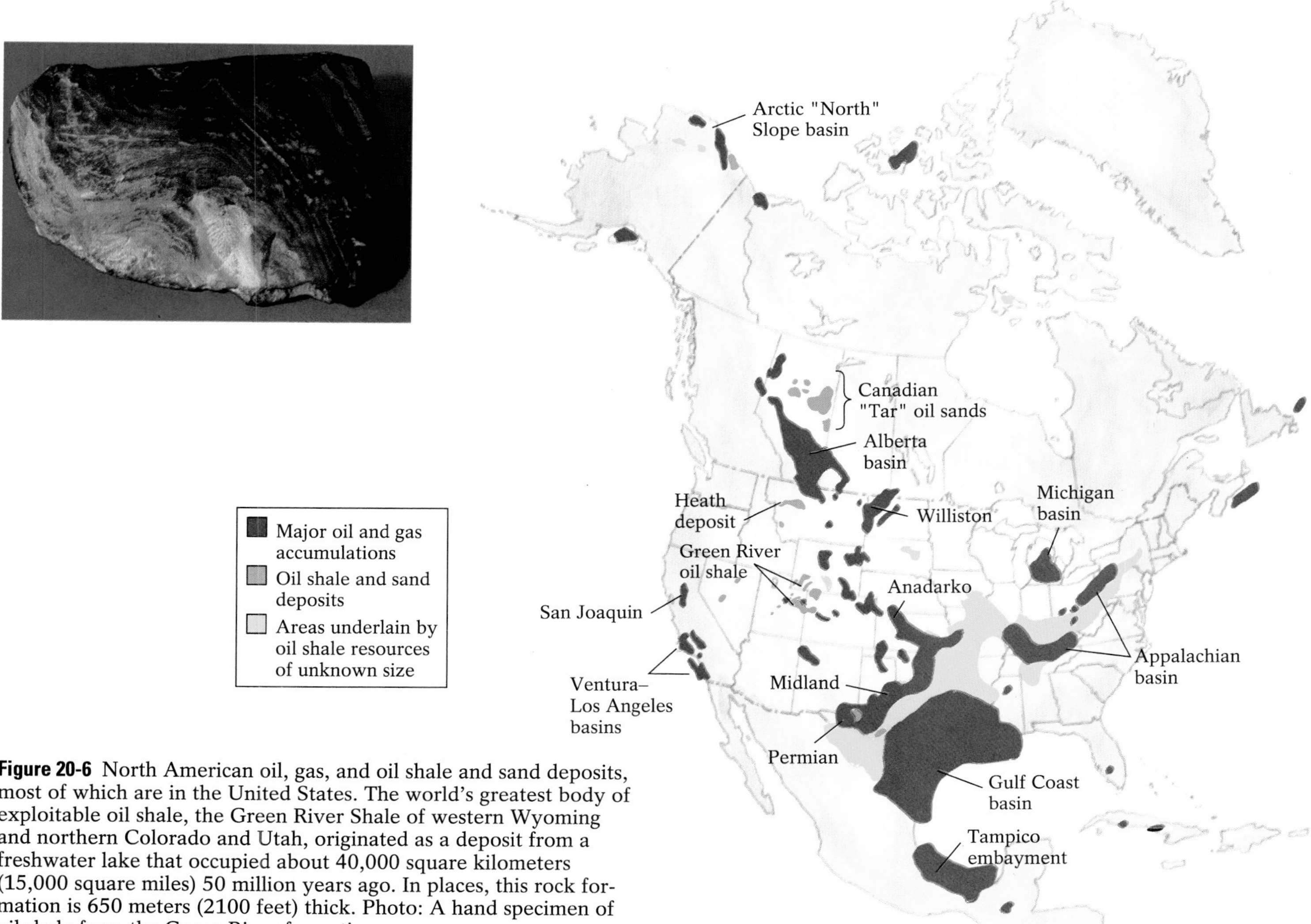

Figure 20-6 North American oil, gas, and oil shale and sand deposits, most of which are in the United States. The world's greatest body of exploitable oil shale, the Green River Shale of western Wyoming and northern Colorado and Utah, originated as a deposit from a freshwater lake that occupied about 40,000 square kilometers (15,000 square miles) 50 million years ago. In places, this rock formation is 650 meters (2100 feet) thick. Photo: A hand specimen of oil shale from the Green River formation.

To develop oil shale, it must be mined, its trapped kerogen must be removed from the rock and then processed, and the waste rock must be disposed of in an environmentally sound fashion. After mining, the shale must be crushed and then heated to more than 500°C (900°F) to vaporize its kerogen, which condenses as crude oil. Like popcorn, which occupies more space than the kernels from which it has popped, fragments of oil shale expand in volume by as much as 20% when heated, so waste rock, if not removed, would overflow the hole from which it was mined. If the United States were to obtain its entire oil supply from oil shale, it would need to dispose of 13 billion tons of waste rock each year, enough to fill 130 million freight-train cars.

Technology may someday provide a solution to this problem; microwaves or radiowaves could perhaps be used to heat kerogen in the shale without removing and crushing it. Given the cost, however, oil shale deposits will be exploited only after more accessible, less expensive sources are exhausted.

Oil sand, a mixture of unconsolidated sand and clay that contains a semi-solid, tar-like hydrocarbon called **bitumen,** poses similar development problems (Fig. 20-7). Bitumen can be refined to produce gasoline, fuel oil, or other commercially viable hydrocarbon products, but it adheres strongly to the mineral grains in oil sand; it does not flow, nor can it be pumped from the ground. Oil sand must be mined, then heated with hot water and steam to release the bitumen in a more fluid state. The development of oil sand, or any other fossil fuel, is viable only if the fuel can provide more energy than is consumed by mining and processing.

The world's largest oil sand deposit, about 400 kilometers (250 miles) north of Edmonton in Alberta, Canada, has been producing crude oil since 1967. Today it produces about 200,000 barrels per day, about 16% of Canada's oil. It is commercially exploitable because it is close enough to the surface to be extracted by surface mining, which involves the removal of overlying sediment layers, a less costly process than underground mining. Surface mining of oil sand, however, produces enormous piles of residual oily sand that can contaminate local surface water and groundwater. To avoid this environmental hazard, Canadian petroleum engineers are developing a process to extract bitumen by injecting hot water or steam directly into underground deposits to soften the bitumen so it can be pumped out.

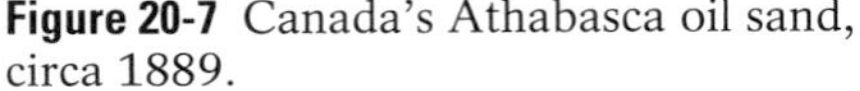

Figure 20-7 Canada's Athabasca oil sand, circa 1889.

Coal and Peat

Coal, the Earth's most abundant fossil fuel, is a combustible chemical sedimentary rock that forms from the highly compressed remains of land plants (see Chapter 6). It contains the energy stored in living plants by photosynthesis, the process through which sunlight, water, and carbon dioxide produce the materials for plant growth. When coal burns, it releases the energy that was stored in plants millions of years ago.

Native Americans used coal thousands of years ago to fire pottery; a thousand years ago, Europeans mined it to heat homes and fuel the fires of smelting industries. Cheap, plentiful coal powered the newly invented steam engine in the Industrial Revolution of nineteenth-century Europe and North America. By 1900, coal supplied 90% of U.S. energy needs for industry and domestic heating.

The use of coal has declined over the last 40 years and coal now provides less than 20% of U.S. energy needs, largely because oil and natural gas have been abundant and could be economically extracted with fewer problems. Coal can be difficult and sometimes dangerous to mine, and is costly to process. It also pollutes the air when burned. But as its replacements become depleted, coal will likely rebound as our principal fossil fuel, especially in electricity-producing power plants. New technology makes it economically feasible to convert coal to liquid and gaseous fuels; in these forms existing world resources could meet energy needs for hundreds of years. The United States has more than 30% of the world's currently accessible coal. However, what may be the world's largest coal field has recently been discovered in Antarctica; it stretches for hundreds of kilometers along the eastern side of the Transantarctic Mountains, and when eventually mined will add substantially to the world's coal inventories.

The Origin of Peat and Coal The luxuriant plant growth that produced the world's coal deposits likely occurred in tropical or semi-tropical swamps that became completely covered either by the growth of more vegetation or by overlying sediments. Once covered, the plant remains could not completely decompose by oxidation. The cumulative weight of overlying deposits squeezed water from the porous mass of incompletely decomposed vegetation. As plant remains became more deeply buried, increasing pressure, geothermal heat, and bacterial reactions removed more water as well as the organic gases, such as CO_2 and CH_4 (methane), produced by bacterial activity and oxidation.

These processes create a variety of fossil fuels categorized by increasing carbon and decreasing gas and water content (see Fig. 6-23). Moderate pressure produces **peat,** the first fuel to form from buried vegetation. A soft brown mass of compressed, largely undecomposed and still recognizable plant structures, peat contains substantial amounts of water, organic acids, and gases such as hydrogen, nitrogen, and oxygen; it is only about 50% carbon. Peat is dried and then burned as fuel for home heating and other domestic uses in rural Ireland, England, and elsewhere in Europe.

Bacterial action continues in peat, eventually breaking it down to form a vegetative muck, rich in kerogen, that becomes compressed as **lignite.** This soft, brown coal is about 70% carbon (plus 20% water and 10% oxygen); the higher carbon content makes it a more efficient heat source than peat. Deep burial of lignite and the accompanying rise in pressure and geothermal heat convert it to lustrous black **bituminous coal** having a carbon content of 80 to 93%, which produces more heat, with much less smoke, than peat or lignite. Metamorphic conditions gradually transform lignite and bituminous coal

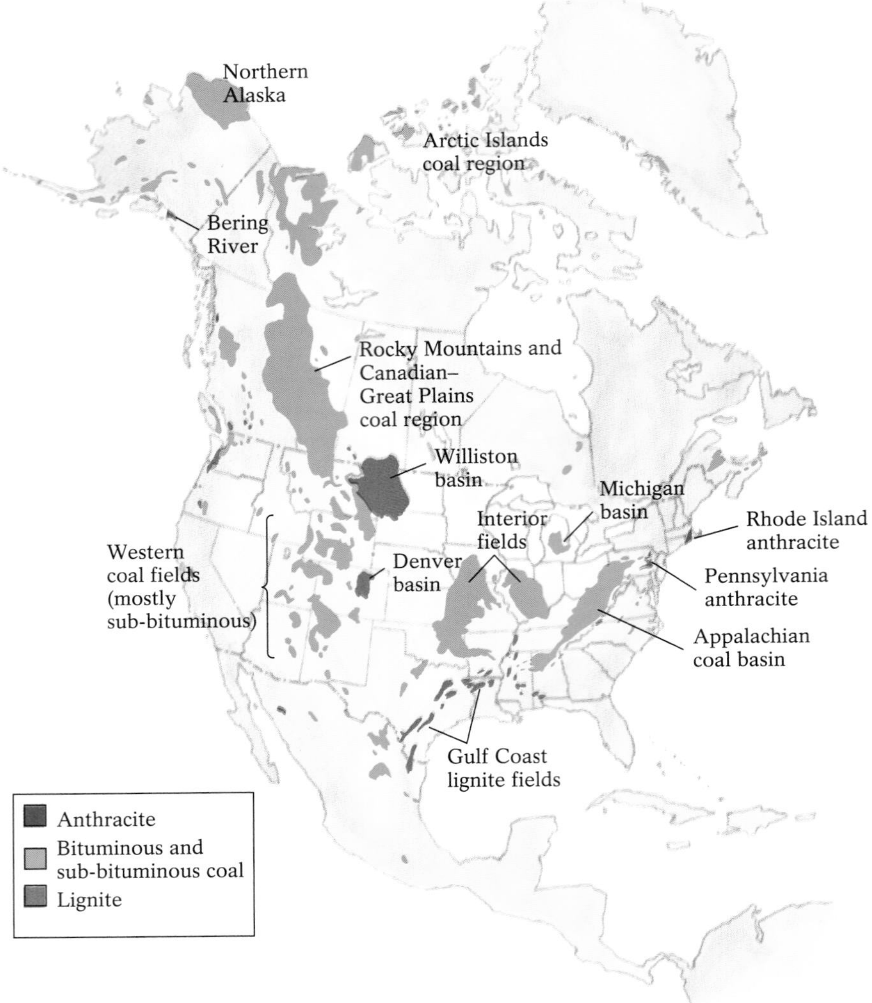

Figure 20-8 Major coal deposits of North America. Above: A coal seam in northeastern Oklahoma.

deposits to **anthracite**, a hard, jet-black coal containing 93 to 98% carbon and very little gas. Anthracite burns with an extremely hot flame and very little smoke. Unlike lignite and bituminous coal, which are widely distributed, anthracite exists only in metamorphic zones in mountainous regions, notably the Appalachians of northeastern Pennsylvania (Fig. 20-8). Anthracite seams are often found within the steeply dipping limbs of folded sedimentary rocks, where coal cannot be strip mined; this requires more difficult, dangerous, and costly deep underground coal-mining processes.

Peat, lignite, and bituminous coal typically occur as distinct seams or beds in sequences of detrital sedimentary rocks, particularly those that accumulated in warm, moist, coastal environments. A series of alternating detrital sediments and coal beds typically represents cycles of rising and falling sea level and their corresponding periods of land submergence (and thus detrital sedimentation) and land emergence (and ensuing swamp development) (see Chapter 6).

Much of North America's vast coal deposits formed during ancient episodes of high worldwide sea level, when shallow seas invaded the continent's interior. The coal of Utah, Montana, and the Dakotas formed at such seas' swampy margins. Coal deposits have also formed on the vegetated flood plains of intermountain rivers. The coals of Wyoming's Powder River basin occur as layers between floodplain sediments. Coal deposits may also accumulate at continental margins, where wide continental shelves are flooded periodically by fluctuating sea levels. The coal of the Pennsylvania and West Virginia Appalachians formed in such an environment.

Peat and coal have probably been accumulating continuously since land plants first appeared 450 million years ago. During certain periods, warmer-than-average climatic conditions and a high proportion of tropical or semi-tropical landmasses promoted more extensive coal formation. During the Mississippian, Pennsylvanian, and Permian periods of the Paleozoic Era (345–225 million years ago), giant ferns and scale trees (gymnosperms) in great swamps and forests covered the tropical equatorial lowlands of the supercontinent Pangaea. This was the Carboniferous Period, when the great bituminous coal beds of western Europe (Great Britain, Germany, and Poland) and the anthracites and bituminous deposits of eastern North America's Appalachians were laid down. A second great period of coal formation occurred about 135 to 30 million years ago, producing lignites and bituminous coals from flowering plants (angiosperms) similar to those of today. Some of these deposits contain well-preserved fossils that identify the types of plants from which they formed. These coals extend from New Mexico to the Great Plains of North Dakota and Saskatchewan. Coal is also being formed today: In the Great Dismal Swamp of coastal Virginia and North Carolina, a 2-meter (7-foot)-thick layer of vegetation has accumulated over a 5000-square-kilometer (2000-square-mile) area during the last 5000 years. The next major rise in sea level that buries these organic deposits with detrital sediment will halt their further decomposition, and eventually provide the pressure and heat that will convert them to peat and coal.

Fossil Fuels and the Environment

Widespread use of coal, oil shale, and other fossil fuels can produce a variety of environmentally damaging side effects. Concerns about acid rain, global warming, and massive marine oil spills have led to increased public awareness and recent legislation to reduce or prevent such effects.

Acid Rain Burning fossil fuels releases secondary materials into the atmosphere, reducing the quality of the air we breathe. When coal is burned, for example, coal ash—fragments of noncombustible silicates and toxic metals—enters the atmosphere. When sulfur- and nitrogen-rich hydrocarbons are burned, sulfur and nitrogen oxides are released and combine with water in the atmosphere to form sulfuric acid (H_2SO_4) and nitric acid (HNO_3), the principal human causes of **acid rain** (Fig. 20-9). Acid rain, along with acid mine water, is widely considered responsible for damaging forests and crops, killing aquatic life in lakes, and accelerating the destructive weathering of human-made structures. Damage from acid rain is most pronounced downwind of major coal-burning industrial regions.

In some settings, environmental damage from acid rain is lessened by the areas' geology. Exposed carbonate bedrock (limestone, dolostone) somewhat neutralizes acid rain by the chemical-weathering process of carbonation

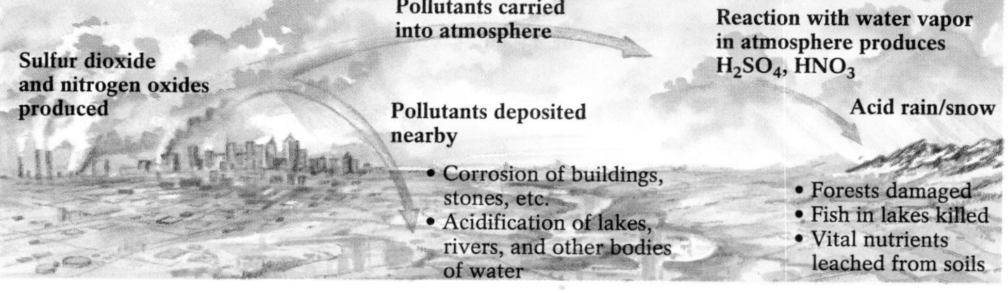

Figure 20-9 Acid rain has two major causes: the release of sulfur dioxide into the atmosphere when sulfur-rich coal is burned, where it combines with water to form sulfuric acid; and the emission of nitrogen oxides in automobile exhaust, which combine with water to produce nitric acid. Once these pollutants enter the atmosphere, their effects can spread to nearby and even distant environments.

(see Chapter 5); thick layers of organic-rich soils absorb acids harmlessly. But where thin, organically poor soils overlie granitic bedrock, as in the lake and forest country of the Canadian Shield area of the northern Great Lakes and adjacent parts of southeastern Canada, acid rain damage to the ecosystem can be severe. Eastern Canada receives acid rain caused by the coal-burning smokestacks of the heavy steel and automotive industries of the American Midwest; attempts to neutralize it feature low-flying aircraft that spread limestone dust across the landscape.

Global Warming Burning all types of fossil fuels increases the volume of carbon dioxide (CO_2) in the atmosphere. Measurements made from 1958 to 1984 at the top of Hawaii's Mauna Loa, far from industrial pollution sources and large population centers, showed a global atmospheric CO_2 increase of about 9%. As discussed in Chapters 5 and 17, atmospheric CO_2 absorbs and traps heat from the Earth's surface, creating what is known as the greenhouse effect and causing warming. Some climatologists predict that if atmospheric CO_2 levels continue to rise at their present rate, atmospheric temperatures could rise 1.5° to 4.5°C (2.5° to 8°F) by the middle of the twenty-first century. This could cause droughts in some regions, increase desertification in others (see Chapter 18), and significantly reduce the agricultural potential of today's marginally cultivatable lands. It would also promote melting of the Earth's ice masses, which along with thermal expansion of warming seawater would cause sea levels to rise worldwide, perhaps by 1 to 2 meters per 100 years, for 200 to 500 years. A rise of 2 to 10 meters (6.5–33 feet) would accelerate coastal erosion and submerge a substantial amount of coastal land, including most major port cities on every continent.

Marine Oil Spills Every so often, an oil tanker becomes disabled and leaks its cargo of crude oil into the sea. Extraordinary environmental damage was caused by a marine oil spill in March 1989, when the supertanker *Exxon Valdez* went aground in Alaska's Prince William Sound. More than 10.2 million gallons of crude oil flowed from its cracked hull into the sea, washing up on shore in layers tens of centimeters thick. The oil killed thousands of birds and marine mammals, halted herring and salmon fishing during peak season, and coated hundreds of kilometers of Alaska's coastline (Fig. 20-10).

Figure 20-10 One problem associated with oil exploration and development is the potential for oil spills, which can cause large-scale environmental damage to oceans and coastal habitats. The negative impact on marine life and habitats can be profound, as dramatized by this oil-covered bird found following the 1989 breakup of the supertanker *Exxon Valdez* in Alaska's Prince William Sound; the effects of the *Valdez* spill are still being felt today.

Oil spills in calm seas can usually be successfully confined within floating barriers placed around the oil, which can then be skimmed from the surface. In Alaska, unfortunately, inclement weather caused rough seas that breached the floating barriers; the Exxon Company recovered only about 500,000 gallons of their oil. Efforts to ignite and burn the spilled oil failed as well, although some of the lightest hydrocarbons evaporated and some were consumed by oil-eating bacteria. But millions of gallons washed onshore, necessitating an enormous cleanup operation. Standing pools of oil were soaked up with peat moss, wood shavings, and even chicken feathers. High-powered steam hoses removed some of the oil coating; workers even scrubbed rocks individually by hand. After several months and some $4 billion in cleanup costs, more than 85% of the spilled oil was gone. The remainder consisted of thick asphalt clumps that could not be scrubbed, did not evaporate in sunlight, and would not decompose by bacterial action. These fouled breeding grounds and other wildlife habitats on land, as well as major fishing grounds.

The *Exxon Valdez* disaster dramatized the need for improved methods to respond to such events, sparked renewed congressional demands for double-hulled, spill-proof tankers, and prompted research that may some day yield a new strain of microbes that voraciously consume petroleum.

Alternative Energy Resources

As fossil fuel reserves dwindle and environmental damage due to their use increases, governments and industries seek alternative ways to meet growing energy needs. Some alternative energy sources already in use are renewable, such as solar and wind power; others rely on nonrenewable natural resources such as uranium.

Renewable Alternative Energy Sources

Renewable alternative energy resources are those that can be used virtually without depletion, or are replenished over a relatively short time span. Such resources include geothermal, hydroelectric, tidal, solar, and wind energy and the energy produced by burning such renewable organic materials as trees and agricultural waste.

Geothermal Energy Reykjavik, the capital of Iceland, is relatively pollution-free because it has a clean, inexpensive source of energy: Heat from shallow magma pooled beneath the surface is used to convert groundwater to hot water and steam. The hot water is circulated through pipes and radiators to heat homes and municipal buildings; the steam drives electric generators (Fig. 20-11a).

Since such *geothermal heat* was first tapped as an energy source in Larderello, Italy, in 1904, about 20 countries have used it, including the United States (Fig. 20-11b). More nations would use this relatively inexpensive, nonpolluting energy if they could, but unlike oil, coal, and natural gas, geothermal energy cannot be exported and must be used close to its source. Every nation using geothermal energy is located on a currently or recently active plate margin or near an intraplate hotspot, where magmatic heat has not yet dissipated.

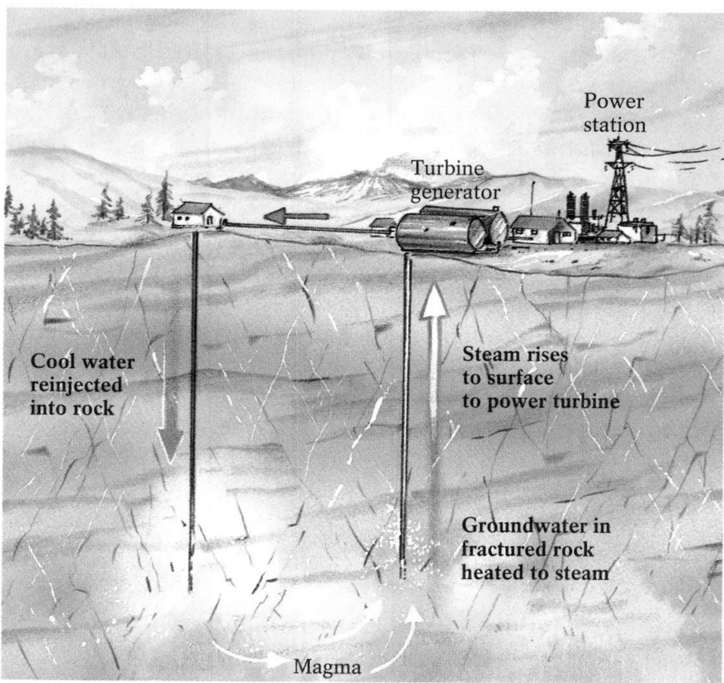

(a)

(b)

Figure 20-11 Geothermal heat provides nonpolluting energy. (**a**) Groundwater heated by shallow magma is converted to steam, which is extracted to drive turbines, generating electricity. Cooled water is generally reinjected into the system to keep the cycle going. (**b**) The Geysers geothermal plant, 140 kilometers (90 miles) north of San Francisco in Sonoma County, is the world's largest geothermal operation. The plant uses heat from subterranean rocks warmed by recent volcanic activity to supply the energy needs of 500,000 homes in the Bay Area.

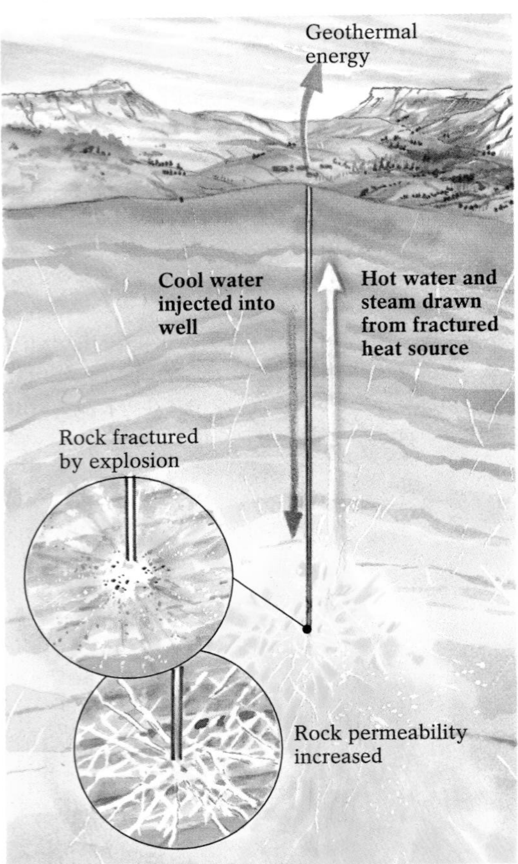

Figure 20-12 Geothermal energy can be harnessed even in arid regions, by injecting imported water into warm rocks fractured by explosives.

In some areas, such as the American Southwest, where there is significant subterranean heat but little water to transfer it to the surface, experimental "dry-rock" geothermal projects are being attempted. Warm, shallow rocks are fractured by explosives to increase their permeability, then imported water is injected into the fractures, creating hot water or steam (Fig. 20-12). In the Jemez Mountains of north-central New Mexico, water is injected to a depth of 10 kilometers (6 miles) into young volcanic rocks whose temperature is about 200°C (400°F); the resulting steam drives turbines and produces electricity.

Hydroelectric Energy Falling water has been used for centuries as an energy source to mill flour, saw logs, and power numerous machines. Today, hydroelectric facilities use falling water to produce electricity. To generate hydroelectric power, a high-discharge stream is impounded by a dam to create enough vertical drop to rotate the blades of large turbines.

Hydroelectric power is widely available; since 1983, nearly one-third of all new electricity-generating plants in the United States have been hydroelectric installations. At present, however, the United States generates only about 15% of its current electricity output this way. The Federal Power Commission estimates that if every sizeable river in the United States were used, hydroelectric power could supply 50% of our total electricity needs. Global hydroelectric development lags even further; only about 6% of the world's hydroelectric potential is being used, and in South America and Africa, where potential is greatest, only 1% has been developed. Canada, at the other extreme, gets 75% of its electricity from this clean resource.

Although hydroelectric power is nonpolluting, dams can disrupt the local ecological balance by altering or destroying wildlife habitats. In addition, they impede natural erosion processes; their reservoirs eventually fill with sediment. Decisions to build dams must balance their environmental costs against their energy yield.

Tidal Power In coastal areas with a high tidal range—the difference in the water surface level between high and low tide—energy from rising and falling water levels can be harnessed by building a dam across a narrow bay or inlet. The dam's gates are opened during rising tides and then closed to trap the water at its maximum height. The elevated water is then channeled seaward through turbines connected to electrical generators, producing renewable, pollution-free energy (Fig. 20-13). The world's largest tide-powered plant, the Rance River project in France, provides virtually all the electricity used by the French province of Brittany.

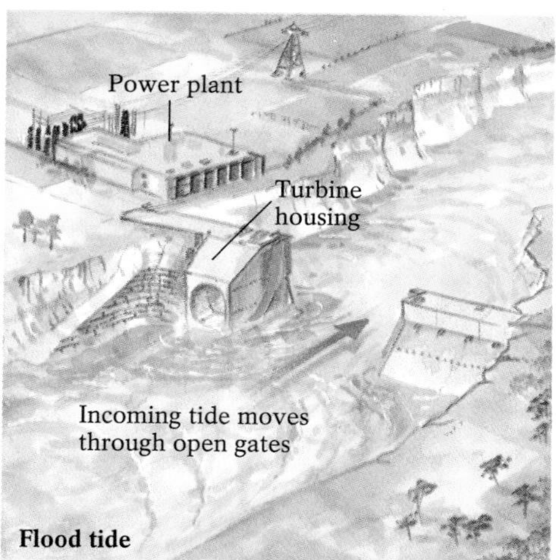

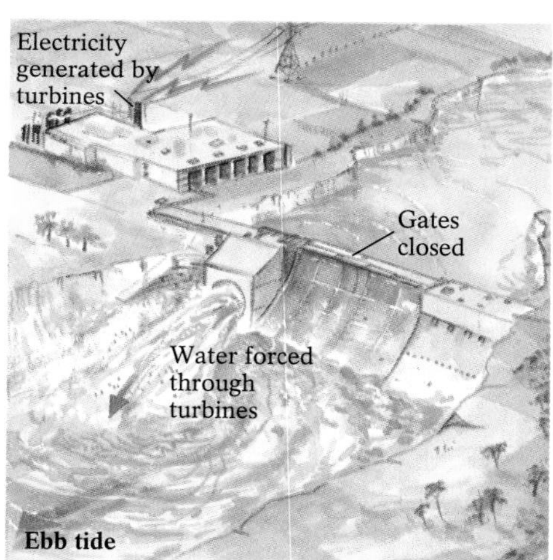

Figure 20-13 Producing electricity through tidal power. Tidal power taps the energy of falling water by trapping water at high, or flood, tide and then releasing it seaward through electricity-producing turbines at low, or ebb, tide.

Tidal power, however, requires a minimum tidal range of 8 meters (26 feet), and disturbs the ecology of estuarine habitats. There are as yet no tidal-power facilities in North America, although Passamaquoddy Bay in northeastern Maine, with a tidal range of 15 meters (50 feet), is a strong candidate for future development, as is the Bay of Fundy in New Brunswick, Canada, with a tidal range of 20 meters (65 feet). Maximum development of the United States' potential tidal power would provide only 1% of our total electricity needs, although it could become a significant supplement to other energy sources in some localities. Worldwide potential, only slightly better, is about 2%.

Solar Energy Solar-powered pocket calculators and wristwatches use an energy source that requires no expensive drilling or destructive strip mining, cannot be monopolized by unfriendly political regimes, and produces no hazardous wastes or air pollution. The Sun, expected to shine for another 5 billion years or so, is a totally renewable energy source. Solar power can be used to heat buildings and living spaces and to generate electricity, energy needs that together account for two-thirds of North America's energy consumption.

Solar heating can be passive or active. Passive solar heating distributes the heat naturally by radiation, conduction, and convection. At mid-northern latitudes, the simplest way to heat spaces passively is to construct buildings with some windows facing south. Sunlight passes through the window glass and heats objects within the room; their heat then radiates to warm the air (Fig. 20-14a, following page). Such an architectural design, coupled with efficient insulation, sharply reduces both air pollution and the cost of heating with fossil fuels.

Active solar heating uses water-filled, roof-mounted panels with black linings to absorb maximum sunlight. The solar-heated water is circulated throughout the building for space heating, or directly to the building's hot-water system (Fig. 20-14b). Solar panels are most productive in mild, sunny climates such as Florida, Texas, the Southwest, and California, where they can provide as much as 90% of a building's heating needs. Even in colder regions, such as northeastern North America, where solar panels are less productive, they can still significantly reduce the need for other energy sources.

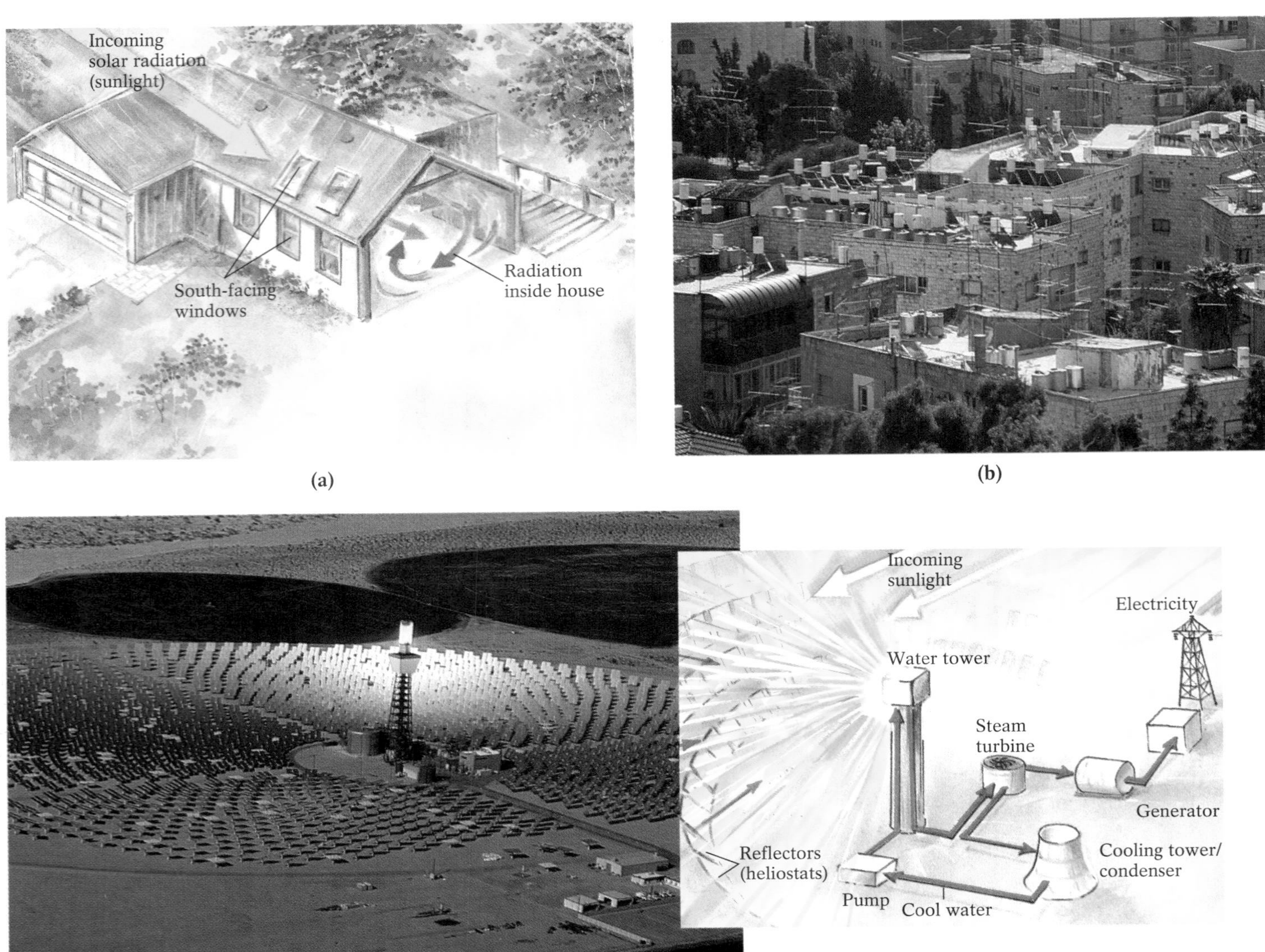

Figure 20-14 There are several ways to use solar power. (**a**) Solar energy passively heats a single dwelling. (**b**) Water-filled panels on rooftops in Jerusalem, Israel, provide hot water and space heating. This is a type of active solar heating. (**c**) This facility near sunny Dagget, California, uses solar energy to generate electricity. Reflecting mirrors focus on a water tower, concentrating the Sun's energy and converting the water to turbine-driving steam.

Solar energy can also generate electricity. An array of many mirrors reflects sunlight onto a large water tower; the water is heated to create steam, which drives turbines attached to electrical generators (Fig. 20-14c).

Electricity can be generated by solar energy more efficiently in some regions than in others, and the technology to maximize its effectiveness at low cost is not yet widely available. Current U.S. electricity needs would require a system of collecting mirrors that would occupy about 25,000 square kilometers—about one-tenth the size of the state of Nevada—which is clearly unfeasible. Substantial utilization of solar energy for the generation of electricity remains decades away.

Wind Power Like falling water, the motion of the tides, and sunlight, wind power is a clean, renewable, nonpolluting energy source that has long been in use. The picturesque windmills of the Netherlands and those of rural midwestern North America have pumped groundwater and powered sawmills and flour mills for centuries. But wind power is not cost-effective on a large scale: Only in a few sparsely populated mountain passes do winds blow constantly, forcefully, and from a consistent direction—all requirements for a practical application.

Figure 20-15 The productivity of this wind farm at Altamont Pass, California, is made possible by sustained westerly winds that buffet California's Sierra Nevada.

During the energy crisis of the 1970s, when fuel costs rose due to curtailed supplies of imported oil, engineers in the Department of Energy studied wind power as a possible way to decrease U.S. energy dependence. In the 1980s, a pilot project at Altamont Pass, east of San Francisco, connected 2000 wind turbines to electrical generators (Fig. 20-15). Altamont has contributed substantially to the Bay Area's electricity needs, and California's government plans to "harvest" as much as 8% of the state's electricity from wind "farms" by the year 2000.

Drawbacks to wind power include its limited geographic application and the extent of the area needed to develop it significantly. In North America, the wind force in the East and industrial Midwest does not meet minimum requirements, nor do those densely populated areas have sufficient available land on which to locate wind-power facilities. The gusty Great Plains of Nebraska, Kansas, Oklahoma, and Colorado are possible candidates, although California, with its large, undeveloped, windy passes remains the best possibility.

Biomass In developing countries, as much as 35% of the energy used for cooking and heating comes from burning wood and animal dung. These fuels, derived from plants and animals, are known collectively as **biomass fuels.** Biomass fuels also include grain alcohol, used as an additive to gasoline, methane gas that rises from decaying garbage in landfills, combustible urban trash, and plant waste from crops such as sugar cane, peanuts, and corn. The most widely used biomass fuel is wood, which today heats about 10% of North America's homes, more than are heated by electricity from nuclear power plants.

Biomass fuel is a renewable resource. Although trees grow slowly, continuous planting and harvesting can produce a steady supply. Many countries have begun to develop biomass energy sources in anticipation of the end of the era of fossil fuels. Burning trash fuels an electrical generator in Rotterdam, the Netherlands, providing efficient disposal of wastes as well as electricity for 250,000 people.

Biomass fuels, however, unlike most other renewable energy resources, can create air pollution and desertification problems when their use is widespread. As with oil and coal, burning biomass introduces noxious gases and particles into the air, reducing air quality and increasing global warming. Moreover, in arid regions (see Chapter 18), overreliance on scrub and trees for energy removes the root systems that help retain water and soil; this contributes to desertification, which consequently also eliminates animals that provide dung for fuel.

Nuclear Energy—A Nonrenewable Alternative

In the mid-twentieth century, physicists harnessed the energy released from the decaying nuclei of radioactive isotopes (nuclear fission). Now they are attempting to tap the energy produced when atomic nuclei are fused together (nuclear fusion). These energy sources may provide a twenty-first–century solution to the inevitable exhaustion of fossil fuels.

Energy from Fission When the nuclei of certain heavy elements' isotopes are bombarded with neutrons, they split into several lighter elements, releasing part of the binding energy of an atom's nucleus, additional neutrons, and an enormous quantity of heat. This is **nuclear fission.** The heat can be used to convert water to steam, driving turbines and producing electricity. The released neutrons can be used, in turn, to bombard other heavy nuclei, setting off a chain reaction.

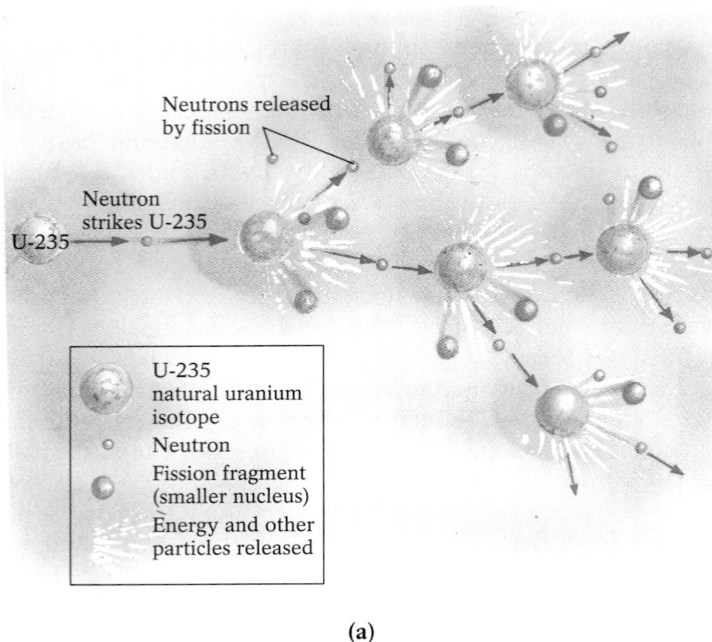

Figure 20-16 (a) A neutron released by the natural decay of a U-235 nucleus initiates a chain reaction. (b) Uranium fuel is contained within 12 meters (40 feet) of water at the Indian Point 2 nuclear power plant in Buchanan, New York.

In commercial nuclear reactors, naturally decaying uranium-235 nuclei release neutrons that bombard other U-235 nuclei, which undergo fission (split), producing heat and, most important, more neutrons, which then bombard and split neighboring U-235 nuclei (Fig. 20-16a). Nuclear reactors are designed to control the flow of neutrons to the fissionable material so that uncontrolled chain reactions (as in a nuclear bomb) do not occur. Neutron-absorbing control rods are moved in and out of the target uranium fuel to regulate the rate of reactions. Reactors and their components are lined with layers of a neutron-absorbing material such as graphite to prevent excess neutrons from reaching the fuel. The enormous amount of heat generated by nuclear reactions is transferred to a coolant, usually water, which is converted to steam to drive turbines (Fig. 20-16b).

Uranium, a relatively uncommon element in the Earth's crust, becomes concentrated in fluids of granitic magmas in the late stages of cooling; thus it may occur in pegmatites (discussed in Chapter 3) as a crystalline uranium oxide, such as the mineral uraninite (UO_2). Uranium is highly soluble in water, so it readily dissolves when igneous rocks are weathered. It is then transported by groundwater into permeable sediments and sedimentary rocks, where it bonds to the surfaces of clay particles and organic matter. These concentrations are the source of most uranium, which is mined from ancient stream sands and gravels, either in granular form as the black, shiny, noncrystalline uranium oxide pitchblende, or as a canary-yellow encrustation on sand grains, the potassium–uranium–vanadium mineral carnotite.

In North America, uranium deposits are found in the Mesozoic stream gravels of Colorado, New Mexico, Wyoming, and Texas, often associated with fossilized plant and wood remains. (The uranium in circulating groundwater bonds with and becomes concentrated in plant cells.) During the uranium boom of the 1950s, western prospectors toting Geiger counters tested tens of thousands of petrified logs. In Canada, near Great Bear Lake in the Northwest Territories, there are important uranium deposits in ancient organic-rich stream gravels.

Uranium-235, the only naturally occurring radioactive isotope that can maintain a nuclear chain reaction, is consumed during fission. Uranium-235 accounts for only 0.7% of natural uranium, whereas nonfissionable U-238 makes up 99.3%. At its current rate of use, recoverable reserves of U-235 may be sufficient to power the world's 575 nuclear power plants for only another 30 years or so.

To address the issue of dwindling U-235 supplies, nuclear physicists are experimenting with ways to produce fissionable plutonium-239 from the abundant nonfissionable U-238. Uranium-238 is placed in a reactor with a small amount of U-235. The U-235 decays spontaneously, and its neutrons bombard the U-238 nuclei, converting them to a form of plutonium, Pu-239. These processes occur in **breeder reactors,** where fission of the manufactured plutonium atoms produces a surplus of neutrons that can then be used to create, or "breed," more fissionable material (Pu-239) than they consume (U-235).

There are today few breeder reactors, primarily because Pu-239 is a weapons-grade nuclear material. Funding for the Clinch River breeder reactor in Tennessee was canceled in the mid-1980s because of public concern about proliferation of nuclear weapons. The Super Phenix breeder reactor in France is the world's largest functional breeder facility.

In 1974, the U.S. Geological Survey predicted that by the year 2000, the United States would get 60% of its electricity from nuclear plants; in 1993, the figure was about 16% and dropping. Meanwhile, western Europe and Japan, with little indigenous fossil fuel, were expanding their nuclear-energy facilities; France, for example, derives 65% of its electricity from its many reactors. There are a number of reasons for the reluctance to develop nuclear energy in the United States. In addition to concern about the spread of weapons-grade fuels, there are technological problems involving reactor safety and radioactive-waste disposal, and numerous other economic, political, and psychological problems associated with the growth of nuclear energy.

At Three Mile Island, Pennsylvania, in 1979, an instrument malfunction led plant operators to conclude that too much water was flushing through the reactor's cooling system. They responded by reducing water flow, leaving the reactor core uncovered for several hours and allowing it to overheat to critical temperatures, although little measurable radiation escaped. At Chernobyl, USSR, in 1986, technicians accidentally allowed a runaway chain reaction to develop. Two small explosions blew the roof off the building, showering the immediate area with radioactive material. Eastern Europe and Scandinavia were covered with a cloud of radioactive steam. Numerous cases of radiation sickness and cancer developed in Chernobyl's immediate vicinity and agriculture and commerce were disrupted over a wide area.

Because radioactive nuclear waste is so toxic that it must be isolated from all life for thousands of years, safe disposal is the industry's greatest technological problem. Spent fuel rods, internal machinery from reactor cores, and the waste products of nuclear-fuel processing must be disposed of in a way that prevents any leakage to the atmosphere or groundwater system. A burial site must be seismically stable—an earthquake could damage the facility or a new fault could breach the repository—and completely isolated from groundwater. In 1991, the U.S. Congress selected Yucca Mountain in southwestern Nevada as the first U.S. nuclear-waste repository (Fig. 20-17).

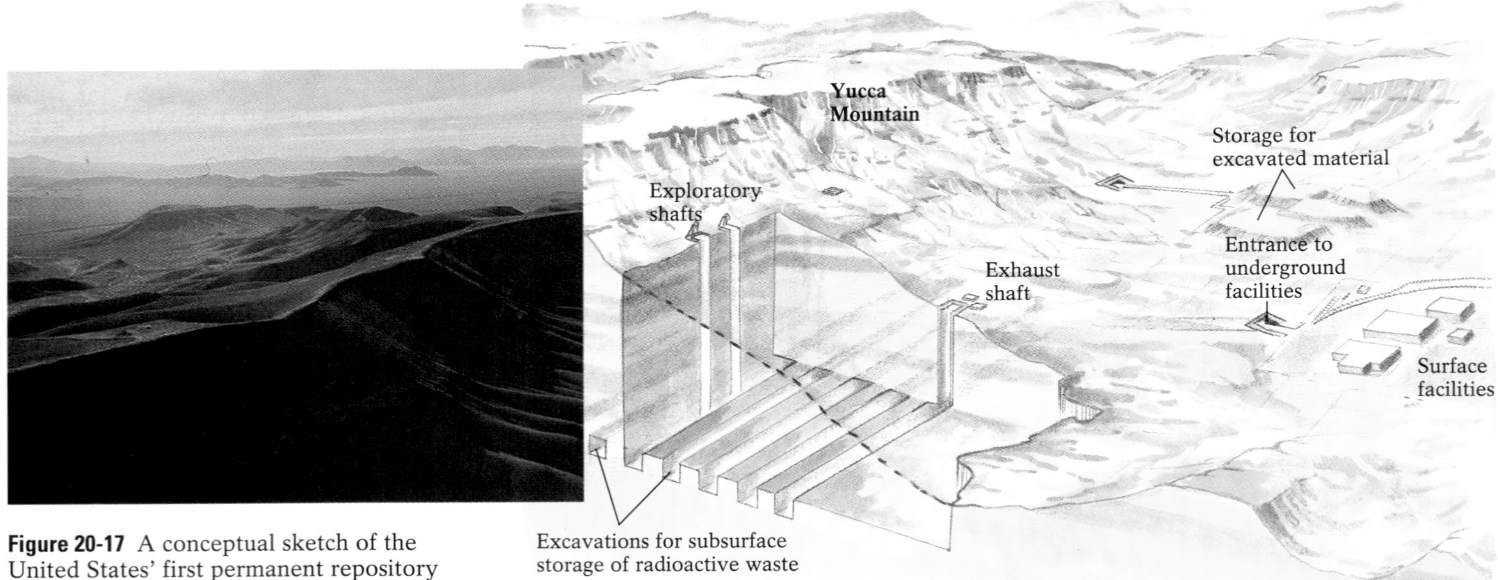

Figure 20-17 A conceptual sketch of the United States' first permanent repository for high-level nuclear waste at Yucca Mountain, Nevada. When preparation of the site is completed in 2003, radioactive waste from around the country will arrive in 3-meter (10-foot)-long containers and be transferred to a cavern excavated into a thick layer of relatively impermeable ash-flow tuff, 300 meters (1000 feet) below the surface. Inset: The ash-flow tuff of Yucca Mountain.

Located within the Nevada Test Site (where the first atomic bomb was tested), the area is largely uninhabited, has been relatively free of earthquakes during recorded time, and is so arid that groundwater flow is very deep. Studies expected to last until the end of the century are being conducted to confirm the safety of this site before it can become operational.

The nuclear-energy industry also faces the very high expense of constructing reactors, a lengthy process for obtaining permits, the potential for theft of weapons-grade plutonium and terrorist attacks on installations, and the possibility of accidents and sabotage when wastes are transported. There is also the psychological issue of public acceptance and confidence in nuclear power, especially since the accidents at Three Mile Island and Chernobyl. After these accidents, strenuous local opposition on Long Island, New York, prevented a nuclear-energy plant in Shoreham, 90 kilometers (55 miles) east of New York City, from coming on line.

Energy from Fusion **Nuclear fusion** occurs when extremely high pressure and temperature cause the atomic nuclei of ultra-light elements to fuse together to form heavier atoms. Like fission, this process releases an enormous amount of heat energy that can be used to convert water to steam, driving turbines to produce electricity. The principal fuel of fusion, however, is hydrogen, one of the most abundant elements, and rather than toxic radioactive waste, the product of fusion is helium, a harmless inert gas.

Fusion's potential is enormous: The energy that could be generated by the hydrogen in one cubic kilometer of seawater exceeds the total energy stored in the world's oil reserves. In recent experiments, heavy isotopes of hydrogen (tritium (H_3) and deuterium (H_2)) were compressed in a powerful magnetic field and then heated by intense laser bursts to about 50 million °C, producing the heavier element helium (He) and an enormous amount of heat. This is the same reaction that powers the Sun's thermal furnace. Sustained small-scale hydrogen fusion is now possible, although it is not yet economically feasible because the amount of energy needed to achieve fusion still exceeds the amount produced by it. Further study and expensive experimentation will be needed to make it commercially viable.

Mineral Resources

Minerals, as we saw in Chapter 2, are naturally occurring inorganic solids consisting of chemical elements in specific proportions whose atoms are arranged in a systematic internal pattern. Of the 3000 or so known minerals, only a few dozen are economically valuable. These include a variety of metallic minerals, such as the oxides and sulfides of iron, copper, and aluminum; certain metals, such as gold, silver, and platinum, that commonly occur uncombined with other elements; and various nonmetallic rock and mineral materials, such as sand and gravel (used as building materials), limestone (for cement), halite (common table salt), and several highly prized gemstones (such as diamonds, rubies, and emeralds).

Metals

Most metals combine readily either with oxygen to form oxides, or with sulfur to form sulfides. Common oxides include those of tin, uranium, iron, and aluminum; common sulfides include those of zinc, lead, iron, copper, and molybdenum. *Native* metals, such as gold, platinum, and silver, do not combine with other elements.

The first known human use of metals, about 9000 years ago in what is now Turkey, was to hammer naturally pure copper into amulets, tools, and weapons. By 6000 years ago, metallurgists in Europe and Asia Minor were separating metals from their host rocks by *smelting;* this was accomplished by heating rocks to the melting points of their incorporated metals and collecting the molten metal for further processing. The earliest smelting operations extracted copper from copper sulfides such as chalcocite (Cu_2S) and chalcopyrite ($CuFeS_2$). By 5000 years ago, lead, tin, and zinc, as well as copper, were being smelted and combined in their molten state. This resulted in **alloys,** or metal mixtures, that were harder than their component metals, and maintained sharper points and edges. Bronze, a mixture of copper and tin, was a common alloy that dominated the tool-and-weapon industries that flourished between 5000 and 2650 years ago; its widespread use characterized the archaeological period known as the Bronze Age. By about 2650 years ago, metallurgists had learned to smelt and work iron, which quickly replaced bronze as the principal metal used in the manufacture of tools, weapons, and armor, ushering in the Iron Age. The demand for metals has grown ever since.

Most metals are scattered thinly through the Earth's crust; extraction is economically feasible only where various rock-forming processes have gathered them into mineable concentrations. These processes include precipitation from hot, metal-rich water, settling of early-forming crystals from crystallizing magma, chemical interactions of hot fluids during metamorphism, separation of dense metallic particles during sedimentation, oxidation and precipitation (of iron), dissolution of surrounding minerals, and precipitation from circulating groundwater.

Processes That Concentrate Metals **Hydrothermal deposits** form by precipitation of metallic ions from hot ion-rich water. Gold, for example, which constitutes only about 0.0000002% of the Earth's crust, can be mined profitably only where circulating hot water has dissolved it from its source rocks and precipitated it elsewhere in highly concentrated form. The rich gold deposits of California's Sierra Nevada and along Cripple Creek in the Colorado Rockies precipitated from hydrothermal solutions, as did the silver deposits of Coeur D'Alene, Idaho, the copper of Butte, Montana, and the Keweenaw Peninsula of northern Michigan, and the lead-zinc deposits of the tri-state area of Missouri, Oklahoma, and Arkansas.

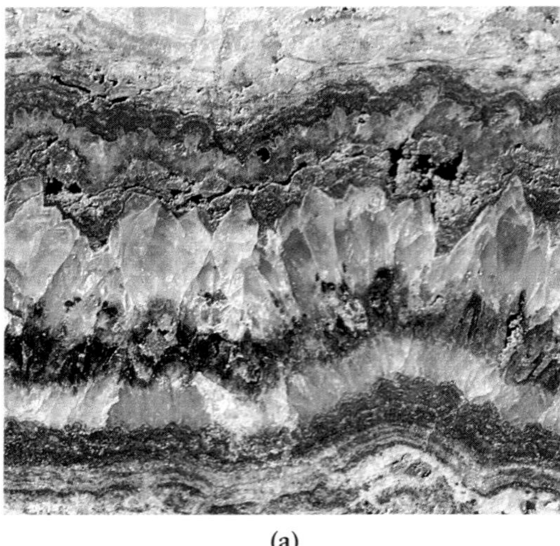

(a)

(b)

Figure 20-18 Hydrothermal processes that concentrate valuable metals. Metals may precipitate from magmatic water or from circulating groundwater or seawater heated by the magma. (**a**) These veins of gold and silver ore were formed when hot magmatic solutions of the metals infiltrated cracks around the cooling magma. (**b**) These massive sulfide deposits in Quebec formed at a mid-ocean ridge when seawater heated by basaltic magma precipitated large quantities of metals on the ocean floor.

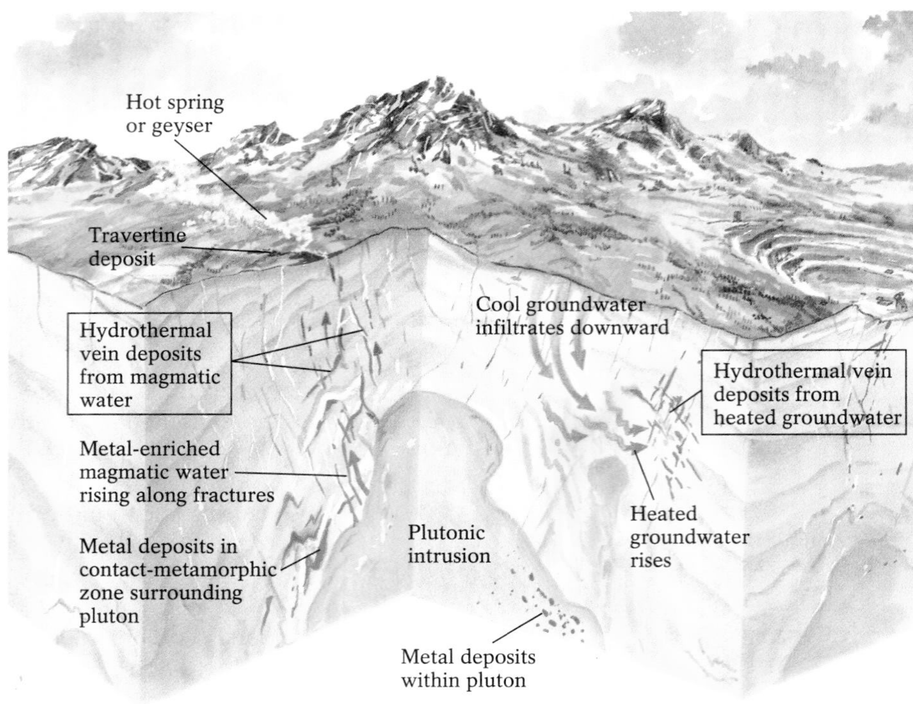

A common source of metal-rich hot water is a cooling magma that contains metallic ions left over after most silicate minerals have crystallized (see Chapter 3). Such hot solutions may contain copper, lead, zinc, gold, silver, platinum, and uranium, none of which bond readily with silicate tetrahedra, as well as some uncrystallized silica (SiO_2). The hot water typically infiltrates faults, fractures, and bedding planes of surrounding rocks, where it cools further and deposits its solutes as silica-rich mineral *veins* that may contain visible masses of metals (Fig. 20-18a). Some of the world's most productive gold and silver mines are hydrothermal vein deposits that cross-cut rocks adjacent to granitic batholiths.

Porphyry copper deposits (which form in plutonic rocks with porphyritic textures, discussed in Chapter 3) form when the water expelled from a cooling body of water-rich felsic magma is converted to pressurized steam. As it expands outward, the steam ruptures the newly crystallized rock at the magma's edge, creating numerous hairline fractures at the margin of the batholith. Hydrothermal solutions then fill these fractures, precipitating copper and other metals as they cool.

Hydrothermal deposits also originate from deep-circulating groundwater that has been heated by close contact with shallow magmas or young, warm, plutonic rocks. As the heated water rises, it dissolves metals from the pluton and surrounding rocks; as it flows into fractures, joints, bedding planes, and faults, it precipitates the metals as concentrated vein deposits. Other hydrothermal solutions pass through porous rocks and cool slowly, precipitating their metals over large areas that are typically not cost-effective to mine.

Massive sulfide deposits form where cold seawater enters the fractures associated with divergent zones, where it is heated by shallow basaltic magma. As the water rises through the fractures, it dissolves some of the trace metals in the basaltic mid-ocean ridge, and also picks up the sulfurous gases typical of basaltic eruptions. The resulting hydrothermal solution contains sulfides of copper, zinc, manganese, lead, and iron (Fig. 20-18b). When the solution erupts at the sea floor, it cools instantaneously, precipitating its metals in a black, jet-like plume called a *black smoker* (see Fig. 12-14). The "smoke" is actually fine, black particles of sulfide minerals such as pyrite (FeS_2), chalcopyrite ($CuFeS_2$), sphalerite (ZnS_2), and galena (PbS_2).

Figure 20-19 Deposits of chromite (black) and platinum (white) in the Bushveld complex of southern Africa. In addition to its vast deposits of chromite, the Bushveld complex contains 70% of the world's platinum, one of the most precious metals. It is a huge sill, 240 kilometers (150 miles) by 480 kilometers (300 miles) in area and 8 kilometers (5 miles) thick, in which gravity settling of heavy early-forming crystals has occurred.

The processes by which minerals crystallize from magma (discussed in Chapter 3) also produce metal deposits. In *gravity settling,* dense, early-crystallizing minerals sink to the bottom of the magma chamber, where they accumulate in layers; in *filter pressing,* large, early-forming crystals remain when tectonic forces compress a magma chamber and force the still-liquid portion of the magma into fractures in adjacent rocks, causing the large crystals to become concentrated against the wall of the magma chamber. Gravity settling and filter pressing of mafic and ultramafic magmas have produced valuable ore bodies of iron (in magnetite, Fe_3O_4), chromium (in chromite, $FeCr_2O_4$), titanium (in ilmenite, $FeTiO_3$), and nickel (in pentlandite, $(Fe,Ni)_9Si_8$)). The rich chromite deposits of the Stillwater intrusive complex of northwestern Montana and the Bushveld complex of southern Africa formed from gravity settling of chromite crystals from mafic and ultramafic magmas (Fig. 20-19). The iron ore of the Kiruna district of northern Sweden (60% iron) is the result of filter pressing of magnetite crystals. The world's largest nickel deposit is at Sudbury, Ontario, near the shores of Lake Superior. It probably formed as droplets of nickel sulfide rose from the Earth's mantle and filtered through a ring-shaped system of fractures believed to have been caused by the impact of an enormous meteorite about 1.9 billion years ago.

Felsic magmas are a potential source of valuable ore bodies as well. Late in the cooling of granitic magma, a substantial amount of water remains along with uncrystallized silica (SiO_2) and some rare elements that have not yet crystallized. These may include lithium, beryllium, boron, uranium, fluorine, and cesium. In this highly fluid residual magma, unbonded ions migrate freely and bond readily to growing crystal structures. The resulting pegmatites typically contain very large crystals (see Chapter 3). Pegmatites at Kings Mountain, North Carolina, contain feldspar crystals the size of two-story houses, and those in the Black Hills of South Dakota contain crystals of the lithium-rich mineral spodumene as large as telephone poles. Some pegmatites also contain precious crystals of beryllium-rich emerald and aquamarine and boron-rich tourmaline.

When hot, ion-rich fluids move through rock bodies, their heat can produce contact-metamorphic mineral alterations in the host rock, and metallic ores often result. For example, when hot, ion-rich fluids enter impure limestones and dolostones containing aluminum-rich clay, chemical reactions in the contact zone release CO_2, which migrates outward, and produce an extensive metamorphic aureole (see Chapter 7) containing corundum (Al_2O_3) and other metals. A large amount of the calcium in limestone may also be replaced by metallic ions from the hot fluids; in the vicinity of a basaltic intrusion, the replacement ions are typically iron. Other metallic deposits produced by contact metamorphism include zinc (in the mineral sphalerite), lead (in galena), copper (in chalcopyrite and bornite), and iron (in magnetite).

Sedimentary processes also produce a range of valuable metal deposits, from the gold nuggets and gem crystals that settle out from the slow-moving segments of rivers to the vast iron deposits precipitated on the floors of ancient oceans. Water-borne heavy materials such as gold, platinum, and tin are sorted out during transport by their density and durability, becoming concentrated as **placer deposits** wherever flowing water slows down. Placer deposits commonly occur in potholes in the stream bed, inside meander bends, downstream from constrictions in the channel, at the confluence of two or more streams, and at coasts, where they are too heavy to be moved by wave action. Extremely durable minerals, such as diamonds, sapphires, rubies, and emeralds, also form placer deposits because they resist abrasion and can survive long journeys that might wear away softer minerals (Fig. 20-20).

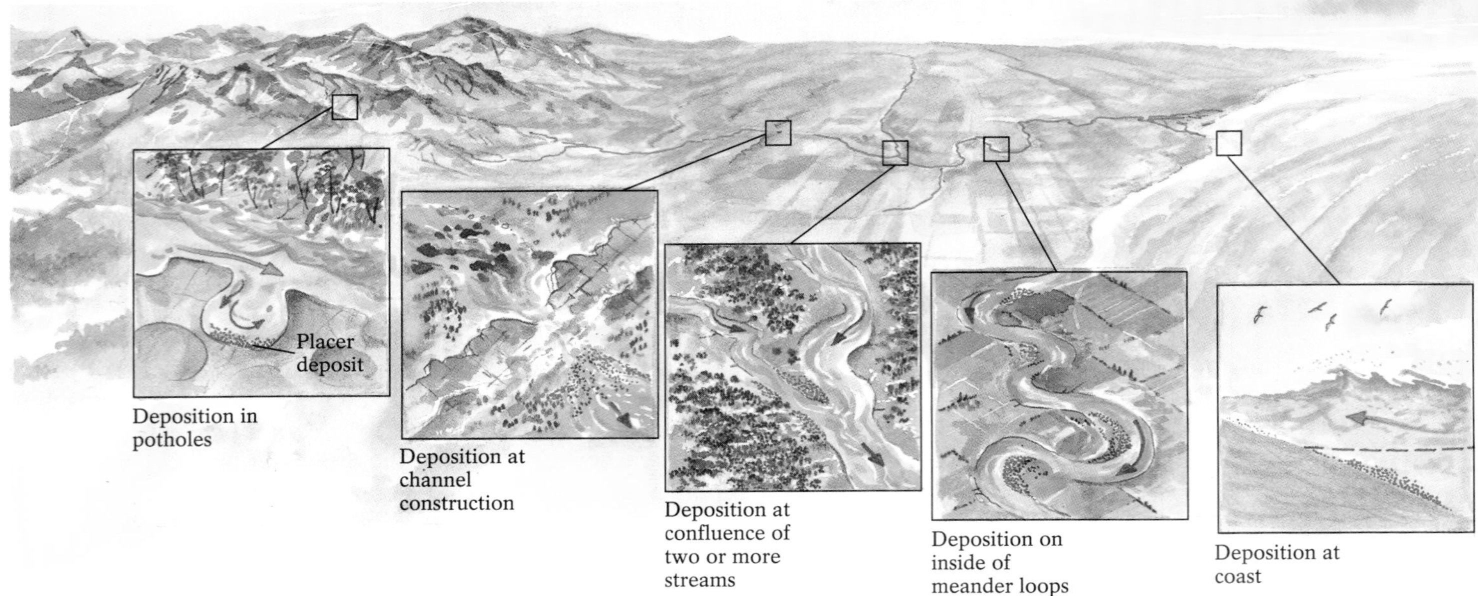

Figure 20-20 Placer deposits develop wherever the velocity of flowing water is significantly reduced, such as at potholes in a stream bed, at the inside of a meander bend, downstream from a constriction in a channel, at the confluence of two or more streams, and at coasts, where dense minerals settle from the water and are reworked by wave action. (**a**) Placer deposits of emeralds being mined in Colombia. (**b**) Emerald merchants in Colombia show their wares.

The most valuable deposits are of gold, the high density of which (19 grams per cubic centimeter) causes it to sink from its transporting stream. Nuggets of gold weathered and eroded from the hydrothermal vein deposits in California's Sierra Nevada batholiths have been carried down the range's numerous streams for hundreds of thousands of years. Placer deposits are generally discovered first; lucky prospectors may then be able to retrace their journey upstream to the "mother lode" from which the placers eroded.

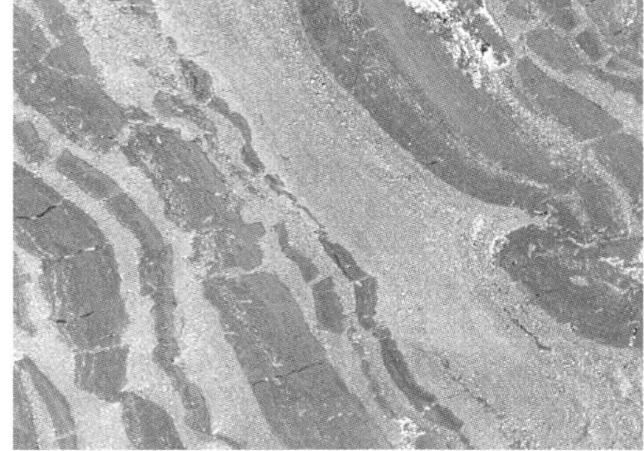

Figure 20-21 This banded iron formation began to form about 2 billion years ago when enough oxygen from photosynthesizing plants accumulated in the Earth's atmosphere to promote oxidation and precipitation of iron.

Most sedimentary iron deposits are found in a unique series of layered Precambrian sediments that were deposited in marine basins on every continent about 2 billion years ago. These **banded iron formations** consist of alternating layers of light-colored recrystallized chert and dark-colored highly concentrated iron oxides (Fig. 20-21), and are found in Labrador, eastern Canada, Brazil, western Australia, Russia, India, the western African nations of Gabon, Mauritania, and Liberia, and around Lake Superior in Minnesota, Wisconsin, and Michigan. Two billion years ago, when these sediments were deposited, vegetation (such as microscopic algae) had become plentiful enough to produce a significant amount of oxygen through photosynthesis. The oxygen content of the Earth's atmosphere and surface waters, which had been very low, began to increase, and iron in solution was able to precipitate in the form of such oxides as magnetite and hematite. Iron is very soluble in an oxygen-poor environment, and thus any iron that weathered from continental rocks prior to this time would have remained in solution while being

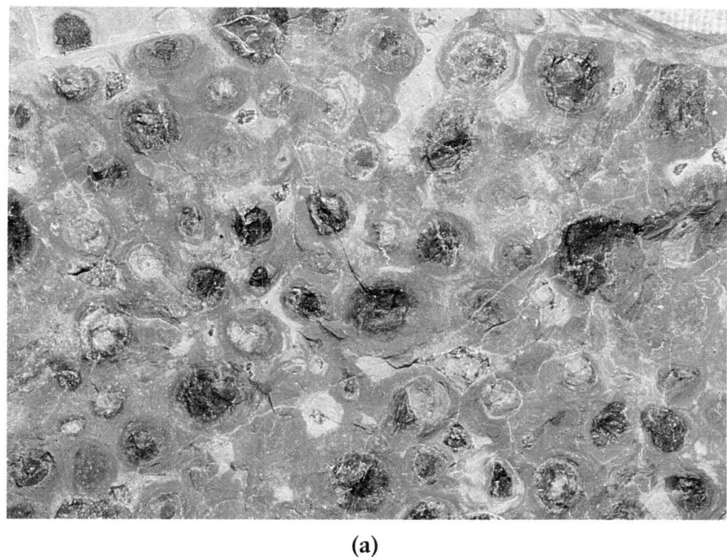

(a)

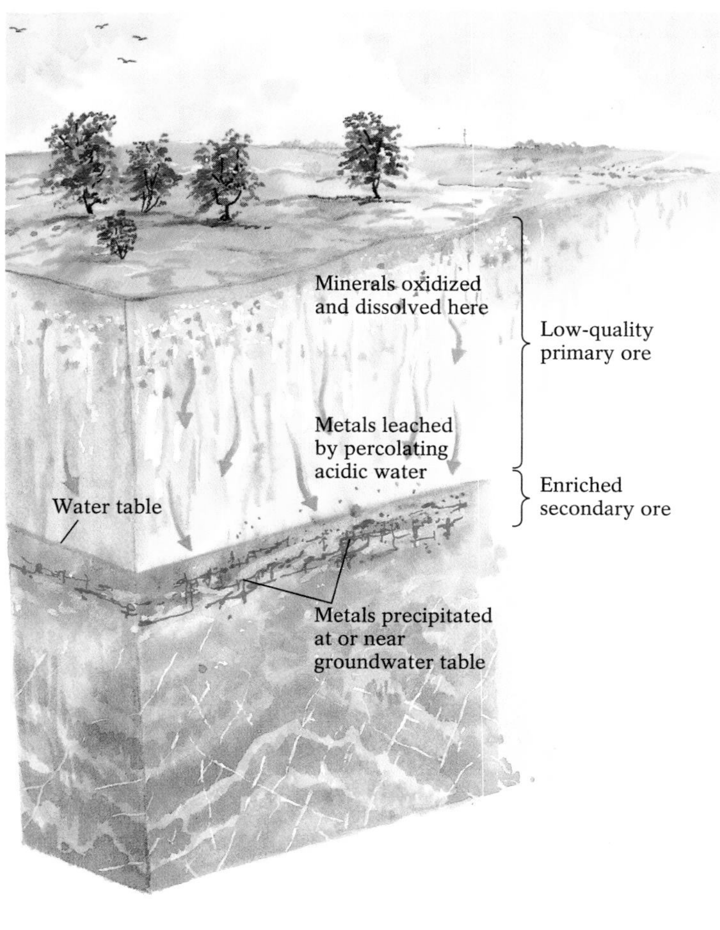

(b)

Figure 20-22 Secondary enrichment occurs when groundwater either (**a**) dissolves the matrix around valuable metals, leaving them as a residual deposit, such as this Arkansas bauxite, or (**b**) dissolves the metals themselves and transports them in solution down to the water table. Copper is particularly susceptible to secondary enrichment.

transported to the oceans. Expansion of flora on Earth added oxygen to the chemical composition of the planet's air and water, producing the widespread oxide deposits that today account for more than 90% of the iron ore mined every year.

Metals can also become concentrated either when other minerals weather and dissolve, leaving a metallic residue at or just below the surface, or when metals dissolved by groundwater are precipitated elsewhere in concentrated form. These **secondary enrichment** processes require a warm climate, abundant atmospheric water, permeable bedrock (to promote groundwater flow), and a soluble matrix surrounding economically valuable materials.

The most prominent weathering-produced ore is bauxite ($Al_2O_3 \cdot nH_2O$), an aluminum oxide that is the most abundant source of aluminum. This valuable metal occurs in sizeable deposits only where the extreme chemical weathering of warm, humid climates breaks down the feldspars in granitic rocks and the aluminum-rich clays in impure limestones. Such residual bauxite is found today in tropical locations such as Jamaica, as well as in nontropical regions, such as France, Australia, and Arkansas (Fig. 20-22a), that were tropical environments in the past.

Soluble metals dissolved by acidic groundwater may be transported downward to the water table, where they may precipitate in concentrated form. For example, when sulfur is released from weathered sulfide minerals such as pyrite (FeS_2), the sulfides oxidize to produce sulfuric acid. The acid dissolves some metals, which are then transported in solution down to the less-acidic water table, where the metals are precipitated. In this way, metals originally scattered throughout a body of rock can become concentrated in the subsurface (Fig. 20-22b).

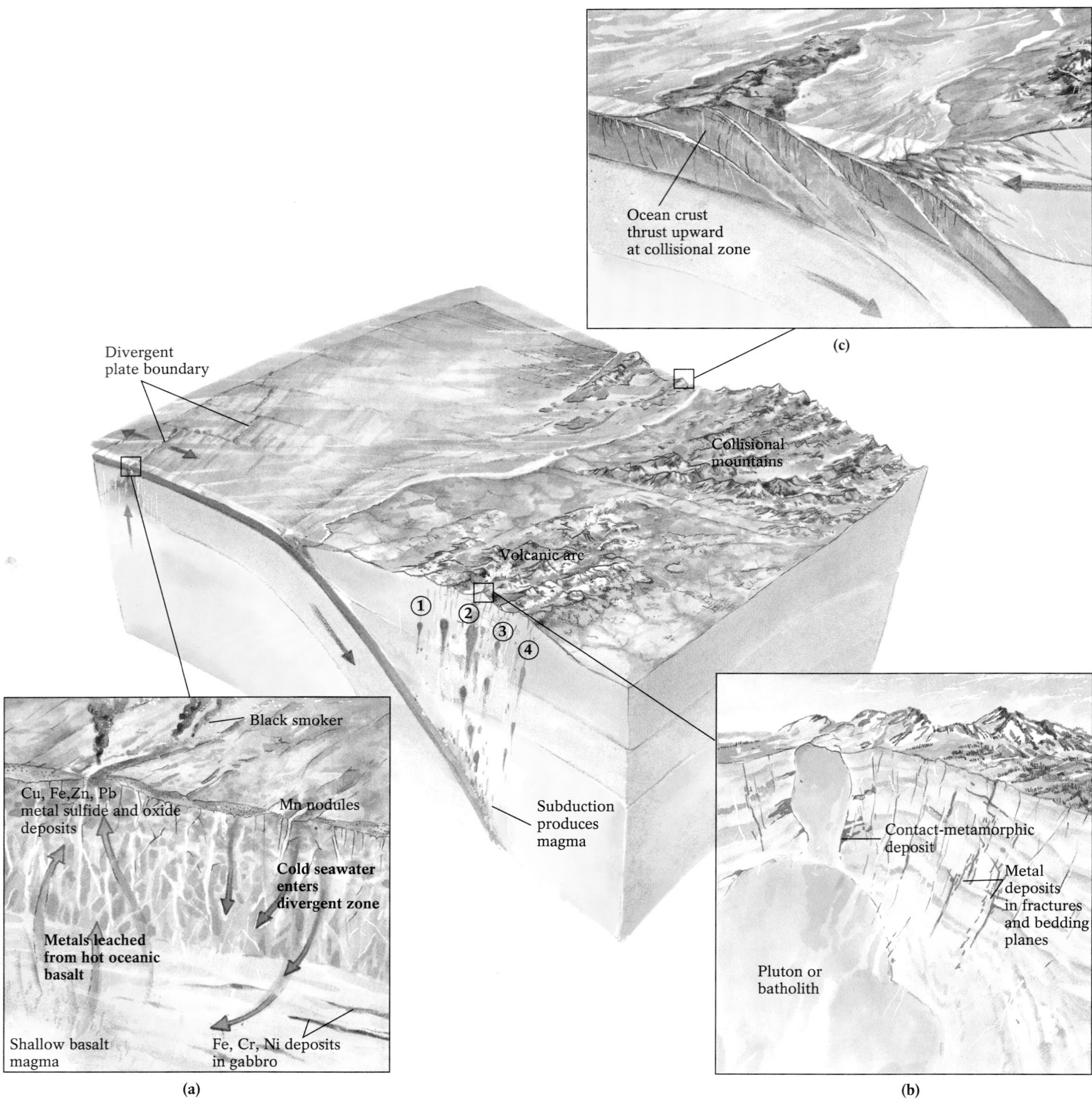

Figure 20-23 The processes that occur at each type of plate boundary (and the available materials) determine the type of metal found at specific locations. (**a**) Divergence produces massive sulfides and oxides rich in copper, lead, zinc, manganese, iron, nickel, and chromium. (**b**) Subduction separates metals by melting point into distinct bands: 1) The first to melt are iron and apatite, which form contact-metamorphic deposits; 2) a little farther inland, gold and porphyry copper deposits, associated with granitic intrusions, are found; 3) in the next band of deposits, melting at somewhat greater depths, are hydrothermal veins of lead, zinc, silver, and copper; 4) tin and molybdenum are the last to melt and are the metals found farthest inland at a subduction zone. (**c**) Deeply buried deposits formed by plate divergence or subduction may be thrust to the surface by a continental collision.

Metals and Plate Boundaries Most of the world's major metal deposits occur at past or present plate boundaries where magmas are generated, hot hydrothermal solutions circulate through rocks, and plate collisions thrust metal-rich, sea-floor rocks upward above sea level.

Massive sulfide and oxide deposits, rich in copper, zinc, lead, chromium, and manganese typically precipitate at or near mid-ocean ridges, where plate divergence opens an extensive network of fractures through which cold seawater circulates into warm oceanic lithosphere (Fig. 20-23a). In the mid-1960s, a hot salty solution mixed with a black powdery substance was collected at the floor of northwestern Africa's Red Sea, which occupies the recent rift between the African and Arabian plates. The black powder contained various sulfides and oxides of iron, zinc, and copper.

The sulfide deposits that form at divergent zones become incorporated into the oceanic lithosphere that spreads out from the mid-ocean ridge. As spreading continues, they may eventually reach a subduction zone and descend to depths where their metals are heated to their melting points. The subduction-produced magmas rise, cool, and deposit their metal content in batholiths. This process has produced the rich metal ores found at the western edges of the Americas from Alaska to Chile. Different metals appear at different positions within subduction-zone batholiths, sometimes as distinct vertical bands, because metals with low melting points melt at shallow depths (and lower temperatures), whereas those with high melting points descend to greater depths before melting (Fig. 20-23b).

After subduction consumes an oceanic plate, the extreme compression that occurs when plates collide may thrust slices of the subducted sea floor back to the Earth's surface (see Chapter 12). For this reason, continental collision zones often contain masses of metal-rich rock that originated at a mid-ocean ridge (Fig. 20-23c). The island of Cyprus, for example, has risen from the floor of the eastern Mediterranean as the African and Eurasian plates have collided. Cyprus's vast copper deposits (from which the island derives its name) have been mined for 5000 years.

Environmental Problems Caused by Metal Use Most major metal deposits are mined from open pits by operations that begin with removal of millions of tons of *overburden,* the rock and regolith that cover ore deposits. This waste material, along with the great volume of waste from ore mills, is typically piled into huge hills and often left without covering vegetation. In this form it is highly susceptible to rapid erosion and mass movement. Surface-water runoff from waste piles may pollute regional streams by clogging them with silt and producing toxicity from dissolved metals. In the mine pits, ore rocks may be in contact with groundwater, which may become contaminated with sulfides. Sulfides oxidize to form sulfuric acid, thus leading to acid groundwater.

These and other problems have prompted the U.S. Congress to pass legislation requiring monitoring and control of mine-water discharge. State and federal regulations also require land reclamation after a mining operation has been completed; this involves isolating the sulfides from the groundwater system, restoring the topography, and replanting vegetation to prevent erosion.

Nonmetals

The search for gold, silver, and platinum has unleashed stampedes of millions of prospectors and altered the histories of entire nations. Although there has been no "gravel rush" or "phosphate rush" to match the gold rushes of nineteenth-century America, such nonmetals are at least as useful and valuable as precious metals. Nonmetal resources range from the building materials derived from common rocks to the natural mineral fertilizers used to enhance our food supply.

Nonmetal Building Materials The combined economic value of natural building materials is exceeded only by that of petroleum. Limestone, the most widely used building material, provides crushed stone for road beds, cut stone for monuments and other stately buildings, and is the key ingredient in Portland cement, a staple of the construction industry. Portland cement is made by mixing finely ground limestone and shale and heating it to 1480°C (2700°F) to drive off limestone's carbon dioxide. The resulting quicklime (CaO) reacts with clays in the shale to produce a calcium slag that is then mixed with a little gypsum ($CaSO_4 \cdot 2H_2O$). Cement hardens slowly when mixed with water.

Gypsum, which is soft and soluble, is rarely used as a building stone. When heated to 177°C (351°F), however, about 75% of its water is driven off, leaving behind a powdery substance known as plaster of Paris (named for the gypsum quarries near Paris that produce high-quality plaster). When water is added to plaster of Paris and the wet mixture congeals, tiny crystals of gypsum form within the plaster, creating a firm solid used to fashion smooth interior walls and for numerous other purposes in such fields as art, dentistry, and orthopedics.

Figure 20-24 A gravel operation in glacial outwash in Ontario, Canada.

Sand and gravel are mixed with cement to form concrete, used principally in building foundations and roads. A single kilometer of four-lane highway, for example, requires about 40 tons of gravel for its concrete. Common sources of sand and gravel are river channel and bar deposits, coastal offshore bars, beach deposits, sand dunes, and glacial eskers and outwash (Fig. 20-24). As urban construction worldwide has increased in the last 25 years, demand for sand and gravel has doubled, depleting many sources. In Scandinavia, entire eskers have been removed to meet the growing need for these materials.

Clay minerals (see Chapter 5) are end-products of the chemical weathering of feldspars, and are therefore relatively stable in the Earth's surface weathering environment. When wet and plastic, clays can be shaped into a variety of building materials, such as decorative terra cotta bricks, tiles, and pipes, that harden when fired.

Various types of stone are quarried for specific building purposes. Slate is used for roofing, flooring, fireplaces, and patios; slabs of polished granite, diorite, gneiss, and other attractive coarse-grained rocks face office and government buildings; retaining walls that support unstable slopes are made of blocks of basalt; and tons of crushed limestone, granite, marble, and schist lie beneath highways as road-bed fill.

Nonmetals for Agriculture and Industry Soon after the year 2000, unless we curb population growth, the number of people on Earth will exceed 7 billion. To feed so many, using soil which has in many areas already been depleted of nutrients by overcultivation, will require a vast supply of chemical fertilizers to hasten plant growth and increase the rate of sugar and starch production. Phosphorus and potassium are two of the principal elements in agricultural fertilizers. Phosphorus is derived primarily from the mineral apatite ($Ca_5(PO_4)_3(OH,F)$), found in certain marine sedimentary rocks, in guano deposits (bird and bat droppings) in tropical caves, and in some calcium-rich igneous rocks. In North America, most phosphorus comes from ancient sedimentary deposits in Montana and Wyoming and from recent sediments along coastal North Carolina and Florida. Potassium is obtained primarily from sylvite (KCl), an evaporite deposited in shallow marine basins during periods of climatic warming. Large sylvite deposits are found in New Mexico, Utah, Colorado, Montana, and Saskatchewan.

Sulfur is used in both agriculture and industry. In the form of sulfuric acid, it is added to alkaline soils to maintain optimal conditions for plant growth. Sulfuric acid is also used to treat rubber, explosives, and wood pulp for the manufacture of paper. Some sulfur is mined in native form from the deposits that crystallize from escaping sulfurous gases around the vents of volcanoes, but most is refined from the mineral pyrite (FeS_2) or from sulfate minerals such as gypsum.

One of the most common nonmetal resources is pure quartz sand, which is melted and then cooled rapidly (quenched) to produce glass. Ancient Egyptians were the first to fabricate glass from melted sand, 4000 years ago. Today, 30 million tons of sand are quarried each year in North America from the purest quartz arenite deposits, such as the St. Peter sandstone of Illinois and Missouri, which is 95% quartz. Slight impurities in sand impart color to glass.

Another nonmetal resource with multiple industrial uses is *asbestos,* a generic term for the white, gray, and green varieties of amphibole. This nonflammable material is used primarily to fireproof clothing, in electrical insulation, and in the linings of automobile brakes. It is no longer used in the construction industry, however, because inhalation of asbestos dust has been linked to several lung and digestive-tract diseases. The asbestiform amphiboles are produced primarily by metamorphism of olivines and pyroxenes in ultramafic mantle rocks of continental collision zones (Chapter 12). There are large asbestos reserves in the high-grade metamorphic rocks of the northern Appalachians of Vermont and Quebec.

Future Use of Natural Resources

The United States has about 50 billion barrels of known oil reserves recoverable by current methods, and may have an additional, as yet undiscovered, 35 billion barrels. The current rate of use is 6 billion barrels per year, about 50% of it imported. Thus, the domestic supply, supplemented with imports, will last only about 30 years. The impending petroleum shortage will surely force many changes in the way we consume this energy source. Many of us will undoubtedly be around when the last drop of American oil trickles from the pumps.

Large new petroleum finds on American soil are unlikely; virtually all potential oil-producing rock formations have already been explored. Indeed, most of the Earth's potential oil-bearing regions have been well explored. Even if large new reserves were discovered, at the current rate of use they would be rapidly exhausted. The most recent significant domestic discovery, about 10 billion barrels at Alaska's North Slope, will cover less than two years of total U.S. oil needs. At the current worldwide petroleum use rate of about 21 billion barrels per year, the world's 700-billion-barrel reserves will last only 35 years or so. Somewhere around the year 2030, cars, buses, and trucks everywhere are likely to stop in place and rust along the side of the road, unless more effort, money, and scientific ingenuity are immediately devoted to finding ways to extend the finite crude oil supply or develop alternatives.

Some countries have already exhausted their supplies of crucial resources and are strictly importers; others are just beginning to feel the effects of critical shortages. What does the future hold for humankind as the Earth's resources dwindle?

Meeting Future Energy Needs

Substantial oil shortages and price increases in developed countries may lead to reduced consumption, thus extending the life of current supplies; widespread use of gasoline additives, such as ethanol (a form of alcohol), may reduce gasoline consumption by as much as 25%.

One way to maximize our oil reserves is to extract as much as possible from known reservoir rocks. In newly drilled wells, oil usually gushes out due to the confining pressure exerted by overlying rocks on the reservoir rocks. Collection of this first gush of oil is called *primary recovery*. As reservoir rocks become partially depleted and are no longer oil-saturated, pressure on the remaining oil is reduced and the gushing gradually stops. In the past, wells were abandoned when they no longer gushed, or were subjected to *secondary recovery* methods, such as injection of water into the reservoir rocks to buoy oil upward. Even after secondary recovery, however, one-half to two-thirds of the oil remains in the ground.

In the United States alone, there may be more than 300 billion barrels in the reservoir rocks of known oil fields, perhaps enough to provide oil for an additional 50 years of use. Old abandoned wells are now being reopened and subjected to *enhanced recovery* methods. In one such method, compressed carbon dioxide gas and superheated steam are injected to make the petroleum less viscous, so it flows more readily. In another, the petroleum is burned in place to produce electricity at the source. Enhanced recovery is more expensive than primary and secondary recovery, but as oil supplies dwindle and prices rise, it may become economically feasible.

Other technological initiatives seek alternatives to petroleum as the primary energy provider for transportation. Liquefied and gasified coal show promise as transportation fuel. Electric cars with long-distance cruising capacity, capable of superhighway speeds, are already in the planning stages at most major automobile manufacturers. Other current attempts to prepare for the inevitable disappearance of petroleum include solar-powered buses and trains propelled by powerful electromagnets (Fig. 20-25).

Figure 20-25 Magnetic-levitation trains are one alternative to petroleum-based transportation systems.

Meeting Future Mineral Needs

The supply of mineral resources remains adequate, although some are located in remote places. Valuable deposits of important metals and nonmetals are still being discovered. Major ore bodies have recently been unearthed in Chile, Australia, and Siberia. Extensive outcrops rich in zinc, lead, copper, nickel, cobalt, and uranium have also been found in Antarctica, along with a 120-kilometer (70-mile)-long, 100-meter (330-feet)-thick deposit of iron, which is large enough to meet the world's iron demand for 200 years.

Mineral prospecting today is no longer a pick-and-shovel operation. To find metals buried beneath more than 150 meters (500 feet) of sediment on the sea floor, gravimeters and magnetometers are used to identify gravity and magnetic anomalies (discussed in Chapter 11). Gravimeters carried in airplanes and helicopters have even detected vast bodies of chromite, magnetite, and nickel beneath the ice of Antarctica. Mass spectrometers analyze the chemistry of soils, and even that of the gases trapped between soil particles, to determine the composition of the underlying bedrock from which the soil developed. Infrared satellite images and satellite-borne radar are being used to map the surface and shallow-subsurface geology of the Earth's most remote regions, identifying previously unknown mineral deposits with mining potential.

We may soon be able to mine the vast deposits on the sea floor, which in some places is covered by billions of nodules of manganese oxide (MnO_2),

Figure 20-26 Nodules of manganese oxide, which are plentiful on the ocean floor, may become a source of valuable minerals in the near future.

some approaching the size of bowling balls, that contain lesser amounts of iron, nickel, copper, zinc, and cobalt. These nodules (Fig. 20-26), which accumulate where the deep-sea sedimentation rate is slow enough to allow them to grow without being buried, may collectively be the Earth's largest mineral deposit. Although their origin is uncertain, they may form much like ooliths (discussed in Chapter 6), by precipitation of dissolved minerals around an organic nucleus such as a shark's tooth or shell fragment. As the technology to gather manganese oxide nodules improves, there will be considerable international debate regarding ownership and mining rights.

Figure 20-27 Beverage cans are processed into high-density bales in a recycling plant in New Jersey.

Most mineral resources can be recycled. Steel from "tin" cans, automobiles, and old bridges, mercury from discarded thermometers, copper from electrical wire, and platinum from the catalytic converters of abandoned automobiles can all be reclaimed for reuse. Recycling has several benefits: It reduces the volume of waste to be disposed of, the land area disturbed by new mining operations, and the use of energy to mine and refine new ores. Scrap aluminum recycling, for example, uses only one-twentieth the energy needed to mine and process an equivalent amount of new aluminum from bauxite (Fig. 20-27).

Of course, some metals are difficult to recycle, such as the lead in batteries, or the aluminum and steel intermingled with other materials in such complex manufactured objects as refrigerators and lawn mowers. Precious metals, such as gold and silver, are easiest to reuse. The gold fashioned today into a new pair of earrings may have been mined thousands of years ago by ancient Romans or the Inca of Peru.

The discarded, seemingly worthless waste-rock piles, or tailings, from old mining operations are today a source of valuable minerals. Acids and other solutions can be flushed through mine tailings to leach metals from them, or to dissolve the surrounding matrix materials. In this way, substantial quantities of valuable minerals are concentrated from the mine waste, which is in turn reduced in volume.

Figure 20-28 This garbage dump near Mt. Everest illustrates that virtually no place on Earth escapes the environmental impact of careless human behavior. This refuse will remain here for decades or centuries, because the rate of decomposition is slow in the cold, dry climate of the region.

Everything we have, or can make, comes from the Earth. Now that you understand the processes that built, moved, and shaped the continents and ocean floors, and the vast amount of time over which the Earth and its resources developed, you can appreciate why the materials that make our modern lives possible must be managed wisely. If they are not, most assuredly, little will remain for future generations. Industry, local governments, and the world community of nations must cooperate to ensure that the search for, development of, and use of the planet's resources do not squander those resources or irreversibly damage our shared environment (Fig. 20-28). With the knowledge you have acquired from your geology training, you are now prepared to contribute to the ongoing debate about the need to balance resource development and environmental protection. By all means, use your training and be heard.

Chapter Summary

Natural resources, generally classified as either energy or mineral resources, include all types of fuels and a variety of metals and nonmetals. The availability of natural resources depends on their concentration, the state of the technology needed to extract and develop them, and the economic forces of supply and demand. **Reserves** are quantities of natural resources that have been discovered and can be exploited profitably with current technology. **Resources** are deposits that we believe to exist, but that are not exploitable today for various technological, economic, or political reasons. **Ores** are mineral deposits that can be mined profitably. Renewable natural resources are either replenished naturally over a short time span, such as trees, or can be used continuously without being depleted, such as sunlight. Nonrenewable resources form so slowly that they are consumed much more quickly than nature can replenish them, such as **fossil fuels,** which are derived from the organic remains of past life.

Most of our energy needs are provided for by the gaseous, liquid, and semi-solid substances known as **petroleum.** These fossil fuels are composed of **hydrocarbons,** complex organic molecules consisting entirely of hydrogen and carbon. Petroleum typically forms in a marine basin where abundant nutrient input and warm, shallow water nourish microscopic organisms that are eventually buried by younger sediments. Pressure and geothermal heat transform the organic remains first into **kerogen,** a solid waxy material, and then into gas, oil, and other hydrocarbons. Petroleum forms in **source rocks**, which are typically marine muds, then migrates into permeable **reservoir rocks. Oil shale** is a relatively impermeable, fine-grained source rock that retains its petroleum. **Oil sand** is unconsolidated clay or sand that contains a semi-solid, tar-like hydrocarbon called **bitumen.**

Coal is a sedimentary rock consisting of the combustible, highly compressed remains of land plants. It begins to form with the accumulation of **peat,** a soft brown mass of compressed, largely undecomposed plant materials. Bacterial action within the sediment gradually breaks down the vegetation to an organic muck. Pressure from the weight of overlying deposits squeezes water and volatile gases from this muck, increasing the proportion of carbon to form **lignite** (70% carbon). Deep burial and increasing geothermal heat convert lignite to **bituminous coal** (80–93% carbon). Metamorphic conditions transform lignite and bituminous coal to **anthracite** (93–98% carbon). The more carbon coal contains, the harder it is and the more energy per unit volume it can produce.

Coal is the most abundant fossil fuel, but its widespread use causes several significant environmental problems. **Acid rain** is one consequence of burning sulfur-rich coal. Coal-burning releases sulfur dioxide gas, which combines with rainwater to produce sulfuric acid. Nitric acid, formed from automobile emissions, also falls as acid rain. Global warming and marine oil spills are other environmental problems associated with the use of fossil fuels.

As fossil fuel supplies dwindle, alternative energy sources must be discovered and developed. Communities and industries today tap the Earth's geothermal energy, dam streams and rising tides for hydroelectric and tidal power, harness the energy of the Sun and the wind, and burn **biomass fuels,** derived from plants and animals, to heat homes and generate electricity. Physicists and engineers have captured the energy released when the nuclei of certain atoms are split, a process known as **nuclear fission. Breeder reactors** can create an unlimited supply of fissionable materials. Safe disposal of radioactive waste is one of the challenges facing the nuclear-energy industry. Attempts are also being made to generate energy by fusing atomic nuclei together, a process called **nuclear fusion.**

The other principal group of resources, minerals, includes metals, such as iron, copper, aluminum, gold, and silver, and nonmetals, such as sand and gravel for building, limestone for cement, potassium and phosphorus for fertilizers, and gemstones, such as diamonds and emeralds. Metals have been used by humans for thousands of years, initially in their native state, then smelted from rocks and combined to form harder metallic **alloys.**

For metals to be economically useful, some geological process must have concentrated them somewhere near the Earth's surface. Metals may be precipitated from magma-derived water or from circulating ground- or seawater to produce concentrated **hydrothermal deposits. Porphyry copper deposits,** found in the granitic rocks of many subduction-zone batholiths, form when water-rich felsic magma cools and expels its water, which is converted to steam. The steam expands rapidly, rupturing newly crystallized rock, and deposits hydrothermal solutions containing copper and other metals in innumerable fractures at the margin of the batholith. **Massive sulfide deposits** form principally at oceanic divergent zones where seawater enters cracks in hot, young basaltic rock, is heated, dissolves metals, then rises to the sea floor, depositing the metals around surface vents.

Sedimentary and weathering processes can also concentrate valuable metal deposits. Minerals accumulate as **placer deposits** where the velocity of surface waters decreases, forcing dense particles of the transported sediment, including precious metals—such as gold and silver and occasionally diamonds, emeralds, and other gemstones—to settle out of the current. On every continent, there are vast **banded iron formations** dating from about 2 billion years ago, when the oxygen content of the atmosphere became great enough to cause oxidation and precipitation of the iron dissolved in surface waters. Dissolution of metals by chemical weathering and their subsequent precipitation from groundwater, or dissolution of a soluble matrix surrounding a metal deposit, concentrate metallic ore bodies by processes of **secondary enrichment.** Most major metal deposits occur near past or present plate boundaries, where magmas and hot hydrothermal solutions are likely to have been generated.

Mining operations typically produce large volumes of waste material and can contaminate groundwater. These environmental problems can be minimized by isolating sulfides from groundwater and carrying out land reclamation when mining has ended.

Nonmetals are widely used as building materials (such as sand and gravel for concrete, limestone for cement, gypsum for plaster of Paris, clay for bricks) and agricultural fertilizers (such as phosphorus and potassium). Among nonmetals commonly used in industry are sulfur, used to treat a variety of products, pure quartz sand, used for glass production, and asbestos, used for fireproofing clothing and electrical insulation.

As the use of energy and other resources increases, future resource needs will be met only through development of alternative energy sources, conservation, and enhanced recovery of known reserves. The search for new mineral supplies will take us to such remote places as the sea floor (for manganese nodules) and Antarctica.

Key Terms

reserves (p. 564)
resources (p. 564)
ore (p. 564)
fossil fuels (p. 565)
petroleum (p. 565)
hydrocarbons (p. 565)
kerogen (p. 566)
source rocks (p. 566)
reservoir rocks (p. 566)
oil shale (p. 567)
oil sand (p. 568)
bitumen (p. 568)
peat (p. 569)
lignite (p. 569)
bituminous coal (p. 569)
anthracite (p. 570)
acid rain (p. 571)
biomass fuels (p. 577)
nuclear fission (p. 577)
breeder reactor (p. 579)
nuclear fusion (p. 580)
alloys (p. 581)
hydrothermal deposits (p. 581)
porphyry copper deposits (p. 582)
massive sulfide deposits (p. 582)
placer deposits (p. 583)
banded iron formations (p. 584)
secondary enrichment (p. 585)

Questions for Review

1. How do resources differ from reserves? Give an example of each.

2. What factors determine whether a metal deposit is an ore?

3. List three fossil fuels and three alternative energy sources. List three metallic and three nonmetallic resources.

4. What are the key physical properties of a petroleum reservoir rock? Is oil shale a source rock or a reservoir rock?

5. Briefly explain how tidal power is harnessed.

6. Describe how a breeder reactor produces its own fuel.

7. Discuss two ways in which dispersed metals in rocks may become concentrated by geological processes.

8. Name three types of locations where placer deposits may be found.

9. How does plate tectonics affect the distribution of the Earth's metal deposits?

10. Name three sources for the sand and gravel used in the building industry.

For Further Thought

1. Using your general knowledge of the geology of North America, explain why aquatic life in the lakes of Yosemite National Park, California (see Chapter 3) would be more susceptible to the effects of acid rainfall than aquatic life in the lakes of Florida and Indiana (see Chapter 16).

2. Referring back to the illustration in Figure 20-20, page 584, in which areas would you search for oil? gold? sand and gravel?

3. Speculate on why copper mining in the United States has been so drastically reduced in recent years. Under what circumstances might U.S. mining companies resume extensive copper mining?

4. Propose a U.S. energy plan for the year 2050. Which energy sources would you develop? Why?

5. Suppose the Earth's natural resources become completely exhausted and it is decided to prospect on neighboring planets. On which planet might we find extensive iron, chromium, and platinum deposits? On which might we find sand and gravel? Could any of our planetary neighbors provide us with limestone or gypsum?

Appendix A

Conversion Factors for English and Metric Units

Length	1 centimeter	=	0.3937 inches
	1 inch	=	2.54 centimeters
	1 meter	=	3.2808 feet
	1 foot	=	0.3048 meters
	1 yard	=	0.9144 meters
	1 kilometer	=	0.6214 miles (statute)
	1 kilometer	=	3281 feet
	1 mile (statute)	=	1.6093 kilometers
Velocity	1 kilometer/hour	=	0.2778 meters/second
	1 mile/hour	=	0.4471 meters/second
Area	1 square centimeter	=	0.16 square inches
	1 square inch	=	6.45 square centimeters
	1 square meter	=	10.76 square feet
	1 square meter	=	1.20 square yards
	1 square foot	=	0.093 square meters
	1 square kilometer	=	0.386 square miles
	1 square mile	=	2.59 square kilometers
	1 acre (U.S.)	=	4840 square yards
Volume	1 cubic centimeter	=	0.06 cubic inches
	1 cubic inch	=	16.39 cubic centimeters
	1 cubic meter	=	35.31 cubic feet
	1 cubic foot	=	0.028 cubic meters
	1 cubic meter	=	1.31 cubic yards
	1 cubic yard	=	0.76 cubic meters
	1 liter	=	1000 cubic centimeters
	1 liter	=	1.06 quarts (U.S. liquid)
	1 gallon (U.S. liquid)	=	3.79 liters
Mass	1 gram	=	0.035 ounces
	1 ounce	=	28.35 grams
	1 kilogram	=	2.205 pounds
	1 pound	=	0.45 kilograms
Pressure	1 kilogram/square centimeter	=	0.97 atmospheres
	1 kilogram/square centimeter	=	14.22 pounds/square inch
	1 kilogram/square centimeter	=	0.98 bars
	1 bar	=	0.99 atmospheres
Temperature	°F (degrees Fahrenheit)	=	°C(9/5) + 32
	°C (degrees Celsius)	=	(°F – 32)(5/9)

Appendix B

A Statistical Portrait of Planet Earth

Surface Areas

Landmasses	150,142,300 kilometers^2 (57,970,000 miles^2)
Oceans and Seas	362,032,000 kilometers^2 (138,781,000 miles^2)
Entire Earth	512,175,090 kilometers^2 (197,751,500 miles^2)

Continental Areas

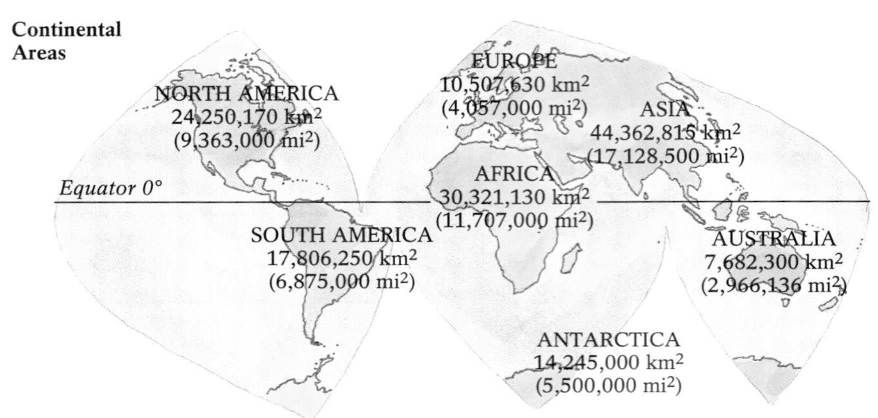

Areas of Oceans

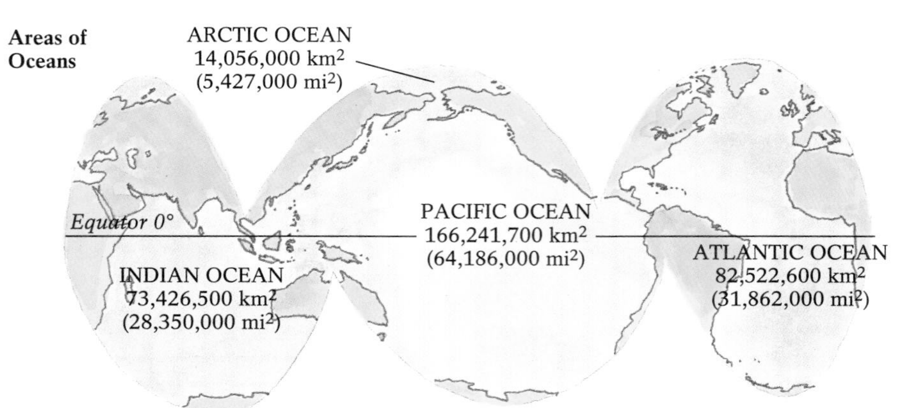

Distribution of Water, by Volume

Oceans and Seas	1.37×10^9 kilometers^3 (3.3×10^8 miles^3)
Glaciers	2.5×10^7 kilometers^3 (7×10^6 miles^3)
Groundwater	8.4×10^6 kilometers^3 (2×10^6 miles^3)
Lakes	1.25×10^5 kilometers^3 (3×10^4 miles^3)
Rivers	1.25×10^3 kilometers^3 (3×10^2 miles^3)

Elevations, Depths, and Distances

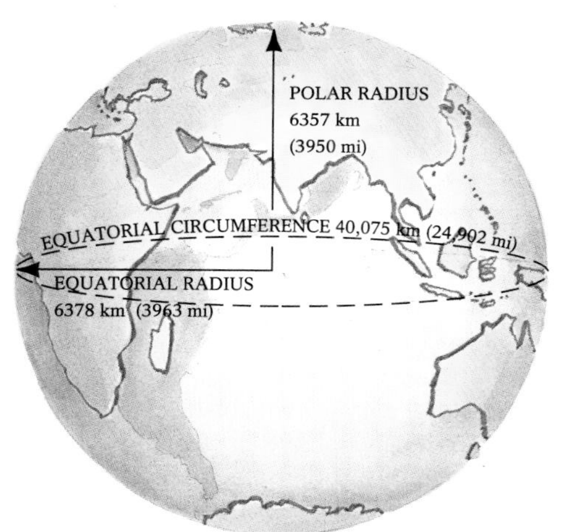

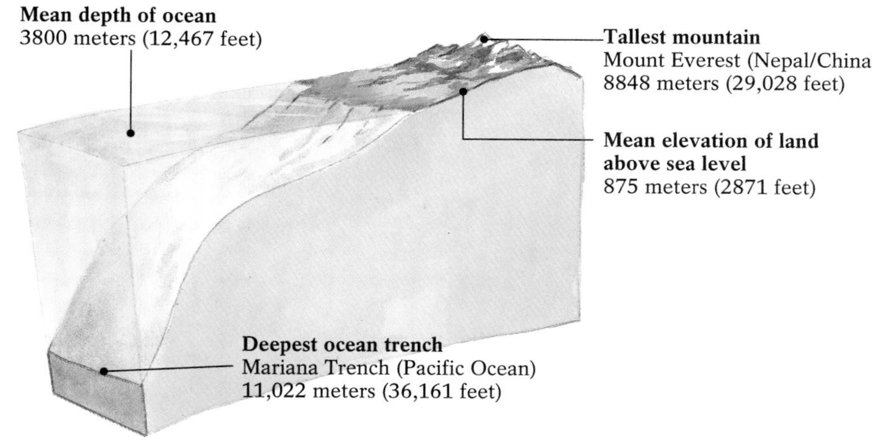

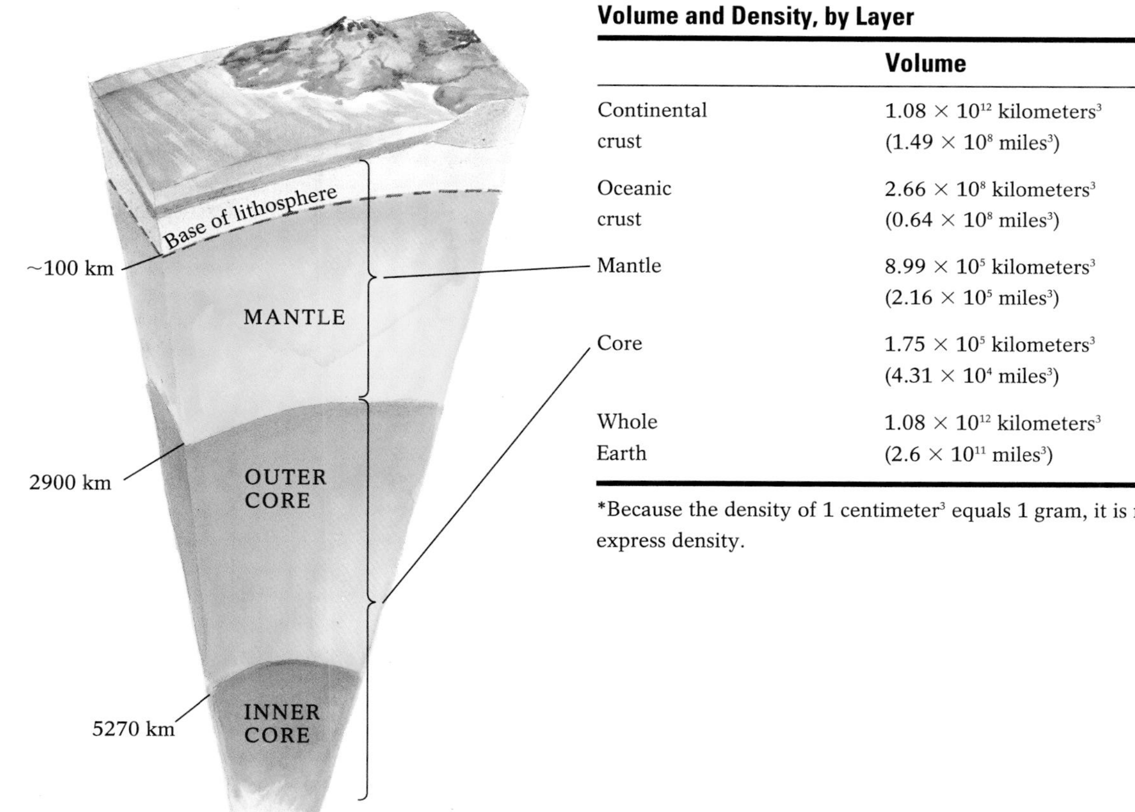

Volume and Density, by Layer

	Volume	Density*
Continental crust	1.08×10^{12} kilometers^3 (1.49×10^8 miles^3)	2.7 grams/centimeter^3
Oceanic crust	2.66×10^8 kilometers^3 (0.64×10^8 miles^3)	3.0 grams/centimeter^3
Mantle	8.99×10^5 kilometers^3 (2.16×10^5 miles^3)	4.5 grams/centimeter^3
Core	1.75×10^5 kilometers^3 (4.31×10^4 miles^3)	10.7 grams/centimeter^3
Whole Earth	1.08×10^{12} kilometers^3 (2.6×10^{11} miles^3)	5.5 grams/centimeter^3 (mean density)

*Because the density of 1 centimeter^3 equals 1 gram, it is most convenient to use metric units to express density.

Appendix C

Reading Topographic and Geologic Maps

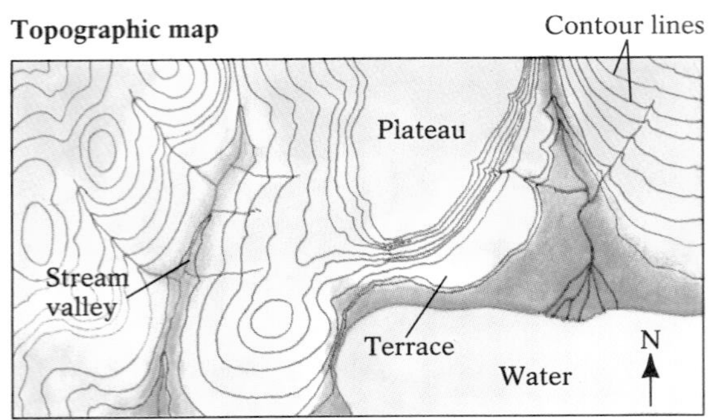

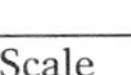

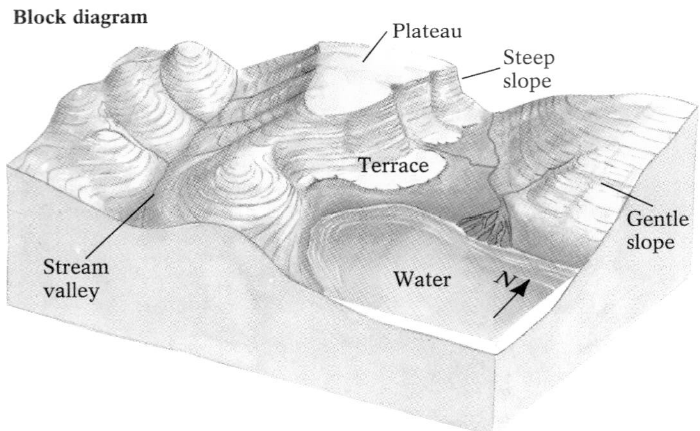

Throughout this textbook, maps have been used to show the locations of and relationships among various features. Geologists use maps for soil management, flood control, environmental planning, finding such resources as ores and groundwater, and determining optimal locations for fuel pipelines, highways, recreational areas, and the like.

Most maps show all or part of the Earth's surface, drawn to scale. A map's scale is usually shown at the bottom of the map. It may feature miles, kilometers, or any other convenient unit of measurement. Most maps from the United States Geological Survey are drawn to a scale of 1:24,000, meaning that 1 centimeter = 240 meters or 1 inch = 2000 feet; the latter would be read as "one inch on this map is equal to two thousand feet in real-world distance." The U.S. Geological Survey has been phasing in metric topographic maps using a scale of 1:100,000.

Most maps show two dimensions, marked by longitude (vertical) and latitude (horizontal) lines. A third dimension–elevation or depth–can be shown on *topographic maps,* which show relief using contour lines that mark specific heights above or depths below sea level. Every point on a contour line is at the same elevation. Every contour line closes upon itself—delineating the border of a discrete area—although its entire length may not be visible within the margins of a given map. A map's contour interval, or the distance between adjacent contour lines, is critical to both the usefulness and the appearance of the map: Contour lines that occur close together represent a steep slope; those that are farther apart represent a more gentle one. In an area of low relief, intervals designating elevation differences of only 5 or 10 feet are appropriate, whereas steeper terrain may require intervals of 50 feet or more. On a given map, all contour intervals are the same.

In the past, accurate map-making depended predominantly upon actual measurements made on site by surveyors, geologists, and others. The advent of aerial photography, and then of space-based satellite photography, has made accurate standardized revisions possible.

The U.S. Geological Survey uses color to indicate specific features:

black and/or red solid lines	: major roads
brown lines	: contour lines
black	: human-made structures, names
light red lines	: town limits
blue	: water features
green	: wooded areas
white	: open fields, deserts, and other nonvegetated areas

Dotted or dashed lines are used for temporary features (that is, solid blue represents a lake, whereas a dashed blue line marks a seasonal stream). Any special symbols used on a map are explained in its legend, which appears with the scale in the bottom margin.

A region's underlying geology can be represented on a *geologic map.* Symbols representing various types of rock have been standardized; such commonly accepted rock symbols have been used throughout this book. Standardized colors are used to show rock ages. The key to a geologic map's labeling, colors, and symbols usually appears at its side margins.

Columbus Quadrangle, New Jersey

Scale 1:24,000
Contour interval 10 feet

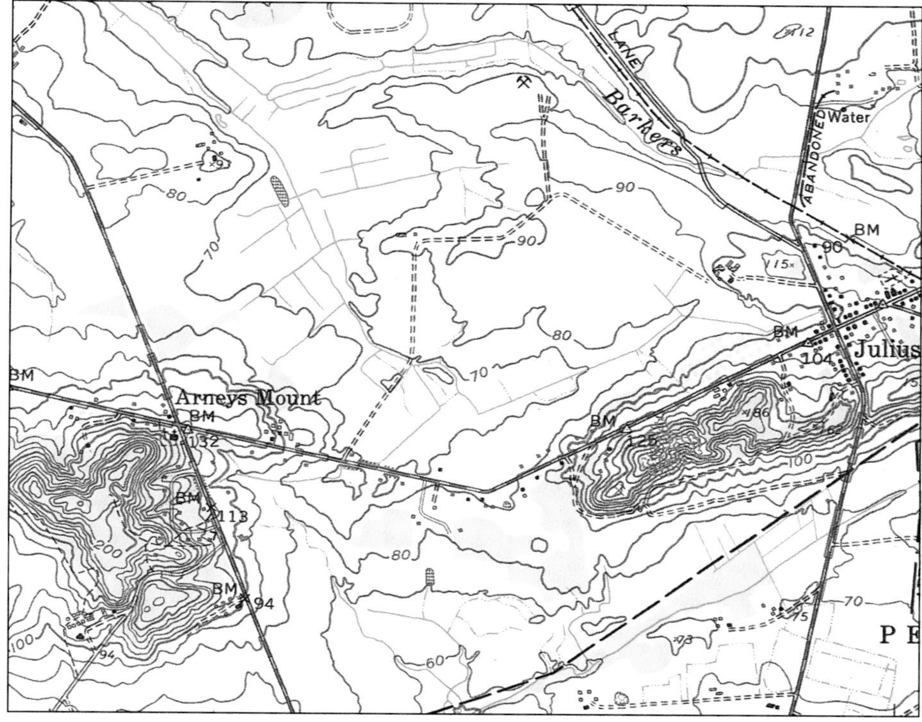

Topographic map

Geologic map

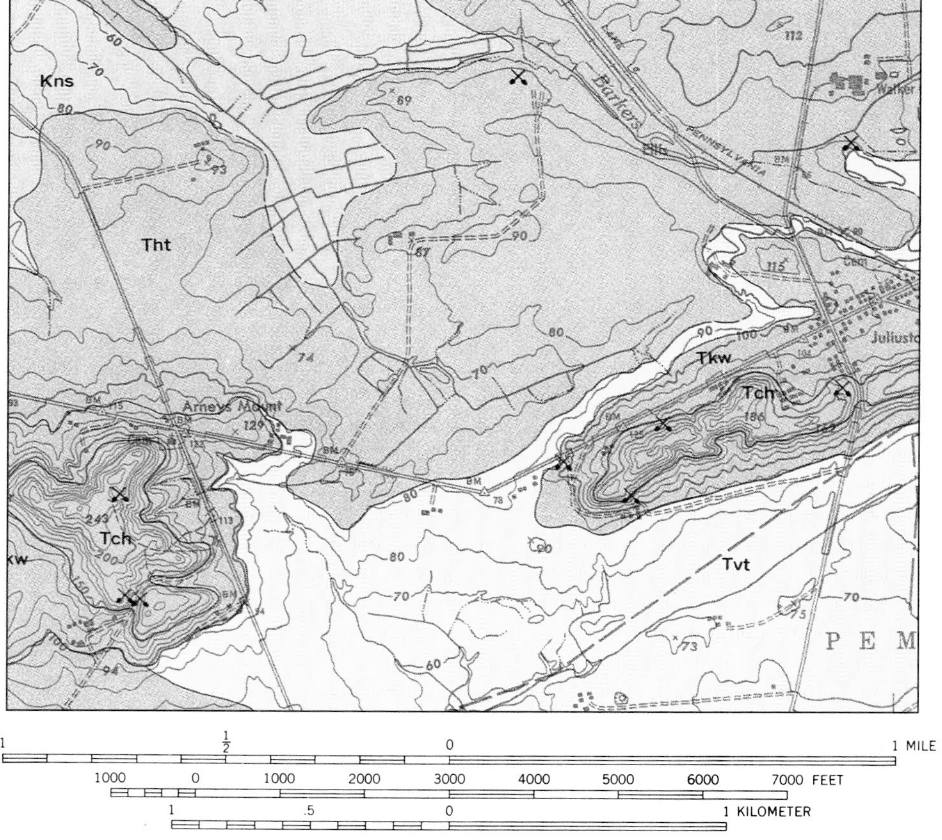

Tertiary Period Deposits

Tch Cohansey sand
Sand, quartz, light-gray to yellow-brown, medium- to coarse-grained, pebbly, ilmenitic, micaceous, stratified

Tkw Kirkwood formation
Sand, quartz, light-gray to tan, fine- to very fine-grained, clayey, micaceous, ilmenitic, kaolinitic, sparingly lignitic, massive-bedded

Tmq Manasquan formation
Sand, quartz, dark green-gray, medium- to coarse-grained, glauconitic, clayey

Tvt Vincentown formation
Upper member—calcarenite, quartz and glauconite, dusky-yellow to pale-olive, clayey; lower member—sand, quartz, dark-gray, poorly sorted, fine to coarse, clayey, glauconitic, entire formation very fossiliferous

Tht Hornerstown sand
Sand, glauconite, dusky-green, medium- to coarse-grained, clayey, massive-bedded

Cretaceous Period Deposits

Krb Red Bank sand
Lower member—sand, glauconite, dark grayish-black, coarse-grained, very clayey, micaceous, lignitic, massive-bedded. (Upper member not present.)

Kns Navesink formation
Sand, glauconite, varying amounts of quartz, greenish-black to brown, medium- to coarse-grained, clayey

Kml Mount Laurel sand
Sand, quartz, reddish-brown to green-gray, poorly sorted, fine- to coarse-grained, glauconitic, massive-bedded

Appendix D

Mineral-Identification Charts

Most Common Rock-Forming Minerals

			MINERAL OR GROUP NAME	COMPOSITION/ VARIETIES	CRYSTAL FORM/OTHER DIAGNOSTIC FEATURES	CLEAVAGE/ FRACTURE	USUAL COLOR/ LUSTER	STREAK	HARDNESS
Light-colored; abundant in all rock types	SILICATES	Framework	Feldspar	Potassium (orthoclase) feldspar ($KAlSi_3O_8$)	Coarse crystals or fine grains	Good cleavage in two directions at 90°; cleavage surfaces striated	White to gray, with pearly luster	White or pink	6
				Sodium, calcium (plagioclase) feldspar Albite: $NaAlSi_3O_8$ Anorthite: $CaAlSi_2O_8$		Good cleavage in two directions at 90°; cleavage surfaces not striated	White to gray, sometimes green- or yellowish	White	
			Quartz	SiO_2	Six-sided crystals; individual or in masses	No cleavage; conchoidal fracture	Colorless or slightly smokey gray, pink or yellow	White	7
		Sheet	Mica	Muscovite $KAl_3Si_3O_{10}(OH)_2$	Thin, disc-shaped crystals	one perfect cleavage plane; splits into very thin sheets	Colorless or slightly gray, green, or brown, with vitreous luster	White	2–2½
				Biotite $K(Mg, Fe)_3AlSi_3O_{10}(OH)_2$	Irregular foliated masses	One perfect cleavage plane; splits into thin sheets	Black, brown, or green, with vitreous luster	White or gray	2½–3
				Chlorite $(Mg, Fe)_5(Al, Fe)_2Si_3O_{10}(OH)_8$			Yellowish, brown, green, or white	White or colorless	2–2½
Dark-colored; abundant in metamorphic and igneous rocks		Double Chain	Amphibole	Actinolite $Ca_2(Mg, Fe)_5Si_8O_{22}(OH)_2$	Long, six-sided crystals, fibrous or in aggregates	Two good cleavage planes at 56° and 124°	Pale to dark green or black, with vitreous luster. Pure actinolite white.	Pale green or white	5–6
				Hornblende $(Ca, Na)_{2\text{-}3}(Mg, Fe, Al)_5Si_6(Si, Al)_2O_{22}(OH)_2$					
		Single Chain	Pyroxene	Augite $Ca(Mg, Fe, Al)(Al, Si_2O_6)$	Short, four- or six-sided crystals	Two good cleavage planes at about 90°	Light to dark green, with vitreous luster	Pale green	5–6
				Diopside $CaMg(Si_2O_6)$			White to light green, with vitreous luster	White or pale green	
				Orthopyroxene $MgSiO_3$			Pale gray, green, brown, or yellow, with vitreous luster	White	
		Single Tetrahedra	Olivine	$(Mg, Fe)_2SiO_4$	Small grains and granular masses	No cleavage; conchoidal fracture	Grayish green or brown, with vitreous, glassy luster	White	6½–7
			Garnet	$(Ca, Mg, Fe, Al)_n(SiO_4)_3$	12- or 24-sided crystals	No cleavage; conchoidal fracture	Deep red, with vitreous to resinous luster	White	6½–7½
Light-colored; abundant in sedimentary rocks	CARBONATES		Calcite	$CaCO_3$	Fine to coarsely crystalline. Effervesces rapidly in HCl	Three oblique cleavage planes, forming rhombohedral cleavage pieces	White or gray, with pearly luster	White	3
			Dolomite	$CaMg(CO_3)_2$	Fine to coarsely crystalline. Effervesces slowly in HCl when powdered		Colorless, white, or pink, and may be tinted by impurities; with pearly luster	White to pale gray	3½–4
	CLAY MINERALS (Hydrous alumino-silicates)		Kaolinite	$Al_2Si_2O_5(OH)_4$	Very fine grains; found as bedded masses in soils and sedimentary rocks; earthy odor	Earthy fracture	White to buff, or tinted gray by impurities	White, off-white, or colorless	1½–2½
			Illite	$K_{0.8}Al_2(Si_{3.2}Al_{0.8})O_{10}(OH)_2$					
			Smectite	$Na_{0.3}Al_2(Si_{3.7}Al_{0.3})O_{10}(OH)_2$					

Accessory or Less-Abundant Rock-Forming Minerals

		MINERAL OR GROUP NAME	COMPOSITION/ VARIETIES	CRYSTAL FORM/OTHER DIAGNOSTIC FEATURES	CLEAVAGE/ FRACTURE	USUAL COLOR/ LUSTER	STREAK	HARDNESS
Light-colored; common in sedimentary rocks	SULFATES	Gypsum	$CaSO_4 \cdot 2H_2O$	Tabular crystals in fine-to-granular masses	One perfect cleavage plane, forming thin sheets; also two other good cleavage planes	Colorless to white, with vitreous luster	White or off-white	1–2½
		Anhydrite	$CaSO_4$	Granular masses	Three good to perfect cleavage planes at 90°	White, gray, or blue-gray, with pearly to vitreous luster	White or off-white	3–3½
		Halite	NaCl	Perfect cubic crystals, soluble in water. Tastes salty.	Three excellent cleavage planes at 90°	White or gray, with pearly luster	White	2½
Light-colored; mainly in metamorphic and igneous rocks	ALUMINO-SILICATES	Kyanite	Al_2SiO_5	Long, bladed or tabular crystals	One perfect cleavage plane, parallel to length of crystals	White to light blue, with vitreous luster	White	5 along cleavage plane, 6½–7 across cleavage plane
		Sillimanite		Long, slender crystals or fibrous masses		White to gray, with vitreous luster	White	6–7
		Andalusite		Coarse, nearly square crystals	Irregular fracture	Red, reddish brown, or green, with vitreous luster	White	5–6
		Serpentine	$Mg_6Si_4O_{10}(OH)_8$	Fibrous or platy masses	Splintery fracture	Light to dark green or brownish yellow, with pearly luster	White	4–6
		Talc	$Mg_3Si_4O_{10}(OH)_2$	Foliated masses. Feels soapy.	Perfect in one direction, forming thin flakes	White to pale green, with pearly or greasy luster	White	1–1½
		Corundum	Al_2O_3	Short, six-sided crystals	Irregular fracture	Usually brown, pink or blue, with adamantine luster	None	9
		Fluorite	CaF_2	Octahedral or cubic crystals	Cleaves easily	White, yellow, green, or purple, with vitreous luster	White	4
Dark-colored; common in metamorphic rocks	SILICATES	Epidote (paired tetrahedra)	$Ca_2(Al,Fe)Al_2Si_3O_{12}(OH)$	Usually granular masses; also slender prisms	One good cleavage direction, one poor	Yellow- to dark green, with vitreous luster	White or gray	6–7
		Staurolite (single tetrahedra)	$Fe_2Al_9Si_4O_{22}(O,OH)_2$	Short crystals, some cross-shaped	One poor cleavage direction	Brown or reddish brown to black, with vitreous luster	Off-white to white	7
		Graphite	C	Scaley, foliated masses. Feels greasy	One direction of cleavage	Steel gray to black, with vitreous or pearly luster	Gray or black	1–2
Dark-colored; common in all rock types		Apatite	$Ca_5(PO_4)_3(OH,F,Cl)$	Granular masses	Poor cleavage	Green, brown, or red, with adamantine or greasy luster	Off-white	5
		Magnetite	Fe_3O_4	Granular masses	Uneven fracture	Black, with metallic luster	Black	5½
		Hematite	Fe_2O_3	Granular masses	Uneven fracture	Brown-red to black, with earthy, dull, to metallic luster	Black	5½
		Limonite	$2Fe_2O_3 \cdot 3H_2O$	Earthy masses	Uneven fracture	Yellowish brown to black	Brownish yellow	5–5½
Metallic luster; common in all rock types	SULFIDES	Pyrite	FeS_2	Cubic crystals or granular masses	Uneven fracture	Pale brass yellow, with metallic luster	Greenish black	6–6½
		Galena	PbS	Cubic crystals or granular masses	Three perfect cleavage planes at 90°	Silver gray with metallic luster	Gray	2½
		Sphalerite	ZnS	Granular masses	Six perfect cleavage planes at 60°	White to green, brown, or black, with resinous to submetallic luster	Reddish brown	3½–4
		Chalcopyrite	$CuFeS_2$	Granular masses	Irregular fracture	Brass yellow	Greenish black	3½–4

Minerals and Elements of Industrial or Economic Importance

MINERAL OR GROUP NAME	COMPOSITION	USUAL COLOR/ LUSTER	STREAK	HARDNESS	OTHER PROPERTIES/ COMMENTS	ORIGIN
Asbestos (Chrysotile)	$Mg_3Si_2O_5(OH)_4$	White to pale green, with pearly luster	White	1–2½	Flexible, nonflammable fibers	A variety of serpentines; found mostly in metamorphic rock
Bauxite	$Al(OH)_3$	Reddish to brown, with dull luster	Pale reddish brown	1½–3½	Found in earthy, clay-like masses. Principal source of commercial aluminum.	Weathering of many rock types
Chalcopyrite	$CuFeS_2$	Yellow, with metallic luster	Greenish black	3½–4½	Uneven fracture; softer than pyrite. Irridescent tarnish. Most common ore of copper.	Hydrothermal veins and porphyry copper deposits
Chromite	$FeCr_2O_4$	Black, with metallic or submetallic luster	Dark brown	5½	Massive; granular; compact. Most common ore of chromium. Used in making steel.	Ultramafic igneous rocks
Copper (native)	Cu	Red, with metallic luster	Red	2½–3	Malleable and ductile. Tarnishes easily.	Mafic igneous rock (basaltic lavas); also in oxidized ore deposits
Galena	PbS	Silver-gray, with metallic luster	Gray	2½	Perfect cubic cleavage. Most important ore of lead; also commonly contains silver.	Hydrothermal veins
Gold (native)	Au	Yellow, with metallic luster	Yellow	2½–3	Malleable and ductile	Sedimentary placer deposits; hydrothermal veins
Hematite	Fe_2O_3	Brown-red to black, with earthy, dull, or metallic luster	Dark red	5½–6½	Granular, massive. Most important source of iron.	Found in all types of rocks. Commonly in contact-metamorphic aureoles around mafic igneous sills.
Magnetite	Fe_3O_4	Black, with metallic luster	Black	5½	Uneven fracture. An important source of iron. Strongly magnetic.	Found in all types of rocks
Platinum (native)	Pt	Steel-gray or silver-white, with metallic luster	Off-white, metallic	4–4½	Malleable and ductile. Occasionally magnetic.	Mafic igneous rocks and sedimentary placer deposits
Silver (native)	Ag	Silver-white, with metallic luster	Silver-white	2½–3	Malleable and ductile. Tarnishes to dull gray or black.	Mostly in hydrothermal veins; also in oxidized ore deposits
Sphalerite	ZnS	Shades of brown and red, with resinous to adamantine luster	Reddish brown	3½–4	Perfect cleavage in six directions at 120°. Most important ore of zinc.	Hydrothermal veins

Precious and Semi-Precious Gems*

MINERAL OR GROUP NAME	COMPOSITION	USUAL COLOR/ LUSTER	STREAK	HARDNESS	OTHER PROPERTIES	ORIGIN
Beryl (aquamarine, emerald)	$Be_3Al_2(SiO_3)_6$	Blue, green, yellow, or pink, with vitreous luster	White	7½–8	Uneven fracture; hexagonal crystals	Cavities in granites and pegmatites, and schist
Corundum (ruby, sapphire)	Al_2O_3	Gray, red (ruby), blue (sapphire), with adamantine luster	None	9	Short, six-sided crystals; irregular, occasionally cleavage-like fracture ("parting")	Metamorphic rocks; some igneous rocks
Diamond	C	Colorless or with pale tints, with adamantine luster	None	10	Octahedral crystals	Peridotite; kimberlite; sedimentary placer deposits
(S) Garnet	$(Ca, Mg, Fe, Al)_n(SiO_4)_3$	Deep red, with vitreous to resinous luster	White	6½–7½	No cleavage; 12- or 24-sided crystals	Contact-metamorphic and regionally metamorphosed rocks, and sedimentary placer deposits
(S) Jadite (jade)	$NaAl(Si_2O_6)$	Green, with vitreous luster	White or pale green	6½–7	Compact fibrous aggregates	High-pressure metamorphic rocks
Olivine (peridot)	$(MgFe)_2SiO_4$	Light to dark green, with vitreous, glassy luster	White	6½–7	Uneven fracture, often in granular masses	Basalt, peridotite
(S) Opal	$SiO_2{\cdot}nH_2O$	White with various other colors, with vitreous, pearly luster	White	5½–6½	Conchoidal fracture; amorphous; tinged with various colors in bands	Low-temperature hot springs; weathered near-surface deposits
(S) Quartz (includes amethyst, citrine, agate, onyx, bloodstone, jasper, etc.)	SiO_2	Colorless, white, or tinted by impurities, with luster depending on variety	White	7	No cleavage; six-sided crystals	Origin specific to gem variety. (Quartz found in all rock types except ultramafic igneous rocks.)
Topaz	$Al_2SiO_4(F, OH)_2$	Colorless, white, or pale pink or blue, with vitreous luster	Colorless	8	Cleavage in one direction; conchoidal fracture	Pegmatite, granite, rhyolite
Tourmaline	$(Na, Ca)(Li, Mg, Al)$-$(Al, Fe, Mn)_6(BO_3)_3$-$(Si_6O_{18})(OH_4)$	Black, brown, green, or pink, with vitreous luster	White to gray	7–7½	Poor cleavage; uneven fracture; striated crystals	Metamorphic rocks; pegmatite; granite
(S) Turquois	$CuAl_6(PO_4)_4$-$(OH)_8{\cdot}5H_2O$	Blue-green, with waxy luster	Blue-green or white	5–6	Massive	Hydrothermal veins
Zircon	$Zr(SiO_4)$	Colorless, gray, green, pink, or light blue, with adamantine luster	White	7½	Poor cleavage	Felsic igneous rocks and sedimentary placer deposits

*Gems are classified as semi-precious *(S)* if they are more accessible than precious gems and/or their properties are somewhat less valued.

Appendix E

How to Identify Some Common Rocks

Rocks are aggregates of minerals, and sometimes of organic materials that occur naturally. They are generally classified according to whether they are of igneous, sedimentary, or metamorphic origin, and then on the basis of their physical properties—primarily texture and mineral composition.

Igneous Rocks

Igneous rocks are formed when a magma cools and crystallizes. The chemical composition of a magma determines the composition of the minerals and rocks that will be formed when it cools. The rate at which a magma cools determines the size of the resulting crystals, which in turn determines the texture of the resulting rocks. Both composition and texture are used to classify igneous rocks.

Color is used only broadly as a diagnostic property for igneous rocks. The descriptions *light, dark,* and *intermediate* are generally agreed on. But many igneous rocks comprise two different-colored components; some blend various grays and pinks. Color, then, can help narrow down the possible identity of a rock, but usually does not cast the deciding vote.

I. *Phaneritic*, or coarse-grained rocks: Composed of individual crystals of about 1 to 5 millimeters in length

A. If the rock contains quartz, it is probably **granite.**

B. If the rock contains no quartz, and from 30% to 60% feldspar, it is **diorite.**

C. If the rock contains no quartz and less than 30% feldspar, it is **gabbro.**

D. If the rock contains neither quartz nor feldspar, it is likely to contain ferromagnesian minerals and be ultramafic **peridotite** (consisting largely of olivine or pyroxene).

II. *Aphanitic* rocks: Individual crystals are fine-grained or microscopic

A. If quartz crystals can be identified in the rock, it is **rhyolite;** if the rock is light gray, white, or light green but too fine-grained to determine whether quartz is present, it is probably rhyolite.

B. If the rock contains no visible quartz crystals but does have approximately equal proportions of white or gray feldspar and ferromagnesian minerals, it is **andesite.**

C. If ferromagnesian minerals can be identified in the rock, it is **basalt;** if the rock is dark gray or black, and too fine-grained to identify any crystals, it is probably a basalt.

III. *Porphyritic* rocks: Contain macroscopic crystals (crystals that can be seen without the aid of a microscope) embedded in a matrix of smaller macroscopic or microscopic crystals. Use the descriptions above, for aphanitic rocks, to identify porphyritic rhyolites, basalts, etc.

IV. *Glassy* rocks: Rocks look like solid glass. The most common is **obsidian,** which is rhyolitic in composition.

Sedimentary Rocks

Sedimentary rocks are made up of particles derived either from preexisting rocks or from organic debris. They are classified on the basis of their texture and mineral composition.

In identifying sedimentary rocks, the first step is to determine whether the rock contains carbonate minerals. This is done by applying a drop of dilute hydrochloric acid to the rock surface. (Note: HCl must be used under close supervision to avoid burning one's skin or clothing.)

I. Rocks that do not effervesce (fizz), even when powdered—but may effervesce in some places, such as the fine cement between grains. Such rocks contain little or no carbonate minerals.

A. Rocks with **clastic** texture: Composed of grains in a cement matrix

1. If the grains are more than 2 millimeters in diameter…
   a. and are angular, the rock is **sedimentary breccia.**
   b. and are rounded, the rock is **conglomerate.**
2. If the grains are from 1/16 to 2 millimeters in diameter, and the rock feels gritty, it is **sandstone.**
   a. If more than 90% of the grains are quartz, the rock is **quartz sandstone.**
   b. If more than 25% of the grains are feldspar, the rock is **arkose.**
   c. If more than 25% of the grains are fine fragments of shale, slate, basalt, and the like, the rock is **lithic sandstone.**
   d. If more than 15% of the rock is fine-grained matrix material, the rock is **graywacke.**
3. If the rock is fine-grained, feels smooth, …
   a. and the grains are visible with a hand lens, the rock is **siltstone.**
   b. and the grains are invisible even with a hand lens, and the rock is …
      i. laminated, or layered, it is **shale.**
      ii. unlayered, it is **mudstone.**

B. Rocks with **crystalline** texture: Composed of invisible, interlocking crystals

1. If the rock dissolves in water, it is **rock salt.**
2. If its crystals are fine to coarse, and have a hardness of 2 on the Mohs scale, the rock is **gypsum.**

C. Rocks with indeterminate texture

1. If the rock is smooth, very fine-grained, and fractures conchoidally, it is **chert;** if it is dark in color it is called **flint.**
2. If the rock is black or dark brown and breaks easily, soiling the fingers, it is **coal.**

II. Rock that effervesces strongly is **limestone.** Limestone may be clastic or crystalline in texture.

A. If the rock is clastic in texture and contains fossils, it is **bioclastic limestone.**

1. If it is composed of whole, recognizable fossils, it is **coquina.**
2. If it is very fine-grained, light-colored, and powdery, it is **chalk.**

B. If the rock is clastic and composed of small spheres, it is **oolitic limestone.**

C. If the rock is coarsely crystalline and contains different-colored layers, it is **travertine.**

III. Rock that effervesces only when hammered to a powder is **dolostone.** Dolostone may be clastic or crystalline in texture.

Metamorphic Rocks

Metamorphic rocks are rocks that have had their composition or structure changed by intense heat or pressure. Two factors determine the nature of a metamorphic rock: the composition of the parent rocks and the combination of metamorphic factors that acted upon them. Different circumstances can produce different textures; thus texture is the primary basis for metamorphic rock identification.

I. Is the rock *foliated* (layered)? If so, identify the type of foliation and if possible the rock's mineral content.

A. If the rock is fine-grained and readily splits into sheets, it is **slate,** and will have an earthy luster.

B. If the rock is slate-like but has a silken luster, it is **phyllite.**

C. If the rock contains flat or needle-like mineral crystals that are virtually parallel to each other, it is **schist.** A schist that primarily contains mica is a mica schist; there are, similarly, garnet mica schists, hornblende schists, and talc schists. A schist that contains serpentine is **serpentinite.**

D. If the rock consists of separate layers of light and dark minerals, it is **gneiss.** Light-colored layers consist of feldspars and perhaps quartz; dark-colored layers are probably biotite, amphibole, or pyroxene.

II. Is the rock nonfoliated? If so…

A. Is quartz its primary constituent? If so, the quartz grains will be interlocking and the rock will be hard enough to scratch glass. The rock is **quartzite.**

B. Does it consist of interlocking, coarse crystals of calcite or dolomite? If so, it is **marble.**

C. Does it consist primarily of dark grains too fine to be seen unaided? If so, it is probably **hornfels;** it may also contain a few larger crystals of less common minerals.

Appendix F

Sources of Geologic Literature, Photographs, and Maps

American Geological Institute (AGI)
4220 King Street
Alexandria, VA 22302
or
AGI Publications Center
P.O. Box 205
Annapolis Junction, MD 20701

Issues the monthly "Bibliography and Index of Geology," which includes worldwide references and contains listing by author and subject.

Canadian Centre for Remote Sensing
2464 Sheffield Road
Ottawa, Ontario K1A-0Y7
CANADA

Source for obtaining aerial photographs and satellite imagery of Canada.

Earth Science Information Center
U.S. Geological Survey
507 National Center
Reston, VA 22092

Publishes topographic and other scale maps of the United States. Indexes listing available map scales are available for all the states.

Earthquake Engineering Research Institute
499 14th Street, Suite 320
Oakland, CA 94612

Provides a valuable field guide for learning about and studying earthquakes.

Federal Emergency Management Agency
P.O. Box 70274
Washington, D.C. 20024
ATTN: Publications

Provides a number of publications aimed at helping citizens prepare for geologic disasters, including "Are you Ready? Your Guide to Disaster Preparedness Checklist." Also issues a general "FEMA Publications Catalog."

Geological Society of America
P.O. Box 9140
Boulder, CO 80301

Publishes a large and diverse number of papers and journals concerning the geology and mineral resources of North America.

National Aerial Photography Library
615 Booth Street
Ottawa, Ontario K1A-0E9
CANADA

Source for obtaining aerial photographs and satellite imagery of Canada.

National Climate Center (NCC)
Federal Building
Asheville, NC 28801

Issues a monthly publication of climatological data such as temperature and precipitation statistics for any given state or region. The NCC maintains up-to-date weather records for the entire United States.

National Geophysical and Solar–Terrestrial Data Center (part of the National Oceanic and Atmospheric Administration)
325 Broadway
Boulder, CO 80303

Maintains worldwide computer files of earthquakes recorded by seismographs and historic earthquake data.

National Oceanic and Atmospheric Administration (NOOA)
6001 Executive Boulevard
Rockville, MD 20852

This federal agency maintains a large collection of aerial photographs of the coastal regions of the United States. A detailed index is available.

State geological surveys

All states maintain a geological survey that produces technical reports, maps, and other publications at the state or county level.

U.S. Department of Agriculture
Aerial Photography Field Office
P.O. Box 30010
Salt Lake City, UT 84130

Can provide aerial photographs of most of the United States. A variety of scales is available.

U.S. Geological Survey
Branch of Distribution
604 South Pickett Street
Alexandria, VA 22304
(also maintains offices across the country for over-the-counter sales of publications)

Produces a large number of maps, reports, circulars, professional papers, bulletins, water resources publications, etc. Circular number 777 of the USGS provides an annually updated guide to obtaining information on the earth sciences.

U.S. Geological Survey
EROS Data Center
Sioux Falls, SD 57198

Provides computer listings of most satellite imagery, high-altitude aerial photography, and photographs obtained during the Apollo, Skylab, and Gemini space missions. Worldwide coverage and indexes are available.

Appendix G

Careers in the Geosciences

Careers in the geosciences are numerous and varied. Several have been alluded to throughout this book. Employers in these fields include the energy industry, which hires about half of all professional geoscientists; the mining industry; federal and state governments; consulting firms, especially in environmental issues and hydrogeology; and academia, which employs more than 10% of all geoscientists.

Each of these broad fields, of course, involves a number of subdisciplines: the energy industry, for example, requires expertise in sedimentology and stratigraphy; mining careers require a background in economic geology, petrology, mineralogy, crystallography, or structural geology. The following is a list of some of the occupations in which people with an interest in the geosciences are employed:

Economic geologists conduct field investigations to determine the locations and economic viability of mineral deposits; they also investigate the genesis of mineral deposits. Mining companies typically employ economic geologists.

Engineering geologists investigate the geologic factors that affect human-made structures such as bridges, dams, and buildings, and the geologic effects of mass wasting.

Environmental geologists work in assessing, solving, and preventing problems associated with the pollution of soil, bedrock, and groundwater. For example, they assist in the selection of suitable locations for municipal- and hazardous-waste facilities such as landfills and waste-storage facilities, and may also help design such facilities.

Geochemists investigate the nature and distribution of chemical elements in the geologic environment.

Geochronologists determine the age of geologic materials, helping to reconstruct the geologic history of the Earth.

Geomorphologists study landforms, the rates and intensity of processes that created them, and the relationship of these landforms to the underlying geologic structures and climates that existed during their evolution.

Geophysicists attempt to determine the internal structure and properties of the Earth. They may focus on specific factors such as seismic waves, geomagnetism, or gravity.

Glaciologists investigate the physical and chemical properties of glacial masses. They also study the development, movement, and decay of glaciers and ice sheets and their deposits.

Hydrogeologists study the production, distribution, movement, and quality of groundwater in the Earth's crust. Many hydrogeologists work in the environmental industry, in cooperation with environmental geologists.

Hydrologists study the distribution, movement, and quality of surface bodies of water.

Marine geologists investigate the topography and sediments of the world's oceans. Marine geologists may work closely with petroleum geologists in off-shore exploration projects. They often work closely with oceanographers.

Mineralogists study the formation, composition, and properties—both physical and chemical—of minerals.

Paleontologists collect fossils and determine their age, reconstruct past environments, and regionally correlate rocks by determining the evolutionary sequence of fossil assemblages found in the rocks. Many paleontologists work in the oil industry.

Petroleum geologists work in the oil industry, and are involved with the exploration and production of petroleum and natural gas. They may also research unconventional sources such as oil shales and sands.

Petrologists study the mineralogical relationships of rocks, and specialize in determining the genesis of rocks; they may focus on either sedimentary, igneous, or metamorphic petrology.

Planetary geologists study the planets and their satellites in order to better understand the evolution of the solar system. Most planetary geologists are employed by universities or advanced research organizations such as NASA (National Aeronautics and Space Administration).

Sedimentologists study the formation of sedimentary rocks by assessing the processes of transportation, erosion, and deposition. Many sedimentologists work in the petroleum industry.

Stratigraphers decipher the sequence of rocks using the principle of superposition. They study the time and space relationships of rock sequences in an effort to determine the geologic history of an area.

Structural geologists investigate the phenomena and structures produced by the deformation of the Earth's crust, such as faults and folds. Most of the data used in structural geology are collected during detailed field work.

An academic background that includes courses in math and other sciences as well as in the specific geological fields mentioned here would be practical preparation for a geoscience career.

Glossary

ablation The loss of snow and ice from a *glacier*, caused primarily by melting.

abrasion A form of *mechanical weathering* that occurs when loose fragments or particles of rocks and minerals that are being transported, as by water or air, collide with each other or scrape the surfaces of stationary rocks.

absolute dating The fixing of a geological structure or event in time, as by counting tree rings.

accretionary wedge A mass of *sediment* and oceanic *lithosphere* that is transferred from a *subducting* plate to the less dense, overriding plate with which it *converges.*

accumulation The increase in a *glacier's* volume, caused primarily by snowfall.

acid rain Rain that contains such acidic compounds as sulfuric acid and nitric acid, which are produced by the combination of atmospheric water with oxides released when *hydrocarbons* are burned. Acid rain is widely considered responsible for damaging forests, crops, and human-made structures, and for killing aquatic life.

aeration zone See *zone of aeration.*

aftershock A ground tremor caused by the repositioning of rocks after an *earthquake.* Aftershocks may continue to occur for as long as two years after the initial earthquake. The intensity of an earthquake's aftershocks decreases over time.

aggradation The process by which a stream's *gradient* steepens due to increased deposition of *sediment.*

alloy A metal that is manufactured by combining two or more molten metals. An alloy is always harder than its component metals. Bronze is an alloy of copper and tin.

alluvial fan A triangular deposit of *sediment* left by a stream that has lost velocity upon entering a broad, relatively flat valley.

alluvium A deposit of *sediment* left by a stream on the stream's channel or *flood plain.*

alpine glacier A mountain *glacier* that is confined by highlands.

andesite The dark, *aphanitic, extrusive rock* that has a *silica* content of about 60% and is the second most abundant volcanic rock. Andesites are found in large quantities in the Andes Mountains.

andesite line The geographic boundary between the *basalts* and *gabbros* of the Pacific Ocean basin and the *andesites* at the subductive margins of the surrounding continents.

angle of repose The maximum angle at which a pile of unconsolidated material can remain stable.

anthracite A hard, jet-black coal that develops from *lignite* and *bituminous coal* through *metamorphism,* has a carbon content of 92% to 98%, and contains little or no gas. Anthracite burns with an extremely hot, blue flame and very little smoke, but it is difficult to ignite and both difficult and dangerous to mine.

anticline A convex *fold* in rock, the central part of which contains the oldest section of rock. See also *syncline.*

aphanitic Of or being an *igneous rock* containing grains that are so small as to be barely visible to the naked eye.

aquiclude An impermeable body of rock that may absorb water slowly but does not transmit it.

aquifer A *permeable* body of rock or *regolith* that both stores and transports groundwater.

aquitard A layer of rock having low permeability that stores groundwater but delays its flow.

arête A sharp ridge of erosion-resistant rock formed between adjacent *cirque glaciers.*

aridity index The ratio of a region's potential annual evaporation, as determined by its receipt of solar radiation, to its average annual *precipitation.*

arroyo A small, deep, usually dry channel eroded by a short-lived or intermittent desert stream.

artesian Of, being, or concerning an *aquifer* in which water rises to the surface due to pressure from overlying water.

asthenosphere A layer of soft but solid, mobile rock comprising the lower part of the upper *mantle* from about 100 to 350 kilometers beneath the Earth's surface. See also *lithosphere.*

atoll A circular *reef* that encloses a relatively shallow *lagoon* and extends from a very great depth to the sea surface. An atoll forms when an oceanic island ringed by a *barrier reef* sinks below sea level.

atom The smallest particle that retains all the chemical *properties* of a given *element.*

atomic mass 1. The sum of *protons* and *neutrons* in an atom's nucleus. 2. The combined mass of all the particles in a given atom.

atomic number The number of *protons* in the nucleus of a given atom. *Elements* are distinguished from each other by their atomic numbers.

aureole A section of rock that surrounds an *intrusion* and shows the effects of *contact metamorphism.*

backarc basin A depression landward of a *volcanic arc* in a *subduction* zone, which is lined with trapped *sediment* from the volcanic arc and the plate interior. See also *forearc basin.*

backarc spreading The process by which the overriding plate in a *subduction* zone becomes stretched to the point of *rifting*, so that *magma* can then rise into the gap created by the rift. Backarc spreading typically occurs when the subducting plate sinks more rapidly than the overriding plate moves forward.

backshore The portion of a *beach* that extends from the high-tide line inland to the sea cliff or vegetation line. *Swash* reaches the backshore only during major storms.

backswamp The section of a *flood plain* where deposits of fine silts and clays settle after a flood. Backswamps usually lie behind a stream's *natural levees.*

backwash Water that returns into an ocean or large lake after hitting the shore as *swash.*

banded iron formation A rock that is made up of alternating light silica-rich layers and dark-colored layers of iron-rich minerals, which were deposited in marine basins on every continent about 2 billion years ago.

barchan dune A crescent-shaped *dune* that forms around a small patch of vegetation, lies perpendicular to the prevailing wind direction, and has a gentle, convex *windward* slope and a steep, concave *leeward* slope. Barchan dunes typically form in arid, inland deserts with stable wind direction and relatively little sand.

barrier island A ridge of sand that runs parallel to the main coast but is separated from it by a *bay* or *lagoon.* Barrier islands range from 10 to 100 kilometers in length and from 2 to 5 kilometers in width. A barrier island may be as high as 6 meters above sea level.

barrier reef A long, narrow *reef* that runs parallel to the main coast but is separated from it by a wide *lagoon.*

basal sliding The process by which a *glacier* undergoes thawing at its base, producing a film of water along which the glacier then flows. Basal sliding primarily affects glaciers in warm climates or mid-latitude mountain ranges.

basalt The dark, dense, *aphanitic, extrusive rock* that has a silica content of 40% to 50% and makes up most of the ocean floor. Basalt is the most abundant volcanic rock in the Earth's crust.

basaltic Of, containing, or composed of *basalt.*

base level The lowest level to which a *stream* can erode the channel through which it flows, generally equal to the prevailing global sea level.

basin A round or oval depression in the Earth's surface, containing the youngest section of rock in its lowest, central part.

batholith A massive *discordant pluton* with a surface area greater than 100 square kilometers, typically having a depth of about 30 kilometers. Batholiths are generally found in elongated mountain ranges after the *country rock* above them has eroded.

bay A recess in a *shoreline*, or an inlet between two *headlands.*

baymouth bar A narrow ridge of sand that stretches completely across the mouth of a *bay.* (Also called bay bar and bay barrier.)

beach The part of a *coast* that is washed by waves or tides, which cover it with *sediments* of various sizes and composition, such as sand or pebbles.

beach drift 1. The process by which *swash* and *backwash* move *sediments* along a *beach face.* 2. The sediments so moved. Beach drift typically consists of sand, gravel, shell fragments, and pebbles. See also *longshore drift.*

beach face The portion of a *foreshore* that lies nearest to the sea and regularly receives the *swash* of breaking waves. The beach face is the steepest part of the foreshore.

bed A layer of *sediment* or *sedimentary rock* that can be distinguished from the surrounding layers by such features as chemical composition and grain size.

bed load A body of coarse particles that move along the bottom of a *stream.*

bedding The division of *sediment* or *sedimentary rock* into parallel layers (*beds*) that can be distinguished from each other by such features as chemical composition and grain size.

bedrock The solid mass of rock that makes up the Earth's *crust.*

Benioff zone A region where the *subduction* of oceanic plates causes earthquakes, the *foci* of which are deeper the farther inland they are.

berm A low, narrow layer or mound of *sediment* deposited on a *backshore* by storm waves.

biomass fuel A renewable *fuel* derived from a living organism or the byproduct of a living organism. Biomass fuels include wood, dung, methane gas, and grain alcohol.

bitumen Any of a group of solid and semi-solid *hydrocarbons* that can be converted into liquid form by heating. Bitumens can be refined to produce such commercial products as gasoline, fuel oil, and asphalt.

bituminous coal A shiny black coal that develops from deeply buried *lignite* through heat and pressure, and that has a carbon content of 80% to 93%, which makes it a more efficient heating fuel than lignite.

blind valley A valley formed by and containing *sinkholes* and *disappearing streams*, and therefore dry except during periods of such heavy rainfalls that the sinkholes cannot immediately drain the entire accumulation of water.

body wave A type of *seismic wave* that transmits energy from an earthquake's *focus* through the Earth's interior in all directions. See also *surface wave.*

bond To combine, by means of chemical reaction, with another atom to form a compound. When an atom bonds with another, it either loses, gains, or shares electrons with the other atom.

bornhardt A large, smooth, round or dome-shaped *inselberg.*

Bowen's reaction series The sequence of *igneous rocks* formed from a *mafic magma*, assuming mineral crystals that have already formed continue to react with the liquid magma and so evolve into new minerals, thereby creating the next rock in the sequence.

braided stream A network of converging and diverging *streams* separated from each other by narrow strips of sand and gravel.

breakwater A wall built seaward of a coast to intercept incoming waves and so protect a harbor or shore. Breakwaters are typically built parallel to the coast.

breccia A *clastic rock* composed of particles more than 2 millimeters in diameter and marked by the angularity of its component grains and rock fragments.

breeder reactor A nuclear reactor that manufactures more fissionable isotopes than it consumes. Breeder reactors use the widely available, nonfissionable uranium isotope U-238, together with small amounts of fissionable U-235, to produce a fissionable isotope of plutonium, Pu-239.

brittle failure Rupture of rock, a type of permanent *strain* caused by relatively low *stress.*

burial metamorphism A form of *regional metamorphism* that acts on rocks covered by 5 to 10 kilometers of rock or sediment, caused by heat from the Earth's interior and *lithostatic pressure.*

caldera A vast depression at the top of a *volcanic cone*, formed when an eruption substantially empties the reservoir of *magma* beneath the cone's summit. Eventually the summit collapses inward, creating a caldera. A caldera may be more than 15 kilometers in diameter and more than 1000 meters deep.

caliche A white *soil horizon* consisting of calcium carbonate, typical of arid and semi-arid areas. Brief heavy rains dissolve calcium carbonate in the upper layers of soil and transport it downward; the rainwater then evaporates rapidly, leaving the calcium carbonate to form a new, solid layer of soil.

capacity The ability of a given *stream* to carry *sediment*, measured as the maximum quantity it can transport past a given point on the channel bank in a given amount of time. See also *competence.*

capillary fringe The lowest part of the *zone of aeration*, marked by the rising of water from the *water table* due to the attraction of the water molecules to mineral surfaces and other molecules, and to pressure from the *zone of saturation* below.

carbon-14 dating A form of *radiometric dating* that relies on the 5730-year half-life of radioactive carbon-14, which decays into nitrogen-14, to determine the age of rocks in which carbon-14 is present. Carbon-14 dating is used for rocks from 100 to 100,000 years old.

catastrophism The hypothesis that a series of immense, brief, worldwide upheavals changed the Earth's crust greatly and can account for the development of mountains, valleys, and other features of the Earth. See also *uniformitarianism.*

carbonate One of several minerals containing one central carbon atom with strong *covalent bonds* to three oxygen atoms and typically having *ionic bonds* to one or more positive ions.

cave A naturally formed opening beneath the surface of the Earth, generally formed by *dissolution* of carbonate bedrock. Caves may also form by erosion of coastal bedrock, partial melting of glaciers, or solidification of lava into hollow tubes.

cementation The *diagenetic* process by which sediment grains are bound together by *precipitated* minerals originally dissolved during the chemical weathering of preexisting rocks.

Cenozoic Era The latest era of the *Phanerozoic Eon*, following the *Mesozoic Era* and continuing to the present time, and marked by the presence of a wide variety of mammals, including the first hominids.

chemical sediment *Sediment* that is composed of previously dissolved minerals that have either precipitated from evaporated water or been extracted from water by living organisms and deposited when the organisms died or discarded their shells.

chemical weathering The process by which chemical reactions alter the chemical composition of rocks and minerals that are unstable at the Earth's surface and convert them into more stable substances; *weathering* that changes the chemical makeup of a rock or mineral. See also *mechanical weathering.*

chert A member of a group of *sedimentary rocks* that consist primarily of microscopic *silica* crystals. Chert may be either organic or inorganic, but the most common forms are inorganic.

cinder cone A *pyroclastic cone* composed primarily of *cinders.*

cinders Glassy, porous, *pyroclastic* rock fragments.

cirque A deep, semi-circular basin eroded out of a mountain by an *alpine glacier.*

cirque glacier A small *alpine glacier* that forms inside a *cirque*, typically near the head of a valley.

clastic Being or pertaining to a *sedimentary rock* composed pri-marily from fragments of preexisting rocks or fossils.

clay A mineral particle of any composition that is less than 1/256 of a millimeter in diameter. Not to be confused with *clay minerals.*

clay mineral One of a group of hydrous *silicate* minerals, such as kaolinite and smectite, the extremely small particle size of which imparts the ability to adsorb water. Clay minerals are the stable end-products of the *chemical weathering* of *feldspars.*

cleavage The tendency of certain minerals to break along distinct planes in their *crystal structures* where the bonds are weakest. Cleavage is tested by striking or hammering a mineral, and is classified by the number of surfaces it produces and the angles between adjacent surfaces.

coal A member of a group of easily combustible, organic *sedimentary rocks* composed mostly of plant remains and containing a high proportion of carbon.

coast The area of dry land that borders on a body of water.

cockpit karst A *karst* environment marked by numerous closely spaced, irregular depressions and steep, conical hills.

col A high mountain pass that forms when part of an *arête* erodes.

compaction The *diagenetic* process by which the volume or thickness of sediment is reduced due to pressure from overlying layers of sediment.

composite cone See *stratovolcano.*

competence The ability of a given *stream* to carry *sediment,* measured as the diameter of the largest particle that the stream can transport. See also *capacity.*

compound An electrically neutral substance that consists of two or more *elements* combined in specific, constant proportions. A compound typically has physical characteristics different from those of its constituent elements.

compression *Stress* that reduces the volume or length of a rock, as that produced by the *convergence* of plate margins.

concordant Of or being a pluton that lies parallel to the surrounding layers of rock. See also *discordant.*

cone of depression An area in a *water table* along which water has descended into a well to replace water drawn out, leaving a gap shaped like an inverted cone.

confining pressure See *lithostatic pressure.*

conglomerate A *clastic rock* composed of particles more than 2 millimeters in diameter and marked by the roundness of its component grains and rock fragments.

contact metamorphism *Metamorphism* that is caused by heat from a magmatic *intrusion.*

continental collision The *convergence* of two continental plates, resulting in the formation of mountain ranges.

continental drift The hypothesis, proposed by Alfred Wegener, that today's continents broke off from a single supercontinent and then plowed through the ocean floors into their present positions. This explanation of the shapes and locations of Earth's current continents evolved into the theory of *plate tectonics.*

continental ice sheet An unconfined *glacier* that covers much or all of a continent.

convection cell The cycle of movement in the *asthenosphere* that causes the plates of the *lithosphere* to move. Heated material in the asthenosphere becomes less dense and rises toward the solid lithosphere, through which it cannot rise further and therefore begins to move horizontally, dragging the lithosphere along with it and pushing forward the cooler, denser material in its path. The cooler material eventually sinks down lower into the mantle, becoming heated there and rising up again, continuing the cycle. See also *plate tectonics.*

convergence The coming together of two lithospheric plates. Convergence causes *subduction* when one or both plates is oceanic, and mountain formation when both plates are continental. See also *divergence.*

core The innermost layer of the Earth, consisting primarily of pure metals such as iron and nickel. The core is the densest layer of the Earth, and is divided into the outer core, which is believed to be liquid, and the inner core, which is believed to be solid. See also *crust* and *mantle.*

correlation The process of determining that two or more geographically distant rocks or rock strata originated in the same time period.

country rock 1. The preexisting rock into which a *magma* intrudes. 2. The preexisting rock surrounding a *pluton.*

covalent bond The combination of two or more atoms by sharing electrons so as to achieve chemical stability under the *octet rule.* Atoms that form covalent bonds generally have outer energy levels containing three, four, or five electrons. Covalent bonds are generally stronger than other bonds.

crater See *volcanic crater.*

creep The slowest form of *mass movement,* measured in millimeters or centimeters per year and occurring on virtually all slopes.

cross bed A *bed* made up of particles dropped from a moving current, as of wind or water, and marked by a downward slope that indicates the direction of the current that deposited them.

crust The outermost layer of the Earth, consisting of relatively low-density rocks. See also *core* and *mantle.*

crystal A mineral in which the systematic internal arrangement of atoms is outwardly reflected as a latticework of repeated three-dimensional units that form a geometric solid with a surface consisting of symmetrical planes.

crystal structure 1. The geometric pattern created by the systematic internal arrangement of atoms in a mineral. 2. The systematic internal arrangement of atoms in a mineral. See also *crystal.*

crystalline Marked by the systematic internal arrangement of atoms.

current 1. A broad flow of ocean water that maintains a stable direction and differs from the surrounding water in such features as temperature and salinity. 2. The water in such a flow.

daughter isotope An *isotope* that forms from the *radioactive decay* of a *parent isotope.* A daughter isotope may or may not be of the same element as its parent. If the daughter isotope is radioactive, it will eventually become the parent isotope of a new daughter isotope. The last daughter isotope to form from this process will be stable and nonradioactive.

debris avalanche The sudden, extremely rapid *mass movement* downward of entire layers of *regolith* along very steep slopes. Debris avalanches are generally caused by heavy rains.

debris flow 1. The rapid, downward *mass movement* of particles coarser than sand, often including boulders one meter or more in diameter, at a rate ranging from 2 to 40 kilometers per hour. Debris flows occur along fairly steep slopes. 2. The material that descends in such a flow.

deflation The process by which wind erodes *bedrock* by picking up and transporting loose rock particles.

deformation Any of the processes by which a rock changes its shape, form, or volume.

degradation The process by which a stream's *gradient* becomes less steep, due to the *erosion* of *sediment* from the stream bed. Such erosion generally follows a sharp reduction in the amount of sediment entering the stream.

delta An *alluvial fan* having its apex at the mouth of a *stream.*

dendrochronology A method of *absolute dating* that uses the number of tree rings found in a cross section of a tree trunk or branch to determine the age of the tree.

desert A region with an average annual rainfall of 10 inches or less and sparse vegetation, typically having thin, dry, and crumbly soil. A desert has an *aridity index* greater than 4.0.

desert pavement A closely packed layer of rock fragments concentrated in a layer along the Earth's surface by the *deflation* of finer particles.

desert varnish A thin, shiny red-brown or black layer, principally composed of iron manganese oxides, that coats the surfaces of many exposed desert rocks.

desertification The process through which a desert takes over a formerly nondesert area. When a region begins to undergo desertification, the new conditions typically include a significantly lowered *water table*, a reduced supply of surface water, increased salinity in natural waters and soils, progressive destruction of native vegetation, and an accelerated rate of erosion.

detrital sediment *Sediment* that is composed of transported solid fragments of preexisting igneous, sedimentary, or metamorphic rocks.

diagenesis The set of processes that cause physical and chemical changes in sediment after it has been deposited and buried under another layer of sediment. Diagenesis may culminate in *lithification.*

dike A *discordant pluton* that is substantially wider than it is thick. Dikes are often steeply inclined or nearly vertical. See also *sill.*

dilatancy The expansion of a rock's volume caused by *stress* and *deformation.*

diorite Any of a group of dark, *phaneritic, intrusive rocks* that are the *plutonic* equivalents of *andesite.*

dip The angle formed by the inclined plane of a geological structure and the horizontal plane of the Earth's surface.

dip-slip fault A *fault* in which two sections of rock have moved apart vertically, parallel to the *dip* of the fault plane.

directed pressure Force exerted on a rock along one plane, flattening the rock in that plane and lengthening it in the perpendicular plane.

disappearing stream A surface *stream* that drains rapidly and completely into a *sinkhole.*

discordant Of or being a *pluton* that lies perpendicular or oblique to the surrounding layers of rock. See also *concordant.*

dissolution A form of *chemical weathering* in which water molecules, sometimes in combination with acid or another compound in the environment, attract and remove oppositely charged ions or ion groups from a mineral or rock.

dissolved load A body of sediment carried by a *stream* in the form of *ions* that have dissolved in the water.

distributary One of a network of small *streams* carrying water and sediment from a *trunk stream* into an ocean.

divergence The process by which two lithospheric plates separated by *rifting* move farther apart, with soft mantle rock rising between them and forming new oceanic *lithosphere.* See also *convergence.*

dolostone A *sedimentary rock* composed primarily of dolomite, a mineral made up of calcium, magnesium, carbon, and oxygen. Dolostone is thought to form when magnesium ions replace some of the calcium ions in *limestone,* to which dolostone is similar in both appearance and chemical structure.

dome A round or oval bulge on the Earth's surface, containing the oldest section of rock in its raised, central part. See also *basin.*

drainage basin The area from which water flows into a *stream.* Also called a *watershed.*

drainage divide An area of raised, dry land separating two adjacent *drainage basins.*

drainage pattern The arrangement in which a *stream* erodes the channels of its network of *tributaries.*

drumlin A long, spoon-shaped hill that develops when pressure from an overriding *glacier* reshapes a *moraine.* Drumlins range in height from 5 to 50 meters and in length from 400 to 2000 meters. They slope down in the direction of the ice flow.

ductile deformation See *plastic deformation.*

dune A usually asymmetrical mound or ridge of sand that has been transported and deposited by wind. Dunes form in both arid and humid climates.

dynamothermal metamorphism A form of *regional metamorphism* that acts on rocks caught between two *converging* plates and is initially caused by *directed pressure* from the plates, which causes some of the rocks to rise and others to sink, sometimes by tens of kilometers. The rocks that fall then experience further dynamothermal metamorphism, this time caused by heat from the Earth's interior and *lithostatic pressure* from overlying rocks.

earthflow 1. The *flow* of a dry, highly viscous mass of clay-like or silty *regolith*, typically moving at a rate of one or two meters per hour. 2. The material that descends in such a flow.

earthquake A movement within the Earth's *crust* or *mantle*, caused by the sudden rupture or repositioning of underground rocks as they release *stress*.

ebb tide A *tide* that lowers the water surface of an ocean and moves the shoreline farther seaward.

echo-sounding sonar The mapping of ocean topography based on the time required for sound waves to reach the sea floor and return to the research ship that emits them.

elastic deformation A temporary *stress*-induced change in the shape or volume of a rock, after which the rock returns to its original shape and volume.

elastic limit See *yield point*.

electron A negatively charged particle that orbits rapidly around the *nucleus* of an *atom*. See also *proton*.

element A form of matter that cannot be broken down into a chemically simpler form by heating, cooling, or chemical reactions. There are 106 known elements, 92 of them natural and 14 synthetic. Elements are represented by one- or two-letter abbreviations. See also *atom, atomic number*.

energy level The path of a given electron's orbit around a nucleus, marked by a constant distance from the nucleus.

epicenter The point on the Earth's surface that is located directly above the *focus* of an *earthquake*.

equilibrium line The point in a *glacier* where overall gain in volume equals overall loss, so that the net volume remains stable. The equilibrium line marks the border between the *zone of accumulation* and the *zone of ablation*.

erosion The process by which particles of rock and soil are loosened, as by *weathering*, and then transported elsewhere, as by wind, water, ice, or gravity.

esker A ridge of *sediment* that forms under a glacier's *zone of ablation*, made up of sand and gravel deposited by *meltwater*. An esker may be less than 100 meters or more than 500 kilometers long, and may be anywhere from 3 to over 300 meters high.

evaporite An inorganic *chemical sediment* that *precipitates* when the salty water in which it had dissolved evaporates.

extrusive rock An *igneous rock* formed from *lava* that has flowed out onto the Earth's surface, characterized by rapid solidification and grains that are so small as to be barely visible to the naked eye.

fall The fastest form of *mass movement*, occurring when rock or sediment breaks off from a steep or vertical slope and descends at a rate of 9.8 meters per second. A fall can be extremely dangerous.

fault A *fracture* dividing a rock into two sections that have visibly moved relative to each other.

fault block A section of rock separated from other rock by one or more *faults*.

fault-block mountain A mountain containing tall *horsts* interspersed with much lower *grabens* and bounded on at least one side by a high-angle *normal fault*.

fault metamorphism The *metamorphism* that acts on rocks grinding past one another along a fault and is caused by *directed pressure* and frictional heat.

feldspar Any of a group of light-colored, silicate, *rock-forming minerals* most often found in *plutonic igneous rocks* and *metamorphic rocks* and often containing potassium, sodium, or calcium. Feldspar constitutes 60% of the Earth's crust.

felsic Of or being a light-colored, *igneous rock* with a silica content of 70% or higher. Felsic rocks are generally rich in potassium feldspars, aluminum, and quartz.

firn Firmly packed snow that has survived a summer melting season. Firn has a density of about 0.4 grams per cubic centimeter. Ultimately, firn turns into glacial ice.

fission The division of the *nucleus* of a radioactive *atom*, which causes the release of several subatomic particles. The fission of a given element always occurs at a constant rate. (See also *nuclear fission*.)

fission-track dating A form of *absolute dating* that relies on the constant rate of fission to determine the age of a crystal, by counting the *fission tracks* left in a given area of the crystal.

fission tracks Marks left in the latticework of a mineral crystal by subatomic particles released during the *fission* of a *radioactive* atom trapped inside the crystal.

fjord A deep, steep-walled, U-shaped valley formed by erosion by a *glacier* and submerged with seawater.

flood plain The flat land that surrounds a *stream* and becomes submerged when the stream overflows its banks.

flood tide A *tide* that raises the water surface of an ocean and moves the *shoreline* farther inland.

fluorescence Emission of visible light by a substance, such as a mineral, that is currently exposed to ultraviolet light and absorbs radiation from it. The light appears in the form of glowing, distinctive colors. The emission ends when the exposure to ultraviolet light ends.

focus (plural **foci**) The precise point within the Earth's *crust* or *mantle* where rocks begin to rupture or move in an *earthquake*.

fold A bend that develops in an initially horizontal layer of rock, usually caused by *plastic deformation*. Folds occur most frequently in *sedimentary rocks*.

fold-and-thrust mountain A mountain consisting of *folds*, which developed from extremely thick layers of sediment, and *thrust fault* blocks, and containing both *igneous* and *metamorphic rocks*. Fold-and-thrust mountains may be several thousand kilometers high and a few hundred kilometers wide. The Alps, the Appalachians, the Carpathians, the Himalayas, and the Urals are all fold-and-thrust mountains.

foliation The arrangement of a set of minerals in parallel, sheet-like layers that lie perpendicular to the flattened plane of a rock. Occurs in *metamorphic rocks* on which *directed pressure* has been exerted.

footwall The section of rock that lies below the *fault* plane in a *dip-slip fault.* See also *hanging wall.*

forearc basin A depression in the sea floor located between an *accretionary wedge* and a *volcanic arc* in a *subduction* zone, and lined with trapped *sediment.* See also *backarc basin.*

foreshock A minor, barely detectable *earthquake*, generally preceding a full-scale earthquake with approximately the same *focus.* Major quakes may follow a cluster of foreshocks by as little as a few seconds or as much as several weeks.

foreshore The portion of a *beach* that lies nearest to the sea, extending from the low-tide line to the high-tide line.

fossil A remnant, an imprint, or a trace of an ancient organism, preserved in the Earth's crust.

fossil fuel A nonrenewable energy source, such as oil, gas, or coal, that derives from the organic remains of past life. Fossil fuels consist primarily of *hydrocarbons.*

fractional crystallization The process by which a *magma* produces crystals that then separate from the original magma, so that the chemical composition of the magma changes with each generation of crystals, producing *igneous rocks* of different compositions. The *silica* content of the magma becomes proportionately higher after each crystallization.

fracture (*n*) A crack or break in a rock. (*v*) To break in random places instead of *cleaving.* Said of minerals.

fringing reef A *reef* that forms against or near an island or continental *coast* and grows seaward, sloping sharply towards the sea floor. Fringing reefs usually range from 0.5 to 1.0 or more kilometers in width.

frost wedging A form of *mechanical weathering* caused by the freezing of water that has entered a pore or crack in a rock. The water expands as it freezes, widening the cracks or pores and often loosening or dislodging rock fragments. As the ice forms, it attracts more water, increasing the effects of frost wedging.

fuel A source of energy, especially a combustible substance that can be burned for heat or power, or matter used in *nuclear fission.*

gabbro Any of a group of dark, dense, *phaneritic, intrusive rocks* that are the *plutonic* equivalent to *basalt.*

geochronology The study of the relationship between the history of the Earth and time.

geologic time scale The division of all of Earth history into blocks of time distinguished by geologic and evolutionary events, ordered sequentially and arranged into eons made up of eras, which are in turn made up of periods, which are in turn made up of epochs.

geology The scientific study of the Earth, its origins and evolution, the materials that make it up, and the processes that act on it.

geophysics The branch of *geology* that studies the physics of the Earth, using the physical principles underlying such phenomena as *seismic waves*, heat flow, *gravity,* and *magnetism* to investigate planetary properties.

geyser A *natural spring* marked by the intermittent escape of hot water and steam.

glacial Produced by, transported by, or concerning a *glacier.*

glacial abrasion The process by which a glacier erodes the underlying *bedrock* through contact between the bedrock and rock fragments embedded in the base of the glacier. See also *glacial quarrying.*

glacial drift A load of rock material transported and deposited by a glacier. Glacial drift is usually deposited when the glacier begins to melt.

glacial erratic A rock or rock fragment transported by a glacier and deposited on bedrock of different composition. Glacial erratics range from a few millimeters to several yards in diameter.

glacial quarrying The process by which a glacier erodes the underlying *bedrock* by loosening and ultimately detaching blocks of rock from the bedrock and attaching them instead to the glacier, which then bears the rock fragments away. See also *glacial abrasion.*

glacial till *Drift* that is deposited directly from glacial ice and therefore not *sorted.* Also called *till.* See also *glacial drift.*

glacier A moving body of ice that forms on land from the accumulation and compaction of snow, and that flows downslope or outward due to *gravity* and the pressure of its own weight.

gneiss A coarse-grained, *foliated metamorphic rock* marked by bands of light-colored minerals such as quartz and feldspar that alternate with bands of dark-colored minerals. This alternation develops through *metamorphic differentiation.*

graben A block of rock that lies between two *faults* and has moved downward to form a depression between the two adjacent fault blocks. See also *horst.*

graded bed A *bed* formed by the deposition of sediment in relatively still water, marked by the presence of particles that vary in size, density, and shape. The particles settle in a gradual slope with the coarsest particles at the bottom and the finest at the top.

graded stream A stream maintaining an equilibrium between the processes of erosion and deposition, and therefore between *aggradation* and *degradation.*

gradient The vertical drop in a stream's elevation over a given horizontal distance, expressed as an angle.

granite A pink-colored, *felsic, plutonic rock* that contains potassium and usually sodium *feldspars,* and has a quartz content of about 10%. Granite is commonly found on continents but virtually absent from the ocean basins.

gravity 1. The force of attraction exerted by one body in the universe on another. Gravity is directly proportional to the product of the masses of the two attracted bodies. 2. The force of attraction exerted by the Earth on bodies on or near its surface, tending to pull them toward the Earth's center.

gravity anomaly The difference between an actual measurement of *gravity* at a given location and the measurement predicted by theoretical calculation.

groin A structure that juts out into a body of water perpendicular to the *shoreline* and is built to restore an eroding beach by intercepting *longshore drift* and trapping sand.

guyot A *seamount*, the top of which has been flattened by *weathering*, wave action, or stream *erosion*.

half-life The time necessary for half of the atoms of a *parent isotope* to decay into the *daughter isotope*.

hanging wall The section of rock that lies above the fault plane in a *dip-slip fault*. See also *footwall*.

hardness The degree of resistance of a given mineral to scratching, indicating the strength of the bonds that hold the mineral's atoms together. The hardness of a mineral is measured by rubbing it with substances of known hardness.

headland A cliff that projects out from a *coast* into deep water.

historical geology The study of the history, origin, and evolution of the Earth and all of its life forms and geologic structures.

Holocene Epoch The second epoch of the *Quaternary Period*, beginning approximately 10,000 years ago and continuing to the present time. See also *Pleistocene Epoch*.

hook A *spit* that curves sharply at its coastal end.

horn A high mountain peak that forms when the walls of three or more *cirques* intersect.

hornfels A hard, dark-colored, dense *metamorphic rock* that forms from the *intrusion of* magma into *shale* or *basalt*.

horst A block of rock that lies between two *faults* and has moved upward relative to the two adjacent fault blocks. See also *graben*.

hot spot An area in the upper *mantle*, ranging from 100 to 200 km in width, from which magma rises in a *plume* to form *volcanoes*. A hot spot may endure for ten million years or more.

hydraulic conductivity The extent to which a given substance allows water to flow through it, determined by such factors as sorting and grain size and shape.

hydraulic gradient The difference in *potential* between two points, divided by the lateral distance between the points.

hydraulic lifting The *erosion* of a stream bed by water pressure.

hydrocarbon A molecule that is entirely made up of hydrogen and carbon.

hydrogen bond An intermolecular bond formed with hydrogen.

hydrologic cycle The perpetual movement of water among the mantle, oceans, land, and atmosphere of the Earth.

hydrolysis A form of *chemical weathering* in which ions from water replace equivalently charged ions from a mineral, especially a silicate.

hydrothermal deposit A mineral deposit formed by the *precipitation* of metallic ions from water ranging in temperature from 50° to 700°C.

hydrothermal metamorphism The chemical alteration of preexisting rocks that is caused by the action of hot water.

hypothesis A tentative explanation of a given set of data that is expected to remain valid after future observation and experimentation. See also *theory*.

ice age A period during which the Earth is substantially cooler than usual and a significant portion of its land surface is covered by *glaciers*. Ice ages generally last tens of millions of years.

ice cap An *alpine glacier* that covers the peak of a mountain.

igneous rock A rock made from molten (melted) or partly molten material that has cooled and solidified.

index fossil The *fossil* of an organism known to have existed for a relatively short period of time, used to date the rock in which it is found.

index mineral See *metamorphic index mineral*.

inselberg A steep ridge or hill left when a mountain has eroded and found in an otherwise flat, typically desert plain.

intermolecular bonding The act or process by which two or more groups of atoms or molecules combine due to weak positive or negative charges that develop at various points within each group of atoms due to uneven distribution of their electrons. The side of molecule where electrons are more likely to be found will have a slight negative charge, and the side where they are less likely to be found will have a slight positive charge. Such charged regions attract oppositely charged regions of nearby molecules, forming relatively weak bonds.

internal deformation The rearrangement of the planes within ice crystals, due to pressure from overlying ice and snow, that causes the downward or outward flow of a *glacier*.

intrusion The entrance of *magma* into preexisting rock.

intrusive rock An *igneous rock* formed by the entrance of *magma* into preexisting rock.

ion An atom that has lost or gained one or more electrons, thereby becoming electrically charged.

ionic bond The combination of an atom that has a strong tendency to lose electrons with an atom that has a strong tendency to gain electrons, such that the former transfers one or more electrons to the latter and each achieves chemical stability under the *octet rule*. The atom that loses electrons acquires a positive electric charge and the atom that gains electrons acquires a negative electric charge, so that the resulting compound is electrically neutral.

ionic bonding The act or process of forming of an ionic bond.

ionic substitution The replacement of one type of ion in a mineral by another that is similar to the first in size and charge.

iron catastrophe The sequence of events resulting in the separation of the Earth's matter into concentric zones of differing densities. This sequence began when the temperature of the Earth at depths of 400 to 800 kilometers below the surface rose to the melting point of iron. Molten iron then gravitated toward the Earth's center, and its movement raised the Earth's temperature to approximately 2000ºC. This led other substances to start melting. The densest matter then sank toward the Earth's center, while lighter matter rose toward the surface. The iron catastrophe took place between a few hundred million and one billion years after the Earth formed.

isostasy The equilibrium maintained between the gravity tending to depress and the buoyancy tending to raise a given segment of the *lithosphere* as it floats above the *asthenosphere.*

isotope One of two or more forms of a single element; the atoms of each isotope have the same number of protons but different numbers of neutrons in their nuclei. Thus, isotopes have the same *atomic number* but differ in *atomic mass.*

jetty A structure built along the bank of a stream channel or tidal outlet to direct the flow of a stream or tide and keep the sediment moving so that it cannot build up and fill the channel. Jetties are typically built in parallel pairs along both banks of the channel. Jetties that are built perpendicular to a *coast* tend to interrupt *longshore drift* and thus widen *beaches.*

joint A *fracture* dividing a rock into two sections that have not visibly moved relative to each other. See also *fault.*

juvenile water The steam that accompanies *volcanic* eruptions.

karst A topography characterized by *caves, sinkholes, disappearing streams,* and underground drainage. Karst forms when groundwater dissolves pockets of limestone, dolomite, or gypsum in bedrock.

kerogen A solid, waxy, organic substance that forms when pressure and heat from the Earth act on the remains of plants and animals. Kerogen converts to various liquid and gaseous *hydrocarbons* at a depth of seven or more kilometers and a temperature between 50º and 100ºC.

laccolith A large *concordant pluton* that is shaped like a dome or a mushroom. Laccoliths tend to form at relatively shallow depths and are typically composed of granite. The *country rock* above them often erodes away completely.

lagoon A shallow body of water separated from the sea by a *reef* or *barrier island.*

lahar A *flow* of *pyroclastic* material mixed with water. A lahar is often produced when a snow-capped volcano erupts and hot pyroclastics melt a large amount of snow or ice.

lava *Magma* that comes to the Earth's surface through a *volcano* or fissure.

leeward Of, located on, or being the side of a dune, hill, or ridge that is sheltered from the wind. See also *windward.*

levee A protective barrier built along the banks of a stream to prevent flooding. See also *natural levee.*

lichen Plant-like colonies of fungi and algae that grow on the exposed surface of rocks. Lichen grows at a constant rate within a single geographic area.

lichenometry A method of *absolute dating* that uses the size of *lichen* colonies on a rock surface to determine the surface's age. Lichenometry is used for rock surfaces less than about 9000 years old.

lignite A soft, brownish *coal* that develops from *peat* through bacterial action, is rich in *kerogen,* and has a carbon content of 70%, which makes it a more efficient heating fuel than peat.

limestone A *sedimentary rock* composed primarily of calcium carbonate. 10% to 15% of all sedimentary rocks are limestones. Limestone is usually organic, but it may also be inorganic.

liquefaction The conversion of moderately cohesive, unconsolidated *sediment* into a fluid, water-saturated mass.

lithification The conversion of loose *sediment* into solid *sedimentary rock.*

lithosphere A layer of solid, brittle rock comprising the outer 100 kilometers of the Earth, encompassing both the crust and the outermost part of the upper *mantle.* See also *asthenosphere.*

lithostatic pressure The force exerted on a rock buried deep within the Earth by overlying rocks. Because lithostatic pressure is exerted equally from all sides of a rock, it compresses the rock into a smaller, denser form without altering the rock's shape.

loess A load of *silt* that is produced by the erosion of *outwash* and transported by wind. Much loess found in the Mississippi Valley, China, and Europe is believed to have been deposited during the *Pleistocene Epoch.*

longitudinal dune One of a series of long, narrow *dunes* lying parallel both to each other and to the prevailing wind direction. Longitudinal dunes range from 60 meters to 100 kilometers in length and from 3 to 50 meters in height.

longshore current An ocean *current* that flows close and almost parallel to the *shoreline* and is caused by the rush of waves toward the shore.

longshore drift 1. The process by which a *current* moves *sediments* along a *surf zone.* 2. The sediments so moved. Longshore drift typically consists of sand, gravel, shell fragments, and pebbles. See also *beach drift.*

low-velocity zone An area within the Earth's upper *mantle* in which both *P waves* and *S waves* travel at markedly slower velocities than in the outermost part of the upper mantle. The low-velocity zone occurs in the range between 100 and 350 kilometers of depth.

luster 1. The reflection of light on a given mineral's surface, classified by intensity and quality. 2. The appearance of a given mineral as characterized by the intensity and quality with which it reflects light.

magma Molten (melted) rock that forms naturally within the Earth. Magma may be either a liquid or a fluid mixture of liquid, crystals, and dissolved gases.

magnetic field The region within which the *magnetism* of a given substance or particle affects other substances.

magnetic reversal The process by which the Earth's magnetic north pole and its magnetic south pole reverse their positions over time.

magnetism The property, possessed by certain materials, to attract or repel similar materials. Magnetism is associated with moving electricity.

mantle The middle layer of the Earth, lying just below the *crust* and consisting of relatively dense rocks. The mantle is divided into two sections, the upper mantle and the lower mantle; the lower mantle has greater density than the upper mantle. See also *core* and *crust.*

marble A coarse-grained, *nonfoliated metamorphic rock* derived from *limestone* or *dolostone.*

marine magnetic anomaly An irregularity in magnetic strength along the ocean floor that reflects *sea-floor spreading* during periods of *magnetic reversal.*

massive sulfide deposit An unusually large deposit of sulfide minerals.

mass movement The process by which such Earth materials as *bedrock,* loose *sediment,* and *soil* are transported down slopes by *gravity.*

meandering stream A *stream* that traverses relatively flat land in fairly evenly spaced loops and separated from each other by narrow strips of *flood plain.*

mechanical exfoliation A form of *mechanical weathering* in which successive layers of a large *plutonic rock* break loose and fall when the erosion of overlying material permits the rock to expand upward. The thin slabs of rock that break off fall parallel to the exposed surface of the rock, creating the long, broad steps that can be found on many mountains.

mechanical weathering The process by which a rock or mineral is broken down into smaller fragments without altering its chemical makeup; *weathering* that affects only physical characteristics. See also *chemical weathering.*

mélange A body of rock that forms along the inner wall of an *ocean trench* and is made up of fragments of *lithosphere* and oceanic *sediment* that have undergone *metamorphism.*

meltwater Water formed from the melted ice of a *glacier.*

Mercalli intensity scale A scale designed to measure the degree of intensity of *earthquakes*, ranging from I for the lowest intensity to XII for the highest. The classifications are based on human perceptions.

Mesozoic Era The intermediate era of the *Phanerozoic Eon,* following the *Paleozoic Era* and preceding the *Cenozoic Era*, and marked by the dominance of marine and terrestrial reptiles, and the appearance of birds, mammals, and flowering plants.

metallic bonding The act or process by which two or more atoms of electron-donating elements pack so closely together that some of their electrons begin to wander among the nuclei rather than orbiting the nucleus of a single atom. Metallic bonding is responsible for the distinctive properties of metals.

metamorphic differentiation The process by which minerals from a chemically uniform rock separate from each other during *metamorphism* and form individual layers within a new *metamorphic rock.*

metamorphic facies 1. A group of minerals customarily found together in *metamorphic rocks* and indicating a particular set of temperature and pressure conditions at which metamorphism occurred. 2. A set of *metamorphic rocks* characterized by the presence of such a group of minerals.

metamorphic grade A measure used to identify the degree to which a *metamorphic rock* has changed from its *parent rock.* A metamorphic grade provides some indication of the circumstances under which the metamorphism took place.

metamorphic index mineral One of a set of minerals found in *metamorphic rocks* and used as indicators of the temperature and pressure conditions at which the metamorphism occurred. A metamorphic index mineral is stable only within a narrow range of temperatures and pressures and the metamorphism that produces it must take place within that range.

metamorphic rock A *rock* that has undergone chemical or structural changes. Heat, pressure, or a chemical reaction may cause such changes.

metamorphism The process by which conditions within the Earth, below the zone of *diagenesis,* alter the mineral content, chemical composition, and structure of solid rock without melting it. *Igneous, sedimentary,* and *metamorphic rocks* may all undergo metamorphism.

meteoric water The precipitation of condensed water from clouds as rain, snow, sleet, or hail.

microcontinent A section of continental *lithosphere* that has broken off from a larger, distant continent, as by *rifting.*

mid-ocean ridge An underwater mountain range that develops between the margins of two lithospheric plates, formed by *rifting.*

migmatite A rock that incorporates both *metamorphic* and *igneous* materials.

mineral A naturally occurring, usually inorganic, solid consisting of either a single element or a *compound*, and having a definite chemical composition and a systematic internal arrangement of atoms.

mineraloid A naturally occurring, usually inorganic, solid consisting of either a single element or a *compound*, and having a definite chemical composition but lacking a systemic internal arrangement of atoms. See also *mineral.*

mineral zone An area of rock throughout which a given *metamorphic index mineral* is found, presumed to have undergone metamorphism under uniform temperature and pressure conditions.

Moho (abbreviation for Mohorovičić) The *seismic discontinuity* between the base of the Earth's *crust* and the top of the *mantle. P waves* passing through the Moho change their velocity by approximately one kilometer per second, with the higher velocity occurring in the mantle and the lower in the crust.

molecule The smallest particle that retains all the chemical and physical properties of a given *compound,* consisting of a stable group of bonded atoms.

moraine A single, large mass of *glacial till* that accumulates, typically at the edge of a glacier.

mudcrack A fracture that develops at the top of a layer of fine-grained, muddy sediment when it is exposed to the air, dries out, and then shrinks.

mudflow The rapid flow of typically fine-grained *regolith* mixed with water. There may be as much as 60% water in a mudflow.

natural bridge An arch-shaped stretch of *bedrock* remaining in a *karst* region when the surrounding bedrock has dissolved.

natural levee One of a pair of ridges of *sediment* deposited along both banks of a *stream* during successive floods.

natural spring A place where groundwater flows to the surface and issues freely from the ground.

neutron A particle that is found in the *nucleus* of an *atom,* has a mass approximately equal to that of a *proton,* and has no electric charge.

nonfoliated Being a *metamorphic rock* that does not show *foliation.*

normal fault A *dip-slip fault* marked by a generally steep *dip* along which the *hanging wall* has moved downward relative to the *footwall.*

nuclear fission The division of the *nuclei* of *isotopes* of certain heavy *elements,* such as uranium and plutonium, effected by bombardment with *neutrons.* Nuclear fission causes the release of energy, additional neutrons, and an enormous quantity of heat. Nuclear fission is used in nuclear power plants and nuclear weapons. A byproduct of nuclear fission is toxic radioactive waste. See also *nuclear fusion.*

nuclear fusion The combination of the *nuclei* of certain extremely light *elements,* especially hydrogen, effected by the application of high temperature and pressure. Nuclear fusion causes the release of an enormous amount of heat energy, comparable to that released by *nuclear fission.* The principle byproduct of nuclear fusion is helium.

nucleus (plural **nuclei**) The central part of an *atom,* containing most of the atom's mass and having a positive charge due to the presence of *protons.*

nuée ardente A sometimes glowing cloud of gas and *pyroclastics* erupted from a *volcano* and moving swiftly down its slopes. Also called a *pyroclastic flow.*

ocean trench A deep, linear, relatively narrow depression in the sea floor, formed by the *subduction* of oceanic plates.

octet rule A scientific law stating that all atoms, except those of hydrogen and helium, require eight electrons in the outermost *energy level* in order to maintain chemical stability.

oil sand A mixture of unconsolidated sand and clay that contains a semi-solid *bitumen.*

oil shale A brown or black *clastic source rock* containing *kerogen.*

ophiolite suite The group of *sediments, sedimentary rocks,* and mafic and ultramafic *igneous rocks* that make up the oceanic *lithosphere.*

ore A mineral deposit that can be mined for a profit.

orogenesis Mountain formation, as caused by *volcanism, subduction, plate divergence, folding,* or the movement of *fault blocks.* Also called *orogeny.*

oscillatory motion The circular movement of water up and down, with little or no change in position, as a wave passes.

outwash A load of *sediment,* consisting of sand and gravel, that is deposited by *meltwater* in front of a *glacier.*

oxbow lake A crescent-shaped body of standing water formed from a single loop that was cut off from a *meandering stream,* typically by a flood that allowed the stream to flow through its *flood plain* and bypass the loop.

oxidation The process of combining with oxygen ions. A mineral that is exposed to air may undergo oxidation as a form of *chemical weathering.*

oxide One of several minerals containing negative oxygen ions bonded to one or more positive metallic ions.

paleosol An ancient, buried soil whose composition may reflect a climate significantly different from the climate now prevalent in the area where the soil is found.

paleomagnetism 1. The fixed orientation of a rock's crystals, based on the Earth's *magnetic field* at the time of the rock's formation, that remains constant even when the magnetic field changes. 2. The study of such phenomena as indicators of the Earth's magnetic history.

Paleozoic Era The earliest era of the *Phanerozoic Eon,* marked by the presence of marine invertebrates, fish, amphibians, insects, and land plants.

parabolic dune A horseshoe-shaped *dune* having a concave *windward* slope and a convex *leeward* slope. Parabolic dunes tend to form along sandy ocean and lake shores. They may also develop from *transverse dunes* through *deflation.*

parent isotope A *radioactive isotope* that changes into a different isotope when its nucleus decays. See also *daughter isotope.*

parent material The source from which a given soil is chiefly derived, generally consisting of *bedrock* or *sediment.*

parent rock The preexisting rock from which a *metamorphic rock* forms.

partial melting The incomplete melting of a rock composed of minerals with differing melting points. When partial melting occurs, the minerals with higher melting points remain solid while the minerals whose melting points have been reached turn to *magma.*

passive continental margin A border that lies between continental and oceanic *lithosphere,* but is not a *plate* margin. It is marked by lack of seismic and volcanic activity.

peat A soft brown mass of compressed, partially decomposed vegetation that forms in a water-saturated environment and has a carbon content of 50%. Dried peat can be burned as fuel.

pediment A broad surface at the base of a receding mountain. The pediment develops when running water erodes most of the mass of the mountain.

pegmatite A coarse-grained *igneous rock* with exceptionally large *crystals*, formed from a *magma* that contains a high proportion of water.

perched water table A saturated area that lies within a *zone of aeration.*

peridotite An *igneous rock* composed primarily of the iron-magnesium *silicate* olivine and having a silica content of less than 40%.

permafrost Permanently frozen *regolith*, ranging in thickness from 30 centimeters to over 1000 meters.

permeability The capability of a given substance to allow the passage of a fluid. Permeability depends upon the size of and the degree of connection among a substance's pores.

petroleum Any of a group of naturally occurring substances made up of *hydrocarbons.* These substances may be gaseous, liquid, or semisolid.

phaneritic Of or being an *igneous rock* containing components large enough to be seen with the unaided eye.

Phanerozoic Eon The eon that started 570 million years ago, when numerous fossils of sea shells began to be formed, and that continues to the present time.

phosphorescence Emission of visible light by a substance, such as a mineral, that is exposed to ultraviolet light and absorbs radiation from it. The light appears in the form of glowing, distinctive colors. The emission continues after the exposure to ultraviolet light ends.

phyllite A *foliated metamorphic rock* that develops from *slate* and is marked by a silky sheen and medium grain size.

placer deposit A deposit of heavy or durable minerals, such as gold or diamonds, typically found where the flow of water abruptly slows.

plastic deformation A permanent *strain* that entails no rupture.

plate One of the large, thin, rigid units making up the Earth's *lithosphere.* Plates may be continental, oceanic, or both.

plate tectonics The theory that the Earth's *lithosphere* consists of large, rigid plates that move horizontally in response to the flow of the *asthenosphere* beneath them, and that interactions among the plates at their borders cause most major geologic activity, including the creation of oceans, continents, mountains, volcanoes, and earthquakes.

playa A dry lake basin found in a desert.

Pleistocene Epoch The first epoch of the *Quaternary Period*, beginning two to three million years ago and ending approximately 10,000 years ago. See also *Holocene Epoch.*

plume An upward flow of hot material from the Earth's *mantle* into the *crust.*

pluton An *intrusive rock*, as distinguished from the preexisting *country rock* that surrounds it.

plutonic rock An *intrusive rock* formed inside the Earth.

pluvial lake A lake that formed from rainwater falling into a landlocked basin during a *glacial* period marked by greater precipitation than is found in the region in prior or subsequent periods.

point bar A low ridge of *sediment* that forms along the inner bank of a *meandering stream.*

polymorph A mineral that is identical to another mineral in chemical composition but differs from it in *crystal structure*.

porosity The percentage of a soil, rock, or sediment's volume that is made up of pores.

porphyritic Of or being an *igneous rock* containing some large grains within a smaller-grained matrix.

porphyry copper deposit A crystallized rock, typically *porphyritic*, having hairline fractures that contain copper and other metals.

potassium-argon dating A form of *radiometric dating* that relies on the extremely long *half-life* of radioactive isotopes of potassium, which decay into isotopes of argon, to determine the age of rocks in which argon is present. Potassium-argon dating is used for rocks between 100,000 and 4 billion years old.

potential The combined influence of *gravity* and water pressure on groundwater flow at a given depth.

potentiometric surface The level to which the water in an *artesian aquifer* would rise if unaffected by friction with the surrounding rocks and sediments.

precipitate To separate from solution in solid form. Minerals may precipitate because of cooling, evaporation, or loss of acidity.

precipitation 1. The process by which a substance becomes *precipitated.* 2. Water that falls from the atmosphere to Earth's surface in the form of rain, snow, sleet, or hail.

primary coast A *coast* shaped primarily by nonmarine processes, such as *glacial erosion* or biological processes.

principle of cross-cutting relationships The *scientific law* stating that a *pluton* is always younger than the rock that surrounds it.

principle of faunal succession The *scientific law* stating that an organism is always simpler than those that evolved later and more complex than those that evolved earlier.

principle of inclusions The *scientific law* stating that rock fragments contained within a larger body of rock are always older than the surrounding body of rock.

principle of original horizontality The *scientific law* stating that *sediments* settling out from bodies of water are deposited horizontally or nearly horizontally in layers that lie parallel or nearly parallel to the Earth's surface.

principle of superposition The *scientific law* stating that in any unaltered sequence of rock strata, each stratum is younger than the one beneath it and older than the one above it, so that the youngest stratum will be at the top of the sequence and the oldest at the bottom.

principle of uniformitarianism The *scientific law* stating that the geological processes taking place in the present operated similarly in the past and can therefore be used to explain past geologic events.

property A characteristic that distinguishes one substance from another.

proton A positively charged particle that is found in the *nucleus* of an *atom* and has a mass approximately 1836 times that of an *electron.*

P wave (abbreviation for **primary wave**) A *body wave* that causes the *compression* of rocks when its energy acts upon them. When the P wave moves past a rock, the rock expands beyond its original volume, only to be compressed again by the next P wave. P waves are the fastest of all *seismic waves.* See also *S wave.*

P-wave shadow zone The region that extends from 103° to 143° from the *epicenter* of an *earthquake* and is marked by the absence of *P waves.* The P-wave shadow zone is due to the refraction of *seismic waves* in the liquid outer *core.* See also *S-wave shadow zone.*

pyroclastic Being or pertaining to rock fragments formed in a volcanic eruption.

pyroclastic cone A usually steep, conic *volcano* composed almost entirely of an accumulation of loose *pyroclastic* material. Pyroclastic cones are usually less than 450 meters high. Because no *lava* binds the *pyroclastics,* pyroclastic cones erode easily.

pyroclastic eruption A volcanic eruption of *viscous,* gas-rich magma. Pyroclastic eruptions tend to produce a great deal of solid volcanic fragments rather than fluid *lava.*

pyroclastic flow A rapid, extremely hot, downward stream of *pyroclastics,* air, gases, and ash ejected from an erupting *volcano.* A pyroclastic flow may be as hot as 800°C or more and may move at speeds higher than 150 kilometers per hour.

pyroclastics (used only in the plural) Particles and chunks of *igneous rock* ejected from a *volcanic vent* during an eruption.

quake See *earthquake.*

quartzite An extremely durable, *nonfoliated metamorphic rock* derived from pure *sandstone* and consisting primarily of quartz.

Quaternary ice age An *ice age* that began approximately 1.6 million years ago and continues to the present time.

Quaternary Period The second period of the *Cenozoic Era,* beginning two to three million years ago and continuing to the present time.

quick clay *Sediment* that sets off a sudden *mudflow* by changing rapidly from solid to liquid form, as after an earthquake, an explosion, or thunder.

radioactive decay The process of spontaneously emitting *protons* and *neutrons* that transforms one *isotope* into another.

radiometric dating The process of using relative proportions of *parent* to *daughter isotopes* in *radioactive decay* to determine the age of a given rock or rock stratum.

rain shadow effect The result of the process by which moist air on the *windward* side of a mountain rises and cools, causing precipitation and leaving the *leeward* side of the mountain dry.

recrystallization The *diagenetic* process by which unstable minerals in buried sediment are transformed into stable ones.

reef A ridge that forms in clear, moderately salty seawater near the *shoreline* and is composed of the carbonate remains of algae, sponges, and especially corals.

regional metamorphism *Metamorphism* that affects rocks over vast geographic areas stretching for thousands of square kilometers.

regolith The unconsolidated material that covers almost all of the Earth's land surface and is composed of *soil, sediment,* and fragments from the *bedrock* beneath it.

relative dating The fixing of a geologic structure or event in a chronological sequence relative to other geologic structures or events. See also *absolute dating.*

reserve A known *resource* that can be exploited for profit with available technology under existing political and economic conditions.

reservoir rock A permeable rock containing oil or gas.

resource A mineral or fuel deposit, known or not yet discovered, that may be or become available for human exploitation.

reverse fault A *dip-slip fault* marked by a *hanging wall* that has moved upward relative to the *footwall.* Reverse faults are often caused by the *convergence* of lithospheric plates.

rhyolite Any of a group of *felsic igneous rocks* that are the *extrusive* equivalents of *granite.*

Richter scale A logarithmic scale that measures the amount of energy released during an *earthquake* on the basis of the amplitude of the highest peak recorded on a *seismogram.* Each unit increase in the Richter scale represents a 10-fold increase in the amplitude recorded on the seismogram and a 30-fold increase in energy released by the earthquake. Theoretically the Richter scale has no upper limit, but the *yield point* of the Earth's rocks imposes an effective limit between 9.0 and 9.5.

rifting The tearing apart of a *plate* to form a depression in the Earth's *crust* and often eventually separating the plate into two or more smaller plates.

rip current A strong, rapid, and brief *current* that flows out to sea, moving perpendicular to the *shoreline.*

ripple marks A pattern of wavy lines formed along the top of a *bed* by wind, water currents, or waves.

riprap A pile of large, angular boulders built seaward of the *shoreline* in order to prevent erosion by waves or currents. See also *seawall.*

rock A naturally formed aggregate of usually inorganic materials from within the Earth.

rock cycle A series of events through which a *rock* changes, over time, between *igneous*, *sedimentary*, and *metamorphic* forms.

rock-forming mineral One of the twenty or so minerals contained in the rock that composes the Earth's crust and mantle.

rubidium-strontium dating A form of *radiometric dating* that relies on the 47-billion-year half-life of radioactive isotopes of rubidium, which decay into isotopes of strontium, to determine the age of rocks in which strontium is present. Rubidium-strontium dating is used for rocks that are at least 10 million years old, deep-Earth plutonic rocks, and Moon rocks.

sand 1. A particle of rock or mineral material, coarser than *silt*, that has been transported from its place of origin, as by water or wind. A particle of sand is usually between 1/16 and two millimeters in diameter. Sands are frequently composed of quartz. 2. A loosely connected body of such particles.

sandstone A *clastic rock* composed of particles that range in diameter from 1/16 millimeter to 2 millimeters in diameter. Sandstones make up about 25% of all sedimentary rocks.

saturation zone See *zone of saturation.*

scarp The steep cliff face that is formed by a *slump.*

scientific law 1. A natural phenomenon that has been proven to occur invariably whenever certain conditions are met. 2. A formal statement describing such a phenomenon and the conditions under which it occurs. Also called *law.*

scientific methods Techniques that involve gathering all available data on a subject, forming an *hypothesis* to explain the data, conducting experiments to test the hypothesis, and modifying or confirming the hypothesis as necessary to account for the experimental results.

schist A coarse-grained, strongly *foliated metamorphic rock* that develops from *phyllite* and splits easily into flat, parallel slabs.

sea stack A steep, isolated island of rock, separated from a *headland* by the action of waves, as when the overhanging section of a *sea arch* is eroded.

sea-floor spreading The formation and growth of oceans that occurs following *rifting* and is characterized by eruptions along *mid-ocean ridges*, forming new oceanic *lithosphere*, and expanding ocean basins. See also *divergence.*

seamount A conical underwater mountain formed by a *volcano* and rising 1000 meters or more from the sea floor.

seawall A wall of stone, concrete, or other sturdy material, built along the *shoreline* to prevent erosion even by the strongest and highest of waves. See also *riprap.*

secondary coast A *coast* shaped primarily by erosion or deposition by sea currents and waves.

secondary enrichment The process by which a metal deposit becomes concentrated when other minerals are eliminated from the deposit, as through *dissolution, precipitation,* or *weathering.*

sediment A collection of transported fragments or precipitated materials that accumulate, typically in loose layers, as of sand or mud.

sedimentary environment The continental, oceanic, or coastal surroundings in which sediment accumulates.

sedimentary facies 1. A set of characteristics that distinguish a given section of sedimentary rock from nearby sections. Such characteristics include mineral content, grain size, shape, and density. 2. A section of sedimentary rock so characterized.

sedimentary rock A *rock* made from the consolidation of solid fragments, as of other rocks or organic remains, or by *precipitation* of minerals from solution.

sedimentary structure A physical characteristic of a *detrital sediment* that reflects the conditions under which the sediment was deposited.

seismic Of, concerning, subject to, or produced by an *earthquake.*

seismic discontinuity A surface marking the boundary between two layers of the Earth differing in composition. *Seismic waves* passing through a seismic discontinuity undergo an abrupt change in velocity.

seismic gap A locked fault segment that has not experienced seismic activity for a long time. Because *stress* tends to accumulate in seismic gaps, they often become the sites of major *earthquakes.*

seismic profiling The mapping of rocks lying along and beneath the ocean floor by recording the reflections and refractions of *seismic waves.*

seismic tomography The process whereby a computer first synthesizes data on the velocities of *seismic waves* from thousands of recent earthquakes in order to make a series of images depicting successive planes within the Earth, and then uses these images to construct a three-dimensional representation of the Earth's interior.

seismic wave One of a series of progressive disturbances that reverberate through the Earth to transmit the energy released from an *earthquake.*

seismogram A visual record produced by a *seismograph* and showing the arrival times and magnitudes of various *seismic waves.*

seismograph A machine for measuring the intensity of *earthquakes* by recording the *seismic waves* that they generate.

seismology The study of *earthquakes* and the structure of the Earth, based on data from *seismic waves.*

shale A *sedimentary rock* composed of *detrital sediment* particles less than 0.004 millimeters in diameter. Shales tend to be red, brown, black, or gray, and usually originate in relatively still waters.

shearing stress *Stress* that slices rocks into parallel blocks that slide in opposite directions along their adjacent sides. Shearing stress may be caused by *transform motion.*

shield volcano A low, broad, gently sloping, dome-shaped structure that forms over time as repeated eruptions eject *basaltic lava* through one or more *vents* and the lava solidifies in approximately the same volume all around.

shock metamorphism The *metamorphism* that results when a meteorite strikes rocks at the Earth's surface. The meteoric impact generates tremendous pressure and extremely high temperatures that cause minerals to shatter and recrystallize, producing new minerals which cannot arise under any other circumstances.

shoreline The boundary between a body of water and dry land.

silica A *compound* consisting of silicon and oxygen.

silicate One of several rock-forming minerals that contain silicon, oxygen, and usually one or more other common elements.

silicon-oxygen tetrahedron A four-sided geometric form created by the tight bonding of four oxygen atoms to each other, and also to a single silicon atom that lies in the middle of the form.

sill A *concordant pluton* that is substantially wider than it is thick. Sills form within a few kilometers of the Earth's surface. See also *dike.*

silt 1. A particle of rock or mineral material, finer than *sand* but coarser than *clay*, that has been transported from its place of origin, typically by wind or water. A particle of silt is usually between 1/16 and 1/256 of a millimeter in diameter. 2. A loosely connected body of such particles.

sinkhole A circular, often funnel-shaped depression in the ground that forms when soluble rocks dissolve.

skarn A coarse-grained, *nonfoliated metamorphic rock* containing *silicates* that are rich in calcium.

slate A fine-grained, *foliated metamorphic rock* that develops from *shale* and tends to break into thin, flat sheets.

slide The *mass movement* of a single, intact mass of rock, soil, or unconsolidated material along a weak plane, such as a *fault, fracture,* or *bedding* plane. A slide may involve as little as a minor displacement of soil or as much as the displacement of an entire mountainside.

slip face The steep *leeward* slope of a *dune.*

slip plane A weak plane in a rock mass from which material is likely to break off in a *slide.*

slump 1. A downward and outward *slide* occurring along a concave *slip plane.* 2. The material that breaks off in such a slide.

snowline The lowest point at which snow remains year-round.

soil The top few meters of *regolith,* generally including some organic matter derived from plants.

soil horizon A layer of soil that can be distinguished from the surrounding soil by such features as chemical composition, color, and texture.

soil profile A vertical strip of soil stretching from the surface down to the *bedrock* and including all of the successive *soil horizons.*

solifluction A form of *creep* in which soil flows downslope at 0.5 to 15 centimeters per year. Solifluction occurs in relatively cold regions when the brief warmth of summer thaws only the upper meter or two of *regolith,* which becomes waterlogged because the underlying ground remains frozen and therefore the water cannot drain down into it.

source rock A rock in which *hydrocarbons* originate.

sorting The process by which a given *transport medium* separates out certain particles, as on the basis of size, shape, or density.

specific gravity The ratio of the weight of a particular volume of a given substance to the weight of an equal volume of pure water.

speleothem A mineral deposit of calcium carbonate that precipitates from solution in a *cave.*

spheroidal weathering The process by which *chemical weathering,* especially by water, decomposes the angles and edges of a rock or boulder, leaving a rounded form from which concentric layers are then stripped away as the weathering continues.

spit A narrow, fingerlike ridge of sand that extends from land into open water.

stalactite An icicle-like mineral formation that hangs from the ceiling of a *cave* and is usually made up of *travertine,* which precipitates as water rich in dissolved limestone drips down from the cave's ceiling. See also *stalagmite.*

stalagmite A cone-shaped mineral deposit that forms on the floor of a *cave* and is usually made up of *travertine*, which precipitates as water rich in dissolved limestone drips down from the cave's ceiling. See also *stalactite.*

star dune A *dune* with three or four arms radiating from its usually higher center so that it resembles a star in shape. Star dunes form when winds blow from three or four directions, or when the wind direction shifts frequently.

stratification See *bedding.*

stratovolcano A cone-shaped *volcano* built from alternating layers of *pyroclastics* and viscous *andesitic lava.* Stratovolcanos tend to be very large and steep.

stratum (plural **strata**) A layer of *sedimentary rock* that is visibly distinct from the surrounding layers.

streak The color of a mineral in its powdered form. This color is usually determined by rubbing the mineral against an unglazed porcelain slab and observing the mark made by it on the slab.

strain The change in the shape or volume of a rock that results from *stress.*

stream A body of water found on the Earth's surface and confined to a narrow topographic depression, down which it flows and transports rock particles, sediment, and dissolved particles. Rivers, creeks, brooks, and runs are all streams.

stream discharge The volume of water to pass a given point on a stream bank per unit of time, usually expressed in cubic meters of water per second.

stream terrace A level plain lying above and running parallel to a stream bed. A stream terrace is formed when a stream's bed erodes to a substantially lower level, leaving its flood plain high above it.

stress The force acting on a rock or another solid to deform it, measured in kilograms per square centimeter or pounds per square inch.

striation One of a group of usually parallel scratches engraved in bedrock by a *glacier* or other geological agent.

strike 1. The horizontal line marking the intersection between the inclined plane of a solid geological structure and the Earth's surface. 2. The compass direction of this line, measured in degrees from true north.

strike-slip fault A *fault* in which two sections of rock have moved horizontally in opposite directions, parallel to the line of the *fracture* that divided them. Strike-slip faults are caused by *shearing stress.*

structural geology The scientific study of the geological processes that deform the Earth's crust and create mountains.

subduction The sinking of an oceanic *plate* edge as a result of *convergence* with a plate of lesser density. Subduction often causes *earthquakes* and creates *volcano* chains.

subsidence The lowering of the Earth's surface, caused by such factors as compaction, a decrease in groundwater, or the pumping of oil.

sulfate One of several minerals containing positive sulfur ions bonded to negative oxygen ions.

sulfide One of several minerals containing negative sulfur ions bonded to one or more positive metallic ions.

surface wave One of a series of *seismic waves* that transmits energy from an earthquake's *epicenter* along the Earth's surface. See also *body wave.*

surf zone The area running from the *shoreline* to the farthest point in the sea where waves begin to break.

surge To flow more rapidly than usually. Said of a *glacier.*

suspended load A body of fine, solid particles, typically of sand, clay, and silt, that travels with stream water without coming into contact with the stream bed.

suture zone The area where two continental plates have joined together through *continental collision.* Suture zones are marked by extremely high mountain ranges, such as the Himalayas and the Alps.

swash The rush of water onto a beach after a wave breaks.

S wave (abbreviation for **secondary wave**) A *body wave* that causes the rocks along which it passes to move up and down perpendicular to the direction of its own movement. See also *P wave.*

S-wave shadow zone The region within an arc of 154° directly opposite an earthquake's *epicenter* that is marked by the absence of *S waves.* The S-wave shadow zone is due to the fact that S waves cannot penetrate the liquid outer core. See also *P-wave shadow zone.*

syncline A concave *fold*, the central part of which contains the youngest section of rock. See also *anticline.*

talus A pile of rock fragments lying at the bottom of the cliff or steep slope from which they have broken off.

tarn A deep, typically circular lake that forms when a *cirque glacier* melts.

tectonic creep The almost constant movement of certain *fault blocks* that allows *strain* energy to be released without major *earthquakes.*

tension *Stress* that stretches or extends rocks, so that they become thinner vertically and longer laterally. Tension may be caused by *divergence* or *rifting.*

tephra (plural noun) *Pyroclastic* materials that fly from an erupting volcano through the air before cooling, and range in size from fine dust to massive blocks.

terminus The outer margin of a *glacier.*

theory A comprehensive explanation of a given set of data that has been repeatedly confirmed by observation and experimentation and has gained general acceptance within the scientific community but has not yet been decisively proven. See also *hypothesis* and *scientific law.*

thermal expansion A form of *mechanical weathering* in which heat causes a mineral's crystal structure to enlarge.

thermal plume A vertical column of upwelling *mantle* material, 100 to 250 kilometers in diameter, that rises from beneath a continent or ocean and can be perceived at the Earth's surface as a *hot spot.* Thermal plumes carry enough energy to move a plate, and they may be found both at plate boundaries and plate interiors.

thrust fault A *reverse fault* marked by a *dip* of 45° or less.

tidal bore A turbulent, abrupt, wall-like wave that is caused by a *flood tide.*

tide 1. The cycle of alternate rising and falling of the surface of an ocean or large lake, caused by the gravitational pull of the Sun and especially Moon in interaction with the Earth's rotation. Tides occur on a regular basis, twice every day on most of the Earth. 2. A single rise or fall within this cycle.

till See *glacial till.*

topography The set of physical features, such as mountains, valleys, and the shapes of landforms, that characterizes a given landscape.

transition zone The *seismic discontinuity* located in the upper *mantle* just beneath the *asthenosphere* and characterized by a marked increase in the velocity of *seismic waves.*

transform motion The movement of two adjacent lithospheric plates in opposite directions along a parallel line at their common edge. Transform motion often causes *earthquakes.*

translatory Of, concerning, or being the movement of water over a significant distance in the direction of a wave.

transport medium A natural agent, such as water, air, or ice, that moves a particle or particles from one location on the Earth's surface to another.

transverse dune One of a series of *dunes* having an especially steep *slip face* and a gentle *windward* slope and standing perpendicular to the prevailing wind direction and parallel to each other. Transverse dunes typically form in arid and semi-arid regions with plentiful sand, stable wind direction, and scarce vegetation. A transverse dune may be as much as 100 kilometers long, 200 meters high, and three kilometers wide.

travertine *Crystalline* deposits of calcium carbonate precipitated from solution, often found in *caves.*

tributary A *stream* that supplies water to a larger stream.

trunk stream A large stream into which *tributaries* carry water and sediment.

tsunami (plural **tsunami**) A vast sea wave caused by the sudden dropping or rising of a section of the sea floor following an *earthquake.* Tsunami may be as much as 30 meters high and 200 kilometers long, may move as fast as 250 kilometers an hour, and may continue to occur for as long as a few days.

tuff See *volcanic tuff.*

uniformitarianism The hypothesis that current geologic processes, such as the slow erosion of a coast under the impact of waves, have been occurring in a similar manner throughout the Earth's history and that these processes can account for past geologic events. See also *catastrophism.*

unconformity A boundary separating two or more rocks of markedly different ages, marking a gap in the geologic record.

upwarped mountain A mountain consisting of a broad area of the Earth's *crust* that has moved gently upward without much apparent deformation, and usually containing *sedimentary, igneous,* and *metamorphic rocks.*

uranium-lead dating A form of *radiometric dating* that relies on the extremely long *half-life* of radioactive isotopes of uranium, which decay into isotopes of lead, to determine the age of rocks in which uranium and lead are present.

valley glacier An *alpine glacier* that flows through a preexisting stream valley.

van der Waals bond A relatively weak kind of *intermolecular bond* that forms when one side of a *molecule* develops a slight negative charge because a number of *electrons* have temporarily moved to that side of the *molecule*, and this negative charge attracts the *nuclei* of the *atoms* of a neighboring molecule, while the side of the molecule with fewer electrons develops a slight positive charge that attracts the electrons of the atoms of neighboring molecules.

varve A pair of sediment *beds* deposited by a lake on its floor, typically consisting of a thick, coarse, light-colored bed deposited in the summer and a thin, fine-grained, dark-colored bed deposited in the winter. Varves are most often found in lakes that freeze in the winter. The number and nature of varves on the bottom of a lake provides information about the lake's age and geologic events that affected the lake's development.

vent An opening in the Earth's surface through which *lava,* gases, and hot particles are expelled. Also called *volcanic vent* and *volcano.*

ventifact A stone that has been flattened and sharpened by *wind abrasion.* Ventifacts are commonly found strewn across a *desert* floor.

viscosity A fluid's resistance to flow. Viscosity increases as temperatures decrease.

volcanic arc A chain of *volcanoes* fueled by magma that rises from an underlying *subducting* plate.

volcanic cone A cone-shaped mountain that forms around a *vent* from the debris of *pyroclastics* and *lava* ejected by numerous eruptions over time.

volcanic crater A steep, bowl-shaped depression surrounding a *vent.* A volcanic crater forms when the walls of a vent collapse inward following an eruption.

volcanic dome A bulb-shaped solid that forms over a *vent* when *lava* so *viscous* that it cannot flow out of the *volcanic crater* cools and hardens. When a volcanic dome forms, it traps the volcano's gases beneath it. They either escape along a side vent of the volcano or build pressure that causes another eruption and shatters the volcanic dome.

volcanic rock See *extrusive rock.*

volcanic tuff A solid rock made up of *tephra* that have consolidated and become cemented together. Also called *tuff.*

volcanism The set of geological processes that result in the expulsion of *lava, pyroclastics*, and gases at the Earth's surface.

volcano The solid structure created when lava, gases, and hot particles escape to the Earth's surface through *vents.* Volcanoes are usually conical. A volcano is "active" when it is erupting or has erupted recently. Volcanoes that have not erupted recently but are considered likely to erupt in the future are said to be "dormant." A volcano that has not erupted for a long time and is not expected to erupt in the future is "extinct."

watershed See *drainage basin.*

water table The surface that lies between the *zone of aeration* and the underlying *zone of saturation.*

wave refraction The process by which a wave approaching the shore changes direction due to slowing of those parts of the wave which enter shallow water first, causing a sharp decrease in the angle at which the wave approaches until the wave is almost parallel to the coast.

wave-cut bench A relatively level surface formed when waves erode the base of a cliff, causing the overlying rock to fall into the surf. A wave-cut bench stands above the water and extends seaward from what remains of the cliff.

weathering The process by which exposure to atmospheric agents, such as air or moisture, causes *rocks* and *minerals* to break down. This process takes place at or near the Earth's surface. Weathering entails little or no movement of the material that it loosens from the rocks and minerals. See also *erosion.*

wetland A lake, marsh, or swamp that supports wildlife and replenishes the groundwater system.

wind abrasion The process by which wind erodes *bedrock* through contact between the bedrock and rock particles carried by the wind.

windward Of, located on, or being the side of a dune, hill, or ridge facing into the wind. See also *leeward.*

xenolith A preexisting rock embedded in a newer *igneous rock*. Xenoliths are formed when a rising *magma* incorporates the preexisting rock. If the preexisting rock does not melt, it will not be assimilated into the magma and will therefore remain distinct from the new igneous rock that surrounds it.

X-ray diffraction The scattering of X-rays passed through a mineral sample so as to form a pattern peculiar to the given mineral.

yield point The maximum *stress* that a given rock can withstand without becoming permanently deformed.

zone of ablation The part of a *glacier* in which there is greater overall loss than gain in volume. A zone of ablation can be identified in the summer by an expanse of bare ice. See also *zone of accumulation.*

zone of accumulation The part of a *glacier* in which there is greater overall gain than loss in volume. A zone of accumulation can be identified by a blanket of snow that survives summer melting. See also *zone of ablation.*

zone of aeration A region below the Earth's surface that is marked by the presence of both water and air in the pores of rocks and soil. Also called *aeration zone.*

zone of saturation A region that lies below the *zone of aeration* and is marked by the presence of water and the absence of air in the pores of rocks and soil.

Illustration Credits

Chapter 1

Opener p. 2–3 James Balog/Black Star; **Figure 1-2** Matthew Naythans/ Gamma-Liaison; **Figure 1-3 (a)** U.S. Geological Survey; **(b)** James L. Amos/Photo Researchers; **(c)** NASA; **Figure 1-4 (a)** Prof. W. Alvarez/ Science Photo Library/Photo Researchers; **(b)** Runk/Schoenberger/Grant Heilman, specimen from North Museum/Franklin Marshall College; **(c)** G. A. Izett, U.S. Geological Survey; **Figure 1-10 (a)** U.S. Geographical Survey W. 100th, no. 47/U.S. Geological Survey; **(b)** R.M. Turner, no. 17/U.S. Geological Survey; **Figure 1-15** Bruce C. Heesen & Marie Tharp, 1977; **Figure 1-18 (a)** Comstock; **(b)** Craig Tuttle/The Stock Market; **Figure 1-20** John S. Shelton; **Figure 1-22** Library of Congress, print courtesy of TIME, Inc. Picture Collection; **Figure 1-26 (a)** Breck P. Kent; **(b)** Betty Crowell/Faraway Places; **Figure 1-27 (a)** Dan Guravich/Photo Researchers.

Chapter 2

Opener p. 34–35 Jeffrey Scovil; **Figure 2-2 (a)** Breck P. Kent; **(b)** Courtesy of International Business Machine Corporation; **Figure 2-3** Fred Ward/ Black Star; **Figure 2-4** Runk/Schoenberger/Grant Heilman; **Figure 2-12 (a)** *(top and bottom)* Jeffrey Scovil; **(b)** Jeffrey Scovil; **Figure 2-13 (a)** and **(b)** Ken Lucas/Biological Photo Service; **Figure 2-14 (a)** Stuart Cohen/ Comstock; **(b)** M. Claye/Jacana/Photo Researchers; **(c, d, e)** Breck P. Kent; **Figure 2-15** Breck P. Kent; **Figure 2-16** M. Claye/Jacana/Photo Researchers; **Figure 2-17 (a)** Ed Degginger/Bruce Coleman, Inc.; **(b)** Breck P. Kent; **(c)** Barry L. Runk/Grant Heilman; **(d)** and **(e)** Francis Örs Lustwerk-Dudás; **Figure 2-18 (a)** Jeffrey Scovil; **(b)** Manfred Kage/Peter Arnold, Inc.; **Figure 2-19** Breck P. Kent; **Figure 2-20 (a)** E. R. Degginger/ Bruce Coleman, Inc.; **(b)** Breck P. Kent; **(c)** Ken Lucas/Biological Photo Service; **(d)** Jeffrey Scovil; **Figure 2-22 (a, b, c)** Bruce Iverson; **Figure 2-23 (a)** and **(b)** Breck P. Kent; **Figure 2-26 (a)** American Museum of Natural History; **(b, c , d)** Breck P. Kent; **(e)** Jeffrey Scovil; **Figure 2-27 (a)** Breck P. Kent; **(b)** and **(c)** M. Claye/Jacana/Photo Researchers; **Figure 2-28 (a)** John S. Shelton; **(b)** Lunar and Planetary Institute; **p. 63** *(top)* Tom McHugh/Photo Researchers; *(bottom)* Breck P. Kent.

Chapter 3

Opener p. 64–65 Comstock; **Figure 3-3 (a)** May/Biological Photo Service; *(inset a)* William E. Ferguson; **Figure 3-3 (b)** William E. Ferguson; **Figure 3-3 (c)** Biological Photo Service; *(inset c)* William E. Ferguson; **Figure 3-3 (d)** Breck P. Kent; **Figure 3-3 (e)** Breck P. Kent; *(inset e)* Doug Sokell/Visions Unlimited; **Figure 3-4** *(granite and diorite)* Breck P. Kent; *(basalt)* Barry L. Runk/Grant Heilman; *(peridotite)* Breck P. Kent; **Figure 3-12** Francis Örs Lustwerk-Dudás; **Figure 3-15** Breck P. Kent; **Figure 3-16** William E. Ferguson; **Figure 3-18** John S. Shelton; **Figure 3-19 (a)** National Archives; **Figure 3-21** Rob Badger; **Figure 3-22** Timothy Holst; **Figure 3-23** May/Biological Photo Service; **Figure 3-28** Space Imagery Center, Lunar and Planetary Lab, University of Arizona; **Figure 3-29** NASA—Starlight; **Figure 3-30** David L. Brown/The Stock Market; **p. 93** William E. Ferguson.

Chapter 4

Opener p. 94–95 Courtesy of RESTEC and Nihon University, Japan; **Figure 4-2** Comstock; **Figure 4-3** John Reed/Science Library/Photo Researchers; **Figure 4-4 (a)** Gary Braasch; **(b)** Ralph Perry/Black Star; **Figure 4-5** G. Brad Lewis; **Figure 4-6** U.S. Geological Survey; **Figure 4-7** Glenn Oliver/ Visuals Unlimited; **Figure 4-8 (b)** Courtesy of Northern Ireland Tourist Board; **Figure 4-9 (b)** Ramesh Venkatakrishnan; **(c)** Fred Grassle/Woods Hole Oceanographic Institute; **Figure 4-10 (a)** and **(b)** William E. Ferguson; **Figure 4-12** Alberto Garcia/SABA; **Figure 4-13** Anthony Suau/Black Star; **Figure 4-14 (a)** Steve Kaufman/Peter Arnold, Inc.; **(b)** John S. Shelton; **Figure 4-15** Marc Schechter/Photo Resource Hawaii; **Figure 4-16** Gary Braasch; **Figure 4-17** Hydrographic Department of Japan; **Figure 4-18** Leonard Von Matt/Photo Researchers; **Figure 4-20 (a)** Royal Commonwealth Society, London; **Figure 4-20 (b)** © 1902 Underwood & Underwood, courtesy of Library of Congress; **Figure 4-21** Ray Atkeson/The Stock Market; **Figure 4-23** Link/Visuals Unlimited; **Figure 4-24** Georg Gerster/Comstock; **Figure 4-27 (a)** James Sugar/Black Star; **(b)** James Mason/Black Star; **(c)** James Sugar/Black Star; **Figure 4-28** Gary Braasch; **Figure 4-29** William E. Ferguson; **Figure 4-30** *(inset)* NASA/ERTS; **Figure 4-31** Fred M. Bullard; **Figure 4-32, Figure 4-33** and **Figure 4-34** NASA; **p. 125** David Ball/Allstock.

Chapter 5

Opener p. 126–127 David Ball/The Stock Market; **Figure 5-2** Mount Rushmore National Memorial; **Figure 5-5** Richard Thom/Visuals Unlimited; **Figure 5-6 (a)** Central Park Conservancy; **(b)** Runk/Schoenberger/Grant Heilman; **Figure 5-8** William E. Ferguson; **Figure 5-9** Runk/Schoenberger/Grant Heilman; **Figure 5-10** Ramesh Venkatakrishnan; **Figure 5-12** Runk/ Schoenberger/Grant Heilman; **Figure 5-13** and **Figure 5-14** *(left and right)* John S. Shelton; **Figure 5-16** Ramesh Venkatakrishnan; **Figure 5-18** Norman Owen Tomalin/Bruce Coleman, Inc.; **Figure 5-21** Runk/ Schoenberger/Grant Heilman; **Figure 5-22 (b)** F.C. Earney/Visuals Unlimited; **Figure 5-24** Betty Crowell/Faraway Places; **Figure 5-25** George Holton/Photo Researchers; **Figure 5-26** NASA; **Figure 5-27** TASS/Sovfoto; **Figure 5-28** NASA; **p. 149** Patrick Spencer.

Chapter 6

Opener p. 150–151 Francois Gohier/Photo Researchers; **Figure 6-3 (a)** Michael B. E. Bograd/Mississippi Office of Geology; **(b)** Ramesh Venkatakrishnan; **Figure 6-4** John S. Shelton; **Figure 6-6 (a)** Ramesh Venkatakrishnan; **Figure 6-9** Glacier National Park; **Figure 6-12 (a)** and **(b)** Ward's; **(c)** Francis Örs Lustwerk-Dudás; **Figure 6-13 (a)** John S. Shelton; **(b)** Breck P. Kent; **Figure 6-14** Larry Davis, Washington State University; **Figure 6-15** J. Richardson/Woodfin Camp & Associates.; **Figure 6-19** J. Fennell/Bruce Coleman, Inc.; **Figure 6-20 (a)** E.R. Degginger/Bruce Coleman, Inc.; **(b)** Lynn McLaren/Photo Researchers; **Figure 6-21** Alfred Pasieka/Bruce Coleman, Inc.; **Figure 6-22 (a)** and **(b)** Robert S. Garrison, University of California; **Figure 6-23** Georg Gerster/Comstock; **Figure 6-29** M. Lustbader/Photo Researchers; **p. 179** Betty Crowell/Faraway Places.

Chapter 7

Opener p. 180–181 James Cowlin/Image Enterprises; **Figure 7-4** Kenneth Murray/Photo Researchers; **Figure 7-6** Mary M. Thatcher/Photo Researchers; **Figure 7-7** John D. Cunningham/Visuals Unlimited; **Figure 7-8** William Felger/Grant Heilman; **Figure 7-9** Calvin Alexander; **Figure 7-10** and **Figure 7-12 (b)** Betty Crowell/Faraway Places; **Figure 7-15** William E. Ferguson; **Figure 7-21** *(blueschist facies and greenschist facies)* William E. Ferguson; *(granulite facies)* Francis Örs Lustwerk-Dudás; **Figure 7-22** Calvin Alexander; **Figure 7-23** Scott Frances/Esto; **Figure 7-24 (a)** and **(b)** Treë; **Figure 7-26** Betty Crowell/Faraway Places; **p. 205** Ramesh Venkatakrishnan.

Chapter 8

Opener p. 206–207 Betty Crowell/Faraway Places; **Figure 8-3 (a)** P.C. Bateman/U.S. Geological Survey; **(b)** David Weintraub/Photo Researchers; **Figure 8-4** William E. Ferguson; **Figure 8-5 (a)** G. Shanmugam; **Figure 8-6** Martin G. Miller/Visuals Unlimited; **Figure 8-7** William E. Ferguson; **Figure 8-10 (a)** and **(b)** Breck P. Kent; **(c)** Ken Lucas/Biological Photo Service; **(d)** Breck P. Kent; **Figure 8-11 (a)** Jack K. Clark/Comstock; **(b)** Grant Heilman; **(c)** Breck P. Kent; **(d)** David Schwimmer/Bruce Coleman, Inc.; **(e)** William E. Ferguson; **Figure 8-13** Edward A. Hay, De Anza College, Cupertino, CA; **Figure 8-16** Breck P. Kent; **Figure 8-20** Sam Bowring/MIT; **Figure 8-24 (b)** William E. Ferguson; **Figure 8-25 (b)** Grant Heilman; **Figure 8-28 (a)** Breck P. Kent; **(b)** William E. Ferguson; **Figure 8-29** Tom McHugh/California Academy of Sciences/Photo Researchers; **Figure 8-30** NASA; **Figure 8-31** Joy Spurr/Bruce Coleman, Inc.

Chapter 9

Opener p. 238–239 Dan Guravich/Photo Researchers; **Figure 9-2 (a)** Gary Braasch; **(b)** John D. Cunningham/Visuals Unlimited; **Figure 9-6 (a)** A.J. Copley/Visuals Unlimited; **(b)** Breck P. Kent; **(c)** Jeff Lepore/Photo Researchers; **Figure 9-10 (a)** GEOPIC©, Earth Satellite Corporation; **Figure 9-13 (a)** David Muench/Allstock; **(b)** Martin G. Miller/Visuals Unlimited; **Figure 9-13 (c)** John S. Shelton; **Figure 9-16 (b)** Georg Gerster/Wingstock/Comstock; **Figure 9-19** Simon Fraser/Science Photo Library/Photo Researchers; **Figure 9-20** Breck P. Kent; **Figure 9-24 (a)** John S. Shelton; **Figure 9-27** Clyde H. Smith/Allstock; **p. 265** Ramesh Venkatakrishnan.

Chapter 10

Opener p. 266–267 Les Stone/Sygma; **Figure 10-9** U.S. Geological Survey/Science Photo Library/Photo Researchers; **Figure 10-11** John S. Shelton; **Figure 10-12** The Anchorage Museum of History & Art; **Figure 10-13** Art Resource, New York; **Figure 10-16** Dorantes/Sygma; **Figure 10-20** California Institute of Technology; **Figure 10-22** R.E Wallace, no. 311/U.S. Geological Survey; **Figure 10-23** Paula Scully/Gamma Liaison; **Figure 10-24** Peter Vadnai/The Stock Market; **Figure 10-25** UPI/Bettmann; **Figure 10-26** John S. Shelton.

Chapter 11

Opener p. 302–303 American Museum of Natural History; **Figure 11-8** Yu-Shen Zhang, University of California, Santa Cruz; **Figure 11-12 (a)** Ian Worpole. From Jeanloz, Raymond, and Lay, Thorne."The Core-Mantle Boundary", *Scientific American 268(5): 55.* © 1993 by *Scientific American, Inc.* All rights reserved; **(b)** Douglas L. Peck; **Figure 11-13** John S. Shelton; **Figure 11-24** Tom Bean/Allstock.

Chapter 12

Opener p. 330–331 ERIM/Tony Stone Worldwide; **Figure 12-8** N.O.A.A./National Geophysical Data Center; **Figure 12-13** Robert R. Hessler; **Figure 12-14** Dudley Foster/Woods Hole Oceanographic Institute; **Figure 12-17** R. Dietmar Müller, Walter R. Roest, Jean-Yves Royer, Lisa M. Gahagan, John G. Sclater/University of California, San Diego; **Figure 12-18** Betty Crowell/Faraway Places; **Figure 12-20** Comstock.

Chapter 13

Opener p. 362–363 Jeff Foott/Bruce Coleman, Inc.; **Figure 13-2 (b)** Peter L. Kresan; **Figure 13-5** *(inset)* Albert Copley/Visuals Unlimited; **Figure 13-7** David Falconer/Bruce Coleman, Inc.; **Figure 13-8 (b)** Ramesh Venkatak rishnan; **Figure 13-10 (b)** Betty Crowell/Faraway Places; **Figure 13-13** *(top)* John S. Shelton; *(bottom)* Peter L. Kresan; **Figure 13-15** Bill Bachman/Photo Researchers; **Figure 13-16** M. Western/Comstock; **Figure 13-17 (a)** John S. Shelton; **Figure 13-18** W. Cross, no. 765, U.S. Geological Survey; **Figure 13-20** Courtesy of the Geological Survey of Canada (GSC 201669); **Figure 13-21** Fletcher & Bayils/Photo Researchers; **Figure 13-22 (a)** and **(b)** Lloyd S. Cluff; **Figure 13-24** Betty Crowell/Faraway Places; **Figure 13-29 (a)** Siva Bonatti; **(b)** Gary Rogers, Virginia Military Institute; **Figure 13-30** NASA; **Figure 13-31** U.S. Geological Survey; **p. 385** Ramesh Venkatakrishnan.

Chapter 14

Opener p. 386–387 Andrew Holbrooke/Gamma Liaison; **Figure 14-7** Norman Weiser; **Figure 14-8** Peter L. Kresan; **Figure 14-11** Jeff Lepore/Photo Researchers; **Figure 14-12** Martin G. Miller/Visuals Unlimited; **Figure 14-13** Comstock; **Figure 14-14** John S. Flannery/Bruce Coleman, Inc.; **Figure 14-16 (a)** Jeff Lepore/Photo Researchers; **(b)** Andrew J. Martinez/Photo Researchers; **(c)** Art Gingert/Comstock; **Figure 14-17** Betty Crowell/Faraway Places; **Figure 14-19** Ramesh Venkatakrishnan; **Figure 14-21** Andrew Holbrooke/Gamma Liaison; **Figure 14-22** Peter L. Kresan; **Figure 14-23 (b)** G.R. Roberts; **(c)** Ramesh Venkatakrishnan; **Figure 14-26** U.S. Geological Survey Photo Library; **Figure 14-27** ERIM, Ann Arbor, MI; **Figure 14-32** Earth Observation Satellite Co.; **Figure 14-33** Galen Rowell/Peter Arnold, Inc.; **Figure 14-34** NASA; **p. 417** Alex S. Maclean/Peter Arnold, Inc.

Chapter 15

Opener p. 418–419 Norman Weiser; **Figure 15-11 (b)** C.E. Siebenthal/U.S. Geological Survey; **Figure 15-12** Comstock; **Figure 15-14 (b)** Russ Kinne/Comstock; **Figure 15-15 (a)** Farrell Grehan/Science Source/Photo Researchers; **(b)** Gregory D. Dimijian, M.D./Photo Researchers; **Figure 15-16** Peter L. Kresan; **Figure 15-21 (b)** Porterfield-Chickering/Photo Researchers; **Figure 15-23** U.S. Department of Interior, U.S. Geological Survey; **Figure 15-24** Porterfield-Chickering/Photo Researchers.

Chapter 16

Opener p. 444–445 Gene Aherns/Bruce Coleman, Inc.; **Figure 16-2** Indiana Geological Survey; **Figure 16-3 (a)** E.R. Degginger/Bruce Coleman, Inc.; **(b)** D. Hardley/Bruce Coleman, Inc.; **Figure 16-4** Rene Burri/Magnum; **Figure 16-6** Ramesh Venkatakrishnan; **Figure 16-8 (a)** Robert & Linda Mitchell; **(b)** Albert Copley/Visuals Unlimited; **(c)** Jeff Lepore/Photo Researchers; **(d)** Stephen Alvarez/TIME Magazine; **(e)** D. & J. McClurg/

Bruce Coleman, Inc.; (f) Robert & Linda Mitchell; **Figure 16-9** Joy Spurr/Bruce Coleman, Inc.; **Figure 16-11 (b)** Stephen Alvarez/TIME Magazine; **Figure 16-12 (c)** Nancy Simmerman/Bruce Coleman, Inc.; **Figure 16-13** Gregory G. Dimijian/Photo Researchers; **Figure 16-14** Kim Heacox/Peter Arnold, Inc.; **Figure 16-17** Ramesh Venkatakrishnan; **Figure 16-19 (a)** Timothy O'Keefe/Bruce Coleman, Inc.; **(b)** Paul H. Moser, Alabama Geological Survey; **Figure 16-20** A. N. Palmer, Department of Earth Sciences, SUNY Oneonta; **Figure 16-21 (a)** Calvin Alexander/Ramesh Venkatakrishnan; **(b)** Calvin Alexander/Ramesh Venkatakrishnan; **Figure 16-22** Paolo Koch/Photo Researchers; **Figure 16-23** Steve Solum/Bruce Coleman, Inc.; **Figure 16-24** AP/Wide World Photos; **Figure 16-25** Betty Crowell/Faraway Places; **Figure 16-26** Ramesh Venkatakrishnan; **p. 465** U.S.D.A., Soil Conservation Service.

Chapter 17

Opener p. 466–467 Zefa-Herfort/The Stock Market; **Figure 17-6 (a)** Comstock; **(b)** Peter Dunwiddie/Visuals Unlimited; **(c)** Betty Crowell/Faraway Places; **(d)** Charles Krebs/The Stock Market; **Figure 17-7** *(inset)* Tom Bean/Allstock; **Figure 17-9** U.S. Department of Interior, U.S. Geological Survey EROS Data Center, Sioux Falls, SD, Scene ID # AB586003588 ROLL Frame # 2572; **Figure 17-10 (a)** Peter L. Kresan; **(b)** B.J. Spenceley/Bruce Coleman, Inc.; **Figure 17-11** William E. Ferguson; **Figure 17- 12 (b)** Inge King; **(c)** Betty Crowell/Faraway Places; **Figure 17-13 (a)** Advanced Satellite Productions, Inc., 1993 ; **(b)** U.S. Department of Interior, U.S. Geological Survey EROS Data Center, Sioux Falls, SD, Scene ID # E-1272-99CT; **Figure 17-14** Peter Arnold/Peter Arnold, Inc.; **Figure 17-16** *(inset)* Art Gingert/Comstock; **Figure 17-18** John S. Shelton; **Figure 17-19 (b)** John S. Shelton; **Figure 17-20** Alan Busacca, Washington State University; **Figure 17-21** John S. Shelton; **Figure 17-24 (a, b, c)** John S. Shelton; **Figure 17-26** *(inset a)* Glenn Oliver/Visuals Unlimited; *(inset b)* Bernard Hallet, University of Washington; *(inset c)* Dr. Troy L. Péwé, # 3161; *(inset d)* Peter Dunwiddie/Visuals Unlimited; **Figure 17-32** Betty Crowell/Faraway Places; **Figure 17-36** John W. Attig, Wisconsin Geological Survey; **Figure 17-37** Peter L. Kresan; **Figure 17-40** U.S. Geological Survey, Flagstaff, AZ; **p. 501** Ramesh Venkatakrishnan.

Chapter 18

Opener p. 502–503 Peter L. Kresan; **Figure 18-2 (a)** David Barnes/Allstock; **(b)** and **(c)** Jeff Foott/Bruce Coleman, Inc.; **Figure 18-7** Leo Meier/Comstock; **Figure 18-8** Tom Bean/Allstock; **Figure 18-10** David Ball/Allstock; **Figure 18-11 (b)** Betty Crowell/Faraway Places; **Figure 18-12** Norman Weiser; **Figure 18-15** Scott Berner/Visuals Unlimited; **Figure 18-16** J.C. Carton/Bruce Coleman, Inc.; **Figure 18-17** John Gerlach/Visuals Unlimited; **Figure 18-18** Martin G. Miller/Visuals Unlimited; **Figure 18-19** Georg Gerster/Comstock; **Figure 18-21** U.S. Farm Security Administration Collection, Prints and Photographs Division, Library of Congress; **Figure 18-23** Lee Rentz/Bruce Coleman, Inc.; **Figure 18-24** Carl Purcell/Photo Researchers; **Figure 18-26** Georg Gerster/Comstock; **Figure 18-27** Gary Braasch; **Figure 18-28 (a)** Peter L. Kresan; **(b)** Tad Nichols/Peter L. Kresan Photography; **(c)** Michael E. Long, National Geographic Society; **(e)** Wingstock/Comstock; **Figure 18-29** Stephenie S. Ferguson; **Figure 18-31** NASA; **Figure 18-32** M.&R. Borland/Bruce Coleman, Inc.; **Figure 18-33** NASA; **Figure 18-34** Peter Ward/Bruce Coleman, Inc.; **Figure 18-35** Georg Gerster/Comstock; **P. 531** John S. Shelton.

Chapter 19

Opener p. 532–533 NASA; **Figure 19-9 (b)** Clyde H. Smith/Peter Arnold, Inc.; **Figure 19-10 (a)** and **(b)** William E. Ferguson; **Figure 19-11** Dick Poe/Visuals Unlimited; **Figure 19-12** William C. Jorgensen/Visuals Unlimited; **Figure 19-13** Michael J. Chrzastowski, Illinois State Geological Survey; **Figure 19-14 (b)** John Elk/Bruce Coleman, Inc.; **(c)** G. R. Roberts; **Figure 19-15 (a)** Betty Crowell/Faraway Places; **(b)** G. Richard Whittecar, Old Dominion University; **Figure 19-17 (a)** Stewart Farrell, Stockton State College, NJ; **(b)** John S. Shelton; **Figure 19-18** John S. Shelton; **Figure 19-19 (a)** Fairchild Air Photos 0-139 & E-5780, UCLA Department of Geography Aerial Photo Archives; **(b)** John S. Shelton; **Figure 19-20 (inset a)** John S. Shelton; **(inset b)** Townsend P. Dickinson/Comstock; **Figure 19-21 (a)** Bruce W. Heinemann/The Stock Market; **(b)** Peter L. Kresan; **Figure 19-23** GEOPIC, Earth Satellite Corporation; **Figure 19-25** S. J. Krasemann/Peter Arnold, Inc.; **Figure 19-27 (b)** David Ball/The Stock Market; **(c)** William E. Ferguson; **Figure 19-28** Peter L. Kresan; **Figure 19-29** John S. Shelton.

Chapter 20

Opener p. 562–563 M. Rio Branco/Magnum; **Figure 20-2** Novosti/Gamma Liaison; **Figure 20-3** Laurent van der Stockt/Gamma Liaison; **Figure 20-4** Drake Well Museum, Titusville, Pa.; **Figure 20-6** Silvia Dinale/Collected by John Wearing; **Figure 20-7** Geological Survey of Canada Photo Library; **Figure 20-8** Bruce Coleman, Inc.; **Figure 20-10** Karen Jettmar/Allstock; **Figure 20-11** Nicholas deVore III/Bruce Coleman, Inc.; **Figure 20-14 (b)** Norman Weiser; **(c)** William E. Ferguson; **Figure 20-15** Kevin Schafer/Allstock; **Figure 20-16** Courtesy of Consolidated Edison of New York; **Figure 20-17** U.S. Department of Energy; **Figure 20-18 (a)** Peter L. Kresan; **(b)** Donna C. Rona/Bruce Coleman, Inc.; **Figure 20-19** JB Pictures; **Figure 20-20 (a)** Diego Guidice/Contrasto/SABA; **(b)** Timothy Ross/Picture Group; **Figure 20-21** Peter L. Kresan; **Figure 20-22** and **Figure 20-24** William E. Ferguson; **Figure 20-25** Thyssen Henschel/Stadler GMBH; **Figure 20-26** Bruce Dale, National Geographic Society; **Figure 20-27** Hank Morgan/Science Source/Photo Researchers; **Figure 20-28** Seny Norasingh/Light Sensitive.

Index